Lecture Notes in Computer Science 3838

Aart Middeldorp Vincent van Oostrom
Femke van Raamsdonk Roel de Vrijer (Eds.)

Processes, Terms and Cycles: Steps on the Road to Infinity

Essays Dedicated to Jan Willem Klop
on the Occasion of His 60th Birthday

Springer

Volume Editors

Aart Middeldorp
University of Innsbruck, Institute of Computer Science
Technikerstr. 21a, 6020 Innsbruck, Austria
E-mail: aart.middeldorp@uibk.ac.at

Vincent van Oostrom
Utrecht Universiteit, Department of Philosophy
Heidelberglaan 8, 3584 CS Utrecht, The Netherlands
E-mail: Vincent.vanOostrom@phil.uu.nl

Femke van Raamsdonk
Roel de Vrijer
Vrije Universiteit, Department of Computer Science
De Boelelaan 1081a, 1081 HV Amsterdam, The Netherlands
E-mail: {femke,rdv}@cs.vu.nl

Library of Congress Control Number: 2005937699

CR Subject Classification (1998): F.4.1, F.4, I.2.3-4, F.3

ISSN 0302-9743
ISBN-10 3-540-30911-X Springer Berlin Heidelberg New York
ISBN-13 978-3-540-30911-6 Springer Berlin Heidelberg New York

Springer is a part of Springer Science+Business Media

springer.com

Printed in Germany

Typesetting: Camera-ready by author, data conversion by Scientific Publishing Services, Chennai, India
Printed on acid-free paper SPIN: 11601548 06/3142 5 4 3 2 1 0

Jan Willem Klop

Preface

This Festschrift is dedicated to Jan Willem Klop on the occasion of his 60th birthday on December 19, 2005. Its focus is on the lambda calculus, term rewriting and process algebra, the fields where Jan Willem has made fundamental contributions. Without attempting to give a balanced account of Jan Willem's scientific achievements, we recall three accomplishments from the early years of his career that especially stand out.

The first is his counterexample showing that the extension of the lambda calculus with surjective pairing lacks the Church–Rosser property, or, in modern terminology, is not confluent [7, 9]. This settled a famous open problem, which had challenged several researchers in the lambda calculus community for years.

The second is his pioneering work in term rewriting. In his PhD thesis [9], Jan Willem gave a systematic study of orthogonal rewriting in the general setting of combinatory reduction systems (CRSs), thereby putting the areas of higher-order rewriting and orthogonality firmly on the map. Some of the ideas in the thesis trace back to the famous Blue Preprint [2], from the period that Jan Willem and some other students were graduating in mathematics and logic, under the supervision of Dirk van Dalen and Henk Barendregt.

The third feat is the creation, together with Jan Bergstra [23], of the algebra of communicating processes (ACP).

With his early work, Jan Willem provided inspiration for many years of fruitful research, continuing to this day. For decades he has been a creative and stimulating force in the areas of term rewriting and process algebra. Some of his recent interests are infinitary rewriting, graph rewriting and the geometry of processes.

Jan Willem's scientific world is inhabited by objects like processes, streams, terms, cycles and many other puzzling and intriguing phenomena, and in this world he paves the way, guided by his extraordinary intuition and his great care in visualizing that intuition. To his colleagues, Jan Willem's adventures and the way he talks about them are a never-ending source of inspiration.

This Festschrift contains scientific papers by close friends and colleagues of Jan Willem, written specifically for this book. The papers are different in nature: some report on new research, others have the character of a survey, and again others are mainly expository. Every contribution has been thoroughly refereed at least twice. In many cases the first round of referee reports led to significant revision of the original paper, which was again reviewed. This introduction includes a list of Jan Willem Klop's publications, for reference, and as an overview of the development of his scientific interests and achievements over the years. Although this bibliography is quite extensive, we do not claim it to be complete.

We thank all authors for their contribution to the Festschrift for Jan Willem Klop and we are grateful to the referees for their constructive cooperation. We also wish to thank Springer for publishing this book in their LNCS Festschrift series, and for the smooth publishing process.

The Festschrift was presented to Jan Willem on his 60th birthday, during a one-day symposium with the same title as this book: *Processes, terms and cycles: steps on the road to infinity.* The symposium not only celebrated Jan Willem's 60th birthday, but also the 25th anniversary of his connection with the CWI in Amsterdam. His first working day as a post-doc at the Mathematical Centre (MC), the predecessor of the CWI, was December 15, 1980.

We invited Zena Ariola (University of Oregon, USA), Arvind (MIT, USA), Henk Barendregt (Radboud University, The Netherlands), Jan Bergstra (University of Amsterdam and Utrecht University, The Netherlands), Nachum Dershowitz (Tel Aviv University, Israel), Mariangiola Dezani (University of Torino, Italy), Roger Hindley (Swansea University, UK), Jean-Jacques Lévy (INRIA, France), and Ronan Sleep (University of East Anglia, UK) to give a talk at the symposium. We are very pleased that they kindly accepted. All speakers are leading researchers in the fields of term rewriting, lambda calculus and process algebra, and their friendship with Jan Willem dates back many years.

The symposium was organized at and together with the CWI. The collaboration with Susanne van Dam and Jaco van de Pol from the CWI has been a pleasure. We wish to thank them, and more generally the CWI, for all their assistance. We also gratefully acknowledge the generous financial support for the symposium from the Vrije Universiteit, CWI, NWO, and the Radboud University.

Also on behalf of Susanne and Jaco, we would like to conclude by stating that it has been an honour and a pleasure to work on the preparation of this book and the symposium. It gave us the opportunity to experience how much Jan Willem is valued by colleagues all over the world. The willingness to contribute and participate and make the best of the book and symposium has been enormous. For this we are thankful of course, but more importantly, the manifestation of this enthusiasm reflects the esteem that Jan Willem has in the scientific community. Congratulations, Jan Willem! We wish you and all of us a fruitful continuation of our cooperation and sharing of interests for many years to come.

October 2005

Aart Middeldorp
Vincent van Oostrom
Femke van Raamsdonk
Roel de Vrijer

Organization

Sponsors

Centrum voor Wiskunde en Informatica (CWI)
Nederlandse Organisatie voor Wetenschappelijk Onderzoek (NWO)
Radboud University Nijmegen
Vrije Universiteit Amsterdam

Reviewers

Luca Aceto
Giovanna d'Agostino
Jos Baeten
Christel Baier
Twan Basten
Lev Beklemishev
Stefano Berardi
Frédéric Blanqui
Stefan Blom
Luca Cardelli
Andrea Corradini
Mariangiola Dezani
Frank Drewes
Irène Durand
Zoltán Ésik
Azadeh Farzan
Wan Fokkink
Alfons Geser
Rob van Glabbeek
John Glauert
Andrew Gordon
Jean Goubault-Larrecq
Bernhard Gramlich
Annegret Habel
Jan Heering
Rosalie Iemhoff
Emil Jeřábek
Alan Jeffrey
Stefan Kahrs
Karl Trygve Kalleberg
Juhani Karhumäki
Jeroen Ketema
Steven Klusener
Salvador Lucas
Bas Luttik
Ian Mackie
Pasquale Malacaria
Ralph Matthes
Eric Monfroy
Georg Moser
Jun Pang
Alban Ponse
Julian Rathke
Francesca Rossi
Manfred Schmidt-Schauß
Géraud Sénizergues
Jeffrey O. Shallit
Paweł Sobociński
Mark-Oliver Stehr
Marielle Stoelinga
Peter Stuckey
Pawel Urzyczyn
Steven Vickers
Albert Visser
Fer-Jan de Vries
Johannes Waldmann
Nobuko Yoshida

Bibliography of Jan Willem Klop

1. J.W. Klop. On solvability by λI-terms. In C. Böhm, editor, *λ-Calculus and Computer Science Theory*, volume 37 of *Lecture Notes in Computer Science*, pages 342–345. Springer, 1975.
2. H.P. Barendregt, J.A. Bergstra, J.W. Klop, and H. Volken. Degrees, reductions and representability in the lambda calculus. Preprint 22, Department of Mathematics, Utrecht University, 1976. The blue preprint.
3. H.P. Barendregt, J.A. Bergstra, J.W. Klop, and H. Volken. Representability in lambda algebras. *Indagationes Mathematicae*, 38(5):377–387, 1976.
4. J.W. Klop. A counterexample to the Church-Rosser property for λ-calculus + $DMM \to M$. Typed note, Department of Mathematics, Utrecht University, 1976.
5. H.P. Barendregt, J.A. Bergstra, J.W. Klop, and H. Volken. Degrees of sensible lambda theories. *Journal of Symbolic Logic*, 43(1):45–55, 1978.
6. J.A. Bergstra and J.W. Klop. Church-Rosser strategies in the lambda calculus. *Theoretical Computer Science*, 9(1):27–38, 1979.
7. J.W. Klop. A counterexample to the Church-Rosser property for lambda calculus with surjective pairing. In B. Robinet, editor, *Proc. of the École de Printemps d'Informatique Théorique: Lambda Calcul et Sémantique Formelle des Langages de Programmation*. LITP-ENSTA, 1979. Also published as Preprint 102, Department of Mathematics, Utrecht University.
8. J.A. Bergstra and J.W. Klop. Invertible terms in the lambda calculus. *Theoretical Computer Science*, 11(1):19–37, 1980.
9. J.W. Klop. *Combinatory Reduction Systems*. PhD thesis, Utrecht University, 1980. Also published as volume 127 of *Mathematical Centre Tracts*, Mathematisch Centrum, Amsterdam, 1980.
10. J.W. Klop. Reduction cycles in combinatory logic. In J.R. Hindley and J.P. Seldin, editors, *To H.B. Curry: Essays on Combinatory Logic, Lambda Calculus and Formalism*, pages 193–214. Academic Press, 1980.
11. J.W. de Bakker, J.W. Klop, and J.-J.Ch. Meyer. Correctness of programs with function procedures. In D. Kozen, editor, *Proc. of the 1981 Workshop on Logic of Programs*, volume 131 of *Lecture Notes in Computer Science*, pages 94–113. Springer, 1981.
12. J.A. Bergstra and J.W. Klop. Algebraic specifications for parametrized data types with minimal parameter and target algebras. In M. Nielsen and E.M. Schmidt, editors, *Proc. of the 9th International Colloquium on Automata, Languages and Programming (ICALP'82)*, volume 140 of *Lecture Notes in Computer Science*, pages 23–34. Springer, 1982.
13. J.A. Bergstra and J.W. Klop. Fixed point semantics in process algebras. Technical Report IW 206/82, Mathematical Centre, Amsterdam, 1982. Revised version: A convergence theorem in process algebra, CWI Report CS-R8733, Amsterdam, 1987.
14. J.A. Bergstra and J.W. Klop. A formalized proof system for total correctness of while programs. In *Proc. of the 5th International Symposium on Programming*, volume 137 of *Lecture Notes in Computer Science*, pages 26–36. Springer, 1982.
15. J.A. Bergstra and J.W. Klop. Strong normalization and perpetual reductions in the lambda calculus. *Elektronische Informationsverarbeitung und Kybernetik*, 18(7/8):403–417, 1982.
16. J.W. Klop. Extending partial combinatory algebras. *Bulletin of the European Association for Theoretical Computer Science*, 16:472–482, 1982.

17. J.W. de Bakker, J.A. Bergstra, J.W. Klop, and J.-J.Ch. Meyer. Linear time and branching time semantics for recursion with merge. In *Proc. of the 10th International Colloquium on Automata, Languages and Programming (ICALP'83)*, pages 39–51, 1983. Extended version appeared as [21].
18. J.A. Bergstra and J.W. Klop. Formal proof systems for program equivalence. In D. Bjørner, editor, *Proc. of the 2nd IFIP WG 2.2 Conference on Formal Description of Programming Concepts*, pages 289–303. North-Holland, 1983.
19. J.A. Bergstra and J.W. Klop. Initial algebra specifications for parametrized data types. *Elektronische Informationsverarbeitung und Kybernetik*, 19(1/2):17–31, 1983.
20. J.A. Bergstra and J.W. Klop. A proof rule for restoring logic circuits. *Integration, the VLSI Journal*, 1:161–178, 1983.
21. J.W. de Bakker, J.A. Bergstra, J.W. Klop, and J.-J.Ch. Meyer. Linear time and branching time semantics for recursion with merge. *Theoretical Computer Science*, 34(1/2):135–156, 1984. Extended version of [17].
22. J.A. Bergstra and J.W. Klop. The algebra of recursively defined processes and the algebra of regular processes. In J. Paredaens, editor, *Proc. of the 11th International Colloquium on Automata, Languages and Programming (ICALP'84)*, volume 172 of *Lecture Notes in Computer Science*, pages 82–95. Springer, 1984.
23. J.A. Bergstra and J.W. Klop. Process algebra for synchronous communication. *Information and Control*, 60(1-3):109–137, 1984.
24. J.A. Bergstra and J.W. Klop. Proving program inclusion using Hoare's logic. *Theoretical Computer Science*, 30(1):1–48, 1984.
25. J.A. Bergstra, J.W. Klop, and J.V. Tucker. Algebraic tools for system construction. In E. Clarke and D. Kozen, editors, *Proc. of the 1983 Workshop on Logic of Programs*, volume 164 of *Lecture Notes in Computer Science*, pages 34–44. Springer, 1984.
26. J.A. Bergstra and J.W. Klop. Algebra of communicating processes with abstraction. *Theoretical Computer Science*, 37(1):77–121, 1985.
27. J.A. Bergstra and J.W. Klop. Verification of an alternating bit protocol by means of process algebra. In W. Bibel and K.P. Jantke, editors, *Proc. of the Spring School on Mathematical Methods of Specification and Synthesis of Software Systems*, volume 215 of *Lecture Notes in Computer Science*, pages 9–23. Springer, 1985.
28. J.A. Bergstra, J.W. Klop, and J.V. Tucker. Process algebra with asynchronous communication mechanisms. In S.D. Brookes, A.W. Roscoe, and G. Winskel, editors, *Seminar on Concurrency*, volume 197 of *Lecture Notes in Computer Science*, pages 76–95. Springer, 1985.
29. J.W. Klop. Term rewriting systems. Unpublished lecture notes for the seminar on reduction machines, Ustica, 1985.
30. J.C.M. Baeten, J.A. Bergstra, and J.W. Klop. Syntax and defining equations for an interrupt mechanism in process algebra. *Fundamenta Informaticae*, 9(2):127–167, 1986.
31. J.A. Bergstra and J.W. Klop. Algebra of communicating processes. In J.W. de Bakker, M. Hazewinkel, and J.K. Lenstra, editors, *Mathematics and Computer Science*, CWI Monograph 1, pages 89–138. North-Holland, Amsterdam, 1986.
32. J.A. Bergstra and J.W. Klop. Conditional rewrite rules: Confluence and termination. *Journal of Computer and System Sciences*, 32(3):323–362, 1986.
33. J.A. Bergstra and J.W. Klop. Process algebra: Specification and verification in bisimulation semantics. In M. Hazewinkel, J.K. Lenstra, and L.G.L.T. Meertens, editors, *Mathematics and Computer Science II*, CWI Monograph 4, pages 61–94. North-Holland, Amsterdam, 1986.

34. J.C.M. Baeten, J.A. Bergstra, and J.W. Klop. Conditional axioms and α/β calculus in process algebra. In M. Wirsing, editor, *Proc. of the 3rd IFIP WG 2.2 Conference on Formal Description of Programming Concepts*, pages 53–75. North-Holland, 1987.
35. J.C.M. Baeten, J.A. Bergstra, and J.W. Klop. Decidability of bisimulation equivalence for processes generating context-free languages. In J.W. de Bakker, A.J. Nijman, and P.C. Treleaven, editors, *Proc. of the 1st International Conference on Parallel Architectures and Languages Europe (PARLE'87), Vol. I: Parallel Languages*, volume 259 of *Lecture Notes in Computer Science*, pages 94–111. Springer, 1987. Extended version appeared as [64].
36. J.C.M. Baeten, J.A. Bergstra, and J.W. Klop. On the consistency of Koomen's fair abstraction rule. *Theoretical Computer Science*, 51(1&2):129–176, 1987.
37. J.C.M. Baeten, J.A. Bergstra, and J.W. Klop. Ready-trace semantics for concrete process algebra with the priority operator. *The Computer Journal*, 30(6):498–506, 1987.
38. J.C.M. Baeten, J.A. Bergstra, and J.W. Klop. Term rewriting systems with priorities. In P. Lescanne, editor, *Proc. of the 2nd International Conference on Rewriting Techniques and Applications (RTA'87)*, volume 256 of *Lecture Notes in Computer Science*, pages 83–94. Springer, 1987. Extended version appeared as [47].
39. H.P. Barendregt, J.R. Kennaway, J.W. Klop, and M.R. Sleep. Needed reduction and spine strategies for the lambda calculus. *Information and Computation*, 75(3):191–231, 1987.
40. J.A. Bergstra and J.W. Klop. ACP_τ: A universal axiom system for process specification. In M. Wirsing and J.A. Bergstra, editors, *Proc. of the Workshop on Algebraic Methods: Theory, Tools and Applications*, volume 394 of *Lecture Notes in Computer Science*, pages 447–463. Springer, 1987.
41. J.A. Bergstra and J.W. Klop. *Semi-complete termherschrijfsystemen.* Programmatuurkunde. Kluwer, 1987. In Dutch.
42. J.A. Bergstra, J.W. Klop, and E.-R. Olderog. Failures without chaos: a new process semantics for fair abstraction. In M. Wirsing, editor, *Proc. of the 3rd IFIP WG 2.2 Conference on Formal Description of Programming Concepts*, pages 77–103. North-Holland, 1987.
43. J.W. Klop. Term rewriting systems: a tutorial. *Bulletin of the European Association for Theoretical Computer Science*, 32:143–183, 1987.
44. J.W. Klop and E. Kranakis. Lower bounds for a class of Kostka numbers. *Ars Combinatoria*, 23:121–129, 1987.
45. J.A. Bergstra and J.W. Klop. A complete inference system for regular processes with silent moves. In F.R. Drake and J.K. Truss, editors, *Proc. of the 1986 Logic Colloquium*, pages 21–81. North-Holland, 1988.
46. J.A. Bergstra, J.W. Klop, and E.-R. Olderog. Readies and failures in the algebra of communicating processes. *SIAM Journal on Computing*, 17(6):1134–1177, 1988.
47. J.C.M. Baeten, J.A. Bergstra, J.W. Klop, and W.P. Weijland. Term-rewriting systems with rule priorities. *Theoretical Computer Science*, 67(2&3):283–301, 1989. Extended version of [38].
48. J.A. Bergstra and J.W. Klop. Process theory based on bisimulation semantics. In J.W. de Bakker, W.-P. de Roever, and G. Rozenberg, editors, *Proc. of the REX Workshop on Linear Time, Branching Time and Partial Order in Logics and Models for Concurrency*, volume 354 of *Lecture Notes in Computer Science*, pages 50–122. Springer, 1989.
49. J.A. Bergstra, J.W. Klop, and A. Middeldorp. *Termherschrijfsystemen.* Programmatuurkunde. Kluwer, 1989. In Dutch.

50. R.N. Bol, K.R. Apt, and J.W. Klop. On the safe termination of PROLOG programs. In G. Levi and M. Martelli, editors, *Proc. of the 6th International Conference on Logic Programming*, pages 353–368. The MIT Press, 1989.
51. J.W. Klop and R.C. de Vrijer. Unique normal forms for lambda calculus with surjective pairing. *Information and Computation*, 80(2):97–113, 1989.
52. Y. Toyama, J.W. Klop, and H.P. Barendregt. Termination for the direct sum of left-linear term rewriting systems – preliminary draft. In N. Dershowitz, editor, *Proc. of the 3rd International Conference on Rewriting Techniques and Applications (RTA'89)*, volume 355 of *Lecture Notes in Computer Science*, pages 477–491. Springer, 1989. Extended version appeared as [79].
53. J.C.M. Baeten and J.W. Klop, editors. *Proc. of the 1st International Conference on Theories of Concurrency: Unification and Extension (CONCUR'90)*, volume 458 of *Lecture Notes in Computer Science*. Springer, 1990.
54. J.A. Bergstra and J.W. Klop. An introduction to process algebra. In J.C.M. Baeten, editor, *Applications of Process Algebra*, volume 17 of *Cambridge Tracts in Theoretical Computer Science*, pages 1–21. Cambridge University Press, 1990.
55. R.N. Bol, K.R. Apt, and J.W. Klop. On the power of subsumption and context checks. In A. Miola, editor, *Proc. of the International Symposium on Design and Implementation of Symbolic Computation Systems (DISCO'90)*, volume 429 of *Lecture Notes in Computer Science*, pages 131–140. Springer, 1990.
56. J.W. Klop. Term rewriting systems: From Church-Rosser to Knuth-Bendix and beyond. In M.S. Paterson, editor, *Proc. of the 17th International Colloquium on Automata, Languages and Programming (ICALP'90)*, volume 443 of *Lecture Notes in Computer Science*, pages 350–369. Springer, 1990.
57. R.N. Bol, K.R. Apt, and J.W. Klop. An analysis of loop checking mechanisms for logic programs. *Theoretical Computer Science*, 86(1):35–79, 1991.
58. N. Dershowitz, J.-P. Jouannaud, and J.W. Klop. Open problems in rewriting. In Ronald V. Book, editor, *Proc. of the 4th International Conference on Rewriting Techniques and Applications (RTA'91)*, volume 488 of *Lecture Notes in Computer Science*, pages 445–456. Springer, 1991.
59. J.R. Kennaway, J.W. Klop, M.R. Sleep, and F.J. de Vries. Transfinite reductions in orthogonal term rewriting systems (extended abstract). In Ronald V. Book, editor, *Proc. of the 4th International Conference on Rewriting Techniques and Applications (RTA'91)*, volume 488 of *Lecture Notes in Computer Science*, pages 1–12. Springer, 1991. Extended version appeared as [78].
60. J.W. Klop and A. Middeldorp. Sequentiality in orthogonal term rewriting systems. *Journal of Symbolic Computation*, 12:161–195, 1991.
61. J.W. Klop and R.C. de Vrijer. Extended term rewriting systems. In S. Kaplan and M. Okada, editors, *Proc. of the 2nd International Workshop on Conditional and Typed Rewriting Systems (CTRS'90)*, volume 516 of *Lecture Notes in Computer Science*, pages 26–50. Springer, 1991.
62. F.S. de Boer, J.W. Klop, and C. Palamidessi. Asynchronous communication in process algebra. In *Proc. of the 7th Annual Symposium on Logic in Computer Science (LICS'92)*, pages 137–147. IEEE Computer Society Press, 1992.
63. J.W. Klop. Term rewriting systems. In S. Abramsky, D.M. Gabbay, and T.S.E. Maibaum, editors, *Handbook of Logic in Computer Science*, volume 2, pages 1–116. Oxford University Press, 1992.
64. J.C.M. Baeten, J.A. Bergstra, and J.W. Klop. Decidability of bisimulation equivalence for processes generating context-free languages. *Journal of the ACM*, 40(3):653–682, 1993. Extended version of [35].

65. N. Dershowitz, J.-P. Jouannaud, and J.W. Klop. More problems in rewriting. In Claude Kirchner, editor, *Proc. of the 5th International Conference on Rewriting Techniques and Applications (RTA'93)*, volume 690 of *Lecture Notes in Computer Science*, pages 468–487. Springer, 1993.
66. J.R. Kennaway, J.W. Klop, M.R. Sleep, and F.J. de Vries. The adequacy of term graph rewriting for simulating term rewriting. In M.R. Sleep, M.J. Plasmeijer, and M.C.J.D. van Eekelen, editors, *Term Graph Rewriting – Theory and Practice*, pages 157–169. John Wiley & Sons, 1993.
67. J.R. Kennaway, J.W. Klop, M.R. Sleep, and F.J. de Vries. Event structures and orthogonal term graph rewriting. In M.R. Sleep, M.J. Plasmeijer, and M.C.J.D. van Eekelen, editors, *Term Graph Rewriting – Theory and Practice*, pages 141–155. John Wiley & Sons, 1993.
68. J.R. Kennaway, J.W. Klop, M.R. Sleep, and F.J. de Vries. An infinitary Church-Rosser property for non-collapsing orthogonal term rewriting systems. In M.R. Sleep, M.J. Plasmeijer, and M.C.J.D. van Eekelen, editors, *Term Graph Rewriting – Theory and Practice*, pages 47–59. John Wiley & Sons, 1993.
69. J.R. Kennaway, J.W. Klop, M.R. Sleep, and F.J. de Vries. An introduction to term graph rewriting. In M.R. Sleep, M.J. Plasmeijer, and M.C.J.D. van Eekelen, editors, *Term Graph Rewriting - Theory and Practice*, pages 1–12. John Wiley & Sons, 1993.
70. J.W. Klop, V. van Oostrom, and F. van Raamsdonk. Combinatory Reduction Systems: Introduction and survey. *Theoretical Computer Science*, 121(1&2):279–308, 1993. Special issue in honour of Corrado Böhm.
71. Z.M. Ariola, J.R. Kennaway, J.W. Klop, M.R. Sleep, and F.J. de Vries. Syntactic definitions of undefined: on defining the undefined. In M. Hagiya and J.C. Mitchell, editors, *Proc. of the 1st International Conference on Theoretical Aspects of Computer Software (TACS'94)*, volume 789 of *Lecture Notes in Computer Science*, pages 543–554. Springer, 1994.
72. Z.M. Ariola and J.W. Klop. Cyclic lambda graph rewriting. In *Proc. of the 9th Annual Symposium on Logic in Computer Science (LICS'94)*, pages 416–425. IEEE Computer Society Press, 1994.
73. J.R. Kennaway, J.W. Klop, M.R. Sleep, and F.J. de Vries. On the adequacy of graph rewriting for simulating term rewriting. *ACM Transactions on Programming Languages and Systems*, 16(3):493–523, 1994.
74. J.W. Klop, A. Middeldorp, Y. Toyama, and R.C. de Vrijer. Modularity of confluence: a simplified proof. *Information Processing Letters*, 49:101–109, 1994.
75. J.A. Bergstra and J.W. Klop. The algebra of recursively defined processes and the algebra of regular processes. In *Algebra of Communicating Processes*, Workshops in Computing, pages 1–25. Springer, 1995.
76. N. Dershowitz, J.-P. Jouannaud, and J.W. Klop. Problems in rewriting III. In Jieh Hsiang, editor, *Proc. of the 6th International Conference on Rewriting Techniques and Applications (RTA'95)*, volume 914 of *Lecture Notes in Computer Science*, pages 457–471. Springer, 1995.
77. J.R. Kennaway, J.W. Klop, M.R. Sleep, and F.J. de Vries. Infinitary lambda calculus and Böhm models. In Jieh Hsiang, editor, *Proc. of the 6th International Conference on Rewriting Techniques and Applications (RTA'95)*, volume 914 of *Lecture Notes in Computer Science*, pages 257–270. Springer, 1995. Extended version appeared as [87].
78. J.R. Kennaway, J.W. Klop, M.R. Sleep, and F.J. de Vries. Transfinite reductions in orthogonal term rewriting systems. *Information and Computation*, 119(1):18–38, 1995. Extended version of [59].

79. Y. Toyama, J.W. Klop, and H.P. Barendregt. Termination for direct sums of left-linear complete term rewriting systems. *Journal of the ACM*, 42(6):1275–1304, 1995. Extended version of [52].
80. Z.M. Ariola and J.W. Klop. Equational term graph rewriting. *Fundamenta Informaticae*, 26(3/4):207–240, 1996.
81. I. Bethke and J.W. Klop. Collapsing partial combinatory algebras. In G. Dowek, J. Heering, K. Meinke, and B. Möller, editors, *Proc. of the 2nd International Workshop on Higher-Order Algebra, Logic, and Term Rewriting (HOA'95)*, volume 1074 of *Lecture Notes in Computer Science*, pages 57–73. Springer, 1996.
82. I. Bethke, J.W. Klop, and R.C. de Vrijer. Completing partial combinatory algebras with unique head-normal forms. In *Proc. of the 11th Annual Symposium on Logic in Computer Science (LICS'96)*, pages 448–454. IEEE Computer Society Press, 1996.
83. J.R. Kennaway, J.W. Klop, M.R. Sleep, and F.J. de Vries. Comparing curried and uncurried rewriting. *Journal of Symbolic Computation*, 21(1):15–39, 1996.
84. J.W. Klop. Term graph rewriting. In G. Dowek, J. Heering, K. Meinke, and B. Möller, editors, *Proc. of the 2nd International Workshop on Higher-Order Algebra, Logic, and Term Rewriting (HOA'95)*, volume 1074 of *Lecture Notes in Computer Science*, pages 1–16. Springer, 1996.
85. Z.M. Ariola and J.W. Klop. Lambda calculus with explicit recursion. *Information and Computation*, 139(2):154–233, 1997.
86. Z.M. Ariola, J.W. Klop, and D. Plump. Confluent rewriting of bisimilar term graphs. *Electronic Notes in Theoretical Computer Science*, 7, 1997.
87. J.R. Kennaway, J.W. Klop, M.R. Sleep, and F.J. de Vries. Infinitary lambda calculus. *Theoretical Computer Science*, 175(1):93–125, 1997. Extended version of [77].
88. M. Bezem, J.W. Klop, and V. van Oostrom. Diagram techniques for confluence. *Information and Computation*, 141(2):172–204, 1998.
89. I. Bethke, J.W. Klop, and R.C. de Vrijer. Extending partial combinatory algebras. *Mathematical Structures in Computer Science*, 9(4):483–505, 1999.
90. Z.M. Ariola, J.W. Klop, and D. Plump. Bisimilarity in term graph rewriting. *Information and Computation*, 156:2–24, 2000.
91. I. Bethke, J.W. Klop, and R.C. de Vrijer. Descendants and origins in term rewriting. *Information and Computation*, 159:59–124, 2000.
92. J.W. Klop, V. van Oostrom, and R.C. de Vrijer. A geometric proof of confluence by decreasing diagrams. *Journal of Logic and Computation*, 10(3):437–460, 2000.
93. Terese. *Term Rewriting Systems*, volume 55 of *Cambridge Tracts in Theoretical Computer Science*. Cambridge University Press, 2003.
94. J. Ketema, J.W. Klop, and V. van Oostrom. Vicious circles in orthogonal term rewriting systems. In S. Antoy and Y. Toyama, editors, *Proc. of the 4th International Workshop on Reduction Strategies in Rewriting and Programming (WRS'04)*, volume 124(2) of *Electronic Notes in Theoretical Computer Science*, pages 65–77. Elsevier Science, 2005.

Table of Contents

The Spectra of Words

Robin Milner

Cambridge University, Cambridge, UK

Abstract. The k-spectrum of a word is the multiset of its non-contiguous subwords of length k. For given k, how small can n be for a pair of different words of length n to exist, with equal k- spectra? From the Thue-Morse word we find that n is at most 2^k. The construction of this paper decreases this upper bound to θ^k, where $\theta \doteq 1.6$ is the golden ratio; the construction was found, though not published, over thirty years ago. Recently the bound has been further reduced, but remains considerably greater than the greatest known lower bound.

Jan Willem Klop is renowned for his contributions to process algebra, term rewriting, and graphical models of computation; also for his exquisite diagrammatic presentations. Alongside these interests he finds delight in phenomena involving complex illustrations and counterexamples; for example, a single lambda term that takes several pages. He enjoys long words.

One long – indeed infinite – word is the Thue-Morse word. This word, and its finite prefixes of length 2^k, have a rich variety of properties. This paper is about one property that they almost, but not quite, enjoy. There is a sense in which these finite prefixes are not optimal, but appear to be nearly so.

I discovered this phenomenon in 1972, together with a counterexample to the optimality, but did not to publish it. In the last decade or so, the results have been repeated and indeed improved; in my concluding remarks I give a brief survey of some of this more recent work, enough for interested readers to discover the current state of this special problem area. But this festschrift for Jan Willem gives an opportunity to publish the phenomenon as it first appeared to me, in a form which I hope is readily digestible by many who, like me, are not expert in combinatorics.

We are concerned with words a over an alphabet A; for simplicity, we henceforth assume $A = \{0, 1\}$. A word s is a *subword* of a if the members of s occur in a in the correct sequence, not necessarily contiguously. A subword may occur often; for example, 01 occurs once in 010 but five times in 01011. (It is worth noting that there is not universal agreement of terminology; some authors use 'subword' to mean a contiguous occurrence, and in that case 01 appears only twice in 01011.) Denote by $\#(s, a)$ the number of non-contiguous occurrences of s in (i.e. as a subword of) a. Remarkably, the quantities $\#(s, a)$ share many properties with the binomial coefficients; indeed, the latter correspond to the case in which the alphabet A is a singleton. An elegant theory of these quantities $\#(s, a)$ is presented by Sakarovitch and Simon [8]. What little we need of that theory is included here, to make the paper self-contained.

A. Middeldorp et al. (Eds.): Processes... (Klop Festschrift), LNCS 3838, pp. 1–5, 2005.

For any natural number k, the *k-spectrum* of a word a is a function giving, for each word s of length k, the value of $\#(s,a)$. For example, the 2-spectrum of 0110 is

s	$\#(s, 0110)$
00	1
01	2
10	2
11	1

We ask the question: if we know the k-spectrum of a, does this fix a? Certainly not, if a has length at least 2^k. The finite prefixes of the Thue-Morse word provide a counterexample. The prefix a_k of length 2^k is given by

$$a_0, b_0 \stackrel{\text{def}}{=} 0, 1$$
$$a_{k+1}, b_{k+1} \stackrel{\text{def}}{=} a_k\, b_k, b_k\, a_k \ .$$

For example $a_3, b_3 = 01101001, 10010110$. An easy inductive proof shows that a_k and b_k have the same k-spectrum.

One way of presenting this proof depends on a Lemma that will come in useful later. It involves the notion of a *rewriting rule* $u \to v$, which may be used to replace a (contiguous) occurrence of u by v in a larger string. When the rule is understood we shall write $a \to b$, or sometimes $b \leftarrow a$, to mean that a single use of the rule can transform a into b.

Lemma. *Let u, v have the same k-spectrum. Suppose a can be rewritten into b by applying the rewrite rule $u \to v$ once and then applying $v \to u$ once, i.e. $a = cuc'$, $cvc' = dvd'$ and $dud' = b$. Then a, b have the same $k{+}1$-spectrum.*

To see that this provides an inductive proof that the Thue-Morse pair a_k, b_k have the same k-spectrum, assume the property for k, and consider a_{k+1}, b_{k+1}. In the Lemma take a, b to be this pair, and u, v to be a_k, b_k; then the result follows by taking $c = d' = \varepsilon$ (the empty word) and $c' = d = b_k$.

Let us look more closely at this when $k = 3$. The rewriting rule is $0110 \to 1001$, so if we underline the occurrences to be rewritten we have

$$a_3 = \underline{0110}1001 \to 1001\underline{1001} \to 10010110 = b_3 \ .$$

Note that the two rewrites are non-overlapping. Now, consider a sequence of pairs a_k, b_k of unequal strings of equal length ℓ_k, where ℓ_k grow *more slowly* than 2^k; a similar inductive proof that a_k and b_k have the same k-spectrum would require that a_{k+1} can be transformed into b_{k+1} by an application of $a_k \to b_k$ followed by an application of $b_k \to a_k$ in which the second rewrite *overlaps the first*! It seemed unlikely that such pairs a_k, b_k could be defined. And it is not easy to think of constructing a sequence of unequal pairs with equal k-spectra other than inductively, applying something like the Lemma to yield the proof.

Therefore it was reasonable to hope to prove that each Thue-Morse pair a_k, b_k is *optimal*, in the sense that no shorter unequal pair has equal k-spectra.

But this claim is false; we now give a counter-example. Considering the recurrence relation for the Thue-Morse pairs, we naturally attempt to define pairs based instead on a Fibonacci sequence, whose length will grow asymptotically as θ^k, where $\theta = (1+\sqrt{5})/2 \simeq 1.6$ is the golden ratio. The details require care. By analogy with the Thue-Morse pairs, given two initial pairs a_0, c_0 and a_1, c_1 we define for $k \geq 0$:

$$a_{k+2},\ c_{k+2} \stackrel{\mathsf{def}}{=} \begin{cases} c_k\, a_{k+1},\ a_{k+1}\, c_k & (k \text{ even}) \\ a_{k+1}\, c_k,\ c_k\, a_{k+1} & (k \text{ odd}). \end{cases}$$

Then, for all $k \geq 0$ define

$$b_{k+2} \stackrel{\mathsf{def}}{=} \begin{cases} a_{k+1}\, a_k & (k \text{ even}) \\ a_k\, a_{k+1} & (k \text{ odd}). \end{cases}$$

Now let us choose $a_0, c_0 = 010, 100$ and $a_1, c_1 = 01, 10$ (we discuss this choice later). Then the following table shows the first few pairs a_k, b_k:

k	a_k	b_k	c_k	ℓ_k
0	010	–	100	3
1	01	–	10	2
2	10001	01010	01100	5
3	1000110	0110001	1010001	7
4	011001000110	100011010001	100011001100	12
5	0110010001101010001	1000110011001000110	...	19

Theorem 1. *Choose $a_0, c_0 = 010, 100$ and $a_1, c_1 = 01, 10$ and define a_k, b_k, c_k ($k \geq 2$) as above. Then for all $k \geq 2$ the strings a_k, b_k are unequal, with equal lengths and equal k-spectra. Also the lengths ℓ_k of the members of the pairs form a Fibonacci sequence, and $\ell_k < 2\theta^k$ for all $k \geq 2$.*

Proof. It is obvious that the strings a_k, b_k have equal length. To show that they differ, one can prove by induction that (i) their first letters always differ, and (ii) the first letters of b_k and c_k differ iff k is odd.

Now, a simple calculation yields that for $k \geq 0$

$$b_{k+2} = \begin{cases} a_k\, c_{k+1} & (k \text{ even}) \\ c_{k+1}\, a_k & (k \text{ odd}). \end{cases}$$

Using this, we give an inductive proof that, for $k \geq 2$, the strings a_k, b_k have equal k-spectra. For $k = 2$, i.e. for the pair 10001 and 01010, this is easily checked. For the inductive step, we assume the property for k and prove it for $k+1$. Let us use $\rightarrow$ for a rewrite by $a_k \rightarrow b_k$, and $\leftarrow$ for the inverse rewrite. Then for odd $k \geq 2$ we have

$$\begin{aligned} a_{k+1} &= c_{k-1}\, a_k && \text{by the recurrence} \\ &\rightarrow c_{k-1}\, b_k && \text{by the rewrite} \\ &= c_{k-1}\, a_{k-2}\, a_{k-1} && \text{by definition} \\ &= b_k\, a_{k-1} && \text{by the above expression} \end{aligned}$$

$$\begin{aligned} &\leftarrow a_k\, a_{k-1} \text{ by the inverse rewrite} \\ &= b_{k+1} \qquad \text{by definition} \end{aligned}$$

and similarly for even k. But by assumption the strings a_k, b_k have equal k-spectra; thus, by the Lemma, a_{k+1}, b_{k+1} have equal $k+1$-spectra. This completes the inductive proof.

Finally, the lengths ℓ_k form the (non-standard) Fibonacci sequence 3, 2, 5, 7, 12, 19, ..., and using $\ell_{k+2} = \ell_{k+1} + \ell_k$ and $\theta^2 = \theta + 1$ we obtain $\ell_k < 2\theta^k$ for all $k \geq 2$ by a simple induction. □

The first two pairs a_k, b_k $(k = 2, 3)$ in our sequence are an oddity. They appear to be the shortest that initiate the recurrence correctly so that the Lemma can be applied. Readers may like to see if they can throw any light on this matter; at present it appears that we have stumbled by accident on a sequence that works!

Now, define a pair a, b to be *k-spectral* if a and b are unequal, but of equal length and with equal k-spectra; call the pair *k-optimal* if it is k-spectral and there is no shorter k-spectral pair.

In our sequence the pair a_k, b_k is not always k-optimal. For example, the pair $10001, 01010$ is not 2-optimal; the Thue-Morse pair $1001, 0110$ is 2-optimal. Our pairs for $k = 3, 4$ are optimal, and already shorter than the corresponding Thue-Morse pairs – which are therefore non-optimal. Our pair for $k = 5$, with $\ell_5 = 19$, is not 5-optimal; there is a 5-spectral pair of length 16:

$$0110000011100001\;,\;1000011100000110\;.$$

This suggests that later pairs in the sequence will also be not k-optimal.

To summarise, let us define the function

$$opt(k) \stackrel{\text{def}}{=} \text{the length of a } k\text{-optimal pair.}$$

Theorem 1 implies an exponential upper bound to $opt(k)$. When I discussed this problem with David Klarner [3][1] he quickly provided a lower bound; he proved that if a and b have equal length less than $2k$ and equal k-spectra then $a = b$. We can summarise these results by the following:

Theorem 2. $2k \leq opt(k) < 2\theta^k$.

This is a wide gap; to narrow it significantly appears non-trivial. However, recent work has made some progress in this direction.

Recent work. As indicated above, these results have been confirmed and improved in recent years. Apparently the problem of reconstructing a word from subwords was first introduced in the literature by Kalashnik (1973) [4]. The fact that the 2^k prefix of the Thue-Morse word and its complement are k-spectral goes

[1] David Klarner was a specialist in combinatorics. He and I did not meet again after 1972. In May 2000, having found his website, I tried to contact him to see if he had done further work on subwords. Sadly he had died two months before, in March 2000, so I had no answer.

back at least to to Manvel *et al* (1991) [7]. The standard reference for a uniform construction of shorter k-spectral pairs, with length determined by a Fibonacci sequence, is by Choffrut and Karhumäki (1997) [1]; see also Manuch (1999) [6]. At about the same time Krasikov and Roditty (1997) [5] increased the linear lower bound of $2k$ to a quadratic lower bound. Finally, Dudík and Schulman (2003) [2] have found a smaller upper bound of the form $\exp(\Omega(\log^2 k))$. Thus the gap between the known lower and upper bounds on $opt(k)$ is decreasing, but remains wide.

I would like to thank the referees for helping me to improve this paper, and especially for help with the bibliography.

References

1. C. Choffrut and J. Karhumäki (1997), Combinatorics of words. In: G. Rozenberg and A. Salomaa (eds), *Handbook of Formal Languages, Vol 1*, Springer, Berlin, pp 329–438.
2. M. Dudík and L.J. Schulman (2002), Reconstruction from subsequences. In *J. Combin. Theory A* 103, pp 337–348.
3. David Klarner (1972). Private communication.
4. L.O. Kalashnik (1973), The reconstruction of a word from fragments. In *Numerical Mathematics and Computer Technology*, Akad. Nauk. Ukrain. SSR Inst. Mat., Preprint IV: pp56–57.
5. L. Krasikov and Y. Roditty (1997), On the reconstruction problem for sequences. In *J. Combin. Theory A* 77, pp 344–348.
6. J. Manŭch (1999), Characterisation of a word by its subwords. In *Preproceedings of DLT99*, Aachener-Informatik-Berichte 99-5, pp 357-367.
7. B. Manvel, A. Meyerowitz, A. Schwenk, K. Smith and P. Stockmeyer (1991), Reconstruction of sequences. In *Discrete Math* 94, pp 209–219.
8. J. Sakarovitch and I. Simon (1982). Subwords. In *Combinatorics on Words, Encyclopaedia of Mathematics and its Applications*, Addison Wesley.

On the Undecidability of Coherent Logic

Marc Bezem

Department of Computer Science, University of Bergen,
P.O. Box 7800, N-5020 Bergen, Norway
bezem@ii.uib.no

Abstract. Through a reduction of the halting problem for register machines we prove that it is undecidable whether or not a coherent formula is a logical consequence of a coherent theory. We include a simple completeness proof for coherent logic. Although not published in the present form, these results seem to be folklore. Therefore we do not claim originality. Given the undecidability of the halting problem for register machines the presentation is self-contained.

1 Introduction

As far as we know, Skolem [12] was the first who used coherent logic (*avant la lettre*) to solve a decision problem in lattice theory and to prove the independence of Desargues' Axiom from the other axioms of projective plane geometry. Modern coherent logic, also called finitary geometric logic or even simply geometric logic, arose in algebraic geometry, see for example [5–Sect. 16.4], and is actually a fragment of higher-order logic. In this note we define coherent logic (abbreviated by CL) as the fragment of first-order logic (FOL) consisting of implicitly universally quantified implications of the following form:

$$A_1 \wedge \cdots \wedge A_n \to E_1 \vee \cdots \vee E_m$$

Here the A_i are first-order atoms. In contrast to resolution logic [9], where the E_j must also be atoms, they may here be existentially quantified conjunctions of atoms. Thus the general format of a *coherent formula* reads:

$$A_1 \wedge \cdots \wedge A_n \to \exists \boldsymbol{x}_1.C_1 \vee \cdots \vee \exists \boldsymbol{x}_m.C_m \qquad (1)$$

where the C_j are conjunctions of atoms. The special cases $n = 0$, $m = 0$ and no existential quantification, in all possible combinations, are understood to be included. (If the premiss is empty we leave out the $\to$ as well, an empty conclusion is denoted by $\bot$, *falsum*.) A *coherent theory* is a set of coherent formulas. Closed atoms will also be called *facts*.

The fact that first-order logic is semidecidable certainly constitutes an upper bound for the coherent fragment as well. Resolution logic with only constants is decidable, since quantification over finite Herbrand domains can be reduced to propositional logic. Horn clause logic [6] is the format (1) with $m \leq 1$ and E_1 atomic. In the presence of one constant and one unary function symbol, Horn

A. Middeldorp et al. (Eds.): Processes... (Klop Festschrift), LNCS 3838, pp. 6–13, 2005.

clause logic is undecidable [11]. This provides the clue for the undecidability of CL without function symbols, since a unary function $f(x)$ can be replaced by a binary predicate $F(x, y)$ plus coherent axioms $\exists y.F(x, y)$ and $F(x, y) \wedge F(x, z) \rightarrow E(y, z)$ as well as congruence axioms for E, which are also coherent. Hence the undecidability result in itself is not surprising, but we show that one can do without all axioms in which E occurs. One can even do away with the constant by using an extra unary predicate. Undecidability of CL can be obtained in many other ways, for example, as an immediate corollary of the linear translation of FOL to CL given in [1]. The current exposition offers an insightful correspondence between computations and proofs.

There are several reasons why coherent/geometric logic is interesting. See [3] for the relevance to computer science. Reasoning in CL is constructive and can be used for, e.g., the constructivization of classical abstract algebra, see [4]. A substantial number of reasoning problems (e.g., in confluence theory, lattice theory and projective geometry) can be formulated *directly* in CL without any clausification or skolemization. This gives some additional benefits in terms of guiding an automated theorem prover and using the proof objects in other logical frameworks. The automation of CL has been studied in [1], inspired by the system SATCHMO [7] for resolution logic.

2 Proof System

CL has a natural proof system which is based on forward ground reasoning with case distinction. Existential quantifiers are eliminated by introducing witnesses. A *witness* is a new constant witnessing the truth of an existential statement. Witnesses play a similar role as eigenvariables in systems of natural deduction and should be chosen completely fresh, not introduced earlier in the proof, not occurring in the theory, nor in the formula to be proven.

In order to elaborate the proof system a bit more, let T be a coherent theory. Assume we have a set I of witnesses and initial constants. The latter constants are the constants occurring in T and in the goal G, the formula to be proven. A *goal* is a closed formula of the same form as a conclusion in a coherent formula. We first explain how to prove a goal and then generalize this to arbitrary coherent formulas. Let X be a set of facts in which only constants from I occur. Together I and X form a so-called (reasoning) *state*. A conjunction of facts is true in this state if all these facts occur in X. A closed formula of the form $\exists \boldsymbol{x}.C$, with C a conjunction of atoms, is true in this state if there exist witnesses $\boldsymbol{w} \in I$ such that $C[\boldsymbol{x}{:=}\boldsymbol{w}]$ is true in the state. A goal is true in a state if at least one of its disjuncts is true in that state. A reasoning *step* in the state (I, X) consists in picking a closed I-instance $C \rightarrow D$ of an axiom from T that is invalid in the state. This means that the premiss C is true in the state, but the conclusion D is not.

As an example, consider a state with $I = \{0\}$, $X = \{Nat(0)\}$ and an axiom

$$Nat(x) \rightarrow \exists y.(Nat(y) \wedge S(x, y)) \tag{2}$$

Assume we would like to prove $G = \exists xy.(S(0, x) \wedge S(x, y))$. The instance of (2) with $x{:=}0$ is invalid in the current state, since the premiss is true but the conclusion is not. The reasoning step now so to say remedies this failure by making the conclusion of the instance true by adding a witness to I, suggestively denoted as 1, and adding the facts $Nat(1)$ and $S(0,1)$ to X. The reasoning process then continues in the state with $I' = \{0,1\}$ and $X' = \{Nat(0), Nat(1), S(0,1)\}$. In the new state we pick the instance with $x{:=}1$ of the same axiom (2) to arrive at a state with constants $I'' = \{0,1,2\}$ and facts $X'' = \{Nat(0), Nat(1), S(0,1), Nat(2), S(1,2)\}$. Now we can stop since the goal G is true in this state: just take $x{:=}1$ and $y{:=}2$ and observe that $S(0,1), S(1,2) \in X''$. Note that the suggestive names 1 and 2 for the witnesses are inessential.

In the case of a disjunctive conclusion the reasoning process branches and the goal has to be proved in all the branches corresponding to the disjuncts in the conclusion. In the special case of an empty disjunction there are no branches and we are done. This special case corresponds to the Ex Falso rule. Disjunctive conclusions give rise to a *tree* of states in which branches are *closed* in leaf states in which the goal is true or in which an empty disjunction can be derived. If all branches are closed we have a proof. If in some branch the procedure breaks down with all axioms true but the goal still false, then the state in question constitutes a counter model. It is also possible that the reasoning process goes on forever, something which reflects the undecidability to be proved in Section 4.

The above procedure actually allows us to prove the coherent formula

$$Nat(0) \to \exists yz.(S(0,y) \wedge S(y,z))$$

from axiom (2) by assuming the premiss (the state (I,X) above) and then proving the conclusion (the goal G). We can further generalize to a proof of

$$Nat(x) \to \exists yz.(S(x,y) \wedge S(y,z)) \tag{3}$$

from axiom (2) in the empty state, since the constant 0 is fresh with respect to (2) and (3) and the empty state. In this way, by taking a fresh closed instance, assuming the premiss and then proving the conclusion as a goal, the procedure easily generalizes to proving arbitrary coherent formulas. The resulting proof system is sound and complete with respect to Tarskian truth. (We include models with empty domains.) In the next section we sketch a completeness proof.

3 Completeness

Without loss of generality we can restrict completeness to finding a proof of a true fact F in a finite coherent theory T. The idea of the proof is to use the procedure of the previous section to build a (possibly infinite) tree of states by applying in a systematic way all axioms from T with F as goal. If this doesn't yield a proof of F, then a model of T in which F is not true can be read off the tree.

Soundness and completeness can be stated as follows. For all states (I,X), a fact F can be proved in T from X if and only if F is true in all models of X,T (we assume the signature of T to be extended by the witnesses in I).

The only-if part is soundness, and this is obvious (but relies essentially on the freshness of the witnesses). For the if-part, assume F is true in all models of X, T. Since both the state and the theory are finite, there exist at most finitely many closed I-instances of axioms from T that are invalid in X. List these by $C_1 \to D_1, \ldots, C_n \to D_n$. If there are no such instances, then X can be viewed as a model of T and F is by assumption true in X, so $F \in X$ and we have a proof of length zero. If $F \notin X$, take a reasoning step in state (I, X) using $C_1 \to D_1$. This leads to zero or more new states, depending on the length of the disjunction D_1 (if this length is zero we are done by Ex Falso). A crucial observation here is that $C_1 \to D_1$ is valid in any of the new states. In each of these not containing F, if $C_2 \to D_2$ is still invalid, take a reasoning step using $C_2 \to D_2$. In this way we work through the whole list, observing that every $C_i \to D_i$ $(1 \leq i \leq n)$ is valid in any of the resulting states. After having processed the whole list, not all branches have to be closed by Ex Falso or by containing F. Such open branches may end in states in which new instances of axioms from T have become invalid. In each such state we have to list again all invalid closed instances of axioms from T and work through this list as above. Will this yield a proof of F? A finite tree with all branches closed is a proof of F. What if the procedure goes on forever and the tree becomes infinite? We shall show that this conflicts with F being true in all models of X, T. In order to see this, assume the tree is infinite. Observe that the tree is finitely branching. By König's Lemma there exists an infinite branch, say β. Along β we find a strictly increasing sequence of states $(I, X) = \beta_0, \beta_1, \ldots$. We can collect all the witnesses and all the facts along β and view them as a Herbrand universe U and a Herbrand model M, respectively. (A non-empty domain is guaranteed if there is at least one initial constant.) Now we obtain a contradiction by the following three observations. First, F is not in M since β is infinite. Second, X is included in M since X is at the root. Third, M is a model of T. In order to see the latter we remark first that all constants from the signature of T are understood to be included in U. Now take an arbitrary closed U-instance $C \to D$ of an axiom from T such that C is true in M. For some $k \geq 0$, all constants occurring in $C \to D$ as well as all facts in C occur in β_k. In this state β_k some finite list of closed instances of axioms from T is being worked through. If D has not become true in β_l $(l \geq k)$ when this list is finished, then $C \to D$ will be on the next list, and hence D will be true in some later state β_m $(m > l)$. In both cases D is true in M. Since an infinite tree conflicts with F being true in all models of X, T, the tree must be finite and hence a proof of F. This completes the completeness proof.

Completeness for arbitrary coherent formulas follows easily from completeness for facts using the generalized proof procedure given at the end of Section 2.

4 Undecidability

Crucial for the undecidability of CL without function symbols is the existential quantification, which allowed us to formulate the axiom (2) in Section 2. Axiom (2) generates representatives of the natural numbers, with the binary predicate

S expressing the successor *relation*. Of course one needs a starting point for applying (2). In the example this was the fact $Nat(0)$.

Given a representation of the natural numbers, there are a number of Turing complete models of computation available for proving undecidability. One of the most convenient ones is the so-called register machine [10], also called counter machine. This machine model, and in particular the unsolvability of its halting problem already for two registers, goes back to [8].

A register machine is a device with *registers* $x_1, \ldots, x_m$, each capable of storing an arbitrarily large natural number, together with a program. A register machine program is a finite enumeration of *instructions* from the following instruction set: $inc(x_i), dec(x_i), jpz(x_i, l, l')$. These instructions lead to the following respective actions: *inc*rement register x_i, *dec*rement register x_i, jump to instruction l if x_i is zero and to l' otherwise. Decrementing a register which has value 0 is not allowed and can be prevented by using conditional jumps preceeding any decrement instruction.

The execution model for register machines uses one additional register, the so-called *program counter*, which addresses the current instruction. Execution of the program starts at the first instruction. The program counter is incremented after each instruction $inc(x_i), dec(x_i)$, its value is changed to either l or l' in case of a conditional jump. Execution terminates when the program counter gets a value not corresponding to an instruction of the program. The registers used by a register machine are by default $x_1, \ldots, x_m$ with m the highest index of a register occurring in the program.

As an example consider the following program ADD:

0 $jpz(x_2, 4, 1)$
1 $dec(x_2)$
2 $inc(x_1)$
3 $jpz(x_1, 0, 0)$

This program obviously adds the contents of x_2 to x_1 and terminates by jumping to 4, beyond the last instruction. Instruction 3 exhibits an unconditional jump. (The program would also work correctly with instructions 3 $jpz(x_{13}, 0, 0)$ or 3 $jpz(x_2, 13, 1)$ instead of 3 $jpz(x_1, 0, 0)$.)

It is undecidable whether or not a given register machine will terminate when started with all registers initialised with 0. We will reduce the latter problem to provability in CL, thereby showing that provability in CL is undecidable.

In order to illustrate the reduction we translate the example program ADD into the following coherent theory.

$$P_0(x_1, 0) \rightarrow P_{halt}(x_1, 0) \text{ and } P_0(x_1, x_2) \wedge S(y, x_2) \rightarrow P_1(x_1, x_2)$$
$$P_1(x_1, x_2) \wedge S(y, x_2) \rightarrow P_2(x_1, y)$$
$$P_2(x_1, x_2) \wedge S(x_1, y) \rightarrow P_3(y, x_2)$$
$$P_3(x_1, x_2) \rightarrow P_0(x_1, x_2)$$

This translation can be applied to any register machine M operating on registers $x_1, \ldots, x_m$, where the registers are the arguments of atoms $P_l(x_1, \ldots, x_m)$

and the instructions of M's program lead to coherent formulas as in the above example. The conditional jump is the only instruction leading to two coherent formulas, but their closed instances do not overlap, since there is no y such that $S(y,0)$. An unconditional jump can do with only one formula. Any jump out the program is translated into a coherent formula $\cdots \to P_{halt}(\ldots)$.

Starting with registers initialised with 0 is represented by adding $P_0(0,\ldots,0)$ to the initial state. Finally, adding axiom (2) generating the natural numbers completes the coherent theory T_M corresponding to the register machine M. The reduction of the halting problem for register machines to provability in coherent logic can now be made precise in the following theorem.

Theorem 1. *For every register machine M operating on registers $x_1,\ldots,x_m$, when started with all registers initially 0, we have that the execution of M terminates if and only if $\exists x_1 \ldots x_m.P_{halt}(x_1,\ldots,x_m)$ is provable in T_M from an initial state with $I_0 = \{0\}$ and $X_0 = \{Nat(0), P_0(0,\ldots,0)\}$.*

Proof. For convenience, we identify reasoning states with sets of facts, the set of constants in any state consisting of all the constants occurring in the facts of that state. Again we use suggestive names $1,\ldots,n{+}1$ for the constants introduced by axiom (2). Assume M as above is terminating. For every step in the execution of M, take the following two reasoning steps:

1. $X_n \rightsquigarrow X_n \cup \{Nat(n{+}1), S(n,n{+}1)\} = Y_{n+1}$ by axiom (2) instantiated with $x{:=}n$.
2. $Y_{n+1} \rightsquigarrow Y_{n+1} \cup \{P_l(n_1,\ldots,n_m)\} = X_{n+1}$ by the axiom corresponding to the instruction executed in step $n{+}1$ of the computation, instantiated with the register values just before the step. Here we assume that step $n{+}1$ of the computation leads to a state in which the program counter is l and the contents of the registers are $n_1,\ldots,n_m$, respectively.

Note that there are always enough natural numbers to instantiate the right instance of the formula corresponding to each instruction, and that its premiss is true. Also, its conclusion must be false, since $Y_{n+1} = X_{n+1}$ would mean that $P_l(n_1,\ldots,n_m)$ has been inferred already, so that the program is actually looping, contradicting termination. There is no branching, which reflects the fact that the computation is deterministic. When the execution of M terminates, that is, jumps out the program, we have $l = halt$ and we have proved $\exists x_1 \ldots x_m.P_{halt}(x_1,\ldots,x_m)$.

For the converse, assume $\exists x_1 \ldots x_m.P_{halt}(x_1,\ldots,x_m)$ is provable in T_M from the initial state. In any state of this proof *exactly* one instance of axiom (2) is invalid. Moreover *at most* one instance of the axioms corresponding to the instructions of M is invalid: in some cases axiom (2) has to generate a new natural number before an increment can take place. It is quite possible that too many natural numbers are generated, but this does no harm. The proof necessarily follows the execution of M, interleaved with (possibly too many) applications of axiom (2). The terminating execution is easily read off.

The above argument still uses the constant 0. We can eliminate the use of 0 by introducing a unary predicate Z. The initial state is then replaced by the axiom

$$\exists x.Z(x) \wedge Nat(x) \wedge P_0(x, \ldots, x)$$

The constant 0 has also to be eliminated from the axioms corresponding to *jpz* instructions. For example, in the translation of the program ADD, the first axiom should be replaced by $P_0(x_1, x_2) \wedge Z(x_2) \rightarrow P_{halt}(x_1, x_2)$. In this way we obtain a coherent theory T'_M corresponding to the register machine M. Theorem 1 can be proved with T'_M instead of T_M and an empty initial state by almost the same argument as above. In order to shift from 'provable' to 'logical consequence' using only the restricted completeness theorem from Section 3, one should add the axiom $P_{halt}(x_1, \ldots, x_m) \rightarrow halt$ to T'_M. Thus we obtain:

Corollary 1. *The consequence relation of coherent logic without function symbols is undecidable.*

We finish by discussing some closely related results. The key idea is beyond any doubt the Horn clause representation of register machines from [11]. This idea has also been exploited to get undecidablity results in typed lambda calculus [2] and term rewriting systems [14–Section 5.3.2]. The latter two systems allow the use of function symbols. In [13–Ch. First-order logic] the undecidability of the $\forall, \rightarrow, \bot$-fragment of intuitionistic predicate logic is based on the same idea. In this fragment Horn clauses are expressible by iterating $\rightarrow$ and $(\forall y.(S(x, y) \rightarrow \bot)) \rightarrow \bot$ replaces $\exists y.S(x, y)$. The latter weakening of existential quantification is still sufficient to obtain undecidability without the use of function symbols. All these approaches have in common that register machine computations correspond to some kind of normalized proofs.

Acknowledgements

The author is indebted to Dimitri Hendriks and to two anonymous reviewers for constructive criticism on an earlier version of this note.

References

1. M.A. Bezem and T. Coquand, Automating Geometric Logic, *Report 33/04 Research Group on Mathematical Linguistics*, Universitat Rovira i Virgili, Tarragona.
2. M.A. Bezem and J. Springintveld, A simple proof of the undecidability of inhabitation in λP, *Journal of Functional Programming* **6**(5):1–5, 1996.
3. A. Blass, Topoi and computation, *Bulletin of the EATCS* **36**:57–65, 1998.
4. M. Coste, H. Lombardi and M.-F. Roy, Dynamical methods in algebra: effective Nullstellensätze, *Annals of Pure and Applied Logic* **111**(3):203–256, 2001.
5. R. Goldblatt, *Topoi : the categorial analysis of logic*, revised edition, North-Holland, 1984.
6. A. Horn, On sentences which are true of direct unions of algebras, *Journal of Symbolic Logic* **16**(1):14–21, 1951.

7. R. Manthey and F. Bry, SATCHMO: a theorem prover implemented in Prolog. In E. Lusk and R. Overbeek, editors, *Proceedings of the 9-th Conference on Automated Deduction*, Lecture Notes in Computer Science **310**:415–434, Springer-Verlag, 1988.
8. M.L. Minsky, Recursive unsolvability of Post's problem of 'tag' and other topics in theory of Turing machines, *Annals of Mathematics* **74**(3):437–455, 1961.
9. J.A. Robinson, A Machine-Oriented Logic Based on the Resolution Principle, *Journal of the ACM* **12**(1): 23–41, 1965.
10. J.C. Shepherdson and H.E. Sturgis, Computability of recursive functions, *Journal of the ACM*, **10**:217–255, 1963.
11. J.C. Shepherdson, *Undecidability of Horn clause logic and pure Prolog*, unpublished manuscript, 1985.
12. Th. Skolem, *Logisch-kombinatorische Untersuchungen über die Erfüllbarkeit und Beweisbarkeit mathematischen Sätze nebst einem Theoreme über dichte Mengen*, Skrifter I **4**:1–36, Det Norske Videnskaps-Akademi, 1920. Also in: Jens Erik Fenstad, editor, *Selected Works in Logic by Th. Skolem*, pp. 103–136, Universitetsforlaget, Oslo, 1970.
13. M.H. Sørensen and P. Urzyczyn, *Lectures on the Curry-Howard Isomorphism*, to appear.
14. Terese, *Term Rewriting Systems*, CUP, 2003.

Löb's Logic Meets the μ-Calculus

Albert Visser

Department of Philosophy, Utrecht University,
Heidelberglaan 8, 3584 CS Utrecht, The Netherlands
Albert.Visser@phil.uu.nl

This paper is dedicated to Jan Willem Klop on the occasion of his 60th birthday. Jan Willem's way of doing research is a perfect illustration of the saying Vakmanschap is Meesterschap.

Abstract. In this paper, we prove that Löb's Logic is a retract of the modal μ-calculus in a suitable category of interpretations. We show that various salient properties like decidability and uniform interpolation are preserved over retractions. We prove a generalization of the de Jongh-Sambin theorem.

1 Introduction

Fixed points are a central subject of metamathematics. Two flavours are of interest to us here: fixed points of operators that are in some sense guarded and fixed points of monotonic operators. We will study such fixed points in the context of propositional modal logic. The simplicity of propositional modal logic helps us to gain control and overview.

There are two prominent modal logics of fixed points. One is Löb's Logic, aka GL. Löb's Logic is a logic of *guarded fixed points*. It is an important tool in the study of arithmetical self-reference. Löb's Logic has been extensively studied. See the expository papers and books [1], [2], [3], [4], [5].

The other prominent logic is the modal μ-calculus. This logic is a logic for *minimal and maximal fixed points of monotonic operators*. It was designed for applications in Computer Science. This logic was introduced in [6]. See also the survey paper [7].

Both logics are very beautiful and have many desirable properties such as decidability and uniform interpolation.

Johan van Benthem, in his paper [8], showed that Löb's Logic can be faithfully interpreted in the μ-calculus. Moreover, he proved that Löb's Logic has definable fixed points for operators defined by formulas in which the designated variable p occurs only positively. Thus, the μ-calculus can be interpreted in Löb's Logic. Our paper is a commentary on, and an extension of van Benthem's paper. We describe more fully the relationship between both logics: Löb's Logic is a retract of the μ-calculus in a suitable category of interpretations. From this, it follows that properties like decidability and uniform interpolation can be transferred from the μ-calculus to Löb's Logic[1].

[1] Of course, these facts were known already for Löb's Logic. But at least they do receive markedly different proofs via our results.

A. Middeldorp et al. (Eds.): Processes... (Klop Festschrift), LNCS 3838, pp. 14–25, 2005.

Along a different line, van Benthem's arguments are semantical in nature. It is always satisfactory to see semantical arguments replaced by syntactical ones. In this paper we work entirely with syntactical arguments, which are often very simple.

In Section 4, we do a bit more than needed for the rest of the paper. We prove the appropriate generalization of both the de Jongh-Sambin fixed point theorem and van Benthem's theorem that GL has definable minimal fixed points.

1.1 Löb's Logic

Löb's Logic is the logic K4 plus Löb's principle.

LP $\quad \vdash \Box(\Box\phi \to \phi) \to \Box\phi.$

This logic is the logic of upwards well-founded transitive frames.

An occurrence of a variable p in ϕ is *modalized* or *boxed* iff it is in the scope of a necessity operator. Consider a formula $\phi p\boldsymbol{q}$ in which all occurrences of p are boxed. We assume all variables of ϕ are among $p, \boldsymbol{q}$. The de Jongh-Sambin fixed point theorem tells us that there is a formula $\psi\boldsymbol{q}$, with only variables among $\boldsymbol{q}$, such that $\mathsf{GL} \vdash \psi\boldsymbol{q} \leftrightarrow \phi(\psi\boldsymbol{q})\boldsymbol{q}$. Moreover, by the Bernardi-de Jongh-Sambin uniqueness theorem this ψ is unique modulo provable equivalence.

For more information see [1], [2], [3], [4], [5].

1.2 The μ-Calculus

The modal μ-calculus was introduced in [6]. See also the survey paper [7].

For our purposes, the language of the μ-calculus will be the uni-modal language extended with the variable binding operator μp[2]. The formation of $\mu p \cdot \phi$ is allowed precisely if all occurrences of p in ϕ are positive. The μ-calculus is axiomatized by the axioms and rules of K plus the following principle and rule, for ϕ in which all occurrences of p are positive.

min1 $\quad \vdash \mu p \cdot \phi p \leftrightarrow \phi(\mu p \cdot \phi p).$
min2 $\quad \vdash \phi\alpha \to \alpha \;\Rightarrow\; \vdash \mu p \cdot \phi p \to \alpha.$

Frames for the μ-calculus are the usual frames for uni-modal logic. The semantics of $\mu p \cdot \phi p$ is as follows. It gives us the minimal fixed point of the operator naturally associated with the formula ϕp and the designated variable p.

We can define a maximal fixed point operator as follows:

$$\nu p \cdot \phi p := \neg\, \mu p \cdot \neg\, \phi \neg\, p.$$

We can easily verify that ν satisfies the following principle and rule.

max1 $\quad \vdash \nu p \cdot \phi p \leftrightarrow \phi(\nu p \cdot \phi p).$
max2 $\quad \vdash \alpha \to \phi\alpha \;\Rightarrow\; \vdash \alpha \to \nu p \cdot \phi p.$

We will write '$\mu \vdash \phi$' for: ϕ is derivable in the μ-calculus.

[2] Usually, the μ-calculus is formulated for a multi-modal language. However, for our present purposes, the extra modalities would do no work.

2 Interpretations

We need a modest framework of interpretations to implement our comparison of GL with the μ-calculus. Consider modal propositional logics U and V. The languages of U and V are allowed to be different. Specifically, we allow variable binding operators like the μ-operator. A b-interpretation[3] K of U in V is given as a triple $\langle U, \tau, V \rangle$, where τ is a translation-mapping from the formulas of U to the formulas of V. We demand the following of K.

1. The mapping τ commutes with propositional variables and the connectives of propositional logic.
2. The free variables of $\tau\phi$ are among the free variables of ϕ.
3. We have: $U \vdash \phi \Rightarrow V \vdash \tau\phi$.

Par abus de langage, we will also call the translation-mapping: 'K'. We write $K : U \rightarrow_{\mathsf{b}} V$ for: K is a b-interpretation of U in V. We count two b-interpretations $K, M : U \rightarrow_{\mathsf{b}} V$ as *equal* iff $V \vdash K\phi \leftrightarrow M\phi$ for all ϕ. It is easy to see that b-interpretations modulo equality give us a category.

For most of our present purposes, it is sufficient to consider b-interpretations. However, we may wish to consider interpretations that commute with substitutions in an appropriate sense. Such interpretations are intuitively more satisfying. Moreover, we have two results, Theorem 2 and Theorem 3, that essentially use interpretations that commute with substitutions.

A substitution is a mapping from a finite set of propositional variables to formulas of the modal language under consideration. We treat substitutions as the identity mapping outside their domain. Consider a substitution σ for the language of U. We write $\overline{\sigma}$ for the canonical extension of σ to the full language. We employ an implicit mechanism of α-conversion to avoid that variables get bound in subsitution (if our language contains variable-binding operators). We say that K is an *f-interpretation*[4] iff it is a b-interpretation and, for all substitutions σ, we have:

$$V \vdash (K \circ \overline{\sigma})(\phi) \leftrightarrow ((\overline{K \circ \sigma}) \circ K)(\phi).$$

It is easy to see that f-interpretations form a sub-category of the category of b-interpretations. We employ similar conventions for f-interpretations as we do for b-interpretations[5].

[3] Here 'b' stands for *boole*. Our b-interpretations are in fact Boolean morphisms with the extra property of reverse preservation of propositional variables.

[4] Here 'f' stands for *full*.

[5] There is a temptation to either call b-interpretations or f-interpretations simply 'interpretations'. However, I have a doubt. On the one hand, both kinds may be too restrictive: we also would want to allow interpretations that do not preserve propositional variables —that is certainly suggested by what is usually called 'interpretations' in the study of predicate logic. On the other hand, b-interpretations only have a thin claim on being interpretations at all. There is just too little uniformity in the way they treat the modal operators. They are rather Boolean morphisms that reversely preserve propositional variables. Probably, all f-interpretations are interpretations, but even here I feel some hesitation. Did we collect all reasonable properties? More experimentation is needed!

We need the notion of *uniform interpolation*. Uniform interpolation was introduced and studied independently in [9] (for Intuitionistic Propositional Logic) and in [10] (for Löb's Logic). The subject was developed further in [11], [12]. See [13] for an exposition (and a lot of new material). Uniform interpolation for the modal μ-calculus was proved in [14].

Consider a logic Λ, a formula ϕ and a finite set of propositional variables $\boldsymbol{p}$. A formula ψ is a *(uniform) post-interpolant* of ϕ w.r.t. $\boldsymbol{p}$ in Λ iff ψ contains only propositional variables from the intersection of the set of free variables of ϕ with $\boldsymbol{p}$ and, for all χ with only free variables in $\boldsymbol{p}$, we have:

$$\Lambda \vdash \phi \to \chi \iff \Lambda \vdash \psi \to \chi.$$

It is immediate that post-interpolants are unique modulo provable equivalence. We will write $\exists \boldsymbol{p}\, \phi$ for the post-interpolant of ϕ w.r.t. $\boldsymbol{p}$.

Similarly, a formula ψ is a *(uniform) pre-interpolant* of ϕ w.r.t. $\boldsymbol{p}$ in Λ iff ψ contains only propositional variables from the intersection of the set of free variables of ϕ with $\boldsymbol{p}$ and, for all χ with only free variables in $\boldsymbol{p}$, we have:

$$\Lambda \vdash \chi \to \phi \iff \Lambda \vdash \chi \to \psi.$$

We will write $\forall \boldsymbol{p}\, \phi$ for the pre-interpolant of ϕ w.r.t. $\boldsymbol{p}$.

A logic has uniform interpolation iff, for all ϕ and $\boldsymbol{p}$, we have post-interpolants and pre-interpolants. As long as we are in classical logic, it is easy to see that the existence of post-interpolants implies the existence of pre-interpolants and vice versa.

Consider the following situation (in any category). We have two morphisms $K : U \to V$ and $M : V \to U$. Suppose $M \circ K = \mathsf{id} : U \to U$. In this case we say that K is a *split monomorphism* or *co-retraction* and that M is a *split epimorphism* or *retraction*. U will be a *retract* of V.

Theorem 1. *We have, in the category of b-interpretations:*

1. *Split monomorphisms are faithful, that is:* $U \vdash \phi \iff V \vdash K\phi$. *Hence, retractions preserve decidability, if their corresponding coretractions are computable.*
2. *Retractions preserve interpolation.*
3. *Retractions preserve uniform interpolation.*

Proof. Ad (1). Suppose $V \vdash K\phi$. Then, $U \vdash (M \circ K)(\phi)$, so $U \vdash \phi$.

Ad (2). Suppose that V has interpolation. Suppose further that $U \vdash \phi \to \psi$. Say, the shared variables of ϕ and ψ are $\boldsymbol{p}$. It follows that $V \vdash K\phi \to K\psi$. Since, K preserves variables, we find that the shared variables of $K\phi$ and $K\psi$ are among $\boldsymbol{p}$. So, we can find a ι with variables among $\boldsymbol{p}$, such that $V \vdash K\phi \to \iota$ and $V \vdash \iota \to K\psi$. It follows that $U \vdash (M \circ K)(\phi) \to M\iota$ and $U \vdash M\iota \to (M \circ K)(\psi)$. Hence, $U \vdash \phi \to M\iota$ and $U \vdash M\iota \to \psi$. Moreover, the free variables of $M\iota$ are among $\boldsymbol{p}$. So, $M\iota$ is the desired interpolant.

Ad (3). Suppose V has uniform interpolation. Consider a formula ϕ in the language of U and a set of variables $\boldsymbol{p}$. Let $\exists \boldsymbol{p}\, K\phi$ be the uniform post-interpolant w.r.t. $\boldsymbol{p}$ of $K\phi$ in V. We claim that $M(\exists \boldsymbol{p}\, K\phi)$ is the uniform post-interpolant w.r.t. $\boldsymbol{p}$ of ϕ in U. Since $V \vdash K\phi \rightarrow \exists \boldsymbol{p}\, K\phi$, it follows that $U \vdash (M \circ K)(\phi) \rightarrow M(\exists \boldsymbol{p}\, K\phi)$ and, hence, $U \vdash \phi \rightarrow M(\exists \boldsymbol{p}\, K\phi)$. Using this last result, we find, for any ψ not containing variables from $\boldsymbol{p}$,

$$
\begin{aligned}
U \vdash \phi \rightarrow \psi \ &\Rightarrow \ V \vdash K\phi \rightarrow K\psi && (1)\\
&\Rightarrow \ V \vdash \exists \boldsymbol{p}\, K\phi \rightarrow K\psi && (2)\\
&\Rightarrow \ U \vdash M(\exists \boldsymbol{p}\, K\phi) \rightarrow (M \circ K)(\psi) && (3)\\
&\Rightarrow \ U \vdash M(\exists \boldsymbol{p}\, K\phi) \rightarrow \psi && (4)\\
&\Rightarrow \ U \vdash \phi \rightarrow \psi && (5)
\end{aligned}
$$

So, $M(\exists \boldsymbol{p}\, K\phi)$ is the uniform post-interpolant of ϕ in U w.r.t. $\boldsymbol{p}$.

A similar argument works for uniform pre-interpolation. Alternatively, we can define the pre-interpolant from the post-interpolant as $\neg \exists \boldsymbol{p} \neg$. Note that we proved in passing: $M \circ \exists \boldsymbol{p} \circ K \equiv \exists \boldsymbol{p}$. □

Note that the above result is preserved to the subcategory of f-interpretations on trivial grounds. The next result uses f-interpretations essentially. We say that a rule (ϕ/ψ) is *admissible* for a logic Λ iff, for all substitutions σ, $\Lambda \vdash \overline{\sigma}\phi \Rightarrow \Lambda \vdash \overline{\sigma}\psi$.

Theorem 2. *Suppose U and V are logics. Suppose $K : U \rightarrow_{\mathsf{f}} V$ is a faithful f-morphism. Then, (ϕ/ψ) is admissible in U if $(K\phi/K\psi)$ is admissible in V. It follows that admissible rules are preserved by f-isomorphisms.*

Proof. Suppose $(K\phi/K\psi)$ is admissible in V. We have:

$$
\begin{aligned}
U \vdash \overline{\sigma}\phi \ &\Rightarrow \ V \vdash K\overline{\sigma}\phi && (6)\\
&\Rightarrow \ V \vdash (\overline{K \circ \sigma})K\phi && (7)\\
&\Rightarrow \ V \vdash (\overline{K \circ \sigma})K\psi && (8)\\
&\Rightarrow \ V \vdash K\overline{\sigma}\psi && (9)\\
&\Rightarrow \ U \vdash \overline{\sigma}\psi && (10)
\end{aligned}
$$

□

The following result, which is closely related to the previous one, was suggested by one of the referees. Consider a logic U. A formula ϕ is *projective* with projective substitution σ, if we have: $U \vdash \overline{\sigma}\phi$ and $U \vdash \phi \rightarrow (p \leftrightarrow \sigma p)$. Projective formulas were introduced by Silvio Ghilardi in [15]. They play an important role both in the study of unification in logics and in the study of admissible rules.

Theorem 3. *Suppose U and V are logics and $K : U \rightarrow_{\mathsf{f}} V$. Suppose ϕ is a projective formula in U with projective substitution σ. Then, $K\phi$ is a projective formula in V with projective substitution $K \circ \sigma$.*

Proof. Suppose ϕ is projective in U with projective substitution σ. Since, we have $U \vdash \overline{\sigma}\phi$, it follows that $V \vdash K\overline{\sigma}\phi$, and, hence, $V \vdash (\overline{K \circ \sigma})K\phi$. From $U \vdash \phi \to (p \leftrightarrow \sigma p)$, we may infer $V \vdash K\phi \to (p \leftrightarrow K\sigma p)$. □

Remark 1. We may wish to consider *theories* in the propositional language rather than propositional logics (which are closed under substitution). Let's relax our framework, for the moment, by also allowing theories. We have the following. Suppose $U \to_{\mathsf{f}} V$, where V is a logic and U is a theory. It is easy to see that, if K is faithful, then U will also be a logic.

3 Interpreting Löb's Logic in the μ-Calculus

In this section we define our interpretation gm of GL in the μ-calculus. Our interpretation is not quite the same as van Benthem's in [8]. The reason for the divergence is that van Benthem's interpretation does not quite fit our framework. We define the following special fixed points in the μ-calculus.

- $\mathsf{H} := \mu p \cdot \Box p$[6].
- $\Diamond^\star\phi := \mu p \cdot (\Diamond p \vee \Diamond\phi)$[7].

We can easily verify that $\Box^\star\phi$ is $\nu p \cdot (\Box p \wedge \Box\phi)$. Thus we have:

$$\mu \vdash \Box^\star\phi \leftrightarrow (\Box\Box^\star\phi \wedge \Box\phi) \tag{11}$$

and

$$\mu \vdash \alpha \to (\Box\alpha \wedge \Box\phi) \;\Rightarrow\; \mu \vdash \alpha \to \Box^\star\phi \tag{12}$$

We can now derive the $\mathsf{K4}$ axioms and rules for $\Box^\star$ in the usual way. Moreover, we have:

$$\mu \vdash \Box^\star\phi \to \phi \;\Rightarrow\; \mu \vdash \Box(\Box^\star\phi \wedge \phi) \to (\Box^\star\phi \wedge \phi) \tag{13}$$

$$\Rightarrow\; \mu \vdash \mathsf{H} \to (\Box^\star\phi \wedge \phi) \tag{14}$$

We define $\Box^{\mathsf{h}}\phi := \Box^\star(\mathsf{H} \to \phi)$. It is easy to check the $\mathsf{K4}$ axioms and rules for $\Box^{\mathsf{h}}$. We show that $\Box^{\mathsf{h}}$ satisfies Löb's rule over the μ-calculus.

$$\mu \vdash \Box^{\mathsf{h}}\phi \to \phi \;\Rightarrow\; \mu \vdash \Box^\star\phi \to \phi \tag{15}$$

$$\Rightarrow\; \mu \vdash \mathsf{H} \to \phi \tag{16}$$

$$\Rightarrow\; \mu \vdash \Box^{\mathsf{h}}\phi \tag{17}$$

$$\Rightarrow\; \mu \vdash \phi \tag{18}$$

As is well known, Löb's theorem follows from $\mathsf{K4}$ plus Löb's rule.

6 'H' stands for Henkin, since H is a Henkin fixed point in the tradition of provability logic.

7 We use the superscript $\star$ to signal the transitive closure. It is not to be confused with the superscript $*$ which is usually employed to mean the transitive reflexive closure.

The interpretation gm is the interpretation generated by the following clause:

$$\mathsf{gm}(\Box\phi) := \Box^{\mathsf{h}}\,\mathsf{gm}(\phi).$$

It is immediate that gm is a b-interpretation of GL in the μ-calculus.

To see that gm is an f-interpretation, it is sufficient to note that $\Box^{\mathsf{h}}$ commutes, μ-provably, with substitutions.

Using semantical arguments similar to those employed by van Benthem in [8], we can easily show that gm is faithful. This fact will also follow by syntactical arguments from Theorem 1 in combination with the result of Section 6.

4 Fixed Points in Löb's Logic

Johan van Benthem shows in his paper [8] that in Löb's Logic we can give explicit definitions of minimal fixed points of the monotonic operators corresponding to formulas ϕp in which the designated variable p occurs only positively. In this section we prove a strenghening of that result. The strengthening is such that restriction to formulas in which the designated variable p occurs only boxed gives us precisely the de Jongh-Sambin theorem and restriction to formulas in which p occurs only positivily gives us precisely van Benthem's theorem.

For our interpretation of the μ-calculus in GL, we only need van Benthem's result. However, it is nice to see the more general result, which involves not much more effort to prove.

We will say that a formula ϕp is *semi-positive in* p if all non-modalized occurrences of p are positive, or, equivalently, if all negative occurrences of p are modalized. We will show that any formula ϕp that is semi-positive in p has an explicit fixed point that is *locally minimal.*

- The formula χ is a locally minimal fixed point of ψp w.r.t. p iff $\mathsf{GL} \vdash \chi \leftrightarrow \psi\chi$ and $\mathsf{GL} \vdash (\Box(r \leftrightarrow \chi) \wedge (r \leftrightarrow \psi r)) \rightarrow (\chi \rightarrow r)$.

Here r is supposed to be fresh. It is easy to see that all locally minimal fixed points of ψp (if any) are provably equivalent in GL. The notion of *local minimality* replaces *minimality.* As we will see, formulas that are semi-positive in p do not generally have minimal fixed points.

Let $\widetilde{\phi} ps$ be such that (i) all occurrences of p are unboxed and positive, (ii) all occurrences of s are boxed and (iii) $\widetilde{\phi} pp = \phi p$. Now note that, by propositional logic, we have:

$$\mathsf{GL} \vdash \widetilde{\phi}\bot s \leftrightarrow \widetilde{\phi}(\widetilde{\phi}\bot s)s \tag{19}$$

Since, all occurrences of s in $\widetilde{\phi}\bot s$ are boxed, by the de Jongh-Sambin Theorem, we may find a χ such that:

$$\mathsf{GL} \vdash \chi \leftrightarrow \widetilde{\phi}\bot\chi \tag{20}$$

Putting things together, it follows that χ is a fixed point of ϕ:

$$\begin{aligned} \mathsf{GL} \vdash\ \ \chi &\leftrightarrow \widetilde{\phi}\bot\chi && (21)\\ &\leftrightarrow \widetilde{\phi}(\widetilde{\phi}\bot\chi)\chi && (22)\\ &\leftrightarrow \widetilde{\phi}\chi\chi && (23)\\ &\leftrightarrow \phi\chi && (24)\end{aligned}$$

To see that χ is locally minimal, we reason as follows.

$$\begin{aligned} \mathsf{GL} \vdash\ \ (\Box(r \leftrightarrow \chi) \wedge (r \leftrightarrow \phi r)) &\rightarrow (r \leftrightarrow \widetilde{\phi} rr) && (25)\\ &\rightarrow (r \leftrightarrow \widetilde{\phi} r\chi) && (26)\\ &\rightarrow (\widetilde{\phi}\bot\chi \rightarrow r) && (27)\\ &\rightarrow (\chi \rightarrow r) && (28)\end{aligned}$$

Example 1. Here is an example that there need not be a minimal fixed point. Consider the Kripke model on the ordinal 2. Take:

$$\phi p := ((\Box\bot \wedge p) \vee (\neg\Box\bot \wedge \neg\Box p)).$$

In the top-node 1 we have two choices for our fixed point: to make it true or to make it false. The minimal choice is to make it false. In 0, we do not have a choice. If p is true in 1, then p is false in 0, and if p is false in 1, then p is true in 0. Thus, there are only two fixed points of ϕp on this model, to wit $\{0\}$ and $\{1\}$. These fixed points are incomparable. The locally minimal one is $\{0\}$, corresponding to the minimal choice in 1. ▽

Finally, we prove that if all occurrences of p in ϕp are positive, then χ is the minimal fixed point of ϕp. We write $\boxdot\theta$ for: $\theta \wedge \Box\theta$. We have:

$$\begin{aligned} \mathsf{GL} \vdash\ \ (\boxdot(r \leftrightarrow \phi r) \wedge \Box(\chi \rightarrow r)) &\rightarrow (\widetilde{\phi} r\chi \rightarrow \widetilde{\phi} rr) && (29)\\ &\rightarrow (\widetilde{\phi} r\chi \rightarrow r) && (30)\\ &\rightarrow (\widetilde{\phi}\bot\chi \rightarrow r) && (31)\\ &\rightarrow (\chi \rightarrow r) && (32)\end{aligned}$$

The desired result is now immediate by the strengthened Löb's Rule.

Question 1. Can we also prove the above result more abstractly, i.e. directly from local minimality instead of by using the already obtained characterization of the locally minimal fixed point? ▽

We end this section with a theorem saying that the fixed point construction commutes with substitution.

Theorem 4. *Suppose ϕ is semi-positive in p. Let τ be any substitution such that (i) τ only substitutes for variables other than p and (ii) p does not occur in the formulas in the range of τ. Suppose, in* GL*, α represents the locally minimal fixed point of ϕ w.r.t. p and β represents the locally minimal fixed point of $\tau\phi$ w.r.t. p. Then, we have* $\mathsf{GL} \vdash \beta \leftrightarrow \tau\alpha$.

Proof. Suppose first that p occurs only modalized in ϕ. The fixed point property of α tells us that $\mathsf{GL} \vdash \alpha \leftrightarrow [p := \alpha]\phi$, and, hence,

$$\mathsf{GL} \vdash \tau\alpha \leftrightarrow \tau[p := \alpha]\phi \tag{33}$$
$$\leftrightarrow [p := \tau\alpha]\tau\phi \tag{34}$$

So, $\tau\alpha$ is a fixed point of $\tau\phi$. By the Bernardi-de Jongh-Sambin uniqueness theorem, $\tau\alpha$ must be provably equivalent to β.

We turn to the case that all non-modalized occurrences of p are positive. Suppose that p' is not in ϕ and not in the domain and not in the range of τ. We construct $\widetilde{\phi}p'p$ in such a way that $\widetilde{\phi}pp = \phi p$, all occurrences of p' are non-modalized and positive and all occurrences of p are boxed. By our preceding results α is, modulo provable equivalence, the fixed point of $\widetilde{\phi}\bot p$.

Consider $\tau\widetilde{\phi}p'p$. It is easy to see that $[p' := p]\tau\widetilde{\phi}p'p = \tau\widetilde{\phi}pp = \tau\phi p$. Moreover, all occurrences of p' in $\tau\widetilde{\phi}p'p$ are positive and all ocurrences of p are boxed. So, β is, modulo provable equivalence, the fixed point of $[p' := \bot]\tau\widetilde{\phi}p'p = \tau\widetilde{\phi}\bot p$.

We may now apply our preceding result for guarded formulas to $\widetilde{\phi}\bot p$. □

Remark 2. We have shown that, for any ϕp that is semi-positive in p, we can construct an explicit fixed point in GL. Did we capture all possible formulas ϕp that have a fixed point in this way (modulo provable equivalence)? The answer is *no*. E.g., $(\neg p \vee \Box p)$ has the (unique) fixed point $\top$. But, by a simple Kripke-model argument, this formula is not GL-provably equivalent to a formula that is semi-positive in p.

We do have the following. Consider ϕp. Let $\phi'' p$ be the result of replacing all non-modalized occurrences of p by $\top$. Let $\phi' p$ be $(p \wedge \phi'' p)$. Then, $\phi' p$ is semi-positive in p. Moreover,

$$\mathsf{GL} \vdash (p \leftrightarrow \phi p) \to (\phi p \leftrightarrow \phi' p).$$

Hence, $\mathsf{GL} \vdash \boxdot(p \leftrightarrow \phi p) \to \boxdot(\phi p \leftrightarrow \phi' p)$. ▽

Question 2. Is there a syntactical characterization of all formulas (modulo GL-provable equivalence) that have a definable fixed point in GL? ▽

5 Interpreting the μ-Calculus in Löb's Logic

In this section we introduce the interpretation mg. To make the definition work we need a lemma.

Lemma 1. *Consider a formula ϕpq and suppose that p occurs only positively and q occurs only positively (negatively). Then, there is a formula ψq in which q occurs only positively (negatively) such that ψq defines the minimal fixed point of ϕpq, w.r.t. p, in* GL.

As we have seen in the previous section, we may find a minimal fixed point of ϕ by applying the de Jongh-Sambin theorem to a modified formula. The proof of the lemma is by careful inspection of one of the standard fixed point constructions verifying the de Jongh-Sambin theorem. The one I used is the very simple version due to Per Linström. See [3]. We could prove a nicer lemma by showing that the polarity of any *occurrence* of an atom q is preserved to its descendants in the fixed point construction, assuming that the designated variable p occurs only positively. However, this would require the detour of defining what it is to be a descendant in the fixed point construction.

We will use the lemma to ensure that if p occurs only positively (negatively) in ϕ, then p occurs only positively (negatively) in $\mathsf{mg}(\phi)$. The interpretation mg is generated by the following clauses.

- $\mathsf{mg}(\Box\phi) := \Box\, \mathsf{mg}(\phi)$.
- $\mathsf{mg}(\mu p \cdot \phi)$ is a minimal fixed point formula of $\mathsf{mg}(\phi)$ that is chosen in such a way that, for any q, if q occures only positively (negatively) in $\mathsf{mg}(\phi)$, then q occurs only positively (negatively) in $\mathsf{mg}(\mu p \cdot \phi)$.

Using the lemma and induction on ϕ, we can see that our definition works. We can now easily verify that mg is indeed a b-interpretation of the μ-calculus in GL.

Theorem 5. mg *is an f-interpretation.*

Proof. The proof is by induction on μ-formulas ϕ. We treat the induction step for the case of μ. Consider the formula ψp in which p occurs only positively. Let σ be a substitution for variables other than p. Moreover, we assume that the formulas in the range of σ do not contain p. We want to show that:

$$\mathsf{GL} \vdash \mathsf{mg}(\overline{\sigma}(\mu p \cdot \psi p)) \leftrightarrow (\overline{\mathsf{mg} \circ \sigma})(\mathsf{mg}(\mu p \cdot \psi p)).$$

Note that $\overline{\sigma}(\mu p \cdot \psi p) = \mu p \cdot \overline{\sigma}\psi p$. So $\mathsf{mg}(\overline{\sigma}(\mu p \cdot \psi p))$ is the minimal fixed point of $\mathsf{mg}(\overline{\sigma}\psi p)$. Moreover, by the Induction Hypothesis, modulo provable equivalence, $\mathsf{mg}(\overline{\sigma}\psi p)$ is $(\overline{\mathsf{mg} \circ \sigma})(\mathsf{mg}(\psi p))$. Thus, we need that the minimal fixed point of $(\overline{\mathsf{mg} \circ \sigma})(\mathsf{mg}(\psi p))$ is $\overline{\mathsf{mg} \circ \sigma}$ applied to the minimal fixed point of $\mathsf{mg}(\psi p)$. But this is precisely what Theorem 4 tells us with $\mathsf{mg}(\psi p)$ in the role of ϕ and $\overline{\mathsf{mg} \circ \sigma}$ in the role of τ. □

Note that mg is far from faithful, since it yields Löb's logic for the μ-free fragment.

Question 3. Is there a faithful b-interpretation of the μ-calculus in GL? If so, is there is a faithful f-interpretation of the μ-calculus in GL? In the light of the well-known result that between any two countable Boole algebras without atoms there is a Boolean isomorphism, it seems to be very well possible that the answer to our first question is *yes*. I conjecture *no*, for the second question. ▽

6 Habemus Retractionem

We show that gm is a split monomorphism or co-retraction, and hence faithful, and that mg is the corresponding split epimorphism or retraction. What we have to show is that $\mathsf{mg} \circ \mathsf{gm} = \mathsf{id}_{\mathsf{GL}}$. We show, by induction on ϕ, that $\mathsf{GL} \vdash (\mathsf{mg} \circ \mathsf{gm})(\phi) \leftrightarrow \phi$. The only non-trivial step is the case that $\phi = \Box\psi$.

Note that we have, for any χ in the language of the μ-calculus,

$$\mu \vdash \Box^{\star}\chi \leftrightarrow (\Box\chi \wedge \Box\Box^{\star}\chi) \tag{35}$$

So, it follows that,

$$\mathsf{GL} \vdash \mathsf{mg}(\Box^{\star}\chi) \leftrightarrow (\Box\,\mathsf{mg}(\chi) \wedge \Box\,\mathsf{mg}(\Box^{\star}\chi)) \tag{36}$$

So, $\mathsf{mg}(\Box^{\star}\chi)$ is a modalised fixed point of $\Box\,\mathsf{mg}(\chi) \wedge \Box p$, and hence unique. Since $\Box\,\mathsf{mg}(\chi)$ is a solution of the equation, we find that $\mathsf{mg}(\Box^{\star}\chi)$ is GL-provably equivalent to $\Box\,\mathsf{mg}(\chi)$.

We also have: $\mu \vdash \mathsf{H} \leftrightarrow \Box\mathsf{H}$. Hence, $\mathsf{GL} \vdash \mathsf{mg}(\mathsf{H}) \leftrightarrow \Box\,\mathsf{mg}(\mathsf{H})$. So, by Löb's Theorem, $\mathsf{GL} \vdash \mathsf{mg}(\mathsf{H})$.

Thus, we have, using the Induction Hypothesis for ψ,

$$\begin{aligned} \mathsf{GL} \vdash (\mathsf{mg} \circ \mathsf{gm})(\Box\psi) &\leftrightarrow \mathsf{mg}(\Box^{\star}(\mathsf{H} \to \mathsf{gm}(\psi))) && (37)\\ &\leftrightarrow \Box(\mathsf{mg}(\mathsf{H}) \to (\mathsf{mg} \circ \mathsf{gm})(\psi)) && (38)\\ &\leftrightarrow \Box\psi && (39) \end{aligned}$$

By Theorem 1, we may conclude that gm is faithful and that, thus, the decidablility of the μ-calculus implies the decidability of GL. Moreover, mg preserves uniform interpolation. Thus, uniform interpolation for the μ-calculus, proved by d'Agostino and Hollenberg in their [14] proves uniform interpolation for Löb's Logic, which was first proved by Shavrukov in his [10]. By Theorem 2, we see that if $(\mathsf{gm}(\phi)/\mathsf{gm}(\psi))$ is admissible in the μ-calculus, then (ϕ/ψ) is admissible in GL.

The rule $(\Box p/p)$ is admissible in GL. However, the rule $(\mathsf{gm}(\Box p)/\mathsf{gm}(p))$, i.e. $(\Box^{\star}(\mathsf{H} \to p)/p)$ is not admissible in the μ-calculus. So, the converse of the above insight does not hold[8].

Acknowledgements. I thank Giovanna d'Agostino, Lev Beklemishev, Johan van Benthem, Emil Jerabek, Jaap van Oosten, Vincent van Oostrom and Yde Venema for enlightening discussions. I am grateful to Lev Beklemishev and Johan van Benthem for their comments on the penultimate manuscript. I thank the anonymous referees for the improvements they suggested.

References

1. Smoryński, C.: Self-Reference and Modal Logic. Universitext. Springer, New York (1985)
2. Boolos, G.: The logic of provability. Cambridge University Press, Cambridge (1993)

[8] This example is due to one of the referees.

3. Lindström, P.: Provability logic – a short introduction. Theoria **62** (1996) 19–61
4. Japaridze, G., de Jongh, D.: The logic of provability. In Buss, S., ed.: Handbook of proof theory. Amsterdam edn. North-Holland Publishing Co. (1998) 475–546
5. Artemov, S., Beklemishev, L.: Provability logic. In Gabbay, D., Guenthner, F., eds.: Handbook of Philosophical Logic, 2nd ed. Volume 13. Kluwer, Dordrecht (2004) 229–403
6. Kozen, D.: Results on the propositional μ-calculus. Theoretical Computer Science **27** (1983) 234–241
7. Bradfield, J., Stirling, C.: Modal logics and mu-calculi: an introduction. In Bergstra, J., Ponse, A., Smolka, S., eds.: Handbook of Process Algebra. Elsevier, Amsterdam (2001) 293–330
8. van Benthem, J.: Modal frame correspondences generalized. Technical Report DARE electronic archive 148523, Institute for Logic, Language and Computation, University of Amsterdam, Plantage Muidergracht 24 (2005) to appear in Studia Logica, issue dedicated to the memory of Wim Blok.
9. Pitts, A.: On an interpretation of second order quantification in first order intuitionistic propositional logic. Journal of Symbolic Logic **57** (1992) 33–52
10. Shavrukov, V.: Subalgebras of diagonalizable algebras of theories containing arithmetic. Dissertationes mathematicae (Rozprawy matematyczne) **CCCXXIII** (1993)
11. Ghilardi, S., Zawadowski, M.: A sheaf representation and duality for finitely presented Heyting algebras. Journal of Symbolic Logic **60** (1995) 911–939
12. Visser, A.: Uniform interpolation and layered bisimulation. In Hájek, P., ed.: Gödel '96, Logical Foundations of Mathematics, Computer Science and Physics — Kurt Gödel's Legacy, Berlin, Springer (1996) 139–164 reprinted as Lecture Notes in Logic 6, Association of Symbolic Logic.
13. Ghilardi, S., Zawadowski, M.: Sheaves, Games and Model Completions. Volume 14 of Trends In Logic, Studia Logica Library. Kluwer, Dordrecht (2002)
14. d' Agostino, G., Hollenberg, M.: Uniform interpolation, automata and the modal μ-calculus. In Kracht, M., de Rijke, M., Wansing, H., Zakharyaschev, M., eds.: Advances in Modal Logic. Volume 1. CSLI Publications, Stanford (1998) 73–84
15. Ghilardi, S.: Unification in intuitionistic logic. The Journal of Symbolic Logic **64** (1999) 859–880

A Characterisation of Weak Bisimulation Congruence

Rob J. van Glabbeek

National ICT Australia
and School of Computer Science and Engineering,
The University of New South Wales
rvg@cs.stanford.edu

Abstract. This paper shows that weak bisimulation congruence can be characterised as rooted weak bisimulation equivalence, even without making assumptions on the cardinality of the sets of states or actions of the processes under consideration.

Introduction

Weak bisimulation equivalence, also known as observation equivalence [7], is a fundamental semantic equivalence used in system verification, and one of the first proposed in the literature. It upgrades strong bisimulation equivalence by featuring abstraction from internal actions.

In order to allow compositional system verification, semantic equivalence relations need to be *congruences* for the operators under consideration, meaning that the equivalence class of an n-ary operator f applied to arguments $p_1, \ldots, p_n$ is completely determined by the equivalence classes of these arguments. Although strong bisimulation equivalence is a congruence for the operators of CCS, ACP_τ and many other languages found in the literature, weak bisimulation equivalence fails to be a congruence for the *choice* or *alternative composition* operator $+$ of CCS, as well as for the left-merge $\llfloor$ of ACP_τ. To bypass this problem, one uses the coarsest congruence relation for $+$ that is finer than weak bisimulation equivalence, called *weak bisimulation congruence*, and characterised as *rooted weak bisimulation equivalence* in [2]. This equivalence turns out to be a minor variant of weak bisimulation equivalence, and a congruence for all of CCS, ACP_τ and many other languages.

Classical proof sketches arguing that rooted weak bisimulation equivalence is indeed weak bisimulation congruence typically make some cardinally assumptions, such as that there is an infinite alphabet of actions of which each process uses only a finite subset. The current contribution establishes the validity of this characterisation without making such assumptions. It also argues that the *root condition* that turns weak bisimulation into rooted weak bisimulation embodies two properties, one of which is needed to obtain a congruence for the $+$, and one to obtain a congruence for the left-merge.

A. Middeldorp et al. (Eds.): Processes... (Klop Festschrift), LNCS 3838, pp. 26–39, 2005.

1 Process Graphs

Definition 1 ([3]). A *process graph* over an alphabet of actions *Act* is a rooted, directed graph whose edges are labelled by elements of *Act*. Formally, a process graph g is a triple $(\text{NODES}(g), \text{ROOT}(g), \text{EDGES}(g))$, where

- $\text{NODES}(g)$ is a set, of which the elements are called the *nodes* or *states* of g,
- $\text{ROOT}(g) \in \text{NODES}(g)$ is a special node: the *root* or *initial state* of g,
- and $\text{EDGES}(g) \subseteq \text{NODES}(g) \times Act \times \text{NODES}(g)$ is a set of triples (s, a, t) with $s, t \in \text{NODES}(g)$ and $a \in Act$: the *edges* or *transitions* of g.

Normally, one is not interested in the names of the nodes in a process graph. For this reason, process graphs are considered up to isomorphism.

Definition 2. Let g and h be process graphs. A *graph isomorphism* between g and h is a bijective function $f : \text{NODES}(g) \rightarrow \text{NODES}(h)$ satisfying

- $f(\text{ROOT}(g)) = \text{ROOT}(h)$ and
- $(s, a, t) \in \text{EDGES}(g) \;\Leftrightarrow\; (f(s), a, f(t)) \in \text{EDGES}(h)$.

Graphs g and h are *isomorphic*, notation $g \cong h$, if there exists a graph isomorphism between them.

If $g \cong h$ then g and h differ only in the identity of their nodes. Graph isomorphism is an equivalence relation on the class of process graphs.

Further on, process graphs are pictured by using open dots ($\circ$) to denote nodes, and labelled arrows to denote edges. The root is represented by an incoming arrow, not originating from another node. These drawings present process graphs only up to isomorphism.

Let $\mathbb{G}(Act)$ be the class of process graphs over the alphabet of actions *Act* *up to isomorphism.* This means that I am satisfied with a level of precision in describing elements of $\mathbb{G}(Act)$ that fails to distinguish isomorphic process graphs. In the digression below I will indicate how to raise the precision of my definitions to a fully formal level; the digression should also make clear that it is not really worthwhile to maintain this level throughout the paper.

Next I define the most basic process algebraic operations on $\mathbb{G}(Act)$: a constant 0 for *inaction*, a binary infix written operator $+$ for *alternative composition* or *choice*, and unary operators $a.$ for *action prefixing* for each $a \in Act$. For the sake of convenience, in the definition below I will only consider *root-acyclic* process graphs. In Sect. 3 I will extend the definition to arbitrary process graphs.

Definition 3 ([3]). A process graph is *root-acyclic* if it has no incoming edges at the root. Let $\mathbb{G}^\rho(Act)$ be the class of root-acyclic process graphs over *Act* up to isomorphism. The constant 0 and the operators $a.$ and $+_\rho$ are defined on $\mathbb{G}^\rho(Act)$ as follows. (The subscript ρ serves to distinguish this alternative composition from the more general one that will be defined in Sect. 3.)

- 0 is interpreted as the trivial graph, having one node (the root) and no edges;
- $a.g$ is obtained from g by adding a new node, which will be the root of $a.g$, and a new a-labelled edge from the root of $a.g$ to the root of g;
- $g +_\rho h$ is obtained by identifying the root nodes of disjoint copies of g and h.

Digression: Distinguishing Isomorphic Process Graphs

In Def. 3 I have not bothered to tell which node exactly will be the only node of the process graph 0 and which new node will be added in the construction of $a.g$. Moreover, in taking the disjoint union of g and h, no explicit solution is offered for what to do when g and h have nodes in common. Here I provide two possible answers. As it doesn't matter at all which one is chosen, the reader may pick himself, or make up a third.

Making arbitrary choices to resolve ambiguity.

The definitions of 0 and $a.g$ could for instance be given as follows:

- $\text{NODES}(0) = \{*\}$,
- $\text{ROOT}(0) = *$,
- $\text{EDGES}(0) = \emptyset$.

- $\text{NODES}(a.g) = \{*\} \cup \{s' \mid s \in \text{NODES}(g)\}$,
- $\text{ROOT}(a.g) = *$,
- $\text{EDGES}(a.g) = \{(*, a, \text{ROOT}(g)')\} \cup \{(s', a, t') \mid (s, a, t) \in \text{EDGES}(g)\}$.

Here the nodes of g are renamed from s into s', so as to make sure that none of them happens to be the symbol $*$ that is used to name the new root node.

Working modulo isomorphism.

In this approach $\mathbb{G}(Act)$ is the class of process graph *modulo isomorphism*, meaning that the elements of $\mathbb{G}(Act)$ are isomorphism classes of process graphs.

Now $0 \in \mathbb{G}^\rho(Act)$ is defined as the isomorphism class of all trivial process graphs, after observing that all trivial graphs are isomorphic. To obtain the isomorphism class $a.G$, for G an isomorphism class of process graphs, I first pick a representative $g \in G$ and a fresh object $r \notin \text{NODES}(g)$. Then $a.g$ is defined by

- $\text{NODES}(a.g) = \text{NODES}(g) \cup \{r\}$,
- $\text{ROOT}(a.g) = r$ and
- $\text{EDGES}(a.g) = \text{EDGES}(g) \cup \{(r, a, \text{ROOT}(g))\}$.

Finally, $a.G$ is defined to be the isomorphism class containing $a.g$, and the exercise is concluded by showing that the result is independent of the choice of $g \in G$. Likewise, $G +_\rho H$, for $G, H \in \mathbb{G}^\rho(Act)$, is obtained as the isomorphism class of $g +_\rho h$, where $g \in G$ and $h \in H$ are chosen in such a way that $\text{NODES}(g) \cap \text{NODES}(h) = \text{ROOT}(g) = \text{ROOT}(h)$, and $g +_\rho h$ is defined by

- $\text{NODES}(g +_\rho h) = \text{NODES}(g) \cup \text{NODES}(h)$,
- $\text{ROOT}(g +_\rho h) = \text{ROOT}(g) = \text{ROOT}(h)$ and
- $\text{EDGES}(g +_\rho h) = \text{EDGES}(g) \cup \text{EDGES}(h)$.

Again, it must be shown that the result is independent of the choice of g and h.

2 Bisimulation Semantics

To make process graphs into a useful semantic model of calculi that enable system verification, a semantic equivalence coarser than isomorphism needs to be defined. The most popular choices are strong and weak bisimulation equivalence, and some of their variants. Such an equivalence can be used fruitfully when it is a *congruence* for the process algebraic operators that are considered in a particular application. These almost always include the operators 0, $a.$ and $+$. A semantic equivalence $\sim$ is a congruence for these operators if $g \sim h$ implies $a.g \sim a.h$ and $g_1 \sim h_1 \wedge g_2 \sim h_2$ implies $g_1 + g_2 \sim h_1 + h_2$.

2.1 Strong Bisimulation

Definition 4. Let $g, h \in \mathbb{G}(Act)$. The graphs g and h are *(strong) bisimulation equivalent*, notation $g \leftrightarroweq h$, if there exists a binary relation $R \subseteq \text{NODES}(g) \times \text{NODES}(h)$, called a *bisimulation* between g and h, satisfying, for all $a \in Act$:

- $\text{ROOT}(g)\ R\ \text{ROOT}(h)$.
- If sRt and $(s, a, s') \in \text{EDGES}(g)$, then $\exists (t, a, t') \in \text{EDGES}(h)$ such that $s'Rt'$.
- If sRt and $(t, a, t') \in \text{EDGES}(h)$, then $\exists (s, a, s') \in \text{EDGES}(g)$ such that $s'Rt'$.

It is well-known and easy to check that $\leftrightarroweq$ is an equivalence relation indeed. I will now show that it is a congruence relation for $a.$ and $+_\rho$. Because these operators have so-far not been defined outside $\mathbb{G}^\rho(Act)$, for now this result will pertain to root-acyclic process graphs only.

Definition 5. A relation between the nodes of two root-acyclic process graphs is called *rooted* if it relates root nodes with root nodes only.

Lemma 1. *Let $g, h \in \mathbb{G}^\rho(Act)$. If $g \leftrightarroweq h$ then there exist a rooted bisimulation between g and h.*

Proof. Let R be a bisimulation between g and h. A rooted bisimulation is obtained from R by omitting all liaisons between root nodes and non-root nodes.

Proposition 1. *On $\mathbb{G}^\rho(Act)$, bisimulation equivalence is a congruence for $a.$ and $+_\rho$.*

Proof. Suppose R is a bisimulation between g and h. Then

$$R \cup \{(\text{ROOT}(a.g), \text{ROOT}(a.h))\}$$

is a bisimulation between $a.g$ and $a.h$. Moreover, invoking Lemma 1, let R_i be a rooted bisimulation between g_i and h_i for $g_i, h_i \in \mathbb{G}^\rho(Act)$ and $i = 1, 2$ then $R_1 \cup R_2$ is a bisimulation between $g_1 +_\rho g_2$ and $h_1 +_\rho h_2$. □

2.2   Weak Bisimulation

Let $\tau \in Act$ be the *invisible action* or *silent step*. Henceforth, write $s \xrightarrow{a}_g s'$ for $(s, a, s') \in$ EDGES(g) and $s \Longrightarrow_g s'$ when there are $s_0, \ldots, s_n$ in NODES(g) such that $s = s_0 \xrightarrow{\tau}_g s_1 \xrightarrow{\tau}_g \cdots \xrightarrow{\tau}_g s_n = s'$. Moreover, $s \xRightarrow{a}_g s'$ denotes that there are nodes s_1 and s_2 in g such that $s \Longrightarrow_g s_1 \xrightarrow{a}_g s_2 \Longrightarrow_g s'$ and $s \xRightarrow{(a)}_g s'$ is a shorthand for $s \Longrightarrow_g s'$ when $a = \tau$, and $s \xRightarrow{a}_g s'$ when $a \neq \tau$. Thus, $s \xRightarrow{(\tau)}_g s'$ says that in g one can travel from s to s' by performing a sequence of zero or more τ-steps, whereas $s \xRightarrow{\tau}_g s'$ requires at least one τ-step. For $a \neq \tau$ there is no difference between $\xRightarrow{(a)}_g$ and $\xRightarrow{a}_g$.

Definition 6. Let $g, h \in \mathbb{G}(Act)$. The graphs g and h are *weak bisimulation equivalent*, notation $g \underline{\leftrightarrow}_w h$, if there exists a binary relation $R \subseteq$ NODES$(g) \times$ NODES(h), called a *weak bisimulation* between g and h, satisfying, for all $a \in Act$:

- ROOT(g) R ROOT(h).
- If sRt and $s \xrightarrow{a}_g s'$, then there is a t' such that $t \xRightarrow{(a)}_h t'$ and $s'Rt'$.
- If sRt and $t \xrightarrow{a}_g t'$, then there is an s' such that $s \xRightarrow{(a)}_h s'$ and $s'Rt'$.

It is well-known and easy to check that $\underline{\leftrightarrow}_w$ is an equivalence relation indeed. However, $\underline{\leftrightarrow}_w$ fails to be a congruence for the $+$. Namely, $\tau.a.0 \underline{\leftrightarrow}_w a.0$ but $\tau.a.0 +_\rho b.0 \not\underline{\leftrightarrow}_w a.0 +_\rho b.0$. The proof of Prop. 1 does not generalise to $\underline{\leftrightarrow}_w$ because Lemma 1 does not hold for $\underline{\leftrightarrow}_w$: there is no rooted weak bisimulation between $\tau.a.0$ and $a.0$.

2.3   Rooted Weak Bisimulation

Although weak bisimulation equivalence captures the invisible nature of the silent step rather well, in order to obtain a congruence, a finer equivalence relation is needed. Such an equivalence was proposed by Bergstra & Klop in [2]. In fact it is the obvious "fix" in the definition of weak bisimulation equivalence needed to inherit the proof of Prop. 1.

Definition 7 ([2]). Two graphs $g, h \in \mathbb{G}^\rho(Act)$ are *rooted weak bisimulation equivalent*, notation $g \underline{\leftrightarrow}_{rw} h$, if there exists a rooted weak bisimulation between them (recall Def. 5).

Again, it is easy to check that $\underline{\leftrightarrow}_{rw}$ is an equivalence relation. By definition it is finer than $\underline{\leftrightarrow}_w$. It is strictly finer because $\tau.0 \underline{\leftrightarrow}_w 0$ but $\tau.0 \not\underline{\leftrightarrow}_{rw} 0$ (and likewise $\tau.a.0 \underline{\leftrightarrow}_w a.0$ but $\tau.a.0 \not\underline{\leftrightarrow}_{rw} a.0$). Moreover, Lemma 1 implies that $\underline{\leftrightarrow}_{rw}$ is coarser than $\underline{\leftrightarrow}$. It is strictly coarser because $\tau.\tau.0 \underline{\leftrightarrow}_{rw} \tau.0$ but $\tau.\tau.0 \not\underline{\leftrightarrow} \tau.0$.

Proposition 2. *On* $\mathbb{G}^\rho(Act)$, *rooted weak bisimulation equivalence is a congruence for* a. *and* $+_\rho$.

Proof. Suppose R is a rooted weak bisimulation between g and h. Then

$$R \cup \{(\text{ROOT}(a.g), \text{ROOT}(a.h))\}$$

is a rooted weak bisimulation between $a.g$ and $a.h$. Moreover, let R_i be a rooted weak bisimulation between g_i and h_i for $g_i, h_i \in \mathbb{G}^\rho(Act)$ and $i = 1, 2$ then $R_1 \cup R_2$ is a rooted weak bisimulation between $g_1 +_\rho g_2$ and $h_1 +_\rho h_2$. □

3 Root Unwinding

In this section I will generalise the definitions and results of Sections 1 and 2 from root-acyclic to general process graphs.

3.1 The Definition of Alternative Composition

The definition of 0 and $a.$ on $\mathbb{G}(Act)$ is exactly as on $\mathbb{G}^\rho(Act)$ (see Def. 3). However, defining the + on $\mathbb{G}(Act)$ as in Def. 3 would yield counterintuitive results. Namely we would have $g+h=m$, as in the top row of Fig. 1. However, m is able to first do a number of a-actions from g and then a b from h; this is inconsistent with the idea that $g+h$ should from the initial state onwards behave either as g or as h. In fact, strong bisimulation equivalence would fail to be a congruence for this definition of +, for $g \leftrightarroweq g^\rho$, yet $m \not\leftrightarroweq k$.

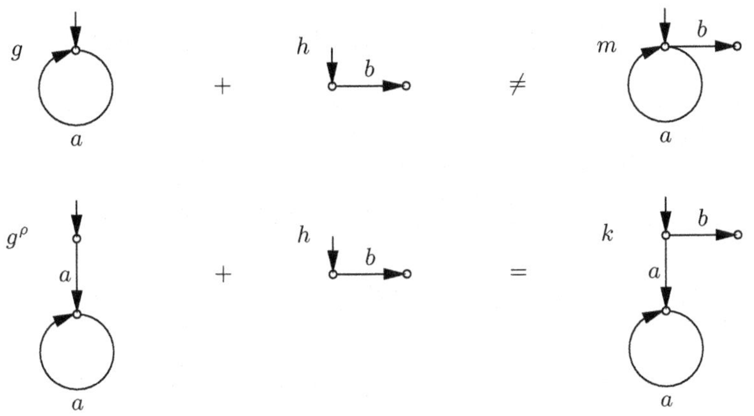

Fig. 1. Alternative composition of process graphs

For most applications we are interested in process graphs only up to strong bisimulation equivalence. Up to strong bisimilarity, the definition of the + on $\mathbb{G}(Act)$ is completely determined by its definition on $\mathbb{G}^\rho(Act)$, because every process graph is strongly bisimilar with a root-acyclic process graph. The following construction is used to establish this.

Definition 8 ([3]). *Root unwinding* is the operator ρ on $\mathbb{G}(Act)$ given by

- $\textsc{nodes}(\rho(g)) = \textsc{nodes}(g) \dot\cup \{*\}$,
- $\textsc{root}(\rho(g)) = \{*\}$ and
- $\textsc{edges}(\rho(g)) = \textsc{edges}(g) \cup \{(*, a, s) \mid (\textsc{root}(g), a, s) \in \textsc{edges}(g)\}$.

Note that $\rho(g) \in \mathbb{G}^\rho(Act)$ for all $g \in \mathbb{G}(Act)$.

Proposition 3 ([3]). *$g \leftrightarroweq \rho(g)$ for every process graph $g \in \mathbb{G}(Act)$.*

Proof. The relation $\{(s,s) \mid s \in \textsc{nodes}(g)\} \cup \{(\textsc{root}(g), *)\}$ is a bisimulation between g and $\rho(g)$. □

Definition 9 ([3]). The definition of the $+$ is extended from $\mathbb{G}^\rho(Act)$ to $\mathbb{G}(Act)$ by $g+h := \rho(g) +_\rho \rho(h)$, where $+_\rho$ is given by Def. 3.

This construction automatically entails that $\leftrightarroweq$ is a congruence for the $+$.

Proposition 4. *On $\mathbb{G}(Act)$, $\leftrightarroweq$ is a congruence for $a.$ and $+$.*

Proof. The case of $a.$ goes exactly as in Prop. 1. Let $g \leftrightarroweq g'$ and $h \leftrightarroweq h'$.
Then $\rho(g) \leftrightarroweq g \leftrightarroweq g' \leftrightarroweq \rho(g')$ and likewise $\rho(h) \leftrightarroweq \rho(h')$.
Hence $g+h = \rho(g) +_\rho \rho(h) \leftrightarroweq \rho(g') +_\rho \rho(h') = g'+h'$ by Prop. 1. □

On $\mathbb{G}^\rho(Act)$ it is in general not the case that $g+h \cong g +_\rho h$. However, up to strong bisimilarity, both versions of the $+$ agree.

Proposition 5. *Let $g,h \in \mathbb{G}^\rho(Act)$. Then $g+h \leftrightarroweq g +_\rho h$.*

Proof. $g+h = \rho(g) +_\rho \rho(h) \leftrightarroweq g +_\rho h$, using Propositions 3 and 1. □

It is also possible to merge the definitions of $+$, $+_\rho$ and root unwinding:

Definition 10. Let $g,h \in \mathbb{G}^\rho(Act)$. Then $g+h$ can alternatively be defined by

- $\text{NODES}(g+h) = \text{NODES}(g) \dot\cup \text{NODES}(h) \dot\cup \{*\}$,
- $\text{ROOT}(g+h) = \{*\}$ and
- $\text{EDGES}(g+h) = \text{EDGES}(g) \cup \text{EDGES}(h) \cup$
  $\{(*,a,s) \mid (\text{ROOT}(g),a,s) \in \text{NODES}(g) \vee (\text{ROOT}(h),a,s) \in \text{NODES}(h)\}$.

It is trivial to check that up to isomorphism this definition yields the same alternative composition operator as Def. 9.

3.2 Rooted Weak Bisimulation

Postulating that rooted weak bisimulation has to be a coarser equivalence than strong bisimulation, Prop. 3 implies that there is a unique extension of $\leftrightarroweq_{rw}$ from $\mathbb{G}^\rho(Act)$ to $\mathbb{G}(Act)$.

Definition 11. Let $g,h \in \mathbb{G}(Act)$. Then $g \leftrightarroweq_{rw} h$ iff $\rho(g) \leftrightarroweq_{rw} \rho(h)$ as per Def. 7.

Proposition 6. *On $\mathbb{G}(Act)$, $\leftrightarroweq_{rw}$ is a congruence for $a.$ and $+$.*

Proof. The case of $a.$ goes exactly as in Prop. 2. Let $g \leftrightarroweq_{rw} g'$ and $h \leftrightarroweq_{rw} h'$.
Then $\rho(g) \leftrightarroweq g \leftrightarroweq_{rw} g' \leftrightarroweq \rho(g')$, so $\rho(g) \leftrightarroweq_{rw} \rho(g')$, and likewise $\rho(h) \leftrightarroweq_{rw} \rho(h')$.
Hence $g+h = \rho(g) +_\rho \rho(h) \leftrightarroweq_{rw} \rho(g') +_\rho \rho(h') = g'+h'$ by Prop. 2. □

The following characterisation of rooted weak bisimulation equivalence was taken as definition in [7]. Let, for $g \in \mathbb{G}(Act)$ and $s \in \text{NODES}(g)$, g_s denote the process graph obtained from g by appointing s as its root.

Proposition 7. *Let $g, h \in \mathbb{G}(Act)$. Then $g \leftrightarroweq_{rw} h$ iff*

- *if* $\text{ROOT}(g) \xrightarrow{a}_g s$ *then there is a* t *such that* $\text{ROOT}(h) \overset{a}{\Longrightarrow}_h t$ *and* $g_s \leftrightarroweq_w h_t$*;*
- *if* $\text{ROOT}(h) \xrightarrow{a}_h t$ *then there is an* s *such that* $\text{ROOT}(g) \overset{a}{\Longrightarrow}_g s$ *and* $g_s \leftrightarroweq_w h_t$*.*

Proof. "If": Let $B = \{(s,t) \in \text{NODES}(g) \times \text{NODES}(h) \mid g_s \leftrightarroweq_w h_t\}$ and

$$B^\rho = B \mathbin{\dot\cup} \{(\text{ROOT}(\rho(g)), \text{ROOT}(\rho(h)))\} \subseteq \text{NODES}(\rho(g)) \times \text{NODES}(\rho(h)),$$

recalling that $\text{NODES}(\rho(g)) = \text{NODES}(g) \mathbin{\dot\cup} \{\text{ROOT}(\rho(g))\}$, and likewise for $\rho(h)$. Assume that both clause above hold. It suffices to show that B^ρ is a rooted weak bisimulation between $\rho(g)$ and $\rho(h)$, for by Def. 11 this implies $g \leftrightarroweq_{rw} h$. The relation B^ρ is rooted by construction. So I need to show it is a weak bisimulation.

Let $sB^\rho t$ and $s \xrightarrow{a}_{\rho(g)} s'$. It suffices to show that there is a t' such that $t \overset{(a)}{\Longrightarrow}_{\rho(h)} t'$ and $s' B^\rho t'$. The other requirement then follows by symmetry.

First assume sBt. In that case, $s \xrightarrow{a}_g s'$ and $g_s \leftrightarroweq_w h_t$, so there is a weak bisimulation R between g_s and h_s. Thus sRt, and there must be a t' such that $t \overset{(a)}{\Longrightarrow}_h t'$ and $s'Rt'$. Hence $g_{s'} \leftrightarroweq_w h_{t'}$, $s'Bt'$ and $s'B^\rho t'$. Def. 8 yields $t \overset{(a)}{\Longrightarrow}_{\rho(h)} t'$.

Now assume $s = \text{ROOT}(\rho(g))$ and $t = \text{ROOT}(\rho(h))$, so $\text{ROOT}(\rho(g)) \xrightarrow{a}_{\rho(g)} s'$. Then, by Def. 8, $\text{ROOT}(g) \xrightarrow{a}_g s'$, so by the first clause of Prop. 7 there is a t' such that $\text{ROOT}(h) \overset{a}{\Longrightarrow}_h t'$ and $g_{s'} \leftrightarroweq_w h_{t'}$. Hence $s'Bt'$ and $s'B^\rho t'$. Def. 8 yields $\text{ROOT}(\rho(h)) \overset{a}{\Longrightarrow}_{\rho(h)} t'$, hence $t = \text{ROOT}(\rho(h)) \overset{(a)}{\Longrightarrow}_{\rho(h)} t'$.

"Only if": Let R be a rooted weak bisimulation between $\rho(g)$ and $\rho(h)$. Suppose $\text{ROOT}(g) \xrightarrow{a}_g s$. Def. 8 yields $\text{ROOT}(\rho(g)) \xrightarrow{a}_{\rho(g)} s$, so there must be a t such that $\text{ROOT}(\rho(h)) \overset{(a)}{\Longrightarrow}_{\rho(h)} t$ and sRt. As R is rooted, the case that $a = \tau$ and $t = \text{ROOT}(\rho(h))$ cannot apply, so $\text{ROOT}(\rho(h)) \overset{a}{\Longrightarrow}_{\rho(h)} t$, hence $\text{ROOT}(h) \overset{a}{\Longrightarrow}_h t$ by Def. 8. Furthermore, sRt implies $g_s \leftrightarroweq_w h_t$. The other clause follows by symmetry.

4 Weak Bisimulation Congruence

A different modification of weak bisimulation equivalence into a congruence for the $+$ was proposed in [6].

Proposition 8. *For every equivalence relation $\sim$ on $\mathbb{G}(Act)$ and every set L of operators on $\mathbb{G}$ there is a coarsest congruence relation $\sim^c$ that is finer than $\sim$.*

Proof. Let a *semantic context* $C[\cdot]$ be an expression build from process graphs $g \in \mathbb{G}(Act)$ and the *hole* $[\cdot]$ through application of operators from L, and in which the hole occurs exactly once. If $C[\cdot]$ is a semantic context and $g \in \mathbb{G}(Act)$, then $C[g]$ denotes the process graph obtained by evaluating the expression $C[\cdot]$ in which g is substituted for the hole $[\cdot]$.

Alternatively, a semantic context can be regarded as a unary operator on process graphs: a *primitive context* is obtained from an operator in L by instantiating all but one of its arguments by process graphs; and a general semantic context is the composition of any number (possibly 0) of primitive contexts.

An equivalence relation $\approx$ on $\mathbb{G}(Act)$ is a congruence for L iff for every n-ary operator f in L one has $g_1 \approx h_1 \wedge \cdots \wedge g_n \approx h_n \Rightarrow f(g_1, \ldots, g_n) \approx f(h_1, \ldots, h_n)$. This is the case iff for every semantic context $C[\cdot]$ one has $g \approx h \Rightarrow C[g] \approx C[h]$.

Given an equivalence relation $\sim$ on $\mathbb{G}(Act)$, define $\sim^c$ by

$$g \sim^c h \text{ iff } C[g] \sim C[h] \text{ for every semantic context } C[\cdot].$$

By construction, $\sim^c$ is a congruence on $\mathbb{G}(Act)$. For if $g \sim^c h$ and $D[\cdot]$ is a semantic context, then for every semantic context $C[\cdot]$ also $C[D[\cdot]]$ is a semantic context, so $\forall C[\cdot](C[D[g]] \sim C[D[h]])$ and hence $D[g] \sim^c D[h]$.

The trivial context guarantees that $g \sim^c h \Rightarrow g \sim h$, so $\sim^c$ is finer than $\sim$.

Finally, $\sim^c$ is the coarsest congruence finer than $\sim$, because if $\approx$ is any congruence finer than $\sim$, then

$$g \approx h \;\Rightarrow\; \forall C[\cdot](C[g] \approx C[h]) \;\Rightarrow\; \forall C[\cdot](C[g] \sim C[h]) \;\Rightarrow\; g \sim^c h. \qquad \square$$

Definition 12. The coarsest congruence w.r.t. the $+$ that is finer than weak bisimulation equivalence is called *weak bisimulation congruence*, notation $\leftrightarrows_w^c$.

Here I address the question whether both approaches coincide, i.e. whether $\leftrightarrows_{rw} = \leftrightarrows_w^c$. Prop. 6 immediately yields $\leftrightarrows_{rw} \subseteq \leftrightarrows_w^c$. The following proof sketch, due to Jan Willem Klop [personal communication], is a first attempt to establish the reverse.

Proof Sketch. It suffices to restrict attention to the graphs in $\mathbb{G}^\rho(Act)$, for when $g \leftrightarrows_w^c h \Rightarrow g \leftrightarrows_{rw} h$ for $g, h \in \mathbb{G}^\rho(Act)$, then by Prop. 3 the same holds for $g, h \in \mathbb{G}(Act)$. [Namely $g \leftrightarrows_w^c h \Rightarrow \rho(g) \leftrightarrows g \leftrightarrows_w^c h \leftrightarrows \rho(h) \Rightarrow \rho(g) \leftrightarrows_w^c \rho(h) \Rightarrow \rho(g) \leftrightarrows_{rw} \rho(h) \Rightarrow g \leftrightarrows \rho(g) \leftrightarrows_{rw} \rho(h) \leftrightarrows h \Rightarrow g \leftrightarrows_{rw} h$.]

So let $g, h \in \mathbb{G}^\rho(Act)$ and assume $g \leftrightarrows_w^c h$. Let $a \neq \tau$ be an action that does not occur in either g or h. Then $g + a.0 \leftrightarrows_w h + a.0$ by the definition of $\leftrightarrows_w^c$. Let R be a weak bisimulation between $g + a.0$ and $h + a.0$.

Claim. R must be rooted.

Proof of Claim. $\textsc{root}(g + a.0) \xrightarrow{a}_{g+a.0} \textsc{root}(0)$, so if $\textsc{root}(g + a.0)\, R\, t$ then $t \overset{a}{\Longrightarrow}_{h+a.0} t'$ for some $t' \in \textsc{nodes}(h{+}a.0)$ with $\textsc{root}(0)\, R\, t'$. This is only possible if $t = \textsc{root}(h + a.0)$. By symmetry, $s\, R\, \textsc{root}(h + a.0)$ is only possible if $s = \textsc{root}(g + a.0)$.

Application of Claim. The restriction of R to the nodes of g and h is a rooted weak bisimulation between g and h, showing that $g \leftrightarrows_{rw} h$. $\square$

The only weak point in this proof sketch is in the choice of an action $a \neq \tau$ that does not occur in g or h. What if such an action does not exists? Below I present two solutions to this problem.

4.1 The Fresh Atom Principle

The *Fresh Atom Principle* (FAP) allows us to use fresh actions in proofs. It was invented and named by Jan Willem Klop [personal communication]. In order to justify the use of FAP, one needs to realise that for any choice of a set of actions *Act* containing τ there exists a class $\mathbb{G}(Act)$ of process graphs over *Act*. Likewise a *parametrised* semantic equivalence $\sim$ has an incarnation in each of these classes, and I write $\sim^{Act}$ to denote the equivalence $\sim$ as it exists in $\mathbb{G}(Act)$. Obviously, $\mathbb{G}(A) \subseteq \mathbb{G}(B)$ for sets of actions $A \subseteq B$. Now say that a parametrised equivalence $\sim$ *satisfies* FAP, if, whenever $A \subseteq B$, $\sim^A$ is the restriction of $\sim^B$ to $\mathbb{G}(A)$, i.e. if for $g, h \in \mathbb{G}(A) \subseteq \mathbb{G}(B)$ one has $g \sim^A h \Leftrightarrow g \sim^B h$. It is immediate from their definitions that $\leftrightarrows$, $\leftrightarrows_w$, $\leftrightarrows_{rw}$ and most other semantic equivalences defined in the literature satisfy FAP. In fact, satisfying FAP looks like a good sanity check for any meaningful equivalence.

In this spirit, one may want to use, instead of $\leftrightarrows_w^c$, the coarsest congruence finer than $\leftrightarrows_w$ that satisfies FAP, notation $\leftrightarrows_w^{fc}$. This congruence is obtained by allowing, in the proof of Prop. 8, contexts C that may involve fresh actions.

Theorem 1. *$\leftrightarrows_w^{fc}$ coincides with $\leftrightarrows_{rw}$.*

Proof. The proof sketch above applies here. After assuming $g \leftrightarrows_w^{fc} h$ in $\mathbb{G}(Act)$, apply FAP to obtain $g \leftrightarrows_w^{fc} h$ in $\mathbb{G}(Act \dot{\cup} \{a\})$. As in the proof sketch, conclude that $g \leftrightarrows_{rw} h$ in $\mathbb{G}(Act \dot{\cup} \{a\})$. Since $\leftrightarrows_{rw}$ satisfies FAP this implies that $g \leftrightarrows_{rw} h$ in $\mathbb{G}(Act)$. □

4.2 Arbitrary Many Non-bisimilar Processes

Even though we now know that $\leftrightarrows_{rw}$ coincides with $\leftrightarrows_w^{fc}$ and thus is the coarsest sane equivalence contained in $\leftrightarrows_w$ that is a congruence for the $+$, the question still remains if also $\leftrightarrows_w^c$ coincides with $\leftrightarrows_{rw}$. In case $Act = \{\tau\}$, this is not the case. For in that case $\leftrightarrows_w$, and hence also $\leftrightarrows_w^c$, is the universal relation, but $\tau.0 \not\leftrightarrows_{rw} 0$. However, as I will show in the following, in all other cases we have in fact that $\leftrightarrows_w^c = \leftrightarrows_{rw}$.

Proposition 9 (Jan Willem Klop). *Provided that there is at least one action $a \in Act - \{\tau\}$, for each infinite cardinal κ there are at least κ bisimulation equivalence classes of τ-free process graphs in $\mathbb{G}(Act)$ with less than κ nodes.*

Proof. For each ordinal λ define $g_\lambda \in \mathbb{G}(Act)$ as follows.

- $g_0 := 0$,
- $g_{\lambda+1} := g_\lambda + a.g_\lambda$ and
- for λ a limit ordinal, $g_\lambda := \sum_{\mu<\lambda} g_\mu$, meaning that g_λ is constructed from all graphs g_μ for $\mu < \lambda$ by identifying their root.

Claim 1. $\textsc{root}(g_\lambda) \xrightarrow{a}_{g_\lambda} s \quad \Leftrightarrow \quad s = \textsc{root}(g_\mu)$ for some $\mu < \lambda$.

Proof of Claim 1. A straightforward transfinite induction on λ.

Claim 2. If $\mu < \lambda$ then $g_\lambda \not\leftrightarrows g_\mu$.

Proof of Claim 2, by transfinite induction on λ: Let $\mu < \lambda$. By Claim 1 we have $\textsc{root}(g_\lambda) \xrightarrow{a}_{g_\lambda} \textsc{root}(g_\mu)$. However, when $\textsc{root}(g_\mu) \xrightarrow{a}_{g_\mu} s$ we have $s = \textsc{root}(g_\rho)$ for some $\rho < \mu$, and by induction $g_\mu \not\leftrightarrows g_\rho$. Thus $g_\lambda \not\leftrightarrows g_\mu$.

Claim 3. For infinite λ, $|\text{NODES}(g_\lambda)| = |\lambda|$.

Proof of Claim 3. A straightforward transfinite induction on λ, using that the cardinality of a disjoint union of $|\lambda|$ sets of cardinality $0 < \kappa \leq |\lambda|$ is $|\lambda|$.

Application of the claims. For each infinite cardinal κ there are κ ordinals λ smaller than κ. Now the proposition follows from Claims 2 and 3. □

Theorem 2. *Provided that there is an $a \in Act - \{\tau\}$, on $\mathbb{G}(Act)$ weak bisimulation congruence coincides with rooted weak bisimulation equivalence.*

Proof. Following the idea from the earlier proof sketch, let $g \leftrightarrows_w^c h$ with $g, h \in \mathbb{G}^\rho(Act)$. Let κ be the smallest infinite cardinal, such that g and h have less than κ nodes. So there are less than κ process graphs g_s with $s \in \text{NODES}(g)$ and h_t with $t \in \text{NODES}(h)$. Hence by Prop. 9 there is a τ-free process graph k with $|\text{NODES}(k)| < \kappa$, such that $k \not\leftrightarrows g_s$ and $k \not\leftrightarrows h_t$ for any such g_s and h_t. Now $g + a.k \leftrightarrows_w h + a.k$ by the definition of $\leftrightarrows_w^c$. Let R be a weak bisimulation between $g + a.k$ and $h + a.k$.

Claim 1. The restriction of R to $\text{NODES}(g) \times \text{NODES}(h)$ must be rooted.

Proof of Claim 1. $\text{ROOT}(g + a.k) \xrightarrow{a}_{g+a.k} \text{ROOT}(k)$, so if $\text{ROOT}(g + a.k)\, R\, t$ for some $t \in \text{NODES}(h) - \{\text{ROOT}(h)\}$, then $t \overset{a}{\Longrightarrow}_{h+a.k} t'$ for some $t' \in \text{NODES}(h)$ with $\text{ROOT}(k)\, R\, t'$ and hence $k \leftrightarrows_w h_{t'}$. By the definition of k this is impossible. Likewise, it is impossible that $s\, R\, \text{ROOT}(h + a.k)$ for $s \in \text{NODES}(g) - \{\text{ROOT}(g)\}$.

Claim 2. The restriction of R to $\text{NODES}(g) \times \text{NODES}(h)$ is a weak bisimulation.

Proof of Claim 2. Let $\text{ROOT}(g) \xrightarrow{b}_g s$. Then $\text{ROOT}(g + a.k) \xrightarrow{b}_{g+a.k} s$, so there must be a $t \in \text{NODES}(h + a.k)$ such that $\text{ROOT}(h + a.k) \overset{(b)}{\Longrightarrow}_{h+a.k} t$ and $s\, R\, t$. The possibility that $b = a$ and $t = \text{ROOT}(k)$ can not occur, for this would imply $g_s \leftrightarrows_w k$. The possibility that $b = \tau$ and $t = \text{ROOT}(h + a.k)$ can not occur by Claim 1. Therefore $t \in \text{NODES}(h) - \{\text{ROOT}(h)\}$ and $\text{ROOT}(h) \overset{b}{\Longrightarrow}_h t$. Likewise, $\text{ROOT}(h) \xrightarrow{b}_h t$ implies that there is an $s \in \text{NODES}(g) - \{\text{ROOT}(g)\}$ with $\text{ROOT}(g) \overset{b}{\Longrightarrow}_g s$ and $s\, R\, t$. It follows that the restriction of R to $\text{NODES}(g) \times \text{NODES}(h)$ is a rooted weak bisimulation between g and h. Hence $g \leftrightarrows_{rw} h$. □

Rather than working with arbitrary process graphs, some people prefer to set a bound on the number of nodes. This happens for instance when one insists that $\mathbb{G}(Act)$ should be set rather than a proper class. That can be achieved by first fixing a set $\mathcal{N}$ of potential nodes, and then allowing only process graphs g with $\text{NODES}(g) \subseteq \mathcal{N}$. For any infinite cardinal κ let $\mathbb{G}_\kappa(Act)$ be the class of process graphs over Act with less than κ nodes. In particular $\mathbb{G}_{\aleph_0}(Act)$ is the class of regular, or finite-state, process graphs. Theorem 2 transfers smoothly from $\mathbb{G}(Act)$ to $\mathbb{G}_\kappa(Act)$.

Theorem 3. *Provided that there is an $a \in Act - \{\tau\}$, on $\mathbb{G}_\kappa(Act)$ weak bisimulation congruence coincides with rooted weak bisimulation equivalence.*

Proof. The proof above goes through because the process k fits in $\mathbb{G}_\kappa(Act)$. □

Another cardinality restriction that is sometimes imposed, is the requirement that each node should have less than κ outgoing edges. In case $\kappa > \aleph_0$, this class of *κ-branching* process graphs is not essentially different from $\mathbb{G}_\kappa(Act)$, and the same proof applies. More interesting is the case $\kappa = \aleph_0$. Up to strong bisimulation equivalence, the *finitely branching* process graphs form a proper superclass of $\mathbb{G}_{\aleph_0}(Act)$, the finite-state process graphs, and a proper subclass of $\mathbb{G}_{\aleph_1}(Act)$, the countable state process graphs. However, as shown in [1], any countable state process graph is weakly bisimilar to a finitely branching process graph. Using this, the proof of Theorem 2 can be adapted to apply to the class of finitely branching process graphs as well.

5 The Left-Merge and Rooted Weak Simulations

In [2] is has been shown that $\leftrightarrows_{rw}$ is a congruence for all operators of ACP_τ. Just as for the $+$, the root condition comes to the rescue in the congruence proof for the left-merge $\lfloor\!\lfloor$. One has $\tau.a.0 \lfloor\!\lfloor b.0 \not\leftrightarrows_w a.0 \lfloor\!\lfloor b.0$, because only the former process can do a b before an a, so $\leftrightarrows_w$ fails to be a congruence for this operator. The reason that the root condition (Def. 5) helps here, is that in $g \lfloor\!\lfloor h$ steps from h can happen in any state of g except the initial one; thus only a weak bisimulation between g_1 and g_2 that does not relate roots with non-roots can be modified into a weak bisimulation between $g_1 \lfloor\!\lfloor h$ and $g_2 \lfloor\!\lfloor h$.

There is a directional difference in the way the root condition solves the congruence problems for $+$ and $\lfloor\!\lfloor$. This can be seen by applying it to weak simulation.

Definition 13. Let $g, h \in \mathbb{G}^\rho(Act)$. The graph g is *weakly simulated* by the graph h if there exists a binary relation $R \subseteq \text{NODES}(g) \times \text{NODES}(h)$, called a *weak simulation* from g to h, satisfying, for all $a \in Act$:

- $\text{ROOT}(g)\ R\ \text{ROOT}(h)$.
- If sRt and $s \xrightarrow{a}_g s'$, then there is a t' such that $t \overset{(a)}{\Longrightarrow}_h t'$ and $s'Rt'$.

A weak simulation from g to h is *source rooted* if $s\, R\, \text{ROOT}(h) \;\Rightarrow\; s = \text{ROOT}(g)$. It is *target rooted* if $\text{ROOT}(g)\, R\, t \;\Rightarrow\; t = \text{ROOT}(h)$.

The relation of "being weakly simulation by" is a *preorder* (transitive and reflexive), called the *weak simulation preorder*. Likewise one obtains the *source rooted weak simulation preorder* and the *target rooted weak simulation preorder*.

Proposition 10. *The target rooted weak simulation preorder is a precongruence for the $+$. This means that if there are target rooted weak simulations from g_1 to h_1 and from g_2 to h_2, then there is a target rooted weak simulation from $g_1 + g_2$ to $h_1 + h_2$.*

Proof. Suppose that R_i is a target rooted weak simulation from g_i to h_i for $g_i, h_i \in \mathbb{G}^\rho(Act)$ and $i = 1, 2$ then $R_1 \cup R_2$ is a target rooted weak simulation from $g_1 +_\rho g_2$ to $h_1 +_\rho h_2$. □

Lemma 2. *If there is a weak simulation from g to h then there is a target rooted weak simulation from g to h.*

Proof. Omit all liaisons of the form $\text{ROOT}(g)\,R\,t$ with $t \neq \text{ROOT}(h)$. □

Corollary 1. *The weak simulation preorder is a precongruence for the +, as well as for action prefixing, even without upgrading it with root conditions.*

A complete axiomatisation of this preorder is given by the usual axioms for strong bisimulation, together with the axioms $x \sqsubseteq x + y$ and $\tau.x = x$, where the latter is a shorthand for $\tau.x \sqsubseteq x \wedge x \sqsubseteq \tau.x$.

Proposition 11. *The source rooted simulation preorder is a precongruence for the* $\|\!\!\lfloor$.

Proof Idea. Suppose that R is a weak simulation from g to h for $g, h \in \mathbb{G}^\rho(Act)$. When adapting R to a weak simulation from $g\|\!\!\lfloor k$ to $h\|\!\!\lfloor k$, one needs to make sure that when $g\|\!\!\lfloor k$ can do a step from k, then so can $h\|\!\!\lfloor k$. The only way this can fail is when a non-root state from g is related to the root of h. In that case $g\|\!\!\lfloor k$ can do a step from k, but $h\|\!\!\lfloor k$ cannot. □

The weak simulation preorder fails to be a precongruence for the $\|\!\!\lfloor$, for $\tau.a.0$ is weakly simulated by $a.0$, but $\tau.a.0\|\!\!\lfloor b.0$ is not weakly simulated by $a.0\|\!\!\lfloor b.0$. This is because the weak simulation from $\tau.a.0$ to $a.0$ is not source rooted. It follows that the source rooted weak simulation preorder is strictly finer than the (target rooted) weak simulation preorder.

It is possible to upgrade the notion of weak simulation by various conditions, such as the following *stability* requirement [4].

Definition 14. A weak simulation R from g to h is *stability respecting* if

– if sRt and $s \overset{\tau}{\not\longrightarrow}_g$ then there is a t' such that $t \Longrightarrow_h t' \overset{\tau}{\not\longrightarrow}_h$ and sRt'.

Again, target rootedness is needed to make the induced preorder into a precongruence for the +, and source rootedness to make it into a precongruence for the $\|\!\!\lfloor$. This time the two rooted variants are incomparable, because there is a target rooted, but not source rooted, stability respecting weak simulation from $\tau.a.0$ to $a.0$, and a source rooted, but not target rooted, stability respecting weak simulation from $a.0$ to $\tau.a.0$.

6 Concluding Remark

The method to turn simulation or bisimulation based equivalences into congruences by insisting on root conditions generalises smoothly to other notions of (bi-)simulation. In the case of branching, delay and η-bisimulation, this is elaborated in [5]. Interestingly, the alternative characterisation of rooted weak bisimulation presented in Prop. 7 takes a rather different shape when applied to rooted branching bisimulation. However, the characterisation with the root condition of Def. 5 remains the same.

Acknowledgement. Many crucial ideas incorporated in this paper originate from Jan Willem Klop, and stem from the mid eighties. Originally, Jan Willem and I planned to write a joint pamphlet on this matter, but as this plan never materialised, I finally collected the material in this paper, dedicated to Jan Willem at the occasion of his sixtieth birthday.

References

1. J.C.M. Baeten, J.A. Bergstra & J.W. Klop (1987): *On the consistency of Koomen's fair abstraction rule.* *Theoretical Computer Science* 51(1/2), pp. 129–176.
2. J.A. Bergstra & J.W. Klop (1985): *Algebra of communicating processes with abstraction.* *Theoretical Computer Science* 37(1), pp. 77–121.
3. J.A. Bergstra & J.W. Klop (1986): *Algebra of communicating processes.* In J.W. de Bakker, M. Hazewinkel & J.K. Lenstra, editors: *Mathematics and Computer Science*, CWI Monograph 1, North-Holland, pp. 89–138.
4. R.J. van Glabbeek (1993): *The linear time – branching time spectrum II; the semantics of sequential systems with silent moves (extended abstract).* In E. Best, editor: Proceedings *CONCUR'93*, 4th International Conference on *Concurrency Theory*, Hildesheim, Germany, LNCS 715, Springer, pp. 66–81.
5. R.J. van Glabbeek & W.P. Weijland (1996): *Branching time and abstraction in bisimulation semantics.* *Journal of the ACM* 43(3), pp. 555–600.
6. M. Hennessy & R. Milner (1985): *Algebraic laws for nondeterminism and concurrency.* *Journal of the ACM* 32(1), pp. 137–161.
7. R. Milner (1990): *Operational and algebraic semantics of concurrent processes.* In J. van Leeuwen, editor: *Handbook of Theoretical Computer Science*, chapter 19, Elsevier Science Publishers B.V. (North-Holland), pp. 1201–1242.

Böhm's Theorem, Church's Delta, Numeral Systems, and Ershov Morphisms

Richard Statman[1] and Henk Barendregt[2]

[1] Department of Mathematics, Carnegie-Mellon University, Pittsburgh PA, USA
[2] Faculty of Science, Radboud University, Nijmegen, The Netherlands

Abstract. In this note we work with untyped lambda terms under β-conversion and consider the possibility of extending Böhm's theorem to infinite RE (recursively enumerable) sets. Böhm's theorem fails in general for such sets $\mathcal{V}$ even if it holds for all finite subsets of it. It turns out that generalizing Böhm's theorem to infinite sets involves three other superficially unrelated notions; namely, Church's delta, numeral systems, and Ershov morphisms. Our principal result is that Böhm's theorem holds for an infinite RE set $\mathcal{V}$ closed under beta conversion iff $\mathcal{V}$ can be endowed with the structure of a numeral system with predecessor iff there is a Church delta (conditional) for $\mathcal{V}$ iff every Ershov morphism with domain $\mathcal{V}$ can be represented by a lambda term.

1 Introduction

We suppose the reader knows some lambda calculus, as e.g. in [1], Chapters 6, 7, 8 and 10.

Definition 1.1. (i) *The set of untyped closed lambda terms is denoted by* $\Lambda^{\emptyset}$. A combinator *is an element of* $\Lambda^{\emptyset}$.

(ii) *We denote congruence under beta conversion by* $=$.

(iii) *We write* $:=$ *for "equal by definition".*

(iv) *We define the following combinators.*

$$\mathbf{c}_n := \lambda fx.f^n x, \qquad \textit{the } \text{Church numerals.}$$
$$\mathsf{U}^n_k := \lambda x_1 \ldots x_n.x_k, \qquad \textit{for } 1 \leq k \leq n, \textit{ the } \text{projections.}$$
$$\Omega := (\lambda x.xx)(\lambda x.xx).$$

(v) *For lambda terms* $\boldsymbol{P} = P_1,\ldots,P_n$ *we write*

$$\langle P_1,\ldots,P_n\rangle := \lambda z.zP_1\ldots P_n.$$

Note that

$$\langle P_1,\ldots,P_n\rangle = \langle Q_1,\ldots,Q_n\rangle \Leftrightarrow P_1 = Q_1 \;\&\; \ldots \;\&\; P_n = Q_n.$$

The classical theorem of Böhm states the following.

A. Middeldorp et al. (Eds.): Processes... (Klop Festschrift), LNCS 3838, pp. 40–54, 2005.

Theorem 1.2 ([3]). *For all combinators M_1 and M_2 having a β-nf (normal form) the following are equivalent.*

(i) *For all combinators N_1, N_2 there exist combinators $\boldsymbol{P}$ such that*

$$M_1\boldsymbol{P} = N_1 \ \&\ M_2\boldsymbol{P} = N_2.$$

(ii) *There exists a combinator F such that*

$$FM_1 = \lambda xy.x \ \&\ FM_2 = \lambda xy.y.$$

(iii) *$M_1 = M_2$ is inconsistent with $\lambda\beta$.*
(iv) *$M_1 = M_2$ is inconsistent with $\lambda\beta\eta$.*
(v) *M_1 and M_2 have distinct $\beta\eta$-nfs (normal forms).*

Proof. (i)⇒(ii) Let $N_i := \lambda x_1 x_2.x_i$, for $1 \leq i \leq 2$. By (i) there are $\boldsymbol{P}$ such that $M_i\boldsymbol{P} = N_i$. Take $F := \lambda m.m\boldsymbol{P}$.

(ii)⇒(iii) From the equation $M_1 = M_2$ one can by (ii) derive $\lambda xy.x = \lambda xy.y$, from which one can derive any equation; all derivations using just $\lambda\beta$.

(iii)⇒(iv) Trivial.

(iv)⇒(v) By the hypothesis that M_1, M_2 have β-nf and [1], Corollary 15.1.5, it follows that M_1, M_2 have $\beta\eta$-nfs. If these were equal, then $M_1 =_{\beta\eta} M_2$ and hence $M_1 = M_2$ would be consistent.

(v)⇒(i) This is the core of Böhm's theorem. A proof can be found in [1], Theorem 10.4.2. □

The equivalences do not hold for arbitrary terms M_1, M_2, not in β-nf.

Remark 1.3. Referring to Theorem 1.2 one has the following.

1. In the list of equivalences on could add (iv^{a}) $M \neq_{\beta\eta} N$. Indeed, (iv) ⇒ (iv^{a}) ⇒ (v).
2. The implications (i)⇒(ii), (ii)⇒(iii), (iii)⇒(iv) and (v)⇒(iv) hold trivially for all M_1, M_2. Also (v)⇒(i) holds (but not trivially), as the condition of normalizability holds by assumption.
3. In general (iv)$\not\Rightarrow$(v). One has $\Omega = \mathsf{I}$ is consistent with $\lambda\beta\eta$, as follows by the technique of [8], but Ω does not have a $\beta\eta$-nf.
4. Similarly (iv)$\not\Rightarrow$(iii). For example the set of equations

$$\{\Omega\mathsf{I} = \mathsf{U}^2_1, \Omega\mathbf{c}_1 = \mathsf{U}^2_2\}$$

is consistent with $\lambda\beta$, see [1], Corollaries 15.3.6 and 15.3.7. But the set is inconsistent with $\lambda\beta\eta$, as $\mathsf{I} =_{\beta\eta} \mathbf{c}_1$. Hence $\langle\Omega\mathsf{I}, \Omega\mathbf{c}_1\rangle = \langle\mathsf{U}^2_1, \mathsf{U}^2_2\rangle$ is consistent with $\lambda\beta$, but not with $\lambda\beta\eta$.
5. As to (iii)$\not\Rightarrow$(ii), the equation $\Omega_3 = \mathsf{I}$, with $\Omega_3 \equiv (\lambda x.xxx)(\lambda x.xxx)$, is inconsistent as shown in [8]. But if $F\Omega_3 = \lambda xy.x$ and $F\mathsf{I} = \lambda xy.y$ for some F, then by [1], Proposition 14.3.24, it follows that either Ω_3 is solvable, which it isn't, or $\forall M.FM = \lambda xy.x$, which contradicts the second equation.

6. (i)$\not\Leftarrow$(v) Let $M_1 = \langle \lambda xy.x, \Omega \rangle, M_2 = \langle \lambda xy.y, \Omega \rangle$. Taking $\boldsymbol{P} := \mathsf{U}_1^2, N_1, N_2$ one has $M_1 \boldsymbol{P} = N_1$ & $M_2 P = N_2$, but M_i has no β-nf.
7. We believe that (ii)$\Rightarrow$(i) also holds in general.

Let $\mathcal{F} = \{M_1, \ldots, M_n\}$ be a finite set of combinators.

Definition 1.4. *$\mathcal{F}$ is called* separable *iff there exists a combinator F such that*

$$FM_1 = \mathsf{U}_1^n \;\&\; \ldots \;\&\; FM_n = \mathsf{U}_n^n.$$

This notion of separability has a different but equivalent definition.

Lemma 1.5. *Let $\mathcal{F} = \{M_1, \ldots, M_n\}$ be a finite set of arbitrary combinators. Then the following are equivalent.*

(i) *There exists an $F \in \Lambda^{\emptyset}$ such that*

$$FM_1 = \mathsf{U}_1^n \;\&\; \ldots \;\&\; FM_n = \mathsf{U}_n^n.$$

(ii) *There exists an $F \in \Lambda^{\emptyset}$ such that*

$$FM_1 = \mathbf{c}_1 \;\&\; \ldots \;\&\; FM_n = \mathbf{c}_n.$$

Proof. (i)$\Rightarrow$(ii) Given F, take $F' := \lambda m.Fm\mathbf{c}_1 \ldots \mathbf{c}_n$.

(ii)$\Rightarrow$(i) Assume (ii) for some F. By induction on n one can show that for some G_n one has

$$1 \leq k \leq n \;\Rightarrow\; G_n \mathbf{c}_k = \mathsf{U}_k^n \qquad (*)$$

For $n = 0$ there is nothing to prove and for $n = 1$ we can take $G_1 = \mathsf{K}\mathsf{U}_1^1$. Suppose that G_n has been defined and satisfies $(*)$. Define

$$\begin{aligned} G_{n+1} &:= \lambda c.\,\texttt{If (Zero? } c\texttt{) then } \mathsf{U}_1^{n+1} \texttt{else } (G_n(P^- c)\mathsf{U}_2^{n+1} \ldots \mathsf{U}_{n+1}^{n+1}) \\ &:= \lambda c.(\texttt{Zero?}\ c)\mathsf{U}_1^{n+1}(G_n(P^- c)\mathsf{U}_2^{n+1} \ldots \mathsf{U}_{n+1}^{n+1}), \end{aligned}$$

where (If B then P else Q)$\equiv BPQ$, $\texttt{Zero?} \equiv \lambda n.n(\lambda x.\mathsf{U}_2^2)\mathsf{U}_1^2$ so that

$$\texttt{Zero?}\mathbf{c}_0 = \mathsf{U}_1^2, \;\; \texttt{Zero?}\mathbf{c}_{k+1} = \mathsf{U}_2^2$$

and $P^- \in \Lambda^{\emptyset}$ is a representation of the predecessor function for the Church numerals. Then G_{n+1} works. Now we can take $F' := \lambda m.G_n(Fm)$. □

For infinite sets the two ways of defining the notions of separability are no longer equivalent. Separability for possibly infinite sets has to be defined as the existence of a definable 1-1 map (modulo β-conversion) to the Church numerals.

Definition 1.6. *$\mathcal{V}$ is said to be* separable *if for some combinator D one has*

(i) $\forall M \in \mathcal{V} \exists n \in \mathbb{N}.DM = \mathbf{c}_n$.
(ii) $\forall M, N \in \mathcal{V}.[DM = DN \Leftrightarrow M = N]$.

For such $\mathcal{F}$ one has the following generalizaton of Theorem 1.2.

Theorem 1.7 ([4]). *For all $\mathcal{F} = M_1, \dots, M_n$, where each M_i has a β-nf, the following are equivalent.*

(i) *For all combinators $N_1, \dots, N_n$ there exist combinators $\boldsymbol{P}$ such that*

$$M_1\boldsymbol{P} = N_1 \; \& \; \dots \; \& \; M_n\boldsymbol{P} = N_n.$$

(ii) *$\mathcal{F}$ is separable.*
(iii) *$M_p = M_q$ is inconsistent with $\lambda\beta$, for $1 \le p, q \le n$ with $p \ne q$.*
(iv) *$M_p = M_q$ is inconsistent with $\lambda\beta\eta$, for $1 \le p, q \le n$ with $p \ne q$.*
(v) *The $M_1, \dots, M_n$ have pairwise distinct $\beta\eta$-nfs.*

Proof. Again, the only non-trivial implication is (v)⇒(i) and is proved in [4], see [1], Corollary 10.4.14. □

In trying to generalize Theorem 1.7 by dropping the requirement that $\mathcal{F}$ is finite or that its elements have a nf, several problems arise.

Remark 1.8. (i) For finite sets $\mathcal{F}$ of combinators having a β-nf the property of consisting of pairwise $\beta\eta$-inconvertible terms is equivalent to separability. Indeed, the Church-Rosser Theorem implies that

$$M \ne_{\beta\eta} N \;\Leftrightarrow\; M, N \text{ have distinct } \beta\eta\text{-nfs},$$

hence Theorem 1.7 applies. If we drop the requirement that the elements of $\mathcal{F}$ have a β-nf, then this is no longer true. For example this is the case with

$$\mathcal{F} = \{\Omega\mathbf{c}_1, \dots, \Omega\mathbf{c}_n\}.$$

This set consists of pairwise β-inconvertible terms, but is not separable, as follows from the Genericity Lemma in [1], Proposition 14.3.24.

(ii) For infinite sets $\mathcal{F}$ of terms having pairwise distinct $\beta\eta$-nf separability does not necessarily hold either. An example is the collection of projections

$$\mathcal{F} = \{\mathsf{U}^n_k \mid n \in \mathbb{N} \;\&\; 1 \le k \le n\}.$$

The set clearly consists of pairwise distinct $\beta\eta$-nfs. This $\mathcal{F}$ is such that each finite subset of it is separable but not the whole set itself. That $\mathcal{F}$ is not separable can be seen as follows. Suppose that F is a combinator which maps the set of projections into the Church numerals injectively. Let U range over the projections. FU is $\beta\eta$-convertible to a Chruch numeral which is a λI-term for all U, except possibly one (that is mapped to $\mathbf{c}_1$). Then for those U it follows, by η-postponement, see [1], Corollary 15.1.6, and the obvious fact that η-conversion does not change the status of being a λI-term, that $FU =_\beta N$ for some λI-term N in nf. Consider a standard reduction $FU \twoheadrightarrow_\beta N$. Then $Fx \twoheadrightarrow_\beta N'$, with $x \in \mathrm{FV}(N')$ (as F is not a constant map) and to the left of the leftmost occurrence of x in N' there is no expression of the form $(\lambda y.P)$. But then for almost all U one has that $FU = N'[x := U]$ is not convertible to a λI-term, a contradiction.

A characterization of separability for finite $\mathcal{F}$, possibly containing terms without normal form is due to [5] and can be found also in [1], Theorem 10.4.13. To give a flavor of that theorem we give some of its consequences, rather than repeating its precise formulation.

1. The set $\left\{\begin{array}{l}\lambda x.x\mathbf{c}_0\Omega, \\ \lambda x.x\mathbf{c}_1\Omega\end{array}\right\}$ is separable.
2. $\left\{\begin{array}{l}\lambda x.x(\lambda y.y\Omega), \\ \lambda x.x(\lambda y.y\mathbf{c}_0)\end{array}\right\}$ is not separable.
3. $\left\{\begin{array}{l}\lambda x.x(\lambda y.y\mathbf{c}_0\Omega(\lambda z.z\Omega), \\ \lambda x.x(\lambda y.y\mathbf{c}_1\Omega(\lambda z.z\mathbf{c}_1), \\ \lambda x.x(\lambda y.y\mathbf{c}_1\Omega(\lambda z.z\mathbf{c}_2)\end{array}\right\}$ is separable.
4. $\left\{\begin{array}{l}\lambda x.x\mathbf{c}_0\mathbf{c}_0\Omega, \\ \lambda x.x\mathbf{c}_1\Omega\mathbf{c}_1, \\ \lambda x.x\Omega\mathbf{c}_2\mathbf{c}_2\end{array}\right\}$ is not separable, although each proper subset is.

Definition 1.9. *A set* $\mathcal{X} \subseteq \Lambda^{\varnothing}$ *is called an* adequate numeral system *iff for some combinators* $O, S, Z_?, P$ *one has*

(i) $\mathcal{X} = \{S^n O \mid n \in \mathbb{N}\}$.
(ii) $P(S^{n+1}O) = S^n O$.
(iii) $Z_? O = \mathsf{U}_1^2$ & $Z_?(S^{n+1}O) = \mathsf{U}_2^2$.

Then all partial computable functions can be represented on $\mathcal{X}$, see [1] Proposition 6.4.3 and the remark following.

Definition 1.10. *Let* $\mathcal{V}$ *be a set of combinators.*

(i) $\mathcal{V}$ *is called* RE (recursively enumerable) *if the set* $\#\mathcal{V} = \{\#M \mid M \in \mathcal{V}\}$ *is an RE set of natural numbers.*
(ii) $\mathcal{V}$ *is* closed under β-conversion *iff* $\forall M, N.(M \in \mathcal{V}$ & $M =_\beta N) \Rightarrow N \in \mathcal{V}$.
(iii) $\mathcal{V}^\beta = \{N \mid \exists M \in \mathcal{V}.M =_\beta N\}$, *the* β-closure of $\mathcal{V}$.
(iv) $\mathcal{V}$ *is a* V-set *iff* $\mathcal{V}$ *is RE and closed under* β*-conversion. These sets are the closed sets in the Visser topology, see [12] or [1], Definition 17.1.12.*

Lemma 1.11. $\mathcal{V} \subseteq \Lambda^{\varnothing}$ *be non-empty . Then* $\mathcal{V}$ *is a V-set iff for some* $F \in \Lambda^{\varnothing}$ *one has* $\mathcal{V} = \{F\mathbf{c}_n \mid n \in \mathbb{N}\}^\beta$.

Proof. ($\Rightarrow$) By assumption $\#\mathcal{V}$ is non-empty and RE. Then $\#\mathcal{V} = \{g(n) \mid n \in \mathbb{N}\}$ for some total computable function g. Then

$$V = \{\mathsf{E}\mathbf{c}_n \mid n \in \#\mathcal{V}\}^\beta.$$

Let $G \in \Lambda^{\varnothing}$ lambda define g. Then

$$\begin{aligned} \mathcal{V} &= \{\mathsf{E}\mathbf{c}_{g(n)} \mid n \in \mathbb{N}\}^\beta, \\ &= \{\mathsf{E}(G\mathbf{c}_n) \mid n \in \mathbb{N}\}^\beta, \\ &= \{F\mathbf{c}_n \mid n \in \mathbb{N}\}^\beta, \qquad \text{with } F = \mathsf{E} \circ G. \end{aligned}$$

($\Leftarrow$) $M \in \{F\mathbf{c}_n \mid n \in \mathbb{N}\}^\beta$ iff $\exists n.N =_\beta F\mathbf{c}_n$, which is RE. Clearly this set is closed under β-conversion. □

In the present paper the following will be proved. See 2.1(iv) for the definition of morphisms. Some of these results have been proved in [9] under stronger hypotheses: (i) $\Leftrightarrow$ (iii), (v) $\Rightarrow$ (i).

Theorem 1.12. *Suppose that $\mathcal{V}$ is an infinite RE set (after coding) of combinators closed under β-conversion. Then the following are equivalent.*

(i) *$\mathcal{V}$ is an adequate numeral system.*

(ii) *For every morphism Φ with dom(Φ) $\subseteq \mathcal{V}$ there is an $F \in \Lambda^{\emptyset}$ such that*

$$\forall M \in \mathcal{V}.\Phi(M) = FM.$$

(iii) *There exists a combinator Δ such that for all $M, N \in \mathcal{V}$*

$$\begin{aligned} \Delta MN &= \mathsf{U}^2_1, \text{ if } M = N; \\ &= \mathsf{U}^2_2, \text{ else.} \end{aligned}$$

(iv) *There is a morphism Φ with dom(Φ) $\subseteq V$ such that for all $M, N \in \mathcal{V}$*

$$\begin{aligned} \Phi(M)N &= \mathsf{U}^2_1, \text{ if } M = N, \\ &= \mathsf{U}^2_2, \text{ else.} \end{aligned}$$

(v) *$\mathcal{V}$ is separable.*

One way to think of Böhm's theorem is that it says that separating morphisms can be realized by terms.

2 Preliminaries

Definition 2.1. (i) *Let $\# : \Lambda^{\emptyset} \to \mathbb{N}$ be an effective surjective Gödel numbering of combinators. We write $\ulcorner M \urcorner$ for $\mathbf{c}_{\#M}$.*

(ii) *There is an inverse* E, *called* Kleene's enumerator, *such that* $\mathsf{E}\ulcorner M \urcorner = M$, *for all combinators M, see [1] Theorem 8.1.6.*

(iii) *For $m, n \in \mathbb{N}$ we write $m \sim n \Leftrightarrow \mathsf{E}\mathbf{c}_m = \mathsf{E}\mathbf{c}_n$.*

(iv) *A* (partial Ershov) morphism *$\Phi : \Lambda^{\emptyset}/= \to \Lambda^{\emptyset}/=$ is a partial map such that for some partial computable function $\varphi : \mathbb{N} \to \mathbb{N}$ and all combinators M*

$$\Phi(M) \cong \mathsf{E}(\mathbf{c}_{\varphi(\#M)}),$$

where $P \cong Q$ means that if one of the two expressions P, Q is defined, then so is the other and $P = Q$. This is implied by $\#\Phi(M) \simeq \varphi(\#M)$, with a similar meaning for $\simeq$: for expressions e_1, e_2 involving partial functions we define

$$e_1 \simeq e_2 \Leftrightarrow [e_1{\downarrow} \Leftrightarrow e_2{\downarrow}] \;\&\; [e_1{\downarrow} \Rightarrow e_1 \sim e_2].$$

See the last section for a discussion about the origin of Ershov morphisms.

(v) *The notion of morphism generalizes naturally to binary maps.* $\Phi : (\Lambda^{\emptyset})^2 \to \Lambda^{\emptyset}$ *is a morphism if for some binary partial computable* φ *one has*

$$\Phi(M,N) \cong \mathbf{Ec}_{\varphi(\#M,\#N)}.$$

(vi) *We write* $\Phi(M)\downarrow, \varphi(m)\downarrow$ *for convergence of the partial functions (being defined); similarly* $\Phi(M)\uparrow, \varphi(m)\uparrow$ *for divergence (being undefined).*

Lemma 2.2. *A partial morphism* Φ *is completely determined by a partial computable* φ *such that*

$$\begin{aligned} \Phi(M) &\cong \mathsf{Ec}_{\varphi(\#M)};\\ n \sim m &\Rightarrow \varphi(n) \simeq \varphi(m). \end{aligned}$$

In this case Φ *is called the morphism corresponding to* φ.

Proof. This is the defining property for morphisms. We emphasize here that if $M =_\beta N$ and $\Phi(M)\downarrow$, then $\Phi(M) = \Phi(N)$, hence $\varphi(\#M) \sim \varphi(\#N)$. □

Although there are partial recursive functions that cannot be made total, this is not the case for partial morphisms, as shown in [10].

Theorem 2.3 (Morphism extension). *Suppose that* Φ *is a partial morphism. Then there exists a total morphism* F *extending* Φ.

Proof. Let Φ correspond to the partial computable function φ. Construct a combinator P such that

$$Px\ulcorner M\urcorner =_\beta \begin{cases} \mathsf{E}x\ulcorner N\urcorner & \text{if } N =_\beta M \;\&\; \#N < \#M \;\&\\ & N \text{ is the first such found in some enumeration}\\ & \text{of the beta converts of } M;\\ \mathsf{Ec}_{\varphi(\#M)} & \text{if } \varphi(\#M) \text{ converges before such } N \text{ is found.} \end{cases}$$

[If $\varphi(\#M)\uparrow$ and $\neg\exists N =_\beta M.(\#N < \#M)$, then the search continues forever; in that case $Px\ulcorner M\urcorner$ will be unspecified and can be arranged to be unsolvable.] By the second fixed-point theorem, see [1], Theorem 6.5.9, there is a combinator Q such that $P\ulcorner Q\urcorner = Q$. Then

$$Q\ulcorner M\urcorner =_\beta \begin{cases} Q\ulcorner N\urcorner & \text{if } N =_\beta M \;\&\; \#N < \#M \;\&\\ & N \text{ is the first such found in some enumeration}\\ & \text{of the beta converts of } M;\\ \mathsf{Ec}_{\varphi(\#M)} & \text{if } \varphi(\#M) \text{ converges before such } N \text{ is found.} \end{cases}$$

If $\varphi(M)\downarrow$, then, using Lemma 2.2, it can be seen that a typical computation for $Q\ulcorner M\urcorner$ is the following

$$\begin{aligned} Q\ulcorner M\urcorner &= Q\ulcorner M_1\urcorner = \ldots = Q\ulcorner M_k\urcorner = \mathsf{Ec}_{\#\varphi(M_k)} =\\ &= \Phi(M_k) = \ldots = \Phi(M_1) = \Phi(M), \end{aligned}$$

with

$$\begin{array}{rrcr} \#M > & \#M_1 > & \ldots > & \#M_k, \\ M =_\beta & M_1 =_\beta & \ldots =_\beta & M_k, \\ \#M \sim & \#M_1 \sim & \ldots \sim & \#M_k, \\ \varphi(\#M) \sim & \varphi(\#M_1) \sim & \ldots \sim & \varphi(\#M_k), \\ \Phi(M) = & \Phi(M_1) = & \ldots = & \Phi(M_k). \end{array}$$

Therefore one has

$$Q\ulcorner M \urcorner =_\beta \begin{cases} \Phi(N) & \text{if } \Phi(M)\downarrow; \\ Q\ulcorner M_0 \urcorner & \text{else,} \end{cases}$$

where $M_0 =_\beta M$ with the smallest Gödel number and $Q\ulcorner M_0 \urcorner$ is unsolvable.

Finally let f be defined by $f(\#M) = \#(Q\ulcorner M \urcorner)$. Then it is easy to see that

1. f is a total computable function;
2. if $M =_\beta N$, then $Q\ulcorner M \urcorner =_\beta Q\ulcorner N \urcorner$, hence $f(\#M) \sim f(\#N)$;
3. if $\varphi(\#M)\downarrow$, then $\varphi(\#M) \sim f(\#M)$.

Thus there is a total morphism F determined by f. Moreover F extends Φ: $FM = Q\ulcorner M \urcorner = \Phi(M)$, if the latter is defined. □

Definition 2.4. *Let $\mathcal{V} \subseteq \Lambda^{\emptyset}$. Then $\mathcal{V}$ is a* V-set *if it is RE and closed under β-conversion.*

Definition 2.5. *Let $\mathcal{V}$ be a V-set.*

(i) *A $\mathcal{V}$-morphism is a partial Ershov morphism whose domain includes $\mathcal{V}$*

(ii) *A $\mathcal{V}$-morphism f is $\mathcal{V}$-representable if there exists an $F \in \Lambda^{\emptyset}$ such that*

$$\forall M \in \mathcal{V}. FM = f(M).$$

A similar definition holds for binary morphisms.

(iii) *Δ is a* Church discriminator (or Church delta) *for $\mathcal{V}$ if for all $M, N \in \mathcal{V}$ one has*

$$\Delta MN = \mathsf{U}_1^2, \ \textit{if } M = N;$$
$$\Delta MN = \mathsf{U}_2^2, \ \textit{if } M \neq N.$$

(iv) *Let $M \in \mathcal{V}$ and $\Phi = \Phi_M$ be a $\mathcal{V}$-morphism. Then Φ is a $\mathcal{V}$-*equality test *for M if for all $N \in \mathcal{V}$*

$$\Phi(N) = \mathsf{U}_1^2, \ \textit{if } M = N,$$
$$\Phi(N) = \mathsf{U}_2^2, \ \textit{if } M \neq N.$$

Lemma 2.6. *Let $\mathcal{V}$ be a V-set. If every unary morphism on $\mathcal{V}$ is $\mathcal{V}$-representable, then the same is true for binary morphisms.*

Proof. Given a binary morphism Φ, define

$$\Psi(M) := \Phi(M\,\mathsf{U}_1^2, M\,\mathsf{U}_2^2).$$

Let Ψ be $\mathcal{V}$-representable by ψ. Construct a binary partial computable function φ such that $\varphi(\#M, \#N) = \psi(\#\langle M, N\rangle)$. Then Φ is represented by φ:

$$\begin{aligned}\Phi(M,N) &= \Psi(\langle M,N\rangle)\\ &= \mathsf{E}\mathbf{c}_{\psi(\#\langle M,N\rangle)}\\ &= \mathsf{E}\mathbf{c}_{\varphi(\#M,\#N)}.\end{aligned}$$

□

Fact 2.7. The following statements are equivalent for a V-set $\mathcal{V}$.

(i) There is a $\mathcal{V}$-morphism Φ such that

$$\forall M,N \in \mathcal{V}.[\Phi(M) = \Phi(N) \;\Rightarrow\; M = N] \;\&$$
$$\forall M \in \mathcal{V}\exists n \in \mathbb{N}.\Phi(M) = \mathbf{c}_n.$$

(ii) $\{\langle M,N\rangle \mid M,N \in \mathcal{V} \;\&\; M \neq N\}$ is RE.

Hence if $\mathcal{V}$ is a separable V-set, then $\{\langle M,N\rangle \mid M,N \in \mathcal{V} \;\&\; M \neq N\}$ is RE.

3 Böhm's Theorem for V-Sets

Theorem 3.1. *For $\mathcal{V}$ an infinite V-set the following are equivalent.*

(i) *$\mathcal{V}$ is an adequate numeral system.*
(ii) *Every $\mathcal{V}$-morphism is $\mathcal{V}$-representable.*
(iii) *There is a Church discriminator for $\mathcal{V}$.*
(iv) *There is a $\mathcal{V}$-morphism Φ such that*

$$\forall M \in \mathcal{V}.\Phi(M) \textit{ is a } \mathcal{V}\textit{-equality test for } M.$$

(v) *$\mathcal{V}$ is separable.*

Proof. We shall prove (i) ⇒ (ii) ⇒ (iii) ⇒ (iv) ⇒ (v) ⇒ (i).

(i) ⇒ (ii). Write $\mathbf{v}_n := S^n O$. By [1] Lemma 6.4.5 there exists an H such that

$$H(\mathbf{v}_n) = \mathbf{c}_n. \tag{1}$$

The function $q(n) = \#\mathbf{v}_n$ (reminiscent of quoting) is total computable, so by [1] Theorem 6.4.3, it is lambda definable w.r.t. $(V, S, P, O, Z_?)$ by, say, Q, i.e.

$$Q\mathbf{v}_n = \mathbf{v}_{q(n)} = \mathbf{v}_{\#\mathbf{v}_n}. \tag{2}$$

Now suppose Φ is a partial morphism whose domain contains the set $\mathcal{V}$. By definition there is a partial computable function φ, such that

$$\Phi(M) = \mathsf{E}\mathbf{c}_{\varphi(\#M)}. \tag{3}$$

This f is lambda definable on $(V, S, P, O, Z_?)$ by, say, F. This means that

$$F\mathbf{v}_n = \mathbf{v}_{\varphi(n)}. \tag{4}$$

Let E be Kleene's enumerator and set $J := \lambda x.\mathsf{E}(H(F(Qx)))$. Then

$$\begin{aligned} J(\mathbf{v}_n) &= E(H(F(Q(\mathbf{v}_n)))) \\ &= E(H(F(\mathbf{v}_{\#\mathbf{v}_n}))), && \text{by (2)}, \\ &= E(H(\mathbf{v}_{\varphi(\#\mathbf{v}_n)})), && \text{by (4)}, \\ &= E\mathbf{c}_{\varphi(\#\mathbf{v}_n)}, && \text{by (1)}, \\ &= \Phi(\mathbf{v}_n), && \text{by (3)}. \end{aligned}$$

Thus Φ is $\mathcal{V}$-represented by J.

(ii) $\Rightarrow$ (iii). Let $M, N \in \mathcal{V}$ with $M \neq N$. Define a partial morphism Φ by

$$\begin{aligned} L = M &\Rightarrow \Phi(L) = \mathsf{U}^2_1 \\ L = N &\Rightarrow \Phi(L) = \mathsf{U}^2_2. \end{aligned}$$

Then this partial morphism extends to a total morphism by Theorem 2.3, which is *a fortiori* a $\mathcal{V}$-morphism. By hypothesis this morphism is $\mathcal{V}$-representable and thus for some F one has

$$FM = \mathsf{U}^2_1 \;\&\; FN = \mathsf{U}^2_2.$$

Hence, in particular, the set $\{(M, N) \in \mathcal{V}^2 \mid M \neq N\}$ is RE. Thus the partial function Φ on $\mathcal{V}^2$ such that

$$\begin{aligned} \Phi(M, N) &= \mathsf{U}^2_1, \text{ if } M = N, \\ \Phi(M, N) &= \mathsf{U}^2_2, \text{ if } M \neq N, \end{aligned}$$

is a $\mathcal{V}$-morphism, which by hypothesis and Lemma 2.6 is representable. In conclusion, there is a Church discriminator for $\mathcal{V}$.

(iii) $\Rightarrow$ (iv). Immediate, taking $\Phi(N) := \Delta N$.

(iv) $\Rightarrow$ (v). Let Φ be as in (iv) correspond to the partial computable φ which is λ-defined by F. Let G be an enumeration of $\mathcal{V}$, i.e. $\mathcal{V} = \{G\mathbf{c}_n \mid n \in \mathbb{N}\}$, possible by the definition of V-set. We want to define D such that

$$\begin{aligned} DM &= \mathbf{c}_{\mu y.[G\mathbf{c}_y = M]} && (5) \\ &= \mathbf{c}_{\mu y.[\Phi((G\mathbf{c}_y))M = \mathsf{U}^2_1]} \\ &= \mathbf{c}_{\mu y.[\mathsf{E}(F(G\mathbf{c}_y))M = \mathsf{U}^2_2]}. \end{aligned}$$

Now the right-hand-side can be defined as $HM\mathbf{c}_0$ if

$$HM = \lambda y.\mathsf{E}(F(Gy))My(HM(S^+y)),$$

where $S^+$is the successor for the Church numerals. This is the case if we take

$$H := \mathsf{Y}(\lambda hmy.\mathsf{E}(F(Gy))my(hm(S^+y)),$$

where Y is the Fixed-point combinator. Then (v) via $D := \lambda m.Hm\mathbf{c}_0$, by (5).

(v) ⇒ (i). Let $\mathcal{V} = \{F\mathbf{c}_e \mid e \in \mathbb{N}\}$ and let $d : \mathcal{V} \to \mathbb{N}$ be an injection definable by lambda term D . Then the set $\{(M, N) \in \mathcal{V}^2 \mid M \neq N\}$ is (after coding) RE. Define

$$\begin{aligned} O &:= F\mathbf{c}_0 \\ S &:= \lambda x.F(\mu y.[\forall z \leq (\mu n.[F\mathbf{c}_n = x]).Fy \neq Fz]) \\ P &:= \lambda x.F(\mu z.[S(Fz) = x]) \\ Z_? &:= \mathrm{Eq}\,(DO)(Dx), \end{aligned}$$

where Eq is the test for equality on the Church numerals. Then

$$\mathcal{V} = \{S^n O \mid n \in \mathbb{N}\},\ P(S^{n+1}O) = S^n O,\ Z_? O = \mathsf{U}^2_1 \text{ and } Z_?(S^{n+1}O) = \mathsf{U}^2_2. \ \square$$

Corollary 3.2. *Not every total Ershov morphism on $\Lambda^{\emptyset}$ is representable.*

Proof. Indeed, by Theorem 3.1 it would follow that there is a Church discriminator Δ for $\Lambda^{\emptyset}$, but then $\lambda x.\Delta x \mathsf{U}^2_2$ has no fixed-point, contradiction. □

We end with some examples showing that there are various ways in which one can have equality tests.

Proposition 3.3. (i) *Let $\mathcal{V}$ be a V-set with a $\mathcal{V}$-equality test for each member of $\mathcal{V}$ (but not uniformly so). This means that for all $M \in \mathcal{V}$ there exists a $\mathcal{V}$-morphism Φ_M such that for all $N \in \mathcal{V}$*

$$\begin{aligned} \Phi_M(N) &= \mathsf{U}^2_1, \ \textit{if } M = N; \\ &= \mathsf{U}^2_2, \ \textit{else}. \end{aligned}$$

Then it does not follow that there is a Church's discriminator for $\mathcal{V}$.

(ii) *Let $\mathcal{V}$ be a V-set with a $\mathcal{V}$-equality test for each member of $\mathcal{V}$ that is $\mathcal{V}$-representable. This means that for all $M \in \mathcal{V}$ there exists an $F_M \in \Lambda^{\emptyset}$ such that for all $N \in \mathcal{V}$*

$$\begin{aligned} F_M N &= \mathsf{U}^2_1, \text{ if } M = N; \\ &= \mathsf{U}^2_2, \text{ else}. \end{aligned}$$

Suppose moreover that $\{\langle M, N\rangle \mid M, N \in \mathcal{V} \ \&\ M \neq N\}$ is RE. Even then $\mathcal{V}$ does not necessarily have a Church discriminator.

Proof. (i) We will construct $\mathcal{V}$ with $\{\langle M, N\rangle \mid M, N \in \mathcal{V} \ \&\ M \neq N\}$ is not RE. This suffices by Fact 2.7. Define the following partial computable function

$$\begin{aligned} \psi(e, x) &= x, && \text{if } \{e\}(0) \text{ converges in exactly } x \text{ steps}, \\ &= 1 + \psi(e, x+1), && \text{else}. \end{aligned}$$

Here $\{e\}(x)$ is the result of the partial computable function with code (program) e and input x. By Kleene's theorem ψ is represented by a lambda term G and set $F := \lambda n.Gn\mathbf{c}_0$. Thus $F\mathbf{c}_e$ has finite Böhm tree $\mathrm{BT}(\mathbf{c}_n)$ if $\{e\}(0)$ converges in n steps and it can be arranged that otherwise $F\mathbf{c}_e$ has the infinite Böhm tree

$$\infty := \quad \begin{array}{c} \lambda xy.\,x \\ | \\ x \\ | \\ x \\ | \\ \ldots \end{array}$$

For each e there are many e^* such that $\{e^*\} = \{e\}$. In case $e(0)\uparrow$ for such e, e^*, one has

$$e \neq e^* \;\Rightarrow\; F\mathbf{c}_e \neq F\mathbf{c}_{e^*}, \tag{1}$$

even if $\mathrm{BT}(F\mathbf{c}_e) = \mathrm{BT}(F\mathbf{c}_{e^*})$ (the difference is 'pushed to infinity', hence the trees are equal). Take $\mathcal{V} = \{F\mathbf{c}_e \mid e \in \mathbb{N}\}/=_\beta$. We show that this V-set works. For each fixed combinator $M \in \mathcal{V}$, say $M = F\mathbf{c}_e$, we have to decide whether for a given combinator N one has $N = M$.

Case 1. $\{e\}(0)\downarrow$ in n steps. Then $M = \mathbf{c}_n$. Given $N \in \mathcal{V}$ we develop its Böhm tree, which will be one of $\{\mathrm{BT}(\mathbf{c}_0), \mathrm{BT}(\mathbf{c}_1), \ldots; \infty\}$. If $\mathrm{BT}(N) = \mathbf{c}_k$ with $k < n$, then $N \neq M$ and the output should be (the Gödel number of) U^2_2. If $\mathrm{BT}(N) = \mathbf{c}_n$, then the output should be U^2_1. Finally, if $\mathrm{BT}(N)$ keeps growing beyond $\mathrm{BT}(\mathbf{c}_n)$, then the output is again U^2_2.

Case 2. $\{e\}(0)\uparrow$. Then for $N \in \mathcal{V}$ one can check $N = M$ as follows. Find an e' such that $N = F\mathbf{c}_{e'}$. Then $N = M \;\Leftrightarrow\; F\mathbf{c}_e \neq F\mathbf{c}_{e^*} \;\Leftrightarrow\; e' = e$, by (1).

(ii) We will construct such a $\mathcal{V}$. First let T be Kleene's T predicate, i.e. $T(e, x, y)$ iff y is the code of a terminating computation for $\{e\}(x)$, the result of the partial computable function with program e on input x. Define the following total computable function.

$$\begin{aligned} f(e,n) &= 0, && \text{if } n \leq e \;\&\; \neg\exists k \leq n.T(e,0,k); \\ &= 1, && \text{if } n \leq e \;\&\; \exists k \leq n.T(e,0,k); \\ &= f(e,e), && \text{else.} \end{aligned}$$

Write $\mathbf{e}_n := \mathbf{c}_{f(e,n)}$. Define the following lambda terms for $n \in \mathbb{N}$:

$$n_\infty = \mathsf{Y}(\lambda p.\langle \mathbf{c}_n, p\rangle).$$

Then $n_\infty = \langle \mathbf{c}_n, n_\infty \rangle$ and e.g. 0_∞ has as Böhm-tree

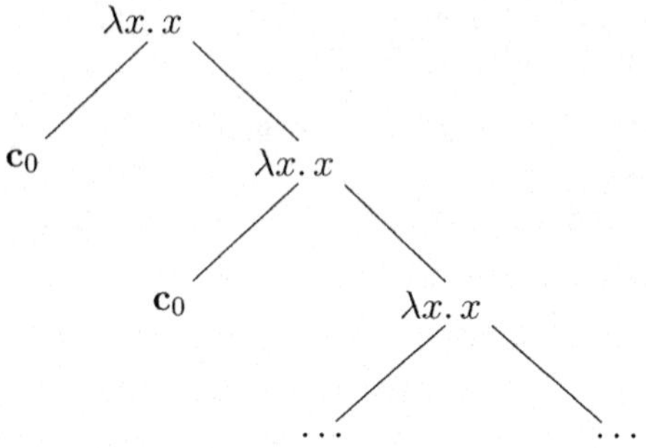

It is not hard to construct terms M_e such that

$$\begin{aligned} M_e &:= H\mathbf{c}_e\mathbf{c}_0; \\ H\mathbf{c}_e\mathbf{c}_n &:= \text{If } (n \le e \ \& \ \neg T(e,0,n)) \text{ then } \langle \mathbf{c}_0, H\mathbf{c}_e\mathbf{c}_{n+1} \rangle \\ &\qquad \text{else [If } (n \le e \ \& \ T(e,0,n)) \text{ then } 1_\infty \text{ else } 0_\infty]. \end{aligned}$$

having as Böhm-Trees

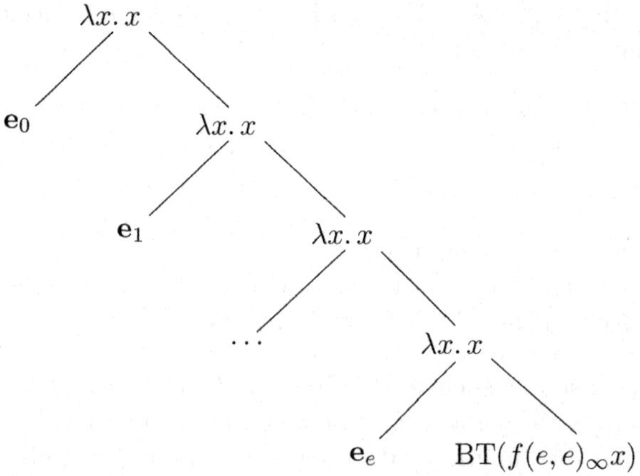

This tree has as its short left branches the trees of $\mathbf{c}_0, \mathbf{c}_0, \ldots$ ad infinitum, unless a computation of $\{e\}(0)$ converges and this happens in $k \le e$ steps, in which case the $k+p$-th branches become $\mathbf{c}_1$ for all p. Let

$$\mathcal{V} = \{M_e \mid e \in \mathbb{N}\}.$$

One has the following.

(1) $\forall k. f(e,k) = f(e',k) \ \Leftrightarrow\ \forall k \le \max\{e,e'\}. f(e,k) = f(e',k)$
$\Leftrightarrow\ M_e =_\beta M_{e'}$.

(2) $\{(e,e') \mid M_e \ne_\beta M_{e'}\}$ is RE.

(3) $\forall M \in \mathcal{V} \exists D_M \in \Lambda^{\emptyset} \forall N \in \mathcal{V}.\ D_M N = \mathsf{U}^2_1$, if $M = N$,

$D_M N = \mathsf{U}^2_2$, else.

Given M one can determine the e such that $M := M_e$. Then $M = N$, for $N \in \mathcal{V}$, iff up to level e one has $\mathrm{BT}(M) = \mathrm{BT}(N)$.

(4) $\neg \exists \Delta \in \Lambda^{\emptyset} \forall M, N \in \mathcal{V}.\ \Delta M N = \mathsf{U}^2_1$, if $M = N$,

$\Delta M N = \mathsf{U}^2_2$, else.

If such a Δ would exist, then by the continuity of application with respect to the tree topology, see [1], Theorem 14.3.22, the value of $\Delta M N$ is determined by a fixed finite approximation of the Böhm-trees of $M, N \in \mathcal{V}$. But there are always terms in $\mathcal{V}$ that start to be different at deeper levels. □

4 Discussion

1. In [6], see also [12], the notion of *numbered set* is intyroduced. This is a pair (S, γ) with $\gamma : \mathbb{N} \to S$ a surjection. An n such that $\gamma(n) = s$ is called a *code* of s. A (Ershove) morphism between numbered sets $(S_1, \gamma_1), (S_2, \gamma_2)$ is a map $\mu : S_1 \to S_2$ such that for some total computable function $f : \mathbb{N} \to \mathbb{N}$ one has

$$\forall n \in \mathbb{N}. \mu(\gamma_1(n)) = \gamma_2(f(n)).$$

That is, a morphism is determined by a computable map on the codes. Given a numbered set (S, γ) one defines an equivalence relation on $\mathbb{N}$ by $n \sim m \Leftrightarrow \gamma(n) = \gamma(m)$. This numbered set is called *pre-complete* iff every partial computable function on $\mathbb{N}$ can be made total modulo $\sim$:

$$\forall \psi \text{ partial computable} \exists f \text{ total computable} \forall n \in \mathbb{N}. \psi(n)\downarrow \ \Rightarrow \ \psi(n) \sim f(n).$$

One of the first results in the the theory of numbered sets is that each morphism from a pre-complete numbered set to itself has a fixed point. An example of a pre-complete numbered set is $(\Lambda^{\emptyset}, \gamma_{\mathsf{E}})$, with $\gamma_{\mathsf{E}}(n) = \mathsf{E}\mathbf{c}_n$.

2. There are numeral systems on which all partial computable functions can be represented, without there being a test for zero $Z_?$, see [7] or [2]. Such numeral systems are not separable.
3. In [1] one uses the notation $\ulcorner M \urcorner = \ulcorner \# M \urcorner$, for M a lambda term. Here one uses a different system of numerals, denoted by $\ulcorner n \urcorner$, for $n \in \mathbb{N}$. This does not matter, as the $\mathbf{c}_n$ and the $\ulcorner n \urcorner$ are equivalent in the sense that for some combinators G, H one has $G\mathbf{c}_n = \ulcorner n \urcorner$ & $H \ulcorner n \urcorner = \mathbf{c}_n$.

References

1. H. Barendregt. *The Lambda Calculus, its Syntax and Semantics*, volume 103 of *Studies in Logic and the Foundations of Mathematics*. North-Holland Publishing Co., Amsterdam, revised edition, 1984.
2. E. Barendsen. Theoretical pearls: an unsolvable numeral system in lambda calculus. *J. Funct. Programming*, 1(3):367–372, 1991.

3. C. Böhm. Alcune proprietà delle forme $\beta\eta$-normali nel λK-calcolo. Technical Report 696, Istituto per le Applicazioni del Calcolo (IAC), Viale del Policlinico 137, 00161 Rome, Italy, 1968.
4. C. Böhm, M. Dezani-Ciancaglini, P. Peretti, and S. Ronchi. A discrimination algorithm inside $\lambda\beta$-calculus. *Theoret. Comput. Sci.*, 8(3):271–291, 1979.
5. M. Coppo, M. Dezani-Ciancaglini, and S. Ronchi. (Semi-)separability of finite sets of terms in Scott's D_∞-models of the λ-calculus. In *Automata, languages and programming (Fifth Internat. Colloq., Udine, 1978)*, volume 62 of *Lecture Notes in Comput. Sci.*, pages 142–164. Springer, Berlin, 1978.
6. Y.L. Ershov. Theorie der Numerierungen I, II, III. *Zeitschr. math. Logik Grundl. Math.*, 19,21,23:289–388, 473–584, 289–371, 1973,1975,1977.
7. B. Intrigila. Some results on numerical systems in λ-calculus. *Notre Dame J. Formal Logic*, 35(4):523–541, 1994.
8. G. Jacopini. A condition for identifying two elements of whatever model of combinatory logic. In *λ-calculus and computer science theory (Proc. Sympos., Rome, 1975)*, pages 213–219. Lecture Notes in Comput. Sci., Vol. 37. Springer, Berlin, 1975.
9. S. Ronchi della Rocca. Discriminability of infinite sets of terms in the D_∞-models of the λ-calculus. In *CAAP '81 (Proc. Sixth Colloq., Genoa, 1981)*, volume 112 of *Lecture Notes in Comput. Sci.*, pages 350–364. Springer, Berlin, 1981.
10. R. Statman. Morphisms and partitions of V-sets. In *Computer science logic (Brno, 1998)*, volume 1584 of *Lecture Notes in Comput. Sci.*, pages 313–322. Springer, Berlin, 1999.
11. R. Statman and H. Barendregt. Applications of Plotkin-terms: partitions and morphisms for closed terms. *J. Funct. Programming*, 9(5):565–575, 1999.
12. A. Visser. Numerations, lambda calculus, and arithmetic. 1980. In: [?], pp. 259-284.

Explaining Constraint Programming

Krzysztof R. Apt[1,2,3]

[1] School of Computing, National University of Singapore
[2] CWI, Amsterdam
[3] University of Amsterdam, The Netherlands

Abstract. We discuss here constraint programming (CP) by using a proof-theoretic perspective. To this end we identify three levels of abstraction. Each level sheds light on the essence of CP.

In particular, the highest level allows us to bring CP closer to the computation as deduction paradigm. At the middle level we can explain various constraint propagation algorithms. Finally, at the lowest level we can address the issue of automatic generation and optimization of the constraint propagation algorithms.

1 Introduction

Constraint programming is an alternative approach to programming which consists of modelling the problem as a set of requirements (constraints) that are subsequently solved by means of general and domain specific methods.

Historically, constraint programming is an outcome of a long process that has started in the seventies, when the seminal works of Waltz and others on computer vision (see, e.g.,[30]) led to identification of constraint satisfaction problems as an area of Artificial Intelligence. In this area several fundamental techniques, including constraint propagation and enhanced forms of search have been developed.

In the eighties, starting with the seminal works of Colmerauer (see, e.g., [16]) and Jaffar and Lassez (see [21]) the area constraint logic programming was founded. In the nineties a number of alternative approaches to constraint programming were realized, in particular in ILOG solver, see e.g., [20], that is based on modeling the constraint satisfaction problems in C++ using classes. Another, recent, example is the Koalog Constraint Solver, see [23], realized as a Java library.

This way constraint programming eventually emerged as a distinctive approach to programming. In this paper we try to clarify this programming style and to assess it using a proof-theoretic perspective considered at various levels of abstraction. We believe that this presentation of constraint programming allows us to more easily compare it with other programming styles and to isolate its salient features.

2 Preliminaries

Let us start by introducing the already mentioned concept of a constraint satisfaction problem. Consider a sequence $X = x_1, \ldots, x_m$ of variables with respective

A. Middeldorp et al. (Eds.): Processes... (Klop Festschrift), LNCS 3838, pp. 55–69, 2005.

domains $D_1, \ldots, D_n$. By a ***constraint*** on X we mean a subset of $D_1 \times \ldots \times D_m$. A ***constraint satisfaction problem (CSP)*** consists of a finite sequence of variables $x_1, \ldots, x_n$ with respective domains $D_1, \ldots, D_n$ and a finite set $\mathcal{C}$ of constraints, each on a subsequence of X. We write such a CSP as

$$\langle \mathcal{C} \; ; \; x_1 \in D_1, \ldots, x_n \in D_n \rangle.$$

A ***solution*** to a CSP is an assignment of values to its variables from their domains that satisfies all constraints. We say that a CSP is ***consistent*** if it has a solution, ***solved*** if each assignment is a solution, and ***failed*** if either a variable domain is empty or a constraint is empty. Intuitively, a failed CSP is one that obviously does not have any solution. In contrast, it is not obvious at all to verify whether a CSP is solved. So we introduce an imprecise concept of a '***manifestly solved***' CSP which means that it is computationally straightforward to verify that the CSP is solved. So this notion depends on what we assume as 'computationally straightforward'.

In practice the constraints are written in a first-order language. They are then atomic formulas or simple combinations of atomic formulas. One identifies then a constraint with its syntactic description. In what follows we study CSPs with finite domains.

3 High Level

At the highest level of abstraction constraint programming can be seen as a task of formulating specifications as a CSP and of solving it. The most common approach to solving a CSP is based on a ***top-down search*** combined with ***constraint propagation***.

The top-down search is determined by a *splitting strategy* that controls the splitting of a given CSP into two or more CSPs, the 'union' of which (defined in the natural sense) is ***equivalent*** to (i.e, has the same solutions as) the initial CSP. In the most common form of splitting a variable is selected and its domain is partitioned into two or more parts. The splitting strategy then determines which variable is to be selected and how its domain is to be split.

In turn, constraint propagation transforms a given CSP into one that is equivalent but *simpler*, i.e, easier to solve. Each form of constraint propagation determines a notion of ***local consistency*** that in a loose sense approximates the notion of consistency and is computationally efficient to achieve. This process leads to a search tree in which constraint propagation is alternated with splitting, see Figure 1.

So the nodes in the tree are CSPs with the root (level 0) being the original CSP. At the even levels the constraint propagation is applied to the current CSP. This yields exactly one direct descendant. At the odd levels splitting is applied to the current CSP. This yields more than one descendant. The leaves of the tree are CSPs that are either failed or manifestly solved. So from the leaves of the trees it is straightforward to collect all the solutions to the original CSP.

The process of tree generation can be expressed by means of proof rules that are used to express transformations of CSPs. In general we have two types of

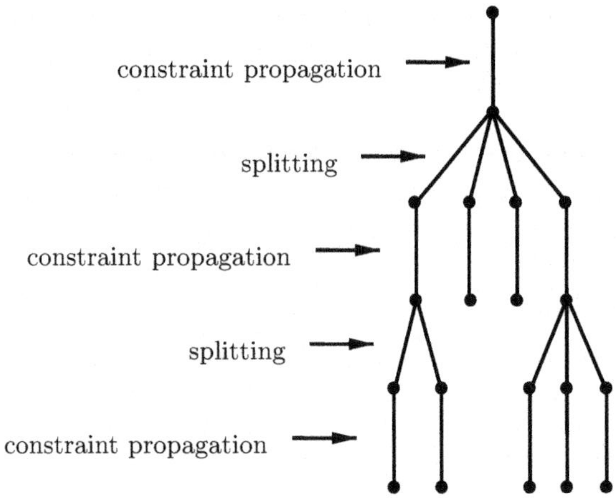

Fig. 1. A search tree for a CSP

rules. The ***deterministic*** rules transform a given CSP into another one. We write such a rule as:

$$\frac{\phi}{\psi}$$

where ϕ and ψ are CSPs.

In turn, the ***splitting*** rules transform a given CSP into a sequence of CSPs. We write such a rule as:

$$\frac{\phi}{\psi_1 \mid \ldots \mid \psi_n}$$

where ϕ and $\psi_1, \ldots, \psi_n$ are CSPs.

It is now easy to define the notion of an ***application of a proof rule*** to a CSP. In the case of a deterministic rule we just replace (after an appropriate renaming) the part that matches the premise of the rule by the conclusion. In the case of a splitting rule we replace (again after an appropriate renaming) the part that matches the premise of the rule by *one* of the CSPs ψ_i from the rule conclusion.

We now say that a deterministic rule

$$\frac{\phi}{\psi}$$

is ***equivalence preserving*** if ϕ and ψ are equivalent and that a splitting rule

$$\frac{\phi}{\psi_1 \mid \ldots \mid \psi_n}$$

is ***equivalence preserving*** if the union of ψ_i's is equivalent to ϕ.

In what follows all considered rules will be equivalence preserving. In general, the deterministic rules are more 'fine grained' than the constraint propagation step that is modeled as a single 'step' in the search tree. In fact, our intention is to model constraint propagation as a *repeated* application of deterministic rules. In the next section we shall discuss how to schedule these rule applications efficiently.

The search for solutions can now be described by means of ***derivations***, just like in logic programming. In logic programming we have in general two types of finite derivations: successful and failed. In the case of proof rules as defined above a new type of derivations naturally arises.

Definition 1. *Assume a finite set of proof rules.*

- *By a* ***derivation*** *we mean a sequence of CSPs such that each of them is obtained from the previous one by an application of a proof rule.*
- *A finite derivation is called*
  - ***successful*** *if its last element is a first manifestly solved CSP in this derivation,*
  - ***failed*** *if its last element is a first failed CSP in this derivation,*
  - ***stabilizing*** *if its last element is a first CSP in this derivation that is closed under the applications of the considered proof rules.* □

The search for a solution to a CSP can now be described as a search for a successful derivation, much like in the case of logic programming. A new element is the presence of stabilizing derivations.

One of the main problems constraint programming needs to deal with is how to limit the size of a search tree. At the high level of abstraction this matter can be addressed by focusing on the derivations in which the applications of splitting rules are postponed as long as possible. This bring us to a consideration of stabilizing derivations that involve only deterministic rules. In practice such derivations are used to model the process of constraint propagation. They do not lead to a manifestly solved CSP but only to a CSP that is closed under the considered deterministic rules. So solving the resulting CSP requires first an application of a splitting rule. (The resulting CSP can be solved but to determine it may be computationally expensive.)

This discussion shows that at a high level of abstraction constraint programming can be viewed as a realization of the ***computation as deduction*** paradigm according to which the computation process is identified with a constructive proof of a formula from a set of axioms. In the case of constraint programming such a constructive proof is a successful derivation. Each such derivation yields at least one solution to the initial CSP.

Because so far no specific rules are considered not much more can be said at this level. However, this high level of abstraction allows us to set the stage for more specific considerations that belong to the middle level.

4 Middle Level

The middle level is concerned with the form of derivations that involve only deterministic rules. It allows us to explain the ***constraint propagation algorithms*** which are used to enforce constraint propagation. In our framework these algorithms are simply efficient schedulers of appropriate deterministic rules. To clarify this point we now introduce examples of specific classes of deterministic rules. In each case we discuss a scheduler that can be used to schedule the considered rules.

Example 1: Domain Reduction Rules

These are rules of the following form:

$$\frac{\langle \mathcal{C} \ ; \ x_1 \in D_1, \ldots, x_n \in D_n \rangle}{\langle \mathcal{C}' \ ; \ x_1 \in D'_1, \ldots, x_n \in D'_n \rangle}$$

where $D'_i \subseteq D_i$ for all $i \in [1..n]$ and $\mathcal{C}'$ is the result of restricting each constraint in $\mathcal{C}$ to $D'_1, \ldots, D'_n$.

We say that such a rule is ***monotonic*** if, when viewed as a function f from the original domains $D_1, \ldots, D_n$ to the reduced domains $D'_1, \ldots, D'_n$, i.e.,

$$f(D_1, \ldots, D_n) := (D'_1, \ldots, D'_n),$$

it is monotonic:

$$D_i \subseteq E_i \text{ for all } i \in [1..n] \text{ implies } f(D_1, \ldots, D_n) \subseteq f(E_1, \ldots, E_n).$$

That is, smaller variable domains yield smaller reduced domains.

Now, the following useful result shows that a large number of domain reduction rules are monotonic.

Theorem 1. *([10]) Suppose each D'_i is obtained from D_i using a combination of*

- *union and intersection operations,*
- *transposition and composition operations applied to binary relations,*
- *join operation $\bowtie$,*
- *projection functions, and*
- *removal of an element.*

Then the domain reduction rule is monotonic.

This repertoire of operations is sufficient to describe typical domain reduction rules considered in various constraint solvers used in constraint programming systems, including solvers for Boolean constraints, linear constraints over integers, and arithmetic constraints over reals, see, e.g., [10].

Monotonic domain reduction rules are useful for two reasons. First, we have the following observation.

Note 1. Assume a finite set of monotonic domain reduction rules and an initial CSP $\mathcal{P}$. Every stabilizing derivation starting in $\mathcal{P}$ yields the same outcome.

Second, monotonic domain reduction rules can be scheduled more efficiently than by means of a naive round-robin strategy. This is achieved by using a

generic iteration algorithm which in its most general form computes the least common fixpoint of a set of functions F in an appropriate partial ordering. This has been observed in varying forms of generality in the works of [12], [28], [17] and [7]. This algorithm has the following form. We assume here a finite set of functions F, each operating on a given partial ordering with the least element $\bot$.

GENERIC ITERATION algorithm

```
d := ⊥;
G := F;
WHILE G ≠ ∅ DO
  choose g ∈ G;
  IF d ≠ g(d) THEN
    G := G ∪ update(G, g, d);
    d := g(d)
  ELSE
    G := G − {g}
  END
END
```

where for all G, g, d

A $\{f \in F - G \mid f(d) = d \wedge f(g(d)) \neq g(d)\} \subseteq update(G, g, d)$.

The intuition behind the assumption **A** is that $update(G, g, d)$ contains at least all the functions from $F - G$ for which d is a fixpoint but $g(d)$ is not. So at each loop iteration if $d \neq g(d)$, such functions are added to the set G. Otherwise the function g is removed from G.

An obvious way to satisfy assumption **A** is by using the following *update* function:

$$update(G, g, d) := \{f \in F - G \mid f(d) = d \wedge f(g(d)) \neq g(d)\}.$$

The problem with this choice of *update* is that it is expensive to compute because for each function f in $F - G$ we would have to compute the values $f(g(d))$ and $f(d)$. So in practice, we are interested in some approximations from above of this *update* function that are easy to compute. We shall return to this matter in a moment.

First let us clarify the status of the above algorithm. Recall that a function f on a partial ordering $(D, \sqsubseteq)$ is called ***monotonic*** if $x \sqsubseteq y$ implies $f(x) \sqsubseteq f(y)$ for all x, y and ***inflationary*** if $x \sqsubseteq f(x)$ for all x.

Theorem 2. *([7]) Suppose that $(D, \sqsubseteq)$ is a finite partial ordering with the least element $\bot$. Let F be a finite set of monotonic and inflationary functions on D. Then every execution of the* GENERIC ITERATION *algorithm terminates and computes in d the least common fixpoint of the functions from F.*

In the applications we study the iterations carried out on a partial ordering that is a Cartesian product of the component partial orderings. More precisely, given n partial orderings $(D_i, \sqsubseteq_i)$, each with the least element $\bot_i$, we assume that each considered function g is defined on a 'partial' Cartesian product $D_{i_1} \times \ldots \times D_{i_l}$. Here $i_1, \ldots, i_l$ is a subsequence of $1, \ldots, n$ that we call the *scheme* of g. Given $d \in D_1 \times \cdots \times D_n$, where $d := d_1, \ldots, d_n$, and a scheme $s := i_1, \ldots, i_l$ we denote by $d[s]$ the sequence $d_{i_1}, \ldots, d_{i_l}$.

The corresponding instance of the above GENERIC ITERATION algorithm then takes the following form.

GENERIC ITERATION FOR COMPOUND DOMAINS algorithm

$d := (\bot_1, \ldots, \bot_n)$;
$d' := d$;
$G := F$;
WHILE $G \neq \emptyset$ DO
  choose $g \in G$;
  $d'[s] := g(d[s])$, where s is the scheme of g;
  IF $d'[s] \neq d[s]$ THEN
    $G := G \cup \{f \in F \mid \text{scheme of } f \text{ includes } i \text{ such that } d[i] \neq d'[i]\}$;
    $d[s] := d'[s]$
  ELSE
    $G := G - \{g\}$
  END
END

So this algorithm uses an *update* function that is straightforward to compute. It simply checks which components of d are modified and selects the functions that depend on these components. It is a standard scheduling algorithm used in most constraint programming systems.

Example 2: Arc Consistency

Arc consistency, introduced in [24], is the most popular notion of local consistency considered in constraint programming. Let us recall the definition.

Definition 2.

- *Consider a binary constraint C on the variables x, y with the domains D_x and D_y, that is $C \subseteq D_x \times D_y$. We call C* **arc consistent** *if*
  - $\forall a \in D_x \exists b \in D_y \ (a, b) \in C$,
  - $\forall b \in D_y \exists a \in D_x \ (a, b) \in C$.
- *We call a CSP* **arc consistent** *if all its binary constraints are arc consistent.*

So a binary constraint is arc consistent if every value in each domain has a ***support*** in the other domain, where we call b a support for a if the pair (a, b) (or, depending on the ordering of the variables, (b, a)) belongs to the constraint.

In the literature several arc consistency algorithms have been proposed. Their purpose is to transform a given CSP into one that is arc consistent without losing any solution. We shall now illustrate how the most popular arc consistency algorithm, AC-3, due to[24], can be explained as a specific scheduling of the appropriate domain reduction rules. First, let us define the notion of arc consistency in terms of such rules.

Assume a binary constraint C on the variables x, y. We introduce the following two rules.

ARC CONSISTENCY 1

$$\frac{\langle C\ ;\ x \in D_x, y \in D_y \rangle}{\langle C\ ;\ x \in D'_x, y \in D_y \rangle}$$

where $D'_x := \{a \in D_x \mid \exists\, b \in D_y\ (a, b) \in C\}$.

ARC CONSISTENCY 2

$$\frac{\langle C\ ;\ x \in D_x, y \in D_y \rangle}{\langle C\ ;\ x \in D_x, y \in D'_y \rangle}$$

where $D'_y := \{b \in D_y \mid \exists\, a \in D_x\ (a, b) \in C\}$.

So in each rule a selected variable domain is reduced by retaining only the supported values. The following observation characterizes the notion of arc consistency in terms of the above two rules.

Note 2 (Arc Consistency). A CSP is arc consistent iff it is closed under the applications of the *ARC CONSISTENCY* rules 1 and 2.

So to transform a given CSP into an equivalent one that is arc consistent it suffices to repeatedly apply the above two rules for all present binary constraints. Since these rules are monotonic, we can schedule them using the GENERIC ITERATION FOR COMPOUND DOMAINS algorithm. However, in the case of the above rules an *improved* generic iteration algorithm can be employed that takes into account commutativity and idempotence of the considered functions, see [8].

Recall that given two functions f and g on a partial ordering we say that f is ***idempotent*** if $f(f(x)) = f(x)$ for all x and say that f and g ***commute*** if $f(g(x)) = g(f(x))$ for all x. The relevant observation concerning these two properties is the following.

Note 3. Suppose that all functions in F are idempotent and that for each function g we have a set of functions $Comm(g)$ from F such that each element of $Comm(g)$ commutes with g. If $update(G, g, d)$ satisfies the assumption **A**, then so does the function $update(G, g, d) - Comm(g)$.

In practice it means that in each iteration of the generic iteration algorithm less functions need to be added to the set G. This yields a more efficient algorithm.

In the case of arc consistency for each binary constraint C the functions corresponding to the *ARC CONSISTENCY* rules 1 and 2 referring to C commute. Also, given two binary constraints that share the first (resp. second) variable, the corresponding *ARC CONSISTENCY* rules 1 (resp. 2) for these two constraints commute, as well. Further, all such functions are idempotent. So, thanks to the above Note, we can use an appropriately 'tighter' *update* function. The resulting algorithm is equivalent to the `AC-3` algorithm.

Example 3: Constructive Disjunction

One of the main reasons for combinatorial explosion in search for solutions to a CSP are ***disjunctive constraints***. A typical example is the following constraint used in scheduling problems:

$$\texttt{Start[task}_1\texttt{]} + \texttt{Duration[task}_1\texttt{]} \leq \texttt{Start[task}_2\texttt{]} \vee$$
$$\texttt{Start[task}_2\texttt{]} + \texttt{Duration[task}_2\texttt{]} \leq \texttt{Start[task}_1\texttt{]}$$

stating that either $\texttt{task}_1$ is scheduled before $\texttt{task}_2$ or vice versa. To deal with a disjunctive constraint we can apply the following splitting rule (we omit here the information about the variable domains):

$$\frac{C_1 \vee C_2}{C_1 \mid C_2}$$

which amounts to a case analysis.

However, as already explained in Section 3 it is in general preferable to postpone an application of a splitting rule and try to reduce the domains first. ***Constructive disjunction***, see [29], is a technique that occasionally allows us to do this. It can be expressed in our rule-based framework as a domain reduction rule that uses some auxiliary derivations as side conditions:

CONSTRUCTIVE DISJUNCTION

$$\frac{\langle C_1 \vee C_2 \ ; \ x_1 \in D_1, \ldots, x_n \in D_n \rangle}{\langle C_1' \vee C_2' \ ; \ x_1 \in D_1' \cup D_1'', \ldots, x_n \in D_n' \cup D_n'' \rangle} \quad \text{where} \quad der_1, der_2$$

with

$$der_1 := \langle C_1 \ ; \ x_1 \in D_1, \ldots, x_n \in D_n \rangle \vdash \langle C_1' \ ; \ x_1 \in D_1', \ldots, x_n \in D_n' \rangle,$$
$$der_2 := \langle C_2 \ ; \ x_1 \in D_1, \ldots, x_n \in D_n \rangle \vdash \langle C_2' \ ; \ x_1 \in D_1'', \ldots, x_n \in D_n'' \rangle,$$

and where C_1' is the result of restricting the constraint in C_1 to $D_1', \ldots, D_n'$ and similarly for C_2'.

In words: assuming we reduced the domains of each disjunct separately, we can reduce the domains of the disjunctive constraint to the respective unions of the reduced domains. As an example consider the constraint

$$\langle |x - y| = 1 \ ; \ x \in [4..10], y \in [2..7] \rangle.$$

We can view $|x - y| = 1$ as the disjunctive constraint $(x - y = 1) \vee (y - x = 1)$. In the presence of the *ARC CONSISTENCY* rules 1 and 2 rules we have then

$$\langle x - y = 1 \ ; \ x \in [4..10], y \in [2..7] \rangle \vdash \langle x - y = 1 \ ; \ x \in [4..8], y \in [3..7] \rangle$$

and

$$\langle y - x = 1 \ ; \ x \in [4..10], y \in [2..7] \rangle \vdash \langle y - x = 1 \ ; \ x \in [4..6], y \in [5..7] \rangle.$$

So using the *CONSTRUCTIVE DISJUNCTION* rule we obtain

$$\langle |x - y| = 1 \ ; \ x \in [4..8], y \in [3..7] \rangle.$$

If each disjunct of a disjunctive constraint is a conjunction of constraints, the auxiliary derivations in the side conditions can be longer than just one step. Once the rules used in these derivations are of an appropriate format, their applications can be scheduled using one of the discussed generic iteration algorithms. Then the single application of the *CONSTRUCTIVE DISJUNCTION* rule consists in fact of two applications of the appropriate iteration algorithm.

It is straightforward to check that if the auxiliary derivations involve only monotonic domain reduction rules, then the *CONSTRUCTIVE DISJUNCTION* rule is itself monotonic. So the GENERIC ITERATION FOR COMPOUND DOMAINS algorithm can be applied both within the side conditions of this rule and for scheduling this rule together with other monotonic domain reduction rules that are used to deal with other, non-disjunctive, constraints.

In this framework it is straightforward to formulate some strengthenings of the constructive disjunction that lead to other modification of the constraints C_1 and C_2 than C_1' and C_2'.

Example 4: Propagation Rules

These are rules that allow us to add new constraints. Assuming a given set $\mathcal{A}$ of 'allowed' constraints we write such rules as

$$\frac{\mathcal{B}}{\mathcal{C}}$$

where $\mathcal{B}, \mathcal{C} \subseteq \mathcal{A}$.

This rule states that in presence of all constraints in $\mathcal{B}$ the constraints in $\mathcal{C}$ can be added, and is a shorthand for a deterministic rule of the following form:

$$\frac{\langle \mathcal{B} \ ; \ x_1 \in D_1, \ldots, x_n \in D_n \rangle}{\langle \mathcal{B}, \mathcal{C} \ ; \ x_1 \in D_1, \ldots, x_n \in D_n \rangle}$$

An example of such a rule is the transitivity rule:

$$\frac{x < y, y < z}{x < z}$$

that refers to a linear ordering $<$ on the underlying domain (for example natural numbers).

In what follows we focus on another example of propagation rules, ***membership rules***. They have the following form:

$$\frac{y_1 \in S_1, \ldots, y_k \in S_k}{z_1 \neq a_1, \ldots, z_m \neq a_m}$$

where $y_i \in S_i$ and $z_j \neq a_j$ are unary constraints with the obvious meaning.

Below we write such a rule as:

$$y_1 \in S_1, \ldots, y_k \in S_k \;\rightarrow\; z_1 \neq a_1, \ldots, z_m \neq a_m.$$

The intuitive meaning of this rule is: if for all $i \in [1..k]$ the domain of each y_i is a subset of S_i, then for all $j \in [1..m]$ remove the element a_j from the domain of z_j.

The membership rules allow us to reason about constraints given explicitly in a form of a table. As an example consider the three valued logic of Kleene. Let us focus on the conjunction constraint $\texttt{and3}(x, y, z)$ defined by the following table:

	t	f	u
t	t	f	u
f	f	f	f
u	u	f	u

That is, and3 consists of 9 triples. Then the membership rule $y \in \{u, f\} \rightarrow z \neq t$, or more precisely the rule

$$\frac{\langle \texttt{and3}(x, y, z), y \in \{u, f\} \;;\; x \in D_x, y \in D_y, z \in D_z\rangle}{\langle \texttt{and3}(x, y, z), y \in \{u, f\}, z \neq t \;;\; x \in D_x, y \in D_y, z \in D_z\rangle}$$

is equivalence preserving. This rule states that if y is either u or f, then t can be removed from the domain of z.

We call a membership rule is *minimal* if it is equivalence preserving and its conclusions cannot be established by either removing from its premise a variable or by expanding a variable range. For example, the above rule $y \in \{u, f\} \rightarrow z \neq t$ is minimal, while neither $x \in \{u\}, y \in \{u, f\} \rightarrow z \neq t$ nor $y \in \{u\} \rightarrow z \neq t$ is. In the case of the and3 constraint there are 18 minimal membership rules.

To clarify the nature of the membership rules let us mention that, as shown in [9], in the case of two-valued logic the corresponding set of minimal membership rules entails a form of constraint propagation that is equivalent to the unit propagation, a well-known form of resolution for propositional logic. So the membership rules can be seen as a generalization of the unit propagation to the explicitly given constraints, in particular to the case of many valued logics.

Membership rules can be alternatively viewed as a special class of monotonic domain reductions rules in which the domain of each z_i variable is modified by removing a_i from it. So we can schedule these rules using the Generic Iteration for Compound Domains algorithm.

However, the propagation rules, so in particular the membership rules, satisfy an important property that allows us to schedule them using a more efficient, fine-tuned, scheduler. We call this property ***stability***. It states that in each derivation the rule needs to be applied at most once: if it is applied, then it does not need to be applied again. So during the computation the applied rules that are stable can be *permanently removed* from the initial rule set. The resulting scheduler for the membership rules and its further optimizations are discussed in [14].

5 Low Level

The low level allows us to focus on matters that go beyond the issue of rule scheduling. At this level we can address matters concerned with further optimization of the constraint propagation algorithms. Various improvements of the `AC-3` algorithm that are concerned with specific choices of the data structures used belong here but cannot be explained by focusing the discussion on the corresponding *ARC CONSISTENCY 1* and *2* rules.

On the other hand some other optimization issues can be explained in proof-theoretic terms. In what follows we focus on the membership rules for which we worked out the details. These rules allow us to implement constraint propagation for explicitly given constraints. We explained above that they can be scheduled using a fine-tuned scheduler. However, even when an explicitly given constraint is small, the number of minimal membership rules can be large and it is not easy to find them all.

So a need arises to generate such rules automatically. This is what we did in [11]. We also proved there that the resulting form of constraint propagation is equivalent to ***hyper-arc consistency***, a natural generalization of arc consistency to n-ary constraints introduced in [25].

A further improvement can be achieved by removing some rules *before* scheduling them. This idea was pursued in [14]. Given a set of monotonic domain reduction rules $\mathcal{R}$ we say that a rule r is ***redundant*** if for each initial CSP $\mathcal{P}$ the unique outcome of a stabilizing derivation (guaranteed by Note 1) is the same with r removed from $\mathcal{R}$. In general, the iterated removal of redundant rules does not yield a unique outcome but in the case of the membership rules some useful heuristics can be used to appropriately schedule the candidate rules for removal.

We can summarize the improvements concerned with the membership rules as follows:

- For explicitly given constraints all minimal membership rules can be automatically generated.
- Subsequently redundant rules can be removed.
- A fine-tuned scheduler can be used to schedule the remaining rules.
- This scheduler allows us to remove permanently some rules which is useful during the top-down search.

To illustrate these matters consider the 11-valued `and11` constraint used in the automatic test pattern generation (ATPG) systems. There are in total 4656 minimal membership rules. After removing the redundant rules only 393 remain. This leads to substantial gains in computing. To give an idea of the scale of the improvement here are the computation times in seconds for three schedulers used to find all solutions to a CSP consisting of the `and11` constraint and solved using a random variable selection, domain ordering and domain splitting:

	Fine-tuned	Generic	CHR
all rules	1874	3321	7615
non-redundant rules	157	316	543

CHR stands for the standard CHR scheduler normally used to schedule such rules. (CHR is a high-level language extension of logic programming used to write user-defined constraints, for an overview see [18].) So using this approach a 50 fold improvement in computation time was achieved. In general, we noted that the larger the constraint the larger the gain in computing achieved by the above approach.

6 Conclusions

In this paper we assessed the crucial features of constraint programming (CP) by means of a proof-theoretic perspective. To this end we identified three levels of abstraction. At each level proof rules and derivations played a crucial role. At the highest level they allowed us to clarify the relation between CP and the computation as deduction paradigm. At the middle level we discussed efficient schedulers for specific classes of rules. Finally, at the lowest level we explained how specific rules can be automatically generated, optimized and scheduled in a customized way.

This presentation of CP suggests that it has close links with the rule-based programming. And indeed, several realizations of constraint programming through some form of rule-based programming exist. For example, constraint logic programs are sets of rules, so constraint logic programming can be naturally seen as an instance of rule-based programming. Further, the already mentioned CHR language is a rule-based language, though it does not have the full capabilities of constraint programming. In practise, CHR is available as a library of a constraint programming system, for example ECL^iPS^e (see [1]) or SICStus Prolog (see [3]). In turn, ELAN, see [2], is a rule-based programming language that can be naturally used to explain various aspects of constraint programming, see for example [22] and [15].

In our presentation we abstracted from specific constraint programming languages and their realizations and analyzed instead the principles of the corresponding programming style. This allowed us to isolate the essential features of constraint programming by focusing on proof rules, derivations and schedulers. This account of constraint programming draws on our work on the subject carried out in the past seven years. In particular, the high level view was introduced in [6]. In turn, the middle level summarizes our work reported in [7, 8]. Both levels are discussed in more detail in [10]. Finally, the account of propagation rules and of low level draws on [11, 14].

This work was pursued by others. Here are some representative references. Concerning the middle level, [26] showed that the framework of Section 4 allows us to parallelize constraint propagation algorithms in a simple and uniform way, while [13] showed how to use it to derive constraint propagation algorithms for soft constraints. In turn, [19] explained other arc consistency algorithms by slightly extending this framework.

Concerning the lowest level, [27] considered rules in which parameters (i.e., unspecified constants) are allowed. This led to a decrease in the number of generated rules. In turn, [4] presented an algorithm that generates more general and more expressive rules, for example with variable equalities in the conclusion. Finally, [5] considered the problem of generating the rules for constraints defined intensionally over infinite domains.

Acknowledgments

The work discussed here draws partly on a joint research carried out with Sebastian Brand and with Eric Monfroy. In particular, they realized the implementations discussed in the section on the low level. We also acknowledge useful comments of the referees.

References

1. The ECL^{i}PS^e Constraint Logic Programming System. `http://www-icparc.doc.ic.ac.uk/eclipse/`.
2. ELAN, Version 3.3. `http://www.iist.unu.edu/~alumni/software/other/inria/elan/elan-presentation.html`.
3. SICStus Prolog. `http://www.sics.se/isl/sicstuswww/site/index.html`.
4. S. Abdennadher and Ch. Rigotti. Automatic generation of rule-based constraint solvers over finite domains. *ACM Transactions on Computational Logic*, 5(2):177–205, 2004.
5. S. Abdennadher and Ch. Rigotti. Automatic generation of CHR constraint solvers. *Theory and Practice of Logic Programming*, 2005. To appear.
6. K. R. Apt. A proof theoretic view of constraint programming. *Fundamenta Informaticae*, 33(3):263–293, 1998. Available via `http://arXiv.org/archive/cs/`.
7. K. R. Apt. The essence of constraint propagation. *Theoretical Computer Science*, 221(1–2):179–210, 1999. Available via `http://arXiv.org/archive/cs/`.
8. K. R. Apt. The role of commutativity in constraint propagation algorithms. *ACM Transactions on Programming Languages and Systems*, 22(6):1002–1036, 2000. Available via `http://arXiv.org/archive/cs/`.
9. K. R. Apt. Some remarks on Boolean constraint propagation. In K. R. Apt, A. C. Kakas, E. Monfroy, and F. Rossi, editors, *New Trends in Constraints*, volume 1865 of *Lecture Notes in Artificial Intelligence*, pages 91 – 107. Springer-Verlag, 2000. Available via `http://arXiv.org/archive/cs/`.
10. K. R. Apt. *Principles of Constraint Programming*. Cambridge University Press, 2003.
11. K. R. Apt and E. Monfroy. Constraint programming viewed as rule-based programming. *Theory and Practice of Logic Programming*, 1(6):713–750, 2001. Available via `http://arXiv.org/archive/cs/`.
12. F. Benhamou. Heterogeneous constraint solving. In M. Hanus and M. Rodriguez-Artalejo, editors, *Proceeding of the Fifth International Conference on Algebraic and Logic Programming (ALP 96)*, Lecture Notes in Computer Science 1139, pages 62–76, Berlin, 1996. Springer-Verlag.
13. S. Bistarelli, R. Gennari, and F. Rossi. Constraint propagation for soft constraint satisfaction problems: Generalization and termination conditions. In Rina Dechter, editor, *Proceedings of Constraint Programming 2000 (CP2000)*, Lecture Notes in Computer Science 1894, pages 83–97, Berlin, 2000. Springer-Verlag.

14. S. Brand and K. R. Apt. Schedulers and redundancy for a class of constraint propagation rules. *Theory and Practice of Logic Programming*, 2005. To appear.
15. C. Castro. Building constraint satisfaction problem solvers using rewrite rules and strategies. *Fundamenta Informaticae*, 33(3):263–293, 1998.
16. Alain Colmerauer. Opening the PROLOG-III universe. *BYTE Magazine*, 12(9), August 1987.
17. F. Fages, J. Fowler, and T. Sola. Experiments in reactive constraint logic programming. *Journal of Logic Programming*, 37(1–3):185–212, 1998.
18. T. Frühwirth. Theory and practice of Constraint Handling Rules. *Journal of Logic Programming*, 37(1–3):95–138, October 1998. Special Issue on Constraint Logic Programming (P. J. Stuckey and K. Marriot, Eds.).
19. R. Gennari. Arc consistency via subsumed functions. In John Lloyd, editor, *Proceedings of Computational Logic 2000 (CL2000)*, Lecture Notes in Artificial Intelligence 1861, pages 358–372, Berlin, 2000. Springer-Verlag.
20. ILOG. Ilog white papers, 2003. Available via `http://www.ilog.com/products/optimization/papers.cfm`.
21. J. Jaffar and J.-L. Lassez. Constraint logic programming. In *POPL'87: Proceedings 14th ACM Symposium on Principles of Programming Languages*, pages 111–119. ACM, 1987.
22. C. Kirchner and C. Ringeissen. Rule-based constraint programming. *Fundamenta Informaticae*, 34(3):225–262, September 1998.
23. Koalog. `http://www.koalog.com`, 2005.
24. A. Mackworth. Consistency in networks of relations. *Artificial Intelligence*, 8(1):99–118, 1977.
25. R. Mohr and G. Masini. Good old discrete relaxation. In Y. Kodratoff, editor, *Proceedings of the 8th European Conference on Artificial Intelligence (ECAI)*, pages 651–656. Pitman Publishers, 1988.
26. E. Monfroy and J.-H. Réty. Chaotic iteration for distributed constraint propagation. In J. Carroll, H. Haddad, D. Oppenheim, B. Bryant, and G. Lamont, editors, *Proceedings of the 14th ACM Symposium on Applied Computing, ACM SAC'99, Scientific Computing Track*, pages 19–24, San Antonio, Texas, USA, March 1999. ACM Press.
27. C. Ringeissen and E. Monfroy. Generating propagation rules for finite domains: a mixed approach. In K. R. Apt, A. C. Kakas, E. Monfroy, and F. Rossi, editors, *New Trends in Constraints*, volume 1865 of *Lecture Notes in Artificial Intelligence*, pages 150–172. Springer-Verlag, 2000.
28. V. Telerman and D. Ushakov. Data types in subdefinite models. In J. A. Campbell J. Calmet and J. Pfalzgraf, editors, *Artificial Intelligence and Symbolic Mathematical Computations*, Lecture Notes in Computer Science 1138, pages 305–319, Berlin, 1996. Springer-Verlag.
29. P. Van Hentenryck, V. Saraswat, and Y. Deville. Design, implementation, and evaluation of the constraint language cc(fd). *Journal of Logic Programming*, 37(1–3):139–164, 1998. Special Issue on Constraint Logic Programming (P. J. Stuckey and K. Marriot, Eds.).
30. D. L. Waltz. Generating semantic descriptions from drawings of scenes with shadows. In P. H. Winston, editor, *The Psychology of Computer Vision*, pages 19–91. McGraw Hill, 1975.

Sharing in the Weak Lambda-Calculus

Tomasz Blanc, Jean-Jacques Lévy, and Luc Maranget

INRIA - Rocquencourt
{Tomasz.Blanc, Luc.Maranget, Jean-Jacques.Levy}@inria.fr
http://moscova.inria.fr/~{tblanc, levy, maranget}

Abstract. Despite decades of research in the λ-calculus, the syntactic properties of the weak λ-calculus did not receive great attention. However, this theory is more relevant for the implementation of programming languages than the usual theory of the strong λ-calculus. In fact, the frameworks of weak explicit substitutions, or computational monads, or λ-calculus with a `let` statement, or super-combinators, were developed for adhoc purposes related to programming language implementation. In this paper, we concentrate on sharing of subterms in a confluent variant of the weak λ-calculus. We introduce a labeling of this calculus that expresses a confluent theory of reductions with sharing, independent of the reduction strategy. We finally state that Wadsworth's evaluation technique with sharing of subterms corresponds to our formal setting.

1 Introduction

In the terminology of the λ-calculus, a *strong* calculus validates the following ξ-rule

$$(\xi)\ \frac{M \to N}{\lambda x.M \to \lambda x.N}$$

A *weak* calculus does not validate this rule. One easily shows that the weak λ-calculus is not confluent. In [18], an extension of the weak λ-calculus was introduced. It is strongly inspired from the one of Çağman and Hindley [6] for Combinatory Logic. In this calculus, a restricted version of the ξ-rule is valid; this new ξ′-rule is intuitively defined by

$$(\xi')\ \frac{M \stackrel{R}{\to} N \quad x \notin R}{\lambda x.M \stackrel{R}{\to} \lambda x.N}$$

meaning that the ξ-rule is valid when the bound variable x is not free in the redex R contracted between M and N (This rule will be presented in Section 2 in a form slightly different from — but equivalent to — the σ-rule used in [18]). The resulting new weak λ-calculus is confluent as shown in [18].

The theory of optimal reductions in the λ-calculus [5] has been extensively studied by Abadi, Asperti, Coppola, Gonthier, Guerrini, Lamping, Lawall, Lévy, Mairson, Martini et al [3, 4, 9, 13, 15, 17]. These authors represent λ-terms as graphs with shared subcontexts. For instance, in $(\lambda x.xa(xb))(\lambda y.Iy)$ where $I = \lambda x.x$, it is necessary to share the redex Iy independently of the value

A. Middeldorp et al. (Eds.): Processes... (Klop Festschrift), LNCS 3838, pp. 70–87, 2005.

of y. Therefore the subcontext $I[\ ]$ has to be shared. But after reduction of the external redex, we get $(\lambda y.Iy)a((\lambda y.Iy)b)$ and further $Ia((\lambda y.Iy)b)$, where the shared subcontext $I[\ ]$ is instantiated with two different terms. Technically, in these graphs, a shared subcontext can be referenced through a *fan-in* node to multiplex incoming arcs from terms using it; and context holes are filled through *fan-out* nodes to demultiplex outgoing arcs pointing to terms filling these holes. For instance, Lamping's graphs [13] operate on fans and *brackets*, which can be decomposed into more elementary fans, brackets and *croissants* obeying to reduction rules directed by the *context semantics* defined in [9].

In the weak λ-calculus, reductions are not performed under λ-abstractions. In the above example, the subterm Iy in $(\lambda x.xa(xb))(\lambda y.Iy)$ cannot be reduced since inside the abstraction $\lambda y.Iy$. Thus the subcontext $I[\ ]$ needs not be shared. The λ-terms with sharing can be represented by directed acyclic graphs (*dags*) instead of the (cyclic) Lamping's graph structures required to implement shared subcontexts. In this paper, we present a weak labeled λ-calculus that expresses a confluent theory of sharing within the weak λ-calculus corresponding to the shared evaluation strategy by Wadsworth [25] defined with dags in 1971!

The weak λ-calculus corresponds to runtime systems in functional languages, since runtimes just pass arguments to functions and never compute function bodies, i.e. under λ-abstractions. At compile-time, inlining or partial evaluation are feasible; but the weak λ-calculus just corresponds to the execution phase. However a runtime of a functional language usually implements a particular reduction strategy such as call-by-name, call-by-need or call-by-value. We prefer to model these runtimes by a confluent calculus which allows to consider mixed strategies alternating call-by-need and call-by-value. Moreover, a confluent calculus makes independent sharing and reduction strategy, which are two independent concepts. In runtimes of lazy functional languages (Haskell, LML, G-machine) [20, 22], the call-by-need strategy will correspond to a (weak) leftmost outermost reduction with some amount of sharing.

In most functional runtime systems (See e.g. [16]), functions are implemented by closures (See e.g. [16]), i.e. a pairs of a λ-abstraction (program) and a substitution (environment). The theory of explicit substitutions is related to the notion of closure but does not restrict reduction strategies [1]. This theory is not simple. It uses de Bruijn indices and is not confluent for open terms: Klop's counterexample [11] for surjective pairing can be adapted to the calculus of explicit substitutions [1]. However, confluence (on open terms) can be recovered at the price of either considering a much more complex calculus of explicit substitutions [7], or a theory of weak explicit substitutions as in [8, 18]. In the latter case, a theory of sharing was sketched, through the definition of a weak labeled calculus of explicit substitutions.

Closures and explicit substitutions are not necessary to express sharing of λ-terms. For instance, call-by-need strategies by Launchbury, Odersky or Ariola et al. [14, 19, 2] use a λ-calculus with a new `let` construct. However, these calculi have often critical pairs (in the sense of term rewriting) and extra rules to handle the `let` construct. Term rewriting systems (TRS) can also be used by lifting free variables and transforming each λ-abstraction into the application of several of its free variables to a (super)combinator. In this paper we want to stay with the classical set of λ-terms and subtheories of the classical λ-calculus.

In Section 2, we recall the definitions and basic properties of the weak λ-calculus introduced in [18]. In Section 3, we consider the weak labeled λ-calculus and prove its confluence. In Section 4, we relate labels and sharing. In Section 5, we discuss several differences between the weak labeled λ-calculus and dag implementations. In Section 6, we comment on the relation between TRSs and our formal setting. We conclude in Section 7.

2 The Weak λ-Calculus

The weak λ-calculus is defined in [18]. As usual, the set of λ-terms is defined by

$$M, N ::= x \mid MN \mid \lambda x.M$$

and β-reduction is

$$(\beta) \qquad (\lambda x.M)N \rightarrow M[\![x \backslash N]\!]$$

where $M[\![x \backslash N]\!]$ is recursively defined by

$$\begin{aligned} x[\![x \backslash P]\!] &= P \\ y[\![x \backslash P]\!] &= y \quad (x \neq y) \\ (MN)[\![x \backslash P]\!] &= M[\![x \backslash P]\!]\, N[\![x \backslash P]\!] \\ (\lambda y.M)[\![x \backslash P]\!] &= \lambda y.M[\![x \backslash P]\!] \quad (x \neq y,\ y \text{ not free in } P) \end{aligned}$$

In the last case, the substitution must not bind free variables in P. We keep α-conversion implicit, and freely use renaming of bound variables. Equality on terms is defined up to the renaming of bound variables (α-conversion). In the (strong) λ-calculus, every context is active, since any subterm may be reduced at any time. By contrast, in the λ-calculus without the ξ-rule, no reduction inside a λ-abstraction occurs. As a consequence, the Church-Rosser property (confluence) does not hold: when $N \rightarrow N'$ we have

$$\begin{array}{ccc} (\lambda x.\lambda y.M)N & \longrightarrow & (\lambda x.\lambda y.M)N' \\ \downarrow & & \downarrow \\ (\lambda y.M[\![x \backslash N]\!]) & & (\lambda y.M[\![x \backslash N']\!]) \end{array}$$

The term $(\lambda y.M[\![x \backslash N]\!])$ is in normal form and cannot be reduced, and the previous diagram does not commute. The problem has been known for a long time in combinatory logic [10], although often kept as a "folk theorem". In [6], it is specifically stated, and shown as being relevant when translating the λ-calculus into combinatory logic. In [18], it is proved that confluence is recovered by adding the new inference rule

$$(\sigma)\ \frac{N \to N'}{M[\![x\backslash N]\!] \to M[\![x\backslash N']\!]}\ (M \text{ linear in } x)$$

where the free variable M is linear in x if x has exactly one occurrence in M. The use of substitution aims at reflecting the fact that N does not contain variables bound outside N. Intuitively, it means that N does not depend on an enclosing λ-abstraction or its argument. Therefore N may be reduced in such a context. The linearity condition avoids to consider parallel reduction steps.

Here, instead of the σ-rule, we use a related, more direct approach to recover confluence in the weak λ-calculus. Let us write $x \notin M$, when x is not a free variable in M. Formally:

$$\frac{}{x \notin y}\ (x \neq y) \qquad \frac{x \notin M}{x \notin \lambda y.M} \qquad \frac{x \notin M \quad x \notin N}{x \notin MN}$$

Then our alternative presentation of the weak λ-calculus adds a new ξ'-rule to the classical μ and ν-rules.

$$(\beta)\ R = (\lambda x.M)N \overset{R}{\to} M[\![x\backslash N]\!] \qquad (\nu)\ \frac{M \overset{R}{\to} M'}{MN \overset{R}{\to} M'N} \qquad (w)\ \frac{M \overset{R}{\to} N}{M \to N}$$

$$(\xi')\ \frac{M \overset{R}{\to} M' \quad x \notin R}{\lambda x.M \overset{R}{\to} \lambda x.M'} \qquad (\mu)\ \frac{N \overset{R}{\to} N'}{MN \overset{R}{\to} MN'}$$

The reduction step relation $\overset{R}{\to}$ is annotated with the contracted redex R. A λ-abstraction to the left of the reduction step relation cannot bind a variable in R. Therefore α-conversion does not change the redex annotating the reduction step relation. Like the σ-rule, the ξ'-rule allows the reduction of a subterm located under a λ-abstraction if the contracted redex does not contain a variable bound by the λ-abstraction. As usual, we write $\twoheadrightarrow$ for the transitive and reflexive closure of $\to$. So $M \twoheadrightarrow N$ iff M can reduce in several steps (maybe none) to N.

Lemma 1. *In the weak λ-calculus, one has*

(i) $N \to N' \Rightarrow M[\![x\backslash N]\!] \twoheadrightarrow M[\![x\backslash N']\!]$
(ii) $M \to M' \Rightarrow M[\![x\backslash N]\!] \to M'[\![x\backslash N]\!]$
(iii) $M[\![x\backslash N]\!][\![y\backslash N']\!] = M[\![y\backslash N']\!][\![x\backslash N[\![y\backslash N']\!]]\!] \quad (x \notin N')$

Theorem 1. *The weak λ-calculus is confluent.*

Proof: Straightforward application of previous lemma. □

In comparison with the strong λ-calculus, there is an additional way of creating a redex. After contracting $M = (\lambda x.Ix)y$, we obtain a redex Iy which is neither a residual of a redex of M nor a created redex of the strong λ-calculus. In fact, although the subterm Ix in M is a redex for the strong λ-calculus, it is not a redex for the weak λ-calculus: it is *frozen* in M by the occurrence of x. The contraction of the enclosing λ-abstraction *activates* Iy.

The weak λ-calculus enjoys syntactic properties simpler than the strong λ-calculus. In the weak λ-calculus, the finite developments theorem [5] is easy to prove, since residuals of disjoint redexes cannot be nested. In the strong λ-calculus, residuals of two disjoint redexes may be nested. The typical example is: $M = (\lambda x.Ix)(Jy)$ with $I = J = \lambda x.x$. Then, the two disjoint Ix and Jy redexes have nested residuals:

$$(\lambda x.Ix)(Jy) \rightarrow I(Jy)$$

But in the weak λ-calculus this case does not occur since the subterm Ix in M contains the bound variable x and is not considered as a redex of the weak λ-calculus. In this calculus, residuals of disjoint redexes are disjoint redexes.

This proposition allows now to state another interesting theorem of the weak λ-calculus, namely Curry's standardization theorem. A standard reduction is usually defined as a reduction contracting redexes in an outside-in and left-to-right way. Precisely a reduction of the form

$$M = M_0 \rightarrow M_1 \rightarrow \ldots M_n = N \quad (n \geq 0)$$

is standard when for all i and j such that $0 \leq i < j < n$, the R_j-redex contracted at step j in M_{j-1} is not a residual of a redex external to or to the left of the R_i-redex contracted at step i in M_{i-1}. We write $M \underset{\mathrm{st}}{\twoheadrightarrow} N$ for the existence of a standard reduction from M to N. Notice that the leftmost outermost reduction is a standard reduction (in the usual λ-calculus), but standard reductions may be more general.

Theorem 2. *If $M \twoheadrightarrow M'$, then $M \underset{st}{\twoheadrightarrow} M'$.*

Proof: One follows the proof scheme in [11] or checks the axioms of [21]. The basic step of the proof follows from the observation that when $M \xrightarrow{R} M'$ and S' is a residual of S (in the usual sense of the strong λ-calculus) in M not inside R, then S is a redex of the weak λ-calculus. Then assume that $M \rightarrow M' \rightarrow N'$ by contracting R in M and S' in M'. Suppose R and S' are not in the standard ordering. Then S' is residual of a redex S in M to the left of or outside R. Thus, we know that S is a redex of the weak λ-calculus and we may contract it getting N. By confluence (actually by finite developments), we converge to N' (by a finite development of the residuals of R in N). □

A normal form is a term without redex. For instance $I = \lambda x.x$ or $\lambda y.Iy$ are normal forms. Let $\xrightarrow[\mathrm{norm}]{}\!\!\!\!\rightarrow$ be the reduction contracting, at each step, the leftmost outermost redex.

Theorem 3. *If $M \twoheadrightarrow N$ and N is a normal form, then $M \underset{norm}{\twoheadrightarrow} N$.*

Proof: By persistence of the leftmost outermost redex and the use of the previous theorem. □

3 The Weak Labeled λ-Calculus

In this section, a calculus for sharing in the weak λ-calculus is developed with the help of a confluent, labeled calculus. Labels are used to name subterms; subterms which are copies along reductions keep their labels, but new subterms must have new labels. We want the weak labeled λ-calculus to be confluent, as sharing and reduction strategy are independent concepts. Therefore an adequate naming scheme should be invariant through reductions equivalent by permutations of reduction steps. The labeled calculus is different from the one used in the labeled calculus of weak explicit substitutions [18], since we have no longer closures and explicit substitutions, and we have to take care of variable binders.

We base our labeling scheme on the labeling of the strong λ-calculus [17].

$U, V ::= \alpha : X$		labeled term
$X, Y ::= S \mid U$		clipped or labeled term
$S, T ::= x \mid UV \mid \lambda x.U$		clipped term
$\alpha, \beta ::= a \mid \lceil \alpha' \rceil \mid \lfloor \alpha' \rfloor \mid [\alpha', \beta] \mid \langle \alpha', \beta \rangle$		labels
$\alpha', \beta' ::= \alpha_1 \alpha_2 \cdots \alpha_n \quad (n > 0)$		compound labels

The labeled term $\alpha : X$ is said to have label α. Labels can be stacked as in $\alpha_1 : \alpha_2 : \cdots \alpha_n : X$. Compound labels, used in the definition of the ℓ-reduction, are sequences of labels. An (atomic) label can be a simple letter, or formed by overlining $\lceil \alpha' \rceil$, or underlining $\lfloor \alpha' \rfloor$; it can also be a pair $[\alpha', \beta]$ or $\langle \alpha', \beta \rangle$ of compound label α' and label β. In the pair with square brackets, we say that α' *tags* β; for angle brackets, then α' *marks* β.

The labeled reduction ℓ-rule is defined as

$$(\ell) \qquad R = (\alpha' \cdot \lambda x.U)V \overset{R}{\rightarrow} \lceil \alpha' \rceil : (\alpha' \circledast U)[\![x \backslash \lfloor \alpha' \rfloor : V]\!]$$

where

$$\alpha_1 \alpha_2 \cdots \alpha_n \cdot S = \alpha_1 : \alpha_2 : \cdots \alpha_n : S$$

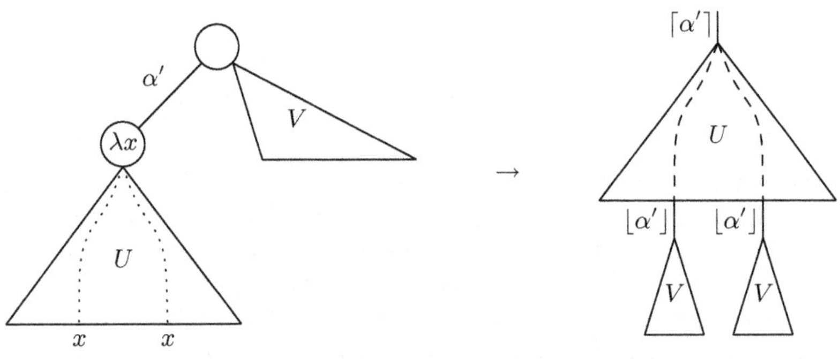

Fig. 1. A reduction step in the weak labeled λ-calculus (dotted lines represent paths from the root of U to occurrences of the free variable x in U; dashed lines represent the paths on which diffusion operates)

The *name* of R is α'; we write $name(R) = \alpha'$. We assume that substitution has a higher precedence than labeling. Hence $\lceil\alpha'\rceil : U[\![x\backslash V]\!]$ is read as $\lceil\alpha'\rceil : (U[\![x\backslash V]\!])$. As in the strong labeled λ-calculus, we sandwich the body of the function part of the redex with its name overlined and underlined as shown in Fig 1. The diffusion $\alpha' \textcircled{x} U$ creates new labels along paths from the root of U to free occurrences of x, as illustrated in Fig 1. Therefore new labels appeared for every subterm of U containing a free occurrence of x. Formally substitution and diffusion are defined as follows:

$$
\begin{aligned}
x[\![x\backslash W]\!] &= W \\
y[\![x\backslash W]\!] &= y \\
(UV)[\![x\backslash W]\!] &= U[\![x\backslash W]\!]\, V[\![x\backslash W]\!] \\
(\lambda y.U)[\![x\backslash W]\!] &= \lambda y.U[\![x\backslash W]\!] \\
(\beta : X)[\![x\backslash W]\!] &= \beta : X[\![x\backslash W]\!]
\end{aligned}
$$

$$
\begin{aligned}
\alpha' \textcircled{x} X &= X \ \text{ if } x \notin X \\
\alpha' \textcircled{x} x &= x \\
\alpha' \textcircled{x} UV &= (\alpha' \textcircled{x} U \ \ \alpha' \textcircled{x} V) \ \text{ if } x \in U \\
\alpha' \textcircled{x} UV &= (\langle \alpha', U\rangle \ \ \alpha' \textcircled{x} V) \ \text{ if } x \notin U \text{ and } x \in V \\
\alpha' \textcircled{x} \lambda y.U &= \lambda y.\, \alpha' \textcircled{x} U \ \text{ if } x \in \lambda y.U \\
\alpha' \textcircled{x} \beta : X &= [\alpha', \beta] : \alpha' \textcircled{x} X \ \text{ if } x \in X
\end{aligned}
$$

$$\langle \alpha', \beta : X\rangle = \langle \alpha', \beta\rangle : X$$

An example of reduction is illustrated in Fig 2.

During diffusion, we mark the left subterm of an application when it does not contain the bound variable x, since the application may become a redex of which we want to keep the history. Take for instance, the unlabeled $(\lambda x.Ix)y$, where $I = \lambda u.u$. Then Ix is not a redex in $(\lambda x.Ix)y$ since its argument contains x, but it becomes the redex Iy after contracting the enclosing redex. We want to mark the name of redex Iy with the name of the enclosing redex that activated it. In the weak labeled λ-calculus, this example becomes:

$$
\begin{aligned}
& a : (b : (\lambda x.\, c : (d : \lambda u.\, e : u)\ f : x)\ g : y) \\
\to\ & a : \lceil b\rceil : [b, c] : (\langle b, d\rangle : (\lambda u.e : u)\ \ [b, f] : \lfloor b\rfloor : g : y)
\end{aligned}
$$

where the name $\langle b, d\rangle$ of Iy contains the name b of $(\lambda x.Ix)y$, whereas if the diffusion did not mark the left part I of the application Ix, the name d of Iy would not have contained the name b of the redex contracted to create Iy.

As for the theory of the strong labeled λ-calculus, the label of a labeled subterm reflects its history. A simple letter stands for a labeled term existing in the initial term (empty history). Overlining $\lceil\alpha'\rceil$, underlining $\lfloor\alpha'\rfloor$, marks $\langle\alpha', \beta\rangle$ and tags $[\alpha', \beta]$ indicate past contraction of a redex of name α'. Overlined and underlined labels are kept to relate the weak labeled λ-calculus with the strong case.

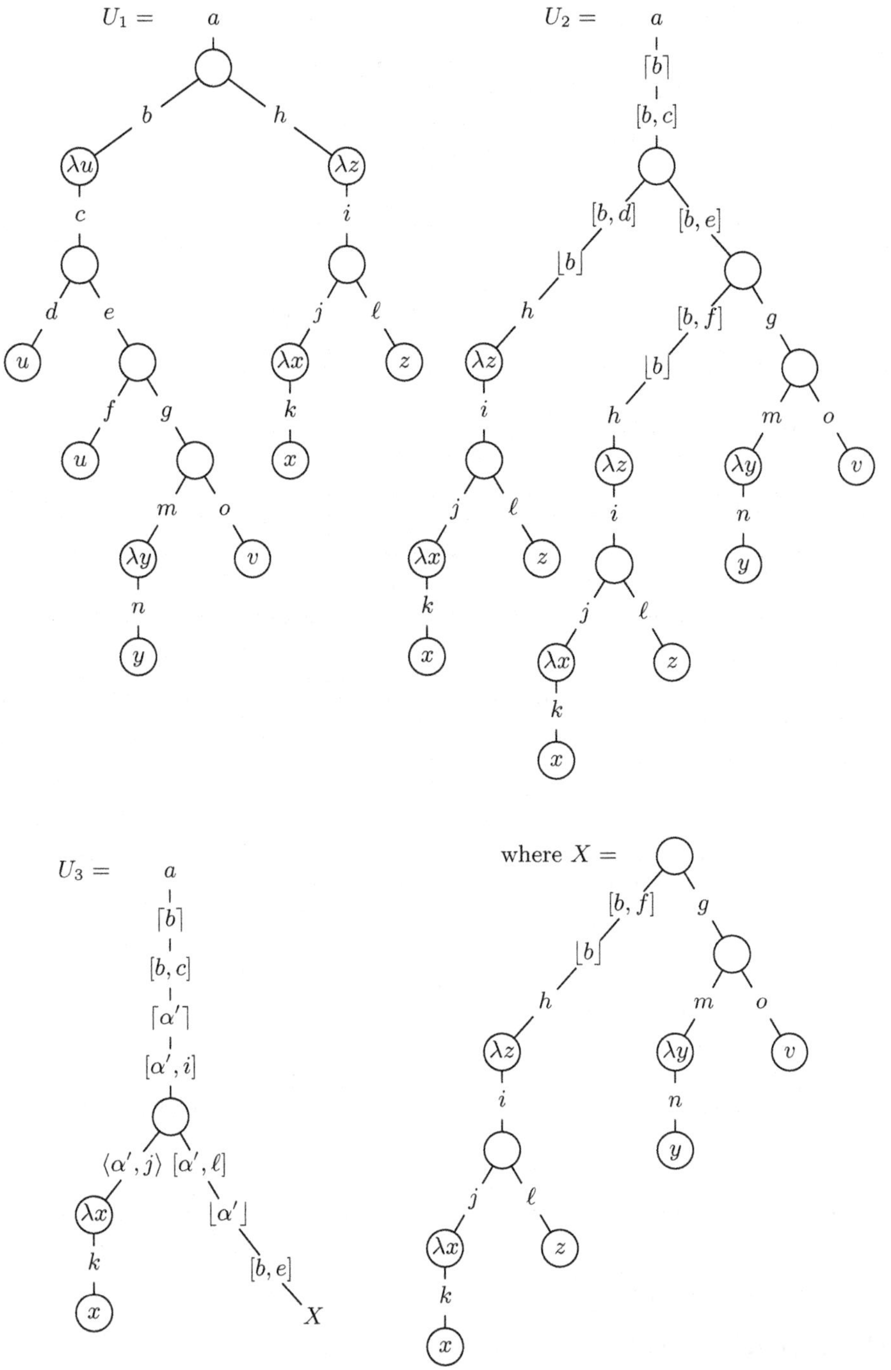

Fig. 2. An example of reduction in the weak labeled λ-calculus where $U_1 \to U_2 \to U_3$ and $\alpha' = [b,d]\lfloor b \rfloor h$. (The redex $(\lambda y.y)v$ could also be contracted in U_1, U_2 and U_3; however $(\lambda x.x)z$ with name j is not a redex in U_1 and U_2; but its "residual" in U_3 is a created redex in U_3, therefore its name is marked by the name α' of the contracted redex.)

Free variables are defined as in the unlabeled λ-calculus. The context rules are defined as follows for labeled reduction:

$$(\nu)\ \frac{U \xrightarrow{R} U'}{UV \xrightarrow{R} U'V} \qquad (\lambda)\ \frac{X \xrightarrow{R} X'}{\alpha\!:\!X \xrightarrow{R} \alpha\!:\!X'} \qquad (w)\ \frac{X \xrightarrow{R} X'}{X \rightarrow X'}$$

$$(\mu)\ \frac{V \xrightarrow{R} V'}{UV \xrightarrow{R} UV'} \qquad (\xi')\ \frac{U \xrightarrow{R} U' \quad x \notin R}{\lambda x.U \xrightarrow{R} \lambda x.U'}$$

We also write $\twoheadrightarrow$ for the transitive and reflexive closure of $\rightarrow$.

Lemma 2. *If $X \xrightarrow{R} X'$ and $x \notin R$, then $\alpha' \textcircled{x} X \rightarrow \alpha' \textcircled{x} X'$*

Proof: We first remark that as $x \notin R$ and $X \xrightarrow{R} X'$, we have $x \in X$ if and only if $x \in X'$. Clearly, $x \notin X$ implies $x \notin X'$ since the set of free variables of the contractum of a redex is a subset of the free variables of the contracted redex. Conversely, a free variable disappears in the contractum iff the redex has all occurrences of this free variable in its argument and the redex erases its argument. But as $x \notin R$, this case is not possible.

Now we proceed by cases on the definition of diffusion and by induction on the size of X:

1. If $x \notin X$, then $x \notin X'$ and the lemma is obvious, since $\alpha' \textcircled{x} X = X$ and $\alpha' \textcircled{x} X' = X'$.
2. If $x \in X$, we have again several cases.
   (a) Let R be in U and $X = UV \rightarrow U'V = X'$.
      i. If we have $x \in U$, one gets by induction $\alpha' \textcircled{x} U \rightarrow \alpha' \textcircled{x} U'$. Then the first remark yields $x \in U'$. And, by the definition of diffusion, we have $\alpha' \textcircled{x} X = (\alpha' \textcircled{x} U\ \alpha' \textcircled{x} V)$ and $\alpha' \textcircled{x} X' = (\alpha' \textcircled{x} U'\ \alpha' \textcircled{x} V)$. By the ν-rule, we obtain $\alpha' \textcircled{x} X \rightarrow \alpha' \textcircled{x} X'$.
      ii. If we have $x \notin U$, the first remark yields $x \notin U'$. Then, we have $\alpha' \textcircled{x} X = (\langle \alpha', U\rangle\ \alpha' \textcircled{x} V)$ and $\alpha' \textcircled{x} X' = (\langle \alpha', U'\rangle\ \alpha' \textcircled{x} V)$. By the λ-rule for context, we get $\langle \alpha', U\rangle \rightarrow \langle \alpha', U'\rangle$, and therefore, by the ν-rule $\alpha' \textcircled{x} X \rightarrow \alpha' \textcircled{x} X'$.
   (b) If R is in V, we have a simpler but similar argument.
   (c) The other cases when $X = \lambda y.U$ and $X = \beta\!:\!Y$ are similar.

□

Lemma 3. *If $U \rightarrow U'$ and $X \rightarrow X'$, then $X[\![x\backslash U]\!] \twoheadrightarrow X[\![x\backslash U']\!]$ and $X[\![x\backslash U]\!] \rightarrow X'[\![x\backslash U]\!]$.*

Proof: Straightforward by induction on the size of X. □

Theorem 4. *The weak labeled λ-calculus is confluent.*

Proof: By the Tait–Martin-Lof method, see [5]. The only interesting cases are the two commuting diagrams, when $x \notin R$, and $U \xrightarrow{R} U'$, $V \rightarrow V'$:

$$\begin{array}{ccc}
(\alpha' \cdot \lambda x.U)V & \xrightarrow{\;R\;} & (\alpha' \cdot \lambda x.U')V \\
\downarrow & & \downarrow \\
\lceil \alpha' \rceil : (\alpha' \textcircled{x} U)[\![x \backslash \lfloor \alpha' \rfloor : V]\!] & \longrightarrow & \lceil \alpha' \rceil : (\alpha' \textcircled{x} U')[\![x \backslash \lfloor \alpha' \rfloor : V]\!]
\end{array}$$

$$\begin{array}{ccc}
(\alpha' \cdot \lambda x.U)V & \longrightarrow & (\alpha' \cdot \lambda x.U)V' \\
\downarrow & & \downarrow \\
\lceil \alpha' \rceil : (\alpha' \textcircled{x} U)[\![x \backslash \lfloor \alpha' \rfloor : V]\!] & \twoheadrightarrow & \lceil \alpha' \rceil : (\alpha' \textcircled{x} U)[\![x \backslash \lfloor \alpha' \rfloor : V']\!]
\end{array}$$

which can be proved with the help of previous lemmas. □

4 Sharing of Subterms

In the labeled λ-calculus, the name of a redex records its history. Namely if a redex S is the residual of a redex R, their names are equal; and if a redex S is created by the contraction of a redex R, the name of R is contained in the name of S. We formalize this central property of the weak labeled λ-calculus with the following $\prec$ relation. We say that α' is strictly smaller than β', we write $\alpha' \prec \beta'$, for the following cases:

$$\alpha' \prec \lceil \alpha' \rceil \qquad \alpha' \prec \lfloor \alpha' \rfloor \qquad \alpha' \prec [\alpha', \beta] \qquad \alpha' \prec \langle \alpha', \beta \rangle$$

$$\alpha' \prec \beta_i \Rightarrow \alpha' \prec \beta_1 \cdots \beta_n \qquad \alpha' \prec \beta' \prec \gamma' \Rightarrow \alpha' \prec \gamma'$$

Hence the name α' of a redex is smaller than any label where it is overlined or underlined. It is also smaller than a pair of which it is the first component. If it is smaller than some β_i, it is smaller than a word for which β_i is a subcomponent. Moreover we close this relation by transitivity. This relation is clearly a strict ordering. This relation expresses the intuitive interpretation of labels we gave when defining labels: $\alpha' \prec \beta'$ if the contraction of a redex of name α' participates to the "creation" of β'. We recall that a reduction $X \twoheadrightarrow Y$ creates redex S in Y if S is not a residual (along this reduction) of a redex R in X.

Lemma 4. *If $X \xrightarrow{R} Y$ and redex S in Y is created in this reduction step, then $name(R) \prec name(S)$.*

Proof: Straightforward by case inspection. □

The goal of this section is to prove that two subterms with the same label are equal in the weak labeled λ-calculus, provided that we start reductions from a term with distinct letters on its subterms. This will mean that two subterms labeled with the same label can be shared in a dag representation of terms. To prove this property, we first show that several invariants are preserved by reductions.

Invariant 1. *$\mathcal{Q}(W)$ holds iff we have $\alpha' \not\prec \beta$ for every redex R with name α' and any subterm $\beta : X$ in W.*

A *complete labeled reduction* step $U \stackrel{\alpha'}{\Longrightarrow} V$ is the finite development of all redexes with name α in U. We first prove that invariant $\mathcal{Q}$ is preserved by complete labeled reduction steps. Intuitively, invariant $\mathcal{Q}$ means that the names of redexes are maximal. It guarantees that when performing complete labeled reduction steps, the names of the contracted redexes are all distinct.

Lemma 5. *If $\mathcal{Q}(W)$ and $W \stackrel{\gamma'}{\Longrightarrow} W'$, then $\mathcal{Q}(W')$.*

Proof: Let R with name α' be a redex in W'. Let $\beta : U$ be a subterm of W' such that $\alpha' \prec \beta$.

1. R is a residual of a redex R' of W. Then $name(R) = name(R') = \alpha'$.
   (a) β is a label in W. Impossible since $\mathcal{Q}(W)$.
   (b) β is created by the reduction from W to W'.
   By cases on the label's creation:
      i. $\beta = \lceil \gamma' \rceil$. Then $\alpha' \prec \lceil \gamma' \rceil$. There are again two cases:
         A. $\alpha' = \gamma'$. Impossible by definition of $\stackrel{\gamma'}{\Longrightarrow}$.
         B. $\alpha' \prec \gamma'$. If $\gamma' = \gamma_1 \dots \gamma_n$, there exists γ_i such that $\alpha' \prec \gamma_i$. Therefore there is a subterm $\gamma_i : V$ in W with $\alpha' \prec \gamma_i$. This contradicts $\mathcal{Q}(W)$.
      ii. $\beta = \lfloor \gamma' \rfloor$ or $\beta = [\gamma', \beta_1]$ or $\beta = \langle \gamma', \beta_1 \rangle$. These cases are similar to the previous one.
2. R is created by the reduction from W to W'. Then $\gamma' \prec \alpha' \prec \beta$.
   (a) β is a label in W. We have $\gamma' \prec \beta$. Impossible by $\mathcal{Q}(W)$.
   (b) β is created by the reduction from W to W'.
   By case on the label's creation:
      i. $\beta = \lceil \gamma' \rceil$. Thus $\gamma' \prec \beta$ As $\alpha' \prec \lceil \gamma' \rceil$, we have $\alpha' \preceq \gamma'$ and therefore $\alpha' \prec \alpha'$, since $\gamma' \prec \alpha'$. Contradiction.
      ii. $\beta = \lfloor \gamma' \rfloor$ or $\beta = [\gamma', \beta_1]$ or $\beta = \langle \gamma', \beta_1 \rangle$. These cases are similar to the previous one.

□

To limit the number of cases to inspect, we consider a simple technical invariant on left subterms of application nodes and show that they cannot be labeled by an overlined or an underlined label.

Invariant 2. *$\mathcal{R}(W)$ holds iff for any clipped subterm UV in W, we have either $U = a : X$, or $U = [\alpha', \beta] : X$, or $U = \langle \alpha', \beta \rangle : X$.*

Lemma 6. *If $\mathcal{R}(W)$ and $W \rightarrow W'$, then $\mathcal{R}(W')$.*

Proof: Let $U'V'$ be a clipped subterm in W'. This application node comes from an application node $T = UV$ in W. Let $R = (\alpha' \cdot \lambda x.A)B$ be the contracted redex.

1. If R is inside T, the only interesting case is when U contains R. Then we have $U = \beta : X$, with R inside X. Therefore β is unchanged by the reduction step, and $\mathcal{R}(W)$ gives the form a, $[\gamma', \beta_1]$, or $\langle \gamma', \beta_1 \rangle$ of β.

2. If R contains T in its argument B. Then $U' = U$ and they have same labels.
3. Otherwise U' comes from an U in the function body part A of R. Then, if $x \notin U'V'$, then $U' = U$ and this case is again straightforward. Otherwise, there can be a diffusion inside U', but then the label of U' is a tagged or marked pair. In all cases, $\mathcal{R}(W')$ holds.

□

Marked labels of the form $\langle \alpha', \beta \rangle$ record histories of reductions. A subterm with a marked label indicates that the application node just above has been activated by a redex with name α'. Although marked labels are necessary for recording histories of reduction (see lemma 5), these labels have no impact on the definition of sharing. The following sharing relation $\simeq$ means that two labels are equal by erasing marks (in marked labels). It is defined inductively by:

$$
\begin{array}{lll}
a \simeq a & \lceil \alpha' \rceil \simeq \lceil \alpha' \rceil & \lfloor \alpha' \rfloor \simeq \lfloor \alpha' \rfloor \\
\beta \simeq \gamma \;\Rightarrow\; [\alpha', \beta] \simeq [\alpha', \gamma] & \beta \simeq \gamma \;\Rightarrow\; \langle \alpha', \beta \rangle \simeq \langle \alpha', \gamma \rangle & \\
\beta \simeq \gamma \;\Rightarrow\; \beta \simeq \langle \alpha', \gamma \rangle & \beta \simeq \gamma \;\Rightarrow\; \langle \alpha', \beta \rangle \simeq \gamma &
\end{array}
$$

Thus, if $\alpha \simeq \beta$, the labeled terms $\alpha : X$ and $\beta : Y$ are shared. This is expressed by the following invariant.

Invariant 3. *$\mathcal{P}(W)$ holds iff, for any pair of subterms $\alpha : X$ and $\beta : Y$ such that $\alpha \simeq \beta$, we have $X = Y$.*

To prove this invariant, we need two extra invariants. The first one states that labels of free and bound variables cannot be equal up to the sharing relation.

Invariant 4. *$\mathcal{S}(W)$ holds iff, for any pair of subterms $\alpha : x$ and $\beta : y$ such that $\alpha \simeq \beta$, we have x free in W iff y free in W.*

Lemma 7. *If $\mathcal{S}(W)$ and $W \to W'$, then $\mathcal{S}(W')$.*

Proof: Firstly, a free (resp. bound) variable x in W' can only come from a free (resp. bound) variable x in W. Secondly, a labeled subterm $\alpha : x$ in W' can only come from a labeled subterm $\alpha_1 : x$ in W where $\alpha \simeq \alpha_1$. Combining these two remarks gives the full proof. □

In the weak labeled λ-calculus, the name of a redex is the compound label found on the left part of an application towards the corresponding abstraction. In a dag representation of terms, we only share subterms. Hence both application and abstraction nodes of a shared redex are shared. This means that for a given redex name α', there must be sharing of the application subterm making the redex. In the terminology of the weak labeled λ-calculus, it means that we share the atomic label of the application node. To be more precise, we state the following invariant.

Invariant 5. *$\mathcal{T}(W)$ holds iff, for any application subterms $\beta : (\alpha : X)U$ and $\gamma : (\alpha : Y)V$, we have $\beta \simeq \gamma$.*

Lemma 8. *If* $\mathcal{P}(W) \wedge \mathcal{Q}(W) \wedge \mathcal{R}(W) \wedge \mathcal{T}(W)$ *and* $W \stackrel{\gamma'}{\Longrightarrow} W'$, *then* $\mathcal{T}(W')$.

Proof: Take $\beta\!:\!(\alpha\!:\!X')U'$ and $\gamma\!:\!(\alpha\!:\!Y')V'$ in W'.

1. α exists in W. Then $\gamma' \not\prec \alpha$ since $\mathcal{Q}(W)$. Therefore, there are two labeled terms $U_1 = \beta_1 : (\alpha : X)U$ and $V_1 = \gamma_1 : (\alpha : Y)V$ in W, with a possible diffusion marking U_1 and/or V_1. Therefore, $\beta_1 \simeq \beta$ and $\gamma_1 \simeq \gamma$. By $\mathcal{T}(W)$, we have $\beta_1 \simeq \gamma_1$. Thus $\beta \simeq \gamma$.
2. $\alpha = \lceil \gamma', \alpha_1 \rceil$. Then $\beta = \lceil \gamma', \beta_1 \rceil$. Then $\gamma = \lceil \gamma', \gamma_1 \rceil$, since both applications received a single diffusion, since all γ' redexes are disjoint by $\mathcal{P}(W)$. In W, the corresponding applications are $\beta_1 : (\alpha_1 : XU)$ and $\gamma_1 : (\alpha_1 : YV)$. By $\mathcal{T}(W)$, we get $\beta_1 \simeq \gamma_1$. Thus $\beta \simeq \gamma$.
3. $\alpha = \langle \gamma', \alpha_1 \rangle$. Case similar to previous one.
4. Other cases, impossible since $\mathcal{R}(W)$.

□

Lemma 9. *If* $\mathcal{P}(W) \wedge \mathcal{Q}(W) \wedge \mathcal{R}(W) \wedge \mathcal{S}(W) \wedge \mathcal{T}(W)$ *and* $W \stackrel{\gamma'}{\Longrightarrow} W'$, *then* $\mathcal{P}(W')$.

Proof: By $\mathcal{T}(W)$, the labels of all redexes with name γ' have the same value γ modulo $\simeq$. Therefore, by $\mathcal{P}(W)$, all redexes with name γ' are the same $R = (\gamma' \cdot \lambda x.A)B$, and their contractums are identical $R' = \lceil \gamma' \rceil : (\gamma' ⓧ A)[\![x \backslash \lfloor \gamma' \rfloor : B]\!]$. As these redexes are identical, they are all disjoint. Conversely, every subterm R is a redex by $\mathcal{S}(W)$ if at least one R is a redex in W.

Let $U' = \alpha\!:\!X'$ and $V' = \beta\!:\!Y'$ two subterms of W' such that $\alpha \simeq \beta$.

1. Both U' and V' come from subterms $U = \alpha\!:\!X$ and $V = \beta\!:\!Y$ in W, with same label as U' and V'. By $\mathcal{P}(W)$, we know that $X = Y$. There are two cases.
   (a) U' contains a contractum. Then either U contains x replaced by $\lfloor \gamma' \rfloor\!:\!B$. Either U contains R. The first alternative is impossible since then R would contain R in its argument. Since every R in X and Y is a redex in W, then $X = Y$, $X \stackrel{\gamma'}{\Longrightarrow} X'$ and $Y \stackrel{\gamma'}{\Longrightarrow} Y'$. Thus $X' = Y'$.
   (b) U' does not contain a contractum of R.
      i. U' is disjoint from the contractums. Thus U is a subterm of W disjoint from redexes R. Then $U = U'$.
      ii. U' is in the argument part of a contractum. Thus, U is a subterm of W in the argument part of a redex R.
      iii. U' is in the body part of the contractum. Thus, U is a subterm of W in the function body part of a redex R, but U does not contain the variable x bound in R, since otherwise its label α would not be equal to the one of U'.

      In any of these cases, $U' = \alpha : X' = U = \alpha : X$. Therefore $X' = X$. Moreover, as $X = Y$, then $V = \beta\!:\!X$ cannot contain R (since X would do) or an occurrence of the bound variable of the function part in R, which means that $Y = Y'$. Thus $X' = Y'$.
2. Both labels α and β are new in W'. Thus $\gamma' \prec \alpha$ and $\gamma' \prec \beta$.
   (a) $\alpha = \lceil \gamma', \alpha_1 \rceil$. As $\alpha \simeq \beta$, we have $\beta = \lceil \gamma', \beta_1 \rceil$ and $\alpha_1 \simeq \beta_1$.
      In W, there exist two labeled subterms $\alpha_1\!:\!X_1$ and $\beta_1\!:\!Y_1$ such that

$\alpha{:}X' = (\gamma' ⓧ \alpha_1 : X_1)[\![x\backslash\lfloor\gamma'\rfloor : B]\!]$ and $\beta{:}Y' = (\gamma' ⓧ \beta_1 : Y_1)[\![x\backslash\lfloor\gamma'\rfloor : B]\!]$. By $\mathcal{P}(W)$, we have $X_1 = Y_1$ which implies $X' = Y'$.

(b) $\alpha = \lceil\gamma'\rceil$. By $\mathcal{Q}(W)$, this label is absent from W. It is created at the top of contractums of γ'-redexes, which are all equal by $\mathcal{P}(W)$.

(c) $\alpha = \lfloor\gamma'\rfloor$. By $\mathcal{Q}(W)$, this label is absent from W. It is created at the top of copies of arguments of R in the contractums. These copies are all equal.

(d) $\alpha = \langle\gamma', \alpha_1\rangle$. As $\alpha \simeq \beta$ and β is new, we can only have $\beta = \langle\gamma', \beta_1\rangle$ and $\alpha_1 \simeq \beta_1$. Both U' and V' come from U and V to the left of an application after a diffusion, when $x \notin U$ and $x \notin V$. So $U = \alpha_1 : X$ and $V = \beta_1 : Y$. Moreover $X = X'$ and $Y = Y'$, since $x \notin X$ and $x \notin Y$. By $\mathcal{P}(W)$, we get $X = Y$. Hence $X' = Y'$.

3. α is created, but β already exists in W. By $\mathcal{Q}(W)$, we know that $\gamma' \not\prec \beta$. As $\alpha \simeq \beta$, we can only have $\alpha = \langle\gamma', \alpha_1\rangle$ and $\alpha_1 \simeq \beta$. Therefore, there is a term $U = \alpha_1 : X$ to the left of an application subterm in W, with $x \notin X$, and $U' = \langle\gamma', \alpha_1\rangle : X'$, with $X = X'$. As $V' = \beta : Y'$ comes from $\beta : Y$ in W, we know by $\mathcal{P}(W)$ that $X = Y$. If Y is modified from W to W', this can only be because Y contains a γ'-redex, which is impossible since X is contained in a γ'-redex; or because it has been substituted after a diffusion, but then it is unchanged since $x \notin X = Y$. Therefore $Y = Y'$, and $X' = X = Y = Y'$.
4. α exists in W, but β is created. Analogous to previous case.

□

We can now state our sharing theorem, characterizing a dag implementation for evaluating terms in the weak λ-calculus. We need to have a notation for terms without sharing.

Notation 1. *Let Init(U) holds when every subterm of U is labeled with a distinct letter.*

Theorem 5. *Let Init(U) and $U \Longrightarrow\!\!\!\!\!\!\!\!\;\;\; V$, then $\mathcal{P}(V)$.*

Proof: We first notice that, if *Init*(U), then $\mathcal{P}(U) \wedge \mathcal{Q}(U) \wedge \mathcal{R}(U) \wedge \mathcal{S}(U) \wedge \mathcal{T}(U)$. Thanks to previous lemmas, this invariant remains valid along reduction steps $\Longrightarrow$. □

5 Dag Implementation

The weak labeled λ-calculus provides a formal setting for a dag implementation of the weak λ-calculus. If we start from a term labeled with distinct letters, each label identifies the address of a shared subterm in the dag implementation. There is a subtlety: first components of marked labels must be skipped, because they are only relevant for storing the history of redexes. The goal of our labeled calculus is to coincide with the dag implementation in Wadsworth [25].

Wadsworth has two implementations with sharing. The first one shares the argument of the contracted redex at each reduction step. When there are n occurrences of x in A, instead of having n copies of the argument B in the contractum of $(\lambda x.A)B$, Wadsworth proposes to replace x by a reference to a single subterm B. However his method demands to copy the function part $\lambda x.A$ of the contracted redex if the reference counter of $\lambda x.A$ is strictly greater than 1. This disallows the substitution of x in other subterms using A. In his second dag implementation, Wadsworth notices that it is unnecessary to copy subterms in A that do not contain occurrences of the free variable x. This variant of the dag implementation avoids unnecessary copies.

However finding if a term does not contain an occurrence of the x variable is a global operation. The occurrences of x in a redex $(\lambda x.A)B$ cannot disappear in the weak λ-calculus, since there cannot be redexes containing x inside A. Therefore, one may compile any abstraction $\lambda x.A$ in the initial term to mark each subterm of A with free variables bound outside $\lambda x.A$. This approach is taken by supercombinators and λ-lifting [22], where each abstraction is λ-lifted into a combinator to which are applied the free variables contained in this abstraction. In this context of the calculus of supercombinator, which is is not a subcalculus of the λ-calculus, Peyton-Jones and Hughes introduced the idea of *full lazyness* which we believe to be related to our diffusion operator. More precisely they create combinators by abstracting over *maximal free subterms*. It would be interesting to connect our sharing theory and their compilation scheme.

Recently, Shivers and Wand [23] provide a realistic implementation of weak reduction with sharing where, in the abstract tree corresponding to any abstraction $\lambda x.A$, the top node points to the list of occurrences of the variable x bound in this abstraction. In their representation of terms, any subterm points to the subterm directly containing it. Thus, terms are represented by graphs with a double-linked connection between vertices (i.e. nodes of the abstract trees). This representation facilitates access to terms and subterms in the usual top-down way, but makes also possible bottom-up traversals of terms from binders of abstractions through the bound occurrences of variables towards the root of an abstraction, or the root of a term. Therefore during a reduction step contracting redex $(\lambda x.A)B$, one first duplicates the paths from top of A to x in a bottom-up traversal from every occurrence of x towards the top node of A. We thus copy nodes corresponding to subterms containing x by performing local operations. However, as on the path from an occurrence of x to the top node of A, one may encounter an other binder, it is also necessary to make a recursive call to duplicate paths to its bound variable. This technique is an efficient way of implementing the second method in Wadsworth [25].

A closer look at our labeled λ-calculus demonstrates several differences between Shivers and Wand's method and our calculus. The creation of nodes along paths to occurrences of the bound variable x corresponds to the new tagged labels created by diffusion. But we can meet other binders λy on these paths. Then we do not copy the paths from such a λy to the free variable y during diffusion. Therefore we can have several binders for the same occurrence of the bound variable y. However no confusion occurs in our calculus, since sharing is virtual and is only represented through labels. But this situation in an actual shared implementation is unsafe, since it would hardly respect alpha renaming.

Extending the diffusion of our calculus could address this problem but it remains to prove that the calculus is still confluent, which looks feasible. Such a modification of the weak λ-calculus would be one way of providing a formal theory to the implementation in [23].

There are other subtle differences between our labeled λ-calculus and standard dag implementations. One is the introduction of new nodes, since, in a reduction step, we create new nodes, with $\lceil\alpha\rceil : U$ and $\lfloor\alpha\rfloor : U$. These nodes guarantee that we do not lose the history of the creation of redexes along a reduction. However, they do not seem necessary with respect to sharing. Similarly, it is quite debatable to have a label on a bound variable. We have labeled variables since the structure looks then more regular. But a calculus without these labels seems also feasible, without destroying sharing.

We can notice that the weak labeled λ-calculus makes a confluent theory of dag implementation. Usually, confluence proofs with graphs are rather delicate unless strong restrictions as in interaction nets [12]. But in the latter case, the theory only considers $\Longrightarrow$ reduction steps. Here we can reason about terms with induction as in the classical λ-calculus. This is what we gain with labeled calculi, since we have a textual representation of dags by handling the node address, the atomic label, and the λ-term.

Finally, the call-by-need strategies $\underset{\text{norm}}{\Longrightarrow}$ of the labeled calculus of weak explicit substitutions correspond to complete labeled reduction steps in the weak labeled λ-calculus. If *Init*(U) holds in the initial labeled term, one can show that the number of steps to get a normal form with this reduction is always minimal. The proof technique follows the one for the classical λ-calculus [17].

6 Relation with Term Rewriting Systems

One referee pointed out that the weak λ-calculus could be considered as a first-order term rewriting system (TRS) [24]. He mentioned that, for instance, the pattern of the weak β-redex $(\lambda x.Ix(\lambda y.Ixy))S$ is $(\lambda x.Z_1x(\lambda y.Z_2xZ_3))X$ giving rise to the first-order rule $(\lambda x.Z_1x(\lambda y.Z_2xZ_3))X \rightarrow Z_1X(\lambda y.Z_2XZ_3)$, where any terms can be substituted for the meta variables Z_1, Z_2, Z_3 and X. One could then substitute I for both Z_1 and Z_2, y for Z_3 and S for X to yield the step $(\lambda x.Ix(\lambda y.Ixy))S \rightarrow IS(\lambda y.ISy)$. With this remark, most of the properties of residuals hold for TRSs and thus might look more natural. However, the corresponding labeled TRS (see [24, 20]) differs from our weak labeled λ-calculus. As mentioned in the introduction, we think that our calculus has the advantage of staying in a subtheory of the classical λ-calculus.

7 Conclusion

In this paper, we recalled the definitions and basic properties of the weak λ-calculus introduced in [18]. This calculus follows the standard presentations of the λ-calculus, since the classical (strong) λ-calculus is an extension of it.

We presented a weak labeled λ-calculus which is a confluent theory of sharing for the weak λ-calculus. We stated that it corresponds to the representation of terms in Wadsworth's dag implementations but a very precise connection has still to be formalized.

The correspondence between the labeled λ-calculus and the calculus of sharing with explicit substitutions could also be studied as in [18]. This would make a bridge with the call-by-need strategies used in the evaluation of functional programs. One can also relate supercombinators and the shared structures for evaluating them in lazy languages to our calculus. A similar approach with respect to frameworks with a `let` statement could also be studied, although these research directions are not focused on basic syntactic properties such as confluence. (Notice that our calculus has no critical pairs.)

The theory of optimal reductions inside the weak λ-calculus is missing in this article, although one might easily guess it following the one of the classical λ-calculus. It remains to achieve it.

Finally, many of the results of this paper were considered as folk theorems, rather easy to prove. We hope to have shown that some of the proofs deserve attention. In fact, some of them are not easy at all. This paper also showed that the theory of the weak λ-calculus still needs more research.

Acknowledgements

We have benefited from numerous comments by the referees.

References

1. Martín Abadi, Luca Cardelli, Pierre-Louis Curien, and Jean-Jacques Lévy. Explicit substitutions. *Journal of Functional Programming*, 6(2):299–327, 1996.
2. Zena M. Ariola, Matthias Felleisen, John Maraist, Martin Odersky, and Philip Wadler. A call-by-need lambda calculus. In *Proc. 22nd ACM Symposium on Principles of Programming Languages*, pages 2330–246, January 1995.
3. Andrea Asperti, Paolo Coppola, and Simone Martini. (Optimal) duplication is not elementary recursive. *Information and Computation*, 193/1:21–56, 2004.
4. Andrea Asperti and Stefano Guerrini. *The Optimal Implementation of Functional Programming Languages.* Cambridge University Press, 1999.
5. Henk P. Barendregt. *The Lambda Calculus, Its Syntax and Semantics.* North-Holland, 1981.
6. Naim Çağman and J.R̃oger Hindley. Combinatory weak reduction in lambda calculus. *Theoretical Computer Science*, 198:239–249, 1998.
7. Pierre-Louis Curien, Thér'ese Hardin, and Jean-Jacques Lévy. Confluence properties of weak and strong calculi of explicit substitutions. *Journal of the ACM*, 43(2):362–397, 1996.
8. Maribel Fernández and Ian Mackie. Closed reductions in the lambda-calculus. In Jörg Flum and Mario Rodríguez-Artalejo, editors, *Computer Science Logic*, volume 1683 of *Lecture Notes in Computer Science*, pages 220–234. Springer, 1999.

9. Georges Gonthier, Martín Abadi, and Jean-Jacques Lévy. The geometry of optimal lambda reduction. In *Proc. of the 19th Conference on Principles of Programming Languages*, pages 15–26. ACM Press, 1992.
10. J.Roger Hindley. Combinatory reductions and lambda reductions compared. *Zeit. Math. Logik*, 23:169–180, 1977.
11. Jan Willem Klop. *Combinatory Reduction Systems.* PhD thesis, Rijksuniversiteit Utrecht, 1980.
12. Y. Lafont. Interaction nets. In *Principles of Programming Languages*, pages 95–108. ACM Press, 1990.
13. John Lamping. An algorithm for optimal lambda-calculus reduction. In *Proceedings of the 17th Annual ACM Symposium on Principles of Programming Languages*, pages 16–30, 1990.
14. John Launchbury. A natural semantics for lazy evaluation. In *Proc. of the 1993 conference on Principles of Programming Languages*, pages 144–154. ACM Press, 1993.
15. Julia L. Lawall and Harry G. Mairson. On the global dynamics of optimal graph reduction. In *ACM International Conference on Functional Programming*, pages 188–195, 1997.
16. Xavier Leroy. The ZINC experiment: an economical implementation of the ML language. Technical report 117, INRIA, 1990.
17. Jean-Jacques Lévy. *Réductions correctes et optimales dans le lambda-calcul.* PhD thesis, Univ. of Paris 7, Paris, 1978.
18. Jean-Jacques Lévy and Luc Maranget. Explicit substitutions and programming languages. In *Proc. 19th Conference on the Foundations of Software Technology and Theoretical Computer Science, Electronic Notes in Theoretical Computer Science*, volume 1738, pages 181–200, 1999.
19. John Maraist, Martin Odersky, David N. Turner, and Philip Wadler. Call-by-name, call-by-value, call-by-need, and the linear lambda calculus. In *Electronic Notes in Theoretical Computer Science*, pages 41–62, March 1995.
20. Luc Maranget. *La stratégie paresseuse.* PhD thesis, Univ. of Paris 7, Paris, 1992.
21. Paul-André Melliès. *Description Abstraite des Systèmes de Réécriture.* PhD thesis, Univ. of Paris 7, december 1996.
22. Simon L. Peyton-Jones. *The implementation of Functional Programming Languages.* Prentice-Hall, 1987.
23. Olin Shivers and Mitchell Wand. Bottom-up beta-substitution: uplinks and lambda-dags. Technical report, BRICS RS-04-38, DAIMI, Department of Computer Science, University of Århus, Århus, Denmark, 2004.
24. Terese. *Term Rewriting Systems.* Cambridge University Press, 2003.
25. Christopher P. Wadsworth. *Semantics and pragmatics of the lambda-calculus.* PhD thesis, Oxford University, 1971.

Term Rewriting Meets Aspect-Oriented Programming

Paul Klint, Tijs van der Storm, and Jurgen Vinju

Centrum voor Wiskunde en Informatica (CWI),
Kruislaan 413, NL-1098 SJ Amsterdam, The Netherlands
{Paul.Klint, T.van.der.Storm, Jurgen.Vinju}@cwi.nl

Dedicated to Jan Willem Klop on the occasion of his 60th anniversary.

Term rewriting is in the intersection of our interests and physical distance has never been large. Nonetheless we seem to be living at opposite ends of the term rewriting galaxy. Here is a story from the other side of that galaxy.

Abstract. We explore the connection between term rewriting systems (TRS) and aspect-oriented programming (AOP). Term rewriting is a paradigm that is used in fields such as program transformation and theorem proving. AOP is a method for decomposing software, complementary to the usual separation into programs, classes, functions, etc. An aspect represents code that is scattered across the components of an otherwise orderly decomposed system. Using AOP, such code can be modularized into aspects and then automatically weaved into a system.

Aspect weavers are available for only a handful of languages. Term rewriting can offer a method for the rapid prototyping of weavers for more languages. We explore this claim by presenting a simple weaver implemented as a TRS.

We also observe that TRS can benefit from AOP. For example, their flexibility can be enhanced by factoring out hardwired code for tracing and logging rewrite rules. We explore methods for enhancing TRS with aspects and present one application: automatically connecting an interactive debugger to a language specification.

1 Introduction

Software engineering is about conquering the complexity of real life software systems. Can large systems be organized such that they remain manageable? Many solutions have been tried from structured programming to abstract data types, modules, objects, components and agents. In specific areas some of these approaches have been successful but the problem of structuring and organizing software remains mostly open for research. In more recent years, *aspects*, *concerns* or *dimensions* of software systems have been investigated [19]. These approaches aim at encapsulating functionality that cuts across boundaries of conventional modularization. In this way, software would become composable along different

A. Middeldorp et al. (Eds.): Processes... (Klop Festschrift), LNCS 3838, pp. 88–105, 2005.

axes and the desired flexibility and composability could be achieved. While providing potential solutions to the software composition problem, they pose new problems as well: how can such new methods of modularization be combined with existing languages and how can they be supported by tools?

Term rewriting [29] is a well-known paradigm used in program transformation, and thus a natural candidate for developing language-oriented tool support. We first give quick introductions to aspect-oriented programming (Sect. 1.1) and applications of term rewriting (Sect. 1.2) and then we explain why it is interesting to explore connections between these two fields and how they can benefit from each other (Sect. 1.3).

The contributions of the paper can be summarized as follows:

- It raises the awareness that term rewriting techniques can be relevant for the implementation of aspect-oriented programming (Sect. 2).
- It explores the application of aspect-oriented techniques to term rewriting systems themselves (Sect. 3 & 4).
- It formulates research questions in the field of term rewriting that are brought forward by the previous two points (Sect. 5).

1.1 Aspect-Oriented Programming

One of the most important principles in software engineering is the principle of *separation of concerns.* Separating concerns in modules (e.g., functions, classes etc.) promotes maintainability and reuse, because the dependencies between modules are loose and explicit.

There are, however, concerns that cannot be adequately modularized using conventional mechanisms. Typical examples of these so-called crosscutting concerns are profiling, tracing, debugging, error handling, origin tracking, caching, and transaction management. In all these cases, the code to implement these concerns occurs in many modules since all these modules are affected by the concern in question. This situation is referred to as *code scattering.*

Aspect-Oriented Programming (AOP) [19] is an approach to ameliorate this situation by introducing a new modularization concept: *aspects.* An important characteristic of AOP is *quantification* [11]. For example, "whenever condition C arises in program P, do X" is a quantified statement over program P. The scattering of code for crosscutting concerns is avoided by automatically *weaving* the aspect code X in places where condition C holds.

In many AOP implementations, quantification over a program is achieved by specifying *pointcuts.* A pointcut is an addressing mechanism for the static or dynamic identification of execution points in the base program. These execution points are called *joinpoints* since at these points in the source code, the aspect code is joined with the base code.

To illustrate the notion of a pointcut, consider the example in the upper left part of Fig. 1 expressed in AspectJ [18], the aspect language for Java. The pointcut `creatingFoo` captures all calls to constructors of class `Foo`, disregarding the argument signature.

Advice specifications describe how the aspect code should be weaved at the joinpoints captured by a certain pointcut. There are three kinds of advice: before, after or around. Figure 1 contains an example of around advice. The directive **proceed**() is used to continue the delayed execution of the joinpoint, in this case constructing a new Foo-object. Figure 1 also shows on the left an example program and on the right the result of applying the given pointcut and advice to it.

Pointcut specification

```
pointcut creatingFoo():
  call (Foo.new(..))
```

Advice code

```
around (): creatingFoo {
if (flag)
  proceed ();
else
  throw
    new Exception("No flag!");
}
```

Initial Program

```
import foo.*;
public class Bar {
 private boolean flag;
 public void doBar() {
  Foo foo = new Foo();
 }
}
```

Result of aspect weaving

```
import foo.*;
public class Bar {
 private boolean flag;
 public void doBar() {
  Foo foo;
  if (flag) {
    foo = new Foo();
  } else {
    throw new
      Exception("No flag!");
  }
 }
}
```

Fig. 1. Pointcut and advice are used to weave code into a small program

1.2 Applications of Term Rewriting

In this paper we consider term rewriting to be a programming paradigm. The concept of term rewriting systems is used in many application areas, such as functional programming, program transformation, theorem proving, and language semantics [14,29]. From the viewpoint of these application areas term rewriting systems are *programs* [21].

To cater for different kinds of applications, most rewriting implementations extend basic rewrite rules with additional features. Such features include, for instance, concrete (mixfix) syntax [22], conditional rewrite rules [4], ordered rules [2], list matching and AC matching [10], traversal functions [30], strategies [6] and more. For this paper, the conditional rewrite rules are essential, while concrete syntax and traversal functions are practical utilities for the application we present.

A conditional rewrite rule is a normal rewrite rule, extended with a list of predicates. Now a redex must also satisfy all such predicates, before it can be contracted. Different kinds of predicates are allowed. For example, in our system we only allow (in)equality between two terms that are first normalized, or an (un)successful match between a normalized term and an open term.

A full account of the theoretical foundations of term rewriting and of systems implementing it is given in [29]. We will give examples using the notation found in the ASF+SDF META-ENVIRONMENT [7,22].

1.3 Connections Between the Two Fields

Why is it interesting to explore the connections between AOP and term rewriting?

Aspect-Oriented Programming Can Profit from Term Rewriting. The ideas for AOP contribute to software composition and maintenance and have been applied to mainstream programming languages (Java [18], C/C++ [28], C# [20], SmallTalk [15], Cobol [24]). In all these cases language-specific tool support has been developed. It is worthwhile to wonder whether AOP is applicable in the context of other languages, such as Perl, PHP, legacy languages, or even domain specific languages. The question, then, is how to develop tool support to evaluate such hypotheses. Term rewriting is used in other kinds of program transformation, so it would be natural to apply it to aspect weaving as well. In Sect. 2 we will investigate whether term rewriting is a good choice for rapidly implementing aspect weavers for new languages.

Term Rewriting Can Profit from Aspect-Oriented Programming. In many applications the *side-effects* of term rewriting systems are important. However, such side-effects are usually hardwired into a particular term rewriting engine. For example, each engine typically implements one kind of reduction tracing or debugging. We propose to separate these hard-wired aspects from the engine, and promote them to programmable aspects on the term rewrite system level. The result is that the application of existing engines can be made much more flexible and reusable. In Sect. 3 we explore a way to add reusable side-effects to an ASF+SDF term rewriting system, by employing aspect-oriented programming.

2 Aspect Weaving Implemented by Term Rewriting Systems

Since term rewriting is well equipped to deal with program transformation, aspect weaving is also a natural application area. This section provides an example of how to use term rewriting to implement aspect weavers.

The view that aspect weaving is a kind of program transformation is not new. For example, in [12] a case is made for a term rewriting approach to weaving aspects. Using special pattern-matching operators the authors are able to succinctly specify how aspects should be weaved. Graph writing for aspect weaving is discussed in [1]. It is argued that graph rewriting is more suitable than term rewriting because the base language consists of class graphs. Semantic information can be stored naturally in graphs, so more complex pointcuts may be expressible.

In most approaches that use rewriting for aspect weaving, rewrite rules directly function as the aspect weaving language. In [13] the authors present an aspect-oriented programming language for ObjectPascal which is implemented on top of DMS [3]. DMS contains a term rewriting component that forms the basis of their weaving algorithm, but this fact is hidden from the user. The strong motivation for using term rewriting is rapid development of aspect weavers for legacy languages.

In this section we adopt the latter approach, and demonstrate the implementation of a simple aspect-oriented programming language called μAspectJ. It consists of a very simple weaver written in ASF+SDF. It is able to weave advice similar to the example of Fig. 1. After this example, we will evaluate the fitness of term rewriting in this application area.

2.1 Implementing a Weaver for μ AspectJ

The expected behavior of μAspectJ on the example from the introduction is displayed in Fig. 2. The function `weave` is applied to two arguments: a Java compilation unit and an advice specification (shown on the left). The result is a new Java compilation unit that is the result from weaving the advice in the original Java code (shown on the right).

This weaver can be defined in one module of ASF+SDF code, which we present here. ASF+SDF modules consist of two parts. The syntactic part defines the

Input term

```
weave(
 import foo.*;
 public class Bar {
  private boolean flag;
  public void doBar() {
   Foo foo = new Foo();
  }
 }
,
 around(): Foo.new(..) {
  if (flag) {
    proceed();
  } else {
    throw new
      Exception("No flag!");
  }
 }
)
```

Output term

```
import foo.*;
public class Bar {
 private boolean flag;
 public void doBar() {
  Foo foo;
  if (flag) {
    foo = new Foo();
  } else {
    throw new
      Exception("No flag!");
  }
 }
}
```

Fig. 2. Weaving is rewriting: a μAspectJ program is weaved by reducing the `weave` symbol

```
module MuAspectJ
imports Java Substitute[Statement Statements]
exports context-free syntax
  AdviceKind "(" ")" PointCut ":" Block → Advice
  "before" | "after" | "around"         → AdviceKind
  "call" "(" Signature ")"              → PointCut
  Class "." "new" "(" ".." ")"          → Signature
```

Fig. 3. μAspectJ: syntax of pointcuts and advices

```
  weave(CompilationUnit, Advice)  → CompilationUnit {traversal}
```

Fig. 4. μAspectJ: signature of the weave traversal function

```
equations
[1] weave(S*1 Type Id = new Class(Param*); S*2,
          before() : call(Class.new(..)) { S* }) =
          S*1 S* Type Id = new Class(Param*); S*2

[2] weave(S*1 Type Id = new Class(Param*); S*2,
          after() : call(Class.new(..)) { S* }) =
          S*1 Type Id = new Class(Param*); S* S*2

[3] S*' := substitute(proceed();, Id = new Class(Param*);, S*)
    ================================================
    weave(S*1 Type Id = new Class(Param*); S*2,
          around() : call(Class.new(..)) { S* }) =
          S*1 Type Id; S*' S*2
```

Fig. 5. Equations defining μAspectJ weaving

types of input terms and functions. These types are defined using context-free syntax productions. The semantic part contains equations between terms expressed in concrete syntax. These equations are rewrite rules when read from left to right.

The definition of μAspectJ consists of three parts: the syntax of pointcuts and advice (Fig. 3), the signature of the weave traversal function (Fig. 4), and the equations defining this function (Fig. 5). The module `MuAspectJ` imports the syntax of Java and a generic parameterized module for substitution. In the syntax section, the syntax of very simple advices is defined. For the sake of brevity, we only allow pointcuts that capture arbitrary constructor invocations. The pointcuts follow the syntax of AspectJ. Advice consists of a kind (before, after or around), a pointcut and a Java statement block.

The `weave` function maps a compilation unit together with an advice specification to a compilation unit with the advice weaved in. To avoid writing a lot of boiler-plate code for traversing a compilation unit, the weave function is defined as traversal function [30]. Weave matches lists of statements containing a constructor invocation for classes that match the pointcut contained in the advice. For each kind of advice the associated advice code is weaved in accordingly. In these equations, pattern variables such as `S*`, `Class`, and `Type` are used to capture lists of statements, class names and type identifiers, respectively.

Consider equation `[1]` in Fig. 5 which matches Java statements of the form "`Type Id = new Class(Param*)`" in the context of a list of statements and applies a **before** advice to it. Note that the name of the class (variable `Class`) in the statement and the advice should be the same (non-left-linearity).

In the resulting code on the right-hand side of the rewrite rule, the body of the advice, which is matched by the variable `S*`, is placed just before the original statement. The other equations work in a similar fashion.

2.2 The Fitness of Term Rewriting for Aspect Weaving

The example shows some characteristics of aspect weaving. It needs at least complex pattern matching, pattern construction, tree traversal, and non-local information (the advice code) at redex positions. The first two features are provided by basic term rewrite rules, the last two features emerge from using traversal functions [30]. As an alternative, traversal strategies and dynamic rewrite rules could be used to implement the same behavior [32]. Matching modulo associativity (list matching) in ASF+SDF, makes the weaver deal easily with lists of statements.

The above example shows how term rewriting provides a large number of primitive operations that are necessary for implementing aspect weavers. Because these operations are readily available in term rewriting, we expect to be able to develop an aspect weaver more rapidly.

The example does not show how to scale up to more advanced pointcut specifications, such as exist in AspectJ. Such specifications need name space resolution, or even control flow information. Any aspect weaver implemented using term rewriting will therefore have to be preceded by a static semantic analysis phase to collect additional context information for matching pointcuts, and take this as an extra argument to the weave function. Note that static semantic analyses can be implemented as term rewriting systems too. Alternatively, code could be generated to dynamically resolve semantic issues, but this may have significant repercussions on run-time efficiency.

We have discussed a method for rapidly creating prototype aspect weavers using term rewriting. Depending on the requirements, these term rewriting systems may immediately be used as full implementations of aspect weavers.

3 Aspects in Term Rewriting Systems

We now shift perspective and explore whether aspect-orientation can contribute to term rewriting. There are at least two directions in which aspect-oriented term rewriting can be considered:

- *Crosscutting concerns in rewrite rules.* Starting from existing pure term rewriting systems we can identify the crosscutting concerns in large sets of rewrite rules and factor them out as programmable rewrite rule aspects.
- *Side-effects in term rewriting engines.* Starting from implementations of term rewriting engines, we notice that crosscutting concerns already occur naturally in the form of side-effects. Usually the implementation of these side-effects is hardwired in the rewriting engine that is used.

The first direction is particularly interesting in the field of language definitions. In this field we try to define rewriting based semantics for programming languages. The combination of aspect orientation and programming language definitions has received some attention. This has mainly focused on making language definitions more modular and extensible.

For example, in [31] the idea of implementing language extensions as aspects is explored in the context of attribute grammars. A similar strategy is explored in [23]. In this paper, declarative language definitions (e.g., in SOS) are evolved by transforming the semantic rules. This allows for the incremental addition of language facets (e.g., state, input/output, exception handling, etc.) to some base language. Primitives to achieve this include adding parameters to semantic functions, adding conditions to conditional rewrite rules, and the like.

The second direction is interesting in many applications of term rewriting. Figure 6 (Step 1) displays the process of rewriting a term. An engine takes a TRS and a term, and produces a normal form and some side-effects. Typical side-effects of the rewriting process include:

- Tracing: an exported trace of a reduction sequence represents a proof. Each reduction step contained in this trace is an equational deduction step.
- Profiling: measure the execution behavior (frequency, call graph, timings) of the term rewriting system.
- Debugging: instrument the term rewriting system with debugging information and interaction.

For some applications of term rewriting, the side-effects are even more important than the normal form. Depending on the domain, or on a specific application, in which a term rewriting engine is applied, these side-effects are specialized in different ways.

For example, a term rewriting engine that is used in concert with a proof assistant (e.g., Elan with Coq [25,16]) generates a trace that communicates with a specific deduction process. Specializing the rewriting engine to emit such a trace makes an otherwise generic term rewriting system less applicable in a different context, let alone in another domain. For example, if we want to connect the same engine to a different brand of proof assistant, a large part of its implementation must be adapted. In [25], this problem is also recognized, and attacked by providing a separate translation scheme from a canonical representation to the specific proof term syntax of the proof assistant. In this paper, we try to generalize this separation of concerns, and make it available for other aspects of term rewriting besides tracing.

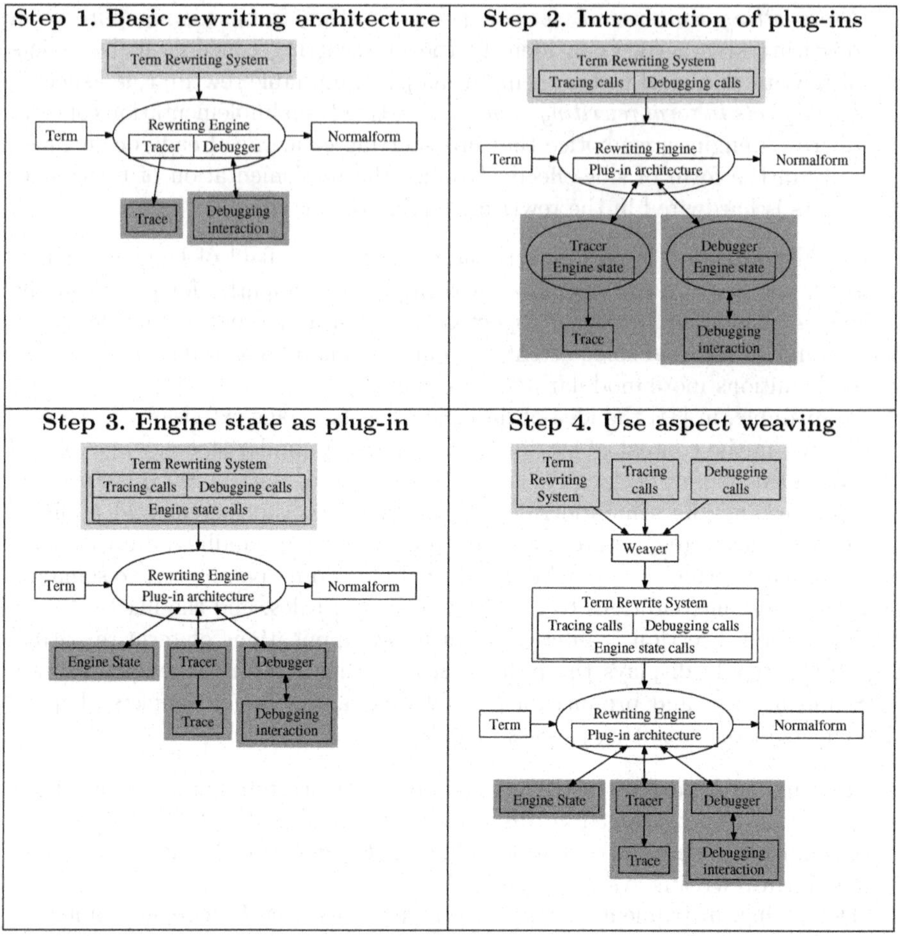

Fig. 6. From a rewriting engine with hardwired side-effects, to a completely decoupled engine with pluggable side-effects

Another problem with hardwired side-effects is the lack of control available to the user. For example, execution traces can be huge, but the user might be only interested in a localized section of a term rewriting system. In terms of space efficiency, such lack of control can lead to a lack of scalability. The possibilities that open up when the user could specialize side-effects for specific term rewriting applications are significant. In Sect. 4 we will present how a language specific debugger is obtained by adding debugging side-effects to a term rewriting system that implements the operational semantics of a language (i.e., a language interpreter).

We know of one instance of weaving side-effects into a language definition. In [33] a debugging aspect is weaved into an ANTLR [27] language definition of a domain specific language. Our example also shows how to weave in debugging support, but we use a dedicated aspect language instead of a general purpose transformation language (the authors of [33] use DMS). Note that a debugging feature does not change the semantics of a programming language, so there is no need to evolve the signatures of semantic functions.

3.1 From Hardwired Side-Effects to Programmable Aspects

Figure 6 depicts how a term rewriting engine can be refactored from an architecture with hardwired side-effects to a flexible plug-in architecture with programmable side-effects. We will discuss each step in turn.

Step 1 represents a basic implementation of term rewriting. Side-effects such as debugging, profiling and tracing are hardwired into the implementation. All side-effects are initiated by the term rewriting engine itself.

Step 2 represents the case that a TRS can be extended with explicit calls to library functions that might have side-effects. In line with the convention in other frameworks, we call these library functions *plug-ins* and add a plug-in architecture to the rewrite engine. This is achieved by adding a plug-in API (Application Programming Interface) to the engine to communicate a large part of the engine's state information to a plug-in. The engine will now be able to communicate with arbitrary plug-ins and achieve arbitrary side-effects. Unlike the situation in Step 1, side-effects are no longer initiated by the engine, but explicit calls to plug-ins must be added to the original term rewriting system.

Step 3 deals with the fact that interaction between different plug-ins is largely determined by the engine state. Therefore, we separate the engine state as a new plug-in. The engine state can now be queried from a term rewriting system dynamically. This information can then be used to trigger tracing, debugging or any other side-effect in a user-defined manner.

Step 4 introduces an aspect language and separates all calls to plug-ins into separate aspects. Note that the original term rewriting system of Step 1 is back as a separate input. By weaving in several aspects we automatically obtain the complete rewriting system of Step 3. From now on, libraries of reusable side-effects can be provided for the general use-cases of a term rewriting engine, while at the same time user-defined specializations can be created without much effort.

We have now separated hardwired functionality, and replaced it by plug-ins and aspect weaving. By doing this we have gained flexibility and reusability. For example, the syntax of a reduction trace can be adapted to the input syntax of a particular proof assistant, and a language semantics can be used with and without debugging support. In both examples, the original term rewriting system does not have to be changed, but only aspects have to be weaved in.

3.2 Introducing AspectAsf

The aspect weaver in Fig. 6 has been introduced for a reason. Side-effects often represent crosscutting concerns: many, if not all, rules of a TRS should be modified to include calls to the library of plug-ins. Take for instance the tracing of a reduction sequence: all firing rewriting rules have to communicate context information to the tracing plug-in. This would mean adding plug-in calls to all rules.

It would be advantageous if one could quantify over the set of rules and thus declaratively specify which rules should be adapted to incorporate the side-effect in question. For the tracing functionality, one would then say something like "add a call to trace to every rule". The actual adaptation of rewrite rules is subsequently enacted by automatically transforming the TRS.

This section introduces a simple aspect language for aspect-oriented programming in ASF+SDF. Using this language, the calls to the plug-ins can be specified separately. Invasive modifications of the TRS are not needed. Moreover, aspects have the additional advantage that we can now use a TRS with and without side-effects, or even add side-effects to parts of the TRS.

The next paragraphs will focus on how the declaration of these aspects would look like in a language called AspectASF. Firstly we will define pointcut patterns that are used for identifying sets of equations. Secondly, we show how these patterns are used in specifying pointcuts and advice.

AspectStratego [17] is another experiment in adding aspects to rewriting. That language allows much more complex pointcut specifications for inserting code along a reduction sequence. AspectASF, however, aims to primarily illustrate the viability of combining TRS with aspect-oriented techniques.

Pointcut Patterns. The pointcut pattern language is a pattern matching language on the structure of equations[1]. The examples in Fig. 7 illustrate the approach. The _ pattern functions as a placeholder for concrete terms that are not of interest. The * pattern is a wild-card for quantifying over parts of literals.

```
[_] _                        captures all equations
[_] eval(_,_)                ... with outermost symbol eval
[_] eval(_, Env)             ... with 2nd arg an Env variable
[int*] _                     ... with label like int..
[int*] _ or [real*] _        ... with label like int.. or real..
```

Fig. 7. A number of example pointcut patterns in AspectASF

For the sake of exposition we only allow pattern matching on labels of equations and left-hand sides. Patterns are expressed in concrete syntax. They do not contain any meta-variables. So, in the third example, Env matches with a regular ASF variable named Env; no binding is taking place at weaving time.

Note that this pointcut pattern language can be made more expressive by adding, for example, higher order matching, meta-variables, associative matching on conditions,sort assertions etc. For this presentation, we restrict ourselves

[1] Pointcut patterns are referred to as *signatures* in the AspectJ community. We avoid this term for clarity.

to first-order matching and simple boolean connectives (**and** and **or**) for constructing composite patterns.

Pointcuts and Advice. Pointcut patterns are used in the definition of pointcuts. We identify two kinds of pointcuts in AspectAsf: entering an equation (after a successful match of the left-hand side), and exiting an equation (just before returning the right-hand side). In an equation without conditions, **entering** and **exiting** are equivalent. The pointcut

entering `[_] eval(_)`

captures the points in the reduction sequence where the left-hand side of an equation with outermost function symbol `eval` has been matched successfully against a redex. The **exiting** pointcut is interpreted similarly.

Pointcuts are used in advice specifications. The kinds of advice that are allowed, exactly correspond to the kinds of pointcuts: **after entering** an equation, and **before exiting** an equation. Note that **before entering** and **after exiting** are meaningless, since such expressions do not correspond to identifiable points in the code.

As is common in aspect languages, the host language is reused for specifying advice code. In our case this language is the language of Asf conditions. To weave conditions **after entering** means prepending the advice conditions to the list of conditions of the equation that is matched by the pointcut. Similarly, advice **before exiting** corresponds to appending the advice conditions to the list of conditions of the subject equation.

Advice code can benefit from access to the context of the equation that it is weaved in. This information is provided in part by the weaver (e.g., the equation name), in part by calls to the library of plug-ins in the advice conditions itself (e.g., to obtain the depth of the evaluation stack). In the next section we will see an example of a call to such a library plug-in.

4 Applying Aspect-Oriented Term Rewriting

In this section we apply aspect-oriented programming to an Asf+Sdf-specification of a small programming language. As a small case-study a plug-in for debugging side-effects was constructed. This plug-in sends and receives messages from a generic visual debugging tool called Tide [26]. With this plug-in, we can instrument term rewriting systems such that they stop at certain points in the execution and allow inspection of the current state. To avoid the pollution of the term rewriting system with calls to the debugging plug-in, our goal is to automatically add such calls to the specification.

Starting from a term rewriting system that implements the semantics of the toy programming language Pico, we can now easily obtain an interactive debugger. We are interested in two functions of this semantics definition: `evs(Series, Env) -> Env` (used for evaluating series of statements) and `evst(Stat, Env) -> Env` (used for evaluating a single statement). For the sake of brevity we omit some of the equations that define these functions. The equations are displayed in Fig. 8.

```
[1] Env' := evst(Stat, Env),
    Env'' := evs(Stat*, Env')
    ================================
    evs(Stat ; Stat*, Env) =  Env''

[2] eve(Exp, Env) != 0
    ================================================
    evst(if Exp then Series1 else Series2 fi, Env) =
                                    evs(Series1, Env)

[3] eve(Exp, Env) == 0
    ================================================
    evst(if Exp then Series1 else Series2 fi, Env) =
                                    evs(Series2, Env)
```

Fig. 8. Fragment of the original Pico interpreter

```
pointcut statementStep: entering [_] evs(Stat ; Stat*, Env)
pointcut conditionStep:
  entering [_] evst(if Exp then Series1 else Series2 fi, Env)
        or [_] evst(while Exp do Series od, Env)
after: statementStep tide-step(get-location(Stat))
after: conditionStep tide-step(get-location(Exp))
```

Fig. 9. A debugging aspect for the Pico semantics

Recall that ASF+SDF allows matching on terms in concrete syntax. So in the first equation, the string `evs(Stat ; Stat*, Env)` is the redex pattern. Variables in these patterns start with an uppercase character (`Stat`, `Stat*`, `Env`, etc.).

Communication with the Tide debugger occurs via one library function, called `tide-step`. It receives one argument, which represents the source location of the active element of the Pico program. The source code location of the active element is obtained by calling the function `get-location`. This function is implemented as a separate plug-in. In a similar fashion one could query for other aspects of the term rewriting engine state.

Setting a breakpoint on a statement or conditional expression (in while- and if-statements) has the effect of pausing the execution. To obtain this functionality for the Pico language, the equation for statement sequencing, as well as the equations dealing with if and while statements, should include a call to `tide-step`.

The aspect declaration in Fig. 9 is used to weave in calls to the debugging side-effect at the points of interest. The pointcut `statementStep` captures the points of entry of the equation for statement sequencing. Pointcut `conditionStep` matches the `evst` equations defining the if- and while-statements. Both pointcuts are used in advice specifications, which ensure that a `tide-step` is executed at the appropriate places. Note how the advice uses some of the variables from the pointcut definitions.

```
[1] tide-step(get-location(Stat)),
    Env' := evst(Stat, Env),
    Env'' := evs(Stat*, Env')
    ================================
    evs(Stat ; Stat*, Env) =  Env''

[2] tide-step(get-location(Exp)),
    eve(Exp, Env) != 0
    ================================================
    evst(if Exp then Series1 else Series2 fi, Env) =
                                   evs(Series1, Env)

[3] tide-step(get-location(Exp)),
    eve(Exp, Env) == 0
    ================================================
    evst(if Exp then Series1 else Series2 fi, Env) =
                                   evs(Series2, Env)
```

Fig. 10. Pico interpreter fragment with debugging support automatically weaved in

The result of weaving this aspect into the Pico interpreter TRS is shown in Fig. 10. Each equation is appended with a special (tautological) condition that communicates with the Tide debugger. Any previously existing condition is evaluated after the debugger is informed. A screen-shot of the resulting interactive debugger is shown in Fig. 11.

We conclude that the original Pico interpreter (Fig. 8) remains completely separated from the debugging aspect (Fig. 9) and that the Pico interpreter with debugging support (Fig. 10) can be generated fully automatically. The example clearly shows the benefits of aspect-oriented programming. Firstly, the concerns for evaluation (base TRS) and debugging (aspect) are completely separated. Secondly, the scattering of calls is avoided: two aspects affect five equations. In larger specifications this ratio (2/5) is expected to be even better.

5 Discussion and Further Research

What can we conclude from these explorations of the connection between term rewriting and aspect-oriented programming? We have shown that term rewriting is a natural choice for implementing aspect weavers as has been illustrated in the μAspectJ case (Sect. 2). This regards primarily the transformations carried out by a weaver. Another source of complexity in weavers is the amount of type information that is needed for the weaving process, like, for instance name resolution. Here, term rewriting has no particular advantage over other techniques.

Another conclusion is that the application of aspect-orientation to term rewriting itself opens up several new possibilities and research questions as has been illustrated by the AspectAsf case (Sect. 3).

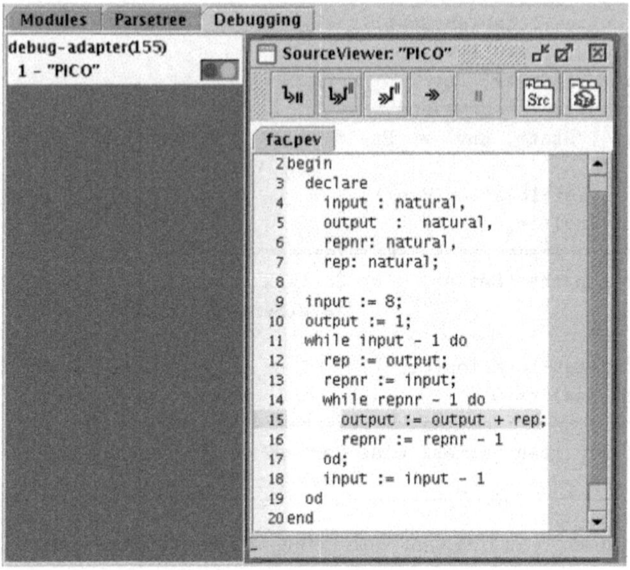

Fig. 11. A screen-shot of the generated Pico debugger in action

On the theoretical side, the composition of term rewriting systems has been studied for the relatively simple case of taking the union of rule sets. Aspect weaving, however, introduces a composition operation with more computational power thus decreasing the chances of predicting properties of the weaving result. Questions are:

- Is it possible to impose restrictions on the weaving TRS such that properties of the weaving result can be guaranteed? For instance, if the original TRS has a certain property (e.g. confluence, termination) how should the weaving TRS be restricted in order to guarantee certain properties of the weaving result?
- Can AOP be helpful to restructure an existing TRS in such a way that it becomes easier to prove properties of the AOP version?

On the practical side, other questions abound:

- Is it possible to design a sufficiently flexible aspect language for term rewriting systems or should one resort to full meta-programming as provided in, for instance, Maude [8]? Instead of the pointcuts and advices used in this paper, one can then use the full expressive power of term rewriting by transforming complete (collections of) rewrite rules.
- What are the implications of static versus dynamic weaving? In the former case, the initial TRS is changed by the weaver before rewriting. In the latter case, the weaving is done during rewriting. The meaning and implications of weaving during rewriting are unexplored.

- In the examples given in this paper, side-effects play a crucial role in the definition of the various aspects. Is it the case that AOP is *the* manner of introducing side-effects in a TRS without completely disturbing the underlying rewriting semantics, or are there alternatives?
- In most AOP implementations, origin information is lost. That is, if a weaved program is compiled or executed and there is an error, finding the source of this error is hard: is it in the base code, the advice code, or the interaction of the two? Origin tracking for term rewriting is a well-researched subject [9,5] and provides a solution for this problem. Aspect weaving may, however, require specialization or extension of that technique.
- Another question is related to origin tracking as well. As the authors of [9] state, different applications of origin tracking, such as program animation, error handling and debugging, require different notions of origin tracking. It would be interesting to investigate whether the code to propagate origins could be weaved in the TRS using aspect-oriented techniques.

All these questions show that cross-fertilization between the areas of term rewriting and aspect-oriented programming is possible and desirable.

Acknowledgments

We would like to thank Pieter Olivier and Bas Cornelissen for their work on debuggers for term rewriting systems and Mark van den Brand, Jan Heering and Ralf Laemmel for comments on drafts of this paper.

References

1. U. Aßmann and A. Ludwig. Aspect weaving with graph rewriting. In *Proceedings of the First International Symposium on Generative and Component-Based Software Engineering*, pages 24–36. Springer-Verlag, 2000.
2. J.C.M. Baeten, J.A. Bergstra, J.W. Klop, and W.P. Weijland. Term rewriting systems with rule priorities. *Theoretical Computer Science*, 67:283–301, 1989.
3. I. Baxter, C. Pidgeon, and M. Mehlich. DMS: program transformation and practical scalable software evolution. In *International Conference on Software Engineering*, May 2004.
4. J.A. Bergstra and J.W. Klop. Conditional rewrite rules: confluence and termination. *Journal of Computer and Systems Sciences*, 32:323–362, 1986.
5. I. Bethke, J.W. Klop, and R. de Vrijer. Descendants and origins in term rewriting. *Information and Computation*, 159:59–124, 2000.
6. P. Borovansky, C. Kirchner, H. Kirchner, P. Moreau, and C. Ringeissen. An overview of ELAN. In C. Kirchner and H. Kirchner, editors, *Second Intl. Workshop on Rewriting Logic and its Applications*, volume 15 of *Electronic Notes in Theoretical Computer Science*, 1998.
7. M.G.J. van den Brand, A. van Deursen, J. Heering, H.A. de Jong, M. de Jonge, T. Kuipers, P. Klint, L. Moonen, P.A. Olivier, J. Scheerder, J.J. Vinju, E. Visser, and J. Visser. The ASF+SDF Meta-Environment: a Component-Based Language Development Environment. In R. Wilhelm, editor, *Compiler Construction*, volume 2027 of *Lecture Notes in Computer Science*, pages 365–370. Springer-Verlag, 2001.

8. M. Clavel, F. Durán, S. Eker, P. Lincoln, N. Martí-Oliet, J. Meseguer, and J. F. Quesada. Maude: Specification and programming in rewriting logic. *Theoretical Computer Science*, 2001.
9. A. van Deursen, P. Klint, and F. Tip. Origin tracking. *Journal of Symbolic Computation*, 15:523–545, 1993.
10. S.M. Eker. Associative-commutative matching with bipartite graph matching. *Computer Journal*, 38(5):381–399, 1995.
11. R. Filman and D. Friedman. Aspect-Oriented Programming is quantification and obliviousness. In *Workshop on Advanced Separation of Concerns at OOPSLA*, 2000.
12. P. Fradet and M. Südholt. AOP: towards a generic framework using program transformation and analysis. In *ECOOP Workshop Reader*, volume 1543 of *LNCS*, pages 394–397. Springer-Verlag, July 1998.
13. J. Gray and S. Roychoudhury. A technique for constructing aspect weavers using a program transformation engine. In *Proceedings of the 3rd international conference on Aspect-oriented software development*, pages 36–45, Lancaster, UK, 2004. ACM Press.
14. J. Heering and P. Klint. Semantics of programming languages: A tool-oriented approach. *ACM Sigplan Notices*, 35(3):39–48, March 2000.
15. R. Hirschfeld. AspectS – aspect-oriented programming with Squeak. In M. Aksit, M. Mezini, and R. Unland, editors, *Objects, Components, Architectures, Services, and Applications for a Networked World*, number 2591 in LNCS, pages 216–232. Springer, 2003.
16. G. Huet, G. Kahn, and Ch. Paulin-Mohring. *The* Coq *Proof Assistant - A tutorial - Version 8.0*, April 2004. `http://coq.inria.fr`.
17. K. T. Kalleberg and E. Visser. Combining aspect-oriented and strategic programming. In H. Cirstea and N. Marti-Oliet, editors, *Workshop on Rule-Based Programming (RULE'05)*, Electronic Notes in Theoretical Computer Science, Nara, Japan, April 2005. Elsevier Science Publishers.
18. G. Kiczales, E. Hilsdale, J. Hugunin, M. Kersten, J. Palm, and W. G. Griswold. An overview of AspectJ. In *Proceedings of the 15th European Conference on Object-Oriented Programming*, pages 327–353. Springer-Verlag, 2001.
19. G. Kiczales, J. Lamping, A. Menhdhekar, C. Maeda, C. Lopes, J. Loingtier, and J. Irwin. Aspect-oriented programming. In M. Akşit and S. Matsuoka, editors, *Proceedings European Conference on Object-Oriented Programming*, volume 1241, pages 220–242, Berlin, Heidelberg, and New York, 1997. Springer-Verlag.
20. H. Kim. AspectC#: An AOSD implementation for C#. Master's thesis, Trinity College, November 2002.
21. H. Kirchner and P.-E. Moreau. Promoting Rewriting to a Programming Language: A Compiler for Non-Deterministic Rewrite Programs in Associative-Commutative Theories. *Journal of Functional Programming (JFP)*, 11(2):207–251, March 2001.
22. P. Klint. A meta-environment for generating programming environments. *ACM Transactions on Software Engineering and Methodology*, 2(2):176–201, April 1993.
23. R. Lämmel. Declarative aspect-oriented programming. In O. Danvy, editor, *Proceedings of the ACM SIGPLAN Workshop on Partial Evaluation and Semantics-Based Program Manipulation (PEPM)*, volume NS-99-1 of *BRICS Notes Series*, pages 131–146, San Antonio, Texas, January 1999.
24. R. Lämmel and K. De Schutter. What does aspect-oriented programming mean to Cobol? In *Proceedings of Aspect-Oriented Software Development.* ACM Press, March 2005. 12 pages.

25. Q-H. Nguyen, C. Kirchner, and H. Kirchner. External rewriting for skeptical proof assistants. volume 29(3–4), pages 309–336. Kluwer Academic Publishers, 2002.
26. P. A. Olivier. *A Framework for Debugging Heterogeneous Applications.* PhD thesis, Universiteit van Amsterdam, 2000.
27. T. J. Parr and R. W. Quong. ANTLR: A predicated-$LL(k)$ parser generator. *Software – Practice & Experience*, 7(25):789–810, 1995.
28. O. Spinczyk, A. Gal, and W. Schröder-Preikschat. AspectC++: An aspect-oriented extension to C++. In *Proceedings of the 40th International Conference on Technology of Object-Oriented Languages and Systems*, February 2002.
29. Terese. *Term Rewriting Systems*, volume 55 of *Cambridge Tracts in Theoretical Computer Science.* Cambridge University Press, 2003.
30. M. G. J. van den Brand, P. Klint, and J. J. Vinju. Term rewriting with traversal functions. *ACM Trans. Softw. Eng. Methodol.*, 12(2):152–190, 2003.
31. E. Van Wyk. Aspects as modular language extensions. In *Proc. of Language Descriptions, Tools and Applications (LDTA)*, volume 82.3 of *Electronic Notes in Theoretical Computer Science.* Elsevier Science, 2003.
32. E. Visser. Program transformation with Stratego/XT: Rules, strategies, tools, and systems in StrategoXT-0.9. In C. Lengauer, editor, *Domain-Specific Program Generation*, volume 3016 of *Lecture Notes in Computer Science*, pages 216–238. Springer-Verlag, June 2004.
33. H. Wu, J. Gray, S. Roychoudhury, and M. Mernik. Weaving a debugging aspect into domain-specific language grammars. In *ACM Symposium for Applied Computing (SAC) – Programming for Separation of Concerns Track*, Santa Fe NM, March 2005.

Observing Reductions in Nominal Calculi *Via* a Graphical Encoding of Processes*

Fabio Gadducci and Ugo Montanari

Dipartimento di Informatica, Università di Pisa

Abstract. The paper introduces a novel approach to the synthesis of labelled transition systems for calculi with name mobility. The proposal is based on a graphical encoding: Each process is mapped into a (ranked) graph, such that the denotation is fully abstract with respect to the usual structural congruence (i.e., two processes are equivalent exactly when the corresponding encodings yield the same graph).

Ranked graphs are naturally equipped with a few algebraic operations, and they are proved to form a suitable (bi)category of cospans. Then, as proved by Sassone and Sobocinski, the synthesis mechanism based on *relative pushout*, originally proposed by Milner and Leifer, can be applied. The resulting labelled transition system has ranked graphs as both states and labels, and it induces on (encodings of) processes an observational equivalence that is reminiscent of early bisimilarity.

Keywords: Nominal calculi, reduction semantics, synthesised labelled transition systems, relative pushouts, graph transformations.

1 Introduction

The dynamics of many computational devices is often defined in terms of *reduction relations*. Let us consider for example the paradigmatic functional language, the λ-calculus. Its operational semantics is aptly provided by the *β-reduction rule* $(\lambda x.M)N \Rightarrow M[N/x]$ that models the application of a functional process $\lambda x.M$ to the actual argument N. The reduction relation is then obtained by freely instantiating and contextualising the rule. This is quite typical in many calculi, since such a rule represents an *internal reduction* of a system component.

Moving towards calculi for interaction, let us consider now the reduction rule $a.P \mid \bar{a} \Rightarrow P$ for *asynchronous CCS-like communication*. The metavariable P actually denotes any possible process, let it be $P = \bar{b}$, and the rule can be contextualised in unary contexts such as $C[\_] = b.0 \mid [\_]$. Under those assumptions, the mechanism yields the rewriting step $b.0 \mid a.\bar{b} \mid \bar{a} \Rightarrow b.0 \mid \bar{b}$.

Reduction semantics have the advantage of conveying the semantics of calculi with relatively few compact rules. Its main drawback is poor compositionality, in

* Partly supported by the EU within the project HPRN-CT-2002-00275 SegraVis (*Syntactic and Semantic Integration of Visual Modelling Techniques*); and within the FETPI Global Computing, project IST-2004-16004 SEnSOria (*Software Engineering for Service-Oriented Overlay Computers*).

A. Middeldorp et al. (Eds.): Processes... (Klop Festschrift), LNCS 3838, pp. 106–126, 2005.

the sense that the dynamic behaviour of arbitrary stand alone terms (like $a.P$ in the example above) can be interpreted only by inserting them in the appropriate context (i.e., $[\_] \mid \bar{a}$), where a reduction may take place.

In different terms, reduction semantics is often less suitable whenever specific behaviours other than confluence (termination, reachability) are of interest. In fact, simply using the reduction relation for defining equivalences between components (e.g. in terms of *bisimulation*) fails to obtain a compositional framework, and in order to recover a suitable notion of equivalence it is often necessary to verify the behaviour of single components under any viable execution context. This is the way leading from the research on termination-under-context-closure equivalences for the λ-calculus to barbed and dynamic equivalences for the π-calculus. In these approaches, though, proofs of equivalence are often tedious as well as involuted, and they are left to the ingenuity of the researcher.

A standard way out of the empasse, reducing the complexity of such analyses, is to express the behaviour of a computational device by a *labelled transition system* (LTS). Should the label associated to a component evolution faithfully express how that component might interact with the whole of the system, it would be possible to analyse *in vitro* the behaviour of a single component, without considering all contexts. Thus, a "well-behaved" LTS represents a fundamental step towards a compositional semantics of the computational device.

Milner's proposal for an alternative semantics for the π-calculus [18] based on reactive rules modulo a suitable structural congruence, inspired by the CHAM paradigm [4], has been the source of an ongoing stream of research focussing on the investigation of the relationship between the LTS based semantics for nominal calculi and their more abstract reduction semantics.

Early attempts by Sewell [24] devised a strategy for obtaining an LTS from a reduction relation by adding contexts as labels on transitions. The technique was refined by Leifer and Milner [16] who introduced *relative pushouts* (RPOs) in order to capture the notion of *minimal context* activating a reduction. The generality of this proposal (and its bicategorical formulation due to Sassone and Sobocinski [22]) allows it to be applied to a large class of formalisms. More importantly, such attempts share the basic property of synthesising a congruent bisimulation equivalence, thus ensuring that the resulting LTS semantics is compositional. However, for the time being there are few case studies which either involve rich calculi, or succeed in making comparisons with standard behavioural equivalences. To tackle a full-fledged case study is the main aim of this paper.

Our starting point for the synthesis of an LTS are the graphical techniques proposed by the authors for modelling the reduction semantics of nominal calculi [11]. There is a long tradition in the use of graphical formalisms for describing the operational semantics of a computational device. They are often biased towards an implementation view, ranging from the functional paradigm (culminating on the works on *optimal implementation* [17]) to the imperative one (using *term graph rewriting* as an efficient technique for equational deduction [2]).

Only recent years have seen proposals concerning the use of graphical techniques for simulating reduction in process calculi, in particular for their mobile extensions. Typically, the use of graphs allows for getting rid of the problems concerning the implementation of reduction over the structural equivalence, such as e.g. the α-conversion of bound names. Most of these proposals (among them one of the better known formalisms, Milner's *bigraphs* [19]) follow the same pattern: At first, a suitable graphical syntax is introduced, and its operators used for implementing processes. After that, usually ad-hoc graph rewriting techniques are developed for simulating the reduction semantics. Most often, the resulting graphical structures are eminently hierarchical (that is, roughly, each node/edge is itself a structured entity, and possibly a graph). From a practical point of view, this is unfortunate, since the restriction to standard graphs would allow for the reuse of already existing theoretical techniques and practical tools.

In a recent series of papers the authors pursed instead the use of standard tools from graph transformation theory for modelling a large class of these calculi, ranging from mobile ambients to fusion [8, 11]. The use of unstructured (that is, non hierarchical) graphs allows for the reuse of standard graph transformation theory and tools for simulating the reduction semantics of a calculus, such as the double-pushout (DPO) approach and the associated concurrent semanticss [1]. The relevant bit here, however, is that these coding techniques can be successfully employed for the synthesis of suitable LTSs for nominal calculi. This is possible thanks to general results concerning the presentation of graph transformations as suitable reductions over so-called *cospan categories* [9].

Summing up, our paper is then to be considered a combination of the graphical techniques of encoding proposed by the authors for modelling nominal calculi, and of the categorical tools used by Sassone and Sobocinski for obtaining suitable LTS semantics out of graph transformation systems, presented according to the DPO style [23]. Even if for the sake of presentation the present work focuses on the finite, deterministic fragment of the π-calculus, it could be easily extended to recursive processes. We thus believe that it may offer novel insights on the synthesis of LTSs, as well as offering further evidence of the adequacy of graph-based formalisms for system design and verification.

The structure of the paper follows. Section 2 presents the finite, deterministic fragment of the π-calculus, and its reduction semantics. Section 3 recalls some definitions concerning ranked graphs, whilst Section 4 illustrates their use in an encoding of π-calculus processes. Finally, Section 5 presents our use of the graphical encoding for providing an alternative labelled transition system semantics for the π-calculus. The final section outlines future research avenues, while the Appendix contains most of the categorical notions used in the paper.

2 Synchronous (Finite) π-Calculus

We now introduce the finite, deterministic fragment of synchronous π-calculus.

Definition 1 (processes). *Let $\mathcal{N}$ be a set of* names, *ranged over by $a, b, c, \ldots$; and let $\Delta = \{a(b), \bar{a}b \mid a, b \in \mathcal{N}\}$ be the set of* prefix operators, *ranged over by δ. A* process *P is a term generated by the syntax*

$$P ::= \mathsf{0} \mid (\nu a)P \mid P \mid P \mid \delta.P$$

We let $P, Q, R, \ldots$ range over the set $\mathcal{P}$ of processes.

The standard definitions for the sets of free and bound names of a process P, denoted by $\mathtt{fn}(P)$ and $\mathtt{bn}(P)$ respectively, are assumed. Similarly for α-conversion with respect to the *restriction* operators $(\nu a)P$ and the *input* operators $b(a).P$: In both cases, the name a is bound in P, and it can be freely α-converted.

Using the definitions above, the behavior of a process P is described as a relation over *abstract processes*, i.e., a relation obtained by closing a set of basic rules under structural congruence.

Definition 2 (structural congruence). *The* structural congruence *for processes is the relation $\equiv \subseteq \mathcal{P} \times \mathcal{P}$, closed under process construction and α-conversion, inductively generated by the following set of axioms*

$$P \mid Q = Q \mid P \qquad P \mid (Q \mid R) = (P \mid Q) \mid R \qquad P \mid \mathsf{0} = P \qquad (\nu a)\mathsf{0} = \mathsf{0}$$

$$(\nu a)(\nu b)P = (\nu b)(\nu a)P \qquad (\nu a)(P \mid Q) = P \mid (\nu a)Q \text{ for } a \notin \mathtt{fn}(P)$$

$$(\nu a)\delta.P = \delta.(\nu a)P \text{ for } a \notin \mathtt{fn}(\delta) \cup \mathtt{bn}(\delta)$$

Definition 3 (reduction semantics). *The* reduction relation *for processes is the relation $R_\pi \subseteq \mathcal{P} \times \mathcal{P}$, closed under the structural congruence $\equiv$, inductively generated by the following set of axioms and inference rules*

$$\frac{}{a(b).P \mid \overline{a}c.Q \rightarrow P\{^c/_b\} \mid Q} \qquad \frac{P \rightarrow Q}{(\nu a)P \rightarrow (\nu a)Q} \qquad \frac{P \rightarrow Q}{P \mid R \rightarrow Q \mid R}$$

where $P \rightarrow Q$ means that $(P, Q) \in R_\pi$.

The first rule denotes the communication between two processes: Process $\overline{a}c.Q$ is ready to communicate the (possibly global) name c along the channel a; it then synchronizes with process $a(b).P$, and the local name b is substituted by c on the residual process P, denoting the resulting process with $P\{^c/_b\}$. The latter rules state the closure of the reduction relation with respect to the operators of restriction and parallel composition.

There are a few differences with respect to the standard syntax and operational semantics for the π-calculus, as proposed e.g. in the initial chapter of [21] (see Definition 1.1.1, Table 1.1 and Table 1.3). First of all, the lack of the prefix operator $\tau.P$ and of the choice operator $P_1 + P_2$. They are both simplifying assumptions, and see [8] for a graphical encoding of the calculus with these two operators. Instead, the axioms concerning the distributivity of the restriction operators with respect to the two prefix operators are not standard, even if they have been already considered in the literature, see e.g. [7]. These equalities do not change substantially the reduction semantics, and they indeed hold in all the observational equivalences we are aware of. Moreover, they allow for a simplified presentation of the graphical encoding: We refer the reader to [11] for a more articulate analysis of the resulting structural congruence.

Example 1. We introduce now a very simple example, the process **race**, defined as $(\nu c)\bar{a}c.\bar{c}c \mid a(b).\bar{b}d$, which seems to us well-suited for illustrating the reduction semantics of the calculus, as well as the graphical encoding of processes in the next sections. The sub-process on the left is ready to send a bound name c via a channel a. The sent name will then used by both component processes as output in their respective continuations. After a scope extension of the restriction operator, a possible commitment of **race** thus consists of a synchronization on b: **race** $\rightarrow$ $(\nu c)(\bar{c}c \mid \bar{c}d)$. The residual process is deadlocked, since the restriction forbids c to be observed.

3 Graphs and Their Ranked Extension

We recall a few definitions concerning (labeled hyper-)graphs, and their *ranked* extension, referring to [5] for a detailed introduction and a comparison with the standard presentation [20]. In the following we assume a chosen signature (Σ, S), for Σ a set of operators (edge labels), and S a set of sorts (node labels), such that the *arity* of an operator in Σ is a pair (s, ω), for $\omega \in S^*$ and $s \in S$.

Definition 4 (graphs). *A* graph *d (over (Σ, S)) is a tuple $d = \langle N, E, l, s, t\rangle$, where N, E are the sets of* nodes *and* edges*; l is the pair of* labeling *functions $l_e : E \rightarrow \Sigma$, $l_n : N \rightarrow S$; $s : E \rightarrow N$ and $t : E \rightarrow N^*$ are the* source *and* target *functions; and such that for each edge $e \in E$, the arity of $l_e(e)$ is $(l_n(s(e)), l_n^*(t(e)))$, i.e., each edge preserves the arity of its label.*

Let d, d' be graphs. A graph morphism *$f : d \rightarrow d'$ is a pair of functions $f_n : N \rightarrow N'$, $f_e : E \rightarrow E'$ that preserves the labeling, source and target functions.*

With an abuse of notation, in the definition above we let l_n^* stand for the extension of the function l_n from nodes to strings of nodes; sometimes, we use l as a shorthand for l_n and l_e. In the following, we denote the components of a graph d by N_d, E_d, l_d, s_d and t_d, dropping the subscript if clear from the context.

In order to inductively define the encoding for processes, we need operations over graphs. The first step is to equip them with suitable "handles" for interacting with an environment, built out of other graphs.

Definition 5 (ranked graphs). *Let d_r, d_v be graphs with no edges. A (d_r, d_v)-*ranked graph *(a* graph of rank *(d_r, d_v)) is a triple $G = \langle r, d, v\rangle$, for d a graph and $r : d_r \rightarrow d$, $v : d_v \rightarrow d$ the* root *and* variable *morphisms.*

Let G, G' be ranked graphs of the same rank. A ranked graph morphism *$f : G \rightarrow G'$ is a graph morphism $f_d : d \rightarrow d'$ between the underlying graphs that preserves the root and variable morphisms.*

We let $d_r \stackrel{r}{\Rightarrow} d \stackrel{v}{\Leftarrow} d_v$ denote a (d_r, d_v)-ranked graph. With an abuse of notation, we sometimes refer to the image of the root and variable morphisms as roots and variables, respectively. More importantly, in the following we will often refer implicitly to a ranked graph as the representative of its isomorphism class, still using the same symbols to denote it and its components.

Definition 6 (two composition operators). *Let* $G = d_r \overset{r}{\Rightarrow} d \overset{v}{\Leftarrow} d_i$ *and* $H = d_i \overset{r'}{\Rightarrow} d' \overset{v'}{\Leftarrow} d_v$ *be ranked graphs. Then, their* sequential composition *is the ranked graph* $G \circ H = d_r \overset{r''}{\Rightarrow} d'' \overset{v''}{\Leftarrow} d_v$*, for* d'' *the disjoint union* $d \uplus d'$*, modulo the equivalence on nodes induced by* $v(x) = r'(x)$ *for all* $x \in N_{d_i}$*, and* $r'' : d_r \to d''$*,* $v'' : d_v \to d''$ *the uniquely induced arrows.*

Let $G = d_r \overset{r}{\Rightarrow} d \overset{v}{\Leftarrow} d_v$ *and* $H = d'_r \overset{r'}{\Rightarrow} d' \overset{v'}{\Leftarrow} d'_v$ *be ranked graphs. Then, their* parallel composition *is the ranked graph* $G \otimes H = (d_r \cup d'_r) \overset{r''}{\Rightarrow} d'' \overset{v''}{\Leftarrow} (d_v \cup d'_v)$*, for* d'' *the disjoint union* $d \uplus d'$*, modulo the equivalence on nodes induced by* $r(x) = r'(x)$ *for all* $x \in N_{d_r} \cap N_{d'_r}$ *and* $v(y) = v'(y)$ *for all* $y \in N_{d_v} \cap N_{d'_v}$*, and* $r'' : d_r \cup d'_r \to d'', v'' : d_v \cup d'_v \to d''$ *the uniquely induced arrows.*

Intuitively, the sequential composition $G \circ H$ is obtained by taking the disjoint union of the graphs underlying G and H, and glueing the variables of G with the corresponding roots of H. Similarly, the parallel composition $G \otimes H$ is obtained by taking the disjoint union of the graphs underlying G and H, and glueing the roots (variables) of G with the corresponding roots (variables) of H. Note that the two operations are defined on "concrete" graphs. Nevertheless, the result is clearly independent of the choice of the representative, up-to isomorphism.[1]

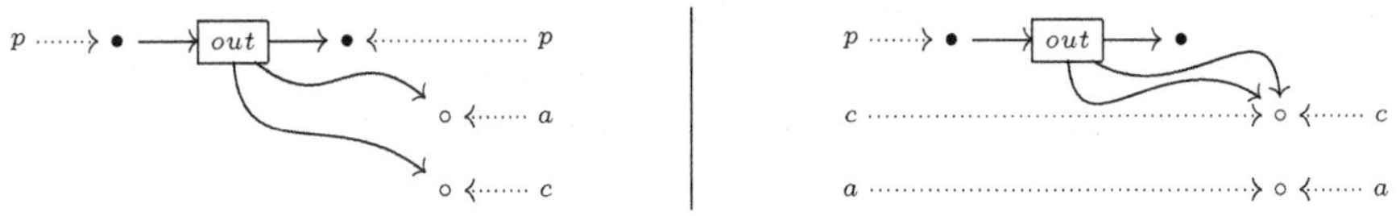

Fig. 1. Ranked graphs $out_{a,c}$ (left) and $\llbracket \bar{c}c \rrbracket \otimes id_{\{a,c\}}$ (right)

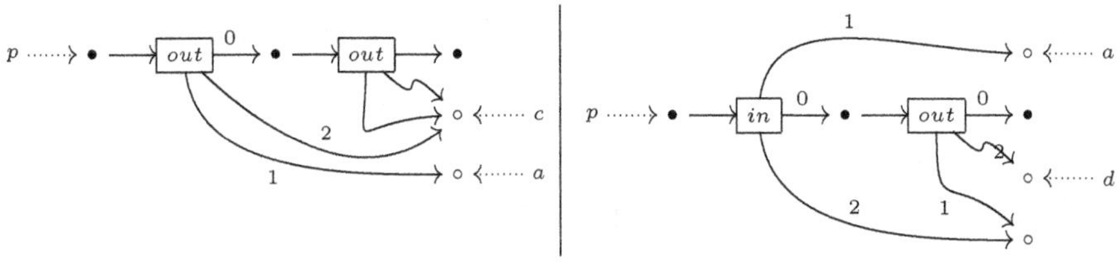

Fig. 2. Ranked graphs $out_{a,c} \circ (\llbracket \bar{c}c \rrbracket \otimes id_{\{a,c\}})$ (left) and $\llbracket a(b).\bar{b}d \rrbracket$ (right)

Example 2 (sequential and parallel composition). Fig. 1 depicts two ranked graphs: As we shall see, they are part of the encoding of our running example, and with an abuse of notation we denote them by using still to be defined

[1] While the sequential operator precisely corresponds to categorical composition, the parallel operator does not coincide with tensor product of monoidal categories [3]. A more standard definition for the latter operator is e.g. in [5]. Our choice, though, allows for a compact presentation of the graphical encoding in the following sections.

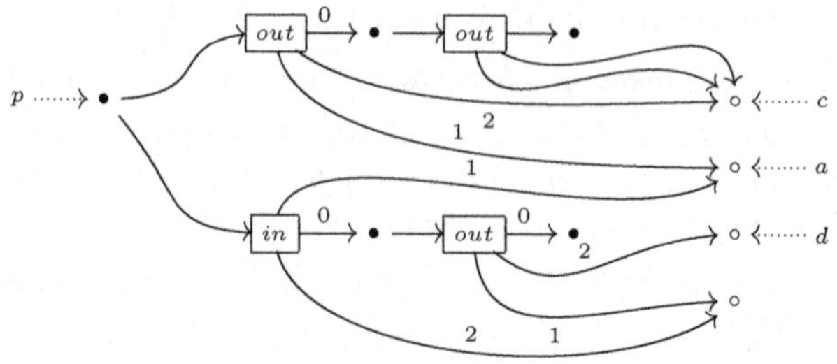

Fig. 3. The ranked graph $\llbracket \bar{a}c.\bar{c}c \rrbracket \otimes \llbracket a(b).\bar{b}d \rrbracket$

symbols. Their sequential composition is depicted in Fig. 2 (left), while the parallel composition of the graphs of Fig. 2 is represented in Fig. 3.

The nodes in the domain of the root (variable) morphism are depicted as a vertical sequence on the left (right, resp.); the variable and root morphisms are represented by dotted arrows, directed from right-to-left and left-to-right, respectively. Edges are represented by a boxed label, from where arrows pointing to the target nodes leave, and to where the arrow from the source node arrive; the sequence of target nodes is usually the clockwise order of the start points of the tentacles, even if sometimes it is indicated by a numbering on the tentacles: For the edge of the leftmost graph of Fig. 1 the sequence is $(v(p), v(a), v(c))$.

The leftmost graph of Fig. 1 has rank $(\{p\}, \{p, a, c\})$, four nodes and one edge labeled by *out*; the rightmost graph has rank $(\{p, a, c\}, \{a, c\})$, four nodes of two different sorts (for graphical convenience, in the underlying graph nodes of different sorts are denoted differently) and one edge labeled by *out*.

A *graph expression* is a term over the syntax containing all ranked graphs as constants, and parallel and sequential composition as binary operators. An expression is *well-formed* if all occurrences of the parallel and sequential operators are defined for the rank of the argument sub-expressions, according to Definition 6; its rank is computed inductively from the rank of the graphs occurring in it, and its *value* is the graph obtained by evaluating all operators in it.

4 From Processes to Graphs

We now present the encoding of π-calculus processes into ranked graphs, inspired to [8]. It is based on a signature (Σ_π, S_π), and it preserves structural congruence. The set of sorts S_π is $\{s_p, s_n\}$: Intuitively, a graph reachable from a node of sort s_p corresponds to a process, while each node of sort s_n represents a name. The set Σ_π contains the operators $\{in, out\}$ of sort $(s_p, s_p s_n s_n)$, clearly simulating the input and output prefixes, respectively. There is no operator for simulating either the restriction operators or the parallel composition of processes.

The second step is the characterization of a class of graphs, such that all processes can be encoded into an expression containing only those graphs as constants, and parallel and sequential composition as binary operators. Let $p \notin \mathcal{N}$: Our choice of graphs as constants is depicted in Fig. 4, for all $a, b \in \mathcal{N}$.

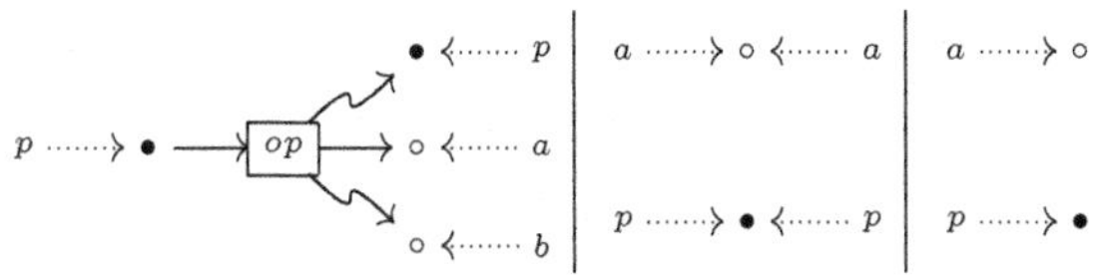

Fig. 4. Ranked graphs $op_{a,b}$ (for $op \in \{in, out\}$), id_a and id_p, 0_a and 0_p

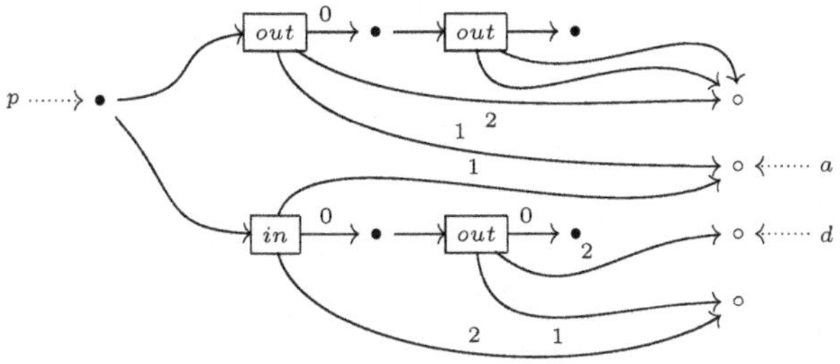

Fig. 5. The ranked graph $\llbracket (\nu c)\bar{a}c.\bar{c}c \mid a(b).\bar{b}d \rrbracket$

Finally, let us denote id_Γ as a shorthand of $\bigotimes_{x \in \Gamma} id_x$, for a set Γ of names (since the ordering is immaterial). The encoding of processes into ranked graphs, mapping each finite process into a graph expression, is presented below.

Definition 7 (encoding for processes). *Let P be a process. The* encoding *$\llbracket P \rrbracket$, mapping a process P into a ranked graph, is defined by structural induction according to the following rules*

$$
\begin{aligned}
\llbracket (\nu a)P \rrbracket &= \begin{cases} \llbracket P \rrbracket & \textit{if } a \notin \mathtt{fn}(P) \\ \llbracket P \rrbracket \circ (0_a \otimes id_{\mathtt{fn}(P)\setminus\{a\}}) & \textit{otherwise} \end{cases} \\
\llbracket P \mid Q \rrbracket &= \llbracket P \rrbracket \otimes \llbracket Q \rrbracket \\
\llbracket 0 \rrbracket &= 0_p \\
\llbracket \bar{a}b.P \rrbracket &= out_{a,b} \circ (\llbracket P \rrbracket \otimes id_{\{a,b\}}) \\
\llbracket a(b).P \rrbracket &= in_{a,b} \circ (\llbracket P \rrbracket \otimes id_{\{a,b\}}) \circ (0_b \otimes id_{\mathtt{fn}(P)\setminus\{b\}})
\end{aligned}
$$

The mapping is well-defined, since the resulting graph expression is well-formed; moreover, the encoding $\llbracket P \rrbracket$ is a graph of rank $(\{p\}, \mathtt{fn}(P))$.

Example 3 (mapping a process). In order to give some intuition about the intended meaning of the previous rules, we show the construction of the encoding for the process $\bar{a}c.\bar{c}c$ (a subprocess of our running example) whose graphical representation is depicted in Fig. 2 (left)

$$
\llbracket \bar{a}c.\bar{c}c \rrbracket \;=\; out_{a,c} \circ (\llbracket \bar{c}c \rrbracket \otimes id_{\{a,c\}}) \;=\; out_{a,c} \circ ((out_{c,c} \circ (0_p \otimes id_c)) \otimes id_{\{a,c\}})
$$

The denotation of $(\llbracket \bar{c}c \rrbracket \otimes id_{\{a,c\}})$ coincides with $(out_{c,c} \otimes id_{\{a,c\}}) \circ (0_p \otimes id_{\{a,c\}})$, and the latter is clearly matched by its graphical representation, see Fig. 1 (right). The graphical representation of $\llbracket \textbf{race} \rrbracket$ is depicted in Fig. 5.

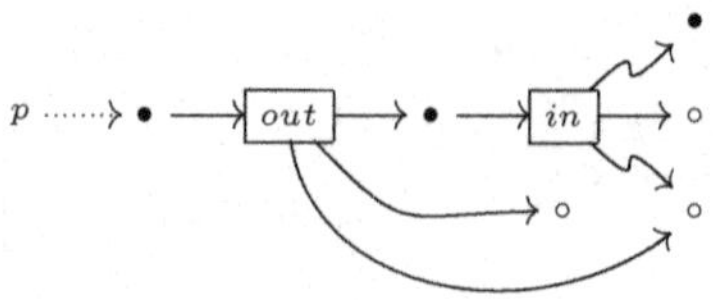

Fig. 6. A ranked graph with a forbidden name-sharing situation

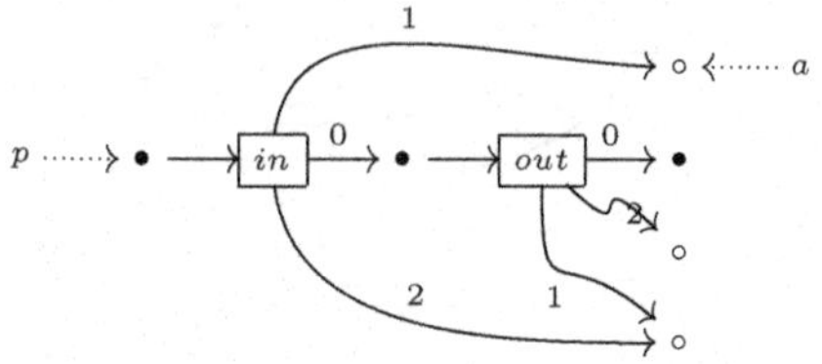

Fig. 7. Ranked graph encoding for both $(\nu d)a(b).\bar{b}d$ and $a(b).(\nu d)\bar{b}d$

The mapping $\llbracket \cdot \rrbracket$ is not surjective, since there are graphs of rank $(\{p\}, \Gamma)$ that do not belong to the image of any process. As an example, let us consider the graph in Fig. 6: It represents a name-sharing situation which is not allowed in the process construction, where a name that is local to the process below the input prefix is made visible globally.

Nevertheless, let us assume that we restrict our attention to processes verifying a mild syntactical condition, namely, forbidding the occurrences of input prefixes such as $a(a)$. Then, our encoding is sound and complete, as stated by the proposition below (adapted from [8]).

Proposition 1. *Let P, Q be processes. Then, $P \equiv Q$ if and only if $\llbracket P \rrbracket = \llbracket Q \rrbracket$.*

Note in particular how the lack of restriction operators is dealt with by manipulating the rank of the interface, even if the price to pay is the presence of "floating" axioms for prefixes, as shown by Fig. 7.

5 Reductions Via Sequential Composition

A recent series of papers advocated the use of graph transformation for modelling the reduction semantics of nominal calculi. In particular, the authors proposed the use of tools and techniques from the double-pushout (DPO) approach for obtaining an implementable, concurrent semantics for these calculi [8, 11, 12].

This section follows a parallel path. The aim is to obtain an algebraic mechanism for specifying graphs, thus presenting their transformation *via* a suitable rewriting system. The technical trick is the recasting of DPO derivations as cells on a suitable bicategory on cospan categories. The fact has been originally noted in [9, 10]. It has been further refined in recent work by Sassone and Sobocinski [23], where the construction has been exploited for obtaining a labelled transition system using Milner and Leifer's *relative pushouts* [16].

In order to simplify our presentation, we plan to recast most of the categorical machinery in terms of the set-theoretic definitions used for ranked graphs. The drawback is that sometimes the statements are going to be loose, and the reasoning mostly driven by examplifications. Nevertheless, all the relevant underlying notions and theorems are provided in the Appendix.

5.1 Completeness of the Specification

Let us consider again the graphs in Fig. 4. Whilst sufficient for encoding processes, there exists ranked graphs that are not described by a graph expression containing only those graphs as constants.

Let us then consider the ranked graphs below, which can be used to either hide roots or performing a renaming on the interfaces.

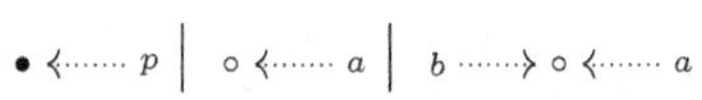

Fig. 8. Ranked graphs ν_p, ν_a and $\sigma_{b,a}$

There has been in recent years a research thread on the algebraic presentation of (ranked) graphs, see e.g. [13, 14]. These approaches differ in the choice of the alternative sets of constants and inference rules for characterizing graph expressions. Variants of the result below are thus frequent in the literature: The present statement is adapted from [5–Theorem 9].

Proposition 2. *Let G be a graph of rank (I, J), for I, J finite subsets of $\{p\} \cup \mathcal{N}$. Then, G can be denoted by a graph expression, possibly containing the graphs in Fig. 4 and Fig. 8 as constants.*

5.2 Encoding the Rules

Despite its appealing simplicity, the DPO approach to graph transformation still lacks suitable proof and analysis techniques, differently from e.g. classical term rewriting. This state of affairs seems mostly due, as argued in [5], to the lack of alternative presentations of the formalism based on structural induction. As pointed out in [9, 10], and confirmed by the recent [23], DPO graph transformation systems can be recast as suitable rewriting systems, obtaining an inductive characterization for the formalism by exploiting this presentation. This section rephrases those results in terms of ranked graphs and their composition.

First of all, though, since we would also like to describe open terms, we consider a set $\{p\} \cup \mathcal{V}$ of metavariables, ranged over by $U, V, \ldots$, and we assume the constants ν_V, id_V, 0_V and $\sigma_{U,V}$, defined as expected; the mapping $\llbracket V \rrbracket = \sigma_{p,V}$; and the encoding $\llbracket P \rrbracket_V = \sigma_{V,p} \circ \llbracket P \rrbracket$.

Now, exploiting the presentation of reductions as graph rewrites [8], and considering the encoding of DPO rules as cospans [9], the reduction rule of the calculus, namely $a(b).P \mid \bar{a}c.Q \rightarrow P\{^c/_b\} \mid Q$, can be simulated as a pair of ranked

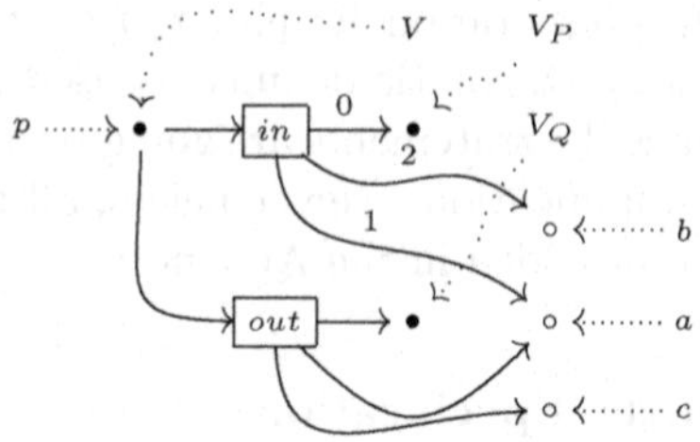

Fig. 9. The encoding R_l of the left-hand side

graphs, with the singleton $\{p\}$ as unique root for both. The graph denoting the left-hand side of the rule is presented in Fig. 9.

Informally, note that V_P and V_Q are the placeholders for the continuation of the processes to which the rule is applied; similarly, V indicate the possible context $[\_] \mid R$ into which the pair of communicating processes can be inserted.

A similar graph R'_l is actually needed for simulating $a(b).P \mid \overline{a}a.Q$ (even if its graphical depiction is not presented here). This corresponds to considering only injective matches in the DPO derivations; or, as we shall see, to sequentially compose (the graphical encoding of) the rule and a graph with injective variable morphism. As we argued in [12–Section 5.3], this is a reasonable restriction when dealing with calculi showing a complex name matching.

On the positive side, please note that only one rule is needed. In fact, the three (different) meta-variables do occur as nodes in the graph, whilst they represent concrete process instances in the corresponding reduction rule of the π-calculus. Similarly, there is no need for rules representing the closure of the reduction with respect to the restriction and parallel operators, since these operators are now embedded into the graph context in which the rule occurs.

The right-hand side of the rule is depicted in Fig. 10. The three nodes of sort s_p are merged, indicating that the continuations occur now at the top of the process; similarly, also the nodes for the variables b and c are coalesced.

The following result, an adaptation of [8–Theorem 1], explains how the graphical encoding of the rules may actually simulate a reduction between processes.

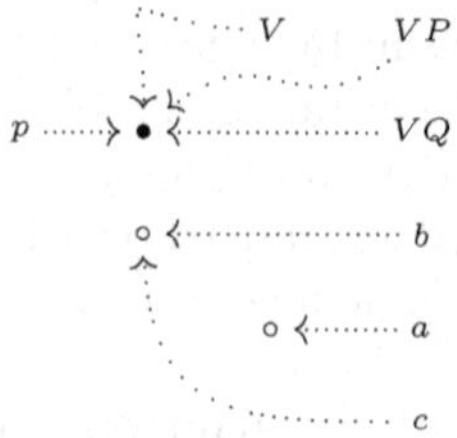

Fig. 10. The encoding R_r of the right-hand side

Proposition 3 (encoding preserves reductions). *Let P, Q be processes. If $P \rightarrow Q$, then there exists a ranked graph G with injective variable morphism such that $\llbracket P \rrbracket$ coincides with either $R_l \circ G$ or $R'_l \circ G$ and $\llbracket Q \rrbracket \otimes \nu_{\mathtt{fn}(P)}$ coincides with $R_r \circ G$ ($R'_r \circ G$, respectively).*

Intuitively, the graph G is built by considering the context into which the rule has to be mapped, in order to capture the encoding of the process. The key point is that any such context can be expressed as a suitable graph expression.

In order to exemplify the construction, we round up the section with a more detailed example. Let us consider again the derivation $(\nu c)(\overline{a}c.\overline{c}c \mid a(b).\overline{b}d) \rightarrow (\nu c)(\overline{c}c \mid \overline{c}d)$. The starting process can be simulated by the sequential composition of the left-hand side R_l of the rule, depicted in Fig. 9, with the graph $G_{race} = 0_V \otimes G_P \otimes G_Q$, for the graph expressions $G_P = (\llbracket \overline{b}d \rrbracket_{V_P} \otimes id_b) \circ (\nu_b \otimes id_d)$ and $G_Q = (\llbracket \overline{c}c \rrbracket_{V_Q} \otimes id_{\{a,c\}}) \circ (\nu_c \otimes id_a)$ depicted in Fig. 11. Note also that the latter coincides with the graph on the right of Fig. 1, modulo the obvious renaming of the root and the hiding of the variable c.

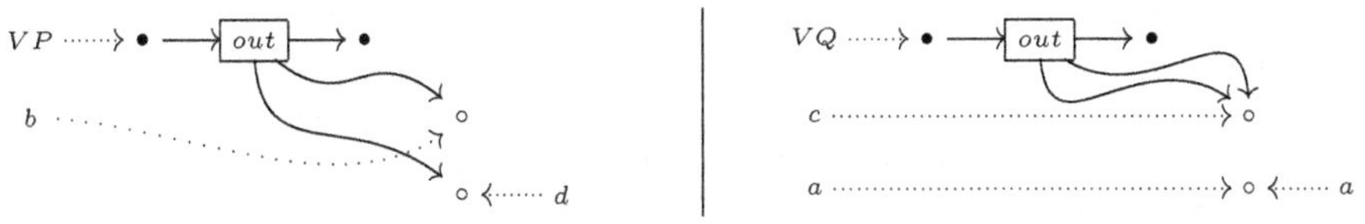

Fig. 11. Ranked graphs G_P (left) and G_Q (right)

The ranked graph $R_r \circ G_{race}$, the sequential composition of the right-hand side R_r of the rule with the "context" G_{race}, is presented in Fig. 12. It coincides with the denotation of $\llbracket (\nu c)(\overline{c}c \mid \overline{c}d) \rrbracket \otimes \nu_a$.

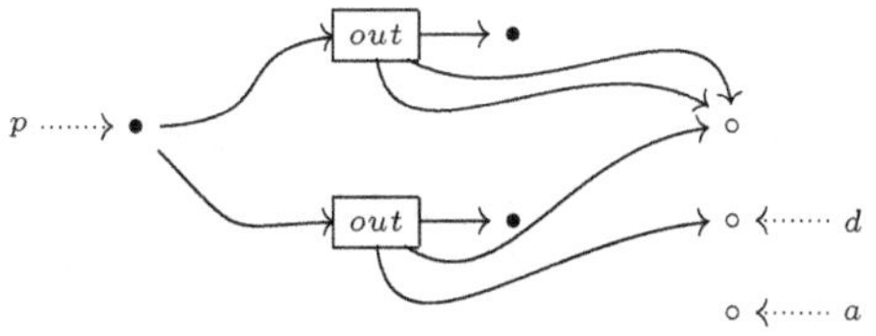

Fig. 12. The ranked graph $\llbracket (\nu c)\overline{c}c \mid \overline{c}d \rrbracket \otimes \nu_a$

5.3 Observing Reductions

Exploiting the results sketched above, this last section presents a labelled transition system for the π-calculus, with graphical encodings of processes as states.

The mechanism to be followed for obtaining the labels is suggested by relative pushouts. Its formal construction is provided in Definition 15. Roughly, its states are (isomorphic classes of) ranked graphs G, and its labels are those "minimal" ranked graphs C such that $G \circ C$ can perform a reduction.

Later in this section we try to exemplify the minimality of a context, referring to the Appendix for its categorical construction. In order to provide a set-theoretic presentation, let us first consider the composition $\alpha \circ R$ of an isomorphism $\alpha : G \rightarrow H$ between graphs of rank (d_r, d_i) and a graph R of rank (d_i, d_v) as the uniquely induced isomorphism from $G \circ R$ into $H \circ R$.

Definition 8 (minimal context). *Let us consider a graph $G \circ C$, isomorphic to $R_l \circ D$ for an isomorphim α, of rank $(\{p\}, d_v)$.*

Moreover, let us consider a triple $\langle C', D'E \rangle$ of ranked graphs and three isomorphisms $\beta : G \circ C' \rightarrow R_l \circ D'$, $\gamma : C' \circ E \rightarrow C$, and $\delta : D' \circ E \rightarrow D$ such that α coincides with functional composition of $C \circ \gamma$, $\beta \circ E$ and $R_l \circ D$.

Then, the context C is minimal *with respect to G and D if whenever the two conditions above hold, there exists a unique ranked graph L (up-to a unique isomorphism) and three compatible isomorphisms $\gamma' : C \circ L \rightarrow C'$, $\delta' : D \circ L \rightarrow D'$, and $\xi : id_{d_v} \rightarrow L \circ E$ (such that e.g. the functional composition of $C \circ \xi$, $\gamma' \circ E$, and γ coincides with the identity on C).*

In the above definition we let id_{d_v} denote the ranked graph with the identity on d_v as both the root and the variable morphims.

Note that, by construction, if C is minimal then E must be discrete. In other terms, the requirement of minimality boils down to ensure that, whenever a graph $G \circ C$ is decomposed as $R_l \circ D$, then the decomposition is unique, up-to renaming of the variables in the interface.

Definition 9 (labelled transitions for graphical encodings). *The labelled transition system $LTS(\mathbb{C}_\pi)$ is given by*

1. *the states of $LTS(\mathbb{C}_\pi)$ are (isomorphic classes of) graphs of rank $(\{p\}, d_v)$;*
2. *there exists a transition $G \xrightarrow{C} R_r \circ D$ iff C is a minimal graph with respect to G and D.*

Hence, a transition $G \xrightarrow{C} R_r \circ D$ can be performed if the ranked graph $G \circ C$, obtained by the sequential composition of the initial state of the transition with the label, can be decomposed as $R_l \circ D$ and C is minimal with respect to G and D. Spelled out, the definition above coincides with the construction in Definition 15, as generated by the bireactive system $\mathbb{C}_\pi$ specified in Definition 17.

Example 4. This final part of the section provides some examples of labelled transitions. First, let us consider the derived encoding $\llbracket P \rrbracket^p = \llbracket P \rrbracket \otimes id_p$, intuitively allowing for a graph to be inserted into a larger context via sequential composition. Let us consider the term $\overline{a}c$, obtained as a sub-process of the left-hand side of the reduction rule, where the process Q is istantiated to $\mathsf{0}$. The graph $\llbracket \overline{a}c \rrbracket^p$ of rank $(\{p\}, \{p, a, c\})$ reduces to $\llbracket V_P \rrbracket \otimes \nu_a \otimes \hat{\nu}_{\{b,c\}}$ (the latter being the derived operator $\nu_b \circ (id_b \otimes \sigma_{b,c})$), and the label $\hat{in}_{a,b} \otimes id_{\{a,c\}}$ (the former being the derived operator $in_{a,b} \circ (\llbracket V_p \rrbracket \otimes id_{\{a,b\}})$ represents the minimal ranked graph (up-to renaming of the metavariable) allowing for the corresponding process reduction to be performed. The transition is depicted in Fig. 13.

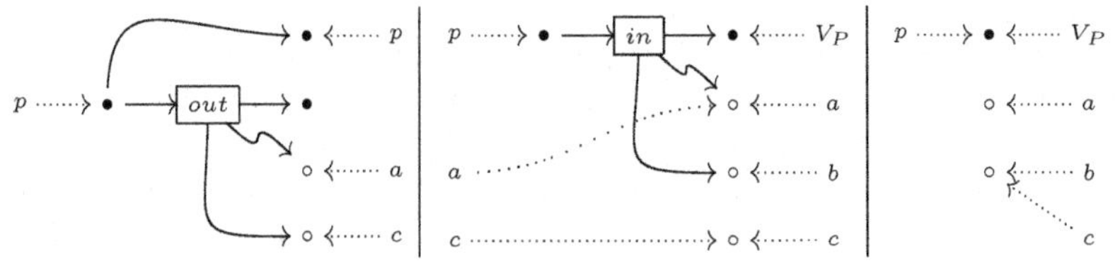

Fig. 13. Components of transition $\llbracket \overline{a}c \rrbracket^p \xrightarrow{\hat{in}_{a,b} \otimes id_{\{a,c\}}} \llbracket V_P \rrbracket \otimes \nu_a \otimes \hat{\nu}_{\{b,c\}}$

Even if b is a bound name, it has to appear, possibly modulo a renaming, among the variables of the label, in order for the latter to be a minimal context.

Let us then elaborate on the previous example.

- Reactions can be applied to open processes. Let us consider the encoding $\llbracket \overline{a}c.V_Q \rrbracket^p$, for a metavariable $V_Q \neq V_P$: Its has rank $(\{p\}, \{p, V_Q, a, c\})$, and via the observation $\hat{in}_{a,b} \otimes id_{\{V_Q,a,c\}}$ can be reduced to $\llbracket V_P \mid V_Q \rrbracket \otimes \nu_a \otimes \hat{\nu}_{\{b,c\}}$.
- Reactions can be applied to restricted processes. Let us consider the encoding $\llbracket (\nu c)\overline{a}c \rrbracket^p$: Its has rank $(\{p\}, \{p, a\})$, and via the observation $\hat{in}_{a,b} \otimes id_a$ can be reduced to $\llbracket V_P \rrbracket \otimes \nu_{\{a,b\}}$.

Perhaps more interestingly, let us consider the encoding $\llbracket a(b).V_P \rrbracket^p$, for a metavariable $V_P \neq V_Q$. Now b is bound in the source state, so its identity should be irrelevant in the computation. In fact, the graph reduces to $\llbracket V_P \mid V_Q \rrbracket \otimes \nu_{\{a,d\}}$ with observation $\llbracket \overline{a}d.V_Q \rrbracket \otimes id_{\{V_P,a\}}$ for *any* name d. The resulting labelled transition is depicted in Fig. 14.

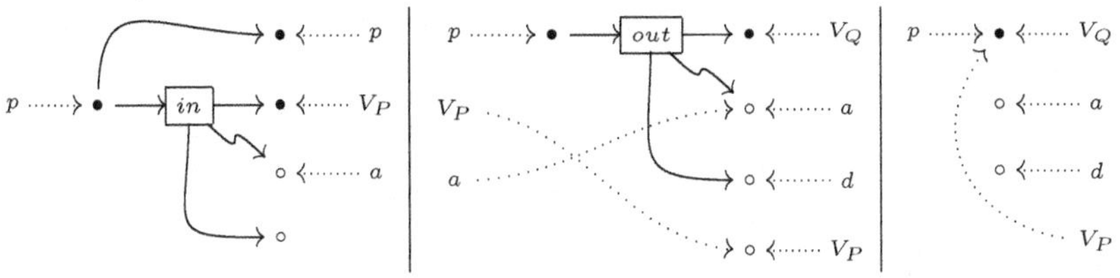

Fig. 14. Components of transition $\llbracket a(b).V_P \rrbracket^p \xrightarrow{\llbracket \overline{a}d.V_Q \rrbracket \otimes id_{\{V_P,a\}}} \llbracket V_P \mid V_Q \rrbracket \otimes \nu_{\{a,d\}}$

Finally, consider the deadlocked $(\nu c)(\overline{c}c \mid \overline{c}d)$. All transitions in $LTS(\mathbb{C}_\pi)$ departing from that process must include R_l in their label: In fact, there is no context such that the node denoting the name c can be linked to an edge labelled *in*, since that node is not referred to in the interface of $\llbracket (\nu c)(\overline{c}c \mid \overline{c}d) \rrbracket^p$.

6 Conclusions and Further Work

The aim of our paper is quite straightforward: To synthesise a labelled transition system for the π-calculus, out of a graphical encoding of its reduction system.

We highlight five different contributions with a pivotal role in the development of our work. We first considered a well-known approach to the synthesis of a labelled transition system out a of reactive system, namely, Leifer and Milner's relative pushouts [16]. We then took into account its generalisation to groupoidal relative pushouts due to Sassone and Sobocinski [22], and its application on the category of cospans [23]. We further included our own proposal for encoding the reduction semantics for nominal calculi using DPO tools [11] (in particular its application to the π-calculus [8]), and the description of graph transformation systems as suitable reactive systems on the bicategory of cospans [10].

The present paper thus comes out as a case study in the growing field of synthesised labelled transition systems: An important one, though, since it is one of the very few examples concerning a rich calculus. We envision a few possible extensions of this work. First of all, however, we would like to make precise the correspondence between the synthesised bisimulation congruence and a more standard observational equivalence: Possibly early bisimulation [21–Table 1.5, Section 2.2], as the transition depicted in Fig. 14 seems to suggest.

References

1. P. Baldan, A. Corradini, H. Ehrig, M. Löwe, U. Montanari, and F. Rossi. Concurrent semantics of algebraic graph transformation. In H. Ehrig, H.-J. Kreowski, U. Montanari, and G. Rozenberg, editors, *Handbook of Graph Grammars and Computing by Graph Transformation*, volume 3, pages 107–187. World Scientific, 1999.
2. H.P. Barendregt, M.C.J.D. van Eekelen, J.R.W. Glauert, J.R. Kennaway, M.J. Plasmeijer, and M.R. Sleep. Term graph reduction. In J.W. de Bakker, A.J. Nijman, and P.C. Treleaven, editors, *Parallel Architectures and Languages Europe*, volume 259 of *Lect. Notes in Comp. Sci.*, pages 141–158. Springer, 1987.
3. M. Barr and C. Wells. *Category Theory for Computing Science.* Les Publications CMR, 1999.
4. G. Berry and G. Boudol. The chemical abstract machine. *Theor. Comp. Sci.*, 96:217–248, 1992.
5. A. Corradini and F. Gadducci. An algebraic presentation of term graphs, via gs-monoidal categories. *Applied Categorical Structures*, 7:299–331, 1999.
6. H. Ehrig, A. Habel, J. Padberg, and U. Prange. Adhesive high-level replacement categories and systems. In G. Engels and F. Parisi-Presicce, editors, *Graph Transformation*, Lect. Notes in Comp. Sci. Springer, 2004.
7. J. Engelfriet and T. Gelsema. Multisets and structural congruence of the π-calculus with replication. *Theor. Comp. Sci.*, 211:311–337, 1999.
8. F. Gadducci. Term graph rewriting and the π-calculus. In A. Ohori, editor, *Programming Languages and Semantics*, volume 2895 of *Lect. Notes in Comp. Sci.*, pages 37–54. Springer, 2003.
9. F. Gadducci and R. Heckel. An inductive view of graph transformation. In F. Parisi-Presicce, editor, *Recent Trends in Algebraic Development Techniques*, volume 1376 of *Lect. Notes in Comp. Sci.*, pages 219–233. Springer, 1997.
10. F. Gadducci, R. Heckel, and M. Llabrés. A bi-categorical axiomatisation of concurrent graph rewriting. In M. Hofmann, D. Pavlovič, and G. Rosolini, editors, *Category Theory and Computer Science*, volume 29 of *Electr. Notes in Theor. Comp. Sci.* Elsevier Science, 1999.

11. F. Gadducci and U. Montanari. A concurrent graph semantics for mobile ambients. In S. Brookes and M. Mislove, editors, *Mathematical Foundations of Programming Semantics*, volume 45 of *Electr. Notes in Theor. Comp. Sci.* Elsevier Science, 2001.
12. F. Gadducci and U. Montanari. Graph processes with fusions: concurrency by colimits, again. In H.-J. Kreowski *et al.*, editor, *Formal Methods (Ehrig Festschrift)*, volume 3393 of *Lect. Notes in Comp. Sci.*, pages 84–100. Springer, 2005.
13. M. Hasegawa. *Models of Sharing Graphs.* PhD thesis, University of Edinburgh, Department of Computer Science, 1997.
14. A. Jeffrey. Premonoidal categories and a graphical view of programs. Technical report, School of Cognitive and Computing Sciences, University of Sussex, 1997.
15. S. Lack and P. Sobociński. Adhesive and quasiadhesive categories. *Informatique Théorique et Applications/Theoretical Informatics and Applications*, 39:511–545, 2005.
16. J. Leifer and R. Milner. Deriving bisimulation congruences for reactive systems. In C. Palamidessi, editor, *Concurrency Theory*, volume 1877 of *Lect. Notes in Comp. Sci.*, pages 243–258. Springer, 2000.
17. J.-J. Lévy. Optimal reductions in the lambda-calculus. In J.P. Seldin and J.R. Hindley, editors, *Combinatory Logic, Lambda Calculus and Formalism: Essays in honour of Haskell B. Curry*, pages 159–191. Academic Press, 1980.
18. R. Milner. The polyadic π-calculus: A tutorial. In F.L. Bauer, W. Brauer, and H. Schwichtenberg, editors, *Logic and Algebra of Specification*, volume 94 of *Nato ASI Series F*, pages 203–246. Springer, 1993.
19. R. Milner. Bigraphical reactive systems. In K.G. Larsen and M. Nielsen, editors, *Concurrency Theory*, volume 2154 of *Lect. Notes in Comp. Sci.*, pages 16–35. Springer, 2001.
20. D. Plump. Term graph rewriting. In H. Ehrig, G. Engels, H.-J. Kreowski, and G. Rozenberg, editors, *Handbook of Graph Grammars and Computing by Graph Transformation*, volume 2, pages 3–61. World Scientific, 1999.
21. S. Sangiorgi and D. Walker. *The π-calculus: A Theory of Mobile Processes.* Cambridge University Press, 2001.
22. V. Sassone and P. Sobociński. Deriving bisimulation congruences using 2-categories. *Nordic Journal of Computing*, 10:163–183, 2003.
23. V. Sassone and P. Sobociński. Reactive systems over cospans. In *Logic in Computer Science*, pages 311–320. IEEE Computer Society Press, 2005.
24. P. Sewell. From rewrite rules to bisimulation congruences. *Theor. Comp. Sci.*, 274:183–230, 2004.

Appendix A: Some Categorical Notions

On Adhesive Categories

We recall here the definition of adhesive categories [15]. We do not provide any introduction to basic categorical constructions such as products, pullbacks and pushouts, referring the reader to Sections 5 and 9 of [3].

Definition 10 (adhesive categories). *A category is called* adhesive *if*

- *it has pushouts along monos;*
- *it has pullbacks;*
- *pushouts along monos are* Van Kampen *(*VK*) squares.*

Referring to Fig. 15, a VK *square is a pushout like (i), such that for each commutative cube like (ii) having (i) as bottom face and the back faces of which are pullbacks, the front faces are pullbacks if and only if the top face is a pushout.*

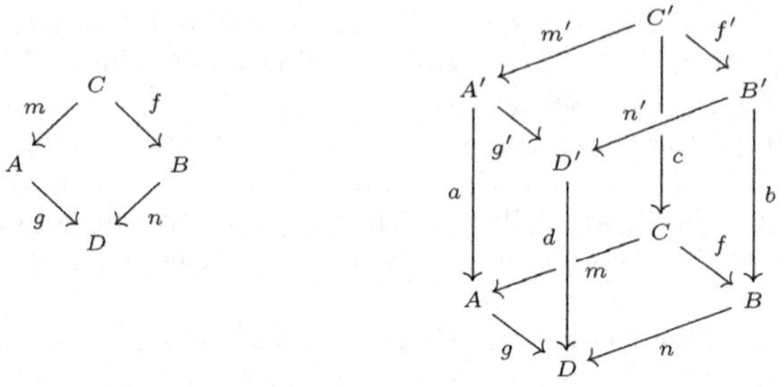

Fig. 15. A pushout square (*i*), left, and a commutative cube (*ii*), right

There are at least two properties of interest for adhesive categories. The first is that adhesive categories subsume many properties of HLR categories [6]. This ensures that several results about parallelism are also valid for DPO rewriting in adhesive categories, if the rules are given by spans of monos [15].

The second fact is concerned with the associated category of *input-linear cospans* (i.e., pairs of arrows with common target, where the first is a mono). As already suggested in [9], any DPO rule can be represented by a pair of cospans, and the bicategory freely generated from the rules represents faithfully all the derivations obtained using monos as matches [10]. Furthermore, the resulting bicategory has relative pushouts [16], hence it is possible to derive automatically a well-behaved behavioral equivalence [23], namely, a bisimulation equivalence which is also a congruence with respect to the closure under (suitable) contexts.

On Bicategories

A *bicategory* $\mathcal{C}$ is described concisely as a category where every homset (the collections of arrows between any pair of objects a and b) is the class of objects of some category $\mathcal{C}(a,b)$ and, correspondingly, whose composition "functions" $\mathcal{C}(a,b) \times \mathcal{C}(b,c) \to \mathcal{C}(a,c)$ are functors.

Definition 11 (bicategories). *A bicategory* $\mathcal{C}$ *consists of*

1. *a class of* objects $a, b, c, \ldots$;
2. *for each* $a, b \in \mathcal{C}$ *a category* $\mathcal{C}(a,b)$;
3. *for each* $a, b, c \in \mathcal{C}$ *a functor* $* : \mathcal{C}(a,b) \times \mathcal{C}(b,c) \to \mathcal{C}(a,c)$.

The objects of $\mathcal{C}(a,b)$ are called *1-cells*, or simply arrows, and denoted by $f : a \to b$. Its morphisms are called *2-cells*, and are written $\alpha : f \Rightarrow g : a \to b$. Composition in $\mathcal{C}(a,b)$ is denoted by $\bullet$ and referred to as *vertical* composition. Identity 2-cells are denoted by $1_f : f \Rightarrow f$.

Actually, a bicategory is also equipped with a family of coherence cells, and the *horizontal* composition $*$ must additionally satisfy a weak associative law, also admitting 1_{id_a} as identities. We refer the reader to [10–Section 4], where the link between bicategories and cospan categories is made explicit.

Definition 12 (2- and groupoidal categories). *A* 2-category *is a bicategory such that horizontal composition is associative. A* groupoidal category *(or G-category) is a 2-category where all 2-cells are invertible.*

On Reactive Systems

Reactive systems were proposed by Leifer and Milner as a general framework for the study of simple formalisms equipped with a reduction semantics [16]. The setting was extended by Sassone and Sobocinski [22] in order to deal with contexts of a formalism that is equipped with a structural congruence relation. For instance, in examples which contain a parallel composition operator, it is usually not satisfactory to simply quotient out terms with respect to its commutativity—intuitively, it is important to know the precise location within the term where the reaction occurs. This information is expressed as a 2-dimensional structure, where the 2-cells are isomorphisms which "permute" the structure of the term.

Definition 13 (reactive system). *A* (bi)reactive system $\mathbb{C}$ *consists of*

1. *a bicategory $\mathcal{C}$ of* contexts*;*
2. *an object $\iota \in \mathcal{C}$;*
3. *a composition-reflecting, 2-full sub-bicategory $\mathcal{E}$ of* evaluation contexts[2]*;*
4. *a set $R \subseteq \bigcup_{a \in \mathcal{E}} \mathcal{C}(\iota, a) \times \mathcal{C}(\iota, a)$ of* reaction rules.

Reaction rules are closed with respect to evaluation contexts in order to obtain the reaction relation on the closed terms (arrows with domain ι) of $\mathcal{C}$.

On Groupoidal Relative Pushouts as Labels

We briefly introduce now *groupoidal* relative pushouts, a bicategorical version of pushouts in slice categories. They can be considered as a way for quotienting out the common context shared between terms, described as arrows in a category.

Definition 14 (GRPOs [23–Definition 3.2]). *Let $\mathcal{C}$ be a bicategory with isomorphic 2-cells. Referring to Fig. 16, a candidate for a cell $\alpha : c; a \Rightarrow d; b$ like (i) is a tuple $\langle E, e, o, h, \beta, \gamma, \delta \rangle$ like (ii) such that its cells past up (taking into account the associativity morphims) to give α.*

A GRPO is a candidate which satisfies a universal property, i.e., such that for any other candidate $\langle E', e', o', h', \beta', \gamma', \delta' \rangle$ there must be a unique (up-to unique isomorphic cell) arrow $l : E \to E'$ and cells $\phi : e; l \Rightarrow e'$, $\phi : o; l \Rightarrow o'$, and $\xi : l; h' \Rightarrow h$ making the two candidate compatible.

Finally, a diagram above on the left is a groupoidal-idem pushout (GIPO) if its GRPO is the tuple $\langle D, g, n, id_D, \alpha, 1_g, 1_n \rangle$.

[2] $\mathcal{E}$ is full on the two-dimensional structure and $e_1; e_2 \in \mathcal{E}$ implies $e_1 \in \mathcal{E}$ and $e_2 \in \mathcal{E}$.

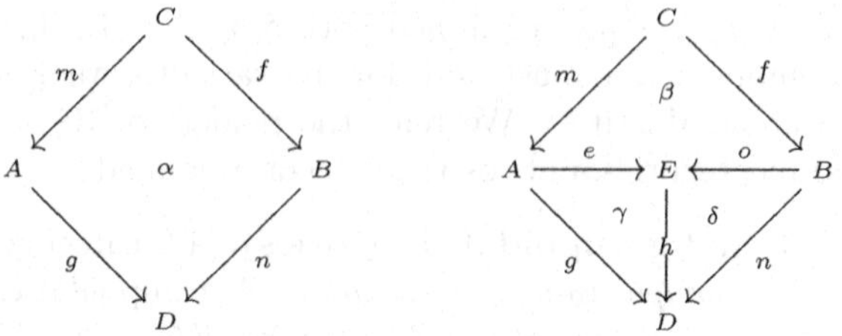

Fig. 16. A cell (i), left, and a candidate GRPO (ii), right

Now, GRPOs can be fruitfully used to define a labelled transtition system. The basic idea, originally due to Sewell [24], is that the labels represent the smallest contexts which allow a reaction to occur. This is obtained by labelling a transitions with those arrows precisely arising from GIPOs.

Definition 15 (a labelled transition system). *Let $\mathbb{C}$ be a bireactive system. The associated labelled transition system $LTS(\mathbb{C})$ is given by*

1. *the states of $LTS(\mathbb{C})$ are (isomorphic classes of) arrows $[s] : \iota \to a$ in $\mathcal{C}$*

2. *there is a transition $[s] \xrightarrow{[f]} [r; t]$ iff there exists $\langle l, r \rangle \in R$, $t \in \mathcal{E}$ and 2-cell $\alpha : s; f \Rightarrow l; t$ such that the square below is a GIPO.*

$$
\begin{array}{ccc}
\iota & \xrightarrow{s} & a \\
l \downarrow & \alpha & \downarrow f \\
b & \xrightarrow{t} & c
\end{array}
$$

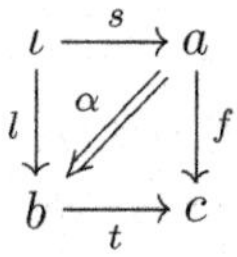

Note that the states and the transitions of the LTS are obtained by quotienting arrows and cells with respect to isomorphism—in other words, the 2-dimensional structure is no longer necessary and may be discarded.

One of the main results that holds for such an LTS is that when the underlying bicategory $\mathcal{C}$ has enough (G)RPOs, then bisimilarity is a congruence (i.e., it is closed with respect to left-composition for each arrow in $\mathcal{C}$).

Proposition 4 (observational congruence). *Let $\mathbb{C}$ be a bireactive system, and let $f, g \in \mathcal{C}(\iota, a)$ be arrows of the underlying bicategory. If f and g are (strong) bisimilar in $\mathbb{C}$, then so are $f; h$ and $g; h$ for all arrows $h \in \mathcal{C}(a, b)$.*

This was originally shown by Leifer and Milner [16–Theorem 1] and extended to the bicategorical setting by Sassone and Sobocinski [22–Theorem 1].

On Cospan Categories

We close the Appendix with a result ensuring the relevant properties for ranked graphs, as stated in Proposition 7 below.

Definition 16 (bicategories of cospans). *Let $\mathcal{C}$ be a category with chosen binary pushouts. Then, the bicategory of input-linear cospans is given by the* *triple* $\langle Ob_{\mathbb{C}}, CoSpan(\mathbb{C}), *\rangle$, *where* $Ob_{\mathbb{C}}$ *is the set of objects of* $\mathbb{C}$; *the arrows of* $CoSpan(\mathbb{C})(a, b)$ *are the triples* $\langle f, c, g\rangle$ *for* $f : a \to c$ *a mono and* $g : b \to c$ *an arrow in* $\mathbb{C}$; *the cells* $l : \langle f, c, g\rangle \Rightarrow \langle h, d, i\rangle$ *are those arrows* $l : c \to d$ *in* $\mathbb{C}$ *making the diagrams commute; the horizontal (i.e., cospan) composition is the family of functors* $*_{a,b,c} : CoSpan(\mathbb{C})(a, b) \times CoSpan(\mathbb{C})(b, c) \to CoSpan(\mathbb{C})(a, c)$, *defined by the chosen pushouts.*

We do not explicitly mention here all the relevant isomorhism cells that are induced by the universal property of pushouts. Note that ranked graphs thus coincide with the category of cospans over typed graphs, or better, its sub-bicategory obtained by restricting to those objects which are discrete graphs.

Proposition 5 (cospans and GRPOs). *Let* $\mathbb{C}$ *be a adhesive category with chosen binary pushouts. Then, the associated bicategory of input-linear cospans and isomorphic cells has GRPOs.*

The previous proposition is the main result obtained in [23] (see Theorem 4.1): It is instantiated to Proposition 7 below, thus allowing for our presentation of a *π-reactive system* and its labelled transition system semantics.

Some Results on Graphs as an Adhesive Category

The aim of this section is to present some easy technical lemma, characterizing the category of ranked graphs as a bicategory of cospans, hence enabling the previous mechanism to be instantiated to our graphical encoding for processes.

Proposition 6 (on adhesiveness). *Graphs and their morphisms (see Definition 4) form an adhesive category.*

The proof is rather straightforward. The category laws clearly hold. Concerning adhesiveness, hyper-graphs form an adhesive category, as proved in [15–Corollary 3.6]; moreover, labelled (hyper-)graphs clearly correspond to typed (hyper-)graphs, for the obvious graph associated to a signature (see e.g. Fig. 17 for the graph associated to the π-calculus[3]), and adhesiveness is closed under the slice construction, as proved again in [15–Proposition 3.5].

Proposition 7 (on groupoidal relative pushouts). *Ranked graphs with injective variable morphism and their isomorphisms (see Definition 5) form a bicategory with groupoidal relative pushouts.*

Note that a ranked graph is just a cospan over the category of typed graphs, see Definition 16. The latter category can be equipped with a choice of pushouts which is compatible with the notion of sequential composition we have given for

[3] Remember that, for graphical convenience, the nodes are represented either by an hollow or as a full circle, in order to distinguish those nodes used for names (the former) from the nodes denoting a (sub-)process in the encoding (the latter); similar considerations hold for the labels *in* and *out* inside the edges.

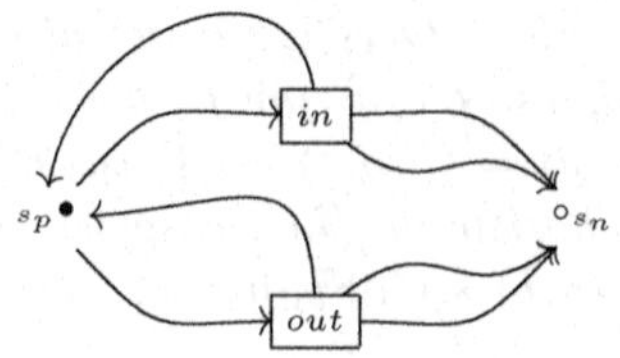

Fig. 17. The type graph for π-calculus

ranked graphs. Thus, ranked graphs are just a 2-full sub-bicategory of the category of input-linear cospans for typed graphs, as defined e.g. according to [23–Definition 2.5], restricted to *discrete* interfaces (i.e., graphs with no edges). The existence of groupoidal relative pushouts (GRPOs) in the sub-bicategory is confirmed by the analysis of the construction of the candidate in [23–Algorithm 4.2].

Definition 17 (π-reactive system). *The (bi)reactive system $\mathbb{C}_\pi$ consists of*

1. *the bicategory of ranked graphs;*
2. *the object $\{p\}$;*
3. *the π-reaction rule $\langle R_l, R_r \rangle$.*

Primitive Rewriting *

Nachum Dershowitz

School of Computer Science, Tel Aviv University,
Ramat Aviv 69978, Israel
nachum.dershowitz@cs.tau.ac.il

For Jan Willem with admiration

Abstract. Undecidability results in rewriting have usually been proved by reduction from undecidable problems of Turing machines or, more recently, from Post's Correspondence Problem. Another natural candidate for proofs regarding term rewriting is Recursion Theory, a direction we promote in this contribution.

We present some undecidability results for "primitive" term rewriting systems, which encode primitive-recursive definitions, in the manner suggested by Klop. We also reprove some undecidability results for orthogonal and non-orthogonal rewriting by applying standard results in recursion theory.

1 Introduction

Indeed, if general recursive function
is the formal equivalent of effective calculability,
its formulation may play a role
in the history of combinatory mathematics
second only to that of the formulation of natural number.
— Emil Post (1944)

A number of models of computation vie for the rôle of "most basic" mechanism for defining effective computations. These include: semi-Thue systems, Markov's normal algorithms, Church's lambda calculus, Schönfinkel's combinatory logic, Turing's "logical computing" machines, and Gödel's recursive functions. Although they operate over different domains (strings, terms, numbers), they are all of equivalent computational power.[1]

First-order term rewriting makes for a very natural symbolic programming paradigm based on subterm replacement, without bound variables or built-in operations. The two most basic properties a rewrite system may possess are termination (a.k.a. strong normalization) and confluence (the famous Church-Rosser property). Variations on these include (weak) normalization, unique normal forms, and ground confluence. For a comprehensive text on rewriting, see the recent volume by the "Terese" group, based in Amsterdam [38].

* This research was supported by the Israel Science Foundation (grant no. 250/05).

[1] See [4, 5] for a discussion of problems pertaining to comparisons of computational power of models operating over diverse domains.

A. Middeldorp et al. (Eds.): Processes... (Klop Festschrift), LNCS 3838, pp. 127–147, 2005.

It comes as no surprise that rewrite systems have the same computational power as the other basic models.[2] Moreover, rewrite systems may be restricted in various ways, including left-linearity, orthogonality, and constructor-basedness, without weakening the model from the point of view of computability.

To quote Klop [31–p. 356]: "As is to be expected, most of the properties of TRSs [term rewriting systems] are undecidable. Consider only TRSs with finite signature and finitely many reduction rules. Then it is undecidable whether confluence holds, and also whether termination holds." Early undecidability results in string and term rewriting were proved by reduction from undecidable problems of Turing machines (e.g. [9, 25]). More recently, Post's Correspondence Problem [47] has been used: for string rewriting by Book [6]; for term rewriting in [28, 36] (see also [14, 15] and [38–Sect. 5.3.3]). The most natural candidate for proofs regarding term rewriting, however, is recursion theory, a direction we promote here.

Recursive function theory is a uniquely suitable candidate for demonstrating, by means of suitable reductions, that various properties of members of classes of rewrite systems are undecidable. Standard works on recursion theory include [41, 45, 48, 53]. The encoding of recursive functions as term-rewriting systems is part of the field's age-old "folklore", and is mentioned by Klop as an exercise in his 1992 survey [32].

This paper present some undecidability results for "primitive" term rewriting systems, which encode primitive-recursive definitions. Primitive rewriting is defined in Sect. 3. Section 4 shows how to also faithfully encode partial-recursive functions. Kleene's computation predicate—which is central to the undecidability results—is coded as a primitive rewrite system in the Appendix, and its properties are discussed in Sect. 5. In Sects. 6 and 7, we reprove (and improve) some undecidability results for orthogonal and non-orthogonal rewriting (see [38–Chap. 5]) by applying standard results in recursion theory. The concluding section lists what we believe to be new by way of sufficient conditions for undecidability obtained in this way.

2 Background

A total function f over the natural numbers is *primitive recursive* if it is the constant $\lambda.0$, a projection function $\lambda x_1, \ldots, x_k.x_i$, the successor function $\lambda x.x{+}1$, the composition of other primitive-recursive functions, or else is itself definable by primitive recursion of the form:

$$f(n, \ldots, x_i, \ldots) \quad := \quad \begin{cases} g(\ldots, x_i, \ldots) & n = 0 \\ h(f(n-1, \ldots, x_i, \ldots), n-1, \ldots, x_i, \ldots) & \text{otherwise}, \end{cases}$$

where g and h are already known to be primitive recursive.

A partial function f over the natural numbers is *partial recursive* if it is primitive recursive, or if it can be defined by composition or primitive recursion

[2] Of course, the classical Church-Turing Thesis asserts that these sets of functions are exactly what are mechanistically computable. See [3].

from other partial-recursive functions, or if it can be defined by minimization (μ-recursion):

$$f(\ldots, x_i, \ldots) \quad := \quad \mu_{n \in \mathbb{N}} \{q(n, \ldots, x_i, \ldots)\} ,$$

where q is a partial-recursive predicate.[3] Recursive functions are computed leftmost-innermost [48–Sect. 1.2], which is the computation rule that goes into an infinite loop whenever any computation rule can (see, e.g., [37]). In other words, the result is always the least-defined partial function possible.[4]

A partial-recursive function is *(general) recursive* if it is total (always defined). It is well-known that the class of general recursive functions coincides with the Turing-computable (total) functions over (encodings of) the naturals.

Kleene's Normal-Form Theorem [48–Thm. 1-X] states that there exist primitive recursive functions U and T^K such that

$$\lambda \bar{x}.\ U(\mu z. T^K(j, \bar{x}, z)) \tag{1}$$

enumerates all the partial-recursive functions ($\bar{x}$ is a sequence of variables). The computation predicate $T^K(j, \bar{a}, z)$ ("Kleene's T"; see [29]), checks whether $z \in \mathbb{N}$ is (a numerical encoding of) a list beginning with the term $f_j(\bar{a})$, with arguments $a_i \in \mathbb{N}$, continuing step-by-step as in a valid computation, and ending with a natural number for the value of $f_j(\bar{a})$; U extracts that number. In modern parlance, we would say that the partial-recursive function

$$\lambda j. \lambda \bar{x}.\ U(\mu z. T^K(j, \bar{x}, z)) \tag{2}$$

is an "interpreter" for partial recursion, analogous to the Universal Turing Machine.

Two basic undecidability results in recursion theory (due to Kleene [29]) follow from the existence of this partial-recursive function:

> DEF(f, n). Given a definition of a partial-recursive function $f : \mathbb{N} \rightharpoonup \mathbb{N}$ and a natural number $n \in \mathbb{N}$, it is undecidable whether $f(n)$ is defined [48–Thm. 1-VII]. By "definition", we mean here the index of an enumeration or the Gödel number of a program.
>
> TOT(f). Given a definition of a partial-recursive function $f : \mathbb{N} \rightharpoonup \mathbb{N}$, it is undecidable whether f is recursive [48–Thm. 1-VIII]. The TOT problem, like its analogue for Turing machines, is not even semi-decidable.

In the appendix, we define an injection $\sharp : f \mapsto f^\sharp$ from the definition of a partial-recursive function f (that is, from the sequence of compositions, recursions, and minimizations that define f) into the naturals. The rewrite program

---

[3] By *predicate* we mean any function, but with 0 interpreted as false and non-zero (usually 1) signifying truth.

[4] For example, with definitions $\kappa(x) := 1$ and $\omega := \mu\{0\}$, $\kappa(\omega)$ is undefined.

TK given there defines a primitive-recursive function $\mathcal{T}$ such that (restricting to one-argument functions):

$$\mathrm{DEF}(f,n) \quad\Leftrightarrow\quad \exists y \in \mathbb{N}.\ \mathcal{T}(f^\sharp, n, y) = 1\,. \tag{3}$$

For $j \in \mathbb{N}$ that do not correspond to any program, $\mathcal{T}(j,n,y) = 0$, for all n and y. This suffices for our purposes.

3 Primitive Rewriting

In [32–Ex. 2.2.9], Klop mentioned that the primitive-recursive functions can be directly programmed as a terminating, orthogonal, constructor-based term-rewriting system, where the two constructors are the constant 0 and the unary successor function s. There are collapsing rules

$$\pi_i^k(x_1,\ldots,x_k) \quad\to\quad x_i\,,$$

for each k and i, $1 \le i \le k$, corresponding to projections. All other functions f are defined by rules that are either recursion-free compositions of the following form:

$$f(\ldots,x_i,\ldots) \quad\to\quad g(h_1(\ldots,x_i,\ldots),\ldots,h_k(\ldots,x_i,\ldots))\,,$$

or else primitive recursions of the form

$$\begin{array}{rcl} f(\mathsf{0},\ldots,x_i,\ldots) & \to & g(\ldots,x_i,\ldots) \\ f(\mathsf{s}(n),\ldots,x_i,\ldots) & \to & h(f(n,\ldots,x_i,\ldots),n,\ldots,x_i,\ldots)\,. \end{array}$$

More conveniently, one can allow arbitrary compositions of previously defined functions on right-hand sides of defining rules. Thus, primitive-recursive functions can be defined either by composition:

$$f(\ldots,x_i,\ldots) \quad\to\quad G[\ldots,x_i,\ldots]\,,$$

or by non-nested recursion over the naturals:

$$\begin{array}{rcl} f(\mathsf{0},\ldots,x_i,\ldots) & \to & G[\ldots,x_i,\ldots] \\ f(\mathsf{s}(n),\ldots,x_i,\ldots) & \to & H[f(n,\ldots,x_i,\ldots),n,\ldots,x_i,\ldots]\,, \end{array}$$

where G and H are "contexts" (terms with holes) built from already-defined functions, and the recursive call and the arguments n and x_i may appear any number of times in H. (One can also allow the recursive decrease to be in any one, fixed position.) Such definitions can be easily deconstructed into a sequence of composed functions, preserving derivability, $\to^+$.

Numbers $n \in \mathbb{N}$ are represented by terms $\widehat{n} = \mathsf{s}^n(\mathsf{0})$ in standard unary successor notation. Let $\widehat{\mathbb{N}} = \{\widehat{n} : n \in \mathbb{N}\}$ be the set of these "tally" numbers. Factorial, for example, is defined as follows (using standard infix notation):

$$\begin{array}{rcl} \mathsf{0}+x & \to & x \\ \mathsf{s}(z)+x & \to & \mathsf{s}(z+x) \\ \mathsf{0}\times x & \to & \mathsf{0} \\ \mathsf{s}(z)\times x & \to & (z\times x)+x \\ \mathsf{0}! & \to & \mathsf{s}(\mathsf{0}) \\ (\mathsf{s}(z))! & \to & \mathsf{s}(z)\times(z!)\,. \end{array}$$

Such *primitive* rewriting systems can be made non-erasing (sometimes called "regular"), in the sense that all variables on the left of a rule also appear on the right, and non-collapsing—no right side just a variable, by using the following primitive functions:

$$\begin{array}{rcl} \iota(\mathsf{0}) & \rightarrow & \mathsf{0} \\ \iota(\mathsf{s}(n)) & \rightarrow & \mathsf{s}(\iota(n)) \\ \varepsilon(\mathsf{0}, n) & \rightarrow & \iota(n) \\ \varepsilon(\mathsf{s}(m), n) & \rightarrow & \varepsilon(m, n) \,. \end{array}$$

Then the right side x_i of each projection rule π_i^k can be enveloped with calls to ε for each of the irrelevant variables:

$$\pi_i^k(x_1, \ldots, x_k) \quad \rightarrow \quad \varepsilon(x_1, \ldots \varepsilon(x_{i-1}, \varepsilon(x_{i+1}, \ldots, \varepsilon(x_n, x_i) \cdots)) \cdots) \,.$$

To reduce the depth of right sides, one can use a sequence of "erasure" rules, instead:

$$\begin{array}{rcll} \pi_1^1(x) & \rightarrow & x & \\ \pi_1^k(x_1, \ldots, x_k) & \rightarrow & \varepsilon(x_k, \pi_1^{k-1}(x_1, \ldots, x_{k-1})) & k > 1 \\ \pi_i^k(x_1, \ldots, x_k) & \rightarrow & \varepsilon(x_1, \pi_{i-1}^{k-1}(x_2, \ldots, x_k)) & i, k > 1 \,. \end{array}$$

So massaged, every primitive rewrite system possesses the following properties:

1. it is terminating;
2. it is what we will call *definitional*, that is,
   (a) orthogonal—left-linear with no critical pairs, hence
   (b) confluent,
   (c) constructor-based—all but the outermost symbols on the left are constructors (either $\mathsf{0}$ or s) or variables,
   (d) constructor complete—every non-constructor variable-free term is reducible,[5]
   (e) non-erasing, and
   (f) non-collapsing;
3. and it is 3-deep (having maximum nesting, on the left and on the right, of 3), with only variables occuring at depth 3 (that is, below at most 2 symbols).

Plaisted [46] noted that every primitive-recursive function written as a rewrite system (as above) is provably terminating with his simple path ordering (of order type ω^{ω^N}). Likewise, they can be shown terminating with a lexicographic or multiset path ordering (see [11]). In the other direction, Hofbauer [23] (taking the

[5] Sufficient completeness (the "no junk" condition) means that every ground (variable-free) term is equal (convertible) to a constructor-only term. In a rewriting context, one normally asks that ground non-constructor terms actually normalize to constructor-only terms (incorporating a degree of ground confluence). Since we already have a separate termination property, we only ask that every ground non-constructor term be rewritable. Combined with termination, this yields the usual sufficient-completeness property. (I am borrowing the term "constructor completeness" from notes by Heinrich Hußmann.)

exponential termination functions of Iturriaga [27] a few steps further) showed that any rewrite system that can be proved terminating using a recursive path ordering must have primitive-recursive derivation length. For some recent related results, see [1, 8].

4 Partial Rewriting

Algebraic rewriting does not have bound variables, so to simulate general recursion we employ a trick, namely, separate minimization functions for each predicate:

$$\begin{array}{rcl} \mu_q(z, \mathsf{s}(y), \ldots, x_i, \ldots) & \rightarrow & z \\ \mu_q(z, 0, \ldots, x_i, \ldots) & \rightarrow & \mu_q(\mathsf{s}(z), q(\mathsf{s}(z), \ldots, x_i, \ldots), \ldots, x_i, \ldots)\,, \end{array}$$

where q is a partial-recursive predicate. Better yet, we can let an arbitrary expression serve as test, with any non-zero value signifying truth:

$$\begin{array}{rcl} \mu_Q(z, \mathsf{s}(y), \ldots, x_i, \ldots) & \rightarrow & z \\ \mu_Q(z, 0, \ldots, x_i, \ldots) & \rightarrow & \mu_Q(\mathsf{s}(z), Q[\mathsf{s}(z), \ldots, x_i, \ldots], \ldots, x_i, \ldots)\,. \end{array}$$

Then, to compute a function f defined by minimization vis-à-vis Q, we start off with

$$f(\ldots, x_i, \ldots) \quad \rightarrow \quad \mu_q(0, Q[0, \ldots, x_i, \ldots], \ldots, x_i, \ldots)\,.$$

To avoid introducing spurious cases of nontermination, q (or Q) must be *monotonic*, in the sense that $q(k, \bar{x}) > 0$ implies $q(m, \bar{x}) > 0$ for all $m > k$. Were we to allow non-monotonic predicates q, then the computation of $\mu_q(\widehat{N}, 0, \ldots, K_i, \ldots)$ might diverge for large N, even as f itself never does, since $q(z, \ldots, x_i, \ldots)$ may yield false for all but one z.

Luckily, with no loss of generality, any ordinary predicate q' can be recast monotonically as

$$q(n, \ldots, x_i, \ldots) \quad := \quad \sum_{i=0}^{n} q'(i, \ldots, x_i, \ldots)\,,$$

where the sum serves as disjunction, and is primitive recursive when q is. The minima μ_q and $\mu_{q'}$ satisfying the two tests (when starting from 0) are the same.

By extension, such rewrite systems, built from primitive recursion and monotonic minimization, will be called *partial recursive*. When they terminate they are *general recursive*.

For example, natural-number division (which is actually primitive recursive) may be defined as follows:

$$\begin{array}{rcl} \mathsf{p}(0) & \rightarrow & 0 \\ \mathsf{p}(\mathsf{s}(z)) & \rightarrow & z \\ x \dot{-} 0 & \rightarrow & x \\ x \dot{-} \mathsf{s}(z) & \rightarrow & p(x \dot{-} z) \\ \mu(z, \mathsf{s}(v), x, y) & \rightarrow & z \\ \mu(z, 0, x, y) & \rightarrow & \mu(\mathsf{s}(z),\ (y \times \mathsf{s}(\mathsf{s}(z))) \dot{-} x,\ x,\ y) \\ x \div y & \rightarrow & \mu(0, y \dot{-} x, x, y)\,. \end{array}$$

$$\begin{array}{c}\text{primitive-recursive system}\\ \cap \\ \text{general-recursive system}\\ \cap \\ \text{partial-recursive system}\\ \cap \\ \text{definitional system}\end{array}$$

Fig. 1. Hierarchy of recursive rewriting

Rules for the base case of minimization can be made non-erasing, like we did for projections. That done, a term reduces to a numeral *only* if it is a *ground* term built from the constructors, 0and s, and from functions defined according to the above schemata.

Partial-recursive rewrite systems are definitional, and, as such, they are confluent. General-recursive systems are definitional and terminating. These inclusions are summarized in Fig. 1.

It is important to take note of the fact that partial-recursive rewrite systems terminate regardless of strategy (in other words, they are strongly normalizing) if, and only if, they terminate via innermost rewriting, since partial systems are orthogonal [42] ([38–Thm. 4.8.7]).[6] Furthermore, as they are also non-erasing, a partial-recursive rewriting system terminates if, and only if, it is (weakly) normalizing [7] ([38–Thm. 4.8.5]). These observations remain true for questions regarding specific initial terms, too [38–p. 128].

Our rewriting implementation of primitive and partial recursion is sound:

Proposition 1. *For all partial-recursive functions* $f : \mathbb{N}^k \rightharpoonup \mathbb{N}$, *implemented as described above by a symbol* f *in rewrite system* F,

$$f(a_1, \ldots, a_k) = n \qquad \Leftrightarrow \qquad \mathsf{f}(\widehat{a_1}, \ldots, \widehat{a_k}) \xrightarrow[F]{!} \widehat{n}\ ,$$

for all $n, a_1, \ldots, a_k \in \mathbb{N}$, *where* $\widehat{n}$ *is the (normal form) term representing the number* n *and* $\to^!$ *is reduction to normal form—using any arbitrary rewriting strategy.*

Proof. If $f(a_1, \ldots, a_k) = n$, then leftmost-innermost rewriting with F will mimic the recursive computation and yield $\widehat{n}$. Since orthogonal systems are confluent, $\widehat{n}$ is the only normal form. The strategy does not matter, since, as just pointed out, non-erasing orthogonal systems terminate regardless of strategy for a given term if they normalize using any strategy.[7] □

[6] Left-linearity is inessential [16]; the non-overlapping condition can also be weakened [21, 13].

[7] To explicate: The non-erasing version of κ from footnote 4 would be $\kappa(x) \to \varepsilon(x, \hat{1})$. Since the rules for ε would not apply to the term ω or any of its descendants, all computations of $\kappa(\omega)$ diverge. With the erasing version, on the other hand, $\kappa(\omega) \to^! \hat{1}$.

The undecidability of definedness (DEF) for partial-recursive rewriting follows directly, by a standard diagonalization argument: We have that $\mathrm{DEF}(F,n)$, for rewrite system F and $n \in \mathbb{N}$, if $\exists m \in \widehat{\mathbb{N}}.\ \mathsf{f}(\widehat{n}) \rightarrow^!_F m$. Were there a recursive system defining a recursive function D for deciding DEF, then the following system X, with the system for D, would be partial recursive (cf. [54]):

$$\begin{array}{rcl} \neg x & \rightarrow & \mathsf{s}(0) \dot{-} x \\ \mu(z, \mathsf{s}(y), f) & \rightarrow & z \\ \mu(z, 0, f) & \rightarrow & \mu(\mathsf{s}(z), \neg\mathsf{D}(f,f), f) \\ \mathsf{X}(g) & \rightarrow & \mu(0, \neg\mathsf{D}(g,g), g)\ . \end{array}$$

But then

$$\begin{array}{ccccc} \mathsf{X}(X^\sharp) \xrightarrow[X]{!} 0 & \Leftrightarrow & \neg\mathsf{D}(X^\sharp, X^\sharp) \xrightarrow[X]{!} \mathsf{s}(0) & \Leftrightarrow \\ \mathsf{D}(X^\sharp, X^\sharp) \xrightarrow[X]{!} 0 & \Leftrightarrow & \forall m \in \widehat{\mathbb{N}}.\ \mathsf{X}(X^\sharp) \not\xrightarrow[X]{!} m\ , \end{array}$$

a contradiction. The first biconditional derives from the definition of X; the last, from the presumption that D decides DEF.

5 Computations

A primitive rewrite system **TK** for the computation predicate $\mathcal{T}$ (implementing T^K), for functions of any arity, is given in the Appendix. Soundness of this implementation of T^K means the following:

Proposition 2. *For all partial-recursive functions* $f : \mathbb{N}^k \rightharpoonup \mathbb{N}$,

$$f(\ldots, a_i, \ldots) = n \quad \Leftrightarrow \quad \exists y \in \widehat{\mathbb{N}}.\ \mathcal{T}(f^\sharp, \ldots, \widehat{a_i}, \ldots, y) \xrightarrow[\mathbf{TK}]{!} \hat{1} \ \wedge\ \mathcal{U}(y) \xrightarrow[\mathbf{TK}]{!} \widehat{n}$$

and

$$f(\ldots, a_i, \ldots) = \bot \quad \Leftrightarrow \quad \forall y \in \widehat{\mathbb{N}}.\ \mathcal{T}(f^\sharp, \ldots, \widehat{a_i}, \ldots, y) \xrightarrow[\mathbf{TK}]{!} 0\ ,$$

where $\bot$ *denotes undefined,* $\mathcal{T}$ *is the symbol for the computation predicate* T^K *in rewrite system* **TK**, $\mathcal{U}$ *is the symbol for the last-element function, and* $f^\sharp$ *is the numeral representing the rewrite program for* f.

The monotonic version of predicate $\mathcal{T}$ is the following:

$$\begin{array}{rcl} \mathcal{T}^*(j, \ldots, x_i, \ldots, 0) & \rightarrow & \mathcal{T}(j, \ldots, x_i, \ldots, 0) \\ \mathcal{T}^*(j, \ldots, x_i, \ldots, \mathsf{s}(y)) & \rightarrow & \mathcal{T}^*(j, \ldots, x_i, \ldots, y) + \mathcal{T}(j, \ldots, x_i, \ldots, \mathsf{s}(y))\ . \end{array}$$

Then, by the Normal Form Theorem, to compute any partial-recursive function f, one can use primitive-recursive $\mathcal{T}^*$ along with the following general-recursive rules:

$$\begin{array}{rcl} \mu(z, \mathsf{s}(y), \ldots, x_i, \ldots) & \rightarrow & z \\ \mu(z, 0, \ldots, x_i, \ldots) & \rightarrow & \mu(\mathsf{s}(z), \mathcal{T}^*(f^\sharp, \ldots, x_i, \ldots, \mathsf{s}(z)), \ldots, x_i, \ldots) \\ \mathsf{f}(\ldots, x_i, \ldots) & \rightarrow & \mathcal{U}(\mu(0, \mathcal{T}^*(f^\sharp, \ldots, x_i, \ldots, 0), \ldots, x_i, \ldots))\ . \end{array}$$

Call this system (including **TK**, and made non-erasing) $\mathbf{R}_f$.

Proposition 3. *For all partial-recursive functions $f : \mathbb{N}^k \rightharpoonup \mathbb{N}$, implemented by a symbol* f *in rewrite system* $\mathbf{R}_f$*,*

$$f(a_1, \ldots, a_k) = n \qquad \Leftrightarrow \qquad \mathsf{f}(\widehat{a_1}, \ldots, \widehat{a_k}) \xrightarrow[\mathbf{R}_f]{!} \widehat{n} \,.$$

As previously mentioned, it is also undecidable whether a partial-recursive system is actually general-recursive. This TOT problem for rewriting can be shown to be non-semi-decidable by using **TK**, with no need to rely on results for Turing machines or recursive functions.

6 Word Problems

In this and the following section, we restrict attention to properties of unary functions. As a corollary of Proposition 2, we have

Proposition 4.

$$\mathrm{DEF}(f, n) \qquad \Leftrightarrow \qquad \exists y \in \widehat{\mathbb{N}}.\ \mathcal{T}(f^\sharp, \widehat{n}, y) \xrightarrow[\mathbf{TK}]{!} \hat{1} \,.$$

Proof. This is due to the fact that **TK** (see Appendix) is designed so as to reduce all (ground) terms headed by $\mathcal{T}$ to either $\mathsf{0}$ (for false) or $\hat{1}$ (for true). □

The (rewrite) matching problem, MATCH, is

$$\mathrm{MATCH}(R, t, N) \qquad := \qquad \exists \sigma.\ t\sigma \xrightarrow[R]{!} N \,,$$

where R is a rewrite system, t is a term containing variables, N is a ground (that is, variable-free) normal form, and σ is a (ground) substitution.

Theorem 1. *Matching of primitive rewriting is undecidable.*

The proof of undecidability of matching for terminating confluent systems in [22–Cor. 3.11] uses a non-erasing, but overlapping system. The simpler proof in [2], based on the unsolvability of Diophantine equations, uses a system for addition and multiplication of integers that is overlapping, non-left-linear, collapsing, erasing, and non-constructor-based. It has recently been shown that matching (as well as unification) is decidable for confluent systems if no variables on the right appear below the root [40].

Undecidability can be shown from the older recursion theory results—without recourse to the difficult resolution of Hilbert's Tenth Problem, as follows:

Proof. The reduction is from undecidable $\mathrm{DEF}(f, n)$ to the instance $\mathrm{MATCH}(\mathbf{TK}, \mathcal{T}(f^\sharp, \widehat{n}, y), \hat{1})$. We have

$$\begin{aligned} \mathrm{DEF}(f, n) \qquad &\Leftrightarrow \qquad \exists y \in \mathcal{G}.\ \mathcal{T}(f^\sharp, \widehat{n}, y) \xrightarrow[\mathbf{TK}]{!} \hat{1} \\ &\Leftrightarrow \qquad \exists \sigma.\ \mathcal{T}(f^\sharp, \widehat{n}, y)\sigma \xrightarrow[\mathbf{TK}]{!} \hat{1} \\ &\Leftrightarrow \qquad \mathrm{MATCH}(\mathbf{TK}, \mathcal{T}(f^\sharp, \widehat{n}, y), \hat{1}) \,, \end{aligned}$$

where $\mathcal{G}$ is the set of ground (variable-free) terms over the vocabulary of **TK**. The first equivalence is Proposition 4, except that we need the fact that **TK** is constructor complete to ensure that any ground y that satisfies $\mathcal{T}$ reduces to a numeral, since it must be a numeral for **TK** to reduce the term to normal form. The second step is simply because y is the sole variable in the initial term. □

It is similarly undecidable if two terms have the same normal form.

The ground confluence problem GCR (for terminating systems) is

$$\mathrm{GCR}(R) \quad := \quad \forall s,t,u \in \mathcal{G}.\ u \xrightarrow[R]{!} s \wedge u \xrightarrow[R]{!} t \Rightarrow s = t\,,$$

where $\mathcal{G}$ is the set of variable-free terms over the vocabulary of R.

Theorem 2. *Ground confluence of terminating left-linear constructor-based non-erasing non-collapsing constructor-complete rewrite systems is undecidable.*

Ground confluence of terminating systems was shown undecidable in [28], both for terminating string systems (which are left- and right-linear, non-erasing, and non-collapsing) and for terminating left- *or* right-linear systems with right-side depth limited to 2. One can't have orthogonality (absence of critical pairs, in addition to left-linearity) here, since orthogonal systems are confluent.

Proof. The reduction is

$$\mathrm{DEF}(f,n) \quad \Leftrightarrow \quad \neg\mathrm{GCR}(\mathbf{TK} \cup K_f^n)\,,$$

where K_f^n contains the (non-erasing, non-collapsing) rule

$$\mathcal{T}(f^\sharp, \widehat{n}, y) \quad \to \quad \varepsilon(y, \mathsf{0})\,.$$

Note that this rule overlaps rules of **TK**. If, and only if, $f(n)$ is defined, is there a (ground) numeral $\widehat{y} \in \widehat{\mathbb{N}}$ such that

$$\mathcal{T}(f^\sharp, \widehat{n}, \widehat{y}) \quad \xrightarrow[\mathbf{TK}]{!} \quad \hat{1}\,,$$

making for two normal forms $(\mathsf{0}, \hat{1})$ for $\mathcal{T}(f^\sharp, \widehat{n}, \widehat{y})$. □

The confluence problem CR is

$$\mathrm{CR}(R) \quad := \quad \forall s,t,u. \quad u \xrightarrow[R]{*} s \wedge u \xrightarrow[R]{*} t \quad \Rightarrow \quad \exists v.\ s \xrightarrow[R]{*} v \wedge t \xrightarrow[R]{*} v\,.$$

Theorem 3. *Confluence of non-overlapping constructor-based non-erasing non-collapsing rewriting is undecidable.*

Undecidability of confluence of nonterminating systems is claimed in [26]: "The property of confluence is undecidable for arbitrary rewriting systems, since a confluence test could be used to decide the equivalence, for instance, of recursive program schemes." Standard proofs (e.g. [38–Thm. 5.2.1]) are based on overlapping, but left-linear, constructions. Since orthogonal systems are always confluent, one can't have both left-linearity and non-overlappingness.[8]

Confluence is known to be undecidable (in fact, not even semi-decidable or co-r.e.) even if the rules are left- and right-linear, constructor-complete, and constructor-based, and all critical pairs obtained from overlaps resolve (i.e. the system is locally, or weakly, confluent) [15–Sect. 4]. The terminating case is decidable, even in the presence of overlapping left sides, by the famous Critical Pair Lemma of Knuth [33]; see [38–Thm. 2.7.16]. It has also recently been shown that confluence is decidable for right-linear systems if no variables appear below depth 1 [18] (extending earlier decidability results [44, 20]).

Proof. We cannot use the same K_f^n as in the previous proof, since its left side overlaps rules of **TK**. Instead let B_f^n be

$$\mathsf{B}(\hat{1}, y) \quad \to \quad \mathsf{s}(\mathsf{B}(\mathcal{T}(f^\sharp, \widehat{n}, y), y)) ,$$

which has the property that

$$\mathsf{B}(\hat{1}, \widehat{y}) \quad \xrightarrow{!} \quad \mathsf{s}(\mathsf{B}(\hat{1}, \widehat{y})) ,$$

for some numeral $\widehat{y}$, if, and only if, $\mathcal{T}(f^\sharp, \widehat{n}, \widehat{y}) \to^* \hat{1}$. Now

$$\begin{array}{lcl} \mathrm{DEF}(f, n) & \Leftrightarrow & \exists y \in \widehat{\mathbb{N}}.\ \mathcal{T}(f^\sharp, \widehat{n}, y) \xrightarrow[\mathbf{TK}]{!} \hat{1} \\ & \Leftrightarrow & \neg \mathrm{CR}(\mathbf{TK} \cup B_f^n \cup H) , \end{array}$$

where non-linear system H is

$$\begin{array}{rcl} \mathsf{H}(x, x) & \to & \varepsilon(x, a) \\ \mathsf{H}(\mathsf{s}(x), x) & \to & \varepsilon(x, c) . \end{array}$$

The rules for B and H are akin to Huet's [24] example of non-terminating non-overlapping non-confluence. So, if $f(n)$ is defined, then $\mathsf{H}(\mathsf{B}(\hat{1}, \widehat{y}), \mathsf{B}(\hat{1}, \widehat{y}))$ rewrites to a term containing a by the first rule of H and to a term containing c in two stages, applying B_f^n, followed by the second H rule. Since there is no other way for a term t to rewrite to $\mathsf{s}(t)$, non-confluence is a perfect indication that $f(n)$ is defined. □

The modular (shared-constructor) confluence problem CR2 is

$$\mathrm{CR2}(R, S) \quad := \quad \mathrm{CR}(R \cup S) ,$$

where R and S are confluent systems with only constructors in common.

Since H shares no defined symbols with **TK** or B:

Corollary 1. *Modular confluence of constructor-based rewriting is undecidable.*

That constructor-sharing combinations need not preserve confluence was pointed out in [35] (just consider H together with $\mathsf{b} \to \mathsf{s}(\mathsf{b})$).

[8] Recursive program schemes [38–Def. 3.4.7] are orthogonal, but two schemes together are not.

7 Halting Problems

The normalizability problem NORM is

$$\mathrm{NORM}(R,t) \quad := \quad \exists z.\, t \xrightarrow[R]{!} z\,.$$

Theorem 4. *Normalizability of definitional rewriting is undecidable.*

That normalizability is undecidable for non-constructor-based rewriting is obvious from the rewrite system CL for combinatory logic in Klop's monograph [30–Sect. 2.2].

Proof. DEF(f,n) reduces to NORM$(F, f(\widehat{n}))$, where F is the partial recursive rewrite system for f. This is basically just faithfulness of F, as stated in Proposition 1. Thus, normalizability for partial-recursive rewriting is undecidable, and, *a fortiori*, for arbitrary definitional rewriting. □

For the (weak) normalization problem WN,

$$\mathrm{WN}(R) \quad := \quad \forall t.\ \mathrm{NORM}(R,t)\,,$$

the situation is the same:

Theorem 5. *Normalization of definitional rewriting is not semi-decidable.*

Proof. The reduction is

$$\mathrm{TOT}(f) \quad \Leftrightarrow \quad \mathrm{WN}(\mathbf{R}_f)\,.$$

By Proposition 3, f is total if, and only if, $\mathbf{R}_f$ normalizes all terms $\mathsf{f}(\widehat{a})$. So, if f is not total, there is a term that $\mathbf{R}_f$ fails to normalize.

For the other direction, when a term (ground or not) is non-terminating for non-overlapping $\mathbf{R}_f$, there must be—by standard techniques in rewriting [10, 16]—an infinite innermost derivation,

$$\mu(z,0,a) \xrightarrow[\mathbf{R}_f]{} \mu(\mathsf{s}(z), \mathcal{T}^*(f^\sharp, a, \mathsf{s}(z)), a) \xrightarrow[\mathbf{R}_f]{+} \mu(\mathsf{s}(z),0,a) \xrightarrow[\mathbf{R}_f]{+} \cdots,$$

punctuated by instances of the main μ-rule (the only potentially non-terminating rule), all subterms of which are already in normal-form. Considering how $\mathbf{R}_f$ looks, that means that

$$\mathcal{T}^*(f^\sharp, a, \mathsf{s}^i(z)) \xrightarrow[\mathbf{R}_f]{+} 0\,,$$

for all $i > 0$. Since $\mathbf{R}_f$ is non-erasing and constructor-based, this can only be if the culprits z and a reduce to numerals. Since $\mathcal{T}^*$ is monotonic, we have

$$\mathcal{T}^*(f^\sharp, \widehat{n}, y) \xrightarrow[\mathbf{R}_f]{+} 0\,,$$

for all $y \in \widehat{\mathbb{N}}$, where $\widehat{n}$ is the normal form of a. In other words, $f(n)$ admits no finite computation.

We needed to use Kleene Normal Form here (that is, $\mathbf{R}_f$ instead of F) to preclude the possibility that f is total, whereas a function g that it uses is not, in which case some term containing g would never terminate. □

The (uniform/strong) termination problem SN is

$$\mathrm{SN}(R) \quad := \quad \neg\exists t.\; t \xrightarrow[R]{} \bot\;,$$

where the postfixed notation $\rightarrow \bot$ indicates the existence of a divergent derivation: $t \rightarrow_R \bot$ for system R if there are terms $\{t_i\}_i$ such that $t \rightarrow_R t_0 \rightarrow_R t_1 \rightarrow_R \cdots$.

Corollary 2. *Termination of definitional rewriting is not semi-decidable.*

Proof. As mentioned above, non-erasing orthogonal systems like $\mathbf{R}_f$ terminate (SN) if and only if they are normalizing (WN). □

Termination (and normalization) of (overlapping) string rewriting was proved undecidable in a technical report by Huet and Lankford [25], using a Turing-machine construction for the semi-Thue word problem. (See [38–Sect. 5.3.1] for a similar proof; such a reduction was given by Davis in [9]; see also [53].) Lescanne's proof in [36] is left-linear and constructor-based and can be made non-overlapping. It has recently been shown that termination is decidable for right-linear systems if no variables appear on the right below depth 1 [19] (extending earlier decidability results [25, 10]).

The (disjoint) modular termination problem SN2 is

$$\mathrm{SN2}(R, S) \quad := \quad \mathrm{SN}(R \cup S)\;,$$

where R and S are terminating systems with no common function symbols or constants *at all.*

Theorem 6 ([39]). *Modular termination of left-linear rewriting is undecidable.*

Modular termination of left-linear rewriting was found undecidable in [39, 50].[9] Were the systems also locally confluent, their disjoint union would perforce be terminating [52].

Proof. The reduction is

$$\begin{aligned} \mathrm{DEF}(f, n) \quad &\Leftrightarrow \quad \exists y \in \widehat{\mathbb{N}}.\; \mathcal{T}(f^\sharp, \widehat{n}, y) \xrightarrow[\mathbf{TK}]{!} \hat{1} \\ &\Leftrightarrow \quad \neg\mathrm{SN}(\mathbf{TK} \cup G_f^n \cup Z) \\ &\Leftrightarrow \quad \neg\mathrm{SN2}(\mathbf{TK} \cup G_f^n, Z)\;, \end{aligned}$$

[9] Middeldorp [39–p. 65] credits the author of this paper with simultaneity.

where G^n_f is

$$\begin{array}{rcl} \mathsf{G}(0,x,y) & \to & x \\ \mathsf{G}(\hat{1},x,y) & \to & y \\ \mathsf{G}(\mathcal{T}(f^\sharp,\widehat{n},z),x,y) & \to & x\,, \end{array}$$

and Z (cf. [51]) is

$$\mathsf{Z}(\mathsf{a},\mathsf{b},z) \quad \to \quad \mathsf{Z}(z,z,z)\,.$$

The point here is that Z alone is terminating, but turns nonterminating when combined with (terminating) rules that rewrite some term to both a and b. System G^n_f, though not containing those constants, does just that if, but only if, $\mathcal{T}(f^\sharp,\widehat{n},z)$ reduces to $\hat{1}$, for some computation z. (The first G rule is only there to keep one of the systems sufficiently complete.) □

The undecidability of termination for various forms of hierarchically combined systems [12, 34, 43] follows directly from the strictly hierarchical form of partial-recursive rewriting, whose right-hand sides only refer to previously defined symbols and whose left sides are, by nature, non-overlapping.

8 Conclusion

The relation between the lambda calculus, combinatory logic, and recursion theory is classical. In 1980, Klop [30] forged the link between combinators and rewriting. Here, we have tried to flesh out the "missing" connection, bridging recursive function theory and term rewriting.

As a consequence of this connection, and the tool added to our arsenal, we have made the following small improvements over well-known undecidability results in rewriting:

- Matching is undecidable for convergent (that is, confluent and terminating), left-linear, non-erasing, constructor-based systems, even when they are non-overlapping and sufficiently complete (Theorem 1).
- Matching is undecidable for convergent, sufficiently-complete systems, even when they are left-linear, non-overlapping, constructor-based, non-erasing, and non-collapsing (Theorem 1).
- Confluence is undecidable for non-erasing, non-collapsing, constructor-based systems, even if they are non-overlapping (Theorem 3).
- Ground confluence is undecidable for terminating, left-linear, non-erasing, non-collapsing, sufficiently-complete systems, even if they are constructor-based (Theorem 2).
- Modular (shared-constructor) confluence is undecidable—even for non-erasing, non-collapsing, constructor-based systems (Corollary 1).
- Normalizability is not decidable for orthogonal, constructor-based systems, even if they are constructor complete and non-erasing (Theorem 4).
- Termination and (weak) normalization are not semi-decidable for orthogonal, constructor-based systems, even if they are constructor complete and non-erasing (Theorem 5; Corollary 2).

Acknowledgement

I thank the referees for their critical reading and constructive suggestions.

References

1. Amir M. Ben-Amram. General size-change termination and lexicographic descent. In Torben Mogensen, David Schmidt, and I. Hal Sudborough, editors, *The Essence of Computation: Complexity, Analysis, Transformation. Essays Dedicated to Neil D. Jones*, volume 2566 of *Lecture Notes in Computer Science*, pages 3–17. Springer-Verlag, 2002.
2. Alexander Bockmayr. A note on a canonical theory with undecidable unification and matching problem. *J. Automated Reasoning*, 3(4):379–381, 1987.
3. Udi Boker and Nachum Dershowitz. A formalization of the Church-Turing Thesis. Available at `http://www.cs.tau.ac.il/ nachum/papers/ChurchTuringThesis.pdf`.
4. Udi Boker and Nachum Dershowitz. Comparing computational power. *Logic Journal of the IGPL*, 2006. To appear; available at: `http://www.cs.tau.ac.il/ nachum/papers/ComparingComputationalPower.pdf`.
5. Udi Boker and Nachum Dershowitz. A hypercomputational alien. *J. of Applied Mathematica & Computation*, 2006. To appear; available at `http://www.cs.tau.ac.il/~nachum/papers/HypercomputationalAlien.pdf`.
6. Ronald V. Book. Thue systems as rewriting systems. *J. Symbolic Computation*, 3(1&2):39–68, February/April 1987.
7. Alonzo Church. *The Calculi of Lambda Conversion*, volume 6 of *Ann. Mathematics Studies*. Princeton University Press, Princeton, NJ, 1941.
8. E. Adam Cichon and Elias Tahhan-Bittar. Strictly orthogonal left linear rewrite systems and primitive recursion. *Ann. Pure Appl. Logic*, 108(1–3):79–101, 2001.
9. Martin Davis. *Computability and Unsolvability*. McGraw-Hill, New York, 1958.
10. Nachum Dershowitz. Termination of linear rewriting systems (Preliminary version). In *Proceedings of the Eighth International Colloquium on Automata, Languages and Programming (Acre, Israel)*, volume 115 of *Lecture Notes in Computer Science*, pages 448–458, Berlin, July 1981. European Association of Theoretical Computer Science, Springer-Verlag.
11. Nachum Dershowitz. Termination of rewriting. *J. Symbolic Computation*, 3(1&2):69–115, February/April 1987. Corrigendum: *4*, 3 (December 1987), 409–410.
12. Nachum Dershowitz. Hierarchical termination. In N. Dershowitz and N. Lindenstrauss, editors, *Proceedings of the Fourth International Workshop on Conditional and Typed Rewriting Systems (Jerusalem, Israel, July 1994)*, volume 968 of *Lecture Notes in Computer Science*, pages 89–105, Berlin, 1995. Springer-Verlag.
13. Nachum Dershowitz and Charles Hoot. Natural termination. *Theoretical Computer Science*, 142(2):179–207, May 1995.
14. Alfons Geser, Aart Middeldorp, Enno Ohlebusch, and Hans Zantema. Relative undecidability in term rewriting: I. The termination hierarchy. *Information and Computation*, 178(1):101–131, 2002.
15. Alfons Geser, Aart Middeldorp, Enno Ohlebusch, and Hans Zantema. Relative undecidability in term rewriting: II. The confluence hierarchy. *Information and Computation*, 178(1):132–148, 2002.

16. Oliver Geupel. Overlap closures and termination of term rewriting systems. Technical report, Universität Passau, Passau, West Germany, July 1989.
17. Kurt Gödel. Über formal unentscheidbare Sätze der Principia Mathematica und verwandter Systeme I. *Monatshefte für Mathematik und Physik*, 38:173–198, 1931. Translated as "On Formally Undecidable Propositions of Principia Mathematica and Related Systems. I", for Basic Books (New York, 1962) and in M. Davis (ed.), *The Undecidable*, Raven Press, Hewlett, NY, 1965. Available at `http://home.ddc.net/ygg/etext/godel/godel3.htm` (viewed September 2005).
18. Guillem Godoy and Ashish Tiwari. Confluence of shallow right-linear rewrite systems. In L. Ong, editor, *Computer Science Logic, 14th Annual Conf.*, volume 3634 of *LNCS*, pages 541–556. Springer-Verlag, August 2005.
19. Guillem Godoy and Ashish Tiwari. Termination of rewrite systems with shallow right-linear, collapsing, and right-ground rules. In R. Nieuwenhuis, editor, *20th Intl. Conf. on Automated Deduction*, volume 3632 of *LNCS*, pages 164–176. Springer-Verlag, July 2005.
20. Guillem Godoy, Ashish Tiwari, and Rakesh Verma. Characterizing confluence by rewrite closure and right ground term rewrite systems. *Applied Algebra on Engineering, Communication and Computer Science*, 15(1):13–36, June 2004.
21. Bernhard Gramlich. On proving termination by innermost termination. In H. Ganzinger, editor, *Proceedings of the 7th International Conference on Rewriting Techniques and Applications (RTA-96, New Brunswick, NJ)*, volume 1103 of *Lecture Notes in Computer Science*, pages 93–107. Springer-Verlag, July 1996.
22. Stephan Heilbrunner and Steffen Hölldobler. The undecidability of the unification and matching problem for canonical theories. *Acta Informatica*, 24(2):157–171, April 1987.
23. Dieter Hofbauer and Clemens Lautemann. Termination proofs and the length of derivations (Preliminary version). In N. Dershowitz, editor, *Proceedings of the International Conference on Rewriting Techniques and Applications*, volume 355 of *Lecture Notes in Computer Science*, pages 167–177, Berlin, April 1989. Springer-Verlag.
24. Gérard Huet. Confluent reductions: Abstract properties and applications to term rewriting systems. *J. of the Association for Computing Machinery*, 27(4):797–821, October 1980.
25. Gérard Huet and Dallas S. Lankford. On the uniform halting problem for term rewriting systems. Rapport laboria 283, Institut de Recherche en Informatique et en Automatique, Le Chesnay, France, March 1978.
26. Gérard Huet and Derek C. Oppen. Equations and rewrite rules: A survey. In R. Book, editor, *Formal Language Theory: Perspectives and Open Problems*, pages 349–405. Academic Press, New York, 1980.
27. Renato Iturriaga. *Contributions to Mechanical Mathematics*. PhD thesis, Department of Computer Science, Carnegie-Mellon University, Pittsburgh, PA, 1967.
28. Deepak Kapur, Paliath Narendran, and Friedrich Otto. On ground confluence of term rewriting systems. *Information and Computation*, 86(1):14–31, May 1990.
29. Stephen C. Kleene. General recursive functions of natural numbers. *Mathematische Annales*, 112:727–742, 1936.
30. Jan Willem Klop. *Combinatory Reduction Systems*, volume 127 of *Mathematical Centre Tracts*. Mathematisch Centrum, Amsterdam, 1980.
31. Jan Willem Klop. Term rewriting systems: from Church-Rosser to Knuth-Bendix and beyond. In M. S. Paterson, editor, *Proceedings of the 17th International Colloquium on Automata, Languages, and Programming (Warwick, England)*, Lecture Notes in Computer Science, pages 350–369. Springer-Verlag, July 1990.

32. Jan Willem Klop. Term rewriting systems. In S. Abramsky, D. M. Gabbay, and T. S. E. Maibaum, editors, *Handbook of Logic in Computer Science*, volume 2, chapter 1, pages 1–117. Oxford University Press, Oxford, 1992.
33. Donald E. Knuth and P. B. Bendix. Simple word problems in universal algebras. In J. Leech, editor, *Computational Problems in Abstract Algebra*, pages 263–297. Pergamon Press, Oxford, U. K., 1970. Reprinted in *Automation of Reasoning 2*, Springer-Verlag, Berlin, pp. 342–376 (1983).
34. Madala R. K. Krishna Rao. Modular proofs for completeness of hierarchical term rewriting systems. *Theoretical Computer Science*, 151:487–512, 1995.
35. M. Kurihara and A. Ohuchi. Modularity of simple termination of term rewriting systems with shared constructors. *Theoretical Computer Science*, 103:273–282, 1992.
36. Pierre Lescanne. On termination of one rule rewrite systems. *Theoretical Computer Science*, 132(1–2):395–401, 1994.
37. Zohar Manna. *Mathematical Theory of Computation.* McGraw-Hill, New York, 1974.
38. "Terese" (Marc Bezem, Jan Willem Klop, and Roel de Vrijer, eds.). *Term Rewriting Systems.* Cambridge University Press, 2002.
39. Aart Middeldorp. *Modular Properties of Term Rewriting Systems.* PhD thesis, Vrije Universiteit, Amsterdam, 1990.
40. Ichiro Mitsuhashi, Michio Oyamaguchi, Yoshikatsu Ohta, and Toshiyuki Yamada. The joinability and unification problems for confluent semi-constructor TRSs. In Vincent van Oostrom, editor, *Proceedings of the 15th International Conference on Rewriting Techniques and Applications, Aachen, Germany, June 2004*, volume 3091 of *Lecture Notes in Computer Science*, pages 285–300. Springer-Verlag, 2004.
41. Piergiorgio Odifreddi. *Classical Recursion Theory*, volume 125 of *Studies in Logic and the Foundations of Mathematics.* North-Holland, Amsterdam, 1989.
42. Michael J. O'Donnell. *Computing in Systems Described by Equations*, volume 58 of *Lecture Notes in Computer Science.* Springer-Verlag, Berlin, 1977.
43. Enno Ohlebusch. *Modular Properties of Composable Term Rewriting Systems.* PhD thesis, Abteilung Informationstechnik, Universität Bielefeld, Bielefeld, Germany, 1994.
44. Michio Oyamaguchi. The Church-Rosser property for ground term rewriting systems is decidable. *Theoretical Computer Science*, 49(1):43–79, 1987.
45. Rózsa Péter. *Recursive Functions.* Academic Press, 1967.
46. David A. Plaisted. Well-founded orderings for proving termination of systems of rewrite rules. Report R-78-932, Department of Computer Science, University of Illinois, Urbana, IL, July 1978.
47. Emil L. Post. Recursive unsolvability of a problem of Thue. *J. of Symbolic Logic*, 13:1–11, 1947.
48. Hartley Rogers, Jr. *Theory of Recursive Functions and Effective Computability.* McGraw-Hill, New York, 1966.
49. Kai Salomaa. Confluence, ground confluence, and termination of monadic term rewriting systems. *Journal of Information Processing and Cybernetics EIK*, 28:279–309, 1992.
50. Kai Salomaa. On the modularity of decidability of completeness and termination. *Journal of Automata, Languages and Combinatorics*, 1:37–53, 1996.
51. Yoshihito Toyama. Counterexamples to termination for the direct sum of term rewriting systems. *Information Processing Letters*, 25:141–143, 1987.

52. Yoshihito Toyama, Jan Willem Klop, and Hendrik Pieter Barendregt. Termination for direct sums of left-linear complete term rewriting systems. *J. of the Association for Computing Machinery*, 42(6):1275–1304, November 1995.
53. Ann Yasuhara. *Recursive Function Theory and Logic*. Academic Press, 1971.
54. Doron Zeilberger. A 2-minute proof of the 2nd-most important theorem of the 2nd millennium. At `http://www.math.rutgers.edu/ zeilberg/mamarim/mamarimhtml/halt.html` (viewed September 2005).

Appendix: The Computation Predicate

Most descriptions of computation predicates are non-algorithmic. (An exception is [41–Chap. I.7].) To fill this lacuna, we give here a complete primitive rewrite program **TK** for T^K.[10]

Lists are encoded as pairs (of pairs) in the customary fashion, due to Gödel:

$$\langle m, n\rangle \mapsto 2^m(2n+1)\ .$$

The empty list is 0. Lists $\langle x_1, \ldots, x_k\rangle$ are just nested pairs of the form $\langle x_1, \langle x_2, \langle \cdots x_k\rangle \cdots\rangle\rangle$, etc. Terms are encoded as lists. With this encoding, it is easy to show by induction that $\lg \ell$ is an upper bound on the length of a list ℓ ($\lg \mathsf{nil} = \lg 0 = 0$), and $1 + \lg \ell$ on its list-nesting depth.

Table 1 recapitulates definitions of the standard logical and arithmetic operations, including binary logarithm ($\lg x := \lfloor \log_2 x\rfloor$)[11] and a conditional if-then-else that evaluates all three of its arguments.

The primitive-recursive Lisp operations are as follows:

$$\begin{array}{rcl}
\mathsf{nil} & \to & 0 \\
x\mathbin{:}y & \to & 2^x(2y+1) \\
\mathsf{car}_2(0,y) & \to & \mathsf{nil} \\
\mathsf{car}_2(\mathsf{s}(x),y) & \to & \mathsf{if}\ 2^{x+1} \nmid y\ \mathsf{then}\ \mathsf{car}_2(x,y) \\
 & & \qquad\qquad \mathsf{else}\ \mathsf{car}_2(x,y)+1 \\
\mathsf{car}\ x & \to & \mathsf{car}_2(\lg x, x) \\
\mathsf{cdr}\ x & \to & x \div 2^{(\mathsf{car}\ x)+1}
\end{array}$$

For readability we use a colon for Lisp's list constructor, **cons**. We will need the fact that, as programmed, $\mathsf{car}(\mathsf{nil}) = \mathsf{cdr}(\mathsf{nil}) = \mathsf{nil}$.

Other standard list operations are easy:

$$\begin{array}{rcl}
\mathsf{cadr}\ x & \to & \mathsf{car}(\mathsf{cdr}\ x) \\
\mathsf{cddr}\ x & \to & \mathsf{cdr}(\mathsf{cdr}\ x) \\
\mathsf{nthcdr}(0,y) & \to & y \\
\mathsf{nthcdr}(\mathsf{s}(n),y) & \to & \mathsf{cdr}(\mathsf{nthcdr}(n,y)) \\
\mathsf{nth}(n,y) & \to & \mathsf{car}(\mathsf{nthcdr}(n,y)) \\
\mathsf{length}_2(0,y) & \to & y
\end{array}$$

[10] The rules are also online at `www.cs.tau.ac.il/~nachum/TK.r`, and a faithful coding in Lisp at `www.cs.tau.ac.il/~nachum/TK.l`.

[11] The lg notation for $\log_2$ was suggested by Ed Reingold and popularized by Don Knuth.

Table 1. Basic arithmetic and logic. Primitive rules for predecessor (p), addition (+), natural subtraction ($\dot{-}$), and multiplication ($\times$ or juxtaposition) were given in Sects. 3 and 4 of the text.

1	$\rightarrow$	$\mathsf{s}(0)$	2	$\rightarrow$	$\mathsf{s}(1)$
3	$\rightarrow$	$\mathsf{s}(2)$	4	$\rightarrow$	$\mathsf{s}(3)$
5	$\rightarrow$	$\mathsf{s}(4)$			
2^0	$\rightarrow$	1	$2^{\mathsf{s}(n)}$	$\rightarrow$	$2 \cdot 2^n$
$\neg x$	$\rightarrow$	$1 \dot{-} x$	$\delta(x)$	$\rightarrow$	$\neg\neg x$
$m > n$	$\rightarrow$	$\delta(m \dot{-} n)$	true	$\rightarrow$	1
$x \vee y$	$\rightarrow$	$\delta(x+y)$	$x \wedge y$	$\rightarrow$	xy
$m \neq n$	$\rightarrow$	$(m > n) \vee (n > m)$	$m = n$	$\rightarrow$	$\neg(m \neq n)$
$\mathsf{if}\ x\ \mathsf{then}\ y\ \mathsf{else}\ z$	$\rightarrow$	$(x > 0)y + (x = 0)z$			
$\mathsf{div}(0, m, n)$	$\rightarrow$	0			
$\mathsf{div}(\mathsf{s}(k), m, n)$	$\rightarrow$	$\mathsf{if}\ m+1 > (k+1)n\ \mathsf{then}\ k+1\ \mathsf{else}\ \mathsf{div}(k, m, n)$			
$m \div n$	$\rightarrow$	$\mathsf{div}(m, m, n)$	$m \nmid n$	$\rightarrow$	$n > (n \div m)m$
$\lg 0$	$\rightarrow$	0			
$\lg \mathsf{s}(n)$	$\rightarrow$	$\mathsf{if}\ 2 \nmid (n+1)\ \mathsf{then}\ \lg n\ \mathsf{else}\ 1 + \lg n$			

$$
\begin{array}{rcl}
\mathsf{length}_2(\mathsf{s}(x), y) & \rightarrow & \mathsf{if}\ \mathsf{nthcdr}(\lg y \dot{-} (x+1), y) = \mathsf{nil}\ \mathsf{then}\ \lg y \dot{-} (x+1) \\
 & & \qquad \mathsf{else}\ \mathsf{length}_2(x, y) \\
|x| & \rightarrow & \mathsf{length}_2(\lg x, x) \\
\mathsf{append}_2(0, x, y) & \rightarrow & y \\
\mathsf{append}_2(\mathsf{s}(n), x, y) & \rightarrow & \mathsf{nth}(|x| \dot{-} (n+1), x)\mathsf{:}\mathsf{append}_2(n, x, y) \\
x * y & \rightarrow & \mathsf{append}_2(|x|, x, y)
\end{array}
$$

The function $|x|$ gives the list length of x; we use an asterisk $*$ for the list append function (as did Gödel [17]).

We will have recourse to a few additional functions for lists and terms:

$$
\begin{array}{rcl}
\mathsf{nthcadr}(0, x) & \rightarrow & x \\
\mathsf{nthcadr}(\mathsf{s}(n), x) & \rightarrow & \mathsf{cadr}(\mathsf{nthcadr}(n, x)) \\
\mathsf{prefix}(0, x) & \rightarrow & \mathsf{nil} \\
\mathsf{prefix}(\mathsf{s}(n), x) & \rightarrow & \mathsf{prefix}(n, x) * (\mathsf{nth}(n, x)\mathsf{:nil}) \\
\mathsf{pos}_2(0, p, x) & \rightarrow & x \\
\mathsf{pos}_2(\mathsf{s}(n), p, x) & \rightarrow & \mathsf{nth}(\mathsf{nth}(n, p), \mathsf{pos}(n, p, x)) \\
\mathsf{pos}(p, x) & \rightarrow & \mathsf{pos}_2(|p|, p, x) \\
\mathcal{U}(z) & \rightarrow & \mathsf{nth}(|z| \dot{-} 1, z)
\end{array}
$$

Function nthcadr digs down first arguments; pos returns a subterm at a given Dewey decimal position; $\mathcal{U}$ is the last element in a sequence, used in Proposition 2.

Primitive programs are enumerated (into the naturals) in the following manner:

- 0 is the constant (function) 0;
- 1 is the unary successor function s;

– $1{:}i$ is the ith projection rule π_i for any arity;
– $2{:}g{:}\bar{h}$ is the composition $g(\ldots, h_i, \ldots)$ of g with h_i;
– $3{:}g{:}h$ is primitive recursion, with base case g and recursive case h;
– $4{:}q$ is minimization over predicate q;
– $5{:}q$ is an auxiliary function for minimization.

A function application $f(x_1, \ldots, x_k)$ is encoded as a list $\langle f, x_1, \ldots, x_k\rangle$. Accordingly, the successor of numeral n is the pair $\langle 1, n\rangle$, while its predecessor is cadr n. So, normal forms look like $\langle 1, \langle 1, \langle \cdots \langle 1, 0\rangle \cdots\rangle\rangle$. The auxiliary function, $5{:}q$, is for minimization starting from some given lower bound. That is, $(5{:}q)(k, c, \bar{x})$ is k when c is true; otherwise, it is the smallest $i > k$ such that $q(i, \bar{x})$.

To find the next (leftmost innermost) redex, we use a test nf for normal form (i.e. a numeral) and a function next to find the first non-normal-form in a list:

$$
\begin{array}{rcl}
\mathsf{nf}_2(0, x) & \to & \mathsf{true} \\
\mathsf{nf}_2(\mathsf{s}(n), x) & \to & (\mathsf{nthcadr}(n, x) = 0 \vee \mathsf{car}(\mathsf{nthcadr}(n, x)) = 1) \wedge \mathsf{nf}_2(n, x) \\
\mathsf{nf}(x) & \to & \mathsf{nf}_2(\lg x, x) \\
\mathsf{next}_2(0, x) & \to & |x| \\
\mathsf{next}_2(\mathsf{s}(n), x) & \to & \mathsf{if}\ \mathsf{nf}(\mathsf{nth}(|x| \dot{-} (n+1), x))\ \mathsf{then}\ \mathsf{next}_2(n, x) \\
 & & \qquad\qquad\qquad\qquad\qquad\quad \mathsf{else}\ |x| \dot{-} (n+1)
\end{array}
$$

$$
\begin{array}{rcl}
\mathsf{next}(x) & \to & \mathsf{next}_2(|x|, x) \\
\mathsf{redex}_2(0, x) & \to & \mathsf{nil} \\
\mathsf{redex}_2(\mathsf{s}(n), x) & \to &
\end{array}
$$

$$
\begin{array}{l}
\mathsf{if}\ \mathsf{next}(\mathsf{cdr}(\mathsf{pos}(\mathsf{redex}_2(n, x), x))) = |\mathsf{cdr}(\mathsf{pos}(\mathsf{redex}_2(n, x), x))| \\
\quad \mathsf{then}\ \mathsf{redex}_2(n, x) \\
\quad \mathsf{else}\ \mathsf{redex}_2(n, x) * (1 + \mathsf{next}(\mathsf{cdr}(\mathsf{pos}(\mathsf{redex}_2(n, x), x)))){:}\mathsf{nil} \\
\qquad\quad \mathsf{redex}(x) \quad \to \quad \mathsf{redex}_2(1 + \lg x, x)
\end{array}
$$

To apply the function definition at that point, we proceed by case analysis on the different ways of building partial-recursive functions:

$$
\begin{array}{rcl}
\mathsf{succ}(x) & \to & 1{:}x{:}\mathsf{nil} \\
\mathsf{dist}(0, g, x) & \to & \mathsf{nil} \\
\mathsf{dist}(\mathsf{s}(n), g, x) & \to & (\mathsf{nth}(|g| \dot{-} (n+1), g){:}x){:}\mathsf{dist}(n, g, x) \\
\mathsf{apply}(f, x) & \to & \mathsf{if}\ \mathsf{car}\ f = 0\ \mathsf{then}\ 0 \\
 & & \mathsf{else\ if}\ \mathsf{car}\ f = 1\ \mathsf{then}\ \mathsf{nth}(\mathsf{cdr}\ f, x) \\
 & & \mathsf{else\ if}\ \mathsf{car}\ f = 2\ \mathsf{then}\ \mathsf{cadr}\ f{:}\mathsf{dist}(|\mathsf{cddr}\ f|, \mathsf{cddr}\ f, x) \\
 & & \mathsf{else\ if}\ \mathsf{car}\ f = 3 \wedge \mathsf{car}\ x = 0\ \mathsf{then}\ \mathsf{cadr}\ f{:}\mathsf{cdr}\ x \\
 & & \mathsf{else\ if}\ \mathsf{car}\ f = 3\ \mathsf{then}\ \mathsf{cddr}\ f{:}(f{:}\mathsf{cadr}(\mathsf{car}\ x){:}\mathsf{cdr}\ x){:}x \\
 & & \mathsf{else\ if}\ \mathsf{car}\ f = 4\ \mathsf{then}\ (5{:}\mathsf{cdr}\ f){:}0{:}(\mathsf{cdr}\ f{:}0{:}x){:}x \\
 & & \mathsf{else\ if}\ \mathsf{car}\ f = 5 \wedge \mathsf{cadr}\ x\ \mathsf{then}\ \mathsf{car}\ x \\
 & & \mathsf{else\ if}\ \mathsf{car}\ f = 5 \\
 & & \qquad \mathsf{then}\ f{:}(\mathsf{cdr}\ f{:}\mathsf{succ}(\mathsf{car}\ x){:}x){:}\mathsf{succ}(\mathsf{car}\ x){:}x \\
 & & \mathsf{else}\ f{:}x
\end{array}
$$

The helper function dist distributes a list of functions over shared arguments.

Finally, to check the validity of a computation, we check that the first element is the function call in question, that each step is an application of one of the above function applications, and that the final element is a numeral:

$$
\begin{array}{rcl}
\mathsf{change}(n,x,y) & \to & \mathsf{prefix}(n,x) * (y{:}\mathsf{nthcdr}(n+1,x)) \\
\mathsf{term}(0,p,x) & \to & \mathsf{apply}(\mathsf{car}\ x, \mathsf{cdr}\ x) \\
\mathsf{term}(\mathsf{s}(n),p,x) & \to & \mathsf{change}(\mathsf{nth}(|p| \dot{-} (n+1), p), x, \mathsf{term}(n,p,x)) \\
\mathsf{step}(x) & \to & \mathsf{term}(|\mathsf{redex}(x)|, \mathsf{redex}(x), x) \\
\mathsf{steps}(0,z) & \to & \mathsf{true} \\
\mathsf{steps}(\mathsf{s}(n),z) & \to & \mathsf{step}(\mathsf{nth}(n,z)) = \mathsf{nth}(n+1,z) \wedge \mathsf{steps}(n,z) \\
\mathcal{T}(f,x,y) & \to & (\mathsf{car}(y) = f{:}x) \wedge \mathsf{steps}(|y| \dot{-} 1, y) \wedge \mathsf{nf}(\mathcal{U}(y)) \\
\mathcal{T}^*(f,x,0) & \to & \mathcal{T}(f,x,0) \\
\mathcal{T}^*(f,x,\mathsf{s}(y)) & \to & \mathcal{T}^*(f,x,y) + \mathcal{T}(f,x,\mathsf{s}(y))
\end{array}
$$

The function $\mathsf{change}(n,x,y)$ replaces the nth element x_n in a sequence x with y; term substitutes a reduct for redex at a given position; $\mathcal{T}$ is Kleene's computation predicate; $\mathcal{T}^*$ is its monotonic counterpart.

With the above definitions, but sans the auxiliary functions, **TK** boils down to 42 convergent, orthogonal, constructor-based rules. As explained in the text, the system should be transformed into one that is non-erasing and non-collapsing.

Recursive rules are of nesting depth 3 (with only variables on level three) on the left and on the right, and compositions also have right-depth 3. Were all right sides non-nested and linear (they are not), then confluence and termination would be decidable [49]. In fact, one needs only two rules of left nesting-depth 3 to encode all of partial-recursive rewriting and bring on the undecidability results of the previous sections. To that end, one would systematically replace each non-shallow recursive definition f, other than the predecessor function p (defined in the text), with left-shallow rules

$$
\begin{array}{rcl}
f(z,\ldots,x_i,\ldots) & \to & f'(\mathsf{t}(z), \mathsf{p}(z), \ldots, x_i, \ldots) \\
f'(\mathsf{F}, z, \ldots, x_i, \ldots) & \to & g(\ldots, x_i, \ldots) \\
f'(\mathsf{T}, z, \ldots, x_i, \ldots) & \to & h(f(z,\ldots,x_i,\ldots), z, \ldots, x_i, \ldots)\,,
\end{array}
$$

and then add the following:

$$
\begin{array}{rcl}
\mathsf{t}(0) & \to & \mathsf{F} \\
\mathsf{t}(\mathsf{s}(x)) & \to & \mathsf{T}\,.
\end{array}
$$

Infinitary Rewriting: From Syntax to Semantics*

Richard Kennaway[1], Paula Severi[2], Ronan Sleep[1], and Fer-Jan de Vries[2]

[1] School of Computing Sciences, University of East Anglia, U.K.
[2] Department of Computer Science, University of Leicester, U.K.

1 Introduction

Rewriting is the repeated transformation of a structured object according to a set of rules. This simple concept has turned out to have a rich variety of elaborations, giving rise to many different theoretical frameworks for reasoning about computation. Aside from its theoretical importance, rewriting has also been a significant influence on the design and implementation of real programming languages, most notably the functional and logic programming families of languages. For a theoretical perspective on the place of rewriting in Computer Science, see for example [14]. For a programming language perspective, see for example [16].

Much of the interest in rewriting paradigms for programming arises from the possibility of a dual reading of a rewrite rule. On the one hand, a rule can be read as a syntactic transformation on a structure. On the other hand, a rule can be read as an equation. For example, the rule:

$$\mathit{fibs} = f(0,1) \quad \text{where} \quad f(m,n) = \mathit{Cons}(m, f(n, m+n))$$

can be read either as an equational definition of a structure which is the infinite list of Fibonacci numbers, or alternatively as instructions for a rewriting machine to construct increasingly better approximations to this infinite list. No real machine can compute the whole of an infinite structure, but by defining suitable finite selectors, we can write programs for rewriting machines which define finite structures in terms of infinite ones. Thus giving the command:

$$\mathit{print}(\mathit{nth}(15, \mathit{fibs}))$$

to a suitable rewriting machine will result in the printing of the 15th Fibonacci number. The rewriting machine has to be careful about how it uses the definitions if it is to achieve a result. From a purely rewriting perspective, the problem is to find a sequence of reductions which is normalising, for example the famous normal order reduction for the lambda calculus [4]. Solutions to this problem are the basis of lazy functional languages. Using implementations of such languages, it is possible to program by devising a suitable set of equations over infinite data structures which can be read as syntactic rewrite rules which deliver an effective means of computing the solution.

---

* Dedicated in friendship to Jan Willem Klop on the occasion of his 60th birthday.

A. Middeldorp et al. (Eds.): Processes... (Klop Festschrift), LNCS 3838, pp. 148–172, 2005.

However, it is rather easy to write down things that look as if they have both equational and rewrite interpretations, but which do not do what one might expect. Here is an example:

$$primesthenfibs = append(primes, fibs)$$

where *append* appends one list to another, and *primes* and *fibs* are the infinite lists of prime and Fibonacci numbers. One can write this program in a lazy functional language such as Haskell, but the result is just the infinite list of primes — the Fibonacci numbers disappear. The problem here is that the first list does not have an end to attach the second list to, so the *append* function seeks forever.

It is clear that some styles of building infinite terms can be computationally useful, whilst others are not. This raises an interesting question for the underlying theory of term rewriting, which is: what happens to various standard results for term rewriting if we allow infinite terms and infinite rewriting sequences, and what should those infinitary concepts be? Do the standard confluence and related results still hold for orthogonal infinitary systems?

We give an account of a theory of infinitary rewriting, beginning with the initial work done with and inspired by Jan Willem Klop, and ending with some recent work on lambda calculus which derives model theoretic notions from the kind of infinite terms which obstruct some traditional theorems of finitary rewriting.

2 Infinite Term Rewriting Systems

In this section we will introduce the basic concepts of infinite term and reduction sequence of transfinite length. We introduce the notion of a strongly convergent reduction sequences for the more general setting of abstract reduction systems. Then we will describe some of the basic theorems that hold for infinite extensions of term rewriting systems. Detailed proofs can be found in [7, 9, 10].

2.1 Infinite Terms

By interpreting finite terms as trees, infinite terms can be defined as trees having infinite branches as in Figure 1. There is a decision to be made about whether an infinite path in such a tree may be allowed to have a symbol at its end, which could then have further descendants, allowing paths from the root of a term to its leaves to have any ordinal length. We have taken the view that such terms have no computational meaning. Although we might imagine the limit of a reduction sequence $A \to B(A) \to B(B(A)) \to \ldots$ to be the term $B(B(B(...(A))))$ with infinitely many occurrences of B, there is no corresponding infinite process by which the symbol at the end of such a branch might be brought back up to the root.

We shall also require trees to be finitely branching, that is, that every operator symbol have finite arity. It is not clear whether allowing infinite arities

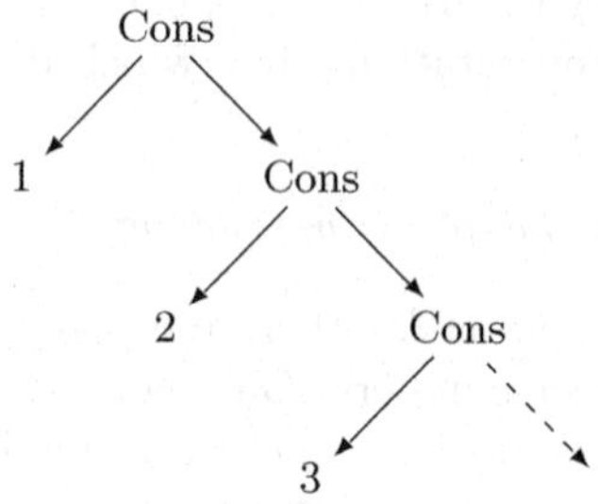

Fig. 1. An infinite term

would significantly change the resulting theory, but it would significantly complicate the exposition for little gain, and the restriction is reasonable on intuitive computational grounds.

Definition 1. *Let $\mathcal{V}$ be a set of variables and let $\Sigma = \bigcup \Sigma^n$ be a signature, i.e. a disjoint union of sets Σ^n of function symbols of arity $n \geq 0$. The set $\mathcal{T}^\infty(\Sigma, \mathcal{V})$ (abbreviated as $\mathcal{T}^\infty$) of finite and infinite terms is defined by coinduction from the grammar:*

$$t := x \mid f(t_1, \ldots, t_n) \text{ for } x \in X \text{ and } f \in \Sigma^n$$

To talk about subterms in a precise way, we introduce the concept of an address. An address is a finite sequence of positive integers $\alpha = i_1 \ldots i_n$, where n is called the *depth* of α. Given a term t, the subterm of t at α (if it exists) is denoted by $t|_\alpha$ and defined as follows. If α is empty then $t|_\alpha$ is t. If $\alpha = i\beta$, $t = f(t_1, \ldots, t_m)$, and $i \leq n$, then $t|_\alpha = t_i|_\beta$. The result of replacing the subterm of t at α by a term t' is denoted by $t[\alpha := t']$.

A term in $\mathcal{T}^\infty$ can equivalently be defined as a function from a set of addresses to function symbols and variables. The set of addresses must satisfy three conditions: it must be prefix-closed; if αi is in the set, so is αj for $j < i$; and for each α there is an upper bound to the i for which αi is in the set. This is Rosen's definition of a "tree domain" [18], generalised to infinite terms with finite arities. Variables may only occur at leaves, i.e. at addresses α for which no αi is in the set.

Note that addresses are always finite, even for infinite terms. In a term such as that of Figure 1, there is nothing at the end of the infinite branch, and no infinite address $222\ldots$. However, we do not require any sort of regularity or computability. Given the symbols *Cons*, 0, and 1, all infinite lists of 0 and 1 are included in the set of infinite terms, even lists which are not recursively enumerable.

2.2 Reduction on Infinite Terms

A term rewrite rule is defined as usual, that is, a pair of terms, written $t \to t'$, such that t is not a variable and contains every free variable of t'. We allow t' to be infinite, but require t to be finite. The rule is called *left-linear* if no variable occurs more than once in t.

A single reduction step is defined in the same way as for finitary term rewriting. A *substitution* is a function from a set of variables to terms. A substitution σ is applied to a term t by replacing every occurrence in t of every variable x in the domain of σ by $\sigma(x)$. The result is denoted by $\sigma(t)$. Given a rule $p \to q$ and a term t, if $t|_\alpha = \sigma(p)$ for some substitution σ of terms for variables, then t can be reduced to $t[\alpha := \sigma(q)]$. The concept of an orthogonal set of rules is also identical to that for finitary term rewriting: all rules must be left-linear, and for any two rules $p \to q$ and $r \to s$, there is no address α such that $p|_\alpha$ exists and is not a variable and r and $p|_\alpha$ are unifiable (excluding the trivial case where the two rules are the same rule and α is the empty address).

On computational grounds we might restrict the sets of function symbols and of rewrite rules to be finite, but none of our results depend on such a restriction.

2.3 Reduction Sequences of Transfinite Length

To define transfinite rewriting sequences, we must have some notion of the limit of a sequence of terms. The natural notion is one which arises from the standard metric on trees. For distinct terms t and t', define the *tree distance* $d(t,t') = 2^{-n}$, where n is the length of the shortest address α for which t and t' have different symbols at α, or for which α is in the tree domain of one but not the other. For example, $d(x,y) = 1$, $d(A(B),A(C)) = \frac{1}{2}$, and $d(A(B(C)),A(B)) = \frac{1}{4}$.

For an infinite rewriting sequence to be considered to converge to a limit, we might simply require that the sequence of its terms converge in the metric. This type of convergence (called *weak* or *Cauchy* convergence) was first substantially studied in [6].

Example 2. With the rules $A(x,y) \to A(y,x)$ and $B \to C$, the term $A(B,B)$ reduces to itself infinitely often. If one occurrence of B is reduced to C, the standard construction of a confluent diagram for finite orthogonal rewriting cannot be completed, because the bottom row of the diagram does not converge:

$$\begin{array}{ccccccccc} A(B,B) & \to & A(B,B) & \to & A(B,B) & \to & A(B,B) & \to \dots & A(B,B) \\ \downarrow & & \downarrow & & \downarrow & & \downarrow & & \downarrow \\ A(B,C) & \to & A(C,B) & \to & A(B,C) & \to & A(C,B) & \to \dots & ? \end{array}$$

Example 3. With the rule $I(x) \to x$, we can reduce $I^\omega = I(I(I(\dots)))$ to itself infinitely often, reducing at the root of the term each time:

$$I(I(I(\dots))) \to I(I(I(\dots))) \to I(I(I(\dots))) \to \dots \; I(I(I(\dots)))$$

But if we track the identity of the occurrences of I throughout the sequence, we observe something strange:

$$I_1(I_2(I_3(\dots))) \to I_2(I_3(I_4(\dots))) \to I_3(I_4(I_5(\dots))) \to \dots \; I_?(I_?(I_?(\dots)))$$

Every redex in the original term is reduced, yet we still have infinitely many in the final term, none of which derive from any part of the initial term.

Example 4. With the rule $A(x) \to A(B(x))$ we have the following convergent sequence:

$$A(C) \to A(B(C)) \to A(B(B(C))) \to \dots\ A(B(B(\dots)))$$

If we again track the identities of subterms, we see that an endless stream of Bs flows down from the root, but none of those occurring in the final term derive from them:

$$A_1(C) \to A_2(B_1(C)) \to A_3(B_2(B_1(C))) \to \dots\ A_?(B_?(B_?(\dots)))$$

Example 5. (Due to Simonsen [24].) With an infinite set of rules:

$$\begin{array}{rll} A \to B & & \\ F(G^k(C), x, y) \to F(G^{k+1}(C), y, y) & & \text{for even } k \\ F(G^k(C), x, y) \to F(G^{k+1}(C), A, y) & & \text{for odd } k \end{array}$$

we can construct the following weakly convergent reduction:

$$F(C, A, A) \to F(G(C), A, A) \to F(G(G(C)), A, A) \to \dots\ F(G^\omega, A, A)$$

We also have $F(C, A, A) \to F(C, A, B)$. However, $F(C, A, B)$ and $F(G^\omega, A, A)$ have no common reduct. Thus although the system is orthogonal, it is not confluent.

In all of these examples, what goes wrong is that although the terms of the sequence have larger and larger prefixes in common, rewriting always continues at the root of the term. In order to be able to relate the structure of the limiting term to the structures of the terms of the sequence, we require a stronger notion of convergence, according to which not only must the terms of the sequence converge, but the depths at which rewrites take place must increase without bound, so that larger and larger prefixes of the term remain "stable".

We can capture the essentials of the situation by considering abstract reduction systems equipped with a measure of the depth of a reduction.

Definition 6. *An* abstract reduction system *is a set $\mathcal{A}$ of objects called terms, and a function from a set $\mathcal{L}$ to $\mathcal{A} \times \mathcal{A}$. We write $a \xrightarrow{l} b$ if $l \in \mathcal{L}$ is mapped to (a, b), and call this a* reduction step. *Note that there can be more than one reduction step from a to b, of different sizes.*

A metric abstract reduction system *in addition has a metric on $\mathcal{A}$ and a measure of size s mapping $\mathcal{L}$ to positive real numbers.*

In such a system, a strongly convergent reduction sequence *of length α, for an ordinal α, consists of:*

1. *a sequence of terms t_β for all $\beta \leq \alpha$, and*
2. *for each $\beta < \alpha$, a reduction step $t_\beta \xrightarrow{s_\beta} t_{\beta+1}$,*

such that for every limit ordinal $\lambda \leq \alpha$, the sequence $\{s_\beta | \beta < \lambda\}$ tends to zero.

We write $t \to t'$ *for a single reduction step,* $t \twoheadrightarrow t'$ *for a finite sequence of reductions, and* $t \twoheadrightarrow\!\!\!\!\!\rightarrow t'$ *for a possibly infinite strongly convergent sequence.* $t \to^{\alpha} t'$ *denotes a strongly convergent sequence of length* α*.*

The equality relation generated by the transfinite rewrite relation is the equivalence closure of $\twoheadrightarrow\!\!\!\!\!\rightarrow$*, that is,* $(\twoheadrightarrow\!\!\!\!\!\rightarrow \cup \leftarrow\!\!\!\!\!\twoheadleftarrow)^*$*.*

A term rewriting system forms a metric abstract reduction system in an obvious way: the size of a reduction step is 2^{-d} where d is the length of the address of the redex, and the metric is the tree distance.

Metric abstract reduction systems on their own, however, have too little structure to produce interesting theorems. For that we depend on the term structure.

Example 7. With the rule $I(x) \to x$ and the term I^{ω} as in Example 3, we can reduce every other redex of the initial term, and obtain a limiting term whose subterms all arise from subterms of terms earlier in the sequence:

$$I_1(I_2(I_3(I_4(\ldots)))) \to I_2(I_3(I_4(I_5(\ldots)))) \to I_2(I_4(I_5(I_6(\ldots)))) \to$$

$$I_2(I_4(I_6(I_7(\ldots)))) \to^{\omega} I_2(I_4(I_6(I_8(\ldots))))$$

Example 8. With the rule $A \to B(A)$, we can generate an infinite term in a way similar to Example 4:

$$A_1 \to B_1(A_1) \to B_1(B_2(A_1)) \to^{\omega} B_1(B_2(B_3(\ldots)))$$

However in this case the place where reductions happen moves down the term instead of staying at the root.

The movement of reductions to deeper and deeper levels is the crucial property that allows the structure of the limiting term to be related to that of the earlier terms in the sequence.

Note that when a rewrite system is "top-terminating" (having no reduction sequences performing infinitely many reductions at the root), a condition introduced by Dershowitz *et al.* [6], weak convergence and strong convergence are equivalent. However, many systems of interest are not top-terminating.

2.4 Compression of Transfinite Sequences to Length ω

Once we have the concept of an infinite rewriting sequence that converges to a limit term, we cannot avoid opening the door to rewriting sequences of any ordinal length. If the limit term after ω steps contains redexes, we can continue to rewrite, to generate a reduction sequence of length $\omega + \omega$, ω^2, or longer.

From Example 3 we can see that a reduction of at least any countable ordinal length can be constructed. This is in fact the maximum: because arities and addresses are finite, there are no uncountably long strongly convergent sequences.

Theorem 9. *Every strongly convergent sequence has countable length.*

Proof. In a strongly convergent sequence, there can be only finitely many reductions of depth n, for any given finite n. Therefore the total number of steps must be countable. □

For left-linear systems we can prove a much stronger result, which helps to give computational meaning to sequences longer than ω: they are all equivalent to sequences of length at most ω.

Theorem 10 (Compression Lemma). *In a left-linear term rewriting system, for every strongly convergent sequence $t \to^{\alpha} t'$, there is a reduction from t to t' of length at most ω.*

Proof. This is proved by induction on α.

If $\alpha = \lambda + 1$ for a limit ordinal λ, then the redex reduced by the final step must, by strong convergence and the finiteness of left hand sides, already exist at some point before λ. One can show that it is possible to reduce it at such an earlier point, and to carry out the remainder of the original reduction sequence in no more than λ steps. By repetition, this proves the theorem for $\lambda + n$ for all finite n.

If α is a limit ordinal greater than ω, then we proceed by considering the minimum depth d of any step in the sequence. One can reorder the sequence so as to perform all of the steps at depth d within some finite initial segment of an equivalent sequence no longer than α. The remainder of the sequence performs reductions only at depth at least $d + 1$. Repeating the argument generates a sequence consisting of at most ω finite subsequences, in which the nth subsequence performs reductions only at depth at least n. This sequence must converge to the limit of the original sequence. □

Example 11. The Compression Lemma does not hold for weakly convergent reduction in left-linear systems. Consider the rules $G(x, B) \to G(F(x), B)$ and $B \to C$. $G(A, B)$ reduces by weakly convergent reduction to $G(F^\omega, C)$ in $\omega + 1$ steps but not in any smaller number:

$$G(A, B) \to G(F(A), B) \to G(F(F(A)), B) \to^{\omega} G(F^\omega, B) \to G(F^\omega, C)$$

Example 12. The Compression Lemma does not hold for strongly converging reductions in non-left-linear systems. Consider the rules $A \to G(A)$, $B \to G(B)$, and $F(x, x) \to C$. Then $F(A, B) \to^{\omega} F(G^\omega, G^\omega) \to C$, but $F(A, B)$ does not reduce to C in fewer than $\omega + 1$ steps.

2.5 Confluence

One of the fundamental properties of finite rewriting in orthogonal systems is confluence. Surprisingly, this turns out to not quite hold for strongly convergent reductions. A limited version does hold, called the Strip Lemma.

Theorem 13 (Strip Lemma). *If $t_0 \to t_1$ and $t_0 \twoheadrightarrow t_2$, then for some t_3, $t_1 \twoheadrightarrow t_3$ and $t_2 \twoheadrightarrow t_3$.*

Proof. The proof is essentially the same as for finitary term rewriting. We consider the set of residuals of the redex $t_0 \to t_1$ in each term in the reduction of t_0 to t_2. Because the residuals of a subterm are always disjoint from each other (that is, none of them is a subterm of any other), each of these sets of residuals has a strongly convergent complete development. It is a straightforward matter to show that the resulting construction of a tiling diagram can be carried through, and that its bottom side is strongly convergent. □

But confluence fails.

Example 14. Consider the rules $A(x) \to x$ and $B(x) \to x$. In the term $A(B(A(B(\ldots))))$, if we reduce all of the A redexes, we obtain $B(B(B(\ldots)))$ but if we reduce all of the B redexes, we obtain $A(A(A(\ldots)))$. These two terms reduce only to themselves, and have no common reduct.

Example 15. By adding the rule $C \to A(B(C))$ to the previous example, we obtain an example in which all the terms in the two sequences except for the limiting terms are finite.

$$C \to A(B(C)) \to A(C) \to A(A(B(C))) \to A(A(C)) \to^{\omega} A(A(A(\ldots)))$$
$$C \to A(B(C)) \to B(C) \to B(A(B(C))) \to B(B(C)) \to^{\omega} B(B(B(\ldots)))$$

However, the situation is not lost. Examples similar to the above are essentially the only way in which an orthogonal transfinite term rewriting system can fail to be confluent.

Definition 16. *A* collapsing rule *is a rewrite rule whose right hand side is a variable. A* hyper-collapsing term *is a term whose every reduct reduces to a redex of a collapsing rule. A* collapsing tower *is a term of the form* $t_1[\alpha_1 := t_2[\alpha_2 := t_3[\alpha_3 := \ldots]]]$, *where each term* $t_i[\alpha_i := x]$ *is a redex of a collapsing rule* $t \to x$ *such that* $t|_{\alpha_i} = x$.

Theorem 17. *Strongly convergent reduction in an orthogonal term rewriting system is confluent if and only if it contains at most one collapsing rule, and the left hand side of that rule contains only one variable.*

We can also prove restricted versions of confluence for systems not covered by the above theorem, to the effect that if collapsing towers do not arise in the construction of a particular tiling diagram, its construction can be completed. For details we refer to [8].

The types of orthogonal rewriting system that are used to model functional languages almost always contain multiple collapsing rules, for example, to implement selectors for data structures:

$$Head(Cons(x, y)) \to x \qquad Tail(Cons(x, y)) \to y$$

These rules immediately give counterexamples to confluence like that of Example 14.

Instead of proving exact confluence for restricted situations, we can prove approximate versions of confluence for all orthogonal systems. Such theorems can be found by further consideration of the meaning of hyper-collapsing terms.

2.6 A More General Way of Restoring Confluence

The collapsing towers which obstruct confluence do not have an obvious meaning. In domain-theoretic terms, a term such as I^ω with the rewrite rule $I(x) \to x$ suggests the least fixed point of the identity function, which is undefined. The same is true of all the hyper-collapsing terms. If we regard these terms as meaningless, and identify them all with each other, it turns out that the confluence property is recovered for orthogonal systems.

Definition 18. *Given a class of terms $\mathcal{U}$, rewriting is* confluent modulo $\mathcal{U}$ *if, whenever $t_0 \twoheadleftarrow t_1 \xleftrightarrow{\mathcal{U}} t_2 \twoheadrightarrow t_3$, there exist t_4 and t_5 such that $t_0 \twoheadrightarrow t_4 \xleftrightarrow{\mathcal{U}} t_5 \twoheadleftarrow t_3$.*

Theorem 19. *An orthogonal term rewriting system is confluent modulo $\mathcal{HC}$.*

The next theorem assures us that the identification of all hyper-collapsing terms with each other introduces no new equalities, since they are already provably equal.

Theorem 20. *Any two hyper-collapsing terms t and t' are interconvertible. Specifically, there exist terms t'', s, and s' such that $t \twoheadrightarrow s \twoheadleftarrow t'' \twoheadrightarrow s' \twoheadleftarrow t'$.*

Proof. Since t and t' are hyper-collapsing, they reduce to collapsing towers $C_0[C_1[C_2[\ldots]]]$ and $D_0[D_1[D_2[\ldots]]]$. The term $C_0[D_0[C_1[D_1[C_2[D_2[\ldots]]]$ reduces to each of these towers. □

2.7 Axiomatic Treatment of Undefinedness

Theorem 19 was proved for orthogonal term rewriting systems in [8], but later work has shown that it does not depend on the details of this particular set of terms. Instead, we can state a set of axioms which any set of "undefined" terms should satisfy, and derive confluence modulo undefinedness from these axioms. Some preliminary definitions are necessary:

Definition 21. *For any set $\mathcal{U}$ of terms, define $t \xleftrightarrow{\mathcal{U}} t'$ if t' can be obtained from t by replacing some (finite or infinite) set of subterms of t in $\mathcal{U}$ by terms in $\mathcal{U}$. The transitive closure of $\xleftrightarrow{\mathcal{U}}$ is denoted by $\stackrel{\mathcal{U}}{=}$.*

Let t contain a redex by a rule $p \to q$ at address α, and a subterm at address β. That subterm overlaps *the redex if $\beta = \alpha\gamma$ for some nonempty γ such that $p|_\gamma$ exists and is not a variable.*

A term t is root-active *if every reduct of t can be reduced to a redex. The set of root-active terms is denoted by $\mathcal{RA}$.*

Definition 22. *A set $\mathcal{U}$ satisfying the following axioms will be called* a set of undefined terms.

1. *Closure. For all $s \twoheadrightarrow t$, $s \in \mathcal{U}$ if and only if $t \in \mathcal{U}$.*
2. *Overlap. If t is a redex, and some subterm of t overlapping the redex is in $\mathcal{U}$, then $t \in \mathcal{U}$.*

3. *Activeness.* $\mathcal{U}$ *includes* $\mathcal{RA}$.
4. *Indiscernability.* *If* $t \stackrel{\mathcal{U}}{\longleftrightarrow} t'$ *then* $t \in \mathcal{U}$ *if and only if* $t' \in \mathcal{U}$.

These axioms were first stated in [10], except that we have here strengthened the Closure axiom, which originally required only that $\mathcal{U}$ be closed under reduction. This extra condition ensures that the Compression Lemma continues to hold for an extended form of reduction we shall introduce in Section 2.8. In most cases, the Indiscernability axiom is the only axiom requiring any significant effort to prove. An equivalent way of stating it is that the $\stackrel{\mathcal{U}}{\longleftrightarrow}$ and $\stackrel{\mathcal{U}}{=}$ relations are identical.

In [10, 7] it is proved that for any set $\mathcal{U}$ satisfying enough of these axioms, transfinite reduction is confluent modulo $\mathcal{U}$, and also possesses the following genericity property:

Definition 23. *Call a term* totally meaningful *if none of its subterms is in* $\mathcal{U}$. $\mathcal{U}$ *is* generic *if for every* $s \in \mathcal{U}$ *and every term* t, *if* $t[x := s]$ *reduces to a totally meaningful term* t', *then for every term* r, $t[x := r]$ *also reduces to* t'.

Theorem 24. *In an orthogonal sytem, if* $\mathcal{U}$ *satisfies all the axioms except possibly Activeness, and includes* $\mathcal{HC}$, *then reduction is confluent modulo* $\mathcal{U}$. *If* $\mathcal{U}$ *satisfies Closure and Overlap, it is generic.*

The root-active terms are themselves a class satisfying all the axioms, and the hyper-collapsing terms satisfy all but the Activeness axiom. In the next section we will give some other concrete examples.

2.8 Syntactic Domain Models from Sets of Undefined Terms

Another way of looking at the concept of reduction modulo undefinedness is to identify all undefined terms with each other by introducing a new symbol $\bot$. Terms which may contain $\bot$ are called *partial terms*, and form the set $\mathcal{T}_\bot^\infty$. $\mathcal{T}_\bot^\infty$ is partially ordered by the prefix order $\preceq$, defined as the least partial order for which $\bot$ is the bottom element and all the function symbols are monotonic. The rewrite relation of the original rewrite system $\mathcal{R}$ extends immediately to partial terms. A set $\mathcal{U}$ of undefined terms can be extended to a set $\mathcal{U}_\bot \subseteq \mathcal{T}_\bot^\infty$ by defining $t \in \mathcal{U}_\bot$ if there is a way of replacing all occurrences of $\bot$ in t by terms in $\mathcal{U}$ to obtain a term in $\mathcal{U}$. (Note that by the Indiscernability property, if one such substitution yields a term in $\mathcal{U}$, then every substitution does.) We then add an additional rule called $\bot_\mathcal{U}$-reduction, which allows any undefined subterm to be replaced by $\bot$. Let $\mathcal{R}_{\bot_\mathcal{U}}^\infty$ denote this extension of the original system $\mathcal{R}$. We write $\rightarrow_\mathcal{R}$ for rewriting by the original rules, $\rightarrow_{\bot_\mathcal{U}}$ for rewriting by the new rule, and $\rightarrow_{\mathcal{R}\bot_\mathcal{U}}$ for the combination.

For any set $\mathcal{U}$ of undefined terms in an orthogonal term rewriting system, the following statements hold.

1. $\mathcal{R}_{\bot_\mathcal{U}}^\infty$ satisfies the Compression Lemma.
2. $\mathcal{R}_{\bot_\mathcal{U}}^\infty$ is confluent.

3. $\bot_{\mathcal{U}}$-reductions can always be postponed after ordinary reductions. That is, if $t \twoheadrightarrow_{\mathcal{R}\bot_{\mathcal{U}}} t'$, then for some t'', $t \twoheadrightarrow_{\mathcal{R}} t'' \twoheadrightarrow_{\bot} t'$. This fact serves to connect reductions in the augmented system with plain reductions of ordinary terms.
4. Every term t has a unique normal form $\mathsf{NF}(t)$ by strongly converging $\twoheadrightarrow_{\mathcal{R}\bot_{\mathcal{U}}}$ reduction.

This allows the construction of models of a term rewriting system. The normal forms of $\mathcal{R}^{\infty}_{\bot_{\mathcal{U}}}$ are the values, with the semantics given by the mapping NF of terms to their unique normal forms.

The properties of this interpretation depend on the choice of $\mathcal{U}$. In some pathological cases the normal form function is not monotonic, and the set of normal forms may not be a complete partial order with respect to the prefix order. As a somewhat contrived counterexample, consider a term rewriting system with two unary symbols s and p, and no rewrite rules. Any set of the form $\{t\}$, where t is any term which is not a proper subterm of itself (as infinite terms can be) satisfies all of the axioms of undefinedness. (Closure, Overlap, and Activeness are trivial when there are no rewrite rules.) Take $\mathcal{U} = \{s(p^{\omega})\}$. Then $s(\bot)$ is a normal form for $\rightarrow_{\mathcal{R}\bot_{\mathcal{U}}}$, but $s(\bot) \preceq s(p^{\omega}) \rightarrow_{\bot_{\mathcal{U}}} \bot$, and so $\mathsf{NF}(s(\bot)) \not\preceq \mathsf{NF}(s(p^{\omega}))$. Furthermore, NF is not continuous at limit points, since every finite prefix of $s(p^{\omega})$ is its own normal form.

2.9 Sets of Undefined Terms

There may be many different sets of undefined terms in an orthogonal term rewriting system. The set of root-active terms is the smallest set of undefined terms. Trivially, the set of all terms is the largest. We call sets which are smaller than the set of all terms *consistent*. The intersection of any set of sets of undefined terms is a set of undefined terms. This does not necessarily hold for unions: Closure, Overlap, and Activeness all hold for unions, but Indiscernability may not.

An interesting set of undefined terms is the *opaque* terms. A term t is opaque if no reduct of that term can overlap any redex. This is proved to be a set of undefined terms in [10] for the axioms used there. Our stronger form of the Closure axiom can be ensured by extending the set to include every term that reduces to an opaque term, and we shall use this as our definition of opaqueness here. Orthogonality immediately implies that all root-active terms are opaque, but in general there are many others. For example, $\mathit{Head}(\mathit{Nil})$ is opaque in a term rewriting system with just the rule $\mathit{Head}(\mathit{Cons}(x, y)) \rightarrow x$.

Of more interest is the concrete term rewriting system for calculating Fibonacci numbers in Figure 2. The opaque terms in this TRS are the terms

$$\begin{array}{lll} 0 + y \rightarrow y & \mathit{nth}(0, y\!:\!z) \rightarrow y & \mathit{fibs} \rightarrow f(0, s(0)) \\ s(x) + y \rightarrow s(x + y) & \mathit{nth}(s(x), y\!:\!z) \rightarrow \mathit{nth}(x, z) & f(x, y) \rightarrow x\!:\!f(y, x + y) \end{array}$$

Fig. 2. The orthogonal Fibonacci TRS

that cannot reduce to any instance of 0, $s(x)$, or $x:y$. This includes root-active terms like $0+(0+(\dots))$, but also some normal forms such as $nth(0,0)$. The normal forms of $\twoheadrightarrow_{\mathcal{R}\bot_{opaque}}$ reduction are all the terms built from the constructor symbols, i.e. 0, s, and :, together with $\bot$.

3 Infinite Lambda Calculus

The theory of infinitary term rewriting can be extended to lambda calculus in a straightforward way, allowing us to prove confluence modulo a similar notion of undefined term, and to derive models from sets of undefined terms. The collection of all sets of undefined lambda terms turns out to be much richer than our original collection of three such sets in [9]. We will explain and extend some of the recent developments in [21, 22, 23].

3.1 Infinite λ-Terms

The concept of an infinite term can be defined for lambda calculus in the same way as for terms, interpreting application as a binary operator and λx as a unary operator for each x.

Definition 25. *The set of $\Lambda^{\infty}_{\bot}$ of finite and infinite λ-terms is defined by coinduction from the grammar:*

$$M ::= \bot \mid x \mid \lambda x.M \mid MM$$

The set Λ^{∞} consists of the terms in $\Lambda^{\infty}_{\bot}$ which do not contain $\bot$.

We ignore the identity of bound variables and do not distinguish alpha-equivalent terms, considering $(\lambda x.x)(\lambda x.x)$ and $(\lambda y.y)(\lambda z.z)$ to be the same term. In particular, the distance between two terms is defined to be the minimum tree distance between any members of their alpha-equivalence classes.

Definition 26. *We will need the following abbreviations of λ-terms:*

1. $\mathbf{\Delta} = \lambda x.xx$, $\mathbf{\Omega} = \mathbf{\Delta\Delta}$, $\mathbf{K} = \lambda x \lambda y.x$, $\mathbf{I} = \lambda x.x$ *and the fixed point combinator* $\mathbf{Y} = \lambda f.(\lambda x.f(xx))(\lambda x.f(xx))$
2. *The normal form of the fixed point* $\mathbf{YK}$ *of* $\mathbf{K}$ *is* $\mathbf{O} = \lambda x_1 \lambda x_2 \lambda x_3 \dots$*, also known as the* ogre.[1]

3.2 Reduction on Infinite λ-Terms

The rule of β-reduction extends in the obvious way to $\Lambda^{\infty}_{\bot}$. The concept of a strongly convergent reduction sequence in Definition 6 applies to the set $\Lambda^{\infty}_{\bot}$. Since beta reduction is a collapsing rule, it is not surprising that the confluence property fails, for the same reason it fails for term rewriting. In fact, even the

[1] Because it eats an unlimited number of arguments.

Strip Lemma fails. This is because in lambda calculus, unlike term rewriting, the residuals of a redex can be nested within each other, and in the Strip Lemma diagram, it is possible to find examples in which the set of residuals of the initial redex by an infinite sequence form a collapsing tower.

Example 27. We show a simple counterexample to the Strip Lemma which can be found in [2]. Define $W = \lambda x.\mathbf{I}(xx)$. Then the term $\mathbf{\Delta}W$ has a one-step reduction to $\mathbf{\Omega} = \mathbf{\Delta\Delta}$ and an infinite reduction to $\mathbf{I}(\mathbf{I}(\ldots)))$, namely

$$\mathbf{\Delta}W \to_\beta WW \to_\beta \mathbf{I}(WW) \to_\beta \mathbf{I}(\mathbf{I}(WW)) \twoheadrightarrow_\beta \mathbf{I}(\mathbf{I}(\mathbf{I}(\ldots)))$$

Both $\mathbf{\Delta\Delta}$ and $\mathbf{I}(\mathbf{I}(\mathbf{I}(\ldots)))$ reduce only to themselves, and have no common reduct. Note that both terms are examples of root-active terms in the sense of Definition 21 applied to lambda calculus.

Despite the counterexample there are several useful restricted forms of Strip Lemma [7]. For instance:

Theorem 28 ([7]). *If $M_0 \to_\beta M_1$ and $M_0 \twoheadrightarrow_\beta M_2$ then for some M_3, $M_1 \to_\beta M_3$ and $M_2 \twoheadrightarrow_\beta M_3$ provided $M_0 \to M_1$ is a head β-reduction.*

It is interesting to note that although root-active terms may not have common reducts, they are all interconvertible.

Theorem 29. *For every root-active term M, there is a term which reduces to both M and $\mathbf{I}^\omega$.*

Proof. For any term M, define $M^\mathbf{I}$ to be the term resulting from replacing every application PQ in M by $\mathbf{I}(PQ)$. Clearly $M^\mathbf{I} \twoheadrightarrow M$. We also have $(P[x := Q])^\mathbf{I} = P^\mathbf{I}[x := Q^\mathbf{I}]$ (which is immediate by considering the introduced copies of $\mathbf{I}$ as labels attached to the applications, and applying the technique of labelled reduction [17]). Hence also

$$((\lambda x.P)Q)^\mathbf{I} = \mathbf{I}(\lambda x.P^\mathbf{I})Q^\mathbf{I} \to \mathbf{I}(P^\mathbf{I}[x := Q^\mathbf{I}]) = \mathbf{I}(P[x := Q])^\mathbf{I} \to (P[x := Q])^\mathbf{I}$$

This lets us mimic for $M^\mathbf{I}$ any reduction of M: if $M \twoheadrightarrow M'$ then $M^\mathbf{I} \twoheadrightarrow M'^\mathbf{I}$.

If, however, we modify this construction by omitting the reduction of $\mathbf{I}$ whenever it occurs at the root, then we instead reduce $M^\mathbf{I}$ to $\mathbf{I}^n(M'^\mathbf{I})$ when the reduction of M to M' performs n reductions at the root. This transforms a reduction of M which performs infinitely many such reductions to a strongly convergent reduction of $M^\mathbf{I}$ to $\mathbf{I}^\omega$. □

Note that the proof of Theorem 20 does not work for lambda calculus, since a root-active term in lambda calculus (for example, $\mathbf{\Omega}$) need not be reducible to a collapsing tower. This is because in term rewriting, a reduction at the root cannot create new redexes, whereas in lambda calculus it can.

3.3 Undefinedness in Lambda Calculus

The remedies for the failure of confluence are the same as for term rewriting: we can identify a set of terms as undefined and define rewriting modulo this set,[2] or extend reduction with a $\bot$ rule reducing undefined terms to $\bot$, and prove that these forms of rewriting are confluent.

The Closure, Activeness, and Indiscernability axioms carry over unchanged. Note that since the β rule is a collapsing rule, all root-active terms are hyper-collapsing. The Overlap rule is also unchanged, but can be stated in a simpler and more explicit form. There is also an additional axiom requiring closure under substitution. This last axiom was not necessary for term rewriting, because the variables in a term behave like constant symbols, and are never substituted for by the process of reduction.

Definition 30. *A set $\mathcal{U} \subseteq \Lambda^\infty$ will be called a* set of undefined terms *if it satisfies the Axioms of Closure, Activeness and Indiscernability of Definition 22 and the following two axioms:*

1. *Overlap. If $(\lambda x.P) \in \mathcal{U}$ then $(\lambda x.P)Q \in \mathcal{U}$.*
2. *Substitution. $\mathcal{U}$ is closed under substitution.*

Now let $\mathcal{U}$ be a set of terms of Λ^∞ satisfying the axioms. We add the following rewrite rule:

$$\frac{M[\bot := \mathbf{\Omega}] \in \mathcal{U} \quad M \neq \bot}{M \to \bot} \; (\bot_\mathcal{U})$$

(Note that by Indiscernability, there is nothing special about the use of the term $\mathbf{\Omega}$ — any other member of $\mathcal{U}$ could be used.) The infinitary lambda calculus over $\Lambda^\infty_\bot$ with the β and $\bot_\mathcal{U}$ rules is denoted $\lambda^\infty_{\beta\bot_\mathcal{U}}$, and the combined reduction relation written simply $\to$. Reductions using only one or other of the rules will be denoted $\to_\beta$ or $\to_{\bot_\mathcal{U}}$

The $\bot_\mathcal{U}$ rule is of course not computable (since $\mathcal{U}$ is not recursively enumerable unless $\mathcal{U} = \Lambda^\infty$), but it provides a mathematically convenient way of talking about terms modulo undefinedness. The postponement property in the next theorem serves to connect reductions in $\lambda^\infty_{\beta\bot_\mathcal{U}}$ with plain beta reduction.

Theorem 31. *Let $\mathcal{U}$ be a set of undefined terms.*

1. *Strongly converging reduction in $\lambda^\infty_{\beta\bot_\mathcal{U}}$ is confluent.*
2. *Every term M has a strongly converging reduction to normal form, which by the first part is unique and will be denoted by $\mathsf{NF}_\mathcal{U}(M)$.*

---

[2] Recently Ketema and Simonsen [12, 13] have shown that strongly converging reduction is confluent modulo $\mathcal{HC}$ in fully-extended orthogonal infinitary combinatory term rewriting systems with rules with finite right hand sides. Since the notions root-active and hyper-collapsing coincide in lambda calculus (because the beta rule is hyper-collapsing) their result generalises our results on confluence modulo $\mathcal{HC}$ for orthogonal term rewriting and confluence modulo $\mathcal{RA}$ for lambda calculus with the beta rule.

3. *$\bot$-reduction can be postponed after β-reduction. That is, if $M \twoheadrightarrow N$, then for some term L, $M \twoheadrightarrow_\beta L \twoheadrightarrow_{\bot_\mathcal{U}} N$.*
4. *The Compression Lemma holds for strongly converging reduction in $\lambda^\infty_{\beta\bot_\mathcal{U}}$.*

Thus $\lambda^\infty_{\beta\bot_\mathcal{U}}$ is a complete (normalising and confluent) extension of the finite lambda calculus λ_β.

Theorem 32. *Let $\mathcal{U}$ be a set of undefined terms. For any term M in $\lambda^\infty_{\beta\bot_\mathcal{U}}$ we have $\mathsf{NF}_\mathcal{U}(M) = \bot$ iff $M[\bot := \mathbf{\Omega}] \in \mathcal{U}$.*

Proof. "If" is trivial. "Only if": suppose that for a term M in $\Lambda^\infty_\bot$ we have that its normal form in $\lambda^\infty_{\beta\bot_\mathcal{U}}$ is equal to $\bot$. Hence there is a reduction $M \twoheadrightarrow_{\beta\bot} \bot$. By Theorem 31 this factors as $M \twoheadrightarrow_\beta K \twoheadrightarrow_\bot \bot$. Hence $M[\bot := \mathbf{\Omega}] \twoheadrightarrow_\beta K[\bot := \mathbf{\Omega}] \twoheadrightarrow_\bot K \twoheadrightarrow_\bot \bot$. By definition of $\bot_\mathcal{U}$-reduction and indiscernability it follows that $K[\bot := \mathbf{\Omega}] \twoheadrightarrow_\bot \bot$ implies $K[\bot := \mathbf{\Omega}] \to_\bot \bot$. Hence $K[\bot := \mathbf{\Omega}] \in \mathcal{U}$. Since $\mathcal{U}$ is closed under β expansion we find that $M[\bot := \mathbf{\Omega}] \in \mathcal{U}$. □

3.4 Sets of Undefined Lambda Terms

In this section we will study the collection $\mathbf{U}$ of all sets of undefined terms. Since $\mathbf{U}$ is closed under intersections (though not under unions), it forms a complete lattice under set inclusion. The top and bottom elements are Λ^∞ and $\overline{\mathcal{TN}}$, and the meet operation is intersection. The join of a set of sets of undefined terms is the intersection of all sets of undefined terms containing their union.

Let us now give some concrete examples of such sets. For a while the following three sets were the only known sets of undefined lambda terms (cf. [8, 1, 10, 7]).

Definition 33. 1. *A term $M \in \Lambda^\infty$ is a* head normal form *if M is of the form $\lambda x_1 \ldots x_n.y P_1 \ldots P_k$. $\overline{\mathcal{HN}}$ is the set of terms without a finite β-reduction to head normal form.*
2. *A term $M \in \Lambda^\infty$ is a* weak head normal form *if M is a head normal form or $M = \lambda x.N$. $\overline{\mathcal{WN}}$ is the set of terms without a finite β-reduction to weak head normal form.*
3. *A term $M \in \Lambda^\infty$ is a* top normal form *if it is either a weak head normal for or an application (NP) if there is no Q such that $N \twoheadrightarrow_\beta \lambda x.Q$. $\overline{\mathcal{TN}}$ is the set of terms without a finite β-reduction to top normal form.*

Lemma 34. *$\overline{\mathcal{HN}}$, $\overline{\mathcal{WN}}$ and $\overline{\mathcal{TN}}$ satisfy the axioms for undefined terms.*

Proof. Apart from closure under expansion all the axioms have been proved to hold for $\overline{\mathcal{HN}}$, $\overline{\mathcal{WN}}$ and $\overline{\mathcal{TN}}$ in [10]. We show the expansion property for $\overline{\mathcal{HN}}$. Suppose N is a term in Λ^∞ without a β-reduction to head normal form. Suppose also that $M_1 \twoheadrightarrow_\beta N$ and M_1 has a head normal form M_2. Without loss of generality we may assume that there is a finite head reduction from M_1 to M_2. Repeated application of the Restricted Strip Lemma 28 then gives us a common reduct M_4 of both N and M_2. The term M_4 is a head normal form because it is a reduct of the head normal form M_2. This contradicts the assumption that N has no head normal form. Hence M_1 has no head normal form either. Closure under expansion for the other two sets can be proved in a similar way. □

If we now apply the Main Theorem 31 of the previous section to these three sets we find that $\lambda^\infty_{\beta\bot_{\overline{\mathcal{HN}}}}$, $\lambda^\infty_{\beta\bot_{\overline{\mathcal{WN}}}}$ and $\lambda^\infty_{\beta\bot_{\overline{\mathcal{TN}}}}$ are confluent and normalising extensions of finite lambda calculus. The normal form of a term M in $\lambda^\infty_{\beta\bot_{\overline{\mathcal{HN}}}}$ (respectively $\lambda^\infty_{\beta\bot_{\overline{\mathcal{WN}}}}$ and $\lambda^\infty_{\beta\bot_{\overline{\mathcal{TN}}}}$) corresponds to the Böhm tree (respectively the Lévy-Longo tree and Berarducci tree) of M. As a useful corollary of the confluence and normalisation property of $\lambda^\infty_{\beta\bot_{\overline{\mathcal{TN}}}}$ we obtain a useful refinement of the old observation of Wadsworth [4] that finite lambda terms are either of the form $\lambda x_1 \ldots \lambda x_n.yM_k \ldots M_1$ or $\lambda x_1 \ldots \lambda x_n.(\lambda y.P)QM_k \ldots M_1$ where $n, k \geq 0$.

Lemma 35 ([23]). *A term in $\Lambda^\infty_\bot$ has one of the following five forms:*

1. $\lambda x_1 \ldots \lambda x_n.yM_k \ldots M_1$
2. $\lambda x_1 \ldots \lambda x_n.(\lambda y.P)QM_k \ldots M_1$
3. $\lambda x_1 \ldots \lambda x_n.\bot M_k \ldots M_1$
4. $\lambda x_1 \ldots \lambda x_n.(((\ldots M_3)M_2)M_1$
5. $\lambda x_1 \lambda x_2 \lambda x_3 \ldots = \mathbf{O}$

Of course, the third option does not apply to terms in Λ^∞.

Now the key to constructing other sets of undefined terms lies in finding a definition of these sets in terms of what they include, rather than what they exclude [22, 23]. For doing so we need some terminology.

Definition 36. 1. A term $M \in \Lambda^\infty_\bot$ is *root-active* (with respect to β) if for all $M \twoheadrightarrow\!\!\!\!\twoheadrightarrow_\beta N$ there exists a redex $(\lambda x.P)Q$ such that $N \twoheadrightarrow_\beta (\lambda x.P)Q$.
2. A term $M \in \Lambda^\infty_\bot$ is a *head active form* if $M = \lambda x_1 \ldots x_n.RP_1 \ldots P_k$ and R is root-active.
3. A term $M \in \Lambda^\infty_\bot$ is a *strong active form* if $M = RP_1 \ldots P_k$ and R is root-active.
4. A term $M \in \Lambda^\infty_\bot$ is a *strong active form relative to X* if $M = RP_1 \ldots P_k$ and R is root-active and $P_1, \ldots, P_k \in X$.
5. A term $M \in \Lambda^\infty_\bot$ is an *infinite left spine form* if $M = \lambda x_1 \ldots x_n.((\ldots)P_2)P_1$.
6. A term $M \in \Lambda^\infty_\bot$ is a *strong infinite left spine form* if $M = ((\ldots)P_2)P_1$. A term $M \in \Lambda^\infty_\bot$ is a *strong infinite left spine form relative to X* if $M = ((\ldots)P_2)P_1$ and $P_i \in X$ for all i.

Example 37. 1. The term $\mathbf{\Omega}$ is a finite root-active term. The fixed point $\mathbf{YI}$ reduces to the infinite root-active term $\mathbf{I}(\mathbf{I}(\mathbf{I}(\ldots)))$.
2. $\mathbf{\Omega}xyz$, $(\mathbf{YI})xyz$ and $\mathbf{I}(\mathbf{I}(\mathbf{I}(\ldots)))xyz$ are strong active terms.
3. The finite term $\mathbf{\Omega}_3 = (\lambda x.xxx)(\lambda x.xxx)$ reduces to the strong infinite left spine form $((\ldots)\omega_3)\omega_3$, where $\omega_3 = \lambda x.xxx$.

We can now redefine the sets $\overline{\mathcal{HN}}$, $\overline{\mathcal{WN}}$ and $\overline{\mathcal{TN}}$.

Lemma 38. *1. A term $M \in \Lambda^\infty$ has no top normal form if and only if M is root-active.*
2. A term $M \in \Lambda^\infty$ has no weak head normal form if and only if M reduces to a strong head active form, or a strong infinite left spine form.
3. A term $M \in \Lambda^\infty$ has no head normal form if and only if M reduces to a head active form, an infinite left spine form, or the ogre.

The reformulation of the set $\overline{\mathcal{WN}}$ reveals that the terms without weak head normal form in the lambda calculus are precisely the strong zero terms, terms of which no instance can reduce to an abstraction. Strong zero terms can also be characterised as those terms no reduct of which can overlap any redex, which are exactly the terms that we called *opaque* in Section 2.9.

Before we can define a partition of Λ^∞ we need to define some notation.

Definition 39. We define the following subsets of Λ^∞.

$$\begin{aligned}
\mathcal{HA} &= \{M \in \Lambda^\infty \mid M \twoheadrightarrow_\beta N \text{ and } N \text{ is head active}\}\\
\mathcal{IL} &= \{M \in \Lambda^\infty \mid M \twoheadrightarrow_\beta N \text{ and } N \text{ is an infinite left spine form}\}\\
\mathcal{O} &= \{M \in \Lambda^\infty \mid M \twoheadrightarrow_\beta \mathbf{O}\}\\
\mathcal{RA} &= \{M \in \Lambda^\infty \mid M \text{ is root-active}\}\\
\mathcal{SA} &= \{M \in \Lambda^\infty \mid M \twoheadrightarrow_\beta N \text{ and } N \text{ is strong active}\}\\
\mathcal{SIL} &= \{M \in \Lambda^\infty \mid M \twoheadrightarrow_\beta N \text{ and } N \text{ is a strong infinite left spine form}\}
\end{aligned}$$

Theorem 40. *Λ^∞ is the disjoint union of $\mathcal{HN}$, $\mathcal{HA}$, $\mathcal{IL}$ and $\mathcal{O}$.*

With these components we can make the sets of undefined terms of Figure 3.

Theorem 41 ([23]). *All eight sets of Figure 3 are sets of undefined terms.*

Proof. The proofs for the three sets defined in Definition 33 can be found in [10]. The proofs for all other sets can be found in [23]. □

There are many more sets of undefined terms besides the eight depicted in Figure 3. Although we do not have a complete classification, we can say where these other sets can be found in relation to those eight sets. In the figure we use solid arrows $X \longrightarrow Y$ to express that $X \supset Y$ and there are NO other sets of undefined terms in between X and Y. Dashed arrows $X \dashrightarrow Y$ indicate that $X \supset Y$ and that there are at least $2^{\mathfrak{c}}$ many other sets of undefined terms in between X and Y, where $\mathfrak{c}$ is the cardinality of the continuum. To prove the correctness of these arrows we will first prove a useful lemma.

Lemma 42. *Let $\mathcal{U}$ be a set of undefined terms.*

1. *If $\lambda x.M \in \mathcal{U}$ then $M \in \mathcal{U}$.*
2. *If $\lambda x.M \in \mathcal{U}$ for some M then $\mathcal{HA} \subseteq \mathcal{U}$.*
3. *If $\mathbf{O} \in \mathcal{U}$ then $\mathcal{HA} \subseteq \mathcal{U}$.*
4. *If $\lambda x.M \in \mathcal{U}$ and $\mathcal{U} \subseteq \mathcal{SA} \cup \mathcal{SIL}$ then $\mathcal{U} \subseteq \mathcal{HA} \cup \mathcal{IL}$.*
5. *If $\mathcal{SIL} \subseteq \mathcal{U}$ then $\mathcal{SA} \subseteq \mathcal{U}$.*

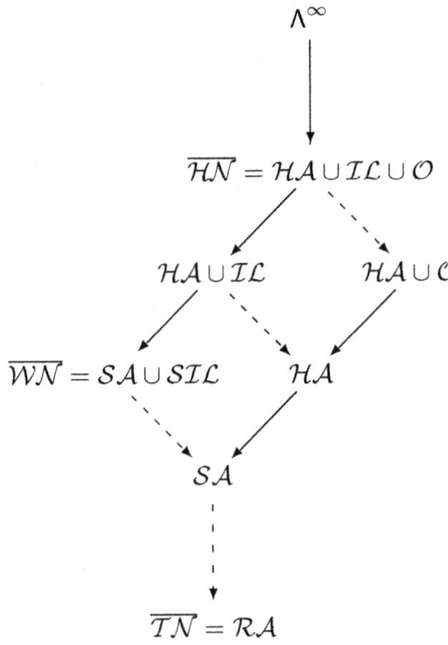

Fig. 3. The eight main sets of undefined terms

6. *If* $\mathcal{IL} \subseteq \mathcal{U}$ *then* $\mathcal{HA} \subseteq \mathcal{U}$.
7. *If a head normal form is in* $\mathcal{U}$ *then* $\mathcal{U} = \Lambda^\infty$.

Proof. Proofs as in [23]. These are straightforward deductions from the axioms: for example, to prove (1), $\lambda x.M \in \mathcal{U}$ implies $(\lambda x.M)x \in \mathcal{U}$ by indiscernability, therefore $M \in \mathcal{U}$ by Closure. □

Theorem 43. *The set* $\overline{\mathcal{HN}}$ *is the largest set of undefined terms which is a proper subset of* Λ^∞.

Proof. The first statement follows from Lemma 42(7). The second statement follows directly from the axioms of undefinedness. □

Definition 44. *Let* $A \subseteq B$ *be two sets of undefined* λ*-terms. The* (open) interval $\langle A, B\rangle$ *is the set* $\{C \mid A \subsetneq C \subsetneq B$ *and* C *is a set of undefined* λ*-terms*$\}$.

Theorem 45. *1. The interval* $\langle \mathcal{SA}, \mathcal{HA} \cup \mathcal{O}\rangle$ *contains only the element* $\mathcal{HA}$.
2. The interval $\langle \mathcal{SA} \cup \mathcal{SIL}, \mathcal{HA} \cup \mathcal{IL} \cup \mathcal{O}\rangle$ *contains only the element* $\mathcal{HA} \cup \mathcal{IL}$.

Proof. These follow from Lemma 42, parts (2) and (4) respectively. □

Theorems 43 and 45 imply that the solid arrows in Figure 3 are correctly drawn. From Theorem 45(2) it follows also that the collection of sets of undefined terms is not closed under unions. The union of the sets of undefined terms $\mathcal{SA} \cup \mathcal{SIL}$ and $\mathcal{HA}$ is $\mathcal{SIL} \cup \mathcal{HA}$, which is not a set of undefined terms. The next theorem will imply the correctness of the dashed arrows in Figure 3.

Definition 46. *Let $X \subset \Lambda^\infty$.*

1. *We say that a term M is a* strongly head active term relative to X *if M reduces to a term of the form $RX_1 \ldots X_n$, where $R \in \mathcal{RA}$ and $X_1, \ldots, X_n \in X$. We denote the set of strongly head active terms relative to X by $\mathcal{SA}_X$.*
2. *We say that a term M is a* strong infinite left spine term relative to X *if M reduces to a term of the form $\ldots X_3 X_2 X_1$ where all $X_i \in X$. We denote the set of strong infinite left spine terms relative to X by $\mathcal{SIL}_X$.*
3. *We say that a term M is an* almost strong infinite left spine term relative to X *if M reduces to a term of the form $\ldots X_3 X_2 X_1 N_k \ldots N_1$ where all $X_i \in X$ and the $N_i \in \Lambda^\infty_\perp$. We denote the set of almost strong infinite left spine terms relative to X by $\mathcal{SIL}^+_X$.*
4. *The set $\mathcal{IL}^+_X$ of* almost infinite left spine terms relative to X *is defined similarly.*

Lemma 47. *If X is a subset of closed normal forms in Λ^∞ then $\mathcal{SA}_X$ is a set of undefined terms. Moreover if X does not contain subterms which are infinite left spine forms. then also $\mathcal{SA}_X \cup \mathcal{SIL}_X$, $\mathcal{SA}_\cup \mathcal{SIL}^+_X$, $\mathcal{HA} \cup \mathcal{IL}^+_X$ and $\mathcal{HA} \cup \mathcal{IL}^+_X \cup \mathcal{O}$ are sets of undefined terms*

Proof. In [23] we have shown that under their respective conditions $\mathcal{SA}_X$ and $\mathcal{SA}_X \cup \mathcal{SIL}_X$ are sets of undefined terms. The proofs of the similar statements for the other three sets are similar. □

Theorem 48. *The cardinality of each of the open intervals $\langle \mathcal{RA}, \mathcal{SA} \rangle$, $\langle \mathcal{SA}, \mathcal{SA} \cup \mathcal{SIL} \rangle$, $\langle \mathcal{HA}, \mathcal{HA} \cup \mathcal{IL} \rangle$, and $\langle \mathcal{HA} \cup \mathcal{O}, \mathcal{HA} \cup \mathcal{IL} \cup \mathcal{O} \rangle$ is at least $2^\mathfrak{c}$.*

Proof. There are at least $2^\mathfrak{c}$ subsets X of closed normal forms in Λ^∞ that do not contain subterms which are infinite left spine forms. The sets $\mathcal{SA}_X$, $\mathcal{SA}_\cup \mathcal{SIL}^+_X$, $\mathcal{HA} \cup \mathcal{IL}^+_X$ and $\mathcal{HA} \cup \mathcal{IL}^+_X \cup \mathcal{O}$ are element of the respective intervals listed in the theorem. □

3.5 Normal form Models of the Lambda Calculus

By Theorem 31, each set $\mathcal{U}$ of undefined terms gives rise to a complete extension $\lambda^\infty_{\beta\perp\mathcal{U}}$ of the finite lambda calculus λ_β. From each $\lambda^\infty_{\beta\perp\mathcal{U}}$ we can construct a generalised Böhm model $\mathfrak{M}_\mathcal{U}$ of the finite lambda calculus as follows. As underlying set we take the set $\mathsf{NF}_\mathcal{U}(\Lambda^\infty_\perp)$ of normal forms of terms in $\lambda^\infty_{\beta\perp\mathcal{U}}$. Let $\mathsf{NF}_\mathcal{U} : \Lambda^\infty_\perp \to \mathsf{NF}_\mathcal{U}(\Lambda^\infty_\perp)$ be the function that maps each M in $\Lambda^\infty_\perp$ to its normal form. On $\mathsf{NF}_\mathcal{U}(\Lambda^\infty_\perp)$ we define application simply by:

$$\mathsf{NF}_\mathcal{U}(M_1) \bullet \mathsf{NF}_\mathcal{U}(M_2) = \mathsf{NF}_\mathcal{U}(M_1 M_2)$$

The applicative structure $\mathfrak{M}_\mathcal{U} = \langle \mathsf{NF}_\mathcal{U}, \bullet \rangle$ extends readily to a syntactic model of the finite lambda calculus along the lines of Definition 5.3.2 in [4]. The construction works because of normalisation and confluence properties of $\lambda^\infty_{\beta\perp\mathcal{U}}$.

We will call these models *normal form models.* The three well-known models of the Böhm trees [3, 4], the Lévy-Longo [15] trees and the Berarducci trees [5, 9] can be seen as examples of this construction and correspond respectively to $\mathfrak{M}_{\overline{\mathcal{HN}}}$, $\mathfrak{M}_{\overline{\mathcal{WN}}}$ and $\mathfrak{M}_{\overline{\mathcal{TN}}}$.[3] There are many different sets of undefined terms, and so there are also many different normal form models. Note that $\mathfrak{M}_{\Lambda_\perp^\infty}$ degenerates to the single element $\perp$. The construction provides non-trivial models for all other sets of undefined terms. We will now examine some properties of these models.

Definition 49. Let $M, N \in \Lambda_\perp^\infty$. We say that M is a prefix of N (we write $M \preceq N$) if M is obtained from N by replacing some subterms of N by $\perp$.

The pair $(\Lambda_\perp^\infty, \preceq)$ is an algebraic cpo and its compact elements are the finite λ-terms. As for term rewriting, the pair $(\mathsf{NF}_\mathcal{U}(\Lambda_\perp^\infty), \preceq)$ may not be a cpo:

Counterexample 50 ([22]). *Consider the term* $(((\ldots \mathbf{K})\mathbf{K})\mathbf{I}$. *The set* $\mathcal{U} = \mathcal{RA} \cup \{M \in \Lambda^\infty \mid M \twoheadrightarrow_\beta ((\ldots \mathbf{K})\mathbf{K})\mathbf{I})\}$ *is a set of undefined terms. The term* $((\ldots \mathbf{K})\mathbf{K})\mathbf{I}$ *is a redex in* $\lambda^\infty_{\beta\perp_\mathcal{U}}$ *but none of its prefixes* $(((\perp\mathbf{K})\ldots \mathbf{K})\mathbf{K})\mathbf{I}$ *contains a redex. Let X be the set of prefixes of* $((\ldots \mathbf{K})\mathbf{K})\mathbf{I}$. *Clearly* $\bigcup X \neq ((\ldots \mathbf{K})\mathbf{K})\mathbf{I}$. *Hence* $(\mathsf{NF}_\mathcal{U}, \preceq)$ *is not a cpo.*

Notwithstanding such counterexamples it is not hard to show that the eight main sets of undefined terms give rise to models whose underlying set is a cpo.

Theorem 51. $(\mathsf{NF}_\mathcal{U}, \preceq)$ *is a cpo for any $\mathcal{U}$ chosen from the main sets of undefined terms of Figure 3.*

Next we consider the properties continuity and monotony.

Definition 52. *1. Let $\sqsubseteq$ be a partial order on $\Lambda_\perp^\infty$. A function $F : \Lambda_\perp^\infty \to \Lambda_\perp^\infty$ is called* monotone *in $(\Lambda_\perp^\infty, \sqsubseteq)$, if $F(M) \sqsubseteq F(N)$ for all $M, N \in \Lambda_\perp^\infty$ such that $M \sqsubseteq N$.*
2. Let $(\Lambda_\perp^\infty, \sqsubseteq)$ be a cpo. A function $F : \Lambda_\perp^\infty \to \Lambda_\perp^\infty$ is called continuous *in $(\Lambda_\perp^\infty, \sqsubseteq)$, if $\bigcup_{i\in I} F(M_i) = F(\bigcup_{i \in I} M_i)$ for any directed set $\{M_i \mid i \in I\} \subseteq \Lambda_\perp^\infty$, where a subset X of $\Lambda_\perp^\infty$ is* directed *if for any two elements $M_1, M_2 \in X$ there exists an $M_3 \in X$ such that $M_1 \sqsubseteq M_3$ and $M_2 \sqsubseteq M_3$.*

The function $\mathsf{NF}_\mathcal{U} : \Lambda_\perp^\infty \to \Lambda_\perp^\infty$ is not always continuous, or even monotone:

Counterexample 53. The map $\mathsf{NF}_\mathcal{U} : \Lambda_\perp^\infty \to \Lambda_\perp^\infty$ is not continuous in the cpo $(\Lambda_\perp^\infty, \preceq)$ in the following cases:

1. Case $\mathcal{U} = \overline{\mathcal{TN}}$: the Berarducci trees are not monotone in $(\Lambda_\perp^\infty, \preceq)$. Take $M = \perp y$, $N = (\lambda x.\perp)y$. Then $M \preceq N$ but $\mathsf{NF}_{\overline{\mathcal{TN}}}(M) \not\preceq \mathsf{NF}_{\overline{\mathcal{TN}}}(N)$.
2. Case $\mathcal{U} = \mathcal{HA} \cup \mathcal{IL}$. Now $\mathsf{NF}_{\mathcal{HA}\cup\mathcal{IL}}$ is monotone but not continuous. This can be seen as follows. The infinite sequence of abstractions $\mathbf{O} = \lambda x_1 x_2 \ldots$ is in normal form but the truncations $\mathbf{O}^n = \lambda x_1 \ldots x_n.\perp$ reduce to $\perp$ for all n. Hence $\bigcup_{n\in\omega} \mathbf{O}^n = \mathbf{O} = \mathsf{NF}(\mathbf{O}) \neq \bigcup_{n \in \omega} \mathsf{NF}(\mathbf{O}^n) = \perp$.

[3] The concept of a Berarducci tree also applies to orthogonal term rewriting, since it is based on the concept of root-active term. Ketema asks in [11] whether the concepts of Böhm tree and Lévy-Longo tree also apply to term rewriting. Sections 2.9 and 3.4 answer this affirmatively for Lévy-Longo trees, because in lambda calculus, the opaque terms are exactly the terms without weak head normal form.

The prefix relation behaves well with respect to continuity only for the cases of Böhm and Lévy-Longo trees:

Theorem 54 ([22]). $\mathsf{NF}_{\mathcal{U}} : \Lambda^{\infty}_{\perp} \to \Lambda^{\infty}_{\perp}$ *is continuous in* $(\Lambda^{\infty}_{\perp}, \preceq)$ *if and only if* $\mathcal{U} = \overline{\mathcal{HN}}$ *or* $\mathcal{U} = \overline{\mathcal{WN}}$.

Recall that Barendregt's proof [4] of the fact that the Böhm trees form a model for lambda calculus depends heavily on continuity. The previous theorem implies that this proof technique does not generalise to models other than the Lévy-Longo model.

Theorem 55 ([23]). $\mathsf{NF}_{\mathcal{U}} : \Lambda^{\infty}_{\perp} \to \Lambda^{\infty}_{\perp}$ *is monotone in* $(\Lambda^{\infty}_{\perp}, \preceq)$ *for any* $\mathcal{U}$ *chosen among the following:* $\mathcal{SA}$, $\mathcal{HA}$, $\mathcal{HA} \cup \mathcal{O}$, $\mathcal{SA} \cup \mathcal{SIL}$, $\mathcal{HA} \cup \mathcal{IL}$ *and* $\mathcal{HA} \cup \mathcal{IL} \cup \mathcal{O}$.

3.6 Another Proof of Incompleteness of the Finite Lambda Calculus

In [23] we have shown that there are at least $2^{\mathfrak{c}}$ many sets $\mathcal{U}$ of undefined terms such that $\mathfrak{M}_{\mathcal{U}}$ cannot be ordered by a partial order with a least element and for which application and abstraction are monotone. The idea was to use sets of undefined terms of the form $\mathcal{U} = \mathcal{SA}_{X \cup \{\mathbf{O}\}}$ for suitable X. Here we will improve this result and use it to obtain another proof of incompleteness of the finite lambda calculus.

Definition 56. We say that $\langle \mathfrak{M}_{\mathcal{U}}, \sqsubseteq \rangle$ is a po$^{\bullet}$ model if $\sqsubseteq$ is a partial order on $\mathsf{NF}(\Lambda^{\infty}_{\perp})$ with a least element (which may be different from $\perp$), and application is monotone wrt $\sqsubseteq$, i.e. whenever $M_1 \sqsubseteq N_1$ and $M_2 \sqsubseteq N_2$ then $M_1 \bullet M_1 \sqsubseteq N_1 \bullet N_2$.

Theorem 57. *If* $\langle \mathfrak{M}_{\mathcal{U}}, \sqsubseteq \rangle$ *is a* po$^{\bullet}$ *model then:*

1. *Either* $\perp$ *is the least element of* $\sqsubseteq$ *and* $\perp P \to_{\perp} \perp$ *for all* $P \in \Lambda^{\infty}_{\perp}$, *or*
2. $\mathbf{O}$ *is the least element of* $\sqsubseteq$.

Proof. Suppose that $M \in \mathsf{NF}(\Lambda^{\infty}_{\perp})$ is the least element. Then $M \sqsubseteq \lambda x.M$ for some x free in M. If application is monotone then $M \bullet P \sqsubseteq (\lambda x.M) \bullet P =_{\mathsf{NF}} M$ and hence $MP =_{\mathsf{NF}} M$ for all P for all $P \in \mathsf{NF}(\Lambda^{\infty}_{\perp})$. Now either $M = \perp$ in which case $\perp P \to_{\perp} \perp$ for all $P \in \Lambda^{\infty}_{\perp}$. Or $M \neq \perp$ and then $Mx = M$ for all x. Hence M is the solution of the recursive equation $M = \lambda x.M$ and so $M = \mathbf{O}$. □

We can now strengthen Theorem 47 in [23]:

Theorem 58. *The interval* $\langle \mathcal{RA}, \mathcal{SA} \rangle$ *contains at least* $2^{\mathfrak{c}}$ *many sets* $\mathcal{U}$ *of undefined terms for which there exist no partial order such that* $\langle \mathfrak{M}_{\mathcal{U}}, \sqsubseteq \rangle$ *is a* po$^{\bullet}$ *model.*

Proof. Take a non-empty subset X of closed terms in $\mathsf{BerT}(\Lambda^{\infty}_{\perp})$ without $\perp$. Clearly there are $2^{\mathfrak{c}}$ many choices for this X. Let $\mathcal{U}$ be the set of terms in Λ^{∞} with a beta reduction (not necessarily finite) to a term of the form $RN_0N_0N_1N_1 \dots N_kN_k$ where $k \geq 0$, $R \in \mathcal{RA}$ and all $N_i \in X$.

Suppose there exists a partial order $\sqsubseteq$ on $\mathsf{NF}(\mathcal{U})$ such that $\langle \mathfrak{M}_{\mathcal{U}}, \sqsubseteq \rangle$ is a po$^\bullet$ model. By Theorem 57 we have that $\mathbf{O}$ is the least element of $\sqsubseteq$. Choose $M \in X$ such that $\mathbf{O} \neq M$. Then $\mathbf{\Omega} MM \in \mathcal{U}$. Consider also $\mathbf{\Omega OO}$. Since $\mathbf{O}$ is the least element wrt to $\sqsubseteq$ we have $\mathbf{O} \sqsubseteq M$.

On one hand, $\bot\mathbf{OO}$ and $\mathbf{\Omega}MM$ reduce both to $\bot$, as they are elements of $\mathcal{U}$. On the other hand, $\bot M\mathbf{O}$ does not reduce to $\bot$, because $\bot M\mathbf{O} \notin \mathcal{U}$. But $\bot = \bot\mathbf{OO} \sqsubseteq \bot M\mathbf{O} \sqsubseteq \bot MM = \bot$ implying that $\bot = \bot M\mathbf{O}$. Contradiction.

Hence there is no partial order such that $\langle \mathfrak{M}_{\mathcal{U}}, \sqsubseteq \rangle$ is a po$^\bullet$ model. □

For each model $\mathfrak{M}_{\mathcal{U}}$ there is a corresponding lambda theory, namely the collection of pairs of closed finite lambda terms with the same interpretation in the model. As a corollary we obtain an alternative proof for Salibra's theorem that any semantics of lambda calculus given in terms of a partially ordered model with a bottom element is incomplete.

Corollary 59 (Salibra [19]). *There are at least continuum many lambda theories that cannot be ordered with a* po$^\bullet$ *model.*

Proof. Restrict in the previous proof the collection X to closed finite normal forms in Λ. There are continuum many such X. Clearly for any two different such sets, the corresponding lambda theories are different. □

Salibra's proof is different. He considers first the enumerable lambda theory Π axiomatised by $\mathbf{\Omega}xx = \mathbf{\Omega}$ to prove with the help of a nice idea by Plotkin that any semantics of lambda calculus given in terms of po$^\bullet$-models with a bottom element is incomplete (cf. [19]). Then he uses a theorem by Visser [26, 4] to obtain a continuum of distinct unorderable enumerable lambda theories satisfying the conditions: $\mathbf{\Omega}xx = \mathbf{\Omega}$ and $\mathbf{\Omega}(\mathbf{\Omega}\mathbf{KI})\mathbf{\Omega} \neq \mathbf{\Omega}$. Note that in the proof of Theorem 58 none of the constructed models $\mathfrak{M}_{\mathcal{U}}$ is a model of Salibra's theory Π, because they do not validate $\mathbf{\Omega\Omega\Omega} = \mathbf{\Omega}$.

This section demonstrates that infinitary lambda calculus can be a convenient tool for proving facts about finite lambda calculus.

3.7 Extensional Infinite Lambda Calculus

Far less is known about *extensional* lambda calculus. The collection of normal form models of extensional lambda calculus is still waiting to be explored. Our earlier work [9] on infinite lambda calculus depended heavily on the Compression property, which does not hold for extensional lambda calculus. An anonymous referee of this paper suggested us an elegant counterexample, simpler than the one we gave in [9].

Counterexample 60. *Let* M *be* $\lambda x.(\lambda y.\mathbf{K}xy(\mathbf{K}xy(\ldots)))x$. *Then neither* M *nor its finite* β*-reducts contain any* η*-redexes. However,* M *can* β*-reduce in* ω *steps to* $\lambda x.(\lambda y.y(y(\ldots)))x$, *which can* η*-reduce further to* $(\lambda y.y(y(\ldots)))$. *This reduction clearly cannot be compressed to a shorter one.*

The transfinite tiling diagram used in [7] to prove confluence of $\lambda^{\infty}_{\beta\bot\mathcal{U}}$ opens the way to confluence proofs of $\lambda^{\infty}_{\beta\bot\mathcal{U}}$ extended with extensionality for certain $\mathcal{U}$. In [20, 21] we have shown confluence and normalisation of $\lambda^{\infty}_{\beta\bot\eta}$ and $\lambda^{\infty}_{\beta\bot\eta!}$ for $\mathcal{U} = \overline{\mathcal{HN}}$. Here $\eta!$ is a strengthened version of the η rule, defined with the help of the concept of strongly converging η-expansion:

$$\frac{x \notin FV(M)}{\lambda x.Mx \to M}(\eta) \qquad \frac{x \notin FV(M)}{M \to \lambda x.Mx}(\eta^{-1}) \qquad \frac{x \twoheadrightarrow_{\eta^{-1}} N \qquad x \notin FV(M)}{\lambda x.MN \to M}(\eta!)$$

In $\lambda^{\infty}_{\beta\bot\eta}$ and $\lambda^{\infty}_{\beta\bot\eta!}$ we also have that extensionality postpones over both β reduction and $\bot$ reduction. Despite the above counterexample against the Compression lemma, there is still a weaker form of compression: any strongly converging reduction in $\lambda^{\infty}_{\beta\bot\eta}$ and $\lambda^{\infty}_{\beta\bot\eta!}$ can be compressed to a strongly converging reduction of length at most $\omega + \omega$.

We are currently working to extend these results to other sets $\mathcal{U}$ and to other forms of extensionality.

4 Summary and Conclusions

The application of rewriting theory to functional languages leads naturally to the consideration of infinite rewriting sequences and their limits. Our theory of transfinite rewriting puts this intuitive concept on a sound footing through the concept of a strongly convergent rewriting sequence, of which the crucial property is that not only does the sequence of terms tend to a limit, but the sequence of redex positions tends to infinite depth.

This notion allows us to demonstrate some classical results for orthogonal systems. However, it transpires that the most important of these, confluence, fails in certain cases. These cases can be precisely characterised, and confluence can be re-established modulo the equality of the set of terms which obstruct exact confluence. The offending terms are of a form that can reasonably be viewed as representing infinite computations that produce no result, and in this sense are undefined.

Further consideration of this set of terms reveals that any set which satisfies certain natural axioms can serve as the class of undefined terms. Not only does confluence hold relative to any such class, but we can immediately construct a semantic model of the rewrite system in which the undefined terms are all mapped to the same element.

For the lambda calculus this has yielded a new uniform characterisation of several known models, and a construction of several classes of new ones.

References

1. Z. M. Ariola, J. R. Kennaway, J. W. Klop, M. R. Sleep, and F. J. de Vries. Syntactic definitions of undefined: On defining the undefined. In M. Hagiya and J. Mitchell, editors, *Proceedings of the 2nd International Symposium on Theoretical Aspects of Computer Software (TACS '94), Sendai*, volume 789 of *Lecture Notes in Computer Science*, pages 543–554. Springer-Verlag, 1994.

2. Z. M. Ariola and J. W. Klop. Cyclic lambda graph rewriting. In *Proceedings of the 8th IEEE Symposium on Logic in Computer Science*, pages 416–425, 1994.
3. H. P. Barendregt. The type free lambda calculus. In J. Barwise, editor, *Handbook of Mathematical Logic*, pages 1091–1132. North-Holland Publishing Company, Amsterdam, 1977.
4. H. P. Barendregt. *The Lambda Calculus: Its Syntax and Semantics.* North-Holland, Amsterdam, Revised edition, 1984.
5. A. Berarducci. Infinite λ-calculus and non-sensible models. In *Logic and algebra (Pontignano, 1994)*, pages 339–377. Dekker, New York, 1996.
6. N. Dershowitz, S. Kaplan, and D. A. Plaisted. Rewrite, rewrite, rewrite, rewrite, rewrite. *Theoretical Computer Science*, 83:71–96, 1991.
7. J. R. Kennaway and F. J. de Vries. Infinitary rewriting. In Terese [25], pages 668–711.
8. J. R. Kennaway, J. W. Klop, M. R. Sleep, and F. J. de Vries. Transfinite reductions in orthogonal term rewriting systems. *Information and Computation*, 119(1):18–38, 1995.
9. J. R. Kennaway, J. W. Klop, M. R. Sleep, and F. J. de Vries. Infinitary lambda calculus. *Theoretical Computer Science*, 175(1):93–125, 1997.
10. J. R. Kennaway, V. van Oostrom, and F. J. de Vries. Meaningless terms in rewriting. *Journal of Functional and Logic Programming*, 1:35 pp, 1999.
11. J. Ketema. Böhm-like trees. In V. van Oostrom, editor, *Proceedings of the 15th international conference on Rewriting Techniques and Applications (RTA '04)*, volume 3097 of *LNCS*, pages 233–248. Springer-Verlag, 2004.
12. J. Ketema and J. G. Simonsen. Infinitary combinatory reduction systems. In J. Giesl, editor, *Term Rewriting and Applications (RTA '05)*, volume 3467 of *LNCS*, pages 438–452. Springer-Verlag, 2005.
13. J. Ketema and J. G. Simonsen. On conflunece of infinitary combinatory reduction systems. In *Proceedings of the 12th international conference on Logic for programming Artificial Intelligence (LPAR '05)*, LNCS. Springer-Verlag, 2005. To appear.
14. J. van Leeuwen. *Handbook of Theoretical Computer Science, Volume B.* Elsevier, 1990.
15. G. Longo. Set-theoretical models of λ-calculus: theories, expansions, isomorphisms. *Annals of Pure and Applied Logic*, 24(2):153–188, 1983.
16. J. C. Mitchell. *Foundations for Programming Languages.* MIT Press, 1996.
17. V. van Oostrom and R. de Vrijer. Equivalence of reductions. In Terese [25], pages 301–474.
18. B. Rosen. Tree-manipulating systems and Church-Rosser theorems. *Journal of the Association for Computing Machinery*, 20:160–187, 1973.
19. A. Salibra. Topological incompleteness and order incompleteness of the lambda calculus. *ACM Transactions on Computational Logic*, 4(3):379–401, 2003. (Special Issue LICS 2001).
20. P. Severi and F. J. de Vries. A Lambda Calculus for D_∞. Technical Report TR-2002-28, University of Leicester, 2002.
21. P. Severi and F. J. de Vries. An extensional Böhm model. In *Proceedings of the 13th International Conference on Rewriting Techniques and Applications (RTA '02)*, volume 2378 of *LNCS*, pages 159–173. Springer-Verlag, 2002.
22. P. Severi and F. J. de Vries. Continuity and discontinuity in lambda calculus. In P. Urzyczyn, editor, *Typed Lambda Calculus and Applications (TLCA '05)*, volume 3461 of *LNCS*, pages 369–385. Springer-Verlag, 2005.

23. P. Severi and F. J. de Vries. Order structures on Böhm-like models. In L. Ong, editor, *Computer Science Logic (CSL '05)*, volume 3634 of *LNCS*, pages 103–118. Springer-Verlag, 2005.
24. J. G. Simonsen. On confluence and residuals in Cauchy convergent transfinite rewriting. *Inf. Proc. Letters*, 91:141–146, 2004.
25. Terese, editor. *Term Rewriting Systems*, volume 55 of *Cambridge Tracts in Theoretical Computer Science*. Cambridge University Press, 2003.
26. A. Visser. Numerations, λ-calculus and arithmetic. In J. R. Hindley and J. P. Seldin, editors, *To H.B. Curry: Essays on combinatory logic, lambda-calculus and formalism*, pages 259–284. Academic Press, New York and London, 1980.

Reducing Right-Hand Sides for Termination

Hans Zantema

Department of Computer Science, TU Eindhoven,
P.O. Box 513, 5600 MB, Eindhoven, The Netherlands
h.zantema@tue.nl

Abstract. We propose two transformations on term rewrite systems (TRSs) based on reducing right-hand sides: one related to the transformation order and a variant of dummy elimination. Under mild conditions we prove that the transformed system is terminating if and only if the original one is terminating. Both transformations are very easy to implement, and make it much easier to prove termination of some TRSs automatically.

Preface

Before introducing the technical contents of this paper first I want to spend some personal words. Several years ago, around 1990, I was looking for a new research area. At that time I was employed at Utrecht University, and among other things I was responsible for a seminar in algebraic specification. Only vaguely I was aware of the area of term rewriting providing a way for implementation of algebraic specifications. Just before that a nice booklet appeared, in Dutch, about term rewriting. I liked this booklet, and decided to use it for my seminar. This booklet appeared to be the course material of a course by Jan Willem Klop at the Free University in Amsterdam, only 40 kilometers from Utrecht. I heard that the group around Jan Willem Klop was active in research in term rewriting, and that they had meetings every two or three weeks around this research, called TeReSe: term rewriting seminar. Since I liked the topic as I learned it from the booklet, and still was looking for a new research area, I decided to follow these meetings. There I met Jan Willem Klop and the people of his group: Aart Middeldorp, Fer-Jan de Vries, Roel de Vrijer, Vincent van Oostrom and Femke van Raamsdonk. I liked the meetings and the pleasant atmosphere, and very naturally and smoothly inside this area I found challenges that happened to grow out to my own research topics.

Now fifteen years have been passed, and I may look back to (co-)authoring dozens of papers related to this area. Although I have never been a member of Jan Willem's group, I realize that in the way sketched above Jan Willem and his group have played a crucial role in my development as a scientist. I am very grateful for this. To mark one issue, on several places in the present paper the underlying theory is based on completions of diagrams as you may see from the pictures if you browse through the paper. For sure this way of completion of diagrams, preferably in the setting of abstract reduction systems, is inspired by the way Jan Willem propagated to do so in these TeReSe meetings long ago.

A. Middeldorp et al. (Eds.): Processes... (Klop Festschrift), LNCS 3838, pp. 173–197, 2005.

1 Introduction

Developing techniques for proving termination of TRSs is a challenging research area already for a long time. In recent years the emphasis in this area has shifted towards implementation: for new techniques to prove termination it is no longer sufficient that they can be used to prove termination of particular TRSs in theory, but also tools should be able to use these techniques to prove termination fully automatically. Several tools have been developed for this goal, and there is a yearly competition in which all of these tools are applied to an extensive set of examples (TPDB [20], the termination problem data base), and compared, see

`http://www.lri.fr/~marche/termination-competition/.`

In this paper we present two transformations on TRSs for which termination of the original TRS can be concluded from termination of the transformed TRS. Since these transformations are very easy to implement and proving termination of the transformed TRS by standard techniques is often much simpler than proving termination of the original TRS, they are very suitable to be used as preprocessing steps before using any of the tools.

Both transformations do not change left-hand sides, and reduce right-hand sides. In the first transformation, related to the *transformation ordering* [3] this is done by rewriting right hand sides using the same TRS. So here it is assumed that at least one right-hand side of a rule is not in normal form. In the second transformation, a variant of *dummy elimination* [8], the right-hand sides are decomposed with respect to a special symbol (a *dummy symbol*) that occurs in a right-hand side but in no left-hand side.

The technique of rewriting right-hand sides was considered before in [12], but there it was required that the whole TRS is non-overlapping (or a mild weakening of it), while our approach does not have such global restrictions. Our approach is based on the transformation ordering from [3] presented in a more abstract setting in [24]. The first approach to implement this was described in [17]. In order to use this technique for rewriting right-hand sides we had to adjust the underlying theory. In this paper all required theory is included.

For our present variant of dummy elimination the main theorem states that the original TRS is terminating if the transformed TRS is terminating, just as in [8]. However, in case of left-linearity we also have the converse, as we prove in this paper. Therefore the new variant is called *complete dummy elimination*, and is often stronger than the earlier version from [8].

For string rewriting the techniques described in this paper have been implemented in TORPA: Termination of Rewriting Proved Automatically [21], a tool developed by the author, which was the winner in the above mentioned competition in the category of string rewriting, both in 2004 and 2005.

For term rewriting the techniques described in this paper have been implemented in the tool TPA: Termination Proved Automatically, written by Adam Koprowski, [15]. In the above mentioned termination competition in 2005 this tool was third among 6 participants in the category of term rewriting, after AProVE [6] and TTT [13].

The organization of this paper is as follows. First in Section 2 the preliminaries are given, both for abstract rewriting and term rewriting. Then in Section 3 the theory and implementation of the technique of rewriting right-hand sides is presented. Next, Section 4 treats complete dummy elimination, first for term rewriting and then for string rewriting. To derive the result for string rewriting from the result for term rewriting in Subsection 4.2 we apply a general theorem. A TRS having no symbols of arity greater than one is transformed to an SRS simply by ignoring all parentheses and variable symbols. The theorem states that the TRS is terminating if and only if the SRS is terminating. Finally, in Section 5 we give some conclusions.

2 Preliminaries

2.1 Abstract Rewriting

In the following R, S and T are arbitrary binary relations on a fixed set. In the applications they will correspond to rewrite relations of TRSs. We write a dot symbol for relational composition, i.e., one has $t(R \cdot S)t'$ if and only if there exists a t'' such that tRt'' and $t''St'$. We write R^+ for the transitive closure of R and R^* for the reflexive transitive closure of R, and we write R^{-1} for the inverse of R. Further we write $R \subseteq S$ if tRt' implies tSt'. Clearly, if $R \subseteq S$ then $R \cdot T \subseteq S \cdot T$ and $T \cdot R \subseteq T \cdot S$.

Using these notations confluence of a relation R, written as $\mathsf{CR}(R)$, can be expressed shortly as $(R^{-1})^* \cdot R^* \subseteq R^* \cdot (R^{-1})^*$. Similarly, local confluence of a relation R, written as $\mathsf{WCR}(R)$, can be expressed as $R^{-1} \cdot R \subseteq R^* \cdot (R^{-1})^*$.

We write $\infty(t, R)$ if there exists an infinite sequence $tRt_1Rt_2Rt_3R\cdots$. Such an infinite sequence is called an *infinite R-reduction.* A relation R is called *terminating* on t, written as $\mathsf{SN}(t, R)$, if not $\infty(t, R)$. A relation R is called *terminating*, written as $\mathsf{SN}(R)$, if it is terminating on every t, i.e., no infinite R-reduction exists at all.

For a terminating relation R we can apply *induction on R*, i.e. if for all elements t we can prove

$$(\forall t' : (tRt' \Rightarrow P(t'))) \Rightarrow P(t)$$

then we may conclude that the property $P(t)$ holds for all t. The assumption $\forall t' : (tRt' \Rightarrow P(t'))$ is called the *induction hypothesis.*

We write R/S for $S^* \cdot R \cdot S^*$. For instance, $(R/S)^+$ describes a sequence of $R \cup S$-steps containing at least one R-step, so

$$(R/S)^+ \;=\; S^* \cdot R \cdot (R \cup S)^* \;=\; (R \cup S)^* \cdot R \cdot S^*.$$

2.2 Term Rewriting

Write $\mathsf{Var}(t)$ for the set of variables in a term t. A *rewrite rule* is a pair of terms (ℓ, r), written as $\ell \to r$, such that ℓ is not a variable and $\mathsf{Var}(r) \subseteq \mathsf{Var}(\ell)$. The terms ℓ, r are called the *left-hand side* (lhs) and the *right-hand side* (rhs) of the rule $\ell \to r$, respectively. A rule $\ell \to r$ is called *left-linear* if every variable occurs at most once in ℓ. A rule $\ell \to r$ is called *non-erasing* if $\mathsf{Var}(r) = \mathsf{Var}(\ell)$.

A *term rewrite system* (TRS) is defined to be a set of rewrite rules. A TRS is called left-linear if all its rules are left-linear. A TRS is called non-erasing if all its rules are non-erasing.

A term t rewrites to a term u w.r.t. a TRS $\mathcal{R}$, notation $t \to_{\mathcal{R}} u$, if there is a rule $\ell \to r$ in $\mathcal{R}$, a context C and a substitution σ such that $t = C[\ell\sigma]$ and $u = C[r\sigma]$. A TRS $\mathcal{R}$ is said to be terminating, confluent or locally confluent (notation: $\mathsf{SN}(\mathcal{R}), \mathsf{CR}(\mathcal{R}), \mathsf{WCR}(\mathcal{R})$) if the corresponding property holds for the binary relation $\to_{\mathcal{R}}$ on terms. Basic techniques to prove termination of TRSs include recursive path order [5] and polynomial interpretations [4]. More involved techniques in which TRSs are first transformed before basic techniques are applied include semantic labelling [23] and dependency pairs [1, 14, 11]. For an overview of techniques for proving termination of TRSs see [24]. For a general introduction to rewriting see [2, 18].

The TRS $\mathcal{E}mb$ is defined to consist of all rules of the shape $f(x_1, \ldots, x_n) \to x_i$. A rule $\ell \to r$ is called *self-embedding* if $r \to^*_{\mathcal{E}mb} \ell$. A TRS $\mathcal{R}$ is called *simply terminating* if $\mathcal{R} \cup \mathcal{E}mb$ is terminating. It is obvious that a TRS containing a self-embedding rule is not simply terminating. It is well-known ([22, 24]) that termination of a TRS can not be proved by recursive path order or polynomial interpretations if the TRS is not simply terminating.

Two non-variable terms t, u are said to have an overlap if there are substitutions σ, τ such that either $t'\sigma = u\tau$ for a non-variable subterm t' of t, or $t\sigma = u'\tau$ for a non-variable subterm u' of u. Two rules $\ell_1 \to r_1$ and $\ell_2 \to r_2$ are said to have a non-trivial overlap if either the rules are distinct and ℓ_1 and ℓ_2 have an overlap, or the rules are equal and there are substitutions σ, τ such that $t'\sigma = \ell_1\tau$ for a non-variable proper subterm t' of ℓ_1. Here properness is essential to exclude the trivial overlap caused by $\ell_1\sigma = \ell_1\tau$ for $\sigma = \tau$. It is well-known that $\mathsf{WCR}(\mathcal{R})$ holds if no two (possibly equal) rules of $\mathcal{R}$ have a non-trivial overlap.

3 Rewriting Right-Hand Sides

3.1 The Theory

Our theory of rewriting right-hand sides is based on a modification of a commutation property that was the basis of the transformation ordering, [3].

Lemma 1. *Let S, T be binary relations satisfying*

1. *$S \cup T$ is terminating,*
2. *T is locally confluent, and*
3. *$T^{-1} \cdot S \subseteq (S/T)^+ \cdot (T^{-1})^*$.*

Then $(T^{-1})^* \cdot (S/T)^+ \subseteq (S/T)^+ \cdot (T^{-1})^*$.

Proof. We prove by induction on $S \cup T$ that for every t the following holds:

Let $t'(T^{-1})^*t(S/T)^+w$. Then there exists w' satisfying $t'(S/T)^+w'\ (T^{-1})^*w$.

First observe that $(S/T)^+ = T^* \cdot S \cdot (S \cup T)^*$. Since $(S \cup T)^* = T^* \cup (S/T)^+$, we obtain u, v satisfying $tT^*uSv(T^* \cup (S/T)^+)w$. Since T is terminating by *1* and locally confluent by *2*, we have $\mathsf{CR}(T)$ by Newman's Lemma: T is confluent. Since tT^*t', tT^*u and $\mathsf{CR}(T)$ we obtain u' satisfying $t'T^*u'$ and uT^*u'. If $u' = u$ then we may choose $w' = w$ indeed satisfying $t'(S/T)^+w'(T^{-1})^*w$ and we are done. In the remaining case we have u'' satisfying $uTu''T^*u'$. Applying condition *3* to $u''T^{-1}uSv$ yields v'' satisfying $u''(S/T)^+v''(T^{-1})^*v$. Since tT^*uTu'' we may apply the induction hypothesis to u'', yielding v' satisfying $u'(S/T)^+v'(T^{-1})^*v''$. Now we have vT^*v' and either vT^*w or $v(S/T)^+w$. In the first case $\mathsf{CR}(T)$ yields w' satisfying $v'T^*w'(T^{-1})^*w$, in the second case the induction hypothesis applied to v yields w' satisfying $v'(S/T)^+w'(T^{-1})^*w$. In all cases we have $t'T^*u'(S/T)^+v'(T^* \cup (S/T)^+)w'$, so $t'(S/T)^+w'$, and wT^*w', and we are done. Summarized in a picture:

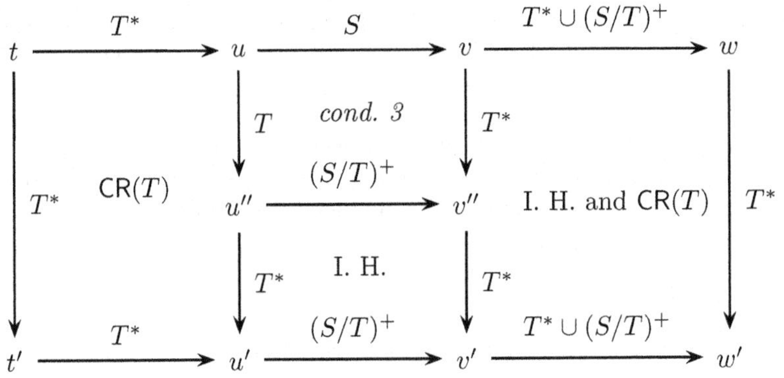

□

Theorem 2. *Let R, S, T be binary relations satisfying*

1. *$S \cup T$ is terminating,*
2. *T is locally confluent,*
3. *$T^{-1} \cdot S \subseteq (S/T)^+ \cdot (T^{-1})^*$, and*
4. *$R \subseteq ((S/T)^+ \cdot (T^{-1})^*) \cup T$.*

Then R is terminating.

Proof. Assume R is not terminating. So there is an infinite R-reduction, i.e., a sequence $t_1, t_2, t_3, \ldots$ such that $t_i R t_{i+1}$ for all $i = 1, 2, 3, \ldots$. Write $R' = (S/T)^+ \cdot (T^{-1})^*$. Let $u_1 = t_1$, and define u_i for $i = 2, 3, 4, \ldots$ satisfying $t_i T^* u_i$ in the following way:

- If $t_{i-1} T t_i$ then choose u_i such that $t_i T^* u_i$ and $u_{i-1} T^* u_i$. This can be done since T is confluent, following from Newman's Lemma and conditions *1*, *2*.
- Otherwise, by condition *4* we have $t_{i-1} R' t_i$, i.e., there exists v_i satisfying $t_{i-1}(S/T)^+ v_i (T^{-1})^* t_i$. By Lemma 1 we may choose u_i such that $u_{i-1}(S/T)^+ u_i$ and $v_i T^* u_i$, by which we obtain $t_i T^* u_i$.

A typical initial part of this construction is sketched in the following picture:

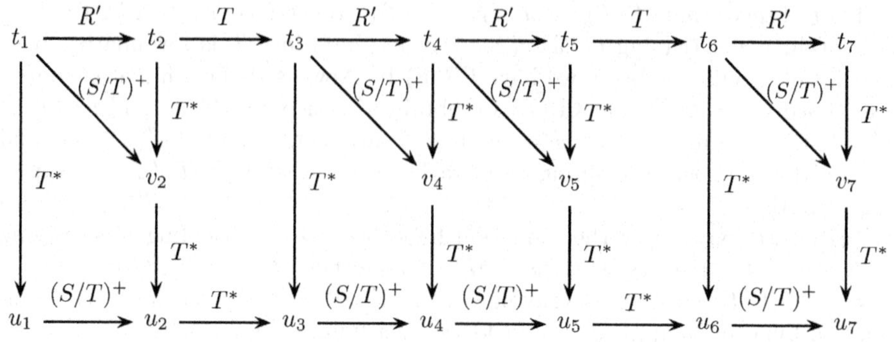

Since T is terminating by condition *1*, the second case $t_{i-1}R't_i$ occurs infinitely often. So we have $u_{i-1}(S/T)^+u_i$ for infinitely many values of i, while for the other values of i we have $u_{i-1}T^*u_i$. Hence $u_1 \to u_2 \to u_3 \to \cdots$ is an infinite $S \cup T$-reduction, contradicting condition *1*, concluding the proof. □

Theorem 2 is closely related to the underlying theory in the transformation ordering, [3]. A generalization of this underlying theory is expressed in Theorem 6.5.16 in [24]. In fact this Theorem 6.5.16 coincides with the present Theorem 2 where conditions *3*, *4* are replaced by

3'. $T^{-1} \cdot S \subseteq T^* \cdot S \cdot (S \cup T \cup T^{-1})^*$,
4'. $R \subseteq (S/(T \cup T^{-1}))^+$,

respectively. Condition *3'* is a strict weakening of condition *3*. However, conditions *4'* and *4* are incomparable, since in condition *4* it is allowed that some R-step is only a single T-step and in condition *4'* it is not. In our application this is essential, therefore this new Theorem 2 was developed rather than applying the earlier result. Later on, it was pointed out by Vincent van Oostrom that by applying Theorem 6.5.16 to $R \setminus T$ rather than R then a simple argument shows that not only $\mathsf{SN}(R \setminus T)$ can be concluded, but also $\mathsf{SN}(R)$. In this way a variant of Theorem 6.5.16 is obtained in which condition *4'* is weakened to $R \subseteq T \cup (S/(T \cup T^{-1}))^+$. Now our new Theorem 2 can be obtained as a corollary of this variant of Theorem 6.5.16. For being self-contained we kept the direct proof of Theorem 2; moreover this proof is slightly simpler than the proof of Theorem 6.5.16.

Next we apply Theorem 2 to term rewriting. In order to do so we first give a lemma analyzing how applications of non-overlapping rewrite rules commute, similar to the well-known critical pair lemma.

Lemma 3. *Let $\ell_i \to r_i$ be rewrite rules for $i = 1, 2$ for which ℓ_1 and ℓ_2 do not have an overlap, and having rewrite relations $\to_1, \to_2$, respectively. Let C be a context and σ, τ be substitutions such that $C[\ell_1\sigma] = \ell_2\tau$. Then $\ell_2 = C_2[x]$ for* *some context* C_2 *and some variable* x*, for which the two reducts* $C[r_1\sigma]$ *and* $r_2\tau$ *of* $C[\ell_1\sigma] = \ell_2\tau$ *satisfy*

$$C[r_1\sigma] \to_1^{n-1} \cdot \to_2 \cdot \leftarrow_1^k r_2\tau,$$

where n, k *are the numbers of occurrences of* x *in* ℓ_2, r_2*, respectively.*

Proof. Since there is no overlap between ℓ_1 and ℓ_2 we can write $C = C_2[D]$ where $\ell_2 = C_2[x]$ for contexts C_2, D and some variable x for which $x\tau = D[\ell_1\sigma]$. Define τ' by $x\tau' = D[r_1\sigma]$, and $y\tau' = y\tau$ for $y \neq x$. By applying the reduction $x\tau \to_1 x\tau'$ to the occurrences of $x\tau$ corresponding to the other $n-1$ occurrences of x in $\ell_2 = C_2[x]$, we obtain $C[r_1\sigma] \to_1^{n-1} \ell_2\tau'$. Conversely we obtain $r_2\tau \to_1^k r_2\tau'$ since x occurs k times in r_2. Combining these observations yields

$$C[r_1\sigma] \to_1^{n-1} \ell_2\tau' \to_2 r_2\tau' \leftarrow_1^k r_2\tau,$$

proving the lemma. □

To sketch a typical example of how Lemma 3 applies, consider $\ell_1 \to r_1$ to be the rule $a \to b$ and $\ell_2 \to r_2$ to be the rule $f(x,x) \to g(x,x,x)$. Let $C = f(\Box, a)$ and $x\tau = a$; since ℓ_1 does not contain variables, σ plays no role. Indeed we have $C[\ell_1\sigma] = f(a,a) = \ell_2\tau$, and

$$C[r_1\sigma] = f(b,a) \to_1 f(b,b) \to_2 g(b,b,b) \leftarrow_1^3 g(a,a,a) = r_2\tau.$$

Now we are ready to give the main theorem.

Theorem 4. *Let* $\mathcal{R}$ *be a TRS for which a rhs is not in normal form, i.e.,* $\mathcal{R}$ *contains a rule* $\ell \to r$ *and a rule of the shape* $\ell' \to C[\ell\sigma]$*. Assume that*

- $\ell \to r$ *is left-linear,*
- $\ell \to r$ *is non-erasing,*
- $\mathsf{WCR}(\{\ell \to r\})$*, and*
- *there is no overlap between* ℓ *and the lhs of any rule of* $\mathcal{R} \setminus \{\ell \to r\}$*.*

Let $\mathcal{R}'$ *be obtained from* $\mathcal{R}$ *by replacing the rule* $\ell' \to C[\ell\sigma]$ *by* $\ell' \to C[r\sigma]$*. Then* $\mathcal{R}$ *is terminating if and only if* $\mathcal{R}'$ *is terminating.*

Proof. The 'only if'-part is immediate from the observation that every $\mathcal{R}'$-step can be mimicked by one or two $\mathcal{R}$-steps: an $\mathcal{R}'$-step applying the rule $\ell' \to C[r\sigma]$ is mimicked by first applying the $\mathcal{R}$-rule $\ell' \to C[\ell\sigma]$ and then the $\mathcal{R}$-rule $\ell \to r$, all other $\mathcal{R}'$-steps are $\mathcal{R}$-steps themselves. It remains to prove the 'if'-part.

First note that the rules $\ell \to r$ and $\ell' \to C[\ell\sigma]$ are distinct, since otherwise the rule $\ell \to C[C[\ell\sigma]\sigma]$ contained in $\mathcal{R}'$ is not terminating.

We apply Theorem 2 for the binary relations

- T being the rewrite relation of the single rule $\ell \to r$,
- S being the rewrite relation of $\mathcal{R}' \setminus \{\ell \to r\}$, and
- R being the rewrite relation of the TRS $\mathcal{R}$.

Termination of the TRS $\mathcal{R}$ is proved by checking all four conditions of Theorem 2.

Condition *1* holds since $S \cup T$ is the rewrite relation of $\mathcal{R}'$ and $\mathcal{R}'$ is assumed to be terminating.

Condition *2* holds by assumption.

For proving condition *3* assume that $uT^{-1}tSv$, i.e., a term t rewrites by T to u and by S to v. We distinguish three cases:

– The T-redex is in parallel with the S-redex. Then we have

$$(u,v) \in S \cdot T^{-1} \subseteq (S/T)^+ \cdot (T^{-1})^*.$$

– The T-redex is above the S-redex. Then we apply Lemma 3 for $\ell_2 \to r_2$ being $\ell \to r$, so $\to_2 = T$, and $\to_1 \subseteq S$. This yields

$$v = C[r_1\sigma] \to_1^{n-1} \cdot \to_2 \cdot \leftarrow_1^k r_2\tau = u,$$

where n, k are the numbers of occurrences of x in ℓ, r, respectively. Since $\ell \to r$ is left-linear and non-erasing, we have $k > 0$ and $n = 1$. So

$$(v,u) \in \to_2 \cdot \leftarrow_1^+,$$

hence $(u,v) \in \to_1^+ \cdot \leftarrow_2 \subseteq S^+ \cdot T^{-1} \subseteq (S/T)^+ \cdot (T^{-1})^*$.

– The S-redex is above the T-redex. Then we apply Lemma 3 for $\ell_1 \to r_1$ being $\ell \to r$, so $\to_1 = T$, and $\to_2 \subseteq S$. This yields

$$u = C[r_1\sigma] \to_1^* \cdot \to_2 \cdot \leftarrow_1^* r_2\tau = v,$$

so $(u,v) \in \to_1^* \cdot \to_2 \cdot \leftarrow_1^* \subseteq T^* \cdot S \cdot (T^{-1})^* \subseteq (S/T)^+ \cdot (T^{-1})^*$.

In all cases we proved $(u,v) \in (S/T)^+ \cdot (T^{-1})^*$, concluding condition *3*.

Condition *4* is verified by considering all three possibilities for an $\mathcal{R}$-rewrite step $t \to u$.

– If $t \to u$ is an application of the rule $\ell \to r$ then tTu.
– If $t \to u$ is an application of the rule $\ell' \to C[\ell\sigma]$ then $tS \cdot T^{-1}u$ where the S-step is an application of the rule $\ell' \to C[r\sigma]$.
– If $t \to u$ is an application of another rule, then this rule is in $\mathcal{R}' \setminus \{\ell \to r\}$, so tSu.

In all three cases we conclude $(t,u) \in (S \cdot (T^{-1})^*) \cup T \subseteq ((S/T)^+ \cdot (T^{-1})^*) \cup T$, concluding condition *4*. □

In the dependency pair framework [1, 14, 11] it often occurs that proving innermost termination is simpler than proving termination, and therefore conditions have been investigated for which termination can be concluded from innermost termination. In a similar way Theorem 4 can be seen as a theorem stating that full termination can be concluded from termination with respect to a particular strategy: R'-rewriting can be seen as R-rewriting following the strategy that the application of a rule for which the rhs is not in normal form is always followed by reduction of this rhs.

One may wonder whether all conditions of Theorem 4 are essential. Indeed they are, as is shown by the following four examples.

In the first example ([12], Example 5) let $\mathcal{R}$ consist of the two rules

$$f(x) \to a, \quad b \to f(b).$$

Let $\ell \to r$ be the first rule, applicable to the rhs of the second rule. Then $\mathcal{R}'$ consisting of the rules $f(x) \to a$, $b \to a$ is terminating, while $\mathcal{R}$ is not, and all conditions of Theorem 4 hold except for non-erasingness of $\ell \to r$.

In the second example let $\mathcal{R}$ consist of the two rules

$$a \to b, \quad a \to a.$$

Let $\ell \to r$ be the first rule, applicable to the rhs of the second rule. Then $\mathcal{R}'$ consisting of two copies of the rule $a \to b$ is terminating, while $\mathcal{R}$ is not, and all conditions of Theorem 4 hold except for the non-overlappingness condition.

In the third example let $\mathcal{R}$ consist of the three rules

$$f(f(x)) \to g(x), \quad h(x) \to f(f(x)), \quad g(f(a)) \to h(h(a)).$$

Let $\ell \to r$ be the first rule, applicable to the rhs of the second rule. Then $\mathcal{R}'$ consisting of the three rules

$$f(f(x)) \to g(x), \quad h(x) \to g(x), \quad g(f(a)) \to h(h(a))$$

is terminating by recursive path order using the precedence $f > h > g$. However, $\mathcal{R}$ is not terminating due to the reduction

$$h(h(a)) \to h(f(f(a))) \to f(f(f(f(a)))) \to f(g(f(a))) \to f(h(h(a))).$$

All conditions of Theorem 4 hold except for $\mathsf{WCR}(\{\ell \to r\})$.

In the last example let $\mathcal{R}$ consist of the four rules

$$f(x, x) \to g(x), \quad a \to b, \quad a \to c, \quad f(b, c) \to f(a, a).$$

Let $\ell \to r$ be the first rule, applicable to the rhs of the last rule. Then $\mathcal{R}'$ is terminating, while $\mathcal{R}$ is not due to the reduction $f(b, c) \to f(a, a) \to f(a, c) \to f(b, c)$. All conditions of Theorem 4 hold except for left-linearity of $\ell \to r$.

A possible generalization of Theorem 4 would be in weakening the restriction of non-overlap to a restriction of critical pairs having a particular kind of common reduct. Moreover, even in case this critical pair condition does not hold one can think of extending T by the normalized versions of the corresponding critical pairs, introducing a kind of completion as in [3, 17]. However, in this paper we want to concentrate on very simple criteria not involving branching choices as is introduced in searching for common reducts of critical pairs in typically non-confluent TRSs.

A variant of Theorem 4 was given by Gramlich in [12]:

Theorem 5. *Let $\mathcal{R}$ be a non-overlapping TRS for which a rhs is not in normal form, i.e., $\mathcal{R}$ contains a rule $\ell \to r$ and a rule of the shape $\ell' \to C[\ell\sigma]$. Assume that $\ell \to r$ is non-erasing. Let $\mathcal{R}'$ be obtained from $\mathcal{R}$ by replacing the rule $\ell' \to C[\ell\sigma]$ by $\ell' \to C[r\sigma]$. Then $\mathcal{R}$ is terminating if and only if $\mathcal{R}'$ is terminating.*

So here we do not have a left-linearity requirement for $\ell \to r$ any more, but the full TRS $\mathcal{R}$ is required to be non-overlapping, while our Theorem 4 only requires non-overlappingness involving the rule $\ell \to r$. In typical applications to TRSs describing arithmetic and having a rule $p(s(x)) \to x$ to be applied to some rhs, Theorem 5 is only applicable if the TRS is non-overlapping. So this approach fails as soon as overlapping combinations of usual rules like

$$x - 0 \to x, \quad s(x) - s(y) \to x - y, \quad x - x \to 0$$

occur. In fact, in Gramlich's paper the requirement of non-overlappingness is slightly weakened, but not overcoming these drawbacks. Our Theorem 4 still applies directly if the TRS contains rules of this shape. Therefore we think that in practice our Theorem 4 is more powerful than Theorem 5.

3.2 Implementation

We propose to use Theorem 4 as a pre-processing phase for any tool for proving termination of TRSs as follows. Let $\mathcal{R}$ be any finite TRS for which termination has to be proved.

Basic procedure:

Check if $\mathcal{R}$ can be written as

$$\mathcal{R} = \mathcal{R}_0 \cup \{\ell \to r, \ \ell' \to C[\ell\sigma]\}$$

where $\ell \to r$ is left-linear and non-erasing, and has no non-trivial overlap with any rule of R.

If so, then replace $\mathcal{R}$ by $\mathcal{R}_0 \cup \{\ell \to r, \ \ell' \to C[r\sigma]\}$, and start again.

Applying this basic procedure is straightforward. From Theorem 4 it follows that for every step the replaced TRS is terminating if and only if the original TRS is terminating; note that local confluence of the single rule $\ell \to r$ follows from the property that $\ell \to r$ has no non-trivial overlap with itself. So for any number of steps the resulting TRS is terminating if and only if the original TRS is terminating.

In case $\mathcal{R}$ is terminating, then the basic procedure is terminating too. This can be seen as follows. Assume that the procedure goes on forever, respectively yielding TRSs $\mathcal{R}_1 = \mathcal{R}, \mathcal{R}_2, \mathcal{R}_3, \mathcal{R}_4, \ldots$. Since every $\mathcal{R}_{i+1}$-step can be mimicked by one or two $\mathcal{R}_i$-steps, as we saw in the 'only if'-part of the proof of Theorem 4, we conclude that $\to_{\mathcal{R}_i} \subseteq \to_{\mathcal{R}}^{+}$ for every $i = 1, 2, 3, \ldots$. Since $\mathcal{R}_{i+1}$ is obtained from $\mathcal{R}_i$ by applying $\to_{\mathcal{R}_i}$ to a rhs of $\mathcal{R}_i$, we can also obtain $\mathcal{R}_{i+1}$ from $\mathcal{R}_i$ by applying $\rightarrow_{\mathcal{R}}^{+}$ to one of the rhs's. Since there are only finitely many rules, but infinitely many steps from $\mathcal{R}_i$ to $\mathcal{R}_{i+1}$, there is some rhs of the original TRS $\mathcal{R}$ on which $\rightarrow_{\mathcal{R}}^{+}$ is applied infinitely often, contradicting termination of $\mathcal{R}$.

Unfortunately, in case $\mathcal{R}$ is not terminating then it can be the case that the basic procedure does not terminate. For instance, if $\mathcal{R}$ consists of the two rules $a \rightarrow a$, $b \rightarrow a$ then the basic procedure can be applied again yielding $\mathcal{R}$ after one step. This process may go on forever. More general, if $\mathcal{R}$ is of the shape $\mathcal{R}_0 \cup \{\ell \rightarrow r\}$ in which r admits an infinite $\mathcal{R}_0$-reduction, then the basic procedure applied to $\mathcal{R}$ may go on forever. A simple way to get a robust implementation of the basic procedure that terminates on every TRS is to put some upper bound on the total number of steps of the basic procedure.

String rewriting can be seen as a particular case of term rewriting in which all symbols have arity 1. Since in this case variables and parentheses are redundant, they are usually omitted.

A tool for automatically proving termination of string rewriting is called TORPA: Termination of Rewriting Proved Automatically [21]. This tool has been developed by the author. After having ideas in mind for years the actual implementation started in July 2003. Earlier versions of TORPA have been described in [25] (version 1.1), in [26] (version 1.2) and in [27] (version 1.3). The extensive paper [27] also contains a full treatment of all the underlying theory.

Our basic procedure for rewriting right-hand sides was implemented for string rewriting in the newest version of TORPA, version 1.4. This version of the TORPA tool participated in the termination competition in 2005, and was the winner among the eight participants in the string rewriting category, see

`http://www.lri.fr/~marche/termination-competition/2005/.`

In the text generated by TORPA our technique is called *transformation order*. As an example we consider the string rewriting system consisting of the following five rules

$$f0 \rightarrow s0, \quad d0 \rightarrow 0, \quad ds \rightarrow ssdps, \quad fs \rightarrow dfps, \quad ps \rightarrow e,$$

where e represents the empty string. This system describes computation of powers of 2: think of s being successor, p being predecessor, d being doubling and f being exponentiation.[1] It is easy to observe that fs^n0 rewrites to its normal form $s^{2^n}0$ for every $n = 0, 1, 2, \ldots$. The normal form of f^n0 has super-exponential size and requires a super-exponential number of steps to be computed. Note that the system is not simply terminating: both the third and the fourth rule are self-embedding. TORPA yields the following termination proof:

[1] The fact that the constant 0 may be treated here as a unary symbol will be justified by Theorem 12.

```
TORPA 1.4 is applied to the string rewriting system
f 0  -> s 0
d 0  -> 0
d s  -> s s d p s
f s  -> d f p s
p s  -> e
Choose polynomial interpretation f: lambda x.x+1, rest identity
remove: f 0  -> s 0
Remaining rules:
    d 0  -> 0
    d s  -> s s d p s
    f s  -> d f p s
    p s  -> e

Transformation order: apply rule 4 on rhs of rule 2, result:
d 0  -> 0
d s  -> s s d
f s  -> d f p s
p s  -> e

Transformation order: apply rule 4 on rhs of rule 3, result:
d 0  -> 0
d s  -> s s d
f s  -> d f
p s  -> e
%

Choose polynomial interpretation p: lambda x.x+1, rest identity
remove: p s  -> e
Remaining rules:
    d 0  -> 0
    d s  -> s s d
    f s  -> d f

Terminating by recursive path order with precedence:
  d>s  f>d
```

For term rewriting our basic procedure has been implemented in the tool TPA, written by Adam Koprowski, [15].

Let $\mathcal{R}$ be any TRS and let $\mathcal{R}'$ be the result of applying the basic procedure to $\mathcal{R}$. From many examples we observe that proving termination of $\mathcal{R}'$ is much simpler than proving termination of $\mathcal{R}$ directly. We should like to have evidence that it is never the other way around. Since the notion of 'simpler' depends on the unspecified set of techniques to be used, it is hard to make this claim solid. However, by construction we have $\to_{\mathcal{R}'} \subseteq \to_{\mathcal{R}}^+$, and the lhs's of $\mathcal{R}$ are equal to the lhs's of $\mathcal{R}'$. Under these conditions for all techniques known by us it is very unlikely that proving termination of $\mathcal{R}'$ may be harder than proving termination of $\mathcal{R}$. In particular, since $\to_{\mathcal{R}' \cup \mathcal{E}mb} \subseteq \to_{\mathcal{R} \cup \mathcal{E}mb}^+$, simple termination of $\mathcal{R}$ implies simple termination of $\mathcal{R}'$.

Therefore applying our basic procedure as a pre-processing before trying any other tool for proving termination will often increase the power of the tool, and probably never decrease it.

One may wonder whether it is natural to have rhs's that are not in normal form. Of course this is hard to answer since there is no precise definition of *natural*. To our knowledge the most extensive list of termination problems in term rewriting is TPDB, the Termination Problem Data Base [20]. This database was used in the above mentioned competition. Restricted to term rewriting (excluding string rewriting, which is a separate category) it contains 773 TRSs for which the problem of termination has been posed. They are from a wide scala of origins and application areas. Therefore we think it makes sense to consider these TRSs to get an impression of the applicability of our technique. It turns out that among these 773 TRSs there are 98 TRSs for which not only a rhs is not in normal form, but also the extra conditions are satisfied. So for these TRSs our basic procedure is applicable. For several of them, proving termination of the transformed system is much simpler than proving termination of the original system. Of course again the meaning of 'simpler' has not been defined precisely, but it sounds reasonable to consider a proof only using basic techniques like recursive path order and linear polynomial interpretations, simpler than a proof using a combination of dependency pairs, argument filtering and the same basic techniques.

For instance, consider the classical TRS describing computation of factorials consisting of the following rules

$$\begin{aligned} p(s(x)) &\to x \\ \mathit{fact}(0) &\to s(0) \\ \mathit{fact}(s(x)) &\to *(s(x), \mathit{fact}(p(s(x)))) \end{aligned} \qquad \begin{aligned} *(0, y) &\to 0 \\ *(s(x), y) &\to +(*(x, y), y) \\ +(x, 0) &\to x \\ +(x, s(y)) &\to s(+(x, y)). \end{aligned}$$

This TRS is `D33/21.trs` in the TRS category of TPDB. In the 2005 competition only two (AProVE [6] and TTT [13]) of the six participating tools were able to prove termination of this TRS. Note that the TRS is not simply terminating since the rule $\mathit{fact}(s(x)) \to *(s(x), \mathit{fact}(p(s(x))))$ is self-embedding, so only using recursive path order and polynomial interpretations will fail. However, by applying our basic procedure this self-embedding rule is replaced by $\mathit{fact}(s(x)) \to *(s(x), \mathit{fact}(x))$, which is replaced again by $\mathit{fact}(s(x)) \to +(*(x, \mathit{fact}(x)), \mathit{fact}(x))$. Termination of the resulting TRS is easily concluded by the recursive path order using the precedence $\mathit{fact} > * > + > s$. This is exactly the proof as it is found automatically by the tool TPA within a fraction of a second.

Using our basic procedure, the tool TPA was able to prove termination of 6 more TRSs in TPDB than without it, including this factorial system.

4 Complete Dummy Elimination

In the basic procedure based on Theorem 4 rhs's are rewritten. So a part of such an rhs matches with an lhs. In this section we consider the opposite: we consider rhs's containing symbols that do not occur in lhs's at all. These symbols are called *dummy symbols.* Again we keep the lhs's and reduce the rhs's, but this reduction is done completely different than before: the dummy symbol now is used to split up the rhs into several smaller rhs's, each generating a rule in the transformed TRS, with its lhs kept unchanged. This approach was studied before in [8] and was called *dummy elimination.* An earlier version already appeared in [22]. The main theorem states that if the TRS after applying dummy elimination is terminating, then the original TRS is terminating too. In general the converse is not true. Here we present a modification of dummy elimination for which the same property holds, but for which in case of left-linearity also the converse holds (the transformed TRS is terminating if and only if the original TRS is). Due to this completeness result our new variant is called *complete dummy elimination.* Moreover, complete dummy elimination is more powerful to be used in tools for automatically proving termination of TRSs or SRSs.

4.1 Complete Dummy Elimination for Term Rewriting

Before giving precise definitions first we give a very simple example sketching the general idea. Consider the TRS $\mathcal{R}$ consisting of the single rule

$$f(g(x)) \;\rightarrow\; f(a(g(x))).$$

Here the symbol a is a dummy symbol: it does not occur in any lhs. Intuitively this means that this dummy symbol does not play an essential role in further reductions of the term, and further reductions can be localized as either affecting the part above the dummy symbol or affecting the part below it. This can be formalized by decomposing the rhs's into smaller terms in which the dummy acts as a separator. In this case this means that the term $f(h(g(x)))$ is decomposed into two terms $f(\diamond)$ and $b(g(x))$, where $\diamond$ is a fresh constant and b is a fresh unary symbol. The lhs's remain the same. The result is the transformed system $DE_a(\mathcal{R})$, in this example consisting of the two rules

$$\begin{array}{l} f(g(x)) \rightarrow f(\diamond) \\ f(g(x)) \rightarrow b(g(x)). \end{array}$$

The main result states that $DE_a(\mathcal{R})$ is terminating if and only if $\mathcal{R}$ is terminating. So termination of $\mathcal{R}$ can be proved by proving termination of $DE_a(\mathcal{R})$, which is straightforward by recursive path order choosing the precedence $f > b$, $g > \diamond$.

In order to give a precise definition for complete dummy elimination we need some auxiliary definitions. We fix one dummy symbol a of a TRS $\mathcal{R}$. Let n be the arity of a. Choose a fresh constant $\diamond_a$ and a fresh unary symbol b_a, i.e., $\diamond_a$ and b_a do not occur in $\mathcal{R}$. As long as a is fixed, we omit the subscripts, simply writing b and $\diamond$. For any term t we define inductively a term $\mathsf{cap}_a(t)$ and a set of terms $\mathsf{dec}_a(t)$:

$$\begin{array}{lll} \mathsf{cap}_a(x) & = x & \text{for all } x \in \mathsf{Var}, \\ \mathsf{cap}_a(f(t_1,\ldots,t_k)) & = f(\mathsf{cap}_a(t_1),\ldots,\mathsf{cap}_a(t_k)) & \text{for all } f,\ f \neq a \\ \mathsf{cap}_a(a(t_1,\ldots,t_n)) & = \diamond & \end{array}$$

$$\begin{array}{lll}\mathsf{dec}_a(x) & = \emptyset & \text{for all } x \in \mathsf{Var},\\ \mathsf{dec}_a(f(t_1,\ldots,t_k)) & = \bigcup_{i=1}^{k} \mathsf{dec}_a(t_i) & \text{for all } f,\ f \neq a\\ \mathsf{dec}_a(a(t_1,\ldots,t_n)) & = \bigcup_{i=1}^{n} (\mathsf{dec}_a(t_i) \cup \{b(\mathsf{cap}_a(t_i))\}). & \end{array}$$

Roughly speaking we decompose a term t by using the symbol a as a separator, where occurrences of a are replaced by $\diamond$ and arguments of a are marked by the symbol b. Now the term $\mathsf{cap}_a(t)$ is the topmost part of this decomposition, while $\mathsf{dec}_a(t)$ is the set of all other parts in this decomposition. Now we define the TRS $DE_a(\mathcal{R})$ for any TRS $\mathcal{R}$ having a as a dummy symbol by

$$DE_a(\mathcal{R}) = \{\ell \to u \mid u = \mathsf{cap}_a(r) \vee u \in \mathsf{dec}_a(r) \text{ for a rule } \ell \to r \in \mathcal{R}\}.$$

The transformation DE_a is called *complete dummy elimination.* For instance, applying DE_a on a rule of the shape

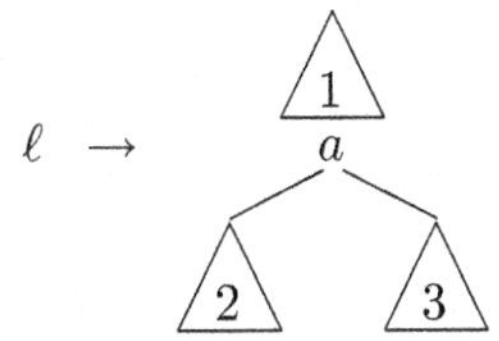

where the binary dummy symbol a occurs only once in the rhs, yields the three rules

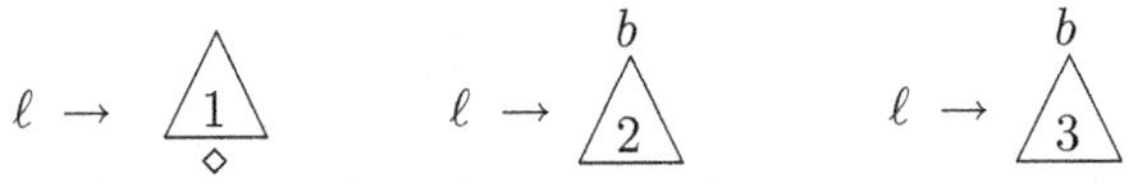

Theorem 6. *Let a be a dummy symbol in a TRS $\mathcal{R}$ for which $DE_a(\mathcal{R})$ is terminating. Then $\mathcal{R}$ is terminating too.*

Before proving this theorem we give an example slightly more complicated than the one given above, and we recall the earlier dummy elimination theorem. Let the TRS $\mathcal{R}$ consist of the two rules

$$\begin{array}{l} f(g(x)) \to f(a(g(a(x,f(x))),g(f(x))))\\ g(f(x)) \to g(g(a(f(x),g(g(x))))). \end{array}$$

Then $DE_a(\mathcal{R})$ consists of the rules

$$\begin{array}{ll} f(g(x)) \to f(\diamond) & f(g(x)) \to b(g(f(x)))\\ f(g(x)) \to b(g(\diamond)) & g(f(x)) \to g(g(\diamond))\\ f(g(x)) \to b(x) & g(f(x)) \to b(f(x))\\ f(g(x)) \to b(f(x)) & g(f(x)) \to b(g(g(x))). \end{array}$$

Indeed Theorem 6 is helpful for proving termination of $\mathcal{R}$: termination of $DE_a(\mathcal{R})$ is easily proved by recursive path order, choosing the precedence $f > g > b > \diamond$.

In the version of dummy elimination from [8] the symbol b was omitted. More precisely, for a TRS $\mathcal{R}$ having a as a dummy symbol the TRS $E(\mathcal{R})$ was defined exactly as $DE_a(\mathcal{R})$, with the only difference that $\mathsf{dec}_a(a(t_1,\ldots,t_n))$ was defined to be $\bigcup_{i=1}^n(\mathsf{dec}_a(t_i)\cup\{\mathsf{cap}_a(t_i)\})$ rather than $\bigcup_{i=1}^n(\mathsf{dec}_a(t_i)\cup\{b(\mathsf{cap}_a(t_i))\})$. As a consequence, the TRS $E(\mathcal{R})$ is obtained from $DE_a(\mathcal{R})$ by removing all symbols b from it. As the main result we recall:

Theorem 7. *Let a be a dummy symbol in a TRS $\mathcal{R}$ for which $E(\mathcal{R})$ is terminating. Then $\mathcal{R}$ is terminating too.*

For a proof of Theorem 7 we refer to [8] or [7], where a slightly more general version has been treated. An alternative proof has been given in [16], where even the restriction of the dummy not occurring in lhs's has been weakened slightly. A generalization of this result to rewriting modulo equations has been given in [9].

Now we give the proof of Theorem 6.

Proof. Let a be a dummy symbol of arity n in a TRS $\mathcal{R}$ for which $DE_a(\mathcal{R})$ is terminating. Assume $\mathcal{R}$ is not terminating, so admits an infinite reduction. We define a transformation Φ on terms and TRSs replacing every a by $a(b(-),\ldots,b(-))$, more precisely:

$$\begin{array}{lll}\Phi(x) & = x & \text{for all } x \in \mathsf{Var},\\ \Phi(f(t_1,\ldots,t_k)) & = f(\Phi(t_1),\ldots,\Phi(t_k)) & \text{for all } f \text{ with } f \neq a,\\ \Phi(a(t_1,\ldots,t_n)) & = a(b(\Phi(t_1)),\ldots,b(\Phi(t_n))), & \\ \Phi(\mathcal{R}) & = \{\Phi(\ell)\to\Phi(r) \mid \ell\to r\in\mathcal{R}\}. & \end{array}$$

From this definition it is straightforwardly proved that if $t \to_{\mathcal{R}} u$, then $\Phi(t) \to_{\Phi(\mathcal{R})} \Phi(u)$. So the assumed infinite $\mathcal{R}$ reduction transforms by Φ to an infinite $\Phi(\mathcal{R})$ reduction.

On the other hand the symbol a is still a dummy symbol in $\Phi(\mathcal{R})$. By construction we have $E(\Phi(\mathcal{R})) = DE_a(\mathcal{R})$, which was assumed to be terminating. Hence by Theorem 7 we conclude termination of $\Phi(\mathcal{R})$, contradiction. □

We want to use complete dummy elimination in proving termination automatically is as follows: if termination of $\mathcal{R}$ has to be proved, and $\mathcal{R}$ has a dummy symbol a, then apply DE_a to $\mathcal{R}$, and proceed with the search for termination proofs on $DE_a(\mathcal{R})$. For this approach to be useful we should also like to have the converse of Theorem 6: $\mathcal{R}$ is terminating only if $DE_a(\mathcal{R})$ is terminating. In other words, apart from soundness Theorem 6, we also want completeness. This is seen as follows: if $\mathcal{R}$ is terminating but $DE_a(\mathcal{R})$ is not, then trying to prove termination of $DE_a(\mathcal{R})$ will fail. For instance, let $\mathcal{R}$ consist of the two rules

$$f(g(x)) \to g(f(f(x))),\quad g(f(x)) \to g(a(g(g(x)))).$$

Then indeed $\mathcal{R}$ is terminating, but trying to prove this by proving termination of $E(\mathcal{R})$ consisting of the three rules

$$f(g(x)) \to g(f(f(x))),\quad g(f(x)) \to g(\diamond),\quad g(f(x)) \to g(g(x))$$

will fail since $E(\mathcal{R})$ is not terminating due to

$$f(f(g(x))) \rightarrow_{E(\mathcal{R})} f(g(f(f(x)))) \rightarrow_{E(\mathcal{R})} f(g(g(f(x)))) \rightarrow_{E(\mathcal{R})} g(\underbrace{f(f(g}(f(x))))).$$

Note that Theorem 4 does not apply here due to an overlap between the rules. We conclude that dummy elimination E rather than DE_a is not complete.

Next we show that in case of left-linearity, the desired completeness, and hence the 'if and only if' property holds for DE_a. First we need two lemmas.

Let $\mathcal{B}l$ be the TRS defined to consist of all rules of the shape $f(x_1, \ldots, x_n) \rightarrow \diamond$, for all symbols f of arity $n \geq 0$. This TRS is used for blocking reductions.

Lemma 8. *Let $\mathcal{R}$ be a left-linear TRS in which the constant $\diamond$ and the unary symbol b do not occur in any lhs.*

1. *If $t \rightarrow_{\mathcal{B}l}^* u$ and $\infty(u, \mathcal{R})$, then $\infty(t, \mathcal{R})$.*
2. *If $\infty(C[b(t)], \mathcal{R})$ for any context C and any term t, then either $\infty(C[\diamond], \mathcal{R})$ or $\infty(t, \mathcal{R})$.*

Proof. Part *1.*

Let t, u, v be terms satisfying $t \rightarrow_{\mathcal{B}l} u \rightarrow_{\mathcal{R}} v$. Then u is obtained from t by replacing any subterm by $\diamond$. So the redex of $u \rightarrow_{\mathcal{R}} v$ is either above or parallel to this occurrence of $\diamond$. Since $\mathcal{R}$ is left-linear and $\diamond$ does not occur in the lhs of the corresponding rule in $\mathcal{R}$, the TRS $\mathcal{R}$ could also be applied directly to t yielding $t \rightarrow_{\mathcal{R}} \cdot \rightarrow_{\mathcal{B}l}^* v$. Hence we conclude $\rightarrow_{\mathcal{B}l} \cdot \rightarrow_{\mathcal{R}} \subseteq \rightarrow_{\mathcal{R}} \cdot \rightarrow_{\mathcal{B}l}^*$. Using this property one easily proves $\rightarrow_{\mathcal{B}l}^* \cdot \rightarrow_{\mathcal{R}} \subseteq \rightarrow_{\mathcal{R}} \cdot \rightarrow_{\mathcal{B}l}^*$, applying induction on the number of $\rightarrow_{\mathcal{B}l}$-steps. Using this inclusion the infinite $\mathcal{R}$-reduction starting in u is transformed to an infinite $\mathcal{R}$-reduction starting in t, as is sketched in the following picture:

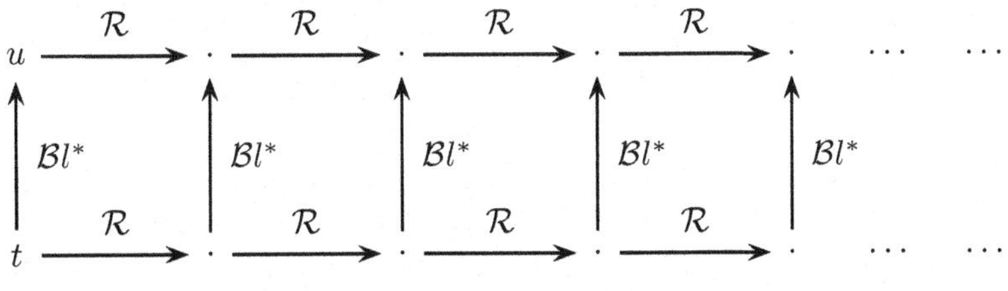

Part *2.* We prove the more general claim for multiple hole contexts:

> Let C be a multi-hole context for which $\mathsf{SN}(C[\diamond, \ldots, \diamond], \mathcal{R})$ and $\infty(C[b(t_1), \ldots, b(t_n)], \mathcal{R})$. Then $\infty(t_i, \mathcal{R})$ for some $i = 1, \ldots, n$.

We prove this for all contexts C satisfying $\mathsf{SN}(C[\diamond, \ldots, \diamond], \mathcal{R})$, by induction on $\rightarrow_{\mathcal{R}}$ restricted to reducts of $C[\diamond, \ldots, \diamond]$. Consider the infinite $\mathcal{R}$-reduction starting in $C[b(t_1), \ldots, b(t_n)]$. If all redex positions are below C then every step is in one of the n displayed subterms $b(t_1), \ldots, b(t_n)$ of $C[b(t_1), \ldots, b(t_n)]$, so at least one of the $b(t_i)$ is rewritten infinitely often. Since b does not occur in any lhs, the same holds for t_i and we are done. In the remaining case the infinite $\mathcal{R}$-reduction is of the shape

$$C[b(t_1),\ldots,b(t_n)] \to_{\mathcal{R}}^* C[b(u_1),\ldots,b(u_n)] \to_{\mathcal{R}} D[b(v_1),\ldots,b(v_k)] \to_{\mathcal{R}}^* \cdots,$$

where $t_i \to_{\mathcal{R}}^* u_i$ for every i, and for every j, $1 \leq j \leq k$, there exists i such that $v_j = u_i$. Since $\mathcal{R}$ is left-linear and b's do not occur in lhs's, we conclude $C[\diamond,\ldots,\diamond] \to_{\mathcal{R}} D[\diamond,\ldots,\diamond]$. Left-linearity is essential here, for instance for $\mathcal{R}$ consisting of $f(x,x) \to a$, $a \to a$ both this claim and the lemma do not hold for $n = 1, t_1 = c, C = f(b(c),\Box)$. Since $C[\diamond,\ldots,\diamond] \to_{\mathcal{R}} D[\diamond,\ldots,\diamond]$ we may apply the induction hypothesis to $D[b(v_1),\ldots,b(v_k)]$, yielding j satisfying $\infty(v_j,\mathcal{R})$. Since there exists i such that $v_j = u_i$ we obtain $t_i \to_{\mathcal{R}}^* u_i = v_j \to_{\mathcal{R}}^\infty$, so $\infty(t_i,\mathcal{R})$. □

Lemma 9. *Let t be any term and a any symbol. Then*

1. $t \to_{\mathcal{B}l}^* \mathsf{cap}_a(t)$, *and*
2. *for every $v \in \mathsf{dec}_a(t)$ the terms t, v can be written as $t = C[t']$ and $v = b(v')$, where $t' \to_{\mathcal{B}l}^* v'$.*

Proof. By induction on the structure of t, straightforward from the definitions of cap_a and dec_a. □

Theorem 10. *Let $\mathcal{R}$ be a left-linear terminating TRS having a dummy symbol a. Then $DE_a(\mathcal{R})$ is terminating.*

Proof. We prove that $\mathsf{SN}(t, DE_a(\mathcal{R}))$ for every term t, by induction on $\mathcal{R}$. So the induction hypothesis states that $\mathsf{SN}(w, DE_a(\mathcal{R}))$ for every term w satisfying $t \to_{\mathcal{R}}^+ w$.

Assume that t admits an infinite $DE_a(\mathcal{R})$-reduction

$$t \to_{DE_a(\mathcal{R})} u \to_{DE_a(\mathcal{R})} \cdot \to_{DE_a(\mathcal{R})} \cdots .$$

For the step $t \to_{DE_a(\mathcal{R})} u$ we distinguish two cases, implied by the definition of $DE_a(\mathcal{R})$.

- $t = C[\ell\sigma]$ and $u = C[\mathsf{cap}_a(r)\sigma]$ for some context C, some substitution σ and some rule $\ell \to r$ in $\mathcal{R}$. Let $v = C[r\sigma]$. By part *1* of Lemma 9 we conclude that $r \to_{\mathcal{B}l}^* \mathsf{cap}_a(r)$, so $v = C[r\sigma] \to_{\mathcal{B}l}^* C[\mathsf{cap}_a(r)\sigma] = u$. Since $\infty(u, DE_a(\mathcal{R}))$ and $DE_a(\mathcal{R})$ is left-linear and has no $\diamond$ or b symbols in lhs's, we may apply part *1* of Lemma 8, yielding $\infty(v, DE_a(\mathcal{R}))$, contradicting the induction hypothesis.
- $t = C[\ell\sigma]$ and $u = C[v\sigma]$ for some context C, some substitution σ, $v \in \mathsf{dec}_a(r)$, and some rule $\ell \to r$ in $\mathcal{R}$. By part *2* of Lemma 9 we obtain $r = C'[r']$ and $v = b(v')$, where $r' \to_{\mathcal{B}l}^* v'$. By part *2* of Lemma 8 we may distinguish two cases based on the infinite $DE_a(\mathcal{R})$-reduction of $u = C[v\sigma] = C[b(v'\sigma)]$:
  - $\infty(C[\diamond], DE_a(\mathcal{R}))$. Since $C[r\sigma] \to_{\mathcal{B}l} C[\diamond]$ we conclude $\infty(C[r\sigma], DE_a(\mathcal{R}))$ from part *1* of Lemma 8.
  - $\infty(v'\sigma, DE_a(\mathcal{R}))$. Since $r' \to_{\mathcal{B}l}^* v'$ we may apply part *1* of Lemma 8, yielding $\infty(r'\sigma, DE_a(\mathcal{R}))$. Since $C[r\sigma] = C[C'\sigma[r'\sigma]]$ we obtain $\infty(C[r\sigma], DE_a(\mathcal{R}))$.

In both cases we obtain $\infty(w, DE_a(\mathcal{R}))$ for $w = C[r\sigma]$ satisfying $t = C[\ell\sigma] \to_{\mathcal{R}} C[r\sigma] = w$, contradicting the induction hypothesis.

□

Left-linearity is essential in Theorem 10 as is shown by the following example. Let $\mathcal{R}$ consist of the single rule

$$f(x, x) \to f(a(c), a(d)).$$

Then $\mathcal{R}$ is terminating, but $DE_a(\mathcal{R})$ is not since it contains the non-terminating rule $f(x, x) \to f(\diamond, \diamond)$.

For proving termination of a TRS $\mathcal{R}$ containing a dummy symbol automatically we propose always to try proving termination of $DE_a(\mathcal{R})$ first. For non-left-linear TRSs this may fail even if $\mathcal{R}$ is terminating as was shown by the above example. However, even then it may be a good strategy first to search for some time for a termination proof of $DE_a(\mathcal{R})$, since often termination proofs for $DE_a(\mathcal{R})$ are substantially simpler than direct termination proofs for $\mathcal{R}$.

Just like we did for the technique of rewriting right-hand sides we investigated on how many problems in the termination problem data base TPDB [20] the technique of complete dummy elimination is directly applicable. It turns out that among the 773 TRSs there are 65 TRSs for which it is. Apart from these 65 it may occur that after some transformation complete dummy elimination is applicable, but this latter figure of course depends on details of the tool.

In case a TRS contains more than one dummy symbol it is a natural question how to proceed. It turns out that just like in earlier versions of dummy elimination the order of applying the corresponding DE operations does not influence the result, e.g., if both a_1 and a_2 are dummy symbols in $\mathcal{R}$, then $DE_{a_1}(DE_{a_2}(\mathcal{R})) = DE_{a_2}(DE_{a_1}(\mathcal{R}))$. In constructing this combined dummy elimination we can apply it for all dummy symbols in one run, introducing a fresh constant $\diamond_a$ and a fresh unary symbol b_a for every dummy symbol a. So in case a TRS contains more than one dummy symbol we propose always to proceed by this simultaneous dummy elimination.

The best tool at the moment for proving TRS termination is AProVE [6]. We give two examples now showing that our DE-strategy is able to enhance the 2005 version of AProVE. The first TRS consists of two rules

$$f(f(g(g(x)))) \to g(g(g(f(f(f(x)))))), \quad f(x) \to a(x, x).$$

AProVE fails to prove termination of this TRS. However, after applying DE_a the resulting TRS consisting of the rules

$$f(f(g(g(x)))) \to g(g(g(f(f(f(x)))))), \quad f(x) \to \diamond, \quad f(x) \to b(x)$$

is proved to be terminating by AProVE in a fraction of a second.

As the second example consider the TRS consisting of the rules

$$\begin{array}{ll} f(g(x)) \to f(h(h(a(h(h(g(f(x)))))))) & f(g(x)) \to g(g(f(h(x)))) \\ f(h(x)) \to h(g(f(x))) & g(h(x)) \to h(g(x)) \\ h(f(x)) \to g(g(h(h(a(f(x)))))) & f(x) \to g(g(h(x))). \end{array}$$

Again AProVE fails to prove termination of this TRS, but by applying DE_a the two rules containing a in their rhs's are replaced by

$$f(g(x)) \to f(h(h(\diamond))), \quad f(g(x)) \to b(h(h(g(f(x))))),$$
$$h(f(x)) \to g(g(h(h(\diamond)))), \quad h(f(x)) \to b(f(x)),$$

resulting in a TRS for which termination is proved easily, e.g., by recursive path order choosing the precedence $f > g > h > b > \diamond$. In the tool TPA complete dummy elimination has been implemented, and indeed for this TRS TPA finds the proof just sketched in a fraction of a second.

In particular this last example is of interest with respect to the following. In [10] it was proved that if $E(\mathcal{R})$ is DP simply terminating then $\mathcal{R}$ is DP simply terminating too. Here roughly speaking DP simple termination means that termination can be proved by the dependency pair technique using argument filtering and a simplification order. As a theorem this is correct, but in [10] it is literally claimed that it implies that *using dummy elimination as a preprocessing step to the dependency pair technique does not have any advantage.* However, our latter example convincingly shows the converse with respect to the present AProVE implementation: here no dependency pair transformation was required, but if the dependency pair transformation had been applied to the resulting system, a straightforward termination proof only using recursive path order would have been found easily too. The difference between $E(\mathcal{R})$ and $DE_a(\mathcal{R})$ is only in the symbol b, and does not play any role: if it is omitted then the same proof holds.

4.2 Complete Dummy Elimination for String Rewriting

For term rewriting we believe that the operation DE_a is the most natural and most powerful variant of dummy elimination, due to the combination of Theorem 6 and Theorem 10. However, for string rewriting there is a drawback: due to the introduction of the constant $\diamond_a$ for a string rewriting system (SRS) $\mathcal{R}$, being a TRS over a signature only containing unary symbols, the transformed system $DE_a(\mathcal{R})$ is not an SRS any more.

This can be solved by defining a variant DE'_a of DE_a, where the only difference is that $\diamond_a$ is a unary symbol rather than a constant. In this way a symmetry between $\diamond_a$ and b_a is introduced. To express this symmetry in the notation, we will write $a_\$$ instead of $\diamond_a$, and $_\$a$ instead of b_a. As usual, we will identify a term $a_1(a_2(\cdots(a_n(x))\cdots))$ with the string $a_1a_2\cdots a_n$, by simply ignoring all parentheses and the variable symbol. So the single variable x in term notation is written as the empty string λ in string notation. Now for a dummy symbol a in an SRS $\mathcal{R}$ we define

$$DE'_a(\mathcal{R}) = \{\ell \to u \mid u = \mathsf{cap}'_a(r) \vee u \in \mathsf{dec}'_a(r) \text{ for a rule } \ell \to r \in \mathcal{R}\},$$

where

$\mathsf{cap}'_a(\lambda) = \lambda$
$\mathsf{cap}'_a(fs) = f\mathsf{cap}'_a(s)$ for all symbols f with $f \neq a$ and all strings s

$$\begin{aligned}
\mathsf{cap}'_a(as) &= a_\$ \\
\mathsf{dec}'_a(\lambda) &= \emptyset \\
\mathsf{dec}'_a(fs) &= \mathsf{dec}'_a(s) \text{ for all symbols } f \text{ with } f \neq a \text{ and all strings } s \\
\mathsf{dec}'_a(as) &= \mathsf{dec}'_a(s) \cup \{{}_\$a(\mathsf{cap}'_a(s))\}.
\end{aligned}$$

Applied on string rewriting, the transformation DE'_a is called *complete dummy elimination*, just as DE_a applied on term rewriting.

First we show how DE'_a acts on a well-known simple standard example. Let the SRS $\mathcal{R}$ consist of the single self-embedding rule $bb \to bab$. Termination of $DE'_a(\mathcal{R})$ consisting of the two rules

$$b\,b \;\to\; b\,a_\$,\quad b\,b \;\to\; {}_\$a\,b,$$

is trivial by counting the number of b-symbols.

The main theorem about DE'_a is the following.

Theorem 11. *Let $\mathcal{R}$ be an SRS having a dummy symbol a. Then $\mathcal{R}$ is terminating if and only if $DE'_a(\mathcal{R})$ is terminating.*

In order to prove Theorem 11 we need a general theorem relating termination of TRSs over constants and unary symbols, and SRSs.

The function ϕ is defined on terms over constants and unary symbols, yielding strings, is defined as follows:

$$\phi(x) = \lambda,\quad \phi(c) = c,\quad \phi(f(t)) = f\phi(t)$$

for all variables x, all constants c and all unary symbols f. A TRS $\mathcal{R}$ over constants and unary symbols is mapped to an SRS $\phi(\mathcal{R})$ as follows:

$$\phi(\mathcal{R}) \;=\; \{\,\phi(\ell) \to \phi(r) \mid \ell \to r \in \mathcal{R}\,\}.$$

Theorem 12. *Let $\mathcal{R}$ be a TRS over constants and unary symbols. Then $\mathcal{R}$ is terminating if and only if $\phi(\mathcal{R})$ is terminating.*

For the proof of this theorem we refer to [19].

The impact of Theorem 12 goes far beyond dummy elimination. In fact Theorem 12 states that proving termination of string rewriting is equivalent to termination of term rewriting as long as no symbols of arity higher than one occur.

Now we are ready to prove Theorem 11.

Proof. (of Theorem 11)

Since an SRS is left-linear, by Theorem 6 and Theorem 10 we conclude that $\mathcal{R}$ is terminating if and only if $DE_a(\mathcal{R})$ is terminating. By Theorem 12 this holds if and only if $\phi(DE_a(\mathcal{R}))$ is terminating. By construction $\phi(DE_a(\mathcal{R}))$ and $DE'_a(\mathcal{R})$ coincide, up to renaming of $\diamond_a$ to $a_\$$ and b_a to ${}_\$a$. □

The transformation DE'_a for string rewriting has been implemented in TORPA, version 1.4. As an example, we give the result of TORPA on the same example we considered before:

```
f g  -> f h h a h h g f
f h  -> h g f
h f  -> g g h h a f
f g  -> g g f h
g h  -> h g
f  -> g g h

Apply dummy elimination, result:
f g  -> f h h a$
f g  -> $a h h g f
f h  -> h g f
h f  -> g g h h a$
h f  -> $a f
f g  -> g g f h
g h  -> h g
f  -> g g h


Choose polynomial interpretation f: lambda x.x+1, rest identity
remove: h f  -> g g h h a$
remove: f  -> g g h
Choose polynomial interpretation
f: lambda x.4x
g: lambda x.x+5
h: lambda x.x+2
a$: lambda x.x+1
$a: lambda x.x+1
remove: f g  -> $a h h g f
remove: f h  -> h g f
remove: h f  -> $a f
remove: f g  -> g g f h
Choose polynomial interpretation g: lambda x.x+1, rest identity
remove: f g  -> f h h a$
Choose polynomial interpretation:
g: lambda x.10x,  rest lambda x.x+1
remove: g h  -> h g
Terminating since no rules remain.
```

5 Conclusions

We described two techniques to transform a given TRS to another one, in such a way that termination of the given TRS can be concluded from termination of the transformed one, and proving termination of the transformed TRS is often easier than proving termination of the given TRS directly.

Both techniques are easy to implement, and have the nice property that no choice has to be made, so never an explosion of the search space will be caused, and no heuristics have to be developed. On the other hand both techniques have a drawback: they are only applicable for a restricted class of TRSs. For rewriting right-hand sides a rhs is required not to be in normal form, and for complete dummy elimination a dummy symbol is required, i.e., a symbol occurring in a rhs but in no lhs. However, both our techniques may be applied not only as a preprocessor, but also in proofs consisting of several transformations of TRSs. If in a proof search the remaining proof obligation is finding a termination proof for some TRS, then both our techniques may be applied, even if they are not applicable to the original TRS.

One may wonder when to apply these techniques. Our proposal is: whenever you can. For rewriting right-hand sides we proved that the original TRS is terminating if and only if the transformed TRS is terminating, and we are not aware of TRSs for which termination of the transformed TRS is harder to prove than termination of the original TRS, while the converse often occurs. For left-linear TRSs the same can be said for complete dummy elimination. So the only situation where the effect may be negative is for complete dummy elimination for non-left-linear TRSs. Indeed for the single rule $f(x, x) \rightarrow f(a(c), a(d))$ we saw that complete dummy elimination should not be applied, since then the transformed TRS is not terminating while the original one is.

Also combinations of both techniques described in this paper make sense: one easily constructs artificial examples on which both rewriting right-hand sides and complete dummy elimination are applicable, and then they can be applied both. We believe that it does not make sense to investigate which order of application of these techniques is preferred, since examples where this makes a difference are really artificial.

Acknowledgments

We want to thank Adam Koprowski for implementing the techniques described here in his tool TPA, and for careful proofreading this paper. We want to thank the anonymous referees for their useful remarks.

References

1. T. Arts and J. Giesl. Termination of term rewriting using dependency pairs. *Theoretical Computer Science*, 236:133–178, 2000.
2. F. Baader and T. Nipkow. *Term Rewriting and All That.* Cambridge University Press, 1998.
3. F. Bellegarde and P. Lescanne. Termination by completion. *Applicable Algebra in Engineering, Communication and Computing*, 1(2):79–96, 1990.
4. A. Ben-Cherifa and P. Lescanne. Termination of rewriting systems by polynomial interpretations and its implementation. *Science of Computer Programming*, 9:137–159, 1987.
5. N. Dershowitz. Orderings for term-rewriting systems. *Theoretical Computer Science*, 17:279–301, 1982.
6. J. Giesl et al. AProVE: Automated program verification environment. `http://aprove.informatik.rwth-aachen.de/`

7. M. Ferreira. *Termination of Term Rewriting – Well-Foundedness, Totality, and Transformations*. PhD thesis, University of Utrecht, 1995.
8. M. Ferreira and H. Zantema. Dummy elimination: Making termination easier. In *Proceedings of the 10th International Conference on Fundamentals of Computation Theory*, volume 965 of *Lecture Notes in Computer Science*, pages 243–252. Springer, 1995.
9. M. C. F. Ferreira. Dummy elimination in equational rewriting. In H. Ganzinger, editor, *Proceedings of the 7th Conference on Rewriting Techniques and Applications*, volume 1103 of *Lecture Notes in Computer Science*, pages 63–77. Springer, 1996.
10. J. Giesl and A. Middeldorp. Eliminating dummy elimination. In *Proceedings of the 17th International Conference on Automated Deduction (CADE)*, volume 1831 of *Lecture Notes in Computer Science*, pages 309–323. Springer, 2000.
11. J. Giesl, R. Thiemann, and P. Schneider-Kamp. The dependency pair framework: Combining techniques for automated termination proofs. In *Proceedings of the 11th International Conference on Logic for Programming, Artificial Intelligence, and Reasoning (LPAR 2004)*, volume 3452 of *Lecture Notes in Computer Science*, pages 301–331. Springer, 2005.
12. B. Gramlich. Simplifying termination proofs for rewrite systems by preprocessing. In Maurizio Gabrielli and Frank Pfenning, editors, *Proceedings of the 2nd International Conference on Principles and Practice of Declarative Programming (PPDP), Montreal, Canada*, pages 139–150, Montreal, Canada, September 2000. ACM Press.
13. N. Hirokawa and A. Middeldorp. TTT: Tyrolean termination tool. `http://cl2-informatik.uibk.ac.at/ttt/`.
14. N. Hirokawa and A. Middeldorp. Dependency pairs revisited. In V. van Oostrom, editor, *Proceedings of the 15th Conference on Rewriting Techniques and Applications (RTA)*, volume 3091 of *Lecture Notes in Computer Science*, pages 249–268. Springer, 2004.
15. A. Koprowski. TPA: Termination proved automatically. `http://www.win.tue.nl/tpa`
16. A. Middeldorp, H. Ohsaki, and H. Zantema. Transforming termination by self-labelling. In *Proceedings of the 13th International Conference on Automated Deduction (CADE)*, volume 1104 of *Lecture Notes in Artificial Intelligence*, pages 373–386. Springer, 1996.
17. J. Steinbach. Automatic termination proofs with transformation orderings. In J. Hsiang, editor, *Proceedings of the 6th Conference on Rewriting Techniques and Applications*, volume 914 of *Lecture Notes in Computer Science*, pages 11–25. Springer, 1995.
18. Terese. *Term Rewriting Systems*. Cambridge University Press, 2003.
19. R. Thiemann, H. Zantema, J. Giesl, and P. Schneider-Kamp. Adding constants to string rewriting. Technical report, RWTH Aachen, 2005. To appear, available via `http://www-i2.informatik.rwth-aachen.de/AProVE/ConstantsSRS.pdf` .
20. TPDB. Termination problem data base. `http://www.lri.fr/~marche/tpdb/`.
21. H. Zantema. TORPA: Termination of rewriting proved automatically. `http://www.win.tue.nl/~hzantema/torpa.html`
22. H. Zantema. Termination of term rewriting: Interpretation and type elimination. *Journal of Symbolic Computation*, 17:23–50, 1994.
23. H. Zantema. Termination of term rewriting by semantic labelling. *Fundamenta Informaticae*, 24:89–105, 1995.

24. H. Zantema. Termination. In *Term Rewriting Systems, by Terese*, pages 181–259. Cambridge University Press, 2003.
25. H. Zantema. Termination of string rewriting proved automatically. Technical Report CS-report 03-14, Eindhoven University of Technology, 2003. Available via `http://www.win.tue.nl/inf/onderzoek/en_index.html`.
26. H. Zantema. TORPA: termination of rewriting proved automatically. In V. van Oostrom, editor, *Proceedings of the 15th Conference on Rewriting Techniques and Applications (RTA)*, volume 3091 of *Lecture Notes in Computer Science*, pages 95–104. Springer, 2004.
27. H. Zantema. Termination of string rewriting proved automatically. *Journal of Automated Reasoning*, 2005. Accepted for publication.

Reduction Strategies for Left-Linear Term Rewriting Systems*

Yoshihito Toyama

RIEC, Tohoku University,
Katahira 2-1-1, Aoba-ku, Sendai 980-8577, Japan
toyama@nue.riec.tohoku.ac.jp

Abstract. Huet and Lévy (1979) showed that needed reduction is a normalizing strategy for orthogonal (i.e., left-linear and non-overlapping) term rewriting systems. In order to obtain a decidable needed reduction strategy, they proposed the notion of strongly sequential approximation. Extending their seminal work, several better decidable approximations of left-linear term rewriting systems, for example, NV approximation, shallow approximation, growing approximation, etc., have been investigated in the literature. In all of these works, orthogonality is required to guarantee approximated decidable needed reductions are actually normalizing strategies. This paper extends these decidable normalizing strategies to left-linear overlapping term rewriting systems. The key idea is the balanced weak Church-Rosser property. We prove that approximated external reduction is a computable normalizing strategy for the class of left-linear term rewriting systems in which every critical pair can be joined with root balanced reductions. This class includes all weakly orthogonal left-normal systems, for example, combinatory logic CL with the overlapping rules $pred \cdot (succ \cdot x) \to x$ and $succ \cdot (pred \cdot x) \to x$, for which leftmost-outermost reduction is a computable normalizing strategy.

1 Introduction

Normalizing reduction strategies of reduction systems, such as leftmost-outermost evaluation of lambda calculus [2, 11], combinatory logic [7, 11], ordinal recursive program schemata [25] and left-normal term rewriting systems [8, 17, 22] guarantee a *safe* evaluation which reduces a given expression to its normal form whenever it exists. Hence, normalizing reduction strategies play an important role in the implementation of functional programming languages based on reduction systems.

Strong sequentiality formalized by Huet and Lévy [8] is a well-known practical criterion guaranteeing an efficiently computable normalizing reduction strategy for orthogonal (i.e., left-linear and non-overlapping) term rewriting systems. They showed that for every strongly sequential orthogonal term rewriting system R, strongly needed reduction is a computable normalizing strategy, that is,

* A part of this paper was published as preliminary version in [24].

A. Middeldorp et al. (Eds.): Processes... (Klop Festschrift), LNCS 3838, pp. 198–223, 2005.

by rewriting a redex called a *strongly needed redex* at each step, every reduction starting with a term having a normal form eventually terminates at the normal from. Here, the strongly needed redex is defined as a *needed redex* concerning an approximation of R which is obtained by analyzing the left-hand sides only of the rewrite rules of R. Moreover, Huet and Lévy [8] proved the decidability of strong sequentiality. A simpler proof by Klop and Middeldorp can be found in [12] and a proof based on second order monadic logic and tree automata by Comon in [3].

Inspired by the seminal work by Huet and Lévy [8], several better decidable approximations of left-linear term rewriting systems, for example, NV approximation [21], shallow approximation [3], growing approximation [9, 15], etc., have been investigated in the literature. Moreover, Durand and Middeldorp [6] presented a simple uniform framework for normalizing reduction strategies based on decidable approximations. In all of these works [6, 9, 10, 15], however, the non-overlapping restriction is still required to guarantee that approximated decidable needed reductions are actually normalizing strategies; hence, they cannot be applied to term rewriting systems with overlapping rules such as

$$\begin{cases} pred(succ(x)) \to x \\ succ(pred(x)) \to x. \end{cases}$$

Though it is known [6, 9, 10, 15] that only the left-linearity restriction is necessary for considering decidability issues, the question whether there exists an approximated decidable normalizing strategy for left-linear overlapping term rewriting systems has received quite a bit of attention.

The main purpose of this paper develops decidable normalizing reduction strategies for left-linear overlapping term rewriting systems. The notion of sequentiality defined by Huet and Lévy [8] is naturally adapted to that of externality. An external term rewriting system R guarantees that every reducible term contains an outer needed redex, called an external redex, which remains at an outer position until it is rewritten. Under this new framework, we show that external reduction is normalizing for the class of external *root balanced joinable* term rewriting systems. A root balanced joinable term rewriting system is defined as a term rewriting system in which every critical pair can be joined with *root balanced reductions.* We also show that for weakly orthogonal left-normal systems, the leftmost-outermost reduction strategy is normalizing. For example, the leftmost-outermost reduction strategy is normalizing for combinatory logic CL $\cup \{pred\cdot(succ\cdot x) \to x,\ succ\cdot(pred\cdot x) \to x\}$. Here, combinatory logic CL [2, 7, 11] is the orthogonal term rewriting system having the following rewrite rules:

$$\text{CL} \quad \begin{cases} ((S\cdot x)\cdot y)\cdot z \to (x\cdot z)\cdot(y\cdot z) \\ (K\cdot x)\cdot y \to x. \end{cases}$$

Moreover, our result can be applied to term rewriting systems not having the Church-Rosser property too. For example, the leftmost-outermost reduction strategy is again normalizing for CL $\cup$

$$\begin{cases} (K \cdot A) \cdot y \to (K \cdot B) \cdot y \\ (K \cdot B) \cdot y \to (K \cdot A) \cdot y \\ A \to A \\ B \to B, \end{cases}$$

though the system is not Church-Rosser since $(K \cdot A) \cdot y$ can be reduced into two constants A and B which cannot be joined.

The approach presented here is more accessible than that based on sequentiality of orthogonal term rewriting systems by Huet and Lévy [8]. The key idea is the balanced weak Church-Rosser property, which was first considered by Toyama [24] for analyzing normalizing reduction strategies of strongly sequential left-linear overlapping term rewriting systems. We first explain this idea in an abstract framework. Section 2 introduces preliminary concepts of abstract reduction systems. In Section 3, we introduce the *balanced weak Church-Rosser property* of abstract reduction systems and explain how this property is related to a normalizing reduction strategy. Our results are carefully partitioned between abstract properties depending solely on the reduction relation and properties depending on term structure. In Section 4 we present preliminary concepts for term rewriting systems and in the next section we introduce the notion of *externality* of (possibly) overlapping term rewriting systems. In Section 6, by using the *balanced weak Church-Rosser property* of *external reduction*, we prove that external reduction of *root balanced joinable* term rewriting systems is normalizing. Section 7 extends external reduction to *quasi-external reduction*. In Section 8, we present *computable normalizing strategies* based on *decidable approximations*. Finally, Section 9 discusses a syntactic characterization of external overlapping term rewriting systems.

2 Reduction Systems

Assuming that the reader is familiar with the basic concepts and notations concerning reduction systems in [1, 18, 22], we briefly present notations and definitions.

A *reduction system* (or an *abstract reduction system*) is a structure $A = \langle D, \to \rangle$ consisting of some set D and some binary relation $\to$ on D (i.e., $\to\ \subseteq D \times D$), called a *reduction relation*. A *reduction* (starting with x_0) in A is a finite or infinite sequence $x_0 \to x_1 \to x_2 \to \cdots$. The identity of elements x, y of D is denoted by $x \equiv y$. $\to^{\equiv}$ is the reflexive closure of $\to$, $\leftrightarrow$ is the symmetric closure of $\to$, $\to^{+}$ is the transitive closure of $\to$, $\to^{*}$ is the transitive reflexive closure of $\to$, and $=$ is the equivalence relation generated by $\to$ (i.e., the transitive reflexive symmetric closure of $\to$). $x {\to}^{m} y$ denotes a reduction of m $(m \geq 0)$ steps from x to y. $x {\leftrightarrow}^{m} y$ denotes a chain $x {\leftrightarrow}^{*} y$ of length m, i.e., there exists a sequence $x = x_0 \leftrightarrow x_1 \leftrightarrow \cdots \leftrightarrow x_m = y$ of m steps.

If $x \in D$ is minimal with respect to $\to$, i.e., $\neg \exists y \in D, [x \to y]$, then we say that x is a *normal form*; let NF be the set of all normal forms. If $x {\to}^{*} y$ and $y \in NF$ then we say x has a normal form y and y is a normal form of x. We say x is reducible if $x \notin NF$.

A reduction system $A = \langle D, \rightarrow \rangle$ ($\rightarrow$ for short) is *strongly normalizing* (or *terminating*) if every reduction in A terminates, i.e., there is no infinite sequence $x_0 \rightarrow x_1 \rightarrow x_2 \rightarrow \cdots$. A is *Church-Rosser* (or *confluent*) if $\forall x, y, z \in D, [x \rightarrow^* y \wedge x \rightarrow^* z \Rightarrow \exists w \in D, y \rightarrow^* w \wedge z \rightarrow^* w]$. A is *weakly Church-Rosser* (or *locally confluent*) if $\forall x, y, z \in D, [x \rightarrow y \wedge x \rightarrow z \Rightarrow \exists w \in D, y \rightarrow^* w \wedge z \rightarrow^* w]$. A is *complete* if A is Church-Rosser (confluent) and strongly normalizing. A has the *normal form property* if $\forall x \in D, \forall y \in NF, [x = y \Rightarrow x \rightarrow^* y]$. A has the *unique normal form property* if $\forall x, y \in NF, [x = y \Rightarrow x \equiv y]$. Note that the normal form property implies the unique normal form property.

The notions of *confluent*, *strongly normalizing*, *complete* on systems are related to the notions on elements. An element $x \in D$ is *confluent* if $\forall y, z \in D, [x \rightarrow^* y \wedge x \rightarrow^* z \Rightarrow \exists w \in D, y \rightarrow^* w \wedge z \rightarrow^* w]$. x is *strongly normalizing* if every reduction starting with x terminates. x is *complete* if x is confluent and strongly normalizing.

Definition 1 (Reduction Strategy). *Let $A = \langle D, \rightarrow \rangle$ and let $\rightarrow_s$ be a sub-relation of $\rightarrow^+$ (i.e., if $x \rightarrow_s y$ then $x \rightarrow^+ y$) such that a normal form concerning $\rightarrow_s$ is also a normal form concerning $\rightarrow$ (i.e., the two binary relations $\rightarrow_s$ and $\rightarrow$ have the same domain). Then, we say that $\rightarrow_s$ is a reduction strategy for A (or for $\rightarrow$). If $\rightarrow_s$ is a sub-relation of $\rightarrow$ then we call it a one step reduction strategy; otherwise $\rightarrow_s$ is called a many step reduction strategy.*

Definition 2 (Normalizing Strategy). *A reduction strategy $\rightarrow_s$ is normalizing iff for each x having a normal form concerning $\rightarrow$, there exists no infinite sequence $x \equiv x_0 \rightarrow_s x_1 \rightarrow_s x_2 \rightarrow_s \cdots$ (i.e., every $\rightarrow_s$ reduction starting with x must eventually terminate at a normal form of x).*

3 Balanced Weak Church-Rosser Property

This section introduces the balanced weak Church-Rosser property. Though in later sections this concept will play an important role for analyzing normalizing strategies of term rewriting systems, our results concerning the balanced weak Church-Rosser property can be presented in an abstract framework depending solely on the reduction relation.

Let $A = \langle D, \rightarrow \rangle$ be an abstract reduction system.

Definition 3. *$A = \langle D, \rightarrow \rangle$ (or $\rightarrow$) is balanced weakly Church-Rosser (BWCR) iff $\forall x, y, z \in D, [x \rightarrow y \wedge x \rightarrow z \Rightarrow \exists w \in D, \exists k \geq 0, y \rightarrow^k w \wedge z \rightarrow^k w]$ (Figure 1).*

Lemma 1 (BWCR Lemma). *Let $A = \langle D, \rightarrow \rangle$ be BWCR. Let $x = y$ and $y \in NF$. Then,*

(1) *x is complete,*
(2) *all the reductions from x to y have the same length (i.e., the same number of reduction steps).*

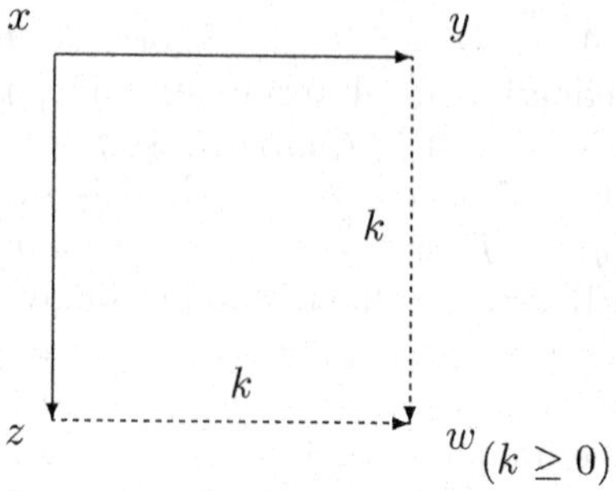

Fig. 1.

Proof. We first prove the following claim: if $x \rightarrow^n y$ and $y \in NF$ then x satisfies the properties (1) and (2).

Proof of the claim. We show the claim by induction on n. The case $n = 0$ is trivial. Let $x \rightarrow x' \rightarrow^{n-1} y \in NF$. Take any one step reduction $x \rightarrow z$ starting with x. By the balanced weak Church-Rosser property, there exists some w and k such that $z \rightarrow^k w$ and $x' \rightarrow^k w$. By the induction hypothesis, the properties (1) and (2) hold at x'; hence $x' \rightarrow^k w \rightarrow^* y$ must have $n - 1$ steps in length. Thus, $w \rightarrow^{n-1-k} y$; see Figure 2. Since $z \rightarrow^k w$, we obtain $z \rightarrow^{n-1} y$. By the induction hypothesis, z satisfies the properties (1) and (2). Therefore, the claim follows.

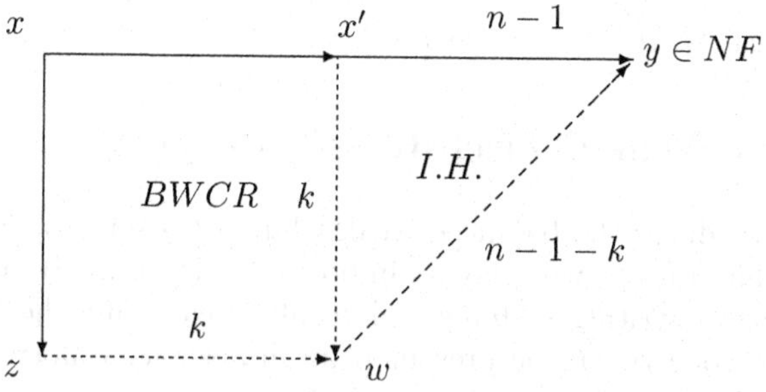

Fig. 2.

We next show that if $x \leftrightarrow^n y$ and $y \in NF$ then $x \rightarrow^* y$. The proof is by induction on n. The case $n = 0$ is trivial. Let $x \leftrightarrow x' \leftrightarrow^{n-1} y$. By the induction hypothesis, we have $x' \rightarrow^* y$. The case $x \rightarrow x'$ is trivial. Let $x \leftarrow x'$. By applying the claim to $x' \rightarrow^* y \in NF$, it is obtained that x' is complete. Thus, $x \rightarrow^* y$.

Therefore, from the claim it follows that if $x = y$ and $y \in NF$ then x satisfies the properties (1) and (2). □

Lemma 1 (BWCR Lemma) is a generalization to Theorem 2 and Corollary 2.1 of Newman [16], which requires the following property instead of BWCR:

$\forall x, y, z \in D, [x \to y \wedge x \to z \wedge y \not\equiv z \Rightarrow \exists w \in D, y \to w \wedge z \to w]$. An extension of BWCR is discussed in Van Oostrom [20].

Corollary 1. *If an abstract reduction system A is BWCR then A has the normal form property.*

Proof. From the BWCR Lemma, it is trivial. □

Next we will explain how the balanced weak Church-Rosser property is related to a normalizing reduction strategy. Let $d(x)$ denote the length of a reduction from x to a normal form if it exists. Note that if $\to$ is balanced weakly Church-Rosser and x has a normal form then $d(x)$ is well-defined according to the BWCR Lemma. We write $x \leftarrow^m y$ if $y \to^m x$.

Lemma 2. *Let* $\to$ *be balanced weakly Church-Rosser. Let* $x \longrightarrow^{m_1} \cdot \longleftarrow^{n_1} \cdot \longrightarrow^{m_2} \cdot \longleftarrow^{n_2} \cdots \longrightarrow^{m_p} \cdot \longleftarrow^{n_p} y$ *for some* $p, m_1, \cdots, m_p, n_1, \cdots, n_p \geq 0$ *and let* x *have a normal form. Then* y *has a normal form and* $d(x) - d(y) = \sum m_i - \sum n_i$.

Proof. By the BWCR Lemma it is clear that y has a normal form. We prove $d(x) - d(y) = \sum m_i - \sum n_i$ by induction on p. The case $p = 0$ is trivial. Let $x \longrightarrow^{m_1} \cdot \longleftarrow^{n_1} \cdot \longrightarrow^{m_2} \cdot \longleftarrow^{n_2} \cdots \longrightarrow^{m_{p-1}} \cdot \longleftarrow^{n_{p-1}} y' \longrightarrow^{m_p} z \longleftarrow^{n_p} y$. By the BWCR Lemma, $d(y')$ and $d(z)$ are well-defined and $d(y') - m_p = d(z) = d(y) - n_p$. Thus, we have $d(y') - d(y) = m_p - n_p$. From the induction hypothesis, $d(x) - d(y') = \sum_{i=1}^{p-1} m_i - \sum_{i=1}^{p-1} n_i$. Therefore, $d(x) - d(y) = \sum m_i - \sum n_i$. □

We write $x \longleftrightarrow\!\!\!\!\rightarrow y$ if there exists a connection $x \longrightarrow^{m_1} \cdot \longleftarrow^{n_1} \cdot \longrightarrow^{m_2} \cdot \longleftarrow^{n_2} \cdots \longrightarrow^{m_p} \cdot \longleftarrow^{n_p} y$ such that $\sum m_i > \sum n_i$. We sometimes write $x \twoheadleftarrow\!\!\!\!\longrightarrow y$ instead of $y \longleftrightarrow\!\!\!\!\rightarrow x$.

Lemma 3. *Let* $\to$ *be balanced weakly Church-Rosser. Let* $x \longleftrightarrow\!\!\!\!\rightarrow y$ *and let* x *have a normal form. Then* y *has a normal form and* $d(x) > d(y)$.

Proof. It is trivial from Lemma 2. □

The following lemma and corollary explain how the BWCR Lemma implies the normalizing property of a reduction strategy $\to_s$ for $\to$ (i.e., $\to_s \subseteq \to$ and the two reduction relations $\to_s$ and $\to$ have the same set of normal forms.)

Lemma 4. *Let* $\to_s$ *be a reduction strategy for* $\to$ *such that:*

(1) $\to_s$ *is balanced weakly Church-Rosser,*
(2) *if* $x \to y$ *then;*
   (i) $x =_s y$ *or,*
   (ii) $x \longleftrightarrow\!\!\!\!\rightarrow_s \cdot \leftrightarrow \cdot \twoheadleftarrow\!\!\!\!\longrightarrow_s y$.

If $x = y$ *and* $y \in NF$ *then we have* $x \to_s^* y$.

Proof. We first show the claim: if $x \leftrightarrow \cdot \rightarrow_s^m y$ and $y \in NF$, then we have $x =_s y$. The proof is by induction on m. For the base step we let $m = 0$. Then $x \leftrightarrow y \in NF$. Suppose that it satisfies the condition (ii), i.e., $x \longleftarrow\!\!\!\twoheadrightarrow_s x' \leftrightarrow y' \twoheadleftarrow\!\!\!\longrightarrow_s y$ holds for some x' and y'. Then by Lemma 3 and $y \in NF$ we have $d(y') < d(y) = 0$; it contradicts $d(y') \geq 0$. Thus $x \leftrightarrow y$ must satisfy the condition (i). Induction Step: Let $x \leftrightarrow z \rightarrow_s^m y \in NF$ $(m > 0)$. Then $x \leftrightarrow z$ must satisfy (i) or (ii) as each condition is symmetric. If $x =_s z$ then $x =_s y$ is trivial. Assume that $x \longleftarrow\!\!\!\twoheadrightarrow_s x' \leftrightarrow z' \twoheadleftarrow\!\!\!\longrightarrow_s z$. By applying Lemma 3 to $z \longleftarrow\!\!\!\twoheadrightarrow_s z'$, we have $z' \rightarrow_s^{m'} y$ with $m' < m$; see Figure 3. Applying the induction hypothesis of the claim to x', we have $x' =_s y$; thus, $x =_s y$ because of $x \longleftarrow\!\!\!\twoheadrightarrow_s x' =_s y$.

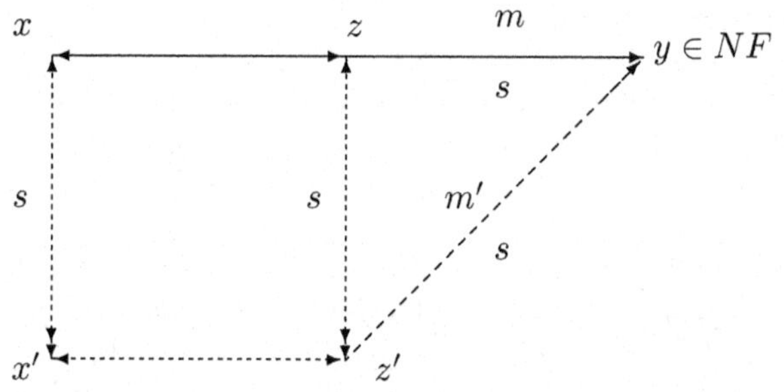

Fig. 3.

We next prove that if $x \leftrightarrow^n y$ and $y \in NF$, then $x \rightarrow_s^* y$. The proof is by induction on n. The case $n = 0$ is trivial. Let $x \leftrightarrow x' \leftrightarrow^{n-1} y \in NF$. From the induction hypothesis, we have $x' \rightarrow_s^* y$. Thus, from the claim, $x =_s y$. From the BWCR Lemma, it follows that $x \rightarrow_s^* y$. □

Corollary 2. *Let $\rightarrow_s$ be a reduction strategy for $\rightarrow$ such that:*

(1) *$\rightarrow_s$ is balanced weakly Church-Rosser,*
(2) *if $x \rightarrow y$ then;*
  (i) *$x =_s y$ or,*
  (ii) *$x \longleftarrow\!\!\!\twoheadrightarrow_s \cdot \leftrightarrow \cdot \twoheadleftarrow\!\!\!\longrightarrow_s y$.*

Then $\rightarrow$ has the normal form property and $\rightarrow_s$ is a normalizing strategy.

Proof. It is trivial from the BWCR Lemma and Lemma 4. □

In Lemma 4 and Corollary 2 we cannot relax the condition (ii) $x \longleftarrow\!\!\!\twoheadrightarrow_s \cdot \leftrightarrow \cdot \twoheadleftarrow\!\!\!\longrightarrow_s y$ to $x \longleftarrow\!\!\!\twoheadrightarrow_s \cdot \leftrightarrow^+ \cdot \twoheadleftarrow\!\!\!\longrightarrow_s y$. Consider the abstract reduction system A with the reduction relation $\rightarrow$ and the reduction strategy $\rightarrow_s$ for $\rightarrow$ presented in Figure 4. Then A does not have the normal form property. Note that $c \rightarrow b$ satisfies $c \longleftarrow\!\!\!\twoheadrightarrow_s \cdot \leftrightarrow^+ \cdot \twoheadleftarrow\!\!\!\longrightarrow_s b$ as $c \rightarrow_s c \rightarrow b \rightarrow a \leftarrow_s b$, and $c \rightarrow d$ satisfies $c \longleftarrow\!\!\!\twoheadrightarrow_s \cdot \leftrightarrow^+ \cdot \twoheadleftarrow\!\!\!\longrightarrow_s d$ as $c \rightarrow_s c \rightarrow d \rightarrow e \leftarrow_s d$.

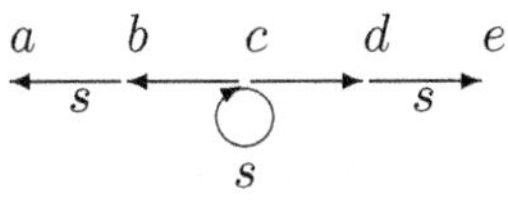

Fig. 4.

In Corollary 2 if we need only to show that $\rightarrow_s$ is a normalizing strategy for $\rightarrow$, we may replace the symmetric condition (ii) $x \longleftrightarrow\!\!\!\!\!\rightarrow_s \cdot \leftrightarrow \cdot \twoheadleftarrow\!\!\longrightarrow_s y$ with an asymmetric weaker condition as follows.

Corollary 3. *Let $\rightarrow_s$ be a reduction strategy for $\rightarrow$ such that:*

(1) *$\rightarrow_s$ is balanced weakly Church-Rosser,*
(2) *if $x \rightarrow y$ then;*
  (i) *$x{=}_s y$ or,*
  (ii) *$x{=}_s \cdot \rightarrow \cdot \twoheadleftarrow\!\!\longrightarrow_s y$.*

Then $\rightarrow_s$ is a normalizing strategy.

Proof. Similarly to the proof of Lemma 4, we can show the claim: if $x{\rightarrow}^* y \in NF$ then $x{\rightarrow}_s^* y$. Thus from the BWCR Lemma the corollary holds. □

In Corollary 3 the normal form property of $\rightarrow$ need not hold. Consider the abstract reduction system A with the reduction relation $\rightarrow$ and the reduction strategy $\rightarrow_s$ for $\rightarrow$ presented in Figure 5. Then A does not have the normal form property though $\rightarrow_s$ is a normalizing strategy for $\rightarrow$. Note that $b{\rightarrow}c$ satisfies $b{=}_s \cdot \rightarrow \cdot \twoheadleftarrow\!\!\longrightarrow_s c$ as $b{\rightarrow}c{\rightarrow}_s c$, and $d{\rightarrow}c$ satisfies $d{=}_s \cdot \rightarrow \cdot \twoheadleftarrow\!\!\longrightarrow_s c$ as $d{\rightarrow}c{\rightarrow}_s c$.

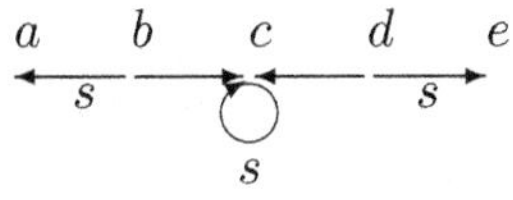

Fig. 5.

4 Term Rewriting Systems

We assume familiarity with the basis of term rewriting systems [1, 18, 22]. Let $\mathcal{F}$ be a set of function symbols denoted by $f, g, h, \cdots$, and let $\mathcal{V}$ be a countably infinite set of variable symbols denoted by $x, y, z, \cdots$ where $\mathcal{F} \cap \mathcal{V} = \varnothing$. By $T(\mathcal{F}, \mathcal{V})$, we denote the set of all terms constructed from $\mathcal{F}$ and $\mathcal{V}$. Terms not containing variables are called *ground* terms. The set of all ground terms built from $\mathcal{F}$ is denoted by $T(\mathcal{F})$. A term t is *linear* if every variable in t occurs only once.

Consider an extra constant $\Box$ called a *hole* and the set $T(\mathcal{F} \cup \{\Box\}, \mathcal{V})$. Then $C \in T(\mathcal{F} \cup \{\Box\}, \mathcal{V})$ is called a *context* over $\mathcal{F}$. We use the notation $C[\ , \ldots, \ ]$

for the context containing n holes ($n \geq 0$), and if $t_1, \ldots, t_n \in T(\mathcal{F}, \mathcal{V})$, then $C[t_1, \ldots, t_n]$ denotes the result of placing $t_1, \ldots, t_n$ in the holes of $C[\ , \ldots, \ ]$ from left to right. In particular, $C[\ ]$ denotes a context containing precisely one hole. A term s is called a *subterm* of t if $t \equiv C[s]$, denoted by $s \trianglelefteq t$. A subterm s of t is *proper*, denoted by $s \triangleleft t$, if $s \not\equiv t$. If a term t has an occurrence of some (function or variable) symbol e, we write $e \in t$.

A *substitution* θ is a mapping from $\mathcal{V}$ to $T(\mathcal{F}, \mathcal{V})$. Substitutions are extended into homomorphisms from $T(\mathcal{F}, \mathcal{V})$ into $T(\mathcal{F}, \mathcal{V})$. We write $t\theta$ instead of $\theta(t)$. We write $s \preceq t$ if $s\theta \equiv t$ for some substitution θ.

A *rewrite rule* over $\mathcal{F}$ is a pair $\langle l, r \rangle$ of terms in $T(\mathcal{F}, \mathcal{V})$ such that $l \notin \mathcal{V}$ and any variable in r also occurs in l. We write $l \to r$ for $\langle l, r \rangle$. A *redex* is a term $l\theta$, where $l \to r$.

A *term rewriting system* (TRS for short) R over $\mathcal{F}$ is a set of rewrite rules over $\mathcal{F}$. (We often simply write R when $\mathcal{F}$ can be inferred from the context.) A TRS R over $\mathcal{F}$ is *finite* if both R and $\mathcal{F}$ are finite. The rewrite rules of R over $\mathcal{F}$ define a *reduction relation* $\to_R$ on $T(\mathcal{F}, \mathcal{V})$ as follows: $t \to_R s$ iff $t \equiv C[l\theta]$ and $s \equiv C[r\theta]$ for some $l \to r \in R$, $C[\ ]$ and θ. When we want to specify the *redex occurrence* $\Delta \equiv l\theta$ of t in this reduction, we write $t \to_R^{\Delta} s$. All the notions defined in the previous sections for abstract reduction systems carry over to TRSs by associating a reduction system $\langle T(\mathcal{F}, \mathcal{V}), \to_R \rangle$ with R. We will simply write $\to$ instead of $\to_R$ when no confusion arises.

Let $l \to r$ and $l' \to r'$ be two rules in R. We assume that they are renamed to have no common variables. Suppose that $s \notin \mathcal{V}$ is a subterm occurrence in l, i.e., $l \equiv C[s]$, such that s and l' are unifiable with a *most general unifier* θ. Then we say that $l \to r$ and $l' \to r'$ are *overlapping*, and that the pair $\langle C[r']\theta, r\theta \rangle$ of terms is *critical* in R [22]. We may choose $l \to r$ and $l' \to r'$ to be the same rule, but in this case we shall not consider the case $s \equiv l$.

If R has a critical pair, then we say that R is *overlapping*; otherwise, *non-overlapping*. We say that R is *left-linear* if for any $l \to r \in R$, l is linear. R is *orthogonal* if R is left-linear and non-overlapping. R is *weakly orthogonal* if R is left-linear and every critical pair $\langle s, t \rangle$ of R is trivial (i.e., $s \equiv t$).

From here on we assume that R is a *finite left-linear* TRS over $\mathcal{F}$ which may have *overlapping* rules. Furthermore, we view R as a TRS over $\mathcal{F} \cup \{\Box\}$ when we consider a reduction relation on $T(\mathcal{F} \cup \{\Box\}, \mathcal{V})$.

5 Externality

The fundamental concept of *neededness* for orthogonal TRSs was introduced by Huet and Lévy [8]. In an orthogonal TRS, every reducible term contains a *needed* redex and needed reduction (i.e., *call-by-need evaluation*) is a normalizing strategy [8]. This section presents a similar framework of *externality* for left-linear overlapping TRSs. An *external* TRS R guarantees that every reducible term contains an outer needed redex, called an *external redex*, which remains at an outer position until R rewrites it. In the next section we shall show that external reduction works as a normalizing strategy for a class of left-linear overlapping TRSs, like needed reduction for orthogonal TRSs.

Consider a left-linear TRS R over $\mathcal{F}$.

Definition 4 (Outer redex). *A context $C[\ ]$ is outer if $C[\ ]$ has no redex occurrence Δ' such that $\square \in \Delta'$. A redex occurrence Δ of $C[\Delta]$ is called outer if $C[\ ]$ is outer. The set of all outer contexts with respect to R is denoted by $OUT(R)$. An outer redex Δ of a term t is outermost if there exists no redex Δ' of t such that $\Delta \lhd \Delta'$.*

Definition 5 (External context). *An outer context $C[\ ]$ is external with respect to R if any s obtained by $C[\ ]\rightarrow_R^* s$ is an outer context. The set of all external contexts with respect to R is denoted by $EXT(R)$.*

If the hole in $C[\ ]$ is deleted or duplicated through a reduction then $C[\ ]$ is not external, since some non-outer context must arise previous to deletion or duplication of the hole.

Definition 6 (External redex). *Let Δ be a redex occurrence in $C[\Delta]$ such that $C[\ ]$ is external. Then the redex occurrence Δ is called external. If Δ is an external redex of $C[\Delta]$ then we write $C[\Delta_E]$; otherwise $C[\Delta_{NE}]$.*

The notion of externality for orthogonal TRSs originates with Huet and Lévy [8]. Externality for non-orthogonal TRSs is presented in Van Oostrom and De Vrier [19], which defines externality as a *reduction step* from a term whose residuals are not nested by other redexes. The definition in Van Oostrom and De Vrier is slightly more abstract than ours, but the two notions are externally same (see 9.2.3 in [19]). The following example is given in [19].

Example 1. Let $R = \{f(x,b)\rightarrow x, a\rightarrow b\}$. Then the context $f(a,\square)$ is external but $f(\square,a)$ is not, since $f(\square,a)\rightarrow f(\square,b) \notin OUT(R)$. Thus, in the term $f(a,a)$ the rightmost redex occurrence a is external but the leftmost occurrence a is not, i.e., $f(a_{NE},a_E)$.

From the definition of external redex it is obvious that in orthogonal TRSs any two external redex occurrences in a term must be disjoint. On the other hand, if a left-linear TRS is overlapping then two external redexes may be overlapping as follows.

Example 2. Let $R = \{p(s(x))\rightarrow x, s(p(x))\rightarrow x\}$. Then we have the overlapping external redexes $f(s(p(s(x))_E)_E)$ since $f(\square)$ and $f(s(\square))$ are external. Thus, external redexes need not be outermost [19].

One might think that overlapping redex occurrences always make overlapping external redexes if one of them is external, but this is not the case from the following example.

Example 3. Let $R = \{b\rightarrow c, f(b)\rightarrow c, g(f(x),c)\rightarrow x\}$. Then we have $g(f(b_{NE})_E, b_E)$. Note that two redex occurrences $f(b)$ and b are overlapping but the redex b occurring in $f(b)$ is not external since the context $g(f(\square),b)$ is not external.

In a left-linear overlapping TRS, external redexes need not exist; for example, in $R = \{a \to b, f(b,x) \to c, f(x,b) \to c\}$ the reducible term $f(a,a)$ has no external redexes [19].

Definition 7 (External TRS). *A reduction $t \to^{\Delta} s$ is external if Δ is an external redex of t. We write $t \to_E s$ if there exists an external reduction $t \to^{\Delta} s$ for some external redex Δ; otherwise $t \to_{NE} s$. We say that R is external if for each term $t \notin NF$, t has an external redex.*

We shortly mention the relationship between *neededness* and *externality* of left-linear TRSs. For details of neededness and externality not treated here we refer to Van Oostrom and De Vrier [19]. The following definition of neededness is due to [6].

Definition 8 (Needed redex). *A context $C[\ ]$ is needed with respect to R if any s obtained by $C[\ ] \to_R^* s$ is not a normal form in $T(\mathcal{F}, \mathcal{V})$. A redex occurrence Δ in a term $C[\Delta]$ is called needed if $C[\ ]$ is needed. A reduction $t \to^{\Delta} s$ is needed if Δ is a needed redex of t.*

As external contexts have no normal forms in $T(\mathcal{F}, \mathcal{V})$, external redexes are (outermost) needed redexes; however, the revers need not hold. In Example 1, the leftmost redex occurrence a of the term $f(a,a)$ is (outermost) needed but not external [19]. For orthogonal TRSs we have the following properties of externality (neededness) [19].

- Any reducible term contains an external redex (a needed redex).
- External (needed) reduction is a normalizing strategy.
- Externality (neededness) of a redex is undecidable.

For a left-linear external TRS R, external reduction is a reduction strategy as every reducible term has an external redex. However, external reduction need not be a normalizing strategy if R is non-orthogonal. (See 9.2.4 in [19] too).

Example 4. Consider $R = \{a \to b, f(x) \to f(x), f(b) \to b\}$. Clearly, R is external. In the term $f(a)$ the outermost redex occurrence $f(a)$ is external but the innermost redex occurrence a not. Then, external reduction starting with $f(a)$ produces an infinite sequence $f(a) \to_E f(a) \to_E f(a) \to_E \cdots$. For normalizing $f(a) \to f(b) \to b$, we need a non-external reduction step $f(a) \to_{NE} f(b)$.

Externality of arbitrary left-linear TRSs is not decidable and external reduction is not computable in general. Hence, in order to obtain computable external reduction, we need to strengthen the notion of externality by decidable approximations. We address this problem in Section 8.

6 Normalization of External Reduction

We will now explain how to prove the normalizing property of external reduction for *overlapping* TRSs by using the BWCR Lemma. We first define *root balanced joinable* TRSs.

Root reduction $t \rightarrow_r s$ is defined as a reduction $t \rightarrow s$ contracted at the *root* position of t (i.e., $t \rightarrow^{\Delta} s$ and $\Delta \equiv t$).

Lemma 5. *Let $C[\Delta_E]$ for some Δ and let $t \rightarrow_r s$. Then $C[t] \rightarrow_E C[s]$.*

Proof. It is trivial from the definition of the root reduction. □

Definition 9. *A critical pair $\langle s, t\rangle$ is root balanced joinable if $s \rightarrow_r^k t'$ and $t \rightarrow_r^k t'$ for some t' and $k \geq 0$. A TRS R is root balanced joinable if every critical pair is root balanced joinable.*

In general it is undecidable whether a critical pair is root balanced joinable. The following example illustrates this problem.

Example 5. Consider a TRS R containing a constant b in normal forms and a ground term s such that reachability of root reduction $s \rightarrow^*{}_r b$ is undecidable. (Such a TRS R and a ground term s exist due to universal computation capability of TRSs; for example, see an encoding of Turing machine to a TRS in [22]). Let R' be $R \cup \{a \rightarrow s, a \rightarrow b, b \rightarrow b\}$ where a is a fresh constant. Then, the critical pair $\langle s, b\rangle$ of R' is root balanced joinable iff $s \rightarrow^*{}_r b$; this is undecidable.

Note that every weakly orthogonal TRS is trivially root balanced joinable since every critical pair is root balanced joinable with $k = 0$. We show that the root balanced joinability is sufficient to guarantee the balanced weak Church-Rosser property of left-linear TRSs.

Definition 10. *Let Δ and Δ' be two redex occurrences in a term t, and let $\Delta \equiv C[x_1\theta, \cdots, x_m\theta]$ where $C[x_1, \cdots, x_m]$ is the left-hand side of a rewrite rule and no variables occur in $C[\ , \cdots,\ ]$. Then Δ and Δ' (or Δ' and Δ) are overlapping if $\Delta' \trianglelefteq \Delta$ and $\Delta' \ntrianglelefteq x_i\theta$ for any subterm occurrence $x_i\theta$.*

Lemma 6. *Let R be left-linear root balanced joinable. Let $t \rightarrow_E^{\Delta} t'$ and $t \rightarrow^{\Delta'} t''$, where $\Delta' \trianglelefteq \Delta$ and Δ and Δ' are overlapping. Then, we have $t' \rightarrow_E^k s$ and $t'' \rightarrow_E^k s$ for some s and $k \geq 0$.*

Proof. Let $t \equiv C[\Delta] \equiv C[C'[\Delta']]$, $t' \equiv C[p]$, $t'' \equiv C[q]$. From the root balanced joinability of the critical pair concerning Δ and Δ', we have $p \rightarrow_r^k s'$ and $q \rightarrow_r^k s'$ for some s' and $k \geq 0$, similarly to the Critical Pair Lemma [1, 18, 22]. Thus, from $C[\Delta_E]$ and Lemma 5, it follows that $C[p] \rightarrow_E^k C[s']$ and $C[q] \rightarrow_E^k C[s']$. □

Lemma 7. *Let $C[\Delta_E, s]$. Then $C[\Delta_E, t]$ for any $s \rightarrow^* t$.*

Proof. Since $C[\Box, s]$ is external and $C[\Box, s] \rightarrow^* C[\Box, t]$, $C[\Box, t]$ is external. Thus we have $C[\Delta_E, t]$. □

Lemma 8. *Let R be left-linear root balanced joinable. Then external reduction $\rightarrow_E$ has the balanced weak Church-Rosser property.*

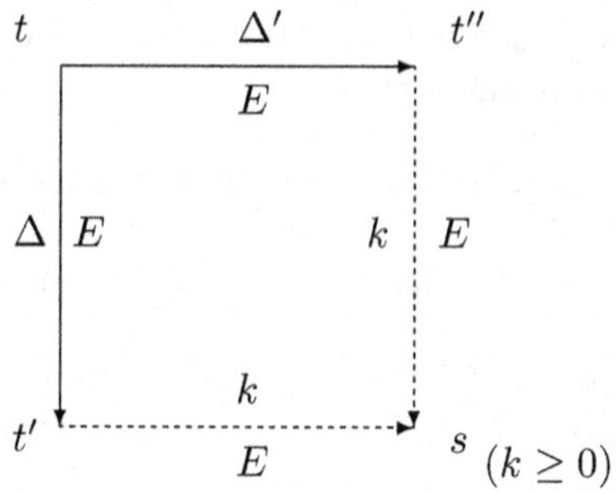

Fig. 6.

Proof. Let $t \rightarrow_E^{\Delta} t'$, $t \rightarrow_E^{\Delta'} t''$. We shall show that $t' \rightarrow_E^{k} s$ and $t'' \rightarrow_E^{k} s$ for some s and $k \geq 0$ (Figure 6). If Δ and Δ' are disjoint, then from Lemma 7 the theorem clearly holds with $k = 1$. Assume that Δ and Δ' are not disjoint, say $\Delta' \trianglelefteq \Delta$. Then, Δ and Δ' must be overlapping as they both are external. Apply Lemma 6. □

We next consider the relation between external reduction $\rightarrow_E$ and arbitrary reduction. Since external reduction $\rightarrow_E$ is balanced weakly Church-Rosser, the normalization of external reduction is obtained if we can apply Corollary 2 by taking $\rightarrow_E$ as $\rightarrow_s$. However, this is impossible as (ii) in the corollary is not satisfied because of duplication of redexes through reduction by a non-right-linear rewrite rule. To overcome this problem, we use *parallel reduction* of disjoint redexes.

Parallel reduction $t \mathrel{\|\!\!\!\!\longrightarrow} s$ is defined by $t \equiv C[\Delta_1, \cdots, \Delta_n] \rightarrow^{\Delta_1} \cdots \rightarrow^{\Delta_n} s$ for some disjoint redexes $\Delta_1, \cdots, \Delta_n$ $(n \geq 0)$. A parallel reduction $t \mathrel{\|\!\!\!\!\longrightarrow} s$ is *proper* if $n > 0$, and we write $t \mathrel{\|\!\!\!\!\longrightarrow}' s$. Since $\rightarrow$ and $\mathrel{\|\!\!\!\!\longrightarrow}'$ have the same set of normal forms and $\rightarrow_E \subseteq \mathrel{\|\!\!\!\!\longrightarrow}'$, it is obvious that $\rightarrow_E$ is a reduction strategy for $\rightarrow$ iff it is a reduction strategy for $\mathrel{\|\!\!\!\!\longrightarrow}'$. In the following lemmas we use $\mathrel{\|\!\!\!\!\longrightarrow}$ instead of $\mathrel{\|\!\!\!\!\longrightarrow}'$ because of technical convenience.

Lemma 9. *Let R be left-linear root balanced joinable and external, and let $t \mathrel{\|\!\!\!\!\longrightarrow} s$. Then $t =_E s$ or $t \longleftrightarrow\!\!\!\!\!\twoheadrightarrow_E \cdot \mathrel{\|\!\!\!\!\longrightarrow} \cdot \twoheadleftarrow\!\!\!\!\!\longleftrightarrow_E s$.*

Proof. Let $t \mathrel{\|\!\!\!\!\longrightarrow}^{\Delta_1 \cdots \Delta_n} s$ $(n \geq 0)$. The proof is by induction on n. The case $n = 0$ is trivial as $t' \equiv t \equiv s \in NF$. Induction Step:

Case 1: Some Δ_i, say without loss of generality Δ_1, is external. We have $t \rightarrow_E^{\Delta_1} t' \mathrel{\|\!\!\!\!\longrightarrow}^{\Delta_2 \cdots \Delta_n} s$. By applying the induction hypothesis to $t' \mathrel{\|\!\!\!\!\longrightarrow}^{\Delta_2 \cdots \Delta_n} s$, we obtain the lemma.

Case 2: No Δ_i is external. From externality there must exist an external redex, say Δ, in t. Let $t \rightarrow_E^{\Delta} t''$ and consider the following two cases.

Case 2-1: Δ and Δ_i $(i = 1 \cdots n)$ are non-overlapping. By using left-linearity of R, we can easily show that $t'' \mathrel{\|\!\!\!\!\longrightarrow} s'$ and $s \rightarrow_E s'$ for some s' (Figure 7).

Case 2-2: Δ and some Δ_i, say without loss of generality Δ_1, are overlapping. Let $t \rightarrow_{NE}^{\Delta_1} t' \mathrel{\|\!\!\!\!\longrightarrow}^{\Delta_2 \cdots \Delta_n} s$. Note that $\Delta_1 \trianglelefteq \Delta$. From Lemma 6, it follows that

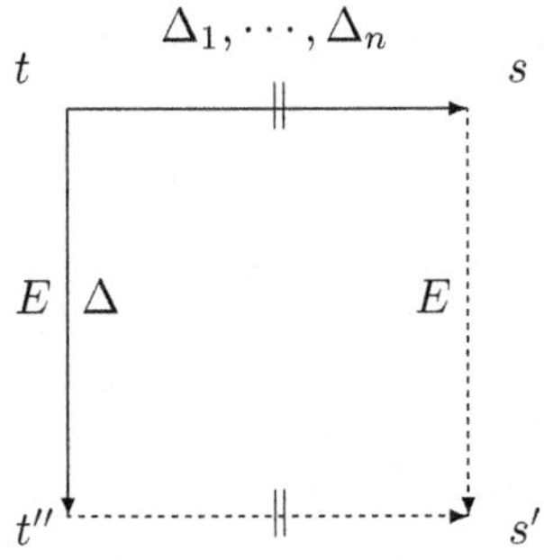

Fig. 7.

$t'' \to^k_E s'$ and $t' \to^k_E s'$ for some s' and $k \geq 0$ (Figure 8). Thus, we can obtain $t \longleftrightarrow\!\!\!\!\twoheadrightarrow_E t'$. Apply the induction hypothesis to $t' \twoheadrightarrow\!\!\!\!\!\!\!+\!\!\!\!+\;^{\Delta_2 \cdots \Delta_n} s$. □

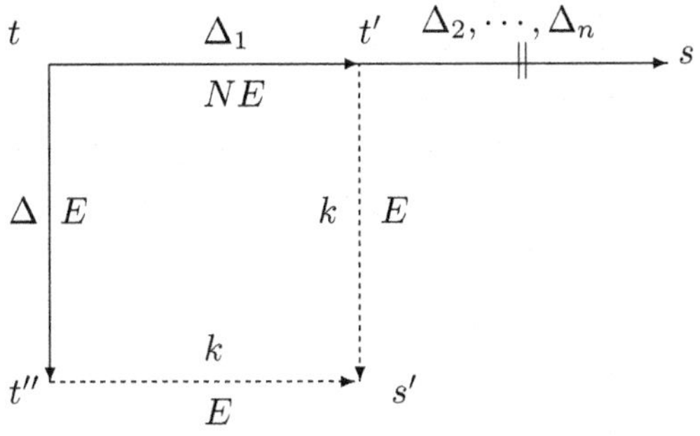

Fig. 8.

Theorem 1. *Let a TRS R be left-linear root balanced joinable and external. Then, R has the normal form property, and external reduction $\to_E$ is a normalizing strategy.*

Proof. Note that by externality, we have $NF = NF_E$ (NF_E denotes the set of the normal forms concerning $\to_E$). Thus, $\to_E$ is a reduction strategy for $\to$, and also for $\twoheadrightarrow\!\!\!\!\!\!\!+\!\!\!\!+\;'$. From Lemma 8 $\to_E$ is BWCR. From Lemma 9 it follows that if $t \twoheadrightarrow\!\!\!\!\!\!\!+\!\!\!\!+\;' s$ then $t =_E s$ or $t \longleftrightarrow\!\!\!\!\twoheadrightarrow_E \cdot \twoheadrightarrow\!\!\!\!\!\!\!+\!\!\!\!+\;' \cdot \twoheadleftarrow\!\!\!\!\longrightarrow_E s$. Taking $\to_E$ as $\to_s$ and $\twoheadrightarrow\!\!\!\!\!\!\!+\!\!\!\!+\;'$ as $\to$ respectively, we can apply Corollary 2. Thus, $\twoheadrightarrow\!\!\!\!\!\!\!+\!\!\!\!+\;'$ has the normal form property and $\to_E$ is a normalizing strategy for $\twoheadrightarrow\!\!\!\!\!\!\!+\!\!\!\!+\;'$. From $\to^+ = \twoheadrightarrow\!\!\!\!\!\!\!+\!\!\!\!+\;'^+$, the theorem follows. □

We remark that in Definition 9 *root reduction* imposed for balanced joinability of critical pairs can be relaxed to *stably external reduction*. The notion of *stable externality* was considered first as *stable index* by Nagaya, Sakai and Toyama [14].

An external context $C[\,]$ in $EXT(R)$ is *stable* if $C'[C[\,]\theta]$ is in $EXT(R)$ for any $C'[\,]$ in $EXT(R)$ and substitution θ. A redex occurrence Δ in $C[\Delta]$ is called *stably external* if $C[\,]$ is stably external. A reduction $t \to^{\Delta} s$ is *stably external* if Δ is a stably external redex of t. (Note that root reduction is clearly stably external as the hole $\square$ is a stably external context; thus, root redaction can be viewed as a special case of stably external reduction.) By replacing *root reduction* in Definition 9 with *stably external reduction*, we define *stable balanced joinability*. Then, all the proofs relied on *root balanced joinability* work also for *stable balanced joinability*. For example, Theorem 1 is improved by replacing "*root balanced joinable*" with "*stable balanced joinable*". Unfortunately, stably external redexes are not decidable; thus, appropriate decidable approximations of them are necessary for computable normalizing strategy. For decidable approximations of stable index reduction, we refer to [14].

7 Normalization of Quasi-External Reduction

This section improves the result of Theorem 1 by extending external reduction to quasi-external reduction; that is, there exist no infinite reduction sequences starting with a term having a normal form in which infinitely many external redexes are contracted. *Quasi-external reduction* (or *hyper-external reduction* [19]) is defined as $\to^*_{NE} \cdot \to_E \cdot \to^*_{NE}$ [22]. We first prove the next lemma.

Lemma 10. *Let R be left-linear root balanced joinable and external. Let $t \to^n_E s \in NF$ for some $n \geq 0$ and $t \to^* t'$. Then, we have $t' \to^m_E s$ for some $m \leq n$ (Figure 9).*

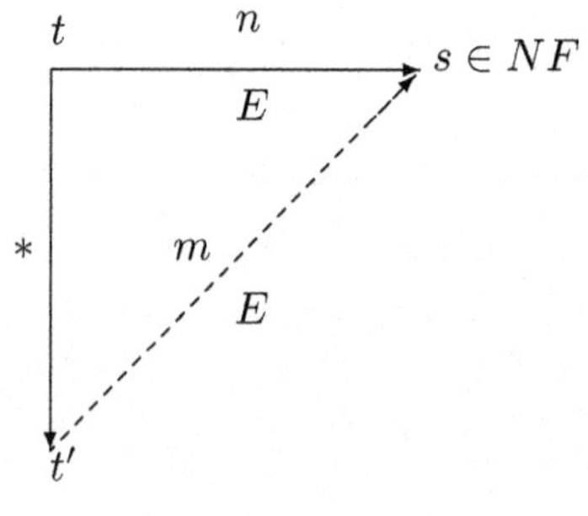

Fig. 9.

Proof. The proof is by induction on n. The case $n = 0$ is trivial as $t \equiv s \in NF$. Induction Step: We first prove the following claim: if $t \to^n_E s \in NF$ and $t \to t'$ then $t' \to^m_E s$ for some $m \leq n$ (Figure 10).

Proof of the claim. Let $t \to^{\Delta}_E t'' \to^{n-1}_E s \in NF$ and $t \to^{\Delta'} t'$ (Figure 11).

Case 1: Δ and Δ' are non-overlapping. By left-linearity of R, we can easily show that $t' \to_E s'$ and $t'' \to^* s'$ for some s'. From the induction hypothesis, it follows that $s' \to^{m'}_E s$ for some $m' \leq n-1$. Thus, we obtain $t' \to^{m'+1}_E s$.

Case 2: Δ and Δ' are overlapping. By using Lemma 6, we have $t' \to_E^i s'$ and $t'' \to_E^i s'$ for some s' and $i \geq 0$. Applying the BWCR Lemma to t'', we have $s' \to_E^{n-1-i} s$. Thus, it holds that $t' \to_E^{n-1} s$. Therefore, the claim follows.

We next prove that if $t \to_E^n s \in NF$ and $t \to^k t'$ then $t' \to_E^m s$ for some $m \leq n$. The proof is by induction on k. The case $k = 0$ is trivial. Induction Step: Let $t \to \hat{t} \to^{k-1} t'$. From the claim we have $\hat{t} \to_E^{n'} s$ for some $n' \leq n$. Thus, from the induction hypothesis with respect to k in case $n' = n$ or with respect to n in case $n' < n$, it follows that $t' \to_E^m s$ for some $m \leq n$. □

Theorem 2. *Let a TRS R be left-linear root balanced joinable and external. Then quasi-external reduction* $\to_{NE}^* \cdot \to_E \cdot \to_{NE}^*$ *is a normalizing strategy.*

Proof. Let t have a normal form s. Then by Theorem 1 we have $t \to_E^n s$ for some n. By using induction on n we prove that every quasi-external reduction starting with t is normalizing. The case $n = 0$ is trivial as $t \equiv s$. Let $t \to_E^n s$ $(n > 0)$. Take any one-step quasi-external reduction starting with t, say $t \to_{NE}^* t' \to_E t'' \to_{NE}^* \hat{t}$. From Lemma 10 we have $t' \to_E^{n'} s$ for some $n' \leq n$. Thus, by applying the BWCR Lemma to t' we obtain $t'' \to_E^{n''} s$ for some $n'' < n$ as $n'' + 1 = n'$. Again from Lemma 10 it holds that $\hat{t} \to_E^m s$ for some $m \leq n''$. From $m < n$ and the induction hypothesis it follows that every quasi-external reduction starting with $\hat{t}$ is normalizing. Therefore the theorem holds. □

8 Decidable Approximations of Externality

In this section we address the problem to find *decidable approximations* of external reduction. Durand and Middeldorp [6] presented a simple framework of decidable approximations to show normalizing strategies of orthogonal TRSs. We adapt this framework to *left-linear overlapping* (i.e., *non-orthogonal*) TRSs, based on the notions of *balanced weak Church-Rosser property* and *externality*. The framework of decidable approximations presented in [6] heavily relies on *tree automata* techniques. We first recall the basic notions concerning tree automata [4].

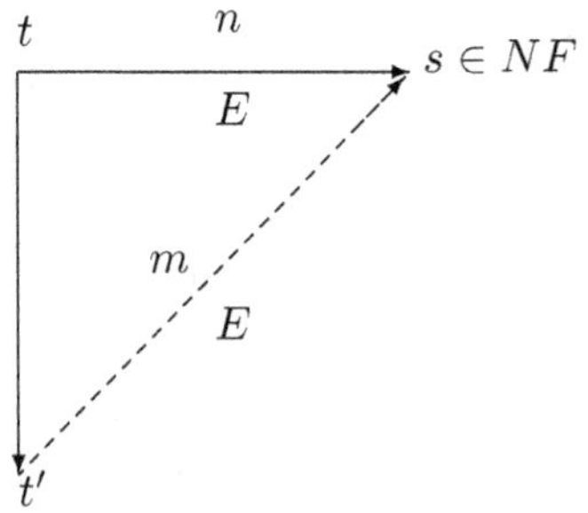

Fig. 10.

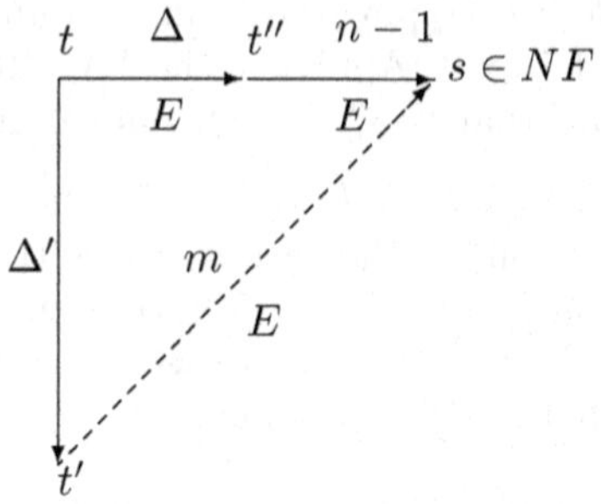

Fig. 11.

A *tree automaton* is a tuple $\mathcal{A} = (\mathcal{F}, Q, Q^f, \Pi)$ where $\mathcal{F}$ is a finite set of function symbols, Q is a finite set of *states*, $Q^f \subseteq Q$ is a set of *final* states and Π is a set of ground rewrite rules of the form $f(q_1, \ldots, q_n) \rightarrow q$ or $q \rightarrow q'$ where $f \in \mathcal{F}$, $q_1, \ldots, q_n, q, q' \in Q$. We use $\rightarrow_{\mathcal{A}}$ for the reduction relation $\rightarrow_{\Pi}$ on $T(\mathcal{F} \cup Q)$. A term $t \in T(\mathcal{F})$ is *accepted* by $\mathcal{A}$ if $t \rightarrow^*_{\mathcal{A}} q$ for some $q \in Q^f$. The tree language $L(\mathcal{A})$ recognized by $\mathcal{A}$ is the set of all terms accepted by $\mathcal{A}$. A set L is *regular* if there exists a tree automaton $\mathcal{A}$ such that $L = L(\mathcal{A})$. The following properties of tree automata are well-known [4]:

- The class of regular languages is effectively closed under union, intersection, difference, and complementation.
- The *membership* and the *emptiness* problems for regular languages are decidable.

Consider a TRS R over $\mathcal{F}$. We denote the set of all normal forms of R in $T(\mathcal{F}, \mathcal{V})$ by $NF(R)$ and the set of all redexes of R in $T(\mathcal{F}, \mathcal{V})$ by $RED(R)$. We introduce a fresh constant $\circ \notin \mathcal{F}$ and let $\mathcal{F}^\circ = \mathcal{F} \cup \{\circ\}$. We view R as a TRS over $\mathcal{F}^\circ$ when a reduction relation on $T(\mathcal{F}^\circ)$ is considered. Note that $NF(R)$ and $RED(R)$ have no terms containing $\circ$ since they are defined as subsets of $T(\mathcal{F}, \mathcal{V})$.

Let t° denote the term in $T(\mathcal{F}^\circ)$ obtained from a term $t \in T(\mathcal{F}, \mathcal{V})$ by replacing each variable in t with the constant $\circ$. We write $T^\circ = \{t^\circ \mid t \in T\}$ for a set $T \subseteq T(\mathcal{F}, \mathcal{V})$. We say a term set $T \subseteq T(\mathcal{F}, \mathcal{V})$ is *variable insensitive* if, for all $t \in T(\mathcal{F}, \mathcal{V})$, $t \in T$ iff $t^\circ \in T^\circ$. These notions are naturally extended to contexts over $\mathcal{F}$. Note that if T is a variable insensitive set of terms (or contexts) over $\mathcal{F}$ and T° is regular then T is decidable.

Lemma 11. *Let R be a left-linear TRS. Then, $RED(R)$, $NF(R)$ and $OUT(R)$ are variable insensitive. Moreover, $NF(R)^\circ$, $RED(R)^\circ$ and $OUT(R)^\circ$ are regular.*

Proof. From left-linearity of R it holds that s is a redex iff s° is a redex for any $s \in T(\mathcal{F}, \mathcal{V})$. Thus, $RED(R)$ is variable insensitive. Similarly we can show that $NF(R)$ and $OUT(R)$ are variable insensitive. From [4] it is clear that $NF(R)^\circ$, $RED(R)^\circ$ and $OUT(R)^\circ$ are regular. □

Definition 11 (Externality approximation mapping). *An externality approximation mapping α is a function mapping from a TRS R to a set of contexts such that $\alpha(R) \subseteq EXT(R)$. We say that α is decidable if $\alpha(R)$ is decidable for all R, and α is regular if, for all R, (i) $\alpha(R)^\circ$ is regular and (ii) $\alpha(R)$ is variable insensitive.*

Note that if an externality approximation mapping α is regular then it is decidable.

Definition 12 (α-external TRS). *We say a context $C[\ ]$ is α-external with respect to R if $C[\ ] \in \alpha(R)$. A redex occurrence Δ in $C[\Delta]$ is called α-external if $C[\ ]$ is α-external. A reduction $t \to^{\Delta} s$ is α-external if Δ is an α-external redex of t. We say that a TRS R is α-external if each term $t \in T(\mathcal{F}, \mathcal{V}) - NF(R)$ has an α-external redex (i.e., an α-external reduction of R is a reduction strategy).*

As an *α-external redex occurrence* Δ is an *external redex* and an *α-external* TRS R is *external*, we have the following theorem.

Theorem 3. *Let α be an externality approximation mapping (resp. an externality regular approximation mapping), and let a TRS R be left-linear root balanced joinable and α-external. Then, R has the normal form property, and α-external reduction is a normalizing strategy (resp. a computable normalizing strategy).*

Proof. From Theorem 1 it is trivial. □

The following theorem shows that the class of α-external TRSs is decidable if α is regular.

Theorem 4. *Let α be an externality regular approximation mapping. Then it is decidable whether a left-linear TRS R is α-external.*

Proof. Let R be a left-linear TRS over $\mathcal{F}$. Let $L = \{\ C[\Delta] \mid \Delta \in RED(R)$ and $C[\,] \in \alpha(R)\}$. Since $RED(R)$ and $\alpha(R)$ are variable insensible, L is variable insensible and we can write $L^\circ = \{\ C'[\Delta'] \mid \Delta' \in RED(R)^\circ$ and $C'[\,] \in \alpha(R)^\circ\}$. Since $RED(R)^\circ$ and $\alpha(R)^\circ$ are regular, there exist two tree automata $\mathcal{A}_{red} = (\mathcal{F}^\circ, Q_{red}, {Q^f}_{red}, \Pi_{red})$ and $\mathcal{A}_\alpha = (\mathcal{F}^\circ \cup \{\Box\}, Q_\alpha, {Q^f}_\alpha, \Pi_\alpha)$, where $Q_{red} \cap Q_\alpha = \varnothing$, which recognize $RED(R)^\circ$ and $\alpha(R)^\circ$ respectively. Without loss of generality we may suppose ${Q^f}_{red} = \{q_{red}\}$ and ${Q^f}_\alpha = \{q_\alpha\}$ [4]. Let $\mathcal{A} = (\mathcal{F}^\circ, Q_{red} \cup Q_\alpha, \{q_\alpha\}, \Pi_L)$ where $\Pi_L = \Pi_{red} \cup (\Pi_\alpha - \{\Box \to p \mid \Box \to p \in \Pi_\alpha\}) \cup \{q_{red} \to p \mid \Box \to p \in \Pi_\alpha\}$. Then it can be shown that $L^\circ = L(\mathcal{A})$; thus L° is regular. From the definition of α-externality, the TRS R is α-external iff $(T(\mathcal{F}, \mathcal{V}) - NF(R)) - L = \varnothing$. Since $T(\mathcal{F}, \mathcal{V})$, $NF(R)$ and L are variable insensitive, we have that $(T(\mathcal{F}, \mathcal{V}) - NF(R)) - L$ is variable insensitive. Thus, it holds that $(T(\mathcal{F}, \mathcal{V}) - NF(R)) - L = \varnothing$ iff $(T(\mathcal{F}^\circ) - NF(R)^\circ) - L^\circ = \varnothing$. Since $(T(\mathcal{F}^\circ) - NF(R)^\circ) - L^\circ$ is regular, the emptiness of $(T(\mathcal{F}^\circ) - NF(R)^\circ) - L^\circ$ is decidable. □

We next consider *decidable approximations* of TRSs to find an *externality regular approximation mapping* α. An *extended* TRS (eTRS for short) R over $\mathcal{F}$ is a finite set of *extended* rewrite rules $l \rightarrow r$ in which the right-hand side r may have *extra variables* not occurring in the left-hand side l. Similarly to TRSs, we can view R as an eTRS over $\mathcal{F}^\circ \cup \{\Box\}$ when a reduction relation on $T(\mathcal{F}^\circ \cup \{\Box\})$ is considered, imposing the restriction that when an extended rewrite rule $l \rightarrow r$ is applied, the extra variables in r are instantiated by arbitrary terms in $T(\mathcal{F}^\circ)$; thus, a reduction does not generate new holes.

Following Durand and Middeldorp [6], we say that an eTRS R over $\mathcal{F}$ is *regularity preserving* if $(R^{-1})^*(L) = \{\, t \mid \exists s \in L\ t \rightarrow_R^* s \,\}$ is regular for every regular tree language L over $\mathcal{F}$ (resp. $\mathcal{F}^\circ \cup \{\Box\}$), where $\rightarrow_R$ is the reduction relation on $T(\mathcal{F})$ (resp. $T(\mathcal{F}^\circ \cup \{\Box\})$) defined by R.

An eTRS R_a over $\mathcal{F}$ is an *approximation* of a TRS R over $\mathcal{F}$ if $\rightarrow_R^* \subseteq \rightarrow_{R_a}^*$. Note that the definition of approximation is slightly different from that in [6] which imposes the extra condition $NF(R) = NF(R_a)$. The following definition is due to [6].

Definition 13 (Regularity preserving approximation mapping). *A regularity preserving approximation mapping τ is a function mapping from a left-linear TRS R to a left-linear eTRS $\tau(R)$ such that (i) $\tau(R)$ is an approximation of R and (ii) $\tau(R)$ is regularity preserving.*

Definition 14 (τ-external context). *Let τ be a regularity preserving approximation mapping. An outer context $C[\ ]$ is τ-external with respect to R if any s obtained by $C[\ ] \rightarrow_{\tau(R)}^* s$ is in $OUT(R)$. The set of all τ-external contexts with respect to R is denoted by $\alpha_\tau(R)$.*

Lemma 12. *Let τ be a regularity preserving approximation mapping and R a left-linear TRS over $\mathcal{F}$. Let $s \in T(\mathcal{F} \cup \{\Box\}, \mathcal{V})$ and $s^\circ \rightarrow_{\tau(R)}^* u$. Then there exists some $t \in T(\mathcal{F} \cup \{\Box\}, \mathcal{V})$ such that $s \rightarrow_{\tau(R)}^* t$ and $t^\circ \equiv u$.*

Proof. Let $s^\circ \rightarrow_{\tau(R)}^k u$. We prove the claim by induction on k. The case $k = 0$ is trivial. Induction Step: Let $s^\circ \rightarrow_{\tau(R)} p \rightarrow_{\tau(R)}^{k-1} u$. From left-linearity of $\tau(R)$, there exists some $q \in T(\mathcal{F} \cup \{\Box\}, \mathcal{V})$ such that $s \rightarrow_{\tau(R)} q$ and $q^\circ \equiv p$. Thus, form $q^\circ \rightarrow_{\tau(R)}^{k-1} u$ and the induction hypothesis, we have $t \in T(\mathcal{F} \cup \{\Box\}, \mathcal{V})$ such that $q \rightarrow_{\tau(R)}^* t$ and $t^\circ \equiv u$. $\Box$

Lemma 13. *Let τ be a regularity preserving approximation mapping and R a left-linear TRS over $\mathcal{F}$. Then $\alpha_\tau(R)$ is variable insensitive.*

Proof. We show that $C[\,] \in \alpha_\tau(R)$ iff $C[\,]^\circ \in \alpha_\tau(R)^\circ$. *If-part*: Let $C[\,]^\circ \in \alpha_\tau(R)^\circ$ and $C[\ ] \rightarrow_{\tau(R)}^* s$. Then we have $C[\ ]^\circ \rightarrow_{\tau(R)}^* s^\circ$. Since $C[\ ]^\circ \in \alpha_\tau(R)^\circ$, it holds that $s^\circ \in OUT(R)^\circ$. As $OUT(R)$ is variable insensitive, $s \in OUT(R)$. Thus, $C[\,] \in \alpha_\tau(R)$. *Only-if-part*: Let $C[\,] \in \alpha_\tau(R)$ and $C[\ ]^\circ \rightarrow_{\tau(R)}^* u$. By Lemma 12 there exists some t such that $C[\ ] \rightarrow_{\tau(R)}^* t$ and $t^\circ \equiv u$. Since $C[\,] \in \alpha_\tau(R)$, it holds that $t \in OUT(R)$. As $OUT(R)$ is variable insensitive, $u \in OUT(R)^\circ$. Thus, $C[\,]^\circ \in \alpha_\tau(R)^\circ$. $\Box$

Lemma 14. *Let τ be a regularity preserving approximation mapping. Then α_τ is an externality regular approximation mapping. Hence, $\alpha_\tau(R)$ is a computable approximation of $EXT(R)$.*

Proof. Since $\rightarrow_R^* \subseteq \rightarrow_{\tau(R)}^*$, it is trivial that $\alpha_\tau(R) \subseteq EXT(R)$. By Lemma 13, $\alpha_\tau(R)$ is variable insensitive. Hence, we shall show that $\alpha_\tau(R)^\circ$ is regular. Let $CONT$ be the set of all contexts over $\mathcal{F}$ containing precisely one hole and let $NOUT(R) = CONT - OUT(R)$. Then we have $NOUT(R)^\circ = CONT^\circ - OUT(R)^\circ$. Since $CONT^\circ$ and $OUT(R)^\circ$ are regular, $NOUT(R)^\circ$ is regular. Let $L_\tau = \{s \in T(\mathcal{F} \cup \{\Box\}, \mathcal{V}) \mid \exists C[\,] \in NOUT(R),\ s \rightarrow_{\tau(R)}^* C[\,]\}$. Then, by Lemma 12 it is easily shown that $s \rightarrow_{\tau(R)}^* C[\,]$ for some $C[\,] \in NOUT(R)$ iff $s^\circ \rightarrow_{\tau(R)}^* C'[\,]$ for some $C'[\,] \in NOUT(R)^\circ$. Thus, we have $L_\tau^\circ = \{s' \in T(\mathcal{F}^\circ \cup \{\Box\}) \mid \exists C'[\,] \in NOUT(R)^\circ,\ s' \rightarrow_{\tau(R)}^* C'[\,]\}$. Since $\tau(R)$ is regularity preserving, we have $L_\tau^\circ = (\tau(R)^{-1})^*(NOUT(R)^\circ)$; thus, L_τ° is regular. From $\alpha_\tau(R)^\circ = CONT^\circ - L_\tau^\circ$, it follows that $\alpha_\tau(R)^\circ$ is regular. □

Definition 15 (τ-external TRS). *We say $C[\,]$ is τ-external with respect to R if $C[\,] \in \alpha_\tau(R)$. A redex occurrence Δ in $C[\Delta]$ is called τ-external if $C[\,]$ is τ-external. A reduction $t \rightarrow^\Delta s$ is τ-external if Δ is a τ-external redex of t. We say that a TRS R is τ-external if each term $t \in T(\mathcal{F}, \mathcal{V}) - NF(R)$ has a τ-external redex.*

Theorem 5. *Let τ be a regularity preserving approximation mapping and let a TRS R be left-linear root balanced joinable and τ-external. Then, R has the normal form property, and τ-external reduction is a computable normalizing strategy.*

Proof. From Theorem 3 and Lemma 14 it is clear. □

The following theorem shows that the class of τ-external TRSs is decidable.

Theorem 6. *Let τ be a regularity preserving approximation mapping. Then it is decidable whether a left-linear TRS R is τ-external.*

Proof. From Theorem 4 and Lemma 14 it is clear. □

The first idea of regularity preserving approximations was proposed by Huet and Lévy [8] as the *strong* approximation of orthogonal TRSs, which is obtained by replacing the right-hand side of every rewrite rule with a fresh variable not occurring in the left-hand side. Oyamaguchi [21] gave a better approximation, the *NV* approximation, which is obtained by replacing all variables in the right-hand side of every rewrite rule with distinct fresh variables. Jacquemard [9], Nagaya and Toyama [15] introduced the *growing* approximation, which is obtained by replacing all variables in the *left-hand* sides of every rewrite rule that occur at a depth greater than 1 with distinct fresh variables [15]. In these approximations, the regularity preserving property depends only on *left-linearity*, but not on *orthogonality* [9, 15, 6]. Thus, we can use them as regularity preserving approximations for arbitrary left-linear TRSs.

An approximation mapping τ is *strong* (resp. *NV*, *growing*) if $\tau(\mathcal{R})$ is a strong (resp. *NV*, *growing*) approximation of $\mathcal{R}$ for every TRS $\mathcal{R}$. Then, from Theorems 5 and 6, the following corollaries hold.

Corollary 4. *Let a TRS R be left-linear root balanced joinable and strong-external (resp. NV-external, growing-external). Then, R has the normal form property, and strong-external (resp. NV-external, growing-external) reduction is a computable normalizing strategy.*

Corollary 5. *It is decidable whether a left-linear TRS R is strong-external (resp. NV-external, growing-external).*

Example 6. Let R be combinatory logic CL $\cup$

$$\begin{cases} f(g(x,K),K) \to (K \cdot ((K \cdot x) \cdot x)) \cdot x \\ f(g(x,K),K) \to ((S \cdot K) \cdot x) \cdot x \\ f(g(K,y),S) \to g(y,y) \\ g(S,S) \to S. \end{cases}$$

As R has overlapping redexes at the root of $f(g(x,K),K)$, we obtain the critical pair $\langle (K\cdot((K\cdot x)\cdot x))\cdot x,\ ((S\cdot K)\cdot x)\cdot x\rangle$. The critical pair meets by root reductions $(K \cdot ((K \cdot x) \cdot x)) \cdot x \to_r (K \cdot x) \cdot x \to_r x$ and $((S \cdot K) \cdot x) \cdot x \to_r (K \cdot x) \cdot (x \cdot x) \to_r x$. Thus R is root balanced joinable. Let the strong approximation of combinatory logic CL be τ(CL):

$$\begin{cases} ((S \cdot x) \cdot y) \cdot z \to w \\ (K \cdot x) \cdot y \to z. \end{cases}$$

Then the strong approximation $\tau(R)$ of R is τ(CL)$\cup$

$$\begin{cases} f(g(x,K),K) \to z \\ f(g(x,K),K) \to z \\ f(g(K,y),S) \to z \\ g(S,S) \to z. \end{cases}$$

Since R is *transitive* [14, 23] (*forward-branching* [5]), it is strong-external. Thus, from Corollary 4, R has the normal form property, and strong-external reduction is a computable normalizing strategy. Consider a term of the form $f(g(\Delta_1,\Delta_2),\Delta_3)$ in R, where Δ_i $(i = 1,2,3)$ are redex occurrences. Then neither Δ_1 nor Δ_2 is a strong-external redex, as $f(g(\square,\Delta_2),\Delta_3) \to_{\tau(R)} f(g(\square,K),\Delta_3) \to_{\tau(R)} f(g(\square,K),K) \notin OUT(R)$ and $f(g(\Delta_1,\square),\Delta_3) \to_{\tau(R)} f(g(K,\square),\Delta_3) \to_{\tau(R)} f(g(K,\square),S) \notin OUT(R)$ respectively. The rightmost redex occurrence Δ_3 is strong-external since one can easily check $f(g(\Delta_1,\Delta_2),\square) \to^*_{\tau(R)} s \in OUT(R)$ for any term s.

Example 7. Let R be combinatory logic CL $\cup$

$$\begin{cases} f(x,S) \to x \cdot S \\ f(S,y) \to S \cdot y \\ f(x,y) \to x \cdot y. \end{cases}$$

Since R is weakly orthogonal, it is trivially root balanced joinable. Let the strong approximation $\tau(R)$ of R be $\tau(\mathrm{CL})\cup$

$$\begin{cases} f(x,S) \to z \\ f(S,y) \to z \\ f(x,y) \to z. \end{cases}$$

Since a term of the form $f(s,t)$ in R certainly gives a redex independent on s and t, one can easily check strong-externality of R, ignoring the first two rules $f(x,S) \to z$ and $f(S,y) \to z$. Thus, from Corollary 4, R has the normal form property, and strong-external reduction is a computable normalizing strategy. Note that if the third rule $f(x,y) \to z$ does not exist, then R is not strong-external as $f(\Delta_1,\Delta_2)$ has no strong-external redexes.

9 Left-Normal Systems

In this section we discuss a syntactic characterization of *external overlapping* TRSs. Such a syntactical characterization was found by O'Donnell [17] for orthogonal TRSs. He proved that if an orthogonal TRS R is *left-normal* then *leftmost-outermost reduction* is normalizing. We show that his result can be naturally extended to root balanced joinable TRSs.

Definition 16. *The set $T_L(\mathcal{F},\mathcal{V})$ of the left-normal terms constructed from $\mathcal{F}$ and $\mathcal{V}$ is inductively defined as follows:*

1. $x \in T_L(\mathcal{F},\mathcal{V})$ **if** $x \in \mathcal{V}$,
2. $f(t_1,\cdots,t_{p-1},s_p,x_{p+1}\cdots,x_n) \in T_L(\mathcal{F},\mathcal{V}) \quad (0 \le p \le n)$
   if $f \in \mathcal{F}$, $t_1,\cdots,t_{p-1} \in T(\mathcal{F})$, $s_p \in T_L(\mathcal{F},\mathcal{V})$, $x_{p+1},\cdots,x_n \in \mathcal{V}$, *and* $f(t_1,\cdots,t_{p-1},s_p,x_{p+1}\cdots,x_n)$ *is linear.*

A TRS R over $\mathcal{F}$ is *left-normal* [8,17,22] if for any rule $l \to r$ in R, l is a left-normal term in $T_L(\mathcal{F},\mathcal{V})$. From the definition of left-normal terms, a left-normal TRS R is left-linear, and it may be overlapping.

Definition 17 (Left-outer context). *A context $C[\,]$ is left-outer if every redex Δ' of $C[\,]$ occurs right of $\Box$ (i.e., $C[\,] \equiv C'[\Box,\Delta']$ for some $C'[\,,\,]$) whenever it exists.*

Definition 18 (Left-outer redex). *A redex occurrence Δ of $C[\Delta]$ is called left-outer if $C[\,]$ is left-outer. A reduction $t \to^{\Delta} s$ is left-outer if Δ is a left-outer redex of t.*

Let $\alpha_L(R)$ be the set of all left-outer contexts with respect to a TRS R. Then the decidability of $\alpha_L(R)$ is trivial. We shall show that $\alpha_L(R)$ is a decidable approximation of $EXT(R)$ if R is left-normal.

Lemma 15. *Let a TRS R be left-normal. If $C[\,]$ is left-outer and $C[\,] \to^* s$, then:*

(1) *s is a left-outer context,*
(2) *for any $t \in T_L(\mathcal{F},\mathcal{V})$, if $t \preceq s$ then $t \preceq C[\,]$.*

Proof. By induction on the size of $C[\,]$, we will prove (1) and (2) simultaneously.

Basic step: $C[\,] \equiv \Box$. Then (1) and (2) are trivial.

Induction step: Since $C[\,] \not\equiv \Box$, we can write $C[\,] \equiv f(t_1, \cdots, t_{p-1}, C_p[\,], t_{p+1}, \cdots, t_n)$, where $t_1, \cdots, t_{p-1}$ are normal forms and $C_p[\,]$ is left-outer.

(1) Suppose that s is not left-outer. As $C[\,]$ is a left-outer context, there exists some non-left-outer context $\tilde{C}[\,] \equiv f(t_1, \cdots, t_{p-1}, \tilde{C}_p[\,], \tilde{t}_{p+1}, \cdots, \tilde{t}_n)$ such that $C[\,] \rightarrow^* \tilde{C}[\,] \rightarrow^* s$ where $C_p[\,] \rightarrow^* \tilde{C}_p[\,]$ and $t_{p+1} \rightarrow^* \tilde{t}_{p+1}, \cdots, t_n \rightarrow^* \tilde{t}_n$. From the induction hypothesis with respect to (1) and $C_p[\,] \rightarrow^* \tilde{C}_p[\,]$, $\tilde{C}_p[\,]$ is left-outer. Since $\tilde{C}[\,]$ is not left-outer, there exists a redex Δ such that $\Box \in \Delta \trianglelefteq \tilde{C}[\,]$. As $\tilde{C}_p[\,]$ is left-outer, $\Delta \ntrianglelefteq \tilde{C}_p[\,]$. Thus, we have $\Delta \equiv \tilde{C}[\,]$. Hence, $l \preceq \tilde{C}[\,] \equiv f(t_1, \cdots, t_{p-1}, \tilde{C}_p[\,], \tilde{t}_{p+1}, \cdots, \tilde{t}_n)$ for some $l \rightarrow r \in R$ such that $l \equiv f(t_1, \cdots, t_{q-1}, s_q, x_{q+1}, \cdots, x_n) \in T_L(\mathcal{F}, \mathcal{V})$. Since $C[\,]$ is left-outer, it holds that $l \not\preceq C[\,]$; thus, $q \geq p$. Since $t_i \not\preceq \tilde{C}_p[\,]$ for any ground term t_i, $q \leq p$ holds. So we have $p = q$. From the induction hypothesis with respect to (2) and $s_p \preceq \tilde{C}_p[\,]$, we have $s_p \preceq C_p[\,]$. Thus, $f(t_1, \cdots, t_{p-1}, s_p, x_{p+1}, \cdots, x_n) \preceq f(t_1, \cdots, t_{p-1}, C_p[\,], t_{p+1}, \cdots, t_n) \equiv C[\,]$; it contradicts to the fact that $C[\,]$ is left-outer. Hence, s must be left-outer.

(2) From (1) it follows that every s' must be left-outer for $C[\,] \rightarrow^* s' \rightarrow^* s$; thus, we can write $s \equiv f(t_1, \cdots, t_{p-1}, C'_p[\,], t'_{p+1}, \cdots, t'_n)$ where $C_p[\,] \rightarrow^* C'_p[\,]$ and $t_{p+1} \rightarrow^* t'_{p+1}, \cdots, t_n \rightarrow^* t'_n$. Let $t \preceq s$ for some $t \equiv f(t_1, \cdots, t_{q-1}, s_q, x_{q+1}, \cdots, x_n) \in T_L(\mathcal{F}, \mathcal{V})$. If $q < p$ then it is clear that $t \preceq C[\,]$. If $q = p$ then $s_q \equiv s_p \preceq C'_p[\,]$. From the induction hypothesis with respect to (2) we have $s_p \preceq C_p[\,]$. Thus, $t \preceq C[\,]$. □

Lemma 16. *Let a TRS R be left-normal. Then* $\alpha_L(R) \subseteq EXT(R)$.

Proof. Note that the left-outer contexts are outer. Thus from Lemma 15 (1) the left-outer contexts are external. □

Thus, α_L is an externality decidable approximation mapping for the class of left-normal TRSs.

Lemma 17. *Let a TRS R be left-normal. Then R is* α_L*-external (i.e., every reducible term has a left-outer redex).*

Proof. Trivial. □

Theorem 7. *Let a TRS R be root balanced joinable and left-normal. Then, R has the normal form property, and left-outer reduction is a computable normalizing strategy.*

Proof. It follows from Theorem 1, Lemmas 16 and 17. □

Definition 19 (Leftmost-outermost redex). *A redex occurrence* Δ *of t is called* leftmost-outermost *if* Δ *is the leftmost of the outermost redexes of t. A reduction* $t \rightarrow^{\Delta} s$ *is leftmost-outermost if* Δ *is a leftmost-outermost redex of t.*

As *leftmost-outermost redexes* are clearly left-outer redexes, we have the following corollary.

Corollary 6. *Let a TRS R be root balanced joinable and left-normal. Then, R has the normal form property, and leftmost-outermost reduction is a computable normalizing strategy.*

Note that every weakly orthogonal left-normal TRS is root balanced joinable. Thus the following corollary holds.

Corollary 7. *Let a TRS R be weakly orthogonal and left-normal. Then, R has the normal form property, and leftmost-outermost reduction is a computable normalizing strategy.*

Example 8. Let R be combinatory logic CL $\cup$

$$\begin{cases} \text{pred} \cdot (\text{succ} \cdot x) \to x \\ \text{succ} \cdot (\text{pred} \cdot x) \to x. \end{cases}$$

It is clear that R is weakly orthogonal and left-normal. Thus, from Corollary 7, R has the normal form property, and leftmost-outermost reduction is a computable normalizing strategy.

Example 9. Let R be combinatory logic CL $\cup$

$$\begin{cases} (A \cdot x) \cdot y \to ((x \cdot K) \cdot x) \cdot y \\ (A \cdot S) \to (S \cdot K) \cdot A. \end{cases}$$

Clearly, R is left-normal and it has overlapping redexes in $(A \cdot S) \cdot y$. Thus, we have the critical pair $\langle((S \cdot K) \cdot A) \cdot y, ((S \cdot K) \cdot S) \cdot y\rangle$. Since the critical pair can join by root reductions of two steps $((S \cdot K) \cdot A) \cdot y \to_r (K \cdot y) \cdot (A \cdot y) \to_r y$ and $((S \cdot K) \cdot S) \cdot y \to_r (K \cdot y) \cdot (S \cdot y) \to_r y$, R is root balanced joinable. Thus, from Corollary 6, R has the normal form property, and leftmost-outermost reduction is a computable normalizing strategy.

Example 10. Let R be combinatory logic CL $\cup$

$$\begin{cases} (K \cdot A) \cdot y \to (K \cdot B) \cdot y \\ (K \cdot B) \cdot y \to (K \cdot A) \cdot y \\ A \to A \\ B \to B. \end{cases}$$

It is clear that R is left-normal and it has the two critical pairs $\langle (K \cdot A) \cdot y, (K \cdot B) \cdot y\rangle$ and $\langle (K \cdot B) \cdot y, (K \cdot A) \cdot y\rangle$. We have root reduction $(K \cdot A) \cdot y \to_r (K \cdot B) \cdot y \to_r B$ and $(K \cdot B) \cdot y \to_r B \to_r B$ for the critical pair $\langle (K \cdot A) \cdot y, (K \cdot B) \cdot y\rangle$, and $(K \cdot B) \cdot y \to_r (K \cdot A) \cdot y \to_r A$ and $(K \cdot A) \cdot y \to_r A \to_r A$ for the critical pair $\langle (K \cdot B) \cdot y, (K \cdot A) \cdot y\rangle$ respectively. Thus, R is root balanced joinable. Therefore, from Corollary 6, R has the normal form property, and leftmost-outermost reduction is a computable normalizing strategy. Note that though R has the unique normal form property due to the normal form property, R is not Church-Rosser as $(K \cdot A) \cdot y$ can be reduced into two constants A and B which cannot be joined.

10 Conclusion

In this paper we have investigated normalizing strategies for left-linear overlapping TRSs. We have introduced the concept of the balanced weak Church-Rosser (BWCR) property and related it to a normalizing strategy based on the BWCR Lemma, which is presented in an abstract framework depending solely on the reduction relation. Applying this abstract framework to TRSs, we have shown that external reduction is a normalizing strategy for the class of left-linear TRSs in which every critical pair can be joined with root balanced reductions and every reducible term has an external redex. Further, we have presented computable normalizing strategies based on decidable approximations of external redexes.

An interesting direction for further research is application to *higher-order* rewriting systems, like Klop's *combinatory reduction system* [11]. We believe the BWCR lemma can provide an accessible means of developing computable normalizing strategies uniformly for various higher-order rewriting systems. Another interesting issue is *root-external* reduction for non-orthogonal TRSs, which is very parallel to *root-needed* reduction for orthogonal TRSs developed by Middeldorp [13]. As root-normalizing strategy is more fundamental and complicated than normalizing strategy, we need to generalize the theoretical framework for dealing with approximated decidable reduction based on root-externality.

Acknowledgments

The first idea of the balanced weak Church-Rosser property was obtained in 1991 during my stay at the CWI, Amsterdam. The author is especially grateful to Jan Willem Klop and Henk Barendregt for offering a stimulating environment for research and for valuable discussions. The author also thanks Aart Middeldorp, Fer-Jan de Vries, Vincent van Oostrom and Jean-Pierre Jouannaud for helpful comments and suggestions on early drafts, and Yuki Chiba for suggesting the term "*variable insensitive*". The author would like to thank the anonymous referees for their very detailed comments and suggestions that improve the presentation of this article significantly.

References

1. Baader, F. and Nipkow ,T. (1998), Term Rewriting and All That, Cambridge University Press.
2. Barendregt, H. P. (1981), The lambda calculus, its syntax and semantics, North-Holland.
3. Comon, H. (2000), Sequentiality, Second Order Monadic Logic and Tree Automata, *Information and Computation* **157**, 25-51.
4. Comon, H., Dauchet, M., Gilleron, R., Lugiez, D., Tison, S. and Tommasi, M., Tree Automata Techniques and Applications, *Preliminary Version*, *http://www.grappa.univ-lille3.fr/tata/*.
5. Durand, I. (1994), Bounded, Strongly Sequential and Forward-Branching Term Rewriting Systems *Journal of Symbolic Computation* **18**, 319-352.

6. Durand, I. and Middeldorp, A. (2005), Decidable Call-by-Need Computations in Term Rewriting, *Information and Computation* **196**, 95-126.
7. Hindley, J. R. and Seldin, J. P. (1986), Introduction to Combinators and λ-Calculus, Cambridge University Press.
8. Huet, G. and Lévy, J. J. (1991), Computations in Orthogonal Term Rewriting Systems, In Lassez, J.-L. and Plotkin, G. editors, Computational Logic: Essays in Honor of Alan Robinson, MIT Press, 396-443.
9. Jacquemard, F. (1996), Decidable Approximations of Term Rewriting Systems, in *Proceedings, 7th International Conference, Rewriting Techniques and Applications, Lecture Notes in Computer Science* **1103**, 362-376.
10. Jouannaud, J. P. and Sadfi W. (1995), Strong sequentiality of left-linear overlapping rewrite systems, in *Proceedings, 4th International Workshop on Conditional Term Rewriting Systems, Jerusalem, Lecture Notes in Computer Science* **968**, 235-246.
11. Klop, J. W. (1980), Combinatory reduction systems, Dissertation, Univ. of Utrecht.
12. Klop, J. W. and Middeldorp, A. (1991), Sequentiality in orthogonal term rewriting systems, *J. Symbolic Comput.* **12**, 161-195.
13. Middeldorp, A. (1997), Call by need computations to root-stable form, in *Proceedings of the 24th ACM Symposium on Principles of Programming Languages*, 94-105.
14. Nagaya, T., Sakai, M. and Toyama, Y. (1998), Index reduction of overlapping strongly sequential systems, *IEICE TRANS. INF. & SYST.* **E-81D**, 419-426.
15. Nagaya, T. and Toyama, Y. (2002), Decidability for left-linear growing term rewriting systems, *Information and Computation* **178**, 499-514.
16. Newman, M. H. A. (1942), On theories with a combinatorial definition of "equivalence", *Ann. Math.* **43**, 223-243.
17. O'Donnell, M. J. (1977), Computing in systems described by equations, *Lecture Notes in Computer Science* **58**, Springer-Verlag.
18. Ohlebusch, E. (2002), Advanced Topics in Term Rewriting, Springer-Verlag.
19. van Oostrom, V. and de Vrijer, R. (2003), Strategies, In Terese, Term Rewriting Systems, Cambridge University Press, 475-547.
20. van Oostrom, V. (2005), Delimiting diagrams, in *Proceedings of 5th International Workshop on Reduction Strategies in Rewriting and Programming*, 17-24.
21. Oyamaguchi, M. (1993), NV-Sequentiality: A Decidable Condition for Call-by-Need Computations in Term Rewriting Systems, *SIAM Journal on Computation* **22**(1), 112-135.
22. Terese (2003), Term Rewriting Systems, Cambridge University Press.
23. Toyama, Y.Smetsers, S. Eeklen, M. Plasmeijer, R. (1993), The functional strategy and transitive term rewriting systems, In Sleep,M.R. etc., editors, Term Graph Rewriting, Wiley, 61-75.
24. Toyama, Y. (1992), Strong Sequentiality of Left-Linear Overlapping Term Rewriting Systems, in *Proceedings of the 7th IEEE Symposium on Logic in Computer Science*, 274-284.
25. Vuillemin, J. (1974), Correct and optimal implementations of recursion in a simple programming language, *J. Comput. System Sci.* **9**, 332-354.

Higher-Order Rewriting: Framework, Confluence and Termination

Jean-Pierre Jouannaud*

LIX/CNRS UMR 7161 & École Polytechnique,
F-91400 Palaiseau
http://www.lix.polytechnique.fr/Labo/Jean-Pierre.Jouannaud/

1 Introduction

Equations are ubiquitous in mathematics and in computer science as well. This first sentence of a survey on first-order rewriting borrowed again and again characterizes best the fundamental reason why rewriting, as a technology for processing equations, is so important in our discipline [10]. Here, we consider *higher-order rewriting*, that is, rewriting higher-order functional expressions at higher-types. Higher-order rewriting is a useful generalization of first-order rewriting: by rewriting higher-order functional expressions, one can process abstract syntax as done for example in program verification with the prover Isabelle [27]; by rewriting expressions at higher-types, one can implement complex recursion schemas in proof assistants like Coq [12].

In our view, the role of higher-order rewriting is to design a type-theoretic framework in which computation and deduction are integrated by means of higher-order rewrite rules, while preserving decidability of typing and coherence of the underlying logic. The latter itself reduces to type preservation, confluence and strong normalization.

It is important to understand why there have been very different proposals for higher-order rewriting, starting with Klop's *combinatory reduction systems* in 1980, Nipkow's *higher-order rewriting* in 1991 and Jouannaud and Okada's *executable higher-order algebraic specifications* in 1991 as well: these three approaches tackle the same problem, in different contexts, with different goals requiring different assumptions.

Jan Willem Klop was mostly interested in generalizing the theory of lambda calculus, and more precisely the confluence and finite developments theorems. Klop does not assume any type structure. As a consequence, the most primitive operation of rewriting, searching for a redex, is already a problem. Because he wanted to encode pure lambda calculus and other calculi as combinatory reduction systems, he could not stick to a pure syntactic search based on first-order pattern matching. He therefore chose to search via finite developments, the only way to base a finite search on beta-reduction in the absence of typing

* Project LogiCal, Pôle Commun de Recherche en Informatique du Plateau de Saclay, CNRS, École Polytechnique, INRIA, Université Paris-Sud.

A. Middeldorp et al. (Eds.): Processes... (Klop Festschrift), LNCS 3838, pp. 224–250, 2005.

assumptions. And because of his interest in simulating pure lambda calculi, he had no real need for termination, hence concentrated on confluence results. Therefore, his theory in its various incarnations is strongly influenced by the theory of residuals initially developed for the pure lambda calculus.

Nipkow was mainly interested in investigating the meta-theory of Isabelle and in proving properties of functional programs by rewriting, goals which are actually rather close to the previous one. Functional programs are typed lambda terms, hence he needed a typed structure. He chose searching via higher-order pattern matching, because plain pattern matching is too poor for expressing interesting transformations over programs with finitely many rules. This choice of higher-order pattern matching instead of finite developments is of course very natural in a typed framework. He assumes termination for which proof methods were still lacking at that time. His (local) confluence results rely on the computation of higher-order critical pairs -because higher-order pattern matching is used for searching redexes- via higher-order unification. Nipkow restricted lefthand sides of rewrite rules to be patterns in the sense of Miller [25]. The main reason for this restriction is that higher-order pattern matching and unification are tractable in this case which, by chance, fits well with most intended applications.

Jouannaud and Okada were aiming at developping a theory of typed rewrite rules that would generalize the notion of recursor in the calculus of inductive constructions, itself generalizing Gödel's system T. This explains their use of plain pattern matching for searching a redex: there is no reason for a more sophisticated search with recursors. In this context, the need for strongly terminating calculi has two origins: ensuring consistency of the underlying logic, that is, the absence of a proof for falsity on the one hand, and decidability of the type system in the presence of dependent types on the other hand. Confluence is needed as well, of course, as is type preservation. The latter is easy to ensure while the former is based on the computation of first-order critical pairs -because first-order pattern matching is used for searching redexes. This explains the emphasis on termination criteria in this work and its subsequent developments.

Our goal in this paper is to present a unified framework borrowed from Jouannaud, Rubio and van Raamsdonk [22] for the most part, in which redexes can be searched for by using either plain or higher-order rewriting, confluence can be proved by computing plain or higher-order critical pairs, and termination can be proved by using the higher-order recursive path ordering of Jouannaud and Rubio [19].

We first present examples showing the need for both search mechanisms based on plain and higher-order pattern matching on the one hand, and for a rich type structure on the other hand. These examples show the need for rules of higher type, therefore contradicting a common belief that application makes rules of higher type unnecessary. They also recall that Klop's idea of variables with arities is very handy. Then, we present our framework in more detail, before we address confluence issues, and finally termination criteria. Missing notations and terminology used in rewriting or type theory can be found in [10, 2].

2 Examples

We present here by means of examples the essential features of the three aforementioned approaches to higher-order rewriting. Rather than comparing their respective expressivity [29], we use our unified framework to show why they are important and how they can be smoothly integrated. Framework and syntax are explained in detail in Section 3, but the necessary information is already provided here to make this section self-contained to anybody who is familiar with typed lambda calculi.

Language. It is our assumption that our examples extend a typed lambda calculus, in which abstraction and application (always written explicitly) are written $\lambda x : \sigma.u$ and $@(u, v)$ respectively where u, v are terms, x is a variable and σ a type. We sometimes drop types in abstractions, and write $\lambda xy.u$ for $\lambda x.(\lambda y.u)$, assuming that the scope of an abstraction extends as far to the right as possible. A variable not in the scope of an abstraction is said to be *free*, otherwise it is *bound*. An expression is *ground* if it has no free variable. The (right-associative) type constructor $\rightarrow$ for functional types is our main type constructor, apart from user-defined ones. We will also need a weak notion of polymorphism, requiring the use of a special (untypable) constant $*$ denoting an arbitrary type. We do not have product types, unless introduced by the user. Recall that lambda calculus is a formal model in which functional computations are described by three (higher-order) rewrite rules, beta-reduction, eta-reduction and alpha-conversion:

beta	$@(\lambda x.u, v) \longrightarrow_\beta u\{x \mapsto v\}$	
eta	$\lambda x.@(u, x) \longrightarrow_\eta u$	where x is not a free variable of u
alpha	$\lambda y.u \longrightarrow_\alpha \lambda x.u\{y \mapsto x\}$	where x is not a free variable of u

In these rules, u, v stand for arbitrary terms, making them rule schemas rather than true rewrite rules. The notation $u\{x \mapsto v\}$ stands for substitution of x by v in u. *Variable capture*, that is, a free variable of v becoming bound after substitution, is disallowed which may force renaming bound variables via alpha-conversion before instantiation can take place. The second rule can also be used from right-to-left, in which case it is called an *expansion*. These rules define an equivalence over terms, the *higher-order equality* $=_{\beta\eta}$, which can be decided by computing normal forms: by using beta and eta both as reductions yielding the beta-eta-normal form; or by using beta as a reduction and eta as an expansion (for terms which are not the left argument of an application, see Section 3 for details) yielding the eta-long beta-normal form. Given two terms u and v, *first-order pattern matching* (resp. *unification*) computes a substitution σ such that $u = v\sigma$. (resp. $u\sigma = v\sigma$). *Plain* and *syntactic* are also used instead of first-order, for qualifying pattern matching, unification or rewriting. *Higher-order pattern matching* (resp. *unification*) computes a substitution σ such that $u =_{\beta\eta} v\sigma$ (resp. $u\sigma =_{\beta\eta} v\sigma$). Of course, such substitutions may not exist. First-order pattern matching and unification are decidable in linear time. Higher-order unification is undecidable, while the exact status of higher-order matching is unknown at orders 5 and up [11].

Our examples come in two parts, a signature for the constants and variables, and a set of higher-order rewrite rules added to the rules of the underlying typed lambda calculus. We use a syntax *à la OBJ*, with keywords introducing successively type constants (`Typ`), type variables (`Tva`), term variables (`Var`), constructors (`Con`), defined function symbols (`Ope`), and rewrite rules (`Prr` and `Hor`). Defined function symbols occur as head operators in lefthand sides of rules, while constructors may not. With some exceptions, we use small letters for constants, greek letters for type variables, capital latin letters for free term variables, and small latin letters for bound term variables. Due to the hierarchical structure of our specifications, and the fact that a type declaration binds whatever comes after, the polymorphism generated by the type variables is weak, as in Isabelle: there is no need for explicit quantifiers which are all external.

Finally, our framework is typed with arities: besides having a type, constants also have an arity which is indicated by writing a double arrow $\Rightarrow$ instead of a single arrow $\rightarrow$ to separate input types from the output type in the typing declarations. The double arrow does not appear when there are no input types. We write $f(u_1, \ldots, u_n)$ when f has arity $n > 0$ and simply f when $n = 0$. In general, the use of arities facilitates the reading. Just like constants, variables also will have arities.

Gödel's System T. We give a polymorphic version of Gödel's system T, a simply typed lambda calculus in which natural numbers are represented in Peano notation. This example has an historical significance: it is the very first higher-order rewrite system *added* to a typed lambda calculus, introduced by Gödel to study the logic of (a fragment of) arithmetic; it plaid a fundamental role in the understanding of the Curry-Howard isomorphism which led to the definition of System F by Girard.

Example 1. **Recursor for natural numbers**

`Tva`	α	$: *$
`Typ`	$\mathbb{N}$	$: *$
`Con`	$\mathsf{0}$	$: \mathbb{N}$
`Con`	s	$: \mathbb{N} \Rightarrow \mathbb{N}$
`Ope`	rec	$: \mathbb{N} \rightarrow \alpha \rightarrow (\mathbb{N} \rightarrow \alpha \rightarrow \alpha) \Rightarrow \alpha$
`Var`	X	$: \mathbb{N}$
`Var`	U	$: \alpha$
`Var`	Y	$: \mathbb{N} \rightarrow \alpha \rightarrow \alpha$

`Prr` $\quad \mathsf{rec}(\mathsf{0}, \mathsf{U}, \mathsf{Y}) \rightarrow \mathsf{U}$

`Prr` $\quad \mathsf{rec}(\mathsf{s}(\mathsf{X}), \mathsf{U}, \mathsf{Y}) \rightarrow @(\mathsf{Y}, \mathsf{X}, \mathsf{rec}(\mathsf{X}, \mathsf{U}, \mathsf{Y}))$

In this example $\mathbb{N}$ is the only type constant and α the only type variable. The constants $\mathsf{0}$ and s are the two constructors for the type $\mathbb{N}$ of natural numbers, as it can be observed from their output type. All variables have arity zero. The rec operator provides us with *higher-order primitive recursion*. It can be used to define new functions, such as addition, multiplication, exponentiation or even the

Ackermann function by choosing appropriate instantiations for the higher-order variables U and X in the rules given in Example 1. For example,

Var	M, N	: $\mathbb{N}$
Prr	plus(N, M)	$\rightarrow$ rec(N, M, $\lambda z_1 z_2$.s(z_2))
Prr	mul(N, M)	$\rightarrow$ rec(N, 0, $\lambda z_1 z_2$.plus(M, z_2))

The precise understanding of the recursor requires some familiarity with the so-called Curry-Howard isomorphism, in which types are propositions in some fragment of intuitionistic logic (here, a quantified propositional fragment), terms of a given type are proofs of the corresponding proposition, and higher-order rules describe proof transformations. In system T, rec can be intuitively interpreted as carrying the proof of a proposition α (the output type of rec) done by induction over the natural numbers. Assuming that α has the form $\forall n.P(n)$ (beware that this proposition is not a type here, we would need a type system à la Coq for that), the variable U is then a proof of $P(0)$, while Y is a function which takes a natural number n and a proof of $P(n)$ as inputs and yields a proof of $P(\mathsf{s}(n))$ as output. It is now easy to see that the first rule equates two proofs of $P(0)$. Further, since rec(X, U, Y) in the righthand side of the second rule is a proof of $P(\mathsf{X})$, that rule equates two proofs of $P(\mathsf{s}(\mathsf{X}))$.

This simple example is already quite interesting in our view.

First, it is based on the use of plain pattern matching. This is always so with recursors for inductive types and is indicated here by using the keyword `Prr`. This comes from the fact that a ground expression of an inductive type (like $\mathbb{N}$) which is in normal form must be headed by a constructor (0 or s for $\mathbb{N}$). Now, pattern matching an expression in normal form (like 0 or s(u) for some normal form u) with respect to the terms 0 and s(X) (in the case of rec) or with respect to the variable N (in the case of plus, mul) does not need higher-order pattern matching since beta- and eta-reductions can only occur inside variables (X or N).

Second, there is no way to define recursors by *a finite number of higher-order rules* in the absence of polymorphism. A description saying that U is a variable of an arbitrary ground type amounts to have one rule for each ground type, which does not fit our purpose: to give a finite specification for system T.

Finally, observe that rewriting expressions for which subexpressions of type $\mathbb{N}$ are ground results in a normal form in which the recursor does not appear anymore. In the OBJ jargon, the operator `rec` is *sufficiently defined*. The fact that all defined operators are sufficiently defined is crucial for encoding recursion by `rec`.

Polymorphic Lists.

Example 2. **Recursors for lists**

Typ	list : $* \Rightarrow *$
Tva	$\alpha, \beta : *$
Con	nil : list(α)
Con	cons : $\alpha \rightarrow$ list(α) $\Rightarrow$ list(α)

```
Ope                    map : list(α) → (α → β) ⇒ list(β)
Var                      H : α
Var                      T : list(α)
Var                      F : α → β

Prr           map(nil, F) → nil
Prr   map(cons(H, T), F) → cons(@(F, H), map(T, F))
```

This example familiar to Lisp programmers shows that the above final remark applies to any inductive type, such as polymorphic lists. Here, list is a type operator of arity one, therefore taking an arbitrary type as input. map returns the list of applications of the function F given as second argument to the elements of the list given as first argument.

Example 3 extends Example 2 with a parametric version of insertion sort (called sort), which takes a list as input, sorts the tail and then inserts the head at the right place by using a function insert, whose second argument is therefore a sorted list. The additional parameters X, Y in both sort and insert stand for functions selecting one of their two arguments with respect to some ordering. Instantiating them yields particular sorting algorithms, as for example ascending − sort. The additional signature declarations are omitted, as well as the keyword `Prr`.

Example 3. **Parametric insertion sort**

$$\begin{array}{rll}
\mathsf{max}(0,X) & \to X & \qquad \mathsf{max}(X,0) \to X \\
\mathsf{max}(\mathsf{s}(X),\mathsf{s}(Y)) & \to \mathsf{s}(\mathsf{max}(X,Y)) & \\
\mathsf{min}(0,X) & \to 0 & \qquad \mathsf{min}(X,0) \to 0 \\
\mathsf{min}(\mathsf{s}(X),\mathsf{s}(Y)) & \to \mathsf{s}(\mathsf{min}(X,Y)) & \\
\mathsf{insert}(N,\mathsf{nil},X,Y) & \to \mathsf{cons}(N,\mathsf{nil}) & \\
\mathsf{insert}(N,\mathsf{cons}(M,T),X,Y) & \to \mathsf{cons}(@(X,N,M),\mathsf{insert}(@(Y,N,M),T,X,Y)) & \\
\mathsf{sort}(\mathsf{nil},X,Y) & \to \mathsf{nil} & \\
\mathsf{sort}(\mathsf{cons}(N,T),X,Y) & \to \mathsf{insert}(N,\mathsf{sort}(T,X,Y),X,Y) & \\
\mathsf{ascending}-\mathsf{sort}(L) & \to \mathsf{sort}(L,\lambda xy.\mathsf{min}(x,y),\lambda xy.\mathsf{max}(x,y)) & \\
\mathsf{descending}-\mathsf{sort}(L) & \to \mathsf{sort}(L,\lambda xy.\mathsf{max}(x,y),\lambda xy.\mathsf{min}(x,y)) &
\end{array}$$

As this example shows, many programs can be defined by first-order pattern matching, a well-known fact exploited by most modern functional programming languages, such as those of the ML family. Again, we could (and should) prove that these new defined operators are sufficiently defined, but the argument is actually the same as before.

Differentiation. We now move to a series of examples showing the need for higher-order pattern matching, which will be indicated by using the second keyword `Hor` for rules. First, we need some more explanations about the typed lambda calculus. $\mathbb{N}$ and list($\mathbb{N}$) are called *data types*, while $\mathbb{N} \to \mathbb{N}$, which is headed by $\to$, is a *functional type*. Constant data types like $\mathbb{N}$ are also called *basic types*. In Nipkow's work, the lefthand and righthand side of higher-order rules must have the same basic type, a condition which is usually enforced when needed by applying a term of a functional type to a sequence of variables of the appropriate types. All terms must be in eta-long beta-normal form, forcing us to write $\lambda x.@(F,x)$ instead of simply F. Lefthand sides of rules must be *patterns* [25], an assumption frequently met in practice which makes both higher-order pattern matching and unification decidable in linear time. Finally, there is no notion of constructor and defined symbol, and no recursors coming along with. We choose arbitrarily to have all symbols as defined. We give only one rule, the others should be easily guessed by the reader.

Example 4. **Differentiation 1**

Typ	$\mathbb{R}$	: $*$
Ope	sin, cos	: $\mathbb{R} \rightarrow \mathbb{R}$
Ope	mul	: $\mathbb{R} \rightarrow \mathbb{R} \rightarrow \mathbb{R}$
Ope	diff	: $(\mathbb{R} \rightarrow \mathbb{R}) \rightarrow \mathbb{R} \rightarrow \mathbb{R}$
Var	y	: $\mathbb{R}$
Var	F	: $\mathbb{R} \rightarrow \mathbb{R}$

$$\text{Hor}\quad @(\mathsf{diff}, \lambda x : \mathbb{R}.\, @(\mathsf{sin}, @(F,x)), y) \rightarrow @(\mathsf{mul}, @(\mathsf{cos}, @(F,y)), @(\mathsf{diff}, \lambda x : \mathbb{R}.@(F,x), y))$$

Note that all function symbols and variables have arity zero, since there is no arity in our sense in Nipkow's framework. Both sides of the rule have type $\mathbb{R}$: diff computes the differential of its first argument at the point given as second argument. Unlike the previous examples, these rules use higher-order pattern matching, and this is necessary here to compute the derivative of the function sin itself. Clearly, $@(\mathsf{diff}, \lambda x.@(\mathsf{sin}, x))$ does not match the lefthand side of rule. Let us instantiate the variable F of the lefthand side of rule by the identity function $\lambda x.x$, resulting in the expression $@(\mathsf{diff}, \lambda x.\, @(\mathsf{sin}, @(\lambda x.x, x)), y)$ which beta-reduces to the expected result $@(\mathsf{diff}, @(\mathsf{sin}, x)), y)$. This shows that the latter expression higher-order matches the lefthand side of rule. Using plain pattern matching would require infinitely many rules, one for each possible instantiation of F requiring a beta-reduction. Incorporating beta-reduction into the matching process allows one to have a single rule for all cases.

Using rules of higher-order type as well as function symbols and variables of non-zero arity is possible thanks to a generalisation of Nipkow's work [22], resulting in the following new version of the same example:

Example 5. **Differentiation 2**

Typ	$\mathbb{R}$	: $*$
Ope	sin, cos	: $\mathbb{R} \Rightarrow \mathbb{R}$
Ope	mul	: $(\mathbb{R} \rightarrow \mathbb{R}) \rightarrow (\mathbb{R} \rightarrow \mathbb{R}) \Rightarrow (\mathbb{R} \rightarrow \mathbb{R})$
Ope	diff	: $(\mathbb{R} \rightarrow \mathbb{R}) \Rightarrow (\mathbb{R} \rightarrow \mathbb{R})$
Var	F	: $\mathbb{R} \Rightarrow \mathbb{R}$

$$\text{Hor}\quad \mathsf{diff}(\lambda x : \mathbb{R}.\, \mathsf{sin}(F(x)) \rightarrow \mathsf{mul}(\lambda x : \mathbb{R}.\, \mathsf{cos}(F(x)), \mathsf{diff}(\lambda x : \mathbb{R}.F(x)))$$

Here, the rule has the higher-order type $\mathbb{R} \Rightarrow \mathbb{R}$, making diff the true differential of the function given as argument. In this format, eta-expansion is controlled by the systematic use of arities. This use of arities for replacing applications by parentheses should be considered here as a matter of style: one can argue that arities are not really needed for that purpose, since, traditionally, the application operator is not explicitly written in the lambda calculus: some write $(M\ N)$ for $@(M, N)$ while others favour $M(N)$ instead. But this is a convention. In both cases, the conventional expression is transformed by the parser into the one with explicit application. Here, the syntax forces us to write $@(M, N)$ when M is not a variable or is a variable with arity zero, and $M(N)$ when M is a variable with arity 1. More convincing advantages are discussed later.

To get the best of our format, we can declare F as a function symbol of arity zero and use the beta-eta-normal form instead of the eta-long beta-normal form. With this last change, the rule becomes:

$$\begin{array}{ll} \texttt{Var} & \mathsf{F} : \mathbb{R} \to \mathbb{R} \\ \texttt{Hor} & \mathsf{diff}(\mathsf{sin}(\mathsf{F})) \to \mathsf{mul}(\mathsf{cos}(\mathsf{F}), \mathsf{diff}(\mathsf{F})) \end{array}$$

Simply Typed Lambda Calculus. We end up this list with Klop's favorite example, showing the need for arities of variables in order to encode the lambda calculus. The challenge here is to have true rewrite rules for beta-reduction and eta-reduction: the usual side condition for the eta-rule must be eliminated. This is made possible by allowing us to control which variables can or cannot be captured through substitution when replacing a variable with arity: a substitute for a variable of arity n must be an abstraction of n different bound variables. For example, a variable of arity one depends upon a single variable, like X which must be written $X(x)$ for some variable x bound above, and replaced by an abstraction $\lambda x.u$ for some term u. The instance of $X(x)$ will then become $@(\lambda x.u, x)$ and beta-reduce to u. This idea due to Klop is indeed very much related to Miller's notion of pattern: in a pattern, every occurrence of X must be of the form $X(x)$ with x bound above if X has arity one, and it becomes then natural to define u as the substitute for $X(x)$ rather than $\lambda x.u$ as a substitute for X which does not exist as a syntactic term. We will later see that this relationship is stronger than anticipated.

Example 6. **Simply typed lambda calculus**

$$\begin{array}{lrl} \texttt{Typ} & \alpha, \beta & : * \\ \texttt{Ope} & \mathsf{app} & : (\alpha \to \beta) \to \alpha \Rightarrow \beta \\ \texttt{Ope} & \mathsf{abs} & : (\alpha \to \beta) \Rightarrow (\alpha \to \beta) \\ \texttt{Var} & \mathsf{U} & : \alpha \Rightarrow \beta \\ \texttt{Var} & \mathsf{V} & : \alpha \\ \texttt{Var} & \mathsf{X} & : \alpha \to \beta \end{array}$$

$$\begin{array}{ll} \texttt{Hor} & \mathsf{app}(\mathsf{abs}(\lambda \mathsf{x} : \alpha.\mathsf{U}(\mathsf{x})), \mathsf{V}) \to \mathsf{U}(\mathsf{V}) \\ \texttt{Hor} & \mathsf{abs}(\lambda \mathsf{x} : \alpha.\mathsf{app}(\mathsf{X}, \mathsf{x})) \to \mathsf{X} \end{array}$$

The beta-rule shows the use of a variable of arity one in order to internalize the notion of substitution, while the eta-rule shows the use of a variable of arity zero in order to eliminate the condition that x does not occur free in an instance of X:

since X has arity zero, it cannot have an abstraction for substitute. And because variable capture is forbidden by the definition of substitution, x cannot occur in the substitute. The use of arity 1 for U is not essential here, since we could also choose the arity zero to the price of replacing the first rule by the variant

$$\texttt{Hor} \quad \mathsf{app}(\mathsf{abs}(\mathsf{U}),\mathsf{V}) \rightarrow @(\mathsf{U},\mathsf{V}).$$

The example is a variation of Klop's, since we actually model a lambda calculus with simple types. It also shows the need of a rich enough type structure for specifying an example even as simple as this one.

3 Polymorphic Higher-Order Algebras

This section introduces the framework of polymorphic algebras [20, 22]. The use of polymorphic operators requires a rather heavy apparatus.

3.1 Types

Given a set $\mathcal{S}$ of *sort symbols* of a fixed arity, denoted by $s : *^n \Rightarrow *$, and a set $\mathcal{S}^\forall$ of *type variables*, the set $\mathcal{T}_{\mathcal{S}^\forall}$ of *polymorphic types* is generated from these sets by the constructor $\rightarrow$ for *functional types*:

$$\begin{array}{c}\mathcal{T}_{\mathcal{S}^\forall} := \alpha \mid s(\mathcal{T}_{\mathcal{S}^\forall}^n) \mid (\mathcal{T}_{\mathcal{S}^\forall} \rightarrow \mathcal{T}_{\mathcal{S}^\forall})\\ \text{for } \alpha \in \mathcal{S}^\forall \text{ and } s : *^n \Rightarrow * \ \in \mathcal{S}\end{array}$$

where $s(\mathcal{T}_{\mathcal{S}^\forall}^n)$ denotes an expression of the form $s(t_1, \ldots, t_n)$ with $t_i \in \mathcal{T}_{\mathcal{S}^\forall}$ for all $i \in [1..n]$. We use $\mathcal{V}ar(\sigma)$ for the set of (type) variables of the type $\sigma \in \mathcal{T}_{\mathcal{S}^\forall}$. When $\mathcal{V}ar(\sigma) \neq \emptyset$, the type σ is said to be *polymorphic* and *monomorphic* otherwise. A type σ is *functional* when headed by the $\rightarrow$ symbol, a *data type* when headed by a sort symbol (*basic* when the sort symbol is a constant). $\rightarrow$ associates to the right.

A *type substitution* is a mapping from $\mathcal{S}^\forall$ to $\mathcal{T}_{\mathcal{S}^\forall}$ extended to an endomorphism of $\mathcal{T}_{\mathcal{S}^\forall}$. We write $\sigma\xi$ for the application of the type substitution ξ to the type σ. We denote by $\mathcal{D}om(\sigma) = \{\alpha \in \mathcal{S}^\forall \mid \alpha\sigma \neq \alpha\}$ the domain of $\sigma \in \mathcal{T}_{\mathcal{S}^\forall}$, by $\sigma|_\mathcal{V}$ its restriction to the domain $\mathcal{D}om(\sigma) \cap \mathcal{V}$, by $\mathcal{R}an(\sigma) = \bigcup_{\alpha \in \mathcal{D}om(\sigma)} \mathcal{V}ar(\alpha\sigma)$ its *range*. By a renaming of the type σ apart from $V \subset \mathcal{X}$, we mean a type $\sigma\xi$ where ξ is a type renaming such that $\mathcal{D}om(\xi) = \mathcal{R}an(\sigma)$ and $\mathcal{R}an(\xi) \cap \mathcal{V} = \emptyset$.

We shall use α, β for type variables, $\sigma, \tau, \rho, \theta$ for arbitrary types, and ξ, ζ to denote type substitutions.

3.2 Signatures

We are given a set of function symbols denoted by the letters f, g, h, which are meant to be algebraic operators equiped with a fixed number n of arguments (called the *arity*) of respective types $\sigma_1 \in \mathcal{T}_{\mathcal{S}^\forall}, \ldots, \sigma_n \in \mathcal{T}_{\mathcal{S}^\forall}$, and an *output type* $\sigma \in \mathcal{T}_{\mathcal{S}^\forall}$ such that $\mathcal{V}ar(\sigma) \subseteq \bigcup_i \mathcal{V}ar(\sigma_i)$ if $n > 0$. We call *aritype* the expression $\sigma_1 \rightarrow \ldots \rightarrow \sigma_n \Rightarrow \sigma$ when $n > 0$ and σ when $n = 0$, which can be seen as a notation for the pair made of a type and an arity. The condition $\mathcal{V}ar(\sigma) \subseteq \bigcup_i \mathcal{V}ar(\sigma_i)$ if $n > 0$ ensures that aritypes encode a logical proposition universally quantified outside, making our type system to come both rich and simple enough. Let $\mathcal{F}$ be the set of all function symbols:

$$\mathcal{F} = \biguplus_{\sigma_1,\ldots,\sigma_n,\sigma} \mathcal{F}_{\sigma_1 \rightarrow \ldots \rightarrow \sigma_n \Rightarrow \sigma}$$

The membership of a given function symbol f to the set $\mathcal{F}_{\sigma_1 \rightarrow \ldots \rightarrow \sigma_n \Rightarrow \sigma}$ is called a *type declaration* and written $f : \sigma_1 \rightarrow \ldots \rightarrow \sigma_n \Rightarrow \sigma$. A type declaration is *first-order* if it uses only sorts, and higher-order otherwise. It is *polymorphic* if it uses some polymorphic type, otherwise, it is *monomorphic*. Polymorphic type declarations are implicitly universally quantified: they can be renamed arbitrarily. Note that type instantiation does not change the arity of a function symbol.

Alike function symbols, variables have an aritype. A variable declaration will therefore take the same form as a function declaration.

3.3 Terms

The set $\mathcal{T}(\mathcal{F}, \mathcal{X})$ of (raw) *terms* is generated from the signature $\mathcal{F}$ and a denumerable set $\mathcal{X}$ of arityped variables according to the grammar:

$$\mathcal{T} := (\lambda \mathcal{X} : \mathcal{T}_{\mathcal{S}^{\forall}}.\mathcal{T}) \mid @(\mathcal{T}, \mathcal{T}) \mid \mathcal{X}(\mathcal{T}, \ldots, \mathcal{T}) \mid \mathcal{F}(\mathcal{T}, \ldots, \mathcal{T}).$$

$\overline{s}$ will ambiguously denote a list, a set or a multiset of terms $s_1, \ldots, s_n$. Terms of the form $\lambda x : \sigma.u$ are called *abstractions*, the type σ being possibly omitted. Because a type is a particular aritype, bound variables have arity zero. $@(u, v)$ denotes the application of u to v. Parentheses are omitted for function or variable symbols of arity zero. The term $@(\overline{v})$ is called a (partial) *left-flattening* of $s = @((\ldots @(v_1, v_2)) \ldots v_n)$, v_1 being possibly an application itself. $\mathcal{V}ar(t)$ is the set of free term variables of t.

Terms are identified with finite labeled trees by considering $\lambda x : \sigma.\_$, for each variable x and type σ, as a unary function symbol taking a term u as argument to construct the term $\lambda x : \sigma.u$. *Positions* are strings of positive integers. Λ and $\cdot$ denote respectively the empty string (root position) and string concatenation. $\mathcal{P}os(t)$ is the set of positions in t. $t|_p$ denotes the *subterm* of t at position p. Replacing $t|_p$ at position p in t by u is written $t[u]_p$. The notation $t[]_p$ stands for a *context* waiting for a term to fill its hole.

3.4 Typing Rules

Definition 1. *An* environment *Γ is a finite set of pairs written as $\{x_1 : \sigma_1, \ldots, x_n : \sigma_n\}$, where x_i is a variable, σ_i is an aritype, and $x_i \neq x_j$ for $i \neq j$. $\mathcal{V}ar(\Gamma) = \{x_1, \ldots, x_n\}$ is the set of variables of Γ. The size $|\Gamma|$ of the environment Γ is the sum of the sizes of its constituents. Given two environments Γ and*

Γ', their composition *is the environment* $\Gamma \cdot \Gamma' = \Gamma' \cup \{x : \sigma \in \Gamma \mid x \notin \mathcal{V}ar(\Gamma')\}$. *Two environments Γ and Γ' are* compatible *if* $\Gamma \cdot \Gamma' = \Gamma \cup \Gamma'$.

Our typing judgments are written as $\Gamma \vdash_{\mathcal{F}} s : \sigma$. A term s has type σ in the environment Γ if and only if the judgment $\Gamma \vdash_{\mathcal{F}} s : \sigma$ is provable in the inference system of Figure 1. Given an environment Γ, a term s is *typable* if there exists a type σ such that $\Gamma \vdash_{\mathcal{F}} s : \sigma$.

Functions:

$$\frac{f : \sigma_1 \to \ldots \to \sigma_n \Rightarrow \sigma \in \mathcal{F} \quad \xi \text{ some type substitution} \quad \Gamma \vdash_{\mathcal{F}} t_1 : \sigma_1\xi \ \ldots \ \Gamma \vdash_{\mathcal{F}} t_n : \sigma_n\xi}{\Gamma \vdash_{\mathcal{F}} f(t_1, \ldots, t_n) : \sigma\xi}$$

Variables:

$$\frac{X : \sigma_1 \to \ldots \to \sigma_n \Rightarrow \sigma \in \Gamma \quad \Gamma \vdash_{\mathcal{F}} t_1 : \sigma_1 \ \ldots \ \Gamma \vdash_{\mathcal{F}} t_n : \sigma_n}{\Gamma \vdash_{\mathcal{F}} X(t_1, \ldots, t_n) : \sigma}$$

Abstraction:

$$\frac{\Gamma \cdot \{x : \sigma\} \vdash_{\mathcal{F}} t : \tau}{\Gamma \vdash_{\mathcal{F}} (\lambda x : \sigma . t) : \sigma \to \tau}$$

Application:

$$\frac{\Gamma \vdash_{\mathcal{F}} s : \sigma \to \tau \quad \Gamma \vdash_{\mathcal{F}} t : \sigma}{\Gamma \vdash_{\mathcal{F}} @(s, t) : \tau}$$

Fig. 1. The type system for polymorphic higher-order algebras with arities

Remember our convention that function and variable symbols having arity zero come without parentheses. When writing a judgement $\Gamma \vdash_{\mathcal{F}} s : \sigma$, we must make sure that σ is a type in the environment defined by the signature. Types are indeed unisorted first-order terms (of sort $*$). Since there is only one sort, and type symbols have a fixed arity, verifying that an expression is a type amounts to check that all symbols occurring in the expression are sort symbols or type variables, and that all sort symbols in the expression have the right number of types as inputs, an easily decidable property usually called *well-formedness*.

This typing system enjoys the *unique typing* property: given an environment Γ and a typable term u, it can be easily shown by induction on u that there exists a unique type σ such that $\Gamma \vdash_{\mathcal{F}} u : \sigma$.

Note that type substitutions apply to types in terms: $x\xi = x$, $(\lambda x : \sigma . s)\xi = \lambda x : \sigma\xi . s\xi$, $@(u, v)\xi = @(u\xi, v\xi)$, and $f(\overline{u})\xi = f(\overline{u}\xi)$.

3.5 Substitutions

Definition 2. *A (term)* substitution $\gamma = \{(x_1 : \sigma_1) \mapsto (\Gamma_1, t_1), \ldots, (x_n : \sigma_n) \mapsto (\Gamma_n, t_n)\}$, *is a finite set of quadruples made of a variable symbol, an aritype, an environment and a term, such that*

(i) Let $\sigma_i = \tau_{i_1} \to \ldots \to \tau_{i_p} \Rightarrow \tau_i$. *Then* $t_i = \lambda y_{i_1} : \tau_{i_1} \ldots y_{i_p} : \tau_{i_p}.u_i$ *for distinct variables* $y_{i_1}, \ldots, y_{i_p}$ *and term* $u_i \neq x_i$ *such that* $\Gamma_i \vdash_{\mathcal{F}} t_i : \sigma_i$.

(ii) $\forall i \neq j \in [1..n]$, $x_i \neq x_j$, *and*

(iii) $\forall i \neq j \in [1..n]$, Γ_i *and* Γ_j *are compatible environments.*

We may omit the aritype σ_i *and environment* Γ_i *in* $(x_i : \sigma_i) \mapsto (\Gamma_i, t_i)$.

The set of (input) variables of the substitution γ *is* $\mathcal{V}ar(\gamma) = \{x_1, \ldots, x_n\}$, *its* domain *is the environment* $\mathcal{D}om(\gamma) = \{x_1 : \sigma_1, \ldots, x_n : \sigma_n\}$ *while its* range *is the environment (by assumption (iii))* $\mathcal{R}an(\gamma) = \bigcup_{i\in[1..n]} \Gamma_i$.

Definition 3. *A substitution* γ *is* compatible *with an environment* Γ *if*

(i) $\mathcal{D}om(\gamma)$ *is compatible with* Γ,

(ii) $\mathcal{R}an(\gamma)$ *is compatible with* $\Gamma \setminus \mathcal{D}om(\gamma)$.

We will also say that γ *is compatible with the judgement* $\Gamma \vdash_{\mathcal{F}} s : \sigma$.

Definition 4. *A substitution* γ *compatible with a judgement* $\Sigma \vdash_{\mathcal{F}} s : \sigma$ *operates as an endomorphism on* s *and yields the* instance $s\gamma$ *defined as:*

If $s = @(u, v)$	*then* $s\gamma = @(u\gamma, v\gamma)$
If $s = \lambda x : \tau.u$	*then* $s\gamma = \lambda z : \tau.(u\{x \mapsto z\})\gamma$ *with* z *fresh.*
If $s = f(u_1, \ldots, u_n)$	*then* $s\gamma = f(u_1\gamma, \ldots, u_n\gamma)$
If $s = X(u_1, \ldots, u_n)$ *and* $X \notin \mathcal{V}ar(\gamma)$	*then* $s\gamma = X(u_1\gamma, \ldots, u_n\gamma)$
If $s = X(u_1, \ldots, u_n)$ *and* $(X : \sigma) \mapsto (\Gamma, \lambda y_1 {:} \tau_1 \ldots y_n {:} \tau_n.u) \in \gamma$	*then* $s\gamma = @(\lambda y_1 {:} \tau_1 \ldots y_n {:} \tau_n.u, u_1\gamma, \ldots, u_n\gamma)$

In the last case, we could also perform the introduced beta-reductions therefore hiding the application operator. Writing $s\gamma$ assumes $\mathcal{D}om(\gamma)$ compatible with $\Gamma \vdash_{\mathcal{F}} s : \sigma$ and it follows that $(\Gamma \setminus \mathcal{D}om(\gamma)) \cup \mathcal{R}an(\gamma) \vdash_{\mathcal{F}} s\gamma : \sigma$. Given γ, the *composition* or *instance* $(\{(x_i : \sigma_i) \mapsto (\Gamma_i, t_i)\}_i)\gamma$ is the substitution $\{(x_i : \sigma_i) \mapsto ((\Gamma_i \setminus \mathcal{D}om(\gamma)) \cup \mathcal{R}an(\gamma), t_i\gamma)\}_i$.

3.6 Higher-Order Rewriting Relations

We now introduce the different variants of higher-order rewriting, categorized by their use of pattern matching. Allowing for variables with arities should be considered as a cosmetic variation of the traditional frameworks, since this is the case for the pattern matching and unification algorithms, as well as for the definition of patterns.

Plain Higher-Order Rewriting.

Definition 5. *A* plain higher-order rewrite system *is a set of higher-order rewrite rules* $\{\Gamma_i \vdash l_i \to r_i : \sigma_i\}_i$ *such that*

(i) $\Gamma_i \vdash_{\mathcal{F}} l_i : \sigma_i$ *and* $\Gamma_i \vdash_{\mathcal{F}} r_i : \sigma_i$,

(ii) $\mathcal{V}ar(r) \subseteq \mathcal{V}ar(l)$.

Plain higher-order rewriting is based on plain pattern matching:

$$\Sigma \vdash u \xrightarrow[\Gamma \vdash l \to r]{p} v \text{ if } \Sigma \vdash_{\mathcal{F}} u : \tau \text{ for some type } \tau, u|_p = l\sigma \text{ and } v = u[r\sigma]_p.$$

Note the need for an environment Σ in which u is typable. Then, it is easy to see that $\Sigma \vdash_{\mathcal{F}} v : \tau$, a property called type preservation, which allows us to actually drop the environment Σ so as to make our notations more readable. We also often take the liberty to drop the type of a rule.

A *higher-order equation* is a pair of higher-order rewrite rules $\{\Gamma \vdash l \to r : \sigma, \Gamma \vdash r \to l : \sigma\}$. We abbreviate such a pair by $\Gamma \vdash l = r : \sigma$, and write $\Sigma \vdash u \longleftrightarrow^*_R v$ or $\Sigma \vdash u =_R v$ if R is a set of equations.

Conversion Rules. Conversion rules in the typed lambda calculus are a particular example of higher-order equations:

$$\begin{array}{rl} \{V : \tau\} \vdash & \lambda x : \sigma.V =_\alpha \lambda y : \sigma.V\{x \mapsto y\} \\ \{U : \sigma, V : \tau\} \vdash & @(\lambda x : \sigma.V, U) =_\beta V\{x \mapsto U\} \\ \{U : \sigma \to \tau\} \vdash & \lambda x : \sigma.@(U, x) =_\eta U \end{array}$$

Traditionally, these equations are schemas, in which U and V stand for arbitrary terms, with $x \notin \mathcal{V}ar(U)$ for the eta-rule and $y \notin (\mathcal{V}ar(V) \setminus \{x\})$ for the alpha-rule. Here, these conditions are ensured by our definition of substitution, hence these equations are indeed true higher-order equations.

Orienting the last two equalities from left to right yields the beta-reduction (resp. eta-reduction) rule. Orienting the third equation from right to left yields the eta-expansion rule. Since this rule may not terminate, its use is restricted by spelling out in which context it is allowed:

$$\begin{array}{c} \{u : \sigma_1 \to \ldots \to \sigma_n \to \sigma\} \vdash \\ s[u]_p \longrightarrow^p_\eta s[\lambda x_1 : \sigma_1, \ldots, x_n : \sigma_n.@(u, x_1, \ldots, x_n)]_p \end{array}$$

$$\textbf{if} \begin{cases} \sigma \text{ is a data type,} \quad x_1, \ldots, x_n \notin \mathcal{V}ar(u), \\ u \text{ is not an abstraction and } s|_q \text{ is not an application in case } p = q \cdot 1 \end{cases}$$

The last condition means that the first argument of an application cannot be recursively expanded on top. A variant requires in addition that $p \neq \Lambda$.

Typed lambda-calculi are confluent modulo alpha-conversion, and terminating with respect to beta-reductions and either eta-expansion or eta-reduction, therefore defining normal forms up to the equivalence generated by alpha-conversion. Our notations for normal forms use down arrows for reductions: $u\downarrow_\beta$ and $u\downarrow_{\beta\eta}$; up arrows for expansions: $u\uparrow^\eta$; and their combined version for eta-long beta-normal forms: $u\updownarrow^\eta_\beta$. We use $u\downarrow$ for a normal form of some unspecified kind.

We can now introduce most general substitutions. Given $\Gamma \vdash_{\mathcal{F}} s, t : \sigma$, a *solution* (resp. *higher-order solution*) of the *equation* $s = t$ is a substitution γ such that $s\gamma = t$ (resp. $s\gamma =_{\beta\eta} t$) for a *plain* (resp. *higher-order*) *pattern-matching* problem, and $s\gamma = t\gamma$ (resp. $s\gamma =_{\beta\eta} t\gamma$) for a *plain* (resp. *higher-order*) *unification* problem. Solutions are called *plain/higher-order matches/unifiers*, depending on which case is considered. A unifier γ is *most general* if any unifier θ satisfies $\theta = \gamma\varphi$ (resp. $\theta =_{\beta\eta} \gamma\varphi$) for some substitution φ.

Normal Higher-Order Rewriting. Normal higher-order rewriting is based upon higher-order pattern matching, rules must be normalized, and their left-hand sides must be patterns. Our definition of patterns specializes to Nipkow's [24] when variables have arity zero:

Definition 6. *A higher-order term u is a pattern if for every variable occurrence of some variable $X : \sigma_1 \rightarrow \ldots \sigma_m \Rightarrow \tau_1 \rightarrow \ldots \tau_n \in \mathcal{V}ar(u)$ with $n > 1$ and τ_n a data type, there exists a position $p \in \mathcal{P}os(u)$ such that*
(i) $u|_p = @(X(x_1, \ldots, x_m), x_{m+1}, \ldots, x_{m+n})$,
(ii) $\forall i, j \in [1..m+n]$, $x_i \neq x_j$ or i=j,
(iii) $\forall i \in [1..m+n]$, there exists a position $q < p$ in $\mathcal{P}os(u)$ and a term v such that $u|_q = \lambda x_i.v$.

Assuming $X : \alpha \Rightarrow \beta \rightarrow \gamma$, $\lambda xy.@(X(x), y)$ is a pattern while $\lambda x.X(x)$, $\lambda x.@(X(x), x)$ and $\lambda xy.Y(@(X(x), y)$ are not.

Computing the normal form of a pattern instance is done by normalizing first the pattern u then the substitution γ, before to reduce every subexpression $@(X\gamma(x_1, \ldots, x_m), x_{m+1}, \ldots, x_{m+n})$. This simple schema in which the third step is a development shows that Nipkow's and Klop's notions of rewriting coincide when lefthand sides of rules are patterns. We believe that this is the very reason why higher-order pattern matching and higher-order unification are decidable for patterns, suggesting a more general notion of pattern. A related observation is made in [13].

Definition 7. *A normal higher-order rewrite system is a set of rules $\{\Gamma_i \vdash l_i \rightarrow r_i : \sigma_i\}_i$ such that conditions (i) and (ii) of Definition 5 are satisfied, (iii) l_i and r_i are in normal form and l_i is a pattern.*

Normal higher-order rewriting operates on terms in normal form:

$$\Sigma \vdash u \xrightarrow[\Gamma \vdash l \rightarrow r]{p} v \text{ if } u = u{\downarrow}, u|_p \xleftrightarrow[\beta\eta]{*} l\sigma \text{ and } v = u[r\sigma]_p{\downarrow}$$

There are several variants of normal higher-order rewriting, which use different kinds of normal forms.

In [26] and [24], eta-long beta-normal forms are used. The higher-order rewrite rules must satisfy the following additional condition:

(iv) type constructors are constants, there are no type variables, function symbols and variables have arity 0, and rules are of basic type.

In [21], the framework is generalized, and condition (iv) becomes:

(v) type constructors may have a non-zero arity, type variables are allowed, function symbols have aritypes, variables have arity 0, terms are assumed to be in eta-long beta-normal form except at the top (eta-expansion is not applied at the top of rules, but beta-normalization is), rules can be of any (possibly polymorphic) type.

In [22], beta-eta-normal forms are used, and condition (iv) becomes:

(vi) type constructors have arities, type variables are allowed, constants and variables have aritypes, lefthand sides of rules are of the form $f(l_1, \ldots, l_n)$ for some function symbol f, and rules are of any type.

The use of beta-eta-normal forms allows us to restrict the lefthand sides of rules by assuming they are headed by a function symbol. Were eta-long beta-normal forms used instead, this assumption would imply Nipkow's restriction that rules are of basic type. We will see some advantage of beta-eta-normal forms for proving confluence and termination.

Mixed Higher-Order Rewriting. There is no reason to restrict ourselves to a single kind of higher-order rules. It is indeed possible to use both, by pairing each rule with the pattern-matching algorithm it uses, as done with the keywords Prr, Hor used in the examples of Section 2. Of course, we need to have all lefthand sides of both kinds of rules to be patterns, and need using the same kind of normal form when rewriting terms. The choice of beta-eta-normal forms is dictated by its advantages.

One can wonder whether there is a real need for two different keywords for higher-order rules. As we have seen, higher-order pattern matching is needed if and only if beta-redexes can be created by instantiation at non-variable positions of the lefthand sides (eta-redexes cannot). This requires a subterm $F(u_1, \ldots, u_m)$ in the lefthand side of the rule, where F is a free variable of aritype $\sigma_1 \to \ldots \sigma_m \Rightarrow \sigma$ with $m > 0$, an easily decidable property.

Mixed higher-order rewriting has not yet been studied in the literature although it is explicitly alluded to in [22]. However, all known results can be easily lifted to the mixed case. In the sections to come, we will present the results known for both plain and higher-order rewriting, and formulate their generalization as a conjecture when appropriate.

4 Confluence

In this section, we restrict our attention to terminating relations. Confluence will therefore be checked via critical pairs, whose kind is imposed by the mechanism for searching redexes.

Definition 8. *Given two rules $l \to r$ and $g \to d$, a non-variable position $p \in \mathcal{P}os(l)$ and a most general unifier (resp. most general higher-order unifier) σ of the equation $l|_p = g$, the pair $(r\sigma, l[d\sigma]_p)$ is called a* critical pair *(resp., a* higher-order critical pair*) of $g \to d$ on $l \to r$ at position p.*

A critical pair (s,t) is joinable *if there exists v such that $s \longrightarrow^*_R v$ and $t \longrightarrow^*_R v$. It is* higher-order joinable *(*joinable *when clear from the context) if there exist v, w such that $s \longrightarrow^*_R v, t \longrightarrow^*_R w$ and $v \longleftrightarrow^*_{\beta\eta} w$.*

Plain Higher-Order Rewriting. Using plain pattern matching leads to plain critical pairs computed with plain unification. The following result follows easily from Newman's Lemma [15]:

Theorem 1. *[7]. Given a set R of higher-order rules such that*
(a) $R \cup \{beta, eta\}$ is terminating;
(a) all critical pairs in $R \cup \{beta, eta\}$ are joinable;
then plain higher-order rewriting with $R \cup \{beta, eta\}$ is confluent.

It can easily be checked that Examples 1, 2 and 3 are confluent because they do not admit any critical pair.

Higher-Order Rewriting. Replacing joinability by higher-order joinability does not allow us to get a similar result for higher-order rewriting [24]. It is indeed well-known, in the first-order case, that the natural generalization of *confluence* of a set of rules R in presence of a set of equations E:

$$\begin{array}{c}\forall s,t,u \text{ in normal form such that } u \longrightarrow_R^* s \text{ and } u \longrightarrow_R^* t\\ \exists v,w \text{ such that } s \longrightarrow_R^* v, t \longrightarrow_R^* w \text{ and } v \longleftrightarrow_E^* w\end{array}$$

is not enough for ensuring the *Church-Rosser property*:

$$\begin{array}{c}\forall s,t \text{ such that } s \longleftrightarrow_{R\cup E}^* t\\ \exists v,w \text{ such that } s \longrightarrow_R^* v, t \longrightarrow_R^* w \text{ and } v \longleftrightarrow_E^* w\end{array}$$

when searching for a redex uses E-matching. An additional *coherence* property is needed:

$$\begin{array}{c}\forall s,t,u \text{ such that } u \longrightarrow_R^* s \text{ and } u \longleftrightarrow_E^* t\\ \exists v \text{ and } w \text{ such that } s \longrightarrow_R^* v, t \longrightarrow_R^* w \text{ and } v \longleftrightarrow_E^* w.\end{array}$$

Coherence can be ensured for an arbitrary equational theory E by using so-called Stickel's extension rules [30, 16]. In the case of higher-order rewriting, E is made of alpha, beta and eta. Then, rules headed by an abstraction operator on the left, that is, of the form

$$\lambda x.l \to r \quad \text{need as beta-extension the rule} \quad l \to @(r,x)\downarrow$$

obtained by putting both sides of $\lambda x.l \to r$ inside $@([],x)$, the minimal context generating a beta redex on top of its lefthand side. Normalizing the result yields the extension. Note that a single extension is generated.

For example, the rule $\lambda x.a \to \lambda x.b$, where a,b are constants has $a \to b$ as extension. Indeed, $a \longleftarrow_\beta^\Lambda @(\lambda x.a, x) \longrightarrow_{\lambda x.a \to \lambda x.b}^1 @(\lambda x.b, x) \longrightarrow_\beta^\Lambda b$. However, $a \neq_{\beta\eta} \lambda x.a$ since a and $\lambda x.a$ have different types. Therefore, a and b are different terms in normal-form, although they are equal in the theory generated by eta, beta, and the equation $\lambda x.a = \lambda x.b$. Adding the extension $a \to b$ solves the problem.

This explains Nipkow's restriction that rules must be of basic type: in this case, no lefthand side of rule can be an abstraction. Generalizing Nipkow's framework when this assumption is not met is not hard: it suffices to close the set of rules with finitely many beta-extensions for those rules whose lefthand side is an abstraction. This covers rules of arbitrary polymorphic functional type. Notice also that no beta-extension is needed for rules whose lefthand side is headed by a function symbol. Finally, because of the pattern condition, it is easy to see that no extension is needed for the eta-rule. We therefore have the following result:

Theorem 2. *[22] Given a set R of higher-order rules satisfying assumptions (i,ii,iii) and (v) such that*

(a) $R \cup \{beta, eta^{-1}\}$ is terminating;
(b) R is closed under the computation of beta-extensions;
(c) irreducible higher-order critical pairs of R are joinable;
then, higher-order rewriting with R is Church-Rosser.

Nipkow's result stating confluence of higher-order rewriting when assumption (iv) is met appears then as a corollary. The fact that it is a corollary is not straightforward, since the termination assumption is different: Nipkow assumes termination of higher-order rewriting with R. The fact that this coincides with assumption (i) is however true under assumptions (i,ii,iii,iv) [22]. We can easily see that Examples 4 and 5 (first version) do not admit higher-order critical pairs, hence are Church-Rosser. Adding the other rules for differentiation would clearly not change this situation.

Adding beta-extensions is not such a burden, but we can even dispense with by using the variant of eta-long beta-normal forms in which eta-expansion does not apply at the top of terms. Then, we can assume that the lefthand side of a higher-order rule is headed by an application or by a function symbol of non-zero arity if they are allowed, and no beta-extension is needed anymore.

We now turn to the second kind of normal form:

Theorem 3. *[22] Given a set R of higher-order rules satisfying assumptions (i,ii,iii) and (vi) such that*
(a) $R \cup \{beta, eta\}$ is terminating;
(b) irreducible higher-order critical pairs of R are joinable;
then, higher-order rewriting with R is Church-Rosser.

Example 5, as modified at the end of its paragraph, has no higher-order critical pairs, hence is Church-Rosser.

Altogether, the framework based on $\beta\eta$-normal forms appears a little bit more appealing.

Mixed Higher-Order Rewriting. We end up with the general case of a set of higher-order rules R split into disjoint subsets R_1 using plain pattern matching, and R_2 using higher-order pattern matching for terms in beta-eta-normal form. We assume that R satisfies assumptions (i,ii), that lefthand sides of rules in R_1 are headed by a function symbol, and that R_2 satisfies assumption (iii,vi). When these assumptions are met, we say that R is a mixed set of higher-order rules. Rewriting then searches for R_1-redexes with plain pattern-matching, and uses beta-eta-normalization before to search for R_2-redexes with higher-order pattern matching.

Our conjecture follows a similar analysis made for the first-order case, with left-linear rules using plain pattern matching, and non-left-linear ones using pattern matching modulo some first-order theory E [16]:

Conjecture. *Given a mixed set $R = R_1 \uplus R_2$ of higher-order rules such that*
(i) $R \cup \{\beta\eta\}$ is terminating;
(ii) irreducible plain critical pairs of R_1 are joinable;

(iii) irreducible higher-order critical pairs of R_2 are joinable;
(iv) irreducible higher-order critical pairs of R_2 with R_1 are joinable;
then, mixed higher-order rewriting with R is Church-Rosser.

5 Termination of Plain Higher-Order Rewriting

Given a rewrite system R, a term t is *strongly normalizing* if there is no infinite sequence of rewrites with R issuing from t. R is *strongly normalizing* or *terminating* if every term is strongly normalizing.

Termination of typed lambda calculi is notoriously difficult. Termination of the various incarnations of higher-order rewriting turns out to be even more difficult. There has been little success in coming up with general methods. The most general results have been obtained by Blanqui [5, 6], as a generalization of a long line of work initiated by Jouannaud and Okada [18, 17, 1, 4, 7]. We will not describe these results here, but base our presentation on the more recent work of Jouannaud and Rubio [19, 20] which is strictly more general in the absence of dependent types. The case of dependent types is investigated in [33], but requires more work.

Higher-Order Reduction Orderings. The purpose of this section is to define the kind of ordering needed for plain higher-order rewriting. To a quasi-ordering $\succeq$, we associate its strict part $\succ$ and equivalence $\simeq$. Typing environments are usually omitted to keep notations simple.

Definition 9. *[20] A* higher-order reduction ordering *is a quasi-ordering $\succeq$ of the set of higher-order terms, which (i) well-founded, (ii) monotonic (i.e., $s \succ t$ implies $u[s] \succ u[t]$ for all contexts $u[]$), (iii) stable (i.e., $s \succ t$ implies $s\gamma \succ t\gamma$ for all compatible substitutions γ) and (iv) includes alpha-conversion (i.e. $=_\alpha \ \subseteq \ \simeq$) and beta-eta-reductions (i.e., $\longrightarrow_{\beta\eta} \ \subset \ \succ$). It is* polymorphic *if $s \succ t$ implies $s\xi \succ t\xi$ for any type instantiation ξ.*

Note that the above definition includes the eta-rule, and not only the beta-rule as originally. This extension does not raise difficulties.

Theorem 4. *[20] Assume that $\succeq$ is a higher-order reduction ordering, and let $R = \{\Gamma_i \ \vdash \ l_i \rightarrow r_i\}_{i \in I}$ be a plain higher-order rewrite system such that $l_i \succ r_i$ for every $i \in I$. Assume in addition that $\succeq$ is polymorphic if so is R. Then the relation $\longrightarrow_R \cup \longrightarrow_{\beta\eta}$ is terminating.*

The proof of the previous result is not difficult: it is based on lifting the property $l_i \succ r_i$ to a rewrite step $u \longrightarrow^p_{l_i \rightarrow r_i} v$ by using the properties of the ordering. A simple induction allows then to conclude.

Higher-Order Recursive Path Ordering. We give here a generalization of Derhowitz's recursive path ordering to higher-order terms. Although the obtained relation is transitive in many particular cases, it is not in general, hence is not an ordering. We will nevertheless call it the *higher order recursive path ordering*, keeping Dershowitz's name for the generalization. We feel free to do so because the obtained relation is well-founded, hence its transitive closure is a well-founded ordering with the same equivalence, taken here equal to alpha-conversion for simplicity. We are given:

1. a partition $Mul \uplus Lex$ of $\mathcal{F} \cup \mathcal{S}$;
2. a quasi ordering $\geq_{\mathcal{FS}}$ on $\mathcal{F} \cup \mathcal{S}$, the *precedence*, such that
   (a) $>_{\mathcal{F}}$ is well-founded ($\mathcal{F} \cup \mathcal{S}$ is not assumed to be finite);
   (b) if $f =_{\mathcal{FS}} g$, then $f \in Lex$ iff $g \in Lex$;
   (c) $>_{\mathcal{FS}}$ is extended to $\mathcal{F} \cup \mathcal{X} \cup \mathcal{S}$ by adding all pairs $x \geq_{\mathcal{FS}} x$ for $x \in \mathcal{X}$ (free variables are only comparable to themselves);
3. a set of terms $\mathcal{CC}(s)$ called the *computability closure* of s: in the coming definition, $\mathcal{CC}(f(\overline{s})) = \overline{s}$, but will become a richer set later on.

The definition compares types and terms in the same recursive manner, using the additional proposition (assuming that s and $f(\overline{t})$ are terms)

$$A = \forall v \in \overline{t} \quad s \underset{horpo}{\succ} v \text{ or } u \underset{horpo}{\succeq} v \text{ for some } u \in \mathcal{CC}(s)$$

Definition 10.

$$s : \sigma \underset{horpo}{\succ} t : \tau \quad \textit{iff} \quad (\sigma = \tau = * \textit{ or } \sigma \underset{horpo}{\succeq} \tau) \textit{ and}$$

1. $s = f(\overline{s})$ *with* $f \in \mathcal{F} \cup \mathcal{S}$, *and* $u \underset{horpo}{\succeq} t$ *for some* $u \in \mathcal{CC}(s)$
2. $s = f(\overline{s})$ *and* $t = g(\overline{t})$ *with* $f >_{\mathcal{FS}} g$, *and* A *(see definition below)*
3. $s = f(\overline{s})$ *and* $t = g(\overline{t})$ *with* $f =_{\mathcal{FS}} g \in Mul$ *and* $\overline{s}(\underset{horpo}{\succ})_{mul}\overline{t}$
4. $s = f(\overline{s})$ *and* $t = g(\overline{t})$ *with* $f =_{\mathcal{FS}} g \in Lex$ *and* $\overline{s}(\underset{horpo}{\succ})_{lex}\overline{t}$, *and* A
5. $s = @(s_1, s_2)$ *and* $u \underset{horpo}{\succeq} t$ *for some* $u \in \{s_1, s_2\}$
6. $s = \lambda x : \sigma.u$, $x \notin \mathcal{V}ar(t)$ *and* $u \underset{horpo}{\succeq} t$
7. $s = f(\overline{s})$, $@(\overline{t})$ *a left-flattening of* t, *and* A
8. $s = f(\overline{s})$ *with* $f \in \mathcal{F}$, $t = \lambda x : \alpha.v$ *with* $x \notin \mathcal{V}ar(v)$ *and* $s \underset{horpo}{\succeq} v$
9. $s = @(s_1, s_2)$, $@(\overline{t})$ *a left-flattening of* t *and* $\{s_1, s_2\}(\underset{horpo}{\succ})_{mul}\overline{t}$
10. $s = \lambda x : \alpha.u$, $t = \lambda x : \beta.v$, $\alpha \underset{horpo}{\simeq} \beta$ *and* $u \underset{horpo}{\succ} v$
11. $s = @(\lambda x.u, v)$ *and* $u\{x \mapsto v\} \underset{horpo}{\succeq} t$

11bis. $s = \lambda x.@(u, x)$ *with* $x \notin \mathcal{V}ar(u)$ *and* $u \underset{horpo}{\succeq} t$

12. $s = \alpha \rightarrow \beta$, *and* $\beta \underset{horpo}{\succeq} t$
13. $s = \alpha \rightarrow \beta$, $t = \alpha' \rightarrow \beta'$, $\alpha \underset{horpo}{\simeq} \alpha'$ *and* $\beta \underset{horpo}{\succ} \beta'$
14. $s = X(\overline{s})$ *and* $t = X(\overline{t})$ *and* $\overline{s}(\succ_{horpo})_{mul}\overline{t}$

We assume known how to extend a relation on a set to n-tuples of elements of the set (*monotonic* extension -comparing terms one by one with at least one strict comparison-, or *lexicographic* extension -comparing terms one by one until a strict comparison is found) or to multisets of elements of the set (*multiset* extension).

The computability closure was called computational closure in [19].

Case 14 does not exist as such in [20], but follows easily from Case 9. Neither does Case 11bis, used for proving the eta-rule. An immediate subterm u of s is *type-decreasing* if the type of u is smaller or equal to the type of s. Condition A differs from its first-order version $\forall v \in \bar{t}\ s \succ_{horpo} v$, but reduces to it when all immediate subterms of s are type-decreasing, because $\succ_{horpo}$ enjoys the subterm property for these subterms. Indeed, the restriction of the ordering to subterm-closed sets of terms whose all immediate subterms are type-decreasing, as are types, is transitive.

A more abstract description of the higher-order recursive path ordering is given in [20], in which the relation on terms and the ordering on types are separated, the latter being specified by properties to be satisfied. The version given here, also taken from [20], shows that the same mechanism can apply to terms and types, a good start for a generalization to dependent type structures.

Theorem 5. *$(\succ_{horpo})*$ is a polymorphic higher-order reduction ordering.*

The proof is based on Tait and Girard's computability predicate technique (Girard's candidates are not needed for weak polymorphism) [20].

We go on checking the examples of Section 2, making appropriate choices when using property A.

The Higher-Order Recursive Path Ordering at Work.

Example 1. Let us first recall the rules for rec:

$$\begin{aligned} \mathsf{rec}(0, \mathsf{U}, \mathsf{Y}) &\to U \\ \mathsf{rec}(\mathsf{s}(\mathsf{X}), \mathsf{U}, \mathsf{Y}) &\to @(Y, X, \mathsf{rec}(\mathsf{X}, \mathsf{U}, \mathsf{Y})) \end{aligned}$$

We assume a multiset status for rec.

Since both sides of the first equation have the same (polymorphic) type α, the comparison $\mathsf{rec}(0, \mathsf{U}, \mathsf{Y}) \succ_{horpo} \mathsf{U}$ proceeds by Case 1 and succeeds easily since Y belongs to the computability closure of $\mathsf{rec}(0, \mathsf{U}, \mathsf{Y})$ as one of its subterms.

As it can be expected, both sides of the second equation have again the same type. The comparison $\mathsf{rec}(\mathsf{s}(\mathsf{X}), \mathsf{U}, \mathsf{Y}) \succ_{horpo} @(\mathsf{Y}, \mathsf{X}, \mathsf{rec}(\mathsf{X}, \mathsf{U}, \mathsf{Y}))$ proceeds this time by Case 7, generating three subgoals (we use here property A, choosing a subterm when possible). The first subgoal $\mathsf{Y} \succeq_{horpo} \mathsf{Y}$ is solved readily since Y is a subterm of the lefthand side. The second subgoal $\mathsf{s}(\mathsf{X}) \succeq_{horpo} \mathsf{X}$ is solved by Case 1. The third subgoal, namely $\mathsf{rec}(\mathsf{s}(\mathsf{X}), \mathsf{U}, \mathsf{Y}) \succ_{horpo} \mathsf{rec}(\mathsf{X}, \mathsf{U}, \mathsf{Y})$ is solved by Case 3, which generates the three easy comparisons $\mathsf{s}(\mathsf{X}) \succ_{horpo} \mathsf{X}$, $\mathsf{U} \succeq_{horpo} \mathsf{U}$ and $\mathsf{Y} \succeq_{horpo} \mathsf{Y}$.

Example 2. Let us recall the rules:

$$\mathsf{map}(\mathsf{nil}, \mathsf{F}) \to \mathsf{nil}$$
$$\mathsf{map}(\mathsf{cons}(\mathsf{H}, \mathsf{T}), \mathsf{F}) \to \mathsf{cons}(@(\mathsf{F}, \mathsf{H}), \mathsf{map}(\mathsf{T}, \mathsf{F}))$$

We assume the precedence $\mathsf{map} >_{\mathcal{FS}} \mathsf{cons}$ and membership of map to Mul. The first rule being easy, let us consider the second. Our goal is
(i) $\mathsf{map}(\mathsf{cons}(\mathsf{H}, \mathsf{T}), \mathsf{F}) \succ_{horpo} \mathsf{cons}(@(\mathsf{H}, \mathsf{F}), \mathsf{map}(\mathsf{T}, \mathsf{F}))$
which first checks types and then reduces to two new goals by Case 2:
(ii) $\mathsf{map}(\mathsf{cons}(\mathsf{H}, \mathsf{T}), \mathsf{F}) \succ_{horpo} @(\mathsf{F}, \mathsf{H})$
(iii) $\mathsf{map}(\mathsf{cons}(\mathsf{H}, \mathsf{T}), \mathsf{F}) \succ_{horpo} \mathsf{map}(\mathsf{T}, \mathsf{F})$
Goal (ii) reduces by Case 7 to three new goals
(iv) $\mathsf{list}(\beta) \succ_{horpo} \beta$
(v) $\mathsf{F} \succeq_{horpo} \mathsf{F}$, which disappears
(vi) $\mathsf{cons}(\mathsf{H}, \mathsf{T}) \succeq_{horpo} \mathsf{H}$
Goal (iv) is taken care of by Case 1 while Goal (vi) reduces by Case 1 to
(vii) $\mathsf{list}(\alpha) \succ_{horpo} \alpha$
(viii) $\mathsf{H} \succ_{horpo} \mathsf{H}$, which disappears.
Goal (vii) is taken care of by Case 2. We are left with goal (iii) which reduces by Case 3 to
(iv) $\{\mathsf{cons}(\mathsf{H}, \mathsf{T}), \mathsf{F}\}(\succ_{horpo})_{mul}\{\mathsf{T}, \mathsf{F}\}$
which yields one (last) goal by definition of $(\succ_{horpo})_{mul}$:
(x) $\mathsf{cons}(\mathsf{H}, \mathsf{T}) \succ_{horpo} \mathsf{T}$, which is taken care of by Case 1.

Example 3. The example of parametric insertion sort raises a difficulty. Let us recall the rules:

$$\begin{array}{rlrl}
\mathsf{max}(0, \mathsf{X}) & \to \mathsf{X} & \mathsf{max}(\mathsf{X}, 0) & \to \mathsf{X} \\
\mathsf{max}(\mathsf{s}(\mathsf{X}), \mathsf{s}(\mathsf{Y})) & \to \mathsf{s}(\mathsf{max}(\mathsf{X}, \mathsf{Y})) & & \\
\mathsf{min}(0, \mathsf{X}) & \to 0 & \mathsf{min}(\mathsf{X}, 0) & \to 0 \\
\mathsf{min}(\mathsf{s}(\mathsf{X}), \mathsf{s}(\mathsf{Y})) & \to \mathsf{s}(\mathsf{min}(\mathsf{X}, \mathsf{Y})) & & \\
\mathsf{insert}(\mathsf{N}, \mathsf{nil}, \mathsf{X}, \mathsf{Y}) & \to \mathsf{cons}(\mathsf{N}, \mathsf{nil}) & & \\
\mathsf{insert}(\mathsf{N}, \mathsf{cons}(\mathsf{M}, \mathsf{T}), \mathsf{X}, \mathsf{Y}) & \to \mathsf{cons}(@(\mathsf{X}, \mathsf{N}, \mathsf{M}), \mathsf{insert}(@(\mathsf{Y}, \mathsf{N}, \mathsf{M}), \mathsf{T}, \mathsf{X}, \mathsf{Y})) & & \\
\mathsf{sort}(\mathsf{nil}, \mathsf{X}, \mathsf{Y}) & \to \mathsf{nil} & & \\
\mathsf{sort}(\mathsf{cons}(\mathsf{N}, \mathsf{T}), \mathsf{X}, \mathsf{Y}) & \to \mathsf{insert}(\mathsf{N}, \mathsf{sort}(\mathsf{T}, \mathsf{X}, \mathsf{Y}), \mathsf{X}, \mathsf{Y}) & & \\
\mathsf{ascending_sort}(\mathsf{L}) & \to \mathsf{sort}(\mathsf{L}, \lambda \mathsf{xy}.\mathsf{min}(\mathsf{x}, \mathsf{y}), \lambda \mathsf{xy}.\mathsf{max}(\mathsf{x}, \mathsf{y})) & & \\
\mathsf{descending_sort}(\mathsf{L}) & \to \mathsf{sort}(\mathsf{L}, \lambda \mathsf{xy}.\mathsf{max}(\mathsf{x}, \mathsf{y}), \lambda \mathsf{xy}.\mathsf{min}(\mathsf{x}, \mathsf{y})) & &
\end{array}$$

The reader can check that all comparisons succeed with appropriate precedence and statuses, but the last two:

$$\mathsf{ascending\_sort}(\mathsf{L}) \succ_{horpo} \mathsf{sort}(\mathsf{L}, \lambda \mathsf{xy}.\mathsf{min}(\mathsf{x}, \mathsf{y}), \lambda \mathsf{xy}.\mathsf{max}(\mathsf{x}, \mathsf{y}))$$
$$\mathsf{descending\_sort}(\mathsf{L}) \succ_{horpo} \mathsf{sort}(\mathsf{L}, \lambda \mathsf{xy}.\mathsf{max}(\mathsf{x}, \mathsf{y}), \lambda \mathsf{xy}.\mathsf{min}(\mathsf{x}, \mathsf{y}))$$

This is because the subterm $\lambda \mathsf{xy}.\mathsf{min}(\mathsf{x}, \mathsf{y})$ occurring in the righthand side has type $\alpha \to \alpha \to \alpha$, which is not comparable to any lefthand side type.

Computability Closure. Our definition of $\succ_{horpo}$ is parameterized by the definition of the computability closure. In order for Theorem 5 to hold, it suffices to make sure that the closure is made of terms which are all computable in the sense of Tait and Girard's computability predicate (termed *reducibility candidate* by Girard). Computability is guaranteed for the immediate subterms of a term by an induction argument in the strong normalization proof, and we can then enlarge the set of computable terms by using computability preserving operations.

Definition 11. *Given a term $t = f(\overline{t})$ of type σ, we define its computable closure $\mathcal{CC}(t)$ as $\mathcal{CC}(t, \emptyset)$, where $\mathcal{CC}(t, \mathcal{V})$, with $\mathcal{V} \cap \mathcal{V}ar(t) = \emptyset$, is the smallest set of well-typed terms containing all variables in $\mathcal{V}$, all terms in $\overline{t}$, and closed under the following operations:*

1. *subterm of minimal data-type: let $s \in \mathcal{CC}(t, \mathcal{V})$ with $f \in \mathcal{F}$, and $u : \sigma$ be a subterm of s such that σ is a data-type minimal in the type ordering and $\mathcal{V}ar(u) \subseteq \mathcal{V}ar(t)$; then $u \in \mathcal{CC}(t, \mathcal{V})$;*
2. *precedence: let g such that $f >_{\mathcal{FS}} g$, and $\overline{s} \in \mathcal{CC}(t, \mathcal{V})$; then $g(\overline{s}) \in \mathcal{CC}(t, \mathcal{V})$;*
3. *recursive call: let $\overline{s}$ be a sequence of terms in $\mathcal{CC}(t, \mathcal{V})$ such that the term $f(\overline{s})$ is well typed and $\overline{t}(\succ_{horpo} \cup \rhd_{\succ})_{stat_f} \overline{s}$; then $f(\overline{s}) \in \mathcal{CC}(t, \mathcal{V})$;*
4. *application: let $s : \sigma_1 \to \ldots \to \sigma_n \to \sigma \in \mathcal{CC}(t, \mathcal{V})$ and $u_i : \sigma_i \in \mathcal{CC}(t, \mathcal{V})$ for every $i \in [1..n]$; then $@(s, u_1, \ldots, u_n) \in \mathcal{CC}(t, \mathcal{V})$;*
5. *abstraction: let $x \notin \mathcal{V}ar(t) \cup \mathcal{V}$ and $s \in \mathcal{CC}(t, \mathcal{V} \cup \{x\})$; then $\lambda x.s \in \mathcal{CC}(t, \mathcal{V})$;*
6. *reduction: let $u \in \mathcal{CC}(t, \mathcal{V})$, and $u \succeq_{horpo} v$; then $v \in \mathcal{CC}(t, \mathcal{V})$;*
7. *weakening: let $x \notin \mathcal{V}ar(u, t) \cup \mathcal{V}$. Then, $u \in \mathcal{CC}(t, \mathcal{V} \cup \{x\})$ iff $u \in \mathcal{CC}(t, \mathcal{V})$.*

(Case 11bis can be removed since it now follows from Case 1)

The new definition of the computability closure uses the relation $\succ_{horpo}$, while the new relation is denoted by $\succ_{chorpo}$. Whether this notational difference introduced in [19] is necessary is doubtful, but allows to define the ordering in a hierarchical way rather than as a fixpoint. The proofs would otherwise be surely harder than they already are. Note that Case 11bis can now be removed since it now follows from Case 1.

We can now show termination of the two remaining rules of Example 3:
(i) $\mathsf{ascending\_sort}(\mathsf{L}) \succ_{chorpo} \mathsf{sort}(\mathsf{L}, \lambda \mathsf{xy}.\mathsf{min}(\mathsf{x}, \mathsf{y}), \lambda \mathsf{xy}.\mathsf{max}(\mathsf{x}, \mathsf{y}))$
(ii) $\mathsf{descending\_sort}(\mathsf{L}) \succ_{chorpo} \mathsf{sort}(\mathsf{L}, \lambda \mathsf{xy}.\mathsf{max}(\mathsf{x}, \mathsf{y}), \lambda \mathsf{xy}.\mathsf{min}(\mathsf{x}, \mathsf{y}))$

Since both proofs are almost identical, we only consider goal (i). We assume the precedence $\mathsf{ascending\_sort} >_{\mathcal{FS}} \mathsf{sort}, \mathsf{min}, \mathsf{max}$. First, note that using rule 2 does not work, since we would generate the goal
$\mathsf{ascending\_sort}(\mathsf{L}) \succ_{chorpo} \lambda xy.max(x, y)$
which fails for typing reason. Instead, we immediately proceed with Case 1 of the ordering definition, showing that the whole righthand side is in the computability closure of the lefthand one:
(iii) $\mathsf{sort}(\mathsf{L}, \lambda \mathsf{xy}.\mathsf{max}(\mathsf{x}, \mathsf{y}), \lambda \mathsf{xy}.\mathsf{min}(\mathsf{x}, \mathsf{y})) \in \mathcal{CC}(\mathsf{ascending\_sort}(\mathsf{L}), \emptyset)$
By precedence Case 2 of the computability closure, we get three goals:
(iv) $\mathsf{L} \in \mathcal{CC}(\mathsf{ascending\_sort}(\mathsf{L}), \emptyset)$
(v) $\lambda \mathsf{xy}.\mathsf{max}(\mathsf{x}, \mathsf{y}) \in \mathcal{CC}(\mathsf{ascending\_sort}(\mathsf{L}), \emptyset)$

(vi) λxy.min(x, y)) $\in \mathcal{CC}$(ascending_sort(L), $\emptyset$)
Goal (iv) is easily done by the basic case of the inductive definition of the computability closure. Goal (v) is similar to goal (vi), which we do now. By weakening Case 7 applied twice, we get the new goal:
(vii) min(x, y)) $\in \mathcal{CC}$(ascending_sort(L), {x, y})
Applying now precedence Case 2 of the closure, we get the new goals
(viii) $x \in \mathcal{CC}$(ascending_sort(L), $\{x, y\}$)
(ix) $y \in \mathcal{CC}$(ascending_sort(L), $\{x, y\}$)
which are both solved by basic case of the inductive definition.

6 Termination of Normal Higher-Order Rewriting

The situation with normal higher-order rewriting is a little bit more delicate because of the definition in which rewritten terms are normalized. Lifting a comparison from a rule $l \succ r$ to an instance $l\sigma\!\downarrow\ \succ r\sigma\!\downarrow$ may not be possible for a given higher-order reduction ordering, but may be for some appropriate subrelation.

Definition 12. *A subrelation $\succ_\beta$ of $\succ$ is said to be* $\downarrow$-stable *if $s \succ_\beta t$ implies $s\gamma\!\downarrow_\beta\ \succ t\gamma\!\downarrow_\beta$ for all normalized terms s, t and normalized substitution γ.*

Theorem 6. *[20] Assume that $\succ$ is a higher-order reduction ordering and that $\succ_\beta$ is a $\downarrow$-stable subrelation of $\succ$. Let $R = \{\Gamma_i \vdash l_i \to r_i\}_{i \in I}$ be a normal higher-order rewrite system such that $l_i \succ_\beta r_i$ for every $i \in I$. Assume in addition that $\succeq$ is polymorphic if so is R. Then normal higher-order rewriting with R is strongly normalizing.*

The proof of this result is quite similar to that of Theorem 4. Let us remark that $\succ_{horpo}$ is not $\downarrow$-stable. The following example shows the kind of problematic term:

Example 8. $@(X, f(a)) \succ_{horpo} @(X, a) \succ_{horpo} a$, while instantiating these comparisons with the substitution $\gamma = \{X \mapsto \lambda y.a\}$ yields $X(f(a))\gamma\!\downarrow\ = X(a)\gamma\!\downarrow\ = a$, contradicting $\downarrow$-stability. □

It turns out that only the beta rule may cause a problem for the higher-order recursive path ordering, hence our notation $\succ_\beta$, by turning a greater than comparison into an greater than or equal to comparison. As a consequence, the coming discussion applies to all kinds of normal higher-order rewriting, using eta either as a reduction or as an expansion, and with or without arities.

Definition 13. *The relation $(\succ_{horpo})_\beta$, called* normal higher-order recursive path ordering *is defined as the relation $\succ_{horpo}$, but restricting Cases 5 and 9 as follows:*

5. *... if s_1 is not an abstraction nor a variable, and $u = s_2$ otherwise.*
9. *... if s_1 is not an abstraction nor a variable, and $\{s_2\}((\succ_{horpo})_\beta\ )_{mul}\bar{t}$ otherwise.*

Although the main argument of the proof is that this definition yields a (well-founded) subrelation of the higher-order recursive path ordering, showing that it is beta-stable happens to be a painful technical task, especially when using eta-expansions. Note that we could think adding two new cases, by incorporating the cases removed from $(\succ_{horpo})_\beta$ in the equivalence part of the new relation:

14. $@(X(\overline{s}), \overline{u})(\succeq_{horpo})_\beta\ @(X(\overline{t}), \overline{v})$ *if* $\overline{s}\,\overline{u}((\succeq_{horpo})_\beta\ )_{mon}\overline{t}\overline{v}$

The study of this variant should be quite easy.

It is unfortunately not enough to modify the ordering as done here in presence of a computational closure which is not reduced to the set of subterms: the definition of the closure itself needs cosmetic modifications to ensure that the ordering is stable. Since they do not impact the example to come, we simply refer to the original article for its precise definition [21]. Let us denote the obtained relation by $(\succ_{chorpo})_\beta$.

Theorem 7. *$((\succ_{horpo})_\beta\ )^*$ and $((\succ_{chorpo})_\beta\ )^*$ are polymorphic $\downarrow$-stable higher-order reduction orderings.*

We now test the ordering $(\succ_{chorpo})_\beta$ against our remaining examples. We shall refer to Definition 10 for the cases of Definition 13 that did not change, and to the modifications for the changed ones.

Example 4. **Differentiation 1.** Let us recall the goal, writing all applications on left, and flattening them all on the right:
(i) $@(@(\mathsf{diff}, \lambda \mathsf{x}.\ @(\mathsf{sin}, (@(\mathsf{F}, \mathsf{x})))), \mathsf{y})\ (\succ_{chorpo})_\beta$
$@(\mathsf{mul}, @(\mathsf{cos}, @(\mathsf{F}, \mathsf{y})), @(\mathsf{diff}, \lambda \mathsf{x}.@(\mathsf{F}, \mathsf{x}), \mathsf{y}))$

We take $D >_{\mathcal{F}} \{mul, sin, cos\}$ for precedence and assume that the function symbols are in Mul. By Case 9 applied to the goal (i), we get
(ii) $\{@(\mathsf{diff}, \lambda \mathsf{x}.\ @(\mathsf{sin}, (@(\mathsf{F}, \mathsf{x})))), \mathsf{y}\}((\succ_{chorpo})_\beta\ )_{mul}$
$\{\mathsf{mul}, @(\mathsf{cos}, @(\mathsf{F}, \mathsf{y})), @(\mathsf{diff}, \lambda \mathsf{x}.@(\mathsf{F}, \mathsf{x}), \mathsf{y})\}$
Because y occurs free in the term $@(\mathsf{cos}, @(\mathsf{F}, \mathsf{y}))$ taken from the multiset on the right and because y is the only term in the left multiset in which y occurs free, there is no possibility to solve the previous goal and the whole comparison fails. There might be a way out by using the closure mechanism for applications, but this variant has not been studied yet.

Example 5. **Differentiation 2.** Let us recall this applications-free goal:
(i) $\mathsf{diff}(\lambda \mathsf{x}.\ \mathsf{sin}(\mathsf{F}(\mathsf{x})))(\succ_{chorpo})_\beta\ \mathsf{mul}(\lambda \mathsf{x}.\ \mathsf{cos}(\mathsf{F}(\mathsf{x})), \mathsf{diff}(\lambda \mathsf{x}.\mathsf{F}(\mathsf{x})))$
We take the precedence $\mathsf{diff} >_{\mathcal{FS}} \mathsf{mul} >_{\mathcal{FS}} \mathsf{sin} =_{\mathcal{FS}} \mathsf{cos}$ and assume that all function symbols are in Mul. By Case 2 applied to goal (i), we get:
(ii) $\lambda \mathsf{x}.\ \mathsf{sin}(\mathsf{F}(\mathsf{x}))(\succeq_{chorpo})_\beta\ \lambda \mathsf{x}.\ \mathsf{cos}(\mathsf{F}(\mathsf{x}))$
(iii) $\mathsf{diff}(\lambda \mathsf{x}.\ \mathsf{sin}(\mathsf{F}(\mathsf{x})))(\succ_{chorpo})_\beta\ \mathsf{diff}(\lambda \mathsf{x}.\mathsf{F}(\mathsf{x}))$
Goal (ii) reduces successively to
(iv) $\mathsf{sin}(\mathsf{F}(\mathsf{x}))(\succeq_{chorpo})_\beta\ \mathsf{cos}(\mathsf{F}(\mathsf{x}))$ by Case 10
(v) $\mathsf{F}(\mathsf{x})(\succeq_{chorpo})_\beta\ \mathsf{F}(\mathsf{x})$ by Case 3, which disappears.
We proceed with goal (iii) which reduces successively to
(vi) $\lambda \mathsf{x}.\ \mathsf{sin}(\mathsf{F}(\mathsf{x}))(\succ_{chorpo})_\beta\ \lambda \mathsf{x}.\mathsf{F}(\mathsf{x})$ by Case 3

(viii) $\mathsf{sin}(\mathsf{F}(\mathsf{x}))(\succ_{chorpo})_\beta\ \mathsf{F}(\mathsf{x})$ by Case 10
(ix) $\mathsf{F}(\mathsf{x}) \in \mathcal{CC}(\mathsf{sin}(\mathsf{F}(\mathsf{x})))$ by Case 1,
which disappears by base case of the closure definition.

With the same precedence, the reader can now try the modified version of Example 5: the computation goes as if using Dershowitz's recursive path ordering. We observe the influence of seemingly equivalent definitions on the computation, suggesting that some work is still necessary to improve the higher-order recursive path ordering in the normalized case.

7 Termination of Mixed Higher-Order Rewriting

Conjecture. *Given a mixed set* $R = R_1 \uplus R_2$ *of higher-order rules, a polymorphic higher-order ordering* $\succ$ *and a* $\downarrow_{\beta\eta}$*-stable subrelation* $\succ_{\beta\eta}$ *of* $\succ$ *such that* $l \succ r$ *for any rule* $l \to r \in R_1$ *and* $l \succ_{\beta\eta} r$ *for any rule* $l \to r \in R_2$*, then mixed rewriting with* R *is terminating.*

The proof follows from the fact $\succ$ decreases along beta-eta-reductions.

8 Conclusion

We have shown here how the various versions of higher-order rewriting relate to each other, and can be enriched so as to yield a new framework in which they coexist. We have also shown how the classical properties of rewriting can be checked with the help of powerful tools, like higher-order critical pairs, beta-extensions, the higher-order recursive path ordering, and the computing closure. We left out sufficient completeness, which has been only recently addressed in the case of plain rewriting [8].

Acknowledgments. This paper would not exist without the pioneering work of Klop on higher-order rewriting and the work I have done myself with my coauthors Femke van Raamsdonk and Albert Rubio, my colleague Mitsuhiro Okada with who I started getting into that subject, and my students Maribel Fernandez, Frederic Blanqui and Daria Walukiewicz who investigated some of these questions within the framework of the calculus of constructions. An anonymous referee and Femke van Raamsdonk deserve special thanks for their help in shaping the initial draft.

References

1. Franco Barbanera, Maribel Fernández, and Herman Geuvers. Modularity of strong normalization and confluence in the algebraic-λ-cube. In *Proc. of the 9th Symp. on Logic in Computer Science*, IEEE Computer Society, 1994.
2. Henk Barendregt. *Handbook of Logic in Computer Science*, chapter Typed lambda calculi. Oxford Univ. Press, 1993. eds. Abramsky et al.

3. Terese. Term Rewriting Systems. Cambridge Tracts in Theoretical Computer Science 55, Marc Bezem, Jan Willem Klop and Roel de Vrijer eds., Cambridge University Press, 2003.
4. Frédéric Blanqui, Jean-Pierre Jouannaud, and Mitsuhiro Okada. The Calculus of Algebraic Constructions. In Narendran and Rusinowitch, Proc. RTA, LNCS 1631, 1999.
5. Frédéric Blanqui. Inductive Types Revisited. Available from the web.
6. Frédéric Blanqui. Definitions by rewriting in the Calculus of Constructions, 2003. To appear in Mathematical Structures in Computer Science.
7. Frédéric Blanqui, Jean-Pierre Jouannaud, and Mitsuhiro Okada. Inductive Data Types. *Theoretical Computer Science* 277, 2001.
8. Jacek Chrzaczsz and Daria Walukiewicz-Chrzaczsz. Consistency and Completeness of Rewriting in the Calculus of Constructions. Draft.
9. Nachum Dershowitz. Orderings for term rewriting systems. *Theoretical Computer Science*, 17(3):279–301, March 1982.
10. Nachum Dershowitz and Jean-Pierre Jouannaud. Rewrite systems. In J. van Leeuwen, editor, *Handbook of Theoretical Computer Science*, volume B, pages 243–309. North-Holland, 1990.
11. Gilles Dowek. Higher-Order Unification and Matching. Handbook of Automated Reasonning, A. Voronkov ed., vol 2, pages 1009–1062.
12. Gilles Dowek, Amy Felty, Hugo Herbelin, Gérard Huet, Christine Paulin-Mohring, and Benjamin Werner. The Coq proof assistant user's guide version 5.6. INRIA Rocquencourt and ENS Lyon.
13. Gilles Dowek, Thérèse Hardin, Claude Kichner and Franck Pfenning. Unification via explicit substitutions: The case of Higher-Order Patterns. In *JICSLP*:259–273, 1996.
14. Jean-Yves Girard, Yves Lafont, and Patrick Taylor. *Proofs and Types.* Cambridge Tracts in Theoretical Computer Science. Cambridge University Press, 1989.
15. Gérard Huet. Confluent reductions: abstract properties and applications to term rewriting systems. In *Journal of the ACM* 27:4(797–821), 1980.
16. Jean-Pierre Jouannaud and Hélène Kirchner. Completion of a Set of Rules Modulo a Set of Equations. In *Siam Journal of Computing* 15:4(1155–1194), 1984.
17. Jean-Pierre Jouannaud and Mitsuhiro Okada. Abstract data type systems. *Theoretical Computer Science*, 173(2):349–391, February 1997.
18. Jean-Pierre Jouannaud and Mitsuhiro Okada. Higher-Order Algebraic Specifications. In *Annual IEEE Symposium on Logic in Computer Science*, Amsterdam, The Netherlands, 1991. IEEE Comp. Soc. Press.
19. Jean-Pierre Jouannaud and Albert Rubio. The higher-order recursive path ordering. In Giuseppe Longo, editor, *Fourteenth Annual IEEE Symposium on Logic in Computer Science*, Trento, Italy, July 1999. IEEE Comp. Soc. Press.
20. Jean-Pierre Jouannaud and Albert Rubio. Higher-order recursive path orderings. Available from the web.
21. Jean-Pierre Jouannaud and Albert Rubio. Higher-order orderings for normal rewriting. Available from the web.
22. Jean-Pierre Jouannaud and Albert Rubio and Femke van Raamsdonk. Higher-order Rewriting with Types and Arities. Available from the web.
23. Jan Willem Klop. Combinatory Reduction Systems. Mathematical Centre Tracts 127. Mathematisch Centrum, Amsterdam, 1980.
24. Richard Mayr and Tobias Nipkow. Higher-order rewrite systems and their confluence. *Theoretical Computer Science*, 192(1):3–29, February 1998.

25. Dale Miller. A Logic Programming Language with Lambda-Abstraction, Function Variables, and Simple Unification. In *Journal and Logic and Computation* 1(4):497–536, 1991.
26. Tobias Nipkow. Higher-order critical pairs. In *6th IEEE Symp. on Logic in Computer Science*, pages 342–349. IEEE Computer Society Press, 1991.
27. Tobias Nipkow, Laurence C. Paulson and Markus Wenzel. Isabelle/HOL — A Proof Assistant for Higher-Order Logic. LNCS 2283, Springer Verlag, 2002.
28. Tobias Nipkow and Christian Prehofer. Higher-Order Rewriting and Equational Reasonning. In *Automated deduction — A basis for Applications. Volume I: Foundations*, Bibel and Schmitt editors. Applied Logic Series 8:399–430, Kluwer, 1998.
29. Femke van Raamsdonk. Higher-order rewriting. In [3].
30. Gerald E. Peterson and Mark E. Stickel. Complete sets of reductions for some equational theories. In *JACM* 28(2):233–264, 1981.
31. Franck Pfenning. Logic Programming in the LF Logical Framework. In *Logical Frameworks*, Gérard Huet and Gordon D. Plotkin eds., Cambridge University Press, 1991.
32. Femke van Raamsdonk. Confluence and Normalization for Higher-Order Rewrite Systems. phd thesis, Vrije Universiteit, Amsterdam, The Netherlands, 1996.
33. Daria Walukiewicz-Chrzaszcz. Termination of rewriting in the Calculus of Constructions. In *Proceedings of the Workshop on Logical Frameworks and Metalanguages, Santa Barbara, California*, 2000.

Timing the Untimed: Terminating Successfully While Being Conservative

J.C.M. Baeten, M.R. Mousavi, and M.A. Reniers

Department of Computer Science,
Eindhoven University of Technology (TU/e),
NL-5600 MB Eindhoven, The Netherlands

Abstract. There have been several timed extensions of ACP-style process algebras with successful termination. None of them, to our knowledge, are equationally conservative (ground-)extensions of ACP with successful termination. Here, we point out some design decisions which were the possible causes of this misfortune and by taking different decisions, we propose a spectrum of timed process algebras ordered by equational conservativity ordering.

1 The Untimed Past

The term "process algebra" was coined by Jan Bergstra and Jan Willem Klop in [8] to denote an algebraic approach to concurrency theory. Their process algebra had uniform atomic actions a_i for $i \in I$ (with I some index set), sequential composition $\_ \cdot \_$, choice (alternative composition) $\_ + \_$ and left merge $\_ \mathbin{\|\!\!\lfloor} \_$ as the basic composition operators.[1]

Much of the core theory of [8] remained intact in the course of more than 20 years of developments in the ACP-school (for Algebra of Communicating Processes) of process algebra. Their theory has however been subject to a number of, rather important, extensions and improvements. Next, we list some of the developments that are most relevant to the subject matter of this paper.

1. A major improvement over the process algebra of [8] was combining the concepts of *communication and concurrency* in the *Algebra of Communicating Processes (ACP)* which was proposed by Bergstra and Klop in [9, 10]. In the process algebra of [8], parallel composition $x \,||\, y$ was a shorthand as defined below.
$$x \,||\, y \doteq (x \mathbin{\|\!\!\lfloor} y) + (y \mathbin{\|\!\!\lfloor} x)$$
There was no possibility for the parallel components to communicate or synchronize. The situation was improved in [9, 10] by introducing a (total) communication function, defining a communication merge operator $|$ and raising the parallel composition operator $||$ to a basic composition operator in the algebra, rather than a defined term.

[1] Sequential composition was called "concatenation" and choice was called "union" in [8].

A. Middeldorp et al. (Eds.): Processes... (Klop Festschrift), LNCS 3838, pp. 251–279, 2005.

2. Another major improvement has been the addition of *identity elements*. Bergstra and Klop in [8] did study the addition of a constant 0 which is an identity element for both nondeterministic choice and sequential composition but then they ruled out this option by observing that the addition of 0 leads to the following counter-intuitive equality:

$$x \cdot y = (x+0) \cdot y = (x \cdot y) + (0 \cdot y) = (x \cdot y) + y$$

   The above equality states that the sequential composition may forget about its first argument which is indeed pathological. A couple of years had to pass to reveal that, as in ordinary rings, two process constants ϵ and δ can be used to give $\_ \cdot \_$ and $\_ + \_$ their identity elements, respectively [16, 22]. (Note that unlike in rings, left-distributivity of choice over sequential composition is still prohibited in the extended process algebra.) Hence, the process algebra PA_δ^ϵ of [22] had two extra constants ϵ and δ. A different proposal for the interplay of ϵ and parallel composition was formulated in [4, 7]. There, a new unary function symbol $\surd(\_)$ is added to the signature in order to capture the possibility of termination for complex terms. ACP of [9, 10] had δ as an identity element for choice but lacked ϵ. In both [16, 22], ϵ is added to ACP resulting in ACP^ϵ. The constant ϵ denotes termination, whereas the action constant encompasses both the action execution and the termination afterwards.
3. The third improvement concerning the subject matter of this paper was the addition of *quantitative time*. Baeten and Bergstra, in [2], proposed a real-time-stamped extension of ACP. In [3], they extend ACP with discrete time using prefix operators $\sigma_{rel}.\_$ and $\sigma_{abs}.\_$ for relative and absolute timing, respectively.

Vereijken tried to extend the result of the first and second improvements with the third aspect in Chapter 6 of his Ph.D. thesis [20]. There, he introduced $ACP_{drt,\epsilon} - ID$ as a discrete time extension of ACP^ϵ (here, $-ID$ denotes the absence of an immediate deadlock constant). However, as it turns out, the above three extensions do not match perfectly: while the extensions in each direction can be interpreted as a conservative one, there is no conservativity result for the extension of ACP^ϵ with timing. In the next section, we review design decisions on the way to timing untimed process algebras. Among the design decisions, we try to find possible cause(s) for this misfortune and will try to improve the situation by redesigning the extensions. This way, we may deviate from the commonly accepted principles of ACP, as we see appropriate. The result will be a lattice of process theories ordered by equational conservativity ordering.

2 Timing the Untimed

The following design decisions have to be taken in order to extend an untimed process algebra with timing information:

1. Delayable vs. urgent actions: When extending an untimed process algebra with timing, a natural question is how to deal with the timing behavior

of untimed basic actions. One choice is to regard them as urgent actions without any timing behavior. Another choice is to allow for an arbitrary timing behavior and introduce new urgent actions. The same decision has to be taken for deadlock and successful termination constants. We believe that taking the untimed actions (deadlock, termination) to be delayable is the more natural choice. The fact that no timing information is given should allow for an arbitrary timing of the implementation, rather than only allowing for the case of the urgent process. Further elaboration of the arguments can be found in [1].

2. Time stamped actions vs. separation of actions and time: Timing can be added to actions in terms of time stamps or alternatively, in terms of a separate delay operator. We choose the second option, as does most of the literature [17, 19, 15].
3. Time (non)determinism: Time determinism means that passage of time cannot make a choice. Usually, in timed extensions of ACP a weaker version of time determinism is used by forcing that time cannot make a choice unless one of the options prevents time to pass. In the latter case, a time transition shall resolve the choice in favor of options that allow for it. This is called weak time-determinism, and allows for a simple description of a time-out mechanism.
4. Time domain: Several decisions can be taken for the appropriate time domain. Existence or absence of a least element, discreteness or denseness of the time domain and having a partial (branching) or full (linear) ordering on the elements of the time domain often lead to different timed theories. Here, we choose for a discrete time domain with a total linear ordering. This choice leads to the simplest theory, and is also the choice taken most often in the literature [17, 19, 15] (see [20] for an overview).
5. Relative vs. absolute time: The timing information may be taken relative to the successful termination of causally preceding actions or alternatively, to a fixed starting point. Here, we take the relative timing approach which is simpler to deal with. Our discussions carry over to the setting with absolute timing. We refer to [5] for more information on relative and absolute timing.

Let us take the above design decisions, and let us consider the timed extension of a simple process algebra, BPA_{δ}^{ϵ}, the theory BPA of [8] extended with constants δ and ϵ (see [20, 1] for an overview of the above design decisions). Operational rules for this theory are given in Table 1.

We fix the notation used for transition system specifications (TSS) in this paper, as follows. The table containing the TSS is labelled with $TSS(Name)$ where $Name$ is the name of the process theory for which we are defining a TSS. For example in Table 1, we define a TSS for BPA_{δ}^{ϵ} and hence the table is labelled by it. Then, we name the TSS which we include in the definition, i.e., the TSS being extended. In an extension, we include the signature, transition relations, predicates and deduction rules from the original TSS and hence do not re-state them in the extended TSS. Table 1 does not extend any previous theory and hence the extension line is empty. Subsequently, we give the signature of the

Table 1. Transition system specification for $BPA_\delta^\epsilon(A)$

$TSS(BPA_\delta^\epsilon(A))$

constant: $\delta, \epsilon, (a)_{a\in A}$; binary: $\_\cdot\_, \_+\_$;

$\_\!\downarrow$; $(\_ \xrightarrow{a} \_)_{a\in A}$;

x, x', y, y';

$$\epsilon\downarrow \qquad a \xrightarrow{a} \epsilon \qquad \frac{x \xrightarrow{a} x'}{x\cdot y \xrightarrow{a} x'\cdot y} \qquad \frac{x\downarrow \quad y \xrightarrow{a} y'}{x\cdot y \xrightarrow{a} y'} \qquad \frac{x\downarrow \quad y\downarrow}{x\cdot y\downarrow}$$

$$\frac{x\downarrow}{x+y\downarrow} \qquad \frac{y\downarrow}{x+y\downarrow} \qquad \frac{x \xrightarrow{a} x'}{x+y \xrightarrow{a} x'} \qquad \frac{y \xrightarrow{a} y'}{x+y \xrightarrow{a} y'}$$

theory in terms of function symbols and their arities. In Table 1, the signature consists of constants δ, ϵ and actions $a \in A$, as well as binary function symbols $\_\cdot\_$ and $\_+\_$ for sequential composition and choice, respectively. The transition relations and predicates being defined by the TSS follow afterwards. In the TSS of Table 1, $\rightarrow$ is the transition relation labelled by $a \in A$ and $\downarrow$ is the termination predicate. Finally, the set of deduction rules is presented. Most of the deduction rules given in Table 1 are quite standard and self-explanatory.

Before we continue with the extension of the basic process algebra to the timed setting, we fix the semantics of TSS's in terms of their induced transition relations and predicates. We write $TSS(A) \models p \xrightarrow{l} q$ and $TSS(A) \models P(p)$, where p and q are closed terms from the signature of $TSS(A)$, and $\xrightarrow{l}$ is a transition relation and P is a predicate, and by that we mean *formulae* $p \xrightarrow{l} q$ and $P(p)$ are *provable* in $TSS(A)$. Due to the presence of negative premises (in the TSS's that are yet to be presented in this paper), it is not clear what is a proof for negative premises and several different interpretations exist in the literature (see [13] for an overview). However, for the purpose of TSS's presented in this paper all the existing interpretations coincide (since they are all *strictly stratified*, see [14, 11] for the definition and details) and hence, it does not matter which interpretation we choose. Next, we define the notion of *stable model* as an intuitive semantics for TSSs with negative premises.

Definition 1 (Stable Model). *We say a positive closed formula ϕ is* provable *from a set of positive formulae T and a TSS tss, denoted by $(T, tss) \vdash \phi$ when there is a well-founded upwardly branching tree with nodes labelled by closed formulae such that:*

- *the root node is labelled by ϕ, and*
- *if the label of a node q, denoted by ψ, is a positive formula and $\{\psi_i \mid i \in I\}$ is the set of labels of the nodes directly above q, then there is a deduction rule*

$\dfrac{\{\chi_i \mid i \in I\}}{\chi}$ *in tss (N.B.* χ_i *can be a positive or a negative formula) and a substitution* σ *such that* $\sigma(\chi) = \psi$ *and for all* $i \in I$, $\sigma(\chi_i) = \psi_i$*;*

- *if the label of a node q, denoted by* $p \stackrel{l}{\nrightarrow}$, *is a negative formula then there exists no* p' *such that* $p \stackrel{l}{\rightarrow} p' \in T$ *(or similarly, if it is of the form* $\neg p\downarrow$ *then* $p \notin \downarrow$*).*

A stable model *defined by tss is a set of formulae T such that for all closed positive formulae* ϕ, $\phi \in T$ *if and only if* $(T, tss) \vdash \phi$.

Using the notion of stable model, we can associate a transition system to each closed term in the signature. To define an equality on transition systems and turn them into model for process algebras, we need a notion of behavioral equivalence. Strong bisimilarity is one such notion of behavioral equivalence which can be efficiently checked in practice and usually leads to elegant theories.

Definition 2 (Bisimilarity). *A symmetric relation R on closed terms is called a (strong) bisimulation relation with respect to a transition relations* $\rightarrow$ *and a predicates P, when for all* $(p, q) \in R$, *and for all labels l and closed terms* p',

- *if* $p \stackrel{l}{\rightarrow} p'$ *then there exists a* q' *such that* $q \stackrel{l}{\rightarrow} q'$ *and* $(p', q') \in R$, *and*
- *if* $p\downarrow$ *then* $q\downarrow$.

We write $TSS(A) \models p \leftrightarrows p'$ *for closed terms p and* p' *when they are (strongly) bisimilar with respect to the stable model of* $TSS(A)$.

To extend BPA_δ^ϵ with timing, we add the unit time transitions $\mapsto$ (which can be considered as an acronym for $\stackrel{1}{\rightarrow}$ where 1 is a fresh label dedicated to time transitions). At the same time, we add additional constants $\underline{\underline{\epsilon}}$ (termination in the current time slice), $\underline{\underline{\delta}}$ (deadlock in the current time slice), $\underline{\underline{a}}$ (action execution in the current time slice, for $a \in A$) and σ (unit delay). The resulting extension is called $BPA_{drt,\delta}^\epsilon$; see Table 2.

The first three deduction rules of Table 2 specify the delayable nature of δ, ϵ and a whereas the next three rules specify the undelayable nature of $\underline{\underline{\delta}}$, $\underline{\underline{\epsilon}}$ and $\underline{\underline{a}}$. Let's next focus on the last three deduction rules: they specify the time-deterministic behavior of choice, i.e., time transitions cannot decide about a choice unless one of the two arguments of choice prohibits time from progressing. The remaining deduction rules specify the behavior of sequential composition. These are rather involved since they have to maintain time-determinism. If a delay can take place in two forms, i.e., by delaying the first argument and by delaying the second argument after the termination of the first, then both options are kept open (the first rule). Otherwise, if exactly one of these two forms is possible then there remains no choice and only possible delay takes place (the second and the third rule).

While this theory can be worked out in full, and indeed has an elegant axiomatization, it does lead to complications. These complications have to do with the fact that the action constants involve both action execution and termination. In

Table 2. Transition system specification for $BPA^{\epsilon}_{drt,\delta}(A)$.

$TSS(BPA^{\epsilon}_{drt,\delta}(A))$

$TSS(BPA^{\epsilon}_{\delta}(A))$

constant: $\underline{\underline{\delta}}, \underline{\underline{\epsilon}}, (\underline{\underline{a}})_{a\in A}, \sigma$;

$\_ \mapsto \_$;

x, x', y, y';

$$\delta \mapsto \delta \qquad \epsilon \mapsto \epsilon \qquad a \mapsto a \qquad \underline{\underline{\epsilon}}\downarrow \qquad \underline{\underline{a}} \xrightarrow{a} \underline{\underline{\epsilon}} \qquad \sigma \mapsto \underline{\underline{\epsilon}}$$

$$\frac{x \mapsto x' \quad x\downarrow \quad y \mapsto y'}{x \cdot y \mapsto x' \cdot y + y'} \qquad \frac{x \mapsto x' \quad x\not\downarrow}{x \cdot y \mapsto x' \cdot y}$$

$$\frac{x \mapsto x' \quad y \not\mapsto}{x \cdot y \mapsto x' \cdot y} \qquad \frac{x \not\mapsto \quad x\downarrow \quad y \mapsto y'}{x \cdot y \mapsto y'}$$

$$\frac{x \mapsto x' \quad y \mapsto y'}{x + y \mapsto x' + y'} \qquad \frac{x \mapsto x' \quad y \not\mapsto}{x + y \mapsto x'} \qquad \frac{x \not\mapsto \quad y \mapsto y'}{x + y \mapsto y'}$$

the timed extension, the immediate and delayable options for action execution and termination lead to four different combinations:

- Action execution after an arbitrary delay, followed by termination after an arbitrary delay (a);
- Immediate action execution followed by immediate termination ($\underline{\underline{a}}$);
- Action execution after an arbitrary delay, followed by immediate termination ($\epsilon \cdot \underline{\underline{a}}$);
- Immediate action execution, followed by termination after an arbitrary delay ($\underline{\underline{a}} \cdot \epsilon$).

In the extension of BPA^{ϵ}_{δ} with timing, we are forced to take the first option. The problem is that this does not match with timed extensions of process algebra without the ϵ constant. In BPA_{δ} [7, 9], the operational rule for action constants is $a \xrightarrow{a} \surd$, where $\surd$ is not a process expression but a special symbol denoting termination. Extending with time, time transitions can be added before action execution, but not afterwards, so we are forced to take the third option.

In [20], the author follows this, so he interprets untimed actions following the third choice. When extending with ϵ he gets $a = \epsilon \cdot \underline{\underline{a}}$. As a result, the extension is not conservative, as the ground equation $a \cdot \epsilon = a$ does not hold any longer in the timed extension.

To combat this mismatch, in [1] it is proposed to separate action execution and termination: by replacing action constants by action prefixing, termination becomes explicit. As we will show further on, separating action execution and termination by abandoning the idea of basic actions as constructors in the signature resolves many of the difficulties.

This involves a deviation of a basic design decision in the untimed process algebra: *action prefixing* is taken as the basic composition operator instead of *sequential composition*, and sequential composition is added later on. However, both in [1] and in [6] conservativity is not maintained for the operators $\lfloor\!\lfloor$ and $|$ that are used to define parallel composition. This inadequacy is solved in the present paper.

Thus, we choose a departure point for the extension of process algebra with termination and timing that has action prefixing instead of action constants. Following [1, 6], we choose a basic theory called MPT (for Minimal Process Theory) with deadlock, action prefixing and choice. Then, we extend this theory with successful termination which results in the theory BSP. Sequential and parallel composition are subsequently added to BSP, resulting in the theories TSP and TCP, respectively. Observe that ACP cannot be considered a conservative (ground-)extension of TCP since the signature of TCP is not contained in the signature of ACP and vice versa TCP is not a conservative (ground-)extension of ACP since the signature of ACP is not contained in the signature of TCP. However, it is possible to embed ACP into TCP by mapping the action constants $a \in A$ on the TCP-terms $a.\epsilon$.

As our goal is to have equationally conservative ground-extensions of process algebras, we extend our theory with timing at each level and establish the conservativity of the extension. In order to make the transition to timed settings smoother, in addition to the time delay operator $\sigma.\_$, we add an arbitrary time delay operator $\sigma^*.\_$ which is very helpful in the axiomatization of complex theories such as TSP and TCP. The result of the extension of theory X with undelayable action prefix, undelayable termination and deadlock, (discrete) time delay and arbitrary delay operators is denoted by X^{drt^*}. The lattice of process theories that we present in this paper (ordered by equational (ground-)conservativity relation) is depicted in Figure 1. Each arrow is labelled by the function symbols introduced in the target of the extension.

To give a formal meaning to the arrows presented in Figure 1, we define a few concepts regarding conservativity. The first definition concerns the traditional notion of operational conservativity.

Definition 3 (Operational Conservativity [12]). *Consider TSS's $TSS(A)$ and $TSS(B)$ defined on signatures Σ_A and Σ_B such that $TSS(B)$ includes $TSS(A)$ in its definition. Also, let $C(\Sigma_x)$ denote the set of closed terms built upon Σ_x. The TSS $TSS(B)$ is an* operationally conservative extension *of $TSS(A)$ when $\forall_{p \in C(\Sigma_A)}\ TSS(A) \vDash p\downarrow \Leftrightarrow TSS(B) \vDash p\downarrow$, and $\forall_{p \in C(\Sigma_A)}\ \forall_{l \in L_B}\ \forall_{p' \in C(\Sigma_B)}$ $TSS(A) \vDash p \xrightarrow{l} p' \Leftrightarrow TSS(B) \vDash p \xrightarrow{l} p'$.*

In the above definition, implicitly it is not allowed to have old relations from old terms ($C(\Sigma_P)$) to a new term ($C(\Sigma_B)$). Note that the transition relations and predicates in the above definition are taken from the extended TSS, i.e., $TSS(B)$ and hence an operationally conservative extension denies any new transition or predicate from the terms from the old syntax, i.e., from $C(\Sigma_A)$. This turns out to be too restrictive for time extensions since we decided to interpret untimed

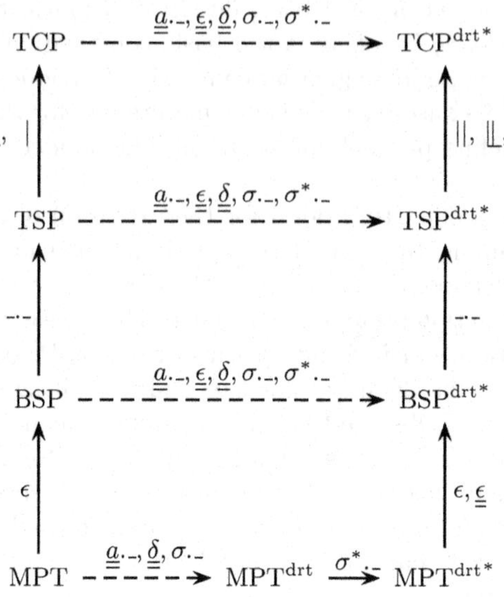

Fig. 1. The Lattice of Process Theories and Their Timed Extensions

basic actions as delayable and hence we have to add timing behavior to them. This has been noted in [18] where the following alternative and more relaxed notion of orthogonality was proposed.

Definition 4 (Orthogonality [18]). *Consider TSS's $TSS(A)$ and $TSS(B)$ defined on signatures Σ_A and Σ_B such that $TSS(B)$ includes $TSS(A)$ in its definition. The TSS $TSS(B)$ is an* orthogonal *extension of $TSS(A)$ when*

1. $\forall_{p \in C(\Sigma_A)}\ TSS(A) \vDash p\downarrow \Leftrightarrow TSS(B) \vDash p\downarrow$ *and* $\forall_{p,p' \in C(\Sigma_A)}\ \forall_{l \in L_A}\ TSS(A) \vDash p \xrightarrow{l} p' \Leftrightarrow TSS(B) \vDash p \xrightarrow{l} p'$ *and*
2. $\forall_{p,p' \in C(\Sigma_A)}\ TSS(A) \vDash p \leftrightarrows p' \Leftrightarrow TSS(B) \vDash p \leftrightarrows p'$.

Note that it follows from Definitions 3 and 4 that an operationally conservative extension is orthogonal [18]. Both operational conservativity and orthogonality are useful means to obtain equational conservativity as defined below.

Definition 5 (Equational Conservativity [18]). *An equational theory B on signature Σ_B is an* equationally conservative ground-extension *of equational theory A on Σ_A if and only if $\Sigma_A \subseteq \Sigma_B$ and for all $p, p' \in C(\Sigma_A)$, $A \vdash p = p' \Leftrightarrow B \vdash p = p'$.*

If the axioms of equational theory A are (syntactically) included among the axioms of equational theory B and B is an equationally conservative ground-extension of A, then B is an equationally conservative extension *of A.*

In the above definition $X \vdash p = p'$ means that $p = p'$ is *derivable* from the equations in X. In our settings, this means that $=$ is the congruence relation induced by the equations in X. We drop the prefixes equational and equationally and simply write conservativity and conservative (ground)-extension in the remainder.

In the outline presented in Figure 1, normal arrows denote equationally conservative extensions and dashed arrows denote equationally conservative ground-extensions. Operational conservativity and orthogonality are mostly used as a means to prove conservative extensions and conservative ground-extensions, respectively. In the present paper, we focus on the process theories and on their interrelationships using the previously mentioned notions of conservativity. We present soundness and ground-completeness theorems for all theories, but omit their proofs altogether.

3 Minimal Process Theory

3.1 MPT

The equational theory Minimal Process Theory (MPT) is specified in Table 3.

Table 3. MPT(A)

MPT(A)		
constant: δ;	unary: $(a.\_)_{a \in A}$;	binary: $\_ + \_$;
x, y, z;		
$x + y = y + x$	A1	
$(x + y) + z = x + (y + z)$	A2	
$x + x = x$	A3	
$x + \delta = x$	A6	

The transition system specification associated to the terms of MPT is given in Table 4.

It is straightforward to check that the equational theory of MPT is a sound and ground-complete axiomatization for its transition system semantics modulo strong bisimulation.

Theorem 1 (Soundness). *Let p and q be two closed* MPT(A)*-terms. If* MPT(A) $\vdash p = q$*, then* TSS(MPT(A)) $\models p \leftrightarrows q$.

Theorem 2 (Ground-completeness). *Let p and q be arbitrary closed* MPT(A)*-terms. If* TSS(MPT(A)) $\models p \leftrightarrows q$*, then* MPT($A$) $\vdash p = q$.

3.2 MPT^{drt}

The equational theory of MPT with discrete relative timing (MPT^{drt}) [1] is specified in Table 5. It adds undelayable action prefixing $\underline{\underline{a}}.\_$, undelayable deadlock

Table 4. Transition system specification for MPT(A)

TSS(MPT(A))

constant: δ;	unary: $(a.\_)_{a\in A}$;	binary: $\_+\_$;
$(\_ \xrightarrow{a} \_)_{a\in A}$		
x, x', y, y';		
$a.x \xrightarrow{a} x$	$\dfrac{x \xrightarrow{a} x'}{x+y \xrightarrow{a} x'}$	$\dfrac{y \xrightarrow{a} y'}{x+y \xrightarrow{a} y'}$

Table 5. MPT$^{\mathrm{drt}}(A)$

MPT$^{\mathrm{drt}}(A)$

constant: $\delta, \underline{\underline{\delta}}$; unary: $(a.\_)_{a\in A}, (\underline{\underline{a}}.\_)_{a\in A}, \sigma.\_$; binary: $\_+\_$;

x, y, z;

$x+y=y+x$	A1	$\delta = \sigma.\delta$	DD1
$x+(y+z)=(x+y)+z$	A2	$a.x = \underline{\underline{a}}.x + \sigma.a.x$	DA1
$x+x=x$	A3	$a.x+\delta = a.x$	A6DD
$x+\underline{\underline{\delta}} = x$	A6DRT	$\sigma.x+\sigma.y=\sigma.(x+y)$	DRTF

$\underline{\underline{\delta}}$ plus a time delay operator $\sigma.\_$ to the signature of MPT and uses no auxiliary operators for the axiomatization. Prefixing is binds stronger than the other operators.

This equational theory is not a conservative extension of MPT since the axiom $x+\delta=x$ of MPT does not hold anymore. The role of deadlock δ (now called delayable deadlock) as a unit element for alternative composition is taken over by undelayable deadlock $\underline{\underline{\delta}}$, see Axiom A6DRT. The behavior of delayable deadlock and action prefix is defined recursively by means of the axioms DD1 (Delayable Deadlock) and DA1 (Delayable Action). Axiom A6DD expresses that delayable deadlock is still a unit element for alternative composition of delayable processes. Finally, axiom DRTF (Discrete Relative Time Factorization) expresses that passage of time by itself cannot determine a choice. Hence, this axiom implements the time-determinism discussed in the previous section.

The transition system specification of MPT^{drt} is given in Table 6. It consists of the deduction rules of MPT (Table 4) as well as new deduction rules defining the time transitions $\mapsto$ for MPT terms as well as action and time transitions for the newly introduced terms.

Theorem 3 (Soundness). *Let p and q be two closed* MPT$^{\mathrm{drt}}(A)$*-terms. If* MPT$^{\mathrm{drt}}(A) \vdash p = q$*, then* *TSS*(MPT$^{\mathrm{drt}}(A)$) $\models p \leftrightarrows q$.

Table 6. Transition system specification for $\mathrm{MPT}^{\mathrm{drt}}(A)$.

$\mathit{TSS}(\mathrm{MPT}^{\mathrm{drt}}(A))$

$\mathit{TSS}(\mathrm{MPT}(A))$

constant: $\underline{\underline{\delta}}$; unary: $(\underline{\underline{a}}.\_)_{a \in A}, \sigma.\_$;

$\_ \mapsto \_$;

x, x', y, y';

$$\delta \mapsto \delta \qquad a.x \mapsto a.x \qquad \frac{x \mapsto x' \quad y \mapsto y'}{x + y \mapsto x' + y'}$$

$$\frac{x \mapsto x' \quad y \not\mapsto}{x + y \mapsto x'} \qquad \frac{x \not\mapsto \quad y \mapsto y'}{x + y \mapsto y'}$$

$$\underline{\underline{a}}.x \xrightarrow{a} x \qquad \sigma.x \mapsto x$$

Theorem 4 (Ground-completeness). *Let p and q be arbitrary closed $\mathrm{MPT}^{\mathrm{drt}}(A)$-terms. If $\mathit{TSS}(\mathrm{MPT}^{\mathrm{drt}}(A)) \models p \leftrightarrow q$, then $\mathrm{MPT}^{\mathrm{drt}}(A) \vdash p = q$.*

As explained before, $\mathrm{MPT}^{\mathrm{drt}}$ cannot be a conservative extension of MPT due to the omission of Axiom A6 ($x + \delta = x$) from $\mathrm{MPT}^{\mathrm{drt}}$. It is a conservative ground-extension, nevertheless, since for closed MPT-terms x, this axiom is derivable from the axioms of $\mathrm{MPT}^{\mathrm{drt}}$.

Theorem 5 (Conservative ground-extension). $\mathrm{MPT}^{\mathrm{drt}}$ *is a conservative ground-extension of* MPT.

Proof. We apply the meta-theorems from [18]. The transition system specification that consists of the deduction rules in the first row of Table 6 is source-preserving and strictly stratified using the number of symbols of terms as a measure, and the sources of the conclusions cover the syntax of MPT (see [18] for definitions of notions used here). Furthermore, the deduction rules in the second row of Table 6 have source-dependent negative time transitions as a premise. Hence $\mathit{TSS}(\mathrm{MPT}(A))$ with deduction rules of the first and the second row of Table 6 is a granting extension of $\mathit{TSS}(\mathrm{MPT}(A))$ and hence an orthogonal extension. The extension of the resulting transition system specification with deduction rules of the third row is conservative, hence also orthogonal. Since orthogonality is a preorder, $\mathit{TSS}(\mathrm{MPT}^{\mathrm{drt}}(A))$ is an orthogonal extension of $\mathit{TSS}(\mathrm{MPT}(A))$. Combined with the facts that both MPT and $\mathrm{MPT}^{\mathrm{drt}}$ are sound and complete, we have that $\mathrm{MPT}^{\mathrm{drt}}$ is an equationally conservative ground-extension of MPT.

3.3 $\mathrm{MPT}^{\mathrm{drt}^*}$

A further discrete relative time extension of MPT, called $\mathrm{MPT}^{\mathrm{drt}^*}$ makes use of an auxiliary operator $\sigma^*\_$ to axiomatize this extension: $\sigma^* p$ denotes that the execution of p can be started in any time slice (present or future). Note that the intuitions of $\underline{\underline{\delta}}$ and $\underline{\underline{a}}.\_$ are in line with the "any-time-slice" interpretation of the untimed constants and action prefix operators. This time iteration operator comes very handy in the axiomatization of delayable actions, particularly in the more involved theories that we encounter in the rest of this paper. The axiomatization of $\mathrm{MPT}^{\mathrm{drt}^*}$ is given in Table 7.

Table 7. $\mathrm{MPT}^{\mathrm{drt}^*}(A)$

$\mathrm{MPT}^{\mathrm{drt}^*}(A)$

constant: $\delta, \underline{\underline{\delta}}$; unary: $(a.\_)_{a \in A}, (\underline{\underline{a}}.\_)_{a \in A}, \sigma.\_, \sigma^*\_$; binary: $\_ + \_$;

x, y, z;

$x + y = y + x$	A1	$\delta = \sigma.\delta$	DD1
$x + (y + z) = (x + y) + z$	A2	$a.x = \underline{\underline{a}}.x + \sigma.a.x$	DA1
$x + x = x$	A3	$\delta = \sigma^*\underline{\underline{\delta}}$	DD2
		$a.x = \sigma^*\underline{\underline{a}}.x$	DA2
$x + \underline{\underline{\delta}} = x$	A6DRT	$\sigma.x + \sigma.y = \sigma.(x + y)$	DRTF
$\sigma^* x = x + \sigma.\sigma^* x$	ATS	$\sigma^* x + \sigma^* y = \sigma^*(x + y)$	DRTIF
$\sigma^* \sigma.x = \sigma.\sigma^* x$	DRTA	$\sigma^* \sigma^* x = \sigma^* x$	TITI

In this equational theory the any time slice constant and action prefix operators are defined in terms of their current time slice counterparts and time iteration by axioms DD2 and DA2. Axiom ATS (Any Time Slice) recursively defines time iteration. Axiom DRTIF (Discrete Relative Time Iteration Factorization) expresses that time factorization also applies to time iteration. Axiom DRTA (Discrete Relative Time Axiom) explains that consecutive occurrences of time iteration and time delay can be switched. It is in line with the observation that both $\sigma.\sigma^* p$ and $\sigma^*\sigma.p$ are solutions to the recursive specification $X = \sigma.p + \sigma.X$. Axiom TITI (Time Iteration Time Iteration) says that two consecutive time iterations are equivalent to only one time iteration. As a consequence, any number of consecutive time iterations is considered to be equivalent to a single one. Note that axioms DD1 and DA1 are derivable from the other axioms (e.g., $\delta = \sigma^*\underline{\underline{\delta}} = \underline{\underline{\delta}} + \sigma\sigma^*\underline{\underline{\delta}} = \sigma\delta$).

It is not the case that the newly introduced operators can all be eliminated. Nevertheless, the newly introduced syntax has some redundancy in the sense that either the time iteration operator or the delayable deadlock and delayable action prefixes can be eliminated from closed terms.

The transition system specification associated to the terms of $\mathrm{MPT}^{\mathrm{drt}^*}$ is given in Table 8. It adds deduction rules defining the behavior of time iteration.

Theorem 6 (Soundness). *Let p and q be two closed $\mathrm{MPT}^{\mathrm{drt}^*}(A)$-terms. If $\mathrm{MPT}^{\mathrm{drt}^*}(A) \vdash p = q$, then $TSS(\mathrm{MPT}^{\mathrm{drt}^*}(A)) \models p \leftrightarrows q$.*

Table 8. Transition system specification for $\mathrm{MPT}^{\mathrm{drt}^*}(A)$

$TSS(\mathrm{MPT}^{\mathrm{drt}^*}(A))$
$TSS(\mathrm{MPT}^{\mathrm{drt}}(A))$
unary: $\sigma^*\_$;
x, x';
$\dfrac{x \xrightarrow{a} x'}{\sigma^* x \xrightarrow{a} x'}$ $\qquad$ $\dfrac{x \mapsto x'}{\sigma^* x \mapsto \sigma^* x + x'}$ $\qquad$ $\dfrac{x \not\mapsto}{\sigma^* x \mapsto \sigma^* x}$

Theorem 7 (Ground-completeness). *Let p and q be arbitrary closed $\mathrm{MPT}^{\mathrm{drt}^*}(A)$-terms. If $TSS(\mathrm{MPT}^{\mathrm{drt}^*}(A)) \models p \leftrightarrows q$, then $\mathrm{MPT}^{\mathrm{drt}^*}(A) \vdash p = q$.*

Theorem 8 (Conservative extension). $\mathrm{MPT}^{\mathrm{drt}^*}$ *is a conservative extension of* $\mathrm{MPT}^{\mathrm{drt}}$.

Proof. The sources of the conclusions of all deduction rules from Table 8 mention a new operator. Therefore, the extension of $TSS(\mathrm{MPT}^{\mathrm{drt}}(A))$ with these deduction rules is conservative. Hence, $TSS(\mathrm{MPT}^{\mathrm{drt}^*}(A))$ is an orthogonal extension of $TSS(\mathrm{MPT}^{\mathrm{drt}}(A))$. Since both theories are sound and complete and the axioms of $\mathrm{MPT}^{\mathrm{drt}}(A)$ are included in $\mathrm{MPT}^{\mathrm{drt}^*}(A)$, it follows from the meta-results of [18] that $\mathrm{MPT}^{\mathrm{drt}^*}$ is an equationally conservative extension of $\mathrm{MPT}^{\mathrm{drt}}$.

4 Successful Termination: Basic Sequential Processes

In this section, we discuss the extension of the theories MPT and $\mathrm{MPT}^{\mathrm{drt}^*}$ from the previous section, with a constant denoting successful termination.

4.1 BSP

The process theory MPT is a *minimal* theory; not much can be expressed in it. One aspect that cannot be addressed is successful termination. The distinction between successful and unsuccessful termination turns out to be essential when sequential composition is introduced. In order to express successful termination, the new constant ϵ, referred to as the empty process or the termination constant, is introduced. The extension of the process theory MPT with the empty process ϵ results in process theory BSP, the theory of Basic Sequential Processes. This section gives the equational theory as well as its term model.

Table 9 defines process theory BSP. The only difference between the signature of MPT and the signature of BSP is the constant ϵ. The axioms of BSP, see Table 9, are exactly the axioms of MPT, given in Table 3.

Table 9. BSP(A)

BSP(A)
MPT(A)
constant: ϵ;
-

Table 10. Transition system specification for BSP(A)

TSS(BSP(A))
TSS(MPT(A))
constant: ϵ;
$\_\!\downarrow$;
x, y;
$\epsilon\!\downarrow \qquad \dfrac{x\!\downarrow}{x+y\!\downarrow} \qquad \dfrac{y\!\downarrow}{x+y\!\downarrow}$

The transition system specification associated to the terms of BSP is given in Table 10. It adds deduction rules defining the termination behavior of the new constant ϵ and of the syntax of MPT.

Theorem 9 (Soundness). *Let p and q be two closed* BSP(A)*-terms. If* BSP(A) $\vdash p = q$*, then* TSS(BSP(A)) $\models p \leftrightarrow q$.

Theorem 10 (Ground-completeness). *Let p and q be arbitrary closed* BSP (A)*-terms. If* TSS(BSP(A)) $\models p \leftrightarrow q$*, then* BSP($A$) $\vdash p = q$.

Theorem 11 (Conservative extension). BSP *is a conservative extension of* MPT.

Proof. Both MPT and BSP are sound and ground-complete equational theories for TSS(MPT(A)) and TSS(BSP(A)). Also, following [12], TSS(BSP(A)) is an operationally conservative (orthogonal) extension of TSS(MPT(A)). Furthermore, axioms of MPT(A) are all included in BSP(A). Thus, we conclude that BSP(A) is an equationally conservative extension of MPT(A).

4.2 BSP$^{\mathrm{drt}^*}$

In this section, the timed process algebra MPT$^{\mathrm{drt}^*}$ from the previous section is extended with the constants *any time slice termination* ϵ and *current time slice termination* $\underline{\underline{\epsilon}}$. The axioms of the process theory BSP$^{\mathrm{drt}^*}$ are given in Table 11.

The axiom DT (Delayable Termination) defines the any time slice constant in terms of its current time slice counterpart and time iteration. Note that $\epsilon = \underline{\underline{\epsilon}} + \sigma\epsilon$ is derivable from the axioms of Table 11.

Table 11. $\mathrm{BSP}^{\mathrm{drt}^*}(A)$

$\mathrm{BSP}^{\mathrm{drt}^*}(A)$
$\mathrm{MPT}^{\mathrm{drt}^*}(A)$
constant: $\epsilon, \underline{\underline{\epsilon}}$;
$\epsilon = \sigma^*\underline{\underline{\epsilon}}$ DT

Table 12. Transition system specification for $\mathrm{BSP}^{\mathrm{drt}^*}(A)$

$TSS(\mathrm{BSP}^{\mathrm{drt}^*}(A))$
$TSS(\mathrm{BSP}(A))$, $TSS(\mathrm{MPT}^{\mathrm{drt}^*}(A))$
constant: $\underline{\underline{\epsilon}}$;
x;
$\underline{\underline{\epsilon}}\downarrow \qquad \epsilon \mapsto \epsilon \qquad \dfrac{x\downarrow}{\sigma^* x\downarrow}$

The transition system specification associated to the terms of $\mathrm{BSP}^{\mathrm{drt}^*}$ is given in Table 12. It adds deduction rules defining the termination behavior of the new constant $\underline{\underline{\epsilon}}$ and the time iteration σ^* and the time behavior of ϵ.

Theorem 12 (Soundness). *Let p and q be two closed $\mathrm{BSP}^{\mathrm{drt}^*}(A)$-terms. If $\mathrm{BSP}^{\mathrm{drt}^*}(A) \vdash p = q$, then $TSS(\mathrm{BSP}^{\mathrm{drt}^*}(A)) \models p \leftrightarroweq q$.*

Theorem 13 (Ground-completeness). *Let p and q be two arbitrary closed $\mathrm{BSP}^{\mathrm{drt}^*}(A)$-terms. If $TSS(\mathrm{BSP}^{\mathrm{drt}^*}(A)) \models p \leftrightarroweq q$, then $\mathrm{BSP}^{\mathrm{drt}^*}(A) \vdash p = q$.*

Theorem 14 (Conservative ground-extension). $\mathrm{BSP}^{\mathrm{drt}^*}$ *is a conservative ground-extension of* BSP.

Proof. This proof is similar to the proof that $\mathrm{MPT}^{\mathrm{drt}^*}$ is a conservative ground-extension of MPT. Add the second deduction rule of 12 to the granting part and the first and third deduction rules of Table 12 to the conservative part.Note that the deduction rules for σ^* that are added to $TSS(\mathrm{MPT}^{\mathrm{drt}}(A))$ to obtain $TSS(\mathrm{MPT}^{\mathrm{drt}^*}(A))$ should be added to the conservative part as well.

Theorem 15 (Conservative extension). $\mathrm{BSP}^{\mathrm{drt}^*}$ *is a conservative extension of* $\mathrm{MPT}^{\mathrm{drt}^*}$.

Proof. Both $\mathrm{MPT}^{\mathrm{drt}^*}$ and $\mathrm{BSP}^{\mathrm{drt}^*}$ are sound and ground-complete equational theories for $TSS(\mathrm{MPT}^{\mathrm{drt}^*}(A))$ and $TSS(\mathrm{BSP}^{\mathrm{drt}^*}(A))$. Also, using the meta-theory from [12], $TSS(\mathrm{BSP}^{\mathrm{drt}^*}(A))$ is an operationally conservative (orthogonal) extension of $TSS(\mathrm{MPT}^{\mathrm{drt}^*}(A))$. Thus, we conclude that $\mathrm{BSP}^{\mathrm{drt}^*}(A)$ is an equationally conservative extension of $\mathrm{MPT}^{\mathrm{drt}^*}(A)$.

5 Sequential Composition

5.1 TSP

This section treats the extension with a *sequential composition* operator. Given two process terms p and q, the term $p \cdot q$ denotes the sequential composition of p and q. The intuition of this operation is that upon the *successful* termination of process p, process q is started. If process p ends in a deadlock, also the sequential composition $p \cdot q$ deadlocks. Thus, a pre-requisite for a meaningful introduction of a sequential composition operator is that successful and unsuccessful termination can be distinguished. As already explained in Section 4, this is not possible in the theory MPT as all processes end in deadlock. Thus, as a starting point the theory BSP of the previous section is used. This theory is extended with sequential composition to obtain the *Theory of Sequential Processes* TSP. It turns out that the empty process is a neutral element for sequential composition: $x \cdot \epsilon = \epsilon \cdot x = x$.

To obtain the axioms of the process theory TSP, the axioms from Table 13 are added to the axioms of the process theory BSP from Table 9. Axiom A5 states that sequential composition is associative. As mentioned before, and now formally captured in the axioms A8 and A9, the empty process is a neutral element with respect to sequential composition. Axiom A7 states that after a deadlock

Table 13. The process theory TSP(A)

TSP(A)			
BSP(A)			
binary: $\_ \cdot \_$;			
$x, y, z : P$;			
$(x+y) \cdot z = x \cdot z + y \cdot z$	A4	$\delta \cdot x = \delta$	A7
$(x \cdot y) \cdot z = x \cdot (y \cdot z)$	A5	$x \cdot \epsilon = x$	A8
$a.x \cdot y = a.(x \cdot y)$	A10	$\epsilon \cdot x = x$	A9

has been reached no continuation is possible. Axiom A4 describes the distribution of sequential composition over alternative composition from the right. Recall that the other distributivity property is not desired as it does not respect the moment of choice. Finally, Axiom A10 describes the relation between sequential composition and action prefixes.

Theorem 16 (Elimination). *For any closed* TSP(A)*-term* p*, there exists a closed* BSP(A)*-term* q *such that* TSP(A) $\vdash p = q$.

Proof. The property is proven by providing a term rewriting system with the same signature as TSP(A) such that

1. each rewrite step transforms a process term into a process term that is derivably equal,
2. the term rewriting system is strongly normalizing, and
3. no closed normal form of the term rewriting system contains a sequential composition operator.

Consider the term rewriting system consisting of the following rewrite rules; for any $a \in A$, and TSP(A)-terms x, y, z:

$$\begin{array}{ll} (x+y)\cdot z \to x\cdot z + y\cdot z & \delta\cdot x \to \delta \\ (x\cdot y)\cdot z \to x\cdot(y\cdot z) & \epsilon\cdot x \to x \\ a.x\cdot y \to a.(x\cdot y) & \end{array}$$

Each of the rewrite rules is obtained directly from an axiom of TSP by replacing $=$ by $\to$. As a consequence, each rewrite step transforms a process term in a process term that is derivably equal.

The second step of the proof, the strong normalization of the given term rewriting system, is standard.

The last part of the proof is to show that no closed normal form of the above term rewriting system contains a sequential composition operator. Thereto, let u be a normal form of the above term rewriting system. Suppose that u contains at least one sequential composition operator. Then, u must contain a subterm of the form $v\cdot w$ for some closed TSP(A)-terms u and v. This subterm can always be chosen in such a way that v is a closed BSP(A)-term. It follows immediately from the structure of closed BSP(A)-terms that one of the above rewrite rules can be applied to $v \cdot w$. As a consequence, u is not a normal form. This contradiction implies that u must be a closed BSP(A)-term.

The transition system specification associated to the terms of TSP is given in Table 14. It adds deduction rules defining action transitions and termination behavior of sequential composition.

Theorem 17 (Soundness). *Let* p *and* q *be two closed* TSP(A)*-terms. If* TSP (A) $\vdash p = q$*, then* *TSS*(TSP(A)) $\models p \leftrightarrows q$.

Theorem 18 (Ground-completeness). *Let* p *and* q *be arbitrary closed* TSP (A)*-terms. If* *TSS*(TSP(A)) $\models p \leftrightarrows q$*, then* TSP($A$) $\vdash p = q$.

Table 14. Transition system specification for TSP(A)

TSS(TSP(A))		
TSS(BSP(A))		
binary: $\_ \cdot \_$;		
x, y, x', y';		
$\dfrac{x \xrightarrow{a} x'}{x \cdot y \xrightarrow{a} x' \cdot y}$	$\dfrac{x\downarrow \quad y \xrightarrow{a} y'}{x \cdot y \xrightarrow{a} y'}$	$\dfrac{x\downarrow \quad y\downarrow}{x \cdot y\downarrow}$

Theorem 19 (Conservative extension). TSP *is a conservative extension of* BSP.

Proof. Both BSP and TSP are sound and ground-complete equational theories for TSS(BSP(A)) and TSS(TSP(A)). Also, following [12], TSS(TSP(A)) is an operationally conservative (orthogonal) extension of TSS(BSP(A)). Furthermore, all axioms of BSP are among the axioms of TSP. Thus, we conclude that TSP is an equationally conservative extension of BSP.

Now that the process theories have been extended with sequential composition, the relationship with the process theory BPA_δ^ϵ can be considered. Syntactically, there is still a mismatch between BPA_δ^ϵ and TSP since the former has constants $a \in A$ and the latter does not have those. There are two (equivalent) ways to overcome this difference. First, we can extend TSP with such constants, using axiom

$$a = a.\epsilon,$$

or second, we can use the notion of embedding to find that the process theory BPA_δ^ϵ can be embedded into TSP taking a as $a.\epsilon$.

5.2 TSP$^{\mathrm{drt}*}$

In this section, the process theory BSP$^{\mathrm{drt}*}$ is extended to the process theory TSP$^{\mathrm{drt}*}$. This extension is obtained by extending the signature of BSP$^{\mathrm{drt}*}$ with the *sequential composition* operator $\_ \cdot \_$. The axioms of the process theory are given in Table 15.

Sequential composition is as before, but here the role of unit that was played by ϵ in the untimed theory, is taken over by the current time slice termination constant $\underline{\underline{\epsilon}}$ (see Axioms A8DR and A9DR). In this setting it can be derived that $\epsilon \cdot x = \sigma^* x$ instead. The sequential composition operator has two left-zero elements: both undelayable deadlock (see Axiom A7DR) and delayable deadlock act as such. The axioms $\delta \cdot x = \delta$ and $a.x \cdot y = a.(x \cdot y)$ from the untimed theory have disappeared since they are derivable from the remaining axioms[2]. Axioms

[2] If these axiom were the only axioms of TSP that disappear, then one could consider keeping them, since it would allow for conservativity instead of ground-conservativity.

Table 15. The process theory $\mathrm{TSP}^{\mathrm{drt}^*}(A)$

$\mathrm{TSP}^{\mathrm{drt}^*}(A)$
$\mathrm{BSP}^{\mathrm{drt}^*}(A)$;
binary: $\_\cdot\_$;
$x, y, z : P$;

$(x+y)\cdot z = x\cdot z + y\cdot z$	A4	$\underline{\underline{\delta}}\cdot x = \underline{\underline{\delta}}$	A7DR
$(x\cdot y)\cdot z = x\cdot(y\cdot z)$	A5	$x\cdot\underline{\underline{\epsilon}} = x$	A8DR
$\underline{\underline{a}}.x\cdot y = \underline{\underline{a}}.(x\cdot y)$	A10DRa	$\underline{\underline{\epsilon}}\cdot x = x$	A9DR
$(\sigma.x)\cdot y = \sigma.(x\cdot y)$	A10DRb		
$\sigma^* x\cdot y = \sigma^*(x\cdot y)$	A10DRc		

Table 16. Transition system specification for $\mathrm{TSP}^{\mathrm{drt}^*}(A)$

$TSS(\mathrm{TSP}^{\mathrm{drt}^*}(A))$

$TSS(\mathrm{TSP}(A))$, $TSS(\mathrm{BSP}^{\mathrm{drt}^*}(A))$

$x,' x', y, y'$;

$$\frac{x \mapsto x' \quad x\downarrow \quad y \mapsto y'}{x\cdot y \mapsto x'\cdot y + y'} \qquad \frac{x \mapsto x' \quad x\not\downarrow}{x\cdot y \mapsto x'\cdot y}$$

$$\frac{x \mapsto x' \quad y \not\mapsto}{x\cdot y \mapsto x'\cdot y} \qquad \frac{x \not\mapsto \quad x\downarrow \quad y \mapsto y'}{x\cdot y \mapsto y'}$$

A10DRb and A10DRc express that the passage of time is measured relative to the previous action and thus has no consequences for the future actions: the timing of y is relative to the last action of x, regardless the time prefix or time iteration operator.

Theorem 20 (Elimination). *For any closed* $\mathrm{TSP}^{\mathrm{drt}^*}(A)$*-term* p*, there exists a closed* $\mathrm{BSP}^{\mathrm{drt}^*}(A)$*-term* q *such that* $\mathrm{TSP}^{\mathrm{drt}^*}(A) \vdash p = q$.

The transition system specification associated to the terms of $\mathrm{TSP}^{\mathrm{drt}^*}$ is given in Table 16. It adds deduction rules defining time transitions of sequential composition.

Theorem 21 (Soundness). *Let* p *and* q *be two closed* $\mathrm{TSP}^{\mathrm{drt}^*}(A)$*-terms. If* $\mathrm{TSP}^{\mathrm{drt}^*}(A) \vdash p = q$*, then* $TSS(\mathrm{TSP}^{\mathrm{drt}^*}(A)) \models p \leftrightarrow q$.

Theorem 22 (Ground-completeness). *Let* p *and* q *be arbitrary closed* $\mathrm{TSP}^{\mathrm{drt}^*}(A)$*-terms. If* $TSS(\mathrm{TSP}^{\mathrm{drt}^*}(A)) \models p \leftrightarrow q$*, then* $\mathrm{TSP}^{\mathrm{drt}^*}(A) \vdash p = q$.

The process theory $\mathrm{TSP}^{\mathrm{drt}^*}$ is not a conservative extension of TSP as was the case for all extensions from untimed to timed process algebra. Besides the untimed identity $x+\delta = x$ that does not hold in the timed setting, also the untimed identity $x \cdot \epsilon = x$ does not hold in the timed extension: $\underline{\underline{\epsilon}} \cdot \epsilon \neq \underline{\underline{\epsilon}}$.

Theorem 23 (Conservative ground-extension). $\mathrm{TSP}^{\mathrm{drt}^*}$ *is a conservative ground-extension of* TSP.

Proof. We show that for all p and p' in the syntax of TSP,

1. $\mathit{TSS}(\mathrm{TSP}(A)) \models p \xrightarrow{a} p' \Leftrightarrow \mathit{TSS}(\mathrm{TSP}^{\mathrm{drt}^*}(A)) \models p \xrightarrow{a} p'$,
2. $\mathit{TSS}(\mathrm{TSP}(A)) \models p\downarrow \Leftrightarrow \mathit{TSS}(\mathrm{TSP}^{\mathrm{drt}^*}(A)) \models p\downarrow$,
3. there exists a closed $\mathrm{TSP}(A)$-term q such that $\mathit{TSS}(\mathrm{TSP}^{\mathrm{drt}^*}(A)) \models p \mapsto q$, and
4. for all closed $\mathrm{TSP}^{\mathrm{drt}^*}(A)$-terms q such that $\mathit{TSS}(\mathrm{TSP}^{\mathrm{drt}^*}(A)) \models p \mapsto q$ then $\mathit{TSS}(\mathrm{TSP}(A)) \models p \leftrightarrows q$ and $\mathit{TSS}(\mathrm{TSP}^{\mathrm{drt}^*}(A)) \models p \leftrightarrows q$.

If we show the above list of items to be true, it follows that $\mathit{TSS}(\mathrm{TSP}^{\mathrm{drt}^*}(A))$ is an orthogonal extension of $\mathit{TSS}(\mathrm{TSP}(A))$: The first condition of orthogonality is item 1 in the above list. So, it only remains to show that $\mathit{TSS}(\mathrm{TSP}^{\mathrm{drt}^*}(A)) \models p \leftrightarrows p' \Leftrightarrow \mathit{TSS}(\mathrm{TSP}(A)) \models p \leftrightarrows p'$, which follows immediately from the above items.

To check the first and the second item, one should note that first, a proof in $\mathrm{TSP}^{\mathrm{drt}^*}$ for an a-transition $\xrightarrow{a}$ with a source term from TSP or a termination predicate $\downarrow$ on a TSP-term only involves deduction rules from TSP. Such deduction rules do not have negative premises and are source-dependent [18]. Hence, all such transition and predicate formulae are included in the stable model of TSP. Second, for the inclusion in the other direction, all proofs in TSP remain valid in $\mathrm{TSP}^{\mathrm{drt}^*}$ and since deduction rules of TSP do not have negative premises, all the proven transitions and predicates of TSP are included in the stable model of $\mathrm{TSP}^{\mathrm{drt}^*}$.

We prove the last two items in one go. To that end, we use a structural induction on closed $\mathrm{TSP}(A)$-term p.

If p is a constant, i.e., δ or ϵ, then it can make a self time transition using one of the following deduction rules (from Table 6 and Table 12 respectively) and these are the only matching rules for such constants to make a time transition.

$$\delta \mapsto \delta \qquad \epsilon \mapsto \epsilon$$

Bisimilarity (w.r.t. both TSS's) is reflexive and hence self-transitions satisfy the criteria of item 4.

If p is of the form $a.p'$ (for some closed $\mathrm{TSP}(A)$-term p') then it can make a self time transition due to the following deduction rule (from Table 6) and this is the only matching deduction rule for p to make a time transition.

$$a.x \mapsto a.x$$

If p is of the form $p'+q'$, then by the induction hypothesis, p' and q' can make time transitions and all their time transitions are to bisimilar terms (w.r.t. both TSS's). Then $p'+q'$ can make time transitions using the following deduction rule (from Table 6).

$$\frac{x \mapsto x' \quad y \mapsto y'}{x + y \mapsto x' + y'}$$

and since bisimilarity is a congruence for both $TSS(\mathrm{TSP}(A))$ and $TSS(\mathrm{TSP}^{\mathrm{drt}^*}(A))$ (both TSS's are in the PANTH format of [21]), time transitions of $p'+q'$ that are due to this rule are to bisimilar terms w.r.t. both TSS's. Furthermore, $p'+q'$ cannot make a time transition using the other two deduction rules for choice in Table 6 since both of them have negative premises denying a time transition from p' or q'. Thus, all transitions of $p'+q'$ are to bisimilar terms w.r.t. both TSS's.

If p is of the form $p' \cdot q'$, then either $TSS(\mathrm{TSP}(A)) \models p'\downarrow$ (hence, following item 2, $TSS(\mathrm{TSP}^{\mathrm{drt}^*}(A)) \models p'\downarrow$) or $\neg TSS(\mathrm{TSP}(A)) \models p'\downarrow$ (hence, $TSS(\mathrm{TSP}^{\mathrm{drt}^*}(A)) \not\models p'\downarrow$). Then, by induction hypothesis, $p' \cdot q'$ can make a time transition due to the first or second deduction rule of Table 16 given below, respectively,

$$\frac{x \mapsto x' \quad x\downarrow \quad y \mapsto y'}{x \cdot y \mapsto x' \cdot y + y'} \quad \frac{x \mapsto x' \quad x\not\downarrow}{x \cdot y \mapsto x' \cdot y}$$

and these are the only possibilities for p to make time transitions as the other two deduction rules of Table 16 deny time transitions of p' or q'. Suppose that $p' \mapsto p''$, $q' \mapsto q''$ and $p' \leftrightarrow p''$ and $q' \leftrightarrow q''$ w.r.t. both TSS's. If $p'\downarrow$ and the time transition of p is due the left-hand-side deduction rule, then it is easy to check that $p' \leftrightarrow p' + \epsilon$ w.r.t. both TSS's. Hence, $p'.q' \leftrightarrow (p'+\epsilon).q' \leftrightarrow p'.q' + \epsilon.q' \leftrightarrow p'.q' + q' \leftrightarrow p''.q'' + q''$ (following the axioms of both $\mathrm{TSP}^{\mathrm{drt}^*}$ and TSP, Theorems 17 and 21 and congruence of bisimilarity for both TSS's). Hence, in this case p makes a time transition to bisimilar terms w.r.t. both TSS's. If $p'\not\downarrow$ and the transition is due to the right-hand-side rule, then $p'.q' \leftrightarrow p''.q'$ and again the transition of p is to a bisimilar term w.r.t. both TSS's.

Theorem 24 (Conservative extension). $\mathrm{TSP}^{\mathrm{drt}^*}$ *is a conservative extension of* $\mathrm{BSP}^{\mathrm{drt}^*}$.

Proof. Both $\mathrm{BSP}^{\mathrm{drt}^*}$ and $\mathrm{TSP}^{\mathrm{drt}^*}$ are sound and ground-complete equational theories for both $TSS(\mathrm{BSP}^{\mathrm{drt}^*}(A))$ and $TSS(\mathrm{TSP}^{\mathrm{drt}^*}(A))$. Also, following [12], $TSS(\mathrm{TSP}^{\mathrm{drt}^*}(A))$ is an operationally conservative (orthogonal) extension of $TSS(\mathrm{BSP}^{\mathrm{drt}^*}(A))$. Thus, we conclude that $\mathrm{TSP}^{\mathrm{drt}^*}$ is an equationally conservative extension of $\mathrm{BSP}^{\mathrm{drt}^*}$.

6 Parallel Composition

6.1 TCP

The formal definition of process theory TCP, the Theory of Communicating Processes, is given in Table 17. The theory includes the encapsulation operator, as this operator is essential to enforce communication between processes. As before, it has as a parameter the set of actions A. Besides this, it has as a second parameter a commutative and associative partial communication function $\gamma : A \times A \to A$. The signature of process theory TCP extends the signature of the process theory TSP with the merge operator $\|$, the left merge operator $\mathbin{⌊\!⌊}$, the synchronization merge operator $\mid$ and the encapsulation operator ∂_H. The four new operators bind stronger than choice but weaker than action prefix.

Following the, by now, standard practice of [9], parallel composition is broken up into three alternatives: the part where the first step comes from x, the part where the first step comes from y and the part where x and y execute together.

$$x \| y = x \mathbin{⌊\!⌊} y + y \mathbin{⌊\!⌊} x + x \mid y$$

To tackle the axiomatization of the left merge operator, the following axioms are used [8].

$$a.x \mathbin{⌊\!⌊} y = a.(x \| y) \qquad (x + y) \mathbin{⌊\!⌊} z = x \mathbin{⌊\!⌊} z + y \mathbin{⌊\!⌊} z.$$

Finally, what remains is the behavior of parallel composition with respect to the termination constants δ and ϵ. As the termination behavior of parallel composition is coded into the communication merge operator, this is of no concern here, and the following laws can be put forward.

$$\delta \mathbin{⌊\!⌊} x = \delta \qquad \epsilon \mathbin{⌊\!⌊} x = \delta$$

Next, the standard axioms of communication merge operator are introduced [9].

$$\begin{array}{ll} (x + y) \mid z = x \mid z + y \mid z & x \mid (y + z) = x \mid y + x \mid z \\ \delta \mid x = \delta & x \mid \delta = \delta \\ a.x \mid b.y = c.(x \| y) & \text{if } \gamma(a, b) = c \\ a.x \mid b.y = \delta & \text{if } \gamma(a, b) \text{ not defined.} \end{array}$$

What remains are the cases where ϵ appears as an argument of the communication merge operator.

$$\begin{array}{ll} \epsilon \mid \epsilon = \epsilon & \\ a.x \mid \epsilon = \delta & \epsilon \mid a.x = \delta. \end{array}$$

The axiom system presented in Table 17 contains an axiom stipulating the commutativity of the communication merge. This allows to save on the number of axioms required for the communication merge operator.

Theorem 25 (Elimination). *For any closed* TCP(A, γ)*-term p, there exists a closed* TSP(A)*-term q such that* TCP$(A, \gamma) \vdash p = q$.

The transition system specification associated to the terms of TCP is given in Table 18. It adds deduction rules defining action transitions and termination behavior of encapsulation and the parallel composition operators.

Theorem 26 (Soundness). *Let p and q be two closed* TCP(A, γ)*-terms. If* TCP$(A, \gamma) \vdash p = q$*, then TSS*(TCP(A, γ)) $\models p \leftrightarrows q$.

Table 17. The process theory $\mathrm{TCP}(A,\gamma)$

$\mathrm{TCP}(A,\gamma)$

$\mathrm{TSP}(A)$;

unary: $(\partial_H)_{H\subseteq A}$; binary: $\_\|\_,\ \_\mathbin{\lfloor\!\lfloor}\_,\ \_\mid\_$;

$x, y, z : P$;

$\partial_H(\epsilon) = \epsilon$	D1		
$\partial_H(\delta) = \delta$	D2		
$\partial_H(a.x) = \delta$ if $a \in H$	D3		
$\partial_H(a.x) = a.\partial_H(x)$ otherwise	D4		
$\partial_H(x + y) = \partial_H(x) + \partial_H(y)$	D5		
$x \| y = x \mathbin{\lfloor\!\lfloor} y + y \mathbin{\lfloor\!\lfloor} x + x \mid y$	M	$x \mid y = y \mid x$	SC1
$\delta \mathbin{\lfloor\!\lfloor} x = \delta$	LM1		
$\epsilon \mathbin{\lfloor\!\lfloor} x = \delta$	LM2	$x \| \epsilon = x$	SC2
$a.x \mathbin{\lfloor\!\lfloor} y = a.(x \| y)$	LM3	$\epsilon \mid x + \epsilon = \epsilon$	SC3
$(x + y) \mathbin{\lfloor\!\lfloor} z = x \mathbin{\lfloor\!\lfloor} z + y \mathbin{\lfloor\!\lfloor} z$	LM4		
$\delta \mid x = \delta$	CM1	$(x \| y) \| z = x \| (y \| z)$	SC4
$(x + y) \mid z = x \mid z + y \mid z$	CM2	$(x \mid y) \mid z = x \mid (y \mid z)$	SC5
$\epsilon \mid \epsilon = \epsilon$	CM3	$(x \mathbin{\lfloor\!\lfloor} y) \mathbin{\lfloor\!\lfloor} z = x \mathbin{\lfloor\!\lfloor} (y \| z)$	SC6
$a.x \mid \epsilon = \delta$	CM4	$(x \mid y) \mathbin{\lfloor\!\lfloor} z = x \mid (y \mathbin{\lfloor\!\lfloor} z)$	SC7
$a.x \mid b.y = c.(x \| y)$ if $\gamma(a,b) = c$	CM5		
$a.x \mid b.y = \delta$ if $\gamma(a,b)$ not defined	CM6	$x \mathbin{\lfloor\!\lfloor} \delta = x \cdot \delta$	SC8

Table 18. Transition system specification for $\mathrm{TCP}(A,\gamma)$

$\mathit{TSS}(\mathrm{TCP}(A,\gamma))$

$\mathit{TSS}(\mathrm{TSP}(A))$;

unary: $(\partial_H)_{H\subseteq A}$; binary: $\_\|\_,\ \_\mathbin{\lfloor\!\lfloor}\_,\ \_\mid\_$;

x, y, x', y';

$$\frac{x\downarrow \quad y\downarrow}{x \| y \downarrow} \qquad \frac{x\downarrow \quad y\downarrow}{x \mid y \downarrow} \qquad \frac{x\downarrow}{\partial_H(x)\downarrow} \qquad \frac{x \xrightarrow{a} x' \quad a \notin H}{\partial_H(x) \xrightarrow{a} \partial_H(x')}$$

$$\frac{x \xrightarrow{a} x'}{x \| y \xrightarrow{a} x' \| y} \qquad \frac{y \xrightarrow{a} y'}{x \| y \xrightarrow{a} x \| y'} \qquad \frac{x \xrightarrow{a} x'}{x \mathbin{\lfloor\!\lfloor} y \xrightarrow{a} x' \| y}$$

$$\frac{x \xrightarrow{a} x' \; y \xrightarrow{b} y' \; \gamma(a,b) = c}{x \| y \xrightarrow{c} x' \| y'} \qquad \frac{x \xrightarrow{a} x' \; y \xrightarrow{b} y' \; \gamma(a,b) = c}{x \mid y \xrightarrow{c} x' \| y'}$$

Theorem 27 (Ground-completeness). *Let p and q be arbitrary closed $\mathrm{TCP}(A,\gamma)$-terms. If $\mathit{TSS}(\mathrm{TCP}(A,\gamma)) \models p \leftrightarroweq q$, then $\mathrm{TCP}(A,\gamma) \vdash p = q$.*

Theorem 28 (Conservative extension). TCP *is a conservative extension of* TSP.

Proof. Both TSP and TCP are sound and ground-complete equational theories for $TSS(\mathrm{TSP}(A))$ and $TSS(\mathrm{TCP}(A,\gamma))$. Also, following [12], $TSS(\mathrm{TCP}(A,\gamma))$ is an operationally conservative (orthogonal) extension of $TSS(\mathrm{TSP}(A))$. Thus, we conclude that TCP is an equationally conservative extension of TSP.

The relationship between the theories ACP^{ϵ} and TCP is similar to the relationship between BPA^{ϵ}_{δ} and TSP. One can either extend TCP with the constants $a \in A$ using axioms $a = a.\epsilon$ or one can embed ACP^{ϵ} into TCP by mapping a onto $a.\epsilon$.

6.2 $\mathrm{TCP}^{\mathrm{drt}^*}$

In this section, the process theory $\mathrm{TSP}^{\mathrm{drt}^*}$ is extended to the process theory $\mathrm{TCP}^{\mathrm{drt}^*}$. This extension is obtained by extending the signature of $\mathrm{TSP}^{\mathrm{drt}^*}$ with the current time slice timeout operator ν, the encapsulation operators ∂_H (for $H \subseteq A$), and the parallel composition operators $\_ \| \_$, $\_ \mathbin{\|\!\!\!\lfloor} \_$, and $\_ \mid \_$.

The axioms of the process theory are given in Table 19.

The current time slice time out operator disallows all initial passage of time. It extracts the part of the behavior that executes an action or performs termination in the current time slice. The encapsulation operator is as defined before: encapsulation disallows the actions that occur in the set H and allows all other behavior including passage of time.

The axioms for parallel composition and the auxiliary operators are such that parallel processes have to synchronize the passage of time until one of the processes can terminate, and within each time slice interleave their actions or communicate. To stay as closely as possible to the interpretation of the axioms in the untimed setting, it is necessary for both left merge and communication merge to synchronize the passage of time as well (axioms LM6DR and CM10DR).

Theorem 29 (Elimination). *For any closed* $\mathrm{TCP}^{\mathrm{drt}^*}(A,\gamma)$*-term* p*, there exists a closed* $\mathrm{TSP}^{\mathrm{drt}^*}(A)$*-term* q *such that* $\mathrm{TCP}^{\mathrm{drt}^*}(A,\gamma) \vdash p = q$.

The transition system specification associated to the terms of $\mathrm{TCP}^{\mathrm{drt}^*}$ is given in Table 20. It adds deduction rules defining the behavior of the "now" operator and it adds deduction rules defining time transitions of encapsulation and the parallel composition operators.

Theorem 30 (Soundness). *Let* p *and* q *be two closed* $\mathrm{TCP}^{\mathrm{drt}^*}(A,\gamma)$*-terms. If* $\mathrm{TCP}^{\mathrm{drt}^*}(A,\gamma) \vdash p = q$*, then* $TSS(\mathrm{TCP}^{\mathrm{drt}^*}(A,\gamma)) \models p \leftrightarrows q$.

Theorem 31 (Ground-completeness). *Let* p *and* q *be arbitrary closed* $\mathrm{TCP}^{\mathrm{drt}^*}(A,\gamma)$*-terms. If* $TSS(\mathrm{TCP}^{\mathrm{drt}^*}(A,\gamma)) \models p \leftrightarrows q$*, then* $\mathrm{TCP}^{\mathrm{drt}^*}(A,\gamma) \vdash p = q$.

Table 19. The process theory $\mathrm{TCP}^{\mathrm{drt}^*}(A,\gamma)$

$\mathrm{TCP}^{\mathrm{drt}^*}(A,\gamma)$

$\mathrm{TSP}^{\mathrm{drt}^*}(A)$;

unary: $\nu, (\partial_H)_{H\subseteq A}$; binary: $\_\|\_$, $\_\lfloor\!\lfloor\_$, $\_\mid\_$;

$x, y, z : P$;

$\partial_H(\underline{\underline{\epsilon}}) = \underline{\underline{\epsilon}}$	D1DR	$\nu(\underline{\underline{\epsilon}}) = \underline{\underline{\epsilon}}$	RTO1
$\partial_H(\underline{\underline{\delta}}) = \underline{\underline{\delta}}$	D2DR	$\nu(\underline{\underline{\delta}}) = \underline{\underline{\delta}}$	RTO2
$\partial_H(\underline{\underline{a}}.x) = \underline{\underline{\delta}}$ if $a \in H$	D3DR	$\nu(\underline{\underline{a}}.x) = \underline{\underline{a}}.x$	RTO3
$\partial_H(\underline{\underline{a}}.x) = \underline{\underline{a}}.\partial_H(x)$ otherwise	D4DR		
$\partial_H(x+y) = \partial_H(x) + \partial_H(y)$	D5	$\nu(x+y) = \nu(x) + \nu(y)$	RTO4
$\partial_H(\sigma.x) = \sigma.\partial_H(x)$	D6DR	$\nu(\sigma.x) = \underline{\underline{\delta}}$	RTO5
$\partial_H(\sigma^* x) = \sigma^* \partial_H(x)$	D7DR	$\nu(\sigma^* x) = \nu(x)$	RTO6
$x \mid y = y \mid x$	SC1	$(x \mid y) \mid z = x \mid (y \mid z)$	SC5
$x \| \underline{\underline{\epsilon}} = x$	SC2DR	$(x \lfloor\!\lfloor y) \lfloor\!\lfloor z = x \lfloor\!\lfloor (y \| z)$	SC6
$\underline{\underline{\epsilon}} \mid x + \underline{\underline{\epsilon}} = \underline{\underline{\epsilon}}$	SC3DR	$(x \mid y) \lfloor\!\lfloor z = x \mid (y \lfloor\!\lfloor z)$	SC7
$(x \| y) \| z = x \| (y \| z)$	SC4	$x \lfloor\!\lfloor \underline{\underline{\delta}} = x.\underline{\underline{\delta}}$	SC8DR
$x \| y = x \lfloor\!\lfloor y + y \lfloor\!\lfloor x + x \mid y$	M		
$\underline{\underline{\delta}} \lfloor\!\lfloor x = \underline{\underline{\delta}}$	LM1DR	$\underline{\underline{\delta}} \mid x = \underline{\underline{\delta}}$	CM1DR
$\underline{\underline{\epsilon}} \lfloor\!\lfloor \underline{\underline{\delta}} = \underline{\underline{\delta}}$	LM2DR	$(x+y) \mid z = x \mid z + y \mid z$	CM2
$\underline{\underline{a}}.x \lfloor\!\lfloor y = \underline{\underline{a}}.(x \| y)$	LM3DR	$\underline{\underline{\epsilon}} \mid \underline{\underline{\epsilon}} = \underline{\underline{\epsilon}}$	CM3DR
$(x+y) \lfloor\!\lfloor z = x \lfloor\!\lfloor z + y \lfloor\!\lfloor z$	LM4	$\underline{\underline{a}}.x \mid \underline{\underline{\epsilon}} = \underline{\underline{\delta}}$	CM4DR
$\sigma.x \lfloor\!\lfloor \underline{\underline{\delta}} = \underline{\underline{\delta}}$	LM5DR	$\underline{\underline{a}}.x \mid \underline{\underline{b}}.y = \underline{\underline{c}}.(x \| y)$ if $\gamma(a,b) = c$	CM5DR
$\sigma.x \lfloor\!\lfloor \underline{\underline{\epsilon}} = \sigma.x$	LM6DR	$\underline{\underline{a}}.x \mid \underline{\underline{b}}.y = \underline{\underline{\delta}}$ if $\gamma(a,b)$ not defined	CM6DR
$\sigma.x \lfloor\!\lfloor (\underline{\underline{a}}.y + z) = \sigma.x \lfloor\!\lfloor z$	LM7DR	$\sigma.x \mid \nu(y) = \underline{\underline{\delta}}$	CM7DR
$\sigma.x \lfloor\!\lfloor (\nu(y) + \sigma.z) = \sigma.(x \lfloor\!\lfloor z)$	LM8DR	$\sigma.x \mid \sigma.y = \sigma.(x \mid y)$	CM8DR
$\sigma^* x \lfloor\!\lfloor \sigma^* \nu(y) = \sigma^*(x \lfloor\!\lfloor \sigma^* \nu(y))$	LM9DR	$\sigma^* x \mid \sigma^* y = \sigma^*(x \mid \sigma^* y + \sigma^* x \mid y)$	CM9DR

Theorem 32 (Conservative ground-extension). $\mathrm{TCP}^{\mathrm{drt}^*}$ *is a conservative ground-extension of* TCP.

Proof. Similar to the proof of Theorem 23. We show that for all p and p' in the syntax of TCP,

1. $\mathit{TSS}(\mathrm{TCP}(A,\gamma)) \models p \xrightarrow{a} p' \Leftrightarrow \mathit{TSS}(\mathrm{TCP}^{\mathrm{drt}^*}(A,\gamma)) \models p \xrightarrow{a} p'$,
2. $\mathit{TSS}(\mathrm{TCP}(A,\gamma)) \models p\downarrow \Leftrightarrow \mathit{TSS}(\mathrm{TCP}^{\mathrm{drt}^*}(A,\gamma)) \models p\downarrow$,
3. there exists a closed $\mathrm{TCP}(A)$-term q such that $\mathit{TSS}(\mathrm{TCP}^{\mathrm{drt}^*}(A,\gamma)) \models p \mapsto q$, and
4. for all closed $\mathrm{TCP}^{\mathrm{drt}^*}(A,\gamma)$-terms q such that $\mathit{TSS}(\mathrm{TCP}^{\mathrm{drt}^*}(A,\gamma)) \models p \mapsto q$ then $\mathit{TSS}(\mathrm{TCP}(A,\gamma)) \models p \leftrightarrows q$ and $\mathit{TSS}(\mathrm{TCP}^{\mathrm{drt}^*}(A,\gamma)) \models p \leftrightarrows q$.

From these, it follows that $\mathit{TSS}(\mathrm{TCP}^{\mathrm{drt}^*}(A,\gamma))$ is an orthogonal extension of $\mathit{TSS}(\mathrm{TCP}(A,\gamma))$.

The first and second items in the above list hold trivially since a proof in $\mathrm{TCP}^{\mathrm{drt}^*}$ for an a-transition $\xrightarrow{a}$ with a source term from TCP or a termination

Table 20. Transition system specification for $\mathrm{TCP}^{\mathrm{drt}^*}(A,\gamma)$

$TSS(\mathrm{TCP}^{\mathrm{drt}^*}(A,\gamma))$

$TSS(\mathrm{TCP}(A,\gamma))$, $TSS(\mathrm{TSP}^{\mathrm{drt}^*}(A))$;

unary: $\nu(\_)$;

x, x', y, y';

$$\frac{x\downarrow}{\nu(x)\downarrow} \qquad \frac{x \xrightarrow{a} x'}{\nu(x) \xrightarrow{a} x'} \qquad \frac{x \mapsto x'}{\partial_H(x) \mapsto \partial_H(x')}$$

$$\frac{x \mapsto x' \quad y \mapsto y'}{x \,\|\, y \mapsto x' \,\|\, y'} \qquad \frac{x \mapsto x' \quad y \mapsto y'}{x \,\lfloor\!\lfloor\, y \mapsto x' \,\lfloor\!\lfloor\, y'} \qquad \frac{x \mapsto x' \quad y \mapsto y'}{x \mid y \mapsto x' \mid y'}$$

$$\frac{x \mapsto x' \quad y \not\mapsto \quad y\downarrow}{x \,\|\, y \mapsto x'} \qquad \frac{x \not\mapsto \quad y \mapsto y' \quad x\downarrow}{x \,\|\, y \mapsto y'} \qquad \frac{x \mapsto x' \quad y \not\mapsto \quad y\downarrow}{x \,\lfloor\!\lfloor\, y \mapsto x'}$$

predicate $\downarrow$ on a TCP-term only involves deduction rules from TCP. Such deduction rules do not have negative premises and are source-dependent [18]. Hence, all such transition and predicate formulae are included in the stable model of TCP. For the inclusion in the other direction, all proofs in TCP remain valid in $\mathrm{TCP}^{\mathrm{drt}^*}$ and since deduction rules of TCP do not have negative premises, all the proven transitions and predicates of TCP are included in the stable model of $\mathrm{TCP}^{\mathrm{drt}^*}$.

For the last two items, we use a structural induction on closed TCP-term p. For function symbols in the signature of TSP, the arguments given in Theorem 23 remain valid. Hence, it only remains to check that for terms of the form $\partial_H(p')$, $p' \| q'$, $p' \lfloor\!\lfloor q'$ and $p' \mid q'$ the last two items hold, assuming that these items hold for p' and q' (if applicable).

If p is of the form $\partial_H(p')$, then it can make a time transition due to the following rule:

$$\frac{x \mapsto x'}{\partial_H(x) \mapsto \partial_H(x')}$$

Using the above deduction rule, one can only prove self time transitions and bisimilarity (w.r.t. both TSS's) is reflexive. Furthermore, the above rule is the only rule using which $\partial_H(p')$, hence p, can make a time transition.

If p is of the form $p' \| q'$, it can make a time transition due to the following rule:

$$\frac{x \mapsto x' \quad y \mapsto y'}{x \,\|\, y \mapsto x' \,\|\, y'}$$

and given that the time transitions of p' and q' are only to bisimilar terms (w.r.t. both TSS's) and bisimilarity is a congruence (since both $TSS(\mathrm{TCP}(A,\gamma))$ and $TSS(\mathrm{TCP}^{\mathrm{drt}^*}(A,\gamma))$ are in the PANTH format of [21]), the transition of $p' \| q'$ or p is also to a bisimilar term (w.r.t. both TSS's). Furthermore, p cannot make a time transition due to any of the other two rules for time transitions of merge since both of them have negative premises denying time transitions from p' or q'.

If p is of the form $p' \mathbin{\|\!\!\_} q'$, it can make a time transition due to the following rule:

$$\frac{x \mapsto x' \quad y \mapsto y'}{x \mathbin{\|\!\!\_} y \mapsto x' \mathbin{\|\!\!\_} y'}$$

and given that the time transitions of p' and q' are only to bisimilar terms (w.r.t. both TSS's) and bisimilarity is a congruence, the transition of $p' \mathbin{\|\!\!\_} q'$ or p is also to a bisimilar term (w.r.t. both TSS's). Furthermore, p cannot make a time transition due to the other rule for time transitions of left merge since it has a negative premise denying time transitions from q'.

If p is of the form $p' \mid q'$, it can make a time transition due to the following rule:

$$\frac{x \mapsto x' \quad y \mapsto y'}{x \mid y \mapsto x' \mid y'}$$

and given that the time transitions of p' and q' are only to bisimilar terms (w.r.t. both TSS's) and bisimilarity is a congruence, the transition of $p' \mid q'$ or p is also to a bisimilar term (w.r.t. both TSS's). There is no other deduction rule using which p can make a time transition and this concludes the proof of the last two items.

Theorem 33 (Conservative extension). $\mathrm{TCP}^{\mathrm{drt}^*}$ *is a conservative extension of* $\mathrm{TSP}^{\mathrm{drt}^*}$.

Proof. Both $\mathrm{TSP}^{\mathrm{drt}^*}$ and $\mathrm{TCP}^{\mathrm{drt}^*}$ are sound and ground-complete equational theories for $\mathit{TSS}(\mathrm{TSP}^{\mathrm{drt}^*}(A))$ and $\mathit{TSS}(\mathrm{TCP}^{\mathrm{drt}^*}(A,\gamma))$. Also, following [12], $\mathit{TSS}(\mathrm{TCP}^{\mathrm{drt}^*}(A,\gamma))$ is an operationally conservative (orthogonal) extension of $\mathit{TSS}(\mathrm{TSP}^{\mathrm{drt}^*}(A))$. Thus, we conclude that $\mathrm{TCP}^{\mathrm{drt}^*}$ is an equationally conservative extension of $\mathrm{TSP}^{\mathrm{drt}^*}$.

7 Concluding Remarks

When we extend untimed process algebra with timing, this extension is usually not a conservative extension, as some axioms of untimed process algebra hold for untimed processes only. However, what can be achieved is a conservative ground-extension, a notion that is introduced here. In this paper, we present timed extensions of an incremental presentation of process algebras, involving termination constants, alternative composition, sequential composition and parallel composition with communication, where in each case it is shown we have a conservative ground-extension. In previous papers, conservativity was always violated in some way.

For the realization of these timed extensions, it was necessary to change a basic design principle of ACP-style process algebra: action constants, involving both action execution and termination, are replaced by action prefixing.

References

1. J. C. M. Baeten. Embedding untimed into timed process algebra: the case for explicit termination. *Mathematical Structures in Computer Science (MSCS)*, 13(4):589–618, 2003.
2. J. C. M. Baeten and J. A. Bergstra. Real Time Process Algebra. *Formal Aspects of Computing*, 3:142–188, 1991.
3. J. C. M. Baeten and J. A. Bergstra. Discrete time process algebra. *Formal Aspects of Computing*, 8(2):188–208, 1996.
4. J. C. M. Baeten and R. J. van Glabbeek. Merge and termination in process algebra. In K. V. Nori, editor, *Proceeding of the Seventh Conference on Foundations of Software Technology and Theoretical Computer Science (FST&TCS'87)*, volume 287 of *Lecture Notes in Computer Science*, pages 153–172. Springer-Verlag, Berlin, Germany, 1987.
5. J. C. M. Baeten and C. A. Middelburg. *Process Algebra with Timing.* EATCS Monographs. Springer-Verlag, Berlin, Germany, 2002.
6. J. C. M. Baeten and M. A. Reniers. Timed process algebra (with a focus on explicit termination and relative-timing). In M. A. Bernardo and F. Corradini, editors, *Proceedings of the International School on Formal Methods for the Design of Real-Time Systems (SFM-RT'04)*, volume 3185 of *Lecture Notes in Computer Science*, pages 59–97. Springer-Verlag, Berlin, Germany, 2004.
7. J. C. M. Baeten and W. P. Weijland. *Process Algebra*, volume 18 of *Cambridge Tracts in Theoretical Computer Science.* Cambrdige University Press, 1990.
8. J. A. Bergstra and J. W. Klop. Fixed point semantics in process algebra. Technical Report IW 206/82, Mathematical Center, Amsterdam, The Netherlands, 1982.
9. J. A. Bergstra and J. W. Klop. Process algebra for synchronous communication. *Information and Control*, 60(1-3):109–137, 1984.
10. J. A. Bergstra and J. W. Klop. Algebra of communicating processes. In J. W. de Bakker, M. Hazewinkel, and J. K. Lenstra, editors, *Proceedings of the CWI Symposium Mathematics and Computer Science*, pages 89–138. North-Holland, Amsterdam, The Netherlands, 1986.
11. R. Bol and J. F. Groote. The meaning of negative premises in transition system specifications. *Journal of the ACM (JACM)*, 43(5):863–914, Sept. 1996.
12. W. J. Fokkink and C. Verhoef. A conservative look at operational semantics with variable binding. *Information and Computation (I&C)*, 146(1):24–54, 1998.
13. R. J. van Glabbeek. The meaning of negative premises in transition system specifications II. *Journal of Logic and Algebraic Programming (JLAP)*, 60-61:229–258, 2004.
14. J. F. Groote. Transition system specifications with negative premises. *Theoretical Computer Science (TCS)*, 118(2):263–299, 1993.
15. M. Hennessy and T. Regan. A process algebra for timed systems. *Information and Computation*, 117(2):221–239, 1995.
16. C. P. J. Koymans and J. L. M. Vrancken. Extending process algebra with the empty process. Technical Report 1, Logic Group Preprint Series, Department of Philosophy, Utrecht University, Utrecht, The Netherlands, 1985. Extended and enhanced version appeared as [22].
17. F. Moller and C. M. N. Tofts. A temporal calculus of communicating systems. volume 458 of *Lecture Notes in Computer Science*, pages 401–415.

18. M. Mousavi and M. A. Reniers. Orthogonal extensions in structural operational semantics. In *Proceedings of the 32nd International Colloquium on Automata, Languages and Programming (ICALP'05)*, volume 3580 of *Lecture Notes in Computer Science*, pages 1214–1225. Springer-Verlag, Berlin, Germany, 2005.
19. X. Nicollin and J. Sifakis. The algebra of timed processes ATP: theory and application. *Information and Computation (I&C)*, 114(1):131–178, Oct. 1994.
20. J. J. Vereijken. *Discrete Time Process Algebra.* PhD thesis, Department of Mathematics and Computer Science, Eindhoven University of Technology, Eindhoven, The Netherlands, 1997.
21. C. Verhoef. A congruence theorem for structured operational semantics with predicates and negative premises. *Nordic Journal of Computing*, 2(2):274–302, 1995.
22. J. L. M. Vrancken. The algebra of communicating processes with empty process. *Theoretical Computer Science*, 177(2):287–328, 1997.

Confluence of Graph Transformation Revisited

Detlef Plump

Department of Computer Science, The University of York,
York YO10 5DD, United Kingdom
det@cs.york.ac.uk

Abstract. It is shown that it is undecidable in general whether a terminating graph rewriting system is confluent or not—in contrast to the situation for term and string rewriting systems. Critical pairs are introduced to hypergraph rewriting, a generalisation of graph rewriting, where it turns out that the mere existence of common reducts for all critical pairs of a graph rewriting system does not imply local confluence. A Critical Pair Lemma for hypergraph rewriting is then established which guarantees local confluence if each critical pair of a system has joining derivations that are compatible in that they map certain nodes to the same nodes in the common reduct.

1 Introduction

To compute efficiently with graph transformation rules requires some way of cutting down the nondeterminism in the derivation spaces of graphs. A common solution to this problem in rule-based formalisms is to rely on *confluent* sets of rules so that all terminating derivations from an initial state will yield the same result, making backtracking unnecessary. For example, confluence properties are important for efficiently recognizing graph classes and executing graph algorithms by graph reduction [2, 5, 7], for parsing graph languages by using the inverse rules of graph grammars [16, 31], and for verifying the deterministic behaviour of programs in graph rewriting languages such as PROGRES [34], AGG [15] and GP [29].

In the setting of term rewriting systems, Knuth and Bendix [22] showed that confluence is decidable for terminating sets of rules. It suffices to compute all *critical pairs* $t \leftarrow s \rightarrow u$ of rewrite steps in which s is the superposition of the left-hand sides of two rules, and to check whether t and u reduce to a common term. This procedure is justified by the Critical Pair Lemma [20]—stating that a term rewriting system is locally confluent if and only if all its critical pairs have common reducts—and by Newman's Lemma which asserts the equivalence of confluence and local confluence in the presence of termination.

In contrast to the situation for term and string rewriting, Theorem 5 below will show that confluence is undecidable for terminating graph rewriting systems. Roughly, the reason is that the embedding of derivations into context is more complicated for graphs than for terms and strings, meaning that the existence of common reducts for all critical pairs need not imply local confluence of the system.

A. Middeldorp et al. (Eds.): Processes... (Klop Festschrift), LNCS 3838, pp. 280–308, 2005.

The second major result of this paper is a Critical Pair Lemma for hypergraph rewriting which provides a sufficient condition for local confluence and hence for confluence of terminating systems (Theorem 7). The result requires each critical pair $T \Leftarrow S \Rightarrow U$ to be joinable by two derivations $T \Rightarrow^* W_1 \cong W_2 \Leftarrow^* U$ such that an isomorphism $W_1 \to W_2$ is compatible with the joining derivations, in that each node in S that is preserved by both $S \Rightarrow T$ and $S \Rightarrow U$ is mapped to the same node in W_2 by $S \Rightarrow^* W_2$ and $S \Rightarrow^* W_1$ followed by $W_1 \to W_2$.

The next section recalls some properties of binary relations and defines labelled and directed hypergraphs. Section 3 reviews the double-pushout approach to graph transformation, adapted to the setting of hypergraphs. Subsections 3.2 and 3.3 provide results about the restriction, extension and independence of derivations which will be needed to prove the Critical Pair Lemma. Section 4 starts by arguing, in Subsection 4.1, that confluence modulo isomorphism rather than confluence is the right notion to consider in graph transformation. Subsection 4.2 then presents a reduction of the Post Correspondence Problem showing that confluence is undecidable for terminating graph rewriting systems. Subsection 4.3 introduces critical pairs to hypergraph rewriting and proves the Critical Pair Lemma. Section 5 concludes by mentioning related work and topics for future work. Finally, the Appendix summarises some properties of hypergraph pushouts that are needed in proofs.

The confluence results of this paper were established in [27], but the undecidability of confluence was only shown for hypergraph rewriting; the current result also covers graph rewriting. In addition, the undecidability proof has been simplified so that the number of rule schemata in the reduction of the Post Correspondence Problem decreased from 21 to 12. Moreover, this paper is rigorous with respect to the role of confluence modulo isomorphism.

2 Preliminaries

This section fixes some terminology for binary relations (see also [3, 4]) and introduces hypergraphs and their morphisms.

2.1 Relations

Let $\to$ be a binary relation on a set A. The inverse relation of $\to$ is denoted by $\leftarrow$. The identity on A is the relation $\to^0 = \{\langle a, a\rangle \mid a \in A\}$. The reflexive closure of $\to$ is $\to^= = \to \cup \to^0$. The compositon of two binary relations $\to_1$ and $\to_2$ on A is $\to_1 \circ \to_2 = \{\langle a, c\rangle \mid a \to_1 b \text{ and } b \to_2 c \text{ for some } b\}$. For every $n > 0$, the n-fold composition of $\to$ is $\to^n = \to \circ \to^{n-1}$. The transitive closure of $\to$ is $\to^+ = \bigcup_{n>0} \to^n$, and the transitive-reflexive closure of $\to$ is $\to^* = \to^+ \cup \to^0$. Two elements a and b have a *common reduct* if $a \to^* c \leftarrow^* b$ for some c. If $a \to^* c$ and there is no d such that $c \to d$, then d is a *normal form* of a.

Definition 1 (Termination and confluence). The relation $\to$ is

(1) *terminating* if there is no infinite sequence of the form $a_1 \to a_2 \to a_3 \to \ldots$,
(2) *confluent* if for all a, b and c with $b \leftarrow^* a \to^* c$, elements b and c have a common reduct (see Figure 1(a)),

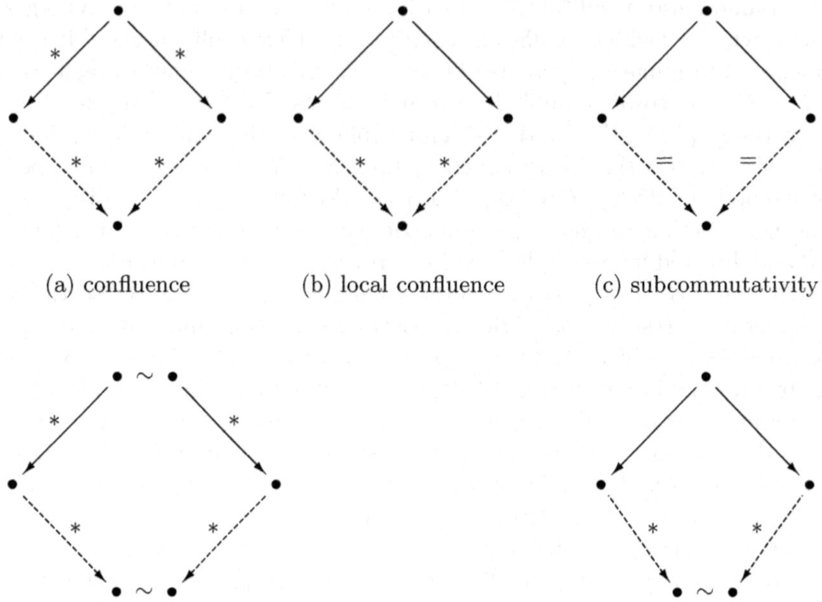

(a) confluence (b) local confluence (c) subcommutativity

(d) confluence modulo $\sim$ (e) local confluence modulo $\sim$

Fig. 1. Confluence properties

(3) *locally confluent* if for all a, b and c with $b \leftarrow a \rightarrow c$, elements b and c have a common reduct (see Figure 1(b)),
(4) *subcommutative* if for all a, b and c with $b \leftarrow a \rightarrow c$ there is some d such that $b \rightarrow^{=} d \leftarrow^{=} c$ (see Figure 1(c)),
(5) *confluent modulo* $\sim$, where $\sim$ is an equivalence relation on A, if for all a, a', b and c with $b \leftarrow^{*} a \sim a' \rightarrow^{*} c$ there are d and d' such that $b \rightarrow^{*} d \sim d' \leftarrow^{*} c$ (see Figure 1(d)),
(6) *locally confluent modulo* $\sim$ if for all a, b and c with $b \leftarrow a \rightarrow c$ there are d and d' such that $b \rightarrow^{*} d \sim d' \leftarrow^{*} c$ (see Figure 1(e)).

By the following well-known result [25], local confluence and confluence are equivalent in the presence of termination.

Lemma 1 (Newman's Lemma). *A terminating relation is confluent if and only if it is locally confluent.*

2.2 Hypergraphs

This paper deals with directed hypergraphs in which nodes and hyperedges carry labels and where the label of a hyperedge can restrict both the number of incident nodes and their possible labels. A *signature* $\Sigma = \langle \Sigma_{\mathrm{V}}, \Sigma_{\mathrm{E}} \rangle$ is a pair of sets of *node labels* and *hyperedge labels* such that each $l \in \Sigma_{\mathrm{E}}$ comes with a set $\mathrm{Type}(l) \subseteq \Sigma_{\mathrm{V}}^{*}$.

(Note the similarity to signatures of many-sorted algebras [17]: sorts correspond to node labels and operation symbols correspond to hyperedge labels.)

Definition 2 (Hypergraph). A *hypergraph* over a signature Σ is a system $G = \langle \mathrm{V}_G, \mathrm{E}_G, \mathrm{mark}_G, \mathrm{lab}_G, \mathrm{att}_G \rangle$ consisting of two finite sets V_G and E_G of *nodes* (or *vertices*) and *hyperedges*, two labelling functions $\mathrm{mark}_G\colon \mathrm{V}_G \to \Sigma_\mathrm{V}$ and $\mathrm{lab}_G\colon \mathrm{E}_G \to \Sigma_\mathrm{E}$, and an attachment function $\mathrm{att}_G\colon \mathrm{E}_G \to \mathrm{V}_G^*$ such that $\mathrm{mark}_G^*(\mathrm{att}_G(e)) \in \mathrm{Type}(\mathrm{lab}_G(e))$ for each hyperedge e.[1]

Hyperedges are said to be *incident* to their attachment nodes. In pictures, nodes and hyperedges are drawn as circles and boxes, respectively, with labels inside. Lines represent the attachment of hyperedges to nodes. If a hyperedge is attached to more than two nodes, the lines are numbered according to the left-to-right order in the attachment string. For hyperedges with just two attachment nodes—ordinary *edges*—, an arrowhead points to the second node; in this case the box may be omitted and the label written next to the arrow. As an example, Figure 2 shows a hypergraph borrowed from [17]. Here Type(PUSH), for instance, contains the string `data stack stack`.

A hypergraph G is a *graph* if each hyperedge e is an ordinary edge, that is, if the attachment sequence $\mathrm{att}_G(e)$ has length two.

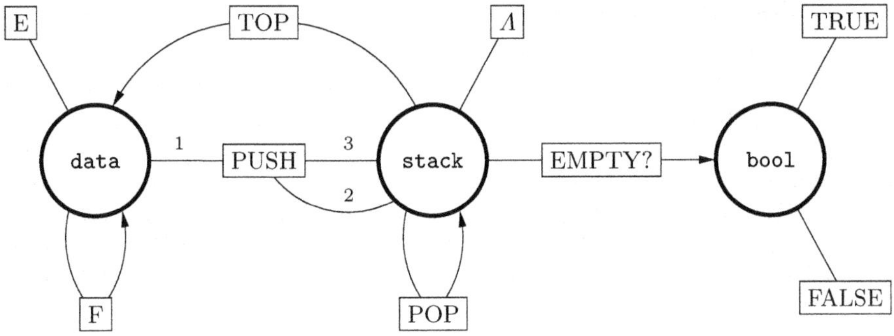

Fig. 2. A hypergraph

Definition 3 (Hypergraph morphism). Given two hypergraphs G and H over Σ, a *hypergraph morphism* $f\colon G \to H$ consists of two functions $f_\mathrm{V}\colon \mathrm{V}_G \to \mathrm{V}_H$ and $f_\mathrm{E}\colon \mathrm{E}_G \to \mathrm{E}_H$ that preserve labels and attachment to nodes, that is, $\mathrm{mark}_H \circ f_\mathrm{V} = \mathrm{mark}_G$, $\mathrm{lab}_H \circ f_\mathrm{E} = \mathrm{lab}_G$ and $\mathrm{att}_H \circ f_\mathrm{E} = f_\mathrm{V}^* \circ \mathrm{att}_G$.

A hypergraph morphism $incl\colon G \to H$ is an *inclusion* if $incl_\mathrm{V}(v) = v$ and $incl_\mathrm{E}(e) = e$ for all $v \in \mathrm{V}_G$ and $e \in \mathrm{E}_G$. In this case G is a *subhypergraph* of H which is denoted by $G \subseteq H$. Every hypergraph morphism $f\colon G \to H$

[1] The extension $f^*\colon A^* \to B^*$ of a function $f\colon A \to B$ maps the empty string to itself and $a_1 \ldots a_n$ to $f(a_1) \ldots f(a_n)$.

induces a subhypergraph of H, denoted by $f(G)$, which has nodes $f_V(V_G)$ and hyperedges $f_E(E_G)$. The *composition* of $f\colon G \to H$ with a morphism $g\colon H \to M$ is the hypergraph morphism $g \circ f\colon G \to M$ consisting of the composed functions $g_V \circ f_V$ and $g_E \circ f_E$. The composition is also written as $G \to H \to M$ if f and g are clear from the context.

The morphism f is *injective* (*surjective*) if f_V and f_E are injective (surjective). Injectivity of f may be indicated by writing $f\colon G \hookrightarrow H$. If f is both injective and surjective, then it is an *isomorphism*. In this case G and H are *isomorphic*, which is denoted by $G \cong H$.

3 Graph Transformation

This section reviews the *double-pushout approach* to graph transformation, where the approach presented in the overviews [9, 6] is generalized in three ways: hypergraphs rather than graphs are considered and rules are matched injectively and can have non-injective right-hand morphisms. Both injective matching and non-injective right-hand morphisms add expressiveness to the double-pushout approach [18].

3.1 Rules and Derivations

From now on an arbitrary but fixed signature is assumed over which all hypergraphs are labelled, unless signatures are explicitly mentioned.

Definition 4 (Rule). A *rule* $r\colon \langle L \hookleftarrow K \to^b R\rangle$ consists of three hypergraphs and two hypergraph morphisms, where $K \hookrightarrow L$ is an inclusion. The hypergraphs L and R are the *left-* and *right-hand side* of r, and K is the *interface*. The rule r is *injective* if $b\colon K \to R$ is injective.

Figure 3 shows a rule which removes two hyperedges and a node, merges two nodes, and creates a hyperedge. The letters x,y,z are node names; they are used to represent the hypergraph morphisms between the interface and the left- and right-hand side. The name x=y indicates that the right-hand morphism identifies node x with node y. The lower half of Figure 3 shows the same rule in a shorthand notation. In this format, only the left- and right-hand sides are depicted while the interface is implicitly given by all the named nodes of the left-hand side.

Definition 5 (Direct derivation). Let G and H be hypergraphs, $r\colon \langle L \hookleftarrow K \to^b R\rangle$ a rule and $f\colon L \hookrightarrow G$ an injective hypergraph morphism. Then G *directly derives* H by r and f, denoted by $G \Rightarrow_{r,f} H$, if there exist two pushouts[2] of the following form:

[2] See the Appendix for the definition and construction of hypergraph pushouts.

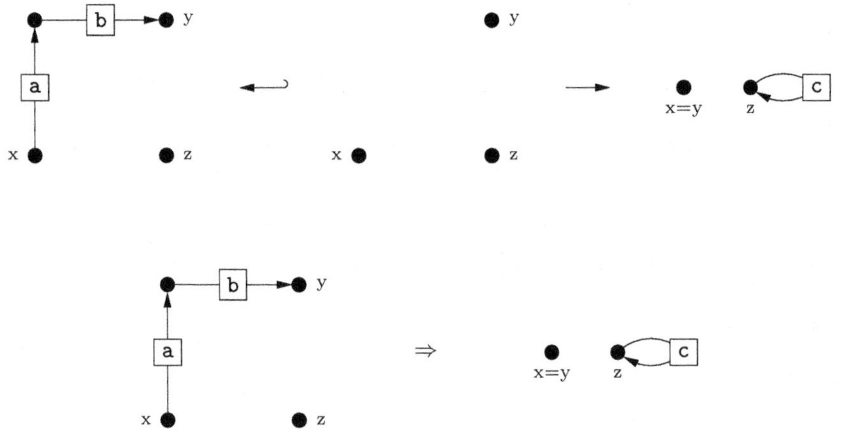

Fig. 3. A rule and its shorthand notation

$$\begin{array}{ccccc} L & \hookleftarrow & K & \xrightarrow{b} & R \\ f\downarrow & & \downarrow & & \downarrow \\ G & \hookleftarrow & D & \longrightarrow & H \end{array} \qquad (1)$$

Intuitively, the left pushout corresponds to the replacement of L with K (equivalently, to the removal of L up to K) and the right to the replacement of K with R. As an example, Figure 4 shows an application of the rule of Figure 3.

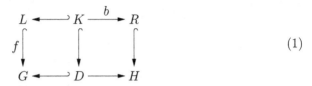

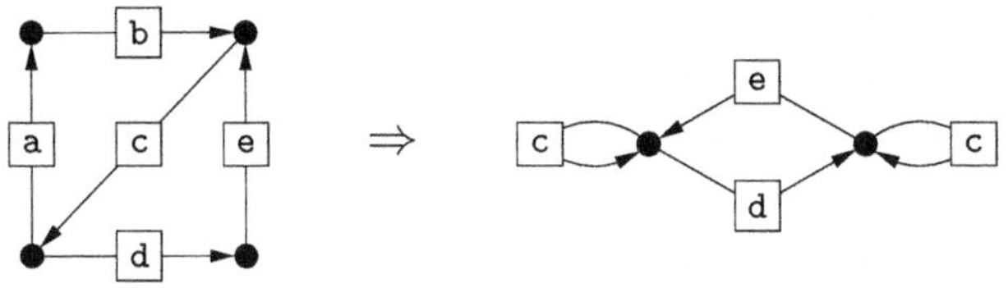

Fig. 4. An application of the rule of Figure 3

A double-pushout as in diagram (1) is called a *direct derivation* from G to H and may be denoted (by abuse of notation) by $G \Rightarrow_{r,f} H$ or just by $G \Rightarrow_r H$ or $G \Rightarrow H$. A *derivation* from G to H is a sequence of direct derivations $G = G_0 \Rightarrow \ldots \Rightarrow G_n = H$ for some $n \geq 0$ and may be denoted by $G \Rightarrow^* H$.

Definition 6 (Dangling condition). Given a rule $r\colon \langle L \hookleftarrow K \rightarrow R\rangle$ and a hypergraph G, a hypergraph morphism $f\colon L \rightarrow G$ satisfies the *dangling condition* if no hyperedge in $\mathrm{E}_G - f_\mathrm{E}(\mathrm{E}_L)$ is incident to a node in $f_\mathrm{V}(\mathrm{V}_L) - f_\mathrm{V}(\mathrm{V}_K)$.

In other words, the dangling condition guarantees that each attachment node of a hyperedge in $\mathrm{E}_G - f_\mathrm{E}(\mathrm{E}_L)$ belongs to either $\mathrm{V}_G - f_\mathrm{V}(\mathrm{V}_L)$ or $f_\mathrm{V}(\mathrm{V}_K)$.

With this condition the existence of a direct derivation can be characterized operationally.

Theorem 1 (Constructing a direct derivation [28]). *Let G and H be hypergraphs, $r\colon \langle L \hookleftarrow K \to^b R\rangle$ a rule and $f\colon L \hookrightarrow G$ an injective hypergraph morphism. Then $G \Rightarrow_{r,f} H$ if and only if f satisfies the dangling condition and H is isomorphic to the hypergraph M constructed as follows:*

- $\mathrm{V}_M = (\mathrm{V}_G - f_\mathrm{V}(\mathrm{V}_L)) + \mathrm{V}_R$ *and* $\mathrm{E}_M = (\mathrm{E}_G - f_\mathrm{E}(\mathrm{E}_L)) + \mathrm{E}_R$,[3]
- $\mathrm{mark}_M(v) = \underline{if}\ v \in \mathrm{V}_R\ \underline{then}\ \mathrm{mark}_R(v)\ \underline{else}\ \mathrm{mark}_G(v)$,
- $\mathrm{lab}_M(e) = \underline{if}\ e \in \mathrm{E}_R\ \underline{then}\ \mathrm{lab}_R(e)\ \underline{else}\ \mathrm{lab}_G(e)$, *and*
- $\mathrm{att}_M(e) = \underline{if}\ e \in \mathrm{E}_R\ \underline{then}\ \mathrm{att}_R(e)\ \underline{else}\ t^*(\mathrm{att}_G(e))$,

where the auxiliary function $t\colon (\mathrm{V}_G - f_\mathrm{V}(\mathrm{V}_L)) \cup f_\mathrm{V}(\mathrm{V}_K) \to \mathrm{V}_M$ *is defined by* $t(v) = \underline{if}\ v = f_\mathrm{V}(v')$ *for some* $v' \in \mathrm{V}_K\ \underline{then}\ b_\mathrm{V}(v')\ \underline{else}\ v$.

With every derivation $\Delta\colon G_0 \Rightarrow^* G_n$ a partial hypergraph morphism[4] can be associated that "follows" the items of G_0 through the derivation: this morphism is undefined for all items in G_0 that are removed by Δ, and maps all other items to the corresponding items in G_n.

Definition 7 (Track morphism). Given a direct derivation $G \Rightarrow H$ as in diagram (1), the *track morphism* $\mathrm{tr}_{G\Rightarrow H}\colon G \to H$ is the partial hypergraph morphism defined by

$$\mathrm{tr}_{G\Rightarrow H}(x) = \begin{cases} c'(c^{-1}(x)) & \text{if } x \in c(D), \\ \text{undefined} & \text{otherwise.} \end{cases}$$

Here $c\colon D \hookrightarrow G$ and $c'\colon D \to H$ are the morphisms in the lower row of (1) and $c^{-1}\colon c(D) \hookrightarrow D$ maps each item $c(x)$ to x.

The track morphism of a derivation $\Delta\colon G_0 \Rightarrow^* G_n$ is defined by $\mathrm{tr}_\Delta = \mathrm{id}_{G_0}$ if $n = 0$ and $\mathrm{tr}_\Delta = \mathrm{tr}_{G_1\Rightarrow^* G_n} \circ \mathrm{tr}_{G_0\Rightarrow G_1}$ otherwise, where id_{G_0} is the identity morphism on G_0.

Definition 8 (Hypergraph rewriting system). A *hypergraph rewriting system* $\langle\Sigma, \mathcal{R}\rangle$ consists of a signature Σ and a set $\mathcal{R}$ of rules over Σ. Such a system is *finite* if Σ_V, Σ_E and $\mathcal{R}$ are finite, and it is *injective* if all rules in $\mathcal{R}$ are injective.

The system $\langle\Sigma, \mathcal{R}\rangle$ is denoted by $\mathcal{R}$ if Σ is irrelevant or clear from the context. Given hypergraphs G and H over Σ such that there is a direct derivation $G \Rightarrow_r H$ with $r \in \mathcal{R}$, this is written $G \Rightarrow_\mathcal{R} H$. The system $\langle\Sigma, \mathcal{R}\rangle$ is a *graph rewriting system* if for each label l in Σ_E, $\mathrm{Type}(l)$ contains only strings of length two.

Example 1. Figure 5 shows a hypergraph rewriting system defining a class of control-flow graphs. A hypergraph belongs to the class if and only if the rules can

[3] "+" denotes the disjoint union of sets.

[4] A *partial hypergraph morphism* $f\colon G \to H$ is a hypergraph morphism from a subhypergraph of G to H.

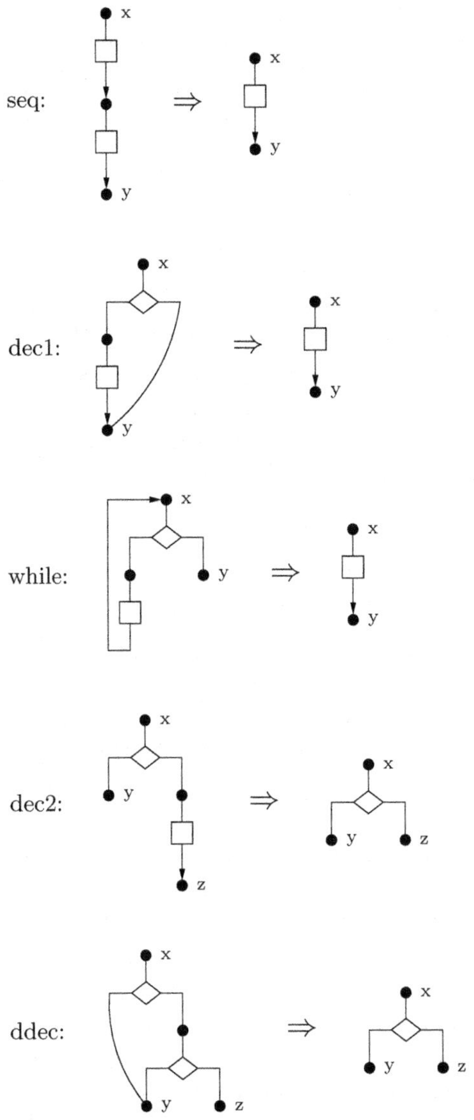

Fig. 5. Hypergraph rewriting system for flow-graph reduction

reduce it to the smallest flow graph, which is the hypergraph on the right-hand side of the rule seq. The underlying signature contains a single node label and two hyperedge labels which are graphically represented by hyperedges formed as squares and rhombs, where the order among the links of a rhomb is "top-left-right". The flow graphs defined in this way correspond to a subset of the so-called semi-structured flow graphs of Farrow, Kennedy and Zucconi [16]. This example will be continued as Example 3 where it is shown that the rewriting system is confluent. □

Since rules are matched injectively, one sometimes wants to include in a hypergraph rewriting system some or all of the homomorphic images of a rule.

Definition 9 (Quotient rule). A rule $r'\colon \langle L' \hookleftarrow K' \to R' \rangle$ is a *quotient* of a rule $r\colon \langle L \hookleftarrow K \to R \rangle$ if there are two pushouts of the form

$$\begin{array}{ccccc} L & \hookleftarrow & K & \longrightarrow & R \\ \downarrow & & \downarrow & & \downarrow \\ L' & \hookleftarrow & K' & \longrightarrow & R' \end{array} \qquad (2)$$

where the vertical hypergraph morphisms are surjective.

For example, Figure 6 shows a rule and its only proper quotient. A rule has—up to isomorphism—only finitely many quotients and hence the double-pushout approach with unrestricted matching morphisms can be simulated in the present setting by replacing each rule with its quotients. However, injective matching allows to select a subset of the quotients and this feature provides additional expressiveness [18].

Fig. 6. A rule and its quotient

3.2 Clipping and Embedding

For proving confluence of graph transformation via critical pairs in Section 4, it will be necessary both to restrict derivations by clipping off context and to extend derivations with context. The technical machinery for these operations is presented next.

Definition 10 (Instance of a derivation). Let the derivation $\Delta\colon G_0 \Rightarrow^* G_n$ be given by the pushouts (1),(1′),…,(n),(n') of Figure 7 and suppose there are pushouts ($\underline{1}$),($\underline{1}'$),…,($\underline{n}$),($\underline{n}'$) whose vertical morphisms are injective. Then the derivation $\Delta'\colon G'_0 \Rightarrow^* G'_n$ consisting of the composed pushouts (1)+($\underline{1}$),…,(n')+($\underline{n}'$)[5] is an *instance* of Δ *based* on the morphism $G_0 \hookrightarrow G'_0$. If moreover $G_0 \hookrightarrow G'_0$ is an isomorphism, then Δ and Δ' are *isomorphic* derivations.[6]

The Clipping Theorem below will show that given a derivation $\Delta'\colon G'_0 \Rightarrow^* G'_n$ and an injective morphism $G_0 \hookrightarrow G'_0$, Δ' can be restricted to a derivation $\Delta\colon G_0 \Rightarrow^* G_n$ if all items in G'_0 that at some stage are used by a rule application in Δ', belong to the image of G_0 in G'_0.

In what follows, given a direct derivation $G \Rightarrow H$ as in diagram (1), the subhypergraph $f(L)$ of G is denoted by Match($G \Rightarrow H$).

[5] See Lemma 14 in the Appendix for the composition of pushouts.

[6] In this case all the morphisms $G_i \hookrightarrow G'_i$ and $D_i \hookrightarrow D'_i$ are isomorphisms.

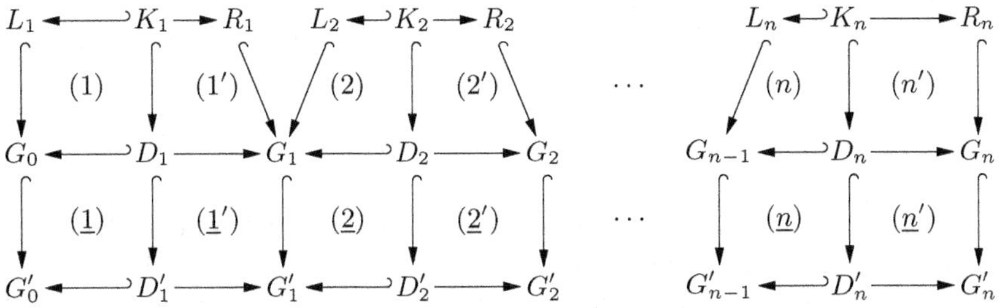

Fig. 7. Derivations $\Delta\colon G_0 \Rightarrow^* G_n$ and $\Delta'\colon G'_0 \Rightarrow^* G'_n$

Definition 11 (Use_Δ)**.** Given a derivation $\Delta\colon G_0 \Rightarrow^* G_n$, the subhypergraph Use_Δ of G_0 consists of all items x such that there is some $i \geq 0$ with $\mathrm{tr}_{G_0 \Rightarrow^* G_i}(x) \in \mathrm{Match}(G_i \Rightarrow G_{i+1})$.

Thus, Use_Δ contains those items of G_0 that at some point will occur in the image of the left-hand side of a rule. It is easily seen that these items constitute a subhypergraph of G_0. The following theorem was first proved in [23], in the setting of injective graph rewriting systems with unrestricted matching.

Theorem 2 (Clipping [28]). *Given a derivation $\Delta'\colon G' \Rightarrow^* H'$ and an injective hypergraph morphism $h\colon G \hookrightarrow G'$ such that $\mathrm{Use}_{\Delta'} \subseteq h(G)$, there exists a derivation $\Delta\colon G \Rightarrow^* H$ such that Δ' is an instance of Δ based on h.*

Next an embedding operation for derivations is considered which is inverse to the clipping operation. The associated theorem will refer to the "persistent" part of the start hypergraph of a derivation.

Definition 12 ($\mathrm{Persist}_\Delta$)**.** Given a derivation $\Delta\colon G \Rightarrow^* H$, $\mathrm{Persist}_\Delta$ is the subhypergraph of G consisting of all items x such that $\mathrm{tr}_{G \Rightarrow^* H}(x)$ is defined.

In other words, $\mathrm{Persist}_\Delta$ is the domain of definition of $\mathrm{tr}_{G \Rightarrow^* H}$. The Embedding Theorem given next allows to extend a derivation with context provided that context edges are not attached to non-persistent nodes. The result was originally established in [8, 23], for injective graph rewriting systems with unrestricted matching.

Theorem 3 (Embedding [28]). *Let $\Delta\colon G \Rightarrow^* H$ be a derivation, $h\colon G \hookrightarrow G'$ an injective hypergraph morphism and Boundary be the discrete subhypergraph of G consisting of all nodes x such that $h(x)$ is incident to a hyperedge in $G' - h(G)$. If Boundary $\subseteq \mathrm{Persist}_\Delta$, then there exists a derivation $\Delta'\colon G' \Rightarrow^* H'$ such that Δ' is an instance of Δ based on h. Moreover, there exists a pushout*

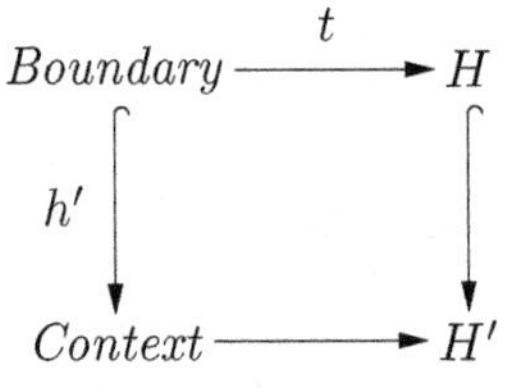

where Context is the subhypergraph $(G' - h(G)) \cup h(Boundary)$ *of* G', h' *is the restriction of* h *to Boundary and Context, and* t *is the restriction of* $\mathrm{tr}_{G \Rightarrow^* H}$ *to Boundary.*

3.3 Independence

Next a basic commutation result is presented showing that independent steps $H_1 \Leftarrow_{r_1} G \Rightarrow_{r_2} H_2$ give rise to steps of the form $H_1 \Rightarrow_{r_2} H \Leftarrow_{r_1} H_2$. Roughly speaking, the given steps are independent if the intersection of the left-hand sides of r_1 and r_2 in G consists of common interface items. If one of the rules is not injective, however, an additional injectivity condition is needed. In the following let r_i denote a rule $\langle L_i \hookleftarrow K_i \rightarrow R_i \rangle$, for $i = 1, 2$.

Definition 13 (Independence). Direct derivations $H_1 \Leftarrow_{r_1} G \Rightarrow_{r_2} H_2$ as in Figure 8 are *independent* if there are hypergraph morphisms $L_1 \rightarrow D_2$ and $L_2 \rightarrow D_1$ such that the following holds:
Commutativity: $L_1 \rightarrow D_2 \hookrightarrow G = L_1 \hookrightarrow G$ and $L_2 \rightarrow D_1 \hookrightarrow G = L_2 \hookrightarrow G$.
Injectivity: $L_1 \rightarrow D_2 \rightarrow H_2$ and $L_2 \rightarrow D_1 \rightarrow H_1$ are injective.

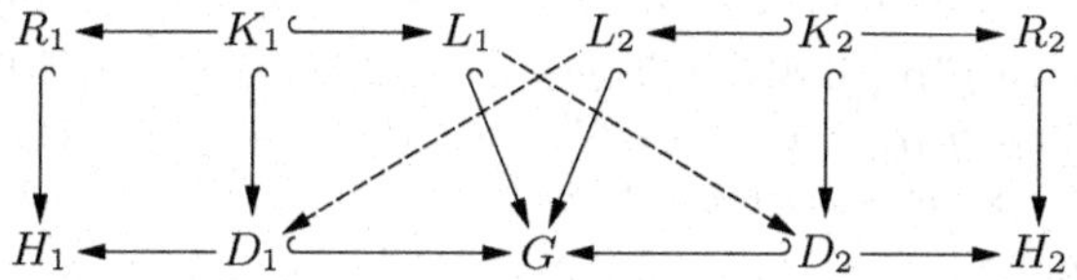

Fig. 8. Independence

If r_1 and r_2 are injective, the direct derivations of Figure 8 are independent if and only if the intersection of the two left-hand sides coincides with the intersection of the two interfaces.

Lemma 2 (Independence for injective rules). *Let* r_1 *and* r_2 *be injective rules. Then direct derivations* $H_1 \Leftarrow_{r_1, g_1} G \Rightarrow_{r_2, g_2} H_2$ *are independent if and only if* $g_1(L_1) \cap g_2(L_2) \subseteq g_1(K_1) \cap g_2(K_2)$.

Lemma 2 does not hold for non-injective rules as the injectivity condition of Definition 13 may be violated [18]. The following result was first proved in [11], for graph rewriting systems with unrestricted matching.

Theorem 4 (Commutativity [28]). *If* $H_1 \Leftarrow_{r_1} G \Rightarrow_{r_2} H_2$ *are independent direct derivations, then there exists an* H *such that* $H_1 \Rightarrow_{r_2} H \Leftarrow_{r_1} H_2$.

4 Confluence

4.1 Rewriting Modulo Isomorphism

In graph transformation, the structure and labelling of (hyper-)graphs matters rather than the identities of nodes and edges. Hence isomorphic hypergraphs are usually considered as equal. This, however, causes a subtle problem as confluence is defined via the reflexive-transitive closure $\Rightarrow^*$ which need not contain isomorphism. For example, consider the rule

$$r\colon \quad \text{(a)} \quad \Rightarrow \quad \text{(b)}$$

and two rewrite steps $R \Leftarrow_r L \Rightarrow_r R'$, where L is the left-hand side of r. Then R and R' are isomorphic but not necessarily equal, so the relation $\Rightarrow_r$ is not confluent in the sense of Definition 1(2). On the other hand, $\Rightarrow_r$ becomes confluent—in fact subcommutative—if it is considered as a relation on isomorphism classes of hypergraphs. A rigorous treatment of confluence requires either to consider the transformation of isomorphism classes of hypergraphs or—equivalently—to replace confluence with confluence modulo isomorphism. In what follows, $[G]$ denotes the isomorphism class $\{G' \mid G' \cong G\}$ of a hypergraph G.

Definition 14 (Rewriting modulo isomorphism). Given a hypergraph rewriting system $\langle \Sigma, \mathcal{R} \rangle$, the relation $\Rightarrow_{\mathcal{R},\cong}$ on isomorphism classes of hypergraphs over Σ is defined by: $[G] \Rightarrow_{\mathcal{R},\cong} [H]$ if there are hypergraphs G' and H' such that $G \cong G' \Rightarrow_{\mathcal{R}} H' \cong H$. The relation $\Rightarrow_{\mathcal{R},\cong}$ is referred to as *hypergraph rewriting modulo isomorphism.*

By the uniqueness of pushouts up to isomorphism (Lemma 13.3), it is clear that $[G] \Rightarrow_{\mathcal{R},\cong} [H]$ if and only if $G \Rightarrow_{\mathcal{R}} H$. But $[G] \Rightarrow^*_{\mathcal{R},\cong} [H]$ need not imply $G \Rightarrow^*_{\mathcal{R}} H$ since G and H may be distinct in the case $[G] = [H]$.

Definition 15 (Confluence of hypergraph rewriting systems). A hypergraph rewriting system $\langle \Sigma, \mathcal{R} \rangle$ is *confluent* (*locally confluent, subcommutative*) if the relation $\Rightarrow_{\mathcal{R},\cong}$ is confluent (locally confluent, subcommutative).

A drawback of hypergraph rewriting modulo isomorphism is that one loses access to nodes and hyperedges. Fortunately, confluence of $\Rightarrow_{\mathcal{R},\cong}$ can be characterized as confluence of $\Rightarrow_{\mathcal{R}}$ modulo isomorphism. (See Definition 1 for (local) confluence modulo isomorphism.)

Lemma 3. *Let $\langle \Sigma, \mathcal{R} \rangle$ be a hypergraph rewriting system.*

(1) *The following are equivalent:*
  - *$\langle \Sigma, \mathcal{R} \rangle$ is confluent.*
  - *The relation $\Rightarrow_{\mathcal{R}}$ is confluent modulo isomorphism.*
  - *For all hypergraphs G, G_1 and G_2 over Σ, $G_1 \Leftarrow^*_{\mathcal{R}} G \Rightarrow^*_{\mathcal{R}} G_2$ implies that there are hypergraphs H_1 and H_2 such that $G_1 \Rightarrow^*_{\mathcal{R}} H_1 \cong H_2 \Leftarrow^*_{\mathcal{R}} G_2$.*

(2) *$\langle \Sigma, \mathcal{R} \rangle$ is locally confluent if and only if $\Rightarrow_{\mathcal{R}}$ is locally confluent modulo isomorphism.*

Proof. By Lemma 13.3, $G \cong G' \Rightarrow_{\mathcal{R}} H' \cong H$ implies $G \Rightarrow_{\mathcal{R}} H$. The above characterizations are easy consequences of this fact. □

Note that—different from confluence—the relation $\Rightarrow_{\mathcal{R},\cong}$ is terminating if and only if $\Rightarrow_{\mathcal{R}}$ is terminating, as $[G] \Rightarrow_{\mathcal{R},\cong} [H]$ if and only if $G \Rightarrow_{\mathcal{R}} H$. A hypergraph rewriting system $\langle \Sigma, \mathcal{R} \rangle$ is said to be *terminating* if the relation $\Rightarrow_{\mathcal{R},\cong}$ (equivalently, $\Rightarrow_{\mathcal{R}}$) is terminating. The following lemma follows directly from Newman's Lemma (Lemma 1).

Lemma 4. *A terminating hypergraph rewriting system is confluent if and only if it is locally confluent.*

4.2 The Decision Problem

This section presents a reduction of the Post Correspondence Problem showing that confluence is undecidable for terminating graph rewriting systems. The precise result is as follows.

Theorem 5. *The following problem is undecidable in general:*

Instance: *A finite, injective and terminating graph rewriting system $\langle \Sigma, \mathcal{R} \rangle$ where Σ_{V} is a singleton.*
Question: *Is $\langle \Sigma, \mathcal{R} \rangle$ confluent?*

The rest of this section is used to prove Theorem 5, with a summary of the proof given at the end of the section. The plan is to encode every instance $\mathcal{I}$ of the Post Correspondence Problem (PCP) as a (finite, injective and) terminating graph rewriting system $\langle \Sigma(\mathcal{I}), \mathcal{R}(\mathcal{I}) \rangle$ that is confluent if and only if $\mathcal{I}$ does not have a solution.

Recall that the PCP is the problem to decide, given a nonempty list

$$\mathcal{I} = \langle (\alpha_1, \beta_1), \ldots, (\alpha_n, \beta_n) \rangle$$

of pairs of words over some finite alphabet Γ, whether there exists a nonempty sequence $i_1, \ldots, i_k$ of indices such that $\alpha_{i_1} \ldots \alpha_{i_k} = \beta_{i_1} \ldots \beta_{i_k}$. The list $\mathcal{I}$ is an *instance* of the PCP, and a sequence $i_1, \ldots, i_k$ with the above property is a *solution* of this instance. It is well-known that it is undecidable in general whether an instance of the PCP has a solution, see for example [33].

The following encoding of the PCP represents strings $a_1 \ldots a_m$ as graphs consisting of m consecutive edges labelled with $a_1, \ldots, a_m$, depicted as

which includes the case $m = 0$ where the graph consists of a single node.

Consider now an arbitrary instance $\mathcal{I} = \langle (\alpha_1, \beta_1), \ldots, (\alpha_n, \beta_n) \rangle$. Let $\Sigma(\mathcal{I})_{\mathrm{V}} = \{\bullet\}$ and $\Sigma(\mathcal{I})_{\mathrm{E}} = \Gamma \cup \{1, \ldots, n\} \cup \{\bowtie, \mathrm{A}, \mathrm{B}\}$ where it is assumed, without loss of generality, that the three latter sets are pairwise disjoint. The rule set $\mathcal{R}(\mathcal{I})$ is partitioned into subsets $\mathcal{R}_1(\mathcal{I})$ to $\mathcal{R}_4(\mathcal{I})$ which are presented by rule schemata in Figure 9 to Figure 12.

The rule schemata s_1 and s_2 of $\mathcal{R}_1$ enable divergent steps $T \Leftarrow_{\mathrm{s}_1} S \Rightarrow_{\mathrm{s}_2} U$ which represent the choice to create a loop labelled with $\bowtie$ or to check a possible

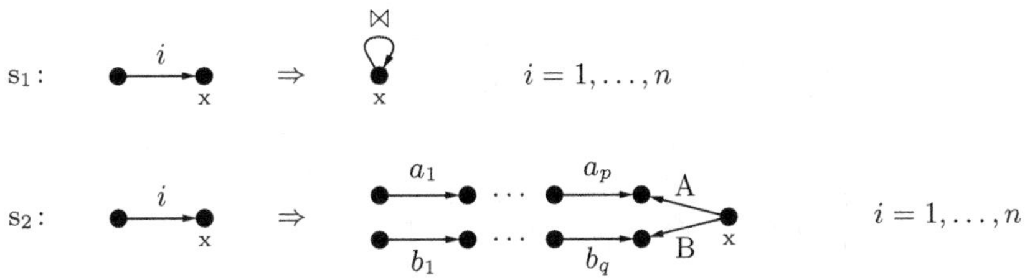

where $\alpha_i = a_1 \dots a_p$ and $\beta_i = b_1 \dots b_q$

Fig. 9. $\mathcal{R}_1(\mathcal{I})$

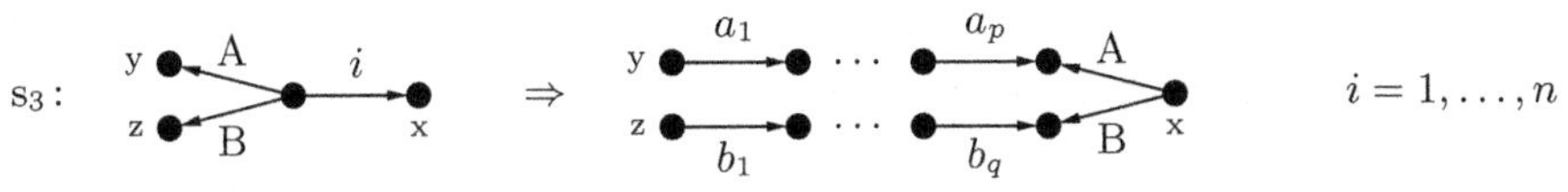

where $\alpha_i = a_1 \dots a_p$ and $\beta_i = b_1 \dots b_q$

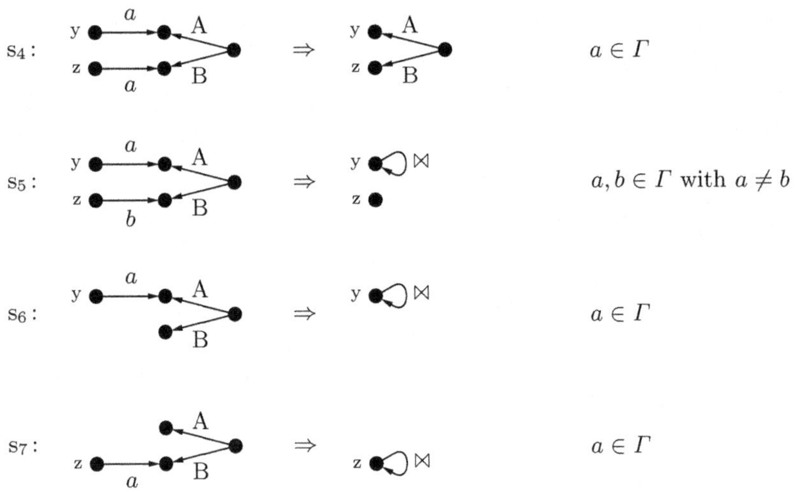

Fig. 10. $\mathcal{R}_2(\mathcal{I})$

solution of $\mathcal{I}$. The rules of $\mathcal{R}_2$ check whether a sequence of indices is a solution of $\mathcal{I}$, $\mathcal{R}_3$ detects ill-shaped graphs, and $\mathcal{R}_4$ performs "garbage collection" in the presence of a $\bowtie$-labelled loop.

We proceed by showing that $\langle \Sigma(\mathcal{I}), \mathcal{R}(\mathcal{I}) \rangle$ is terminating, and that it is confluent if and only if $\mathcal{I}$ does not have a solution.

Lemma 5. *The system $\langle \Sigma(\mathcal{I}), \mathcal{R}(\mathcal{I}) \rangle$ is terminating.*

Proof. Suppose that $\mathcal{R}(\mathcal{I})$ admits an infinite derivation $G_1 \Rightarrow G_2 \Rightarrow \dots$ Since no application of a rule in $\mathcal{R}(\mathcal{I})$ increases the number of edges with label in $\{1, \dots, n\}$, there is some $t \geq 1$ such that the number of these edges is the same

s8 : 

$l \in \Gamma \cup \{\bowtie, \mathrm{A}, \mathrm{B}\}$

s9 : 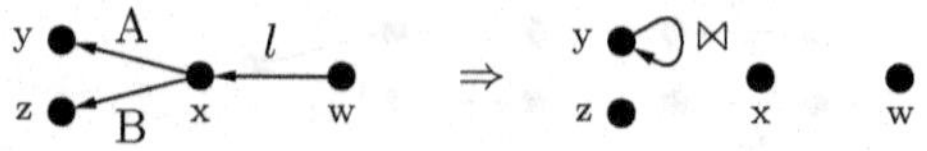

$l \in \Sigma(\mathcal{I})_\mathrm{E}$

including the quotient obtained by merging x and w

s10 : 

$i, j \in \{1, \ldots, n\}$

including the quotient obtained by merging y and z

Fig. 11. $\mathcal{R}_3(\mathcal{I})$

s11 : 

$l \in \Sigma(\mathcal{I})_\mathrm{E}$

including all quotients

s12 : 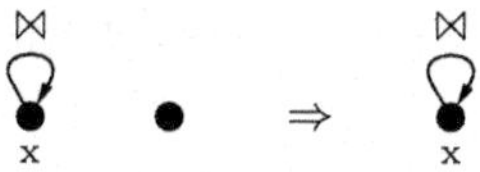

Fig. 12. $\mathcal{R}_4(\mathcal{I})$

in all G_i with $i \geq t$. It follows that the rule schemata s_1 to s_3 are not applied in $G_t \Rightarrow G_{t+1} \Rightarrow \ldots$ But all other rule schemata in $\mathcal{R}(\mathcal{I})$ decrease the sum of the numbers of nodes and edges, hence $G_t \Rightarrow G_{t+1} \Rightarrow \ldots$ cannot be infinite. □

The next four lemmata will show that the instance $\mathcal{I}$ has a solution if and only if $\langle \Sigma(\mathcal{I}), \mathcal{R}(\mathcal{I}) \rangle$ is not confluent. The "only if"-direction follows from the observation that every solution of $\mathcal{I}$—represented as a chain of edges—can be reduced to two non-isomorphic normal forms. Define graphs Join and Success as follows:

Join: 

Success: 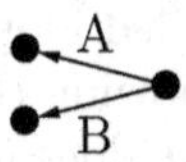

Lemma 6. *Every graph over $\Sigma(\mathcal{I})$ containing a $\bowtie$-labelled loop reduces to* Join.

Proof. Apply the rules in $\mathcal{R}_4(\mathcal{I})$ as long as possible. □

Lemma 7. *If $\mathcal{I}$ has a solution, then $\langle \Sigma(\mathcal{I}), \mathcal{R}(\mathcal{I}) \rangle$ is not confluent.*

Proof. Let $i_1, \ldots, i_k$ be a solution of $\mathcal{I}$. Then the graph

can be reduced to Join by first applying rule schema s_1 and then as long as possible the rules in $\mathcal{R}_4(\mathcal{I})$. On the other hand, by first applying rule schema s_2 and then as long as possible the rule schemata s_3 and s_4, the graph will reduce to Success. Since Join and Success are normal forms, $\langle \Sigma(\mathcal{I}), \mathcal{R}(\mathcal{I}) \rangle$ is not confluent. □

To complete the proof of Theorem 5, it remains to be shown that $\langle \Sigma(\mathcal{I}), \mathcal{R}(\mathcal{I}) \rangle$ is confluent if $\mathcal{I}$ does not have a solution. The next lemma provides a crucial argument for this proof.

Lemma 8. *If $\mathcal{I}$ does not have a solution, then for every graph G over $\Sigma(\mathcal{I})$, $G \Rightarrow_{\mathrm{s}_2} H$ implies $H \Rightarrow^*_{\mathcal{R}(\mathcal{I})}$* Join.

Proof. Let $G \Rightarrow_{\mathrm{s}_2} H$. Call a subgraph C of H an *index chain* of length k, $k \geq 1$, if C has the form

with $i_1, \ldots, i_k \in \{1, \ldots, n\}$, where only the rightmost node may be incident to edges not belonging to C.

Let now C be the longest index chain in G such that the leftmost edge of C is replaced by the step $G \Rightarrow_{\mathrm{s}_2} H$. Let $e_1, \ldots, e_k$ be the edges of C in left-to-right order and $\mathrm{lab}_C(e_j) = i_j$ for $j = 1, \ldots, k$. Then $H \Rightarrow^{k-1}_{\mathrm{s}_3} H'$ for some graph H' such that the jth step in $G \Rightarrow_{\mathrm{s}_2} H \Rightarrow^{k-1}_{\mathrm{s}_3} H'$ replaces e_j by two sequences of edges representing the strings α_{i_j} and β_{i_j}. Let v be the destination of edge e_k in C. In H', v is the source of two edges labelled with A and B which were created in the kth step. If v is incident to any other edges, then $H' \Rightarrow_{\mathcal{R}_3(\mathcal{I})} H''$ for some H'' and hence $H'' \Rightarrow^*$ Join by Lemma 6. Assume that v is incident to no other edges than the two edges labelled with A and B. The generated strings $\alpha_{i_1} \ldots \alpha_{i_k}$ and $\beta_{i_1} \ldots \beta_{i_k}$ are distinct as otherwise $i_1, \ldots, i_k$ would be a solution of $\mathcal{I}$. Let $H' \Rightarrow^* H''$ be a derivation in which the rule schemata s_4, s_5, s_6 and s_7 are applied as long as possible. This derivation must include an application of s_5, s_6 or s_7 because otherwise $\alpha_{i_1} \ldots \alpha_{i_k} = \beta_{i_1} \ldots \beta_{i_k}$. Thus H'' contains a $\bowtie$-edge and, by Lemma 6, reduces to Join. □

By Lemma 3 and Lemma 4, showing that $\langle \Sigma(\mathcal{I}), \mathcal{R}(\mathcal{I}) \rangle$ is confluent if $\mathcal{I}$ has no solution amounts to proving that $\Rightarrow_{\mathcal{R}(\mathcal{I})}$ is locally confluent modulo isomorphism. It will turn out that in every situation $H_1 \Leftarrow_{\mathcal{R}(\mathcal{I})} G \Rightarrow_{\mathcal{R}(\mathcal{I})} H_2$ where H_1 and H_2 are not isomorphic, either the steps are independent and hence can be commuted or H_1 and H_2 reduce to the graph Join.

Lemma 9. *If $\mathcal{I}$ does not have a solution, then $\mathcal{R}(\mathcal{I})$ is confluent.*

Proof. By Lemma 3 and Lemma 4, it suffices to show that $\Rightarrow_{\mathcal{R}(\mathcal{I})}$ is locally confluent modulo isomorphism since $\mathcal{R}(\mathcal{I})$ is terminating. Consider two direct derivations $H_1 \Leftarrow_{r_1,g_1} G \Rightarrow_{r_2,g_2} H_2$ by rules $r_i\colon \langle L_i \hookleftarrow K_i \rightarrow R_i \rangle \in \mathcal{R}(\mathcal{I})$, for $i = 1, 2$, such that $H_1 \not\cong H_2$. Moreover, assume that the two steps are not independent as otherwise they can be commuted by Theorem 4. By the injectivity of r_1 and r_2 and Lemma 2,

$$g_1(L_1) \cap g_2(L_2) \not\subseteq g_1(K_1) \cap g_2(K_2). \tag{3}$$

There are four cases.

Case 1: $r_1, r_2 \in \mathcal{R}_3(\mathcal{I}) \cup \mathcal{R}_4(\mathcal{I})$. Then both H_1 and H_2 contain a $\bowtie$-loop and hence, by Lemma 6, reduce to the graph Join.

Case 2: $r_1 \in \mathcal{R}_1(\mathcal{I}) \cup \mathcal{R}_2(\mathcal{I})$, $r_2 \in \mathcal{R}_3(\mathcal{I}) \cup \mathcal{R}_4(\mathcal{I})$. Then H_2 reduces to Join. If r_1 is an instance of rule schema s_1, s_5, s_6 or s_7, then H_1 contains a $\bowtie$-loop and reduces to Join, too. Let therefore r_1 be an instance of rule schema s_2, s_3 or s_4. By (3) and the dangling condition (Definition 6), four subcases remain. In each of these cases there will be a step $H_1 \Rightarrow_{\mathcal{R}_3(\mathcal{I})} H_1'$ and hence a derivation $H_1 \Rightarrow_{\mathcal{R}_3(\mathcal{I})} H_1' \Rightarrow^*_{\mathcal{R}_4(\mathcal{I})}$ Join.

Case 2.1: r_1 is an instance of s_2 and r_2 is an instance of s_9. Then H_1 contains a node with two outgoing A-edges and two outgoing B-edges, so s_8 is applicable to H_1.

Case 2.2: r_1 is an instance of s_2 and r_2 is an instance of s_{11}. Then H_1 contains a $\bowtie$-loop and hence s_{11} is applicable to H_1.

Case 2.3: r_1 is an instance of s_3 and r_2 is an instance of s_9. Again H_1 contains a node with two outgoing A-edges and two outgoing B-edges, so s_8 is applicable to H_1.

Case 2.4: r_1 is an instance of s_3 and r_2 is an instance of s_{11}. Then H_1 contains a $\bowtie$-loop so that s_{11} is applicable to H_1.

Case 3: $r_1 \in \mathcal{R}_3(\mathcal{I}) \cup \mathcal{R}_4(\mathcal{I})$, $r_2 \in \mathcal{R}_1(\mathcal{I}) \cup \mathcal{R}_2(\mathcal{I})$. This case is symmetric to Case 2.

Case 4: $r_1, r_2 \in \mathcal{R}_1(\mathcal{I}) \cup \mathcal{R}_2(\mathcal{I})$. Then one of the rules, say r_1, is an instance of s_1 and r_2 is an instance of s_2. By Lemma 6 and Lemma 8, both H_1 and H_2 reduce to Join. □

Proof of Theorem 5. For every instance $\mathcal{I}$ of the PCP, the system $\langle \Sigma(\mathcal{I}), \mathcal{R}(\mathcal{I}) \rangle$ is finite and injective and contains only one node label. Moreover, $\mathcal{R}(\mathcal{I})$ is terminating by Lemma 5. By Lemma 7 and Lemma 9, $\langle \Sigma(\mathcal{I}), \mathcal{R}(\mathcal{I}) \rangle$ is confluent if and only if $\mathcal{I}$ does not have a solution. This concludes the proof of Theorem 5 as the PCP is known to be undecidable [33]. □

4.3 Critical Pairs

The concept of a critical pair was introduced by Knuth and Bendix [22] who showed that confluence of a terminating term rewriting system can be tested by checking for each critical pair whether both terms have a common reduct. This subsection explores to what extent this idea can be adopted for the setting of hypergraph rewriting. The motivation is to ensure that arbitrary divergent steps $H_1 \Leftarrow_{r_1} G \Rightarrow_{r_2} H_2$ have a common reduct if this is the case for those steps where G represents a "critical overlap" of the left-hand sides of r_1 and r_2. By Theorem 4 (commutativity), such an overlap is critical only if the two steps are not independent. This suggests the following definition of a critical pair.

Definition 16 (Critical pair). Let $r_i\colon \langle L_i \hookleftarrow K_i \rightarrow R_i\rangle$ be rules, for $i = 1, 2$. A pair of direct derivations $U_1 \Leftarrow_{r_1,g_1} S \Rightarrow_{r_2,g_2} U_2$ is a *critical pair* if

(1) $S = g_1(L_1) \cup g_2(L_2)$ and
(2) the steps are not independent.

Moreover, $g_1 \neq g_2$ has to hold if $r_1 = r_2$.

Two critical pairs $U_1 \Leftarrow S \Rightarrow U_2$ and $U_1' \Leftarrow S' \Rightarrow U_2'$ are *isomorphic* if there is an isomorphism $f\colon S \rightarrow S'$ such that for $i = 1, 2$, $S' \Rightarrow U_i'$ is an instance of $S \Rightarrow U_i$ based on f. In the sequel, isomorphic critical pairs will be equated so that condition (1) guarantees that a finite set of rules has only a finite number of critical pairs.

By Theorem 4, hypergraph rewriting systems without critical pairs enjoy a strong commutation property which implies subcommutativity and hence confluence. This is fundamentally different from the situation for term rewriting systems where the absence of critical pairs guarantees only local confluence [20].

Theorem 6. *If $\langle \Sigma, \mathcal{R}\rangle$ is a hypergraph rewriting system without critical pairs, then $H_1 \Leftarrow_{\mathcal{R}} G \Rightarrow_{\mathcal{R}} H_2$ implies $H_1 \cong H_2$ or that there is a hypergraph H such that $H_1 \Rightarrow_{\mathcal{R}} H \Leftarrow_{\mathcal{R}} H_2$.*

Proof. Let $H_1 \Leftarrow_{r_1,g_1} G \Rightarrow_{r_2,g_2} H_2$. If the two steps are independent, then by Theorem 4 there are two steps of the form $H_1 \Rightarrow_{r_2} H \Leftarrow_{r_1} H_2$. Assume therefore that the given steps are not independent. By Theorem 2 (clipping), there are two restricted steps of the form

$$U_1 \Leftarrow_{r_1} g_1(L_1) \cup g_2(L_2) \Rightarrow_{r_2} U_2$$

such that the given steps are instances of the restricted steps. It is not difficult to show that the latter steps are not independent either. Hence, as there are no critical pairs, $r_1 = r_2$ and $g_1 = g_2$ must hold. It follows $H_1 \cong H_2$ since the result of a rewrite step is determined uniquely up to isomorphism. □

By the proof of the corollary below it follows that in the absence of critical pairs, $[H_1] \Leftarrow_{\mathcal{R},\cong} [G] \Rightarrow_{\mathcal{R},\cong} [H_2]$ with $[H_1] \neq [H_2]$ implies that there is a hypergraph H with $[H_1] \Rightarrow_{\mathcal{R},\cong} [H] \Leftarrow_{\mathcal{R},\cong} [H_2]$. In [4] this property is denoted by CR^1 (for arbitrary binary relations).

Corollary 1. *Hypergraph rewriting systems without critical pairs are subcommutative.*

Proof. Let $\langle \Sigma, \mathcal{R} \rangle$ be a hypergraph rewriting system without critical pairs and consider two steps $[H_1] \Leftarrow_{\mathcal{R},\cong} [G] \Rightarrow_{\mathcal{R},\cong} [H_2]$. Then there are hypergraphs G', G'', H_1' and H_2' such that $H_1 \cong H_1' \Leftarrow_{\mathcal{R}} G' \cong G \cong G'' \Rightarrow_{\mathcal{R}} H_2' \cong H_2$. Hence $H_1 \Leftarrow_{\mathcal{R}} G \Rightarrow_{\mathcal{R}} H_2$ by Lemma 13.3. Thus, by Theorem 6, $[H_1] = [H_2]$ or there is a hypergraph H such that $[H_1] \Rightarrow_{\mathcal{R},\cong} [H] \Leftarrow_{\mathcal{R},\cong} [H_2]$. So, in particular, $\Rightarrow_{\mathcal{R},\cong}$ is subcommutative. □

It follows that hypergraph rewriting systems without critical pairs are confluent, since subcommutative relations are confluent [20]. The next definition adapts the notion of a joinable critical pair from the setting of term rewriting to hypergraph rewriting.

Definition 17 (Joinability). A critical pair $U_1 \Leftarrow S \Rightarrow U_2$ is *joinable* if there exist hypergraphs W_1 and W_2 such that $U_1 \Rightarrow^* W_1 \cong W_2 \Leftarrow^* U_2$.

It turns out that the joinability of all critical pairs of a hypergraph rewriting system does *not* guarantee local confluence. This problem may occur if the track morphisms of the derivations $S \Rightarrow U_1 \Rightarrow^* W_1$ and $S \Rightarrow U_2 \Rightarrow^* W_2$ send some node in S to nodes in W_1 and W_2 that are not related by the isomorphism between W_1 and W_2.

Example 2. Let $\langle \Sigma, \mathcal{R} \rangle$ be a graph rewriting system consisting of the following two rules:

r_1:
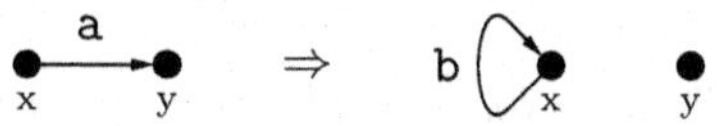

r_2:
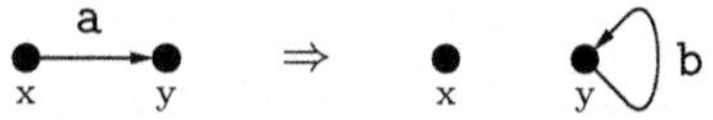

There is only one critical pair, which is clearly joinable:

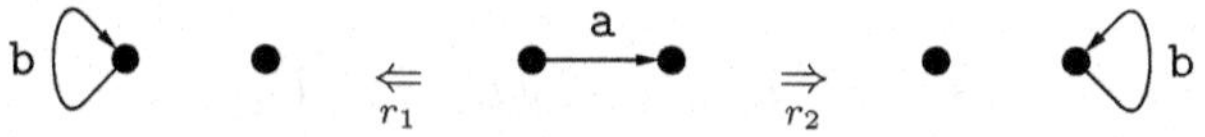

However, $\langle \Sigma, \mathcal{R} \rangle$ is not locally confluent:

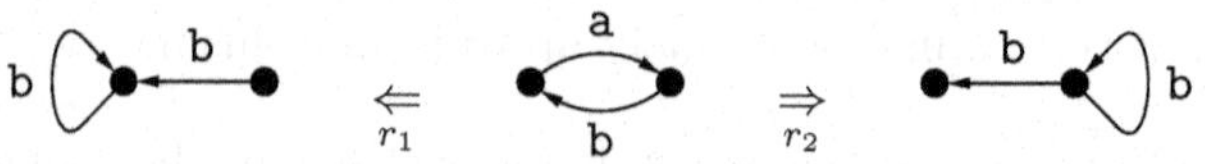

The outer hypergraphs are non-isomorphic normal forms and hence there are no joining derivations. The problem is that the embedding of the critical pair into context destroys the isomorphism between the outer hypergraphs. This is possible because the two steps of the critical pair—although resulting in isomorphic hypergraphs—have incompatible track morphisms.

One could try to overcome this problem by adding the rule

r_3:

which reduces the outer hypergraphs of the critical pair to the empty hypergraph. Adding r_3 does not create new critical pairs, so the only critical pair is "joinable by derivations with identical track morphisms". Still, this is not sufficient for local confluence modulo isomorphism: r_3 cannot be applied to the outer hypergraphs of the latter derivation pair because of the dangling condition for direct derivations. In other words, the joining derivations cannot be embedded into context as r_3 removes the nodes to which the context edge is attached. □

This example suggests that the joining derivations of a critical pair need to preserve certain nodes and possess track morphisms that are compatible with the isomorphism between the resulting hypergraphs.

Definition 18 (Strong joinability). Given a critical pair $\Gamma\colon U_1 \Leftarrow S \Rightarrow U_2$, let $\mathrm{Persist}_\Gamma = \mathrm{Persist}_{S \Rightarrow U_1} \cap \mathrm{Persist}_{S \Rightarrow U_2}$. Then Γ is *strongly joinable* if there are derivations $U_1 \Rightarrow^* W_1$, $U_2 \Rightarrow^* W_2$ and an isomorphism $f\colon W_1 \to W_2$ such that for each node v in $\mathrm{Persist}_\Gamma$,

(1) $\mathrm{tr}_{S \Rightarrow U_1 \Rightarrow^* W_1}(v)$ and $\mathrm{tr}_{S \Rightarrow U_2 \Rightarrow^* W_2}(v)$ are defined and
(2) $f_V(\mathrm{tr}_{S \Rightarrow U_1 \Rightarrow^* W_1}(v)) = \mathrm{tr}_{S \Rightarrow U_2 \Rightarrow^* W_2}(v)$.

So each node that is preserved by both $S \Rightarrow U_1$ and $S \Rightarrow U_2$ has to be preserved by $U_1 \Rightarrow^* W_1$ and $U_2 \Rightarrow^* W_2$ as well, and its descendants in W_1 and W_2 have to be related by the isomorphism f.

Lemma 10 (Critical Pair Lemma). *A hypergraph rewriting system is locally confluent if all its critical pairs are strongly joinable.*

Proof. Let $\langle \Sigma, \mathcal{R} \rangle$ be a hypergraph rewriting system such that all its critical pairs are strongly joinable. By Lemma 3 it suffices to show that $\Rightarrow_\mathcal{R}$ is locally confluent modulo isomorphism. Consider two steps $H_1 \Leftarrow_{r_1,g_1} G \Rightarrow_{r_2,g_2} H_2$ by rules $r_i\colon \langle L_i \hookleftarrow K_i \to R_i \rangle$, for $i = 1, 2$. If the two steps are independent, then Theorem 4 guarantees that there are two steps of the form $H_1 \Rightarrow_{r_2} H \Leftarrow_{r_1} H_2$. Assume therefore that the given steps are not independent. Assume further that $r_1 \neq r_2$ or $g_1 \neq g_2$ as otherwise $H_1 \cong H_2$. Let now $S = g_1(L_1) \cup g_2(L_2)$. Then, for $i = 1, 2$, $\mathrm{Use}_{G \Rightarrow H_i} \subseteq S$ and hence, by Theorem 2 (clipping), there are direct derivations $U_1 \Leftarrow_{r_1, g'_1} S \Rightarrow_{r_2, g'_2} U_2$ such that for $i = 1, 2$, $G \Rightarrow H_i$ is an instance of $S \Rightarrow U_i$ based on the inclusion $S \hookrightarrow G$. It is not difficult to check that since the steps $H_1 \Leftarrow G \Rightarrow H_2$ are not independent, $U_1 \Leftarrow S \Rightarrow U_2$ are not independent either and hence constitute a critical pair Γ. Thus, by assumption, there are derivations $U_1 \Rightarrow^* W_1$, $U_2 \Rightarrow^* W_2$ and an isomorphism $f\colon W_1 \to W_2$ such that for each node v in $\mathrm{Persist}_\Gamma$, $f_V(\mathrm{tr}_{S \Rightarrow U_1 \Rightarrow^* W_1}(v))$ and $\mathrm{tr}_{S \Rightarrow U_2 \Rightarrow^* W_2}(v)$ are defined and equal.

Let *Boundary* be the subhypergraph of S consisting of all nodes that are incident to a hyperedge in $G - S$. Then $Boundary \subseteq \text{Persist}_{G \Rightarrow H_1} \cap \text{Persist}_{G \Rightarrow H_2}$ since $G \Rightarrow H_1$ and $G \Rightarrow H_2$ satisfy the dangling condition. Hence

$$Boundary \subseteq (\text{Persist}_{G \Rightarrow H_1} \cap \text{Persist}_{G \Rightarrow H_2}) \cap S = \text{Persist}_{\Gamma}.$$

Thus $\text{tr}_{S \Rightarrow U_1 \Rightarrow^* W_1}$ and $\text{tr}_{S \Rightarrow U_2 \Rightarrow^* W_2}$ are defined for all nodes in *Boundary*. So $Boundary \subseteq \text{Persist}_{S \Rightarrow U_i \Rightarrow^* W_i}$ for $i = 1, 2$ and hence, by Theorem 3 (embedding), there are derivations $G \Rightarrow_{r_1, g_1''} H_1' \Rightarrow^* M_1$ and $G \Rightarrow_{r_2, g_2''} H_2' \Rightarrow^* M_2$ that are instances of $S \Rightarrow_{r_1, g_1'} U_1 \Rightarrow^* W_1$ and $S \Rightarrow_{r_2, g_2'} U_2 \Rightarrow^* W_2$, respectively. Since both instances are based on the inclusion $S \hookrightarrow G$, g_i'' is the extension of g_i' to G and hence $g_i'' = g_i$, for $i = 1, 2$. It follows $H_1' \cong H_1$ and $H_2' \cong H_2$; thus for $i = 1, 2$, $H_i \Rightarrow^* M_i$ or $H_i \cong M_i$. So it remains to show $M_1 \cong M_2$ for $\Rightarrow_{\mathcal{R}}$ to be locally confluent modulo isomorphism. By Theorem 3, for $i = 1, 2$ there is a pushout

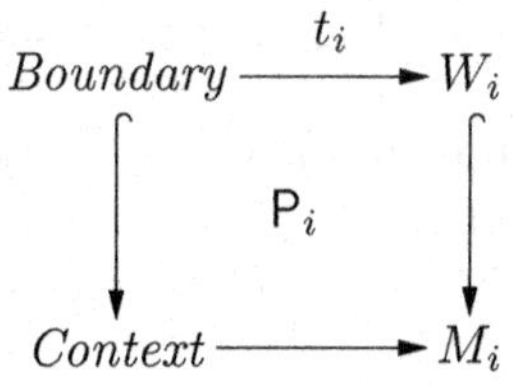

where $Context = (G - S) \cup Boundary$, $Boundary \hookrightarrow Context$ is the inclusion of *Boundary* in *Context*, and t_i is the restriction of $\text{tr}_{S \Rightarrow U_i \Rightarrow^* W_i}$ to *Boundary*. By assumption, $f \circ t_1 = t_2$. So $Boundary \hookrightarrow Context \to M_2 = Boundary \to^{t_2} W_2 \hookrightarrow M_2 = Boundary \to^{t_1} W_1 \to^{f} W_2 \hookrightarrow M_2$. Hence there is a unique morphism $M_1 \to M_2$ such that the diagram in Figure 13 commutes, where the

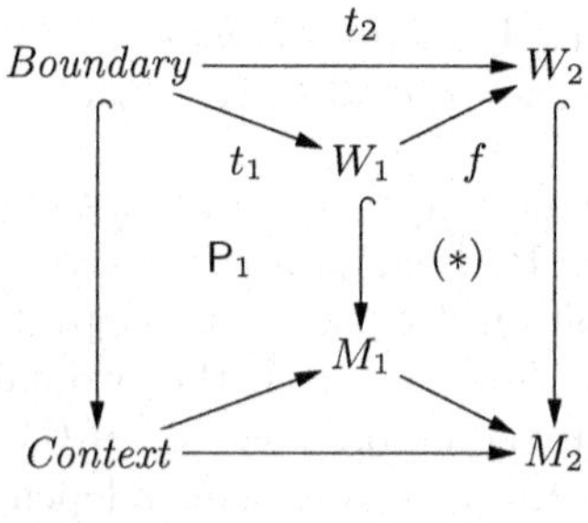

Fig. 13. Decomposing pushout P_2

outer diagram is the pushout P_2. So $(*)$ is a pushout by Lemma 14.2. Since f is an isomorphism, Lemma 13.2 guarantees that $M_1 \to M_2$ is an isomorphism as well. This concludes the proof of Lemma 10. □

Combining the Critical Pair Lemma with Newman's Lemma (Lemma 1) yields a sufficient condition for the confluence of terminating hypergraph rewriting systems.

Theorem 7. *A terminating hypergraph rewriting system is confluent if all its critical pairs are strongly joinable.*

Example 3. This example continues Example 1 by applying Theorem 7 to the hypergraph rewriting system of Figure 5. That system is terminating since each of the rules reduces the size of a hypergraph it is applied to. Figure 14 shows that all critical pairs of the system are strongly joinable (where track morphisms are indicated by node names). Note that for each critical pair $\Gamma\colon U_1 \Leftarrow S \Rightarrow U_2$, $\mathrm{Persist}_\Gamma$ is a proper subset of V_S. For instance, the persistent nodes of the topmost pair are w and z; hence the isomorphism between the outer hypergraphs of this pair is compatible with the track morphisms in the way required by Definition 18.

Thus, by Theorem 7, the hypergraph rewriting system of Example 1 is confluent. As a consequence, membership in the class of flow graphs defined by the system can be checked by a backtracking-free reduction algorithm: an input hypergraph G is reduced to its unique normal form $\mathrm{N}(G)$ by applying the rules in any order; G is a flow graph if and only if $\mathrm{N}(G)$ is the flow graph on the right-hand side of rule seq. □

The converse of Theorem 7 does not hold because if all critical pairs of terminating and confluent systems were strongly joinable, confluence of terminating systems could be checked by testing critical pairs for strong joinability—contradicting Theorem 5. The next example gives two terminating and confluent systems having critical pairs that are not strongly joinable.

Example 4. Let the graph rewriting system $\langle \Sigma, \mathcal{R} \rangle$ consist of singletons Σ_V and Σ_E, and the following rules:

This system is terminating because every rule application decreases the number of edges by one. To see that it is confluent, consider two derivations $H_1 \Leftarrow^*_{\mathcal{R}} G \Rightarrow^*_{\mathcal{R}} H_2$. Then G, H_1 and H_2 have the same number of nodes. Hence $H_1 \Rightarrow^*_{\mathcal{R}} H'_1 \cong H'_2 \Leftarrow^*_{\mathcal{R}} H_2$, where H'_1 and H'_2 consist of $|V_G|$ nodes and either no edges (if G is loop-free) or one loop and no other edges. So the system is confluent. But the following critical pair[7] is not strongly joinable:

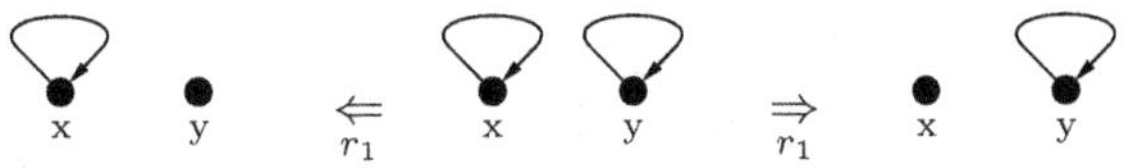

[7] The track morphisms are indicated by node names.

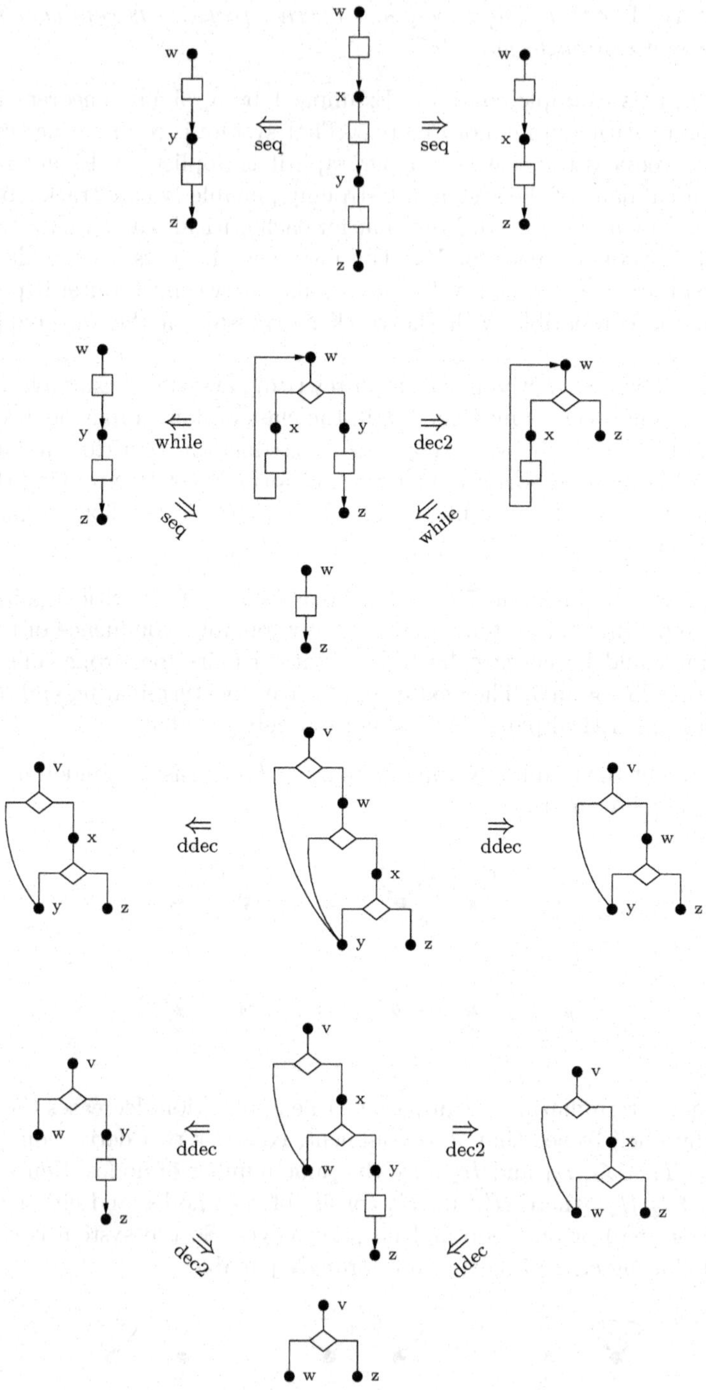

Fig. 14. The critical pairs of the rules of Figure 5

This is because the outer graphs are normal forms and the isomorphism between them is not compatible with the track morphisms of the rewrite steps.

A similar situation can arise in the system $\langle \Sigma(\mathcal{I}), \mathcal{R}(\mathcal{I}) \rangle$ of the previous subsection which encodes the Post Correspondence Problem. That system is confluent if the instance $\mathcal{I}$ of the PCP has no solution. The rule schemata s_1 and s_2 give rise to the critical pair

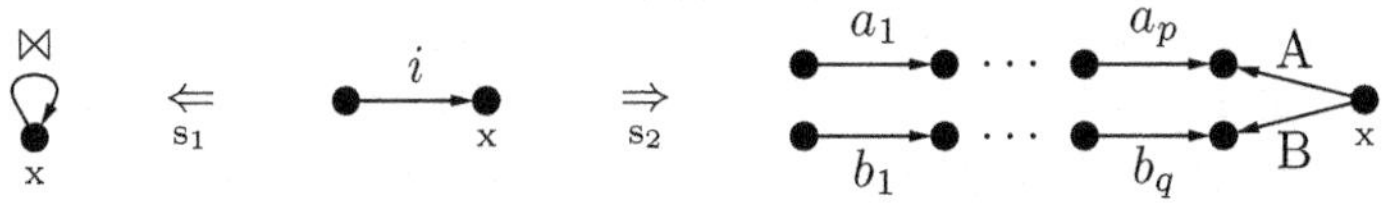

which is joinable if $\mathcal{I}$ has no solution, as then the graph on the right reduces to the graph on the left (see Lemma 9). However, this critical pair is not strongly joinable because the derivation from the right graph to the left graph deletes the node x which is preserved by both steps of the critical pair. □

5 Related Work and Conclusion

This section mentions some related work and a couple of topics for future work.

The proof idea for showing that confluence of terminating systems is undecidable (Theorem 5) was inspired by Kapur's, Narendran's and Otto's proof that ground-confluence is undecidable for terminating term rewriting systems [21].

The phenomenon that the joinability of all critical pairs need not imply local confluence of the rewrite relation refutes the critical pair lemma in [30]. In [26] the problem is avoided by imposing the strong restriction that distinct nodes in a graph must not have the same label.

The critical pair lemma of [27] was adopted to the so-called single-pushout approach to graph transformation in [24]. In [19] a critical pair lemma for a certain kind of attributed graph transformation (in the double-pushout approach) was presented. An abstract critical pair lemma in the setting of so-called adhesive high-level replacement systems was given in [10] and specialised in [14] to a form of attributed graph transformation (different from the aforementioned).

Future research should establish sufficient conditions under which all critical pairs of a confluent (hyper-)graph rewriting system are strongly joinable. For a finite and terminating system satisfying such a condition, confluence can be decided by checking all critical pairs for strong joinability.

Another application of the Critical Pair Lemma could be the completion of non-confluent systems, in analogy to the well-known completion procedure for term rewriting systems invented by Knuth and Bendix [22]. Such a procedure would add rules to a (hyper-)graph rewriting system until all critical pairs are strongly joinable. The procedure should preserve both the equivalence generated by the rewrite relation and termination.

Dedication. This paper is dedicated to Jan Willem Klop on the occasion of his 60th birthday. Since 1997, I have been a regular visitor of Jan Willem at CWI, the University of Nijmegen, and the Free University of Amsterdam. I am grateful for Jan Willem's hospitality and collaboration during all this time. The present paper is not directly concerned with our common topic of interest—term graph rewriting—but confluence of various forms of rewriting plays a prominent role in Jan Willem's scientific work.

Appendix: Pushouts

This appendix presents the definition and construction of hypergraph pushouts and summarises some facts that are used in the proofs of Section 4.

Definition 19 (Pushout). Given two hypergraph morphisms $A \to B$ and $A \to C$, a hypergraph D together with two hypergraph morphisms $B \to D$ and $C \to D$ is a *pushout* of $A \to B$ and $A \to C$ if the following conditions are satisfied:

Commutativity: $A \to B \to D = A \to C \to D$.

Universal property: For all hypergraphs D' and hypergraph morphisms $B \to D'$ and $C \to D'$ such that $A \to B \to D' = A \to C \to D'$, there is a unique morphism $D \to D'$ such that $B \to D \to D' = B \to D'$ and $C \to D \to D' = C \to D'$. (See the right part of Figure 15.)

In this case the diagram on the left of Figure 15 is also called a pushout.

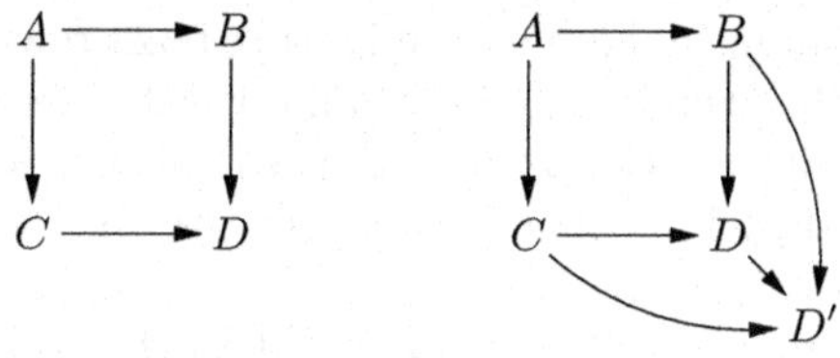

Fig. 15. A pushout diagram (on the left)

Intuitively, hypergraph D is obtained by gluing together B and C in a common part A. In particular, if $A \to C$ is injective, then D can be constructed from C by replacing the image of A with B. The following pushout construction assumes that one of the given morphisms is injective, which is the case for the two pushouts of a direct derivation as defined in Definition 5.

Lemma 11 (Pushout construction). *Let $b: A \to B$ and $c: A \hookrightarrow C$ be hypergraph morphisms such that c is injective. Then a pushout*

$$
\begin{array}{ccc}
A & \xrightarrow{\,b\,} & B \\
{\scriptstyle c}\downarrow & & \downarrow{\scriptstyle f} \\
C & \xrightarrow[\,g\,]{} & D
\end{array}
$$

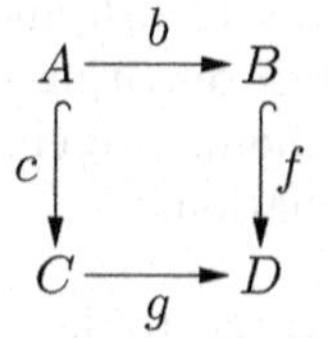

can be constructed as follows:

- $V_D = (V_C - c_V(V_A)) + V_B$ *and* $E_D = (E_C - c_E(E_A)) + E_B$;
  $\mathrm{mark}_D(v) = \underline{\mathrm{if}}\ v \in V_B\ \underline{\mathrm{then}}\ \mathrm{mark}_B(v)\ \underline{\mathrm{else}}\ \mathrm{mark}_C(v)$;
  $\mathrm{lab}_D(e) = \underline{\mathrm{if}}\ e \in E_B\ \underline{\mathrm{then}}\ \mathrm{lab}_B(e)\ \underline{\mathrm{else}}\ \mathrm{lab}_C(e)$;
  $\mathrm{att}_D(e) = \underline{\mathrm{if}}\ e \in E_B\ \underline{\mathrm{then}}\ \mathrm{att}_B(e)\ \underline{\mathrm{else}}\ g_V^*(\mathrm{att}_C(e))$, *where* $g_V\colon V_C \to V_D$ *is defined below.*
- $f(x) = x$, *separately for nodes and hyperedges.*
- $g(x) = \underline{\mathrm{if}}\ x = c(x')$ *for some* x' *in* A $\underline{\mathrm{then}}\ b(x')\ \underline{\mathrm{else}}\ x$, *separately for nodes and hyperedges.*

Proof. Analogous to the corresponding proof for graphs, see [9]. □

Definition 20 (Pushout complement). Given two hypergraph morphisms $A \to B$ and $B \to D$, a hypergraph C together with two morphisms $A \to C$ and $C \to D$ is a *pushout complement* of $A \to B$ and $B \to D$ if diagram (1) in Figure 16 is a pushout.

The following lemma gives a sufficient and necessary condition for the existence of the pushout complement in case $B \to D$ is injective (see [13] for the general case), viz. the dangling condition of Definition 6. A condition for the uniqueness of pushout complements is given in Lemma 13.4.

Lemma 12 (Pushout complement construction). *Let* $b\colon A \to B$ *and* $f\colon B \hookrightarrow D$ *be hypergraph morphisms such that* f *is injective. Then* b *and* f *possess a pushout complement if and only if no hyperedge in* $E_D - f_E(E_B)$ *is incident to a node in* $f_V(V_B) - f_V(b_V(V_A))$*. In this case a pushout complement* $A \to C \hookrightarrow D$ *can be constructed as follows:*

- C *is the subhypergraph of* D *with nodes* $(V_D - f_V(V_B)) \cup f_V(b_V(V_A))$ *and edges* $(E_D - f_E(E_B)) \cup f_E(b_E(E_A))$.
- $C \hookrightarrow D$ *is the inclusion of* C *in* D.
- $A \to C$ *is the restriction of* $A \to B \hookrightarrow D$ *to* C.

Proof. Analogous to the corresponding proof for relational structures in [13]. □

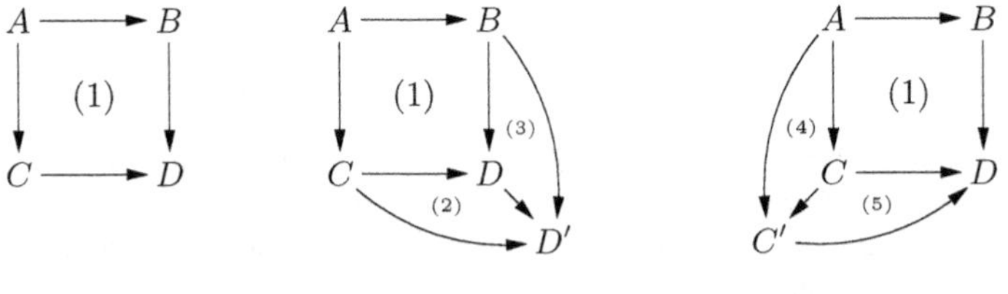

Fig. 16. Uniqueness of pushouts and pushout complements

Lemma 13 (Pushout properties). *Let diagram* (1) *in Figure 16 be a hypergraph pushout. Then the following holds:*

1. Joint surjectivity. *Each item in D has a preimage in B or C.*
2. Injectivity and surjectivity. *If $A \to B$ is injective (surjective), then $C \to D$ is injective (surjective) as well.*
3. Uniqueness of pushouts. *A hypergraph D' together with morphisms $B \to D'$ and $C \to D'$ is a pushout of $A \to B$ and $A \to C$ if and only if there is an isomorphism $D \to D'$ such that the triangles (2) and (3) in Figure 16 commute.*
4. Uniqueness of pushout complements. *If C' together with $A \to C'$ and $C' \to D$ is a pushout complement of $A \to B$ and $B \to D$, and $A \to B$ is injective, then there is an isomorphism $C \to C'$ such that the triangles (4) and (5) in Figure 16 commute.*

Proof. The first and the fourth property are shown (for graphs) in [12] and [32], respectively. The second property holds for set pushouts [1] and hence also for hypergraph pushouts. The third property holds in every category [1]. □

Lemma 14 (Composition and decomposition of pushouts). *Let the diagrams in Figure 17 consist of hypergraph morphisms. Then the following holds:*

1. *If (1) and (2) are pushouts, then (1)+(2) is a pushout.*
2. *If (1)+(2) and (1) are pushouts and (2) commutes, then (2) is a pushout.*
3. *If (1)+(2) and (2) are pushouts, (1) commutes and $B \to E$ is injective, then (1) is a pushout.*

Proof. The first two properties hold in every category, the third is proved (for graphs) in [12]. □

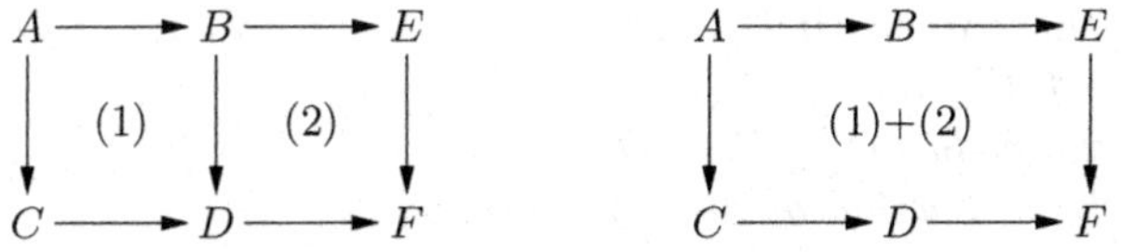

Fig. 17. Pushout composition and decomposition

References

1. Jiří Adámek, Horst Herrlich, and George Strecker. *Abstract and Concrete Categories.* Wiley, 1990.
2. Stefan Arnborg, Bruno Courcelle, Andrzej Proskurowski, and Detlef Seese. An algebraic theory of graph reduction. *Journal of the ACM*, 40(5):1134–1164, 1993.
3. Franz Baader and Tobias Nipkow. *Term Rewriting and All That.* Cambridge University Press, 1998.
4. Marc Bezem, Jan Willem Klop, and Roel de Vrijer, editors. *Term Rewriting Systems.* Cambridge University Press, 2003.
5. Hans L. Bodlaender and Babette de Fluiter. Reduction algorithms for constructing solutions in graphs with small treewidth. In *Proc. Computing and Combinatorics*, volume 1090 of *Lecture Notes in Computer Science*, pages 199–208, 1996.

6. Andrea Corradini, Ugo Montanari, Francesca Rossi, Hartmut Ehrig, Reiko Heckel, and Michael Löwe. Algebraic approaches to graph transformation — Part I: Basic concepts and double pushout approach. In G. Rozenberg, editor, *Handbook of Graph Grammars and Computing by Graph Transformation*, volume 1, chapter 3, pages 163–245. World Scientific, 1997.
7. Babette de Fluiter. *Algorithms for Graphs of Small Treewidth.* Dissertation, Universiteit Utrecht, 1997.
8. Hartmut Ehrig. Embedding theorems in the algebraic theory of graph-grammars. In *Proc. Fundamentals of Computation Theory*, volume 56 of *Lecture Notes in Computer Science*, pages 245–255. Springer-Verlag, 1977.
9. Hartmut Ehrig. Introduction to the algebraic theory of graph grammars. In *Proc. Graph-Grammars and Their Application to Computer Science and Biology*, volume 73 of *Lecture Notes in Computer Science*, pages 1–69. Springer-Verlag, 1979.
10. Hartmut Ehrig, Annegret Habel, Julia Padberg, and Ulrike Prange. Adhesive high-level replacement categories and systems. In *Proc. International Conference on Graph Transformation (ICGT 2004)*, volume 3256 of *Lecture Notes in Computer Science*, pages 144–160. Springer-Verlag, 2004.
11. Hartmut Ehrig and Hans-Jörg Kreowski. Parallelism of manipulations in multidimensional information structures. In *Proc. Mathematical Foundations of Computer Science*, volume 45 of *Lecture Notes in Computer Science*, pages 284–293. Springer-Verlag, 1976.
12. Hartmut Ehrig and Hans-Jörg Kreowski. Pushout-properties: An analysis of gluing constructions for graphs. *Mathematische Nachrichten*, 91:135–149, 1979.
13. Hartmut Ehrig, Hans-Jörg Kreowski, Andrea Maggiolo-Schettini, Barry K. Rosen, and Jozef Winkowski. Transformations of structures: an algebraic approach. *Mathematical Systems Theory*, 14:305–334, 1981.
14. Hartmut Ehrig, Ulrike Prange, and Gabriele Taentzer. Fundamental theory for typed attributed graph transformation. In *Proc. International Conference on Graph Transformation (ICGT 2004)*, volume 3256 of *Lecture Notes in Computer Science*, pages 161–177. Springer-Verlag, 2004.
15. Claudia Ermel, Michael Rudolf, and Gabi Taentzer. The AGG approach: Language and environment. In H. Ehrig, G. Engels, H.-J. Kreowski, and G. Rozenberg, editors, *Handbook of Graph Grammars and Computing by Graph Transformation*, volume 2, chapter 14, pages 551–603. World Scientific, 1999.
16. Rodney Farrow, Ken Kennedy, and Linda Zucconi. Graph grammars and global program data flow analysis. In *Proc. 17th Annual Symposium on Foundations of Computer Science*, pages 42–56. IEEE, 1976.
17. Joseph A. Goguen, Jim W. Thatcher, and Eric G. Wagner. An initial algebra approach to the specification, correctness, and implementation of abstract data types. In R. T. Yeh, editor, *Current Trends in Programming Methodology, Volume 4: Data Structuring*, pages 80–149. Prentice Hall, 1978.
18. Annegret Habel, Jürgen Müller, and Detlef Plump. Double-pushout graph transformation revisited. *Mathematical Structures in Computer Science*, 11(5):637–688, 2001.
19. Reiko Heckel, Jochen Malte Küster, and Gabi Taentzer. Confluence of typed attributed graph transformation systems. In *Proc. International Conference on Graph Transformation (ICGT 2002)*, volume 2505 of *Lecture Notes in Computer Science*, pages 161–176. Springer-Verlag, 2002.
20. Gérard Huet. Confluent reductions: Abstract properties and applications to term rewriting systems. *Journal of the ACM*, 27(4):797–821, 1980.

21. Deepak Kapur, Paliath Narendran, and Friedrich Otto. On ground-confluence of term rewriting systems. *Information and Computation*, 86:14–31, 1990.
22. Donald E. Knuth and Peter B. Bendix. Simple word problems in universal algebras. In J. Leech, editor, *Computational Problems in Abstract Algebras*, pages 263–297. Pergamon Press, 1970.
23. Hans-Jörg Kreowski. Manipulationen von Graphmanipulationen. Doctoral dissertation, Technische Universität Berlin, 1977. In German.
24. Michael Löwe and Jürgen Müller. Critical pair analysis in single-pushout graph rewriting. In Gabriel Valiente Feruglio and Francesc Roselló Llompart, editors, *Proc. Colloquium on Graph Transformation and its Application in Computer Science*, pages 71–77. Technical report B-19, Universitat de les Illes Balears, 1995.
25. M.H.A. Newman. On theories with a combinatorial definition of "equivalence". *Annals of Mathematics*, 43(2):223–243, 1942.
26. Yasuyoshi Okada and Masahiro Hayashi. Graph rewriting systems and their application to network reliability analysis. In *Proc. Graph-Theoretic Concepts in Computer Science*, volume 570 of *Lecture Notes in Computer Science*. Springer-Verlag, 1992.
27. Detlef Plump. Hypergraph rewriting: Critical pairs and undecidability of confluence. In Ronan Sleep, Rinus Plasmeijer, and Marko van Eekelen, editors, *Term Graph Rewriting: Theory and Practice*, chapter 15, pages 201–213. John Wiley, 1993.
28. Detlef Plump. *Computing by Graph Rewriting*. Habilitation thesis, Universität Bremen, Fachbereich Mathematik und Informatik, 1999.
29. Detlef Plump and Sandra Steinert. Towards graph programs for graph algorithms. In *Proc. Int. Conference on Graph Transformation (ICGT 2004)*, volume 3256 of *Lecture Notes in Computer Science*, pages 128–143. Springer-Verlag, 2004.
30. Jean-Claude Raoult. On graph rewritings. *Theoretical Computer Science*, 32:1–24, 1984.
31. Jan Rekers and Andy Schürr. Defining and parsing visual languages with layered graph grammars. *Journal of Visual Languages and Computing*, 8(1):27–55, 1997.
32. Barry K. Rosen. Deriving graphs from graphs by applying a production. *Acta Informatica*, 4:337–357, 1975.
33. Grzegorz Rozenberg and Arto Salomaa. *Cornerstones of Undecidability*. Prentice Hall, 1994.
34. Andy Schürr, Andreas Winter, and Albert Zündorf. The PROGRES approach: Language and environment. In H. Ehrig, G. Engels, H.-J. Kreowski, and G. Rozenberg, editors, *Handbook of Graph Grammars and Computing by Graph Transformation*, volume 2, chapter 13, pages 487–550. World Scientific, 1999.

Compositional Reasoning for Probabilistic Finite-State Behaviors

Yuxin Deng[1,⋆], Catuscia Palamidessi[2,⋆⋆], and Jun Pang[2]

[1] INRIA Sophia-Antipolis and Université Paris 7, France
[2] INRIA Futurs and LIX, École Polytechnique, France

Abstract. We study a process algebra which combines both nondeterministic and probabilistic behavior in the style of Segala and Lynch's simple probabilistic automata. We consider strong bisimulation and observational equivalence, and provide complete axiomatizations for a language that includes parallel composition and (guarded) recursion. The presence of the parallel composition introduces various technical difficulties and some restrictions are necessary in order to achieve complete axiomatizations.

1 Introduction

Process algebras, also known as process calculi, are a powerful mathematical model for the specification and verification of concurrent systems. They provide a formal apparatus for representing and reasoning about the behaviors of distributed systems, algorithms and protocols in a compositional way. Some of the most prominent representants of these formalisms are CCS [27], ACP [8, 6], and CSP [21].

The axiomatic theories of process algebra provide an elegant way for proving properties of systems. Both a system and its desired external behavior can be expressed as process terms. The correctness of the system can then be verified by proving that these two terms are equivalent.

In a process algebra typically there are only a few operators, such as action prefix, summation (nondeterministic choice), recursion and parallel composition. The latter is particularly important for concurrency, as it allows to specify the structural properties of systems composed of several interacting parts. For example, a typical communication protocol for data transferring involves two agents S and R, representing the sender and the receiver, and two lossy channels K and L between them (see Figure 1). The behavior of each of these four components can be described as a process term in a chosen process algebra, and then they are all put together in parallel to form the complete view of the protocol. The parallel composition operator captures both the interleaving behaviors and the possible synchronization of the components. The external behavior of the

⋆ Supported by the EU project PROFUNDIS.
⋆⋆ Partially supported by the projet Rossignol of the ACI Sécurité Informatique (Ministère de la recherche et nouvelles technologies).

A. Middeldorp et al. (Eds.): Processes... (Klop Festschrift), LNCS 3838, pp. 309–337, 2005.

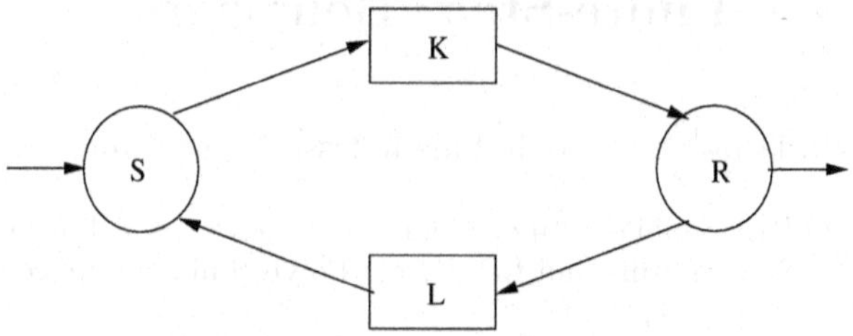

Fig. 1. A communication protocol

protocol can be specified as a FIFO queue. The equivalence proof between the protocol and its external behavior is established by equational reasoning based on axiomatization, hiding internal behavior, using fairness assumption, and the other feasible methods (see e.g. [9, 17]).

Developing a both complete and sound axiomatization for a chosen bisimulation relation over a process algebra expressing finite-state processes has been a research focus for the process algebra community. This led to a wealth of classical results in the literature. Milner [26, 28] gave complete axiomatizations of both strong bisimilarity and observational equivalence for a core CCS (not containing the parallel composition operator) with both unguarded and guarded recursion. Bergstra and Klop [10] axiomatized observational equivalence in an alternative way by using an interesting graph rewriting technique. Hennessy and Milner [20] offered a complete equational axiomatization of strong bisimulation over the recursion free fragment of CCS. To deal with parallel composition, they used the so-called *expansion law*, which is an equation schema with a countably infinite number of instances. Bergstra and Klop [8] gave a finite equational axiomatization of the merge operator (as the parallel composition in CCS) using the auxiliary left merge and communication merge operators. An interesting essay on equational axiomatizations of parallel composition can be found in [2].

Having both recursion and parallel composition in a process algebra complicates the matters to establish a complete axiomatization, mostly because this can give rise to infinite-state systems even with the guardedness condition. For example, let E be the expression $\mu_X(a.(X \mid b))$, then we have the infinite transition graph starting from E in Figure 2. Milner pointed out in [28] that in order to have a complete axiomatization for CCS with both recursion and parallel composition, a sufficient condition is that the parallel composition does not occur in the body of any recursive expression.

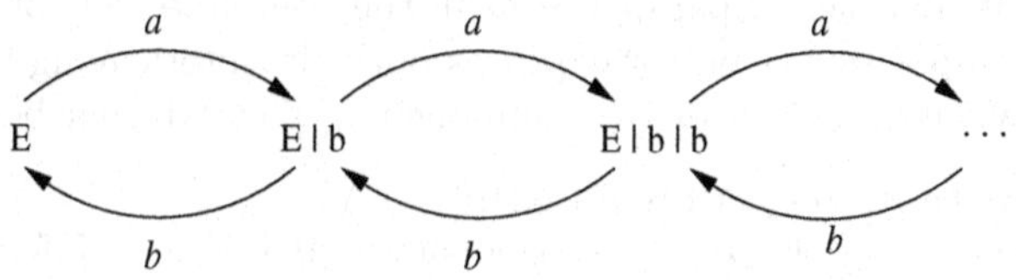

Fig. 2. The transition graph of E

In this paper we relax this restriction by requiring, instead, that free variables do not appear in the scope of parallel composition. A similar restriction was adopted, independently, in [5]. In that paper, Baeten and Bravetti considered a generic process algebra of which CCS, CSP and ACP are subalgebras. Finite-stateness is achieved by requiring that recursion variables do not occur in the scope of *static operators*, which include the parallel composition. Our work and [5] are, in a sense, incomparable, because we consider a probabilistic and nondeterministic framework (as explained in the rest of this introduction) with CCS-like communication, while [5] considers a purely nondeterministic paradigm, but more general than our nondeterministic fragment. The same restriction already appeared in [11], for a nondeterministic process algebra with CSP multiway synchronization.

Recently there has been an increasing interest in the area of formal methods for the specification and analysis of probabilistic behaviors, as exhibited for instance in randomized, distributed and fault-tolerant systems. The notion of probabilistic bisimulation is introduced first by Larsen and Skou [22]. Later many variant behavioural equivalences have been defined for various probabilistic models. A representative model for analyzing probabilistic systems is provided by Segala and Lynch's simple probabilistic automata [30], which take into account both probabilistic and nondeterministic behavior and which have been successfully adopted in the studies of distributed algorithms [24, 29] and practical communication protocols [33]. An axiomatization for the finite sequential fragment of simple probabilistic automata has been provided by Bandini and Segala in [7]. Following this line of research, Deng and Palamidessi [16, 15] have given a sound and complete axiomatization for a larger language, which includes the recursion operator.

In this paper, we improve on [16, 15] by considering also the parallel composition. To our knowledge, it is the first time that an axiomatization for a probabilistic and nondeterministic process algebra with both recursion and parallel operator has been attempted. Similar to the case of classical process algebra, once we have both parallel composition and recursion, the equational axiomatization of strong bisimulation and observational equivalence turns out to be quite complicated to achieve.

To obtain the completeness of the axiomatizations, we develop a probabilistic version of the expansion law to eliminate all occurrences of parallel composition. In order to do that, we heavily rely on the condition that only closed terms are put in parallel (cf. Theorem 3).

Concerning soundness, it turns out to be particularly difficult to prove that strong and weak bisimilarities are closed under the parallel composition operator. Our approach is to manipulate equivalences of distributions on terms. An important property that we exploit in our proofs is Lemma 2, which says that if two distributions are equivalent with respect to an equivalence relation $\mathcal{R}$, then there is a uniform way to extend them so that the resulting distributions in parallel contexts are equivalent with respect to another equivalence relation

$\mathcal{R}^{\mathsf{l}}$. It turns out that if $\mathcal{R}$ is instantiated as strong or weak bisimilarity then $\mathcal{R}^{\mathsf{l}}$ is a subset of $\mathcal{R}$, thus $\mathcal{R}^{\mathsf{l}}$ also relates bisimilar expressions.

Structure of the paper. In the next section we briefly recall some basic concepts and definitions about probabilistic distributions. In Section 3, we present the syntax and operational semantics of a probabilistic process calculus. Next, we give the notions of strong and weak behavioral equivalences in Section 4. We provide complete axiomatizations for strong bisimilarity and observational equivalence in Sections 5 and 6 respectively, restricted to guarded expressions in the second case. In Section 7, we conclude and discuss some related work not yet mentioned in the introduction. Detailed proofs of the main propositions in Section 4 are in the Appendix.

2 Preliminaries

Let S be a set. A function $\eta : S \mapsto [0,1]$ is called a *discrete probability distribution*, or *distribution* for short, on S if the *support* of η, defined as $spt(\eta) = \{x \in S \mid \eta(x) > 0\}$, is finite or countably infinite and $\sum_{x \in S} \eta(x) = 1$. We denote by $\mathcal{P}(S)$ the set of distributions over S. If η is a distribution with finite support and $V \subseteq spt(\eta)$ we use the set $\{s_i : \eta(s_i)\}_{s_i \in V}$ to enumerate the probability associated with each element of V. The constructor $\uplus$ on this kind of sets is defined as follows.

$$\{s_i : p_i\}_{i\in I} \uplus \{s : p\} = \begin{cases} \{s_i : p_i\}_{i\in I\setminus j} \cup \{s_j : (p_j + p)\} & \text{if } s = s_j \text{ for some } j \in I \\ \{s_i : p_i\}_{i\in I} \cup \{s : p\} & \text{otherwise.} \end{cases}$$

$$\{s_i : p_i\}_{i\in I} \uplus \{t_j : p_j\}_{j\in 1..n} = (\{s_i : p_i\}_{i\in I} \uplus \{t_1 : p_1\}) \uplus \{t_j : p_j\}_{j\in 2..n}$$

Given some distributions $\eta_1, ..., \eta_n$ on S and some real numbers $r_1, ..., r_n \in [0,1]$ with $\sum_{i\in 1..n} r_i = 1$, we define the *convex combination* $r_1\eta_1 + ... + r_n\eta_n$ of $\eta_1, ..., \eta_n$ to be the distribution η such that $\eta(s) = \sum_{i\in 1..n} r_i\eta_i(s)$, for each $s \in S$.

A *simple probabilistic automaton* is a tuple $(S, s, \Sigma, \mathcal{T})$, where S is a set of *states*, $s \in S$ is a *start state*, Σ is a set of *actions*, and $\mathcal{T} \subseteq S \times \Sigma \times \mathcal{P}(S)$ is a *transition relation.* Informally, a simple probabilistic automaton is like an ordinary automaton except that a labeled transition leads to a probabilistic distribution over a set of states instead of a single state. Simple probabilistic automata are used in this paper to give operational semantics of our probabilistic process calculus.

3 Probabilistic Process Calculus

We assume a countable set of variables, $Var = \{X, Y, ...\}$, and a countable set of atomic actions, $\mathcal{A} = \{a, b, ...\}$. Given a special action τ not in $\mathcal{A}$, we let $u, v, ...$

range over the set of *actions*, $Act = \mathcal{A} \cup \overline{\mathcal{A}} \cup \{\tau\}$, and let $\alpha, \beta, ...$ range over the set $Var \cup Act$. The class of expressions $\mathcal{E}$ is defined by the following syntax:

$$E, F ::= u.\bigoplus_{i\in 1..n} p_i E_i \;\Big|\; \sum_{i\in 1..m} E_i \;\Big|\; E \mid F \;\Big|\; X \;\Big|\; \mu_X E$$

Here $\bigoplus_{i\in 1..n} p_i E_i$ stands for a *probabilistic choice* operator, where the p_i's represent positive probabilities, i.e., they satisfy $p_i \in (0,1]$ and $\sum_{i\in 1..n} p_i = 1$. When $n = 0$ we abbreviate the probabilistic choice as $\mathbf{0}$; when $n = 1$ we abbreviate it as E_1. Sometimes we are interested in certain branches of the probabilistic choice; in this case we write $\bigoplus_{i\in 1..n} p_i E_i$ as $p_1 E_1 \oplus ... \oplus p_n E_n$ or $(\bigoplus_{i\in 1..(n-1)} p_i E_i) \oplus p_n E_n$ where $\bigoplus_{i\in 1..(n-1)} p_i E_i$ abbreviates (with a slight abuse of notation) $p_1 E_1 \oplus ... \oplus p_{n-1} E_{n-1}$. The second construction $\sum_{i\in 1..m} E_i$ stands for *nondeterministic choice*, and occasionally we may write it as $E_1 + ... + E_m$. As in CCS we let variables range over process expressions. The notation μ_X stands for a recursion which binds the variable X. We shall use $fv(E)$ for the set of free variables (i.e., not bound by any μ_X) in E. As explained in the introduction, we require that only closed expressions are put in parallel composition, i.e., in $E \mid F$ we have $fv(E \mid F) = \emptyset$. As usual we identify expressions which differ only by a change of bound variables. We shall write $E\{F_1, ..., F_n/X_1, ..., X_n\}$ or $E\{\widetilde{F}/\widetilde{X}\}$ for the result of simultaneously substituting F_i for each occurrence of X_i in E ($1 \leq i \leq n$), renaming bound variables if necessary.

Definition 1. *The variable X is* weakly guarded *(resp.* guarded*) in E if every free occurrence of X in E occurs within some subexpression $u.F$ (resp. $a.F$ or $\bar{a}.F$), otherwise X is* weakly unguarded *(resp.* unguarded*) in E.*

The operational semantics of an expression E is defined as a simple probabilistic automaton whose states are the expressions reachable from E and the transition relation is defined by the axioms and inference rules in Table 1, where $E \xrightarrow{\alpha} \eta$ describes a transition that, by performing an action or exposing a free variable, leaves from E and leads to a distribution η over $\mathcal{E}$. The symmetric rules of par and com are omitted.

Finitary weak transitions are defined as in [7]. We abstract away finitely many invisible actions that occur before or after the appearance of a single visible

Table 1. Strong transitions

var	$X \xrightarrow{X} \{\mathbf{0} : 1\}$	psum	$u.\bigoplus_{i\in 1..n} p_i E_i \xrightarrow{u} \biguplus_{i\in 1..n}\{E_i : p_i\}$
rec	$\dfrac{E\{\mu_X E/X\} \xrightarrow{\alpha} \eta}{\mu_X E \xrightarrow{\alpha} \eta}$	nsum	$\dfrac{E_j \xrightarrow{\alpha} \eta}{\sum_{i\in 1..m} E_i \xrightarrow{\alpha} \eta}$ for some $j \in 1..m$
par	$\dfrac{E \xrightarrow{\alpha} \{E_i : p_i\}_i}{E \mid F \xrightarrow{\alpha} \{E_i \mid F : p_i\}_i}$	com	$\dfrac{E \xrightarrow{a} \{E_i : p_i\}_{i\in I} \qquad F \xrightarrow{\bar{a}} \{F_j : q_j\}_{j\in J}}{E \mid F \xrightarrow{\tau} \{E_i \mid F_j : p_i q_j\}_{i\in I, j\in J}}$

Table 2. Weak transitions

$$\textsf{wea1}\ E \Longrightarrow \{E:1\} \qquad \textsf{wea2}\ \frac{E \xrightarrow{\tau} \eta}{E \Longrightarrow \eta} \qquad \textsf{wea3}\ \frac{E \xrightarrow{\alpha} \eta}{E \stackrel{\alpha}{\Longrightarrow} \eta}$$

$$\textsf{wea4}\ \frac{E \stackrel{\alpha}{\Longrightarrow} \{E_i : p_i\}_{i\in I} \quad \forall i \in I : E_i \Longrightarrow \{E_{ij} : p_{ij}\}_{j\in J_i}}{E \stackrel{\alpha}{\Longrightarrow} \{E_{ij} : p_i p_{ij}\}_{i\in I, j\in J_i}}$$

$$\textsf{wea5}\ \frac{E \Longrightarrow \{E_i : p_i\}_{i\in I} \quad \forall i \in I : E_i \stackrel{\alpha}{\Longrightarrow} \{E_{ij} : p_{ij}\}_{j\in J_i}}{E \stackrel{\alpha}{\Longrightarrow} \{E_{ij} : p_i p_{ij}\}_{i\in I, j\in J_i}}$$

action or a variable. It is easy to see that if $E \stackrel{X}{\Longrightarrow} \eta$ then $\eta = \{\mathbf{0} : 1\}$. We use the notation $\stackrel{\hat{\alpha}}{\Longrightarrow}$ to stand for $\stackrel{\alpha}{\Longrightarrow}$ if $\alpha \neq \tau$, for $\Longrightarrow$ otherwise. We also define a *weak combined transition*: $E \stackrel{\hat{\alpha}}{\Longrightarrow}_c \eta$ if there exists a collection $\{\eta_i, r_i\}_{i\in 1..n}$ of distributions and probabilities such that $\sum_{i\in 1..n} r_i = 1$, $\eta = r_1\eta_1 + ... + r_n\eta_n$ and $E \stackrel{\hat{\alpha}}{\Longrightarrow} \eta_i$ for each $i \in 1..n$. Similarly we write $E \stackrel{\alpha}{\Longrightarrow}_c \eta$ if every component is a "normal" (i.e., non-virtual) weak transition, namely, $E \stackrel{\alpha}{\Longrightarrow} \eta_i$ for all $i \leq n$.

4 Behavioral Equivalences

To define behavioral equivalences in probabilistic process algebra, it is customary to consider equivalence of distributions with respect to equivalence relations on expressions.

4.1 Equivalence of Distributions

If η is a distribution on S and $V \subseteq S$, we write $\eta(V)$ for $\sum_{s\in V} \eta(s)$. We lift an equivalence relation on $\mathcal{E}$ to an equivalence relation between distributions over $\mathcal{E}$ in the following way.

Definition 2. *Given two distributions η_1 and η_2 over $\mathcal{E}$, we say that they are equivalent w.r.t. an equivalence relation $\mathcal{R}$ on $\mathcal{E}$, written $\eta_1 \equiv_{\mathcal{R}} \eta_2$, if*

$$\forall V \in \mathcal{E}/\mathcal{R} : \eta_1(V) = \eta_2(V).$$

The following property is simple but important as it underpins many other results in the rest of the paper.

Lemma 1. *If $\eta_1 \equiv_{\mathcal{R}_1} \eta_2$ and $\mathcal{R}_1 \subseteq \mathcal{R}_2$ then $\eta_1 \equiv_{\mathcal{R}_2} \eta_2$.*

Given an equivalence relation $\mathcal{R}$, we construct two relations:

$$\mathcal{R}_G \stackrel{\text{def}}{=} \{(E \mid G,\ F \mid G) \bigm| E\ \mathcal{R}\ F\}$$
$$\mathcal{R}^{\mid} \stackrel{\text{def}}{=} \bigcup\{\mathcal{R}_G \bigm| G \in \mathcal{E}\}.$$

Clearly $\mathcal{R}_G$ and $\mathcal{R}^{|}$ are also equivalence relations. If $V \in \mathcal{E}/_{\mathcal{R}_G}$ then we write $V^{\backslash G}$ for the set $\{E \mid E \mid G \in V\}$. It is easy to see that if $V \in \mathcal{E}/_{\mathcal{R}^{|}}$ then there exists some expression G such that $V \in \mathcal{E}/_{\mathcal{R}_G}$. Furthermore, we observe that $V \in \mathcal{E}/_{\mathcal{R}_G}$ iff $V^{\backslash G} \in \mathcal{E}/_{\mathcal{R}}$. Suppose $\theta_1 = \{E_i : p_i\}_{i \in I}$ and $\theta_2 = \{F_j : q_j\}_{j \in J}$, we introduce the following notation:

$$\theta_1 \mid \theta_2 \stackrel{\text{def}}{=} \{E_i \mid F_j : p_i q_j\}_{i \in I, j \in J}.$$

The following lemma is crucial for showing the congruence property of strong bisimilarity and observational equivalence (cf. Section 4.4). It says that if two distributions θ_1 and θ_2 are equivalent w.r.t. an equivalence relation $\mathcal{R}$, then there is a uniform way to extend the two distributions so that the resulting distributions on composed terms are equivalent w.r.t. another equivalence relation $\mathcal{R}^{|}$.

Lemma 2. *If $\theta_1 \equiv_{\mathcal{R}} \theta_2$ then $(\theta_1 \mid \theta) \equiv_{\mathcal{R}^{|}} (\theta_2 \mid \theta)$.*

Proof. Let $\theta = \{G_k : p_k\}_{k \in K}$. Without loss of generality, we assume that if $i, j \in K$ and $i \neq j$ then $G_i \neq G_j$. For any $V \in \mathcal{E}/_{\mathcal{R}^{|}}$ there exists some expression G such that $V \in \mathcal{E}/_{\mathcal{R}_G}$. There are two cases:

1. if $G \neq G_k$ for all $k \in K$, then $(\theta_1 \mid \theta)(V) = 0 = (\theta_2 \mid \theta)(V)$;
2. if $G = G_k$ for some $k \in K$, then $(\theta_1 \mid \theta)(V) = r_k \theta_1(V^{\backslash G_k}) = r_k \theta_2(V^{\backslash G_k}) = (\theta_2 \mid \theta)(V)$.

In summary, $(\theta_1 \mid \theta)(V) = (\theta_2 \mid \theta)(V)$ for any $V \in \mathcal{E}/_{\mathcal{R}^{|}}$, i.e., $(\theta_1 \mid \theta) \equiv_{\mathcal{R}^{|}} (\theta_2 \mid \theta)$, which is the required result. □

Corollary 1. *If $\theta_1 \equiv_{\mathcal{R}} \theta_2$, $\theta_1' \equiv_{\mathcal{R}} \theta_2'$ and $\mathcal{R}$ is closed under parallel composition, then $(\theta_1 \mid \theta_1') \equiv_{\mathcal{R}} (\theta_2 \mid \theta_2')$.*

Proof. If $\mathcal{R}$ is closed under parallel composition, then $\mathcal{R}^{|} \subseteq \mathcal{R}$. By Lemma 1, we can state Lemma 2 as: if $\theta_1 \equiv_{\mathcal{R}} \theta_2$ then $(\theta_1 \mid \theta) \equiv_{\mathcal{R}} (\theta_2 \mid \theta)$. Similarly we can establish a symmetric property: if $\theta_1 \equiv_{\mathcal{R}} \theta_2$ then $(\theta \mid \theta_1) \equiv_{\mathcal{R}} (\theta \mid \theta_2)$. As a consequence we have $(\theta_1 \mid \theta_1') \equiv_{\mathcal{R}} (\theta_2 \mid \theta_1') \equiv_{\mathcal{R}} (\theta_2 \mid \theta_2')$. □

4.2 Behavioral Equivalences

Strong bisimulation is defined by requiring equivalence of distributions at every step. Because of the way equivalence of distributions is defined, we need to restrict to bisimulations which are equivalence relations.

Definition 3. *An equivalence relation $\mathcal{R} \subseteq \mathcal{E} \times \mathcal{E}$ is a* strong bisimulation *if $E \; \mathcal{R} \; F$ implies:*

– *whenever $E \xrightarrow{\alpha} \eta_1$, there exists η_2 such that $F \xrightarrow{\alpha} \eta_2$ and $\eta_1 \equiv_{\mathcal{R}} \eta_2$.*

Two expressions E, F are strong bisimilar, *written $E \sim F$, if there exists a strong bisimulation $\mathcal{R}$ s.t. $E \; \mathcal{R} \; F$.*

We have shown in [16, 15] that to define weak equivalences it is necessary to use weak combined transitions[1], so weak probabilistic bisimulation is given in the following way.

Definition 4. *An equivalence relation* $\mathcal{R} \subseteq \mathcal{E} \times \mathcal{E}$ *is a* weak probabilistic bisimulation *if* $E\ \mathcal{R}\ F$ *implies:*

- *whenever* $E \xrightarrow{\alpha} \eta_1$, *there exists* η_2 *such that* $F \overset{\hat{\alpha}}{\Longrightarrow}_c \eta_2$ *and* $\eta_1 \equiv_{\mathcal{R}} \eta_2$.

We write $E \approx F$ *whenever there exists a weak probabilistic bisimulation* $\mathcal{R}$ *s.t.* $E\ \mathcal{R}\ F$.

As usual, observational equivalence is defined in terms of weak probabilistic bisimulation.

Definition 5. *Two expressions* E, F *are* observationally equivalent, *written* $E \simeq F$, *if*

1. *whenever* $E \xrightarrow{\alpha} \eta_1$, *there exists* η_2 *such that* $F \overset{\alpha}{\Longrightarrow}_c \eta_2$ *and* $\eta_1 \equiv_{\approx} \eta_2$.
2. *whenever* $F \xrightarrow{\alpha} \eta_2$, *there exists* η_1 *such that* $E \overset{\alpha}{\Longrightarrow}_c \eta_1$ *and* $\eta_1 \equiv_{\approx} \eta_2$.

One can check that all the relations defined above are indeed equivalence relations and we have the inclusion ordering: $\sim \subsetneq \simeq \subsetneq \approx$.

Example 1. Consider the following expressions:

$$\begin{aligned} E_1 &\stackrel{\text{def}}{=} \mu_X(a.X + X) \\ E_2 &\stackrel{\text{def}}{=} \mu_X(\tfrac{1}{2}X \oplus \tfrac{1}{2}(X + X)) \\ F_1 &\stackrel{\text{def}}{=} a.b + \tau.c \\ F_2 &\stackrel{\text{def}}{=} F_1 + \tau.(\tfrac{1}{3}F_1 \oplus \tfrac{2}{3}c) \end{aligned}$$

It can be checked that $E_1 \sim E_2$, $F_1 \approx F_2$, and $\tau.F_1 \simeq \tau.F_2$. Note that $F_1 \not\simeq F_2$ because the transition $F_2 \xrightarrow{\tau} \{F_1 : \frac{1}{3},\ c : \frac{2}{3}\}$ cannot be matched up by the transition $F_1 \xrightarrow{\tau} \{c : 1\}$, which is the only normal transition from F_1 with action τ. □

4.3 Probabilistic "Bisimulation up to" Techniques

A natural way for showing $E \sim F$ in a probabilistic process calculus is to construct an equivalence relation $\mathcal{R}$ which includes the pair (E, F), and then to check that $\mathcal{R}$ is a bisimulation. However, it is often difficult to ensure that the relation $\mathcal{R}$ one constructs is indeed an equivalence relation. In this case we use "bisimulation up to" techniques. The idea is that we extend $\mathcal{R}$ to be $\mathcal{R}'$ such that $\mathcal{R} \subseteq \mathcal{R}'$ and $\mathcal{R}'$ is easily shown to be a bisimulation.

Given a binary relation $\mathcal{R}$ we denote by $\mathcal{R}_{\sim}$ the relation $(\mathcal{R} \cup \sim)^*$, the equivalence closure of $\mathcal{R} \cup \sim$. Similarly for the notation $\mathcal{R}_{\approx}$.

[1] The example given in [16, 15] for supporting this argument is built in probabilistic automata [30], but it is easy to write a similar example in simple probabilistic automata.

Definition 6. *A binary relation* $\mathcal{R}$ *is a* strong bisimulation up to $\sim$ *if* $E \mathcal{R} F$ *implies:*

1. *whenever* $E \xrightarrow{\alpha} \eta_1$, *there exists* η_2 *such that* $F \xrightarrow{\alpha} \eta_2$ *and* $\eta_1 \equiv_{\mathcal{R}_\sim} \eta_2$.
2. *whenever* $F \xrightarrow{\alpha} \eta_2$, *there exists* η_1 *such that* $E \xrightarrow{\alpha} \eta_1$ *and* $\eta_1 \equiv_{\mathcal{R}_\sim} \eta_2$.

A strong bisimulation up to $\sim$ is not necessarily an equivalence relation. It is just an ordinary binary relation included in $\sim$, as shown by the next proposition.

Proposition 1. *If* $\mathcal{R}$ *is a strong bisimulation up to* $\sim$, *then* $\mathcal{R} \subseteq\sim$.

For weak probabilistic bisimulation, the "up to" relation can be defined as well, but we need to be careful.

Definition 7. *A binary relation* $\mathcal{R}$ *is a* weak probabilistic bisimulation up to $\approx$ *if* $E \mathcal{R} F$ *implies:*

1. *whenever* $E \stackrel{\alpha}{\Longrightarrow} \eta_1$, *there exists* η_2 *such that* $F \stackrel{\hat{\alpha}}{\Longrightarrow}_c \eta_2$ *and* $\eta_1 \equiv_{\mathcal{R}_\approx} \eta_2$.
2. *whenever* $F \stackrel{\alpha}{\Longrightarrow} \eta_2$, *there exists* η_1 *such that* $E \stackrel{\hat{\alpha}}{\Longrightarrow}_c \eta_1$ *and* $\eta_1 \equiv_{\mathcal{R}_\approx} \eta_2$.

In the above definition, we are not able to replace the first double arrow in each clause by a simple arrow. Otherwise, the resulting relation would not be included in $\approx$.

Proposition 2. *If* $\mathcal{R}$ *is a weak probabilistic bisimulation up to* $\approx$, *then* $\mathcal{R} \subseteq\approx$.

In a way similar to Definition 7, we introduce an "up to $\simeq$" relation.

Definition 8. *A binary relation* $\mathcal{R}$ *is an* observational equivalence up to $\simeq$ *if* $E \mathcal{R} F$ *implies:*

1. *whenever* $E \stackrel{\alpha}{\Longrightarrow} \eta_1$, *there exists* η_2 *such that* $F \stackrel{\alpha}{\Longrightarrow}_c \eta_2$ *and* $\eta_1 \equiv_{\mathcal{R}_\approx} \eta_2$.
2. *whenever* $F \stackrel{\alpha}{\Longrightarrow} \eta_2$, *there exists* η_1 *such that* $E \stackrel{\alpha}{\Longrightarrow}_c \eta_1$ *and* $\eta_1 \equiv_{\mathcal{R}_\approx} \eta_2$.

As expected, observational equivalence up to $\simeq$ is useful because of the following property.

Proposition 3. *If* $\mathcal{R}$ *is an observational equivalence up to* $\simeq$, *then* $\mathcal{R} \subseteq\simeq$.

4.4 Some Properties of Behavioral Equivalences

By using the "bisimulation up to" techniques introduced in the previous section, together with Lemma 2, we can prove the following results. Their detailed proofs are in Appendices 7 and 7, respectively.

Proposition 4 (Properties of $\sim$).

1. $\sim$ *is a congruence relation;*
2. $\mu_X E \sim E\{\mu_X E/X\}$;
3. $\mu_X(E + X) \sim \mu_X E$;
4. *If* $E \sim F\{E/X\}$ *and* X *is weakly guarded in* F, *then* $E \sim \mu_X F$.

Proposition 5 (Properties of $\simeq$).

1. $\simeq$ *is a congruence relation;*
2. *If* $\tau.E \simeq \tau.E + F$ *and* $\tau.F \simeq \tau.F + E$ *then* $\tau.E \simeq \tau.F$;
3. *If* $E \simeq F\{E/X\}$ *and* X *is guarded in* F *then* $E \simeq \mu_X F$.

5 Axiomatizing Strong Bisimilarity

We present in this section the axiom system $\mathcal{A}_s$ for $\sim$, which includes all axioms and rules displayed in Table 3. We assume the usual rules for equality (reflexivity, symmetry, transitivity and substitutivity), and the alpha-conversion of bound variables. If we omit all the axioms involving probabilities, we obtain the system composed by **S1-3** and **R1-3**, which characterizes exactly the class of nonprobabilistic finite-state behaviors studied in [26]. The two axioms **S4-5** allow us to permute and merge probabilistic branches in a probabilistic choice. **E** is a probabilistic version of the expansion law in CCS.

Table 3. The axiom system $\mathcal{A}_s$

S1 $E + \mathbf{0} = E$
S2 $E + E = E$
S3 $\sum_{i\in I} E_i = \sum_{i\in I} E_{\rho(i)}$ ρ is any permutation on I
S4 $u.\bigoplus_{i\in I} p_i E_i = u.\bigoplus_{i\in I} p_{\rho(i)} E_{\rho(i)}$ ρ is any permutation on I
S5 $u.((\bigoplus_i p_i E_i) \oplus pE \oplus qE) = u.((\bigoplus_i p_i E_i) \oplus (p+q)E)$

R1 $\mu_X E = E\{\mu_X E/X\}$
R2 If $E = F\{E/X\}$, X weakly guarded in F, then $E = \mu_X F$
R3 $\mu_X(E + X) = \mu_X E$

E Assume $E \equiv \sum_i u_i.\bigoplus_j p_{ij} E_{ij}$ and $F \equiv \sum_k v_k.\bigoplus_l q_{kl} F_{kl}$. Then infer:

$$E \mid F = \sum_i u_i.\bigoplus_j p_{ij}(E_{ij} \mid F) + \sum_k v_k.\bigoplus_l q_{kl}(E \mid F_{kl}) + \sum_{u_i\ opp\ v_k} \tau.\bigoplus_{j,l}(p_{ij}q_{kl})(E_{ij} \mid F_{kl})$$

where u_i *opp* v_k means that u_i and v_k are complementary actions, i.e., $\bar{u}_i = v_k$.

The notation $\mathcal{A}_s \vdash E = F$ (and $\mathcal{A}_s \vdash \widetilde{E} = \widetilde{F}$ for a finite sequence of equations) means that the equation $E = F$ is derivable by applying the axioms and rules from $\mathcal{A}_s$. The following theorem shows that $\mathcal{A}_s$ is sound with respect to $\sim$.

Theorem 1 (Soundness of $\mathcal{A}_s$). *If $\mathcal{A}_s \vdash E = E'$ then $E \sim E'$.*

Proof. The soundness of the recursion axioms **R1-3** is shown in Section 4.4; the soundness of **S1-4** and **E** is obvious, and **S5** is a consequence of Definition 2. □

For the completeness proof, the basic points are: (1) if two expressions are bisimilar then we can construct an equation set in a certain format (standard format) that they both satisfy; (2) if two expressions satisfy the same standard equation set, then they can be proved equal by $\mathcal{A}_s$. This schema is inspired by [26, 32], but in our case the definition of standard format and the proof itself are more complicated due to the presence of both probabilistic and nondeterministic dimensions.

Definition 9. *Let* $\widetilde{X} = \{X_1, ..., X_m\}$ *and* $\widetilde{W} = \{W_1, W_2, ...\}$ *be disjoint sets of variables. Let* $\widetilde{H} = \{H_1, ..., H_m\}$ *be expressions with free variables in* $\widetilde{X} \cup \widetilde{W}$. *In the equation set* $S : \widetilde{X} = \widetilde{H}$, *we call* $\widetilde{X}$ formal variables *and* $\widetilde{W}$ free variables. *We say* S *is* standard *if each* H_i *takes the form* $\sum_j E_{f(i,j)} + \sum_l W_{h(i,l)}$ *where* $E_{f(i,j)} = u_{f(i,j)}.\bigoplus_k p_{f(i,j,k)} X_{g(i,j,k)}$. *We call* S weakly guarded *if there is no* H_i *s.t.* $H_i \xrightarrow{X_i} \{\mathbf{0} : 1\}$. *We say that* E provably satisfies S *if there are expressions* $\widetilde{E} = \{E_1, ..., E_m\}$, *with* $E_1 \equiv E$ *and* $fv(\widetilde{E}) \subseteq \widetilde{W}$, *such that* $\mathcal{A}_s \vdash \widetilde{E} = \widetilde{H}\{\widetilde{E}/\widetilde{X}\}$.

We first recall the theorem of unique solution of equations originally appeared in [26]. Adding probabilistic choice does not affect the validity of this theorem.

Theorem 2 (Unique solution of equations I). *If* S *is a weakly guarded equation set with free variables in* $\widetilde{W}$, *then there is an expression* E *which provably satisfies* S. *Moreover, if* F *provably satisfies* S *and has free variables in* $\widetilde{W}$, *then* $\mathcal{A}_s \vdash E = F$.

Proof. Exactly as in [26]. □

Below we give an extension of Milner's equational characterization theorem by accommodating probabilistic choice.

Theorem 3 (Equational characterization I). *For any expression* E, *with free variables in* $\widetilde{W}$, *there exist some expressions* $\widetilde{E} = \{E_1, ..., E_m\}$, *with* $E_1 \equiv E$ *and* $fv(\widetilde{E}) \subseteq \widetilde{W}$, *satisfying* m *equations*

$$\mathcal{A}_s \vdash E_i = \sum_{j \in 1..n(i)} E_{f(i,j)} + \sum_{j \in 1..l(i)} W_{h(i,j)} \qquad (i \leq m)$$

where $E_{f(i,j)} \equiv u_{f(i,j)}.\bigoplus_{k \in 1..o(i,j)} p_{f(i,j,k)} E_{g(i,j,k)}$.

Proof. By induction on the structure of E. We only consider the case that $E \equiv F \mid F'$; all other cases are similar to the proof in [26]. By definition F and F' are closed terms. By induction we have closed terms $F_1, .., F_m$ satisfying m equations

$$\mathcal{A}_s \vdash F_i = \sum_{j \in 1..n(i)} F_{f(i,j)} \qquad (i \leq m)$$

where $F_{f(i,j)} \equiv u_{f(i,j)}.\bigoplus_{k \in 1..o(i,j)} p_{f(i,j,k)} F_{g(i,j,k)}$. Similarly we have closed expressions $F'_1, ..., F'_{m'}$ satisfying m' equations

$$\mathcal{A}_s \vdash F'_{i'} = \sum_{j' \in 1..n'(i')} F'_{f'(i',j')} \qquad (i \leq m')$$

where $F'_{f'(i',j')} \equiv u'_{f'(i',j')}.\bigoplus_{k' \in 1..o'(i',j')} p'_{f'(i',j',k')} F'_{g'(i',j',k')}$. Now set $E_{i,i'} \equiv F_i \mid F'_{i'}$. By the expansion law $\mathbf{E}$ we obtain the equations

$$\begin{aligned}\mathcal{A}_s \vdash E_{i,i'} = &\textstyle\sum_{j \in 1..n(i)} u_{f(i,j)}.\bigoplus_{k \in 1..o(i,j)} p_{f(i,j,k)} E_{g(i,j,k),i'} \\ &+ \textstyle\sum_{j' \in 1..n'(i')} u'_{f'(i',j')}.\bigoplus_{k' \in 1..o'(i',j')} p'_{f'(i',j',k')} E_{i,g'(i',j',k')} \\ &+ \textstyle\sum_{u_{f(i,j)}\ opp\ u'_{f'(i',j')}} \tau.\bigoplus_{k \in 1..o(i,j), k' \in 1..o'(i',j')} (p_{f(i,j,k)} p'_{f'(i',j',k')}) \\ &E_{f(i,j,k),f'(i',j',k')}\end{aligned}$$

where $i \leq m$, $i' \leq m'$ and $u_{f(i,j)}$ *opp* $u'_{f'(i',j')}$ means that $u_{f(i,j)}$ and $u'_{f'(i',j')}$ are complementary actions, i.e., they are a and $\bar{a}$ respectively, for some a, or the inverse.

Moreover, we have $E \equiv F_1 \mid F'_1 \equiv E_{1,1}$. □

The following completeness proof is closely analogous to that of [32]. It is complicated somewhat by the presence of nondeterministic choice. For example, to construct the formal equations, we need to consider a more refined relation $L_{iji'j'}$ underneath the relation $K_{ii'}$ while in [26, 32] it is sufficient to just use $K_{ii'}$.

Theorem 4 (Completeness of $\mathcal{A}_s$). *If $E \sim E'$ then $\mathcal{A}_s \vdash E = E'$.*

Proof. Let E and E' have free variables in $\widetilde{W}$. By Theorem 3 there are provable equations such that $E \equiv E_1$, $E' \equiv E'_1$ and

$$\mathcal{A}_s \vdash E_i = \sum_{j \in 1..n(i)} E_{f(i,j)} + \sum_{j \in 1..l(i)} W_{h(i,j)} \qquad (i \leq m)$$

$$\mathcal{A}_s \vdash E'_{i'} = \sum_{j' \in 1..n'(i')} E'_{f'(i',j')} + \sum_{j' \in 1..l'(i')} W_{h'(i',j')} \qquad (i' \leq m')$$

with

$$E_{f(i,j)} \equiv u_{f(i,j)}.\bigoplus_{k \in 1..o(i,j)} p_{f(i,j,k)} E_{g(i,j,k)}$$

$$E'_{f'(i',j')} \equiv u'_{f'(i',j')}.\bigoplus_{k' \in 1..o'(i',j')} p'_{f'(i',j',k')} E'_{g'(i',j',k')}.$$

Let $I = \{\langle i, i' \rangle \mid E_i \sim E'_{i'}\}$. By hypothesis we have $E_1 \sim E'_1$, so $\langle 1, 1 \rangle \in I$. Moreover, for each $\langle i, i' \rangle \in I$, the following holds, by the definition of strong bisimilarity:

1. There exists a total surjective relation $K_{ii'}$ between $\{1, ..., n(i)\}$ and $\{1, ..., n'(i')\}$, given by

   $$K_{ii'} = \{\langle j, j' \rangle \mid \langle f(i,j), f'(i',j') \rangle \in I\}.$$

   Furthermore, for each $\langle j, j' \rangle \in K_{ii'}$, we have $u_{f(i,j)} = u'_{f'(i',j')}$ and there exists a total surjective relation $L_{iji'j'}$ between $\{1, ..., o(i,j)\}$ and $\{1, ..., o'(i',j')\}$, given by

   $$L_{iji'j'} = \{\langle k, k' \rangle \mid \langle g(i,j,k), g'(i',j',k') \rangle \in I\}.$$

2. $\mathcal{A}_s \vdash \sum_{j \in 1..l(i)} W_{h(i,j)} = \sum_{j' \in 1..l'(i')} W_{h'(i',j')}$.

Now, let $L_{iji'j'}(k)$ denote the image of $k \in \{1, ..., o(i,j)\}$ under $L_{iji'j'}$ and $L^{-1}_{iji'j'}(k')$ the preimage of $k' \in \{1, ..., o'(i',j')\}$ under $L_{iji'j'}$. We write $[k]_{iji'j'}$ for the set $L^{-1}_{iji'j'}(L_{iji'j'}(k))$ and $[k']_{iji'j'}$ for $L_{iji'j'}(L^{-1}_{iji'j'}(k'))$. It follows from the definitions that

1. If $\langle i, i'_1\rangle \in I$, $\langle i, i'_2\rangle \in I$, $\langle j, j'_1\rangle \in K_{ii'_1}$ and $\langle j, j'_2\rangle \in K_{ii'_2}$, then $[k]_{iji'_1j'_1} = [k]_{iji'_2j'_2}$.
2. If $q_1 \in [k]_{iji'j'}$ and $q_2 \in [k]_{iji'j'}$, then $E_{g(i,j,q_1)} \sim E_{g(i,j,q_2)}$.

Define $\nu_{ijk} = \sum_{q\in[k]_{iji'j'}} p_{f(i,j,q)}$ for any i', j' such that $\langle i, i'\rangle \in I$ and $\langle j, j'\rangle \in K_{ii'}$; define $\nu'_{i'j'k'} = \sum_{q'\in[k']_{iji'j'}} p'_{f'(i',j',q')}$ for any i, j such that $\langle i, i'\rangle \in I$ and $\langle j, j'\rangle \in K_{ii'}$. It is easy to see that whenever $\langle i, i'\rangle \in I$, $\langle j, j'\rangle \in K_{ii'}$ and $\langle k, k'\rangle \in L_{iji'j'}$ then $\nu_{ijk} = \nu'_{i'j'k'}$.

We now consider the formal equations, one for each $\langle i, i'\rangle \in I$:

$$X_{i,i'} = \sum_{\langle j,j'\rangle\in K_{ii'}} H_{f(i,j),f'(i',j')} + \sum_{j\in 1..l(i)} W_{h(i,j)}$$

where

$$H_{f(i,j),f'(i',j')} \equiv u_{f(i,j)}. \bigoplus_{\langle k,k'\rangle\in L_{iji'j'}} \left(\frac{p_{f(i,j,k)}p'_{f'(i',j',k')}}{\nu_{ijk}}\right)X_{g(i,j,k),g'(i',j',k')}.$$

These equations are provably satisfied when each $X_{i,i'}$ is instantiated to E_i, since $K_{ii'}$ and $L_{iji'j'}$ are total and the right-hand side differs at most by repeated summands from that of the already proved equation for E_i. Note that each probabilistic branch $p_{f(i,j,k)}E_{g(i,j,k)}$ in the subterm $E_{f(i,j)}$ of E_i becomes the probabilistic summation of several branches like

$$\bigoplus_{q'\in[k']_{iji'j'}} \left(\frac{p_{f(i,j,k)}p'_{f'(i',j',q')}}{\nu_{ijk}}\right)E_{g(i,j,k)}$$

in $H_{f(i,j),f'(i',j')}\{E_i/X_{i,i'}\}_i$, where $\langle i, i'\rangle \in I$, $\langle j, j'\rangle \in K_{ii'}$ and $\langle k, k'\rangle \in L_{iji'j'}$. But they are provably equal because

$$\begin{aligned}\textstyle\sum_{q'\in[k']_{iji'j'}} \left(\frac{p_{f(i,j,k)}p'_{f'(i',j',q')}}{\nu_{ijk}}\right) &= \frac{p_{f(i,j,k)}}{\nu_{ijk}} \cdot \textstyle\sum_{q'\in[k']_{iji'j'}} p'_{f'(i',j',q')}\\ &= \frac{p_{f(i,j,k)}}{\nu_{ijk}} \cdot \nu'_{i'j'k'} = p_{f(i,j,k)}\end{aligned}$$

and then the axiom **S5** can be used. Symmetrically, the equations are provably satisfied when each $X_{i,i'}$ is instantiated to $E'_{i'}$; this depends on the surjectivity of $K_{ii'}$ and $J_{iji'j'}$.

Finally, we note that each $X_{i,i'}$ is weakly guarded in the right-hand sides of the formal equations. It follows from Theorem 2 that $\vdash E_i = E'_{i'}$ for each $\langle i, i'\rangle \in I$, and hence $\vdash E = E'$. □

6 Axiomatizing Observational Equivalence

In this section we axiomatize the observational equivalence $\simeq$. We are not able to give a complete axiomatization for the whole set of expressions (and we conjecture that it is not possible), so we restrict to the subset of $\mathcal{E}$ consisting of *guarded expressions* only. An expression is guarded if for each of its subexpression of the form $\mu_X F$, the variable X is guarded in F (cf. Definition 1).

First let us analyze the system $\mathcal{A}_s$. All axioms except for **R2-3** are still valid for $\simeq$. **R3** is not needed because it deals with unguarded expressions. We can reuse **R2** by requiring X to be (strongly) guarded, so we get **R2′** in Table 4. To establish the system $\mathcal{A}_o$ for $\simeq$, we use five τ-laws, **T1-5** in Table 4, to abstract away invisible actions. Note that **T1** and **T2** together constitute the probabilistic version of Milner's second τ-law ([28] page 231). **T3** and **T4** are the probabilistic extensions of Milner's third and first τ-laws, respectively. The extra rule **T5** has no nonprobabilistic counterpart in CCS, but it plays an important role in the proof of Theorem 8. As in [7] the axiom **C** is needed because we use combined transitions when defining observational equivalence.

Table 4. Some laws for the axiom system $\mathcal{A}_o$

T1 $\tau.\bigoplus_i p_i(E_i + X) = X + \tau.\bigoplus_i p_i(E_i + X)$
T2 $\tau.\bigoplus_i p_i(E_i + u.\bigoplus_j p_{ij}.E_{ij}) + u.\bigoplus_{i,j} p_i p_{ij}.E_{ij}$
$= \tau.\bigoplus_i p_i(E_i + u.\bigoplus_j p_{ij}.E_{ij})$
T3 $u.\bigoplus_i p_i(E_i + \tau.\bigoplus_j p_{ij}.E_{ij}) + u.\bigoplus_{i,j} p_i p_{ij}.E_{ij}$
$= u.\bigoplus_i p_i(E_i + \tau.\bigoplus_j p_{ij}.E_{ij})$
T4 $u.(p\tau.E \oplus \bigoplus_i p_i E_i) = u.(pE \oplus \bigoplus_i p_i E_i)$
T5 If $\tau.E = \tau.E + F$ and $\tau.F = \tau.F + E$ then $\tau.E = \tau.F$.

R2′ If $E = F\{E/X\}$, X guarded in F, then $E = \mu_X F$

C $\sum_{i\in 1..n} u.\bigoplus_j p_{ij} E_{ij} = \sum_{i\in 1..n} u.\bigoplus_j p_{ij} E_{ij} + u.\bigoplus_{i\in 1..n} \bigoplus_j r_i p_{ij} E_{ij}$
with $\sum_{i\in 1..n} r_i = 1$.

Theorem 5 (Soundness of $\mathcal{A}_o$). *If $\mathcal{A}_o \vdash E = F$ then $E \simeq F$.*

Proof. The rules **R2′** and **T5** are proved to be sound in Proposition 5 (its proof is detailed in Appendix 7). The soundness of **C** and **T1-4** is straightforward. □

For the completeness proof, it is convenient to use the following saturation property, which relates operational semantics to term transformation, and which can be shown by using the probabilistic τ-laws **T1-4** and the axiom **C**.

Lemma 3 (Saturation). *Suppose there is no parallel composition in E.*

1. *If $E \stackrel{u}{\Longrightarrow} \eta$ with $\eta = \{E_i : p_i\}_i$, then $\mathcal{A}_o \vdash E = E + u.\bigoplus_i p_i E_i$;*
2. *If $E \stackrel{u}{\Longrightarrow}_c \eta$ with $\eta = \{E_i : p_i\}_i$, then $\mathcal{A}_o \vdash E = E + u.\bigoplus_i p_i E_i$;*
3. *If $E \stackrel{X}{\Longrightarrow} \{\mathbf{0} : 1\}$ then $\mathcal{A}_o \vdash E = E + X$.*

Proof. The first and third clauses are proved by transition induction on the inference $E \stackrel{u}{\Longrightarrow} \eta$; the second clause is a corollary of the first one. □

Below we state two simple properties of weak combined transitions. They will be used in proving Theorem 8.

Lemma 4. *1. If $E \stackrel{\hat{u}}{\Longrightarrow}_c \eta$ then $\tau.E \stackrel{u}{\Longrightarrow}_c \eta$;*
2. If $E \stackrel{X}{\Longrightarrow}_c \{\mathbf{0} : 1\}$ then $E \stackrel{X}{\Longrightarrow} \{\mathbf{0} : 1\}$.

Proof. Trivial. □

Lemma 5. *If $E \stackrel{\hat{u}}{\Longrightarrow}_c \{E_i : p_i\}_i$ then $\mathcal{A}_o \vdash \tau.E = \tau.E + u.\bigoplus_i p_i E_i$.*

Proof. It follows from Lemma 4 and Lemma 3. □

To show the completeness of $\mathcal{A}_o$, we need some notations. Given a standard equation set $S : \widetilde{X} = \widetilde{H}$, which has free variables $\widetilde{W}$, we define the relations $\stackrel{\alpha}{\longrightarrow}_S \subseteq \widetilde{X} \times \mathcal{P}(\widetilde{X})$ (recall that the notation $\mathcal{P}(V)$ represents all distributions on V) as $X_i \stackrel{\alpha}{\longrightarrow}_S \eta$ iff $H_i \stackrel{\alpha}{\longrightarrow} \eta$. From $\stackrel{\alpha}{\longrightarrow}_S$ we can define the weak transition $\stackrel{\alpha}{\Longrightarrow}_S$ in the same way as in Section 3. We shall call S *guarded* if there is no X_i s.t. $X_i \stackrel{X_i}{\Longrightarrow}_S \{\mathbf{0} : 1\}$. The variable W is *guarded* in S if it is not the case that $X_1 \stackrel{W}{\Longrightarrow}_S \{\mathbf{0} : 1\}$.

For guarded expressions, the equational characterization theorem and the unique solution theorem given in last section can now be refined, as done in [28].

Theorem 6 (Equational characterization II). *Each guarded expression E with free variables in $\widetilde{W}$ provably satisfies a standard guarded equation set S with free variables in $\widetilde{W}$. Moreover, if W is guarded in E then W is guarded in S.*

Proof. By induction on the structure of E. Consider the case that $E \equiv u.\bigoplus_{i\in I} p_i E_i$. For each $i \in I$, let X_i be the distinguished variable of the equation set S_i for E_i. We can define S as $\{X = u.\bigoplus_{i\in I} p_i X_i\} \cup \bigcup_{i\in I} S_i$, with the new variable X distinguished. All other cases are the same as in [28]. For the case that $E \equiv F \mid F'$, the arguments are similar to those in Theorem 3. □

Theorem 7 (Unique solution of equations II). *If S is a guarded equation set with free variables in $\widetilde{W}$, then there is an expression E which provably satisfies S. Moreover, if F provably satisfies S and has free variables in $\widetilde{W}$, then $\mathcal{A}_o \vdash E = F$.*

Proof. Nearly the same as the proof of Theorem 2, just replacing the recursion rule **R2** with **R2′**. □

The following theorem plays a crucial role in proving the completeness of $\mathcal{A}_o$.

Theorem 8. *Let E provably satisfy S and F provably satisfy T, where both S and T are standard, guarded equation sets, and let $E \simeq F$. Then there is a standard, guarded equation set U satisfied by both E and F.*

Proof. Suppose that $\widetilde{X} = \{X_1, ..., X_m\}$, $\widetilde{Y} = \{Y_1, ..., Y_n\}$ and $\widetilde{W} = \{W_1, W_2, ...\}$ are disjoint sets of variables. Let

$$S : \widetilde{X} = \widetilde{H}$$

$$T : \widetilde{Y} = \widetilde{J}$$

with $fv(\widetilde{H}) \subseteq \widetilde{X} \cup \widetilde{W}$, $fv(\widetilde{J}) \subseteq \widetilde{Y} \cup \widetilde{W}$, and that there are expressions $\widetilde{E} = \{E_1, ..., E_m\}$ and $\widetilde{F} = \{F_1, ..., F_n\}$ with $E_1 \equiv E$, $F_1 \equiv F$, and $fv(\widetilde{E}) \cup fv(\widetilde{F}) \subseteq \widetilde{W}$, so that

$$\mathcal{A}_o \vdash \widetilde{E} = \widetilde{H}\{\widetilde{E}/\widetilde{X}\}$$
$$\mathcal{A}_o \vdash \widetilde{F} = \widetilde{J}\{\widetilde{F}/\widetilde{Y}\}.$$

Consider the least equivalence relation $\mathcal{R} \subseteq (\widetilde{X} \cup \widetilde{Y}) \times (\widetilde{X} \cup \widetilde{Y})$ such that

1. whenever $(Z, Z') \in \mathcal{R}$ and $Z \xrightarrow{\alpha} \eta$, then there exists η' s.t. $Z' \stackrel{\hat{\alpha}}{\Longrightarrow}_c \eta'$ and $\eta \equiv_{\mathcal{R}} \eta'$;
2. $(X_1, Y_1) \in \mathcal{R}$ and if $X_1 \xrightarrow{\alpha} \eta$ then there exists η' s.t. $Y_1 \stackrel{\alpha}{\Longrightarrow}_c \eta'$ and $\eta \equiv_{\mathcal{R}} \eta'$.

Clearly $\mathcal{R}$ is a weak probabilistic bisimulation on the transition system over $\widetilde{X} \cup \widetilde{Y}$, determined by $\rightarrow \stackrel{\text{def}}{=} \rightarrow_S \cup \rightarrow_T$. Now for two given distributions $\eta = \{X_i : p_i\}_{i \in I}$, $\eta' = \{Y_j : q_j\}_{j \in J}$, with $\eta \equiv_{\mathcal{R}} \eta'$, we introduce the following notations:

$$\begin{aligned} K_{\eta,\eta'} &= \{(i,j) \mid i \in I,\ j \in J, \text{ and } (X_i, Y_j) \in \mathcal{R}\} & \\ \nu_i &= \textstyle\sum\{p_{i'} \mid i' \in I, \text{ and } (X_i, X_{i'}) \in \mathcal{R}\} & \text{for } i \in I \\ \nu_j &= \textstyle\sum\{p_{j'} \mid j' \in J, \text{ and } (Y_j, Y_{j'}) \in \mathcal{R}\} & \text{for } j \in J \end{aligned}$$

Since $\eta \equiv_{\mathcal{R}} \eta'$ it follows by definition that if $(i,j) \in K_{\eta,\eta'}$, for some η, η', then $\nu_i = \nu_j$. Thus we can define the expression

$$G_{\eta,\eta'} \stackrel{\text{def}}{=} \bigoplus_{(i,j) \in K_{\eta,\eta'}} \frac{p_i q_j}{\nu_i} Z_{ij}$$

which will play the same role as the expression $H_{f(i,j),f'(i',j')}$ in the proof of Theorem 4.

Based on the above $\mathcal{R}$ we choose a new set of variables $\widetilde{Z}$ such that

$$\widetilde{Z} = \{Z_{ij} \mid X_i \in \widetilde{X},\ Y_j \in \widetilde{Y} \text{ and } (X_i, Y_j) \in \mathcal{R}\}.$$

Furthermore, for each $Z_{ij} \in \widetilde{Z}$ we construct three auxiliary finite sets of expressions, denoted by A_{ij}, B_{ij} and C_{ij}, by the following procedure.

1. Initially the three sets are empty.
2. For each η with $X_i \xrightarrow{\alpha} \eta$, arbitrarily choose one (and only one — the same principle applies in other cases too) η' (if it exists) satisfying $\eta \equiv_{\mathcal{R}} \eta'$ and $Y_j \stackrel{\alpha}{\Longrightarrow}_c \eta'$. If $\alpha \in Act$ then we construct the expression $G_{\eta,\eta'}$ and update A_{ij} to be $A_{ij} \cup \{\alpha.G_{\eta,\eta'}\}$; if $\alpha = X$ for some X then we update A_{ij} to be $A_{ij} \cup \{X\}$. Similarly for each η' with $Y_j \xrightarrow{\alpha} \eta'$, arbitrarily choose one η (if it exists) satisfying $\eta \equiv_{\mathcal{R}} \eta'$ and $X_i \stackrel{\alpha}{\Longrightarrow}_c \eta$. If $\alpha \in Act$ then we construct the expression $G_{\eta,\eta'}$ and update A_{ij} to be $A_{ij} \cup \{\alpha.G_{\eta,\eta'}\}$; if $\alpha = X$ for some X then we update A_{ij} to be $A_{ij} \cup \{X\}$.
3. For each η with $X_i \xrightarrow{\tau} \eta$, arbitrarily choose one η' (if it exists) satisfying $\eta \equiv_{\mathcal{R}} \eta'$, $Y_j \Longrightarrow_c \eta'$ but not $Y_j \stackrel{\tau}{\Longrightarrow}_c \eta'$, construct the expression $G_{\eta,\eta'}$ and update B_{ij} to be $B_{ij} \cup \{\tau.G_{\eta,\eta'}\}$.

4. For each η' with $Y_j \xrightarrow{\tau} \eta'$, arbitrarily choose one η (if it exists) satisfying $\eta \equiv_{\mathcal{R}} \eta'$, $X_i \Longrightarrow_c \eta$ but not $X_i \stackrel{\tau}{\Longrightarrow}_c \eta$, construct $G_{\eta,\eta'}$ and update C_{ij} to be $C_{ij} \cup \{\tau.G_{\eta,\eta'}\}$.

Clearly the three sets constructed in this way are finite. Now we build a new equation set

$$U : \widetilde{Z} = \widetilde{L}$$

where U_{11} is the distinguished variable and

$$L_{ij} = \begin{cases} \sum_{G \in A_{ij}} G & \text{if } B_{ij} \cup C_{ij} = \emptyset \\ \tau.(\sum_{G \in A_{ij} \cup B_{ij} \cup C_{ij}} G) & \text{otherwise.} \end{cases}$$

We assert that E provably satisfies the equation set U. To see this, we choose expressions

$$G_{ij} = \begin{cases} E_i & \text{if } B_{ij} \cup C_{ij} = \emptyset \\ \tau.E_i & \text{otherwise} \end{cases}$$

and verify that $\mathcal{A}_o \vdash G_{ij} = L_{ij}\{\widetilde{G}/\widetilde{Z}\}$.

In the case that $B_{ij} \cup C_{ij} = \emptyset$, all those summands of $L_{ij}\{\widetilde{G}/\widetilde{Z}\}$ which are not variables are of the form:

$$u. \bigoplus_{(i,j) \in K_{\eta,\eta'}} \frac{p_i q_j}{\nu_i} E'_i$$

where $E'_i = E_i$ or $E'_i = \tau.E_i$ for each i. By **T4** we can prove that

$$u. \bigoplus_{(i,j) \in K_{\eta,\eta'}} \frac{p_i q_j}{\nu_i} E'_i = u. \bigoplus_{(i,j) \in K_{\eta,\eta'}} \frac{p_i q_j}{\nu_i} E_i.$$

Then by some arguments similar to those in Theorem 4, together with Lemma 3, we can show that

$$\mathcal{A}_o \vdash L_{ij}\{\widetilde{G}/\widetilde{Z}\} = H_i\{\widetilde{E}/\widetilde{X}\} = E_i.$$

On the other hand, if $B_{ij} \cup C_{ij} \neq \emptyset$, we let $C_{ij} = \{D_1, ..., D_o\}$ ($C_{ij} = \emptyset$ is a special case of the following argument) and $D = \sum_{l \in 1..o} D_l\{\widetilde{G}/\widetilde{Z}\}$. As in last case we can show that

$$\mathcal{A}_o \vdash L_{ij}\{\widetilde{G}/\widetilde{Z}\} = \tau.(H_i\{\widetilde{E}/\widetilde{X}\} + D).$$

For any l with $1 \leq l \leq o$, let $D_l\{\widetilde{G}/\widetilde{Z}\} = \tau.\bigoplus_k p_k E_k$. It is easy to see that $E_i \Longrightarrow_c \eta$ with $\eta = \{E_k : p_k\}_k$. So by Lemma 5 it holds that

$$\mathcal{A}_o \vdash \tau.E_i = \tau.E_i + D_l\{\widetilde{G}/\widetilde{Z}\}.$$

As a result we can infer

$$\mathcal{A}_o \vdash \tau.E_i = \tau.E_i + D = \tau.E_i + (E_i + D).$$

by Lemma 3. Similarly,

$$\mathcal{A}_o \vdash \tau.(E_i + D) = \tau.(E_i + D) + E_i.$$

Consequently it follows from **T5** that

$$\mathcal{A}_o \vdash \tau.E_i = \tau.(E_i + D) = \tau.(H_i\{\widetilde{E}/\widetilde{X}\} + D) = L_{ij}\{\widetilde{G}/\widetilde{Z}\}.$$

In the same way we can show that F provably satisfies U. At last U is guarded because S and T are guarded. □

Theorem 9 (Completeness of $\mathcal{A}_o$). *If E and F are guarded expressions and $E \simeq F$, then $\mathcal{A}_o \vdash E = F$.*

Proof. A direct consequence by combining Theorems 6, 8 and 7. □

In the axiom system $\mathcal{A}_o$ the rule **T5** deserves more explanations. This rule holds also in the non-probabilistic setting, but usually it is not part of the axiomatization because it is subsumed by other axioms. Here we need it, for instance to derive $\tau.F_1 = \tau.F_2$ for the two expressions F_1, F_2 of Example 1 in Section 4.2. Alternatively, we could use the following equality

$$\tau.E = \tau.(E + \tau.((1-p)E \oplus \textstyle\bigoplus_i pp_iE_i)) \qquad \text{where } E = \tau.(\tau.\textstyle\bigoplus_i p_iE_i + F)$$

which is sound and indeed derivable from **T5**. In fact, we could have introduced the above equality as an axiom in place of **T5** in the axiomatization for $\simeq$ — we would still be able to prove Theorem 8 and the completeness of the alternative axiomatization. In this paper we have chosen **T5** instead merely because it looks more elegant than the above axiom.

7 Conclusion and Related Work

We have proposed a probabilistic process calculus which combines both nondeterministic and probabilistic behavior in the style of Segala and Lynch's simple probabilistic automata. The calculus also admits a restricted form of parallel composition to allow for compositional reasoning of finite-state behaviors. We have presented sound and complete axiomatizations for two behavioral equivalences: strong bisimilarity and observational equivalence.

In CCS there are other static operators such as restriction and relabeling that are not studied in this paper. As with parallel composition, these operators should be treated carefully. For example, the expression $\mu_X((a.X \mid \bar{a})\backslash a)$ appears to be guarded (cf. Definition 1), but actually it is strongly bisimilar to $\mu_X(\tau.X)$ thus should be deemed unguarded. When considering axiomatizations one tends to disallow this kind of expressions by imposing the constraint that free variables do not occur in the scope of static operators [11,5].

As we said before, in this paper many concepts and proof techniques are inherited from [16,15]. The main differences are as follows: (i) in this paper we have added a parallel composition operator to our probabilistic process calculus; (ii) to define the operational semantics of this operator we restrict ourself to simple probabilistic automata, while the results of [16, 15] are valid for all probabilistic automata; (iii) besides strong bisimilarity and observational equivalence, in [16, 15] we also axiomatized two other equivalences: a strong probabilistic bisimilarity and a divergency-sensitive equivalence. We think that it should be possible to adapt those results to the framework of this paper.

In [26] and [28] Milner gave complete axiomatizations for strong bisimilarity and observational equivalence, respectively, for a core CCS [27]. Our results in Section 5 and Section 6 extend [26] and [28] (for guarded expressions) respectively, to a strictly larger language with a probabilistic choice and a parallel composition operator.

The first work to consider (strong) bisimulation for probabilistic processes was [22]. They considered the so-called reactive model, in which at each step the probabilistic choice ranges over the next state, while the action is fixed. In a sequel paper, Larsen and Skou also gave a complete axiomatization for the finite case [23].

Bandini and Segala [7] axiomatized two strong and two weak equivalences for a language similar to the fragment of our calculus without recursion and parallelism. They considered two types of semantics. In both cases, their completeness proofs are done by structural induction on processes, which is, of course, impossible in our setting because of recursion.

Giacalone, Jou and Smolka [18] axiomatized strong bisimulation for a fully probabilistic (i.e. without nondeterminism) extension of Milner's SCCS [25], where parallel composition is synchronous. In contrast, we consider an asynchronous parallel composition and we admit nondeterminism.

Baeten, Bergstra and Smolka [4] proposed a probabilistic ACP by introducing a parameterized composition. They considered generative models, which are fully probabilistic, and axiomatized strong probabilistic bisimilarity for finite processes (without recursion).

Andova [3] studied a different version of probabilistic ACP by allowing nondeterminism and a parallel composition which is not parameterized. She provided a sound and complete axiomatization for strong probabilistic bisimilarity in the case of finite processes. She also gave some sound verification rules for probabilistic branching bisimilarity in a fully probabilistic model without parallelism.

Strong probabilistic bisimilarity was also axiomatized by Stark and Smolka in [32]. They gave a probabilistic version of the results of [26]. However, neither nondeterminism nor parallelism is considered. Later the same calculus was studied in [1], which uses some axioms from iteration algebra to characterize recursion.

In the nonprobabilistic setting, Bergstra and Klop [10] established a sound and complete axiomatization for regular processes with τ-steps and free merge (which allows arbitrary interleaving but no communication). They required that free merge should not appear in the body of any recursive expression. To give a linearization algorithm for pCRL, Groote, Ponse and Usenko adopted a similar restriction for parallel composition [19]. Usenko extended this result to μCRL in his thesis [34]. In this paper our parallel composition operator allows communication and it can appear in the body of a recursive expression, though only in a restricted way. For example, the expression

$$\mu_X(a.X + a.\mu_Y(b.Y) \mid \bar{a}.\mu_Z(c.Z))$$

is a legal expression in our calculus and we are able to manipulate it in our axiom systems.

Baeten and Bravetti [5] axiomatized observational equivalence in a generic process algebra. Their restriction enforced to parallel composition is the same as ours in spirit. Interestingly, they reduced two of Milner's axioms for unguarded recursion [28] to just a single axiom. It remains open whether their results can be adapted to a probabilistic setting. Similarly, it might be interesting to extend van Glabbeek's axiomatization for branching congruence [35] to a probabilistic setting. We believe that the general proof schema laid out in this paper could be reused for branching congruence, but the soundness proof of some axioms such as **R2**′ would be very complicated because, besides the probabilistic and nondeterministic features, we need to consider the branching structure of processes, which is ignored in observational congruence.

Christensen, Hirshfeld and Moller studied a class of standard form CCS [13] where open expressions are allowed to be put in parallel composition. In that language, strong bisimulation is decidable and they obtained a sound and complete sequent based equational theory, but observational equivalence is semi-decidable [12]. In this paper we follow [26, 28] and characterize recursion by laws concerning the explicit fixed point operator μ, while we capture by τ-laws the difference between observational equivalence and strong bisimulation.

Several works in the literature address the problem of how to define appropriate parallel composition operators on various probabilistic models, see [14] for more discussions and [31] for a good survey. In this paper, we work at simple probabilistic automata where parallel composition is easy to define (cf. Table 1).

References

1. L. Aceto, Z. Ésik, and A. Ingólfsdóttir. Equational axioms for probabilistic bisimilarity (preliminary report). Technical Report RS-02-6, BRICS, 2002.
2. L. Aceto and W. J. Fokkink. The quest for equational axiomatizations of parallel composition: Status and open problems. In *Proceedings of the Workshop on Algebraic Process Calculi: The First Twenty Five Years and Beyond*, BRICS Notes Series, 2005. To appear.
3. S. Andova. *Probabilistic Process Algebra*. PhD thesis, Eindhoven University of Technology, 2002.
4. J. C. M. Baeten, J. A. Bergstra, and S. A. Smolka. Axiomatizing probabilistic processes: ACP with generative probabilities. *Information and Computation*, 121(2):234–255, 1995.
5. J. C. M. Baeten and M. Bravetti. A ground-complete axiomatization of finite state processes in process algebra. In *Proceedings of the 16th International Conference on Concurrency Theory*, Lecture Notes in Computer Science. Springer, 2005. To appear.

6. J. C. M. Baeten and W. P. Weijland. *Process Algebra*, volume 18 of *Cambridge Tracts in Theoretical Computer Science*. Cambridge University Press, 1990.
7. E. Bandini and R. Segala. Axiomatizations for probabilistic bisimulation. In *Proceedings of the 28th International Colloquium on Automata, Languages and Programming*, volume 2076 of *Lecture Notes in Computer Science*, pages 370–381. Springer, 2001.
8. J. A. Bergstra and J. W. Klop. Process algebra for synchronous communications. *Information and Control*, 60:109–137, 1984.
9. J. A. Bergstra and J. W. Klop. Verification of an alternating bit protocol by means of process algebra. In *Proceedings of the International Spring School on Mathematical Methods of Specification and Synthesis of Software Systems*, volume 215 of *Lecture Notes in Computer Science*, pages 9–23. Springer, 1986.
10. J. A. Bergstra and J. W. Klop. A complete inference system for regular processes with silent moves. In *Proceedings of Logic Colloquium 1986*, pages 21–81. North Holland, Amsterdam, 1988.
11. M. Bravetti and R. Gorrieri. Deciding and axiomatizing weak ST bisimulation for a process algebra with recursion and action refinement. *ACM Transactions on Computational Logic*, 3(4):465–520, 2002.
12. S. Christensen. *Decidability and Decomposition in Process Algebras.* PhD thesis, University of Edinburgh, 1993.
13. S. Christensen, Y. Hirshfeld, and F. Moller. Decidable subsets of ccs. *Computer Journal*, 37(4):233–242, 1994.
14. P. R. D'Argenio, H. Hermanns, and J.-P. Katoen. On generative parallel composition. *Electronic Notes in Theoretical Computer Science*, 22, 1999.
15. Y. Deng. *Axiomatisations and types for probabilistic and mobile processes.* PhD thesis, Ecole des Mines de Paris, 2005.
16. Y. Deng and C. Palamidessi. Axiomatizations for probabilistic finite-state behaviors. In *Proceedings of the 8th International Conference on Foundations of Software Science and Computation Structures*, volume 3441 of *Lecture Notes in Computer Science*, pages 110–124. Springer, 2005.
17. W. J. Fokkink, J. F. Groote, J. Pang, B. Badban, and J. C. van de Pol. Verifying a sliding window protocol in μCRL. In *10th Conference on Algebraic Methodology and Software Technology, Proceedings*, volume 3116 of *Lecture Notes in Computer Science*, pages 148–163. Springer, 2004.
18. A. Giacalone, C.-C. Jou, and S. A. Smolka. Algebraic reasoning for probabilistic concurrent systems. In *Proceedings of IFIP WG 2.2/2.3 Working Conference on Programming Concepts and Methods*, pages 453–459, 1990.
19. J. F. Groote, A. Ponse, and Y. S. Usenko. Linearization in parallel pCRL. *Journal of Logic and Algebraic Programming*, 48(1-2):39–72, 2001.
20. M. Hennessy and R. Milner. Algebraic laws for nondeterminism and concurrency. *Journal of ACM*, 32:137–161, 1985.
21. C. A. R. Hoare. *Communicating Sequential Processes.* Prentice Hall, 1985.
22. K. G. Larsen and A. Skou. Bisimulation through probabilistic testing. *Information and Computation*, 94(1):1–28, 1991.
23. K. G. Larsen and A. Skou. Compositional verification of probabilistic processes. In W. R. Cleaveland, editor, *CONCUR '92: Third International Conference on Concurrency Theory*, volume 630 of *Lecture Notes in Computer Science*, pages 456–471, Stony Brook, New York, 24–27Aug. 1992. Springer-Verlag.
24. N. A. Lynch, I. Saias, and R. Segala. Proving time bounds for randomized distributed algorithms. In *Proceedings of the 13th Annual ACM Symposium on the Principles of Distributed Computing*, pages 314–323, 1994.

25. R. Milner. Calculi for synchrony and asynchrony. *Theoretical Computer Science*, 25:267–310, 1983.
26. R. Milner. A complete inference system for a class of regular behaviours. *Journal of Computer and System Science*, 28:439–466, 1984.
27. R. Milner. *Communication and Concurrency.* Prentice-Hall, 1989.
28. R. Milner. A complete axiomatisation for observational congruence of finite-state behaviours. *Information and Computation*, 81:227–247, 1989.
29. A. Pogosyants, R. Segala, and N. A. Lynch. Verification of the randomized consensus algorithm of Aspnes and Herlihy: a case study. *Distributed Computing*, 13(3):155–186, 2000.
30. R. Segala and N. A. Lynch. Probabilistic simulations for probabilistic processes. In *Proceedings of the 5th International Conference on Concurrency Theory*, volume 836 of *Lecture Notes in Computer Science*, pages 481–496. Springer, 1994.
31. A. Sokolova and E. P. de Vink. Probabilistic automata: system types, parallel composition and comparison. In *Validation of Stochastic Systems: A Guide to Current Research*, volume 2925 of *Lecture Notes in Computer Science*, pages 1–43. Springer, 2004.
32. E. W. Stark and S. A. Smolka. A complete axiom system for finite-state probabilistic processes. In *Proof, language, and interaction: essays in honour of Robin Milner*, pages 571–595. MIT Press, 2000.
33. M. Stoelinga and F. Vaandrager. Root contention in IEEE 1394. In *Proceedings of the 5th International AMAST Workshop on Formal Methods for Real-Time and Probabilistic Systems*, volume 1601 of *Lecture Notes in Computer Science*, pages 53–74. Springer, 1999.
34. Y. S. Usenko. *Linearization in μCRL.* PhD thesis, Edindhoven University of Technology, 2002.
35. R. J. van Glabbeek. A complete axiomatization for branching bisimulation congruence of finite-state behaviours. In *Proceedings of the 18th International Symposium on Mathematical Foundations of Computer Science*, volume 711 of *Lecture Notes in Computer Science*, pages 473–484. Springer, 1993.

Appendix

A Proof of Proposition 4

Lemma 6. *If* $fv(E) \subseteq \{\widetilde{X}, Z\}$ *and* $Z \notin fv(\widetilde{F})$, *then*

$$E\{E'/Z\}\{\widetilde{F}/\widetilde{X}\} \equiv E\{\widetilde{F}/\widetilde{X}\}\{E'\{\widetilde{F}/\widetilde{X}\}/Z\}.$$

Proof. By induction on the structure of E. □

Lemma 7. *Let* $\eta = r_1\eta_1 + ... + r_n\eta_n$ *and* $\eta' = r_1\eta'_1 + ... + r_n\eta'_n$ *with* $\sum_{i\in 1..n} r_i = 1$. *If* $\eta_i \equiv_{\mathcal{R}} \eta'_i$ *for each* $i \leq n$, *then* $\eta \equiv_{\mathcal{R}} \eta'$.

Proof. For any $V \in \mathcal{E}/\mathcal{R}$, we have

$$\eta(V) = \sum_{i\in 1..n} r_i\eta_i(V) = \sum_{i\in 1..n} r_i\eta'_i(V) = \eta'(V).$$

Therefore $\eta \equiv_{\mathcal{R}} \eta'$ by definition. □

Proposition 6. *If $E \sim F$ then $E \mid G \sim F \mid G$.*

Proof. We show that the relation $\sim^{|}$ is a strong bisimulation. There are four cases, among which we consider two of them, the others are similar.

Case 1: Suppose $\eta_1 = \{E_i \mid G : p_i\}_i$ and $E \mid G \xrightarrow{\alpha} \eta_1$ is derived from the transition $E \xrightarrow{\alpha} \theta_1 = \{E_i : p_i\}_i$. Since $E \sim F$, there exists θ_2 such that $F \xrightarrow{\alpha} \theta_2$ and $\theta_1 \equiv_\sim \theta_2$. Let $\theta_2 = \{F_j : q_j\}_j$, by rule par we have the transition $F \mid G \xrightarrow{\alpha} \{F_j \mid G : q_j\}_j = \eta_2$. Let $\theta = \{G : 1\}$, then we have $\eta_1 = \theta_1 \mid \theta$ and $\eta_2 = \theta_2 \mid \theta$. By Lemma 2 it follows that $\eta_1 \equiv_{\sim^|} \eta_2$.

Case 2: Suppose $E \xrightarrow{a} \theta_1$, $G \xrightarrow{\bar{a}} \theta$, and $E \mid G \xrightarrow{\tau} \eta_1$ with $\eta_1 = \theta_1 \mid \theta$. Since $E \sim F$, there exists θ_2 such that $F \xrightarrow{a} \theta_2$ and $\theta_1 \equiv_\sim \theta_2$. By rule com we have the transition $F \mid G \xrightarrow{\tau} \eta_2$ with $\eta_2 = \theta_2 \mid \theta$. By Lemma 2 it follows that $\eta_1 \equiv_{\sim^|} \eta_2$. □

Proposition 7. *If $E \sim F$ then $E\{G/X\} \sim F\{G/X\}$ for any $G \in \mathcal{E}$.*

Proof. Similar to the proof of Proposition 13, which is detailed in next section. □

Proposition 8. *If $E \sim F$ then $\mu_X E \sim \mu_X F$.*

Proof. Let $\rho \stackrel{\text{def}}{=} \{\mu_X E/X\}$ and $\sigma \stackrel{\text{def}}{=} \{\mu_X F/X\}$. We show that the relation

$$\mathcal{R} = \{(G\rho, G\sigma) \mid fv(G) \subseteq \{X\}\}$$

is a strong bisimulation up to $\sim$. Because of symmetry we only show the assertion:

"if $G\rho \xrightarrow{\alpha} \eta_1$ then there exists η_2 s.t. $G\sigma \xrightarrow{\alpha} \eta_2$ and $\eta_1 \equiv_{\mathcal{R}_\sim} \eta_2$"

by induction on the depth of the inference $G\sigma \to \eta_1$. There are several cases, depending on the structure of G.

1. $G \equiv X$: Then $G\rho \equiv \mu_X E \xrightarrow{\alpha} \eta_1$ and there is a shorter inference $E\rho \xrightarrow{\alpha} \eta_1$. By induction hypothesis there is some θ s.t. $E\sigma \xrightarrow{\alpha} \theta$ and $\eta_1 \equiv_{\mathcal{R}_\sim} \theta$. Since $E \sim F$ we know that $E\sigma \sim F\sigma$ by Proposition 7. Hence there exists some η_2 s.t. $F\sigma \xrightarrow{\alpha} \eta_2$ and $\theta \equiv_\sim \eta_2$. By Lemma 1 and the transitivity of $\equiv_{\mathcal{R}_\sim}$ it follows that $\eta_1 \equiv_{\mathcal{R}_\sim} \eta_2$.
2. $G \equiv u.\bigoplus_i p_i G_i$: Then we have $G\rho \xrightarrow{u} \eta_1 \equiv \{G_i\rho : p_i\}_i$ and $G\sigma \xrightarrow{u} \eta_2 \equiv \{G_i\sigma : p_i\}_i$. Since $G_i\rho \; \mathcal{R} \; G_i\sigma$, it is easy to see that $\eta_1 \equiv_{\mathcal{R}_\sim} \eta_2$.
3. $G \equiv \sum_{i \in 1..m} G_i$: If $G\rho \xrightarrow{\alpha} \eta_1$, then $G_j\rho \xrightarrow{\alpha} \eta_1$ for some $j \in 1..m$, by a shorter inference. By induction hypothesis we have that $G_j\sigma \xrightarrow{\alpha} \eta_2$ such that $\eta_1 \equiv_{\mathcal{R}_\sim} \eta_2$.
4. $G \equiv \mu_Y G'$: If $G\rho \xrightarrow{\alpha} \eta_1$ then $G'\rho\{G\rho/Y\}$ by a shorter inference. Since $G'\rho\{G\rho/Y\} \equiv (G'\{G/Y\})\rho$ we have that $(G'\{G/Y\})\rho \xrightarrow{\alpha} \eta_1$. By induction hypothesis it follows that $(G'\{G/Y\})\sigma \xrightarrow{\alpha} \eta_2$ with $\eta_1 \equiv_{\mathcal{R}_\sim} \eta_2$. Thus $G'\sigma\{G\sigma/Y\} \xrightarrow{\alpha} \eta_2$, which implies $G\sigma \xrightarrow{\alpha} \eta_2$ by the rule rec.

5. $G \equiv G_1 \mid G_2$: Suppose $G\rho \xrightarrow{\alpha} \eta_1$. Depending on the last rule used for deriving the transition, there are four cases. We consider one typical case where the last rule used is com. So we have the transitions $G_1\rho \xrightarrow{a} \theta_1$, $G_2\rho \xrightarrow{\bar{a}} \theta_1'$ and $G\rho \xrightarrow{\tau} \eta_1$ with $\eta_1 = \theta_1 \mid \theta_1'$. By induction hypothesis we have the simulating transitions $G_1\sigma \xrightarrow{a} \theta_2$ and $G_2\sigma \xrightarrow{\bar{a}} \theta_2'$ such that $\theta_1 \equiv_{\mathcal{R}_\sim} \theta_2$ and $\theta_1' \equiv_{\mathcal{R}_\sim} \theta_2'$. By rule com we infer that $G\sigma \xrightarrow{\tau} \eta_2$ with $\eta_2 = \theta_2 \mid \theta_2'$. It is easy to see that $\mathcal{R}$ is closed under parallel composition (here we need the condition of composing closed expressions). By Proposition 6 we know that $\sim$ is also closed under parallel composition. It follows that $\mathcal{R}_\sim$ is closed under parallel composition as well. Therefore by Corollary 1 we can derive that $\eta_1 \equiv_{\mathcal{R}_\sim} \eta_2$. □

Proposition 9 (Congruence). *If $\widetilde{E} \sim \widetilde{F}$ then*

1. $u.\bigoplus_i p_i E_i \sim u.\bigoplus_i p_i F_i$;
2. $\sum_i E_i \sim \sum_i F_i$;
3. $E_1 \mid E_2 \sim F_1 \mid F_2$;
4. $\mu_X E_1 \sim \mu_X F_1$.

Proof. The first two clauses are easy to prove; the last two follow from Proposition 6 and Proposition 8 respectively. □

Proposition 10. $\mu_X E \sim E\{\mu_X E/X\}$.

Proof. Observe that $\mu_X E \xrightarrow{\alpha} \eta$ iff $E\{\mu_X E/X\} \xrightarrow{\alpha} \eta$. □

Proposition 11. $\mu_X(E + X) \sim \mu_X E$

Proof. Let $\rho \stackrel{\text{def}}{=} \{\mu_X(E+X)/X\}$ and $\sigma \stackrel{\text{def}}{=} \{\mu_X E/X\}$. We show that the relation

$$\mathcal{R} = \{(G\rho, G\sigma \mid fv(G \subseteq \{X\}))\}$$

is a strong bisimulation up to $\sim$. We prove the following two assertions:

1. If $G\rho \xrightarrow{\alpha} \eta_1$ then $G\sigma \xrightarrow{\alpha} \eta_2$ and $\eta_1 \equiv_{\mathcal{R}_\sim} \eta_2$;
2. If $G\sigma \xrightarrow{\alpha} \eta_2$ then $G\rho \xrightarrow{\alpha} \eta_1$ and $\eta_1 \equiv_{\mathcal{R}_\sim} \eta_2$.

The proof is carried out by induction on transitions, similar to the proof of Proposition 8. Here we only consider the case that $G \equiv X$.

1. If $G\rho \equiv X\rho \xrightarrow{\alpha} \eta_1$ then $(E+X)\rho \xrightarrow{\alpha} \eta_1$ by a shorter inference. By induction hypothesis it follows that $(E + X)\sigma \xrightarrow{\alpha} \eta_2$ and $\eta_1 \equiv_{\mathcal{R}_\sim} \eta_2$. Then either $E\sigma \xrightarrow{\alpha} \eta_2$ or $X\sigma \xrightarrow{\alpha} \eta_2$. From the first case we can also obtain $X\sigma \xrightarrow{\alpha} \eta_2$ by rule rec. Therefore in both cases we have $G\sigma \xrightarrow{\alpha} \eta_2$.
2. If $G\sigma \equiv X\sigma \xrightarrow{\alpha} \eta_2$ then $E\sigma \xrightarrow{\alpha} \eta_2$ by a shorter inference. By induction hypothesis it follows that $E\rho \xrightarrow{\alpha} \eta_1$ with $\eta_1 \equiv_{\mathcal{R}_\sim} \eta_2$. By the rule nsum we derive $(E+X)\rho \xrightarrow{\alpha} \eta_1$. By rec we get the required result that $G\rho \equiv X\rho \xrightarrow{\alpha} \eta_1$.

□

Lemma 8. *Suppose $fv(G) \subseteq \{X\}$ and all free occurrences of X in G are weakly guarded. If $G\{E/X\} \xrightarrow{\alpha} \eta_1$ with $\eta_1 \equiv \{G_i : p_i\}_i$ then G_i takes the form $G'_i\{E/X\}$; Moreover, for any F, $G\{F/X\} \xrightarrow{\alpha} \eta_2$ with $\eta_2 \equiv \{G'_i\{F/X\} : p_i\}_i$ and $\eta_1 \equiv_{\mathcal{R}_\sim} \eta_2$ where*

$$\mathcal{R} = \{(G\{E/X\}, G\{F/X\}) \mid G \in \mathcal{E} \textit{ and } fv(G) \subseteq \{X\}\}.$$

Proof. By transition induction. □

Proposition 12. *If $E \sim F\{E/X\}$, where all occurrences of X in F are weakly guarded, then $E \sim \mu_X F$.*

Proof. Similar to the proof of Proposition 8. Now we take $\mathcal{R}$ as:

$$\mathcal{R} = \{(G\{E/X\}, G\{\mu_X F/X\}) \mid G \in \mathcal{E} \text{ and } fv(G) \subseteq \{X\}\}$$

Let us consider the case that $G \equiv X$. Suppose $E \xrightarrow{\alpha} \eta_1$. Since $E \sim F\{E/X\}$, there exists θ s.t. $F\{E/X\} \xrightarrow{\alpha} \theta$ and $\eta_1 \equiv_\sim \theta$. By Lemma 8 there exists η_2 s.t. $F\{\mu_X F/X\} \xrightarrow{\alpha} \eta_2$ and $\theta \equiv_{\mathcal{R}_\sim} \eta_2$. By rule rec we have $\mu_X F \xrightarrow{\alpha} \eta_2$. By Lemma 1 and the transitivity of $\equiv_{\mathcal{R}_\sim}$, we have $\eta_1 \equiv_{\mathcal{R}_\sim} \eta_2$. With similar reasoning, one can show that if $\mu_X F \xrightarrow{\alpha} \eta_2$ there exists η_1 s.t. $E \xrightarrow{\alpha} \eta_1$ and $\eta_1 \equiv_{\mathcal{R}_\sim} \eta_2$. □

At last Proposition 4 is proved by collecting all the results in Propositions 9-12.

B Proof of Proposition 5

Lemma 9. *1. If $E \xrightarrow{u} \{E_i : p_i\}_i$ then $E\{G/X\} \xrightarrow{u} \{E_i\{G/X\} : p_i\}_i$;*
2. If $E \xRightarrow{u} \{E_i : p_i\}_i$ then $E\{G/X\} \xRightarrow{u} \{E_i\{G/X\} : p_i\}_i$;
3. If $E \xRightarrow{u}_c \{E_i : p_i\}_i$ then $E\{G/X\} \xRightarrow{u}_c \{E_i\{G/X\} : p_i\}_i$;
4. If $E \xRightarrow{\hat{u}}_c \{E_i : p_i\}_i$ then $E\{G/X\} \xRightarrow{\hat{u}}_c \{E_i\{G/X\} : p_i\}_i$.

Proof. Straightforward by induction on inference. □

Lemma 10. *1. If $E \xrightarrow{X} \{\mathbf{0} : 1\}$ and $G \xrightarrow{\alpha} \eta$ then $E\{G/X\} \xrightarrow{\alpha} \eta$.*
2. If $E \xRightarrow{X} \{\mathbf{0} : 1\}$ and $G \xrightarrow{\alpha} \eta$ then $E\{G/X\} \xRightarrow{\alpha} \eta$.

Proof. Straightforward by examining the structure of E. □

Lemma 11. *If $E\{G/X\} \xrightarrow{\alpha} \eta$ then one of the following two cases holds.*

1. $E \xrightarrow{X} \{\mathbf{0} : 1\}$ and $G \xrightarrow{\alpha} \eta$;
2. $\eta = \{E_i\{G/X\} : p_i\}_i$ and $E \xrightarrow{\alpha} \{E_i : p_i\}_i$.

Proof. By induction on the depth of the inference of $E\{G/X\} \xrightarrow{\alpha} \eta$. □

Proposition 13. *If $E \approx F$ then $E\{G/X\} \approx F\{G/X\}$ for any $G \in \mathcal{E}$.*

Proof. Consider the relation $\mathcal{R} = \{(E\{G/X\}, F\{G/X\}) \mid E, F \in \mathcal{E} \text{ and } E \approx F\}$. Since $\approx$ is an equivalence relation, it follows that $\mathcal{R}$ is also an equivalence relation. So if we can show the assertion:

"If $E\{G/X\} \xrightarrow{\alpha} \eta_1$ then there exists η_2 s.t. $F\{G/X\} \stackrel{\hat{\alpha}}{\Longrightarrow}_c \eta_2$ and $\eta_1 \equiv_{\mathcal{R}} \eta_2$" then it follows from Definition 4 that $\mathcal{R}$ is a weak probabilistic bisimulation.

We now prove the above assertion. From Lemma 11 we know that there are two possibilities:

1. $E \xrightarrow{X} \{\mathbf{0} : 1\}$ and $G \xrightarrow{\alpha} \eta_1$. Thus $F \stackrel{X}{\Longrightarrow}_c \{\mathbf{0} : 1\}$ because $E \approx F$. From Lemma 4 we know that $F \stackrel{X}{\Longrightarrow} \{\mathbf{0} : 1\}$. By Lemma 10 it follows that $F\{G/X\} \stackrel{\alpha}{\Longrightarrow} \eta_1$. We can simply take η_1 as η_2 and finish this case.
2. $\eta_1 = \{E_i\{G/X\} : p_i\}$ and $E \xrightarrow{\alpha} \theta_1 = \{E_i : p_i\}_i$. Since $E \approx F$ there exists $\theta_2 = \{F_j : q_j\}_j$ s.t. $F \stackrel{\hat{\alpha}}{\Longrightarrow}_c \theta_2$ and $\theta_1 \equiv_{\approx} \theta_2$. By Lemma 9 we can derive $F\{G/X\} \stackrel{\hat{\alpha}}{\Longrightarrow}_c \eta_2 = \{F_j\{G/X\} : q_j\}_j$. Observe that for any $E', F' \in \{E_i\}_i \cup \{F_j\}_j$ it holds that $E' \approx F'$ iff $E'\{G/X\} \ \mathcal{R} \ F'\{G/X\}$. Hence it follows from $\theta_1 \equiv_{\approx} \theta_2$ that $\eta_1 \equiv_{\mathcal{R}} \eta_2$ and we complete the proof of this case. □

Proposition 14. *If $E \simeq F$ then $E\{G/X\} \simeq F\{G/X\}$ for any $G \in \mathcal{E}$.*

Proof. Due to symmetry, it suffices to verify that if $E\{G/X\} \xrightarrow{\alpha} \eta_1$ then there exists η_2 s.t. $F\{G/X\} \stackrel{\alpha}{\Longrightarrow}_c \eta_2$ and $\eta_1 \equiv_{\approx} \eta_2$. From Lemma 11 we know that there are two possibilities:

1. $E \xrightarrow{X} \{\mathbf{0} : 1\}$ and $G \xrightarrow{\alpha} \eta_1$. Thus $F \stackrel{X}{\Longrightarrow}_c \{\mathbf{0} : 1\}$ because $E \simeq F$. From Lemma 4 we know that $F \stackrel{X}{\Longrightarrow} \{\mathbf{0} : 1\}$. By Lemma 10 it follows that $F\{G/X\} \stackrel{\alpha}{\Longrightarrow} \eta_1$. We we can simply take η_1 as η_2 and finish this case.
2. $\eta_1 = \{E_i\{G/X\} : p_i\}$ and $E \xrightarrow{\alpha} \theta_1 = \{E_i : p_i\}_i$. Since $E \simeq F$ there exists $\theta_2 = \{F_j : q_j\}_j$ s.t. $F \stackrel{\alpha}{\Longrightarrow}_c \theta_2$ and $\theta_1 \equiv_{\approx} \theta_2$. By Lemma 9 we can derive $F\{G/X\} \stackrel{\alpha}{\Longrightarrow}_c \eta_2 = \{F_j\{G/X\} : q_j\}_j$. By Proposition 13 it holds that for any $E', F' \in \{E_i\}_i \cup \{F_j\}_j$ if $E' \approx F'$ then $E'\{G/X\} \approx F'\{G/X\}$. Hence it follows from $\theta_1 \equiv_{\approx} \theta_2$ that $\eta_1 \equiv_{\approx} \eta_2$ and we complete the proof of this case. □

Lemma 12. *1. The following rules are derivable:*

$$\text{D1} \ \frac{E_j \stackrel{\alpha}{\Longrightarrow}_c \eta}{\sum_{i \in 1..n} E_i \stackrel{\alpha}{\Longrightarrow}_c \eta} \ \textit{for some } j \in 1..n \qquad \text{D2} \ \frac{E\{\mu_X E/X\} \stackrel{\alpha}{\Longrightarrow}_c \eta}{\mu_X E \stackrel{\alpha}{\Longrightarrow}_c \eta}$$

$$\text{D3} \ \frac{E \stackrel{\hat{\alpha}}{\Longrightarrow}_c \{E_i : p_i\}_i}{E \mid F \stackrel{\hat{\alpha}}{\Longrightarrow}_c \{E_i \mid F : p_i\}_i}$$

$$\text{D4} \ \frac{E \stackrel{a}{\Longrightarrow}_c \{E_i : p_i\}_{i \in I} \qquad F \xrightarrow{\bar{a}} \{F_j : q_j)\}_{j \in J}}{E \mid F \stackrel{\tau}{\Longrightarrow}_c \{E_i \mid F_j : p_i q_j)\}_{i \in I, j \in J}}$$

2. *If* $\sum_{i\in 1..n} E_i \stackrel{\alpha}{\Longrightarrow} \eta$ *then* $E_j \stackrel{\alpha}{\Longrightarrow} \eta$ *for some* $j \in 1..n$*, with a shorter inference.*
3. *If* $\mu_X E \stackrel{\alpha}{\Longrightarrow} \eta$ *then* $E\{\mu_X E/X\} \stackrel{\alpha}{\Longrightarrow} \eta$*, with a shorter inference.*

Proof. Straightforward by induction on inference. □

Lemma 13. *1. Let* $\mathcal{R}$ *be a weak probabilistic bisimulation. If* $E\ \mathcal{R}\ F$ *then whenever* $E \stackrel{\hat{\alpha}}{\Longrightarrow}_c \eta$*, there exists* η' *such that* $F \stackrel{\hat{\alpha}}{\Longrightarrow}_c \eta'$ *and* $\eta \equiv_{\mathcal{R}} \eta'$*.*
2. Suppose $E \simeq F$*. If* $E \stackrel{\alpha}{\Longrightarrow}_c \eta$ *then there exists* η' *s.t.* $F \stackrel{\alpha}{\Longrightarrow}_c \eta'$ *and* $\eta \equiv_{\approx} \eta'$*.*

Proof. By transition induction. □

Lemma 14. *If* $E \approx F$ *then* $E \mid G \approx F \mid G$*.*

Proof. We show that the relation $\approx^{|}$ is a weak probabilistic bisimulation. There are four cases, among which we consider two of them, the others are similar.

Case 1: Suppose $\eta_1 = \{E_i \mid G : p_i\}_i$ and $E \mid G \stackrel{\alpha}{\longrightarrow} \eta_1$ is derived from the transition $E \stackrel{\alpha}{\longrightarrow} \theta_1 = \{E_i : p_i\}_i$. Since $E \approx F$, there exists θ_2 such that $F \stackrel{\hat{\alpha}}{\Longrightarrow}_c \theta_2$ and $\theta_1 \equiv_{\approx} \theta_2$. Let $\theta_2 = \{F_j : q_j\}_j$, by rule D3 we have the transition $F \mid G \stackrel{\hat{\alpha}}{\Longrightarrow}_c \{F_j \mid G : q_j\}_j = \eta_2$. Let $\theta = \{G : 1\}$, then we have $\eta_1 = \theta_1 \mid \theta$ and $\eta_2 = \theta_2 \mid \theta$. By Lemma 2 it follows that $\eta_1 \equiv_{\approx^|} \eta_2$.

Case 2: Suppose $E \stackrel{a}{\longrightarrow} \theta_1$, $G \stackrel{\bar{a}}{\longrightarrow} \theta$, and $E \mid G \stackrel{\tau}{\longrightarrow} \eta_1$ with $\eta_1 = \theta_1 \mid \theta$. Since $E \approx F$, there exists θ_2 such that $F \stackrel{a}{\Longrightarrow}_c \theta_2$ and $\theta_1 \equiv_{\approx} \theta_2$. By rule D4 we have the transition $F \mid G \stackrel{\tau}{\Longrightarrow}_c \eta_2$ with $\eta_2 = \theta_2 \mid \theta$. By Lemma 2 it follows that $\eta_1 \equiv_{\approx^|} \eta_2$. □

Proposition 15. *If* $E \simeq F$ *then* $E \mid G \simeq F \mid G$*.*

Proof. Similar to the proof of Lemma 14. We need to use the above proved result that $\approx^{|} \subseteq \approx$. □

Proposition 16. *If* $E \simeq F$ *then* $\mu_X E \simeq \mu_X F$*.*

Proof. Let $\rho = \{\mu_X E/X\}$ and $\sigma = \{\mu_X F/X\}$. We show that the relation

$$\mathcal{R} = \{(G\rho, G\sigma) \mid E, F, G \in \mathcal{E} \text{ and } E \simeq F\}$$

is an observational equivalence up to $\simeq$. Because of symmetry we only need to show that if $G\rho \stackrel{\alpha}{\Longrightarrow} \eta$ there exists η' s.t. $G\sigma \stackrel{\alpha}{\Longrightarrow}_c \eta'$ and $\eta \equiv_{\mathcal{R}_\approx} \eta'$. The proof is carried out by induction on the depth of the inference of $G\rho \stackrel{\alpha}{\Longrightarrow} \eta$. There are several cases depending on the structure of G. We consider three typical ones.

- $G \equiv X$: Then $G\rho \equiv \mu_X E \stackrel{\alpha}{\Longrightarrow} \eta$. By Lemma 12 we have a shorter inference with the conclusion $E\rho \stackrel{\alpha}{\Longrightarrow} \eta$. By induction hypothesis there exists θ s.t. $E\sigma \stackrel{\alpha}{\Longrightarrow}_c \theta$ and $\eta \equiv_{\mathcal{R}_\approx} \theta$. Since $E \simeq F$ we have $E\sigma \simeq F\sigma$ by Proposition 14. By Lemma 13 (2) there exists η' s.t. $F\sigma \stackrel{\alpha}{\Longrightarrow}_c \eta'$ and $\theta \equiv_{\approx} \eta'$. By rule D2 it holds that $\mu_X F \stackrel{\alpha}{\Longrightarrow}_c \eta'$. At last it follows from Lemma 1 and the transitivity of $\equiv_{\mathcal{R}_\approx}$ that $\eta \equiv_{\mathcal{R}_\approx} \eta'$.

- $G \equiv \sum_{i \in 1..n} G_i$: If $G\rho \stackrel{\alpha}{\Longrightarrow} \eta$ then by Lemma 12, $G_j\rho \stackrel{\alpha}{\Longrightarrow} \eta$ for some $j \in 1..n$ with a shorter inference. By induction hypothesis there exists η' s.t. $G_j\sigma \stackrel{\alpha}{\Longrightarrow}_c \eta'$ and $\eta \equiv_{\mathcal{R}_\approx} \eta'$. By rule D1 it holds that $G\sigma \stackrel{\alpha}{\Longrightarrow}_c \eta'$.
- $G \equiv G_1 \mid G_2$: Then $fv(G) = \emptyset$ and $G = G\rho = G\sigma$. Clearly if $G\rho \stackrel{\alpha}{\Longrightarrow} \eta$ then $G\sigma \stackrel{\alpha}{\Longrightarrow} \eta$. □

Proposition 17. *$\simeq$ is a congruence relation.*

Proof. Given $\widetilde{E} \simeq \widetilde{F}$, we need to show the following three clauses:

1. $u.\bigoplus_i p_i E_i \simeq u.\bigoplus_i p_i F_i$;
2. $\sum_{i \in 1..n} E_i \simeq \sum_{i \in 1..n} F_i$;
3. $E_1 \mid E_2 \simeq F_1 \mid F_2$;
4. $\mu_X E_1 \simeq \mu_X F_1$.

Among them, the first two clauses are easy to prove; the last two are shown in Proposition 15 and Proposition 16 respectively. □

Proposition 18. *1. $E \approx F$ iff $\tau.E \simeq \tau.F$;*
2. If $\tau.E \simeq \tau.E + F$ and $\tau.F \simeq \tau.F + E$ then $\tau.E \simeq \tau.F$.

Proof. The first clause is straightforward. For the second one, it suffices to prove that $E \approx F$. Consider the relation

$$\mathcal{R} = \{(E, F) \mid E, F \in \mathcal{E}, \tau.E \simeq \tau.E + F \text{ and } \tau.F \simeq \tau.F + E\}.$$

We show that $\mathcal{R}$ is a weak probabilistic bisimulation up to $\approx$. Suppose that $E \stackrel{\alpha}{\Longrightarrow} \eta$. By the condition $E + \tau.F \simeq \tau.F$ and Lemma 13 (2), there exists η' s.t. $\tau.F \stackrel{\alpha}{\Longrightarrow}_c \eta'$ and $\eta \equiv_\approx \eta'$. Since $\tau.F \approx F$, by Lemma 13 (1) there exists η'' s.t. $F \stackrel{\hat{\alpha}}{\Longrightarrow}_c \eta''$ and $\eta' \equiv_\approx \eta''$. Then it is easy to see that $\eta \equiv_{\mathcal{R}_\approx} \eta''$. Similar result holds when E and F exchange their roles. □

We use a measure $d_X(E)$ to count the depth of guardedness of the free variable X in expression E.

$$\begin{aligned} d_X(X) &= 0 \\ d_X(Y) &= 0 \\ d_X(E \mid F) &= 0 \\ d_X(a.E) &= d_X(E) + 1 \\ d_X(\tau.E) &= d_X(E) \\ d_X(\textstyle\bigoplus_i p_i E_i) &= min\{d_X(E_i)\}_i \\ d_X(\textstyle\sum_i E_i) &= min\{d_X(E_i)\}_i \\ d_X(\mu_Y E) &= d_X(E) \end{aligned}$$

Note that $d_X(E \mid F) = 0$ because $fv(E \mid F) = \emptyset$. If $d_X(E) > 0$ then X is guarded in E.

Lemma 15. *Let $d_X(G) = n$ and $\eta = \{G_i : p_i\}_{i \in I}$. Suppose $G\{E/X\} \stackrel{\alpha}{\Longrightarrow} \eta$. For all $i \in I$, it holds that*

1. *If $n > 0$ and $\alpha = \tau$ then $G_i = G_i'\{E/X\}$ and $d_X(G_i') \geq n$;*
2. *If $n > 1$ and $\alpha \neq \tau$ then $G_i = G_i'\{E/X\}$ and $d_X(G_i') \geq n - 1$.*

Proof. By induction on the depth of the inference of $G\{E/X\} \stackrel{\alpha}{\Longrightarrow} \eta$. □

Lemma 16. *Suppose $d_X(G) > 1$, $\eta = \{G_i : p_i\}_{i \in I}$ and $G\{E/X\} \stackrel{\alpha}{\Longrightarrow} \eta$. Then $G_i = G_i'\{E/X\}$ for each $i \in I$. Moreover, $G\{F/X\} \stackrel{\alpha}{\Longrightarrow} \eta'$ and $\eta \equiv_{\mathcal{R}^*} \eta'$, where $\eta' = \{G_i'\{F/X\} : p_i\}_{i \in I}$ and $\mathcal{R} = \{(G\{E/X\}, G\{F/X\}) \mid$ for any $G \in \mathcal{E}\}$.*

Proof. A direct consequence of Lemma 15. □

The following Lemma is a counterpart of Lemma 8.

Lemma 17. *Let $d_X(G) > 1$. If $G\{E/X\} \stackrel{\alpha}{\Longrightarrow}_c \eta$ then $G\{F/X\} \stackrel{\alpha}{\Longrightarrow}_c \eta'$ such that $\eta \equiv_{\mathcal{R}^*} \eta'$ where $\mathcal{R} = \{(G\{E/X\}, G\{F/X\}) \mid$ for any $G \in \mathcal{E}\}$.*

Proof. Let $\eta = r_1\eta_1 + ... + r_n\eta_n$ and $G\{E/X\} \stackrel{\alpha}{\Longrightarrow} \eta_i$ for each $i \leq n$. By Lemma 16, for each $i \leq n$, there exists η_i' s.t. $G\{F/X\} \stackrel{\alpha}{\Longrightarrow} \eta_i'$ and $\eta_i \equiv_{\mathcal{R}^*} \eta_i'$. Now let $\eta' = r_1\eta_1' + ... + r_n\eta_n'$, thus $G\{F/X\} \stackrel{\alpha}{\Longrightarrow}_c \eta'$. By lemma 7 it follows that $\eta \equiv_{\mathcal{R}^*} \eta'$. □

Proposition 19. *If $E \simeq F\{E/X\}$ and X is guarded in F then $E \simeq \mu_X F$.*

Proof. We show that the relation $\mathcal{R} = \{(G\{E/X\}, G\{\mu_X F/X\}) \mid$ for any $G \in \mathcal{E}\}$ is an observational equivalence up to $\simeq$. That is, we need to show the following assertions:

1. if $G\{E/X\} \stackrel{\alpha}{\Longrightarrow} \eta$ then there exists η' s.t. $G\{\mu_X F/X\} \stackrel{\alpha}{\Longrightarrow}_c \eta'$ and $\eta \equiv_{\mathcal{R}_\approx} \eta'$;
2. if $G\{\mu_X F/X\} \stackrel{\alpha}{\Longrightarrow} \eta'$ then there exists η s.t. $G\{E/X\} \stackrel{\alpha}{\Longrightarrow}_c \eta$ and $\eta \equiv_{\mathcal{R}_\approx} \eta'$;

We concentrate on the first clause since the second one is similar. The proof follows closely the arguments in proving Proposition 16, thus we only consider the case that $G \equiv X$.

We write $G(E)$ for $G\{E/X\}$ and $G^2(E)$ for $G(G(E))$. Since $E \simeq F(E)$, we have $E \simeq F^2(E)$ since $\simeq$ is an congruence relation by Proposition 17. If $E \stackrel{\alpha}{\Longrightarrow} \eta$ then by Lemma 13 (2) there exists θ_1 s.t. $F^2(E) \stackrel{\alpha}{\Longrightarrow}_c \theta_1$ and $\eta \equiv_\approx \theta_1$. Since X is guarded in F, i.e., $d_X(F) > 0$, then it follows that $d_X(F^2(X)) > 1$. By Lemma 17, there exists θ_2 s.t. $F^2(\mu_X F) \stackrel{\alpha}{\Longrightarrow}_c \theta_2$ and $\theta_1 \equiv_{\mathcal{R}^*} \theta_2$. From Proposition 10 we have $\mu_X F \sim F^2(\mu_X F)$, thus $\mu_X F \simeq F^2(\mu_X F)$. By Lemma 13 (2) there exists η' s.t. $\mu_X F \stackrel{\alpha}{\Longrightarrow}_c \eta'$ and $\theta_2 \equiv_\approx \eta'$. From Lemma 1 and the transitivity of $\equiv_{\mathcal{R}_\approx}$ it follows that $\eta \equiv_{\mathcal{R}_\approx} \eta'$. □

Finally Proposition 5 is proved by collecting all the results in Propositions 17-19.

Finite Equational Bases in Process Algebra: Results and Open Questions

Luca Aceto[1,4], Wan Fokkink[2,5], Anna Ingolfsdottir[1,4], and Bas Luttik[2,3]

[1] **BRICS** (**B**asic **R**esearch in **C**omputer **S**cience), Centre of the Danish National Research Foundation, Department of Computer Science, Aalborg University, Fr. Bajersvej 7B, 9220 Aalborg Ø, Denmark
{luca, annai}@cs.aau.dk

[2] CWI, Department of Software Engineering, PO Box 94079, 1090 GB Amsterdam, The Netherlands

[3] Department of Mathematics and Computer Science, Eindhoven Technical University, P.O. Box 513, 5600 MB Eindhoven, The Netherlands
luttik@win.tue.nl

[4] Department of Computer Science, Reykjavík University, Ofanleiti 2, 103 Reykjavík, Iceland
{luca, annai}@ru.is

[5] Vrije Universiteit Amsterdam, Department of Computer Science, Section Theoretical Computer Science, De Boelelaan 1081a, 1081 HV Amsterdam, The Netherlands
wanf@cs.vu.nl

Abstract. Van Glabbeek (1990) presented the linear time/branching time spectrum of behavioral equivalences for finitely branching, concrete, sequential processes. He studied these semantics in the setting of the basic process algebra BCCSP, and tried to give finite complete axiomatizations for them. Obtaining such axiomatizations in concurrency theory often turns out to be difficult, even in the setting of simple languages like BCCSP. This has raised a host of open questions that have been the subject of intensive research in recent years. Most of these questions have been settled over BCCSP, either positively by giving a finite complete axiomatization, or negatively by proving that such an axiomatization does not exist. Still some open questions remain. This paper reports on these results, and on the state-of-the-art in axiomatizations for richer process algebras with constructs like sequential and parallel composition.

1 Introduction

One of Jan Willem Klop's main contributions to the theory of concurrency is the development of the ACP family of process algebras in collaboration with Jan Bergstra—see the original papers [8, 9, 10, 11, 12], the textbooks [6, 18] and the historical paper [5]. Process algebras in the ACP style are defined, following the tradition of the algebraic specification of abstract data types, relying on tools from universal algebra and equational logic. More specifically, languages in the

A. Middeldorp et al. (Eds.): Processes... (Klop Festschrift), LNCS 3838, pp. 338–367, 2005.

ACP family are defined by specifying their signature—that is, the collection of algebraic operations that can be used to build new descriptions of reactive systems in terms of ones that we have already constructed—together with a collection of equational axioms that implicitly define the expected semantic properties of processes. This is an application of the classic axiomatic method, on which the development of modern algebra rests, to concurrency theory.

An example of a typical axiom that holds for all of the classic algebras in the ACP family, and is familiar from the theory of regular languages [16,33], is

$$(x + y) \cdot z \approx (x \cdot z) + (y \cdot z) \ .$$

In the above equation, the operation symbols $+$ and $\cdot$ stand for "alternative composition" (or nondeterministic choice) and "sequencing", respectively. Intuitively, this axiom states that a process that can initially choose to behave either like x or like y, and then proceeds to behave like z, is "equivalent" to one that initially chooses to behave either like $x \cdot z$ or like $y \cdot z$.

On the other hand, the right-distributivity axiom of alternative composition over sequencing familiar from formal language theory, namely

$$x \cdot (y + z) \approx (x \cdot y) + (x \cdot z) \ ,$$

is usually *not* considered part of the axiom systems for process algebras since the left- and right-hand sides of the above equation may exhibit different deadlock potential, and should not be equated as descriptions of reactive systems.

Axiom systems arise from the desire of isolating the features that are common to a collection of algebraic structures—namely, their *models*. Early examples of models of the axiom systems for ACP style process algebras were the "projective limit" model—as employed in, e.g., [8]—, and the "graph model" adopted in [11].

Given a language in the ACP family, one may define intuitively appealing models of its axiom system as quotients of the collection of labelled transition systems modulo some behavioural congruence. *Labelled transition systems* (LTSs) [32] are a fundamental formalism for the description of concurrent computation, which is widely used in light of its flexibility and applicability. In particular, they underlie Plotkin's Structural Operational Semantics [41,42] and, following Milner's pioneering work on CCS [36], are by now the standard formalism for describing the semantics of various process description languages.

LTSs model processes by explicitly describing their states and their transitions from state to state, together with the actions that produced them. Since this view of process behaviours is very detailed, several notions of behavioural equivalence and preorder have been proposed for LTSs. The aim of such behavioural semantics is to identify those (states of) LTSs that afford the same "observations", in some appropriate technical sense. The lack of consensus on what constitutes an appropriate notion of observable behaviour for reactive systems has led to a large number of proposals for behavioural equivalences for concurrent processes. (See the study [24], where van Glabbeek presents the linear time/branching time spectrum—a lattice of known behavioural equivalences and preorders over LTSs, ordered by inclusion.)

Having defined a model of an axiom system for a process algebra in terms of LTSs, it is natural to study the connection between the equations that are valid in the chosen model, and those that are derivable from the axioms using the rules of equational logic. The key questions here are:

- Is the axiom system complete? That is, can all of the equations that hold in the LTS model modulo the chosen notion of behavioural equivalence be derived from the axiom system using the rules of equational logic? (A complete axiom system is also referred to as a *basis* for the algebra it axiomatizes.) Researchers in concurrency theory often restrict themselves to studying axiom systems that are complete with respect to the collection of valid equations that do not contain occurrences of variables.
- Does the algebra of LTSs modulo the chosen notion of behavioural equivalence afford a finite equational axiomatization?

A complete axiomatization of a behavioural congruence yields a purely syntactic characterization, independent of LTSs and of the actual details of the definition of the chosen behavioural equivalence, of the semantics of the process algebra. This bridge between syntax and semantics plays an important role in both the practice and the theory of process algebras. From the point of view of practice, these proof systems can be used to perform system verifications in a purely syntactic way using general purpose theorem provers or proof checkers, and form the basis of purpose built axiomatic verification tools like, e.g., PAM [34]. A positive answer to the first basic question raised above is therefore not just theoretically pleasing, but has potential practical applications. From the theoretical point of view, complete axiomatizations of behavioural equivalences capture the essence of different notions of semantics for processes in terms of a basic collection of identities, and this often allows one to compare semantics which may have been defined in very different styles and frameworks. A review of existing complete equational axiomatizations for many of the behavioural semantics in van Glabbeek's spectrum is offered in [24]. The equational axiomatizations offered *ibidem* are over the language BCCSP, a common fragment of Milner's CCS [36] and Hoare's CSP [31] suitable for describing finite synchronization trees, and characterize the differences between behavioural semantics in terms of a few revealing axioms.

If the answer to the second basic question mentioned above is negative, then one may resort to expanding the signature with auxiliary operations, thus adding expressive power for the purpose of axiomatizing the equational theory. Bergstra and Heering [7] have proved that every algebra with a recursively enumerable equational theory has a finite complete equational axiomatization if it may involve a hidden sort and some auxiliary hidden functions. That the auxiliary functions are declared *hidden* means in particular that they themselves need not be completely axiomatized. So then the question remains whether it is possible to expand the algebra with (visible) auxiliary operations, preferably with an intuitive interpretation of their own, in such a way that the equational theory of the expansion has a finite axiomatization.

A classic example of this line of research, which can again be traced back to Jan Willem Klop's work in concurrency theory, is offered by the paper [10]. There Bergstra and Klop showed how to give a finite axiomatization of the language ACP using the auxiliary left and communication merge operators to characterize parallel composition. As shown by Moller [38, 39], auxiliary operators are needed to obtain a finite basis for that language because the process algebras CCS and ACP without the auxiliary left merge operator from [8] do not have a finite equational axiomatization modulo bisimulation equivalence.

An axiom system E is *ω-complete* when an equation can be derived from E if, and only if, all of its closed instantiations can be derived from E. In theorem proving applications, it is often convenient to work with axiomatizations that are ω-complete. In fact, using an ω-complete axiomatization one can avoid proofs by (structural) induction in favour of purely equational reasoning. Moreover, as argued by Heering in [26], ω-completeness of an axiom system is desirable in the partial evaluation of programs. A classic example of an axiom system that is *not* ω-complete is that for the lambda-calculus—see [40].

Many of the existing axiomatizations of behavioural equivalences over expressive process description languages studied in concurrency theory are powerful enough to prove all of the valid equalities between terms that contain no occurrences of variables, but are *not* ω-complete. In fact, obtaining ω-complete axiomatizations in concurrency theory often turns out to be a difficult question, even in the setting of simple languages like BCCSP. This has raised a host of open questions that have been the subject of intensive investigation by process algebraists in recent years. Most of these questions have been settled over BCCSP and other simple process algebras, either positively by giving a finite ω-complete axiomatization, or negatively by proving that such an axiomatization does not exist. Still some open questions remain—especially for process description languages and behavioural equivalences that, like observation equivalence [29, 36], abstract, in some formal sense, from events in process behaviours that are deemed to be unobservable.

In this paper, we report on positive and negative results pertaining to the existence of (finite) complete axiomatizations for BCCSP and richer process algebras, containing constructs like sequential composition and interleaving. We hope that this survey of results will contribute to their dissemination in our research community, and will stimulate further investigations leading to the solution of the challenging open problems that are left.

The paper is organized as follows. We begin by presenting in Section 2 some basic background on universal algebra and equational logic that will be useful for the remainder of this study. In this general setting, we describe a collection of proof techniques that can be used to establish positive and negative results pertaining to the existence of finite, complete axiomatizations for algebras of processes. Section 3 reports on results and open problems on axiomatizations of behavioural equivalences over the language BCCSP studied by van Glabbeek in [24]. The paper concludes with a survey of the state-of-the-art in the equational theory of extensions of that language with more complex operators such as parallel composition and sequential composition (Sections 4 and 5).

2 General Techniques

Our aim in this section is to present some general techniques that can be used to establish results pertaining to the existence or non-existence of finite equational axiomatizations for behavioural equivalences and preorders over process description languages. A suitable general framework within which these techniques can be described is given by the classic fields of *universal algebra* and *equational logic*. We therefore begin by introducing the basic notions from these areas of mathematical research that will be used throughout this paper. We state at the outset that we shall not need very deep results or constructions from universal algebra in what follows, and that much more on it may be found in, e.g., the classic reference [15]. A self-contained presentation from a computer science perspective of the topics we now proceed to introduce may be found in [27].

2.1 Preliminaries

Σ-Algebras. We start from a countably infinite set V of *variables* with typical elements x, y, w, z. A *signature* Σ consists of a set of *operation symbols*, disjoint from V, together with a function *arity* that assigns a natural number to each operation symbol. The set of *terms* over Σ is the least set such that

- Each $x \in V$ is a term.
- If f is an operation symbol of arity n, and $t_1, \ldots, t_n$ are terms, then $f(t_1, \ldots, t_n)$ is also a term.

An operation symbol f of arity 0 will be often called a *constant* symbol, and the term $f()$ will be abbreviated as f.

We write $\mathbb{T}(\Sigma)$ for the set of all terms over Σ and use t, u, v, possibly subscripted and/or superscripted, to range over terms. A term is *closed* (or *ground*) if it contains no occurrences of variables. We denote by $\mathrm{T}(\Sigma)$ the set of closed terms over Σ. A substitution is a mapping from variables to terms. A substitution is closed if it maps variables to closed terms. For every term t and substitution σ, the term obtained by replacing every occurrence of a variable x in t with the term $\sigma(x)$ will be written $\sigma(t)$. Note that $\sigma(t)$ is closed if σ is. Throughout this paper, we use the symbol "=" to stand for (syntactic) equality.

Example 1. A signature for the natural numbers with the operation max yielding the maximum of two numbers might contain a constant 0, a unary successor operation S and the binary operation symbol $\vee$. We shall use this signature as our running example throughout this section, and use $\vee$ in its customary infix notation for the sake of clarity.

Example 2. A process algebra that will be discussed extensively in Section 3 is BCCSP. Its signature consists of the constant $\mathbf{0}$, the binary operator $\_+\_$ called *alternative composition*, and unary *prefix* operators $a\_$, where a ranges over a nonempty set A of actions.

The collection of terms over a signature Σ yields a language. The semantics of this language can be defined canonically once we equip the set of intended denotations with the structure of a Σ-algebra. A *Σ-algebra* is a structure

$$\mathcal{A} = (\mathbf{A}, \{f^{\mathcal{A}} \mid f \in \Sigma\}) ,$$

where $\mathbf{A}$ is a non-empty set (often called the *carrier* of the algebra), and

$$f^{\mathcal{A}} : \mathbf{A}^n \to \mathbf{A}$$

for each operation symbol $f \in \Sigma$ of arity n. Note that if f is a constant symbol, then $f^{\mathcal{A}}$ can be viewed as an element of $\mathbf{A}$.

In order to interpret terms in $\mathbb{T}(\Sigma)$ in a *Σ-algebra* $\mathcal{A} = (\mathbf{A}, \{f^{\mathcal{A}} \mid f \in \Sigma\})$ we need the notion of an environment. An *environment* is a function ρ mapping variables to elements of $\mathbf{A}$. The mapping ρ can be extended homomorphically to $\mathbb{T}(\Sigma)$ in a unique way by stipulating that

$$\rho(f(t_1, \ldots, t_n)) = f^{\mathcal{A}}(\rho(t_1), \ldots, \rho(t_n))$$

for each operation symbol f of arity n and terms $t_1, \ldots, t_n$. Note that $\rho(t)$ is independent of ρ whenever t is closed. For each closed term t, we write $t^{\mathcal{A}}$ for the element of $\mathcal{A}$ that is the interpretation of t in the algebra $\mathcal{A}$. An element of the carrier set of $\mathcal{A}$ is *denotable* if it is the interpretation of some closed term.

Example 3. A suitable algebra $\mathcal{N}$ in which to interpret the collection of terms over the signature introduced in Example 1 has the set of natural numbers $\mathbb{N}$ as carrier set. The constant symbol 0 is interpreted as the natural number 0, the unary function symbol S is interpreted as the successor function—that is, the function mapping each natural number n to $n+1$—and the binary function symbol $\vee$ is interpreted as the function mapping each pair of natural numbers to the largest of the two.

It is easy to see that each element of $\mathcal{N}$ is denotable. Indeed, the natural number n is the interpretation of the term t_n defined thus:

$$\begin{aligned} t_0 &= 0 \quad \text{and} \\ t_{n+1} &= S(t_n) \ . \end{aligned}$$

The interpretation of the language $\mathbb{T}(\Sigma)$ in a Σ-algebra $\mathcal{A} = (\mathbf{A}, \{f^{\mathcal{A}} \mid f \in \Sigma\})$ naturally induces a congruence relation $=_{\mathcal{A}}$ over $\mathbb{T}(\Sigma)$. This is defined thus:

$$t =_{\mathcal{A}} u \ \text{ if, and only if, } \ \rho(t) = \rho(u), \ \text{ for each environment } \rho \ .$$

Example 4. Examples of identities that hold with respect to the congruence relation $=_{\mathcal{N}}$ induced by the interpretation of the language of terms over the signature for the natural numbers in our running example are

$$x \vee 0 =_{\mathcal{N}} x$$
$$0 \vee x =_{\mathcal{N}} x \quad \text{and}$$
$$S(x) \vee S(y) =_{\mathcal{N}} S(x \vee y) \ .$$

The results reviewed in this paper all aim at using the classic logic of equality to offer a syntactic characterization of the relation $=_{\mathcal{A}}$ for algebras of processes. The study of such axiomatic characterizations of semantic equivalences falls therefore within the realm of equational logic, whose basics we now proceed to present.

Equational Logic. An *axiom system* is a collection E of equations $t \approx u$ over the language $\mathbb{T}(\Sigma)$. (The equations in E are often referred to as *axioms*.) An equation $t \approx u$ is derivable from an axiom system E, notation $E \vdash t \approx u$, if it can be proven from the axioms in E using the rules of equational logic (viz. reflexivity, symmetry, transitivity, substitution and closure under Σ-contexts):

$$t \approx t \qquad \frac{t \approx u}{u \approx t} \qquad \frac{t \approx u \quad u \approx v}{t \approx v} \qquad \frac{t \approx u}{\sigma(t) \approx \sigma(u)}$$

$$\frac{t_i \approx u_i \ \ (1 \leq i \leq n)}{f(t_1, \ldots, t_n) \approx f(u_1, \ldots, u_n)} \ .$$

(The first three rules above state that $\approx$ is an equivalence relation, whereas the latter two state that $\approx$ is closed under substitutions, and is a congruence.) Formally, a proof of an equation $t \approx u$ from E is a sequence $t_i \approx u_i$ $(1 \leq i \leq n)$ of equations such that

- $t_n = t$ and $u_n = u$, and
- for each $1 \leq i \leq n$, the equation $t_i \approx u_i$ is obtained by applying one of the aforementioned inference rules using equations in E or some of the equations that precede it in the sequence as premises.

Without loss of generality one may assume that the substitution rule is only applied to axioms, i.e., that the rule

$$\frac{t \approx u}{\sigma(t) \approx \sigma(u)}$$

may only be used when $(t \approx u) \in E$. In this case, the equation $\sigma(t) \approx \sigma(u)$ is called a *substitution instance* of an axiom in E.

Moreover, by postulating that for each axiom in E also its symmetric counterpart is present in E, one may assume that there are no applications of the symmetry rule in equational proofs.

It is well-known (see, e.g., Sect. 2 in [25]) that if an equation relating two closed terms can be proven from an axiom system E, then there is a closed proof for it.

Definition 1 (Soundness). *Let $\mathcal{A}$ be a Σ-algebra. An equation $t \approx u$ is sound with respect to $=_{\mathcal{A}}$ iff $t =_{\mathcal{A}} u$. An axiom system is sound with respect to $=_{\mathcal{A}}$ iff so is each of its equations.*

The collection of all equations that are sound with respect to $=_{\mathcal{A}}$ is called the equational theory of $\mathcal{A}$.

In other words, an axiom system is sound with respect to $=_{\mathcal{A}}$ if it can only be used to prove equations that are valid in the algebra $\mathcal{A}$. This is, of course, a most natural requirement on an axiom system. However, ideally an axiom system should also allow us to prove all of the equations that hold in a given algebra. This is captured by the technical requirement of *completeness*.

Definition 2 (Completeness). *Let $\mathcal{A}$ be a Σ-algebra. An axiom system E is ground complete with respect to $=_{\mathcal{A}}$ iff $E \vdash t \approx u$ whenever $t =_{\mathcal{A}} u$, for all closed terms t, u.*

E is complete with respect to $=_{\mathcal{A}}$ iff $E \vdash t \approx u$ whenever $t =_{\mathcal{A}} u$, for all terms t, u.

Definition 3 (Equational Bases and Finitely Based Algebras). *An equational basis for an algebra $\mathcal{A}$ is a sound axiom system E that is complete with respect to $=_{\mathcal{A}}$. We say that an algebra $\mathcal{A}$ is finitely based if it has a finite equational basis.*

The notion of completeness of an axiom system relates the proof-theoretic notion of derivability using the rules of equational logic with the model-theoretic one of "validity in a model". From a proof-theoretic perspective, a useful property of an axiom system E is that, for all terms $t, u \in \mathbb{T}(\Sigma)$,

$$E \vdash t \approx u \;\text{ iff }\; E \vdash \sigma(t) \approx \sigma(u), \;\text{ for each closed substitution } \sigma \;. \tag{1}$$

An axiom system with the above property is called *ω-complete*. In theorem proving applications, it is convenient if an axiomatization is ω-complete, because this means that proofs by (structural) induction can be avoided in favour of purely equational reasoning. In fact, suppose that $\sigma(t) \approx \sigma(u)$ is provable from an axiom system E, for each closed substitution σ. If E is ω-complete, then we know that an equational proof of the actual equation $t \approx u$ from E exists. In general, the equation $t \approx u$ might not be derivable from E if E is just ground complete. In that case, we might have to content ourselves with showing that all closed instantiations of that equation are derivable from E, and this is usually done by induction on the structure of the closed terms that can be substituted for the variables occurring in t and u.

Example 5. The collection of equations corresponding to the congruences listed in Example 4 is easily seen to be ground complete with respect to $=_{\mathcal{N}}$. That axiom system is, however, neither complete nor ω-complete. For example, the equation

$$x \vee x \approx x \tag{2}$$

is valid in the algebra $\mathcal{N}$, and all of its closed instantiations are provable from the three equations in Example 4. However, the above equation itself is *not* derivable from the axioms in Example 4. (See Examples 7 and 8 for proofs of this claim.)

A finite basis for the algebra $\mathcal{N}$ is given by the following axiom system

$$\begin{aligned} x \vee 0 &\approx x \\ S(x) \vee S(y) &\approx S(x \vee y) \\ S(x) \vee x &\approx S(x) \\ x \vee x &\approx x \\ x \vee y &\approx y \vee x \quad \text{and} \\ x \vee (y \vee z) &\approx (x \vee y) \vee z \ . \end{aligned}$$

It turns out that completeness and ω-completeness are closely related properties of an axiom system. Indeed, assume that $\mathcal{A}$ is a Σ-algebra each of whose elements is denotable. Suppose that E is sound and complete with respect to $=_{\mathcal{A}}$. It is not hard to argue that, in this case, E is also ω-complete.

Remark 1. For the aforementioned connection between the model-theoretic notion of completeness and the proof-theoretic one of ω-completeness to hold, it is crucial that each element in the algebra $\mathcal{A}$ be denotable. To see this, consider the signature consisting of the constant $\perp$ and the unary function symbol P. Interpret this language over the algebra having $\{0, 1\}$ as carrier set, where $\perp$ is interpreted as 0, and P is interpreted as the constant function 0. We claim that no basis for this algebra can be ω-complete. To see that this holds, note, first of all, that each closed term over the aforementioned signature denotes the element 0. Therefore each closed instantiation of the equation $P(x) \approx x$ holds in the algebra, and is provable from the chosen basis. However, the equation $P(x) \approx x$ is itself *not* provable. This follows because E is sound, and that equation does not hold in the algebra, as can be seen by setting the variable x to 1.

Consider the Σ-algebra obtained by quotienting the set of closed terms $\mathrm{T}(\Sigma)$ with respect to the congruence relation that equates two closed terms t, u iff the equation $t \approx u$ is provable from an axiom system E. As a corollary of the aforementioned observation, we have that an equational basis for that algebra is also ω-complete.

Remark 2. Let $\mathcal{A}$ be a Σ-algebra. It is not hard to see that an axiom system that is both ω-complete and ground complete with respect to $=_{\mathcal{A}}$ is also complete with respect to $=_{\mathcal{A}}$.

One of the classic topics in the field of equational logic, and in its applications in process algebra, is the study of results pertaining to the existence or non-existence of finite bases for algebras. In the realm of concurrency theory, van Glabbeek presented in [23, 24] the linear time/branching time spectrum of behavioral equivalences for finitely branching, concrete, sequential processes. He studied these semantics in the setting of the basic process algebra BCCSP, and tried to give finite ω-complete axiomatizations for them. In many cases this turns out to be a difficult question. Most of these finite basis questions have been settled, either positively by giving a finite ω-complete axiomatization, or negatively by proving that such an axiomatization does not exist. But some open questions remain. The main aim of this paper is to survey such results. Before doing so, however, we give a brief overview of some of the general proof techniques that have been developed in the literature on universal algebra, and more specifically within process algebra, to show that certain algebras afford a finite equational basis, or that no such basis exists. These strategies will then be used in Sections 3–5 to establish positive and negative results on the existence of finite bases for behavioural congruences over several process description languages.

2.2 Methods for Establishing Positive Results

Assume that we have an algebra $\mathcal{A}$ and a (finite) axiom system E that is sound with respect to $=_{\mathcal{A}}$. How can we show that E is complete or ground complete? There are a few general proof techniques that have been applied in the literature to answer this question, and we review some of those in the remainder of this section.

Normal Forms. A classic strategy for showing that an axiom system is complete or ground complete that has had a wealth of applications in process algebra relies on the following two steps:

- **Isolation of normal forms.** In this step one finds a collection of terms, the so-called *normal forms*, with the property that each term t can be proved equal to a normal form using the equations in E. In other words, the set of normal forms is as expressive as the whole collection of terms modulo the equational theory generated by E. (If we are aiming at showing that our axiom system E is ground complete, then the normal forms are closed terms, and it suffices only to prove that each closed term is provably equal to a normal form using the equations in E.)
- **Distinctness of normal forms.** In this second step, one argues that two normal forms are related by $=_{\mathcal{A}}$ if, and only if, they are "identical". This is often done by showing that, for each pair of different normal forms, it is possible to construct an environment ρ distinguishing them.

In applications of this method in process algebra, the former step in this proof strategy is often carried out with the use of term rewriting techniques. In that case, the normal forms are precisely those of the term rewriting system, and the analysis is complicated by the need to consider rewriting modulo commutativity and associativity of certain operators like alternative composition. Moreover, the isolation of a suitable notion of normal form often requires considerable ingenuity, and is a difficult art.

Example 6. The aforementioned strategy based upon the isolation of suitable normal forms for terms can be used to show that the axiom system presented in Example 5 is, as claimed there, a finite basis for the algebra $\mathcal{N}$. Indeed, a suitable set of normal forms for terms over the signature of that algebra is given by the collection of terms of the form

$$\bigvee_{i \in I} S^{n_i}(x_i) \; [\vee S^n(0)] ,$$

where

- I is a finite index set,
- $n_i \geq 0$, for each $i \in I$, and
- the variables x_i ($i \in I$) are all different.

The notation $[\vee\{S^n(0)\}]$ used in defining normal forms means that the term $S^n(0)$ is optional. If that term is present then n must be larger than each of the n_i ($i \in I$). Moreover, for an index set $J = \{j_1, \ldots, j_k\}$ ($k \geq 0$) and collection of terms t_j ($j \in J$), we have used the notation $\bigvee_{j \in J} t_j$ as a short-hand for

$$t_{j_1} \vee \cdots \vee t_{j_k} .$$

(That term stands for 0 if J is empty.)

It is not too hard to argue that

1. each term can be proven equal to a normal form using the equations in Example 5 and
2. if t and u are different normal forms, then there is an environment ρ mapping variables to natural numbers such that $\rho(t) \neq \rho(u)$.

Therefore, as claimed in Example 5, that axiom system is a finite basis for the algebra $\mathcal{N}$. Since each element of $\mathcal{N}$ is denotable (Example 3), E is also ω-complete.

Inverted Substitutions. A proof technique that can be used to prove the ω-completeness of an axiom system, and that originates from research in process algebra, was offered by Groote in [25]. Groote's strategy is based on proof transformations, and proceeds as follows. Assume that we have an axiom system E, and an arbitrary equation $t \approx u$ all of whose closed instantiations are provable from E. The first step in Groote's "inverted substitutions" strategy is to find a closed substitution σ such that a proof of the equation $\sigma(t) \approx \sigma(u)$ from E can be transformed uniformly to a proof of the equation $t \approx u$. This proof transformation is achieved by means of a mapping $\hat{\sigma} : \mathrm{T}(\Sigma) \to \mathbb{T}(\Sigma)$ that intuitively maps each closed term representing a variable to the variable itself. This transformation yields the desired proof of the equation $t \approx u$ from E, provided that the technical conditions stated in the following theorem are met.

Theorem 1 (Groote [25]). *Let E be an axiom system over signature Σ. Assume that, for each equation $t \approx u$ all of whose closed instantiations can be proven from E, there exist a closed substitution σ and a mapping $\hat{\sigma} : T(\Sigma) \to \mathbb{T}(\Sigma)$, satisfying the following conditions:*

1. *E proves the equations* $\hat{\sigma}(\sigma(t)) \approx t$ *and* $\hat{\sigma}(\sigma(u)) \approx u$,
2. *for each operation symbol* f *and terms* $u_1, \ldots, u_n, u'_1, \ldots, u'_n$, *where* n *is the arity of* f, *the equation* $\hat{\sigma}(f(u_1, \ldots, u_n)) \approx \hat{\sigma}(f(u'_1, \ldots, u'_n))$ *is provable from those in* E *and the equations* $u_i \approx u'_i$ *and* $\hat{\sigma}(u_i) \approx \hat{\sigma}(u'_i)$ $(1 \leq i \leq n)$ *and*
3. *the equation* $\hat{\sigma}(\sigma'(t_1)) \approx \hat{\sigma}(\sigma'(t_2))$ *is provable from* E *for each* $(t_1 \approx t_2) \in E$ *and closed substitution* σ'.

Then E *is* ω*-complete.*

The strategy for proving the ω-completeness of axiom systems offered by the above result has been applied with success by Groote and other researchers in the field of process algebra, and, when applicable, often leads to simpler proofs than the standard one based on normal forms. As remarked by Groote in [25], the ω-completeness of the finite basis for the algebra $\mathcal{N}$ given in Example 5 *cannot* be shown using the technique in Theorem 1.

Giving Semantics to All Terms. The algebras that are used in the field of process description languages to interpret terms over some signature Σ are often obtained by taking the quotient $\mathrm{T}(\Sigma)/\!\sim$ of the algebra of closed terms over Σ modulo some notion of congruence $\sim$. The interpretation of a closed term in this algebra is its congruence class with respect to $\sim$, and two arbitrary terms are congruent if, and only if, so are all of their closed instantiations.

Another technique that has been developed in the field of process algebra to establish ω-completeness results for axiom systems relies on the following steps:

- **Define the congruence relation $\sim$ over all terms in $\mathbb{T}(\Sigma)$ directly.** The relation $\sim$ should be defined over $\mathbb{T}(\Sigma)$ in such a way that two terms are related by $\sim$ if, and only if, so are all of their closed instantiations. This means, in particular, that an equation $t \approx u$ is sound in the quotient algebra $\mathrm{T}(\Sigma)/\!\sim$ exactly when $t \sim u$ holds. (This step usually involves giving an operational semantics to open terms, and possibly adapting the definition of the congruence relation $\sim$.)
- **Completeness over terms.** In this second step, one proves that the candidate axiom system E is a basis for the quotient algebra of terms $\mathbb{T}(\Sigma)$ modulo $\sim$, and hence for the quotient algebra of *closed* terms $\mathrm{T}(\Sigma)$ modulo $\sim$. Since each element of the algebra $\mathrm{T}(\Sigma)/\!\sim$ is denotable, it follows that E is also ω-complete.

To the best of our knowledge, this technique was first applied in [35] by Milner to show completeness of his inference system for bisimulation equivalence over the regular fragment of the Calculus of Communicating Systems (CCS) [36].

Cover Equations. This technique from Fokkink and Nain [20] is tailored to BCCSP. The aim is to obtain an explicit description of the equational theory for a particular semantics. The central idea is that if an equation $t \approx u$ is sound for BCCSP modulo some semantics in the linear time/branching time spectrum, then $u + t \approx t$ and $t + u \approx u$ are sound as well; and from the last two equations one can derive $t \approx u$. This implies that it is sufficient to only consider sound equations of the form $at + u \approx u$ (where a denotes an action and t, u are BCCSP terms). These are called the *cover equations*.

When the cover equations have been classified, one can proceed in two ways. Either one can determine an infinite family of cover equations that obstructs a finite basis, or one can determine a finite basis among the cover equations.

2.3 Methods for Establishing Negative Results

To prove that a set of equations cannot be derived from a given, possibly finite, subset of this set, we usually point out one specific equation in the superset, and prove that it is not derivable from the subset. To show that an equational theory—that is, the set of equations that hold in a given algebra—is not finitely based, we extend this reasoning by proving that for *each* finite subset of the theory, there is an equation that cannot be derived from this finite set. Often we obtain this result by establishing a stronger result: we identify a particular countably infinite sequence of equations in the theory with some suitable properties, and show that no finite subset of the theory can prove all of the equations in that sequence.

The proof techniques used for this purpose can roughly be divided into two categories: the model-theoretic techniques and the proof-theoretic ones. In what follows we will try to describe the essence of these two main methodologies.

Model-theoretic Techniques. If a set of equations E is sound in an algebra $\mathcal{A}$, we say that $\mathcal{A}$ is a model for E. By Birkhoff's completeness theorem for equational logic [13], each equation that is derived from E holds in $\mathcal{A}$, if $\mathcal{A}$ is a model for E. Thus, to prove that an equation $t \approx u$ is *not* derivable from E it is sufficient to find an algebra that is a model for E but not of the equation $t \approx u$.

Example 7. In Example 5 we claimed that equation (2) is not derivable from the axioms in Example 4. As argued above, this can be proven by exhibiting a model of the axioms in Example 4 where $\vee$ is not idempotent. A simple example of such a model consists of the collection of all finite strings over the symbol a, where 0 is interpreted as the empty string, the unary operation symbol S is interpreted as the identity function, and $\vee$ is used to stand for concatenation.

In light of the previous observations, to prove that an equational theory is not finitely based, one may therefore proceed as follows:

- isolate a countably infinite collection of equations e_n $(n \geq 0)$ in the equational theory,
- for each finite subset E of the equational theory, construct an algebra $\mathcal{A}_E$ that is a model of E, but in which some of the equations e_n fail.

Examples of the application of this strategy may be found in, e.g., [1, 2, 16, 22].

Proof-theoretic Techniques. Recall that an equation $t \approx u$ is derivable from a set of equations E if there is a sequence $t_i \approx u_i$ $(1 \leq i \leq n)$ of equations such that

- $t_n = t$ and $u_n = u$, and
- for each $1 \leq i \leq n$, the equation $t_i \approx u_i$ is obtained by applying one of the aforementioned inference rules using equations in E or some of the equations that precede it in the sequence as premises.

Proof-theoretic techniques aim at showing that $t \approx u$ is *not* derivable from E, by establishing that no such proof sequence exists. This is often done by finding a property of equations that

- holds true for each instantiation of the axioms in E,
- is preserved by the rules of equational logic—that is, if all of the equations that are premises of the rule have the property, then so does the conclusion of the rule—, and
- fails for the equation $t \approx u$.

This contradicts the existence of a proof for the equation $t \approx u$ from E, showing that $t \approx u$ is not derivable from that axiom system.

Example 8. The aforementioned proof-theoretic strategy can be used to give an alternative proof that the idempotence of $\vee$ is not derivable from the axioms in Example 4. To this end, observe that the left- and right-hand sides of each axiom in Example 4 contain the same number of occurrences of each variable. It is not hard to see that this property is preserved under equational derivations. On the other hand, the term $x \vee x$ contains two occurrences of the variable x, whereas the term x has only one. It follows that equation (2) is not derivable from the axioms in Example 4.

The proof-theoretic strategy we have just described can be applied to show that an equational theory is not finitely based as follows:

- isolate a countably infinite collection of equations e_n $(n \geq 0)$ in the equational theory,
- for each finite subset E of the equational theory, show that there is a property of equations that is satisfied by all of the equations that can be derived from E, but that is not afforded by some of the equations e_n.

Proof-theoretic techniques have found wide application in establishing that algebras of processes do not afford a finite basis. In particular, all of the known proofs of the negative results we survey in Sections 4 and 5 are based on applications of the aforementioned proof-theoretic strategy.

Remark 3. An observation that can sometimes be used to show that an equational theory does not afford a finite equational axiomatization relies on the *compactness theorem* (see, e.g., [15]). Assume that we have an infinite axiomatization E for an equational theory T. If T had a finite axiomatization, then, by the compactness theorem, some finite subset of E would be a complete axiomatization for the theory T. Namely, since E is complete, each axiom in the finite axiomatization for T could be derived from E, and each of these derivations uses only finitely many axioms in E. To prove that T does not have a finite axiomatization, it therefore suffices to show that, for each finite subset E' of E, there is an equation in E that is not provable from E'. This can be achieved using either of the two general proof strategies described above. Applications of this proof methodology may be found in, e.g., [16, 17].

3 On Finite Bases for BCCSP

3.1 The Linear Time/Branching Time Spectrum

Van Glabbeek presented in [23, 24] the linear time/branching time spectrum of behavioural equivalences for finitely branching, concrete processes. In this section, for the sake of completeness, we define the semantics in this spectrum.

A *labelled transition system* contains a set of *states*, with typical element s, and a set of transitions $s \xrightarrow{a} s'$, where a ranges over some set of labels. The set $\mathcal{I}(s)$ consists of those labels a for which there exists a transition $s \xrightarrow{a} s'$.

First we define four semantics based on simulation.

Definition 4 (Simulations). *Assume a labelled transition system.*

- *A binary relation $\mathcal{R}$ on states is a* simulation *if $s_0 \; \mathcal{R} \; s_1$ and $s_0 \xrightarrow{a} s_0'$ imply $s_1 \xrightarrow{a} s_1'$ with $s_0' \; \mathcal{R} \; s_1'$.*
- *A simulation $\mathcal{R}$ is a* ready simulation *if $s_0 \; \mathcal{R} \; s_1$ and $a \notin \mathcal{I}(s_0)$ imply $a \notin \mathcal{I}(s_1)$.*
- *A simulation $\mathcal{R}$ is a* 2-nested simulation *if $\mathcal{R}^{-1}$ is included in a simulation.*
- *A* bisimulation *is a symmetric simulation.*

Next we define six semantics based on decorated versions of execution traces.

Definition 5 (Decorated Traces). *Assume a labelled transition system.*

- *A sequence $a_1 \cdots a_n$, with $n \geq 0$, is a* trace *of a state s_0 if there is a sequence of transitions $s_0 \xrightarrow{a_1} s_1 \xrightarrow{a_2} \cdots s_{n-1} \xrightarrow{a_n} s_n$. It is a* completed trace *of s_0 if moreover $\mathcal{I}(s_n) = \emptyset$.*
- *A pair $(a_1 \cdots a_n, X)$, with $n \geq 0$ and $X \subseteq A$, is a* ready pair *of a state s_0 if there is a sequence of transitions $s_0 \xrightarrow{a_1} s_1 \xrightarrow{a_2} \cdots s_{n-1} \xrightarrow{a_n} s_n$ with $\mathcal{I}(s_n) = X$. It is a* failure pair *of s_0 if $\mathcal{I}(s_n) \cap X = \emptyset$.*
- *A sequence $X_0 a_1 X_1 \ldots a_n X_n$, with $n \geq 0$ and $X_i \subseteq A$, is a* ready trace *of a state s_0 if there is a sequence of transitions $s_0 \xrightarrow{a_1} s_1 \xrightarrow{a_2} \cdots s_{n-1} \xrightarrow{a_n} s_n$ with $\mathcal{I}(s_i) = X_i$ for $i = 0, \ldots, n$. It is a* failure trace *of s_0 if $\mathcal{I}(s_i) \cap X_i = \emptyset$ for $i = 0, \ldots, n$.*

Finally, we define two semantics based on possible futures and on possible worlds.

Definition 6 (Possible Futures/Worlds). *Assume a labelled transition system.*

- *A pair $(a_1 \cdots a_n, X)$, with $n \geq 0$ and $X \subseteq A^*$, is a* possible future *of a state s_0 if there is a sequence of transitions $s_0 \xrightarrow{a_1} s_1 \xrightarrow{a_2} \cdots s_{n-1} \xrightarrow{a_n} s_n$ where X is the set of traces of s_n.*

– *A state s is deterministic if for each $a \in \mathcal{I}(s)$ there is exactly one state s' such that $s \xrightarrow{a} s'$, and moreover s' is deterministic.*
  A state s is a possible world of a state s_0 if s is deterministic and $s\ \mathcal{R}\ s_0$ for some ready simulation $\mathcal{R}$.

Two states s and s' are *simulation*, *ready simulation*, or *2-nested simulation equivalent* if there exist simulations, ready simulations, or 2-nested simulations $\mathcal{R}_1$ and $\mathcal{R}_2$, respectively, with $s\,\mathcal{R}_1\,s'$ and $s'\,\mathcal{R}_2\,s$. They are *bisimilar* if there is a bisimulation that relates them. They are *possible futures*, *possible worlds*, *ready trace*, *failure trace*, *ready*, *failure*, *completed trace*, or *trace equivalent* if they have the same possible futures, possible worlds, ready traces, failure traces, ready pairs, failure pairs, completed traces, or traces, respectively.

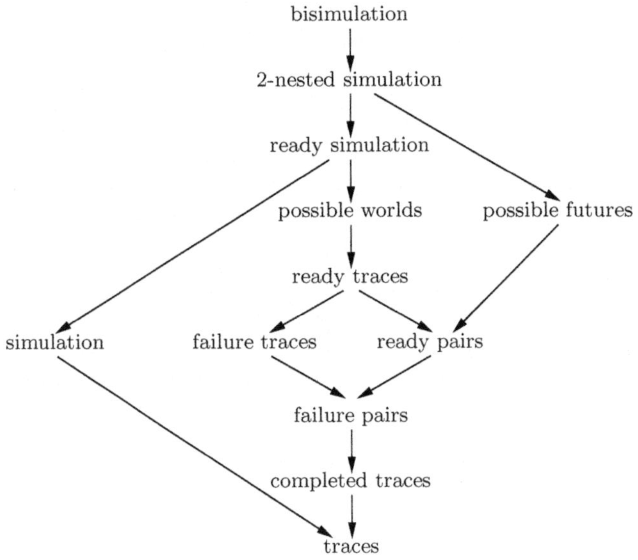

Fig. 1. The Linear Time/Branching Time Spectrum

The linear time/branching time spectrum is depicted in Figure 1, where a directed edge from one semantics to another means that the source of the edge is included in the target.

3.2 BCCSP

BCCSP is a basic process algebra for expressing finite process behaviour. Its signature consists of the constant $\mathbf{0}$, the binary operator $\_ + \_$ called *alternative composition*, and unary *prefix* operators $a\_$, where a ranges over a nonempty set A of actions, called the *alphabet* (with typical elements a, b, c, d). Intuitively, closed BCCSP terms represent finite process behaviour, where $\mathbf{0}$ does not exhibit any behaviour, $p+q$ is the nondeterministic choice between the behaviours of p and q, and ap executes action a to transform into p. This intuition is captured by the transition rules below, in which a ranges over A. They give rise to A-labelled transitions between BCCSP terms.

$$\frac{}{ax \xrightarrow{a} x} \qquad \frac{x \xrightarrow{a} x'}{x+y \xrightarrow{a} x'} \qquad \frac{y \xrightarrow{a} y'}{x+y \xrightarrow{a} y'}$$

We use *summation* $\sum_{i=1}^{n} t_i$, with $n \geq 0$, to denote $t_1 + \cdots + t_n$, where the empty sum denotes $\mathbf{0}$.

The semantics in the linear time/branching time spectrum all constitute a *congruence* for BCCSP, meaning that $p_1 \sim q_1$ and $p_2 \sim q_2$ imply $ap_1 \sim aq_1$ for $a \in A$ and $p_1 + p_2 \sim q_1 + q_2$, where $\sim$ ranges over the semantics in the spectrum.

3.3 Positive and Negative Results for BCCSP

In this section we will survey positive and negative results, and open questions, on the existence of a finite basis for the equational theories of BCCSP modulo the equivalences in the spectrum above. The axiomatizations that we will present for the different semantics in the spectrum were mostly taken from [24].

In case of an infinite alphabet, occurrences of action names in axioms are interpreted as variables (or action schemes).

Bisimulation. The core axioms in Table 1 are sound and ground complete for BCCSP modulo bisimulation. Moller [37] proved using normal forms that this axiomatization is ω-complete; Groote provided an alternative proof of this result in [25] using inverted substitutions.

Table 1. The axioms for bisimulation

A1	$x + y$	$\approx y + x$
A2	$(x+y)+z$	$\approx x + (y+z)$
A3	$x + x$	$\approx x$
A6	$x + \mathbf{0}$	$\approx x$

2-Nested Simulation and Possible Futures. Aceto, Fokkink, van Glabbeek and Ingolfsdottir [4] proved that BCCSP modulo any semantics no coarser than possible futures and no finer than 2-nested simulation does not possess a finite sound and ground complete axiomatization. The infinite family of equations that they used to prove this negative result is defined as follows. Let E be any finite axiomatization for BCCSP that is sound modulo possible futures. Let the *depth* of a BCCSP term t be the largest number of transitions in sequence that t can exhibit. Pick an m such that

$$m > \max\{\mathit{depth}(t), \mathit{depth}(u) \mid (t \approx u) \in E\} \ .$$

For $n \geq 0$, let p_n and q_n be defined inductively as follows, for some $a \in A$:

$$\begin{array}{llll} p_0 &= a^{2m-1}\mathbf{0} & q_0 &= a^{m-1}\mathbf{0} \\ p_{n+1} &= ap_n + aq_n & q_{n+1} &= ap_n \ . \end{array}$$

The equations $p_n \approx q_n$ for $n \geq 2$ are sound modulo 2-nested simulation. However, they cannot be derived from E.

Ready Simulation. Van Glabbeek presented a conditional axiom for ready simulation equivalence: $\mathcal{I}(x) = \mathcal{I}(y) \Rightarrow a(x+y) \approx a(x+y) + ay$. Blom, Fokkink and Nain [14] showed that a sound and ground complete finite equational axiomatization for BCCSP modulo ready simulation exists. It is obtained by extending the four core axioms with

$$a(bx + by + z) \approx a(bx + by + z) + a(bx + z) \ ,$$

where a, b range over A. When A is infinite, Groote's technique of inverted substitutions can be applied to show that this axiomatization is ω-complete. When A is finite, it remains an open question whether BCCSP modulo ready simulation is finitely based.

Simulation. A sound and ground complete axiomatization for BCCSP modulo simulation is obtained by extending the four core axioms with

$$a(x + y) \approx a(x + y) + ay \ .$$

When A is infinite, Groote's technique of inverted substitutions can be applied to show that this axiomatization is ω-complete. When $1 < |A| < \infty$, it remains an open question whether BCCSP modulo simulation is finitely based. When $|A| = 1$, simulation equivalence coincides with trace equivalence, and we will see that in this case a finite basis does exist.

Possible Worlds. A sound and ground complete axiomatization for BCCSP modulo possible worlds is obtained by extending the four core axioms with

$$a(bx + by + z) \approx a(bx + z) + a(by + z) \ .$$

When A is infinite, Groote's technique of inverted substitutions can be applied to show that this axiomatization is ω-complete. Fokkink and Nain [20] showed that when $1 < |A| < \infty$, BCCSP modulo any semantics no coarser than ready equivalence and no finer than possible worlds equivalence does not possess a finite basis. (Note that ready traces are within this semantic range.) Their proof of this negative result, which uses cover equations and applies the compactness theorem to the equational theory for terms of depth 1, is based on the following infinite family of equations:

$$a(\sum_{i=1}^{|A|-1} x_i) + \sum_{j=1}^{|A|-1} a(\sum_{i=1}^{j-1} x_i + \sum_{i=j+1}^{n} x_i) + \sum_{j=|A|}^{n} a(\sum_{i=1}^{|A|-1} x_i + x_j + y_j) \approx$$

$$a(\sum_{i=1}^{|A|-1} x_i) + \sum_{j=1}^{|A|-1} a(\sum_{i=1}^{j-1} x_i + \sum_{i=j+1}^{n} x_i) + \sum_{j=|A|}^{n} a(\sum_{i=1}^{|A|-1} x_i + x_j + y_j) + a(\sum_{i=1}^{n} x_i).$$

These equations are sound modulo possible worlds for $n \geq |A|$. However, any finite axiomatization that is sound for BCCSP modulo ready pairs cannot derive them all. When $|A| = 1$, possible worlds equivalence coincides with completed trace equivalence, and we will see that in this case a finite basis does exist.

Ready Traces. Van Glabbeek presented a conditional axiom for ready trace equivalence: $\mathcal{I}(x) = \mathcal{I}(y) \Rightarrow ax + ay \approx a(x + y)$. Blom, Fokkink and Nain [14] showed that when A is finite, a sound and ground complete finite equational axiomatization for BCCSP modulo ready traces exists. It is obtained by extending the four core axioms with

$$a(\sum_{i=1}^{|A|}(b_i x_i + b_i y_i) + z) \approx a(\sum_{i=1}^{|A|} b_i x_i + z) + a(\sum_{i=1}^{|A|} b_i y_i + z) \ .$$

When A is infinite, they showed using the compactness theorem that a finite sound and ground complete axiomatization does not exist. Their proof is based on the following equations, for $n > 0$:

$$a(\sum_{i=1}^{n}(b_i c\mathbf{0} + b_i d\mathbf{0})) \approx a(\sum_{i=1}^{n} b_i c\mathbf{0}) + a(\sum_{i=1}^{n} b_i d\mathbf{0}) \ .$$

When $1 < |A| < \infty$, the aforementioned negative result from [20] (see the paragraph on possible worlds) implies that BCCSP modulo ready traces does not possess a finite basis. When $|A| = 1$, ready trace equivalence coincides with completed trace equivalence, and we will see that in this case a finite ω-complete axiomatization does exist.

Failure Traces. Van Glabbeek presented a conditional axiom for failure traces (the same one as for ready traces). Blom, Fokkink and Nain [14] showed using normal forms that a sound and ground complete finite equational axiomatization for BCCSP modulo failure traces exists. It is obtained by extending the four core axioms with

$$a(bx + by + z) \approx a(bx + by + z) + a(by + z)$$
$$ax + ay \approx ax + ay + a(x + y) \ .$$

When A is infinite, Groote's technique of inverted substitutions can be applied to show that this axiomatization is ω-complete. When $1 < |A| < \infty$, it remains an open question whether BCCSP modulo failure traces is finitely based. When $|A| = 1$, failure trace equivalence coincides with completed trace equivalence, and we will see that in this case a finite basis does exist.

Ready Pairs. A sound and ground complete axiomatization for BCCSP modulo ready pairs is obtained by extending the four core axioms with

$$a(bx + by + z) \approx a(bx + by + z) + a(by + z) \ .$$

When A is infinite, Groote's technique of inverted substitutions can be applied to show that this axiomatization is ω-complete. When $1 < |A| < \infty$, the aforementioned negative result from [20] (see the paragraph on possible worlds) implies that BCCSP modulo ready pairs does not possess a finite basis. When $|A| = 1$, ready equivalence coincides with completed trace equivalence, and we will see that in this case a finite basis does exist.

Failure Pairs. A sound and ground complete axiomatization for BCCSP modulo failure pairs is obtained by extending the four core axioms with

$$a(bx + by + z) \approx a(bx + by + z) + a(bx + z)$$
$$ax + a(y + z) \approx ax + a(y + z) + a(x + y) \ .$$

Fokkink and Nain [21] proved using cover equations that when A is infinite, this axiomatization is ω-complete. They also proved that when A is finite, one extra axiom is needed to obtain an ω-complete axiomatization:

$$a(\sum_{i=1}^{|A|} b_i x_i + y + z) \approx a(\sum_{i=1}^{|A|} b_i x_i + y + z) + a(\sum_{i=1}^{|A|} b_i x_i + y) \ .$$

Completed Traces. A sound and ground complete axiomatization for BCCSP modulo completed traces is obtained by extending the four core axioms with

$$a(bw + y) + a(cx + z) \approx a(bw + cx + y + z) \ .$$

Groote [25] proved using normal forms that in order to obtain an ω-complete axiomatization, one extra axiom is needed:

$$ax + a(y + z) \approx ax + a(y + z) + a(x + y) \ .$$

Traces. A sound and ground complete axiomatization for BCCSP modulo traces is obtained by extending the four core axioms with

$$ax + ay \approx a(x + y) \ .$$

Groote [25] proved using normal forms that this axiomatization is ω-complete when $|A| > 1$. When $|A| = 1$, it is not hard to see that one extra axiom, $ax + x \approx ax$, suffices to make the axiomatization ω-complete. Indeed, in that case, the algebra of closed BCCSP terms modulo trace equivalence is isomorphic to the algebra $\mathcal{N}$ in Example 3. (To the best of our knowledge this is the first time this last observation appears in print.)

3.4 Overview

Concluding, BCCSP has a finite sound and ground complete axiomatization for most of the semantics in the linear time/branching time spectrum. Only for 2-nested simulation and possible futures, and for ready traces in case of an infinite alphabet, such an axiomatization does not exist.

Regarding ω-completeness, matters are more mixed, especially when $1 < |A| < \infty$. The table below presents an overview, where + means that there a finite basis, – means that there is no finite basis, and ? means that it is unknown whether a finite basis exists. We distinguish between an infinite alphabet, a finite alphabet with more than one element, and a singleton alphabet.

	$\|A\| = 1$	$1 < \|A\| < \infty$	$\|A\| = \infty$
bisim	+	+	+
2-nes sim	–	–	–
poss futu	–	–	–
ready sim	?	?	+
sim	+	?	+
poss worl	+	–	+
ready tr	+	–	–
failure tr	+	?	+
ready	+	–	+
failure	+	+	+
compl tr	+	+	+
traces	+	+	+

4 Parallelism

In this section we discuss extensions of BCCSP with a binary operation $\|$ for parallel composition. We only consider bisimulation semantics. The intuition is that $p \| q$ does a move from either component, or establishes some kind of synchronization between its components. The synchronization mechanism differs from one process description language to another. For the sake of generality, we make use of the mechanism incorporated in ACP, and show how it can be instantiated, e.g., to the synchronization mechanism of CCS.

ACP's synchronization mechanism presupposes a *communication function* γ, i.e., a partial function

$$\gamma : A \times A \rightharpoonup A$$

such that for all $a, b, c \in A$:

(i) if $\gamma(a,b)$ is defined, then so is $\gamma(b,a)$ and moreover $\gamma(a,b) = \gamma(b,a)$; and
(ii) $\gamma(a,\gamma(b,c))$ is defined iff $\gamma(\gamma(a,b),c)$ is defined, and if both are defined, then $\gamma(a,\gamma(b,c)) = \gamma(\gamma(a,b),c)$.

The operational semantics of $\|$ is then given by the following transition rules:

$$\frac{x \xrightarrow{a} x'}{x \| y \xrightarrow{a} x' \| y} \qquad \frac{y \xrightarrow{a} y'}{x \| y \xrightarrow{a} x \| y'} \qquad \frac{x \xrightarrow{a} x',\ y \xrightarrow{b} y',\ \gamma(a,b) = c}{x \| y \xrightarrow{c} x' \| y'}$$

By additional assumptions on γ we can obtain the different versions of parallel composition that are encountered in the literature; we give three examples:

1. The assumption $\gamma = \emptyset$ expresses that there is no communication at all, i.e., the operation $\|$ models *pure interleaving*.

2. The assumption that $\gamma(a, \gamma(b, c))$ is always undefined expresses that there is only *handshaking* communication.
3. We get the operation for parallel composition of CCS by assuming that
   (a) A contains a special action τ;
   (b) there is a bijection $\bar{\cdot}$ on $A - \{\tau\}$ such that $\bar{\bar{a}} = a$ and $\bar{a} \neq a$ for all $a \in A - \{\tau\}$;
   (c) $\gamma(a, \bar{a}) = \gamma(\bar{a}, a) = \tau$ for all $a \in A - \{\tau\}$, and γ is undefined otherwise.

Let $\text{BCCSP}_{\|}$ be the extension of BCCSP with $\|$. A ground complete axiomatization for $\text{BCCSP}_{\|}$ modulo bisimulation equivalence is obtained by adding to the axioms A1–3,6 in Table 1 the equations generated by the so-called *Expansion Law*: for all $t = \sum_{i \in I} a_i x_i$ and $u = \sum_{j \in J} b_j y_j$:

$$t \parallel u \approx \sum_{i \in I} a_i\,(x_i \parallel u) + \sum_{j \in J} b_j\,(t \parallel y_j) + \sum_{i \in I}\sum_{j \in J} \gamma(a_i, b_j)(x_i \parallel y_j) \ , \qquad (3)$$

with, for $i \in I$ and $j \in J$, the summand $\gamma(a_i, b_j)(t_i \parallel u_j)$ only present when $\gamma(a_i, b_j)$ is defined. The result was first established by Hennessy and Milner [29].

Since the Expansion Law generates infinitely many equations, the aforementioned ground complete axiomatization is infinite. If the set of actions A contains at least one element a such that $\gamma(a, a)$ is undefined, then a finite ground complete axiomatization is not possible, as shown by Moller [37, 39]. He establishes that there does not exist a finite set of $\text{BCCSP}_{\|}$-equations, sound with respect to bisimulation equivalence, from which all equations of the form

$$a\mathbf{0} \parallel \varphi_n \approx a\varphi_n + \sum_{i=1}^{n} a a^i \quad (\text{with } \varphi_n = \sum_{i=1}^{n} a^i,\ n \geq 1) \qquad (4)$$

are equationally derivable. Moller carries out his proof in a pure interleaving setting (i.e., $\gamma = \emptyset$), but it is easy to see that the assumption can be relaxed to: $\gamma(a, a)$ is undefined. First note that, with the relaxed requirement, the equations in (4) are still sound with respect to bisimulation equivalence. Now, suppose there does exist a finite basis E for $\text{BCCSP}_{\|}$ modulo bisimulation equivalence. Then, since the equations in (4) are sound with respect to bisimulation equivalence, they are all derivable from E. Let $E' \subseteq E$ be the set of equations in E that are involved in the derivations of the equations in (4). Then E' consists of equations in which no actions other than a occur (for if p and q are bisimulation equivalent closed $\text{BCCSP}_{\|}$-terms, then p and q contain the same actions). Obviously, the equations in E' are all sound for $\text{BCCSP}_{\|}$ with $A = \{a\}$ and $\gamma = \emptyset$, contradicting Moller's result.

Moller's result shows that for a finite axiomatization of parallel composition auxiliary operators are indispensable. Three such auxiliary operators have been proposed in the literature: Bergstra and Klop introduced the *left merge* ($\llfloor$) in [8] and the *communication merge* ($|$) in [10], and Hennessy [28] introduced an operation that we call *Hennessy's merge* ($|\!\!/$). In the remainder of this section we discuss these auxiliary operators in the context of BCCSP. In the next section we examine the consequences of replacing action prefixing in BCCSP by a binary operation for sequential composition.

4.1 Left Merge

First we consider the special case of axiomatizing parallel composition under the pure interleaving assumption ($\gamma = \emptyset$). In that case, as can be seen from the transition rules for $\|$, a parallel composition $p \| q$ either does a move $p \xrightarrow{a} p'$ from its left component p and proceeds as $p' \| q$, or it does a move $q \xrightarrow{a} q'$ from its right component q and proceeds as $p \| q'$. So, intuitively, it is an alternative composition of two subprocesses. The auxiliary operation left merge is a device for expressing these subprocesses in terms of p and q; its operational semantics is given by the following transition rule:

$$\frac{x \xrightarrow{a} x'}{x \mathbin{|\!\!\lfloor} y \xrightarrow{a} x' \| y}$$

Using the left merge the intuition with respect to the behaviour of a parallel composition can be captured in a single equation:

$$\text{M} \quad x \| y \approx x \mathbin{|\!\!\lfloor} y + y \mathbin{|\!\!\lfloor} x \ .$$

The axiom M and the axioms L1–3 in Table 2 allow the elimination of all occurrences of $\|$ and $\mathbin{|\!\!\lfloor}$ from closed terms. (Bergstra and Klop [10] established a similar result in the more general setting of ACP.) Hence, together with the axioms of BCCSP in Table 1, those equations constitute a ground complete axiomatization of $\text{BCCSP}_{\|,\mathbin{|\!\!\lfloor}}$.

Table 2. The axioms for left merge

L1	$\mathbf{0} \mathbin{|\!\!\lfloor} x$	$\approx \mathbf{0}$
L2	$ax \mathbin{|\!\!\lfloor} y$	$\approx a(x \| y)$
L3	$(x + y) \mathbin{|\!\!\lfloor} z$	$\approx x \mathbin{|\!\!\lfloor} z + y \mathbin{|\!\!\lfloor} z$
L4	$(x \mathbin{|\!\!\lfloor} y) \mathbin{|\!\!\lfloor} z$	$\approx x \mathbin{|\!\!\lfloor} (y \| z)$
L5	$x \mathbin{|\!\!\lfloor} \mathbf{0}$	$\approx x$

An ω-complete axiomatization is obtained by adding the axioms L4 and L5 in Table 2. Moller [37] proved this assuming that A is infinite. He used the technique based on normal forms: first he showed that every term is provably equal to a normal form, and then he argued that for distinct normal forms there is a distinguishing environment. Moller used a distinguishing environment that substitutes a special action (not already occurring in either normal form) for every variable, which is only possible if there are infinitely many actions. Both the proof that every term is provably equal to a normal form and the proof that normal forms can be distinguished are quite involved. It turns out that Groote's inverted substitutions technique also applies (see [25]), and the application is in fact quite straightforward.

The requirement that A is infinite seems essential for the application of Groote's technique. However, the authors have recently established that Moller's proof can be adapted with a distinguishing environment that only requires one action. So, if $\gamma = \emptyset$, then the axioms of BCCSP together with the axioms in Table 2 constitute a basis for $\mathrm{BCCSP}_{\|,\llcorner\!\llcorner}$ modulo bisimulation equivalence, for each non-empty set of actions A. Hence, if A is finite, then $\mathrm{BCCSP}_{\|,\llcorner\!\llcorner}$ modulo bisimulation equivalence is finitely based.

4.2 Communication Merge

If $p \xrightarrow{a} p'$ and $q \xrightarrow{b} q'$ and $\gamma(a,b) = c$, then the parallel composition $p \| q$ has the extra option to perform the synchronization move $p \| q \xrightarrow{c} p' \| q'$. The communication merge provides notation for this part of the behaviour of a parallel composition; its operational semantics is given by the following transition rule:

$$\frac{x \xrightarrow{a} x',\ y \xrightarrow{b} y',\ \gamma(a,b) = c}{x \mid y \xrightarrow{c} x' \| y'}$$

Of course, if synchronization is possible, then the axiom M is not sound and needs to be replaced by:

$$\text{M'} \quad x \| y \approx (x \llcorner\!\llcorner y + y \llcorner\!\llcorner x) + x \mid y \ .$$

Using the axiom M', the axioms L1–3 in Table 2 and the axioms C1–5 in Table 3 all occurrences of $\|$, $\llcorner\!\llcorner$ and $\mid$ can be eliminated from closed terms. Hence, together with the axioms of BCCSP in Table 1, those equations constitute a ground complete axiomatization of $\mathrm{BCCSP}_{\|,\llcorner\!\llcorner,\mid}$.

Let us now consider ω-completeness. Note that it critically depends on γ whether certain equations between terms with variables are sound. For instance, the equation

$$x \mid y \approx \mathbf{0}$$

is sound if $\gamma = \emptyset$, but if there exist actions a and b such that $\gamma(a,b)$ is defined, then it is clearly not sound.

Groote [25] proved that if A is a commutative semigroup under γ (which means that γ is an associative and commutative total function on A), and A is moreover freely generated by some infinite subset, then the axioms of BCCSP in Table 1 together with M' and the axioms in Tables 2 and 3 constitute an ω-complete axiomatization. (Of course, since γ is total, the axiom C3 is superfluous

Table 3. The axioms for communication merge

C1	$\mathbf{0} \mid x$	$\approx \mathbf{0}$	
C2	$ax \mid by$	$\approx c(x \| y)$	if $\gamma(a,b) = c$
C3	$ax \mid by$	$\approx \mathbf{0}$	if $\gamma(a,b)$ is undefined
C4	$(x+y) \mid z$	$\approx x \mid z + y \mid z$	
C5	$x \mid y$	$\approx y \mid x$	
C6	$(x \mid y) \mid z$	$\approx x \mid (y \mid z)$	
C7	$x \mid (y \llcorner\!\llcorner z)$	$\approx (x \mid y) \llcorner\!\llcorner z$	

in this axiomatization.) It is an open problem whether it is necessary to require A to be generated by an *infinite* subset.

If γ satisfies the requirement that $\gamma(a, \gamma(b, c))$ is undefined for all $a, b, c \in A$ (i.e., there is only handshaking communication), then the axiom

$$\text{H} \quad x \mid y \mid z \approx \mathbf{0}$$

is sound. We conjecture that if A is non-empty and γ implements the CCS communication mechanism, then the axioms M' and H together with the axioms in Tables 1, 2 and 3 constitute an ω-complete axiomatization.

4.3 Hennessy's Merge

In [28], Hennessy proposed another auxiliary operator, using it in his axiomatizations of observation congruence and timed congruence. *Hennessy's merge*, as we call it, combines the behaviour of the left merge and the communication merge. Its operational semantics is given by the following transition rules:

$$\frac{x \xrightarrow{a} x'}{x \not| \, y \xrightarrow{a} x' \parallel y} \qquad \frac{x \xrightarrow{a} x',\ y \xrightarrow{b} y',\ \gamma(a,b) = c}{x \not| \, y \xrightarrow{c} x' \parallel y'}$$

Note that with Hennessy's merge, parallel composition is definable with the following equation:

$$x \parallel y \approx x \not| \, y + y \not| \, x \ .$$

This may seem promising for the existence of a finite axiomatization of parallel composition that only uses Hennessy's merge as auxiliary operation. However, as was already conjectured by Bergstra and Klop in [10], it turns out that the operation itself cannot be finitely axiomatized. Assuming the CCS synchronization mechanism (see the beginning of Section 4), the authors recently proved in [3] that there does not exist a finite set of sound $\text{BCCSP}_{\parallel, \not|}$-equations from which all equations of the form

$$a\mathbf{0} \not| \, \psi_n \approx a\psi_n + \sum_{i=0}^{n} \tau a^i \quad (\text{with } \psi_n = \sum_{i=0}^{n} \bar{a}a^i,\ n \geq 0)$$

are equationally derivable.

4.4 Overview

In the table below we summarize the results and open problems discussed in this section. A + in the first (respectively, second) column means that there exists a *finite* ground complete (respectively, ω-complete) axiomatization, a − means that such an axiomatization does not exist, and a ? means that it is unknown whether such an axiomatization exists.

	ground complete	ω-complete
$\text{BCCSP}_{\parallel}$	−	−
$\text{BCCSP}_{\parallel, \parallel\!\!\_}$	+	+
$\text{BCCSP}_{\parallel, \parallel\!\!\_, \mid}$ (handshaking)	+	?
$\text{BCCSP}_{\parallel, \not|}$	−	−

Moller's result shows that $\mathrm{BCCSP}_{\|}$ has no finite ground complete axiomatization. With the two auxiliary binary operations $\|\!\!\_$ and $|$ of Bergstra and Klop a finite ground complete axiomatization becomes possible. If one assumes pure interleaving, then adding only $\|\!\!\_$ suffices, and then there even exists a finite ω-complete axiomatization. It remains an open problem whether it is possible to axiomatize $\mathrm{BCCSP}_{\|}$ with arbitrary handshaking or the CCS synchronization mechanism adding only one auxiliary binary operation.

5 Sequential Composition

In this section we discuss the consequences of having sequential composition instead of action prefixing. We remove the constant $\mathbf{0}$ and the unary prefixes $a\_$ from BCCSP, and replace them by a binary operation $\cdot$ for sequential composition, treating the actions in A as constant symbols. Thus we get the signature of BPA [10]. The transition rules for actions and sequential composition are as follows:

$$\frac{}{a \xrightarrow{a} \surd} \qquad \frac{x \xrightarrow{a} x'}{x \cdot y \xrightarrow{a} x' \cdot y}$$

Note the special state $\surd$ that we use to write the transition rules; it signals successful termination. To make the rules work, we stipulate that $\surd \cdot x = x$ and $\surd \parallel x = x \parallel \surd = x$. We also require that bisimulations relate $\surd$ only to $\surd$.

Table 4. The axioms for alternative and sequential composition

A1	$x + y$	$\approx y + x$
A2	$x + (y + z)$	$\approx (x + y) + z$
A3	$x + x$	$\approx x$
A4	$(x + y) \cdot z$	$\approx x \cdot z + y \cdot z$
A5	$(x \cdot y) \cdot z$	$\approx x \cdot (y \cdot z)$

The axioms of BPA are obtained by taking the first three axioms of BCCSP, adding that $\cdot$ distributes from the right over $+$ and that $\cdot$ is associative. For the sake of clarity, we list them all in Table 4. It is folklore that they constitute a ground complete axiomatization of BPA. Moreover, the axiomatization is ω-complete. As far as we know, this latter result does not explicitly appear in print, but a proof can be extracted from the ω-completeness proof for PA [19] that we discuss below.

In [38], Moller adapted his proof that $\mathrm{BCCSP}_{\|}$ is not finitely based to the setting with sequential composition. The infinite family of equations he uses to establish this result is obtained from the equations in (4) by the obvious translation (replace action prefixes by actions and sequential compositions, and omit all occurrences of $\mathbf{0}$).

The signature of PA combines that of BPA with $\|$ and $\|\!\!\_$. Parallel composition in PA stands for pure interleaving (i.e., $\gamma = \emptyset$), so the relation between $\|$ and $\|\!\!\_$

Table 5. The axioms for merge and left merge

M1	$x \parallel y$	$\approx x \mathbin{⌊\!⌊} y + y \mathbin{⌊\!⌊} x$
M2	$a \mathbin{⌊\!⌊} x$	$\approx a \cdot x$
M3	$a \cdot x \mathbin{⌊\!⌊} y$	$\approx a \cdot (x \parallel y)$
M4	$(x + y) \mathbin{⌊\!⌊} z$	$\approx x \mathbin{⌊\!⌊} z + y \mathbin{⌊\!⌊} z$
M5	$(x \mathbin{⌊\!⌊} y) \mathbin{⌊\!⌊} z$	$\approx x \mathbin{⌊\!⌊} (y \parallel z)$
M6	$(x \cdot \alpha) \mathbin{⌊\!⌊} \alpha$	$\approx (x \mathbin{⌊\!⌊} \alpha) \cdot \alpha$

is expressed by the axiom M1 in Table 5. For a ground complete axiomatization of PA it suffices to add the first four axioms in Table 5 to the axioms of BPA. Fokkink and Luttik proved in [19] that if M5 and M6 are added too, then the axiomatization is ω-complete. The α in M6 ranges over finite sums of distinct actions. The equation that results by replacing both occurrences of $⌊\!⌊$ in M6 with $\parallel$ is also sound (it is an instructive exercise to derive it using M6 and the other axioms). It is an example of a so-called *mixed equation*, equating a parallel composition and a sequential composition. There is a deep theory of mixed equations developed by Hirshfeld and Jerrum [30] for the benefit of their proof that bisimulation equivalence is decidable for normed PA. The proof in [19] that the presented axiomatization is ω-complete partly relies on their theory.

Incorporation of synchronization can be done by adding a communication merge and replacing the axiom M1 by M'. It is not difficult to adapt the axioms in Table 3 in such a way that all communication merges can be eliminated from closed terms; thus a ground complete axiomatization can be obtained. Note that this does require the addition of a special constant δ that will assume the rôle of **0**; it satisfies

$$\delta \cdot x \approx \delta$$
$$x + \delta \approx x \ .$$

There are no known ω-completeness results pertaining to the extension of BPA with $\parallel$, $|$ and δ. It would again be necessary to make some assumptions about the synchronization mechanism. If there is only handshaking communication, then the equations

$$(\cdots(((x_1 \mid x_2) \cdot y_1 \mathbin{⌊\!⌊} z_1) \cdot y_2 \mathbin{⌊\!⌊} z_2) \cdot \cdots \cdot y_n \mathbin{⌊\!⌊} z_n) \mid x_3 \approx \delta \qquad (n \geq 0)$$

are sound. We conjecture that there does not exist a finite set of sound equations from which they are all derivable.

References

1. L. Aceto, Z. Ésik, and A. Ingolfsdottir. The max-plus algebra of the natural numbers has no finite equational basis. *Theoretical Computer Science*, 293(1):169–188, 2003.
2. L. Aceto, W.J. Fokkink, and A. Ingolfsdottir. A menagerie of non-finitely based process semantics over BPA*—from ready simulation to completed traces. *Mathematical Structures in Computer Science*, 8(3):193–230, 1998.

3. L. Aceto, W.J. Fokkink, A. Ingolfsdottir, and S.P. Luttik. CCS with Hennessy's merge has no finite equational axiomatization. *Theoretical Computer Science*, 330(3):377–405, 2005.
4. L. Aceto, W.J. Fokkink, R.J. van Glabbeek, and A. Ingolfsdottir. Nested semantics over finite trees are equationally hard. *Information and Computation*, 191(2):203–232, 2004.
5. J.C.M. Baeten. A brief history of process algebra. *Theoretical Computer Science*, 335(2–3):131–146, 2005.
6. J.C.M. Baeten and W.P. Weijland. *Process Algebra*. Cambridge Tracts in Theoretical Computer Science 18. Cambridge University Press, 1990.
7. J.A. Bergstra and J. Heering. Which data types have omega-complete initial algebra specifications? *Theoretical Computer Science*, 124(1):149–168, 1994.
8. J.A. Bergstra and J.W. Klop. Fixed point semantics in process algebras. Report IW 206, Mathematisch Centrum, Amsterdam, 1982.
9. J.A. Bergstra and J.W. Klop. The algebra of recursively defined processes and the algebra of regular processes. In J. Paredaens, editor, *Proceedings 11th Colloquium on Automata, Languages and Programming*, Antwerp, LNCS 172, pages 82–95. Springer-Verlag, 1984.
10. J.A. Bergstra and J.W. Klop. Process algebra for synchronous communication. *Information and Control*, 60(1/3):109–137, 1984.
11. J.A. Bergstra and J.W. Klop. Algebra of communicating processes with abstraction. *Theoretical Computer Science*, 37(1):77–121, 1985.
12. J.A. Bergstra and J.W. Klop. Algebra of communicating processes. In J.W. de Bakker, M. Hazewinkel, and J.K. Lenstra, editors, *Mathematics and Computer Science*, CWI Monograph 1, pages 89–138. North-Holland, 1986.
13. G. Birkhoff. On the structure of abstract algebras. *Proceedings Cambridge Philosophical Society*, 31:433–454, 1935.
14. S.C.C. Blom, W.J. Fokkink, and S. Nain. On the axiomatizability of ready traces, ready simulation and failure traces. In J.C.M. Baeten, J.K. Lenstra, J. Parrow, and G.J. Woeginger, editors, *Proceedings 30th Colloquium on Automata, Languages and Programming*, Eindhoven, LNCS 2719, pages 109–118. Springer-Verlag, 2003.
15. S.N. Burris and H.P. Sankappanavar. *A Course in Universal Algebra*. Graduate Texts in Mathematics. Springer-Verlag, 1981. The Millennium Edition of this book is available at `http://www.math.uwaterloo.ca/~snburris/htdocs/ualg.html`.
16. J.H. Conway. *Regular Algebra and Finite Machines*. In R. Brown and J. De Wet, editors, *Mathematics Series*. Chapman and Hall, 1971.
17. Z. Ésik and S. Okawa. Series and parallel operations on pomsets. In *Proceedings 19th Conference on Foundations of Software Technology and Theoretical Computer Science*, Chennai, LNCS 1738, pages 316–328. Springer-Verlag, 1999.
18. W.J. Fokkink. *Introduction to Process Algebra*. Texts in Theoretical Computer Science, An EATCS Series. Springer-Verlag, January 2000.
19. W.J. Fokkink and S.P. Luttik. An omega-complete equational specification of interleaving. In U. Montanari, J. Rolinn, and E. Welzl, editors, *Proceedings 27th Colloquium on Automata, Languages and Programming*, Geneva, LNCS 1853, pages 729–743. Springer-Verlag, 2000.
20. W.J. Fokkink and S. Nain. On finite alphabets and infinite bases: From ready pairs to possible worlds. In I. Walukiewicz, editor, *Proceedings 7th Conference on Foundations of Software Science and Computation Structures*, Barcelona, LNCS 2897, pages 182–194. Springer-Verlag, 2004.

21. W.J. Fokkink and S. Nain. A finite basis for failure semantics. In L. Caires, G.F. Italiano, L. Monteiro, C. Palamidessi, and M. Yung, editors, *Proceedings 32nd Colloquium on Automata, Languages and Programming*, Lisbon, LNCS 3580, pages 755–765. Springer-Verlag, 2005.
22. J.L. Gischer. The equational theory of pomsets. *Theoretical Computer Science*, 61:199–224, 1988.
23. R.J. van Glabbeek. The linear time-branching time spectrum. In J.C.M. Baeten and J.W. Klop, editors, *Proceedings 1st Conference on Concurrency Theory: Unification and Extension*, Amsterdam, LNCS 458, pages 278–297. Springer-Verlag, 1990.
24. R.J. van Glabbeek. The linear time-branching time spectrum I. The semantics of concrete, sequential processes. In J.A. Bergstra, A. Ponse and S.A. Smolka, editors, *Handbook of Process Algebra*, pages 3–99. North-Holland, 2001.
25. J.F. Groote. A new strategy for proving ω-completeness with applications in process algebra. In J.C.M. Baeten and J.W. Klop, editors, *Proceedings 1st Conference on Concurrency Theory: Unification and Extension*, Amsterdam, LNCS 458, pages 314–331. Springer-Verlag, 1990.
26. J. Heering. Partial evaluation and ω-completeness of algebraic specifications. *Theoretical Computer Science*, 43(2-3):149–167, 1986.
27. M.C.B. Hennessy. *Algebraic Theory of Processes.* MIT Press, 1988.
28. M.C.B. Hennessy. Axiomatising finite concurrent processes. *SIAM Journal on Computing*, 17(5):997–1017, 1988.
29. M.C.B. Hennessy and R. Milner. Algebraic laws for nondeterminism and concurrency. *Journal of the ACM*, 32(1):137–161, 1985.
30. Y. Hirshfeld and M. Jerrum. Bisimulation equivalence is decidable for normed process algebra. In J. Wiedermann, P. van Emde Boas, and M. Nielsen, editors, *Proceedings 26th Colloquium on Automata, Languages and Programming*, Prague, LNCS 1644, pages 412–421. Springer-Verlag, 1999.
31. C.A.R. Hoare. *Communicating Sequential Processes.* Prentice-Hall, 1985.
32. R.M. Keller. Formal verification of parallel programs. *Communications of the ACM*, 19(7):371–384, 1976.
33. S.C. Kleene. Representation of events in nerve nets and finite automata. In C.E. Shannon and J. McCarthy, editors, *Automata Studies*, pages 3–41. Princeton University Press, 1956.
34. Huimin Lin. An interactive proof tool for process algebras. In *Proceedings 9th Symposium on Theoretical Aspects of Computer Science*, Cachan, LNCS 577, pages 617–618. Springer-Verlag, 1992.
35. R. Milner. A complete inference system for a class of regular behaviours. *Journal of Computer and System Sciences*, 28(3):439–466, 1984.
36. R. Milner. *Communication and Concurrency.* Prentice-Hall, 1989.
37. F. Moller. *Axioms for Concurrency.* PhD thesis, Department of Computer Science, University of Edinburgh, July 1989. Report CST-59-89. Also published as ECS-LFCS-89-84.
38. F. Moller. The importance of the left merge operator in process algebras. In M. Paterson, editor, *Proceedings 17th Colloquium on Automata Languages and Programming*, Warwick, LNCS 443, pages 752–764. Springer-Verlag, 1990.
39. F. Moller. The nonexistence of finite axiomatisations for CCS congruences. In *Proceedings 5th Symposium on Logic in Computer Science*, Philadelphia, pages 142–153. IEEE Computer Society Press, 1990.

40. G.D. Plotkin. The λ-calculus is ω-incomplete. *Journal of Symbolic Logic*, 39:313–317, 1974.
41. G.D. Plotkin. A structural approach to operational semantics. Report DAIMI FN-19, Computer Science Department, Aarhus University, 1981.
42. G.D. Plotkin. A structural approach to operational semantics. *Journal of Logic and Algebraic Programming*, 60–61:17–139, 2004. This is a revised version of the original DAIMI memo [41].

Skew and ω-Skew Confluence and Abstract Böhm Semantics

Zena M. Ariola[1] and Stefan Blom[2]

[1] University of Oregon
[2] University of Innsbruck

Abstract. Skew confluence was introduced as a characterization of non-confluent term rewriting systems that had unique infinite normal forms or Böhm like trees. This notion however is not expressive enough to deal with all possible sources of non-confluence in the context of infinite terms or terms extended with letrec. We present a new notion called ω-skew confluence which constitutes a sufficient and necessary condition for uniqueness. We also present a theory that can lift uniqueness results from term rewriting systems to rewriting systems on terms with letrec. We present our results in the setting of Abstract Böhm Semantics, which is a generalization of Böhm like trees to abstract reduction systems.

1 Introduction

For term rewriting systems, it is well-known that uniqueness of normal forms follows from confluence and that given termination, confluence and uniqueness of normal forms are equivalent [27, 32]. Because of this, the normal form of a term is a natural candidate for the semantics of a term. However, there are many term rewriting systems in which there are interesting terms that do not have a normal form. For example, according to the rewriting rule given below:

$$\mathsf{F}\ x \to \mathsf{Cons}(x, \mathsf{F}\ x)$$

the term F 1 does not have a normal form. More precisely, it doesn't have a *finite* normal form. The term does have an *infinite* normal form: the infinite list of ones.

There are several ways to define infinite normal forms on terms. An obvious way to define them is by means of infinitary rewriting [23, 24, 32, 16, 26]. Another way is to use a definition similar to that of the Böhm Tree in the lambda calculus [13] or Böhm like trees for term rewriting systems [25]. These infinite normal forms are closely related to the notion of observational equivalence. A detailed study of the relation between different notions of infinite normal form and contextual or observational equivalence is given in [19].

An advantage of the Böhm Tree approach or Böhm semantics over infinitary rewriting is that it allows one to deal in a simple manner with rewrite rules that remove unused definitions (garbage collection). For example, consider the rewrite rule

$$\mathsf{let}\ x = M\ \mathsf{in}\ N \to N, \text{ if } x \text{ does not occur free in } N\ .$$

A. Middeldorp et al. (Eds.): Processes... (Klop Festschrift), LNCS 3838, pp. 368–403, 2005.

To define Böhm semantics for a rewrite system containing this rule, it suffices to rewrite to normal form with respect to this rule. It should also be possible to deal with this in the setting of infinitary rewriting. The rule can be seen as a rewrite rule in a combinatory reduction system and although the latest results ([26]) cover fully extended CRS's only, it is known how to deal with non fully extended rules such as the η-rule ([30]), so the only thing which is needed is a combination of these two results.

We also prefer the Böhm Tree approach because in programming languages it is very important if a result can be reached in finitely many steps or not. This is immediately clear in the Böhm Tree approach and needs a compression lemma in the infinitary rewriting case.

The notions of infinite normal form and Böhm semantics are related to the notion of information content, also called instant semantics [36] or direct approximation [28, 34]. These notions determine a prefix of the term, which can never be changed by reduction. If we rewrite term graphs represented as terms with letrec then we can follow the same intuition, but we must be careful by how we interpret the prefix. In the style of calculus used by Ariola and Klop, the letrec bindings will remain at the top of the term and keep changing during the entire reduction. Strictly speaking, the only stable prefix is Ω, our constant for undefined. However, if we forget about the syntax and look at the picture of the graph then we will see stable prefixes as usual. Effectively, we have to ignore the letrec's when we determine the stable prefix. The same situation occurs when we consider a term rewriting system in which a strategy has been encoded as a symbol which keeps traveling up and down the term. In that case the administrative symbol(s) have to be ignored. If we consider abstract rewriting systems rather than term rewriting then the information content of a term is simply an observation about that object. The combination of the information contents of all reachable objects is what we refer to as the *Abstract Böhm Semantics.*

An important property is *uniqueness* (also referred to as soundness) of Böhm semantics. This property is similar to uniqueness of normal forms and states that every two convertible terms have the same Böhm semantics. Confluence implies uniqueness of Böhm semantics. However, confluence is not necessary for guaranteeing uniqueness. Skew confluence[1], as introduced in [4, 14, 6], characterizes rewrite systems that have unique Böhm semantics with respect to a notion of direct approximant or notion of finite information content. The idea behind skew confluence is that if there exists a computation that develops a certain information content, then any other computation can be extended to develop more detailed information content. The theory of skew confluence works well for term rewriting and certain forms of term graph rewriting. However, there are problems in applying the notion to other forms of term graph rewriting and infinitary rewriting. These problems are due to the fact that the information content is not really a single observation. It actually is a set of observations. For example, when we observe a term we actually observe the finite prefixes of that term. If the term is finite, the set of observations is finite and skew confluence works.

[1] The name was suggested to us by Jan Willem Klop.

If the term is infinite, the set of observations can become infinite and problems arise with skew confluence. For example, consider the infinite term M below:

$$M \equiv (\lambda f. f\, x\, (f\, x\, (f\, x\, (\cdots))))(I\, g) \ ,$$

where $I \equiv \lambda x.x$. On one side, M rewrites to an infinite normal form in two steps:

$$\begin{aligned}(\lambda f. f\, x\, (f\, x\, (f\, x\, (\cdots))))(I\, g) &\underset{\beta}{\to} (\lambda f. f\, x\, (f\, x\, (f\, x\, (\cdots))))g \\ &\underset{\beta}{\to} g\, x\, (g\, x\, (g\, x\, (\cdots)))) \ .\end{aligned}$$

This infinite normal form has itself as its information content. On the other side, M rewrites to M_1 which does not have a finite normalizing sequence:

$$M \equiv (\lambda f. f\, x\, (f\, x\, (f\, x\, (\cdots))))(I\, g) \underset{\beta}{\to} I\, g\, x\, (I\, g\, x\, (I\, g\, x\, (\cdots))) \equiv M_1 \ .$$

Moreover, in each reduct of M_1 only finitely many of the $(I\, g)$ redexes have been reduced, which means that each reduct has finite information content. For example, the information content of the terms in the sequence

$$I\, g\, x\, (I\, g\, x\, (\cdots)) \underset{\beta}{\to} g\, x\, (I\, g\, x\, (I\, g\, x\, (\cdots))) \underset{\beta}{\to} g\, x\, (g\, x\, (I\, g\, x\, (I\, g\, x\, (\cdots)))) \underset{\beta}{\to} \cdots$$

is

$$\Omega, g\, x\, \Omega, g\, x\, (g\, x\, \Omega), \cdots \ .$$

Because the information content of any reduct of M_1 is finite and the information content of the infinite normal form is infinite, it is impossible that the information content of any reduct of M_1 exceeds that of the normal form. Hence, we do not have skew confluence.

To solve the problem, we introduce a new variant of skew confluence, called *ω-skew confluence*, which is more suitable to the case of infinite information content. The idea behind ω-skew confluence is that if an object (a) reduces to two other objects (a_1, a_2) then for any observation that can be made about the first reduct (a_1) there exists a reduct (a_2') of the second reduct (a_2) about which the same observation can be made. We call this reduct the covering reduct. For example, given any prefix of the infinite normal form of M, we can find a reduct of M_1 whose information content exceeds the given prefix.

Note that for every observation, we may have a different covering reduct (a_2'). If it is possible to find a reduct that covers all observations then we have skew confluence. Such a reduct will always exist if the set of observations is finite, but as we have seen in the example it might not exist if the set of observations is infinite.

Later in this paper, we will consider term graphs represented by cyclic terms or terms with letrec. Cyclic graphs/terms can represent infinite terms. For example, the infinite list of ones can be represented as the graph in Fig. 1. Thus, the information content of a cyclic term can be infinite as well. Because examples of infinite information content for term graphs are lengthy, we will delay them until Sect. 5.

Proving (ω-)skew confluence of a non-confluent rewrite system can be rather tedious and non-confluence often complicates matters. In the case of cyclic calculi the source of non-confluence is often a subset of the rewrite rules which deals

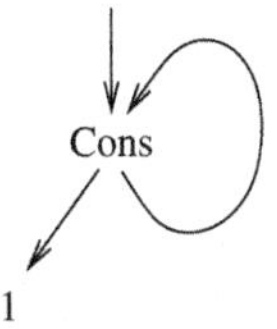

Fig. 1. Finite representation of an infinite list of ones

with substitution and/or unwinding. The idea behind the lifting theory in [15] is to partition the problem into a problem of the substitution rules and a problem of the other rules. The intention is that only the part dealing with substitution is non-confluent. Thus, one deals with non-confluence in a simplified setting and in the more complicated setting one can use confluence. In this paper we present an abstract version of this lifting theory. We start with an abstract rewrite system (ARS for short) equipped with a notion of information content which yields unique Abstract Böhm Semantics. We assume the ARS is equipped with a partial order. (E.g. finite terms with the prefix order.) Next, we consider an extension of the rewrite system where we use the ideal completion of the original set of objects as the semantics of the objects in the extension. (E.g. term graphs using unwinding of a term graph to an infinite term as semantics.) Finally, we define a construction that lifts the notion of information content from the original ARS to the extension and provide conditions under which we have uniqueness of Abstract Böhm Semantics with respect to the lifted notion of information content.

The notions and results about skew confluence are taken from [6]. The notion of abstract Böhm semantics is a modification of the notions in [14, 6]. The results about lifting are abstract versions of the results in [15]. The notion of ω-skew confluence is introduced explicitly for the first time in this paper. It was implicitly present in earlier work, but not identified as a notion.

The paper is organized as follows: We start in Sect. 2 with a few preliminaries. The next section contains an informal description of the different notions of confluence with both abstract examples and naive rewriting examples. In Sect. 4, we formalize skew and ω-skew confluence and the notion of abstract Böhm semantics. In Sect. 5, we present a counterexample to confluence that arises from unwinding a graph in different ways. We discuss how confluence modulo bisimilarity provides a solution. We also explain the need of skew and ω-skew confluence to cope with the loss of confluence when the substitution rules are extended with other rewrite rules. In Sect. 6, we use these notions to show the consistency of the call-by-name and call-by-need cyclic calculi. In Sect. 7, we present an abstract version of the lifting theory. We conclude in Sect. 8.

2 Preliminaries

We will briefly state a few definitions and introduce our notation.

Definition 1. *A partial order is a pair $(S, \leq)$, where $\leq$ is a transitive, reflexive and anti-symmetric binary relation over S. An upper bound of a set $S' \subseteq S$ is* *an element $s \in S$, such that $\forall s' \in S' : s' \leq s$. An element $s \in S$ is the least upper bound of S' (denoted as $\mathrm{lub}\, S'$) if and only if s is an upper bound and $s \leq s'$ for all upper bounds s' of S'. A non-empty set $D \subseteq S$ is a directed set if for every finite subset D' of D there exists $d \in D$ such that d is an upper bound of D'. A partial order $(S, \leq)$ is complete if there exists a least element and every directed subset has a least upper bound. A complete partial order is referred to as a CPO. A set $D \subseteq S$ is downward closed if $\forall d \in S, d' \in D : d \leq d' \Rightarrow d \in D$. A non-empty set $I \subseteq S$ is an ideal if I is downward closed and directed. The set of all ideals over S is denoted by $\mathcal{I}(S)$. The set of all ideals over S ordered by inclusion $(\mathcal{I}(S), \subseteq)$ is called the ideal completion of S, denoted $\mathcal{I}(S, \leq)$. The downward closure of $S' \subseteq S$ is given by*

$$\downarrow S' = \{s \in S \mid \exists s' \in S' : \; s \leq s'\} \; .$$

Definition 2. *Given a CPO $(A, \leq)$. An element $a \in A$ is* finite *if for every directed set $D \subseteq A$, such that $a \leq \mathrm{lub}\, D$, we have that there exists $d \in D$, such that $a \leq d$. The set of all finite elements in A is denoted by $\mathcal{F}(A)$. The CPO is algebraic if $\forall a \in A : a = \mathrm{lub}\{a' \in \mathcal{F}(A) \mid a' \leq a\}$.*

In other words, in an algebraic CPO, each element is a directed limit of its "finite" approximations. For an extensive treatment of the use of partial orders in semantics see [21].

Proposition 1. *If $\mathcal{A} \equiv (A, \leq)$ is a partial order with a least element then the ideal completion $\mathcal{A}_{\mathcal{I}} \equiv (\mathcal{I}(A), \subseteq)$ is an algebraic complete partial order.*

Definition 3. *An ARS is a structure $(A, \rightarrow)$, where A is a set of objects and $\rightarrow \subseteq A \times A$ is a relation, called the reduction relation.*
The transitive, reflexive closure of $\rightarrow$ is denoted $\twoheadrightarrow$.
The equivalence relation generated by $\rightarrow$, also called conversion, is denoted by $\overset{*}{\longleftrightarrow}$ rather than the usual $=$, to avoid overloading of the symbol $=$.

In the lambda calculus, the compatible closure of a relation R is the least relation such that $M\ R\ N \Rightarrow C[M]\ R\ C[N]$ for any context C (see [13]). The constant Ω stands for an undefined term. By replacing an Ω with a larger term you get a "more defined term".

Definition 4. *Let Λ_Ω be the set of lambda calculus terms extended with the constant Ω. We define the order $\leq_\Omega$ as the transitive, reflexive and compatible closure of*

$$\Omega \leq M \;,$$

where $M \in \Lambda_\Omega$.

Note that $(\Lambda_\Omega, \leq_\Omega)$ is a partial order with a least element (Ω). Hence, its ideal completion is an algebraic CPO. Moreover, the ideal completion is one of the representation of infinite lambda terms.

3 Confluence, Skew Confluence and ω-Skew Confluence

In the following, we give an informal description of the properties of *skew* and *ω-skew* confluence through a series of simple examples. To better understand these new properties, we review the well-established notion of confluence and a version of confluence modulo. We start by defining the set of objects A as the set consisting of the bottom element $\bot$, two copies of the set of natural numbers and infinity:

$$A = \{\bot\} \cup \mathbb{N} \cup \{\underline{n} \mid n \in \mathbb{N}\} \cup \{\infty\} \ .$$

By using possibly underlined numbers, we have both a natural equivalence and a natural order on our set of objects. Moreover, the number functions as the information content as well. For example, the numbers $\underline{2}$ and 2 have the same information content: 2.

For each abstract example, we will give a matching example in term rewriting or infinitary term rewriting. These term rewriting examples are derived from graph rewriting examples given in Sect. 5.

3.1 Confluence

Confluence is an important property since it guarantees the consistency of the rewriting theory. If rewriting formalizes execution, then confluence guarantees that execution of a program has a unique result. In other words, diverging computations with the same starting point can always converge on the same intermediate result. We define the reduction relation $\to$ on A as follows:

$$\bot \to 0,\ \ \bot \to \underline{0},\ \ n \to n+1,\ \ \underline{n} \to \underline{n+1},\ \ 2n \to \underline{2n},\ \ \underline{2n+1} \to 2n+1\ .$$

That is, we rewrite each number in $\mathbb{N}$ and its copy to its successor. Moreover, in addition to replacing $\bot$ with 0 and $\underline{0}$ we have rules to relate the numbers and their copies. We then have that $(A, \to)$ is confluent, which means that divergent computations can always be brought together, as shown in the following commuting diagram:

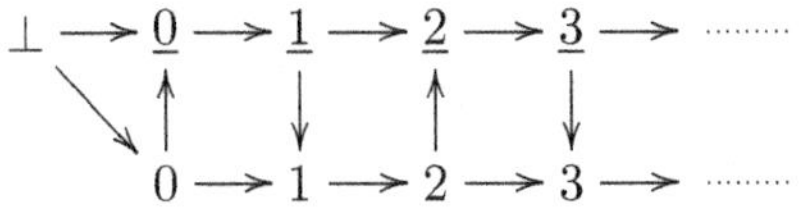

A matching concrete example can be found in term rewriting. The reduction graph of the term A in the TRS:

$$\begin{array}{l} A \to B \\ A \to C \\ C \to B \\ B \to F(C) \\ C \to F(B) \end{array}$$

is:

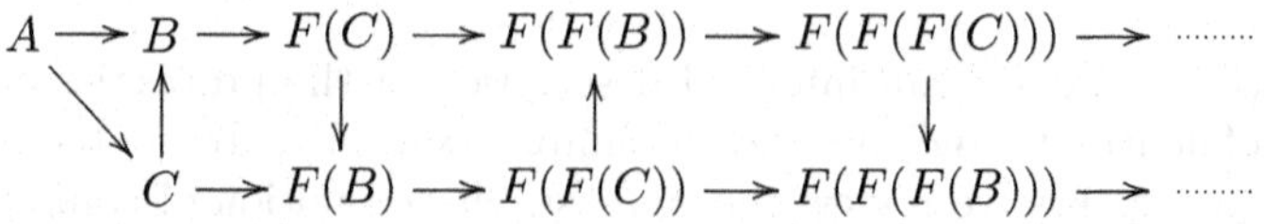

3.2 Confluence Modulo

Confluence could fail for reasons that do not impact the end result. For example, if you optimize computations in different ways, you might not get exactly the same intermediate result, but as long as you perform a single unit of work in a single step you should get equivalent results. Similarly, in modeling execution one often reasons about a program modulo the names of bound variables. To continue with our example, we define the reduction relation $\rightarrow$ on A as follows:

$$\bot \rightarrow 0, \quad \bot \rightarrow \underline{0}, \quad n \rightarrow n+1, \quad \underline{n} \rightarrow \underline{n+1} \ .$$

Unlike before, there are no reduction rules connecting the two copies of $\mathbb{N}$. This causes confluence to fail. To cope with the situation, one defines an equivalence (*i.e.*, reflexive, symmetric and transitive) relation $\sim$ on A and then shows that divergent reductions can always reach *equivalent* terms, as opposed to the *same* term. This is called confluence modulo $\sim$. For our running example, we define $\sim$ on A as follows:

$$a \sim a', \text{ if } |a| = |a'| \ ,$$

where $|.| : A \rightarrow \mathbb{N}$ is defined as:

$$|\bot| = 0, |n| = n+1, |\underline{n}| = n+1 \ .$$

In other words, the difference between the two different copies of n is not essential; we can regard it as "syntactic noise". We then obtain that $(A, \rightarrow)$ is confluent modulo $\sim$. Pictorially:

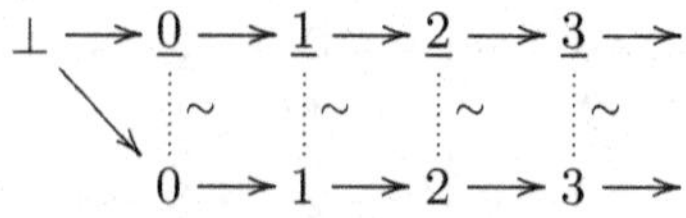

In the context of term graph rewriting, an interesting notion of confluence modulo is confluence modulo bisimilarity [11].

A matching concrete example can be found in term rewriting. The reduction graph of the term A in the TRS:

$$\begin{array}{lcl} A & \rightarrow & B_1 \\ A & \rightarrow & B_2 \\ B_1 & \rightarrow & F(B_1) \\ B_2 & \rightarrow & F(B_2) \end{array}$$

is:

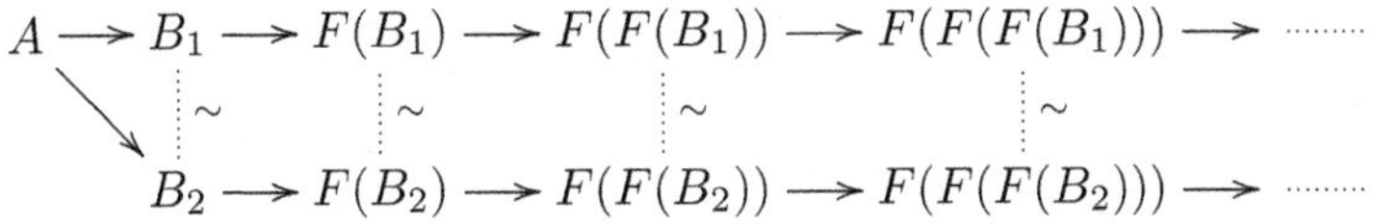

where $\sim$ is the equivalence relation generated by $B_1 \sim B_2$.

3.3 Skew Confluence

The goal of optimization is of course to do more than one unit of work in a single step. But if you do two units of work in a single step then you can easily get an out-of-sync phenomenon. For example, define the reduction relation $\rightarrow$ on A as follows:

$$\bot \rightarrow 0, \quad \bot \rightarrow \underline{1}, \quad n \rightarrow n+2, \quad \underline{n} \rightarrow \underline{n+2} \ .$$

We have that $(A, \rightarrow)$ is neither confluent nor confluent modulo $\sim$. The situation is depicted below:

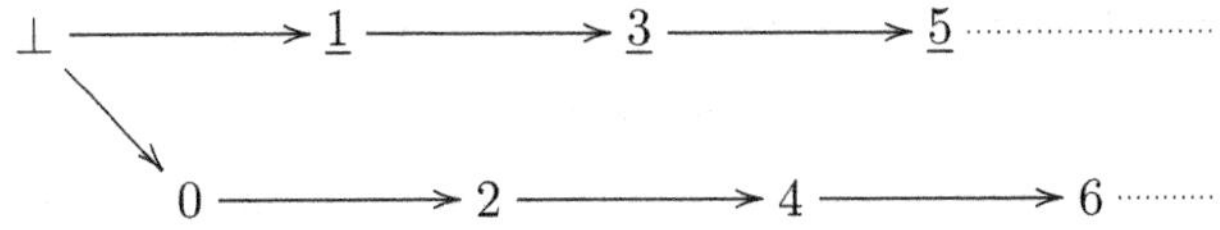

On the top reduction we will always obtain an odd number and on the bottom we will always obtain an even number. However, notice that for every number $\underline{n}$ reached with the top reduction one can always reach a number greater than $\underline{n}$ with the bottom reduction, and vice-versa. Intuitively, it seems that both reductions converge to the same result. That result is infinity and we call it the *abstract Böhm semantics*. In this example, the uniqueness of the abstract Böhm semantics is guaranteed by the notion of skew confluence. Instead of requiring that divergent computations lead to the *same* or *equivalent* term, skew confluence requires that divergent computations reach a result which is *better*. In other words, instead of reasoning up to an equivalence relation we reason up to a quasi-order (*i.e.*, a reflexive and transitive relation). We define a quasi order $\preceq$ on A as follows:

$$a \preceq a', \text{ if } |a| \leq |a'| \ ,$$

where $|.| : A \rightarrow \mathbb{N}$ is defined as before. We say that a' is better than a. We then have that $(A, \rightarrow)$ is skew confluent:

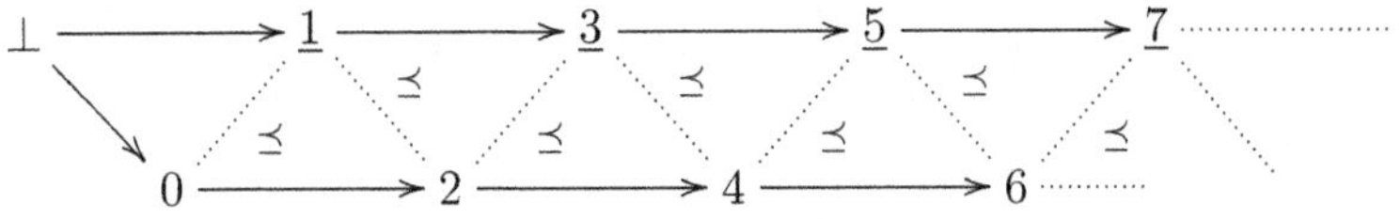

A matching concrete example can be found in term rewriting. The reduction graph of the term A in the TRS:

$$\begin{aligned} A &\to B \\ A &\to F(B) \\ B &\to F(F(B)) \end{aligned}$$

is:

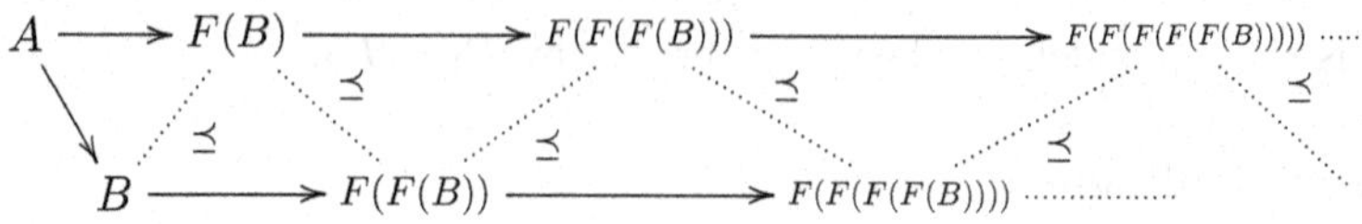

where $s \preceq t$ if t begins with at least as many F symbols as s.

3.4 ω-Skew-Confluence

Another possible outcome of optimization might be that you are suddenly able to do infinitely many units of work in a single step or the opposite where you throw away the possibility of doing more than finitely many units of work in a single step. Skew confluence is not expressive enough to deal with this situation. For example, define the reduction relation $\to$ on A as follows:

$$\bot \to \infty, \bot \to 0, n \to n+1 \ .$$

For this ARS skew confluence fails: $\bot \twoheadrightarrow \infty$ and $\bot \twoheadrightarrow 60$ and there does not exist an n, such that $60 \twoheadrightarrow n$ and $\infty \preceq n$. Pictorially:

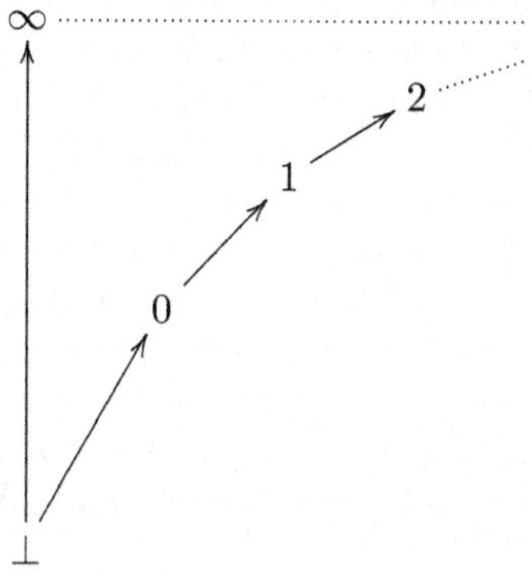

However, for each approximation m of ∞, we have that $60 \twoheadrightarrow m'$ such that $m \preceq m'$. We say that $(A, \to)$ is ω-skew confluent.

A matching concrete example can be found in infinitary rewriting. The reduction graph of the term A in the infinitary TRS:

$$\begin{aligned} A &\to F(G(F(G(\cdots)))) \\ A &\to F(F(G(F(G(\cdots))))) \\ F(F(x)) &\to F(G(x)) \\ G(G(x)) &\to G(F(x)) \end{aligned}$$

is (redexes are underlined):

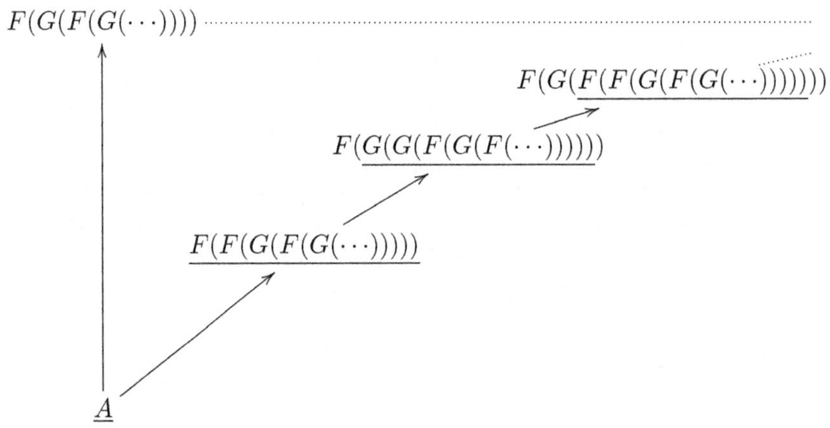

4 Skew and ω-Skew Confluence and Abstract Böhm Semantics

In this section we develop the theory of *Abstract Böhm Semantics*, which is a generalization of the theory of Böhm trees to abstract reduction systems. This generalization is based on the Böhm tree definition of Lévy [28].

4.1 Abstract Böhm Semantics

Seen from an abstract point of view, the first step in the Böhm tree construction of Lévy is to define a notion of information content. Given an ARS $(A, \rightarrow)$ and a partial order $(A, \leq)$, we would need to define a monotonic function $\omega : A \rightarrow A$, such that $\omega(a) \leq a$. Ketema followed this approach in his paper on Böhm like trees [25], but for us this doesn't work because we also want to compute Böhm semantics for terms with letrec. The problem is that terms with letrec can represent infinite terms. Hence, it is possible that the information content of a term with letrec is an infinite term. This is why one wants the domain of the rewrite system to be different from the domain of the information content.

The difference between finite and infinite information content is important. This difference needs to be reflected in our models. An obvious choice is to model finite objects as a partial order and infinite objects as ideals over finite objects. This however forces us to distinguish between finite and infinite objects, which makes for a cluttered presentation. Hence, we have chosen to use an algebraic complete partial order. In an algebraic CPO we can distinguish between finite and infinite elements as a property on the elements. Moreover, the infinite elements are completely defined by their sets of finite approximations.

Thus, we get the following abstraction of a rewrite systems with information content:

Definition 5. *A structure* $\mathcal{A} \equiv ((A, \rightarrow), \omega, (B, \leq))$ *is an* ARS with information content *(ARSI) if* $(A, \rightarrow)$ *is an ARS,* $(B, \leq)$ *is an algebraic complete partial* *order and* $\omega : A \to B$ *is monotonic with respect to* $\to$. *We say that* $\mathcal{A}$ *has finite information content if for every* $a \in A$ *we have that* $\omega(a)$ *is finite.*

Given an ARSI $((A, \to), \omega, (B, \leq))$ we refer to $\omega(a)$ as the information content of a or as the direct approximation of a. The function ω induces a quasi order $\leq_\omega$ on A, defined by $a \leq_\omega a'$ if $\omega(a) \leq \omega(a')$.

In the introduction we viewed the information content of an object as a set of observations to get some intuition. This intuition is consistent with our formalization of ARSI due to the fact that the power set of any set, ordered by inclusion is an algebraic CPO. Moreover, the finite elements in this CPO are the finite subsets of the original set.

The second step in the Böhm tree construction of Lévy is to define the actual tree or in our case the abstract Böhm semantics. The abstract Böhm semantics of an element a is supposed to be the set of all information that can be found in reducts of that element. Following the definition of Lévy we would formalize that set as:

$$\downarrow \{\omega(b) \mid a \twoheadrightarrow b\} \ .$$

The set $\{\omega(b) \mid a \twoheadrightarrow b\}$ is called the reachable information of a. Lévy already found that it is necessary to take its downward closure because otherwise there would be gaps in the set. For example, the reachable information content of

$$(\lambda x.f\,(I\,y)\,(x\,x))\,(\lambda x.f\,(I\,y)\,(x\,x))$$

is

$$\{\Omega, f\,\Omega\,\Omega, f\,y\,\Omega, f\,\Omega\,(f\,\Omega\,\Omega), \cdots\} \ .$$

But when we rewrite the two $I\,y$ redexes then the reachable information content of the result is

$$\{\Omega, f\,y\,\Omega, f\,y\,(f\,y\,\Omega), \cdots\} \ .$$

These sets are different, but their downward closures are the same.

In our case this is not enough. As we have seen in the introduction and Sect. 4, it is possible that a term allows two sequences: one sequence in which an infinite information content (e.g. ∞) is reached in a few steps and one sequence in which the information content is built up in finite pieces (e.g. $1, 2, 3, \cdots$). The set of finite pieces doesn't contain the infinite result so the downward closures will not be the same. To flatten this difference, we introduce the notion of finite element downward closure:

Definition 6. *Given a complete partial order* $(A, \leq)$, *we define*

$$\begin{aligned} \downarrow_{\mathcal{F}} s &= \{a \in \mathcal{F}(A) \mid a \leq s\} &&, \forall s \in A \ ; \\ \downarrow_{\mathcal{F}} S &= \cup_{s \in S} \downarrow_{\mathcal{F}} (s) &&, \forall S \subseteq A \ . \end{aligned}$$

We refer to $\downarrow_{\mathcal{F}} S$ as the finite element downward closure of S and to $\downarrow_{\mathcal{F}} s$ as the set of finite approximations of s. For example:

$$\downarrow_{\mathcal{F}} \{\infty\} = \downarrow_{\mathcal{F}} \{0, 2, 4, \cdots\} = \{0, 1, 2, \cdots\} \ .$$

Note that the finite element downward closure not only fills the gaps between the even numbers like the downward closure would, but also breaks ∞ up into it's finite approximations. Thus, our definition of abstract Böhm semantics is:

Definition 7. *Given an ARSI* $((A, \rightarrow), \omega, (B, \leq))$. *The Abstract Böhm Semantics* $\mathrm{ABS}(a)$ *of an element* $a \in A$ *is defined by*

$$\mathrm{ABS}(a) = \downarrow_{\mathcal{F}} \{\omega(a') \mid a \twoheadrightarrow a'\} \ .$$

The ARSI has unique abstract Böhm semantics if

$$a \xleftrightarrow{*} a' \Rightarrow \mathrm{ABS}(a) = \mathrm{ABS}(a') \ .$$

In the running text, we will often omit the word abstract and just talk about Böhm semantics. As an example of an ARSI, let us define an ARSI whose Böhm semantics is the Böhm tree from the lambda calculus.

Example 1. We consider the ARS $(\Lambda, \underset{\beta}{\rightarrow})$. The function ω_{BT} from lambda terms to possibly infinite lambda terms is defined recursively as follows:

$$\omega_{\mathrm{BT}}(M) = \begin{cases} \lambda x_1 \cdots x_n . x \, \omega_{\mathrm{BT}}(M_1) \cdots \omega_{\mathrm{BT}}(M_k) & \text{, if } M \equiv \lambda x_1 \cdots x_n . x \, M_1 \cdots M_k \\ \Omega & \text{, otherwise} \end{cases}$$

This function is a notion of information content. That is,

$$((\Lambda, \underset{\beta}{\rightarrow}), \omega_{\mathrm{BT}}, \mathcal{I}(\Lambda_\Omega, \leq_\Omega))$$

is an ARSI. Moreover, for all lambda terms M, we have

$$\mathrm{BT}(M) = \mathrm{ABS}_{\omega_{\mathrm{BT}}}(M) \ ,$$

where $\mathrm{BT}(M)$ stands for the Böhm Tree of M.

The notion of uniqueness for abstract Böhm semantics for ARSI's is related to uniqueness of normal forms in ARS's in the following sense. Consider an ARS $(A, \rightarrow)$. We can build an order by adding a bottom element, leaving the original elements incomparable. Next, we define

$$\omega(a) = \begin{cases} a & \text{, if } a \text{ is a normal form} \\ \bot & \text{, otherwise} \end{cases}$$

This gives us an ARSI for any ARS. Moreover, the ARSI has unique abstract Böhm semantics if and only if the ARS has unique normal forms.

Next, we consider sufficient and necessary conditions for uniqueness.

4.2 Skew Confluence

Skew confluence is a sufficient condition for uniqueness. To define skew confluence we need a way of telling if an object is better than another object. We formalize this by considering an ARS and a quasi order.

Definition 8 (skew confluence). *Given an ARS* $\mathcal{A} \equiv (A, \rightarrow)$ *and a quasi order* $(A, \preceq)$. *The ARS* $\mathcal{A}$ *is skew confluent with respect to* $\preceq$ *if*

$$\forall a_1, a_2, a_3 \in A : a_1 \twoheadrightarrow a_2 \wedge a_1 \twoheadrightarrow a_3 \Rightarrow \exists a_4 : a_2 \preceq a_4 \wedge a_3 \twoheadrightarrow a_4 \ .$$

The commutative diagram for skew confluence is

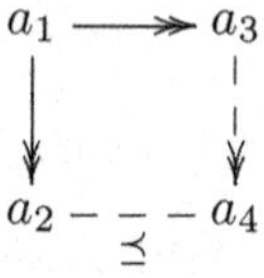

Confluence implies skew confluence. More precisely, if the reduction relation is increasing in a quasi order then confluence implies skew confluence with respect to that quasi order:

Proposition 2. *Given an ARS* $\mathcal{A} \equiv (A, \rightarrow)$ *and a quasi order* $(A, \preceq)$. *If* $\rightarrow \subseteq \preceq$ *and* $\mathcal{A}$ *is confluent then* $\mathcal{A}$ *is skew confluent with respect to* $\preceq$.

The definitions of confluence and skew confluence are easily extended to ARSI's. We say that an ARSI $((A, \rightarrow), \omega, (B, \leq))$ is confluent, if $(A, \rightarrow)$ is confluent and we say that it is skew confluent if $(A, \rightarrow)$ is skew confluent with respect to $\leq_\omega$.

4.3 ω-Skew Confluence

In [6], we defined abstract Böhm semantics (called infinite normal forms in that paper) in a setting where information content was always finite. In that setting, skew confluence is a necessary and sufficient condition for uniqueness of abstract Böhm semantics. In the current setting, it still is a sufficient condition, but it is not necessary. We will now define ω-skew confluence which is a necessary and sufficient condition in the presence of infinite information content. Later, we will show that for finite information content the two properties coincide. Hence, the result in this paper can be seen as an extension of the earlier result.

The definition of ω-skew confluence follows the intuition in the introduction based on observations. An ARSI is ω-skew confluent if given two diverging computations and an observation about the first result, we can find a reduct of the second result which allows the same observation. To make that formal, an observation about an object is defined as a finite element less than or equal to the information content of the object. This results in the following definition:

Definition 9 (ω-skew confluence). *The ARSI* $\mathcal{A} \equiv ((A, \rightarrow), \omega, (B, \leq))$ *is* ω-*skew confluent if*

$$\begin{aligned} &\forall a_1, a_2, a_3 \in A, d_1 \in \mathcal{F}(B) : \\ &\qquad a_1 \twoheadrightarrow a_2 \wedge a_1 \twoheadrightarrow a_3 \wedge d_1 \leq \omega(a_2) \Rightarrow \exists a_4 \in A : \ a_3 \twoheadrightarrow a_4 \wedge d_1 \leq \omega(a_4) \ . \end{aligned}$$

To be able to draw diagrams about ω-skew confluence, we introduce two arrows: $\xrightarrow[\omega]{}$ and $\xrightarrow[\omega]{\mathcal{F}}$. The former computes ω, the latter selects a finite element less than the information content:

Definition 10. *Given an ARSI* $((A, \rightarrow), \omega, (B, \leq))$, *we define the relations* $\xrightarrow[\omega]{} \subseteq A \times B$ *and* $\xrightarrow[\omega]{\mathcal{F}} \subseteq A \times \mathcal{F}(B)$ *as*

$$\begin{array}{ll} \forall a \in A : & a \xrightarrow[\omega]{} \omega(a) \ ; \\ \forall a \in A \ \forall b \in \downarrow_{\mathcal{F}} (\omega(a)) : a \xrightarrow[\omega]{\mathcal{F}} b \ . & \end{array}$$

Based on this definition, we can draw diagrams of skew confluence (SC) and ω-skew confluence (ωSC):

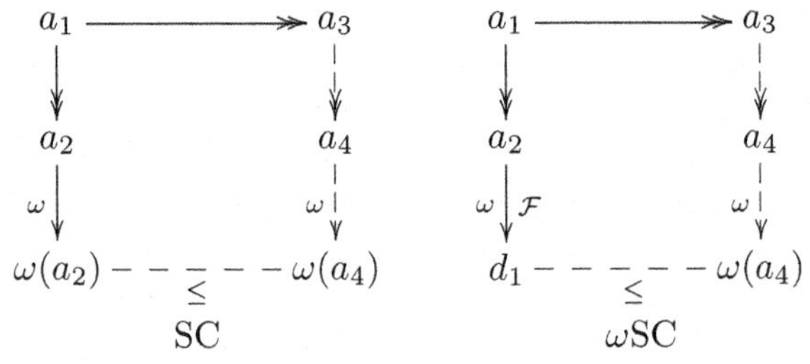

Unfortunately, the diagram of ω-skew confluence is not suitable for diagram proofs using tiling. The reason is that the assumption on the left-hand side involves selecting a finite element and the conclusion on the right-hand side might not yield such an element. Hence, we include the following characterization.

Proposition 3. *For any ARSI* $((A, \rightarrow), \omega, (B, \leq))$ *the following diagram equivalence holds:*

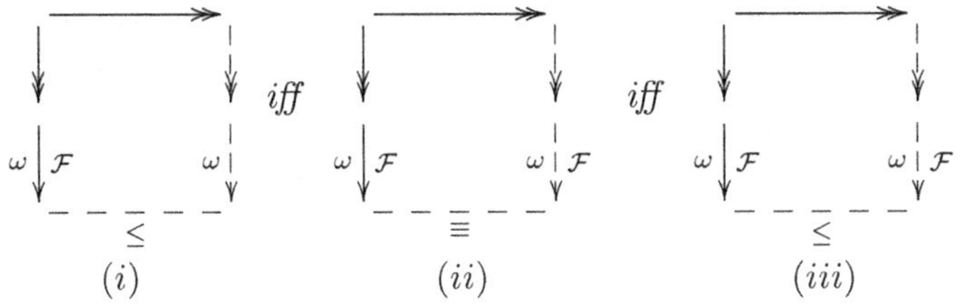

(1)

Proof. **(i)⇒(iii)** Assume that $d \leq \omega(a)$ for some finite $d \in B$ and some $a \in A$. Because the CPO is algebraic, we have $\omega(a) = \text{lub} \downarrow_{\mathcal{F}} (\omega(a))$. Because d is finite, there exists $d' \in \downarrow_{\mathcal{F}} (\omega(a))$ such that $d \leq d'$. For that d', we have $a \xrightarrow[\omega]{\mathcal{F}} d'$ by definition.

(iii)⇒(ii) If $d \leq d'$ and $a \xrightarrow[\omega]{\mathcal{F}} d'$ then $a \xrightarrow[\omega]{\mathcal{F}} d$ must hold as well.

(ii)⇒(i) If $a \xrightarrow[\omega]{\mathcal{F}} d$ then $d \leq \omega(a)$ because $\omega(a) = \text{lub} \downarrow_{\mathcal{F}} (\omega(a))$.

Next, we prove that ω-skew confluence is a necessary and sufficient condition for uniqueness of abstract Böhm semantics.

Theorem 1. *Given an ARSI* $\mathcal{A} \equiv ((A, \rightarrow), \omega, (B, \leq))$. *The ARSI* $\mathcal{A}$ *has unique abstract Böhm semantics iff* $\mathcal{A}$ *is* ω*-skew confluent.*

Proof. From the arrow notation for ω, we can also derive the following equation:

$$\mathrm{ABS}(a) = \{b \mid a \twoheadrightarrow\xrightarrow[\omega]{\mathcal{F}} b\} \ ; \tag{2}$$

$\Rightarrow$ Assume that $\mathcal{A}$ has unique abstract Böhm semantics. Let $a_1, a_2, a_3 \in A$ and $d \in B$ be given such that $a_1 \twoheadrightarrow a_2$, $a_1 \twoheadrightarrow a_3$ and $a_2 \xrightarrow[\omega]{\mathcal{F}} d$.
In particular, we have $a_1 \twoheadrightarrow\xrightarrow[\omega]{\mathcal{F}} d$, so because of Eq. 2, we have that $d \in \mathrm{ABS}(a_1)$. Because of the uniqueness, we have that $\mathrm{ABS}(a_1) = \mathrm{ABS}(a_3)$, so $d \in \mathrm{ABS}(a_3)$. Again by Eq. 2, it follows that $a_3 \twoheadrightarrow\xrightarrow[\omega]{\mathcal{F}} d$. This proves diagram 1.(ii) and hence by the previous proposition ω-skew confluence.

$\Leftarrow$ Assume that $\mathcal{A}$ is ω-skew confluent. Let $a_1, a_2 \in A$ be given. It suffices to show that if $a_1 \to a_2$ then $\mathrm{ABS}(a_1) = \mathrm{ABS}(a_2)$.
From the definition of abstract Böhm semantics it is obvious that $\mathrm{ABS}(a_2) \subseteq \mathrm{ABS}(a_1)$, so the part we need to show is $\mathrm{ABS}(a_1) \subseteq \mathrm{ABS}(a_2)$. Let $d \in \mathrm{ABS}(a_1)$ be given. Then by Eq. 2, it follows that $a_1 \twoheadrightarrow\xrightarrow[\omega]{\mathcal{F}} d$. Because of ω-skew confluence and the previous proposition, diagram 1.(ii) holds. From this diagram it follows that $a_2 \twoheadrightarrow\xrightarrow[\omega]{\mathcal{F}} d$ and hence by Eq. 2 $d \in \mathrm{ABS}(a_2)$.

The definition of ω-skew confluence allows us to select a different matching reduct for every observation that must be matched. If we can match every observation with the same reduct then this is called uniform ω-skew confluence. This is however not a new property because it is equivalent to skew confluence:

Proposition 4. *Given an ARSI $\mathcal{A} \equiv ((A, \to), \omega, (B, \leq))$. We say that $\mathcal{A}$ is uniformly ω-skew confluent if*

$$\begin{aligned}&\forall a_1, a_2, a_3 \in A\ \exists a_4 \in A\ \forall d_1 \in \mathcal{F}(B):\\ &\qquad a_1 \twoheadrightarrow a_2 \wedge a_1 \twoheadrightarrow a_3 \wedge d_1 \leq \omega(a_2) \Rightarrow a_3 \twoheadrightarrow a_4 \wedge d_1 \leq \omega(a_4) \ .\end{aligned}$$

We have that $\mathcal{A}$ is uniformly ω-skew confluent iff $\mathcal{A}$ is skew confluent.

Proof. Follows from the claim that

$$\forall a, a' \in A: \ (\forall d \in \mathcal{F}(B): \ d \leq \omega(a) \Rightarrow d \leq \omega(a')) \Leftrightarrow a \leq_\omega a' \ .$$

The claim follows from the fact that $\omega(a) = \mathrm{lub} \downarrow_{\mathcal{F}} (\omega(a))$ and $\omega(a') = \mathrm{lub} \downarrow_{\mathcal{F}} (\omega(a'))$ because $(B, \leq)$ is an algebraic CPO.

Hence, it is not surprising that we can prove skew confluence implies ω-skew confluence. We also prove that in the case of finite information content the two notions are equivalent.

Proposition 5. *Given an ARSI $\mathcal{A} \equiv ((A, \to), \omega, (B, \leq))$.*

(i) If $\mathcal{A}$ is skew confluent then $\mathcal{A}$ is ω-skew confluent.
(ii) If $\mathcal{A}$ is ω-skew confluent and $\mathcal{A}$ has finite information content then $\mathcal{A}$ is skew confluent.

Proof. (i) By the previous proposition skew confluence implies uniform ω-skew confluence. It is obvious that uniform ω-skew confluence implies ω-skew confluence.

(ii) Because the information content is finite, we have

$$\forall a \in A : \; a \xrightarrow[\omega]{\mathcal{F}} \omega(a)$$

From this fact and ω-skew confluence, skew confluence follows easily.

This completes the presentation of the basic theory of abstract Böhm semantics. We continue with an example of an application area: term graph rewriting based on terms with letrec.

5 Lack of Confluence in Term Graph Rewriting

The need for a less restrictive notion of confluence arises in practice if one wants to provide a more accurate foundation of programming languages. To reason about either execution or optimizations one has to deal with the notion of sharing and cyclic structures [12, 31, 2]. As pointed out by Wadsworth [35], these concerns can be accommodated by considering term graph rewriting as opposed to term (or tree) rewriting.

As pointed out in [9, 10], term graphs can be nicely represented as terms with the letrec [2] construct:

$$\langle M \mid x_1 = M_1, \cdots, x_n = M_n \rangle \; .$$

We sometimes refer to the variables $x_1, \cdots, x_n$ as the recursion variables, to the equations and to M as the internal and external part of the letrec construct, respectively. Because of the capability of the letrec to represent graphs with cycles, we refer to terms with letrec's as cyclic terms.

The cyclic structure depicted in Fig. 1 is represented as

$$\langle x \mid x = \mathsf{Cons}(1, x) \rangle \; .$$

The advantage of this representation is that one can apply existing term rewrite rules directly to the cyclic term. However, the old rewrite rules are not enough: we must also use rules that modify the letrec structure to make potential redexes visible [9, 10]. For example, with respect to the rule

$$\mathsf{F}(1) \to \mathsf{G}(1)$$

the terms:

$$\langle \mathsf{F}(x) \mid x = 1 \rangle \quad \text{and} \quad \langle x \mid x = \mathsf{F}(y), y = 1 \rangle$$

[2] We use the Ariola/Klop notation for letrec ($\langle M \mid E \rangle \equiv$ letrec E in M).

are in normal form. Whereas, their corresponding graphs contain a redex. To cope with this situation, the following two rules for external and internal substitution are introduced:

$$\langle C[x] \mid x = M, E\rangle \underset{\text{es}}{\rightarrow} \langle C[M] \mid x = M, E\rangle$$
$$\langle M \mid x = C[y], y = N, E\rangle \underset{\text{is}}{\rightarrow} \langle M \mid x = C[N], y = N, E\rangle$$

where $C[x]$ stands for a one-hole context filled with variable x, and E for a collection of unordered equations. According to these rules we have:

$$\langle \mathsf{F}(x) \mid x = 1\rangle \underset{\text{es}}{\rightarrow} \langle \mathsf{F}(1) \mid x = 1\rangle$$
$$\langle x \mid x = \mathsf{F}(y), y = 1\rangle \underset{\text{is}}{\rightarrow} \langle x \mid x = \mathsf{F}(1), y = 1\rangle$$

Both right-hand sides contain the redex $\mathsf{F}(1)$. One more substitution rule is needed. Consider the following rule:

$$\mathsf{F}(\mathsf{F}(x)) \rightarrow \mathsf{G}(x)$$

and the term

$$\langle x \mid x = \mathsf{F}(x)\rangle \ .$$

To make the redex explicit in the internal part, one needs a substitution applied to the equation itself. We call it cyclic substitution:

$$\langle M \mid x = C[x], E\rangle \underset{\text{cs}}{\rightarrow} \langle M \mid x = C[C[x]], E\rangle$$

We have:

$$\langle x \mid x = \mathsf{F}(x)\rangle \underset{\text{cs}}{\rightarrow} \langle x \mid x = \mathsf{F}(\mathsf{F}(x))\rangle \rightarrow \langle x \mid x = \mathsf{G}(x)\rangle \ .$$

The problem with these three substitution rules is that confluence is lost. The classical example is:

$$\begin{array}{ccccc}
M & & & & \\
\equiv & & & & \\
\langle x \mid x = \mathsf{F}(x)\rangle & \xrightarrow[\text{es}]{} & \langle \mathsf{F}(x) \mid x = \mathsf{F}(x)\rangle & \xrightarrow[\text{cs}]{} & \langle \mathsf{F}(x) \mid x = \mathsf{F}(\mathsf{F}(x))\rangle \\
\Big\downarrow \text{cs} & & & & \equiv \\
 & & & & M_o \\
\langle x \mid x = \mathsf{F}(\mathsf{F}(x))\rangle & & & & \\
\equiv & & & & \\
M_e & & & &
\end{array}$$

The cyclic terms M_o and M_e do not have a common reduct because any reduct of M_o will contain an odd number of F symbols and any reduct of M_e an even number.

The fact that the three substitution rules aren't confluent is not in itself a big problem. Not only are these rewrite rules confluent modulo bisimulation, but it is also possible to add rewrite rules to regain confluence. Even if we add an orthogonal TRS we can keep the rewrite systems confluent. (See [29]).

However, if we consider combinatory reduction systems or non-orthogonal TRS's then interaction between the rewrite rules and substitution rules becomes a real problem. For example, consider the TRS

$$\begin{aligned} \mathsf{F}(\mathsf{F}(x)) &\to \mathsf{F}(\mathsf{G}(x)) \\ \mathsf{G}(\mathsf{G}(x)) &\to \mathsf{G}(\mathsf{F}(x)) \end{aligned} \tag{3}$$

This TRS is confluent and terminating, but not orthogonal. When we apply these rewrite rules to M_o and M_e we get:

$$\begin{aligned} M_e &\to \langle x \mid x = \mathsf{F}(\mathsf{G}(x))\rangle \equiv M_e' \ ; \\ M_o &\to \langle \mathsf{F}(x) \mid x = \mathsf{F}(\mathsf{G}(x))\rangle \equiv M_o' \ . \end{aligned}$$

As before, a count of the symbols leads to the conclusion that these M_e' and M_o' do not have a common reduct.

All reducts of M_e' will be of the form

$$\langle (\mathsf{FG})^n(x) \mid x = (\mathsf{FG})^m(x)\rangle \ ,$$

where $(\mathsf{FG})^0(x) = x$ and $(\mathsf{FG})^{n+1}(x) = \mathsf{F}(\mathsf{G}((\mathsf{FG})^n(x)))$. Note that a term of this form has exactly the same number of F's and G's and that there it will not contain a redex of the TRS rules from Eq. 3.

All reducts of M_o' will be of the form

$$\langle (\mathsf{FG})^n(\mathsf{G}((\mathsf{FG})^k(x))) \mid x = (\mathsf{FG})^m(x)\rangle \text{ or } \langle (\mathsf{FG})^n(\mathsf{F}((\mathsf{FG})^k(x))) \mid x = (\mathsf{FG})^m(x)\rangle \ .$$

Note that a term of one of these forms has one more F than G's or one more G than F's. More importantly, a term of these forms always has a redex with respect to the TRS rules.

The fundamental difference between the reduction sequences from M to M_e' and from M to M_o' is that in the first sequence the "correct" redex is exposed and contracted. In the second sequence the "wrong" redex is exposed and contracted. The result of that is that a redex remains, which will create a new redex whenever it is contracted. Because a new redex is created in every step, it is impossible to reduce M_0' to normal form in finitely many steps.

We will now define a notion of information content for this TRS and show that the resulting ARSI is ω-skew confluent. The function ω from cyclic terms to infinite terms is defined as follows:

$$\omega(M) = \mathrm{lub}\{N \mid M \twoheadrightarrow_{\omega} N \text{ and } N \text{ is a normal form}\} \ , \tag{4}$$

where $\twoheadrightarrow_{\omega}$ is defined as follows:

$$\begin{aligned} &\langle C[x] \mid x = M, E\rangle && \to_{\omega} \langle C[M] \mid x = M, E\rangle \\ &\mathsf{F}(\mathsf{F}(x)) && \to_{\omega} \Omega \\ &\mathsf{G}(\mathsf{G}(x)) && \to_{\omega} \Omega \\ &\langle M \mid x_1 = M_1, \cdots, x_n = M_n\rangle && \to_{\omega} M[x_1 := \Omega, \cdots, x_n := \Omega] \end{aligned}$$

For example:

$$\begin{aligned} M'_o &\equiv \langle \mathsf{F}(x) \mid x = \mathsf{F}(\mathsf{G}(x))\rangle \\ &\underset{\omega}{\rightarrow} \langle \mathsf{F}(\mathsf{F}(\mathsf{G}(x))) \mid x = \mathsf{F}(\mathsf{G}(x))\rangle \\ &\underset{\omega}{\rightarrow} \langle \Omega \mid x = \mathsf{F}(\mathsf{G}(x))\rangle \\ &\underset{\omega}{\rightarrow} \Omega \ . \end{aligned}$$

Because this normal form is unique, we have that

$$\omega(M'_o) = \mathrm{lub}\{\Omega\} = \Omega \ .$$

The normal form of M'_e with respect to $\underset{\omega}{\rightarrow}$ is not unique. For every n we have

$$M'_e \equiv \langle x \mid x = \mathsf{F}(\mathsf{G}(x))\rangle \underset{\omega}{\twoheadrightarrow} (\mathsf{FG})^n(\Omega) \ .$$

This means that

$$\omega(M'_e) = \mathrm{lub}\{(\mathsf{FG})^n(\Omega) \mid n \in \mathbb{N}\} = (\mathsf{FG})^\omega \ .$$

There does not exist any reduct of M'_o, whose information content is $(\mathsf{FG})^\omega$ so the rewrite system is not skew confluent. However, for every n, we can find a reduct of M'_o, such that the information content of the reduct is $(\mathsf{FG})^n(\Omega)$:

$$M^n = \langle (\mathsf{FG})^n(\mathsf{F}(x)) \mid x = \mathsf{F}(\mathsf{G}(x))\rangle \ .$$

These reducts show that the example is ω-skew confluent.

6 Cyclic Lambda Calculi

In this section we consider extensions of the call-by-name lambda calculus and the call-by-need lambda calculus [8, 7] with cyclic structures. These extensions contain a large number of rules. This is in order to be able to address the correctness of compilation by transformation [18]. But before we give the calculi let us start with the basic principles.

We are interested in cyclic lambda terms. That is, lambda terms extended with letrec.

Definition 11. *The set of cyclic lambda terms $\Lambda\circ$ is defined as follows:*

$$\begin{array}{ll} \textit{Terms} & M ::= x \mid \lambda x.M \mid M\,N \mid \langle M \mid E\rangle \ ; \\ \textit{Equations} & E ::= x_1 = M_1, \ldots, x_n = M_n \ . \end{array}$$

where the variables $x_1, \cdots, x_n$, are distinct from each other and the order of the equations does not matter. Terms are taken up to α-conversion.

Because it is not really important where the definitions are placed, we base our lambda calculi on a rewrite system that brings any cyclic term into a standard form. This standard form is:

$$\begin{array}{l} ST ::= x \mid \langle x \mid SE\rangle \ ; \\ SE ::= x = x \mid x = \lambda y.ST \mid x = x_1\,x_2 \mid SE, SE \ . \end{array}$$

Table 1. A rewrite system for normalizing the representation of a graph

Variable substitution:			
$\langle M \mid x = y, E\rangle$	$\xrightarrow[\text{vs}]{}$	$\langle M[x := y] \mid E[x := y]\rangle$	$x \not\equiv y$
Lift:			
$\langle M \mid E\rangle\ N$	$\xrightarrow[\text{lift}]{}$	$\langle M\ N \mid E\rangle$	
$M\ \langle N \mid E\rangle$	$\xrightarrow[\text{lift}]{}$	$\langle M\ N \mid E\rangle$	
$\lambda x.\langle M \mid E_1, E_2\rangle$	$\xrightarrow[\text{lift}]{}$	$\langle \lambda x.\langle M \mid E_1\rangle \mid E_2\rangle$	C1
Merge:			
$\langle\langle M \mid E_1\rangle \mid E_2\rangle$	$\xrightarrow[\text{em}]{}$	$\langle M \mid E_1, E_2\rangle$	
$\langle M \mid x = \langle N \mid E_1\rangle, E_2\rangle$	$\xrightarrow[\text{im}]{}$	$\langle M \mid x = N, E_1, E_2\rangle$	
Garbage collection:			
$\langle M \mid E_1, E_2\rangle$	$\xrightarrow[\text{gc}]{}$	$\langle M \mid E_1\rangle$	C2
$\langle M \mid\rangle$	$\xrightarrow[\text{gc}]{}$	M	
Naming:			
$C_{\text{safe}}[\lambda y.M]$	$\xrightarrow[\text{name}]{}$	$C_{\text{safe}}[\langle x \mid x = \lambda y.M\rangle]$	C3
$C_{\text{safe}}[M\ N]$	$\xrightarrow[\text{name}]{}$	$C_{\text{safe}}[\langle x \mid x = M\ N]\rangle$	C3

C1: E_2 is non-empty and neither x nor a variable defined in E_1 occurs free in E_2;
C2: E_2 is non-empty and none of the variables defined in E_2 occur free in E_1 or M;
C3: x is a fresh variable and the rule is *not* closed under contexts;

$$\begin{aligned} C_{\text{safe}} &::= C' \mid C[\lambda x.C'] \mid C[C'\ M] \mid C[M\ C'] \ ; \\ C' &::= \Box \mid \langle C' \mid E\rangle \ . \end{aligned}$$

That is, a standard term is either a variable or a letrec with a non-empty list of standard definitions. A standard definition can be a black hole definition ($x = x$), a function definition ($x = \lambda y.ST$) or an application definition ($x = x_1\, x_2$). In Table 1 we present a confluent and terminating rewrite system for computing standard representations. Apart from the usual conditions on lifting and garbage collection it contains a special condition on the naming rules. Unlike the other rules which can be applied in any context these rules are only applicable in safe contexts. This restriction is necessary to guarantee termination.

Representing the same graph is one equivalence. Another equivalence is having the same unwinding. The unwinding of a cyclic term is the unique (infinite) term represented by the cyclic term. The substitution rules compute the unwinding in the sense that we can define a notion of information content such that the unwinding of a cyclic term is the Böhm semantics of the term:

$$\begin{aligned} \omega_{\text{es}}(x) &= x \\ \omega_{\text{es}}(M_1\, M_2) &= \omega_{\text{es}}(M_1)\, \omega_{\text{es}}(M_2) \\ \omega_{\text{es}}(\lambda x.M) &= \lambda x.\omega_{\text{es}}(M) \\ \omega_{\text{es}}(\langle M \mid x_1 = M_1, \cdots, x_n = M_n\rangle) &= \omega_{\text{es}}(M)[x_1 := \Omega, \cdots, x_n := \Omega] \end{aligned}$$

We have used the label es because external substitution is actually the only rule needed to compute the unwinding.

Definition 12. *The unwinding of a cyclic term M, denoted $[\![M]\!]$, is the Böhm semantics of M with respect to the ARSI $((\Lambda\circ, \underset{\text{es}}{\rightarrow}), \omega_{\text{es}}, \mathcal{I}(\Lambda_\Omega, \leq_\Omega))$:*

$$[\![M]\!] = \text{ABS}_{\text{es}}(M) \ .$$

What we want is a set of rewrite rules such that terms with the same unwinding are convertible. To that end, we introduce a rewrite rule for copying. What copying means is that one duplicates definitions and for every reference to a duplicated variable one can choose to refer to the original definition or to one of the copies. In the following definition, we define the copy rewrite rule on graphs by means of a meta rewrite system.

Definition 13. *On cyclic terms extended with the binary symbol $+$, we define the rewrite relation $\underset{+}{\rightarrow}$ as follows:*

$$\begin{array}{ll} \langle M \mid x = N, E\rangle \underset{+}{\rightarrow} \langle M\sigma \mid y = N\sigma, z = N\sigma, E\sigma\rangle & \textit{where } \sigma = [x := y + z] \\ & \textit{for fresh variables } y \textit{ and } z \\ y + z \qquad \underset{+}{\rightarrow} y & \\ y + z \qquad \underset{+}{\rightarrow} z & \end{array}$$

If $M \underset{+}{\twoheadrightarrow} N$ and M nor N contain on occurrence of $+$ then

$$M \underset{\text{cp}}{\rightarrow} N \ .$$

For example:

$$\begin{aligned} \lambda z.\langle u \mid u = z\,u\rangle &\underset{+}{\rightarrow} \lambda z.\langle (x + y) \mid x = z\,(x + y), y = z\,(x + y)\rangle \\ &\underset{+}{\rightarrow}^3 \lambda z.\langle x \mid x = z\,y, y = z\,x\rangle \ . \end{aligned}$$

For a more precise discussion of the issues of representation see [3].

The principle of cyclic lambda calculi is simple. In the beta-rule, instead of a substitution one uses a letrec:

$$(\lambda x.M)\ N \underset{\beta\circ}{\rightarrow} \langle M \mid x = N\rangle \ .$$

To simulate the normal β-rule, we must obviously include substitution. But this is not enough. Consider the term (I stands for the term $\lambda x.x$):

$$\langle \lambda y.x\ y \mid x = I\rangle\ I \ .$$

It contains a potential redex "$(\lambda y.x\ y)\ I$" which needs to be made explicit by moving the equation "$x = I$" around. This is made possible by the first lift rule:

$$\begin{aligned} \langle \lambda y.x\ y \mid x = I\rangle\ I &\underset{\text{lift}}{\rightarrow} \langle (\lambda y.x\ y) I \mid x = I\rangle \\ &\underset{\beta\circ}{\rightarrow} \langle\langle x\ y \mid y = I\rangle \mid x = I\rangle \\ &\underset{\text{es}}{\rightarrow} \langle\langle I\ y \mid y = I\rangle \mid x = I\rangle \\ &\underset{\beta\circ}{\rightarrow} \langle\langle\langle z \mid z = y\rangle \mid y = I\rangle \mid x = I\rangle \\ &\underset{\text{es}}{\twoheadrightarrow} \langle\langle\langle I \mid z = y\rangle \mid y = I\rangle \mid x = I\rangle \ . \end{aligned}$$

Table 2. The cyclic call-by-name lambda calculus $\lambda\circ$

$\beta\circ$:			
$(\lambda x.M)\ N$	$\xrightarrow[\beta\circ]{}$	$\langle M \mid x = N\rangle$	
Substitution:			
$\langle C[x] \mid x = M, E\rangle$	$\xrightarrow[\text{es}]{}$	$\langle C[M] \mid x = M, E\rangle$	
$\langle M \mid x = C[y], y = N, E\rangle$	$\xrightarrow[\text{is}]{}$	$\langle M \mid x = C[N], y = N, E\rangle$	
Lift:			
$\langle M \mid E\rangle\ N$	$\xrightarrow[\text{lift}]{}$	$\langle M\ N \mid E\rangle$	
$M\ \langle N \mid E\rangle$	$\xrightarrow[\text{lift}]{}$	$\langle M\ N \mid E\rangle$	
$\lambda x.\langle M \mid E_1, E_2\rangle$	$\xrightarrow[\text{lift}]{}$	$\langle \lambda x.\langle M \mid E_1\rangle \mid E_2\rangle$	C1
Merge:			
$\langle\langle M \mid E_1\rangle \mid E_2\rangle$	$\xrightarrow[\text{em}]{}$	$\langle M \mid E_1, E_2\rangle$	
$\langle M \mid x = \langle N \mid E_1\rangle, E_2\rangle$	$\xrightarrow[\text{im}]{}$	$\langle M \mid x = N, E_1, E_2\rangle$	
Garbage collection:			
$\langle M \mid E_1, E_2\rangle$	$\xrightarrow[\text{gc}]{}$	$\langle M \mid E_1\rangle$	C2
$\langle M \mid\rangle$	$\xrightarrow[\text{gc}]{}$	M	
Copy:			
M	$\xrightarrow[\text{cp}]{}$	N	

C1: E_2 is non-empty and neither x nor a variable defined in E_1 occurs free in E_2;
C2: E_2 is non-empty and none of the variables defined in E_2 occur free in E_1 or M.

Our entire call-by-name cyclic lambda calculus in given in Table. 2. Basically, it consists of the $\beta\circ$-rule, the substitution rules, the representation rules and copying. However, superfluous rules have been removed. For example, the cyclic substitution is derivable from copying, internal substitution and garbage collection:

$$\begin{aligned}\langle M \mid x = C[x], E\rangle &\xrightarrow[\text{cp}]{} \langle M \mid x = C[y], y = C[x], E\rangle\\ &\xrightarrow[\text{is}]{} \langle M \mid x = C[C[x]], y = C[x], E\rangle\\ &\xrightarrow[\text{gc}]{} \langle M \mid x = C[C[x]], E\rangle\ .\end{aligned}$$

Naming is not included since, due to substitution, it is not needed to equate different representations of a graph.

We explained the third lift rule as a rule needed to normalize terms representing graphs. This is not the only explanation of this rule. It is also needed to be able to capture different kinds of evaluation, such as, full laziness [34]. The rule lifts declarations, that do not contain occurrences of the bound variable, out of a lambda body. As an example, consider the following term:

$$\langle f\,I\,(f\,I) \mid f = \lambda x.\langle w\,x \mid w = \underline{(I\,I)}\rangle\rangle\ .$$

If we do not lift the redex $I\,I$ (*i.e.*, the one underlined) out of the lambda body, that redex will be reduced twice. We have:

$$\begin{aligned}&\langle f\,I\,(f\,I) \mid f = \lambda x.\langle w\,x \mid w = (I\,I)\rangle\rangle\\ &\quad\xrightarrow[\text{lift}]{} \langle f\,I\,(f\,I) \mid f = \langle \lambda x.\langle w\,x \mid\rangle \mid w = (I\,I)\rangle\rangle\ .\end{aligned}$$

To complete the work done by the lift rule we apply the internal merge:

$$\begin{aligned}&\langle f\, I\,(f\, I) \mid f = \langle \lambda x.\langle w\, x \mid\rangle \mid w = (I\, I)\rangle\rangle \\ \xrightarrow[\text{im}]{} &\langle f\, I\,(f\, I) \mid f = \lambda x.\langle w\, x \mid\rangle, w = (I\, I)\rangle \\ \xrightarrow[\text{es}]{} &\langle (\lambda x.\langle w\, x \mid\rangle) I\,(f\, I) \mid f = \lambda x.\langle w\, x \mid\rangle, w = (I\, I)\rangle\ .\end{aligned}$$

Note that the substitution of f did not cause the duplication of the redex $I\,I$.

In Sect. 5, we already pointed out that the substitution rules lead to non-confluence. However, the cyclic call-by-name lambda calculus is skew-confluent with respect to a notion of finite information content, which returns a lambda calculus term extended with a constant Ω. As usual in the field of programming languages, the information content for the call-by-name lambda calculus is derived from that for the Lévy-Longo tree rather than the Böhm tree.

Definition 14. *Given the ARS $(\Lambda_\circ, \xrightarrow[\lambda\circ]{})$ and the partial order $(\Lambda_\Omega, \leq_\Omega)$. The finite information content $\omega_{\lambda\circ}(M)$ of a term $M \in \Lambda\circ$ is the normal form of M with respect to the following rules:*

$$\begin{array}{llll} (\lambda x.M)N & \xrightarrow[\omega_{\lambda\circ}]{} \Omega & & \beta\omega \\ \langle C[x] \mid x = M, E\rangle & \xrightarrow[\omega_{\lambda\circ}]{} \langle C[\Omega] \mid x = M, E\rangle & & es\omega \\ \Omega M & \xrightarrow[\omega_{\lambda\circ}]{} \Omega & & @\omega \\ \langle M \mid E\rangle & \xrightarrow[\omega_{\lambda\circ}]{} M & C & \mathrm{gc}\omega \end{array}$$

C: none of the variables defined in E occurs free in M.

Examples: $\omega_{\lambda\circ}(\langle \lambda x.y\, z \mid y = I\rangle) = \lambda x.\Omega$, $\omega_{\lambda\circ}(\langle x \mid x = x\rangle) = \Omega$, $\omega_{\lambda\circ}(\langle x\, y \mid y = I\rangle\, x) = (x\,\Omega)\, x$, and $\omega_{\lambda\circ}(\langle x\, x \mid x = I\rangle) = \Omega$. Note that even though $\langle x\, y \mid y = I\rangle\, x$ is a lift redex, its information content is not Ω.

Theorem 2. *The ARSI $((\Lambda_\circ, \xrightarrow[\lambda\circ]{}), \omega_{\lambda\circ}, \mathcal{I}(\Lambda_\Omega, \leq_\Omega))$ is skew confluent.*

This theorem guarantees uniqueness of Böhm semantics. A direct proof can be found in [6]. In the next section, we will develop a theory which allows us to prove uniqueness of Böhm semantics from a list of other properties. First, we present a call-by-need calculus.

One of the features of the call-by-need calculus is that duplication of terms is restricted to the class of values. Thus, we need a version of copying which only duplicates a certain class of terms.

Definition 15. *Let C be a set of terms. On cyclic terms extended with the binary symbol $+$, we define the rewrite relation $\xrightarrow[+C]{}$ as follows:*

$$\begin{array}{ll} \langle M \mid x = N, E\rangle \xrightarrow[+C]{} \langle M\sigma \mid y = N\sigma, z = N\sigma, E\sigma\rangle & \textit{where } N \in C \textit{ and} \\ & \sigma = [x := y + z] \textit{ for fresh variables } y \textit{ and } z \\ y + z \quad \xrightarrow[+C]{} y & \\ y + z \quad \xrightarrow[+C]{} z & \end{array}$$

Table 3. The cyclic call-by-need lambda calculus $\lambda\circ_{\text{need}}$

$\beta\circ$:			
$(\lambda x.M)N$	$\xrightarrow[\beta\circ]{}$	$\langle M \mid x = N\rangle$	
Value Substitutions:			
$\langle C[x] \mid x = V, E\rangle$	$\xrightarrow[\text{esv}]{}$	$\langle C[V] \mid x = V, E\rangle$	
$\langle M \mid x = C[x_1], x_1 = V, E\rangle$	$\xrightarrow[\text{isv}]{}$	$\langle M \mid x = C[V], x_1 = V, E\rangle$	
Lift:			
$\langle M \mid E\rangle N$	$\xrightarrow[\text{lift}]{}$	$\langle MN \mid E\rangle$	
$M\langle N \mid E\rangle$	$\xrightarrow[\text{lift}]{}$	$\langle MN \mid E\rangle$	
$\lambda x.\langle M \mid E, VE\rangle$	$\xrightarrow[\text{lift}]{}$	$\langle \lambda x.\langle M \mid E\rangle \mid VE\rangle$	C1
Merge:			
$\langle\langle M \mid E\rangle \mid E'\rangle$	$\xrightarrow[\text{em}]{}$	$\langle M \mid E, E'\rangle$	
$\langle M \mid x = \langle N \mid E\rangle, E_1\rangle$	$\xrightarrow[\text{im}]{}$	$\langle M \mid x = N, E, E_1\rangle$	
Garbage collection:			
$\langle M \mid E, E'\rangle$	$\xrightarrow[\text{gc}]{}$	$\langle M \mid E\rangle$	C2
$\langle M \mid \rangle$	$\xrightarrow[\text{gc}]{}$	M	
Value Copying:			
M	$\xrightarrow[\text{cpv}]{}$	N	
Naming:			
$C_{\text{safe}}[M\ N]$	$\xrightarrow[\text{name}]{}$	$C_{\text{safe}}[\langle x \mid x = M\ N\rangle]$	C3

C1: VE is non-empty and neither x nor a variable defined in E occurs free in VE;
C2: E' is non-empty and none of the variables defined in E' occur free in E or M;
C3: x is a fresh variable and the rule is *not* closed under contexts;

$$\begin{aligned} C_{\text{safe}} &::= C' \mid C[\lambda x.C'] \mid C[C'\ M] \mid C[M\ C'] \ ; \\ C' &::= \Box \mid \langle C' \mid E\rangle \ ; \\ V &::= x \mid \lambda x.M \ ; \\ VE &::= x_1 = V_1, \cdots, x_n = V_n \ . \end{aligned}$$

If $M \twoheadrightarrow_{+C} N$ *and* M *nor* N *contain on occurrence of* $+$ *then*

$$M \xrightarrow[\text{cp}_C]{} N \ .$$

The cyclic call-by-need lambda calculus is defined in Table 3. We can define a notion of information content for it using the information content of the call-by-name calculus:

Definition 16. *Given the ARS* $(\Lambda\circ, \xrightarrow[\lambda\circ_{\text{need}}]{})$ *and the partial order* $(\Lambda_\Omega, \leq_\Omega)$. *The information content* $\omega_{\lambda^\circ_{\text{need}}}$ *of a term* $M \in \Lambda\circ$ *is given as follows:*

$$\omega_{\lambda^\circ_{\text{need}}}(M) = \text{lub}\{\omega_{\lambda\circ}(N) \mid M \twoheadrightarrow_{\text{es}} N\} \ .$$

The Böhm semantics of M *with respect to* $\omega_{\lambda^\circ_{\text{need}}}$ *is denoted* $\text{ABS}_{\text{need}}(M)$.

Due to the fact that the information content is infinite, we do not have skew confluence. Consider the following two reductions:

$$\begin{aligned} M \equiv \langle x \mid x = \lambda z.z\ y, y = \lambda z'.z'\ (x\ z')\rangle \\ \xrightarrow[\lambda\circ_{\text{need}}]{} \langle \lambda z.z\ y \mid y = \lambda z'.z'\ ((\lambda z.z\ y)\ z')\rangle \\ \xrightarrow[\lambda\circ_{\text{need}}]{} \langle \lambda z.z\ y \mid y = \lambda z'.z'\ (z'\ y)\rangle \equiv M_1 \end{aligned}$$

and

$$\begin{aligned} M \equiv \langle x \mid x = \lambda z.z\ y, y = \lambda z'.z'\ (x\ z')\rangle \\ \xrightarrow[\lambda\circ_{\text{need}}]{} \langle x \mid x = \lambda z.z\ (\lambda z'.z'\ (x\ z'))\rangle \equiv M_2\ . \end{aligned}$$

We have that $\omega_{\lambda^\circ_{\text{need}}}(M_1) = \text{ABS}_{\text{need}}(M_1)$, because the only redexes in M_1 and any of its reducts are value substitutions, which are performed as part of the computation of the information content. However, there cannot exist M_3 such that $M_2 \xrightarrow[\lambda\circ_{\text{need}}]{} M_3$ and $\omega_{\lambda^\circ_{\text{need}}}(M_1) \subseteq \omega_{\lambda^\circ_{\text{need}}}(M_3)$ because $\omega_{\lambda^\circ_{\text{need}}}(M_1)$ is infinite whereas the information content of any reduct of M_2 is finite. The reason is that in the unwinding of M we have an infinite number of β-redexes. When we rewrite M into M_1 we do all of those redexes at once and when we rewrite M into M_2 we destroy the opportunity to do them in one step. The consistency of $\lambda^\circ_{\text{need}}$ is guaranteed by the following theorem.

Theorem 3. *The ARSI* $((\Lambda\circ, \xrightarrow[\lambda\circ_{\text{need}}]{}), \omega_{\lambda\circ_{\text{need}}}, \mathcal{I}(\Lambda_\Omega, \leq_\Omega))$ *is* ω*-skew confluent.*

Uniqueness of Böhm semantics follows from this theorem. A direct proof of an equivalent statement can be found in [5].

7 Lifting Abstract Böhm Semantics

The Böhm semantics of both the cyclic call-by-name and call-by-need lambda calculi are closely related to unwinding. The information content for the cyclic call-by-name calculus can be seen as a two step process. First, one computes the normal form with respect to the $es\omega$ and $gc\omega$ rules given in Definition 14. Second, one applies the notion of information content associated to the lambda calculus [34], which consists of computing the normal form with respect to the $\beta\omega$ and $@\omega$ rules. The call-by-need information content of a term is the information content of the unwinding of the term. In this section, we study how to derive these notions of information content in an abstract setting.

We first introduce in Sect. 7.1 the notion of a finite basis and its properties. In Sect. 7.2 we consider extensions consisting of infinite objects over the basis and objects whose semantics are infinite objects. In Sect. 7.3, we consider abstract Böhm semantics of extensions.

7.1 Finite Basis

We start from an ARSI equipped with a partial order on its objects. The partial order should have a least element to ensure that its ideal completion [20] is an algebraic CPO. The finite elements of the ideal completion are the embeddings of the original partial order, so the information content of the finite elements is already defined. To be able to lift the notion of information content to infinite elements, we must require that the notion of information content and the rewrite relation are also monotonic with respect to the partial order. We formalize this starting point with the notion of a *finite basis*, for which we need one auxiliary definition: monotonicity of a rewrite relation with respect to an order.

Definition 17. *Given an ARS* $(A, \rightarrow)$ *and a partial order* $(A, \leq)$ *with a least element, we say that* $\rightarrow$ *is monotonic with respect to* $\leq$ *if*

$$a \rightarrow a' \wedge a \leq a'' \Rightarrow \exists a''' : a' \leq a''' \wedge a'' \rightarrow a''' \ .$$

The diagram of monotonicity is

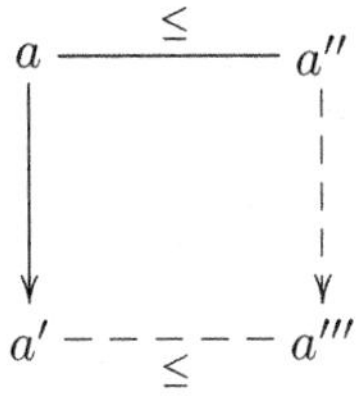

Definition 18 (finite basis). *A tuple* $(A, \rightarrow, \leq_A, \omega, D, \leq_D)$ *is a finite basis if*

- $((A, \rightarrow), \omega, (D, \leq_D))$ *is an ARSI with unique abstract Böhm semantics;*
- ω *is monotonic with respect to* $\leq_A$*:* $a \leq_A a' \Rightarrow \omega(a) \leq_D \omega(a')$*;*
- $\rightarrow$ *is monotonic with respect to* $\leq_A$.

Example 2. Let ω_{LL} stand for the function which given a lambda calculus term M returns the normal form of M with respect to the following ω_{LL}-rules [28]:

$$\begin{array}{ll} (\lambda x.M)\,N & \xrightarrow[\omega_{\text{LL}}]{} \Omega \\ \Omega\,M & \xrightarrow[\omega_{\text{LL}}]{} \Omega \end{array}$$

Then one has that $(\Lambda_\Omega, \xrightarrow[\beta]{}, \leq_\Omega, \omega_{\text{LL}}, \mathcal{I}(\Lambda_\Omega), \subseteq)$ is a finite basis.

Next, we consider rewrite systems, referred to as extensions, whose objects have infinite objects as semantics. Moreover, we want these rewrite systems to mimic the behavior of their finite counterparts. For the cyclic lambda calculi mimicking the finite lambda calculus meant that the rewrite relation induced by the cyclic calculi was contained in the infinitary lambda calculus and that finite reductions in an approximation could be lifted to reductions in the extension. The equivalent of this involves lifting the reduction in a finite basis to a reduction on its ideal completion.

7.2 Extensions

The set of infinite terms can be seen as the ideal completion of the set of finite terms under the prefix order $\leq_\Omega$. Therefore, we treat the ideal completion of a set of objects as infinite objects. We then define a rewrite relation on ideals as follows: we say that an ideal rewrites to another if every sufficiently large element of the first ideal rewrites to an element of the second and every sufficiently large element of the second ideal can be obtained by rewriting an element of the first. This is in a way similar to how Corradini defined complete developments of an infinite set of redexes in an infinite term [17].

Definition 19. *Given a finite basis* $\mathcal{A} = (A, \underset{A}{\rightarrow}, \leq_A, \omega, D, \leq_D)$*. The operator* $[\cdot\rangle : \mathcal{P}(A \times A) \to \mathcal{P}(\mathcal{I}(A) \times \mathcal{I}(A))$ *is defined by* $I_1[R\rangle I_2$ *if*

$$\forall a \in I_1, \exists a' \in I_1, a'' \in I_2 : a \leq_A a' \; R \; a''$$

and

$$\forall a'' \in I_2, \exists a' \in I_2, a \in I_1 : a \; R \; a' \geq_A a'' \; .$$

If for $I \in \mathcal{I}(A)$, we denote $a \in I$ as $I \underset{\alpha}{\rightarrow} a$ then we can phrase this definition with the following two diagrams:

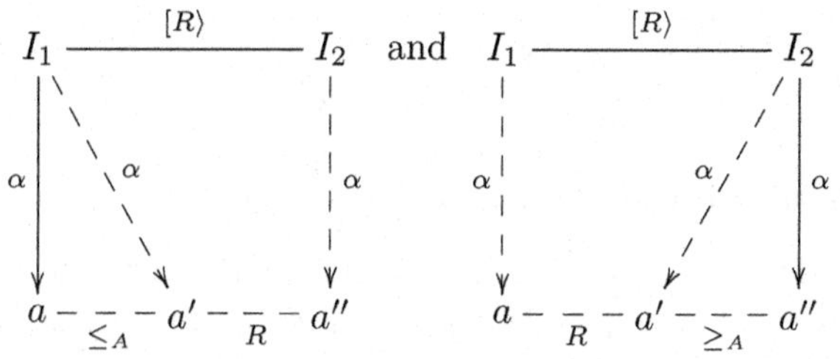

Example 3. Consider the infinitary lambda calculus term

$$(\lambda x.f\,x\,(f\,x\,(f\,x\,(\cdots))))(I\,I) \; .$$

The reduction of $I\,I$ to I can be matched:

$$(\lambda x.f\,x\,(f\,x\,(f\,x\,(\cdots))))(I\,I)\;[\underset{\beta}{\rightarrow}\rangle(\lambda x.f\,x\,(f\,x\,(f\,x\,(\cdots))))I$$

because

$$\begin{array}{rcl} (\lambda x.\Omega)\,(I\,I) & \underset{\beta}{\rightarrow} & (\lambda x.\Omega)\,I \\ (\lambda x.f\,x\,(\Omega))\,(I\,I) & \underset{\beta}{\rightarrow} & (\lambda x.f\,x\,(\Omega))\,I \\ (\lambda x.f\,x\,(f\,x\,(\Omega)))\,(I\,I) & \underset{\beta}{\rightarrow} & (\lambda x.f\,x\,(f\,x\,(\Omega)))\,I \\ \vdots & \vdots & \vdots \end{array}$$

It is obvious that for any single step we can do this. We also have

$$(\lambda x.f\,x\,(f\,x\,(f\,x\,(\cdots))))I\;[\underset{\beta}{\rightarrow}\rangle\; f\,I\,(f\,I\,(f\,I\,(\cdots)))$$

and

$$(\lambda x.f\,x\,(f\,x\,(f\,x\,(\cdots))))(I\,I)\ [\underset{\beta}{\rightarrow}\rangle\ f\,(I\,I)\,(f\,(I\,I)\,(f\,(I\,I)\,(\cdots)))\ .$$

By using $[\underset{\beta}{\twoheadrightarrow}\rangle$ rather than $[\underset{\beta}{\rightarrow}\rangle$, we can also develop infinite sets of redexes. The trick is to develop the finite subset of redexes present in suitable finite prefixes. For example, the fact that

$$f\,(I\,I)\,(f\,(I\,I)\,(f\,(I\,I)\,(\cdots)))\ [\underset{\beta}{\twoheadrightarrow}\rangle\ f\,I\,(f\,I\,(f\,I\,(\cdots)))$$

follows from

$$\begin{array}{rcl} f\,(I\,I)\,\Omega & \underset{\beta}{\twoheadrightarrow} & f\,I\,\Omega \\ f\,(I\,I)\,(f\,(I\,I)\,\Omega) & \underset{\beta}{\twoheadrightarrow} & f\,I\,(f\,I\,(\Omega)) \\ f\,(I\,I)\,(f\,(I\,I)\,(f\,(I\,I)\,(\Omega))) & \underset{\beta}{\twoheadrightarrow} & f\,I\,(f\,I\,(f\,I\,(\Omega))) \\ \vdots & \vdots & \vdots \end{array}$$

Now that we can lift any relation from an order to its ideal completion, it is logical to also extend information content from the order to the ideal completion. Because the information contained in an ideal can be infinite, we define the information content of an ideal as the downward closure of the set of information contents of its elements:

Definition 20. *Given a finite basis $\mathcal{A} = (A, \underset{A}{\rightarrow}, \leq_A, \omega, D, \leq_D)$ and $I \in \mathcal{I}(A)$. Let*

$$\omega_\infty(I) = \text{lub}\{\omega(a) \mid a \in I\}\ .$$

This is well-defined because of the monotonicity of ω with respect to $\leq_A$. Next, we consider the abstract version of an extension which contains objects whose semantics are infinite objects over the basis. Moreover, the reduction relation of the extension should contain a subset that can compute the semantics internally as a normal form. A good example is the call-by-name calculus, where the subset of just external substitution plus ω_{es} as the notion of information content can compute the unwinding. In general, we can always compute the semantics by using the empty (sub)set and the semantics as information content.

Definition 21. *A tuple $\mathcal{B} \equiv (B, \underset{B}{\rightarrow}, \underset{[\![B]\!]}{\rightarrow}, \omega_{[\![\cdot]\!]}, [\![\cdot]\!])$ is an extension of a finite basis $\mathcal{A} \equiv (A, \underset{A}{\rightarrow}, \leq_A, \omega, D, \leq_D)$, if*

- *$(B, \underset{B}{\rightarrow})$ is an ARS;*
- *$\underset{[\![B]\!]}{\rightarrow} \subseteq \underset{B}{\rightarrow}$;*
- *$((B, \underset{[\![B]\!]}{\rightarrow}), \omega_{[\![\cdot]\!]}, \mathcal{I}(A, \leq_A))$ is an ARSI, such that*

$$\forall b \in B:\ \text{ABS}(b) = [\![b]\!]\ .$$

The function $[\![\cdot]\!]$ takes the place of the unwinding. The function $\omega_{[\![\cdot]\!]}$ denotes the visible part of the unwinding. This visible part is used to restrict information

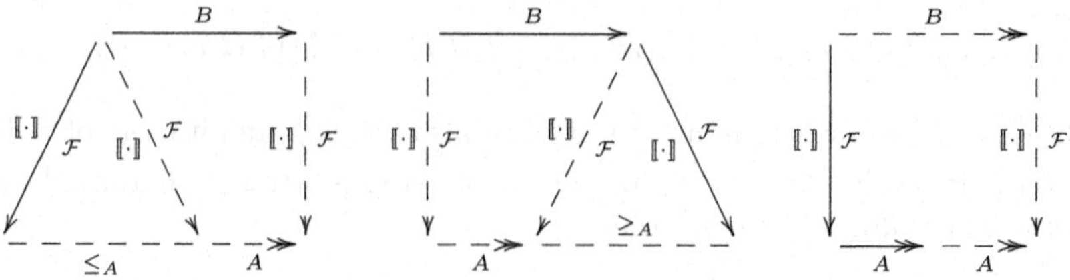

Fig. 2. Soundness and completeness of an extension as commutative diagrams

content. For example, in the call-by-name calculus the visible part would be ω_{es} and in the call-by-need calculus it would be $[\![\cdot]\!]$. There must also be a subset of the rewrite relation, such that the semantics can be computed internally.

The cyclic lambda calculi are extensions of the lambda calculus in this sense because the semantics of a cyclic lambda term is its unwinding, which is an infinite term. For an extension to make sense, we require it to be sound and complete with respect to the basis [22, 1]. In other words, the extension cannot do more than the basis (soundness) and the extension can simulate everything the basis can do (completeness). To define soundness we use the $[\cdot\rangle$ operator. To define completeness we use a simple lifting property:

Definition 22. *Given a finite basis* $\mathcal{A} \equiv (A, \underset{A}{\rightarrow}, \leq_A, \omega, D, \leq_D)$ *and an extension* $\mathcal{B} \equiv (B, \underset{B}{\rightarrow}, \underset{[\![B]\!]}{\longrightarrow}, \omega_{[\![\cdot]\!]}, [\![\cdot]\!])$. *Then,*

- $\mathcal{B}$ *is infinitarily sound with respect to* $\mathcal{A}$ *if*

$$s \underset{B}{\rightarrow} t \Rightarrow [\![s]\!] \; [\underset{A}{\twoheadrightarrow}\rangle \; [\![t]\!] \;;$$

- $\mathcal{B}$ *is infinitarily complete with respect to* $\mathcal{A}$ *if*

$$\forall a, s : a \in [\![s]\!] \wedge a \underset{A}{\twoheadrightarrow} a' \Rightarrow \exists t, a'' : s \underset{B}{\twoheadrightarrow} t \wedge a' \underset{A}{\twoheadrightarrow} a'' \in [\![t]\!] \;.$$

In order to be able to draw diagrams, we use the fact that our notation allows us to denote $a \in [\![s]\!]$ by $s \xrightarrow[{[\![\cdot]\!]}]{\mathcal{F}} a$. Thus, the diagrams for soundness and completeness can be drawn as given in Fig. 2.

7.3 Abstract Böhm Semantics

We can now define an abstract Böhm semantics for extensions. The idea is simple: given an object, we compute the visible part of its semantics and apply the infinite extension of the information content of the basis to it.

Definition 23. *Given a finite basis* $\mathcal{A} \equiv (A, \underset{A}{\rightarrow}, \leq_A, \omega, D, \leq_D)$ *and an extension* $\mathcal{B} \equiv (B, \underset{B}{\rightarrow}, \underset{[\![B]\!]}{\longrightarrow}, \omega_{[\![\cdot]\!]}, [\![\cdot]\!])$. *Define* $\omega_{\mathcal{B}} : B \rightarrow D$ *by*

$$\omega_{\mathcal{B}}(s) = \omega_\infty(\omega_{[\![\cdot]\!]}(s)) \;.$$

The only problem with the above definition is that the visible part and the base information content must fit together to form a proper notion of information content. First, we will give an example that shows that the result might not be an ARSI. Next, we will prove two propositions that help establishing that the result is an ARSI.

Example 4. Consider the cyclic extension

$$\begin{array}{ll} \mathsf{F}(\mathsf{F}(x)) & \to \langle x \mid x = \mathsf{F}(\mathsf{G}(x))\rangle \\ \mathsf{G}(\mathsf{G}(x)) & \to \langle x \mid x = \mathsf{G}(\mathsf{F}(x))\rangle \\ \langle x \mid x = M, E\rangle & \to \langle M \mid x = M, E\rangle \end{array}$$

$$\begin{array}{l} \langle \mathsf{F}(x) \mid x = M, E\rangle \to \mathsf{F}(\langle x \mid x = M, E\rangle) \\ \langle \mathsf{G}(x) \mid x = M, E\rangle \to \mathsf{G}(\langle x \mid x = M, E\rangle) \end{array}$$

and the functions ω, defined in Eq. 4, and ω_{es}, defined below:

$$\begin{array}{ll} \omega_{\mathrm{es}}(x) & = x \\ \omega_{\mathrm{es}}(f(M_1, \cdots, M_n) & = f(\omega_{\mathrm{es}}(M_1), \cdots, \omega_{\mathrm{es}}(M_n)) \\ \omega_{\mathrm{es}}(\langle M \mid x_1 = M_1, \cdots, x_n = M_n\rangle) & = \omega_{\mathrm{es}}(M)[x_1 := \Omega, \cdots, x_n := \Omega] \end{array}$$

The function $\omega \circ \omega_{\mathrm{es}}$ is not a notion of information content:

$$(\omega \circ \omega_{\mathrm{es}})(\mathsf{F}(\mathsf{F}(x))) = \omega(\mathsf{F}(\mathsf{F}(x))) = \mathsf{F}(\Omega)$$

and

$$(\omega \circ \omega_{\mathrm{es}})(\langle x \mid x = \mathsf{F}(\mathsf{G}(x))\rangle) = \omega(\Omega) = \Omega \ .$$

So $\omega \circ \omega_{\mathrm{es}}$ is not monotonic with respect to the reduction relation of the extension.

The following proposition assumes that the visible part of the semantics is the whole semantics.

Proposition 6. *Given a finite basis $\mathcal{A} \equiv (A, \underset{A}{\rightarrow}, \leq_A, \omega, D, \leq_D)$ and an extension $\mathcal{B} \equiv (B, \underset{B}{\rightarrow}, \emptyset, [\![\cdot]\!], [\![\cdot]\!])$. If $\mathcal{B}$ is infinitarily sound with respect to $\mathcal{A}$ then $\mathcal{L} \equiv ((B, \underset{B}{\rightarrow}), \omega_{\mathcal{B}}, (D, \leq_D))$ is an ARSI.*

Proof. We have to establish that $\omega_{\mathcal{B}}$ is monotonic with respect to $\underset{B}{\rightarrow}$. That is, we need to show that if $s \underset{B}{\rightarrow} s'$ then $\omega_{\mathcal{B}}(s) \leq_D \omega_{\mathcal{B}}(s')$. Unfolding definitions we get

$$\omega_{\mathcal{B}}(s) = \omega_{\infty}([\![s]\!]) = \mathrm{lub}\{\omega(a) \mid a \in [\![s]\!]\}$$

and

$$\omega_{\mathcal{B}}(s') = \mathrm{lub}\{\omega(a) \mid a \in [\![s']\!]\} \ .$$

From the soundness of $\mathcal{B}$, we get that

$$\forall a \in [\![s]\!] : \exists a' \in [\![s]\!], a'' \in [\![s']\!] : a \leq_A a' \underset{A}{\twoheadrightarrow} a'' \ .$$

Since $\mathcal{A}$ is a finite basis, we have monotonicity of ω with respect to both $\leq_A$ and $\underset{A}{\rightarrow}$, so

$$\forall a \in [\![s]\!] : \exists a' \in [\![s]\!], a'' \in [\![s']\!] : \omega(a) \leq_D \omega(a') \leq_D \omega(a'') \ .$$

Hence

$$\{\omega(a) \mid a \in [\![s]\!]\} \subseteq \{\omega(a) \mid a \in [\![s']\!]\}$$

and also

$$\omega_{\mathcal{B}}(s) \leq_D \omega_{\mathcal{B}}(s') \ .$$

The problem with the counterexample is that the extended rewrite rules destroy a part of the unwinding, which was already part of the information content. The solution therefore is to require that every rewrite step in the extension preserves enough unwinding to compute at least the information content of the left-hand side:

$$\forall s, t \in B : s \xrightarrow[B]{} t \Rightarrow \exists I \in \mathcal{I}(A) : I \subseteq \omega_{[\![\cdot]\!]}(s) \wedge I \subseteq \omega_{[\![\cdot]\!]}(t) \wedge \omega_\infty(I) = \omega_\infty(\omega_{[\![\cdot]\!]}(s)) \ .$$

So the rewrite step from s to t preserves a part of the unwinding a and a is enough to compute the information content of s.

For the call-by-name calculus this property holds, because if $M \xrightarrow[\lambda\circ]{} N$ then it is either not a $\beta\circ$ step and $\omega_{[\![\cdot]\!]}(M) \leq_\Omega \omega_{[\![\cdot]\!]}(N)$ and we can take $a = \omega_{[\![\cdot]\!]}(M)$ or it is a $\beta\circ$ step $C[(\lambda x.P)\, Q] \xrightarrow[\beta\circ]{} C[\langle P \mid x = Q\rangle]$. In this case we can take $a = \omega_{[\![\cdot]\!]}(C[\Omega])$.

This idea give us our second proposition:

Proposition 7. *Given a finite basis* $\mathcal{A} \equiv (A, \xrightarrow[A]{}, \leq_A, \omega, D, \leq_D)$ *and an extension* $\mathcal{B} \equiv (B, \xrightarrow[B]{}, \xrightarrow[[\![B]\!]]{}, \omega_{[\![\cdot]\!]}, [\![\cdot]\!])$. *If*

1. $\mathcal{B}$ *is infinitarily sound with respect to* $\mathcal{A}$;
2. $\forall s, t \in B : s \xrightarrow[B]{} t \Rightarrow \exists I \in \mathcal{I}(A) : I \subseteq \omega_{[\![\cdot]\!]}(s) \wedge I \subseteq \omega_{[\![\cdot]\!]}(t) \wedge \omega_\infty(I) = \omega_\infty(\omega_{[\![\cdot]\!]}(s))$;

then $\mathcal{L} \equiv ((B, \xrightarrow[B]{}), \omega_{\mathcal{B}}, (D, \leq_D))$ *is an ARSI.*

Proof. We have to establish that $\omega_{\mathcal{B}}$ is monotonic with respect to $\xrightarrow[B]{}$. That is, we need to show that if $s \xrightarrow[B]{} s'$ then $\omega_{\mathcal{B}}(s) \leq_D \omega_{\mathcal{B}}(s')$. By the second condition we can find $I \in \mathcal{I}(A)$, such that

$$I \subseteq \omega_{[\![\cdot]\!]}(s) \wedge I \subseteq \omega_{[\![\cdot]\!]}(s') \wedge \omega_\infty(I) = \omega_\infty(\omega_{[\![\cdot]\!]}(s)) \ .$$

From the fact that ω is monotone w.r.t. $\leq_A$ it is easy to prove that

$$I \subseteq \omega_{[\![\cdot]\!]}(s') \Rightarrow \omega_\infty(I) \leq_D \omega_\infty(\omega_{[\![\cdot]\!]}(s')) \ .$$

Hence

$$\omega_{\mathcal{B}}(s) \leq_D \omega_{\mathcal{B}}(s') \ .$$

Once we have established that the result is an ARSI then we can prove uniqueness of the abstract Böhm semantics. We establish a few lemmas first.

Lemma 1. *Given a finite basis* $\mathcal{A} \equiv (A, \xrightarrow[A]{}, \leq_A, \omega, D, \leq_D)$ *and an extension* $\mathcal{B} \equiv (B, \xrightarrow[B]{}, \xrightarrow[[\![B]\!]]{}, \omega_{[\![\cdot]\!]}, [\![\cdot]\!])$ *such that*

1. *$\mathcal{B}$ is infinitarily sound with respect to $\mathcal{A}$;*
2. *$\mathcal{B}$ is infinitarily complete with respect to $\mathcal{A}$;*
3. *$\mathcal{L} \equiv ((B, \underset{B}{\rightarrow}), \omega_{\mathcal{B}}, (D, \leq_D))$ is an ARSI.*

We have:

$$\forall b_1, b_2 \in B, a \in A : b_1 \underset{B}{\twoheadrightarrow} b_2 \underset{[\![\cdot]\!]}{\overset{\mathcal{F}}{\rightarrow}} a \Rightarrow \exists a', a'' \in A : b_1 \underset{[\![\cdot]\!]}{\overset{\mathcal{F}}{\rightarrow}} a'' \underset{A}{\twoheadrightarrow} a' \geq_A a \ .$$

Proof. Follows by induction on the length of $b_1 \underset{B}{\twoheadrightarrow} b_2$ from soundness and monotonicity of $\underset{A}{\rightarrow}$ with respect to $\leq_A$. In a diagram:

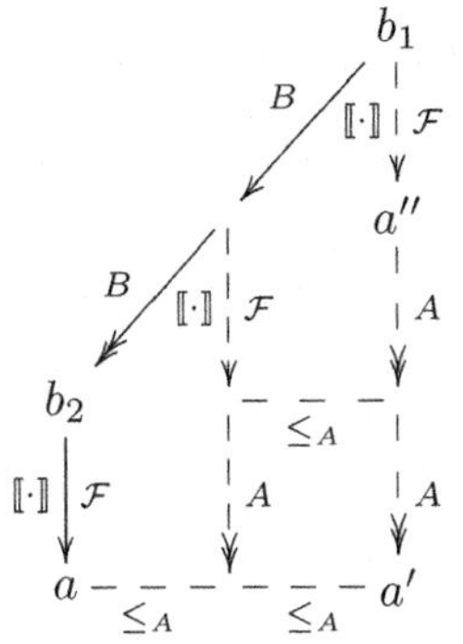

In this diagram, the left part is the induction hypothesis, the top right part is soundness and the bottom right part is monotonicity.

Lemma 2. *Given a finite basis $\mathcal{A} \equiv (A, \underset{A}{\rightarrow}, \leq_A, \omega, D, \leq_D)$ and an extension $\mathcal{B} \equiv (B, \underset{B}{\rightarrow}, \underset{[\![B]\!]}{\rightarrow}, \omega_{[\![\cdot]\!]}, [\![\cdot]\!])$ such that*

1. *$\mathcal{B}$ is infinitarily sound with respect to $\mathcal{A}$;*
2. *$\mathcal{B}$ is infinitarily complete with respect to $\mathcal{A}$;*
3. *$\mathcal{L} \equiv ((B, \underset{B}{\rightarrow}), \omega_{\mathcal{B}}, (D, \leq_D))$ is an ARSI.*

We have:

$$\forall b_1, b_2 \in B, a_1, a_1' \in A, d_1 \in D : b_1 \underset{B}{\twoheadrightarrow} b_2 \wedge b_1 \underset{[\![\cdot]\!]}{\overset{\mathcal{F}}{\rightarrow}} a_1 \underset{A}{\twoheadrightarrow} a_1' \underset{\omega}{\overset{\mathcal{F}}{\rightarrow}} d_1 \Rightarrow$$

$$\exists a_2, a_2' \in A, d_2 \in D : b_2 \underset{[\![\cdot]\!]}{\overset{\mathcal{F}}{\rightarrow}} a_2 \underset{A}{\twoheadrightarrow} a_2' \underset{\omega}{\overset{\mathcal{F}}{\rightarrow}} d_2 \wedge d_1 \leq_D d_2 \ .$$

Proof. By repeating the following diagram:

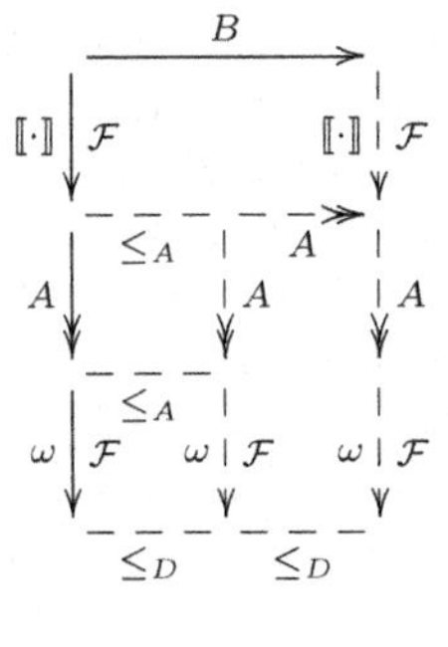

The diagram is built up from an instance of soundness on the top, two instances of monotonicity on the left bottom and an instance of ω-skew confluence of the basis on the right bottom.

Theorem 4. *Given a finite basis* $\mathcal{A} \equiv (A, \underset{A}{\rightarrow}, \leq_A, \omega, D, \leq_D)$ *and an extension* $\mathcal{B} \equiv (B, \underset{B}{\rightarrow}, \underset{[\![B]\!]}{\rightarrow}, \omega_{[\![\cdot]\!]}, [\![\cdot]\!])$. *If*

1. *$\mathcal{B}$ is infinitarily sound with respect to $\mathcal{A}$;*
2. *$\mathcal{B}$ is infinitarily complete with respect to $\mathcal{A}$;*
3. *$\mathcal{L} \equiv ((B, \underset{B}{\rightarrow}), \omega_{\mathcal{B}}, (D, \leq_D))$ is an ARSI;*

then $\mathcal{L}$ yields unique abstract Böhm semantics.

Proof. The following diagram proves ω-skew confluence of $\mathcal{L}$:

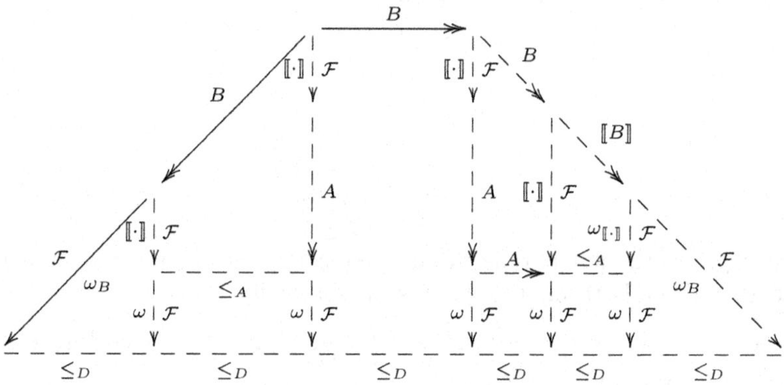

From left to right, we can obtain the sub-diagrams along the top by unfolding the definition of $\omega_{\mathcal{B}}$, the first lemma, the second lemma, completeness, the fact that $[\![b]\!] = \mathrm{ABS}_{[\![\cdot]\!]}(b)$ and again unfolding the definition of $\omega_{\mathcal{B}}$. The bottom rectangles are various applications of monotonicity.

This completes the presentation of the abstract lifting theory. It is a fairly complicated theory, but we think that using all of this machinery is easier than giving a direct proof of uniqueness of Böhm semantics in the extension. With a direct proof, we have to deal with effects of non-confluence for most of the proof. When we use the extension approach, we have a lot more statements to prove, but most of them can be proved in a context where the rewrite system is confluent. For example, in the case of the lambda calculus we can use the fact that the lambda calculus is confluent while proving that the lambda calculus is a basis. Proving that the cyclic lambda calculus is infinitarily sound for the $\beta\circ$ rule is made easy by the fact that the rewrite system consisting of external substitution, external lift and $\beta\circ$ is confluent. Proving soundness for the other rules is somewhat harder, because this involves a property of non-confluent rules. However, it should be noted that what we really have to do is to prove that those rules preserve the unwinding. Hence, the result can be reused for other calculi. Finally, to prove completeness it is sufficient to prove a confluent subset complete.

8 Conclusions

We have given a modified version of the notion of abstract Böhm tree, which is capable of dealing with infinite information content. To guarantee the uniqueness of the new form of abstract Böhm semantics, we have introduced a new member of the confluence family: ω-skew confluence, which is a generalization of the existing notion of skew confluence.

Also, we have developed an abstract framework to be able to construct a new system while deriving properties from the old one. In particular, we have defined the notion of a finite basis, consisting of an ARS and a notion of information content with suitable properties. We have defined how to extend the ARS to its ideal completion. We have defined an extension as an ARS, whose objects have the ideal completion of the basis as their semantics and we have defined when such an extension is sound and complete. We have shown that a finite basis and a sound and complete extension give rise to a notion of information content on the extension.

Acknowledgments. We thank the anonymous referees and the editors for their comments which allowed us to make substantial improvements to the original paper. The first author was supported by National Science Foundation grant number CCR-0204389.

References

1. Z. M. Ariola. Relating graph and term rewriting via Böhm models. *Applicable Algebra in Engineering, Communication and Computing*, 7(5), 1996.
2. Z. M. Ariola and Arvind. Properties of a first-order functional language with sharing. *Theoretical Computer Science*, 146:69–108, 1995.
3. Z. M. Ariola and S. Blom. Lambda calculi plus letrec. Technical Report CIS-TR-97-05, Department of computer and information science, University of Oregon.
4. Z. M. Ariola and S. Blom. Cyclic lambda calculi. In M. Abadi and T. Ito, editors, *Theoretical Aspects of Computer Software*, volume 1281 of *Lecture Notes in Computer Science*, pages 77–106. Springer Verlag, Sept. 1997.
5. Z. M. Ariola and S. Blom. Lambda calculi plus letrec. Technical Report IR-434, Department of Mathematics and Computer Science, Vrije Universiteit Amsterdam, Oct. 1997.
6. Z. M. Ariola and S. Blom. Skew confluence and the lambda calculus with letrec. *Annals of Pure and Applied Logic*, 117(1-3):95–168, october 2002.
7. Z. M. Ariola and M. Felleisen. The call-by-need lambda calculus. *Journal of Functional Programming*, 7(3), 1997.
8. Z. M. Ariola, M. Felleisen, J. Maraist, M. Odersky, and P. Wadler. The call-by-need lambda calculus. In *Proc. ACM Conference on Principles of Programming Languages*, pages 233–246, 1995.
9. Z. M. Ariola and J. W. Klop. Equational term graph rewriting. *Fundamentae Informaticae*, 26(3,4):207–240, 1996. Extended version: CWI Report CS-R9552.
10. Z. M. Ariola and J. W. Klop. Lambda calculus with explicit recursion. *Information and computation*, 139(2):154–233, Dec. 1997.

11. Z. M. Ariola, J. W. Klop, and D. Plump. Bisimilarity in term graphs rewriting. *Information and Computation*, 156(1/2):2–24, 2000.
12. H. Barendregt, T. Brus, M. van Eekelen, J. Glauert, J. Kennaway, M. van Leer, M. Plasmeijer, and M. R. Sleep. Towards an intermediate language based on graph rewriting. In *Proc. Conference on Parallel Architecture and Languages Europe (PARLE '87), Eindhoven, The Netherlands, Springer-Verlag LNCS 259*, June 1987.
13. H. P. Barendregt. *The Lambda Calculus: Its Syntax and Semantics*, volume 103 of *Studies in Logic and the Foundations of Mathematics*. Elsevier, revised edition, 1984.
14. S. Blom. *Term Graph Rewriting - syntax and semantics*. PhD thesis, Vrije Universiteit Amsterdam, 2001.
15. S. Blom. Lifting Infinite Normal Form Definitions from Term Rewriting to Term Graph Rewriting. In *TERMGRAPH 2002 - International Workshop on Term Graph Rewriting*, 2002.
16. S. Blom. An approximation based approach to infinitary lambda calculi. In van Oostrom [33], pages 221–232.
17. A. Corradini. Term rewriting in CT_Σ. In M.-C. Gaudel and J.-P. Jouannaud, editors, *Proc. Colloquium on Trees in Algebra and Programming (CAAP '93), Springer-Verlag LNCS 668*, pages 468–484, 1993.
18. A. L. de Medeiros Santos. *Compilation by Transformation in Non-Strict Functional Languages*. PhD thesis, University of Glasgow, July 1995.
19. M. Dezani-Ciancaglini and E. Giovannetti. From Böhm's theorem to observational equivalences: an informal account. *Electronic Notes in Theoretical Computer Science*, 50(2), 2001.
20. M. Hennessy. *Algebraic Theory of Processes*. MIT Press, 1988.
21. G. Kahn and G. D. Plotkin. Concrete domains. *Theor. Comput. Sci.*, 121(1&2):187–277, 1993.
22. J. R. Kennaway, J. W. Klop, M. R. Sleep, and F. J. de Vries. The adequacy of term graph rewriting for simulating term rewriting. In M. R. Sleep, M. J. Plasmeijer, and M. C. D. J. van Eekelen, editors, *Term Graph Rewriting: Theory and Practice*, pages 157–168. John Wiley & Sons, 1993.
23. J. R. Kennaway, J. W. Klop, M. R. Sleep, and F. J. de Vries. Infinitary lambda calculus. In *Proc. Rewriting Techniques and Applications, Kaiserslautern*, 1995.
24. J. R. Kennaway, J. W. Klop, M. R. Sleep, and F. J. de Vries. Transfinite reductions in orthogonal term rewriting systems. *Information and Computation*, 119(1), 1995.
25. J. Ketema. Böhm-like trees for term rewriting systems. In van Oostrom [33], pages 233–248.
26. J. Ketema and J. G. Simonsen. Infinitary combinatory reduction systems. In J. Giesl, editor, *RTA*, volume 3467 of *Lecture Notes in Computer Science*, pages 438–452. Springer, 2005.
27. J. W. Klop. Term rewriting systems. In S. Abramsky, D. Gabbay, and T. Maibaum, editors, *Handbook of Logic in Computer Science*, volume II, pages 1–116. Oxford University Press, 1992.
28. J.-J. Lévy. *Réductions Correctes et Optimales dans le Lambda-Calcul.* PhD thesis, Universite Paris VII, October 1978.
29. D. Plump. Term graph rewriting. In H. Ehrig, G. Engels, H.-J. Kreowski, and G. Rozenberg, editors, *Handbook of Graph Grammars and Computing by Graph Transformation*, volume 2: Applications, Languages and Tools, chapter 1, pages 3–61. World Scientific, 1999.
30. P. Severi and F.-J. de Vries. An extensional böhm model. In S. Tison, editor, *RTA*, volume 2378 of *Lecture Notes in Computer Science*, pages 159–173. Springer, 2002.

31. M. R. Sleep, M. J. Plasmeijer, and M. C. D. J. van Eekelen, editors. *Term Graph Rewriting: Theory and Practice.* John Wiley & Sons, 1993.
32. Terese. *Term Rewriting Systems.* Cambdrige University Press, 2003.
33. V. van Oostrom, editor. *Rewriting Techniques and Applications, 15th International Conference, RTA 2004, Aachen, Germany, June 3-5, 2004, Proceedings*, volume 3091 of *Lecture Notes in Computer Science.* Springer, 2004.
34. C. Wadsworth. *Semantics And Pragmatics Of The Lambda-Calculus.* PhD thesis, University of Oxford, September 1971.
35. C. Wadsworth. The Relation between Computational and Denotational Properties for Scott's D_∞-Models of the Lambda-Calculus. *Theoretical Computer Science*, 5, 1976.
36. P. Welch. Continuous Semantics and Inside-out Reductions. In *λ-Calculus and Computer Science Theory, Italy (Springer-Verlag LNCS 37)*, March 1975.

A Mobility Calculus with Local and Dependent Types

Mario Coppo[1,⋆], Federico Cozzi[2,⋆⋆], Mariangiola Dezani-Ciancaglini[1,⋆⋆⋆], Elio Giovannetti[1,†], and Rosario Pugliese[3,‡]

[1] Dip. Informatica, Univ. di Torino, corso Svizzera 185, Torino, Italy
[2] Dip. Scienze Matematiche e Informatiche, Univ. di Siena, pian dei Mantellini 44, Siena, Italy
[3] Dip. Sistemi e Informatica, Univ. di Firenze, viale Morgagni 65, Firenze, Italy

Dedicated to Jan Willem Klop on the occasion of his 60th birthday.

Abstract. We introduce an ambient-based calculus that combines ambient mobility with process mobility, uses group names to collect ambients with homologous features, and exploits co-moves and runtime type checking to implement flexible policies for controlling process activities. Types rely on group names and, to support dynamicity, may *depend* on group variables. Policies can dynamically change also through installation of co-moves. The compliance with ambient policies can be checked *locally* to the ambients and requires no global assumptions. We prove that the type assignment system and the operational semantics of the calculus are 'sound', and define a sound and complete type inference algorithm which, when applied to terms whose type decorations only express the desired policies, computes the minimal type annotations required for their execution. As an application of our calculus, we present a couple of examples and linger on the setting up of policies for controlling the activities of the entities involved.

1 Introduction

The foundational research on distributed and mobile computing, driven by the technological advances of the last decades, has produced a number of theoretical models (for example [27,12,29,7,3], to cite just a few), which can be generally assimilated to some form of distributed process calculus or of 'ambient' calculus.

⋆ Partially supported by EU within the FET - Global Computing initiative, project DART IST-2001-33477.

⋆⋆ Partially supported by EU within the project IHP 'Marie Curie DisCo' HPMT-CT-2001-00290.

⋆⋆⋆ Partially supported by EU within the FET - Global Computing initiative, project MIKADO IST-2001-32222 and MURST Cofin'04 project McTafi.

† Partially supported by EU within the FET - Global Computing initiative, project DART IST-2001-33477.

‡ Partially supported by EU within the FET - Global Computing initiative, project MIKADO IST-2001-32222, and FP6-2004-IST-FET Proactive, project SENSORIA proposal contract number 016004.

A. Middeldorp et al. (Eds.): Processes... (Klop Festschrift), LNCS 3838, pp. 404–444, 2005.

All such models rely on (often sophisticated) type systems to express and check behavioural properties concerning mobility, resource access, security, etc. In most of them, a system or component is represented by a term P of a given calculus, a type $\mathtt{V}$ assigned to P and an environment Σ. In the standard view, as is well-known, the term P abstractly describes the implementation, its type $\mathtt{V}$ may express some behavioural properties, and the environment Σ is a set of assumptions on the outside world. There is thus the notion of a global environment, whose corresponding concrete scenario is one where all the interacting parties are known in advance to each other, so that static checks performed before execution ensure the correctness of the whole system.

In particular, type systems for ambient calculi are usually based on the notion of a process/ambient type which describes the kind of communication and the kind of mobility actions a process can perform and, at the same time, the kind of movements and actions an ambient can make because of the activity of its internal processes. Every ambient name is assigned (by a global assumption or by a name restriction) a type which is simply the type of the processes it is allowed to contain. The basic typing rule for such systems is therefore some variant of the rule:

$$\frac{\Sigma,\ m{:}\mathtt{V} \vdash P : \mathtt{V}}{\Sigma,\ m{:}\mathtt{V} \vdash m[P] \text{ is well typed}} \ (\textsc{Amb})$$

When dealing with computing in wide-area distributed and mobile systems, however, static verification is impractical both because of the huge amount of information to be checked and because typing information could be partial, inaccurate or missing. In such 'open' and dynamic systems no global environment can be assumed; on the contrary, there usually exist several different local and autonomous computational environments. Moreover, interaction may take place between parties whose respective properties are unknown or only partially known to each other. If stopping the execution for re-checking is to be avoided, every potentially dangerous component must dynamically carry with it sufficient behavioural information that can be checked at runtime by the other components interacting with it (see, e.g., the approach based on *proof-carrying code* [33]).

To model these scenarios, we propose here an ambient-based calculus which combines ambient mobility with general process mobility (like $\mathbf{M}^3$ [15]) and where there are no global assumptions on ambient names, since there is no static ambient type. Every ambient name m may be used to build an ambient $m[P]$ with any desired content P; on the other hand, process movements are constrained by the presence of co-moves and by runtime type checking. Indeed, following [22], we define an *operational semantics with types* which exploits types to authorize or block reductions but is simpler than a full-fledged *typed operational semantics*, because it only checks that types agree with process movements.

Types rely on group names, sort of 'family names' that group ambients with homologous features. Mobility properties and co-actions (entrance permissions) are expressed in terms of groups. The notion of a group, however, can be as fine-grained as necessary: in principle, each ambient can be in a distinct group. Dynamicity is enhanced by means of group-dependent types: types can contain variables which communication can instantiate to group names. In this way, during computation, processes can acquire knowledge of new group names that can be used either in movement actions or to let processes in.

Our first result shows that the type assignment system and the operational semantics of our calculus are 'sound'. This means that in any reduction sequence, ambient and process movements always comply with the constraints expressed by the types associated to the single ambients, and the types of messages exchanged in communications always match. This is done by first proving a property of *subject reduction*, namely that an ambient's 'policy' – expressed by its inner type – is preserved by reduction, and then by proving a property of *type safety*, i.e., that a process or ambient's behaviour complies, at each reduction step, with the policy of the enclosing ambient.

Since the terms of our calculus are quite heavily decorated with types, one would like to avoid writing all those types explicitly, and to let the system partially infer them *à la* ML: ideally, one would like to write them only for specifying, within an ambient, the rights granted to incoming mobile processes, without having to introduce any extra type annotation.

Also, safety suggests that an ambient, before letting in a process coming from an untrusted ambient, must typecheck it to ascertain that its actual behavioural type agrees with the one it declares, i.e., with the rights it requires (see [23]). More generally, it is useful to know which are, for ambient types, the minimal requirements that ensure the initial consistency of the system.

Both these issues require a type inference algorithm which, when applied to processes whose type decorations only express ambient-access policies, computes the minimal rights needed for running them. Such an algorithm is therefore the second result we present, along with the proofs of its soundness and completeness.

The rest of the paper is organized as follows. In Section 2 the different constructs of the calculus are explained by means of a simple example. In Section 3 the calculus' syntax, type system and operational semantics are formally presented, while in Section 4 the soundness results are stated and proved. Section 5 illustrates an application of our calculus to modelling a public transportation system and to controlling mobility of the entities involved (e.g., trains and passengers). In Section 6, a type inference algorithm is defined and its soundness and completeness are proved. Finally, in Section 7 we draw some short conclusions and touch upon directions for future work and comparisons with related work.

2 An Explanatory Example: The Publisher

The scenario envisaged in the example consists of a publisher PUB that publishes a number of electronic journals to which different institutions may subscribe. Two kinds of institutions are present: those that are fully trusted by the publisher and whose names are known to it, and all the others, whose names are not known in advance to the publisher and which are only partially trusted (in a sense that will be apparent below). The whole system is therefore the parallel composition (via the operator '$|$') of institutions and publisher:

$$(\nu g_{inst})(\nu g_{tdl_1})\dots(\nu g_{tdl_m})(\nu g_{ins}) \\ (\mathrm{INST}_1 \mid \dots \mid \mathrm{INST}_m \mid (\nu g_{dl_1})\mathrm{INS}_1 \mid \dots \mid (\nu g_{dl_n})\mathrm{INS}_n \mid \mathrm{PUB}) \tag{2.1}$$

where the INST_i (with $i = 1, \dots, m$) are the known trusted institutions and the INS_i (with $i = 1, \dots, n$) are those only partially trusted.

Each ambient has a name and a group name: the association of an ambient with a group is not established by a static assumption but is done in the construction of the term. Ambient names and group names respectively act as the ambient's first name and family name; different ambients with homologous features are generally assigned the same group. For example, all the trusted institutions are assigned the group g_{inst}, while all the others belong to the group g_{ins}; the group g_{tdl_i} (for $i = 1, \dots, m$) labels the download processes originated by the trusted institution INST_i, while the group g_{dl_i} (for $i = 1, \dots, n$) labels the download processes originated by the untrusted institution INS_i.

The standard construct $(\nu g)P$ declares that the name g is known only to process P; thus the fact that the group of processes originated by an untrusted institution is not known in advance to the publisher has been modelled in (2.1) by putting the publisher out of the scope of its declaration.

Each of these top-level components is an ambient containing other ambients and processes: the publisher contains the journals and a manager process; an institution is an ambient that sends the publisher a subscription request and hosts a number of download processes; these in turn are represented by mobile ambients moving from their institutions to the publisher (where they access the subscribed journal) and back.

$$\begin{aligned} \mathrm{PUB} &= pub{:}g_{pub}(\langle\!\langle \varnothing, \{g_{jrn}\}\rangle\!\rangle, \mathsf{com}(\mathsf{amb}, \mathsf{group}))[\mathrm{MGR} \mid \mathrm{JRN}_1 \mid \dots \mid \mathrm{JRN}_n] \\ \mathrm{INST}_i &= inst_i{:}g_{inst}(\langle\!\langle \varnothing, \{g_{pub}\}\rangle\!\rangle, \mathsf{shh})[\mathrm{REQ}_i^k \mid \mathrm{TDL}_i^1 \mid \dots \mid \mathrm{TDL}_i^h] \\ \mathrm{INS}_i &= ins_i{:}g_{ins}(\langle\!\langle \varnothing, \{g_{pub}\}\rangle\!\rangle, \mathsf{shh})[\mathrm{RQ}_i^k \mid \mathrm{DL}_i^1 \mid \dots \mid \mathrm{DL}_i^l] \end{aligned}$$

As anticipated in the Introduction, an ambient's properties are purely local and are not committed to global assumptions on the ambient's name or group. So the construct $\alpha{:}\gamma\,\mathtt{V}[P]$ also contains a type $\mathtt{V} \equiv (\langle\!\langle \mathscr{C}, \mathscr{E}\rangle\!\rangle, \mathtt{T})$ which coincides with the type of its inner process P. This type consists of two components, the *mobility* type $\langle\!\langle \mathscr{C}, \mathscr{E}\rangle\!\rangle$ and the *communication* type $\mathtt{T}$, which are sets of group names: $\mathscr{C}$ is the set of ambient groups which α is allowed to *cross* (driven by P) and $\mathscr{E}$ is the set of ambient groups which processes sent by α (and then by P) are allowed to *enter*. On the other hand $\mathtt{T}$ specifies the type of the messages that can be communicated within α by input and output actions of P. It can be shh if no input and output action can be performed by P, or it can be of the form $\mathsf{com}(\mathtt{W})$, where $\mathtt{W}$ is an ambient name, a group name or a capability type. For instance, the ambient *pub* of group g_{pub} can cross no ambients, can send processes only to ambients of group g_{jrn} and internally communicates pairs consisting each of an ambient name and a group name.

The ambient typing rule therefore informally becomes:

$$\frac{\Sigma \vdash \alpha : \mathsf{amb} \qquad \Sigma \vdash \gamma : \mathsf{group} \qquad \Sigma \vdash P : \mathtt{V}}{\Sigma \vdash \alpha{:}\gamma\,\mathtt{V}[P] : \mathtt{V}'} \quad (\textsc{Amb})$$

where V' is any well-formed process type. As usual, the ambient's external process type V', which may be any type, is not to be confused with its inner type V. The global environment Σ is used only for associating types to variables: they can range over different ambients and processes, so global assumptions on their types are unavoidable.

The whole computation starts with a request being sent by an institution (trusted or untrusted) to the publisher for subscribing to a journal JRN_k:

$$\begin{aligned} \text{REQ}_i^k &= \mathsf{to}\ pub{:}g_{pub}\ \mathsf{with}\ (\langle\!\langle \varnothing, \varnothing \rangle\!\rangle, \mathsf{com}(\mathsf{amb}, \mathsf{group}))\,.\,\langle jrn_k, g_{tdl_i}\rangle \\ \text{or}\ \ \text{RQ}_i^k &= \mathsf{to}\ pub{:}g_{pub}\ \mathsf{with}\ (\langle\!\langle \varnothing, \varnothing \rangle\!\rangle, \mathsf{com}(\mathsf{amb}, \mathsf{group}))\,.\,\langle jrn_k, g_{dl_i}\rangle \end{aligned}$$

The request is a simple mobile process that, by performing a to action, moves from the institution to the publisher, where it communicates the journal's name and the institution's 'signature', i.e., the group name of its download ambients. We remark that, although this example employs dyadic communication, the formal definition of the calculus (which will be presented in the next section), for the sake of simplicity, allows only monadic communication.

The construct $\mathsf{to}\,m{:}\gamma\,\mathsf{with}\,\mathsf{V}\,.\,P$ (introduced in a basic form in the calculus $\mathbf{M}^3$) denotes a process that sends its continuation P to a sibling ambient named m of group γ, where it will behave in conformity with the declared type V. Static type checking ensures that the type of P actually equals its declared type V. Here, the request process REQ_i^k announces that, once reached the publisher, it will communicate a pair consisting of an ambient name and a group name, without performing any mobility action.

At runtime, a to action can fire only if the target ambient accepts the incoming process through the consumption of a suitable co-move of the form $\mathsf{co}\ \gamma\ \mathsf{with}\ \mathsf{U}$, where γ is the group of the expected incoming process and U a mobility type. The firing of the to-co transition is subject to the dynamic check that the communication type of the incoming process complies with the one of the entered ambient and that its mobility type is bounded by U (this notion is rendered by a suitable notion of subtyping). Static type checking ensures that U is in turn compatible with the mobility type of the entered ambient. On the other hand, dynamic checking of type compatibility between the incoming process and the entered ambient is necessary since, in the absence of a global type environment for ambient or group names, the type properties of the entered ambient can be known only at runtime. Note that in this way it is possible to allow processes of different groups to enter a same ambient with different mobility rights. Checking the mobility type of a process going from one ambient to another is motivated by the fact that this may be extremely dangerous for the receiving ambient, since the process could take complete control of it.

In our example the control is performed by the manager, which is a process running within the publisher and dedicated to handling incoming requests:

$$\begin{aligned} \text{MGR} = \ & !\,\mathsf{co}\ g_{inst}\ \mathsf{with}\ \langle\!\langle \varnothing, \varnothing \rangle\!\rangle\,.\,(x{:}\mathsf{amb}, y{:}\mathsf{group})\,. \\ & \quad \mathsf{down}\ x{:}g_{jrn}\ \mathsf{with}\ (\langle\!\langle \varnothing, \{y\} \rangle\!\rangle, \mathsf{shh})\,.\,!\mathsf{co}\ y\ \mathsf{with}\ \langle\!\langle \varnothing, \{y\} \rangle\!\rangle \\ \mid\ & !\,\mathsf{co}\ g_{ins}\ \mathsf{with}\ \langle\!\langle \varnothing, \varnothing \rangle\!\rangle\,.\,(x{:}\mathsf{amb}, y{:}\mathsf{group})\,. \\ & \quad \mathsf{down}\ x{:}g_{jrn}\ \mathsf{with}\ (\langle\!\langle \varnothing, \varnothing \rangle\!\rangle, \mathsf{shh})\,.\,!\mathsf{co}\ y\ \mathsf{with}\ \langle\!\langle \varnothing, \varnothing \rangle\!\rangle \end{aligned}$$

The first action performed by the manager is the consumption of the co-move $\mathsf{co}\ g_{inst}\ \mathsf{with}\ \langle\!\langle\varnothing,\varnothing\rangle\!\rangle$, which authorizes a process to enter the publisher only if it cannot move the publisher nor send processes from it. As previously described, subscription requests satisfy such constraint.

The co-move is consumed in the action; the standard replication operator '!' is therefore used to provide authorization for an unlimited number of accesses. After accepting the request, the manager receives from it the journal's name and the institution's signature, goes down into the journal and finally leaves there a co-move that will grant the institution's reading processes the right to access the journal. Depending on whether the request comes from a known trusted institution or from an unknown one, the co-move deposited in the journal will be slightly different: in the second case it grants more restricted rights, which will compel processes from unknown institutions to adopt a more controlled behaviour.

The down primitive is a process mobility primitive analogous to the to: it sends the continuation process down from its ambient into a child ambient. The set of process mobility primitives is completed by the up action, which sends the continuation process to the parent ambient. For the sake of simplicity, there is only one kind of co-move which synchronizes with any of the three kinds of process movements (to, up and down).

The type expression in the co construct may contain a variable bound by an input abstraction. For example, in the case of a request from a fully trusted institution the co-move deposited by the manager in the journal contains the variable y: the journal will thus accept any reading agent coming from a download ambient of group y and going back as a continuation process to a y-ambient (i.e., to an ambient of the same group y). The actual value of y is provided by the input operation performed by the manager prior to dropping the co-move at the journal.

Journals are simple ambients that communicate their contents any number of times:

$$\begin{aligned}\mathrm{JRN}_k = jrn_k{:}g_{jrn}(\langle\!\langle\varnothing,\{g_{tdl_1},\ldots,g_{tdl_m}\}\rangle\!\rangle,\mathsf{com}(\mathsf{paper}))[\\ !\,\langle paper\rangle \mid\ !\,\mathsf{co}\ g_{pub}\ \mathsf{with}\ \langle\!\langle\varnothing,\{g_{tdl_1},\ldots,g_{tdl_m}\}\rangle\!\rangle]\end{aligned}$$

The co-move initially present in the journal enables it to receive from the publisher (through the manager) the co-move that in turn will authorize the journal to accept reader agents from members of an institution, if this has subscribed the journal.

As we have seen, the latter 'dynamic' co-move, in case of a trusted institution, will authorize the reader agent to go back to its originating y-ambient; the former 'static' co-move, to be able to accept the dynamic one, must therefore authorize the movement to any possible value of y, i.e., it must explicitly mention the signatures (i.e., the groups of download ambients) of all the trusted institutions.

Finally, a download process performing an access to a subscribed journal is a mobile ambient going out from its institution into the publisher, where it sends a reading agent to the journal; the agent, represented by a process, reads a paper within the journal and then goes back to the download mobile ambient, which in turn goes back to the institution. In the case of fully trusted institutions such behaviour can be implemented exactly as described. In particular, since the above list of actions has to be performed strictly in the given order, the ambient's main internal component will be a process consisting of a sequence of prefixes, in parallel with the co-move needed to eventually take in again the reading process:

$$
\begin{aligned}
\mathrm{TDL}_i^j = (\nu tdl_i^j)\ tdl_i^j{:}\, g_{tdl_i}(\langle\!\langle \{g_{pub}, g_{inst}\}, \{g_{jrn}\}\rangle\!\rangle, \mathsf{com}(\mathsf{paper}))[&\\
\mathsf{out}\ inst_i{:}\, g_{inst} \,.\, \mathsf{in}\ pub{:}\, g_{pub} \,.&\\
\mathsf{to}\ jrn_k{:}g_{jrn}\ \mathsf{with}\ (\langle\!\langle \varnothing, \{g_{tdl_i}\}\rangle\!\rangle, \mathsf{com}(\mathsf{paper})).&\\
(x{:}\, \mathsf{paper}).&\\
\mathsf{to}\ tdl_i^j{:}g_{tdl_i}\ \mathsf{with}\ (\langle\!\langle \{g_{pub}, g_{inst}\}, \varnothing\rangle\!\rangle, \mathsf{shh})\,.&\\
\mathsf{out}\ pub{:}\, g_{pub} \,.\, \mathsf{in}\ inst_i{:}\, g_{inst} \,.\, !\,(\langle x\rangle \mid \mathsf{co}\ g_{inst}\ \mathsf{with}\ \langle\!\langle \varnothing, \{g_{inst}\}\rangle\!\rangle) \mid&\\
!\mathsf{co}\ g_{jrn}\ \mathsf{with}\ \langle\!\langle \{g_{pub}, g_{inst}\}, \varnothing\rangle\!\rangle]&
\end{aligned}
$$

The main process starts by driving the ambient out of the institution and into the publisher, by means of the usual out and in actions of ambient calculi; the only difference w.r.t. the standard primitives is that in our calculus an action takes as arguments not only an ambient name, but also a group name.

Once the download ambient is in the publisher, the main thread asks to be allowed to jump into the journal, by promising that it will not move the journal itself, will have a continuation coming back to an ambient of group g_{tdl_i}, and will perform an input/output of messages of type paper.

The journal accepts the reader process because it contains the appropriate co-move; then, the process reads the paper in the journal through an input operation and goes back to the download ambient, with the accompanying declaration that it will continue there by driving the ambient across publisher and institution boundaries; which in fact is what it performs in the second-last line of the definition, before finally making available the paper to other processes in the institution.

For the agent to be allowed to return from the journal after reading the paper, the download ambient must contain a suitable co-move, which is shown in the last line of the definition. Its with-component $\langle\!\langle \{g_{pub}, g_{inst}\}, \varnothing\rangle\!\rangle$ exactly matches the one of the returning process.

It is important to observe that an ambient's mobility type $\langle\!\langle \mathscr{C}, \mathscr{E}\rangle\!\rangle$ is an upper bound of the mobilities of the parallel processes it contains and will contain, in the sense that the set $\mathscr{C}$ collects all the group names that are arguments of in or out actions or are found in the $\mathscr{C}$ components of co-actions, while the set $\mathscr{E}$ analogously collects the arguments of to, up or down actions and the members of the $\mathscr{E}$ components of co-actions. The inclusion of the with components of the co-moves guarantees that all the possible future contents of the ambient are taken into account and that therefore the ambient's inner type will not be changed by reduction. This is formally expressed by the theorems relative to soundness, proved in Section 4.

As we have seen, the ambient's inner communication type must be kept at runtime in the ambient construct, to allow the check of incoming processes; on the contrary, the ambient's mobility type is not strictly needed at runtime, since its $\mathscr{C}$ and $\mathscr{E}$ are supersets of the respective components of the co-actions, which are the ones that perform the checks w.r.t. mobility. Nevertheless, we have chosen to also include the mobility type in the ambient syntax, and thus to explicitly attach to each ambient its complete inner type $(\langle\!\langle \mathscr{C}, \mathscr{E} \rangle\!\rangle, \mathtt{T})$, in order to make the calculus more perspicuous and to facilitate the expression of the soundness property.

For example, in the TDL_i^j component, the ambient named tdl_i^j (and of group g_{tdl_i}) is labelled with the type $(\langle\!\langle \{g_{pub}, g_{inst}\}, \{g_{jrn}\} \rangle\!\rangle, \mathsf{com}(\mathsf{paper}))$, corresponding to the fact that the ambient goes across the boundaries of publisher and institutions, and sends a process to a journal.

The download originating from an unknown institution differs from the previous case because the reading process, once read the paper, is not allowed to directly go back from the journal to the download mobile ambient, whose group is not known to the publisher and therefore not mentioned in the authorization to exit the journal. To be able to return to the download ambient, the reader must turn itself into a mobile ambient – in the example, we have named such an ambient with the secret name *box* – which is then able to leave the journal without needing a permission. The ambient *box* acts as a sort of sandbox wherein the otherwise potentially harmful process may run safely, because ambients, differently from processes, in the absence of the open primitive cannot directly act on their surrounding ambients, nor perform input/output operations; they can therefore move around without need of any co-move authorization and runtime check.

Once out of the journal, the reader agent may then go back to the download ambient, which of course knows the family name (i.e., the group) of the sandbox ambient and can therefore authorize the move. The rest of the process is like in the case of known institutions.

$$
\begin{aligned}
\text{DL}_i^j = (\nu\, dl_i^j)\ & dl_i^j\colon g_{dl_i}(\langle\!\langle \{g_{pub}, g_{ins}\}, \{g_{jrn}\} \rangle\!\rangle, \mathsf{com}(\mathsf{paper}))[\\
& \mathsf{out}\ ins_i\colon g_{ins}\,.\ \mathsf{in}\ pub\colon g_{pub}\,.\\
& \mathsf{to}\ jrn_k\colon g_{jrn}\ \mathsf{with}\ (\langle\!\langle \varnothing, \varnothing \rangle\!\rangle, \mathsf{com}(\mathsf{paper})).\\
& (x\colon \mathsf{paper})\\
& (\nu\, box)\ box\colon g_{box}(\langle\!\langle \{g_{jrn}\}\{g_{dl_i}\} \rangle\!\rangle, \mathsf{shh})\\
& \quad [\mathsf{out}\ jrn_k\colon g_{jrn}\,.\ \mathsf{to}\ dl_i^j\colon g_{dl_i}\ \mathsf{with}\ (\langle\!\langle \{g_{pub}, g_{ins}\}, \varnothing \rangle\!\rangle, \mathsf{com}(\mathsf{paper}))\,.\\
& \quad\ \mathsf{out}\ pub\colon g_{pub}\,.\ \mathsf{in}\ ins_i\colon g_{inst}\,.\ !\,(\langle x \rangle \mid \mathsf{co}\ g_{ins}\ \mathsf{with}\ \langle\!\langle \varnothing, \{g_{ins}\} \rangle\!\rangle)] \mid\\
& !\mathsf{co}\ g_{box}\ \mathsf{with}\ \langle\!\langle \{g_{pub}, g_{ins}\}, \varnothing \rangle\!\rangle]
\end{aligned}
$$

A side effect of the above implementation is that at each download access a new *box* ambient is created and then left empty within the publisher. However, the name restriction immediately transforms such ambients into garbage, which can be collected by means of the equivalence $(\nu\ box)\ box\colon g_{box}(\langle\!\langle \{g_{jrn}\}\{g_{dl_i}\} \rangle\!\rangle, \mathsf{shh})$ $[\mathsf{0}] \equiv \mathsf{0}$.

3 The Calculus

This section formally introduces the calculus. The syntax of the pre-terms of the language (where type constraints are ignored) is given in Fig. 1. Processes are

$\mathscr{A}$ denotes the set of *ambient names* and $\mathscr{G}$ denotes the set of *group names*.

α ::= **ambients**
- $m, n, \ldots$ ambient names
- $x, y, \ldots$ ambient variables

γ ::= **groups**
- $g, h, \ldots$ group names
- $x, y, \ldots$ group variables

χ ::= **capabilities**

in $\alpha{:}\gamma$	moves the containing ambient into ambient α of group γ
out $\alpha{:}\gamma$	moves the containing ambient out of ambient α of group γ
co γ with U	lets into the containing ambient a process with mobility rights U coming from an ambient of group γ
$\chi{\cdot}\chi'$	path
$x, y, \ldots$	capability variables

$M,\ N,\ L$ ::= **messages**

α	ambients
γ	groups
χ	capabilities

$P,\ Q$ ::= **processes**

0	null
$\chi\,.\,P$	capability prefix
$\langle M \rangle\,.\,P$	synchronous output
$(x{:}\,\mathtt{W})\,.\,P$	typed input
down $\alpha{:}\gamma$ with V . P	from an ambient, sends the process P, requiring rights V, down to an enclosed ambient α of group γ
up $\alpha{:}\gamma$ with V . P	from an ambient, sends the process P, requiring rights V, up to the enclosing ambient α of group γ
to $\alpha{:}\gamma$ with V . P	from an ambient, sends the process P, requiring rights V, to a sibling ambient α of group γ
$P \mid Q$	parallel composition
$\alpha{:}\gamma\,\mathtt{V}[P]$	ambient
$!\,P$	replication
$(\nu n)P$	name restriction
$(\nu g)P$	group restriction

where U, V and W are defined in Figure 2.

Fig. 1. Syntax

built from the 0 process through the standard constructs of sequential prefixing, parallel composition, ambient formation, replication, name restriction and group restriction. Admissible syntactic prefixes are the capabilities in and out, the co-move co, the input/output actions and the process movement constructs down, up and to. Messages can be ambient names, group names or (sequences of) capabilities; note that process movement actions (to, down, up) are not considered capabilities and cannot be communicated.

In the input construct $(x{:}\,\mathtt{W})\,.\,P$ the prefix $(x{:}\,\mathtt{W})$ binds the variable x in P, while in the name restriction $(\nu n)P$ the binder (νn) binds the name n in P; group re-

	$\mathscr{C}, \mathscr{E}, \ldots$	sets of group names and variables;
$\mathtt{V}$ ::=	$(\mathtt{U}, \mathtt{T})$	process type or *mobcom* type: mobility type and communication type
$\mathtt{U}$ ::=	$\langle\!\langle \mathscr{C}, \mathscr{E} \rangle\!\rangle$	*mobility* type: mobility rights $\mathscr{C}, \mathscr{E}$
$\mathtt{T}$ ::=		*communication* type
	shh	no communication
	com(W)	communication of messages of type W
$\mathtt{W}$ ::=		*message* type
	amb	ambient
	group	group
	cap(U)	capability type
Σ ::=		environment
	$\varnothing$	empty environment
	$\Sigma, x{:}\,\mathtt{W}$	environment containing the typing assumption $x{:}\,\mathtt{W}$

Fig. 2. Types

striction (νg) works analogously. A name or a variable that is not bound is called *free*. The sets of free and bound names/variables of a term are respectively defined in accordance with that. We identify processes modulo renaming of bound names and variables. We will use the notation $\eta \in (\notin) \Upsilon$, where $\eta \in \{\alpha, \gamma\}$ and $\Upsilon \in \{P, \mathscr{C}, \mathtt{V}, \Sigma, \ldots\}$, as short for '$\eta$ does (does not) occur free in Υ'.

The syntax of types is given in Fig. 2. Since types in our system describe communication and mobility properties, the type system is based on the notion of a *mobcom* type (or process type) V which packs the mobility type and the communication type of a process. A *mobility* type U consists of two sets $\mathscr{C}$ and $\mathscr{E}$ whose elements may be both group names and group variables; the intuitive meaning is that $\mathscr{C}$ is the set of the ambient groups which the enclosing ambient is allowed to *cross* when the process executes an in or out action, and $\mathscr{E}$ is the set of the ambient groups which the process may *enter* by means of a down, up or to action. As a matter of notation, if $\mathtt{V} = (\langle\!\langle \mathscr{C}, \mathscr{E} \rangle\!\rangle, \mathtt{T})$, we will write $\mathscr{C}(\mathtt{V})$ and $\mathscr{E}(\mathtt{V})$ to denote $\mathscr{C}$ and $\mathscr{E}$, respectively; an analogous notation will be used for U.

Observe that types may *depend* on group variables, i.e. on parameters of input prefixes; they are therefore affected by reduction, when an input action substitutes a variable with a group name. Types, in turn, may occur in terms: mobcom types appear in the with components of the process mobility primitives (where they characterize process continuations) and in the ambient labels (where they represent the types of the inner processes); mobility types appear in the co-moves, where they represent the rights of the processes which are allowed to enter; message types appear in the input construct. The calculus is therefore a truly typed one, where types play an essential role both in the language definition and in the operational semantics.

The *communication* type T of a process indicates whether the process is silent (type shh) or can engage in the communication of messages of type W. A message type can be the atomic type group of group names, the atomic type amb of ambient names, or a capability type $\mathsf{cap}(\mathtt{U})$ representing the mobility rights required by capabilities.

A type environment Σ is a finite set of pairs $x{:}\,\mathtt{W}$, where x is a variable and $\mathtt{W}$ is its assumed message type. The domain $Dom(\Sigma)$ of the environment Σ is defined as usual:

$$Dom(\varnothing) = \varnothing \qquad Dom(\Sigma, x{:}\,\mathtt{W}) = Dom(\Sigma) \cup \{x\}.$$

A subtyping relation $\leq$ is naturally defined on mobility types by componentwise set inclusion and then trivially extended to capability types, while the other two message types amb and group are not comparable with them. Subtyping on message types does not extend in the same way to communication types, since we do not allow subtyping polymorphism in input/output, except for the particular case of silent processes. Communication types are therefore not comparable with each other, with the exception of the type shh, which is the least element. We keep the same notation ($\leq$) for both relations since there is no risk of confusion. Finally, subtyping on process (mobcom) types is the one trivially induced by subtyping on mobility and communication types.

Definition 1 (Subtyping).

1. $\langle\!\langle \mathscr{C}, \mathscr{E} \rangle\!\rangle \leq \langle\!\langle \mathscr{C}', \mathscr{E}' \rangle\!\rangle$ *if* $\mathscr{C} \subseteq \mathscr{C}'$ *and* $\mathscr{E} \subseteq \mathscr{E}'$*;*
2. $\mathsf{amb} \leq \mathsf{amb}$*;* $\mathsf{group} \leq \mathsf{group}$*;* $\mathsf{cap}(\mathtt{U}) \leq \mathsf{cap}(\mathtt{U}')$ *if* $\mathtt{U} \leq \mathtt{U}'$*;*
3. $\mathtt{T} \leq \mathtt{T}$*;* $\mathsf{shh} \leq \mathtt{T}$*;*
4. $(\mathtt{U}, \mathtt{T}) \leq (\mathtt{U}', \mathtt{T}')$ *if* $\mathtt{U} \leq \mathtt{U}'$ *and* $\mathtt{T} \leq \mathtt{T}'$.

Whenever we write $\Sigma \subseteq \Sigma'$, with Σ and Σ' type environments, we mean the standard set-theoretic inclusion.

We now turn to the typing rules. There are seven kinds of typing judgments:

$\vdash \Sigma$	good environment Σ
$\Sigma \vdash \mathtt{U}$	good mobility type $\mathtt{U}$
$\Sigma \vdash \mathtt{W}$	good message type $\mathtt{W}$
$\Sigma \vdash \mathtt{T}$	good communication type $\mathtt{T}$
$\Sigma \vdash \mathtt{V}$	good mobcom type $\mathtt{V}$
$\Sigma \vdash M : \mathtt{W}$	good message M of type $\mathtt{W}$
$\Sigma \vdash P : \mathtt{V}$	good process P of type $\mathtt{V}$

In the sequel, the generic form $\Sigma \vdash \Gamma$ is intended to range over the last six kinds of judgments.

Figure 3 contains the rules for the well-formedness of environments and types, necessary because the calculus allows variables to occur within mobility types as long as they stand for group names. To simplify the presentation of the rules we rely on the established notational conventions to distinguish between the syntactic categories of objects. So, for instance, $\mathtt{W}$ denotes a message type, $\mathtt{U}$ a mobility type and so on. The rules are all standard; in particular, rule (Mob Type) checks that all the variables contained in a mobility type have been assumed to be of type group.

$$\frac{\forall\gamma \in \mathscr{C} \cup \mathscr{E} \quad \Sigma \vdash \gamma : \mathsf{group}}{\Sigma \vdash \langle\!\langle \mathscr{C}, \mathscr{E} \rangle\!\rangle} \ (\text{Mob Type})$$

$$\frac{\vdash \Sigma}{\Sigma \vdash \mathsf{group}} \ (\text{Group Type}) \qquad \frac{\vdash \Sigma}{\Sigma \vdash \mathsf{amb}} \ (\text{Amb Type}) \qquad \frac{\Sigma \vdash \mathtt{U}}{\Sigma \vdash \mathsf{cap}(\mathtt{U})} \ (\text{Cap Type})$$

$$\frac{\vdash \Sigma}{\Sigma \vdash \mathsf{shh}} \ (\text{Shh Type}) \qquad \frac{\Sigma \vdash \mathtt{W}}{\Sigma \vdash \mathsf{com}(\mathtt{W})} \ (\text{MsgCom Type}) \qquad \frac{\Sigma \vdash \mathtt{U} \quad \Sigma \vdash \mathtt{T}}{\Sigma \vdash (\mathtt{U}, \mathtt{T})} \ (\text{MobCom Type})$$

$$\frac{}{\vdash \varnothing} \ (\text{Empty Env}) \qquad \frac{\Sigma \vdash \mathtt{W} \quad x \notin Dom(\Sigma)}{\vdash \Sigma, x{:}\, \mathtt{W}} \ (\text{Env Constr})$$

Fig. 3. Good Types and Environments

$$\frac{\vdash \Sigma \quad n \in \mathscr{A}}{\Sigma \vdash n : \mathsf{amb}} \ (\text{Amb Const}) \qquad \frac{\vdash \Sigma \quad g \in \mathscr{G}}{\Sigma \vdash g : \mathsf{group}} \ (\text{Grp Const})$$

$$\frac{\Sigma \vdash \mathtt{W}' \quad x{:}\mathtt{W} \in \Sigma \quad \mathtt{W} \leq \mathtt{W}'}{\Sigma \vdash x : \mathtt{W}'} \ (\text{Env})$$

$$\frac{\Sigma \vdash \gamma : \mathsf{group} \quad \Sigma \vdash \mathtt{U}' \quad \mathtt{U} \leq \mathtt{U}'}{\Sigma \vdash \mathsf{co}\,\gamma\,\mathsf{with}\ \mathtt{U} : \mathsf{cap}(\mathtt{U}')} \ (\text{Co}) \qquad \frac{\Sigma \vdash \chi : \mathsf{cap}(\mathtt{U}) \quad \Sigma \vdash \chi' : \mathsf{cap}(\mathtt{U})}{\Sigma \vdash \chi.\chi' : \mathsf{cap}(\mathtt{U})} \ (\text{Path})$$

$$\frac{\Sigma \vdash \alpha : \mathsf{amb} \quad \Sigma \vdash \mathtt{U} \quad \gamma \in \mathscr{C}(\mathtt{U})}{\Sigma \vdash \mathsf{in}\ \alpha{:}\gamma : \mathsf{cap}(\mathtt{U})} \ (\text{In}) \qquad \frac{\Sigma \vdash \alpha : \mathsf{amb} \quad \Sigma \vdash \mathtt{U} \quad \gamma \in \mathscr{C}(\mathtt{U})}{\Sigma \vdash \mathsf{out}\ \alpha{:}\gamma : \mathsf{cap}(\mathtt{U})} \ (\text{Out})$$

Fig. 4. Good Messages

The typing rules for messages and processes are given in Fig. 4 and Fig. 5, respectively. Rules (Amb Const) and (Grp Const) are straightforward.
In the rule (Env), if a variable is assigned a message type $\mathtt{W}$ equal to the atomic type amb or group, its deduced type $\mathtt{W}'$ can only be the same as $\mathtt{W}$. If, on the other hand, a variable is assumed to be of a capability type $\mathtt{W} \equiv \mathsf{cap}(\mathtt{U})$, then subsumption applies, and the deduced type $\mathtt{W}'$ can be any supertype of $\mathtt{W}$.

As usual, subtyping enhances typability by allowing a subterm to have a type that is a subtype of the one required by the construction of the term. For example, the simple process $(x{:}\, \mathsf{cap}(\langle\!\langle \{g\}, \varnothing \rangle\!\rangle)) \,.\, x \,.\, \mathsf{in}\ m{:}g_m \,.\, \mathsf{0}$ is well typed (in the empty environment), with typing:

$$\vdash (x{:}\, \mathsf{cap}(\langle\!\langle \{g\}, \varnothing \rangle\!\rangle)) \,.\, x \,.\, \mathsf{in}\ m{:}g_m \,.\, \mathsf{0} : (\langle\!\langle \{g, g_m\}, \varnothing \rangle\!\rangle, \mathsf{com}(\mathsf{cap}(\langle\!\langle \{g\}, \varnothing \rangle\!\rangle)))$$

$$\frac{\Sigma \vdash \mathtt{V}}{\Sigma \vdash \mathtt{0} : \mathtt{V}} \ (\text{Null}) \qquad \frac{\Sigma \vdash \chi : \mathsf{cap}(\mathtt{U}) \quad \Sigma \vdash P : (\mathtt{U}, \mathtt{T})}{\Sigma \vdash \chi.P : (\mathtt{U}, \mathtt{T})} \ (\text{Cap Prefix})$$

$$\frac{\Sigma \vdash \alpha : \mathsf{amb} \quad \Sigma \vdash P : \mathtt{V} \quad \Sigma \vdash \mathtt{V}' \quad \gamma \in \mathscr{E}(\mathtt{V}')}{\Sigma \vdash \mathsf{down}\ \alpha{:}\gamma\ \mathsf{with}\ \mathtt{V}\,.\,P : \mathtt{V}'} \ (\text{Down})$$

$$\frac{\Sigma \vdash \alpha : \mathsf{amb} \quad \Sigma \vdash P : \mathtt{V} \quad \Sigma \vdash \mathtt{V}' \quad \gamma \in \mathscr{E}(\mathtt{V}')}{\Sigma \vdash \mathsf{up}\ \alpha{:}\gamma\ \mathsf{with}\ \mathtt{V}\,.\,P : \mathtt{V}'} \ (\text{Up})$$

$$\frac{\Sigma \vdash \alpha : \mathsf{amb} \quad \Sigma \vdash P : \mathtt{V} \quad \Sigma \vdash \mathtt{V}' \quad \gamma \in \mathscr{E}(\mathtt{V}')}{\Sigma \vdash \mathsf{to}\ \alpha{:}\gamma\ \mathsf{with}\ \mathtt{V}\,.\,P : \mathtt{V}'} \ (\text{To})$$

$$\frac{\Sigma, x{:}\,\mathtt{W} \vdash P : (\mathtt{U}, \mathsf{com}(\mathtt{W})) \quad x \notin \Sigma \quad x \notin (\mathtt{U}, \mathsf{com}(\mathtt{W}))}{\Sigma \vdash (x{:}\,\mathtt{W})\,.\,P : (\mathtt{U}, \mathsf{com}(\mathtt{W}))} \ (\text{Input})$$

$$\frac{\Sigma \vdash P : (\mathtt{U}, \mathsf{com}(\mathtt{W})) \quad \Sigma \vdash M : \mathtt{W}}{\Sigma \vdash \langle M \rangle\,.\,P : (\mathtt{U}, \mathsf{com}(\mathtt{W}))} \ (\text{Output})$$

$$\frac{\Sigma \vdash \alpha : \mathsf{amb} \quad \Sigma \vdash \gamma : \mathsf{group} \quad \Sigma \vdash P : \mathtt{V} \quad \Sigma \vdash \mathtt{V}'}{\Sigma \vdash \alpha{:}\gamma\ \mathtt{V}[P] : \mathtt{V}'} \ (\text{Amb})$$

$$\frac{\Sigma \vdash P : \mathtt{V} \quad \Sigma \vdash Q : \mathtt{V}}{\Sigma \vdash P \,|\, Q : \mathtt{V}} \ (\text{Par}) \qquad \frac{\Sigma \vdash P : \mathtt{V}}{\Sigma \vdash !P : \mathtt{V}} \ (\text{Repl})$$

$$\frac{\Sigma \vdash P : \mathtt{V}}{\Sigma \vdash (\nu n)P : \mathtt{V}} \ (\text{Amb Res}) \qquad \frac{\Sigma \vdash P : \mathtt{V} \quad g \notin \Sigma \quad g \notin \mathtt{V}}{\Sigma \vdash (\nu g)P : \mathtt{V}} \ (\text{Grp Res})$$

Fig. 5. Good Processes

where the capability that is going to be received in input carries a more restricted mobility than the resulting mobility of the whole process. Though quite natural, that would not be possible without subtyping.

Rule (Co) checks that the mobility type of the co-move is a supertype of the mobility type of the processes the co-move lets in. Rule (Path) checks that the mobility types of the two (sequences of) capabilities are equal.

Rules (In) and (Out) check that the $\mathscr{C}$ component of the mobility type contains the group γ of the ambient across whose border the capability drives its ambient; a similar check is performed in rules (Down), (Up) and (To) for the $\mathscr{E}$ component. Rule (Amb) ensures that the type of the inner process is recorded in the ambient header. The same rule allows an ambient to have an arbitrary type, as does rule (Null) for the 0 process.

Structural Congruence: $(\,|\,, 0)$ is a commutative monoid.

$$(\nu n)(P \,|\, Q) \equiv (\nu n)P \,|\, Q \quad (n \notin Q) \qquad (\nu n)(\nu m)P \equiv (\nu m)(\nu n)P$$
$$n{:}g\,\mathtt{V}[(\nu m)P] \equiv (\nu m)n{:}g\,\mathtt{V}[P] \quad (n \neq m) \qquad (\nu n)(\nu g)P \equiv (\nu g)(\nu n)P$$
$$(\nu g)(P \,|\, Q) \equiv (\nu g)P \,|\, Q \quad (g \notin Q) \qquad (\nu g)(\nu g')P \equiv (\nu g)(\nu g')P$$
$$n{:}g\,\mathtt{V}[(\nu g')P] \equiv (\nu g')n{:}g\,\mathtt{V}[P] \quad (g \neq g' \ \&\ g' \notin \mathtt{V}) \qquad !P \equiv P \,|\, !P$$
$$(\nu n)\mathsf{0} \equiv \mathsf{0} \qquad (\nu g)\mathsf{0} \equiv \mathsf{0} \qquad (\nu n)n{:}g\,\mathtt{V}[\mathsf{0}] \equiv \mathsf{0}$$

Basic reduction rules:

(R-in) $n{:}g_n\,\mathtt{V}_n[\,\mathsf{in}\ m{:}g_m\,.\,P_1 \,|\, P_2\,] \,|\, m{:}g_m\,\mathtt{V}_m[Q]$
$\rightarrow m{:}g_m\,\mathtt{V}_m[\,n{:}g_n\,\mathtt{V}_n[P_1 \,|\, P_2\,] \,|\, Q\,]$

(R-out) $m{:}g_m\,\mathtt{V}_m[\,n{:}g_n\,\mathtt{V}_n[\,\mathsf{out}\ m{:}g_m\,.\,P_1 \,|\, P_2\,] \,|\, Q\,]$
$\rightarrow n{:}g_n\,\mathtt{V}_n[\,P_1 \,|\, P_2\,] \,|\, m{:}g_m\,\mathtt{V}_m[Q]$

(R-down) $n{:}g_n\,\mathtt{V}_n[\mathsf{down}\ m{:}g_m\ \mathsf{with}\ (\mathtt{U},\mathtt{T})\,.\,P_1 \,|\, P_2 \,|\, m{:}g_m\,(\mathtt{U}_m,\mathtt{T}_m)[\mathsf{co}\ g_n\ \mathsf{with}\ \mathtt{U}'.Q_1 \,|\, Q_2]]$
$\rightarrow n{:}g_n\,\mathtt{V}_n[m{:}g_m\,(\mathtt{U}_m,\mathtt{T}_m)[P_1 \,|\, Q_1 \,|\, Q_2] \,|\, P_2]$
if $\mathtt{U} \leq \mathtt{U}'$ and $\mathtt{T} \leq \mathtt{T}_m$

(R-up) $m{:}g_m\,(\mathtt{U}_m,\mathtt{T}_m)[n{:}g_n\,\mathtt{V}_n[\mathsf{up}\ m{:}g_m\ \mathsf{with}\ (\mathtt{U},\mathtt{T})\,.\,P_1 \,|\, P_2] \,|\, \mathsf{co}\ g_n\ \mathsf{with}\ \mathtt{U}'.Q_1 \,|\, Q_2]$
$\rightarrow m{:}g_m\,(\mathtt{U}_m,\mathtt{T}_m)[n{:}g_n\,\mathtt{V}_n[P_2] \,|\, P_1 \,|\, Q_1 \,|\, Q_2]$
if $\mathtt{U} \leq \mathtt{U}'$ and $\mathtt{T} \leq \mathtt{T}_m$

(R-to) $n{:}g_n\,\mathtt{V}_n[\mathsf{to}\ m{:}g_m\ \mathsf{with}\ (\mathtt{U},\mathtt{T})\,.\,P_1 \,|\, P_2] \,|\, m{:}g_m\,(\mathtt{U}_m,\mathtt{T}_m)[\mathsf{co}\ g_n\ \mathsf{with}\ \mathtt{U}'.Q_1 \,|\, Q_2]$
$\rightarrow n{:}g_n\,\mathtt{V}_n[P_2] \,|\, m{:}g_m\,(\mathtt{U}_m,\mathtt{T}_m)[P_1 \,|\, Q_1 \,|\, Q_2]$
if $\mathtt{U} \leq \mathtt{U}'$ and $\mathtt{T} \leq \mathtt{T}_m$

(R-comm) $(x{:}\,\mathtt{W})\,.\,P \,|\, \langle M \rangle\,.\,Q \rightarrow P\{x := M\} \,|\, Q$

Structural reduction rules:

(R-par)	$P \rightarrow Q$	$\Rightarrow$	$P \,|\, R \rightarrow Q \,|\, R$
(R-amb)	$P \rightarrow Q$	$\Rightarrow$	$n{:}g\,\mathtt{V}[P] \rightarrow n{:}g\,\mathtt{V}[Q]$
(R-ν-amb)	$P \rightarrow Q$	$\Rightarrow$	$(\nu n)P \rightarrow (\nu n)Q$
(R-ν-group)	$P \rightarrow Q$	$\Rightarrow$	$(\nu g)P \rightarrow (\nu g)Q$
(R-$\equiv$)	$P' \equiv P',\ P \rightarrow Q,\ Q \equiv Q'$	$\Rightarrow$	$P' \rightarrow Q'$

Fig. 6. Operational Semantics

Rule (Cap Prefix) checks that the mobility type of the process matches the type of its prefix, in the same way as rules (Input) and (Output) check that the communication type of the process matches the type of the message. Rules (Par), (Repl), (Amb Res) and (Grp Res) are standard.

Finally, we present the operational semantics of the calculus. As is common in calculi for distributed computation, the operational semantics relies on a structural congruence and on a reduction relation. The structural congruence equates terms whose syntactical differences should be considered inessential; it is defined as the smallest congruence satisfying the laws in Fig. 6. The only non-standard law is $(\nu n)n{:}g\, \mathsf{V}[\mathsf{0}] \equiv \mathsf{0}$, which – in the absence of the open capability – is handy for disposing of 'garbage' inactive ambients.

The reduction relation is defined only for *closed processes*, i.e., processes with no occurrences of free variables; the rules are given in Fig. 6.

The first five rules deal with mobility and can be divided into two groups: those for ambient mobility and those for process mobility. The former are the standard rules of ambient calculi for the in and out capabilities, which can drive ambients into or out of other ambients. The latter are the ones for the down, up and to actions (already found in $\mathbf{M}^3$), which allow the continuation process to move to a parent, child or sibling ambient, respectively. As discussed earlier, an action of this kind must specify not only the name of the target ambient but also its group, because of the absence of a global mapping from ambients to groups.

In the case of process movements the reduction may take place only if the communication type of the continuation is compatible with the one of the target ambient, and in addition if the ambient contains a suitable co-move letting the process in; both checks make use of the subtyping relations introduced in Definition 1. Note that the co-move is consumed in the reaction and this, in turn, allows the co-move's continuation to synchronize on the entering of a process; moreover, the consumable nature of co-moves allows the implementation of fine-tuned admittance policies.

The last reduction rule is the standard one for communication, local to ambients as in the original ambient calculus.

To get the full reduction relation, the basic reduction rules are then closed under the application of all possible *reduction contexts*, which are those contexts that are built only by parallel composition, ambient formation, and group and name restriction. Note that no reduction rule can be applied within the continuation of an action prefix, in particular within the body of an input process (before the input itself is performed). Thus, if $C[-]$ is a reduction context and $C[P]$ is a good closed process, no free variable can occur in P. Finally, the last rule is standard and states that structurally congruent processes have the same reductions.

4 Soundness

The type system satisfies the usual property of subject reduction (Theorem 1), which here, as in all systems of behavioural types, is particularly meaningful. In fact, joined to type safety (Corollary 1), which states the exact correspondence between types and behavioural properties, subject reduction guarantees that any evolution of a system, i.e., any reduction sequence, satisfies the constraints expressed by types. This represents a form of *soundness*, in the sense that communication and movements of both ambients and processes actually obey the constraints that the type attached to each ambient is supposed to express.

The construction of the proof of the subject reduction theorem starts, as usual, with the statement of generation lemmata, which trivially hold because of the evident syntax-directed character of the typing rules. To simplify the statements we only consider for each syntactic category of terms the corresponding type pattern. It is easy to verify that, in all cases, no other type pattern would be possible.

Lemma 1 (Generation Lemma I).

1. *If* $\Sigma \vdash \Gamma$ *then* $\vdash \Sigma$.
2. *If* $\Sigma \vdash \langle\!\langle \mathscr{C}, \mathscr{E} \rangle\!\rangle$ *then* $\Sigma \vdash \gamma : \mathsf{group}$ *for all* $\gamma \in \mathscr{C} \cup \mathscr{E}$.
3. *If* $\Sigma \vdash \mathsf{cap}(\mathtt{U})$ *then* $\Sigma \vdash \mathtt{U}$.
4. *If* $\Sigma \vdash \mathsf{com}(\mathtt{W})$ *then* $\Sigma \vdash \mathtt{W}$.
5. *If* $\Sigma \vdash (\mathtt{U}, \mathtt{T})$ *then* $\Sigma \vdash \mathtt{U}$ *and* $\Sigma \vdash \mathtt{T}$.
6. *If* $\vdash \Sigma, x{:}\,\mathtt{W}$ *then* $\Sigma \vdash \mathtt{W}$ *and* $x \notin Dom(\Sigma)$.
7. *If* $\Sigma \vdash n : \mathsf{amb}$ *then* $n \in \mathscr{A}$.
8. *If* $\Sigma \vdash g : \mathsf{group}$ *then* $g \in \mathscr{G}$.
9. *If* $\Sigma \vdash x : \mathtt{W}$ *then* $x{:}\,\mathtt{W}' \in \Sigma$ *for some* $\mathtt{W}' \leq \mathtt{W}$.
10. *If* $\Sigma \vdash \mathsf{in}\ \alpha{:}\gamma : \mathsf{cap}(\mathtt{U})$ *then* $\Sigma \vdash \alpha : \mathsf{amb}$ *and* $\Sigma \vdash \mathsf{cap}(\mathtt{U})$ *and* $\gamma \in \mathscr{C}(\mathtt{U})$.
11. *If* $\Sigma \vdash \mathsf{out}\ \alpha{:}\gamma : \mathsf{cap}(\mathtt{U})$ *then* $\Sigma \vdash \alpha : \mathsf{amb}$ *and* $\Sigma \vdash \mathsf{cap}(\mathtt{U})$ *and* $\gamma \in \mathscr{C}(\mathtt{U})$.
12. *If* $\Sigma \vdash \mathsf{co}\ \gamma\ \mathsf{with}\ \mathtt{U} : \mathsf{cap}(\mathtt{U}')$ *then* $\Sigma \vdash \gamma : \mathsf{group}$ *and* $\Sigma \vdash \mathsf{cap}(\mathtt{U}')$ *and* $\mathtt{U} \leq \mathtt{U}'$.
13. *If* $\Sigma \vdash \chi.\chi' : \mathsf{cap}(\mathtt{U})$ *then* $\Sigma \vdash \chi : \mathsf{cap}(\mathtt{U})$ *and* $\Sigma \vdash \chi' : \mathsf{cap}(\mathtt{U})$.

Lemma 2 (Generation Lemma II).

1. *If* $\Sigma \vdash \mathtt{0} : \mathtt{V}$ *then* $\Sigma \vdash \mathtt{V}$.
2. *If* $\Sigma \vdash \chi.P : (\mathtt{U}, \mathtt{T})$ *then* $\Sigma \vdash \chi : \mathsf{cap}(\mathtt{U})$ *and* $\Sigma \vdash P : (\mathtt{U}, \mathtt{T})$.
3. *If* $\Sigma \vdash \mathsf{down}\,\alpha{:}\gamma\,\mathsf{with}\,\mathtt{V}.P : \mathtt{V}'$ *then* $\Sigma \vdash \alpha : \mathsf{amb}$, $\Sigma \vdash \gamma : \mathsf{group}$ *and* $\Sigma \vdash P : \mathtt{V}$ *and* $\Sigma \vdash \mathtt{V}'$ *and* $\gamma \in \mathscr{E}(\mathtt{V}')$.
4. *If* $\Sigma \vdash \mathsf{up}\,\alpha{:}\gamma\,\mathsf{with}\,\mathtt{V}.P : \mathtt{V}'$ *then* $\Sigma \vdash \alpha : \mathsf{amb}$, $\Sigma \vdash \gamma : \mathsf{group}$ *and* $\Sigma \vdash P : \mathtt{V}$ *and* $\Sigma \vdash \mathtt{V}'$ *and* $\gamma \in \mathscr{E}(\mathtt{V}')$.
5. *If* $\Sigma \vdash \mathsf{to}\,\alpha{:}\gamma\,\mathsf{with}\,\mathtt{V}.P : \mathtt{V}'$ *then* $\Sigma \vdash \alpha : \mathsf{amb}$, $\Sigma \vdash \gamma : \mathsf{group}$ *and* $\Sigma \vdash P : \mathtt{V}$ *and* $\Sigma \vdash \mathtt{V}'$ *and* $\gamma \in \mathscr{E}(\mathtt{V}')$.
6. *If* $\Sigma \vdash (x{:}\,\mathtt{W}) \,.\, P : (\mathtt{U}, \mathsf{com}(\mathtt{W}'))$ *then* $\mathtt{W} = \mathtt{W}'$ *and* $\Sigma, x : \mathtt{W} \vdash P : (\mathtt{U}, \mathsf{com}(\mathtt{W}))$ *and* $x \notin \Sigma$ *and* $x \notin (\mathtt{U}, \mathsf{com}(\mathtt{W}))$.
7. *If* $\Sigma \vdash \langle M \rangle \,.\, P : (\mathtt{U}, \mathsf{com}(\mathtt{W}))$ *then* $\Sigma \vdash P : (\mathtt{U}, \mathsf{com}(\mathtt{W}))$ *and* $\Sigma \vdash M : \mathtt{W}$.
8. *If* $\Sigma \vdash \alpha{:}\gamma\,\mathtt{V}[P] : \mathtt{V}'$ *then* $\Sigma \vdash \alpha : \mathsf{amb}$ *and* $\Sigma \vdash \gamma : \mathsf{group}$ *and* $\Sigma \vdash \mathtt{V}'$ *and* $\Sigma \vdash P : \mathtt{V}$.
9. *If* $\Sigma \vdash P \,|\, Q : \mathtt{V}$ *then* $\Sigma \vdash P : \mathtt{V}$ *and* $\Sigma \vdash Q : \mathtt{V}$.
10. *If* $\Sigma \vdash !P : \mathtt{V}$ *then* $\Sigma \vdash P : \mathtt{V}$.
11. *If* $\Sigma \vdash (\nu n)P : \mathtt{V}$ *then* $\Sigma \vdash P : \mathtt{V}$.
12. *If* $\Sigma \vdash (\nu g)P : \mathtt{V}$ *then* $\Sigma \vdash P : \mathtt{V}$, *and* $g \notin \Sigma$ *and* $g \notin \mathtt{V}$.

Note that the lemmata imply that every valid typing judgement has a unique derivation.

We also need a substitution lemma, which in our case handles the substitution of variables by messages. The application of substitutions to types, messages and processes is standard, while environments need some care. The substitution of the variable x by the message M in the environment Σ (denoted by $\Sigma\{x := M\}$) is defined by induction on Σ:

$$\varnothing\{x := M\} = \varnothing$$

$$(\Sigma, y{:}\,\mathtt{W})\{x := M\} = \begin{cases} \Sigma\{x := M\}, y{:}\,\mathtt{W}\{x := M\} & \text{if } x \neq y \\ \Sigma\{x := M\} & \text{if } x = y \text{ and } M \text{ is not a variable} \\ \Sigma\{x := M\}, M{:}\,\mathtt{W}\{x := M\} & \text{otherwise} \end{cases}$$

In the last case, where x is the same as y and M is a variable, the resulting environment might be non-well-formed (if the variable M occurs in Σ); however, that will never happen with substitutions as used in the paper. Also notice that if M is not a variable, then it is an ambient name, a group name or a capability, but in either case no assignment is allowed for it in environments.

Lemma 3 (Substitution Lemma).

1. *If* $\Sigma \vdash \Gamma$ *and* $x \in \Gamma$ *then* $x \in Dom(\Sigma)$.
2. *If* $\vdash \Sigma, x{:}\,\mathtt{W}$ *and* $\Sigma \vdash M : \mathtt{W}$ *then* $\vdash \Sigma\{x := M\}$.
3. *If* $\Sigma, x{:}\,\mathtt{W} \vdash \Gamma$ *and* $\Sigma \vdash M : \mathtt{W}$ *then* $\Sigma\{x := M\} \vdash \Gamma\{x := M\}$.

Proof. The proof of Point (1) by induction on the derivations is standard. Points (2) and (3) can be proved simultaneously by induction on the derivations; we only consider two interesting cases.

The first case is the proof of Point (2) when the last rule applied (in the derivation of the point's first antecedent) is (Env Constr):

$$\frac{\Sigma, x{:}\,\mathtt{W} \vdash \mathtt{W}' \quad y \notin Dom(\Sigma, x{:}\,\mathtt{W})}{\vdash \Sigma, x{:}\,\mathtt{W}, y{:}\,\mathtt{W}'}$$

Then, to the judgement $\Sigma, x{:}\,\mathtt{W} \vdash \mathtt{W}'$ we may apply the inductive hypothesis of Point (3), with Γ instantiated to $\mathtt{W}'$, and we obtain $\Sigma\{x := M\} \vdash \mathtt{W}'\{x := M\}$.

To be able to re-apply the rule (Env Constr) with x substituted by M, and thus to prove the conclusion $\vdash (\Sigma, y{:}\,\mathtt{W}')\{x := M\}$, we only need to show that the side condition $y \notin Dom(\Sigma\{x := M\})$ holds. But this is not hard, since by Lemma 1(1) the well-formedness judgment $\vdash \Sigma, x{:}\,\mathtt{W}$ holds, which in turn by Lemma 1(6) implies $x \notin Dom(\Sigma)$; it follows that $Dom(\Sigma) = Dom(\Sigma\{x := M\})$. From the rule's rightmost premise $y \notin Dom(\Sigma, x{:}\,\mathtt{W})$ in the assumption, we immediately have that $y \notin Dom(\Sigma)$; thus also $y \notin Dom(\Sigma\{x := M\})$.

As a second interesting case we consider the proof of Point (3) when the last rule applied is (Input):

$$\frac{\Sigma, x{:}\,\mathtt{W}, y{:}\,\mathtt{W}' \vdash P : (\mathtt{U}, \mathsf{com}(\mathtt{W}')) \quad y \notin (\Sigma, x{:}\,\mathtt{W}) \quad y \notin (\mathtt{U}, \mathsf{com}(\mathtt{W}'))}{\Sigma, x{:}\,\mathtt{W} \vdash (y{:}\,\mathtt{W}')\,.\,P : (\mathtt{U}, \mathsf{com}(\mathtt{W}'))}$$

By induction on Point (3), with Σ replaced by $\Sigma, y{:}\mathtt{W}'$ and Γ instantiated to $P{:}(\mathtt{U}, \mathsf{com}(\mathtt{W}'))$, we obtain $(\Sigma, y{:}\,\mathtt{W}')\{x := M\} \vdash (P{:}(\mathtt{U}, \mathsf{com}(\mathtt{W}')))\{x := M\}$. As for the two needed side conditions on y, observe that by Point (1) $y \notin (\Sigma, x{:}\mathtt{W})$

and $\Sigma \vdash M{:}\mathtt{W}$ imply $y \notin M$; from $y \notin \Sigma$ and $y \notin M$ it follows that $y \notin \Sigma\{x := M\}$, while the rightmost premise $y \notin (\mathtt{U}, \mathsf{com}(\mathtt{W}'))$, again with the condition $y \notin M$, implies that $y \notin (\mathtt{U}, \mathsf{com}(\mathtt{W}'))\{x := M\}$. We may therefore conclude by applying the rule (INPUT), thus obtaining the consequent $\Sigma\{x := M\} \vdash ((y{:}\,\mathtt{W}').P{:}(\mathtt{U}, \mathsf{com}(\mathtt{W}')))\{x := M\}$. □

The other standard lemmata needed for the proof of subject reduction are those concerning strengthening, weakening, and admissibility of subtyping.

Strengthening is expressed by the first two points of the Lemma below. Point (1) states that if the environment $\Sigma \cup \Sigma'$ is well formed and no x occurring in Σ 'depends' on Σ', i.e., is in the domain of Σ', then Σ is separately well formed. Point (2) says that if the judgment $\Sigma \cup \Sigma' \vdash \Gamma$ holds, and no x occurring (free) in Σ or in Γ is in the domain of Σ', then Σ' may be disposed of, and the strengthened judgment $\Sigma \vdash \Gamma$ holds.

The Lemma's third point expresses the weakening in the usual form.

Lemma 4 (Strengthening and Weakening Lemma).

1. *If* $\vdash \Sigma \cup \Sigma'$ *and* $x \notin \Sigma$ *for all* $x \in Dom(\Sigma')$ *then* $\vdash \Sigma$.
2. *If* $\Sigma \cup \Sigma' \vdash \Gamma$ *and* $x \notin \Sigma$ *and* $x \notin \Gamma$ *for all* $x \in Dom(\Sigma')$ *then* $\Sigma \vdash \Gamma$.
3. *If* $\Sigma \vdash \Gamma$ *and* $\vdash \Sigma'$ *and* $\Sigma \subseteq \Sigma'$, *then* $\Sigma' \vdash \Gamma$.

Proof. Points (1) and (2) can be proved simultaneously by induction on derivations. The proof of Point (3) is also by induction on derivations. We only consider a single interesting case, the one in which the last rule applied is (INPUT):

$$\frac{\Sigma, x{:}\,\mathtt{W} \vdash P : (\mathtt{U}, \mathsf{com}(\mathtt{W})) \quad x \notin \Sigma \quad x \notin (\mathtt{U}, \mathsf{com}(\mathtt{W}))}{\Sigma \vdash (x{:}\,\mathtt{W})\,.\,P : (\mathtt{U}, \mathsf{com}(\mathtt{W}))}$$

and, in addition, $x \in \Sigma'$.

Let x' be a fresh variable; by applying induction to the rule's (leftmost) premise we obtain $\Sigma, x{:}\,\mathtt{W}, x'{:}\,\mathtt{W} \vdash P{:}\,(\mathtt{U}, \mathsf{com}(\mathtt{W}))$, while the rule (ENV) yields $\Sigma, x'{:}\,\mathtt{W} \vdash x'{:}\,\mathtt{W}$. From these two judgments it follows by Lemma 3(3) (since environments are sets) that the typing $\Sigma, x'{:}\,\mathtt{W} \vdash P\{x := x'\} : (\mathtt{U}, \mathsf{com}(\mathtt{W}))$ holds, i.e., is derivable. It is easy to verify that the derivations of $\Sigma, x{:}\,\mathtt{W} \vdash P{:}\,(\mathtt{U}, \mathsf{com}(\mathtt{W}))$ and $\Sigma, x'{:}\,\mathtt{W} \vdash P\{x := x'\} : (\mathtt{U}, \mathsf{com}(\mathtt{W}))$ are isomorphic; then by the induction hypothesis we obtain $\Sigma', x'{:}\,\mathtt{W} \vdash P\{x := x'\} : (\mathtt{U}, \mathsf{com}(\mathtt{W}))$. By the rule (INPUT) this implies that $\Sigma' \vdash (x'{:}\,\mathtt{W})\,.\,P\{x := x'\} : (\mathtt{U}, \mathsf{com}(\mathtt{W}))$, and we are done, since $(x{:}\,\mathtt{W})\,.\,P$ and $(x'{:}\,\mathtt{W})\,.\,P\{x := x'\}$ are α-equivalent. □

Lemma 5 (Admissibility of Subtyping).

1. *If* $\Sigma \vdash M : \mathsf{cap}(\mathtt{U})$ *and* $\mathtt{U} \leq \mathtt{U}'$ *and* $\Sigma \vdash \mathtt{U}'$ *then* $\Sigma \vdash M : \mathsf{cap}(\mathtt{U}')$.
2. *If* $\Sigma \vdash P : \mathtt{V}$ *and* $\mathtt{V} \leq \mathtt{V}'$ *and* $\Sigma \vdash \mathtt{V}'$ *then* $\Sigma \vdash P : \mathtt{V}'$.

Proof. Points (1) and (2) are easily proved simultaneously by induction on the derivations. □

We are now finally able to prove subject reduction. The property is expressed in its most natural form, which only holds for closed processes, i.e., processes without free variables.

Theorem 1 (Subject Reduction).
Let $\vdash P : \mathtt{V}$*. Then*

1. $P \equiv Q$ *implies* $\vdash Q : \mathtt{V}$.
2. $P \to Q$ *implies* $\vdash Q : \mathtt{V}$.

Proof. The proof is standard, by induction on the derivations of $P \equiv Q$ and $P \to Q$ using the Weakening, Substitution, Admissibility of Subtyping and Generation Lemmata. We only explicitly present the case of rule (R-up):

$$m{:}g_m\,(\mathtt{U}_m,\mathtt{T}_m)[n{:}g_n\,\mathtt{V}_n[\mathsf{up}\,m{:}g_m\,\mathsf{with}\,(\mathtt{U},\mathtt{T})\,.\,P_1 \mid P_2] \mid \mathsf{co}\,g_n\,\mathsf{with}\,\mathtt{U}_1.Q_1 \mid Q_2]$$
$$\to m{:}g_m\,(\mathtt{U}_m,\mathtt{T}_m)[n{:}g_n\,\mathtt{V}_n[P_2] \mid P_1 \mid Q_1 \mid Q_2]$$

if $\mathtt{U} \leq \mathtt{U}_1$ and $\mathtt{T} \leq \mathtt{T}_m$.

If

$$\vdash m{:}g_m\,(\mathtt{U}_m,\mathtt{T}_m)[n{:}g_n\,\mathtt{V}_n[\mathsf{up}\,m{:}g_m\,\mathsf{with}\,(\mathtt{U},\mathtt{T})\,.\,P_1 \mid P_2] \mid \mathsf{co}\,g_n\,\mathsf{with}\,\mathtt{U}_1.Q_1 \mid Q_2] : \mathtt{V}$$

then by Lemma 2(8) we get $\vdash \mathtt{V}$ and

$$\vdash n{:}g_n\,\mathtt{V}_n[\mathsf{up}\,m{:}g_m\,\mathsf{with}\,(\mathtt{U},\mathtt{T})\,.\,P_1 \mid P_2] \mid \mathsf{co}\,g_n\,\mathsf{with}\,\mathtt{U}_1.Q_1 \mid Q_2 : (\mathtt{U}_m,\mathtt{T}_m).$$

By Lemma 2(9) we must have

$$\vdash n{:}g_n\,\mathtt{V}_n[\mathsf{up}\,m{:}g_m\,\mathsf{with}\,(\mathtt{U},\mathtt{T})\,.\,P_1 \mid P_2] : (\mathtt{U}_m,\mathtt{T}_m) \tag{4.2}$$
$$\vdash \mathsf{co}\,g_n\,\mathsf{with}\,\mathtt{U}_1.Q_1 : (\mathtt{U}_m,\mathtt{T}_m) \tag{4.3}$$
$$\vdash Q_2 : (\mathtt{U}_m,\mathtt{T}_m). \tag{4.4}$$

From (4.2) by Lemma 2(8) we have $\vdash (\mathtt{U}_m,\mathtt{T}_m)$ and

$$\vdash \mathsf{up}\,m{:}g_m\,\mathsf{with}\,(\mathtt{U},\mathtt{T})\,.\,P_1 \mid P_2 : \mathtt{V}_n,$$

which implies by Lemma 2(9)

$$\vdash \mathsf{up}\,m{:}g_m\,\mathsf{with}\,(\mathtt{U},\mathtt{T})\,.\,P_1 : \mathtt{V}_n \tag{4.5}$$
$$\vdash P_2 : \mathtt{V}_n. \tag{4.6}$$

Now, by Lemma 2(4) we get $\vdash P_1 : (\mathtt{U},\mathtt{T})$ and $\vdash \mathtt{V}_n$ and $g_m \in \mathscr{E}(\mathtt{V}_n)$. Furthermore, by Lemma 5(2) from $\vdash P_1 : (\mathtt{U},\mathtt{T})$ we have $\vdash P_1 : (\mathtt{U}_m,\mathtt{T}_m)$ since $\mathtt{U} \leq \mathtt{U}_m$ and $\mathtt{T} \leq \mathtt{T}_m$. From (4.3) by Lemma 2(2) we get $\vdash Q_1 : (\mathtt{U}_m,\mathtt{T}_m)$.

Rule (Amb) applied to (4.6) gives $\vdash n{:}g_n\,\mathtt{V}_n[P_2] : (\mathtt{U}_m,\mathtt{T}_m)$ since $\vdash (\mathtt{U}_m,\mathtt{T}_m)$. Rule (Par) applied to $\vdash n{:}g_n\,\mathtt{V}_n[P_2] : (\mathtt{U}_m,\mathtt{T}_m)$, $\vdash P_1 : (\mathtt{U}_m,\mathtt{T}_m)$, $\vdash Q_1 : (\mathtt{U}_m,\mathtt{T}_m)$, and (4.4) gives

$$\vdash n{:}g_n\,\mathtt{V}_n[P_2] \mid P_1 \mid Q_1 \mid Q_2 : (\mathtt{U}_m,\mathtt{T}_m).$$

We conclude

$$\vdash m{:}g_m\,\mathtt{V}_m[n{:}g_n\,\mathtt{V}_n[P_2] \mid P_1 \mid Q_1 \mid Q_2] : \mathtt{V}$$

by rule (Amb) since $\vdash \mathtt{V}$. $\square$

Clearly, subject reduction guarantees that in any sequence of reductions every process obeys the constraints imposed by its enclosing ambient. As a matter of fact, a more accurate and stronger property is satisfied: any sub-process of a 'good' process does not only behave in compliance with the policy of its enclosing ambient (if any), but it also complies with the policy against which it has been checked when entering the ambient. Such policy is in general more restrictive than the one of the enclosing ambient, and is expressed by the with component of the co-move that has authorized the movement. Unfortunately this expression completely disappears with its containing co-move as soon as the incoming process is authorized: it thus becomes impossible to formalize the statement that the policy is respected during the whole computation carried out by the process before any further migration.

To overcome this problem, we enrich the process syntax by introducing *tagged processes.* Tags will be used to record, for each process entering an ambient by consuming a to, down, or up capability, the mobility type U associated with it. To accommodate tags in the syntax of Figure 1, we add the production:

$$P ::= \ldots$$
$$\{P\}^{\mathtt{U}} \qquad \text{tagged process}$$

Tagging has no influence on process formation and well-typing rules: the type of a tagged process is the one of its untagged version, obtained by dropping all tags occurring in it.

The operational semantics attaches new tags to processes (when needed), but in reduction it considers tagged processes as ordinary ones; i.e., tags are not exploited to enable or disable reduction steps. We always assume that the initial process of a system is untagged and that all tags are generated during reduction. All properties concerning reduction, notably subject reduction, are then unaffected. Nevertheless, the operational semantics must be extended to represent the generation of tags upon process movements and to allow tagged processes to evolve. The basic reduction rules of Fig. 6 are therefore replaced by those given in Fig. 7, where we convene that a process not explicitly tagged may be either untagged or tagged. In the same figure we also give the additional structural congruence rules for dealing with tagged processes. Notice that the tag disappears when tagging 0 or the ambient formation, and that tags commute with all other process constructs except the ones for mobility (of both ambients and processes). For this reason we need to duplicate the movement reduction rules.

The rules are hopefully self-explanatory. We only remark that after a reduction with rule (R-down), (R-up) or (R-to), a tag is attached to the migrating process to record the policy against which it has been checked. Similarly, in rules (R-down)′, (R-up)′ and (R-to)′ the tag of the migrating process is changed appropriately.

The structural reduction rules of Fig. 6 are still valid, but we need to extend the notion of reduction contexts to also encompass the occurrence of tagged processes. Thus, we will say that $C[-]$ is a reduction context if its untagged version is a reduction context.

Structural congruence (additional rules):

$$\begin{array}{rcl} \{0\}^{\mathtt{U}} & \equiv & 0 \\ \{(x{:}\,\mathtt{W}).P\}^{\mathtt{U}} & \equiv & (x{:}\,\mathtt{W}).\{P\}^{\mathtt{U}} \\ \{P \mid Q\}^{\mathtt{U}} & \equiv & \{P\}^{\mathtt{U}} \mid \{Q\}^{\mathtt{U}} \\ \{(\nu n)P\}^{\mathtt{U}} & \equiv & (\nu n)\{P\}^{\mathtt{U}} \\ \{n{:}g\,\mathtt{V}[P]\}^{\mathtt{U}} & \equiv & n{:}g\,\mathtt{V}[P] \end{array} \qquad \begin{array}{rcl} \{\mathsf{co}\; g \;\mathsf{with}\; \mathtt{U}'.P\}^{\mathtt{U}} & \equiv & \mathsf{co}\; g \;\mathsf{with}\; \mathtt{U}'.\{P\}^{\mathtt{U}} \\ \{\langle M\rangle.P\}^{\mathtt{U}} & \equiv & \langle M\rangle.\{P\}^{\mathtt{U}} \\ \{!P\}^{\mathtt{U}} & \equiv & !\{P\}^{\mathtt{U}} \\ \{(\nu g)P\}^{\mathtt{U}} & \equiv & (\nu g)\{P\}^{\mathtt{U}} \quad (g \notin \mathtt{U}) \\ P \equiv Q & \text{implies} & \{P\}^{\mathtt{U}} \equiv \{Q\}^{\mathtt{U}} \end{array}$$

Basic reduction rules:

(R-in) $n{:}g_n\,\mathtt{V}_n[\,\mathsf{in}\; m{:}g_m\,.\,P_1 \mid P_2\,] \mid m{:}g_m\,\mathtt{V}_m[Q]$
$\rightarrow m{:}g_m\,\mathtt{V}_m[\,n{:}g_n\,\mathtt{V}_n[P_1 \mid P_2\,] \mid Q\,]$

(R-in)′ $n{:}g_n\,\mathtt{V}_n[\,\{\mathsf{in}\; m{:}g_m\,.\,P_1\}^{\mathtt{U}_1} \mid P_2\,] \mid m{:}g_m\,\mathtt{V}_m[Q]$
$\rightarrow m{:}g_m\,\mathtt{V}_m[\,n{:}g_n\,\mathtt{V}_n[\{P_1\}^{\mathtt{U}_1} \mid P_2\,] \mid Q\,]$

(R-out) $m{:}g_m\,\mathtt{V}_m[\,n{:}g_n\,\mathtt{V}_n[\,\mathsf{out}\; m{:}g_m\,.\,P_1 \mid P_2\,] \mid Q\,]$
$\rightarrow n{:}g_n\,\mathtt{V}_n[\,P_1 \mid P_2\,] \mid m{:}g_m\,\mathtt{V}_m[Q]$

(R-out)′ $m{:}g_m\,\mathtt{V}_m[\,n{:}g_n\,\mathtt{V}_n[\,\{\mathsf{out}\; m{:}g_m\,.\,P_1\}^{\mathtt{U}_1} \mid P_2\,] \mid Q\,]$
$\rightarrow n{:}g_n\,\mathtt{V}_n[\,\{P_1\}^{\mathtt{U}_1} \mid P_2\,] \mid m{:}g_m\,\mathtt{V}_m[Q]$

(R-down) $n{:}g_n\,\mathtt{V}_n[\mathsf{down}\; m{:}g_m \;\mathsf{with}\; (\mathtt{U},\mathtt{T})\,.\,P_1 \mid P_2 \mid m{:}g_m\,(\mathtt{U}_m,\mathtt{T}_m)[\mathsf{co}\; g_n \;\mathsf{with}\; \mathtt{U}_1.Q_1 \mid Q_2]]$
$\rightarrow n{:}g_n\,\mathtt{V}_n[m{:}g_m\,(\mathtt{U}_m,\mathtt{T}_m)[\{P_1\}^{\mathtt{U}_1} \mid Q_1 \mid Q_2] \mid P_2]$
if $\mathtt{U} \leq \mathtt{U}_1$ and $\mathtt{T} \leq \mathtt{T}_m$

(R-down)′ $n{:}g_n\mathtt{V}_n[\{\mathsf{down}\; m{:}g_m \;\mathsf{with}(\mathtt{U},\mathtt{T}).P_1\}^{\mathtt{U}'} \mid P_2 \mid m{:}g_m(\mathtt{U}_m,\mathtt{T}_m)[\mathsf{co}\; g_n \mathsf{with}\; \mathtt{U}_1.Q_1 \mid Q_2]]$
$\rightarrow n{:}g_n\,\mathtt{V}_n[m{:}g_m\,(\mathtt{U}_m,\mathtt{T}_m)[\{P_1\}^{\mathtt{U}_1} \mid Q_1 \mid Q_2] \mid P_2]$
if $\mathtt{U} \leq \mathtt{U}_1$ and $\mathtt{T} \leq \mathtt{T}_m$

(R-up) $m{:}g_m\,(\mathtt{U}_m,\mathtt{T}_m)[n{:}g_n\,\mathtt{V}_n[\mathsf{up}\; m{:}g_m \;\mathsf{with}\; (\mathtt{U},\mathtt{T})\,.\,P_1 \mid P_2] \mid \mathsf{co}\; g_n \;\mathsf{with}\; \mathtt{U}_1.Q_1 \mid Q_2]$
$\rightarrow m{:}g_m\,(\mathtt{U}_m,\mathtt{T}_m)[n{:}g_n\,\mathtt{V}_n[P_2] \mid \{P_1\}^{\mathtt{U}_1} \mid Q_1 \mid Q_2]$
if $\mathtt{U} \leq \mathtt{U}_1$ and $\mathtt{T} \leq \mathtt{T}_m$

(R-up)′ $m{:}g_m(\mathtt{U}_m,\mathtt{T}_m)[n{:}g_n\,\mathtt{V}_n[\{\mathsf{up}\; m{:}g_m \;\mathsf{with}\; (\mathtt{U},\mathtt{T})\,.\,P_1\}^{\mathtt{U}'} \mid P_2] \mid \mathsf{co}\; g_n \;\mathsf{with}\; \mathtt{U}_1.Q_1 \mid Q_2]$
$\rightarrow m{:}g_m\,(\mathtt{U}_m,\mathtt{T}_m)[n{:}g_n\,\mathtt{V}_n[P_2] \mid \{P_1\}^{\mathtt{U}_1} \mid Q_1 \mid Q_2]$
if $\mathtt{U} \leq \mathtt{U}_1$ and $\mathtt{T} \leq \mathtt{T}_m$

(R-to) $n{:}g_n\,\mathtt{V}_n[\mathsf{to}\; m{:}g_m \;\mathsf{with}\; (\mathtt{U},T)\,.\,P_1 \mid P_2] \mid m{:}g_m\,(\mathtt{U}_m,\mathtt{T}_m)[\mathsf{co}\; g_n \;\mathsf{with}\; \mathtt{U}_1.Q_1 \mid Q_2]$
$\rightarrow n{:}g_n\,\mathtt{V}_n[P_2] \mid m{:}g_m\,(\mathtt{U}_m,\mathtt{T}_m)[\{P_1\}^{\mathtt{U}_1} \mid Q_1 \mid Q_2]$
if $\mathtt{U} \leq \mathtt{U}_1$ and $\mathtt{T} \leq \mathtt{T}_m$

(R-to)′ $n{:}g_n\mathtt{V}_n[\{\mathsf{to}\; m{:}g_m \;\mathsf{with}\; (\mathtt{U},T)\,.\,P_1\}^{\mathtt{U}'} \mid P_2] \mid m{:}g_m\,(\mathtt{U}_m,\mathtt{T}_m)[\mathsf{co}\; g_n \;\mathsf{with}\; \mathtt{U}_1.Q_1 \mid Q_2]$
$\rightarrow n{:}g_n\,\mathtt{V}_n[P_2] \mid m{:}g_m\,(\mathtt{U}_m,\mathtt{T}_m)[\{P_1\}^{\mathtt{U}_1} \mid Q_1 \mid Q_2]$
if $\mathtt{U} \leq \mathtt{U}_1$ and $\mathtt{T} \leq \mathtt{T}_m$

(R-comm) $(x{:}\,\mathtt{W})\,.\,P \mid \langle M\rangle\,.\,Q \rightarrow P\{x := M\} \mid Q$

Fig. 7. Operational Semantics with Tags

The main property of the tag system is that on the one hand the tag assigned to a process during reduction refines the policy of the enclosing ambient, on the other hand it encompasses all the actions the process can perform while running in that ambient. Formally, it is expressed as follows.

Theorem 2 (Correctness of Tagging). *If P is a good untagged process and $P \rightarrow^* C[m{:}g_m\,(\mathtt{U}_m, \mathtt{T}_m)[\{P_1\}^{\mathtt{U}_1} \mid P_2]]$, then $\mathtt{U}_1 \leq \mathtt{U}_m$ and $\vdash P_1 : (\mathtt{U}_1, \mathtt{T}_m)$.*

Proof. To show that the thesis holds for all the tags generated by the operational semantics, we prove that it holds as soon as a tag is generated and that the property is preserved by reduction. Formally, we reason by induction on the length of the computation that generates the tag. Since by hypothesis the process P is untagged while the process at the right of the arrow has at least one tag, the reduction sequence must be composed of at least one step (the one that introduces the tag). Let then Q be a process such that

$$P \rightarrow^* Q \rightarrow C[m{:}g_m\,(\mathtt{U}_m, \mathtt{T}_m)[\{P_1\}^{\mathtt{U}_1} \mid P_2]] \tag{4.7}$$

The hypotheses imply that Q is a good process (by Theorem 1) and that all the tags occurring within Q have been generated in the computation from P. Therefore, by induction, the thesis holds for all the tags in Q.

To prove that the thesis holds also for the tag $\mathtt{U}_1$, we reason by induction on the depth of the proof of the last reduction step in (4.7). In the inductive case, the last rule applied to infer the reduction step is a structural reduction rule and therefore the property trivially holds. The base case is when only one of the basic reduction rules of Fig. 7 is applied; we distinguish between process movement and ambient movement or communication.

In the case of a process movement, the tag $\mathtt{U}_1$ is generated by the reduction step. We only consider rule (R-down)$'$; the other cases can be dealt with similarly. With reference to the notations of Fig. 7, we have $\vdash P_1 : (\mathtt{U}, \mathtt{T})$ by Lemma 2(3) and $\mathtt{U}_1 \leq \mathtt{U}_m$ by Lemmata 1(12) and 2(2), because Q is a good process. Also, we have $\mathtt{U} \leq \mathtt{U}_1$ and $\mathtt{T} \leq \mathtt{T}_m$, otherwise the rule cannot fire. Hence, by Lemma 5(2), we get $\vdash P_1 : (\mathtt{U}_1, \mathtt{T}_m)$, which proves the thesis.

In the case of a reduction corresponding to an ambient movement or a communication, the tag $\mathtt{U}_1$ was already present in Q. Again, we only consider one significant case, namely when rule (R-in)$'$ is applied. We refer to the notations of Fig. 7, but with n and m exchanged. Let $\mathtt{V}_m = (\mathtt{U}_m, \mathtt{T}_m)$; then we have $\mathtt{U}_1 \leq \mathtt{U}_m$ and $\vdash \mathsf{in}\; n{:}g_n \,.\, P_1 : (\mathtt{U}_1, \mathtt{T}_m)$ by induction, because the tag $\mathtt{U}_1$ was generated previously. Now, $\mathtt{U}_1 \leq \mathtt{U}_m$ still holds because the last reduction step in (4.7) does not change the tag, while $\vdash \mathsf{in}\; n{:}g_n \,.\, P_1 : (\mathtt{U}_1, \mathtt{T}_m)$ implies $\vdash P_1 : (\mathtt{U}_1, \mathtt{T}_m)$ by Lemma 2(2). □

Type safety can now be stated by exploiting tags assigned to processes.

Corollary 1 (Type Safety). *Let Q be a good untagged process, and let $Q \rightarrow^* C[P] \rightarrow C[P']$ where $P \rightarrow P'$ is obtained by applying one of the basic reduction rules of Fig. 7. With reference to the notations of Fig. 7 we have:*

1. *If* $P \to P'$ *by (R-in) or (R-out) and* $\mathtt{V}_n = (\mathtt{U}_n, \mathtt{T}_n)$ *then* $g_m \in \mathscr{C}(\mathtt{U}_n)$.
2. *If* $P \to P'$ *by (R-in)′ or (R-out)′ then* $g_m \in \mathscr{C}(\mathtt{U}_1)$.
3. *If* $P \to P'$ *by (R-up), (R-down) or (R-to) and* $\mathtt{V}_n = (\mathtt{U}_n, \mathtt{T}_n)$ *then* $g_m \in \mathscr{E}(\mathtt{U}_n)$.
4. *If* $P \to P'$ *by (R-up)′, (R-down)′ or (R-to)′ then* $g_m \in \mathscr{E}(\mathtt{U}')$.
5. *If* $P \to P'$ *by (R-comm) and* $P \equiv (x{:}\,\mathtt{W}) . P_1 \mid \langle M \rangle . P_2$ *and* $C[P] \equiv C'[n{:}g_n\,(\mathtt{U}_n, \mathtt{T}_n)[(x{:}\,\mathtt{W}) . P_1 \mid \langle M \rangle . P_2 \mid Q_1]]$ *then* $\mathtt{T}_n = \mathsf{com}(\mathtt{W})$ *and* $\vdash M : \mathtt{W}$.

Proof. The proofs of cases 1, 3 and 5 follow from the observation that by Theorem 1 $C[P]$ is a good process, therefore by Lemma 2 P and all its sub-processes are good processes too. We consider case 5 as an interesting example. In order to type $C[P]$, by Lemma 2(9) we must have $\vdash (x{:}\,\mathtt{W}) . P_1 : (\mathtt{U}_n, \mathtt{T}_n)$ and $\vdash \langle M \rangle . P_2 : (\mathtt{U}_n, \mathtt{T}_n)$ which, by Lemma 2(7) and (8), imply that $\mathsf{com}(\mathtt{W}) = \mathtt{T}_n$ and $\vdash M : \mathtt{W}$.

For cases 2 and 4 notice that by Theorem 2 if $\{P_1\}^{\mathtt{U}_1}$ is a sub-process of P then $\vdash P_1 : (\mathtt{U}_1, \mathtt{T})$ for some $\mathtt{T}$. □

We can therefore conclude that ambient and process moves always respect the mobility rights represented by types, and that the types of messages exchanged in communications always match.

As an example, we apply the previous results to the scenario described in Section 2 and prove some behavioural and security properties of the journals.

Proposition 1. *The ambient* jrn_k *is immobile and can send processes only to ambients of groups* $\{g_{tdl_1}, \ldots, g_{tdl_m}\}$. *Also, whenever pub is the only ambient of group* g_{pub}, *a process entering* jrn_k *from an ambient of group* g_{tdl_i} *is allowed to send a process to an ambient of the same group, while a process from an ambient of group* g_{dl_i} *cannot send processes.*

Proof. The ambient jrn_k is immobile since its internal process is well typed with type $(\langle\!\langle \varnothing, \{g_{tdl_1}, \ldots, g_{tdl_m}\} \rangle\!\rangle, \mathsf{com}(\mathsf{paper}))$. Therefore by Theorem 1 and Corollary 1(1) and (2) no action driving it into or out of another ambient is possible. Similarly, by Theorem 1 and Corollary 1(3) and (4), each process inside jrn_k can send processes only to ambients belonging to the set of groups $\{g_{tdl_1}, \ldots, g_{tdl_m}\}$.

Notice that the initial co-moves in the ambient jrn_k only let in processes coming from ambients of group g_{pub}, while the co-moves in the ambient *pub* only let in 'idle' processes, i.e., processes that can be tagged by $\langle\!\langle \varnothing, \varnothing \rangle\!\rangle$. So, the ambient jrn_k can receive from the ambient *pub* only two kinds of co-moves (see process MGR): $!\mathsf{co}g_{tdl_i}$ with $\langle\!\langle \varnothing, \{g_{tdl_i}\} \rangle\!\rangle$ and $!\mathsf{co}g_{dl_i}$ with $\langle\!\langle \varnothing, \varnothing \rangle\!\rangle$. Therefore, a process P entering jrn_k from an ambient of group g_{tdl_i} will be tagged by $\langle\!\langle \varnothing, \{g_{tdl_i}\} \rangle\!\rangle$ and so by Theorem 2 we get $\vdash P : (\langle\!\langle \varnothing, \{g_{tdl_i}\} \rangle\!\rangle, \mathsf{com}(\mathsf{paper}))$: this, by Theorem 1 and Corollary 1(3) and (4), means that P can send processes to ambients of group g_{tdl_i}. On the other hand, a process entering jrn_k from an ambient of group g_{dl_i} will be tagged by $\langle\!\langle \varnothing, \varnothing \rangle\!\rangle$ and so it cannot send processes. □

5 An Exemplifying Application: The Train Scenario

In this section we focus on the modelling of a public transportation system, the *train* introduced by [9], as a nice illustration of the issues related to the control of mobility.

We want to represent a railway network connecting a set of different places (e.g., cities) in the world. Trains move between stations, travellers may get into and off trains only at stations and cannot drive them (no hijacking is possible). The number of passengers in a train at any given instant cannot exceed the number of seats; a passenger takes a seat on boarding and releases it on getting off. Each train has a fixed route.

For the sake of simplicity, we assume that:

- There is a top-level untrusted ambient *world*, which includes stations, travellers, and some other unspecified process S (e.g. other means of transport); it has mobcom type $\mathtt{V}_w$, but no assumption can be made on it.
- In our intended representation different stations should be found within different cities or localities, and moving from one city to another would only be possible by train. The presence of cities would however increase the size of the example in a trivial manner, without providing more insights; we therefore place stations directly within *world*, although in this way travellers appear to use a train to end up in the same ambient *world* which they started from.
- There are only two stations *stA* and *stB*, and one train TRAIN commuting between them. Initially, the train is within *stA*.

Stations and trains are represented by ambient processes; travellers are represented by simple processes; the number of free seats in a train is represented by the multiplicity of the co-actions allowing to get into the train.

Since there is no communication in any considered ambient except at most in the *world*, we will write for all other ambients mobcom types $\langle\!\langle \mathscr{C}, \mathscr{E} \rangle\!\rangle$ instead of $(\langle\!\langle \mathscr{C}, \mathscr{E} \rangle\!\rangle, \mathsf{shh})$.

Stations are immobile ambients of group g_{st}, and can have travellers both going down into the trains (of group g_{tr}) or up into the world (of group g_w); they can be crossed by trains, and can receive travellers with different rights both from train and from the outside world. Stations contain also an ambient *checkOut* instrumental to preserving the condition of having at most n passengers on the train. Each station thus always contains the process:

$$\text{STATP} \triangleq\ !\,\mathsf{co}\ g_w\ \mathsf{with}\ \mathtt{U}_{from} \mid\ !\,\mathsf{co}\ g_{tr}\ \mathsf{with}\ \mathtt{U}_{to} \mid \mathit{checkOut}{:}g_{check}\,\mathtt{U}_{from}[\,!\,\mathsf{co}\ g_{st}\ \mathsf{with}\ \mathtt{U}_{from}]$$

where $\mathtt{U}_{from} = \langle\!\langle \varnothing, \{g_{tr}\} \rangle\!\rangle$, $\mathtt{U}_{to} = \langle\!\langle \varnothing, \{g_w, g_{check}\} \rangle\!\rangle$. Therefore the process STATP can be typed by the mobility type $\mathtt{U}_{st} = \langle\!\langle \varnothing, \{g_{tr}, g_w, g_{check}\} \rangle\!\rangle$.

The mobility types $\mathtt{U}_{from}$ and $\mathtt{U}_{to}$ specify the behaviours of a passenger respectively in the departure station, when going to board a train, and in the arrival station, when going to exit the station into the outside world or city and to send a notification (through the ambient *checkOut*) that a seat becomes free.

As an immediate application of Theorem 2 and Lemmata 1 and 2 passenger behaviours must respect this policy, so for example a passenger entering the station from the world cannot send a notification to the ambient *checkOut*.

The train is an ambient which can cross stations, send traveller processes into stations and receive at most n passengers from stations, provided they behave as good passengers (and not, for example, as drivers). The train can also receive the notification of free seats from the ambient *checkOut*.

$$\text{TRAIN} \triangleq tr{:}g_{tr}(\langle\!\langle \{g_{st}\}, \{g_{st}\} \rangle\!\rangle)[\underbrace{\mathsf{co}\ g_{st}\ \mathsf{with}\ \mathtt{U}_{psng} \mid \ldots \mid \mathsf{co}\ g_{st}\ \mathsf{with}\ \mathtt{U}_{psng}}_{\mathsf{n}} \mid$$
$$!\,\mathsf{co}\ g_{check}\ \mathsf{with}\ \mathtt{U}_{psng} \mid !\,\mathsf{out}\ stA{:}g_{st}\,.\,\mathsf{in}\ stB{:}g_{st}\,.\,\mathsf{out}\ stB{:}g_{st}\,.\,\mathsf{in}\ stA{:}g_{st}]$$

where $\mathtt{U}_{psng} = \langle\!\langle \varnothing, \{g_{st}\} \rangle\!\rangle$.

A traveller is represented by a parametric process TRAVELLER(*src*, *dst*) which from some unspecified place in the world enters the station *src* to become a passenger of a train that takes him to the station *dst*:

$$\begin{aligned}\text{TRAVELLER}(src, dst) \triangleq\ & \mathsf{down}\ src{:}g_{st}\ \mathsf{with}\ \mathtt{U}_{from}\,.\,\mathsf{down}\ tr{:}g_{tr}\ \mathsf{with}\ \mathtt{U}_{psng}\,.\,\mathsf{up}\ dst{:}g_{st}\ \mathsf{with}\ \mathtt{U}_{to}\,.\\ & (\mathsf{up}\ world{:}g_w\ \mathsf{with}\ \mathtt{U}_w\,.\,P \mid\\ & \ \mathsf{down}\ checkOut{:}g_{check}\ \mathsf{with}\ \mathtt{U}_{from}\,.\,\mathsf{to}\ tr{:}g_{tr}\ \mathsf{with}\ \mathtt{U}_{psng}\,.\,\mathsf{co}\ g_{st}\ \mathsf{with}\ \mathtt{U}_{psng})\end{aligned}$$

where $\mathtt{U}_w$ are the rights of passengers in the world. The traveller after exiting the train sends a co-move to the train using the ambient *checkOut*.

The initial configuration is:

$$\begin{aligned}(\nu stA, stB)(world{:}g_w(\mathtt{V}_w)[\,!\,\mathsf{co}\ g_{st}\ \mathsf{with}\ \mathtt{U}_w \mid S \mid \text{TRVLRS}(stA, stB)\\ \mid stA{:}g_{st}(\mathtt{U}_{st})[\text{TRAIN} \mid \text{STATP}] \mid stB{:}g_{st}(\mathtt{U}_{st})[\text{STATP}]\\ \mid \text{TRVLRS}(stB, stA)\,])\end{aligned}$$

where S is unknown (the world can be dangerous!) and TRVLRS(*src*, *dst*) is a parallel composition of TRAVELLER(*src*, *dst*) processes. The world accepts passengers from stations, since it contains $!\,\mathsf{co}\ g_{st}\ \mathsf{with}\ \mathtt{U}_w$.

A *bad* passenger willing to get off the train when this is not in a station, though it may be statically well-typed, is dynamically not allowed to do so. Suppose the bad passenger is represented by the process

$$\text{BADPSNG} \triangleq \mathsf{down}\ tr{:}g_{tr}\ \mathsf{with}\ \mathtt{U}_{bad}\,.\,\mathsf{up}\ world{:}g_w\ \mathsf{with}\ \mathtt{U}_w\,.\,\text{BP}$$

By assuming $\Sigma \vdash \text{BP} : \mathtt{U}_w$ one may derive the typing

$$\Sigma \vdash \mathsf{up}\ world{:}g_w\ \mathsf{with}\ \mathtt{U}_w\,.\,\text{BP} : \mathtt{U}_{bad} \quad \text{with} \quad \mathtt{U}_{bad} \triangleq \langle\!\langle \varnothing, \{g_w\} \rangle\!\rangle$$

Observe that the mobility type $\mathtt{U}_{bad}$ characterizes a process that, once boarded the train, wants to go from it directly into the world. From the above, we may infer the typing $\Sigma \vdash \text{BADPSNG} : \mathtt{U}_{st}$, since for that it is enough that $\mathtt{U}_{st}$ allows the process to get into the train, i.e., $g_{tr} \in \mathscr{E}(\mathtt{U}_{st})$.

The process BADPSNG is therefore statically allowed to stay within a station, as for example in the well-typed term $stA{:}g_{st}\,\mathtt{U}_{st}[\text{BADPSNG} \mid \text{TRAIN}]$. Nevertheless, when trying at runtime to get into the train, the process is blocked. As a matter of fact, for the action down $tr{:}g_{tr}$ with $\mathtt{U}_{bad}$ to fire, it is required that $\mathtt{U}_{bad} \leq \mathtt{U}_{psng}$, which is not the case since $\mathtt{U}_{bad} = \langle\!\langle \varnothing, \{g_w\} \rangle\!\rangle$ while $\mathtt{U}_{psng} = \langle\!\langle \varnothing, \{g_{st}\} \rangle\!\rangle$: the type $\mathtt{U}_{bad}$ allows going into the world while $\mathtt{U}_{psng}$ does not.

This should have been somehow expected, because in our calculus the dynamic checks, performed when co-moves are consumed, are assigned the very task of controlling that mobile processes either respect some given policies expressed through types, or are blocked. Notice that all the previous properties are guaranteed by exploiting in the operational semantics only information local to the involved processes.

A similar scenario has already been modelled in [9,19,14]. In the first two cases, the mobility control is implemented by informing the passenger when the train has reached the station at which he wants to get off. More specifically, in [9] a new primitive for *ambient renaming* is exploited. Intuitively, the train ambient takes a suitable name to implicitly inform the passengers when it has arrived at a certain station and to allow them to get in or off, while it takes a name unknown to passengers when it is moving (in this way passengers cannot get in or off the train). In [19], mobility policies are implemented through *guardians*, i.e., components attached to ambients for monitoring inner activities and interaction with the external environment. When the train arrives at a station, the attached guardian allows passengers to get in; in addition, the train generates a suitable ambient called *announcement* that informs the passengers of the arrival at a certain station and guides the passengers willing to get off. In [14], the mobility control is performed by exploiting dynamic checks to ensure that mobile processes willing to get in an ambient do respect some fixed policies expressed through types.

6 Type Inference

An inference algorithm for a typed calculus takes a *raw* term, i.e. a simplified form of term with no or only partial type annotations, and reconstructs an ordinary typed term along with a valid typing judgment for it. The first decision to be taken when designing an algorithm therefore concerns the syntax for raw terms.

In our calculus the naive approach of erasing all type annotations does not work, because in this way a typing cannot always be sensibly reconstructed; we have to leave in the term some type information, which – by taking part in the reduction rules – actually 'implements' some specified behaviour of the modelled system.

In particular, we have chosen to erase the type in the input binder and to eliminate as many mobcom annotations as reasonable (those in the with component of the prefixes up, down and to, and in the ambient construct); on the other hand, the group assigned to an ambient occurrence and the group and the mobility type within a co-move are kept, because these annotations define the policies and the mobility constraints established by the designer of the applica-

....		...
R ::= **raw processes**		
....		...
	$(x) \,.\, R$	untyped input
	down $\alpha{:}\gamma \,.\, R$	moves process R out from its ambient down to an enclosed ambient α of group γ
	up $\alpha{:}\gamma \,.\, R$	moves process R out from its ambient up to the enclosing ambient α of group γ
	to $\alpha{:}\gamma \,.\, R$	moves process R out from its ambient to a sibling ambient α of group γ
	$\alpha{:}\gamma[R]$	ambient
....		...

Fig. 8. Raw Processes

tion. The formal definition of raw terms is given in Fig. 8, where the missing parts are as in Fig. 1. We omit the obvious definition of the untyped version $|P|$ of a typed term P.

An analysis of the typing rules shows that they do not directly provide an algorithm for inferring the typing of a term, because of three distinct problems. The first is the implicit weakening present in the rule (ENV), coupled with the fact that the different premises of a single rule must share the same environment. This is well known: in such cases, a more algorithmic system can be obtained by delaying the application of weakening until the end of the inference process and by admitting different environments in different premises.

The second problem arises from the fact that there is no uniqueness of typing: in the rule (NULL), for instance, the type of 0 has no relationship with the term. The standard approach consists in the introduction of type variables; this alone, however, is not sufficient here, since in some rules the type variables occurring in the conclusion are limited in their range by conditions in the premises. For instance, in the rule (IN-I) (corresponding to rule (IN)) the variable u of the capability type scheme $\mathsf{cap}(u)$ that types the conclusion must satisfy the constraint of being greater than or equal to the mobility type $\langle\!\langle\{\gamma\}, \varnothing\rangle\!\rangle$. We cope with this difficulty by employing, jointly with type variables, the technique of delaying the solution by simply recording the constraints: these will then be solved at the very end (see e.g. Chapter 22 of [34]).

The third problem comes from having dependent types, i.e., possible occurrences of group variables and group names within type expressions: when a variable or name γ is being bound, as in rules (INPUT) and (GRP RES), one must check that such γ does not occur in the current environment and type. In the type inference this condition is expressed by adding *non-occurrence constraints*, i.e., constraints of the form $\gamma \notin \mathcal{I}$ where $\mathcal{I}$ is a set of group variables or group names. A substitution satisfies the constraint $\gamma \notin \mathcal{I}$ when, applied to all the variables in $\mathcal{I}$, returns types without occurrences of γ.

Fig. 9 defines *type schemes*, i.e. a type syntax augmented with type variables (denoted by lowercase letters): more specifically, we introduce three distinct sets

...	...	...
V ::=	(U, T)	*mobcom* (or process) type scheme
U ::=		*mobility* type scheme
	U	mobility type
	u	mobility type variable
T ::=		*communication* type scheme
	$\mathsf{com}(W)$	communication of messages of type scheme W
	shh	no communication
	t	communication type variable
W ::=		*message* type scheme
	$\mathsf{cap}(U)$	capability type scheme
	$\mathsf{amb}, \mathsf{group}$	ambient and group types
	w	message type variable
Θ ::=		*environment* scheme
	$\varnothing$	empty environment scheme
	$\Theta, x{:}\ W$	environment scheme containing the assumption $x{:}\ W$

Fig. 9. Type Schemes

of variables respectively for mobility types, message types and communication types. We use ξ to range over communication, message and mobility type variables and Ξ to range over communication, message and mobility type schemes.

We remark that types in raw terms can belong only to the original type syntax, while the inference algorithm can infer (and annotate terms with) type schemes. More precisely, the inference algorithm builds process schemes, environment schemes and sets of constraints, defined respectively as follows.

Definition 2. *1. A* process scheme *is defined by the same syntax as the one of a process in Fig. 1, except that all type decorations are type schemes instead of types.*
2. An environment scheme *is defined by the same syntax as the one of an environment in Fig. 2, except that all predicates are type schemes instead of types.*
3. A set of constraints *is a set whose elements may be equalities and inequalities (w.r.t. the relation $\leq$) between type schemes, and non-occurrence constraints.*

We use S to range over process schemes and $\mathcal{C}$ to range over sets of constraints.

The input of the algorithm is either a message M or a raw term R. In the first case the output consists of: an environment scheme Θ, the message M, a message type scheme W and a set of constraints $\mathcal{C}$. In the second case the output is given by: an environment scheme Θ, a process scheme S, a mobcom type scheme V and a set of constraints $\mathcal{C}$. We respectively use the notations:

$$M \Longrightarrow \langle \Theta \vdash_I M : W \,\|\, \mathcal{C} \rangle \qquad \text{and} \qquad R \Longrightarrow \langle \Theta \vdash_I S : V \,\|\, \mathcal{C} \rangle$$

It is easy to verify, by looking at the inference rules in Fig. 10, 11 and 12, that the elements of a set of constraints are of the form $\Xi = \Xi'$ or $U \leq u$ or $\gamma \notin \mathcal{I}$.

To relate type inference with type assignment, type variables must be replaced by types of the respective kinds. More precisely, a solution of a set of constraints is defined as follows.

Definition 3. *1. A* ground substitution *is a total mapping from mobility variables to mobility types, from message variables to message types and from communication variables to communication types.*

2. A solution of a set of constraints $\mathcal{C}$ *is a ground substitution* ς *such that*
  - *if* $\Xi = \Xi' \in \mathcal{C}$ *then* $\varsigma(\Xi) = \varsigma(\Xi')$*;*
  - *if* $U \leq u \in \mathcal{C}$ *then* $\varsigma(U) \leq \varsigma(u)$*;*
  - *if* $\gamma \notin \mathcal{I} \in \mathcal{C}$ *then* $\gamma \notin \varsigma(\xi)$ *for all* $\xi \in \mathcal{I}$.

As usual, the application of ς to Υ, written $\varsigma(\Upsilon)$, with $\Upsilon \in \{\xi, \Xi, U, \ldots\}$, denotes the term obtained by replacing all the variables ξ occurring in Υ with $\varsigma(\xi)$.

The handling of constraints will be explained later; suffice it now to say that we will present an algorithm solve which accepts as input a set of constraints $\mathcal{C}$ and either fails or outputs a solution, as stated in Proposition 4.

For the description of the inference rules some preliminary definitions are necessary. We need to eliminate from sets of typing assumptions those whose subjects are not variables: let then Ω be a set of typing assumptions whose subjects are names or variables: $\Omega\Downarrow$ is defined as the maximum subset containing only assumptions whose subjects are variables, i.e., $\Omega\Downarrow = \{x{:}\ W \mid x{:}\ W \in \Omega\}$.

In defining the combination of two environment schemes our choice is that whenever they respectively contain two (generally different) assumptions with the same subject, we take the assumption found in the leftmost environment and add to the set of constraints the equality between the two predicates. More precisely we define:

- the *combination* $\Theta \triangleright \Theta'$ of two environment schemes Θ and Θ':

$$\Theta \triangleright \Theta' = \{x{:}\ W \mid x{:}\ W \in \Theta\} \cup \{x{:}\ W \mid x \notin Dom(\Theta) \ \&\ x{:}\ W \in \Theta'\}$$

- the *set of constraints* $\Theta\Diamond\Theta'$ generated by the combination of two environment schemes Θ and Θ':

$$\Theta\Diamond\Theta' = \{W = W' \mid x{:}\ W \in \Theta \ \&\ x{:}\ W' \in \Theta'\}.$$

Clearly, combination might be defined with the roles of Θ and Θ' exchanged; in any case the constraints obtained would be the same, as is obvious from the symmetric nature of the operator $\Diamond$. It is also easy to verify that every solution of the set of constraints, when applied to the combination of the two originating environments, is the same as the union of its application to the environments themselves, as stated by the following proposition.

Proposition 2. *If* ς *is a solution of* $\Theta\Diamond\Theta'$ *then* $\varsigma(\Theta \triangleright \Theta') = \varsigma(\Theta) \cup \varsigma(\Theta')$.

The inference rules are given in Fig. 10, 11 and 12. Type variables that do not occur in the premises are assumed to be fresh; in this way, when different

$$\frac{n \in \mathscr{A}}{n \Longrightarrow \langle \varnothing \vdash_I n : \mathsf{amb} \parallel \varnothing \rangle} \text{(Amb Const-I)} \qquad \frac{g \in \mathscr{G}}{g \Longrightarrow \langle \varnothing \vdash_I g : \mathsf{group} \parallel \varnothing \rangle} \text{(Grp Const-I)}$$

$$\frac{}{x \Longrightarrow \langle \{x \colon w\} \vdash_I x : w' \parallel w \leq w' \rangle} \text{(Env-I)}$$

$$\frac{}{\mathsf{in}\ \alpha{:}\gamma \Longrightarrow \langle \{\alpha \colon \mathsf{amb}, \gamma \colon \mathsf{group}\}{\Downarrow} \vdash_I \mathsf{in}\ \alpha{:}\gamma : \mathsf{cap}(u) \parallel \{\langle\!\langle \{\gamma\}, \varnothing \rangle\!\rangle \leq u\} \rangle} \text{(In-I)}$$

$$\frac{}{\mathsf{out}\ \alpha{:}\gamma \Longrightarrow \langle \{\alpha \colon \mathsf{amb}, \gamma \colon \mathsf{group}\}{\Downarrow} \vdash_I \mathsf{out}\ \alpha{:}\gamma : \mathsf{cap}(u) \parallel \{\langle\!\langle \{\gamma\}, \varnothing \rangle\!\rangle \leq u\} \rangle} \text{(Out-I)}$$

$$\frac{}{\mathsf{co}\ \gamma\ \mathsf{with}\ \mathtt{U} \Longrightarrow \langle \{\gamma \colon \mathsf{group}\}{\Downarrow} \cup \{x \colon \mathsf{group} \mid x \in \mathtt{U}\} \vdash_I \mathsf{co}\ \gamma\ \mathsf{with} \mathtt{U} : \mathsf{cap}(u) \parallel \{\mathtt{U} \leq u\} \rangle} \text{(Co-I)}$$

$$\frac{\chi \Longrightarrow \langle \Theta \vdash_I \chi : W \parallel \mathcal{C} \rangle \qquad \chi' \Longrightarrow \langle \Theta' \vdash_I \chi' : W' \parallel \mathcal{C}' \rangle}{\chi \cdot \chi' \Longrightarrow \langle \Theta \triangleright \Theta' \vdash_I \chi . \chi' : \mathsf{cap}(u) \parallel \mathcal{C} \cup \mathcal{C}' \cup \{W = \mathsf{cap}(u), W' = \mathsf{cap}(u)\} \cup \Theta \Diamond \Theta' \rangle} \text{(Path-I)}$$

Fig. 10. Type Inference Rules for Messages

$$\frac{}{\mathsf{0} \Longrightarrow \langle \varnothing \vdash_I \mathsf{0} : (u, t) \parallel \varnothing \rangle} \text{(Null-I)}$$

$$\frac{\chi \Longrightarrow \langle \Theta \vdash_I \chi : W \parallel \mathcal{C} \rangle \qquad R \Longrightarrow \langle \Theta' \vdash_I S : (u, t) \parallel \mathcal{C}' \rangle}{\chi . R \Longrightarrow \langle \Theta \triangleright \Theta' \vdash_I \chi . S : (u, t) \parallel \mathcal{C} \cup \mathcal{C}' \cup \{W = \mathsf{cap}(u)\} \cup \Theta \Diamond \Theta' \rangle} \text{(Cap Prefix-I)}$$

$$\frac{R \Longrightarrow \langle \Theta \vdash_I S : V \parallel \mathcal{C} \rangle}{\mathsf{down}\ \alpha{:}\gamma . R \Longrightarrow \langle \{\alpha \colon \mathsf{amb}, \gamma \colon \mathsf{group}\}{\Downarrow} \triangleright \Theta \vdash_I \mathsf{down}\ \alpha{:}\gamma\ \mathsf{with}\ V . S : (u, t) \parallel \mathcal{C}' \rangle} \text{(Down-I)}$$
$$\text{where } \mathcal{C}' = \mathcal{C} \cup \{\langle\!\langle \varnothing, \{\gamma\} \rangle\!\rangle \leq u\} \cup \{\alpha \colon \mathsf{amb}, \gamma \colon \mathsf{group}\}{\Downarrow} \Diamond \Theta$$

$$\frac{R \Longrightarrow \langle \Theta \vdash_I S : V \parallel \mathcal{C} \rangle}{\mathsf{up}\ \alpha{:}\gamma . R \Longrightarrow \langle \{\alpha \colon \mathsf{amb}, \gamma \colon \mathsf{group}\}{\Downarrow} \triangleright \Theta \vdash_I \mathsf{up}\ \alpha{:}\gamma\ \mathsf{with}\ V . S : (u, t) \parallel \mathcal{C}' \rangle} \text{(Up-I)}$$
$$\text{where } \mathcal{C}' = \mathcal{C} \cup \{\langle\!\langle \varnothing, \{\gamma\} \rangle\!\rangle \leq u\} \cup \{\alpha \colon \mathsf{amb}, \gamma \colon \mathsf{group}\}{\Downarrow} \Diamond \Theta$$

$$\frac{R \Longrightarrow \langle \Theta \vdash_I S : V \parallel \mathcal{C} \rangle}{\mathsf{to}\ \alpha{:}\gamma . R \Longrightarrow \langle \{\alpha \colon \mathsf{amb}, \gamma \colon \mathsf{group}\}{\Downarrow} \triangleright \Theta \vdash_I \mathsf{to}\ \alpha{:}\gamma\ \mathsf{with}\ V . S : (u, t) \parallel \mathcal{C}' \rangle} \text{(To-I)}$$
$$\text{where } \mathcal{C}' = \mathcal{C} \cup \{\langle\!\langle \varnothing, \{\gamma\} \rangle\!\rangle \leq u\} \cup \{\alpha \colon \mathsf{amb}, \gamma \colon \mathsf{group}\}{\Downarrow} \Diamond \Theta$$

Fig. 11. Type Inference Rules for Raw Processes I

derivations are pasted together in a multiple-premise rule, their variables are all distinct.

Rules (Amb Const-I) and (Grp Const-I) are very similar to (Amb Const) and (Grp Const): the only difference is in the inferred environment, which is empty as is the constraint set. Similarly in (Env-I) the minimal environment is

$$\frac{R \Longrightarrow \langle \Theta \vdash_I S : (u,t) \parallel \mathcal{C} \rangle}{(x).R \Longrightarrow \langle \Theta{\downarrow}x \vdash_I (x{:}\,w).S : (u,t) \parallel \mathcal{C} \cup \{t = \mathsf{com}(w), x \notin \mathcal{I}\} \cup \{x{:}\,w\} \Diamond \Theta \rangle} \quad (\text{Input-I})$$
where $\mathcal{I} = \{w \mid w \in \Theta{\downarrow}x\} \cup \{u, t\}$

$$\frac{R \Longrightarrow \langle \Theta \vdash_I S : (u,t) \parallel \mathcal{C} \rangle \quad M \Longrightarrow \langle \Theta' \vdash_I M : W \parallel \mathcal{C}' \rangle}{\langle M \rangle .R \Longrightarrow \langle \Theta \triangleright \Theta' \vdash_I \langle M \rangle .S : (u,t) \parallel \mathcal{C} \cup \mathcal{C}' \cup \{t = \mathsf{com}(W)\} \cup \Theta \Diamond \Theta' \rangle} \quad (\text{Output-I})$$

$$\frac{R \Longrightarrow \langle \Theta \vdash_I S : V \parallel \mathcal{C} \rangle}{\alpha{:}\gamma[R] \Longrightarrow \langle \{\alpha{:}\,\mathsf{amb}, \gamma{:}\,\mathsf{group}\}{\Downarrow} \triangleright \Theta \vdash_I \alpha{:}\gamma V[S] : (u,t) \parallel \mathcal{C} \cup \{\alpha{:}\,\mathsf{amb}, \gamma{:}\,\mathsf{group}\}{\Downarrow} \Diamond \Theta \rangle} \quad (\text{Amb-I})$$

$$\frac{R \Longrightarrow \langle \Theta \vdash_I S : (u,t) \parallel \mathcal{C} \rangle \quad R' \Longrightarrow \langle \Theta' \vdash_I S' : (u',t') \parallel \mathcal{C}' \rangle}{R \mid R' \Longrightarrow \langle \Theta \triangleright \Theta' \vdash_I S \mid S' : (u,t) \parallel \mathcal{C} \cup \mathcal{C}' \cup \{u = u', t = t'\} \cup \Theta \Diamond \Theta' \rangle} \quad (\text{Par-I})$$

$$\frac{R \Longrightarrow \langle \Theta \vdash_I S : V \parallel \mathcal{C} \rangle}{!\,R \Longrightarrow \langle \Theta \vdash_I !\,S : V \parallel \mathcal{C} \rangle} \quad (\text{Repl-I}) \qquad \frac{R \Longrightarrow \langle \Theta \vdash_I S : V \parallel \mathcal{C} \rangle}{(\nu n)R \Longrightarrow \langle \Theta \vdash_I (\nu n)S : V \parallel \mathcal{C} \rangle} \quad (\text{Amb Res-I})$$

$$\frac{R \Longrightarrow \langle \Theta \vdash_I S : V \parallel \mathcal{C} \rangle}{(\nu g)R \Longrightarrow \langle \Theta \vdash_I (\nu g)S : V \parallel \mathcal{C} \cup g \notin \{\xi \mid \xi \in \Theta \vee \xi \in V\} \rangle} \quad (\text{Grp Res-I})$$

Fig. 12. Type Inference Rules for Raw Processes II

inferred; the inferred type scheme is a fresh message type variable which must be greater than or equal to the message type variable assumed for the term variable in the environment.

Rules (In-I) and (Out-I) are the first interesting cases, because they show how the typing rules are adapted to the inference framework. The condition $\gamma \in \mathscr{C}$ in rules (In) and (Out) becomes the constraint $\langle\!\langle \{\gamma\}, \varnothing \rangle\!\rangle \leq u$, where u is the fresh mobility type variable such that $\mathsf{cap}(u)$ is the type scheme inferred for the capability. The assumptions $\alpha{:}\,\mathsf{amb}$ and $\gamma{:}\,\mathsf{group}$ are added to the environment scheme only if α, γ are variables. To avoid writing several different rules we build the set $\{\alpha{:}\,\mathsf{amb}, \gamma{:}\,\mathsf{group}\}$ and then obtain an environment scheme by filtering it through the operator $\Downarrow$.

In rule (Co-I) we add to the environment (scheme) all the variables that occur in the mobility type $\mathtt{U}$ and we assume for them the type group. The resulting mobility type scheme is $\mathsf{cap}(u)$, where u is a fresh variable constrained to be greater than or equal to $\mathtt{U}$.

In rule (Path-I) it is easy to verify that W and W' can be either message type variables or capability type schemes containing only mobility variables. Besides, W (respectively W') is a message type variable iff χ (respectively χ') is a variable. In both cases it is necessary that the mobility rights of the path are equal to those of the components. This is accomplished by requiring that $W = W' = \mathsf{cap}(u)$, where the fresh variable u represents the mobility rights of χ, χ' and $\chi.\chi'$. We keep track of the previous constraints $\mathcal{C}$ and $\mathcal{C}'$ by adding them to the new constraint set. In addition, the environment schemes of the premises are combined using the operator $\triangleright$ and the equalities generated by the application of $\Diamond$ to these environments are added to the set of constraints.

For the $\mathsf{0}$ process we derive the parametric mobcom type scheme (u, t) from the empty environment and we do not require any constraint.

Rule (Cap Prefix-I) is similar to rule (Path-I): the message type scheme variable representing the mobility rights of the capability must be equal to $\mathsf{cap}(u)$, where the mobility variable u represents the mobility rights of the whole process.

In the rules for the process-moving prefixes (Down-I)–(To-I) we decorate the prefix (after the keyword with) with the mobcom type scheme of the process to be sent. For the type scheme of the whole process we only require that it expresses the right to send a process to an ambient of group γ: this is ensured by the constraint $\langle\!\langle \varnothing, \{\gamma\} \rangle\!\rangle \leq u$. As in rules (In-I) and (Out-I), we add to the environment scheme the premises $\alpha{:}\, \mathsf{amb}$ and $\gamma{:}\, \mathsf{group}$ only if α, γ are variables.

In rule (Input-I) the communication type scheme t is equated with $\mathsf{com}(w)$ where w is a fresh message variable, the input binder is annotated with w, and x is removed from the environment. In fact $\Theta{\downarrow}x$ is defined as $\{y{:}\, W \mid y{:}\, W \in \Theta \;\&\; y \neq x\}$. We add to the set of constraints the equality (if any) generated by combining the assumption $x{:}\, w$ with the environment scheme of the premise. In order to take into account the conditions $x \notin \Sigma$ and $x \notin (\mathtt{U}, \mathsf{com}(\mathtt{W}))$ of rule (Input) we add the constraint $x \notin \{w \mid w \in \Theta{\downarrow}x\} \cup \{u, t\}$.

Rule (Output-I) is similar but simpler: we equate the communication type scheme t of the process scheme with $\mathsf{com}(W)$ where W is the message type scheme of the output. We deal with the environment schemes and the constraints of the premises as in rule (Cap Prefix-I).

In rule (Amb-I) the mobcom type scheme of the ambient is the mobcom type scheme of the enclosed process scheme and the mobcom type scheme of the whole process scheme is just (u, t) with u and t fresh.

Rule (Par-I) needs to ensure that the two process schemes have the same mobcom type scheme: this is achieved by adding to the constraint set the equality between the mobility variables and the communication type schemes of these process schemes.

Finally, in rule (Grp Res-I) we need to ensure that g will not appear in the environment or in the process type: this is accomplished by adding the constraint $g \notin \{\xi \mid \xi \in \Theta \vee \xi \in V\}$.

The following properties of the inference rules can be easily proved by inspection of the rules themselves.

Proposition 3. *Let $R \Longrightarrow \langle \Theta \vdash_I S : V \parallel \mathcal{C} \rangle$. Then:*

1. *$x{:}\, W \in \Theta$ implies $W \in \{w, \mathsf{amb}, \mathsf{group}\}$ for some w;*
2. *$V = (u, t)$ for some u, t;*
3. *$\mathcal{C} = \mathcal{C}_= \cup \mathcal{C}_\leq \cup \mathcal{C}_\notin$ where:*
   (a) *$\mathcal{C}_=$ is a set of equalities of the following forms (equalities symmetric of the listed ones are omitted):*

$$\begin{array}{lllll} \mathsf{amb} = \mathsf{amb} & \mathsf{group} = \mathsf{group} & \mathsf{amb} = \mathsf{group} & \mathsf{amb} = \mathsf{cap}(u) \\ \mathsf{group} = \mathsf{cap}(u) & t = t' & w = w' & u = u' \end{array}$$

$$t = \mathsf{com}(\mathsf{amb}) \quad t = \mathsf{com}(\mathsf{group}) \quad t = \mathsf{com}(w) \quad t = \mathsf{com}(\mathsf{cap}(u))$$
$$w = \mathsf{amb} \quad w = \mathsf{group} \quad w = \mathsf{cap}(u) \quad \mathsf{cap}(u) = \mathsf{cap}(u')$$

(b) $\mathcal{C}_{\leq}$ *is a set of subtyping judgments of the form* $w \leq w'$ *and* $\mathtt{U} \leq u$*;*
(c) $\mathcal{C}_{\not\in}$ *is a set of constraints of the form* $\gamma \not\in \{w_1, \ldots, w_k, u, t\}$ *where* $k \geq 0$*.*

We are now going to describe a procedure solve which, applied to a set of constraints, checks if it is solvable. If the set of constraints is solvable solve generates a particular solution of it, otherwise fails. The procedure makes use of three sub-procedures, executed in sequence: $\mathsf{solve}_=$, which solves the equality constraints; $\mathsf{solve}_{\leq}$, which solves the inequalities; and $\mathsf{solve}_{\not\in}$, which checks the non-occurrence conditions. Any of the three may fail; in this case the set of constraints has no solution, the algorithm stops and the given term is not typable. If all succeed, then solve provides a solution.

The procedure $\mathsf{solve}_=$ applies the standard unification algorithm to $\mathcal{C}_=$. By Proposition 3(3a), $\mathcal{C}_=$ is a set of equations, therefore $\mathsf{solve}_=$ either fails or returns a substitution $\sigma_=$ which is a most general unifier. It is easy to verify that $\sigma_=$ maps communication type variables to communication type schemes different from shh, message type variables to message type schemes and mobility type variables to mobility type variables.

If $\mathcal{C}_=$ is solvable, then the main procedure solve applies the substitution $\sigma_=$ to both sides of the subtyping judgments in $\mathcal{C}_{\leq}$, thus obtaining a set of subtyping judgments $\mathcal{C}^1_{\leq}$. Since the subtyping judgments in $\mathcal{C}_{\leq}$ are of the forms $\mathtt{U} \leq u$ and $w \leq w'$, then $\mathcal{C}^1_{\leq}$ contains only judgments of the forms $\mathtt{U} \leq u$ and $W \leq W'$, where W and W' are either amb, or group, or capability type schemes $\mathsf{cap}(u)$, or message type variables w.

The procedure $\mathsf{solve}_{\leq}$ starts by defining a substitution σ_W from message type variables that occur in $\mathcal{C}^1_{\leq}$ to message type schemes. To this end we build the transitive closure of the relation $\leq$ in $\mathcal{C}^1_{\leq}$, i.e., we add $W_1 \leq W_3$ to $\mathcal{C}^1_{\leq}$ whenever $W_1 \leq W_2$ and $W_2 \leq W_3$ are in it, for some W_2. Let $\mathcal{C}^2_{\leq}$ be the resulting set of inequalities.

Assume initially σ_W as the trivial identity substitution and repeat the following steps (which transform both σ_W and $\mathcal{C}^2_{\leq}$) until possible:

1. for all w such that $\mathsf{amb} \leq w$ or $w \leq \mathsf{amb}$ set $\sigma_W(w) = \mathsf{amb}$ and replace w with amb in $\mathcal{C}^2_{\leq}$;
2. for all w such that $\mathsf{group} \leq w$ or $w \leq \mathsf{group}$ set $\sigma_W(w) = \mathsf{group}$ and replace w with group in $\mathcal{C}^2_{\leq}$;
3. for all w such that $\mathsf{cap}(u') \leq w$ or $w \leq \mathsf{cap}(u')$ set $\sigma_W(w) = \mathsf{cap}(u)$, where u is fresh, and replace w with $\mathsf{cap}(u)$ in $\mathcal{C}^2_{\leq}$.

At the end, define $\sigma_W(w) = \mathsf{group}$ for all remaining message type variables w and replace them accordingly in $\mathcal{C}^2_{\leq}$. Let $\mathcal{C}^3_{\leq}$ be the so obtained set of constraints.

For instance, if $\mathcal{C}^2_{\leq} = \{w_1 \leq w_2, w_1 \leq \mathsf{group}\}$, after the first iteration we get $\sigma_W = id\{w_1 := \mathsf{group}\}$ (where id is the identity substitution) and $\mathcal{C}^2_{\leq} = \{\mathsf{group} \leq w_2, \mathsf{group} \leq \mathsf{group}\}$ and after the second iteration $\sigma_W = id\{w_1 := \mathsf{group}, w_2 :=$

group} and $\mathcal{C}^2_{\leq} = \mathcal{C}^3_{\leq} = \{\mathsf{group} \leq \mathsf{group}\}$. Note the need, in defining σ_{W}, to iterate the previous steps more than once.

If $\mathcal{C}^3_{\leq}$ is inconsistent (i.e., if it contains inequalities involving amb and group, or amb and $\mathsf{cap}(u)$, or $\mathsf{cap}(u)$ and group) then $\mathsf{solve}_{\leq}$ fails. Otherwise the non-trivial subtyping judgments in $\mathcal{C}^3_{\leq}$ are of the forms $\mathsf{cap}(u) \leq \mathsf{cap}(u')$ and $\mathtt{U} \leq u$. The procedure $\mathsf{solve}_{\leq}$ applies the following transformations to $\mathcal{C}^3_{\leq}$:

- replace $\langle\!\langle \mathscr{C}, \mathscr{E} \rangle\!\rangle \leq u$ and $\langle\!\langle \mathscr{C}', \mathscr{E}' \rangle\!\rangle \leq u$ with $\langle\!\langle \mathscr{C} \cup \mathscr{C}', \mathscr{E} \cup \mathscr{E}' \rangle\!\rangle \leq u$
- add $\mathtt{U} \leq u$ whenever $\mathsf{cap}(u') \leq \mathsf{cap}(u)$ and $\mathtt{U} \leq u'$

until a fixed point $\mathcal{C}^4_{\leq}$ is reached. Now define for all mobility variables u:

$$\sigma_{\mathrm{U}}(u) = \begin{cases} \mathtt{U} & \text{if } \mathtt{U} \leq u \in \mathcal{C}^4_{\leq}, \\ \langle\!\langle \varnothing, \varnothing \rangle\!\rangle & \text{otherwise.} \end{cases}$$

Finally, the main procedure solve calls $\mathsf{solve}_{\not\in}$, which checks the satisfiability of the constraints in $\mathcal{C}_{\not\in}$. If there are γ and ξ such that the constraint $\gamma \notin \mathcal{I}$ is in $\mathcal{C}_{\not\in}$, and $\xi \in \mathcal{I}$ and $\gamma \in \sigma_{\mathrm{U}} \circ \sigma_{\mathrm{W}} \circ \sigma_{=}(\xi)$, then the result of $\mathsf{solve}_{\not\in}$ is failure; otherwise $\mathsf{solve}_{\not\in}$ returns success.

For the output of solve to be a ground substitution, the remaining communication type variables must be replaced by types; then, only in the case $\mathsf{solve}_{\not\in}$ returned success, define:

$$\sigma_{\mathrm{T}}(t) = \mathsf{shh}$$

The output of the whole procedure solve is finally defined as

$$\mathsf{solve}(\mathcal{C}) = \begin{cases} \sigma_{\mathrm{T}} \circ \sigma_{\mathrm{U}} \circ \sigma_{\mathrm{W}} \circ \sigma_{=} & \text{if all substitutions are defined,} \\ \textit{failure} & \text{otherwise.} \end{cases}$$

The proof of the following proposition is straightforward:

Proposition 4. *If the set of constraints $\mathcal{C}$ is solvable then* $\mathsf{solve}(\mathcal{C})$ *is a solution of $\mathcal{C}$. Vice versa, if $\mathcal{C}$ is not solvable then* $\mathsf{solve}(\mathcal{C})$ *fails.*

In the rest of this section we will prove soundness and completeness of our inference algorithm. Crucial here is the treatment of free and bound variables. While the type assignment system considers processes modulo renaming of bound variables and names, the inference procedure cannot, since bound variables and names may appear in the set of constraints. For instance, we get

$$(x)\,.\,\mathsf{to}\ n{:}g\,.\,\mathsf{co}\ g'\ \mathsf{with}\ \langle\!\langle \{x\}, \varnothing \rangle\!\rangle\,.\,0 \Longrightarrow \langle\ \vdash_I S_0 : (u,t) \parallel \mathcal{C}_0 \rangle$$

where

$$S_0 = (x{:}\,w)\,.\,\mathsf{to}\ n{:}g\ \mathsf{with}\ (u',t')\,.\,\mathsf{co}\ g'\ \mathsf{with}\ \langle\!\langle \{x\}, \varnothing \rangle\!\rangle\,.\,0$$

and

$$\mathcal{C}_0 = \{\,\langle\!\langle \{x\}, \varnothing \rangle\!\rangle \leq u'', \mathsf{cap}(u'') = \mathsf{cap}(u'), \langle\!\langle \varnothing, \{g\} \rangle\!\rangle \leq u, t = \mathsf{com}(w), w = \mathsf{group}, \\ x \notin \{u,t\}\}.$$

Note that $\mathcal{C}_0$ contains constraints involving the bound variable x. The substitution $\varsigma_0 = \mathsf{solve}(\mathcal{C}_0)$ for the current variables is defined by $\varsigma_0(u'') = \varsigma_0(u') = \langle\!\langle \{x\}, \varnothing \rangle\!\rangle$, $\varsigma_0(u) = \langle\!\langle \varnothing, \{g\} \rangle\!\rangle$, $\varsigma_0(t) = \mathsf{com}(\mathsf{group})$, $\varsigma_0(w) = \mathsf{group}$, $\varsigma_0(t') = \mathsf{shh}$. As will follow from Theorem 3, applying ς_0 to the inferred process scheme S_0 gives a good process; however, to be able to apply ς_0, we are forced to introduce an occurrence of x in the scope of the binder, because u' must be replaced with $\langle\!\langle \{x\}, \varnothing \rangle\!\rangle$. We conclude that:

> when we apply a substitution to a process scheme in the scope of a binder, we need to allow capturing of group variables and names.

Therefore, following the 'nomenclature' of [16], we say that type variables are *replaced* by types.

As regards soundness, note that if the statement $M \Longrightarrow \langle \Theta \vdash_I M : W \parallel \mathcal{C} \rangle$ is derivable from the type inference rules and ς is a solution of $\mathcal{C}$, then $\varsigma(\Theta)$ may not be a good environment because the mobility types appearing in it might contain variables which either are assigned a type different from group or are not in its domain. To obtain a deducible typing judgment we therefore have to require that the enlarged environment $\varsigma(\Theta) \cup \Sigma$ is a good environment, where the domain of Σ contains the missing group variables. The case of statements of the form $R \Longrightarrow \langle \Theta \vdash_I S : V \parallel \mathcal{C} \rangle$ is analogous.

Theorem 3 (Soundness of Inference).

1. *If $M \Longrightarrow \langle \Theta \vdash_I M : W \parallel \mathcal{C} \rangle$ holds and ς is a solution of $\mathcal{C}$ such that $\varsigma(\Theta) \cup \Sigma$ is good, with $\Sigma = \{x{:}\,\mathsf{group} \mid \exists \xi \,.\, (\xi \in \Theta \vee \xi \in W) \wedge x \in \varsigma(\xi)\}$, then the typing $\varsigma(\Theta) \cup \Sigma \vdash \varsigma(M) : \varsigma(W)$ is derivable.*
2. *If $R \Longrightarrow \langle \Theta \vdash_I S : V \parallel \mathcal{C} \rangle$ holds, then $|S| = R$; also, if ς is a solution of $\mathcal{C}$ such that $\varsigma(\Theta) \cup \Sigma$ is good, with $\Sigma = \{x{:}\,\mathsf{group} \mid \exists \xi \,.\, (\xi \in \Theta \vee \xi \in S \vee \xi \in V) \wedge x \in \varsigma(\xi)\}$, then the typing $\varsigma(\Theta) \cup \Sigma \vdash \varsigma(S) : \varsigma(V)$ is derivable.*

Proof. The proofs of (1) and (2) are respectively by induction on the derivations of $M \Longrightarrow \langle \Theta \vdash_I M : W \parallel \mathcal{C} \rangle$ and $R \Longrightarrow \langle \Theta \vdash_I S : V \parallel \mathcal{C} \rangle$. We only consider two representative cases.

Let the last rule applied be (Cap Prefix-I):

$$\frac{\chi \Longrightarrow \langle \Theta \vdash_I \chi : W \parallel \mathcal{C} \rangle \qquad R \Longrightarrow \langle \Theta' \vdash_I S : (u,t) \parallel \mathcal{C}' \rangle}{\chi \,.\, R \Longrightarrow \langle \Theta \triangleright \Theta' \vdash_I \chi \,.\, S : (u,t) \parallel \mathcal{C} \cup \mathcal{C}' \cup \{W = \mathsf{cap}(u)\} \cup \Theta \Diamond \Theta' \rangle}$$

The theorem's first conclusion $|\chi \,.\, S| = \chi \,.\, R$ is immediate since $|\chi \,.\, S| = \chi \,.\, |S|$ and by induction $|S| = R$. As for the second and more important conclusion, first of all observe that a solution ς of the set of constraints $\mathcal{C} \cup \mathcal{C}' \cup \{W = \mathsf{cap}(u)\} \cup \Theta \Diamond \Theta'$ is also a solution of $\mathcal{C}$, $\mathcal{C}'$, and $\Theta \Diamond \Theta'$. A first consequence is that by Proposition 2 one has $\varsigma(\Theta \triangleright \Theta') = \varsigma(\Theta) \cup \varsigma(\Theta')$. Now put

$$\begin{aligned}
\Sigma &= \{x{:}\,\mathsf{group} \mid \exists \xi \,.\, (\xi \in \Theta \vee \xi \in W) \wedge x \in \varsigma(\xi)\};\\
\Sigma' &= \{x{:}\,\mathsf{group} \mid \exists \xi \,.\, (\xi \in \Theta' \vee \xi \in S \vee \xi \in (u,t)) \wedge x \in \varsigma(\xi)\};\\
\Sigma'' &= \{x{:}\,\mathsf{group} \mid \exists \xi \,.\, (\xi \in (\Theta \triangleright \Theta') \vee \xi \in (\chi \,.\, S) \vee \xi \in (u,t)) \wedge x \in \varsigma(\xi)\}.
\end{aligned}$$

The constraint $W = \mathsf{cap}(u)$ implies $\varsigma(W) = \varsigma(\mathsf{cap}(u))$ and then $\Sigma'' = \Sigma \cup \Sigma'$. If we assume that $\Sigma'' \cup (\Theta \triangleright \Theta')$ is a good environment, then so are $\Sigma \cup \Theta$ and $\Sigma' \cup \Theta'$; then by (1) we get $\varsigma(\Theta) \cup \Sigma \vdash \varsigma(\chi) \colon \varsigma(W)$, and by induction on (2) we get $\varsigma(\Theta') \cup \Sigma' \vdash \varsigma(S) \colon \varsigma((u,t))$.

By weakening (Lemma 4(3)) we have $\varsigma(\Theta \triangleright \Theta') \cup \Sigma'' \vdash \varsigma(\chi) \colon \varsigma(W)$ and also $\varsigma(\Theta \triangleright \Theta') \cup \Sigma'' \vdash \varsigma(S) \colon \varsigma((u,t))$. Being $\varsigma(W) = \varsigma(\mathsf{cap}(u))$ we can conclude by applying the rule (Cap Prefix).

Consider now the case where the last rule applied is (Input-I):

$$\frac{R \Longrightarrow \langle \Theta \vdash_{\mathcal{I}} S : (u,t) \parallel \mathcal{C} \rangle}{(x) \,.\, R \Longrightarrow \langle \Theta{\downarrow}x \vdash_{\mathcal{I}} (x \colon w) \,.\, S : (u,t) \parallel \mathcal{C} \cup \{t = \mathsf{com}(w), x \notin \mathcal{I}\} \cup \{x \colon w\} \Diamond \Theta \rangle}$$

where $\mathcal{I} = \{w \mid w \in \Theta{\downarrow}x\} \cup \{u,t\}$. Since by hypothesis ς solves $x \notin \mathcal{I}$, we get $x \notin \varsigma(\Theta{\downarrow}x)$ and $x \notin \varsigma((u,t))$. As in the previous case, by induction, by Proposition 2 and by Lemma 4(3) we get $\varsigma(\{x \colon w\} \triangleright \Theta) \cup \Sigma \vdash \varsigma(S) \colon \varsigma((u,t))$, with Σ properly defined following the type inference rule. By definition $\{x \colon w\} \triangleright \Theta = \{x \colon w\} \cup \Theta{\downarrow}x$ and the constraint $\{t = \mathsf{com}(w)\}$ implies $\varsigma(t) = \varsigma(\mathsf{com}(w))$. So we conclude by applying the rule (Input). □

It is not surprising that the output of solve turns out to produce a derivable statement:

Corollary 2. *If $R \Longrightarrow \langle \Theta \vdash_{\mathcal{I}} S : V \parallel \mathcal{C} \rangle$ and $\varsigma = \mathsf{solve}(\mathcal{C})$, then $\varsigma(\Theta) \vdash \varsigma(S) : \varsigma(V)$ is derivable.*

Proof. By inspection of the type inference rules and of the solve procedure it is easy to check that if x occurs in the range of $\mathsf{solve}(\mathcal{C})$ then R contains a co-move $\mathsf{co}\,\gamma\,\mathsf{with}\ \mathtt{U}$ for some γ, $\mathtt{U}$ such that $x \in \mathtt{U}$. By rule (Co-I) the assumption $x \colon \mathsf{group}$ is in Θ; therefore $\{x \colon \mathsf{group} \mid \exists \xi \,.\, (\xi \in \Theta \vee \xi \in S \vee \xi \in V) \wedge x \in \varsigma(\xi)\}$ is a subset of Θ, and we can conclude by Theorem 3. □

We can state and prove completeness as expected.

Theorem 4 (Completeness of Inference).

1. *If $\Sigma \vdash M : \mathtt{W}$, then $M \Longrightarrow \langle \Theta \vdash_{\mathcal{I}} M : W \parallel \mathcal{C} \rangle$ and there is a solution ς of $\mathcal{C}$ such that $\varsigma(\Theta) \subseteq \Sigma$ and $\varsigma(W) = \mathtt{W}$.*
2. *If $\Sigma \vdash P : \mathtt{V}$, then $|P| \Longrightarrow \langle \Theta \vdash_{\mathcal{I}} S : V \parallel \mathcal{C} \rangle$ and there is a solution ς of $\mathcal{C}$ such that $\varsigma(\Theta) \subseteq \Sigma$ and $\varsigma(S) = P$ and $\varsigma(V) = \mathtt{V}$.*

Proof. The proofs of (1) and (2) are respectively by induction on the derivations of $\Sigma \vdash M : \mathtt{W}$ and $\Sigma \vdash P : \mathtt{V}$. We only consider some representative cases.

If the last rule applied is (Cap Prefix):

$$\frac{\Sigma \vdash \chi : \mathsf{cap}(\mathtt{U}) \qquad \Sigma \vdash P : (\mathtt{U},\mathtt{T})}{\Sigma \vdash \chi.P : (\mathtt{U},\mathtt{T})}$$

then

- by (1), $\chi \Longrightarrow \langle \Theta \vdash_I \chi : W \parallel \mathcal{C} \rangle$ and there is a solution ς_1 of $\mathcal{C}$ such that $\varsigma_1(\Theta) \subseteq \Sigma$ and $\varsigma_1(W) = \mathsf{cap}(\mathtt{U})$;
- by induction on (2), $|P| \Longrightarrow \langle \Theta' \vdash_I S : (u,t) \parallel \mathcal{C}' \rangle$ and there is a solution ς_2 of $\mathcal{C}'$ such that $\varsigma_2(\Theta') \subseteq \Sigma$ and $\varsigma_2(S) = P$ and $\varsigma_2((u,t)) = (\mathtt{U}, \mathtt{T})$.

By rule (Cap Prefix-I) we get:

$$\frac{\chi \Longrightarrow \langle \Theta \vdash_I \chi : W \parallel \mathcal{C} \rangle \quad |P| \Longrightarrow \langle \Theta' \vdash_I S : (u,t) \parallel \mathcal{C}' \rangle}{\chi \,.\, |P| \Longrightarrow \langle \Theta \triangleright \Theta' \vdash_I \chi \,.\, S : (u,t) \parallel \mathcal{C} \cup \mathcal{C}' \cup \{W = \mathsf{cap}(u)\} \cup \Theta \Diamond \Theta' \rangle}$$

We can assume that the sets of type variables which occur free in $\mathcal{C}$ and $\mathcal{C}'$ are disjoint and define for all type variables ξ:

$$\varsigma(\xi) = \begin{cases} \varsigma_1(\xi) & \text{if } \xi \in \mathcal{C}, \\ \varsigma_2(\xi) & \text{otherwise.} \end{cases}$$

By construction ς is a solution of both $\mathcal{C}$ and $\mathcal{C}'$. Moreover, since $\varsigma(W) = \mathsf{cap}(\mathtt{U})$ and $\varsigma((u,t)) = (\mathtt{U}, \mathtt{T})$, the substitution ς solves the constraint $W = \mathsf{cap}(u)$. Lastly, $\varsigma(\Theta) \subseteq \Sigma$ and $\varsigma(\Theta') \subseteq \Sigma$ imply that for all term variables x if $x\colon w \in \Theta$ and $x\colon w' \in \Theta'$ then $\varsigma(w) = \varsigma(w')$, i.e., ς is also a solution of $\Theta \Diamond \Theta'$. By Proposition 2 this implies $\varsigma(\Theta \triangleright \Theta') = \varsigma(\Theta) \cup \varsigma(\Theta')$ and then we get $\varsigma(\Theta \triangleright \Theta') \subseteq \Sigma$.

If the last rule applied is (Input):

$$\frac{\Sigma, x\colon \mathtt{W} \vdash P : (\mathtt{U}, \mathsf{com}(\mathtt{W})) \quad x \notin \Sigma \quad x \notin (\mathtt{U}, \mathsf{com}(\mathtt{W}))}{\Sigma \vdash (x\colon \mathtt{W}) \,.\, P : (\mathtt{U}, \mathsf{com}(\mathtt{W}))}$$

then by induction $|P| \Longrightarrow \langle \Theta \vdash_I S : (u,t) \parallel \mathcal{C} \rangle$ and there is a solution ς of $\mathcal{C}$ such that $\varsigma(\Theta) \subseteq \Sigma, x\colon \mathtt{W}$ and $\varsigma(S) = P$ and $\varsigma((u,t)) = (\mathtt{U}, \mathsf{com}(\mathtt{W}))$. Since $\varsigma(t) = \mathsf{com}(\mathtt{W})$ and $\varsigma(\Theta) \subseteq \Sigma, x\colon \mathtt{W}$, the substitution ς is also a solution of $\{t = \mathsf{com}(w)\} \cup \{x\colon w\}\Diamond\Theta$. The condition $x \notin \Sigma$ implies $x \notin \varsigma(\Theta{\downarrow}x)$; moreover $x \notin (\mathtt{U}, \mathsf{com}(\mathtt{W}))$. Therefore ς satisfies also the constraint $x \notin \{w \mid w \in \Theta{\downarrow}x\} \cup \{u, t\}$. □

7 Conclusions and Related Work

We have introduced a variant of the Calculus of Mobile Ambients (MA) that combines ambient and process mobility and allows the expression of flexible policies for controlling process activities. The calculus exploits co-move actions and runtime type checking to require the agreement between a moving process and the target ambient (similar mechanisms could also be used for ambient movements, but we have omitted them for the sake of simplicity). Policies can dynamically change due to further co-moves being added to an ambient, either by means of process movements or indirectly through communication. The operational semantics and the type assignment system ensure that an incoming process conforms to the policy of the target ambient. The compliance with ambient policies can be checked locally and requires no global assumption. We have defined a sound and complete type inference algorithm and illustrated a few applications of our framework to examples.

As future work, we are considering ways of increasing the expressive power of our type system so as to be able to express stronger properties. For instance, incoming processes are only checked against mobility capabilities. No check about their input/output behaviour is currently done, except that their communication type must comply with the communication type of the entered ambient. Instead, one could need a more strict control as, for example, in the scenario of Section 2 where reader processes entering a journal could be allowed only to read (i.e., input) papers, not to write (i.e., output) them. This would require distinguishing between input and output, as is usually done in calculi with channel-based communication (see, e.g., [35]). One could also obtain more informative policies by asking that the operation of ambient creation be subject to authorization. This would be especially significant when creating ambients of known groups, in which case, in general, one may expect they behave in a controlled way.

Related Work

Modelling wide-area distributed systems requires that the space of locations and the mobility in such space are taken into account as new dimensions of computing. Most foundational languages proposed in the literature model this space either as an evolving graph of fully connected locations, like the Dπ [27] and the language KLAIM [17], or as an evolving forest of trees of locations, like MA [12] and its variants. Some recent proposals explicitly handling the underlying network topology are tKLAIM [18] and Dπ_F [20]. An interesting *core model* generalizing many of the available calculi and languages has been developed within the Mikado project [5].

Many variants of MA have been defined: for a survey see [21]. A crucial choice in all these calculi is the form of *interaction* between processes in different ambients. In the original calculus [12] interaction is only local to an ambient, and for processes in different ambients to communicate, at least one of the ambients' boundaries has to be dissolved by means of the open capability. This approach has also been used in [12,29,6,31,1]. However, the open capability has been considered by many researchers as potentially dangerous, because it could be inadvertently or maliciously used to destroy an ambient's individuality (by dissolving its boundary). Therefore, several variants of MA have been proposed which either are equipped with additional constructs for controlling the execution of open, like the co-moves of Safe Ambients [29] (used with modifications also in [6,31,32,8]), or replace it with other interaction-enabling mechanisms: among them, we mention communication between nested ambients in Boxed Ambients [7,32,8] and in Seal calculus [13], and process (*objective*) mobility in [15,14]. To ensure and enforce behavioural properties, in particular those concerning resource access, communication, mobility and security, ambient calculi are usually typed [10,29,11,6,1,7,31,32,8,30].

The calculus presented in this paper is derived from the variant of $\mathbf{M}^3$ presented in [14]. However, the two calculi use different authorization mechanisms.

In [14], rights to cross or enter an ambient are recorded as passive components (i.e., multisets of rights attached to the ambient); here, authorization relies on co-moves and rights, which allow more flexible policies. Also the mechanisms used to pass permits are different: in [14] specific primitives are used to add permissions to the multisets attached to ambients; the mechanism used in this paper, on the contrary, has been somehow inspired by [24], where policies to access network resources can dynamically change due to communication of permissions.

Before the present calculus, only the calculi of [26,28,6,24,4,23], to our knowledge, considered *type information local to computational environments*, while in the other proposals there is a global environment containing all typing assumptions. In [26,24,28,23] local type information is sufficient because processes are dynamically checked whenever they migrate, which prevents processes not complying with the policies of a locality to get in. This is similar to our approach, though their computational environments are not hierarchically structured. To reduce the amount of dynamic controls, in [28] a relation of trust among nodes is exploited; thus, a process coming from a trusted node is never dynamically type-checked. In [23] each location comes equipped with a *membrane* that controls access by type-checking the incoming processes. The presence of the open capability requires in [6] a careful updating of the local type information of ambients when they migrate. The aim of types in [4] is dual to ours: the type system ensures a *liberal but safe communication policy*, so that ambient movements are only allowed when this does not break the soundness of data exchanges. Also, the calculus of [4] is a variant of Boxed Ambients and therefore communication may cross one ambient boundary.

Dependent types have been widely used in the framework of the calculus Dπ (see, e.g., [25]) to restrict capabilities of processes launched by incoming code. To our knowledge, the type system of [30], where types directly depend on ambient variables, is the only with dependent types for variants of MA. Our type system is simpler but less precise than the one of [30] in the specification of ambient behaviours, since in our approach all ambients of the same group share the same "passive" behaviour.

The first type inference algorithm for MA was presented in [36]. Other algorithms for variants of typed ambient calculi can be found in [2] and [15]. The main challenge in the design of the inference algorithm presented in this paper has been the handling of dependent types, an issue not addressed by any of the above.

References

1. T. Amtoft, A. J. Kfoury, and S. M. Pericas-Geertsen. What are Polymorphically-Typed Ambients? In D. Sands, editor, *Proc. of ESOP'01*, volume 2028 of *LNCS*, pages 206–220. Springer, 2001.
2. F. Barbanera, M. Dezani-Ciancaglini, I. Salvo, and V. Sassone. A type inference algorithm for secure ambients. In M. Lenisa and M. Miculan, editors, *Proc. of TOSCA'01*, volume 62 of *ENTCS*. Elsevier Science, 2002.

3. L. Bettini, V. Bono, R. D. Nicola, G. Ferrari, D. Gorla, M. Loreti, E. Moggi, R. Pugliese, E. Tuosto, and B. Venneri. The KLAIM Project: Theory and Practice. In C. Priami, editor, *Global Computing: Programming Environments, Languages, Security and Analysis of Systems*, volume 2874 of *LNCS*, pages 88–151. Springer, 2003.
4. E. Bonelli, A. Compagnoni, M. Dezani-Ciancaglini, and P. Garralda. Boxed Ambients with Communication Interfaces. In V. Fiala, Jiríand Koubek and K. Jan, editors, *Proc. of MFCS '04*, volume 3153 of *LNCS*, pages 119–148. Springer, 2004.
5. G. Boudol. A Parametric Model of Migration and Mobility, Release 1. Mikado Deliverable D1.2.1, available at http://mikado.di.fc.ul.pt/repository/D1.2.1.pdf, 2003.
6. M. Bugliesi and G. Castagna. Behavioral Typing for Safe Ambients. *Computer Languages*, 28(1):61 – 99, 2002.
7. M. Bugliesi, G. Castagna, and S. Crafa. Access Control for Mobile Agents: The Calculus of Boxed Ambients. *ACM Transactions on Programming Languages and Systems*, 26(1):57–124, 2004.
8. M. Bugliesi, S. Crafa, M. Merro, and V. Sassone. Communication and Mobility Control in Boxed Ambients. *Information and Computation*, 202(1): 39–86, 2005.
9. L. Cardelli. Abstractions for Mobile Computation. In J. Vitek and C. Jensen, editors, *Secure Internet Programming: Security Issues for Mobile and Distributed Objects*, volume 1603 of *LNCS*, pages 51–94. Springer, 1999.
10. L. Cardelli, G. Ghelli, and A. D. Gordon. Mobility Types for Mobile Ambients. In J. Wiedermann, P. van Emde Boas, and M. Nielsen, editors, *Proc. of ICALP'99*, volume 1644 of *LNCS*, pages 230–239. Springer, 1999.
11. L. Cardelli, G. Ghelli, and A. D. Gordon. Types for the Ambient Calculus. *Information and Computation*, 177(2):160–194, 2002.
12. L. Cardelli and A. D. Gordon. Mobile Ambients. *Theoretical Computer Science*, 240(1):177–213, 2000. Special Issue on Coordination, Daniel Le Métayer Editor.
13. G. Castagna, J. Vitek, and F. Z. Nardelli. The Seal Calculus. *Information and Computation*, 201(1):1–54, 2005.
14. M. Coppo, M. Dezani-Ciancaglini, E. Giovannetti, and R. Pugliese. Dynamic and Local Typing for Mobile Ambients. In *Proc. of TCS'04*, pages 583–596. Kluwer, 2004.
15. M. Coppo, M. Dezani-Ciancaglini, E. Giovannetti, and I. Salvo. M3: Mobility Types for Mobile Processes in Mobile Ambients. In J. Harland, editor, *Proc. of CATS'03*, volume 78 of *ENTCS*. Elsevier, 2003.
16. H. B. Curry and R. Feys. *Combinatory Logic*, volume I of *Studies in Logic and the Foundations of Mathematics*. North-Holland, Amsterdam, 1958.
17. R. De Nicola, G. Ferrari, and R. Pugliese. Klaim: a Kernel Language for Agents Interaction and Mobility. *IEEE Transactions on Software Engineering*, 24(5):315–330, 1998.
18. R. De Nicola, D. Gorla, and R. Pugliese. Basic observables for a calculus for global computing. In L. Caires, G. F. Italiano, L. Monteiro, C. Palamidessi, and M. Yung, editors, *Proc. of ICALP'05*, volume 3580 of *LNCS*, pages 1226–1238. Springer, 2005.
19. G. Ferrari, E. Moggi, and R. Pugliese. Guardians for Ambient-based Monitoring. In V. Sassone, editor, *Proc. of F-WAN*, volume 66 of *ENTCS*. Elsevier, 2002.
20. A. Francalanza and M. Hennessy. A theory of system behaviour in the presence of node and link failures. In M. Abadi and L. de Alfaro, editors, *Proc. of CONCUR'05*, volume 3653 of *LNCS*, pages 368–382. Springer, 2005.

21. E. Giovannetti. Ambient Calculi with Types: a Tutorial. In C. Priami, editor, *Global Computing - Programming Environments, Languages, Security and Analysis of Systems*, volume 2874 of *LNCS*, pages 151–191. Springer, 2003.
22. H. Goguen. Typed Operational Semantics. In M. Dezani-Ciancaglini and G. Plotkin, editors, *Proc. of TLCA'95*, volume 902 of *LNCS*, pages 186–200. Springer, 1995.
23. D. Gorla, M. Hennessy, and V. Sassone. Security Policies as Membranes in Systems for Global Computing. In J. Rathke, editor, *Proc. of FGUC'04*, ENTCS. Elsevier, 2004.
24. D. Gorla and R. Pugliese. Resource Acces and Mobility Control with Dynamic Privileges Acquisition. In J. Parrow, editor, *Proc. of ICALP'03*, volume 2719 of *LNCS*, pages 119–132. Springer, 2003.
25. M. Hennessy, J. Rathke, and N. Yoshida. SafeDpi: A language for controlling mobile code (extended abstract). In I. Walukiewicz, editor, *Proc. of FOSSACS'04*, volume 2987 of *LNCS*, pages 241–256, 2004. Extended and revised version to appear in *Acta Informatica.*
26. M. Hennessy and J. Riely. Type-Safe Execution of Mobile Agents in Anonymous Networks. In J. Vitek and C. Jensen, editors, *Secure Internet Programming: Security Issues for Distributed and Mobile Objects*, number 1603 in LNCS, pages 95–115. Springer, 1999.
27. M. Hennessy and J. Riely. Resource Access Control in Systems of Mobile Agents. *Information and Computation*, 173:82–120, 2002.
28. M. Hennessy and J. Riely. Trust and Partial Typing in Open Systems of Mobile Agents. *Journal of Automated Reasoning*, 31(3-4):335–370, 2003.
29. F. Levi and D. Sangiorgi. Controlling Interference in Ambients. *Transactions on Programming Languages and Systems*, 25(1):1–69, 2003.
30. C. Lhoussaine and V. Sassone. A Dependently Typed Ambient Calculus. In D. Schmidt, editor, *Proc. of ESOP'04*, volume 2986 of *LNCS*, pages 171–187. Springer, 2004.
31. M. Merro and M. Hennessy. Bisimulation Congruences in Safe Ambients. In N. D. Jones and X. Leroy, editors, *Proc. of POPL'02*, pages 71–80, New York, 2002. ACM Press.
32. M. Merro and V. Sassone. Typing and Subtyping Mobility in Boxed Ambients. In L. Brim, P. Jancar, M. Kretinsky, and A. Kucera, editors, *Proc. of CONCUR'02*, volume 2421 of *LNCS*, pages 304–320. Springer, 2002.
33. G. C. Necula. Proof-Carrying Code. In N. D. Jones, editor, *Proc. of POPL'97*, pages 106–119. ACM Press, 1997.
34. B. C. Pierce. *Types and Programming Languages.* MIT Press, 2002.
35. B. C. Pierce and D. Sangiorgi. Typing and Subtyping for Mobile Processes. *Mathematical Structures in Computer Science*, 6(5):409–454, 1996. An extract appeared in *Proc. of LICS '93*: 376–385.
36. P. Zimmer. Subtyping and typing algorithms for mobile ambients. In J. Tiuryn, editor, *Proc. of FOSSACS'00*, volume 1784 of *LNCS*, pages 375–390. Springer, 2000.

Model Theory for Process Algebra

Jan A. Bergstra[1,2] and C.A. (Kees) Middelburg[3]

[1] Programming Research Group, University of Amsterdam,
P.O. Box 41882, 1009 DB Amsterdam, The Netherlands
janb@science.uva.nl

[2] Department of Philosophy, Utrecht University,
P.O. Box 80126, 3508 TC Utrecht, The Netherlands
janb@phil.uu.nl

[3] Computing Science Department, Eindhoven University of Technology,
P.O. Box 513, 5600 MB Eindhoven, The Netherlands
keesm@win.tue.nl

Abstract. We present a first-order extension of the algebraic theory about processes known as ACP and its main models. Useful predicates on processes, such as deadlock freedom and determinism, can be added to this theory through first-order definitional extensions. Model theory is used to analyse the discrepancies between identity in the models of the first-order extension of ACP and bisimilarity of the transition systems extracted from these models, and also the discrepancies between deadlock freedom in the models of a suitable first-order definitional extension of this theory and deadlock freedom of the transition systems extracted from these models. First-order definitions are material to the formalization of an interpretation of one theory about processes in another. We give a comprehensive example of such an interpretation too.

1 Introduction

Model theory is for some time now a very active branch of mathematical logic. Therefore, it looks to be worthwhile to introduce various techniques from model theory into the field of process algebra. This forms the greater part of our motivation to take up the work presented in this paper. With great pleasure, we contribute this paper to the Liber Amicorum in honor of the 60th birthday of Jan Willem Klop.

Usually, theories about processes such as ACP [1, 2] and CCS [3, 4] are equationally axiomatized. However, it is also possible to give first-order theories. An important advantage of a first-order approach is that it makes available the tool of first-order definition of predicates and operations on processes.

In this paper, we present a first-order extension of ACP and its main models. The first-order extension concerned includes a binary reachability predicate on processes with an associated first-order axiom schema for subprocess induction. The reachability predicate can be used to give first-order definitions of many general properties of processes, such as deadlock freedom and determinism, and the axiom schema for subprocess induction can then be used to verify whether

A. Middeldorp et al. (Eds.): Processes... (Klop Festschrift), LNCS 3838, pp. 445–495, 2005.

processes have these properties. This is one of the interesting applications of first-order definitions of predicates on processes.

First-order definitions of predicates and operations on processes are generally indispensable for the formalization of an interpretation of one theory about processes in another. For example, a first-order definition of the deadlock freedom predicate permits the formalization of the interpretation of BPA in BPA_δ [2] (both are subtheories of ACP). By first-order definitions of operations on processes, we are able to formalize more complicated interpretations, such as the interpretation of BPPA [5, 6] in the first-order extension of ACP. If one theory is interpretable in another theory, then a model of the former theory can be obtained from each model of the latter theory by taking a submodel of a restriction of an expansion by definitions. The expansion concerns the first-order definable operations on processes needed in the formalization of the interpretation concerned; and the first-order definable predicate on processes needed in the formalization of the interpretation determines the domain of the submodel. This technique to construct models can be regarded as a first-order generalization of the SRM-technique from [7].

In this paper, we analyse the discrepancies between identity in the models of the first-order extension of ACP and external bisimilarity, i.e. bisimilarity of the transition systems extracted from these models. Besides external bisimilarity, we pay attention to observational equivalence; and we have a look at other related issues such as bisimilarity based on structural operational semantics and modal characterization of external bisimilarity. We also analyse the discrepancies between deadlock freedom in the models of a suitable first-order definitional extension of the first-order extension of ACP and external deadlock freedom, i.e. deadlock freedom of the transition systems extracted from these models. Additionally, we briefly consider the comparable discrepancies for determinism.

It happens that the first-order extension of BPA_δ, which is a subtheory of the first-order extension of ACP, gets great expressive power in case it is extended with restricted reachability predicates. Even the first-order extension of ACP can be interpreted in it. In this paper, we formalize the interpretation concerned. Thus, we provide a comprehensive example of the formalization of an interpretation of one theory about processes in another.

The structure of this paper is as follows. First of all, we introduce BPA_δ^{fo}, the (finitary) first-order extension of an important subtheory of ACP, to wit BPA_δ (Sect. 2). Next, we consider some useful infinitary and second-order axioms (Sect. 3). After that, we introduce transition systems, bisimilarity of transition systems (Sect. 4) and full bisimulation models, the main models of BPA_δ^{fo} (Sect. 5). Thereupon, we analyse the discrepancies between external bisimilarity and identity in models of BPA_δ^{fo} (Sect. 6) and investigate the related external equivalence known as observational equivalence (Sect. 7). Following this, we have a closer look at bisimilarity based on structural operational semantics (Sect. 8) and the modal characterization of external bisimilarity (Sect. 9). Then, we extend BPA_δ^{fo} with a deadlock freedom predicate and analyse the discrepancies between external deadlock freedom and internal deadlock freedom in models of the extension of $\mathrm{BPA}_\delta^{\mathrm{fo}}$ concerned (Sect. 10). We also briefly consider the extension with a determinism predicate (Sect. 11). After that, we consider the addition of restricted reachability predicates to $\mathrm{BPA}_\delta^{\mathrm{fo}}$ (Sect. 12). Next, we introduce $\mathrm{ACP}^{\mathrm{fo}}$, the first-order extension of ACP (Sect. 13) and the full bisimulation models of $\mathrm{ACP}^{\mathrm{fo}}$ (Sect. 14). Thereupon, we consider interpretations of one theory in another (Sect. 15) and give as an example the interpretation of $\mathrm{ACP}^{\mathrm{fo}}$ in the extension of $\mathrm{BPA}_\delta^{\mathrm{fo}}$ with restricted reachability predicates (Sect. 16). Finally, we make some concluding remarks (Sect. 17).

Some familiarity with model theory is required. The desirable background can be found in [8, 9, 10].

2 The First-Order Theory $\mathrm{BPA}_\delta^{\mathrm{fo}}$

In this section, we present $\mathrm{BPA}_\delta^{\mathrm{fo}}$, a first-order extension of an important subtheory of ACP, being known as BPA_δ. In $\mathrm{BPA}_\delta^{\mathrm{fo}}$, it is assumed that there is a fixed but arbitrary finite set of *actions* A with $\delta \notin \mathsf{A}$.

The first-order theory $\mathrm{BPA}_\delta^{\mathrm{fo}}$ has the following nonlogical symbols:

- the *deadlock* constant δ;
- for each $a \in \mathsf{A}$, the *action* constant a;
- the binary *alternative composition* operator $+$;
- the binary *sequential composition* operator $\cdot$;
- the binary *summand inclusion* predicate symbol $\sqsubseteq$;
- for each $a \in \mathsf{A}$, the unary *action termination* predicate symbol $\xrightarrow{a}\surd$;
- for each $a \in \mathsf{A}$, the binary *action step* predicate symbol $\xrightarrow{a}$;
- the binary *reachability* predicate symbol $\twoheadrightarrow$.

We use infix notation for the binary operators, postfix notation for the unary predicate symbols and infix notation for the binary predicate symbols. The following precedence conventions are used to reduce the need for parentheses. Operators bind stronger than predicate symbols, and predicate symbols bind stronger than logical connectives and quantifiers. Moreover, the operator $\cdot$ binds stronger than the operator $+$, the logical connective $\neg$ binds stronger than the logical connectives $\wedge$ and $\vee$, and the logical connectives $\wedge$ and $\vee$ bind stronger than the logical connectives $\Rightarrow$ and $\Leftrightarrow$. Quantifiers are given the smallest possible scope. We often use $t \neq t'$, where t and t' are terms of $\mathcal{L}(\mathrm{BPA}_\delta^{\mathrm{fo}})$, as a shorthand for $\neg\, t = t'$.

The constants and operators of $\mathrm{BPA}_\delta^{\mathrm{fo}}$ are the same as the constants and operators of BPA_δ. The additional nonlogical symbols of $\mathrm{BPA}_\delta^{\mathrm{fo}}$ are all predicate symbols. In the context of BPA_δ, the summand inclusion predicate symbol is sometimes used in abbreviations for equations expressing summand inclusions. The action termination and action step predicate symbols are used in the description of the structural operational semantics of BPA_δ. That usage is related to the usage in the theory $\mathrm{BPA}_\delta^{\mathrm{fo}}$, but the one should not be mistaken for the other. A similar remark applies to the reachability predicate symbol.

Let t and t' be closed terms of $\mathcal{L}(\mathrm{BPA}_\delta^{\mathrm{fo}})$. Intuitively, the constants and operators can be explained as follows:

- δ cannot perform any action;
- a first performs action a and then terminates successfully;
- $t + t'$ behaves either as t or as t', but not both;
- $t \cdot t'$ first behaves as t, but when t terminates succesfully it continues by behaving as t'.

Intuitively, the predicates can be explained as follows:

- $t \sqsubseteq t'$ means that t' is capable of behaving as t;
- $t \xrightarrow{a} \surd$ means that t is capable of performing action a and then terminating successfully;
- $t \xrightarrow{a} t'$ means that t is capable of performing action a and then behaving as t';
- $t \twoheadrightarrow t'$ means that t is capable of performing a number of actions and then behaving as t'.

Before we give the axioms of $\mathrm{BPA}_\delta^{\mathrm{fo}}$, we introduce an important notational convention which will be used throughout this paper. If we introduce a term t as $t(x_1, \ldots, x_n)$, where $x_1, \ldots, x_n$ are distinct variables, this indicates that all variables that have occurrences in t are among $x_1, \ldots, x_n$. In the same context, $t(t_1, \ldots, t_n)$ is the term obtained by simultaneously replacing in t all occurrences of x_1 by t_1 and ... and all occurrences of x_n by t_n. Similarly, if we introduce a formula ϕ as $\phi(x_1, \ldots, x_n)$, where $x_1, \ldots, x_n$ are distinct variables, this indicates that all variables that have free occurrences in ϕ are among $x_1, \ldots, x_n$. In the same context, $\phi(t_1, \ldots, t_n)$ is the formula obtained by simultaneously replacing in ϕ all free occurrences of x_1 by t_1 and ... and all free occurrences of x_n by t_n. Bound variables are first renamed if needed to avoid free occurrences of variables in the replacing terms becoming bound.

The axioms of $\mathrm{BPA}_\delta^{\mathrm{fo}}$ are given in Table 1. Many axioms in this table are actually axiom schemas. RDPf and RSPf are axiom schemas where $t_1(x_1, \ldots, x_n), \ldots, t_n(x_1, \ldots, x_n)$ are terms of $\mathcal{L}(\mathrm{BPA}_\delta^{\mathrm{fo}})$ in which all occurrences of variables are guarded. We call an occurrence of a variable x in a term t *guarded* if t has a subterm of the form $a \cdot t'$ with t' containing this occurrence of x. BS and RS are axiom schemas where $\phi(x, y)$ is a formula of $\mathcal{L}(\mathrm{BPA}_\delta^{\mathrm{fo}})$. SI2–SI9, TR1–TR2 and R2 are axiom schemas where a and b are action constants. The instances of axiom schema SI4 are restricted by a side condition to those in which a is not (syntactically) identical to b.

Axioms A1–A7 are the axioms of BPA_δ. So $\mathrm{BPA}_\delta^{\mathrm{fo}}$ imports the (equational) axioms of BPA_δ. Axiom schemas RDPf and RSPf are relevant to the use of recursion for describing (potentially) non-terminating processes. They will be explained separately below. Axiom SI1 is the defining axiom of the summand inclusion predicate. Axiom schemas SI2–SI9 exclude models that identify processes that cannot be related by a bisimulation (a precise definition of bisimulation is given in Sect. 4). Axiom SI10 is an extensionality axiom for summand inclusion. The instances of axiom schema TR1 are the defining axioms of the action termination predicates and the instances of axiom schema TR2 are the defining

Table 1. Axioms of $\mathrm{BPA}_\delta^{\mathrm{fo}}$ (in $t_1, \ldots, t_n$ all occurrences of variables must be guarded)

$x + y = y + x$		A1
$(x + y) + z = x + (y + z)$		A2
$x + x = x$		A3
$(x + y) \cdot z = x \cdot z + y \cdot z$		A4
$(x \cdot y) \cdot z = x \cdot (y \cdot z)$		A5
$x + \delta = x$		A6
$\delta \cdot x = \delta$		A7
$\exists x_1, \ldots, x_n \bullet \bigwedge_{1 \leq i \leq n} x_i = t_i(x_1, \ldots, x_n)$		RDPf
$\bigwedge_{1 \leq i \leq n} x_i = t_i(x_1, \ldots, x_n) \wedge \bigwedge_{1 \leq i \leq n} y_i = t_i(y_1, \ldots, y_n) \Rightarrow \bigwedge_{1 \leq i \leq n} x_i = y_i$		RSPf
$x \sqsubseteq y \Leftrightarrow x + y = y$		SI1
$\neg\, a \sqsubseteq \delta$		SI2
$\neg\, a \cdot x \sqsubseteq \delta$		SI3
$\neg\, a \sqsubseteq b$	if $a \not\equiv b$	SI4
$\neg\, a \cdot x \sqsubseteq b$		SI5
$\neg\, a \sqsubseteq x \cdot y$		SI6
$a \cdot x \sqsubseteq y \cdot z \Rightarrow (a \sqsubseteq y \wedge x = z) \vee \exists y' \bullet (a \cdot y' \sqsubseteq y \wedge x = y' \cdot z)$		SI7
$a \sqsubseteq x + y \Rightarrow a \sqsubseteq x \vee a \sqsubseteq y$		SI8
$a \cdot x \sqsubseteq y + z \Rightarrow a \cdot x \sqsubseteq y \vee a \cdot x \sqsubseteq z$		SI9
$\bigwedge_{a \in \mathsf{A}}((a \sqsubseteq x \Rightarrow a \sqsubseteq y) \wedge \forall z \bullet (a \cdot z \sqsubseteq x \Rightarrow a \cdot z \sqsubseteq y)) \Rightarrow x \sqsubseteq y$		SI10
$x \xrightarrow{a} \surd \Leftrightarrow a \sqsubseteq x$		TR1
$x \xrightarrow{a} y \Leftrightarrow a \cdot y \sqsubseteq x$		TR2
$\phi(x, y) \wedge$ $\forall x', y' \bullet (\phi(x', y') \Rightarrow$ $\quad \bigwedge_{a \in \mathsf{A}}((x' \xrightarrow{a} \surd \Leftrightarrow y' \xrightarrow{a} \surd) \wedge$ $\qquad \forall x'' \bullet (x' \xrightarrow{a} x'' \Rightarrow \exists y'' \bullet (y' \xrightarrow{a} y'' \wedge \phi(x'', y''))) \wedge$ $\qquad \forall y'' \bullet (y' \xrightarrow{a} y'' \Rightarrow \exists x'' \bullet (x' \xrightarrow{a} x'' \wedge \phi(x'', y''))))) \Rightarrow x = y$		BS
$x \twoheadrightarrow x$		R1
$x \xrightarrow{a} y \wedge y \twoheadrightarrow z \Rightarrow x \twoheadrightarrow z$		R2
$x \twoheadrightarrow y \wedge$ $\forall x', y', z' \bullet (\phi(x', x') \wedge \bigwedge_{a \in \mathsf{A}}(x' \xrightarrow{a} y' \wedge \phi(y', z') \Rightarrow \phi(x', z'))) \Rightarrow \phi(x, y)$		RS

axioms of the action step predicates. Axiom schema BS, called the *bisimilarity axiom schema*, excludes models that do not identify processes that can be related by a first-order definable bisimulation. Axiom R1 and axiom schemas R2 and RS concern the reachability predicate. Axiom schema RS is an induction schema, called the *subprocess induction schema*. It is unknown to us whether the reachability predicate is implicitly defined by $\mathrm{BPA}_\delta^{\mathrm{fo}}$.

We do not claim that the axioms of $\mathrm{BPA}_\delta^{\mathrm{fo}}$ are independent. For example, axiom SI2 is derivable from axioms A7 and SI6. Axiom SI10 and axiom schema BS are dependent in a weak sense: extensionality for equality, i.e.

$$\bigwedge_{a \in \mathsf{A}}((a \sqsubseteq x \Leftrightarrow a \sqsubseteq y) \land \forall z \bullet (a \cdot z \sqsubseteq x \Leftrightarrow a \cdot z \sqsubseteq y)) \Rightarrow x = y \,,$$

is not only derivable from SI10 and SI1, but also from BS, TR1 and TR2.

The axiom schemas RDPf and RSPf are called the *recursive definition principle* and the *recursive specification principle* for finite guarded recursive specifications. A *guarded recursive specification* (over $\mathrm{BPA}_\delta^{\mathrm{fo}}$) is a set of equations $E = \{x = t_x \mid x \in V\}$ where V is a set of variables and each t_x is a term of $\mathcal{L}(\mathrm{BPA}_\delta^{\mathrm{fo}})$ in which only the variables in V may have occurrences and all those occurrences are guarded. There is an instance of RDPf and an instance of RSPf for each finite guarded recursive specification E. We write RDPf^E for the instance of RDPf for E and RSPf^E for the instance of RDPf for E. RDPf^E expresses that E has at least one solution and RSPf^E expresses that E has at most one solution.

Because the implications from right to left are derivable, the (outmost) occurrence of "$\Rightarrow$" in SI7–SI10 and BS can be replaced by "$\Leftrightarrow$". The equivalences

$$x = y \Leftrightarrow x \sqsubseteq y \land y \sqsubseteq x \,,$$

$$x + y \sqsubseteq z \Leftrightarrow x \sqsubseteq z \land y \sqsubseteq z \,.$$

are easily derived from axiom SI1 and axiom SI10, respectively. Both equivalences are used in the proof of Theorem 1 (see below).

Using the reachability predicate, we can give explicit definitions of other properties of processes. For example, deadlock freedom, absence of termination, and determinism can be explicitly defined as follows:

$$\begin{aligned}
\mathrm{dlf}(x) \;&\Leftrightarrow \neg\, x \twoheadrightarrow \delta \,,\\
\mathrm{perp}(x) \;&\Leftrightarrow \neg\, x \twoheadrightarrow \delta \land \bigwedge_{a \in \mathsf{A}} \neg\, \exists y \bullet (x \twoheadrightarrow y \land y \xrightarrow{a} \surd) \,,\\
\mathrm{det}(x) \;&\Leftrightarrow \forall y \bullet \Big(x \twoheadrightarrow y \Rightarrow \bigwedge_{a \in \mathsf{A}} \big((y \xrightarrow{a} \surd \Rightarrow \forall z \bullet \neg\, y \xrightarrow{a} z\big) \land \\
&\qquad \forall z, z' \bullet (y \xrightarrow{a} z \land y \xrightarrow{a} z' \Rightarrow z = z'))\Big) \,.
\end{aligned}$$

Using the subprocess induction schema, we can derive a formula according to which case distinction with respect to reachability can be made.

Proposition 1 (Case distinction for reachability). *The following formula is derivable from* $\mathrm{BPA}_\delta^{\mathrm{fo}}$*:*

$$\begin{aligned}
&x \twoheadrightarrow y \Rightarrow \\
&\quad x = y \lor \bigvee_{a \in \mathsf{A}} x \xrightarrow{a} y \lor \exists z \bullet \Big(z \neq x \land \bigvee_{a \in \mathsf{A}} (x \xrightarrow{a} z \land z \twoheadrightarrow y)\Big) \,.
\end{aligned}$$

Proof. We use $\mathrm{cdr}(x, y)$ as an abbreviation for the right-hand side of the above implication. We will apply RS, taking $x \twoheadrightarrow y \land \mathrm{cdr}(x, y)$ for $\phi(x, y)$. When we have shown that $x \twoheadrightarrow y \Rightarrow (x \twoheadrightarrow y \land \mathrm{cdr}(x, y))$, we can immediately conclude that $x \twoheadrightarrow y \Rightarrow \mathrm{cdr}(x, y)$ and we are done.

It remains to be shown by means of RS that $x \twoheadrightarrow y \Rightarrow (x \twoheadrightarrow y \wedge \mathrm{cdr}(x, y))$. First of all, we conclude from R1, because obviously $\mathrm{cdr}(x, x)$, that

$$\forall x' \bullet (x' \twoheadrightarrow x' \wedge \mathrm{cdr}(x', x'))\ .$$

Moreover, we easily derive the following implications:

$$\begin{aligned}
&x' \xrightarrow{a'} y' \wedge y' \twoheadrightarrow z' \Rightarrow x' \twoheadrightarrow z'\ ,\\
&x' \xrightarrow{a'} y' \wedge y' \twoheadrightarrow z' \wedge y' = z' \Rightarrow \bigvee_{a \in \mathsf{A}} x' \xrightarrow{a} z'\ ,\\
&x' \xrightarrow{a'} y' \wedge y' \twoheadrightarrow z' \wedge \bigvee_{a \in \mathsf{A}} y' \xrightarrow{a} z' \Rightarrow\\
&\quad \bigvee_{a \in \mathsf{A}} x' \xrightarrow{a} z' \vee \exists z \bullet \Big(z \neq x' \wedge \bigvee_{a \in \mathsf{A}} (x' \xrightarrow{a} z \wedge z \twoheadrightarrow z')\Big)\ ,\\
&x' \xrightarrow{a'} y' \wedge y' \twoheadrightarrow z' \wedge \exists z \bullet \Big(z \neq y' \wedge \bigvee_{a \in \mathsf{A}} (y' \xrightarrow{a} z \wedge z \twoheadrightarrow z')\Big) \Rightarrow\\
&\quad \exists z \bullet \Big(z \neq x' \wedge \bigvee_{a \in \mathsf{A}} (x' \xrightarrow{a} z \wedge z \twoheadrightarrow z')\Big)\ .
\end{aligned}$$

The first implication is derived using R2, the second implication is derived by elementary logical reasoning, the third implication is derived using R1 and R2 (with distinction between the cases $x' = y'$, $y' = z'$ and $x' \neq y' \wedge y' \neq z'$), and the fourth implication is derived by elementary logical reasoning (with distinction between the cases $x' = y'$ and $x' \neq y'$). The left-hand sides of the second, third and fourth implication are conjunctions of $x' \xrightarrow{a'} y' \wedge y' \twoheadrightarrow z'$ and one of the disjuncts of $\mathrm{cdr}(y', z')$. The right-hand sides of these implication consists of one or two of the disjuncts of $\mathrm{cdr}(x', z')$. Hence, we also conclude that

$$\begin{aligned}
&\forall x', y', z' \bullet\\
&\quad \bigwedge_{a' \in \mathsf{A}} (x' \xrightarrow{a'} y' \wedge (y' \twoheadrightarrow z' \wedge \mathrm{cdr}(y', z')) \Rightarrow x' \twoheadrightarrow z' \wedge \mathrm{cdr}(x', z'))\ .
\end{aligned}$$

Using the subprocess induction schema, it follows from these conclusions that $x \twoheadrightarrow y \Rightarrow (x \twoheadrightarrow y \wedge \mathrm{cdr}(x, y))$. □

A well-known subtheory of BPA_δ is BPA, which is BPA_δ without the deadlock constant and consequently without axioms A6 and A7. Analogously, we have a subtheory of $\mathrm{BPA}_\delta^{\mathrm{fo}}$, to wit $\mathrm{BPA}^{\mathrm{fo}}$. As to be expected, the first-order theory $\mathrm{BPA}^{\mathrm{fo}}$ is $\mathrm{BPA}_\delta^{\mathrm{fo}}$ without the deadlock constant and without axioms A6, A7, SI2 and SI3. In other words, the possibility that a process gets into a deadlock is not covered by $\mathrm{BPA}^{\mathrm{fo}}$.

To prove a statement for all closed terms of $\mathcal{L}(\mathrm{BPA}_\delta^{\mathrm{fo}})$, it is sufficient to prove it for all basic terms over $\mathrm{BPA}_\delta^{\mathrm{fo}}$. The set $\mathcal{B}$ of *basic terms* over $\mathrm{BPA}_\delta^{\mathrm{fo}}$ is inductively defined by the following rules:

- $\delta \in \mathcal{B}$;
- if $a \in \mathsf{A}$, then $a \in \mathcal{B}$;

– if $a \in \mathsf{A}$ and $t \in \mathcal{B}$, then $a \cdot t \in \mathcal{B}$;
– if $t_1, t_2 \in \mathcal{B}$, then $t_1 + t_2 \in \mathcal{B}$.

We can prove that all closed terms of $\mathcal{L}(\mathrm{BPA}_\delta^{\mathrm{fo}})$ are derivably equal to a basic term over $\mathrm{BPA}_\delta^{\mathrm{fo}}$.

Proposition 2 (Elimination). *For all closed terms t of $\mathcal{L}(\mathrm{BPA}_\delta^{\mathrm{fo}})$ there exists a basic term $t' \in \mathcal{B}$ such that $\mathrm{BPA}_\delta^{\mathrm{fo}} \vdash t = t'$.*

Proof. This follows immediately from the elimination property for BPA_δ: the closed terms of $\mathcal{L}(\mathrm{BPA}_\delta^{\mathrm{fo}})$ are the same as the closed terms of $\mathcal{L}(\mathrm{BPA}_\delta)$, and the equational axioms of $\mathrm{BPA}_\delta^{\mathrm{fo}}$ are the same as the axioms of BPA_δ. □

For closed equations, $\mathrm{BPA}_\delta^{\mathrm{fo}}$ is a complete theory.

Theorem 1 (Complete theory for closed equations). *For all closed terms t_1, t_2 of $\mathcal{L}(\mathrm{BPA}_\delta^{\mathrm{fo}})$, we have either $\mathrm{BPA}_\delta^{\mathrm{fo}} \vdash t_1 = t_2$ or $\mathrm{BPA}_\delta^{\mathrm{fo}} \vdash \neg\, t_1 = t_2$, but not both.*

Proof. In Sect. 5, we will show that there exist models of $\mathrm{BPA}_\delta^{\mathrm{fo}}$. From this, it follows by the Extended Completeness Theorem (see e.g. [9]) that there are no closed terms t_1, t_2 of $\mathcal{L}(\mathrm{BPA}_\delta^{\mathrm{fo}})$ such that both $t_1 = t_2$ and $\neg\, t_1 = t_2$ are derivable. Moreover, the equivalence $x = y \Leftrightarrow x \sqsubseteq y \wedge y \sqsubseteq x$ is derivable. For these reasons, and Proposition 2, it is sufficient to prove that for all basic terms $t_1, t_2 \in \mathcal{B}$, either $\mathrm{BPA}_\delta^{\mathrm{fo}} \vdash t_1 \sqsubseteq t_2$ or $\mathrm{BPA}_\delta^{\mathrm{fo}} \vdash \neg\, t_1 \sqsubseteq t_2$. This is easily proved by induction on the sum of the lengths of t_1 and t_2. All cases follow immediately from axioms SI1–SI9, sometimes using the induction hypothesis, except the cases $a \cdot t_1' \sqsubseteq b \cdot t_2'$ and $t_1' + t_1'' \sqsubseteq t_2$. Those cases follow immediately from the derivable equivalences $a \cdot x \sqsubseteq b \cdot y \Leftrightarrow a \sqsubseteq b \wedge x \sqsubseteq y \wedge y \sqsubseteq x$ and $x + y \sqsubseteq z \Leftrightarrow x \sqsubseteq z \wedge y \sqsubseteq z$, using the induction hypothesis. □

For arbitrary closed formula, $\mathrm{BPA}_\delta^{\mathrm{fo}}$ is not a complete theory. This follows from the fact that there are models of $\mathrm{BPA}_\delta^{\mathrm{fo}}$ that are not elementary equivalent (see Theorems 4 and 8).

3 Infinitary and Second-Order Axioms

It appears to be of use to add certain infinitary and second-order axioms to $\mathrm{BPA}_\delta^{\mathrm{fo}}$. In this section, we consider those axioms.

The recursive definition principle and recursive specification principle for finite guarded recursive specifications (RDPf and RSPf) do not exclude models in which there are countably infinite guarded recursive specifications without a unique solution. The infinitary axiom schemas RDP and RSP from Table 2 would exclude all such models. Like in the case of axiom schemas RDPf and RSPf, we write RDP^E and RSP^E for the instances of RDP and RSP, respectively, for guarded recursive specification E.

The instances of axiom schema RSP are formulas of $\mathcal{L}_{\omega_1\omega}(\mathrm{BPA}_\delta^{\mathrm{fo}})$, the first-order language of $\mathrm{BPA}_\delta^{\mathrm{fo}}$ with conjunctions and disjunctions of countable sets of

Table 2. Infinitary first-order axioms

$\exists x_1, x_2, \ldots \bullet \bigwedge_{i \geq 1} x_i = t_i(x_1, x_2, \ldots)$	RDP
$\bigwedge_{i \geq 1} x_i = t_i(x_1, x_2, \ldots) \wedge \bigwedge_{i \geq 1} y_i = t_i(x_1, x_2, \ldots) \Rightarrow \bigwedge_{i \geq 1} x_i = y_i$	RSP

formulas. The instances of axiom schema RDP are formulas of $\mathcal{L}_{\omega_1 \omega_1}(\mathrm{BPA}_\delta^{\mathrm{fo}})$, the first-order language of $\mathrm{BPA}_\delta^{\mathrm{fo}}$ with conjunctions and disjunctions of countable sets of formulas and quantification on countable sets of variables. RDP and RSP are not axiomatizable in the usual finitary first-order language $\mathcal{L}(\mathrm{BPA}_\delta^{\mathrm{fo}})$.

Theorem 2 (RDP and RSP are not axiomatizable in $\mathcal{L}(\mathrm{BPA}_\delta^{\mathrm{fo}})$). *There does not exist a finitary first-order extension of* $\mathrm{BPA}_\delta^{\mathrm{fo}}$ *of which all models satisfy* RDP *and there does not exist a finitary first-order extension of* $\mathrm{BPA}_\delta^{\mathrm{fo}}$ *of which all models satisfy* RSP.

Proof. First, we show that there does not exist a finitary first-order extension of $\mathrm{BPA}_\delta^{\mathrm{fo}}$, say $\mathrm{BPA}_\delta^{\mathrm{fo}} \cup H$, such that $\mathrm{BPA}_\delta^{\mathrm{fo}} \cup H \models \mathrm{RDP}$. Suppose that $\mathrm{BPA}_\delta^{\mathrm{fo}} \cup H \models \mathrm{RDP}$. A contradiction is found as follows. By the Downward Löwenheim-Skolem Theorem (see e.g. [10]), there exists a countable model of $\mathrm{BPA}_\delta^{\mathrm{fo}} \cup H$. Take a countable model $\mathfrak{A} \models \mathrm{BPA}_\delta^{\mathrm{fo}} \cup H$. Let a and b be different actions. Consider the guarded recursive specifications $E_V = \{X_i = a \cdot X_{i+1} \mid i \in V\} \cup \{X_i = b \cdot X_{i+1} \mid i \notin V\}$ for $V \subseteq \mathbb{N}$. E_V encodes the characteristic function of V. Because $\mathrm{BPA}_\delta^{\mathrm{fo}} \cup H \models \mathrm{RDP}$ by our supposition, and $\mathfrak{A} \models \mathrm{BPA}_\delta^{\mathrm{fo}} \cup H$, there exists a solution p_V of E_V for X_0 in $\mathfrak{A}$ for each $V \subseteq \mathbb{N}$. There exist uncountably many V such that $V \subseteq \mathbb{N}$; and it is easily proved by induction on the smallest i such that $i \in V \Leftrightarrow i \notin V'$ that $V \neq V'$ implies $p_V \neq p_{V'}$. Hence, $\mathfrak{A}$ must be an uncountable model, which contradicts the fact that $\mathfrak{A}$ is a countable model.

Next, we show that there does not exist a finitary first-order extension of $\mathrm{BPA}_\delta^{\mathrm{fo}}$, say $\mathrm{BPA}_\delta^{\mathrm{fo}} \cup H$, such that $\mathrm{BPA}_\delta^{\mathrm{fo}} \cup H \models \mathrm{RSP}$. Suppose that $\mathrm{BPA}_\delta^{\mathrm{fo}} \cup H \models \mathrm{RSP}$. A contradiction is found as follows. Let $c_0, c_1, c_2, \ldots$ and $d_0, d_1, d_2, \ldots$ be different new constants; and let a, a', a'' be different actions. Consider the following sets of formulas:

$$\begin{aligned} H' &= \{c_0 \neq d_0\} \cup \{c_i = a \cdot c_{i+1} \mid i \geq 0\} \cup \{d_i = a \cdot d_{i+1} \mid i \geq 0\}\,, \\ H'_n &= \{c_0 \neq d_0\} \cup \{c_i = a \cdot c_{i+1} \mid 0 \leq i < n\} \cup \{d_i = a \cdot d_{i+1} \mid 0 \leq i < n\} \\ &\quad \cup \{c_n = a', d_n = a''\} \\ &(\text{for } n \geq 0)\,. \end{aligned}$$

Take an arbitrary model $\mathfrak{A}$ of $\mathrm{BPA}_\delta^{\mathrm{fo}} \cup H$. It follows easily from the axioms of $\mathrm{BPA}_\delta^{\mathrm{fo}}$ that, for each $n \geq 0$, H'_n is satisfied in the definitional expansion of $\mathfrak{A}$ determined by the definitional extension of $\mathrm{BPA}_\delta^{\mathrm{fo}} \cup H$ with the constants $c_0, \ldots, c_n$, $d_0, \ldots, d_n$ and the equations $c_i = a^{n-i} \cdot a'$ for $0 \leq i < n$, $c_n = a'$, $d_i = a^{n-i} \cdot a''$ for $0 \leq i < n$, $d_n = a''$.[1] Hence, for each $n \geq 0$, H'_n is consistent

[1] For each action a and each $n \geq 1$, the term a^n is defined by induction on n as follows: a^1 is a and a^{n+1} is $a \cdot a^n$.

with $\mathrm{BPA}_\delta^{\mathrm{fo}} \cup H$. Each finite $H'' \subseteq H'$ is consistent with $\mathrm{BPA}_\delta^{\mathrm{fo}} \cup H$ because there is an $n \geq 0$ for which $H'' \subseteq H'_n$. From this, it follows by the Compactness Theorem (see e.g. [9]) that H' is consistent with $\mathrm{BPA}_\delta^{\mathrm{fo}} \cup H$. Now consider an arbitrary model $\mathfrak{A}'$ of $\mathrm{BPA}_\delta^{\mathrm{fo}} \cup H \cup H'$. Because $\mathfrak{A}'$ satisfies H', we have $c_0^{\mathfrak{A}'} \neq d_0^{\mathfrak{A}'}$. Both $c_0^{\mathfrak{A}'}$ and $d_0^{\mathfrak{A}'}$ are solutions of the guarded recursive specification $E = \{X_i = a \cdot X_{i+1} \mid i \in \mathbb{N}\}$ for X_0. Hence, by RSP, it must be the case that $c_0^{\mathfrak{A}'} = d_0^{\mathfrak{A}'}$, which contradicts the fact that $c_0^{\mathfrak{A}'} \neq d_0^{\mathfrak{A}'}$. □

If we restrict ourselves to recursively enumerable theories, we can even give an instance of RDP that is not axiomatizable.

Theorem 3 (Instance of RDP is not axiomatizable in $\mathcal{L}(\mathrm{BPA}_\delta^{\mathrm{fo}})$). *Let T be a finitary first-order extension of $\mathrm{BPA}_\delta^{\mathrm{fo}}$ that is recursively enumerable, let a, b be different actions, let V be a subset of $\mathbb{N}$ that is not recursively enumerable, and let E_V be the guarded recursive specification $\{X_i = a \cdot X_{i+1} \mid i \in V\} \cup \{X_i = b \cdot X_{i+1} \mid i \notin V\}$. Then $T \not\models \mathrm{RDP}^{E_V}$.*

Proof. Let $\psi_n(x)$, for each $n \geq 0$, be the following formula:

$$\exists y \bullet \exists z_0, \ldots, z_n \bullet \Big(x = z_0 \cdot \ldots \cdot z_n \cdot y \;\land \bigwedge_{i \leq n, i \in V} z_i = a \;\land \bigwedge_{i \leq n, i \notin V} z_i = b\Big) .$$

Let Ψ be the set of formulas $\{\psi_n(x) \mid n \in \mathbb{N}\}$. It is easy to see that there does not exist a solution of E_V in a model of $\mathrm{BPA}_\delta^{\mathrm{fo}}$ iff that model omits Ψ. Moreover, by the Omitting Types Theorem (see e.g. [9]), there exists a model that omits Ψ if T or some consistent extension of T locally omits Ψ. Thus, when we have shown that T or some consistent extension of T locally omits Ψ, we can immediately conclude that $T \not\models \mathrm{RDP}^{E_V}$ and we are done.

We prove that some consistent extension of T locally omits Ψ by constructing such an extension of T. Let $\phi_0(x), \phi_1(x), \phi_2(x), \ldots$ be an enumeration of all formulas of $\mathcal{L}(\mathrm{BPA}_\delta^{\mathrm{fo}})$ in which no variable other than x has free occurrences. We start to construct a non-decreasing sequence $T^0, T^1, T^2, \ldots$ of consistent extensions of T as follows:

$$\begin{array}{lll}
T^0 & = T\,, & \\[1ex]
T^{2k+1} & = T^{2k} \cup \{\phi_k(x)\} & \text{if not } T^{2k} \vdash \neg\, \phi_k(x)\,, \\
T^{2k+1} & = T^{2k} \cup \{\neg\, \phi_k(x)\} & \text{otherwise}\,, \\[1ex]
T^{2k+2} & = T^{2k+1} & \text{if } T^{2k+1} \vdash \neg\, \exists x \bullet \phi_k(x)\,, \\
T^{2k+2} & = T^{2k+1} \cup \{\exists x \bullet (\phi_k(x) \;\land\; \neg\, \psi_n(x))\} & \text{otherwise}\,,
\end{array}$$

$$\text{for some } n \in \mathbb{N} \text{ such that not } T^{2k+1} \vdash \neg\, \exists x \bullet (\phi_k(x) \;\land\; \neg\, \psi_n(x)) .$$

For all k, there exists an n such that not $T^{2k+1} \vdash \neg\, \exists x \bullet (\phi_k(x) \;\land\; \neg\, \psi_n(x))$. This is easily proved by contradiction. If it was not the case for some k, then we would have $T^{2k+1} \vdash \forall x \bullet (\phi_k(x) \Rightarrow \psi_n(x))$. Because of the recursive enumerability of T (and therefore also T^{2k+1}), it would follow that V is recursively enumerable. This contradicts the fact that V is not recursively enumerable.

Table 3. Second-order axioms

$\exists R \bullet (R(x,y) \wedge$ $\quad \forall x', y' \bullet (R(x',y') \Rightarrow$ $\quad\quad \bigwedge_{a \in \mathsf{A}}((x' \xrightarrow{a} \surd \Leftrightarrow y' \xrightarrow{a} \surd) \wedge$ $\quad\quad\quad \forall x'' \bullet (x' \xrightarrow{a} x'' \Rightarrow \exists y'' \bullet (y' \xrightarrow{a} y'' \wedge R(x'',y''))) \wedge$ $\quad\quad\quad \forall y'' \bullet (y' \xrightarrow{a} y'' \Rightarrow \exists x'' \bullet (x' \xrightarrow{a} x'' \wedge R(x'',y''))))))) \Rightarrow$ $x = y$	B
$\forall R \bullet (x \twoheadrightarrow y \wedge$ $\quad \forall x', y', z' \bullet (R(x',x') \wedge \bigwedge_{a \in \mathsf{A}}(x' \xrightarrow{a} y' \wedge R(y',z') \Rightarrow R(x',z'))) \Rightarrow$ $\quad R(x,y))$	R

For each $k \in \mathbb{N}$, T^k is consistent by construction. Let $T^\infty = \bigcup_{k \in \mathbb{N}} T^k$. Then T^∞ is also consistent by construction. Moreover, T^∞ locally omits Ψ by construction. □

The bisimilarity axiom schema (BS) from Table 1 does not exclude all models that distinguish between processes that can be related by a bisimulation. It only excludes models that distinguish between processes that can be related by a first-order definable bisimulation. The second-order axiom B from Table 3 would exclude all such models. Axiom B is called the *bisimilarity axiom*. It is a second-order axiom because of the existential quantification on R, which is a variable ranging over binary relations on processes instead of a variable ranging over processes.

The subprocess induction schema (RS) from Table 1 does not exclude all models in which there are processes that have more reachable processes than needed to satisfy axiom R1 and the instances of axiom schema R2. The second-order axiom R from Table 3 would exclude all such models. Axiom R is called the *subprocess induction axiom*.

Let $\mathfrak{A}$ be a model of $\mathrm{BPA}_\delta^{\mathrm{fo}}$, i.e. $\mathfrak{A} \models \mathrm{BPA}_\delta^{\mathrm{fo}}$. Then $\mathfrak{A}$ is a *bisimulation model* if $\mathfrak{A} \models \mathrm{B}$; and $\mathfrak{A}$ is a model with *standard reachability* if $\mathfrak{A} \models \mathrm{R}$.

4 Transition Systems and Bisimilarity

In this section, we introduce transition systems and bisimilarity of transition systems. In Sect. 5, we will make use of transition systems and bisimilarity of transition systems to construct the main models of $\mathrm{BPA}_\delta^{\mathrm{fo}}$.

A *transition system* T consists of the following:

- a set S of *states*;
- a set $\xrightarrow{a} \subseteq S \times S$, for each $a \in \mathsf{A}$;
- a set $\xrightarrow{a}\surd \subseteq S$, for each $a \in \mathsf{A}$;
- an *initial state* $s^0 \in S$.

If $(s, s') \in \xrightarrow{a}$ for some $a \in \mathsf{A}$, then we say that there is a *transition* from state s to state s'. We usually write $s \xrightarrow{a} s'$ instead of $(s, s') \in \xrightarrow{a}$ and $s \xrightarrow{a}\surd$ instead of $s \in \xrightarrow{a}\surd$. Furthermore, we write $\rightarrow$ for the family of sets $(\xrightarrow{a})_{a \in \mathsf{A}}$ and $\rightarrow\surd$ for the family of sets $(\xrightarrow{a}\surd)_{a \in \mathsf{A}}$.

A transition system may have states that are not reachable from its initial state by a number of transitions. Unreachable states, and the transitions between them, are not relevant to the behaviour represented by the transition system. We exclude transition systems with unreachable states as follows.

Let $T = (S, \rightarrow, \rightarrow\surd, s^0)$ be a transition system. Then the *reachability* relation of T is the smallest relation $\twoheadrightarrow \subseteq S \times S$ such that:

- $s \twoheadrightarrow s$;
- if $s \xrightarrow{a} s'$ and $s' \twoheadrightarrow s''$, then $s \twoheadrightarrow s''$.

We write $\mathrm{RS}(T)$ for $\{s \in S \mid s^0 \twoheadrightarrow s\}$. T is called a *connected* transition system if $S = \mathrm{RS}(T)$. Henceforth, we will only consider connected transition systems. However, this often calls for extraction of the connected part of a transition system that is composed of connected transition systems.

Let $T = (S, \rightarrow, \rightarrow\surd, s^0)$ be a transition system that is not necessarily connected. Then the *connected part* of T, written $\Gamma(T)$, is defined as follows:

$$\Gamma(T) = (S', \rightarrow', \rightarrow\surd', s^0) ,$$

where

$$S' = \mathrm{RS}(T) ,$$

and for every $a \in \mathsf{A}$:

$$\begin{aligned} \xrightarrow{a}' &= \xrightarrow{a} \cap (S' \times S') , \\ \xrightarrow{a}\surd' &= \xrightarrow{a}\surd \cap S' . \end{aligned}$$

It is assumed that for each infinite cardinal κ a fixed but arbitrary set $\mathcal{S}_\kappa$ with the following properties has been given:

- the cardinality of $\mathcal{S}_\kappa$ is greater than or equal to κ;
- if $S_1, S_2 \subseteq \mathcal{S}_\kappa$, then $S_1 \uplus S_2 \subseteq \mathcal{S}_\kappa$ and $S_1 \times S_2 \subseteq \mathcal{S}_\kappa$.[2]

Let κ be an infinite cardinal number. Then $\mathbb{TS}_\kappa$ is the set of all connected transition systems $T = (S, \rightarrow, \rightarrow\surd, s^0)$ such that $S \subset \mathcal{S}_\kappa$ and the branching degree of T is less than κ, that is, for all $s \in S$, the cardinality of the set $\{(a, s') \in \mathsf{A} \times S \mid s \xrightarrow{a} s'\} \cup \{a \in \mathsf{A} \mid s \xrightarrow{a}\surd\}$ is less than κ.

The condition $S \subset \mathcal{S}_\kappa$ guarantees that $\mathbb{TS}_\kappa$ is indeed a set.

A connected transition system is said to be *finitely branching* if its branching degree is less than $\aleph_0$. Otherwise, it is said to be *infinitely branching.*

The identity of the states of a connected transition system is not relevant to the behaviour represented by it. Connected transition systems that differ only with respect to the identity of the states are isomorphic.

---

2 We write $A \uplus B$ for the disjoint union of sets A and B, i.e. $A \uplus B = (A \times \{\emptyset\}) \cup (B \times \{\{\emptyset\}\})$. We write μ_1 and μ_2 for the associated injections $\mu_1 : A \rightarrow A \uplus B$ and $\mu_2 : B \rightarrow A \uplus B$, defined by $\mu_1(a) = (a, \emptyset)$ and $\mu_2(b) = (b, \{\emptyset\})$.

Let $T_1 = (S_1, \rightarrow_1, \rightarrow\surd_1, s_1^0)$ and $T_2 = (S_2, \rightarrow_2, \rightarrow\surd_2, s_2^0)$ be connected transition systems. Then T_1 and T_2 are *isomorphic*, written $T_1 \cong T_2$, if there exists a bijective function $b : S_1 \rightarrow S_2$ such that

- $b(s_1^0) = s_2^0$;
- $s_1 \xrightarrow{a}_1 s_1'$ iff $b(s_1) \xrightarrow{a}_2 b(s_1')$;
- $s \xrightarrow{a}\surd_1$ iff $b(s) \xrightarrow{a}\surd_2$.

Henceforth, we will always consider two connected transition systems essentially the same if they are isomorphic.

Remark 1. The set $\mathbb{TS}_\kappa$ is independent of $\mathcal{S}_\kappa$. By that we mean the following. Let $\mathbb{TS}_\kappa$ and $\mathbb{TS}'_\kappa$ result from different choices for $\mathcal{S}_\kappa$. Then there exists a bijection $b : \mathbb{TS}_\kappa \rightarrow \mathbb{TS}'_\kappa$ such that for all $T \in \mathbb{TS}_\kappa$, $T \cong b(T)$.

Bisimilarity of transition systems from $\mathbb{TS}_\kappa$ is defined as follows.

Let $T_1 = (S_1, \rightarrow_1, \rightarrow\surd_1, s_1^0) \in \mathbb{TS}_\kappa$ and $T_2 = (S_2, \rightarrow_2, \rightarrow\surd_2, s_2^0) \in \mathbb{TS}_\kappa$ ($\kappa \geq \aleph_0$). Then a *bisimulation* B between T_1 and T_2 is a binary relation $B \subseteq S_1 \times S_2$ such that $B(s_1^0, s_2^0)$ and for all s_1, s_2 such that $B(s_1, s_2)$:

- $s_1 \xrightarrow{a}\surd_1$ iff $s_2 \xrightarrow{a}\surd_2$;
- if $s_1 \xrightarrow{a}_1 s_1'$, then there is a state s_2' such that $s_2 \xrightarrow{a}_2 s_2'$ and $B(s_1', s_2')$;
- if $s_2 \xrightarrow{a}_2 s_2'$, then there is a state s_1' such that $s_1 \xrightarrow{a}_1 s_1'$ and $B(s_1', s_2')$.

Two transition systems $T_1, T_2 \in \mathbb{TS}_\kappa$ are *bisimilar*, written $T_1 \leftrightarrows T_2$, if there exists a bisimulation B between T_1 and T_2. Let B be a bisimulation between T_1 and T_2. Then we say that B is a bisimulation *witnessing* $T_1 \leftrightarrows T_2$.

Note that $\leftrightarrows$ is an equivalence on $\mathbb{TS}_\kappa$. Let $T \in \mathbb{TS}_\kappa$. Then we write $[T]$ for $\{T' \in \mathbb{TS}_\kappa \mid T \leftrightarrows T'\}$, i.e. the $\leftrightarrows$-equivalence class of T. We write $\mathbb{TS}_\kappa/\!\leftrightarrows$ for the set of equivalence classes $\{[T] \mid T \in \mathbb{TS}_\kappa\}$.

In Sect. 5, we will use $\mathbb{TS}_\kappa/\!\leftrightarrows$ as the domain of a structure that is a model of $\mathrm{BPA}_\delta^{\mathrm{fo}}$. As the domain of a structure, $\mathbb{TS}_\kappa/\!\leftrightarrows$ must be a set. That is the case because $\mathbb{TS}_\kappa$ is a set. The latter is guaranteed by considering only connected transition systems of which the set of states is a subset of $\mathcal{S}_\kappa$.

Remark 2. The question arises whether $\mathcal{S}_\kappa$ is large enough if its cardinality is greater than or equal to κ. This question can be answered in the affirmative. Let $T = (S, \rightarrow, \rightarrow\surd, s^0)$ be a connected transition system of which the branching degree is less than κ. Then there exists a connected transition system $T' = (S', \rightarrow', \rightarrow\surd', s^{0\prime})$ of which the branching degree is less than κ such that $T \leftrightarrows T'$ and the cardinality of S' is less than κ.

It is easy to see that, if we would consider transition systems with unreachable states as well, each transition system would be bisimilar to its connected part. This justifies the choice to consider only connected transition systems. It is easy to see that isomorphic transition systems are bisimilar. This justifies the choice to consider transition systems essentially the same if they are isomorphic.

In the construction of the main models of $\mathrm{BPA}_\delta^{\mathrm{fo}}$ in Sect. 5, we also make use of subsystems of transition systems.

Let $T = (S, \rightarrow, \rightarrow\surd, s^0) \in \mathbb{TS}_\kappa$ and $s \in S$. Then the *subsystem* of T with initial state s, written $(T)_s$, is defined as follows:

$$(T)_s = \Gamma(S, \rightarrow, \rightarrow\surd, s) .$$

5 Full Bisimulation Models of $\text{BPA}_\delta^{\text{fo}}$

In this section, we introduce the full bisimulation models of $\text{BPA}_\delta^{\text{fo}}$. They are models of which the domain consists of equivalence classes of connected transition systems modulo bisimilarity. The qualification "full" will be explained later on.

The models of $\text{BPA}_\delta^{\text{fo}}$ are structures that consist of the following:

- a non-empty set $\mathcal{D}$, called the *domain* of the model;
- for each constant of $\text{BPA}_\delta^{\text{fo}}$, an element of $\mathcal{D}$;
- for each n-ary operator of $\text{BPA}_\delta^{\text{fo}}$, an n-ary operation on $\mathcal{D}$;
- for each n-ary predicate symbol of $\text{BPA}_\delta^{\text{fo}}$, an n-ary relation on $\mathcal{D}$.

In the full bisimulation models of $\text{BPA}_\delta^{\text{fo}}$ that are introduced in this section, the domain is $\mathbb{TS}_\kappa/\leftrightarrows$ for some $\kappa \geq \aleph_0$. We obtain the models concerned by associating certain elements of $\mathbb{TS}_\kappa/\leftrightarrows$, certain operations on $\mathbb{TS}_\kappa/\leftrightarrows$ and certain relations on $\mathbb{TS}_\kappa/\leftrightarrows$ with the constants, operators and predicate symbols of $\text{BPA}_\delta^{\text{fo}}$. We begin by associating elements of $\mathbb{TS}_\kappa$ and operations on $\mathbb{TS}_\kappa$ with the constants and operators, and a binary relation on $\mathbb{TS}_\kappa$ with the reachability predicate symbol. The result of this is subsequently lifted to $\mathbb{TS}_\kappa/\leftrightarrows$.

It is assumed that for each infinite cardinal κ a fixed but arbitrary function $\text{ch}_\kappa : (\mathcal{P}(\mathcal{S}_\kappa) \setminus \emptyset) \rightarrow \mathcal{S}_\kappa$ such that for all $S \in \mathcal{P}(\mathcal{S}_\kappa) \setminus \emptyset$, $\text{ch}_\kappa(S) \in S$ has been given.

We associate with each constant c of $\text{BPA}_\delta^{\text{fo}}$ an element $\widehat{c}$ of $\mathbb{TS}_\kappa$ and with each operator f of $\text{BPA}_\delta^{\text{fo}}$ an operation $\widehat{f}$ on $\mathbb{TS}_\kappa$ as follows.

- $$\widehat{\delta} = (\{s^0\}, \emptyset, \emptyset, s^0) .$$
  where
  $$s^0 = \text{ch}_\kappa(\mathcal{S}_\kappa) .$$
- $$\widehat{a} = (\{s^0\}, \emptyset, \rightarrow\surd, s^0) ,$$
  where
  $$\begin{aligned} s^0 &= \text{ch}_\kappa(\mathcal{S}_\kappa) , \\ \xrightarrow{a}\surd &= \{s^0\} , \end{aligned}$$
  and for every $a' \in \mathsf{A}$ such that $a' \neq a$:
  $$\xrightarrow{a'}\surd = \emptyset .$$

– Let $T_i = (S_i, \rightarrow_i, \rightarrow\surd_i, s_i^0) \in \mathbb{TS}_\kappa$ for $i = 1, 2$. Then

$$T_1 \mathbin{\hat{+}} T_2 = \Gamma(S, \rightarrow, \rightarrow\surd, s^0)\;,$$

where

$$s^0 = \mathrm{ch}_\kappa(\mathcal{S}_\kappa \setminus (S_1 \uplus S_2))\;,$$

$$S = \{s^0\} \cup (S_1 \uplus S_2)\;,$$

and for every $a \in \mathsf{A}$:

$$\begin{aligned}\xrightarrow{a} &= \{(s^0, \mu_1(s)) \mid s_1^0 \xrightarrow{a}_1 s\} \cup \{(s^0, \mu_2(s)) \mid s_2^0 \xrightarrow{a}_2 s\} \\ &\cup \{(\mu_1(s), \mu_1(s')) \mid s \xrightarrow{a}_1 s'\} \cup \{(\mu_2(s), \mu_2(s')) \mid s \xrightarrow{a}_2 s'\}\;,\end{aligned}$$

$$\begin{aligned}\xrightarrow{a}\surd &= \{s^0 \mid s_1^0 \xrightarrow{a}\surd_1\} \cup \{s^0 \mid s_2^0 \xrightarrow{a}\surd_2\} \\ &\cup \{\mu_1(s) \mid s \xrightarrow{a}\surd_1\} \cup \{\mu_2(s) \mid s \xrightarrow{a}\surd_2\}\;.\end{aligned}$$

– Let $T_i = (S_i, \rightarrow_i, \rightarrow\surd_i, s_i^0) \in \mathbb{TS}_\kappa$ for $i = 1, 2$. Then

$$T_1 \mathbin{\hat{\cdot}} T_2 = \Gamma(S, \rightarrow, \rightarrow\surd, s^0)\;,$$

where

$$S = S_1 \uplus S_2\;,$$

$$s^0 = \mu_1(s_1^0)\;,$$

and for every $a \in \mathsf{A}$:

$$\begin{aligned}\xrightarrow{a} &= \{(\mu_1(s), \mu_1(s')) \mid s \xrightarrow{a}_1 s'\} \cup \{(\mu_1(s), \mu_2(s_2^0)) \mid s \xrightarrow{a}\surd_1\} \\ &\cup \{(\mu_2(s), \mu_2(s')) \mid s \xrightarrow{a}_2 s'\}\;,\end{aligned}$$

$$\xrightarrow{a}\surd = \{\mu_2(s) \mid s \xrightarrow{a}\surd_2\}\;.$$

We associate with the reachability predicate symbol $\twoheadrightarrow$ a relation $\widehat{\twoheadrightarrow}$ on $\mathbb{TS}_\kappa$ as follows.

– Let $T_i = (S_i, \rightarrow_i, \rightarrow\surd_i, s_i^0) \in \mathbb{TS}_\kappa$ for $i = 1, 2$. Then

$$T_1 \mathrel{\widehat{\twoheadrightarrow}} T_2 \text{ iff } \exists s \in S_1 \bullet (T_1)_s = T_2\;.$$

In the definition of alternative composition on $\mathbb{TS}_\kappa$, the connected part of a transition system is extracted because the initial states of the transition systems T_1 and T_2 may be unreachable from the new initial state. The new initial state is introduced because, in T_1 and/or T_2, there may exist a transition back to the initial state. In the definition of sequential composition on $\mathbb{TS}_\kappa$, the connected part of a transition system is extracted because the initial state of the transition system T_2 may be unreachable from the initial state of the transition system T_1 – due to absence of termination in T_1.

We do not associate relations on $\mathbb{TS}_\kappa$ with the summand inclusion, action termination and action step predicate symbols. They have defining axioms, which explicitly define them in terms of the other nonlogical symbols of $\mathrm{BPA}_\delta^{\mathrm{fo}}$. Therefore, it is known how to obtain the relations on $\mathbb{TS}_\kappa/\!\leftrightarroweq$ to be associated with these predicate symbols from the elements of $\mathbb{TS}_\kappa/\!\leftrightarroweq$, operations on $\mathbb{TS}_\kappa/\!\leftrightarroweq$ and relations on $\mathbb{TS}_\kappa/\!\leftrightarroweq$ to be associated with the other nonlogical symbols of $\mathrm{BPA}_\delta^{\mathrm{fo}}$.

Remark 3. The elements of $\mathbb{TS}_\kappa$ and the operations on $\mathbb{TS}_\kappa$ defined above are independent of ch_κ. Different choices for ch_κ lead for each constant of $\mathrm{BPA}^{\mathrm{fo}}_\delta$ to isomorphic elements of $\mathbb{TS}_\kappa$ and lead for each operator of $\mathrm{BPA}^{\mathrm{fo}}_\delta$ to operations on $\mathbb{TS}_\kappa$ with isomorphic results.

We can easily show that bisimilarity is a congruence with respect to alternative composition and sequential composition.

Proposition 3 (Congruence). *For all $T_1, T_2, T'_1, T'_2 \in \mathbb{TS}_\kappa$ $(\kappa \geq \aleph_0)$, $T_1 \mathrel{\underline{\leftrightarrow}} T'_1$ and $T_2 \mathrel{\underline{\leftrightarrow}} T'_2$ imply $T_1 \mathbin{\hat{+}} T_2 \mathrel{\underline{\leftrightarrow}} T'_1 \mathbin{\hat{+}} T'_2$ and $T_1 \mathbin{\hat{\cdot}} T_2 \mathrel{\underline{\leftrightarrow}} T'_1 \mathbin{\hat{\cdot}} T'_2$.*

Proof. Let $T_i = (S_i, \rightarrow_i, \rightarrow\!\surd_i, s^0_i)$ and $T'_i = (S'_i, \rightarrow'_i, \rightarrow\!\surd'_i, s^{0\prime}_i)$ for $i = 1, 2$. Let R_1 and R_2 be bisimulations witnessing $T_1 \mathrel{\underline{\leftrightarrow}} T'_1$ and $T_2 \mathrel{\underline{\leftrightarrow}} T'_2$, respectively. Then we construct relations $R_{\hat{+}}$ and $R_{\hat{\cdot}}$ as follows:

- $R_{\hat{+}} = (\{(s^0, s^{0\prime})\} \cup \mu_1(R_1) \cup \mu_2(R_2)) \cap (S \times S')$, where S and S' are the sets of states of $T_1 \mathbin{\hat{+}} T_2$ and $T'_1 \mathbin{\hat{+}} T'_2$, respectively, and s^0 and $s^{0\prime}$ are the initial states of $T_1 \mathbin{\hat{+}} T_2$ and $T'_1 \mathbin{\hat{+}} T'_2$, respectively;
- $R_{\hat{\cdot}} = (\mu_1(R_1) \cup \mu_2(R_2)) \cap (S \times S')$, where S and S' are the sets of states of $T_1 \mathbin{\hat{\cdot}} T_2$ and $T'_1 \mathbin{\hat{\cdot}} T'_2$, respectively.

Here, we write $\mu_i(R_i)$ for $\{(\mu_i(s), \mu_i(s')) \mid R_i(s, s')\}$, where μ_i is used to denote both the injection of S_i into $S_1 \uplus S_2$ and the injection of S'_i into $S'_1 \uplus S'_2$. Given the definitions of alternative composition and sequential composition, it is easy to see that $R_{\hat{+}}$ and $R_{\hat{\cdot}}$ are bisimulations witnessing $T_1 \mathbin{\hat{+}} T_2 \mathrel{\underline{\leftrightarrow}} T'_1 \mathbin{\hat{+}} T'_2$ and $T_1 \mathbin{\hat{\cdot}} T_2 \mathrel{\underline{\leftrightarrow}} T'_1 \mathbin{\hat{\cdot}} T'_2$, respectively. □

The *full bisimulation models* $\mathfrak{P}_\kappa$, one for each $\kappa \geq \aleph_0$, consist of the following:[3]

- a set $\mathcal{P}$, called the domain of $\mathfrak{P}_\kappa$;
- for each constant c of $\mathrm{BPA}^{\mathrm{fo}}_\delta$, an element $\tilde{c}$ of $\mathcal{P}$;
- for each n-ary operator f of $\mathrm{BPA}^{\mathrm{fo}}_\delta$, an n-ary operation $\tilde{f}$ on $\mathcal{P}$;
- for each n-ary predicate symbol R of $\mathrm{BPA}^{\mathrm{fo}}_\delta$, a n-ary relation $\tilde{R}$ on $\mathcal{P}$;

where those ingredients are defined as follows:

$$
\begin{array}{lll}
\mathcal{P} = \mathbb{TS}_\kappa / \underline{\leftrightarrow}\,, & & \\
\tilde{\delta} = [\,\hat{\delta}\,]\,, & [T_1] \mathrel{\tilde{\sqsubseteq}} [T_2] & \text{iff } [T_1] \mathbin{\tilde{+}} [T_2] = [T_2]\,, \\
\tilde{a} = [\,\hat{a}\,]\,, & [T_1] \mathrel{\widetilde{\xrightarrow{a}}} \surd & \text{iff } \tilde{a} \mathrel{\tilde{\sqsubseteq}} [T_1]\,, \\
[T_1] \mathbin{\tilde{+}} [T_2] = [T_1 \mathbin{\hat{+}} T_2]\,, & [T_1] \mathrel{\widetilde{\xrightarrow{a}}} [T_2] & \text{iff } \tilde{a} \mathbin{\tilde{\cdot}} [T_2] \mathrel{\tilde{\sqsubseteq}} [T_1]\,, \\
[T_1] \mathbin{\tilde{\cdot}} [T_2] = [T_1 \mathbin{\hat{\cdot}} T_2]\,, & [T_1] \mathrel{\tilde{\twoheadrightarrow}} [T_2] & \text{iff } \exists T \in [T_2] \bullet T_1 \mathrel{\hat{\twoheadrightarrow}} T\,.
\end{array}
$$

Alternative composition and sequential composition on $\mathbb{TS}_\kappa / \underline{\leftrightarrow}$ are well-defined because $\underline{\leftrightarrow}$ is a congruence with respect to the corresponding operations on $\mathbb{TS}_\kappa$. Reachability on $\mathbb{TS}_\kappa / \underline{\leftrightarrow}$ is well-defined because $\underline{\leftrightarrow}$ preserves reachability on $\mathbb{TS}_\kappa$ up to $\underline{\leftrightarrow}$: if $T_1 \mathrel{\underline{\leftrightarrow}} T'_1$ and $T_1 \mathrel{\hat{\twoheadrightarrow}} T_2$, then there exists a T'_2 such that $T_2 \mathrel{\underline{\leftrightarrow}} T'_2$ and $T'_1 \mathrel{\hat{\twoheadrightarrow}} T'_2$.

The structures $\mathfrak{P}_\kappa$ are models of $\mathrm{BPA}^{\mathrm{fo}}_\delta$.

[3] $\mathfrak{P}$ is the Gothic capital P.

Theorem 4 (Soundness of $\mathbf{BPA}_\delta^{\mathrm{fo}}$). *For all $\kappa \geq \aleph_0$, we have $\mathfrak{P}_\kappa \models \mathrm{BPA}_\delta^{\mathrm{fo}}$.*

Proof. The soundness of all axioms, except RDPf and RSPf, follows easily from the definitions of the ingredients of $\mathfrak{P}_\kappa$. The soundness of RDPf and RSPf follows immediately from Theorem 5 (see below), which states the soundness of RDP and RSP. □

All finite and countably infinite guarded recursive specifications have a unique solution in the full bisimulation models.

Theorem 5 (Soundness of RDP and RSP). *For all $\kappa \geq \aleph_0$, we have $\mathfrak{P}_\kappa \models \mathrm{RDP}$ and $\mathfrak{P}_\kappa \models \mathrm{RSP}$.*

Proof. This is essentially the proof of soundness of RDP and RSP in the graph models of ACP_τ given in [11] adapted to the case without silent steps. □

Moreover, B and R are valid in the full bisimulation models.

Theorem 6 (Soundness of B and R). *For all $\kappa \geq \aleph_0$, we have $\mathfrak{P}_\kappa \models \mathrm{B}$ and $\mathfrak{P}_\kappa \models \mathrm{R}$.*

Proof. The soundness of B follows easily from the definitions of $\widetilde{\xrightarrow{a}\surd}$ and $\widetilde{\xrightarrow{a}}$, the definition of bisimilarity of transition systems and Proposition 4. The soundness of R follows easily from the definitions of $\widetilde{\xrightarrow{a}}$ and $\widetilde{\twoheadrightarrow}$, the definition of the reachability relation of a transition system and Corollary 2.[4] □

As to be expected, the full bisimulation models are related by isomorphic embeddings.

Theorem 7 (Isomorphic embedding). *Let $\aleph_0 \leq \kappa < \kappa'$. Then $\mathfrak{P}_\kappa$ is isomorphically embedded in $\mathfrak{P}_{\kappa'}$.*

Proof. It follows immediately from the definitions of $\mathbb{TS}_\kappa$, $\mathbb{TS}_{\kappa'}$ and $\leftrightarroweq$ that for each $p \in \mathbb{TS}_\kappa/{\leftrightarroweq}$, there exists a unique $p' \in \mathbb{TS}_{\kappa'}/{\leftrightarroweq}$ such that $p \subseteq p'$. Now consider the function $h : \mathbb{TS}_\kappa/{\leftrightarroweq} \to \mathbb{TS}_{\kappa'}/{\leftrightarroweq}$ where for each $p \in \mathbb{TS}_\kappa/{\leftrightarroweq}$, $h(p)$ is the unique $p' \in \mathbb{TS}_{\kappa'}/{\leftrightarroweq}$ such that $p \subseteq p'$. It follows immediately from the definition of h that h is injective. Moreover, it follows easily from the definitions of the operations and relations on $\mathbb{TS}_\kappa/{\leftrightarroweq}$ and $\mathbb{TS}_{\kappa'}/{\leftrightarroweq}$ that h is a homomorphism from $\mathfrak{P}_\kappa$ to $\mathfrak{P}_{\kappa'}$. □

In Sect. 6, we will show that every bisimulation model with standard reachability, i.e. every model that additionally satisfies the second-order axioms B and R, is isomorphically embedded in the models $\mathfrak{P}_\kappa$ from some $\kappa \geq \aleph_0$. This explains why the models $\mathfrak{P}_\kappa$ are called full bisimulation models: within the bound on the branching degree set by κ, $\mathfrak{P}_\kappa$ is full.

The question whether all full bisimulation models are elementary equivalent must be answered in the negative.

[4] Proposition 4 and Corollary 2 are in Sect. 6 and Sect. 10, respectively, because they need definitions of auxiliary notions which are better in place in there.

Theorem 8 (No elementary equivalence). *We have* $\mathfrak{P}_{\aleph_0} \not\equiv \mathfrak{P}_{2^{\aleph_0}}$, $\mathfrak{P}_{\aleph_0} \not\equiv \mathfrak{P}_{2^{2^{\aleph_0}}}$ *and* $\mathfrak{P}_{2^{\aleph_0}} \not\equiv \mathfrak{P}_{2^{2^{\aleph_0}}}$.

Proof. $\mathfrak{P}_{\aleph_0} \not\equiv \mathfrak{P}_{2^{\aleph_0}}$ and $\mathfrak{P}_{\aleph_0} \not\equiv \mathfrak{P}_{2^{2^{\aleph_0}}}$ are proved as follows. Let a be an action. Let ϕ be the following formula of $\mathcal{L}(\mathrm{BPA}_\delta^{\mathrm{fo}})$:

$$\exists x \bullet (x \xrightarrow{a} \delta \wedge \forall y \bullet (x \xrightarrow{a} y \Rightarrow \exists z \bullet (z \neq y \wedge x \xrightarrow{a} z \wedge z \xrightarrow{a} y))) .$$

Clearly, $\mathfrak{P}_{\aleph_0} \not\models \phi$, but $\mathfrak{P}_{2^{\aleph_0}} \models \phi$ and $\mathfrak{P}_{2^{2^{\aleph_0}}} \models \phi$.

$\mathfrak{P}_{2^{\aleph_0}} \not\equiv \mathfrak{P}_{2^{2^{\aleph_0}}}$ is proved as follows. Let a, a', b, b' be different actions. Let $\phi(x)$ be the following formula of $\mathcal{L}(\mathrm{BPA}_\delta^{\mathrm{fo}})$:

$$\forall y \bullet (x \twoheadrightarrow y \Rightarrow \exists! z \bullet y \xrightarrow{a} z \wedge \neg (y \xrightarrow{a'} \surd \Leftrightarrow y \xrightarrow{b'} \surd)) .$$

For all $\kappa \geq \aleph_0$, there exist $2^{\aleph_0}$ different x in the domain of $\mathfrak{P}_\kappa$ for which $\phi(x)$. Let ψ be the following formula of $\mathcal{L}(\mathrm{BPA}_\delta^{\mathrm{fo}})$:

$$\exists w \bullet (\forall x \bullet (\phi(x) \Rightarrow w \xrightarrow{b} x)) .$$

Clearly, $\mathfrak{P}_{2^{\aleph_0}} \not\models \psi$ and $\mathfrak{P}_{2^{2^{\aleph_0}}} \models \psi$. □

We conjecture that there exists a countably infinite set of infinite cardinal numbers $\mathcal{U}$ such that, for $\kappa, \kappa' \in \mathcal{U}$, $\mathfrak{P}_\kappa \not\equiv \mathfrak{P}_{\kappa'}$ if $\kappa \neq \kappa'$.

We can summarize the state of affairs as follows. The full bisimulation models $\mathfrak{P}_\kappa$ are models of $\mathrm{BPA}_\delta^{\mathrm{fo}}$ in which RDP, RSP, B and R are valid. If $\kappa < \kappa'$, then $\mathfrak{P}_\kappa$ is essentially included in $\mathfrak{P}_{\kappa'}$. Moreover, not all full bisimulation models satisfy exactly the same formulas of $\mathcal{L}(\mathrm{BPA}_\delta^{\mathrm{fo}})$. In subsequent sections, we will see that the full bisimulation models have many more interesting properties.

6 External Bisimilarity

Each model of $\mathrm{BPA}_\delta^{\mathrm{fo}}$ induces a transition system for each element of its domain.

Let $\mathfrak{A}$ be a model of $\mathrm{BPA}_\delta^{\mathrm{fo}}$ with domain P, a binary relation $\xrightarrow{a}'$ on P for each predicate symbol $\xrightarrow{a}$, and a unary relation $\xrightarrow{a}\surd'$ on P for each predicate symbol $\xrightarrow{a}\surd$. Moreover, let $p \in P$. Then the transition system of p *induced by* $\mathfrak{A}$, written $\mathrm{TS}(\mathfrak{A}, p)$, is defined as follows:

$$\mathrm{TS}(\mathfrak{A}, p) = \Gamma(P, \rightarrow', \rightarrow\surd', p) .$$

In each of the full bisimulation models, every element of the domain is an equivalence class of transition systems. The transition system of an element induced by the model is (up to isomorphism) a representative of that element.

Lemma 1 ($\mathfrak{P}_\kappa$ induces representatives). *Let* $p \in \mathbb{TS}_\kappa/{\leftrightarrows}$ *for some* $\kappa \geq \aleph_0$. *Then* $\mathrm{TS}(\mathfrak{P}_\kappa, p) \in p$.

Proof. Let $\mathrm{TS}(\mathfrak{P}_\kappa, p) = (P, \rightarrow', \rightarrow\surd', p)$. Take an arbitrary transition system $T = (S, \rightarrow'', \rightarrow\surd'', s^0) \in \mathbb{TS}_\kappa$ such that $[T] = p$. Consider the relation $B \subseteq P \times S$ defined as follows:

$$B = \{([(T)_s], s) \mid s \in S\} .$$

It is easy to see that B is a bisimulation between $\mathrm{TS}(\mathfrak{P}_\kappa, p)$ and T. Hence, $\mathrm{TS}(\mathfrak{P}_\kappa, p) \in [\, T\,] = p$. □

Let $\mathfrak{A}$ be a model of $\mathrm{BPA}_\delta^{\mathrm{fo}}$ with domain P. Then bisimilarity on P is defined as follows:

$$p_1 \leftrightarroweq_{\mathfrak{A}} p_2 \quad \text{iff} \quad \mathrm{TS}(\mathfrak{A}, p_1) \leftrightarroweq \mathrm{TS}(\mathfrak{A}, p_2)\ .$$

Bisimilarity on the domain of a model of $\mathrm{BPA}_\delta^{\mathrm{fo}}$ as defined above is called *external bisimilarity*. In each of the full bisimulation models, external bisimilarity coincides with identity.

Proposition 4 (External bisimilarity is identity in $\mathfrak{P}_\kappa$). *Let $p_1, p_2 \in \mathbb{TS}_\kappa/\!\leftrightarroweq$ for some $\kappa \geq \aleph_0$. Then $p_1 \leftrightarroweq_{\mathfrak{P}_\kappa} p_2$ iff $p_1 = p_2$.*

Proof. Follows immediately from Lemma 1. □

There does not exist a consistent extension of $\mathrm{BPA}_\delta^{\mathrm{fo}}$ with first-order axioms that has only models in which external bisimilarity coincides with identity.

Theorem 9 (Undefinability of external bisimilarity). *Each first-order consistent extension of* $\mathrm{BPA}_\delta^{\mathrm{fo}}$ *has a model in which external bisimilarity is not identity.*

Proof. Suppose that there exists a first-order consistent extension of $\mathrm{BPA}_\delta^{\mathrm{fo}}$, say $\mathrm{BPA}_\delta^{\mathrm{fo}} \cup H$, that has only models in which external bisimilarity is identity. A contradiction is found as follows. Let $c_0, c_1, c_2, \ldots$ and $d_0, d_1, d_2, \ldots$ be different new constants; and let a, a', a'' be different actions. Consider the following sets of formulas:

$$\begin{aligned} H' &= \{c_0 \neq d_0\} \cup \{c_i = a \cdot c_{i+1} \mid i \geq 0\} \cup \{d_i = a \cdot d_{i+1} \mid i \geq 0\}\ , \\ H'_n &= \{c_0 \neq d_0\} \cup \{c_i = a \cdot c_{i+1} \mid 0 \leq i < n\} \cup \{d_i = a \cdot d_{i+1} \mid 0 \leq i < n\} \\ &\quad \cup \{c_n = a', d_n = a''\} \\ &(\text{for } n \geq 0)\ . \end{aligned}$$

Take an arbitrary model $\mathfrak{A}$ of $\mathrm{BPA}_\delta^{\mathrm{fo}} \cup H$. It follows easily from the axioms of $\mathrm{BPA}_\delta^{\mathrm{fo}}$ that, for each $n \geq 0$, H'_n is satisfied in the definitional expansion of $\mathfrak{A}$ determined by the definitional extension of $\mathrm{BPA}_\delta^{\mathrm{fo}} \cup H$ with the constants $c_0, \ldots, c_n$, $d_0, \ldots, d_n$ and the equations $c_i = a^{n-i} \cdot a'$ for $0 \leq i < n$, $c_n = a'$, $d_i = a^{n-i} \cdot a''$ for $0 \leq i < n$, $d_n = a''$. Hence, for each $n \geq 0$, H'_n is consistent with $\mathrm{BPA}_\delta^{\mathrm{fo}} \cup H$. Each finite $H'' \subseteq H'$ is consistent with $\mathrm{BPA}_\delta^{\mathrm{fo}} \cup H$ because there is an $n \geq 0$ for which $H'' \subseteq H'_n$. From this, it follows by the Compactness Theorem that H' is consistent with $\mathrm{BPA}_\delta^{\mathrm{fo}} \cup H$. Now consider an arbitrary model $\mathfrak{A}'$ of $\mathrm{BPA}_\delta^{\mathrm{fo}} \cup H \cup H'$. Because $\mathfrak{A}'$ satisfies H', we have $c_0^{\mathfrak{A}'} \neq d_0^{\mathfrak{A}'}$. Since $\mathrm{TS}(\mathfrak{A}', c_0^{\mathfrak{A}'})$ and $\mathrm{TS}(\mathfrak{A}', d_0^{\mathfrak{A}'})$ are isomorphic transition systems, we have $c_0^{\mathfrak{A}'} \leftrightarroweq_{\mathfrak{A}'} d_0^{\mathfrak{A}'}$. Hence, because external bisimilarity is identity, it must be the case that $c_0^{\mathfrak{A}'} = d_0^{\mathfrak{A}'}$, which contradicts the fact that $c_0^{\mathfrak{A}'} \neq d_0^{\mathfrak{A}'}$. □

We can summarize the state of affairs as follows. It is obvious that equality derivable from $\mathrm{BPA}_\delta^{\mathrm{fo}}$ implies external bisimilarity in each model of $\mathrm{BPA}_\delta^{\mathrm{fo}}$. In the full bisimulation models, external bisimilarity coincides with identity. However, there also exist models of which the domain contains pairs of different elements that are externally bisimilar. Moreover, those models cannot be excluded by extending $\mathrm{BPA}_\delta^{\mathrm{fo}}$ with first-order axioms.

The above-mentioned discrepancy can for the greater part be eliminated in second-order logic, as indicated below by Theorem 10. This theorem states that each bisimulation model with standard reachability is isomorphic to a substructure of one of the full bisimulation models.

Theorem 10 (Isomorphic embedding). *Let* $\mathfrak{A}$ *be a model of* $\mathrm{BPA}_\delta^{\mathrm{fo}}$ *such that* $\mathfrak{A} \models \mathrm{R}$. *Then* $\mathfrak{A} \models \mathrm{B}$ *iff* $\mathfrak{A}$ *is isomorphically embedded in* $\mathfrak{P}_\kappa$ *for some* $\kappa \geq \aleph_0$.

Proof. The implication from left to right is proved as follows. Let P be the domain of $\mathfrak{A}$, κ' be the cardinality of P, and $\kappa > \kappa'$. It follows immediately from the definitions of TS and $\mathbb{TS}_\kappa$ that for each $p \in P$, $\mathrm{TS}(\mathfrak{A}, p) \in \mathbb{TS}_\kappa$. Now consider the function $h{:}P \to \mathbb{TS}_\kappa/{\leftrightarrows}$ such that for each $p \in P$, $h(p) = [\,\mathrm{TS}(\mathfrak{A}, p)\,]$. Because $\mathfrak{A} \models \mathrm{B}$, it follows immediately that h is injective. Because the implications from right to left are derivable, the occurrence of "$\Rightarrow$" in axioms SI7–SI9 (Table 1) can be replaced by "$\Leftrightarrow$". It follows easily from these equivalences and the definitions of alternative composition and sequential composition on $\mathbb{TS}_\kappa/{\leftrightarrows}$ (Sect. 5) that h is a homomorphism with respect to these operations. From this, it follows immediately by axioms SI1, TR1 and TR2 that h is also a homomorphism with respect to the summand inclusion, action termination and action step relations. Because $\mathfrak{A} \models \mathrm{R}$, it follows immediately that h is a homomorphism with respect to the reachability relation. The implication from right to left is trivial. □

Models of $\mathrm{BPA}_\delta^{\mathrm{fo}}$ other than bisimulation models with standard reachability are to $\mathrm{BPA}_\delta^{\mathrm{fo}}$ as nonstandard models of number theory are to number theory.

7 Observational Equivalence

In this section, we have a closer look at observational equivalence as defined in [12]. This equivalence on the domain of models of $\mathrm{BPA}_\delta^{\mathrm{fo}}$ is closely related to external bisimilarity. Observational equivalence is defined in the following way.

Let $\mathfrak{A}$ be a model of $\mathrm{BPA}_\delta^{\mathrm{fo}}$ with domain P, a binary relation $\xrightarrow{a}'$ on P for each predicate symbol $\xrightarrow{a}$, and a unary relation $\xrightarrow{a}\surd'$ on P for each predicate symbol $\xrightarrow{a}\surd$. Then equivalences $\sim_n \subseteq P \times P$ for each $n \geq 0$ are defined as follows:

- $p_1 \sim_0 p_2$ for all $p_1, p_2 \in P$;
- $p_1 \sim_{n+1} p_2$ if
  - $p_1 \xrightarrow{a}\surd'$ iff $p_2 \xrightarrow{a}\surd'$;
  - if $p_1 \xrightarrow{a}' p_1'$, then there is a $p_2' \in P$ such that $p_2 \xrightarrow{a}' p_2'$ and $p_1' \sim_n p_2'$;
  - if $p_2 \xrightarrow{a}' p_2'$, then there is a $p_1' \in P$ such that $p_1 \xrightarrow{a}' p_1'$ and $p_1' \sim_n p_2'$.

Table 4. Approximation induction principle

$x \sim_0 y$	$\mathrm{OBS_0}$
$x \sim_{n+1} y \Leftrightarrow$ $\bigwedge_{a \in \mathsf{A}}((x \xrightarrow{a} \surd \Leftrightarrow y \xrightarrow{a} \surd) \wedge$ $\forall x' \bullet (x \xrightarrow{a} x' \Rightarrow \exists y' \bullet (y \xrightarrow{a} y' \wedge x' \sim_n y')) \wedge$ $\forall y' \bullet (y \xrightarrow{a} y' \Rightarrow \exists x' \bullet (x \xrightarrow{a} x' \wedge x' \sim_n y')))$	$\mathrm{OBS_{n+1}}$
$x \sim y \Leftrightarrow \bigwedge_{n \geq 0} x \sim_n y$	OBS
$x \sim y \Rightarrow x = y$	AIP

Now, p_1 and p_2 are *observationally equivalent* in $\mathfrak{A}$, written $p_1 \sim_{\mathfrak{A}} p_2$, if $p_1 \sim_n p_2$ for all $n \geq 0$.

If all transition systems that can be extracted from a model are finitely branching, then observational equivalence and external bisimilarity coincide.

Theorem 11 (Observational equivalence vs external bisimilarity). *Let $\mathfrak{A}$ be a model of* $\mathrm{BPA}_\delta^{\mathrm{fo}}$ *with domain P. Then $\sim_{\mathfrak{A}} = \leftrightarroweq_{\mathfrak{A}}$ if $\mathrm{TS}(\mathfrak{A}, p) \in \mathbb{TS}_{\aleph_0}$ for all $p \in P$.*

Proof. The proof is analogous to the proof of the corresponding property for process graphs given in [13]. □

An interesting extension of $\mathrm{BPA}_\delta^{\mathrm{fo}}$ is obtained as follows. We add to the non-logical symbols of $\mathrm{BPA}_\delta^{\mathrm{fo}}$, for each $n \geq 0$, a binary *observational equivalence up to depth n* predicate symbol $\sim_n$ and a binary *observational equivalence* predicate symbol $\sim$. Moreover, we add the axioms given in Table 4 to the axioms of $\mathrm{BPA}_\delta^{\mathrm{fo}}$. $\mathrm{OBS_{n+1}}$ is actually an axiom schema with an instance for each $n \geq 0$.

Axiom $\mathrm{OBS_0}$ is the defining axiom of the observational equivalence up to depth 0 predicate; and $\mathrm{OBS_{n+1}}$ is an axiom schema whose instances are the defining axioms of the observational equivalence up to depth $n+1$ predicates. Axiom OBS is the defining axiom of the observational equivalence predicate. Axiom AIP is called the *approximation induction principle.*

We write $\mathfrak{P}_\kappa^\sim$ ($\kappa \geq \aleph_0$) for the unique definitional expansion of $\mathfrak{P}_\kappa$ determined by the definitional extension of $\mathrm{BPA}_\delta^{\mathrm{fo}}$ with the binary predicate symbols $\sim_0$, $\sim_1$, $\sim_2$, ... and $\sim$ and axioms $\mathrm{OBS_0}$, $\mathrm{OBS_1}$, ... and OBS. AIP is valid in $\mathfrak{P}_{\aleph_0}^\sim$, but not in $\mathfrak{P}_\kappa^\sim$ with $\kappa \geq \aleph_1$.

Theorem 12 (Soundness of AIP). *We have $\mathfrak{P}_\kappa^\sim \models$ AIP iff $\kappa = \aleph_0$.*

Proof. It follows immediately from Proposition 4 and Theorem 11 that $\mathfrak{P}_\kappa^\sim \models$ AIP if $\kappa = \aleph_0$. For $\kappa > \aleph_0$, we have the following counterexample. Fix an $a \in \mathsf{A}$. Consider the transition systems $T_1 = (S_1, \rightarrow_1, \emptyset, 0)$ and $T_2 = (S_2, \rightarrow_2, \emptyset, 0)$ where

$$S_1 = \{0\} \cup \{(i,j) \mid i,j \in \mathbb{N}, i \geq j \geq 1\} ,$$

$$\xrightarrow{a}_1 = \{(0,(i,1)) \mid i \in \mathbb{N}, i \geq 1\} \cup \{((i,j),(i,j+1)) \mid i,j \in \mathbb{N}, i > j \geq 1\} ,$$

$$\xrightarrow{a'}_1 = \emptyset \quad \text{for every } a' \in \mathsf{A} \text{ such that } a' \neq a,$$

and

$$S_2 = S_1 \cup \mathbb{N}\,,$$
$$\xrightarrow{a}_2 = \xrightarrow{a}_1 \cup \{(i, i+1) \mid i \in \mathbb{N}\}\,,$$
$$\xrightarrow{a'}_2 = \emptyset \quad \text{for every } a' \in \mathsf{A} \text{ such that } a' \neq a.$$

Clearly, $T_1, T_2 \notin \mathbb{TS}_{\aleph_0}$. Because T_1 has no infinite branch and T_2 has an infinite branch, $T_1 \not\leftrightarrow_{\mathfrak{P}^{\sim}_{\kappa}} T_2$. However, $T_1 \sim_{\mathfrak{P}^{\sim}_{\kappa}} T_2$. □

All models of $\mathrm{BPA}_{\delta}^{\mathrm{fo}} \cup \mathrm{AIP}$ satisfy B. Here, and in Theorem 23, we abuse the name AIP for the set of axioms $\{\mathrm{OBS}_n \mid n \geq 0\} \cup \{\mathrm{OBS}, \mathrm{AIP}\}$.

Proposition 5 (AIP implies B). *We have* $\mathrm{BPA}_{\delta}^{\mathrm{fo}} \cup \mathrm{AIP} \models \mathrm{B}$.

Proof. Take a model $\mathfrak{A}$ of $\mathrm{BPA}_{\delta}^{\mathrm{fo}} \cup \mathrm{AIP}$ with domain P. Let $p, p' \in P$. It is easily proved by induction on n that $p \leftrightarrow_{\mathfrak{A}} p'$ implies $p \sim_n p'$ (in $\mathfrak{A}$) for each $n \geq 0$. Because AIP is satisfied, it follows immediately that B is satisfied. □

We can summarize the state of affairs as follows. In the models of $\mathrm{BPA}_{\delta}^{\mathrm{fo}}$ from which only finitely branching transition systems can be extracted, observational equivalence coincides with external bisimilarity. It happens that observational equivalence can be used to formulate AIP. The strength of AIP is witnessed by the fact that $\mathfrak{P}^{\sim}_{\aleph_0}$ is the only full bisimulation model in which AIP is valid. Moreover, in all models in which AIP is valid, B is also valid.

AIP was first formulated in [14]. To the best of our knowledge, the formulation given here is the first one using observational equivalence explicitly. In [15, 16], more can be found on bisimulation models in which AIP is valid. However, in those papers, only bisimulation models of PA, i.e. ACP without communication (see also Sect. 13), are considered.

Note that the defining axiom of observational equivalence is a formula of $\mathcal{L}_{\omega_1\omega}(\mathrm{BPA}_{\delta}^{\mathrm{fo}})$. Observational equivalence is not definable in $\mathcal{L}(\mathrm{BPA}_{\delta}^{\mathrm{fo}})$. It is shown in [17] that external bisimilarity is not even definable in $\mathcal{L}_{\omega_1\omega}(\mathrm{BPA}_{\delta}^{\mathrm{fo}})$.

8 SOS-Based Bisimilarity

It is customary to associate transition systems with closed terms of the language of an ACP-like theory about processes by means of structural operational semantics and to identify closed terms if their associated transition systems are bisimilar. In this section, we briefly dwell on this approach.

In the presence of recursion the approach requires a special provision, namely constants for the solutions of recursive specifications.

We add to the nonlogical symbols of the first-order theory $\mathrm{BPA}_{\delta}^{\mathrm{fo}}$, for each finite guarded recursive specification E and each variable X that occurs as the left-hand side of an equation in E, a constant standing for the unique solution of E for X. This constant is denoted by $\langle X|E\rangle$. Moreover, we add the axiom (schema) given in Table 5 to the axioms of $\mathrm{BPA}_{\delta}^{\mathrm{fo}}$. We write $\mathrm{BPA}_{\delta c}^{\mathrm{fo}}$ for the

Table 5. Axiom schema for the constants $\langle X|E\rangle$

$\bigwedge_{1\leq i\leq n}\langle X_i|E\rangle = t_i(\langle X_1|E\rangle,\ldots,\langle X_n|E\rangle)$ if $E=\{X_i=t_i(X_1,\ldots,X_n)\mid 1\leq i\leq n\}$	RDPc

Table 6. Structural operational semantics of $\mathrm{BPA}_{\delta\mathrm{c}}^{\mathrm{fo}}$

$$\frac{}{a \xrightarrow{a} \surd}$$

$$\frac{x \xrightarrow{a} \surd}{x+y \xrightarrow{a} \surd} \quad \frac{y \xrightarrow{a} \surd}{x+y \xrightarrow{a} \surd} \quad \frac{x \xrightarrow{a} x'}{x+y \xrightarrow{a} x'} \quad \frac{y \xrightarrow{a} y'}{x+y \xrightarrow{a} y'}$$

$$\frac{x \xrightarrow{a} \surd}{x\cdot y \xrightarrow{a} y} \quad \frac{x \xrightarrow{a} x'}{x\cdot y \xrightarrow{a} x'\cdot y}$$

$$\frac{t_i(\langle X_1|E\rangle,\ldots,\langle X_n|E\rangle) \xrightarrow{a} \surd}{\langle X_i|E\rangle \xrightarrow{a} \surd}\ E=\{X_i=t_i(X_1,\ldots,X_n)\mid 1\leq i\leq n\}$$

$$\frac{t_i(\langle X_1|E\rangle,\ldots,\langle X_n|E\rangle) \xrightarrow{a} x'}{\langle X_i|E\rangle \xrightarrow{a} x'}\ E=\{X_i=t_i(X_1,\ldots,X_n)\mid 1\leq i\leq n\}$$

resulting theory. RDPc is an axiom schema with an instance for each guarded recursive specification E. Note that the models of $\mathrm{BPA}_{\delta\mathrm{c}}^{\mathrm{fo}}$ are simply the expansions of the models of $\mathrm{BPA}_{\delta}^{\mathrm{fo}}$ obtained by associating with each constant $\langle X|E\rangle$ the unique solution in the model concerned of E for X.

The structural operational semantics of $\mathrm{BPA}_{\delta\mathrm{c}}^{\mathrm{fo}}$ is described by the transition rules given in Table 6. It determines a transition system for each process that can be denoted by a closed term of $\mathcal{L}(\mathrm{BPA}_{\delta\mathrm{c}}^{\mathrm{fo}})$. These transition systems are special in the sense that their states are closed terms of $\mathcal{L}(\mathrm{BPA}_{\delta\mathrm{c}}^{\mathrm{fo}})$.

Let t be a closed term of $\mathcal{L}(\mathrm{BPA}_{\delta\mathrm{c}}^{\mathrm{fo}})$. Then the transition system of t *induced by* the structural operational semantics of $\mathrm{BPA}_{\delta\mathrm{c}}^{\mathrm{fo}}$, written $\mathrm{TS}(t)$, is the connected transition system $\Gamma(S,\rightarrow,\rightarrow\surd,s^0)$, where:

- S is the set of closed terms of $\mathcal{L}(\mathrm{BPA}_{\delta\mathrm{c}}^{\mathrm{fo}})$;
- the sets $\xrightarrow{a}\subseteq S\times S$ and $\xrightarrow{a}\surd\subseteq S$ for each $a\in\mathsf{A}$ are the smallest subsets of $S\times S$ and S, respectively, for which the transition rules from Table 6 hold;
- $s^0\in S$ is t.

Clearly, the structural operational semantics does not give rise to infinitely branching transition systems. In other words, for each closed term t of $\mathcal{L}(\mathrm{BPA}_{\delta\mathrm{c}}^{\mathrm{fo}})$, we have $\mathrm{TS}(t)\in\mathbb{TS}_{\aleph_0}$.

Let t_1 and t_2 be closed terms of $\mathcal{L}(\mathrm{BPA}_{\delta\mathrm{c}}^{\mathrm{fo}})$. Then we say that t_1 and t_2 are *bisimilar*, written $t_1 \leftrightarrows_{\mathrm{sos}} t_2$, if $\mathrm{TS}(t_1)\leftrightarrows\mathrm{TS}(t_2)$.

We have the following relationship between bisimilarity of terms, which is based on structural operational semantics, and validity of equations in models of $\mathrm{BPA}_{\delta\mathrm{c}}^{\mathrm{fo}}$.

Theorem 13 (SOS-based bisimilarity and validity of equations).

1. *Let* t_1, t_2 *be closed terms of* $\mathcal{L}(\mathrm{BPA}_\delta^{\mathrm{fo}})$. *Then* $t_1 \leftrightarrows_{\mathrm{sos}} t_2$ *implies* $\mathfrak{A} \models t_1 = t_2$ *for all models* $\mathfrak{A}$ *of* $\mathrm{BPA}_{\delta\mathrm{c}}^{\mathrm{fo}}$.
2. *Let* t_1, t_2 *be closed terms of* $\mathcal{L}(\mathrm{BPA}_{\delta\mathrm{c}}^{\mathrm{fo}})$. *Then* $t_1 \not\leftrightarrows_{\mathrm{sos}} t_2$ *implies* $\mathfrak{A} \models t_1 \neq t_2$ *for all models* $\mathfrak{A}$ *of* $\mathrm{BPA}_{\delta\mathrm{c}}^{\mathrm{fo}}$.

Proof.
Proof of part 1. It follows easily from the structural operational semantics of $\mathrm{BPA}_{\delta\mathrm{c}}^{\mathrm{fo}}$ that, for all closed terms t_1, t_2 of $\mathcal{L}(\mathrm{BPA}_\delta^{\mathrm{fo}})$, $t_1 \leftrightarrows_{\mathrm{sos}} t_2$ iff $\mathrm{BPA}_{\delta\mathrm{c}}^{\mathrm{fo}} \vdash t_1 = t_2$ (see also [18]). From this, it follows immediately that, for all closed terms t_1, t_2 of $\mathcal{L}(\mathrm{BPA}_\delta^{\mathrm{fo}})$, $t_1 \leftrightarrows_{\mathrm{sos}} t_2$ implies $\mathfrak{A} \models t_1 = t_2$ for all models $\mathfrak{A}$ of $\mathrm{BPA}_{\delta\mathrm{c}}^{\mathrm{fo}}$.
Proof of part 2. It follows easily from the structural operational semantics of $\mathrm{BPA}_{\delta\mathrm{c}}^{\mathrm{fo}}$ that, for all closed terms t_1, t_2 of $\mathcal{L}(\mathrm{BPA}_{\delta\mathrm{c}}^{\mathrm{fo}})$, $t_1 \leftrightarrows_{\mathrm{sos}} t_2$ iff $\mathrm{BPA}_{\delta\mathrm{c}}^{\mathrm{fo}} \cup \{\mathrm{OBS_n} \mid n \geq 0\} \vdash t_1 \sim_n t_2$ for all $n \geq 0$ (see also [18]). Moreover, for all closed terms t_1, t_2 of $\mathcal{L}(\mathrm{BPA}_{\delta\mathrm{c}}^{\mathrm{fo}})$ and $n \geq 0$, either $\mathrm{BPA}_{\delta\mathrm{c}}^{\mathrm{fo}} \cup \{\mathrm{OBS_n} \mid n \geq 0\} \vdash t_1 \sim_n t_2$ or $\mathrm{BPA}_{\delta\mathrm{c}}^{\mathrm{fo}} \cup \{\mathrm{OBS_n} \mid n \geq 0\} \vdash \neg\, t_1 \sim_n t_2$, but not both. This is easily proved by induction on n. As a consequence, for all closed terms t_1, t_2 of $\mathcal{L}(\mathrm{BPA}_{\delta\mathrm{c}}^{\mathrm{fo}})$, $t_1 \not\leftrightarrows_{\mathrm{sos}} t_2$ iff $\mathrm{BPA}_{\delta\mathrm{c}}^{\mathrm{fo}} \cup \{\mathrm{OBS_n} \mid n \geq 0\} \vdash \neg\, t_1 \sim_n t_2$ for some $n \geq 0$. From this, it follows easily that, for all closed terms t_1, t_2 of $\mathcal{L}(\mathrm{BPA}_{\delta\mathrm{c}}^{\mathrm{fo}})$, $t_1 \not\leftrightarrows_{\mathrm{sos}} t_2$ implies $\mathrm{BPA}_{\delta\mathrm{c}}^{\mathrm{fo}} \vdash \neg\, t_1 = t_2$. From this, it follows immediately that, for all closed terms t_1, t_2 of $\mathcal{L}(\mathrm{BPA}_{\delta\mathrm{c}}^{\mathrm{fo}})$, $t_1 \not\leftrightarrows_{\mathrm{sos}} t_2$ implies $\mathfrak{A} \models \neg\, t_1 = t_2$ for all models $\mathfrak{A}$ of $\mathrm{BPA}_{\delta\mathrm{c}}^{\mathrm{fo}}$.
□

This theorem implies that, for closed equations of $\mathcal{L}(\mathrm{BPA}_\delta^{\mathrm{fo}})$, validity in all models coincides with (SOS-based) bisimilarity of the closed terms concerned.

We could have introduced constants for the solutions of unguarded recursive specifications as well. In that case, the structural operational semantics would have given rise to countably branching transition systems. Moreover, it would have fixed a particular solution for each unguarded recursive specification. In this paper, we do not consider unguarded recursion.

The following remark on fixing a particular solution in the case of unguarded recursion is in order. Suppose that we also add to the nonlogical symbols of the first-order theory $\mathrm{BPA}_\delta^{\mathrm{fo}}$ a constant, denoted by $\langle X|E\rangle$, for each finite unguarded recursive specification E and each variable X that occurs as the left-hand side of an equation in E. Consider the two unguarded recursive specifications $X = a \cdot X + X$ and $Y = b \cdot Y + Y$, where a and b are different actions. The structural operational semantics of $\mathrm{BPA}_{\delta\mathrm{c}}^{\mathrm{fo}}$ described in Table 6 fixes the obvious solution for each of these unguarded recursive specifications. However, as usual with unguarded recursive specifications, both have more than one solution. The problem is not so much that they have more than one solution, but that the sets of solutions are not disjoint. For example, the solution of the guarded recursive specification $Z = a \cdot Z + b \cdot Z$ is a common solution of $X = a \cdot X + X$ and $Y = b \cdot Y + Y$. The common solutions exclude any possibility to achieve that $\mathfrak{A} \models \langle X|\{X = a \cdot X + X\}\rangle \neq \langle Y|\{Y = b \cdot Y + Y\}\rangle$ for all models $\mathfrak{A}$, although $\langle X|\{X = a \cdot X + X\}\rangle \not\leftrightarrows_{\mathrm{sos}} \langle Y|\{Y = b \cdot Y + Y\}\rangle$.

9 A Modal Fragment of $\mathcal{L}(\mathrm{BPA}_\delta^{\mathrm{fo}})$

In this section, we have a closer look at a modal fragment of $\mathcal{L}(\mathrm{BPA}_\delta^{\mathrm{fo}})$. This fragment corresponds to a variant of HML (Hennessy-Milner Logic), a simple modal logic introduced in [12] to give a modal characterization of bisimilarity.

The set $\mathcal{M}$ of *modal fragment formulas* of $\mathcal{L}(\mathrm{BPA}_\delta^{\mathrm{fo}})$ is inductively defined as follows:

- if x is a variable, then $x = x \in \mathcal{M}$;
- if $\phi \in \mathcal{M}$, then $\neg\, \phi \in \mathcal{M}$;
- if $\phi_1, \phi_2 \in \mathcal{M}$, then $\phi_1 \;\wedge\; \phi_2 \in \mathcal{M}$;
- if $a \in \mathsf{A}$ and x is a variable, then $x \xrightarrow{a} \surd \in \mathcal{M}$;
- if $a \in \mathsf{A}$, x, y are different variables and $\phi \in \mathcal{M}$, then $\exists y \bullet (x \xrightarrow{a} y \;\wedge\; \phi) \in \mathcal{M}$.

We write $\mathcal{M}_1$ for the subset of $\mathcal{M}$ that contains all formulas from $\mathcal{M}$ in which exactly one variable occurs free. The set $\mathcal{M}_1$ of one-variable modal fragment formulas has an interesting property: $\mathcal{M}_1$ is essentially the set of formulas of $\mathcal{L}(\mathrm{BPA}_\delta^{\mathrm{fo}})$ that are invariant for external bisimulation.

Theorem 14 (Invariance for external bisimilarity). *Let $\mathfrak{A}$ be a model of $\mathrm{BPA}_\delta^{\mathrm{fo}}$ with domain P, and let ϕ be a formula of $\mathcal{L}(\mathrm{BPA}_\delta^{\mathrm{fo}})$. Then the following are equivalent:*

- *$\mathfrak{A} \models \phi[p_1]$ iff $\mathfrak{A} \models \phi[p_2]$ for all $p_1, p_2 \in P$ such that $p_1 \leftrightarrows_{\mathfrak{A}} p_2$;*
- *there exists a formula $\phi' \in \mathcal{M}_1$ such that $\phi \Leftrightarrow \phi'$.*

Proof. The proof is analogous to the proof of the corresponding property for first-order formulas that correspond to HML-like modal formulas given in [17]. □

We have the following corollary of Theorem 14.

Corollary 1 (External bisimilarity implies indistinguishability). *Let $\mathfrak{A}$ be a model of $\mathrm{BPA}_\delta^{\mathrm{fo}}$ with domain P, and let $p_1, p_2 \in P$. If $p_1 \leftrightarrows_{\mathfrak{A}} p_2$, then for all $\phi \in \mathcal{M}_1$ we have $\mathfrak{A} \models \phi[p_1]$ iff $\mathfrak{A} \models \phi[p_2]$.*

In general, we do not have the converse of Corollary 1. The transition systems from the counterexample used in the proof of Theorem 12 provide a counterexample here as well. However, we do have the converse in the case of finite branching.

Theorem 15 (Indistinguishability implies external bisimilarity). *Let $\mathfrak{A}$ be a model of $\mathrm{BPA}_\delta^{\mathrm{fo}}$ with domain P, and let $p_1, p_2 \in P$. If for all $\phi \in \mathcal{M}_1$ we have $\mathfrak{A} \models \phi[p_1]$ iff $\mathfrak{A} \models \phi[p_2]$ and moreover $\mathrm{TS}(\mathfrak{A}, p_1), \mathrm{TS}(\mathfrak{A}, p_2) \in \mathbb{TS}_{\aleph_0}$, then $p_1 \leftrightarrows_{\mathfrak{A}} p_2$.*

Proof. The proof is analogous to the proof of the corresponding property for HML-like modal formulas given in [19]. □

Now we come back to the variant of HML of which the formulas correspond to the formulas in $\mathcal{M}$. HML is a modal logic introduced in [12] to be used in a setting where no distinction is made between successful termination and deadlock. The variant of HML considered here is adapted to a setting where distinction is made between successful termination and deadlock. This variant is henceforth also called HML. The set $\mathcal{H}$ of *HML formulas* is inductively defined as follows:

- $\mathsf{T} \in \mathcal{H}$;
- if $\psi \in \mathcal{H}$, then $\neg\, \psi \in \mathcal{H}$;
- if $\psi_1, \psi_2 \in \mathcal{H}$, then $\psi_1 \wedge \psi_2 \in \mathcal{H}$;
- if $a \in \mathsf{A}$, then $\langle a \rangle \surd \in \mathcal{H}$;
- if $a \in \mathsf{A}$ and $\psi \in \mathcal{H}$, then $\langle a \rangle\, \psi \in \mathcal{H}$.

There is a "standard translation" from HML formulas to formulas of $\mathcal{L}(\mathrm{BPA}_\delta^{\mathrm{fo}})$. Let x be a fixed but arbitrary variable. Then the translation of HML formulas is defined as follows:

$$
\begin{aligned}
\mathsf{T}^\bullet &= x = x\,, \\
(\neg\, \psi)^\bullet &= \neg\, (\psi^\bullet)\,, \\
(\psi_1 \wedge \psi_2)^\bullet &= {\psi_1}^\bullet \wedge {\psi_2}^\bullet\,, \\
\langle a \rangle \surd^\bullet &= x \xrightarrow{a} \surd\,, \\
(\langle a \rangle\, \psi)^\bullet &= \exists y \bullet \big(x \xrightarrow{a} y \wedge \psi^\bullet(y)\big) \qquad \text{where } y \text{ is a fresh variable.}
\end{aligned}
$$

This translation is justified by the fact that satisfaction for HML formulas ψ is defined such that $\mathfrak{A} \models \psi$ iff $\mathfrak{A} \models \forall x \bullet \psi^\bullet$.

Clearly, the image of the translation from HML formulas to formulas of $\mathcal{L}(\mathrm{BPA}_\delta^{\mathrm{fo}})$ consists of all formulas from $\mathcal{M}_1$ of which the free variable is x. HML is a modal logic which has been devised to complement the process algebra CCS [3, 4] with a formalism that allows one to express and verify properties of processes which are definable directly in terms of the action steps that are possible at any stage. Apparently, $\mathrm{BPA}_\delta^{\mathrm{fo}}$ can be considered to include a process algebra and a variant of HML as fragments.

10 Deadlock Freedom

In this section, we add a deadlock freedom predicate to $\mathrm{BPA}_\delta^{\mathrm{fo}}$. In Sect. 2, we demonstrated that the deadlock freedom predicate can be explicitly defined by using the reachability predicate. Here, the deadlock freedom predicate will be implicitly defined without using the reachability predicate.

We add to $\mathrm{BPA}_\delta^{\mathrm{fo}}$ the unary *deadlock freedom* predicate symbol dlf and the axioms given in Table 7. We write DLF for this set of axioms. DLFS is an axiom schema where $\psi(x)$ is a formula of $\mathcal{L}(\mathrm{BPA}_\delta^{\mathrm{fo}} \cup \mathrm{DLF})$. Axiom schema DLFS is an induction schema.

The deadlock freedom predicate that is implicitly defined by DLF is equivalent to the one that is explicitly defined by using the reachability predicate.

Theorem 16 (Explicit definability of deadlock freedom). *We have* $\mathrm{BPA}_\delta^{\mathrm{fo}} \cup \mathrm{DLF} \vdash \mathsf{dlf}(x) \Leftrightarrow \neg\, x \twoheadrightarrow \delta$.

Table 7. Axioms for deadlock freedom

$\neg\, \mathsf{dlf}(\delta)$	DLF1
$\bigwedge_{a \in \mathsf{A}} \forall x, y \bullet (\mathsf{dlf}(x) \wedge x \xrightarrow{a} y \Rightarrow \mathsf{dlf}(y))$	DLF2
$\neg\, \psi(\delta) \wedge \bigwedge_{a \in \mathsf{A}} \forall x, y \bullet (\psi(x) \wedge x \xrightarrow{a} y \Rightarrow \psi(y)) \Rightarrow \forall x \bullet (\psi(x) \Rightarrow \mathsf{dlf}(x))$	DLFS

Proof. We will apply RS, taking $\mathsf{dlf}(x) \Rightarrow y \neq \delta$ for $\phi(x, y)$, to prove the implication $\mathsf{dlf}(x) \Rightarrow \neg\, x \twoheadrightarrow \delta$. When we have shown that $x \twoheadrightarrow y \Rightarrow (\mathsf{dlf}(x) \Rightarrow y \neq \delta)$, we can first conclude by substitution of δ for y that $x \twoheadrightarrow \delta \Rightarrow \neg\, \mathsf{dlf}(x)$, and then by contraposition that $\mathsf{dlf}(x) \Rightarrow \neg\, x \twoheadrightarrow \delta$.

It remains to be shown by means of RS that $x \twoheadrightarrow y \Rightarrow (\mathsf{dlf}(x) \Rightarrow y \neq \delta)$. First of all, we immediately conclude from DLF1 that

$$\forall x' \bullet (\mathsf{dlf}(x') \Rightarrow x' \neq \delta)\,.$$

Moreover, we conclude from DLF2, using substitutivity of implication, that

$$\forall x', y', z' \bullet \bigwedge_{a' \in \mathsf{A}} \left(x' \xrightarrow{a'} y' \wedge (\mathsf{dlf}(y') \Rightarrow z' \neq \delta) \Rightarrow (\mathsf{dlf}(x') \Rightarrow z' \neq \delta)\right).$$

Using the subprocess induction schema, it follows from these conclusions that $x \twoheadrightarrow y \Rightarrow (\mathsf{dlf}(x) \Rightarrow y \neq \delta)$.

We will apply DLFS, taking $\neg\, x \twoheadrightarrow \delta$ for $\psi(x)$, to prove the reverse implication $\neg\, x \twoheadrightarrow \delta \Rightarrow \mathsf{dlf}(x)$.

First of all, we immediately conclude from R1 that

$$\neg\,(\neg\, \delta \twoheadrightarrow \delta)\,.$$

Moreover, we conclude from R2, because $(x \xrightarrow{a} y \wedge y \twoheadrightarrow z \Rightarrow x \twoheadrightarrow z) \Leftrightarrow (\neg\, x \twoheadrightarrow z \wedge x \xrightarrow{a} y \Rightarrow \neg\, y \twoheadrightarrow z)$, that

$$\bigwedge_{a \in \mathsf{A}} \forall x, y \bullet \left(\neg\, x \twoheadrightarrow \delta \wedge x \xrightarrow{a} y \Rightarrow \neg\, y \twoheadrightarrow \delta\right).$$

Using DLFS, it follows from these conclusions that $\forall x \bullet (\neg\, x \twoheadrightarrow \delta \Rightarrow \mathsf{dlf}(x))$. □

Using Proposition 1 and Theorem 16, we can easily prove that, for example, the solution of the guarded recursive specification $X = a \cdot X$ is deadlock free.

Proposition 6 (Solution of $X = a \cdot X$ is deadlock free). *We have* $\mathrm{BPA}_\delta^{\mathrm{fo}} \cup \mathrm{DLF} \vdash X = a \cdot X \Rightarrow \mathsf{dlf}(X)$.

Proof. Suppose $\neg\, \mathsf{dlf}(X)$. By Theorem 16, then also $X \twoheadrightarrow \delta$. We distinguish three cases according to Proposition 1:

- $X = \delta$. Then, because $X = a \cdot X$, also $\delta = a \cdot \delta$. This is equivalent to $a \cdot \delta \sqsubseteq \delta$, which contradicts axiom SI3.
- $X \xrightarrow{a} \delta$ for some $a \in \mathsf{A}$. Then $a \cdot \delta \sqsubseteq X$. Because $X = a \cdot X$, this is equivalent to $a \cdot \delta \sqsubseteq a \cdot X$. This in turn implies $\delta = X$, which contradicts the conclusion of the previous case that $X \neq \delta$.

- $X \xrightarrow{a} z$ for some $a \in \mathsf{A}$ and $z \neq X$ with $z \twoheadrightarrow \delta$. Then $a \cdot z \sqsubseteq X$. Because $X = a \cdot X$, this is equivalent to $a \cdot z \sqsubseteq a \cdot X$. This in turn implies $z = X$, which contradicts the fact that $z \neq X$.

So $\neg\, \mathsf{dlf}(X)$ leads in all cases to contradiction. From this, we conclude that $\mathsf{dlf}(X)$. □

Let $\mathfrak{A}$ be a model of $\mathrm{BPA}_\delta^{\mathrm{fo}} \cup \mathrm{DLF}$ with domain P. Then reachability and deadlock freedom on P are defined as follows:

$$p_1 \twoheadrightarrow_{\mathfrak{A}} p_2 \quad \text{iff} \quad p_1 \twoheadrightarrow p_2\ ,$$

where $\twoheadrightarrow$ is the reachability relation of $\mathrm{TS}(\mathfrak{A}, p_1)$;

$$\mathsf{dlf}_{\mathfrak{A}}(p) \quad \text{iff} \quad \text{not } p \twoheadrightarrow_{\mathfrak{A}} \delta^{\mathfrak{A}}\ .$$

Reachability and deadlock freedom on the domain of a model of $\mathrm{BPA}_\delta^{\mathrm{fo}} \cup \mathrm{DLF}$ as defined above are called *external reachability* and *external deadlock freedom*, respectively.

We write $\mathfrak{P}_\kappa^{\mathsf{dlf}}$ ($\kappa \geq \aleph_0$) for the unique definitional expansion of $\mathfrak{P}_\kappa$ determined by the definitional extension of $\mathrm{BPA}_\delta^{\mathrm{fo}}$ with the unary predicate symbol dlf and the formula $\mathsf{dlf}(x) \Leftrightarrow \neg\, x \twoheadrightarrow \delta$. In the proof of Proposition 8 (see below), we will use the next lemma. It states that in the models $\mathfrak{P}_\kappa^{\mathsf{dlf}}$, external reachability coincides with internal reachability.

Lemma 2 (External reachability is internal reachability in $\mathfrak{P}_\kappa^{\mathsf{dlf}}$). *Let $p_1, p_2 \in \mathbb{TS}_\kappa/\underline{\leftrightarrow}$ for some $\kappa \geq \aleph_0$. Then $p_1 \twoheadrightarrow_{\mathfrak{P}_\kappa^{\mathsf{dlf}}} p_2$ iff $p_1 \,\widetilde{\twoheadrightarrow}\, p_2$.*

Proof. By Lemma 1, $\mathrm{TS}(\mathfrak{P}_\kappa, p_1) \in p_1$ and $\mathrm{TS}(\mathfrak{P}_\kappa, p_2) \in p_2$. Hence, $p_1 \,\widetilde{\twoheadrightarrow}\, p_2$ iff $[\,\mathrm{TS}(\mathfrak{P}_\kappa, p_1)\,] \,\widetilde{\twoheadrightarrow}\, [\,\mathrm{TS}(\mathfrak{P}_\kappa, p_2)\,]$. It is easy to see that p is a state of $\mathrm{TS}(\mathfrak{P}_\kappa, p_1)$ iff $p_1 \twoheadrightarrow p$ where $\twoheadrightarrow$ is the reachability relation of $\mathrm{TS}(\mathfrak{P}_\kappa, p_1)$; and also that, if p is a state of $\mathrm{TS}(\mathfrak{P}_\kappa, p_1)$, $(\mathrm{TS}(\mathfrak{P}_\kappa, p_1))_p = \mathrm{TS}(\mathfrak{P}_\kappa, p)$. From this, and the definitions of $\widetilde{\twoheadrightarrow}$ and $\widehat{\twoheadrightarrow}$, it follows that $[\,\mathrm{TS}(\mathfrak{P}_\kappa, p_1)\,] \,\widetilde{\twoheadrightarrow}\, [\,\mathrm{TS}(\mathfrak{P}_\kappa, p_2)\,]$ iff there exists a p such that $p_1 \twoheadrightarrow_{\mathfrak{P}_\kappa} p$ and $\mathrm{TS}(\mathfrak{P}_\kappa, p) \in [\,\mathrm{TS}(\mathfrak{P}_\kappa, p_2)\,]$. Moreover, by Lemma 1, $\mathrm{TS}(\mathfrak{P}_\kappa, p) \in [\,\mathrm{TS}(\mathfrak{P}_\kappa, p_2)\,]$ iff $p = p_2$. Thus, we conclude that $p_1 \,\widetilde{\twoheadrightarrow}\, p_2$ iff $p_1 \twoheadrightarrow_{\mathfrak{P}_\kappa} p_2$. Because $\mathfrak{P}_\kappa^{\mathsf{dlf}}$ is a definitional expansion of $\mathfrak{P}_\kappa$, it follows immediately that also $p_1 \,\widetilde{\twoheadrightarrow}\, p_2$ iff $p_1 \twoheadrightarrow_{\mathfrak{P}_\kappa^{\mathsf{dlf}}} p_2$. □

A useful corollary of the proof of Lemma 2 is the following.

Corollary 2 (External reachability is internal reachability in $\mathfrak{P}_\kappa$). *Let $p_1, p_2 \in \mathbb{TS}_\kappa/\underline{\leftrightarrow}$ for some $\kappa \geq \aleph_0$. Then $p_1 \twoheadrightarrow_{\mathfrak{P}_\kappa} p_2$ iff $p_1 \,\widetilde{\twoheadrightarrow}\, p_2$.*

In the models of $\mathrm{BPA}_\delta^{\mathrm{fo}} \cup \mathrm{DLF}$, internal deadlock freedom implies external deadlock freedom.

Proposition 7 (Internal deadlock freedom implies external deadlock freedom). *Let $\mathfrak{A}$ be a model of $\mathrm{BPA}_\delta^{\mathrm{fo}} \cup \mathrm{DLF}$ with domain P and let $p \in P$. Then $\mathsf{dlf}^{\mathfrak{A}}(p)$ implies $\mathsf{dlf}_{\mathfrak{A}}(p)$.*

Proof. By Theorem 16, $\mathsf{dlf}^{\mathfrak{A}}(p)$ iff not $p \twoheadrightarrow' \delta^{\mathfrak{A}}$, where $\twoheadrightarrow'$ is the binary relation on P associated with the predicate symbol $\twoheadrightarrow$ in $\mathfrak{A}$. By the definition of external deadlock freedom, $\mathsf{dlf}_{\mathfrak{A}}(p)$ iff not $p \twoheadrightarrow'' \delta^{\mathfrak{A}}$, where $\twoheadrightarrow''$ is the reachability relation of $\mathrm{TS}(\mathfrak{A}, p)$. It follows immediately from axioms R1, R2 and RS of $\mathrm{BPA}_{\delta}^{\mathrm{fo}}$ (Table 1) and the definition of reachability relation of a transition system (Sect. 4) that for all $p', p'' \in P$, $p' \twoheadrightarrow'' p''$ implies $p' \twoheadrightarrow' p''$. Hence, $p \twoheadrightarrow'' \delta^{\mathfrak{A}}$ implies $p \twoheadrightarrow' \delta^{\mathfrak{A}}$; and by the above-mentioned equivalences $\mathsf{dlf}^{\mathfrak{A}}(p)$ implies $\mathsf{dlf}_{\mathfrak{A}}(p)$. □

In the full bisimulation models $\mathfrak{P}_{\kappa}^{\mathsf{dlf}}$, external deadlock freedom coincides with internal deadlock freedom.

Proposition 8 (External deadlock freedom is internal deadlock freedom in $\mathfrak{P}_{\kappa}^{\mathsf{dlf}}$). *Let $p \in \mathbb{TS}_{\kappa}/\leftrightarrows$ for some $\kappa \geq \aleph_0$. Then $\mathsf{dlf}_{\mathfrak{P}_{\kappa}^{\mathsf{dlf}}}(p)$ iff $\widetilde{\mathsf{dlf}}(p)$.*

Proof. By Lemma 2, $p \twoheadrightarrow_{\mathfrak{P}_{\kappa}^{\mathsf{dlf}}} \widetilde{\delta}$ iff $p \mathrel{\widetilde{\twoheadrightarrow}} \widetilde{\delta}$. Hence, $\mathsf{dlf}_{\mathfrak{P}_{\kappa}^{\mathsf{dlf}}}(p)$ iff not $p \mathrel{\widetilde{\twoheadrightarrow}} \widetilde{\delta}$. By Theorem 16, also $\widetilde{\mathsf{dlf}}(p)$ iff not $p \mathrel{\widetilde{\twoheadrightarrow}} \widetilde{\delta}$. From this, it follows immediately that $\mathsf{dlf}_{\mathfrak{P}_{\kappa}^{\mathsf{dlf}}}(p)$ iff $\widetilde{\mathsf{dlf}}(p)$. □

There does not exist a consistent extension of $\mathrm{BPA}_{\delta}^{\mathrm{fo}} \cup \mathrm{DLF}$ with first-order axioms that has only models in which external deadlock freedom coincides with internal deadlock freedom.

Theorem 17 (Undefinability of external deadlock freedom). *Each first-order consistent extension of* $\mathrm{BPA}_{\delta}^{\mathrm{fo}} \cup \mathrm{DLF}$ *has a model in which external deadlock freedom is not internal deadlock freedom.*

Proof. Suppose that there exists a first-order consistent extension of $\mathrm{BPA}_{\delta}^{\mathrm{fo}} \cup \mathrm{DLF}$, say $\mathrm{BPA}_{\delta}^{\mathrm{fo}} \cup \mathrm{DLF} \cup H$, that has only models in which external deadlock freedom is internal deadlock freedom. A contradiction is found as follows. Let $c_0, c_1, c_2, \ldots$ be different new constants; and let a be an action. Consider the following sets of formulas:

$$H' = \{\neg\, \mathsf{dlf}(c_0)\} \cup \{c_i = a \cdot c_{i+1} \mid i \geq 0\}\,,$$

$$H'_n = \{\neg\, \mathsf{dlf}(c_0)\} \cup \{c_i = a \cdot c_{i+1} \mid 0 \leq i < n\} \cup \{c_n = \delta\}\,.$$

Take an arbitrary model $\mathfrak{A}$ of $\mathrm{BPA}_{\delta}^{\mathrm{fo}} \cup \mathrm{DLF} \cup H$. It follows easily from the axioms of $\mathrm{BPA}_{\delta}^{\mathrm{fo}} \cup \mathrm{DLF}$ that, for each $n \geq 0$, H'_n is satisfied in the definitional expansion of $\mathfrak{A}$ determined by the definitional extension of $\mathrm{BPA}_{\delta}^{\mathrm{fo}} \cup \mathrm{DLF} \cup H$ with the constants $c_0, \ldots, c_n$ and the equations $c_i = a^{n-i} \cdot \delta$ for $0 \leq i < n$ and $c_n = \delta$. Hence, for each $n \geq 0$, H'_n is consistent with $\mathrm{BPA}_{\delta}^{\mathrm{fo}} \cup \mathrm{DLF} \cup H$. Each finite $H'' \subseteq H'$ is consistent with $\mathrm{BPA}_{\delta}^{\mathrm{fo}} \cup \mathrm{DLF} \cup H$ because there is an $n \geq 0$ for which $H'' \subseteq H'_n$. From this, it follows by the Compactness Theorem that H' is consistent with $\mathrm{BPA}_{\delta}^{\mathrm{fo}} \cup \mathrm{DLF} \cup H$. Now consider an arbitrary model $\mathfrak{A}'$ of $\mathrm{BPA}_{\delta}^{\mathrm{fo}} \cup \mathrm{DLF} \cup H \cup H'$. Because $\mathfrak{A}'$ satisfies H', not $\mathsf{dlf}^{\mathfrak{A}'}(c_0^{\mathfrak{A}'})$. It is easy to see that the reachability relation $\twoheadrightarrow$ of $\mathrm{TS}(\mathfrak{A}', c_0^{\mathfrak{A}'})$ is such that not $c_0^{\mathfrak{A}'} \twoheadrightarrow \delta^{\mathfrak{A}'}$. This means that $\mathsf{dlf}_{\mathfrak{A}'}(c_0^{\mathfrak{A}'})$. Hence, because external deadlock freedom is internal deadlock freedom, $\mathsf{dlf}^{\mathfrak{A}'}(c_0^{\mathfrak{A}'})$, which contradicts the fact that not $\mathsf{dlf}^{\mathfrak{A}'}(c_0^{\mathfrak{A}'})$. □

Apparently, there is a discrepancy in relation to deadlock freedom which is similar to the discrepancy in relation to bisimilarity found in Sect. 6.

We can summarize the state of affairs as follows. Deadlock freedom derivable from $\mathrm{BPA}_\delta^{\mathrm{fo}} \cup \mathrm{DLF}$ implies external deadlock freedom in each model of $\mathrm{BPA}_\delta^{\mathrm{fo}} \cup \mathrm{DLF}$. In the full bisimulation models $\mathfrak{P}_\kappa^{\mathsf{dlf}}$, external deadlock freedom coincides with internal deadlock freedom. However, there also exist models of which the domain contains elements that are externally deadlock free, but not internally deadlock free. Moreover, those models cannot be excluded by extending $\mathrm{BPA}_\delta^{\mathrm{fo}} \cup \mathrm{DLF}$ with first-order axioms.

11 Determinism

In the previous section, the relation between external deadlock freedom and internal deadlock freedom in models of $\mathrm{BPA}_\delta^{\mathrm{fo}} \cup \mathrm{DLF}$ was analysed in detail. It is obvious that there are other properties of processes of which the relation between the external version and the internal version can be analysed. In this section, we briefly consider one other property, namely determinism.

The *determinism* predicate symbol det is explicitly defined in terms of $\mathcal{L}(\mathrm{BPA}_\delta^{\mathrm{fo}})$ by

$$\mathsf{det}(x) \Leftrightarrow \forall y \bullet \Big(x \twoheadrightarrow y \Rightarrow \bigwedge_{a \in \mathsf{A}} \big((y \xrightarrow{a} \surd \Rightarrow \forall z \bullet \neg\, y \xrightarrow{a} z) \wedge \forall z, z' \bullet (y \xrightarrow{a} z \wedge y \xrightarrow{a} z' \Rightarrow z = z')\big)\Big) .$$

External determinism can be defined in the same vein as external deadlock freedom.

In this case, it is easy to see that there exist models of the extension of $\mathrm{BPA}_\delta^{\mathrm{fo}}$ with determinism in which external determinism does not coincide with internal determinism. We know from Theorem 9 that each first-order extension of $\mathrm{BPA}_\delta^{\mathrm{fo}}$ has a model of which the domain contains pairs of different elements that are externally bisimilar. Let $\mathfrak{A}$ be such a model, and let p and p' be elements from the domain of $\mathfrak{A}$ such that $p \leftrightarrows_{\mathfrak{A}} p'$ and not $p = p'$. Clearly, the element $a^{\mathfrak{A}} \cdot^{\mathfrak{A}} p +^{\mathfrak{A}} a^{\mathfrak{A}} \cdot^{\mathfrak{A}} p'$ is externally deterministic, but not internally deterministic.

It is also easy to see that external determinism coincides with internal determinism in the unique expansions of the full bisimulation models $\mathfrak{P}_\kappa$ determined by the explicit definition of det. We know from Proposition 4 that external bisimilarity coincides with identity in those models; and we know from Corollary 2 that external reachability coincides with internal reachability in those models. From this, it is clear that external determinism coincides with internal determinism in those models.

12 Restricted Reachability

In this section, we present an interesting extension of $\mathrm{BPA}_\delta^{\mathrm{fo}}$, called $\mathrm{BPA}_{\delta\mathrm{rr}}^{\mathrm{fo}}$. It is obtained as follows. We add to the nonlogical symbols of $\mathrm{BPA}_\delta^{\mathrm{fo}}$, for each $a \in \mathsf{A}$,

Table 8. First-order and second-order axioms for restricted reachability

$x \xrightarrow{a}\!\!\twoheadrightarrow x$		RR1
$x \xrightarrow{a} y \wedge y \xrightarrow{a}\!\!\twoheadrightarrow z \Rightarrow x \xrightarrow{a}\!\!\twoheadrightarrow z$		RR2
$\exists! y \bullet \psi^{a,b}(x,y)$	if $a \not\equiv b$	RR3
$x \xrightarrow{a}\!\!\twoheadrightarrow y \wedge \forall x', y', z' \bullet (\phi(x',x') \wedge (x' \xrightarrow{a} y' \wedge \phi(y',z') \Rightarrow \phi(x',z'))) \Rightarrow \phi(x,y)$		RRS
$\forall R \bullet (x \xrightarrow{a}\!\!\twoheadrightarrow y \wedge \forall x', y', z' \bullet (R(x',x') \wedge (x' \xrightarrow{a} y' \wedge R(y',z') \Rightarrow R(x',z'))) \Rightarrow R(x,y))$		RR

a binary *reachability by a-steps* predicate symbol $\xrightarrow{a}\!\!\twoheadrightarrow$. Moreover, we add the axioms given in Table 8, with the exception of RR, to the axioms of $\mathrm{BPA}_\delta^{\mathrm{fo}}$. In axiom RR3 and henceforth, $\psi^{a,b}(x,y)$, where a and b are different actions, stands for the formula

$$
\begin{aligned}
& y \xrightarrow{b} x \wedge \forall \overline{x} \bullet \left(y \xrightarrow{b} \overline{x} \Rightarrow x = \overline{x}\right) \wedge \\
& \forall y' \bullet \Big(y \xrightarrow{a}\!\!\twoheadrightarrow y' \Rightarrow \exists x', x'' \bullet \Big(y' \xrightarrow{b} x' \wedge \bigvee_{a' \in \mathsf{A}} x' \xrightarrow{a'} x'' \Rightarrow \\
& \qquad\qquad \exists! y'' \bullet \left(y' \xrightarrow{a} y'' \wedge y'' \xrightarrow{b} x''\right)\Big) \wedge \\
& \qquad \exists x', y'' \bullet \Big(y' \xrightarrow{b} x' \wedge y' \xrightarrow{a} y'' \Rightarrow \\
& \qquad\qquad \exists! x'' \bullet \Big(\bigvee_{a' \in \mathsf{A}} x' \xrightarrow{a'} x'' \wedge y'' \xrightarrow{b} x''\Big)\Big)\Big) .
\end{aligned}
$$

RR1–RR3 are axiom schemas where a and b are action constants. RRS is an axiom schema where a is an action constant and $\phi(x,y)$ is a formula of $\mathcal{L}(\mathrm{BPA}_{\delta\mathrm{rr}}^{\mathrm{fo}})$. The differences of RR1, RR2 and RRS with R1, R2 and RS reflect that $\xrightarrow{a}\!\!\twoheadrightarrow$ is the restricted kind of reachability in which only action a is involved. We will return to the additional axiom schema RR3 below. Axiom schema RRS is called the *restricted subprocess induction schema*.

Similar to RS, the first-order axiom schema RRS does not exclude all models in which there are processes that have more processes reachable by a-steps than needed to satisfy the instances of axiom schemas RR1 and RR2. Similar to R, the second-order axiom schema RR from Table 8, where a is an action constant, would exclude all such models.

One can think of $\psi^{a,b}(x,y)$ as a formula expressing that y produces an indexing of the processes reachable from x with a set of processes reachable from y by a-steps only. Axiom schema RR3 excludes models in which such an indexing cannot be produced for all processes. This looks to be indispensable to establish that the (unrestricted) reachability predicate is explicitly definable by means of a restricted reachability predicate. It is unknown to us whether RR3 is derivable from the other axioms of $\mathrm{BPA}_{\delta\mathrm{rr}}^{\mathrm{fo}}$.

Note further that axiom schema RR3 induces the existence of an indexing operator for each pair of different actions a and b. The formula

$$\chi_{a,b}(x) = y \Leftrightarrow \psi^{a,b}(x,y)$$

is an explicit definition of this operator in terms of $\mathcal{L}(\mathrm{BPA}^{\mathrm{fo}}_{\delta\mathrm{rr}})$. Thus, a definitional extension of $\mathrm{BPA}^{\mathrm{fo}}_{\delta\mathrm{rr}}$ is obtained. Hence, every model of $\mathrm{BPA}^{\mathrm{fo}}_{\delta\mathrm{rr}}$ can be expanded in a unique way with an indexing operation that satisfies this formula. Using an auxiliary operator $\overline{\chi}_{a,b}$, we can equationally characterize $\chi_{a,b}$ as follows:

$$\begin{aligned}
&\chi_{a,b}(x) = b \cdot x + \overline{\chi}_{a,b}(x)\ , \\
&\overline{\chi}_{a,b}(c) = \delta\ , \\
&\overline{\chi}_{a,b}(c \cdot x) = a \cdot \chi_{a,b}(x)\ , \\
&\overline{\chi}_{a,b}(x + y) = \overline{\chi}_{a,b}(x) + \overline{\chi}_{a,b}(x)\ ,
\end{aligned}$$

where c stands for an arbitrary constant of $\mathrm{BPA}^{\mathrm{fo}}_{\delta\mathrm{rr}}$ (i.e. $c \in \mathsf{A} \cup \{\delta\}$).

Now we come back to the explicit definability of unrestricted reachability.

Theorem 18 (Explicit definability of unrestricted reachability). *We have* $\mathrm{BPA}^{\mathrm{fo}}_{\delta\mathrm{rr}} \vdash x \twoheadrightarrow y \Leftrightarrow \mathrm{P}_{\twoheadrightarrow}(x, y)$, *where* $\mathrm{P}_{\twoheadrightarrow}(x, y)$ *stands for the following formula of* $\mathcal{L}(\mathrm{BPA}^{\mathrm{fo}}_{\delta\mathrm{rr}})$:

$$\exists z \bullet \left(\psi^{a,b}(x, z) \ \wedge\ \exists z' \bullet \left(z \xrightarrow{a}\!\!\!\!\twoheadrightarrow z' \ \wedge\ z' \xrightarrow{b} y\right)\right)\ ,$$

with a *and* b *different actions.*

Proof. We will apply the subprocess induction schema RS, taking $\mathrm{P}_{\twoheadrightarrow}(x, y)$ for $\phi(x, y)$, to prove the implication $x \twoheadrightarrow y \Rightarrow \mathrm{P}_{\twoheadrightarrow}(x, y)$.

First of all, we conclude from RR1 and RR3, because $\psi^{a,b}(x, z) \Rightarrow z \xrightarrow{b} x$, that

$$\forall x' \bullet \mathrm{P}_{\twoheadrightarrow}(x', x')\ .$$

Moreover, using SI1, SI9, TR2 and RR2, we easily derive the following:

$$\begin{aligned}
&x' \xrightarrow{a'} y' \ \wedge\ \psi^{a,b}(y', u') \ \wedge\ \exists u'' \bullet \left(u' \xrightarrow{a}\!\!\!\!\twoheadrightarrow u'' \ \wedge\ u'' \xrightarrow{b} z'\right) \Rightarrow \\
&\quad \psi^{a,b}(x', a \cdot u' + b \cdot x') \ \wedge\ \exists u'' \bullet \left(a \cdot u' + b \cdot x' \xrightarrow{a}\!\!\!\!\twoheadrightarrow u'' \ \wedge\ u'' \xrightarrow{b} z'\right)\ .
\end{aligned}$$

Hence, we conclude from RR3, using existential generalization, that

$$\forall x', y', z' \bullet \bigwedge_{a' \in \mathsf{A}} \left(x' \xrightarrow{a'} y' \ \wedge\ \mathrm{P}_{\twoheadrightarrow}(y', z') \Rightarrow \mathrm{P}_{\twoheadrightarrow}(x', z')\right)\ .$$

Using the subprocess induction schema, it follows from these conclusions that $x \twoheadrightarrow y \Rightarrow \mathrm{P}_{\twoheadrightarrow}(x, y)$.

In the proof of the implication $\mathrm{P}_{\twoheadrightarrow}(x, y) \Rightarrow x \twoheadrightarrow y$ given below, $\rho(u, u')$ stands for the formula

$$\begin{aligned}
&\exists x \bullet \psi^{a,b}(x, u) \Rightarrow \\
&\quad \exists! x \bullet \left(\psi^{a,b}(x, u) \ \wedge\ u \xrightarrow{a}\!\!\!\!\twoheadrightarrow u' \ \wedge\ \exists! y \bullet \left(u' \xrightarrow{b} y \ \wedge\ x \twoheadrightarrow y\right)\right)\ .
\end{aligned}$$

We will apply the restricted subprocess induction schema RRS, taking $\rho(x, y)$ for $\phi(x, y)$. When we have shown in this manner that $u \xrightarrow{a}\!\!\!\!\twoheadrightarrow u' \Rightarrow \rho(u, u')$, we can conclude that $\mathrm{P}_{\twoheadrightarrow}(x, y) \Rightarrow x \twoheadrightarrow y$ as follows. Assume $\mathrm{P}_{\twoheadrightarrow}(x, y)$. Then there exist

u and u' such that $\psi^{a,b}(x,u) \wedge u \overset{a}{\twoheadrightarrow} u' \wedge u' \overset{b}{\to} y$. Because $u \overset{a}{\twoheadrightarrow} u' \Rightarrow \rho(u,u')$, also $\rho(u,u')$. This immediately gives $x \twoheadrightarrow y$.

It remains to be shown by means of RRS that $u \overset{a}{\twoheadrightarrow} u' \Rightarrow \rho(u,u')$. First of all, we conclude from RR1 and R1, because $\psi^{a,b}(x,u) \Rightarrow u \overset{b}{\to} x$, that

$$\forall u \bullet \rho(u,u) .$$

Moreover, using RR2 and R2, we easily derive from the hypothesis $\exists x \bullet \psi^{a,b}(x,u)$ the following implications:

$$u \overset{a}{\to} u' \wedge \psi^{a,b}(x',u') \Rightarrow \exists! x \bullet \left(\psi^{a,b}(x,u) \wedge \bigvee_{a' \in \mathsf{A}} x \overset{a'}{\to} x'\right),$$
$$u \overset{a}{\to} u' \wedge u' \overset{a}{\twoheadrightarrow} u'' \Rightarrow u \overset{a}{\twoheadrightarrow} u'',$$
$$\bigvee_{a' \in \mathsf{A}} x \overset{a'}{\to} x' \wedge \exists! y \bullet (u'' \overset{b}{\to} y \wedge x' \twoheadrightarrow y) \Rightarrow \exists! y \bullet (u'' \overset{b}{\to} y \wedge x \twoheadrightarrow y) .$$

The left-hand sides of the first and second implication are conjunctions of $u \overset{a}{\to} u'$ and (an instance of) one of the first two conjuncts occurring in the right-hand side of $\rho(u',u'')$. The left-hand side of the third implication is a conjunction of the second conjunct occurring in the right-hand side of the first implication and (an instance of) the third conjunct occurring in the right-hand side of $\rho(u',u'')$. Hence, we also conclude that

$$\forall u, u', u'' \bullet (u \overset{a}{\to} u' \wedge \rho(u',u'') \Rightarrow \rho(u,u'')) .$$

Using the restricted subprocess induction schema, it follows from these conclusions that $u \overset{a}{\twoheadrightarrow} u' \Rightarrow \rho(u,u')$. □

The following is a corollary of the proof of Theorem 18.

Corollary 3 (RRS implies RS). *We have* $\mathrm{BPA}^{\mathrm{fo}}_{\delta\mathrm{rr}} \setminus \mathrm{RS} \models \mathrm{RS}$.

Moreover, in the models of $\mathrm{BPA}^{\mathrm{fo}}_{\delta\mathrm{rr}}$, R holds if RR holds.

Theorem 19 (RR implies R). *We have* $\mathrm{BPA}^{\mathrm{fo}}_{\delta\mathrm{rr}} \cup \mathrm{RR} \models \mathrm{R}$.

Proof. Suppose $\forall x', y', z' \bullet (R(x',x') \wedge \bigwedge_{a \in \mathsf{A}} (x' \overset{a}{\to} y' \wedge R(y',z') \Rightarrow R(x',z')))$. Then we must show that $\mathrm{BPA}^{\mathrm{fo}}_{\delta\mathrm{rr}} \cup \mathrm{RR} \models x \twoheadrightarrow y \Rightarrow R(x,y)$. By Theorem 18, it is sufficient to show that $\forall u, u' \bullet (\psi^{a,b}(x,u) \wedge u \overset{a}{\twoheadrightarrow} u' \wedge u' \overset{b}{\to} y \Rightarrow R(x,y))$. This is done by induction on the number of steps, say k, required for $u \overset{a}{\twoheadrightarrow} u'$. If $k = 0$, then we immediately have $R(x,y)$. If $k = n+1$, then there exists a u'' such that $u \overset{a}{\to} u''$ and $u'' \overset{a}{\twoheadrightarrow} u'$. It follows from $\psi^{a,b}(x,u)$, that there exists a unique x'' such that $x \overset{a'}{\to} x''$ for some action a' and $u'' \overset{b}{\to} x''$. By the induction hypothesis, $R(x'',y)$. From $x \overset{a'}{\to} x''$ and $R(x'',y)$, it follows that $R(x,y)$. □

For each $\kappa \geq \aleph_0$, $\mathfrak{P}^{\mathrm{rr}}_\kappa$ is the expansion of $\mathfrak{P}_\kappa$ that additionally has for each predicate symbol $\overset{a}{\twoheadrightarrow}$ a binary relation $\widetilde{\overset{a}{\twoheadrightarrow}}$ on $\mathbb{TS}_\kappa/\leftrightarrows$ defined as follows:

$$[T_1] \widetilde{\overset{a}{\twoheadrightarrow}} [T_2] \text{ iff } \exists T \in [T_2] \bullet T_1 \widehat{\overset{a}{\twoheadrightarrow}} T ,$$

where $\widehat{\xrightarrow{a}\!\!\!\!\rightarrow}$ is a binary relation on $\mathbb{TS}_\kappa$ which will be defined below. However, we first introduce an auxiliary notion. Let $T = (S, \rightarrow, \rightarrow\surd, s^0)$ be a transition system. Then, for each $a \in \mathsf{A}$, the *reachability by a-steps* relation of T is the smallest relation $\xrightarrow{a}\!\!\!\!\rightarrow \subseteq S \times S$ such that:

- $s \xrightarrow{a}\!\!\!\!\rightarrow s$;
- if $s \xrightarrow{a} s'$ and $s' \xrightarrow{a}\!\!\!\!\rightarrow s''$, then $s \xrightarrow{a}\!\!\!\!\rightarrow s''$.

We write $\mathrm{RS}_a(T)$ for $\{s \in S \mid s^0 \xrightarrow{a}\!\!\!\!\rightarrow s\}$. Now the relation $\widehat{\xrightarrow{a}\!\!\!\!\rightarrow}$ on $\mathbb{TS}_\kappa$ is defined as follows. Let $T_1, T_2 \in \mathbb{TS}_\kappa$. Then

$$T_1 \widehat{\xrightarrow{a}\!\!\!\!\rightarrow} T_2 \text{ iff } \exists s \in \mathrm{RS}_a(T_1) \bullet (T_1)_s = T_2 \ .$$

Reachability by a-steps on $\mathbb{TS}_\kappa/\underline{\leftrightarrow}$ is well-defined because $\underline{\leftrightarrow}$ preserves reachability by a-steps on $\mathbb{TS}_\kappa$ up to $\underline{\leftrightarrow}$.

The structures $\mathfrak{P}^{\mathrm{rr}}_\kappa$ are models of $\mathrm{BPA}^{\mathrm{fo}}_{\delta\mathrm{rr}}$.

Theorem 20 (Soundness of $\mathbf{BPA}^{\mathrm{fo}}_{\delta\mathrm{rr}}$). *For all $\kappa \geq \aleph_0$, we have $\mathfrak{P}^{\mathrm{rr}}_\kappa \models \mathrm{BPA}^{\mathrm{fo}}_{\delta\mathrm{rr}}$.*

Proof. Because $\mathfrak{P}^{\mathrm{rr}}_\kappa$ is an expansion of $\mathfrak{P}_\kappa$, it is sufficient to show that the additional axioms for restricted reachability are sound. The soundness of all additional axioms for restricted reachability follows easily from the definitions of the ingredients of $\mathfrak{P}^{\mathrm{rr}}_\kappa$. □

The extension of $\mathrm{BPA}^{\mathrm{fo}}_\delta$ to $\mathrm{BPA}^{\mathrm{fo}}_{\delta\mathrm{rr}}$ may seem at first sight rather far-fetched. However, unrestricted reachability is explicit definable in $\mathrm{BPA}^{\mathrm{fo}}_{\delta\mathrm{rr}}$. Moreover, in all models of $\mathrm{BPA}^{\mathrm{fo}}_{\delta\mathrm{rr}}$, the validity of RS is implied by the validity of RRS and the validity of R is implied by the validity of RR. All this strongly suggests that restricted reachability is more basic than unrestricted reachability. In addition, we will see in Sect. 16 that $\mathrm{ACP}^{\mathrm{fo}}$, i.e. the first-order extension of ACP presented in Sect. 13, can be interpreted in $\mathrm{BPA}^{\mathrm{fo}}_{\delta\mathrm{rr}}$.

It is unknown to us whether the restricted reachability predicates $\xrightarrow{a}\!\!\!\!\rightarrow$ are definable in terms of $\mathcal{L}(\mathrm{BPA}^{\mathrm{fo}}_\delta)$ in $\mathrm{BPA}^{\mathrm{fo}}_\delta$. In any case, the extension turns out to have great expressive power. Consider the following formula of $\mathcal{L}(\mathrm{BPA}^{\mathrm{fo}}_{\delta\mathrm{rr}})$:

$$\begin{aligned}
\exists z \bullet \Big(\forall u \bullet \Big(z \xrightarrow{a}\!\!\!\!\rightarrow u \Rightarrow {} & \exists! v \bullet u \xrightarrow{a} v \ \land \\
& \exists! u' \bullet u \xrightarrow{b} u' \ \land \bigwedge_{a' \in \mathsf{A}, a' \neq a, b} \neg\, \exists w \bullet u \xrightarrow{a'} w\Big) \ \land \\
z \xrightarrow{b} x \ \land \qquad & \\
\forall u, v \bullet \Big(z \xrightarrow{a}\!\!\!\!\rightarrow u \land u \xrightarrow{a} v \Rightarrow {} & \\
& \exists u', v' \bullet \Big(u \xrightarrow{b} u' \ \land\ v \xrightarrow{b} v' \ \land \bigvee_{a' \in \mathsf{A}} u' \xrightarrow{a'} v'\Big)\Big)\Big) \ ,
\end{aligned}$$

where a and b are different actions. We use $\infty(x)$ as an abbreviation of the above formula. Let $\mathfrak{A}$ be a model of $\mathrm{BPA}^{\mathrm{fo}}_{\delta\mathrm{rr}}$ with domain P, and let $p \in P$. Then $\mathfrak{A} \models \neg\,\infty(x)\ [p]$ only if p has no infinite path in $\mathrm{TS}(\mathfrak{A}, p)$. If $\mathfrak{A}$ is one of the full

Table 9. Bar induction schema

$$\bigwedge_{a\in\mathsf{A}}\psi(a) \wedge \forall x \bullet (\neg\,\infty(x) \Rightarrow (\forall y \bullet \bigwedge_{a\in\mathsf{A}}(x \xrightarrow{a} y \Rightarrow \psi(y)) \Rightarrow \psi(x))) \Rightarrow \forall x \bullet (\neg\,\infty(x) \Rightarrow \psi(x)) \quad \text{BAR}$$

bisimulation models $\mathfrak{P}^{\mathrm{rr}}_{\kappa}$, "only if" can be replaced by "if and only if". It looks to be that there is no formula of $\mathcal{L}(\mathrm{BPA}^{\mathrm{fo}}_{\delta})$ with analogous properties.

The axiom schema BAR given in Table 9 can be used to prove properties of all processes that have no infinite path. BAR is an axiom schema where $\psi(x)$ is a formula of $\mathcal{L}(\mathrm{BPA}^{\mathrm{fo}}_{\delta\mathrm{rr}})$. Axiom schema BAR is an induction schema, called the *bar induction schema.*

BAR is valid in the full bisimulation models $\mathfrak{P}^{\mathrm{rr}}_{\kappa}$.

Theorem 21 (Soundness of BAR). *For all $\kappa \geq \aleph_0$, we have $\mathfrak{P}^{\mathrm{rr}}_{\kappa} \models \mathrm{BAR}$.*

Proof. We define an ordinal function $\|\_\|$ on the domain $\mathcal{P}$ of $\mathfrak{P}^{\mathrm{rr}}_{\kappa}$ as follows:

- if $\{p' \mid p \mathrel{\widetilde{\twoheadrightarrow}} p'\} = \emptyset$, then $\|p\| = 0$;
- if $\{p' \mid p \mathrel{\widetilde{\twoheadrightarrow}} p'\} \neq \emptyset$ and $\{\|p'\| \mid p \mathrel{\widetilde{\twoheadrightarrow}} p'\}$ has a maximal element, then $\|p\| = \max\{\|p'\| \mid p \mathrel{\widetilde{\twoheadrightarrow}} p'\} + 1$;
- if $\{p' \mid p \mathrel{\widetilde{\twoheadrightarrow}} p'\} \neq \emptyset$ and $\{\|p'\| \mid p \mathrel{\widetilde{\twoheadrightarrow}} p'\}$ has no maximal element, then $\|p\| = \sup\{\|p'\| \mid p \mathrel{\widetilde{\twoheadrightarrow}} p'\}$.

Because $p \mathrel{\widetilde{\xrightarrow{a}}} p'$ implies $\|p'\| < \|p\|$ if p has no infinite path, it is easily proved by transfinite induction on $\|x\|$ that BAR is valid in $\mathfrak{P}^{\mathrm{rr}}_{\kappa}$. □

13 The First-Order Theory ACP^{fo}

In this section, we present ACP^{fo}, a first-order extension of ACP. Like in BPA^{fo}, it is assumed that there is a fixed but arbitrary finite set of *actions* A with $\delta \notin \mathsf{A}$. We write A_δ for $\mathsf{A} \cup \{\delta\}$. In ACP^{fo}, it is further assumed that there is a fixed but arbitrary commutative and associative *communication* function $| : \mathsf{A}_\delta \times \mathsf{A}_\delta \to \mathsf{A}_\delta$ such that $\delta \mid a = \delta$ for all $a \in \mathsf{A}_\delta$. The function $|$ is regarded to give the result of synchronously performing any two actions for which this is possible, and to be δ otherwise.

The first-order theory ACP^{fo} is an extension of $\mathrm{BPA}^{\mathrm{fo}}_{\delta}$. It has the nonlogical symbols of $\mathrm{BPA}^{\mathrm{fo}}_{\delta}$ and in addition:

- the binary *parallel composition* operator $\|$;
- the binary *left merge* operator $\mathbin{\|\!\_}$;
- the binary *communication merge* operator $|$;
- for each $H \subseteq \mathsf{A}$, the unary *encapsulation* operator ∂_H.

We use infix notation for the additional binary operators as well. The precedence conventions for the binary operators are now as follows. The operator $\cdot$ binds stronger than all other binary operators and the operator $+$ binds weaker than all other binary operators.

Table 10. Additional axioms for $\mathrm{ACP^{fo}}$ $(a, b, c \in \mathsf{A}_\delta)$

$x \parallel y = x \mathbin{⌊\!⌊} y + y \mathbin{⌊\!⌊} x + x \mid y$	CM1	$a \mid b = b \mid a$		C1
$a \mathbin{⌊\!⌊} x = a \cdot x$	CM2	$(a \mid b) \mid c = a \mid (b \mid c)$		C2
$a \cdot x \mathbin{⌊\!⌊} y = a \cdot (x \parallel y)$	CM3	$\delta \mid a = \delta$		C3
$(x + y) \mathbin{⌊\!⌊} z = x \mathbin{⌊\!⌊} z + y \mathbin{⌊\!⌊} z$	CM4			
$a \cdot x \mid b = (a \mid b) \cdot x$	CM5	$\partial_H(a) = a$	if $a \notin H$	D1
$a \mid b \cdot x = (a \mid b) \cdot x$	CM6	$\partial_H(a) = \delta$	if $a \in H$	D2
$a \cdot x \mid b \cdot y = (a \mid b) \cdot (x \parallel y)$	CM7	$\partial_H(x + y) = \partial_H(x) + \partial_H(y)$		D3
$(x + y) \mid z = x \mid z + y \mid z$	CM8	$\partial_H(x \cdot y) = \partial_H(x) \cdot \partial_H(y)$		D4
$x \mid (y + z) = x \mid y + x \mid z$	CM9			

The constants and operators of $\mathrm{ACP^{fo}}$ are the same as the constants and operators of ACP.

Let t and t' be closed terms of $\mathcal{L}(\mathrm{ACP^{fo}})$. Intuitively, the additional operators can be explained as follows:

- $t \parallel t'$ behaves as the process that proceeds with t and t' in parallel;
- $t \mathbin{⌊\!⌊} t'$ behaves the same as $t \parallel t'$, except that it starts with performing an action of t;
- $t \mid t'$ behaves the same as $t \parallel t'$, except that it starts with performing an action of t and an action of t' synchronously;
- $\partial_H(t)$ behaves the same as t, except that it does not perform actions in H.

The axioms of $\mathrm{ACP^{fo}}$ are the axioms of $\mathrm{BPA}_\delta^{\mathrm{fo}}$ and the additional axioms given in Table 10. CM2–CM3, CM5–CM7, C1–C3 and D1–D2 are axiom schemas where a, b and c are constants of $\mathrm{ACP^{fo}}$. In D1–D4, H stands for an arbitrary subset of A. So, D3 and D4 are axiom schemas as well.

Axioms A1–A7, CM1–CM9, C1–C3 and D1–D4 are the axioms of ACP. So $\mathrm{ACP^{fo}}$ imports the (equational) axioms of ACP.

A well-known subtheory of ACP is PA, which is ACP without communication. Likewise, we have a subtheory of $\mathrm{ACP^{fo}}$, to wit $\mathrm{PA^{fo}}$. The first-order theory $\mathrm{PA^{fo}}$ is $\mathrm{ACP^{fo}}$ without the communication merge operator, without axioms CM5–CM9 and C1–C3, and with axiom CM1 replaced by $x \parallel y = x \mathbin{⌊\!⌊} y + y \mathbin{⌊\!⌊} x$ (M1). In other words, the possibility that actions are performed synchronously is not covered by $\mathrm{PA^{fo}}$.

To prove a statement for all closed terms of $\mathcal{L}(\mathrm{ACP^{fo}})$, it is sufficient to prove it for all basic terms over $\mathrm{BPA}_\delta^{\mathrm{fo}}$. The reason for this is that all closed terms of $\mathcal{L}(\mathrm{ACP^{fo}})$ are derivably equal to a basic term over $\mathrm{BPA}_\delta^{\mathrm{fo}}$.

Proposition 9 (Elimination). *For all closed terms t of $\mathcal{L}(\mathrm{ACP^{fo}})$ there exists a basic term t' such that $\mathrm{ACP^{fo}} \vdash t = t'$.*

Proof. This follows immediately from the elimination property for ACP: the closed terms of $\mathcal{L}(\mathrm{ACP^{fo}})$ are the same as the closed terms of $\mathcal{L}(\mathrm{ACP})$, and the equational axioms of $\mathrm{ACP^{fo}}$ are the same as the axioms of ACP. □

For closed equations, $\mathrm{ACP^{fo}}$ is a complete theory.

Theorem 22 (Complete theory for closed equations). *For all closed terms* t_1, t_2 *of* $\mathcal{L}(\mathrm{ACP}^{\mathrm{fo}})$*, we have either* $\mathrm{ACP}^{\mathrm{fo}} \vdash t_1 = t_2$ *or* $\mathrm{ACP}^{\mathrm{fo}} \vdash \neg\, t_1 = t_2$*, but not both.*

Proof. This follows immediately from Proposition 9 and Theorem 1. □

We have not yet investigated the decidability of $\mathrm{ACP}^{\mathrm{fo}}$, but it is to be expected that it is an undecidable theory. By adaptation of the proof of a similar theorem from [20], we can easily establish the undecidability of $\mathrm{ACP}^{\mathrm{fo}} \cup \mathrm{AIP}$.

Theorem 23 (Undecidability). $\mathrm{ACP}^{\mathrm{fo}} \cup \mathrm{AIP}$ *is an undecidable theory.*

Proof. We consider a register machine with three registers, numbered 1, 2 and 3. A program for the register machine is a finite sequence $I_1, \ldots, I_k$ of instructions of the following form:

- (add_i, j): add 1 to the contents of register i and go to instruction j;
- (sub_i, j): if the contents of register i equals 0, then go to instruction j, otherwise subtract 1 from the contents of register i and go to instruction j;
- (zero_i, j, j'): if the contents of register i equals 0, then go to instruction j, otherwise go to instruction j';
- halt: halt;

where $i \in \{1, 2, 3\}$ and $j, j' \in \{1, \ldots, k\}$.

Let K be a recursively enumerable but not recursive subset of $\mathbb{N}$, and let $n \in \mathbb{N}$. Then there exists a program for this register machine such that, if the registers are initialized to n, 0 and 0, the program halts iff $n \in K$ (see e.g. [21]). Let $P = I_1, \ldots, I_l$ be this program. We will show that P can be represented in $\mathrm{ACP}^{\mathrm{fo}} \cup \mathrm{AIP}$.

Let $A = \{a_i, s_i, z_i \mid 1 \le i \le 3\}$ and $\overline{A} = \{\overline{a}_i, \overline{s}_i, \overline{z}_i \mid 1 \le i \le 3\}$. We fix the set of actions A and the communication function $\mid$ as follows: $\mathsf{A} = A \cup \overline{A} \cup \{t, h\}$; and $a \mid b = t$ if either $a \in A$, $b \in \overline{A}$ and $\overline{a} = b$, or $a \in \overline{A}$, $b \in A$ and $a = \overline{b}$, and $a \mid b = \delta$ otherwise.

Let E be the guarded recursive specification that consists of the following equations:

$$
\begin{aligned}
R_i &= \overline{z}_i \cdot R_i + \overline{a}_i \cdot R_i' \cdot R_i && \text{for } i \in \{1,2,3\}\,,\\
R_i' &= \overline{s}_i + \overline{a}_i \cdot R_i' \cdot R_i' && \text{for } i \in \{1,2,3\}\,,\\
X_j &= [\![I_j]\!] && \text{for } j \in \{1,\ldots,l\}\,,\\
T &= t \cdot T\,,
\end{aligned}
$$

where the map $[\![\_]\!]$ from register machine instructions to terms of $\mathcal{L}(\mathrm{ACP}^{\mathrm{fo}})$ is defined as follows:

$$[\![(\mathsf{add}_i, j)]\!] = a_i \cdot X_j \;,$$

$$[\![(\mathsf{sub}_i, j)]\!] = (z_i + s_i) \cdot X_j \;,$$

$$[\![(\mathsf{zero}_i, j, j')]\!] = z_i \cdot X_j + s_i \cdot a_i \cdot X_{j'} \;,$$

$$[\![\mathsf{halt}]\!] = h \;.$$

We introduce for $m \geq 0$ the abbreviation $R_i(m)$ defined by $R_i(0) = R_i$ and $R_i(m+1) = R_i' \cdot R_i(m)$. Note that $R_i(m)$ represents register i in the state where its contents is m.

It is easy to see that P does not halt iff

$$\mathrm{ACP^{fo}} \cup \mathrm{AIP} \;\vdash\; E \Rightarrow \partial_H(X_1 \parallel R_1(n) \parallel R_2(0) \parallel R_3(0)) = T \;,$$

where $H = \mathsf{A} \cup \overline{\mathsf{A}}$. Therefore, the problem whether $n \notin K$ is one to one reducible to the problem whether a given formula of $\mathcal{L}(\mathrm{ACP^{fo}} \cup \mathrm{AIP})$ is derivable. Because the former problem is undecidable, we conclude that the latter problem is undecidable as well. This shows that $\mathrm{ACP^{fo}} \cup \mathrm{AIP}$ is an undecidable theory. □

In this section, $\mathrm{BPA}_\delta^{\mathrm{fo}}$ has been extended to $\mathrm{ACP^{fo}}$. $\mathrm{BPA}_{\delta\mathrm{rr}}^{\mathrm{fo}}$ can be extended with the same nonlogical symbols and axioms as $\mathrm{BPA}_\delta^{\mathrm{fo}}$, resulting in $\mathrm{ACP}_{\mathrm{rr}}^{\mathrm{fo}}$.

14 Full Bisimulation Models of ACP^{fo}

In this section, we expand the full bisimulation models of $\mathrm{BPA}_\delta^{\mathrm{fo}}$ to $\mathrm{ACP^{fo}}$. We will use the abbreviation $s \xrightarrow{a} s' \wr s''$ for $s \xrightarrow{a} s' \lor (s \xrightarrow{a}\surd \land s' = s'')$.

First of all, we associate with each additional operator f of $\mathrm{ACP^{fo}}$ an operation $\widehat{f}$ on $\mathbb{TS}_\kappa$ as follows.

– Let $T_i = (S_i, \rightarrow_i, \rightarrow\surd_i, s_i^0) \in \mathbb{TS}_\kappa$ for $i = 1, 2$. Then

$$T_1 \mathbin{\widehat{\parallel}} T_2 = (S, \rightarrow, \rightarrow\surd, s^0) \;,$$

where

$$s^0 = (s_1^0, s_2^0) \;,$$

$$s^{\surd} = \mathrm{ch}_\kappa(\mathcal{S}_\kappa \setminus (S_1 \cup S_2)) \;,$$

$$S = ((S_1 \cup \{s^{\surd}\}) \times (S_2 \cup \{s^{\surd}\})) \setminus \{(s^{\surd}, s^{\surd})\} \;,$$

and for every $a \in \mathsf{A}$:

$$\begin{aligned}
\xrightarrow{a} \;=\; & \{((s_1, s_2), (s_1', s_2)) \mid (s_1', s_2) \in S \land s_1 \xrightarrow{a}_1 s_1' \wr s^{\surd}\} \\
\cup\; & \{((s_1, s_2), (s_1, s_2')) \mid (s_1, s_2') \in S \land s_2 \xrightarrow{a}_2 s_2' \wr s^{\surd}\} \\
\cup\; & \Big\{((s_1, s_2), (s_1', s_2')) \;\Big|\; (s_1', s_2') \in S \;\land \\
& \bigvee_{a', b' \in \mathsf{A}} (s_1 \xrightarrow{a'}_1 s_1' \wr s^{\surd} \land s_2 \xrightarrow{b'}_2 s_2' \wr s^{\surd} \land a' \mid b' = a)\Big\} \;,
\end{aligned}$$

$$\begin{aligned}
\xrightarrow{a}\surd \;=\; & \{(s_1, s^{\surd}) \mid s_1 \xrightarrow{a}\surd_1\} \cup \{(s^{\surd}, s_2) \mid s_2 \xrightarrow{a}\surd_2\} \\
\cup\; & \Big\{(s_1, s_2) \;\Big|\; \bigvee_{a', b' \in \mathsf{A}} (s_1 \xrightarrow{a'}\surd_1 \land s_2 \xrightarrow{b'}\surd_2 \land a' \mid b' = a)\Big\} \;.
\end{aligned}$$

– Let $T_i = (S_i, \rightarrow_i, \rightarrow_{\surd i}, s_i^0) \in \mathbb{TS}_\kappa$ for $i = 1, 2$. Suppose that $T_1 \mathbin{\widehat{\Vert}} T_2 = (S, \rightarrow, \rightarrow_\surd, s^0)$ where $S = ((S_1 \cup \{s^\surd\}) \times (S_2 \cup \{s^\surd\})) \setminus \{(s^\surd, s^\surd)\}$ and $s^\surd = \mathrm{ch}_\kappa(\mathcal{S}_\kappa \setminus (S_1 \cup S_2))$. Then

$$T_1 \mathbin{\widehat{\lfloor\!\lfloor}} T_2 = \Gamma(S', \rightarrow', \rightarrow_\surd, s^{0\prime}) ,$$

where

$$s^{0\prime} = \mathrm{ch}_\kappa(\mathcal{S}_\kappa \setminus S) ,$$

$$S' = \{s^{0\prime}\} \cup S ,$$

and for every $a \in \mathsf{A}$:

$$\xrightarrow{a}{}' = \left\{(s^{0\prime}, (s, s_2^0)) \mid s_1^0 \xrightarrow{a}_1 s \wr s^\surd\right\} \cup \xrightarrow{a} .$$

– Let $T_i = (S_i, \rightarrow_i, \rightarrow_{\surd i}, s_i^0) \in \mathbb{TS}_\kappa$ for $i = 1, 2$. Suppose that $T_1 \mathbin{\widehat{\Vert}} T_2 = (S, \rightarrow, \rightarrow_\surd, s^0)$ where $S = ((S_1 \cup \{s^\surd\}) \times (S_2 \cup \{s^\surd\})) \setminus \{(s^\surd, s^\surd)\}$ and $s^\surd = \mathrm{ch}_\kappa(\mathcal{S}_\kappa \setminus (S_1 \cup S_2))$. Then

$$T_1 \mathbin{\widehat{\mid}} T_2 = \Gamma(S', \rightarrow', \rightarrow_\surd', s^{0\prime}) ,$$

where

$$s^{0\prime} = \mathrm{ch}_\kappa(\mathcal{S}_\kappa \setminus S) ,$$

$$S' = \{s^{0\prime}\} \cup S ,$$

and for every $a \in \mathsf{A}$:

$$\begin{aligned}\xrightarrow{a}{}' &= \Big\{(s^{0\prime}, (s_1, s_2)) \;\Big|\; (s_1, s_2) \in S \;\wedge \\ &\qquad \bigvee_{a', b' \in \mathsf{A}} \left(s_1^0 \xrightarrow{a'}_1 s_1 \wr s^\surd \;\wedge\; s_2^0 \xrightarrow{b'}_2 s_2 \wr s^\surd \;\wedge\; a' \mid b' = a\right)\Big\} \cup \xrightarrow{a} ,\\ \xrightarrow{a}{}_\surd' &= \Big\{s^{0\prime} \;\Big|\; \bigvee_{a', b' \in \mathsf{A}} \left(s_1^0 \xrightarrow{a'}_{\surd 1} \;\wedge\; s_2^0 \xrightarrow{b'}_{\surd 2} \;\wedge\; a' \mid b' = a\right)\Big\} \cup \xrightarrow{a}_\surd .\end{aligned}$$

– Let $T = (S, \rightarrow, \rightarrow_\surd, s^0) \in \mathbb{TS}_\kappa$. Then

$$\widehat{\partial_H}(T) = \Gamma(S, \rightarrow', \rightarrow_\surd', s^0) ,$$

where for every $a \notin H$:

$$\xrightarrow{a}{}' = \xrightarrow{a} ,$$

$$\xrightarrow{a}{}_\surd' = \xrightarrow{a}_\surd ,$$

and for every $a \in H$:

$$\xrightarrow{a}{}' = \emptyset ,$$

$$\xrightarrow{a}{}_\surd' = \emptyset .$$

We can easily show that bisimilarity is a congruence with respect to parallel composition, left merge, communication merge and encapsulation.

Proposition 10 (Congruence). *For all* $T_1, T_2, T_1', T_2' \in \mathbb{TS}_\kappa$ *(*$\kappa \geq \aleph_0$*),* $T_1 \mathrel{\underline{\leftrightarrow}} T_1'$ *and* $T_2 \mathrel{\underline{\leftrightarrow}} T_2'$ *imply* $T_1 \mathbin{\widehat{\parallel}} T_2 \mathrel{\underline{\leftrightarrow}} T_1' \mathbin{\widehat{\parallel}} T_2'$, $T_1 \mathbin{\widehat{\lfloor\!\lfloor}} T_2 \mathrel{\underline{\leftrightarrow}} T_1' \mathbin{\widehat{\lfloor\!\lfloor}} T_2'$, $T_1 \mathbin{\widehat{\mid}} T_2 \mathrel{\underline{\leftrightarrow}} T_1' \mathbin{\widehat{\mid}} T_2'$ *and* $\widehat{\partial_H}(T_1) \mathrel{\underline{\leftrightarrow}} \widehat{\partial_H}(T_1')$*.*

Proof. Let $T_i = (S_i, \rightarrow_i, \rightarrow\!\surd_i, s_i^0)$ and $T_i' = (S_i', \rightarrow_i', \rightarrow\!\surd_i', s_i^{0\prime})$ for $i = 1, 2$. Let R_1 and R_2 be bisimulations witnessing $T_1 \mathrel{\underline{\leftrightarrow}} T_1'$ and $T_2 \mathrel{\underline{\leftrightarrow}} T_2'$, respectively. Then we construct relations $R_{\widehat{\parallel}}$, $R_{\widehat{\lfloor\!\lfloor}}$, $R_{\widehat{\mid}}$ and $R_{\widehat{\partial_H}}$ as follows:

- $R_{\widehat{\parallel}} = \{((s_1, s_2), (s_1', s_2')) \in S \times S' \mid (s_1, s_1') \in R_1 \cup R^\surd, (s_2, s_2') \in R_2 \cup R^\surd\}$, where S and S' are the sets of states of $T_1 \mathbin{\widehat{\parallel}} T_2$ and $T_1' \mathbin{\widehat{\parallel}} T_2'$, respectively, and $R^\surd = \{(\mathrm{ch}_\kappa(\mathcal{S}_\kappa \setminus (S_1 \cup S_2)), \mathrm{ch}_\kappa(\mathcal{S}_\kappa \setminus (S_1' \cup S_2')))\}$;
- $R_{\widehat{\lfloor\!\lfloor}} = (\{(s^0, s^{0\prime})\} \cup R_{\widehat{\parallel}}) \cap (S \times S')$, where S and S' are the sets of states of $T_1 \mathbin{\widehat{\lfloor\!\lfloor}} T_2$ and $T_1' \mathbin{\widehat{\lfloor\!\lfloor}} T_2'$, respectively, and s^0 and $s^{0\prime}$ are the initial states of $T_1 \mathbin{\widehat{\lfloor\!\lfloor}} T_2$ and $T_1' \mathbin{\widehat{\lfloor\!\lfloor}} T_2'$, respectively;
- $R_{\widehat{\mid}} = (\{(s^0, s^{0\prime})\} \cup R_{\widehat{\parallel}}) \cap (S \times S')$, where S and S' are the sets of states of $T_1 \mathbin{\widehat{\mid}} T_2$ and $T_1' \mathbin{\widehat{\mid}} T_2'$, respectively, and s^0 and $s^{0\prime}$ are the initial states of $T_1 \mathbin{\widehat{\mid}} T_2$ and $T_1' \mathbin{\widehat{\mid}} T_2'$, respectively;
- $R_{\widehat{\partial_H}} = R_1 \cap (S \times S')$, where S and S' are the sets of states of $\widehat{\partial_H}(T_1)$ and $\widehat{\partial_H}(T_1')$, respectively.

Given the definitions of parallel composition, left merge, communication merge and encapsulation, it is easy to see that $R_{\widehat{\parallel}}$, $R_{\widehat{\lfloor\!\lfloor}}$, $R_{\widehat{\mid}}$ and $R_{\widehat{\partial_H}}$ are bisimulations witnessing $T_1 \mathbin{\widehat{\parallel}} T_2 \mathrel{\underline{\leftrightarrow}} T_1' \mathbin{\widehat{\parallel}} T_2'$, $T_1 \mathbin{\widehat{\lfloor\!\lfloor}} T_2 \mathrel{\underline{\leftrightarrow}} T_1' \mathbin{\widehat{\lfloor\!\lfloor}} T_2'$, $T_1 \mathbin{\widehat{\mid}} T_2 \mathrel{\underline{\leftrightarrow}} T_1' \mathbin{\widehat{\mid}} T_2'$ and $\widehat{\partial_H}(T_1) \mathrel{\underline{\leftrightarrow}} \widehat{\partial_H}(T_1')$, respectively. □

The *full bisimulation models* $\mathfrak{P}_\kappa'$ of $\mathrm{ACP}^{\mathrm{fo}}$, one for each $\kappa \geq \aleph_0$, are the expansions of the full bisimulation models $\mathfrak{P}_\kappa$ of $\mathrm{BPA}_\delta^{\mathrm{fo}}$ with an n-ary operation $\tilde{f}$ on the domain of $\mathfrak{P}_\kappa$ ($\mathbb{TS}_\kappa/\mathrel{\underline{\leftrightarrow}}$) for each additional n-ary operator f of $\mathrm{ACP}^{\mathrm{fo}}$. Those additional operations are defined as follows:

$$
\begin{aligned}
[T_1] \mathbin{\widetilde{\parallel}} [T_2] &= [T_1 \mathbin{\widehat{\parallel}} T_2]\,, \\
[T_1] \mathbin{\widetilde{\lfloor\!\lfloor}} [T_2] &= [T_1 \mathbin{\widehat{\lfloor\!\lfloor}} T_2]\,, \\
[T_1] \mathbin{\widetilde{\mid}} [T_2] &= [T_1 \mathbin{\widehat{\mid}} T_2]\,, \\
\widetilde{\partial_H}([T_1]) &= [\widehat{\partial_H}(T_1)]\,.
\end{aligned}
$$

Parallel composition, left merge, communication merge and encapsulation on $\mathbb{TS}_\kappa/\mathrel{\underline{\leftrightarrow}}$ are well-defined because $\mathrel{\underline{\leftrightarrow}}$ is a congruence with respect to the corresponding operations on $\mathbb{TS}_\kappa$.

The structures $\mathfrak{P}_\kappa'$ are models of $\mathrm{ACP}^{\mathrm{fo}}$.

Theorem 24 (Soundness of $\mathrm{ACP}^{\mathrm{fo}}$). *For all* $\kappa \geq \aleph_0$*, we have* $\mathfrak{P}_\kappa' \models \mathrm{ACP}^{\mathrm{fo}}$*.*

Proof. Because $\mathfrak{P}'_\kappa$ is an expansion of $\mathfrak{P}_\kappa$, it is sufficient to show that the additional axioms for $\mathrm{ACP^{fo}}$ are sound. The soundness of all additional axioms for $\mathrm{ACP^{fo}}$ follows easily from the definitions of the ingredients of $\mathfrak{P}'_\kappa$. □

It is easy to see that Theorems 5, 7 and 8 go through for $\mathfrak{P}'_\kappa$.

In this section, the full bisimulation models $\mathfrak{P}_\kappa$ of $\mathrm{BPA}^{\mathrm{fo}}_\delta$ have been expanded to obtain the full bisimulation models $\mathfrak{P}'_\kappa$ of $\mathrm{ACP^{fo}}$. The full bisimulation models $\mathfrak{P}^{\mathrm{rr}}_\kappa$ of $\mathrm{BPA}^{\mathrm{fo}}_{\delta\mathrm{rr}}$ can be expanded in the same way to obtain the full bisimulation models $\mathfrak{P}^{\mathrm{rr}\prime}_\kappa$ of $\mathrm{ACP}^{\mathrm{fo}}_{\mathrm{rr}}$.

15 Interpretation of One Theory in Another

Let T be a first-order theory with non-logical symbols Σ. Then we say that Σ is the *signature* of T. We write $\Sigma(T)$ for the signature of T.

Let T and T' be first-order theories, and $\mathfrak{d} \notin \Sigma(T) \cup \Sigma(T')$. Then an interpretation Θ of T in T' is a family of formulas that consists of the following:

- an explicit definition $\Theta_\mathfrak{d}$ of a unary predicate $\mathfrak{d}$ in terms of $\mathcal{L}(T')$;
- for each $\sigma \in \Sigma(T) \setminus \Sigma(T')$, an explicit definition Θ_σ in terms of $\mathcal{L}(T')$;

such that the following holds for $T'' = T' \cup \{\Theta_\sigma \mid \sigma \in (\Sigma(T) \setminus \Sigma(T')) \cup \{\mathfrak{d}\}\}$:

$$T'' \vdash \exists x \bullet \mathfrak{d}(x) ,$$

$$T'' \vdash \mathfrak{d}(x_1) \wedge \ldots \wedge \mathfrak{d}(x_n) \Rightarrow \mathfrak{d}(f(x_1, \ldots, x_n))$$
$$\text{for each } n\text{-ary operator } f \in \Sigma(T) ,$$

$$T'' \vdash \phi^*$$
$$\text{for each axiom } \phi \text{ of } T ,$$

where ϕ^* is the formula obtained from ϕ by first taking the universal closure of ϕ and then replacing each subformula $\forall x \bullet \phi'$ by $\forall x \bullet (\mathfrak{d}(x) \Rightarrow \phi')$ and each subformula $\exists x \bullet \phi'$ by $\exists x \bullet (\mathfrak{d}(x) \wedge \phi')$.

This notion of interpretation of one theory in another is more general than the corresponding notion from [8], but in line with the notion of interpretability of one theory in another from [8]. It is less general than the corresponding notion in [10]. In the terminology of [10], an interpretation as defined here is an injective one-dimensional interpretation. We believe that higher dimensional interpretations are irrelevant to the case where theories about processes are considered. So long as we only consider bisimilarity as the intended notion of identity, non-injective interpretations are irrelevant as well. Note that the last condition in the definition given above can be replaced by

$$T \vdash \phi \quad \text{implies} \quad T'' \vdash \phi^* \quad \text{for each formula } \phi \text{ of } \mathcal{L}(T) .$$

The following is an important property of interpretations. For each interpretation Θ of a theory T in a theory T', $T'' = T' \cup \{\Theta_\sigma \mid \sigma \in (\Sigma(T) \setminus \Sigma(T')) \cup \{\mathfrak{d}\}\}$

is a definitional extension of T'. This means that, for each model $\mathfrak{A}'$ of T', there is a unique expansion of $\mathfrak{A}'$ that is a model of T''.

Let Θ be an interpretation of theory T in theory T', and let $T'' = T' \cup \{\Theta_\sigma \mid \sigma \in (\Sigma(T) \setminus \Sigma(T')) \cup \{\mathfrak{d}\}\}$. Suppose that $\mathfrak{A}'$ is a model of T'. Then a model $\mathfrak{A}$ of T can be obtained from $\mathfrak{A}'$ in the following steps:

1. take the unique expansion $\mathfrak{A}''$ of $\mathfrak{A}'$ such that $\mathfrak{A}'' \models T''$;
2. take the restriction $\mathfrak{A}''|_{\Sigma(T)\cup\{\mathfrak{d}\}}$ of $\mathfrak{A}''$ to $\Sigma(T) \cup \{\mathfrak{d}\}$;
3. take the unique substructure $\mathfrak{A}^*$ of $\mathfrak{A}''|_{\Sigma(T)\cup\{\mathfrak{d}\}}$ such that $\mathfrak{A}^* \models \forall x \bullet \mathfrak{d}(x)$;
4. take the restriction $\mathfrak{A} = \mathfrak{A}^*|_{\Sigma(T)}$ of $\mathfrak{A}^*$ to $\Sigma(T)$.

The most simple example of this construction is the following: The interpretation of BPA in $\mathrm{BPA}_\delta^{\mathrm{fo}}$ consists only of the explicit definition $\mathfrak{d}(x) \Leftrightarrow \neg\, x \twoheadrightarrow \delta$. That is, $\mathfrak{d}$ is in this case just another symbol for the deadlock freedom predicate. If we apply the construction described above to one of the full bisimulation models of $\mathrm{BPA}_\delta^{\mathrm{fo}}$, then we obtain one of the main models of BPA.

MPA_δ, Minimal Process Algebra with deadlock, provides another simple example. MPA_δ, introduced in [22], differs from BPA_δ by having a unary *action prefixing* operator $a\,.\,$ for each $a \in \mathsf{A}$ instead of the binary sequential composition operator of BPA_δ.[5] The axioms of MPA_δ are axioms A1, A2, A3 and A6 from Table 1. The interpretation of MPA_δ in $\mathrm{BPA}_\delta^{\mathrm{fo}}$ consists of the explicit definition $\mathfrak{d}(x) \Leftrightarrow \bigwedge_{a\in\mathsf{A}} \neg\, \exists y \bullet (x \twoheadrightarrow y \wedge y \xrightarrow{a} \surd)$ and an explicit definition $a\,.\,x = y \Leftrightarrow a \cdot x = y$ for each $a \in A$. If we apply the construction described above to one of the full bisimulation models of $\mathrm{BPA}_\delta^{\mathrm{fo}}$, then we obtain one of the main models of MPA_δ.

It needs no explaining that an interpretation of a theory T in a theory T' includes an explicit definition of each non-logical symbol of T that T does not have in common with T'. The examples given above make clear why it also includes an explicit definition of a special unary predicate symbol $\mathfrak{d}$. BPA is only concerned with processes that are deadlock free and MPA_δ is only concerned with processes that are free of successful termination. In the interpretations of BPA and MPA_δ in $\mathrm{BPA}_\delta^{\mathrm{fo}}$ described above, $\mathfrak{d}$ takes care of the restriction to the processes concerned.

16 Interpretation of $\mathbf{ACP^{fo}}$ in $\mathbf{BPA_{\delta rr}^{fo}}$

In this section, we consider the interpretation of $\mathrm{ACP}^{\mathrm{fo}}$ in $\mathrm{BPA}_{\delta\mathrm{rr}}^{\mathrm{fo}}$. This interpretation consists of explicit definitions of the predicate symbol $\mathfrak{d}$ and the operators $\|$, $\mathbin{⌊\!⌊}$, $\mid$ and ∂_H (one for each $H \subseteq \mathsf{A}$). The explicit definition of $\mathfrak{d}$ is simply $\mathfrak{d}(x) \Leftrightarrow x = x$. The explicit definitions of the operators are quite unusual in the sense that they involve an auxiliary process (u) that is used to represent a bisimulation.

5 For action prefixing and sequential composition different kinds of dot, viz. the low dot and the centered dot, are used. In MPA_δ, we have action prefixing without variable binding. In [7], the semicolon is used for action prefixing with variable binding.

First, we consider the explicit definition of the parallel composition operator. We begin by introducing the abbreviation $\mathrm{P}'_{\|}(x, y, z, u)$, which enables us to formulate the explicit definition of $\|$ as $x \| y = z \Leftrightarrow \exists u \bullet \mathrm{P}'_{\|}(x, y, z, u)$. We fix different actions i, l, r and m. We use $\mathrm{P}'_{\|}(x, y, z, u)$ as an abbreviation of the following formula of $\mathcal{L}(\mathrm{BPA}^{\mathrm{fo}}_{\delta\mathrm{rr}})$:

$$\phi_1(x,y,z,u) \wedge \phi_2(x,y,z,u) \wedge \phi_3(u) \wedge \phi_4(u) \wedge \phi_5(u) \wedge \\ \phi_6(u) \wedge \phi_7(u) \wedge \phi_8(u) \wedge \phi_9(u) \wedge \phi_{10}(u) \wedge \phi_{11}(u)\,;$$

where:
$\phi_1(x,y,z,u)$ is the formula

$$\forall u' \bullet \Big(u \stackrel{i}{\twoheadrightarrow} u' \Rightarrow \\ \exists! x', y', z' \bullet (x \twoheadrightarrow x' \wedge y \twoheadrightarrow y' \wedge z \twoheadrightarrow z' \wedge \\ u' \stackrel{l}{\rightarrow} x' \wedge u' \stackrel{r}{\rightarrow} y' \wedge u' \stackrel{m}{\longrightarrow} z') \wedge \\ \bigwedge_{a' \in \mathsf{A}, a' \neq i,l,r,m} \neg\, \exists v' \bullet u' \stackrel{a'}{\longrightarrow} v'\Big)\,,$$

$\phi_2(x,y,z,u)$ is the formula

$$u \stackrel{l}{\rightarrow} x \wedge u \stackrel{r}{\rightarrow} y \wedge u \stackrel{m}{\longrightarrow} z\,,$$

$\phi_3(u)$ is the formula

$$\forall x', y', z', u', x'' \bullet \\ \bigwedge_{a' \in \mathsf{A}} (u \stackrel{i}{\twoheadrightarrow} u' \wedge u' \stackrel{l}{\rightarrow} x' \wedge u' \stackrel{r}{\rightarrow} y' \wedge u' \stackrel{m}{\longrightarrow} z' \wedge x' \stackrel{a'}{\longrightarrow} x'' \Rightarrow \\ \exists u'', z'' \bullet \\ (u' \stackrel{i}{\rightarrow} u'' \wedge u'' \stackrel{l}{\rightarrow} x'' \wedge u'' \stackrel{r}{\rightarrow} y' \wedge u'' \stackrel{m}{\longrightarrow} z'' \wedge z' \stackrel{a'}{\longrightarrow} z''))\,,$$

$\phi_4(u)$ is the formula

$$\forall x', y', z', u', y'' \bullet \\ \bigwedge_{a' \in \mathsf{A}} (u \stackrel{i}{\twoheadrightarrow} u' \wedge u' \stackrel{l}{\rightarrow} x' \wedge u' \stackrel{r}{\rightarrow} y' \wedge u' \stackrel{m}{\longrightarrow} z' \wedge y' \stackrel{a'}{\longrightarrow} y'' \Rightarrow \\ \exists u'', z'' \bullet \\ (u' \stackrel{i}{\rightarrow} u'' \wedge u'' \stackrel{l}{\rightarrow} x' \wedge u'' \stackrel{r}{\rightarrow} y'' \wedge u'' \stackrel{m}{\longrightarrow} z'' \wedge z' \stackrel{a'}{\longrightarrow} z''))\,,$$

$\phi_5(u)$ is the formula

$$\forall x', y', z', u', x'', y'' \bullet \\ \bigwedge_{a', b' \in \mathsf{A}, a' | b' \neq \delta} (u \stackrel{i}{\twoheadrightarrow} u' \wedge u' \stackrel{l}{\rightarrow} x' \wedge u' \stackrel{r}{\rightarrow} y' \wedge u' \stackrel{m}{\longrightarrow} z' \wedge \\ x' \stackrel{a'}{\longrightarrow} x'' \wedge y' \stackrel{b'}{\longrightarrow} y'' \Rightarrow \\ \exists u'', z'' \bullet \\ (u' \stackrel{i}{\rightarrow} u'' \wedge u'' \stackrel{l}{\rightarrow} x'' \wedge u'' \stackrel{r}{\rightarrow} y'' \wedge u'' \stackrel{m}{\longrightarrow} z'' \wedge \\ z' \stackrel{a'|b'}{\longrightarrow} z''))\,,$$

$\phi_6(u)$ is the formula

$$\forall x', y', z', u' \bullet \bigwedge_{a' \in \mathsf{A}} \left(u \xrightarrow{i}\!\!\!\!\rightarrow u' \wedge u' \xrightarrow{l} x' \wedge u' \xrightarrow{r} y' \wedge u' \xrightarrow{m} z' \wedge x' \xrightarrow{a'} \surd \Rightarrow z' \xrightarrow{a'} y'\right),$$

$\phi_7(u)$ is the formula

$$\forall x', y', z', u' \bullet \bigwedge_{a' \in \mathsf{A}} \left(u \xrightarrow{i}\!\!\!\!\rightarrow u' \wedge u' \xrightarrow{l} x' \wedge u' \xrightarrow{r} y' \wedge u' \xrightarrow{m} z' \wedge y' \xrightarrow{a'} \surd \Rightarrow z' \xrightarrow{a'} x'\right),$$

$\phi_8(u)$ is the formula

$$\forall x', y', z', u', x'' \bullet \bigwedge_{a', b' \in \mathsf{A}, a' | b' \neq \delta} \left(u \xrightarrow{i}\!\!\!\!\rightarrow u' \wedge u' \xrightarrow{l} x' \wedge u' \xrightarrow{r} y' \wedge u' \xrightarrow{m} z' \wedge x' \xrightarrow{a'} x'' \wedge y' \xrightarrow{b'} \surd \Rightarrow z' \xrightarrow{a'|b'} x''\right),$$

$\phi_9(u)$ is the formula

$$\forall x', y', z', u', y'' \bullet \bigwedge_{a', b' \in \mathsf{A}, a' | b' \neq \delta} \left(u \xrightarrow{i}\!\!\!\!\rightarrow u' \wedge u' \xrightarrow{l} x' \wedge u' \xrightarrow{r} y' \wedge u' \xrightarrow{m} z' \wedge x' \xrightarrow{a'} \surd \wedge y' \xrightarrow{b'} y'' \Rightarrow z' \xrightarrow{a'|b'} y''\right),$$

$\phi_{10}(u)$ is the formula

$$\forall x', y', z', u' \bullet \bigwedge_{a', b' \in \mathsf{A}, a' | b' \neq \delta} \left(u \xrightarrow{i}\!\!\!\!\rightarrow u' \wedge u' \xrightarrow{l} x' \wedge u' \xrightarrow{r} y' \wedge u' \xrightarrow{m} z' \wedge x' \xrightarrow{a'} \surd \wedge y' \xrightarrow{b'} \surd \Rightarrow z' \xrightarrow{a'|b'} \surd\right),$$

$\phi_{11}(u)$ is the formula

$$\begin{aligned}
&\forall x', y', z', u', z'' \bullet \\
&\bigwedge_{a' \in \mathsf{A}} \Big(u \xrightarrow{i}\!\!\!\!\rightarrow u' \wedge u' \xrightarrow{l} x' \wedge u' \xrightarrow{r} y' \wedge u' \xrightarrow{m} z' \wedge z' \xrightarrow{a'} z'' \Rightarrow \\
&\quad \exists x'', u'' \bullet \\
&\quad\quad (u' \xrightarrow{i} u'' \wedge u'' \xrightarrow{l} x'' \wedge u'' \xrightarrow{r} y' \wedge u'' \xrightarrow{m} z'' \wedge x' \xrightarrow{a'} x'') \vee \\
&\quad \exists y'', u'' \bullet \\
&\quad\quad (u' \xrightarrow{i} u'' \wedge u'' \xrightarrow{l} x' \wedge u'' \xrightarrow{r} y'' \wedge u'' \xrightarrow{m} z'' \wedge y' \xrightarrow{a'} y'') \vee \\
&\quad \exists x'', y'', u'' \bullet \\
&\quad\quad \Big(u' \xrightarrow{i} u'' \wedge u'' \xrightarrow{l} x'' \wedge u'' \xrightarrow{r} y'' \wedge u'' \xrightarrow{m} z'' \wedge \\
&\quad\quad\quad \bigvee_{b', c' \in \mathsf{A}, a' = b' | c'} (x' \xrightarrow{b'} x'' \wedge y' \xrightarrow{c'} y'')\Big) \vee \\
&\quad (x' \xrightarrow{a'} \surd \wedge z'' = y') \vee (y' \xrightarrow{a'} \surd \wedge z'' = x') \vee \\
&\quad\quad \bigvee_{b', c' \in \mathsf{A}, a' = b' | c'} (x' \xrightarrow{b'} z'' \wedge y' \xrightarrow{c'} \surd) \vee \\
&\quad\quad \bigvee_{b', c' \in \mathsf{A}, a' = b' | c'} (x' \xrightarrow{b'} \surd \wedge y' \xrightarrow{c'} z'')\Big) \wedge
\end{aligned}$$

$$\forall x', y', z', u' \bullet$$
$$\bigwedge_{a' \in \mathsf{A}} \Big(u \xrightarrow{i}\!\!\twoheadrightarrow u' \wedge u' \xrightarrow{l} x' \wedge u' \xrightarrow{r} y' \wedge u' \xrightarrow{m} z' \wedge z' \xrightarrow{a'} \surd \Rightarrow \bigvee_{b', c' \in \mathsf{A}, a' = b' \mid c'} (x' \xrightarrow{b'} \surd \wedge y' \xrightarrow{c'} \surd) \Big) .$$

Formula ϕ_1 expresses that each process reachable by i-steps from u relates a process reachable from x and a process reachable from y to a process reachable from z. Formula ϕ_2 expresses that u relates x and y to z. The conjunction of formulas ϕ_3–ϕ_{10} expresses that, if x' and y' are related to z', then z' is capable of behaving as $x' \parallel y'$: formula ϕ_3 expresses that, if x' and y' are related to z' and $x' \xrightarrow{a'} x''$, then there exists a z'' such that $z' \xrightarrow{a'} z''$ and x'' and y' are related to z''; formula ϕ_4 expresses that, if x' and y' are related to z' and $y' \xrightarrow{a'} y''$, then there exists a z'' such that $z' \xrightarrow{a'} z''$ and x' and y'' are related to z''; formula ϕ_5 expresses that, if x' and y' are related to z', $x' \xrightarrow{a'} x''$, $y' \xrightarrow{b'} y''$ and $a' \mid b' \neq \delta$, then there exists a z'' such that $z' \xrightarrow{a' \mid b'} z''$ and x'' and y'' are related to z''; etc. Formula ϕ_{11} expresses that, if x' and y' are related to z', then $x' \parallel y'$ is capable of behaving as z'. In other words, $\mathrm{P}'_{\parallel}(x, y, z, u)$ expresses that u encodes a bisimulation witnessing the bisimilarity of $x \parallel y$ and z.

The formula $x \parallel y = z \Leftrightarrow \exists u \bullet \mathrm{P}'_{\parallel}(x, y, z, u)$ is only admissible as an explicit definition of $\parallel$ if $\exists! z \bullet \exists u \bullet \mathrm{P}'_{\parallel}(x, y, z, u)$ is derivable. This admissibility condition for $\parallel$ can be split into an *existence condition* $\exists z \bullet \exists u \bullet \mathrm{P}'_{\parallel}(x, y, z, u)$ and a *uniqueness condition* $\exists u \bullet \mathrm{P}'_{\parallel}(x, y, z, u) \wedge \exists u \bullet \mathrm{P}'_{\parallel}(x, y, \overline{z}, u) \Rightarrow z = \overline{z}$. The uniqueness condition for $\parallel$ is derivable in $\mathrm{BPA}^{\mathrm{fo}}_{\delta\mathrm{rr}}$.

Proposition 11 (Uniqueness for parallel composition). *We have* $\mathrm{BPA}^{\mathrm{fo}}_{\delta\mathrm{rr}} \vdash \exists u \bullet \mathrm{P}'_{\parallel}(x, y, z, u) \wedge \exists u \bullet \mathrm{P}'_{\parallel}(x, y, \overline{z}, u) \Rightarrow z = \overline{z}$.

Proof. Assume $\mathrm{P}'_{\parallel}(x, y, z, u)$ and $\mathrm{P}'_{\parallel}(x, y, \overline{z}, \overline{u})$. Then we derive $z = \overline{z}$ by applying the bisimulation axiom schema BS, taking the following formula for $\phi(z, \overline{z})$:

$$\exists x', y', u', \overline{u}' \bullet (u \xrightarrow{i}\!\!\twoheadrightarrow u' \wedge u' \xrightarrow{m} z \wedge u' \xrightarrow{l} x' \wedge u' \xrightarrow{r} y' \wedge \overline{u} \xrightarrow{i}\!\!\twoheadrightarrow \overline{u}' \wedge \overline{u}' \xrightarrow{m} \overline{z} \wedge \overline{u}' \xrightarrow{l} x' \wedge \overline{u}' \xrightarrow{r} y') .$$

□

We will come back to the existence condition for $\parallel$ later on.

As mentioned in Sect. 13, left merge and communication merge are the same as parallel composition except that the actions that can be performed at the start are restricted. As a consequence, the explicit definitions of the left merge operator and the communication merge operator can be formulated as $x \,⌊\!⌊\, y = z \Leftrightarrow \exists u \bullet \mathrm{P}'_{⌊\!⌊}(x, y, z, u)$ and $x \mid y = z \Leftrightarrow \exists u \bullet \mathrm{P}'_{\mid}(x, y, z, u)$, where the formulas for which $\mathrm{P}'_{⌊\!⌊}(x, y, z, u)$ and $\mathrm{P}'_{\mid}(x, y, z, u)$ stand are simply obtained from the formula for which $\mathrm{P}'_{\parallel}(x, y, z, u)$ stands by replacing at appropriate places $u \xrightarrow{i}\!\!\twoheadrightarrow u'$ by $u \xrightarrow{i}\!\!\twoheadrightarrow u' \wedge \neg\, u = u'$. We refrain from giving the precise formulas for which $\mathrm{P}'_{⌊\!⌊}(x, y, z, u)$ and $\mathrm{P}'_{\mid}(x, y, z, u)$ stand. We mention that the uniqueness conditions for $⌊\!⌊$ and $\mid$ are derivable in $\mathrm{BPA}^{\mathrm{fo}}_{\delta\mathrm{rr}}$.

Next, we consider the explicit definition of the encapsulation operators. As in the case of parallel composition, we begin by introducing the abbreviation $\mathrm{P}'_{\partial_H}(x, y, u)$, which enables us to formulate the explicit definition of ∂_H as $\partial_H(x) = y \Leftrightarrow \exists u \bullet \mathrm{P}'_{\partial_H}(x, y, u)$. We fix different actions i, l and e. We use $\mathrm{P}'_{\partial_H}(x, y, u)$ as an abbreviation of the following formula of $\mathcal{L}(\mathrm{BPA}^{\mathrm{fo}}_{\delta\mathrm{rr}})$:

$$\phi_1(x, y, u) \wedge \phi_2(x, y, u) \wedge \phi_3(u) \wedge \phi_4(u) \wedge \phi_5(u) ;$$

where:
$\phi_1(x, y, u)$ is the formula

$$\forall u' \bullet \Big(u \xrightarrow{i}\!\!\!\!\rightarrow u' \Rightarrow \exists! x', y' \bullet (x \twoheadrightarrow x' \wedge y \twoheadrightarrow y' \wedge u' \xrightarrow{l} x' \wedge u' \xrightarrow{e} y') \wedge \bigwedge_{a' \in \mathsf{A}, a' \neq i,l,e} \neg\, \exists v' \bullet u' \xrightarrow{a'} v'\Big) ,$$

$\phi_2(x, y, u)$ is the formula

$$u \xrightarrow{l} x \wedge u \xrightarrow{e} y ,$$

$\phi_3(u)$ is the formula

$$\forall x', y', u', x'' \bullet \bigwedge_{a' \in \mathsf{A} \setminus H} (u \xrightarrow{i}\!\!\!\!\rightarrow u' \wedge u' \xrightarrow{l} x' \wedge u' \xrightarrow{e} y' \wedge x' \xrightarrow{a'} x'' \Rightarrow \exists u'', y'' \bullet (u' \xrightarrow{i} u'' \wedge u'' \xrightarrow{l} x'' \wedge u'' \xrightarrow{e} y'' \wedge y' \xrightarrow{a'} y'')) ,$$

$\phi_4(u)$ is the formula

$$\forall x', y', u' \bullet \bigwedge_{a' \in \mathsf{A} \setminus H} (u \xrightarrow{i}\!\!\!\!\rightarrow u' \wedge u' \xrightarrow{l} x' \wedge u' \xrightarrow{e} y' \wedge x' \xrightarrow{a'} \surd \Rightarrow y' \xrightarrow{a'} \surd) ,$$

$\phi_5(u)$ is the formula

$$\forall x', y', u', y'' \bullet \bigwedge_{a' \in \mathsf{A}} \Big(u \xrightarrow{i}\!\!\!\!\rightarrow u' \wedge u' \xrightarrow{l} x' \wedge u' \xrightarrow{e} y' \wedge y' \xrightarrow{a'} y'' \Rightarrow \exists x'', u'' \bullet \big(u' \xrightarrow{i} u'' \wedge u'' \xrightarrow{l} x'' \wedge u'' \xrightarrow{e} y'' \wedge \bigvee_{b' \in \mathsf{A} \setminus H, a' = b'} x' \xrightarrow{b'} x''\big)\Big) \wedge$$
$$\forall x', y', u' \bullet \bigwedge_{a' \in \mathsf{A}} \Big(u \xrightarrow{i}\!\!\!\!\rightarrow u' \wedge u' \xrightarrow{l} x' \wedge u' \xrightarrow{e} y' \wedge y' \xrightarrow{a'} \surd \Rightarrow \bigvee_{b' \in \mathsf{A} \setminus H, a' = b'} x' \xrightarrow{b'} \surd\Big) .$$

The uniqueness condition for ∂_H is derivable in $\mathrm{BPA}^{\mathrm{fo}}_{\delta\mathrm{rr}}$.

Proposition 12 (Uniqueness for encapsulation). *We have* $\mathrm{BPA}^{\mathrm{fo}}_{\delta\mathrm{rr}} \vdash \exists u \bullet \mathrm{P}'_{\partial_H}(x, y, u) \wedge \exists u \bullet \mathrm{P}'_{\partial_H}(x, \overline{y}, u) \Rightarrow y = \overline{y}$.

Table 11. Existence conditions

$\exists z \bullet \exists u \bullet \mathrm{P}'_{\parallel}(x,y,z,u)$	X1
$\exists z \bullet \exists u \bullet \mathrm{P}'_{\llfloor}(x,y,z,u)$	X2
$\exists z \bullet \exists u \bullet \mathrm{P}'_{\mid}(x,y,z,u)$	X3
$\exists y \bullet \exists u \bullet \mathrm{P}'_{\partial_H}(x,y,u)$	X4

Proof. The proof follows the same line as to the proof of Proposition 11. □

The formulas of $\mathcal{L}(\mathrm{BPA}^{\mathrm{fo}}_{\delta\mathrm{rr}})$ that are given in Table 11 are existence conditions for $\parallel$, $\llfloor$, $\mid$ and ∂_H. We write X for this set of formulas. X4 is actually an axiom schema with an instance for each $H \subseteq \mathsf{A}$. The existence conditions from Table 11 are valid in the full bisimulation models $\mathfrak{P}^{\mathrm{rr}}_{\kappa}$ ($\kappa \geq \aleph_0$). It is unknown to us whether they are derivable from $\mathrm{BPA}^{\mathrm{fo}}_{\delta\mathrm{rr}}$.

Theorem 25 (Interpretation of $\mathbf{ACP}^{\mathrm{fo}}$ in $\mathbf{BPA}^{\mathrm{fo}}_{\delta\mathrm{rr}}$). *The following is an interpretation of* $\mathrm{ACP}^{\mathrm{fo}}$ *in* $\mathrm{BPA}^{\mathrm{fo}}_{\delta\mathrm{rr}} \cup \mathrm{X}$*:*

$$\mathfrak{d}(x) \Leftrightarrow x = x\,,$$

$$x \parallel y = z \Leftrightarrow \exists u \bullet \mathrm{P}'_{\parallel}(x,y,z,u)\,,$$

$$x \llfloor y = z \Leftrightarrow \exists u \bullet \mathrm{P}'_{\llfloor}(x,y,z,u)\,,$$

$$x \mid y = z \Leftrightarrow \exists u \bullet \mathrm{P}'_{\mid}(x,y,z,u)\,,$$

$$\partial_H(x) = y \Leftrightarrow \exists u \bullet \mathrm{P}'_{\partial_H}(x,y,u) \quad \text{for each } H \subseteq \mathsf{A}\,.$$

Proof. Because $\mathfrak{d}(x) \Leftrightarrow x = x$, the first two conditions made in the definition of interpretation are trivially fulfilled. Because $\mathfrak{d}(x) \Leftrightarrow x = x$, the third condition becomes

$$\mathrm{BPA}^{\mathrm{fo}}_{\delta\mathrm{rr}} \cup \mathrm{X} \cup \mathrm{E} \vdash \phi \quad \text{for each axiom } \phi \text{ of } \mathrm{ACP}^{\mathrm{fo}}\,,$$

where E is the set of explicit definitions given above. For each axiom ϕ of $\mathrm{BPA}^{\mathrm{fo}}_{\delta}$, we immediately have $\mathrm{BPA}^{\mathrm{fo}}_{\delta\mathrm{rr}} \cup \mathrm{X} \cup \mathrm{E} \vdash \phi$. Hence, it is sufficient to establish $\mathrm{BPA}^{\mathrm{fo}}_{\delta\mathrm{rr}} \cup \mathrm{X} \cup \mathrm{E} \vdash \phi$ only for each axiom ϕ of $\mathrm{ACP}^{\mathrm{fo}}$ that is not an axiom of $\mathrm{BPA}^{\mathrm{fo}}_{\delta}$.

All axioms in question are atomic formulas of $\mathcal{L}(\mathrm{BPA}^{\mathrm{fo}}_{\delta\mathrm{rr}} \cup \mathrm{E})$. Each atomic formula ϕ of $\mathcal{L}(\mathrm{BPA}^{\mathrm{fo}}_{\delta\mathrm{rr}} \cup \mathrm{E})$ is equivalent in $\mathrm{BPA}^{\mathrm{fo}}_{\delta\mathrm{rr}} \cup \mathrm{X} \cup \mathrm{E}$ to an existential formula ϕ' of $\mathcal{L}(\mathrm{BPA}^{\mathrm{fo}}_{\delta\mathrm{rr}} \cup \mathrm{E})$ in which no other terms occur than terms of $\mathcal{L}(\mathrm{BPA}^{\mathrm{fo}}_{\delta\mathrm{rr}})$ and terms $t_1 \parallel t_2$, $t_1 \llfloor t_2$, $t_1 \mid t_2$ and $\partial_H(t_1)$ of which the subterms t_1 and t_2 are terms of $\mathcal{L}(\mathrm{BPA}^{\mathrm{fo}}_{\delta\mathrm{rr}})$ (see e.g. [10]). Because E contains the explicit definitions for $\parallel$, $\llfloor$, $\mid$ and ∂_H, this existential formula ϕ' is equivalent in $\mathrm{BPA}^{\mathrm{fo}}_{\delta\mathrm{rr}} \cup \mathrm{X} \cup \mathrm{E}$ to a formula ϕ'' of $\mathcal{L}(\mathrm{BPA}^{\mathrm{fo}}_{\delta\mathrm{rr}})$. Because definitional extensions are conservative extensions (see e.g. [8]), $\mathrm{BPA}^{\mathrm{fo}}_{\delta\mathrm{rr}} \cup \mathrm{X} \cup \mathrm{E} \vdash \phi''$ iff $\mathrm{BPA}^{\mathrm{fo}}_{\delta\mathrm{rr}} \cup \mathrm{X} \vdash \phi''$. This suggests the following three-steps approach to establish that $\mathrm{BPA}^{\mathrm{fo}}_{\delta\mathrm{rr}} \cup \mathrm{X} \cup \mathrm{E} \vdash \phi$:

1. eliminate from ϕ all nested terms other than terms of $\mathcal{L}(\mathrm{BPA}^{\mathrm{fo}}_{\delta\mathrm{rr}})$, resulting in ϕ';

2. eliminate from ϕ' all atomic formulas in which $\|$, $\lfloor\!\lfloor$, $|$ or ∂_H occur, resulting in ϕ'';
3. derive ϕ'' from $\mathrm{BPA}^{\mathrm{fo}}_{\delta\mathrm{rr}} \cup \mathrm{X}$.

For each axiom of $\mathrm{ACP}^{\mathrm{fo}}$ that is not an axiom of $\mathrm{BPA}^{\mathrm{fo}}_{\delta}$, the first two steps are short and simple. The last step is generally straightforward, but tedious. We outline the proof for axioms CM3 and CM4.

The first two steps result for CM3 in the formula

$$\exists z \bullet (\exists u \bullet \mathrm{P}'_{\lfloor\!\lfloor}(a \cdot x, y, a \cdot z, u) \wedge \exists u' \bullet \mathrm{P}'_{\|}(x, y, z, u'))$$

and for CM4 in the formula

$$\exists v, w \bullet (\exists u \bullet \mathrm{P}'_{\lfloor\!\lfloor}(x + y, z, v + w, u) \wedge \exists u' \bullet \mathrm{P}'_{\lfloor\!\lfloor}(x, z, v, u') \wedge \exists u'' \bullet \mathrm{P}'_{\lfloor\!\lfloor}(y, z, w, u'')) .$$

The last step for CM3 goes as follows. First of all, it follows from X that $\exists z \bullet \exists u' \bullet \mathrm{P}'_{\|}(x, y, z, u')$. Therefore, it is sufficient to show that $\exists u' \bullet \mathrm{P}'_{\|}(x, y, z, u') \Rightarrow \exists u \bullet \mathrm{P}'_{\lfloor\!\lfloor}(a \cdot x, y, a \cdot z, u)$. This is done as follows. Assume $\mathrm{P}'_{\|}(x, y, z, u')$. Take $l \cdot (a \cdot x) + r \cdot y + m \cdot (a \cdot z) + i \cdot u'$ for u. Then $\mathrm{P}'_{\lfloor\!\lfloor}(a \cdot x, y, a \cdot z, u)$ is easily derived.

The last step for CM4 follows essentially the same line as the last step for CM3. However, there are two complications in the construction of u from u' and u''. The first complication is that four cases have to be distinguished according to the reachability of x from x in one or more steps and the reachability of y from y in one or more steps. The second complication is that u has to be constructed from subprocesses of u' and u'' instead of u' and u'' themselves. Thus, although the construction of u is rather straightforward, it becomes very tedious to express it in $\mathcal{L}(\mathrm{BPA}^{\mathrm{fo}}_{\delta\mathrm{rr}})$ and to derive $\mathrm{P}'_{\lfloor\!\lfloor}(x+y, z, v+w, u)$. We refrain from outlining the last step for CM4 further.

The proofs for CM2, CM5–CM7 and D4 are similar to the proof for CM3. The proofs for CM1, CM8–CM9 and D3 are similar to the proof for CM4. The proofs for C1–C3, D1 and D2 are easy. □

17 Concluding Remarks

In this paper, we build on earlier work on ACP. The algebraic theory ACP was first presented in [1] and RDP, RSP and AIP were first formulated in [14]. Moreover, the full bisimulation models are basically the graph models of ACP, which are most extensively described in [11]. In this paper, we extend ACP to a first-order theory and look into that theory from the point of view of classical model theory. Some open problems that arise from this work are:

- Is the reachability predicate $\twoheadrightarrow$ of $\mathrm{BPA}^{\mathrm{fo}}_{\delta}$ first-order definable in $\mathfrak{P}_{\aleph_0}$ if the cardinality of A is given?
- What are the relations between RDP, RSP (Table 2), B, R (Table 3) and AIP (Table 4) in the presence of $\mathrm{BPA}^{\mathrm{fo}}_{\delta}$? In particular, do all models of $\mathrm{BPA}^{\mathrm{fo}}_{\delta}$ extended with R satisfy B?

- Is it derivable from $\mathrm{BPA}_{\delta}^{\mathrm{fo}}$ or a finitary first-order extension thereof, for all pairs of guarded recursive specifications of which the solutions in $\mathfrak{P}_{\aleph_0}$ are not identical, that their solutions are not equal?
- Is axiom RR3 (Table 8) derivable from the other axioms of $\mathrm{BPA}_{\delta\mathrm{rr}}^{\mathrm{fo}}$?
- Are the restricted reachability predicates $\xrightarrow{a}\!\!\!\!\rightarrow$ of $\mathrm{BPA}_{\delta\mathrm{rr}}^{\mathrm{fo}}$ first-order definable in $\mathfrak{P}_{\aleph_0}$ if the cardinality of A is given (they are if the cardinality of A is 1)?
- Are the existence conditions for $\|$, $\lfloor\!\lfloor$, $|$ and ∂_H (Table 11) derivable from $\mathrm{BPA}_{\delta\mathrm{rr}}^{\mathrm{fo}}$?

To the best of our knowledge there is no related work. Many options for future work remain. We mention:

- Development of extensions of $\mathrm{ACP}^{\mathrm{fo}}$ with additional operators, such as the iteration operators from [23, 24, 25].
- Development of first-order extensions of variants of ACP with timing, such as the ones from [26, 27, 28].
- Re-development of the α/β-calculus [29] in the setting of $\mathrm{ACP}^{\mathrm{fo}}$.
- Further analysis of the relation between external and internal versions of predicates on processes.
- Further investigations into interpretation of existing process algebras in $\mathrm{ACP}^{\mathrm{fo}}$.
- Investigations into interpretation of other related algebraic theories, such as the network algebra from [30], in $\mathrm{ACP}^{\mathrm{fo}}$.
- Exploration of the strong and weak points of $\mathrm{ACP}^{\mathrm{fo}}$ for process specification and verification.

Acknowledgements

We thank Luca Aceto from Reykjavík University and Bas Luttik from Eindhoven University of Technology for carefully reading a preliminary version of this paper, for pointing out flaws in some proof outlines and for suggesting improvements of the presentation of the paper. We also thank an anonymous referee for his/her valuable comments.

References

1. Bergstra, J.A., Klop, J.W.: Process algebra for synchronous communication. Information and Control **60** (1984) 109–137
2. Baeten, J.C.M., Weijland, W.P.: Process Algebra. Volume 18 of Cambridge Tracts in Theoretical Computer Science. Cambridge University Press, Cambridge (1990)
3. Milner, R.: A Calculus of Communicating Systems. Volume 92 of Lecture Notes in Computer Science. Springer-Verlag, Berlin (1980)
4. Milner, R.: Communication and Concurrency. Prentice-Hall, Englewood Cliffs (1989)
5. Bergstra, J.A., Loots, M.E.: Program algebra for sequential code. Journal of Logic and Algebraic Programming **51** (2002) 125–156

6. Bergstra, J.A., Bethke, I.: Polarized process algebra and program equivalence. In Baeten, J.C.M., Lenstra, J.K., Parrow, J., Woeginger, G.J., eds.: Proceedings 30th ICALP. Volume 2719 of Lecture Notes in Computer Science., Springer-Verlag (2003) 1–21
7. Baeten, J.C.M., Bergstra, J.A.: On sequential composition, action prefixes and process prefix. Formal Aspects of Computing **6** (1994) 250–268
8. Shoenfield, J.R.: Mathematical Logic. Addison-Wesley Series in Logic. Addison-Wesley, Reading, MA (1967)
9. Chang, C.C., Keisler, H.J.: Model Theory. Third edn. Volume 73 of Studies in Logic and the Foundations of Mathematics. Elsevier, Amsterdam (1990)
10. Hodges, W.A.: Model Theory. Volume 42 of Encyclopedia of Mathematics and Its Applications. Cambridge University Press, Cambridge (1993)
11. Baeten, J.C.M., Bergstra, J.A., Klop, J.W.: On the consistency of Koomen's fair abstraction rule. Theoretical Computer Science **51** (1987) 129–176
12. Hennessy, M., Milner, R.: Algebraic laws for non-determinism and concurrency. Journal of the ACM **32** (1985) 137–161
13. Bergstra, J.A., Klop, J.W.: Algebra of communicating processes. In de Bakker, J.W., Hazewinkel, M., Lenstra, J.K., eds.: Proceedings Mathematics and Computer Science I. Volume 1 of CWI Monograph., North-Holland (1986) 89–138
14. Bergstra, J.A., Klop, J.W.: Process algebra: Specification and verification in bisimulation semantics. In Hazewinkel, M., Lenstra, J.K., Meertens, L.G.L.T., eds.: Proceedings Mathematics and Computer Science II. Volume 4 of CWI Monograph., North-Holland (1986) 61–94
15. Bergstra, J.A., Klop, J.W.: Process theory based on bisimulation semantics. In de Bakker, J.W., de Roever, W.P., Rozenberg, G., eds.: Linear Time, Branching Time and Partial Order in Logics and Models for Concurrency. Volume 354 of Lecture Notes in Computer Science., Springer-Verlag (1989) 50–122
16. Bergstra, J.A., Klop, J.W.: A convergence theorem in process algebra. In de Bakker, J.W., Rutten, J.J.M.M., eds.: Ten Years of Concurrency Semantics, World Scientific (1992) 164–195
17. van Benthem, J.F.A.K., Bergstra, J.A.: Logic of transition systems. Journal of Logic, Language and Information **3** (1995) 247–283
18. Fokkink, W.J.: Introduction to Process Algebra. Texts in Theoretical Computer Science, An EATCS Series. Springer-Verlag, Berlin (2000)
19. van Benthem, J.F.A.K., van Eijck, D.J.N., Stebletsova, V.: Modal logic, transition systems and processes. Journal of Logic and Computation **4** (1994) 811–855
20. Bergstra, J.A., Klop, J.W.: The algebra of recursively defined processes and the algebra of regular processes. In Paredaens, J., ed.: Proceedings 11th ICALP. Volume 172 of Lecture Notes in Computer Science., Springer-Verlag (1984) 82–95
21. Hopcroft, J.E., Ullman, J.D.: Introduction to Automata Theory, Languages and Computation. Addison-Wesley, Reading, MA (1979)
22. Fokkink, W.J.: A complete equational axiomatization for prefix iteration. Information Processing Letters **52** (1994) 333–337
23. Bergstra, J.A., Bethke, I., Ponse, A.: Process algebra with iteration and nesting. Computer Journal **37** (1994) 243–258
24. Bergstra, J.A., Fokkink, W.J., Ponse, A.: Process algebra with recursive operations. In Bergstra, J.A., Ponse, A., Smolka, S.A., eds.: Handbook of Process Algebra. Elsevier, Amsterdam (2001) 333–389
25. Bergstra, J.A., Ponse, A.: Non-regular iterators in process algebra. Theoretical Computer Science **269** (2001) 203–229

26. Baeten, J.C.M., Bergstra, J.A.: Real time process algebra. Formal Aspects of Computing **3** (1991) 142–188
27. Baeten, J.C.M., Bergstra, J.A.: Discrete time process algebra. Formal Aspects of Computing **8** (1996) 188–208
28. Baeten, J.C.M., Middelburg, C.A.: Process Algebra with Timing. Monographs in Theoretical Computer Science, An EATCS Series. Springer-Verlag, Berlin (2002)
29. Baeten, J.C.M., Bergstra, J.A., Klop, J.W.: Conditional axioms and α/β-calculus in process algebra. In Wirsing, M., ed.: Formal Description of Programming Concepts III, North-Holland (1987) 53–75
30. Bergstra, J.A., Middelburg, C.A., Ştefănescu, G.: Network algebra for asynchronous dataflow. International Journal of Computer Mathematics **65** (1997) 57–88

Expression Reduction Systems and Extensions: An Overview

John Glauert[1], Delia Kesner[2], and Zurab Khasidashvili[3]

[1] School of Computing Sciences, University of East Anglia, Norwich, UK
J.Glauert@uea.ac.uk
[2] PPS, CNRS and Université Paris 7, France
kesner@pps.jussieu.fr
[3] Logic and Validation Technology,
Design Technology Division Intel Development Center, Haifa, Israel
zurabk@iil.intel.com

Abstract. Expression Reduction Systems is a formalism for higher-order rewriting, extending Term Rewriting Systems and the lambda-calculus. Here we give an overview of results in the literature concerning ERSs. We review confluence, normalization and perpetuality results for orthogonal ERSs. Some of these results are extended to orthogonal conditional ERSs. Further, ERSs with patterns are introduced and their confluence is discussed. Finally, higher-order rewriting is translated into equational first-order rewriting. The technique develops an isomorphic model of ERSs with variable names, based on de Bruijn indices.

1 Introduction

Many programming languages and logical systems are modelled by transformations that allow programs or expressions to be rewritten as values that are just simpler expressions.

The more traditional rewriting frameworks are *first-order rewriting systems*, where expressions are represented using term algebras, and the λ-*calculus*, where expressions are modelled by λ-terms. Both languages can be combined naturally into a *higher-order rewriting system*, which is a well-suited formalism to deal simultaneously with algebraic data structures and functions. We thus obtain a framework inheriting the advantages from both the first-order and the functional worlds.

Following the ground-breaking work by J. W. Klop [67] on *Combinatory Reduction Systems* (CRS), many different frameworks for higher-order rewriting have been proposed: *Expression Reduction Systems* (ERS) by Z. Khasidashvili [52], *Higher-Order Rewrite Systems* (HRS) by T. Nipkow [81], *Higher-Order Rewriting Systems* (HORS) by V. van Oostrom and F. van Raamsdonk [98], *Higher-Order Term Rewriting Systems* (HOTRS) by D. Wolfram [101], *The General Scheme family* (GS) or *Algebraic-Functional Systems* (AFS) by Jouannaud and Okada [46].

A. Middeldorp et al. (Eds.): Processes... (Klop Festschrift), LNCS 3838, pp. 496–553, 2005.

In this chapter we review some results concerning *Expression Reduction Systems*. We start this introduction by relating ERS to Pkhakadze's work as well as to other well-known formalisms in the literature.

1.1 A Short History of ERSs

Expression Reduction Systems (ERSs) were introduced by Khasidashvili [52], under the supervision of Sh. Pkhakadze. The syntax of ERSs was influenced by the syntax of the Notation Theory of Pkhakadze [86, 87].

Pkhakadze's work was motivated by a study of formal mathematical theories and, in particular, extensions of formal theories with new function or predicate symbols. When new symbols are *defined* using the original or previously defined symbols, one expects that the new, extended theory is a conservative extension of the original one – no new results can be proven on expressions of the original theory.

Besides function and predicate symbols, one often needs to introduce new quantifiers. Bourbaki [18] devoted a large part of their study of set theory to the introduction of new symbols in formal theories. However, a general syntactic framework for defining new symbols was missing there, which would enable, for all introduced symbols, uniform proofs of syntactic results, such as termination or confluence in current rewriting terminology. The aim of Pkhakadze's work [86] was indeed to define a uniform syntax for defining new symbols, and the hard work went into understanding the binding structure of the rules defining new quantifiers. A famous example of a definition of a quantifier is Hilbert's definition of the existential quantifier using the choice operator τ: $\exists x(A) \to (\tau x(A)/x)A$.

Pkhakadze introduced several syntactic categories for defining new symbols. The most general of them are of the form $\sigma a_1 \ldots a_n(A_1 \ldots A_m) \to B$, where a_i are object-metavariables expressing, after instantiation, the binding variables of the quantifier symbol σ. The A_j are term-metavariables that are instantiated to terms or formulae, and B is a meta-expression written using metavariables occurring in the left-hand side and using meta-substitutions of the form $(A_l/a_k)A_r$. Clearly, for the rewrite relation to be well defined, several syntactic constraints need to be imposed on the right-hand sides of the rules, which was done in [86]. These conditions are described in detail in a more recent summary of Pkhakadze's work (written in English) [87].

It would be fair to say that the alphabet in Pkhakadze's system was typed – there were symbols that could take terms or formulae as arguments, and return terms or formulae, depending on their types. The early versions of ERSs [52] used a similar syntax – symbols of types NAT and $BOOL$ were used, but we cannot see them in later versions of ERSs. Similarly, object-metavariables are no longer used in rewrite rules of ERSs.

The main results in [86] concern termination of rewriting (or elimination of defined symbols) and uniqueness of the normal forms. Most of the results there correspond to first-order rewriting. For more information, we refer to [87].

While ERSs were introduced independently from Klop's work [67], later work of the third author on ERSs, and especially on perpetual strategies, was greatly influenced by [67] and by J.W. Klop himself. Indeed, many results on perpetual reductions in ERSs presented in Section 5 generalise or refine Klop's results.

1.2 ERSs with Respect to Other Higher-Order Formalisms

There are many different ingredients in higher-order formalisms and each of them admits an interesting variety of possibilities.

The first property making substantial differences between all the formalisms for higher-order rewriting is the use of types. While CRS and ERS are untyped languages, HRS, HORS and AFS allow specification of type information and work only with well-typed terms.

Metavariables are also defined differently in all these formalisms. A metavariable Z in CRS has a fixed arity m ($m \geq 0$) which has to be respected in order to construct metaterms, while metavariables in ERS are symbols of an alphabet (without arity) and do not appear *applied* to other symbols. For example, $Z([x]M)$ is a CRS metaterm which can be expressed by the ERS metaterm $(\lambda x.M/y)Z$. Also, metaterms in HRS appearing in rewrite rules are required to be in η-long-normal-form which can be cumbersome. Indeed, a metavariable M representing a unary function has to be written as $\lambda x.Mx$.

Binding operators are reduced to the singleton λ in most of the formalisms, while ERS allows different names for different binders. Thus for example, we can write $f(\mu y.\lambda x.M, N)$, where μ and λ are different binder symbols having different behaviour. This enables ERS to be more natural but it is, of course, just a syntactical issue that can be expressed within a syntax having only one binder symbol.

Another important issue is the definition of metavariable substitution. For example, a CRSs substitution replaces a free metavariable of arity n by an abstraction of n bound variables. The result of such a substitution, as for example when applying $\sigma = \{Z \mapsto \underline{\lambda} x.app(x, a)\}$ to $abs([y]Z(y))$ is obtained by a complete development of the term $abs([y](\underline{\lambda} x.app(x, a))(y))$ in the $\underline{\lambda}$-calculus, thus giving $abs([y]app(y, a))$. The concept of complete development can be replaced by a η-long β-normalisation in the typed framework of HRSs, which is more powerful. However, a simpler definition of substitution is adopted in ERS where substitution of metavariables is just defined as replacement and the semantics of the syntactic substitution operator is given by means of the traditional higher-order substitution on terms and a suitable notion of safeness.

The notion of an instance of a rule is also different in these formalisms. While the CRS mechanism used to instantiate metavariables avoids capture of variables, the ERS replacement mechanism is forced to add a notion of safeness in order to guarantee that no bound variable becomes free during reduction.

The decision of whether or not to include the β-rule at the object level also makes these formalisms different. While CRS and ERS do not assume any object-level operation, GS uses β-reduction in the object-level as part of the reduction relation associated to it and HRS and HORS suppose a given substitution calculus in the meta-level which includes β reduction.

1.3 ERS with Pattern Matching

All the higher-order formalisms mentioned above use a binding mechanism acting only on *variables*. However, most functional languages currently in used, such as [20, 84, 42, 78, 96] and most popular proof assistants, such as, for example,[2, 24, 32, 44, 73, 89], allow definition by cases using pattern-matching mechanisms. Thus, a natural extension of higher-order rewriting consists in the use of binders for patterns so that a projection function like $\lambda f(x,y).x$ would be acceptable.

The *Pattern-Matching Calculus* [50], proposed as a theoretical framework to study pattern-matching within a *pure functional paradigm*, allows precisely this kind of binding mechanism. This calculus was later extended with *explicit operators* [21, 22, 33]; weak reduction was widely studied in [33]. Other languages allowing abstractions on patterns appear in [23, 45].

The formalism we discuss here, *Expression Reduction Systems with Patterns* (ERSP), was introduced by J. Forest and D. Kesner. It provides binding mechanisms for complex patterns. The calculus constitutes a generalisation of the Pattern-Matching Calculus to the case of higher-order rewriting (and not only functional rewriting).

1.4 Outline

We attempt to present all known important results about ERSs and their extensions. We assume the reader is familiar with basic concepts of rewriting – the book [94] is an excellent introduction. We however recall here the most basic concepts that are important in this work:

Definition 1. *Let $\rightarrow_S$ be a* **reduction relation** *defined on a set $\mathcal{T}$ and let $\rightarrow_S^*$ be its reflexive-transitive closure. The relation $\rightarrow_S$ is said to have the* **local confluence** *(resp.* **confluence***) property iff for every $t, s, o \in \mathcal{T}$ such that $t \rightarrow_S s$ and $t \rightarrow_S o$ (resp. $t \rightarrow_S^* s$ and $t \rightarrow_S^* o$), there is $e \in \mathcal{T}$ such that $s \rightarrow_S^* e$ and $o \rightarrow_S^* e$. The relation $\rightarrow_S$ is said to have the* **diamond** *property if for every $t, s, o \in \mathcal{T}$ such that $t \rightarrow_S s$ and $t \rightarrow_S o$ there is $e \in \mathcal{T}$ such that $s \rightarrow_S e$ and $o \rightarrow_S e$.*

The relation $\rightarrow_S$ is said to have the **weak normalisation** *property iff for every $t \in \mathcal{T}$ there is at least one finite $\rightarrow_S$-reduction chain $t \rightarrow_S \ldots \rightarrow_S s$ such that s cannot be further reduced; s is then called a* **normal form** *of t. The relation $\rightarrow_S$ is said to have the* **strong normalisation** *property iff for every $t \in \mathcal{T}$ there is no infinite $\rightarrow_S$-reduction chain starting at t.*

We introduce the basic concepts of ERSs and their extensions, give numerous examples, and discuss the results, often sketching the intuition behind the proofs. Proofs are omitted for brevity but references are provided to appropriate sources. We do not discuss related work in detail but instead refer to the recent survey of higher-order rewriting by F. van Raamsdonk, in [94].

We start by introducing the syntax of context-sensitive conditional ERSs, and discuss their expressive power. We define a concept of orthogonality for them, and study confluence, normalisation and perpetuality results. We then introduce

ERSs with patterns and discuss their confluence. Finally, a de Bruijn setting is proposed for rewriting in (simple) ERSs, which has a translation into equational first-order rewriting. We conclude by suggesting a few research directions for future work on ERSs and their extensions.

2 Context-Sensitive Conditional ERSs

In this section, we will introduce context-sensitive conditional ERSs (CCERSs); conditional ERSs and ERSs will be defined as special cases of CCERSs. Conversely, conditional ERSs are obtained from ERSs by restricting arguments of redexes in rewrite rules, and CCERSs are obtained from conditional ERSs by restricting the context in which reduction steps are allowed. We will demonstrate with concrete examples how typed λ-calculi and process calculi, which are not ERSs, can be encoded as CCERSs. We will also discuss an encoding of ERSs with reduction strategies as CCERSs. The material is taken from [63].

2.1 The Syntax of CCERSs

Terms in CCERSs are built from the alphabet as they are in the first-order case. Like the λ in λ-calculus and the $\int$ in integrals, symbols may have binding power and require some binding variables and terms as arguments, as specified by their *arity*. *Scope indicators* are used to specify which variables have binding power in which arguments. For example, a β-redex in the λ-calculus appears as $Ap(\lambda x\ t, s)$, where Ap is a function symbol of arity 2 and λ is an operator sign of arity $(1,1)$ and scope indicator (1). Integrals such as $\int_s^t f(x)\,dx$ can be represented as $\int x(s,t,f(x))$ by using an operator sign $\int$ of arity $(1,3)$ and scope indicator (3).

Metaterms will be used to write rewrite rules. They are constructed from *metavariables* and meta-expressions for substitutions, called *metasubstitutions*. Instantiation of metavariables in metaterms yields terms. Metavariables play the rôle of variables in the TRS rules and of function variables in other formats of higher-order rewriting such as Higher-Order TRSs (HOTRSs) [101], Higher-Order Rewrite Systems (HRS) [82], and Higher-Order Rewriting Systems (HORSs) [98]. Unlike the function variables in HOTRSs, HRSs, and HORSs, however, metavariables *cannot* be bound. In the current formalism, we do not use the object metavariables used by [86, 87, 52] to express bindings in rewrite rules; instead we express bindings using variables, which makes the formalism slightly simpler.

Definition 2. *Let Σ be an* ***alphabet*** *comprising infinitely many* ***variables****, denoted by x, y, z,..., and* ***symbols (signs)****. A symbol σ can be either a* ***function symbol (simple operator)*** *having an* ***arity*** *$n \in \mathcal{N}$ or an* ***operator sign (quantifier sign)*** *of* ***arity*** *$(m,n) \in \mathcal{N}^+ \times \mathcal{N}^+$. An operator sign needs to be supplied with m* ***binding*** *variables $x_1,\ldots,x_m$ to form a* ***quantifier (compound***

operator) *$\sigma x_1 \ldots x_m$, and it also has a* ***scope indicator*** *specifying in which of the n arguments it has binding power.* [1]

Terms *t, s, e, o are constructed from variables, function symbols, and quantifiers in the usual first-order way, respecting (the second component of the) arities. A predicate AT on terms specifies which terms are* ***admissible****.*

Metaterms *are constructed like terms, but also allowing* ***metavariables*** *A, $B, \ldots$ and* ***metasubstitutions*** *$(t_1/x_1, \ldots, t_n/x_n)t_0$, where each t_i is an arbitrary metaterm and the x_i have a binding effect in t_0. Metaterms without metasubstitutions are called* ***simple****. An* ***assignment*** *θ maps each metavariable to a term. The application of θ to a metaterm t is written $t\theta$ and is obtained from t by replacing metavariables with their values under θ and by replacing metasubstitutions $(t_1/x_1, \ldots, t_n/x_n)t_0$, in right to left order (for example), with the result of substitution of terms $t_1, \ldots, t_n$ for free occurrences of $x_1, \ldots, x_n$ in t_0. The substitution operation may involve a* ***renaming*** *of bound variables to avoid collision, and we assume that the set of variables in Σ comes equipped with an equivalence relation, called renaming, such that any equivalence class of variables is infinite. We also assume that any variable can be renamed by any other variable in the corresponding equivalence class. Unless otherwise specified, the default renaming relation is the total binary relation on variables (that is, there is only one equivalence class). A partial renaming relation may be useful for conditional systems.*

The specification of a CCERS consists of an alphabet (generating a set of terms possibly restricted by the predicate AT as specified above), and a set of rules (generating the rewrite relation possibly restricted by *admissibility* predicates AA and AC as specified below). The predicate AT can be used to express sorting and typing constraints, since sets of admissible terms allowed for arguments of an operator can be seen as terms of certain sorts or types. The predicates AA and AC impose restrictions respectively on arguments of (admissible) redexes and on the contexts in which they can be contracted.

The CCERS syntax is very close to the syntax of the λ-calculus. For example, the β-rule is written as $Ap(\lambda x A, B) \to (B/x)A$, where A and B can be instantiated by any term. The η-rule is written as $\lambda x Ap(A, x) \to A$, where for any assignment $\theta \in AA(\eta)$, $x \notin FV(A\theta)$ (the set of free, i.e., unbound, variables of $A\theta$); otherwise an x occurring free in $A\theta$ and therefore bound in $\lambda x Ap(A\theta, x)$ would become free. A rule like $f(A) \to \exists x(A)$ is also allowed, but in that case the assignment θ with $x \in A\theta$ is not allowed in CCERSs. Such a collision between free and bound variables cannot arise when assignments are restricted by the condition $[vcf]$, described below. For that reason, the η-rule is not conditional, but is an (unconditional) ERS rule, as defined below.

Familiar rules for defining the existential quantifier $\exists x$ and the quantifier $\exists! x$ (there exists exactly one x) are written as $\exists x(A) \to (\tau x(A)/x)A$ and $\exists! x(A) \to \exists x(A) \wedge \forall x \forall y(A \wedge (y/x)A \Rightarrow x = y)$, respectively. For the assignment associating

[1] Scope indicators can be avoided at the expense of side conditions of the form $x \notin FV(s)$. In which case, in order to avoid unintended bindings, such conditions must be imposed on construction of admissible terms.

$x = 5$ to the metavariable A, these rules generate rewrite steps $\exists x(x = 5) \to \tau x(x = 5) = 5$ and $\exists! x(x = 5) \to \exists x(x = 5) \wedge \forall x \forall y(x = 5 \wedge y = 5) \Rightarrow x = y)$. In general, evaluation of a reduction step may involve execution of a number of substitutions corresponding to the metasubstitutions in the right-hand-side of the rule. This will be explained below in examples.

Definition 3. *A* ***Context-sensitive Conditional Expression Reduction System*** *(CCERS) is a pair (Σ, R), where Σ is an* **alphabet** *described in Definition 2 and R is a set of* **rewrite rules** *$r : t \to s$, where t and s are closed metaterms (i.e., metaterms possibly containing 'free' metavariables but not containing free variables).*

Furthermore, each rule r has a set of **admissible assignments** *$AA(r)$ which, to prevent confusion of variable bindings, must satisfy the following condition for* **variable-capture-freeness:**

[vcf] for any assignment $\theta \in AA(r)$, any metavariable A occurring in t or s, and any variable $x \in FV(A\theta)$, either every occurrence of A in r is in the scope of some binding occurrence of x in r or no occurrence is.

*For any $\theta \in AA(r)$, $t\theta$ is an r-***redex** *or an R-***redex** *(and so is any* **variant** *of $t\theta$ obtained by renaming of bound variables), and $s\theta$ is the* **contractum** *of $t\theta$. We call R* **simple** *if the right-hand sides of R-rules are simple metaterms. We call redexes that are instances of the same rule* **weakly similar***.*

Furthermore, each pair (r, θ) with $r \in R$ and $\theta \in AA(r)$ has a set $AC(r, \theta)$ of **admissible contexts** *such that if a context $C[\,]$ is admissible for (r, θ) and o is the contractum of $u = r\theta$ according to r, then $C[u] \to C[o]$ is an R-reduction step. In this case, u is* **admissible** *for r in the term $C[u]$. We require that the set of admissible terms be closed under reduction. We also require that admissibility of terms, assignments, and contexts be closed under the renaming of bound variables. That is, if u is admissible in $C[u]$ and if $C'[u']$ is its variant obtained by a renaming of bound variables in $C[u]$, then u' must be a redex admissible for $C'[\,]$.*

We call a CCERS **context-free***, or simply a* **Conditional Expression Reduction System** *(CERS), if every term is admissible, if every context is admissible for any redex, if the rules $r : t \to s$ are such that t is a simple metaterm and is not a metavariable, and if each metavariable that occurs in s also occurs in t. Moreover if for any rule $r \in R$, $AA(r)$ is the maximal set of variable-capture-free assignments, then we call the CERS an unconditional Expression Reduction System, or simply an* **Expression Reduction System** *(ERS).*

Note that in CCERSs (but not in CERSs or ERSs) we allow metavariable-rules like $\eta^{-1} : A \to \lambda x Ap(A, x)$ and metavariable-introduction-rules like $f(A) \to g(A, B)$, which are usually excluded a priori. This is useful only when the system is conditional. As in the η-rule, the requirement [vcf] forces $x \notin FV(A\theta)$ for every $\theta \in AA(\eta^{-1})$. A metavariable-introduction-rule is used in [65] to model a Hilbert style proof system as a CCERS. Such a rule allows adding any axiom (and not any formula) to a sequence of already proven theorems. A rule with a metasubstitution in the left-hand side (allowed in CCERSs) is used there to model the $\exists$-introduction rule. In a simplified form, such a rule is written as $(B/x)A \to \exists xA$. We consider such a rule as a simple rule since no substitution needs to be performed to apply the rule once the left-hand side has been pattern-matched.

Let $r : t \to s$ be a rule in a CCERS R and let θ be admissible for r. Subterms of a redex $v = t\theta$ that correspond to the metavariables in t are the *arguments* of v, and the rest of v is the *pattern* of v (hence the binding variables of the quantifiers occurring in the pattern also belong to the pattern while the corresponding bound variables belong to the arguments). Subterms of v whose head symbols are in its pattern are called the *pattern-subterms* of v. The pattern of the right-hand side of a simple CCERS rule is defined similarly.

Notation. We use a, b, c, d for constants, use t, s, e, o for terms and metaterms, use u, v, w for redexes, and use N, P, Q for reductions (i.e., reduction paths). We write $s \subseteq t$ if s is a subterm (occurrence) of t. A one-step reduction in which a redexoccurrence $u \subseteq t$ is contracted is written as $t \xrightarrow{u} s$ or $t \to s$ or just u. We write $P : t \twoheadrightarrow s$ or $t \overset{P}{\twoheadrightarrow} s$ if P denotes a reduction (sequence) from t to s, write $P : t \twoheadrightarrow$ if P may be infinite. For finite P, $P + Q$ denotes the concatenation of P and Q.

Below, when we refer to terms and redexes, we always mean admissible terms and admissible redexes except when explicitly mentioned.

2.2 Expressive Power of CCERSs

Here we discuss briefly how to encode conditional TRSs [10] and reduction strategies as CCERSs. We also present an encoding of the π-calculus into a CCERS. For more details refer to [66] where, for example, encodings of Hilbert- and Gentzen-style proof systems into CCERSs are also given.

Conditional TRSs. Conditional term rewriting systems (CTRSs) were introduced by Bergstra and Klop [10]. Their conditional rules have the form $t_1 = s_1 \wedge \cdots \wedge t_n = s_n \Rightarrow t \to s$, where s_i and t_i may contain variables in t and s. According to such a rule, $t\theta$ can be rewritten to $s\theta$ if all the equations $s_i\theta = t_i\theta$ are satisfied. CTRSs were classified depending on how satisfaction is defined ('=' can be interpreted as $\twoheadrightarrow$, $\leftrightarrow^*$, etc.). As Bergstra and Klop remark, this can be generalised by allowing for arbitrary predicates on the variables as conditions (cf. also [29, 95]).

Clearly, all these CTRSs are context-free CCERSs since they allow conditions on the arguments but not on the context of rewrite rules. For this reason results for them are sometimes a special case of general results holding for all CCERSs. In particular, *stable* CTRSs for which the unconditional version is orthogonal as defined in [10] are orthogonal in our sense (to be defined in Subsection 3) and hence are confluent.

Encoding of Strategies. In the literature a strategy for a rewriting system (R, Σ) is often defined as a map $F{:}\,Ter(\Sigma) \to Ter(\Sigma)$, such that $t \to F(t)$ if

t is not a normal form, and $t = F(t)$ otherwise (e.g., [5]). Such strategies are deterministic and do not specify the way in which to obtain $F(t)$ from t.

The above definition of a strategy is unsatisfactory for the following reasons:

- In a term, there may be several redex occurrences yielding the same result if reduced. For example, $I(Ix)$ can be β-reduced in one step to Ix, by contracting either of the two I-redexes.
- A redex occurrence can be an instance of more than one rule. For example, $or(true, true) \to true$ by applying either of the two rules $or(true, x) \to true$ and $or(x, true) \to true$,
- The result of a reduction step is in general not determined uniquely by the redex occurrence and the rule that is applied. For example, applying the variable-introducing rule $a \to A$ to the term a in the empty context may lead to any result, depending on the assignment to A.
- One may want to define $t = F(t)$ even if t is not a normal form (e.g., when one is interested in computing head-normal forms [5]).

In [63], a strategy is defined as a set F of triples $(r, \theta, C[\,])$ specifying that a rule $r : t \to s \in R$ can be used with assignment θ in context $C[\,]$ to rewrite $C[t\theta]$ to $C[s\theta]$. With a strategy F one can associate a CCERS R_F, encoding exactly the same information, by taking $\theta, C[\,]$ admissible for r iff $(r, \theta, C[\,]) \in F$. Obviously, this also holds the other way around; that is, every CCERS can be viewed as a strategy for its unconditional version.

Encoding of the π-Calculus as a CCERS. In this subsection we will encode as a CCERS the version of the π-calculus described by Milner [77] . Recall that the π-calculus agents P, Q, ... are defined as follows:

$$P ::= \overline{x}y.P \mid x(y).P \mid 0 \mid P|P \mid !P \mid (x)P$$

Basic interaction is generated from the rule

$$x(y).P|\overline{x}z.Q \to [z/y]P|Q$$

by closing under unguarded contexts and working modulo structural congruence (where a guard is a prefix of form $\overline{x}y.$ or $x(y).$).

A CCERS (Σ_π, R_π) can be associated to the π-calculus as follows. The alphabet Σ_π consists of the function symbols $0, !, |, O$ with respective arities $0, 1, 2, 3$ and the quantifier symbols I and R with arities $(1, 2)$ and $(1, 1)$. I binds only in its last argument. The map $[\,]$ transforms π-terms into terms in $Ter(\Sigma_\pi)$. The only non-obvious cases are input, output, and restriction:

$$[x(y).P] = Iy(x, [P])\,;\; [\overline{x}z.Q] = O(x, z, [Q])\,;\; [(x)P] = Rx([P])$$

Combining the transformation $[\,]$ with the closing under unguarded contexts and the structural congruence leads to rules R_π of the form

$$C_1[Iy(X, P)] \mid C_2[O(X, Z, Q)] \to C_1[(Z/y)P] \mid C_2[Q], \text{ where}$$

1. P, Q, X, Z are metavariables, and admissible assignments for X, Z are variables.
2. The indicated subterms must be unguarded in $C_1[\,]$ and $C_2[\,]$ and not in the scope of RX (among the symbols above them can occur only the operators $|$, $!$ and Rx with $x \neq X$).
3. For any redex only (all) unguarded contexts are admissible.

3 Orthogonal CCERSs

In this section, we introduce a suitable concept of orthogonality for CCERSs, outline a *strict* and *strong confluence* proof methods for them, and indicate how these confluence results can be used for proving confluence for restricted λ-calculi. These results are taken from [53, 63]. Finally, we will discuss briefly a classification of orthogonal CCERSs according to redex-creation.

3.1 Orthogonality and Confluence

The idea of orthogonality is that contraction of a redex does not destroy other redexes (in whatever way) but instead leaves a number of their residuals. A prerequisite for the definition of residual is the concept of *descendant*, also called *trace*, which allows the tracing of subterms along a reduction. Whereas this concept is pretty simple in the first-order case, CCERSs may exhibit complex behaviour due to the possibility of nested metasubstitutions in the right-hand sides of rules, thereby complicating the definition of descendants. A standard technique in higher-order rewriting [67] (illustrated below on examples) is to *decompose* or *refine* each rewrite step into two parts: a *TRS*-part in which the left-hand side is replaced by the right-hand side without evaluating the (meta)substitutions, and a *substitution*-part in which the delayed substitutions are evaluated. To express substitution we use the S-reduction rules

$$S^{n+1}x_1 \ldots x_n A_1 \ldots A_n A_0 \to (A_1/x_1, \ldots, A_n/x_n)A_0, \;\; n = 1, 2, \ldots,$$

where S^{n+1} is a *substitution operator sign* with arity $(n, n+1)$ and scope indicator $(n+1)$ and where $x_1, \ldots, x_n$ and $A_1, \ldots, A_n, A_0$ are pairwise distinct variables and metavariables. (We assume that the CCERS does not contain symbols S^{n+1}; it can of course contain a renamed variant of S-rules. The collection of all substitution rules, renamed or not, is an ERS itself.) Thus S^{n+1} binds only in the last argument. One can think of S-redexes as (simultaneous) let-expressions. Clearly, using just S^2 would be enough.

If a CCERS R is simple, we define $R_{fS} = R_f = R$; otherwise $R_{fS} = R_f \cup S$, where R_f is obtained from R by adding symbols S^{n+1} to the alphabet and by replacing all metasubstitutions of the form $(t_1/x_1, \ldots, t_n/x_n)t_0$ with $S^{n+1}x_1 \ldots x_n t_1 \ldots t_n t_0$ in the right-hand sides of the rules. Thus the descendant relation of a rewrite step can be obtained by composing the descendant relation of the TRS-step in R_f and the descendant relations of the S-reduction steps.

All known concepts of descendants agree in the cases when the subterm $s \subseteq t$ which is to be traced during a step $t \xrightarrow{u} o$ is (1) in an argument e of the contracted redex u, (2) properly contains u, or (3) does not overlap with u. In case (1), one traces the argument e first; and in every descendant e' of e (if any), the subterm of e' at the same position as that of s in e is a descendant of s in o. A special case of (1) for an S-reduction step is the situation when s is a variable occurrence in the last argument, bound by the S-operator, in which case the descendant of s is the subterm occurrence which instantiates it after the substitution step. In cases (2)-(3), the subterm of o at the same position as that of s in t is the unique descendant of s in o. The known descendant concepts differ when s is a pattern-subterm (i.e., when s is in the contracted redex u but is not in any of its arguments), in which case we define the contractum of u to be the descendant of s. According to many definitions, however, s does not have a u-descendant (*descendant* is often used as a synonym of *residual*, which it is not). In the case of TRSs, our definition coincides with Boudol's [17] and the descendant concept corresponds to Boudol-Khasidashvili labelling as defined in [94]. It differs slightly from Klop's [68] definition where the descendants of a contracted redex, as well as of any of its pattern-subterms, are all subterms whose head-symbols are within the pattern of the contractum.

We explain our descendant concept by using examples (for a formal definition, see [63]). Consider a TRS-step $t = f(g(a)) \to h(b) = s$ performed according to the rule $f(g(x)) \to h(b)$. The descendant of both pattern-subterms $f(g(a))$ and $g(a)$ of t in s is $h(b)$ and a does not have a descendant in s.

The refinement of a β-step $t = Ap(\lambda x(Ap(x,x)), z) \to_\beta Ap(z,z) = e$ would be $t = Ap(\lambda x(Ap(x,x)), z) \to_{\beta_f} o = S^2xzAp(x,x) \to_S Ap(z,z) = e$: the descendant of both t and $\lambda x(Ap(x,x))$ after the TRS-step is the contractum $S^2xzAp(x,x)$, the descendants of $Ap(x,x), z \subseteq t$ are the respective subterms $Ap(x,x), z \subseteq o$, the descendant of both $o = S^2xzAp(x,x)$ and $Ap(x,x)$ after the substitution step is the contractum e, and the descendants of $z \subseteq o$, as well as of the bound occurrence of x in $Ap(x,x)$, are the occurrences of z in e. As another example, consider the S-reduction step $t = S^2xf(a)g(x) \to_S g(f(a)) = s$. Then the descendant of $x \subseteq t$ is $f(a) \subseteq s$, and the descendant of $g(x) \subseteq t$ is s. The descendants of $f(a), a \subseteq t$ are the occurrences $f(a), a \subseteq s$, respectively.

The *descendant* concept extends by transitivity to arbitrary reductions consisting of TRS-steps and S-reduction steps. If P is an R-reduction, then P-*descendants* are defined to be the descendants under the refinement of P. The *ancestor* relation is the inverse of the descendant relation. The descendant concept allows us to define residuals:

Definition 4. *Let $t \xrightarrow{u} s$ be in a CCERS R, let $v \subseteq t$ be an admissible redex, and let $w \subseteq s$ be a u-descendant of v. We call w a u-**residual** of v if (a) the patterns of u and v do not overlap (i.e., the pattern-occurrences do not share an occurrence of a symbol in t), (b) w is a redex weakly similar to v (see Definition 3), and (c) w is admissible. (So u itself does not have u-**residuals** in s.) The concept of a **residual** of redexes extends naturally to arbitrary reductions. A redex in s is called a **new** redex or a **created** redex if it is not a residual of a redex in t. The **predecessor** relation is inverse to that of residual.*

Definition 5. *We call a CCERS* ***orthogonal*** *if:*

- *the left-hand sides of rules are not single metavariables,*
- *the left-hand side of any rule is a simple metaterm and its metavariables contain those of the right-hand side, and*
- *all the descendants of an admissible redex u in a term t under the contraction of any other admissible redex $v \subseteq t$ are residuals of u.*

The second condition ensures that rules exhibit deterministic behaviour when they can be applied. Failure of the last condition may easily lead to non-confluence: For example, consider the rules $a \to b$ and $f(A) \to A$ with the (only) admissible assignment $A\theta = a$. The descendant $f(b)$ of the redex $f(a)$ after contraction of a is not a redex because the assignment $A\theta = b$ is not admissible. Hence the system is not orthogonal, nor is it confluent.

Definition 6. *Reductions starting from the same term are called* ***co-initial****. Recall that co-initial reductions $P : t \twoheadrightarrow s$ and $Q : t \twoheadrightarrow e$ are* ***weakly equivalent*** *or* ***Hindley-equivalent*** *[5], written $P \approx_H Q$, if $s = e$ and the residuals of any redex of t under P and under Q are the same redexes in s. Furthermore, P and Q are* ***strictly equivalent*** *[51], written $P \approx_{st} Q$, if $s = e$ and the descendants of any subterm of t under P and under Q are the same subterms in s.*

Using these equivalencies and the above definition of residuals, we can easily infer *strong* [70, 43] and *strict* [51] forms of the *Church-Rosser* (CR) property for CCERSs.

A standard method of proving the *strong* version of CR is one using termination of developments (FD) and the fact that any pair of redexes u, v in a term *strongly commute*: $u + v/u \approx_H v + u/v$ [70] where v/u denotes a **complete development** of a set of residuals of v after contracting u, and similarly for u/v; that latter property will be called *strong local confluence*. Recall that a *development* of a term t is a reduction in which only residuals of redexes present in t are contracted; and a development of a set U of redexes in t is a reduction in which only residuals of redexes in U are contracted. A *complete development* of U is a development $t \twoheadrightarrow s$ of U such that s does not contain residuals of redexes in U. Below U may also denote a complete development of a set of redexes U; and U is also called a *multi-step*. As in orthogonal TRSs [43], the λ-calculus [71, 5], orthogonal CRSs [67], and orthogonal HRSs [97], one can in orthogonal CCERSs use FD and strong commutativity to define for any co-initial reductions P and Q the *residual of P under Q*, written P/Q. One way to do it is to consider P and Q as multi-step reductions, say $P = U + P'$, where U is the first multi-step in P, and define inductively that $P/Q = U/Q + P'/(Q/U)$. For a correctness proof of this definition, we refer the reader to [71]. We write $P \trianglelefteq_L Q$ if $P/Q = \emptyset$ ($\trianglelefteq_L$ is the *Lévy-embedding* relation); P and Q are called *Lévy-equivalent* or *permutation-equivalent* (written $P \approx_L Q$) if $P \trianglelefteq_L Q$ and $Q \trianglelefteq_L P$. It follows from the definition of / that if $P + P'$ and $Q + Q'$ are co-initial finite reductions in an orthogonal CCERS, then $(P + P')/Q \approx_L P/Q + P'/(Q/P)$ and $P/(Q + Q') \approx_L (P/Q)/Q'$. This is all well known and we do not give more details. The strong Church-Rosser theorem then states that, for any co-initial finite reductions P and Q in an orthogonal ERS, $P \sqcup Q \approx_L Q \sqcup P$, where $P \sqcup Q$ means $P + Q/P$. The *Strict Church-Rosser* theorem states that, for any co-initial finite reductions P and Q in an orthogonal ERS, $P \sqcup Q \approx_{st} Q \sqcup P$. (Thus, $P \approx_L Q$ implies $P \approx_{st} Q$ since P/Q and Q/P are empty reductions.) Like the strong CR property, the strict CR property follows from FD and the following *strict local confluence* property: any two co-initial steps u, v *strictly commute*: $u \sqcup v \approx_{st} v \sqcup u$.

Theorem 1. *([53, 63])* **(Finite Developments)** *All developments of a term t in an orthogonal CCERS R eventually terminate.*

Using this theorem and the last condition in the definition of orthogonality, the next theorem follows from some abstract theory of residuals.

Theorem 2. *([53, 63]) Let P and Q be any co-initial finite reductions in an orthogonal CCERS R. Then*

(1) **(Strong Church-Rosser)** $P \sqcup Q \approx_L Q \sqcup P$.
(2) **(Strict Church-Rosser)** $P \sqcup Q \approx_{st} Q \sqcup P$.

By restricting term formation in the λ-calculus [5], one arrives at a large class of typed lambda calculi. Since the rewrite relation in these calculi is not restricted and typed terms are closed under β-reduction, these CCERSs are orthogonal, and hence confluent. As an example, see [63] for a proof that the *call-by-need λ-calculus* of Ariola et al. [3] is an orthogonal CCERS. Other interesting examples of context-sensitive, conditional orthogonal λ-calculi can also be found in [63].

3.2 A Classification of Orthogonal CCERSs According to Redex Creation

Properties of orthogonal CCERSs depend on the types of redex-creation during reduction. It is therefore natural to classify CCERSs according to redex creation. Here we recall from [55, 57] some of the interesting types of CCERSs where redex creation is restricted. In subsequent sections we discuss more results on subclasses of orthogonal CCERSs.

Definition 7. *1. Let $t \longrightarrow_u s$ in an orthogonal ERS R, let $t \longrightarrow_{u_f} t' \twoheadrightarrow_S s$ be its refinement in R_{fS}, and let v be a new redex in s. Redex v is called* ***generated*** *if it is a residual of a redex $v' \subseteq t'$ whose pattern is in the pattern of the contractum of u_f. If furthermore the redex v' is not inside an S-redex in t', then we call v* ***uniformly generated****.*
2. An orthogonal ERS R is ***persistent*** *(PERS) (resp.* ***uniformly persistent****) if each created redex in R is generated (resp. uniformly generated).*
3. An orthogonal ERS R is ***strongly persistent*** *(SPERS) if R_{fS} is persistent.*

It is easy to see that a non-simple orthogonal ERS (OERS) R is strongly persistent iff the left-hand sides of its rules consist of a single operator: R_{fS} contains S-rules which can create any redex whose pattern contains at least two operators. Recall that if R is simple (see Definition 3), then $R_{fS} = R$.

Higher Order Recursive Program Schemes (HRPSs) are the prime examples of persistent ERSs [55]. In HRPSs, the left-hand side of any rewrite rule contains only one function or quantifier symbol. The rules $\exists x(A) \to (\tau x(A)/x)A$ and $\exists! x(A) \to \exists x(A) \wedge \forall x \forall y(A \wedge (y/x)A \Rightarrow x = y)$ are important examples. In the rewrite step $\sigma x(x) \to f((\sigma x(x)/x)x) = f(\sigma x(x))$ corresponding to rule $\sigma x(A) \to f((\sigma x(x)/x)A)$ and assignment $\theta(A) = x$, the created redex $\sigma x(x)$ is generated but not uniformly generated – for steps corresponding to assignments θ' such that $x \notin FV(\theta'(A))$, no redexes will be generated.

For persistent ERSs, weak and strong normalisation are decidable; and for uniformly persistent ERSs, the lengths of shortest reductions can be computed statically [55]. For strongly persistent ERSs, lengths of longest reductions too can be computed statically, see Section 5.3. These results are obtained by analysing *essential similarity* and *strong similarity* of redexes, respectively, which will be introduced in the next section.

4 Normalising Strategies for Orthogonal CCERSs

If a term t in a rewrite system has a normal form (a term without a redex), then a *normalising strategy*, when applied to t, computes a reduction of t to a normal form. A normalising strategy, called a *needed strategy*, was found for orthogonal TRSs by Huet and Lévy [43]. They defined a redex u in a term t of an orthogonal TRS as *needed* if any normalising reduction of t contracts at least one residual of u; and a needed strategy is defined as a strategy that repeatedly contracts a needed redex in the given term, till a normal form is computed. This strategy was later generalised to many other rewrite systems.

In this section we will review normalising strategies developed for orthogonal CCERSs. We will mainly focus on the concept of *essentiality* which makes sense for all subterms, including free or bound variable occurrences, and for the case of redexes coincides with Maranget's concept of neededness [72] which, unlike Huet-Lévy's concept, is meaningful for terms without a normal form as well. We discuss briefly usage of essentiality where neededness cannot be used. We will also discuss properties of external redexes and similarity of redexes, used for proving many results for orthogonal CCERSs.

The material discussed in this section is taken from the following publications [51, 58, 55, 54, 63]. For more on strategies in orthogonal rewrite systems, see [94].

4.1 Relative Normalisation by Neededness in Orthogonal CCERSs

A theory of normalisation by neededness for orthogonal CCERSs is developed in [35]. It introduces a concept of *stable results*, and develops normalisation with respect to this set of "normal forms". Important examples of stable results are normal forms [43], head-normal forms [7], and weak-head-normal forms in the λ-calculus, constructor-head-normal forms for constructor TRSs [83], and root-stable forms (terms that cannot be rewritten to a redex) in TRSs [76]. A labelling system for orthogonal CCERSs, in the style of Lévy's labelling for optimal β-reduction [71], is also introduced in [35] and relative optimal normalisation is studied. In later works by Glauert and Khasidashvili [36, 59, 61, 60, 62], and others (e.g. [Mel98]), this relativized normalisation theory was generalised to abstract reduction systems with axiomatised residual and family relations, and therefore it is less relevant to review these results here in the context of Orthogonal CCERSs. Instead, we review briefly a normalisation proof method which uses the descendant concept discussed in previous sections.

4.2 External Redexes

In this subsection we will show that every reducible term in an orthogonal *fully-extended* (see Definition 9) CCERS has an *external* redex. External redexes for orthogonal TRSs were introduced by Huet and Lévy [43], who also proved the existence of external redexes in every reducible term. Both the original definition of external redexes and the existence proof are quite lengthy.

With our concept of descendant, external redexes can be defined as follows:

Definition 8. *([58, 63]) A subterm $s \subseteq t$ in an orthogonal CCERS is* ***external*** *if no descendants of s along any reduction $P : t \twoheadrightarrow o$ appear inside redex-arguments.*[2]

In the above definition, s may be a redex, in which case its descendants are in fact its residuals. Any external redex is trivially outermost, but an outermost redex is not necessarily external. Contracting a redex *disjoint from it*, may cause its residual to be non-outermost. For example, consider the orthogonal TRS $\{f(x,b) \to c, a \to b\}$. The first a in $f(a,a)$ is outermost but not external; contracting the second a (which is disjoint from it) creates the redex $f(a,b)$ having the residual of the first a as argument. The second a is external.

In an ERS, there may be another reason why an outermost redex need not be external. Contracting a redex *in one of its arguments* may cause its residual to be non-outermost. This already shows up in the $\lambda\beta\eta$-calculus. Let $I = \lambda x.x$ and $K = \lambda xy.x$, as usual [5]. The redex $u = I(KIx)$ in $\lambda x.I(KIx)x$ is outermost but not external; contracting the redex KIx in its argument creates the η-redex $\lambda x.IIx$ having the residual II of u as argument. This particular example, but not the entire $\lambda\beta\eta$-calculus), can be readily encoded as an orthogonal ERS. We will see later that because of rules like η which test for the absence of variables in subterms (occur check!) even the conservation theorem fails for orthogonal CCERSs in general.

To exclude such rules, we restrict ourselves to *fully extended* CCERSs [63] (defined after [38]).

Definition 9. *We call a CCERS R* ***fully-extended*** *iff for any step $t \xrightarrow{u} s$ in R and any occurrence $w \subseteq t$ of an instance of the left-hand-side (of a rule $r \in R$) such that:*

[2] In [58], external redexes are called *unabsorbed*.

(a) the patterns of w and u in t do not overlap, and
(b) w has a u-descendant $w' \in s$ that is a redex,

w is an admissible redex in t weakly similar to w'.

A simple proof of the following theorem can be found in [63]. A similar method for orthogonal TRS was used in [58]. Even for orthogonal TRSs, there is no algorithm for computing an external redex in every term [43].

Theorem 3. *Every reducible term in an orthogonal fully-extended CCERS has an external redex.*

4.3 Normalisation of the Essential Strategy

Normalisation theory based on tracing subterms rather than only redexes was introduced by Khasidashvili for the λ-calculus [51], independently from the work of Huet and Lévy [43]. The proof method generalises easily to orthogonal TRSs [58]. The same method can be used for orthogonal CCERSs; we will review this method briefly here. We will also discuss some important applications where one needs to reason about essentiality of subterms that are not redexes, in particular, about essentiality of variable occurrences, where the concept of neededness cannot be used.

Definition 10. *Let t be a term in an orthogonal CCERS R. We call a subterm s in t **essential** if s has at least one descendant under any finite reduction starting from t and otherwise call it* inessential.

It is easy to prove that contraction of a non-essential redex cannot create an essential redex; therefore any (normalising) reduction $P : t \twoheadrightarrow o$ can be reorganised into a reduction $Q : t \twoheadrightarrow o$ so that essential redexes are contracted first in Q. And since an inessential redex cannot reduce a term to its normal form, the "essential prefix" Q_e of Q must end in a normal form. Finally, there cannot be an infinite essential reduction P' starting from t: indeed, an essential redex cannot erase another essential redex; thus the reduction P'/Q_e would be an infinite essential reduction if P' was an infinite essential reduction, which is impossible since Q_e ends in a normal form o. Any term not in normal form has an essential redex – an external redex constructed in Section 4.2 is essential since it has a descendent along any reduction. Thus we have proven the theorem:

Theorem 4. *The essential strategy is normalising in fully-extended orthogonal CCERSs.*

For persistent and uniformly persistent orthogonal ERSs one can prove more:

Theorem 5 ([55]). *Let t be a term in a persistent ERS R.*

1. *Weak and strong normalisation of t is decidable.*
2. *All essential redexes of t can be computed statically.*

3. *When R is uniformly persistent and t is normalisable, repeated contraction of innermost essential redexes in t yields a shortest normalizing reduction starting from t.*

The last statement of Theorem 5 does not hold for persistent ERSs in general: For example, all normalizing reductions starting from term $\sigma x(f(x))$ in the following persistent ERS $R = \{\sigma x(A) \to (\epsilon x(A)/x)A, \epsilon x(A) \to (\tau x(A)/x)A, f(x) \to g(x,x)\}$ must contract two copies of at least one redex. Indeed, contracting the f-redex $f(x)$ first would duplicate x, which will cause generating two copies of a ϵ-redex when the σ redex is contracted; while contracting the σ-redex first would cause a duplication of the f-redex. Note that developments in orthogonal ERSs can be encoded as ERSs where there is no redex-creation, thus the above theorem allows one to construct shortest developments in orthogonal ERSs. These results can be generalised to orthogonal CCERSs as well.

4.4 Similarity of Redexes

In this section, we discuss several concepts of similarity of redexes which are used in many proofs, including proofs concerning the shortest as well as longest reductions. These concepts have not been studied for other forms of higher-order rewriting. The idea of *similarity* of redexes [55, 54] u and v is that u and v are weakly similar – that is, they match the same rewrite rule – and quantifiers in the pattern of u and v bind 'similarly' in the corresponding arguments. For example, recall that a β-redex $Ap(\lambda xt, s)$ is an *I-redex* if $x \in FV(t)$ and is a *K-redex* otherwise. Then all I-redexes are similar and all K-redexes are similar, but no I-redexes are similar to a K-redex. Consequently, for any pair of corresponding arguments of u and v, either both are erased after contraction of u and v or none is.

A redex in a CCERS has the form $u = C[t_1, \ldots, t_n]$, where C is the pattern and $t_1, \ldots, t_n$ are the arguments. Sometimes we write u as $u = C[\overline{x_1}t_1, \ldots, \overline{x_n}t_n]$, where $\overline{x_i} = \{x_{i_1}, \ldots, x_{i_{n_i}}\}$ is the subset of binding variables of C such that t_i is in the scope of an occurrence of each x_{i_j}, $j = 1, \ldots, n_i$. Let us call the maximal subsequence $j_1, \ldots, j_k$ of $1, \ldots, n$ such that $t_{j_1}, \ldots, t_{j_k}$ have u-descendants the *main sequence* of u (or the *u-main sequence*), call $t_{j_1}, \ldots, t_{j_k}$ the (u-)*main arguments*, and call the remaining arguments (u)-*erased*. Further, call u *erasing* if $k < n$ and *non-erasing* otherwise.

Now the similarity of redexes can be defined as follows: weakly similar redexes $u = C[\overline{x_1}t_1, \ldots, \overline{x_n}t_n]$ and $v = C[\overline{x_1}s_1, \ldots, \overline{x_n}s_n]$ are *similar* if, for any $1 \le i \le n$, $\overline{x_i} \cap FV(t_i) = \overline{x_i} \cap FV(s_i)$. For example, consider the rule $\sigma x(A, B) \to (\sigma x(f(A), A)/x)B$. Then the redexes $u = \sigma x(x, y)$ and $v = \sigma x(f(x), y)$ are similar, while $w = \sigma x(y, y)$ is not similar to any of them since $x \notin FV(y)$. However, note that the second arguments of all the redexes u, v and w are main and the first arguments are erased. In this presentation it is more convenient to use a slightly relaxed concept of similarity, written $\sim$, such that $u \sim v \sim w$:

Definition 11. *We write $u \sim v$ if the main sequences of u and v coincide and for any main argument t_i of u, $\overline{x_i} \cap FV(t_i) = \overline{x_i} \cap FV(s_i)$.*

The following lemma implies in particular that, indeed, if u and v are similar, then $u \sim v$, and that $\sim$ is an equivalence relation. Its proof involves properties of essentiality of subterms, in particular of variable occurrences, and strict confluence. Roughly, the idea is that bound variable occurrences in erased arguments of a redex u are "inessential" even if we do not contract redexes in the arguments of u. Therefore only bindings in main arguments are relevant.

Below, θ will not only denote assignments but will also denote substitutions assigning terms to variables; when we write $o' = o\theta$ for a substitution θ, we assume that no free variables of the substituted subterms become bound in o' (i.e., we rename bound variables in o when necessary).

Lemma 1. *(Redex Similarity) Let u and v be weakly similar redexes of the form $u = C[\overline{x_1}t_1, \ldots, \overline{x_n}t_n]$ and $v = C[\overline{x_1}s_1, \ldots, \overline{x_n}s_n]$, and let for any main argument s_i of v, $\overline{x_i} \cap FV(t_i) \subseteq \overline{x_i} \cap FV(s_i)$. Then the main sequence of u is a subset of the main sequence of v. Furthermore, if $\overline{x_i} \cap FV(t_i) = \overline{x_i} \cap FV(s_i)$, then $u \sim v$; in particular, if $u = v\theta$, then $u \sim v$.*

For many results for CCERSs, one needs to understand how the erasure of arguments depends on the binding structure of the redex. The similarity lemma above establishes such a relation. This lemma is used essentially in many proofs of perpetuality of redexes, longest reductions, as well as for characterising lengths of shortest reductions for classes of orthogonal CCERSs. The following lemma is an example of application of the similarity lemma. It is used in proving properties of perpetual/safe redexes, in Section 5.2.

Lemma 2. *Let $t \xrightarrow{u} s$ be in a CCERS, let $o \subseteq t$ be either in an argument of u or not overlapping with u, and let $o' \subseteq s$ be a u-descendant of o. Then $o' = o\theta$ for some substitution θ. Moreover, if o is a redex, then so is o' and $o \sim o'$.*

For results concerning the shortest and the longest reductions in orthogonal CCERSs, one needs the concepts of *essential* and *strong similarity* of redexes, respectively:

Definition 12. *Let $u = C[\overline{x_1}e_1, \ldots, \overline{x_n}e_n]$ and $v = C[\overline{x_1}o_1, \ldots, \overline{x_n}o_n]$ be weakly similar redexes in an orthogonal CCERS R.*

- *Redexes u and v are called **R-essentially similar** if, for all i and every $x_{i_j} \in \overline{x_i}$, x_{i_j} has an R-essential occurrence in e_i iff it has one in o_i.*
- *Redexes u and v are called **strongly similar** if, for all i and every $x_{i_j} \in \overline{x_i}$, the numbers of occurrences of x_{i_j} in e_i and o_i coincide.*

Consider for example redexes $u = (\lambda x.x)s$, $v = (\lambda x.(\lambda y.z)x)o$, and $w = (\lambda x.xx)e$. Then u and v are similar and strongly similar, but not β-essentially similar; and u and w are similar and β-essentially similar, but not strongly similar.

The properties of essential bindings are needed to prove Theorem 5. The analysis of essential bindings is also crucial for proving normalisation results for the *hyperbalanced λ-calculus* [64]. There, essentiality of bound variables and of redexes can be determined statically (thus one has a static garbage collection algorithm for the hyperbalanced λ-calculus). Furthermore, the number of superdevelopments [99, 69] needed to normalise a hyperbalanced λ-term can be computed statically as well.

Using the concept of strong similarity, an algorithm is developed in [57] for computing the lengths of longest reductions in persistent ERSs; as a corollary, one gets an algorithm for statically computing lengths of longest developments in orthogonal CCERSs.

5 Perpetual and Longest Reductions in Orthogonal CCERSs

In this section we review results on perpetual and longest reductions in orthogonal CCERSs. For brevity, we will not discuss related results in other formats of higher-order rewriting, or in the λ-calculus, and instead refer to [63, 94]. Most of the results in this section are taken from [63].

5.1 A Minimal Perpetual Strategy

In this subsection we introduce a perpetual strategy F_m^∞ for orthogonal fully-extended CCERSs by generalising the *constricting* perpetual strategies in the literature [88, 93, 37, 74, 100]. We also study properties of F_m^∞ that are used in the next subsection to obtain new criteria for the perpetuality of redexes and of redex occurrences in orthogonal fully-extended CCERSs. A survey on perpetual reductions in the λ-calculus and its extensions can be found in [92, 100].

Recall that a term t is called *weakly normalizing* (WN), written $WN(t)$, if it is reducible to a *normal form* (i.e., a term without a redex), and t is called *strongly normalizing* (SN), written $SN(t)$, if it does not possess an infinite reduction. We call t an ∞*-term* (written ∞t), if $\neg SN(t)$. Clearly, for any term t, $SN(t) \Rightarrow WN(t)$. If the converse is also true, then we call t *uniformly normalizing* (UN). So a UN term t either does not have a normal form or all reductions from t eventually terminate. Correspondingly, a rewrite system R is called WN, SN, or UN if all terms in R are WN, SN, or UN, respectively.

Following [9, 68], we call a rewrite step $t \xrightarrow{u} s$, as well as the redex-occurrence $u \subseteq t$, *perpetual* if $\infty t \Rightarrow \infty s$. Otherwise we call them *critical.* We call a redex (not an occurrence) *perpetual* iff its occurrence in every (admissible) context is perpetual. A *perpetual strategy* in an orthogonal fully-extended CCERS is a (partial) function on terms which in any reducible term selects a perpetual redex-occurrence; the orthogonality of the CCERS implies that the redex-occurrence uniquely determines the rewrite rule (and the corresponding admissible assignment) according to which the redex is to be contracted.

Definition 13. *Let $P : t \twoheadrightarrow$ and $s \subseteq t$. Reduction P is* **internal** *to s if it contracts redexes only in (the descendants of) s.*

Definition 14. *(1) Let t be an ∞-term in an orthogonal fully-extended CCERS and let $s \subseteq t$ be a smallest subterm of t such that $\infty(s)$ (i.e., such that every proper subterm $e \subset s$ is SN). Then we call s a* **minimal perpetual subterm** *of t, and call any external redex of s a* **minimal perpetual redex** *of t.*

(2) Let F^∞_m be a one-step strategy that contracts a minimal perpetual redex in t if ∞t and otherwise contracts any redex. Then we call F^∞_m a **minimal perpetual strategy**. *We call F^∞_m* constricting *if for any F^∞_m-reduction $P : t_0 \xrightarrow{u_0} t_1 \xrightarrow{u_1} \dots$ (i.e., any reduction constructed using F^∞_m) starting from an ∞-term t_0 and for any i, $P^*_i : t_i \xrightarrow{u_i} t_{i+1} \xrightarrow{u_{i+1}} \dots$ is internal to s_i, where $s_i \subseteq t_i$ is the minimal perpetual subterm containing u_i. Constricting minimal perpetual strategies will be denoted F^∞_{cm}.*

(3) F^∞_m is the **leftmost** *minimal perpetual strategy, denoted F^∞_{lm}, if in each term it contracts the leftmost minimal perpetual redex.*

Now we can state the main results on minimal perpetual strategies for fully-extended orthogonal CCERSs [63].

Theorem 6. *1. F^∞_m is a perpetual strategy in any orthogonal fully-extended CCERS.*
2. F^∞_{lm} is a constricting strategy in any orthogonal fully-extended CCERS.

It is interesting to note that the constricting perpetual reductions are minimal w.r.t. Lévy's embedding relation $\trianglelefteq_L$ (see [63] for a proof).

5.2 Two Characterisations of Critical Redexes

In this section we give an intuitive characterisation of critical redex occurrences for orthogonal fully-extended CCERSs, generalising Klop's characterisation of critical redex occurrences for orthogonal TRSs [68], and derive from it a characterisation of perpetual redexes similar to Bergstra and Klop's perpetuality criterion for β-redexes [9].

Definition 15. *(1) Let $P : t_0 \xrightarrow{u_0} t_1 \xrightarrow{u_1} \dots \xrightarrow{u_{k-1}} t_k$, be in an orthogonal CCERS, and let $s_0, s_1, \dots, s_k$ be a chain of descendants of s_0 along P (i.e, s_{i+1} is a u_i-descendant of $s_i \subseteq t_i$). Then, following [9], we call P* **passive** *w.r.t. $s_0, s_1, \dots, s_k$ if the pattern of u_i does not overlap s_i (s_i may be in an argument of u_i or be disjoint from u_i) for $0 \leq i < k$, and we call s_k a* **passive descendant** *of s_0. By Lemma 2, $s_k = s_0\theta$ for some substitution θ, which we call a* **passive substitution**, *or* **P-substitution** *(w.r.t. $s_0, s_1, \dots, s_k$).*

(2) Let t be a term in an orthogonal fully-extended CCERS and let $s \subseteq t$. We call s a **potentially infinite** *subterm of t if s has a passive descendant s' s.t. $\infty(s')$. (Thus $\infty(s\theta)$ for some passive substitution θ.)*

Theorem 7. *Let t be an ∞-term and let $t \xrightarrow{v} s$ be a critical step in an orthogonal fully-extended CCERS. Then v erases a potentially infinite argument o (thus $\infty(o\theta)$ for some passive substitution θ).*

Note in the above theorem that if the orthogonal fully-extended CCERS is an orthogonal TRS, a potentially infinite argument is actually an ∞-term (since passive descendants are all identical), implying Klop's perpetuality lemma [68]. O'Donnell's [85] lemma, stating that any term from which an innermost reduction is normalizing is strongly normalizing, is an immediate consequence of Klop's Lemma.

Corollary 1. *Any redex whose erased arguments are closed SN terms is perpetual in orthogonal fully-extended CCERSs.*

Note that Theorem 7 implies a general (although not computable) perpetual strategy: simply contract a redex u in the term t whose erased arguments (if any) are not potentially infinite w.r.t. at least one ∞-subterm $s \subseteq t$ (although the erased arguments of u may be potentially infinite w.r.t. t). Many known perpetual strategies can be obtained as special cases, such as for example those reported in [5, 6, 9, 56, 54, 57, 74, 100, 93]

We conclude this section with a characterisation of the perpetuality of erasing redexes, a characterisation similar to the perpetuality criterion of β_K-redexes that was given by Bergstra and Klop [9].

Below, a substitution θ will be called SN iff $SN(x\theta)$ for every variable x.

Definition 16. *We call a redex u* ***safe*** *(respectively,* ***SN-safe****) if it is non-erasing or if it is erasing and for any (resp. SN-) substitution θ, if $u\theta$ erases an ∞-argument, then the contractum of $u\theta$ is an ∞-term. (Note that, by Lemma 1, u is erasing iff $u\theta$ is, for any θ, erasing.)*

Theorem 8. *In an orthogonal fully-extended CCERS R, any safe redex v is perpetual.*

The following example demonstrates that non-erasing steps need not be perpetual in orthogonal CCERSs in general, that is, the restriction to fully-extended CCERSs is necessary:

Example 1. Consider the ERS with rules: $\lambda x(A, B) \rightarrow (B/x)A$, $\kappa yz(A) \rightarrow (a/z)A$, and $e(A, B) \rightarrow c$, $f(a) \rightarrow f(a)$, where λ is a partial quantifier symbol binding only in its first argument, and $y \notin FV(A\theta)$ for any assignment θ admissible for the κ-rule. Consider the term $s = \kappa yz(\lambda x(e(x, y), f(z)))$. Note that s is not a redex (yet) due to the occurrence of y. On the one hand, contracting the e-redex yields an infinite reduction $s \rightarrow \kappa yz(\lambda x(c, f(z))) \rightarrow \lambda x(c, f(a)) \rightarrow \ldots$. On the other hand, contracting the (non-erasing) λ-redex yields $s \rightarrow \kappa yz(e(f(z), y)) \rightarrow \kappa yz(c) \rightarrow c$ as the only, and strongly normalizing, reduction. Hence the λ-step is non-erasing but critical.

Corollary 2. *([57]) (***Conservation***) If a term t in a fully-extended OERS R has an infinite reduction and $t \longrightarrow_u s$, where u is a non-erasing redex, then s has also an infinite reduction.*

5.3 The Longest Perpetual Reductions in OERSs

In this section, we introduce the *limit strategy* and show that it is perpetual in OERSs and moreover it constructs the longest reductions. We also give an algorithm for characterising the lengths of longest reductions of strongly normalisable terms [57]. The results are obtained with a well known "memory method", first studied by Nederpelt [80] and Klop [67].

The idea of the memory method is to associate with a rewrite system R a non-erasing system R_μ in such a way that all reductions in R can be simulated by reductions in R_μ. One can show that R_μ is an occur-conditional ERS for any OERS R, meaning that for any rule $r \in R$ and an assignment $\theta \in AA(R)$, if an assignment θ' is such that $FV(A\theta) = FV(A\theta')$ for every metavariable A occurring in r, then $\theta' \in AA(r)$, which implies easily that R_μ is orthogonal. Since R_μ is non-erasing, one hopes to prove that if a term in R is weakly normalisable in R_μ, it is also strongly normalisable in R_μ. Because of the above simulation, it would therefore be strongly normalisable in R too. One needs special memory symbols to keep erasable arguments in the right-hand sides of R_μ-rules (we use μ-symbols for that purpose), and the memory symbols may block the creation of redexes during R_μ-steps. One way to ensure the possibility of simulation is to have special 'restructuring rules' for moving these memory symbols away from undesirable positions such as the *shift* rule introduced by Klop [67]. Another way, and the one used here, is simply to extend the left-hand sides by allowing occurrences of memory symbols in them.

Another difference between our method and the Nederpelt-Klop method is that we do not memorise all arguments in the right-hand sides. Only erasable arguments are memorised; thus there are no extra (unnecessary) copies of non-erasable arguments in the right-hand sides of R_μ-rules. This difference is important in that it allows us to characterise the least upper bounds of the lengths of reductions in OERSs in terms of the number of μ-occurrences in corresponding R_μ-normal forms.

Definition 17. *The* ***μ-extension*** (Σ_μ, R_μ) *of an OERS* (Σ, R) *is a conditional ERS defined as follows:*

1. $\Sigma_\mu = \Sigma \cup \{\mu^n \mid n = 0, 1, \ldots\}$, *where* μ^n *is a fresh* n*-ary function symbol. For any subterm* $s = \mu^{n+1}(t_1, \ldots, t_n, t_0)$ *of a term* t *over* Σ_μ, *the arguments* $t_1, \ldots, t_n$, *as well as subterms and symbols in* $t_1, \ldots, t_n$ *and the head-symbol* μ *itself, are called* ***μ-erased*** *– or more precisely,* ***μ'-erased****, where* μ' *is the occurrence of the head symbol of* s *in* t. *The argument* t_0 *is called* ***μ'-main****. Symbols and subterms in* t *that are not* μ*-erased are called* ***μ-main****. We denote by* $[t]_\mu$ *the term obtained from* t *by removing all* μ*-erased symbols.*
2. R_μ *is the set of* all *rules of the form* $r_\mu : t' \to s'$ *such that*
   (a) *there is a rule* $r : t \to s$ *in* R *such that* $[t']_\mu = t$;
   (b) t' *is linear, its head symbol is not a* μ*-symbol, the* μ*-erased arguments of each* μ*-occurrence* μ' *in* t' *are metavariables, and the* μ'*-main argument is not a metavariable;*

(c) *if $A_1, \ldots, A_n$ are all μ-main metavariables of t', $B_1, \ldots, B_j$ are all μ-erased metavariables of t', and k is the number of occurrences of μ-symbols in t', then*

$$s' = \mu^m(\overbrace{\mu^0, \ldots, \mu^0}^{k}, B_1, \ldots, B_j, A_{i_1}, \ldots, A_{i_l}, s),$$

where $i_1, \ldots, i_l$ is the erased sequence of some r-redex. Furthermore, an assignment θ' is admissible for r_μ iff it is variable-capture-free and $i_1, \ldots, i_l$ is the erased sequence of the r-redex $t\theta'$ (considered as an R-redex). If θ' is admissible, the arguments of the r_μ-redex $t'\theta'$ that correspond to the erased (main) arguments of the r-redex $t\theta'$ are called ***quasi-erased (quasi-main)****.*

For example,

$$Ap(\mu^3(A, B, \mu^2(C, \lambda x D)), E) \rightarrow \mu^7(\mu^0, \mu^0, A, B, C, E, (E/x)D)$$

and

$$Ap(\mu^3(A, B, \mu^2(C, \lambda x D)), E) \rightarrow \mu^6(\mu^0, \mu^0, A, B, C, (E/x)D)$$

are two β_μ-rules with the same left-hand sides and different right-hand sides. The arguments A, B, and C are μ-erased, and E and D are μ-main. An assignment θ is admissible for the first rule iff $x \notin FV(D\theta)$ (since E is kept in its right-hand side as a μ-erased argument) and otherwise is admissible for the second one.

The following result is essential for proving properties of longest reduction; we state it here as an example of application of the Similarity Lemma 1.

Corollary 3. *(1) Let R be an OERS, let u and v be R_μ-redexes such that u is in an argument of v, and let $v \longrightarrow_u w$ in R_μ. Then w is an R_μ-redex similar to v, and the quasi-main sequences of v and w coincide.*

(2) Let u be a redex in an OERS R, and let v be an R_μ-redex such that $[v]_\mu = u$ and the set of free variables of any quasi-main argument of v coincides with that of the corresponding argument of u. Then an argument of v is quasi-erased iff the corresponding argument of u is erased.

(3) Let u be a redex in an OERS R and v be an R_μ-redex such that $[v]_\mu = u$. Then the corresponding argument of any quasi-erased argument of v is u-erased.

Definition 18. *Let u_l be a redex in a term t in a fully extended OCCERS, defined as follows: choose an external redex u_1 in t; choose an erased argument s_1 of u_1 that is not in normal form (if any); choose in s_1 an external redex u_2, and so on as long as possible. Let $u_1, s_1, u_2, \ldots, u_l$ be such a sequence. Then we call u_l a* ***limit redex*** *and call $u_1, s_1, u_2, \ldots, u_l$ a* ***limit sequence*** *of t.*

Thus in any term not in normal form there is a limit redex. We call a reduction **limit** if each contracted redex is a limit redex, and call a strategy **limit** if in any term not in normal form it contracts a limit redex.

Theorem 9. *A limit strategy is perpetual in fully-extended OERSs. Moreover, if a term t in a fully-extended OERS R is strongly normalisable, then a limit strategy constructs a longest normalizing reduction starting from t, and its length coincides with the number of occurrences of the μ-symbols in the R_μ-normal form of t.*

6 Expression Reduction Systems with Patterns

The *Pattern-Matching Calculus* [50] employs an evaluation process given by a generalisation of the standard β-rule:

$$(\beta_{PM}) \quad app(\lambda\mathbb{X}.M, N) \longrightarrow M\{\mathbb{X} \text{ by } N\}$$

where $\mathbb{X}$ denotes a pattern and $\{\mathbb{X} \text{ by } N\}$ denotes a substitution resulting from the pattern-matching operation on the pattern $\mathbb{X}$ and the term N.

Expression Reduction Systems with Patterns (ERSP) were introduced in [34] by J. Forest and D. Kesner as an extension of ERS [52] and *Simplified Expression Reduction Systems* (SERS) [15] to the case of patterns, and a generalisation of the Pattern-Matching Calculus to the case of higher-order rewriting (and not only functional rewriting). ERSP patterns are defined as combinations of standard algebraic structures with special *choice* constructors used to denote different possible syntactic forms for any abstracted argument.

This section gives an overview of ERSP and introduces the key notions to get a confluence result for such a formalism. For a formal development and full proofs we refer the reader to [34].

6.1 Basic Notions of the ERSP Formalism

We consider a set of *usual variables* denoted $x, y, z, \ldots$, a set of *choice variables* denoted $a, b, c, \ldots$, a set of *pattern metavariables* denoted $\mathbb{X}, \mathbb{Y}, \ldots$, a set of *term metavariables* denoted $M, N, \ldots$, a set of *function symbols* equipped with a fixed (possibly zero) arity, denoted $f, g, h, \ldots$, a set of *binder symbols* denoted $\lambda, \mu, \nu, \ldots$. We assume all these sets to be denumerable and disjoint.

When no special distinction is needed for the previous sets of variables and metavariables will use the symbols $\widehat{x}, \widehat{y}, \widehat{z}, \ldots$

Metapatterns (p) and metaterms (t) are generated by the grammars:

$p ::=$	x	usual variable	$t ::=$	x	usual variable
$\mid$	$\mathbb{X}$	pattern metavariable	$\mid$	M	term metavariable
$\mid$	$f(p, \ldots, p)$	algebraic	$\mid$	$f(t, \ldots, t)$	algebraic
$\mid$	$a\langle p, \ldots, p\rangle$	choice	$\mid$	$a\langle t, \ldots, t\rangle$	case
$\mid$	$@(p, \ldots, p)$	layered	$\mid$	$\mu p.t$	abstraction
$\mid$	$\_$	wildcard	$\mid$	$t\{p \text{ by } t\}$	meta pattern-matching

The constructor $@()$ is varyadic, i.e. it has no fixed arity. It will be used as a generalisation of the **as** constructor in functional programming [84]. As we will see later, the wildcard pattern $\_$ can be considered as a special case of @() applied to 0 arguments. The constructor $a\langle\ \rangle$ is also varyadic, but with a non-zero arity. We assume that whenever a choice variable a appears inside t, then all its occurrences have the same arity; thus, a term like $\mu a\langle x\rangle.a\langle x,y\rangle$ is not allowed since for every term t and every choice variable a, the arity of a inside t is unique. The symbol $\{\ \mathtt{by}\ \}$ is called the meta pattern-matching constructor. The metaterms $\mu p.t$ and $t\{p\ \mathtt{by}\ t'\}$ define bindings whose scope is t for all the (usual and choice) variables occurring in p. From now on, we write $\mathcal{FV}(t)$ (resp. $\mathcal{BV}(t)$) the set of free (resp. bound) variables of t. Without any loss of generality we assume these sets to be disjoint by working modulo α-conversion on preterms as for example in $\mu a\langle x,y,z\rangle.a\langle x,x,v\rangle =_\alpha \mu b\langle x',y',z'\rangle.b\langle x',x',v\rangle$. Thus, renaming of bound variables is used when necessary to avoid clashes and capture of free variables.

As an example the term $\mu a\langle f(0,y), f(s(x),y)\rangle.a\langle y, s(addition\ f(x,y))\rangle$, where $f(\_,\_)$ denotes a pair constructor, could be used to denote an addition function.

A metapattern (resp. metaterm) is said to be a **pattern** (resp. **preterm**) if it contains no metavariables. A preterm is said to be a **term** if it contains no pattern-matching constructors.

We denote by $\mathcal{V}ar(p)$ the set of all the variables appearing in a metapattern p. We denote by $\mathcal{MV}(t)$ the set of all the *pattern* and *term* metavariables appearing in t.

Definition 19. *A metapattern is called* ***linear*** *if each variable and metavariable appears at most once in it. We use the notation* $p \in p'$ *to say that the metapattern* p appears *inside the metapattern* p'. *A metaterm* t *is called* ***pattern-linear*** *iff every metapattern* p *in* t *is linear.*

A position is a (possibly empty) word over the alphabet $\mathbb{N}$. We use $\mathcal{POS}(t)$ to denote the *set of positions* of a metaterm t. The *submetaterm* of t at position p is written as $t|_p$. When $t|_p = u$, we will say that p is an *occurrence* of u in t.

The following notion is used to describe the set of variables/metavariables appearing along a given path playing a role of "bound" objects.

Definition 20 (Parameter Path). *Given a metaterm* s *and* $p \in \mathcal{POS}(s)$, *we define the* ***parameter path of*** s ***at position*** q, *written* $\mathcal{PP}(s,q)$, *as the following subset of variables and metavariables of* s:

$$
\begin{aligned}
&\mathcal{PP}(s,\epsilon) &&= \emptyset\\
&\mathcal{PP}(f(s_1,\ldots,s_n), i.q) &&= \mathcal{PP}(s_i,q),\ \textit{for}\ i \in \{1\ldots n\}\\
&\mathcal{PP}(a\langle s_1,\ldots,s_n\rangle, i.q) &&= \mathcal{PP}(s_i,q),\ \textit{for}\ i \in \{1\ldots n\}\\
&\mathcal{PP}(\mu p.s, 1.q) &&= \mathcal{V}ar(p) \cup \mathcal{MV}(p) \cup \mathcal{PP}(s,q)\\
&\mathcal{PP}(u\{p\ \mathtt{by}\ v\}, 1.1.q) &&= \mathcal{V}ar(p) \cup \mathcal{MV}(p) \cup \mathcal{PP}(u,q)\\
&\mathcal{PP}(u\{p\ \mathtt{by}\ v\}, 2.q) &&= \mathcal{PP}(v,q)
\end{aligned}
$$

This notion can be extended to contexts by saying that the parameter path of a context $C[\,]$ *is the parameter path of* $C[\,]$ *at the position of the hole.* □

As an example, if $t = M\{g(\mathbb{X}, x) \texttt{ by } \mu a\langle \mathbb{Y}, s(\mathbb{Y})\rangle.N\}$, then we have $\mathcal{PP}(t, 2) = \emptyset$, $\mathcal{PP}(t, 1.1) = \{\mathbb{X}, x\}$, and $\mathcal{PP}(t, 2.1) = \{\mathbb{Y}, a\}$.

Definition 21 (Well-formed metaterm). *A metaterm t is* ***well-formed*** *iff t has no* free *occurrences of choice/usual variables. Thus, $\mu a\langle x, y\rangle.a\langle x, y\rangle$ is well-formed while $f(a\langle g, g\rangle)$ and $f(x)$ are not.*

Also, we assume that different "non parallel" metapatterns appearing on the same path cannot share (meta)variables. Thus for example, $\mu\mathbb{X}.\lambda\mathbb{X}.M$ or $\lambda x.\mu x.M$ are not well-formed but the metaterm $t = \mu a\langle \mathbb{Y}, s(\mathbb{Y})\rangle.N$ is well-formed since the two occurrences of $\mathbb{Y}$ in t are not nested. This is just a generalisation of what is called "Barendregt's convention on bound variables".

The following notion is used to talk about the free variables of a term which remain *after* a given choice on a choice variable.

Definition 22 (Localised Free Variables). *Let t be a preterm and a be a choice variable having fixed arity k inside t. Given $1 \leq i \leq k$, the set $\mathcal{FV}^i_a(t)$ of* ***localised free variables*** *of t can be defined as follows:*

$$
\begin{array}{ll}
\mathcal{FV}^i_a(x) & = \{x\} \\
\mathcal{FV}^i_a(f(t_1, \ldots, t_n)) & = \mathcal{FV}^i_a(t_1) \cup \ldots \mathcal{FV}^i_a(t_n) \\
\mathcal{FV}^i_a(a\langle t_1, \ldots, t_k\rangle) & = \mathcal{FV}^i_a(t_i) \\
\mathcal{FV}^i_a(b\langle t_1, \ldots, t_n\rangle) & = \mathcal{FV}^i_a(t_1) \cup \ldots \mathcal{FV}^i_a(t_n) \cup \{b\} \\
\mathcal{FV}^i_a(\mu p.u) & = \mathcal{FV}^i_a(u) \setminus \mathcal{V}ar(p) \\
\mathcal{FV}^i_a(t\{p \texttt{ by } u\}) & = (\mathcal{FV}^i_a(t) \setminus \mathcal{V}ar(p)) \cup \mathcal{FV}^i_a(u)
\end{array}
$$

Indeed, $\mathcal{FV}^i_a(b\langle x, y, z\rangle) = \{b, z, x, y\}$ for any i, and $\mathcal{FV}^1_a(a\langle x, y, z\rangle) = \{x\}$. Moreover, as we work modulo α-conversion we have $\mathcal{FV}^1_a(\mu a\langle x, y\rangle.a\langle f(x, z), u\rangle) = \mathcal{FV}^1_a(\mu b\langle x, y\rangle.b\langle f(x, z), u\rangle) = \{z, u\}$.

We remark that when the choice variable a is not free in t then we have $\mathcal{FV}^i_a(t) = \mathcal{FV}(t)$ for every i.

Definition 23 (Acceptable preterms). *Acceptability of preterms is defined by induction as follows:*

- *All variables are acceptable.*
- *If $t_1, \ldots, t_n$ are acceptable, then $f(t_1, \ldots, t_n)$ and $a\langle t_1, \ldots, t_n\rangle$ are acceptable for any choice variable a and any function symbol f.*
- *If t is acceptable and p is a pattern such that for all $a\langle p_1, \ldots, p_n\rangle \in p$, for all $i \in 1 \ldots n$, and for all $j \neq i$ we have $(\mathcal{FV}^j_a(t) \setminus \mathcal{V}ar(p_j)) \cap \mathcal{V}ar(p_i) = \emptyset$, then we have that $\mu p.t$ is an acceptable term.*
- *If $\mu p.t$ and u are acceptable, then $t\{p \texttt{ by } u\}$ is acceptable.*

Indeed, the terms $\mu a\langle x, x\rangle.a\langle x, x\rangle$ and $\mu a\langle x, y\rangle.a\langle x, y\rangle$ are acceptable while $\mu a\langle x, y\rangle.b\langle x, y\rangle$ is not since $\mathcal{FV}^1_a(b\langle x, y\rangle) \setminus \mathcal{V}ar(x) = \{y, b\}$ and $\{y, b\} \cap \mathcal{V}ar(y) = \{y\}$. The role of acceptability is to prevent the creation of new free variables during evaluation: let us consider the term $\mu a\langle x, y\rangle.a\langle y, x\rangle$ and let us suppose that we apply this term to some value, forcing in this way the choice of the first branch x in the pattern $a\langle x, y\rangle$, i.e., the choice variable a takes the value 1. We will then force the choice of the first branch y in the term $a\langle y, x\rangle$, thus the variable y will become a *new* free variable. The example will be more clear after definitions of substitution and reduction.

Definition 24 (Contexts). ***Contexts*** *are preterms with one (and only one) occurrence containing a distinguished constant called a "hole" (and denoted $\Box$) in a non bound position and with no occurrence of the pattern-matching constructor. Thus $\mu\mathbb{X}.\Box$ is a context but $\mu\Box.y$ is not.*

We remark that the notion of acceptability is not closed by contexts as for example the preterm $a\langle x, y\rangle$ is acceptable but $\lambda a\langle y, x\rangle.a\langle x, y\rangle$ is not.

Definition 25 (Metasubstitution and Substitution).

- *A* ***metasubstitution*** *is a denumerable set of pairs of the form $\mathbb{X} \rhd p$ and $M \rhd t$ where p is a pattern and t is a term. Metasubstitutions will be denoted by* `rho`, `delta`, `sigma`, `theta`, *etc.*
- *A* ***substitution*** *is a denumerable set of pairs of the form $x \rhd t$ and $a \rhd i$, where t is a term and i is a natural number. Substitutions will be denoted by ρ, δ, σ, θ, etc.*

When no special distinction is needed between metasubstitutions and substitutions will use the symbols $\widehat{\rho}, \widehat{\delta}, \widehat{\sigma}, \ldots$. We denote by id the empty metasubstitution/substitution.

The ***domain*** *of a metasubstitution (resp. substitution) $\widehat{\sigma}$ is given by $Dom(\widehat{\sigma}) = \{\widehat{x} \mid (\widehat{x} \rhd o) \in \sigma \text{ and } \widehat{x} \neq o\}$. When $\widehat{x} \in Dom(\widehat{\sigma})$ we write $\widehat{\sigma}\widehat{x}$ to denote the object o such that $\widehat{x} \rhd o \in \widehat{\sigma}$. The* ***codomain*** *of $\widehat{\sigma}$ is given by $Codom(\widehat{\sigma}) = \bigcup_{\widehat{x} \in Dom(\widehat{\sigma})} \mathcal{FV}(\widehat{\sigma}\widehat{x})$.*

The **union** of two metasubstitutions (resp. two substitutions) $\widehat{\theta}_1$ and $\widehat{\theta}_2$ is denoted by $\widehat{\theta}_1 \sqcup \widehat{\theta}_2$. This union is *only defined* if for every variable $\widehat{x} \in Dom(\widehat{\theta}_1) \cap Dom(\widehat{\theta}_2)$ we have $\widehat{\theta}_1\widehat{x} = \widehat{\theta}_2\widehat{x}$.

We are now ready to define the notion of *pattern-matching*. This operation is not defined in general as a function from patterns and terms to substitutions but from patterns and terms to *sets* of substitutions. We will see latter how to ensure the uniqueness of this result.

Definition 26 (Pattern-matching). *For each pair (p, t), where p is a pattern and t is a term, we associate a set of substitutions as follows:*

$$\begin{array}{lll}
id & \in \{\!\{ _ \text{ by } t \}\!\} & \\
\{x \rhd t\} & \in \{\!\{ x \text{ by } t \}\!\} & \\
\theta_1 \sqcup \ldots \sqcup \theta_n & \in \{\!\{ @(p_1, \ldots, p_n) \text{ by } t \}\!\} & \text{if } \theta_i \in \{\!\{ p_i \text{ by } t \}\!\} \\
\theta_1 \sqcup \ldots \sqcup \theta_n & \in \{\!\{ f(p_1 \ldots p_n) \text{ by } f(t_1 \ldots t_n) \}\!\} & \text{if } \theta_i \in \{\!\{ p_i \text{ by } t_i \}\!\} \\
\{a \rhd i\} \sqcup \theta_i & \in \{\!\{ a\langle p_1 \ldots p_n\rangle \text{ by } t \}\!\} & \text{if } \theta_i \in \{\!\{ p_i \text{ by } t \}\!\}
\end{array}$$

We remark that in the last three cases the result of $\{\!\{p \text{ by } t\}\!\}$ is defined only if $\sqcup$ is defined. When $\{\!\{p \text{ by } t\}\!\}$ is a singleton we will make an abuse of notation by writing $\{\!\{p \text{ by } t\}\!\}$ to denote the only element of this set.

As an example of the previous definition, the pattern-matching $\{\!\{a\langle 0, x\rangle \text{ by } 0\}\!\}$ has two solutions: $\{a \triangleright 1\}$ and $\{a \triangleright 2, x \triangleright 0\}$. This comes from the fact that the pattern $a\langle 0, x\rangle$ contains two "overlapping" subpatterns 0 and x.

Definition 27 (Acceptable/linear metasubstitution/substitution). *Let $\mathcal{S}$ be a set of term metavariables (resp. usual variables). A metasubstitution (resp. substitution) $\widehat{\theta}$ is said to be* ***acceptable*** *(resp.* ***linear****) w.r.t. $\mathcal{S}$ iff for every metavariable $\widehat{x} \in \mathcal{S}$ we have that $\widehat{\theta}(\widehat{x})$ is an acceptable term (resp. a linear term). A metasubstitution (resp. substitution) $\widehat{\theta}$ is said to be* ***acceptable*** *(resp.* ***linear****) if it is acceptable (resp. linear) w.r.t. $Dom(\widehat{\theta})$.*

Lemma 3. *Let t be an acceptable term and p be a pattern. Then any substitution $\sigma \in \{\!\{p \text{ by } t\}\!\}$ is acceptable.*

It is time to make the point w.r.t. capture of variables in higher-order rewriting.

In CRS [67, 69] for example, a metaterm like $\lambda x.M(x)$ allows the (eventual) capture of the variable x while $\lambda x.M$ does not. In this formalism the β-rule has to be written as $app(\lambda x.M(x), N) \longrightarrow M(N)$ which does not correspond to the traditional way to express the β-rule.

In ERS [52] there is a metasubstitution operator which allows the β-rule to be expressed in a more traditional way as $app(\lambda x.M, N) \longrightarrow (N/x)M$. The instantiation of the metavariable M may or may not capture the variable x. However, we cannot assume α-conversion on metaterms in this formalism: if we suppose $\lambda x.M =_\alpha \lambda y.M$, then the instantiation of M by x will give two non α-equivalent terms $\lambda x.x \neq_\alpha \lambda y.x$.

To allow α-conversion on the level of terms but not on that of metaterms, instantiation of metaterms must be done very carefully: metasubstitution will be *first-order* replacement allowing capture of variables and substitution will be *higher-order substitution* dealing with α-conversion on terms.

Definition 28 (Applying a substitution). *The* ***application*** *of a substitution θ to preterm t (or* ***instantiation*** *of t by θ) yields a set of terms, written $\theta(t)$, which is computed as a higher-order substitution (modulo α-conversion) as follows:*

t	$\in \theta(x)$	*if* $(\{x \triangleright t\}) \in \theta$
x	$\in \theta(x)$	*if* $x \notin Dom(\theta)$
$\mu p.t'$	$\in \theta(\mu p.t)$	*if* $t' \in \theta(t)$ *and no capture of variables occurs*
$f(t'_1, \ldots, t'_n)$	$\in \theta(f(t_1, \ldots, t_n))$	*if* $t'_i \in \theta(t_i)$
t'_i	$\in \theta(a\langle t_1, \ldots, t_n\rangle)$	*if* $\theta a = i$ *and* $t'_i \in \theta(t_i)$
$a\langle t'_1, \ldots, t'_n\rangle$	$\in \theta(a\langle t_1, \ldots, t_n\rangle)$	*if* $t'_i \in \theta(t_i)$ *and* $a \notin Dom(\theta)$
t'	$\in \theta(t\{p \text{ by } u\})$	*if* $u' \in \theta(u), \theta' \in \{\!\{p \text{ by } u'\}\!\}$, $t' \in (\theta' \sqcup \theta)(t)$ *and no capture of variables occurs*

Definition 29 (Applying a metasubstitution). *The* ***application*** *of a replacement* theta *to a metaterm t (or* ***instantiation*** *of t by* theta*) yields a set of terms, written* theta(t)*, given by:*

x	$\in \mathtt{theta}(x)$	
t	$\in \mathtt{theta}(M)$	*if* $(M \triangleright t) \in \mathtt{theta}$
$\mu\mathtt{theta}(p).t'$	$\in \mathtt{theta}(\mu p.t)$	*if* $t' \in \mathtt{theta}(t)$
$f(t'_1,\ldots,t'_n)$	$\in \mathtt{theta}(f(t_1,\ldots,t_n))$	*if* $t'_i \in \mathtt{theta}(t_i)$
$a\langle t'_1,\ldots,t'_n\rangle$	$\in \mathtt{theta}(a\langle t_1,\ldots,t_n\rangle)$	*if* $t'_i \in \mathtt{theta}(t_i)$
t'	$\in \mathtt{theta}(t\{p\ \mathtt{by}\ u\})$	*if* $u' \in \mathtt{theta}(u), \theta' \in \{\!\{\mathtt{theta}(p)\ \mathtt{by}\ u'\}\!\}, t' \in \theta'(\mathtt{theta}(t))$

We remark that if a metaterm t has no pattern-matching constructor, then theta(t) is at most a singleton.

Let us see how the application of a metasubstitution works on an example. Consider $\mathtt{theta} = \{\mathbb{X}/a\langle x, f(z,y)\rangle, M/a\langle g(x,x),z\rangle, N/f(x',x')\}$ and let us compute $\mathtt{theta}(M\{\mathbb{X}\ \mathtt{by}\ N\})$. We first compute the set $\mathtt{theta}(N) = \{f(x',x')\}$, then the set

$$\{\!\{\mathtt{theta}(\mathbb{X})\ \mathtt{by}\ \mathtt{theta}(N)\}\!\} = \{\!\{a\langle x, f(z,y)\rangle\ \mathtt{by}\ f(x',x')\}\!\} = \{\rho_1,\rho_2\}$$

where $\rho_1 = \{a \triangleright 1, x \triangleright f(x',x')\}$ and $\rho_2 = \{a \triangleright 2, z \triangleright x', y \triangleright x'\}$. Finally, we conclude with $\rho_1(\mathtt{theta}(M)) = \rho_1(a\langle g(x,x),z\rangle) = g(f(x',x'),f(x',x'))$ and $\rho_2(a\langle g(x,x),z\rangle) = x'$ so that $\mathtt{theta}(M\{\mathbb{X}\ \mathtt{by}\ N\}) = \{g(f(x',x'),f(x',x')),x'\}$.

In order to see how α-conversion takes part in the substitution procedure let us take another example given by $\mathtt{theta} = \{\mathbb{X}/x, \mathbb{Y}/y, M/y, N/x\}$. Let $t = (\mu\mathbb{X}.M)\{\mathbb{Y}\ \mathtt{by}\ N\}$ and $t' = \mu\mathbb{X}.M\{\mathbb{Y}\ \mathtt{by}\ N\}$. We have $\mathtt{theta}(t) = \{(\mu x.y)\{y \triangleright x\}\} = \{\mu z.x\}$ and $\mathtt{theta}(t') = \{\mu x.y\{y \triangleright x\}\} = \{\mu x.x\}$.

Lemma 4. *Let t be an acceptable preterm and let θ be an acceptable substitution w.r.t. $\mathcal{FV}(t)$. Then, the set $\theta(t)$ has only acceptable terms.*

6.2 Rewrite Rules and Reduction Relation

This section introduces the precise syntax used to specify rewrite rules in the ERSP formalism as well as the reduction relation associated to them.

Definition 30. *An* ***Expression Reduction System with Patterns (ERSP)*** *is a set of rewrite rules of the form $l \longrightarrow r$ (written also (l,r)) such that:*

- *l and r are well-formed metaterms,*
- *l contains no occurrence of the pattern-matching constructor and the* head symbol *of l is a function symbol or a binder symbol,*
- *$\mathcal{MV}(r) \subseteq \mathcal{MV}(l)$,*

Thus for example, the rule $app(\lambda\mathbb{X}.M, N) \longrightarrow M\{\mathbb{X} \texttt{ by } N\}$ given in the introduction, which generalises the classical β-rule to the case of patterns, belongs to our framework.

In order to be able to guarantee that no free variable is "generated" during reduction the following notion will be necessary.

Definition 31 (Path condition). *Let M be a term metavariable in a metaterm t. We consider all the occurrences $p_1, \ldots, p_n$ of M in t and their corresponding parameter paths $l_1, \ldots, l_n$. A metasubstitution* `theta` *is said to have the* ***path condition*** *property for M in t iff:*

$$\forall \widehat{x} \in \mathcal{FV}(\texttt{theta}(M)), (\forall 1 \leq i \leq n, \widehat{x} \in \texttt{theta}(l_i)) \vee (\forall 1 \leq i \leq n, \widehat{x} \notin \texttt{theta}(l_i))$$

where the notation $\texttt{theta}(l)$ *denotes the set* $\bigcup_{\widehat{x} \in l} \mathcal{FV}(\texttt{theta}(\widehat{x}))$.

This notion is extended to rewrite rules by saying that `theta` *has the* path condition *for M in (l, r) iff it has the path condition for M in $\mapsto (l, r)$, where $\mapsto$ is* a fresh binary *function symbol. This trick is used to consider a rule as a unique "tree".*

The classical example where the path condition is not satisfied for a rewrite rule is given by the η-rule of the λ-calculus (see for example [52, 15]). Another rule in the same spirit but using patterns is $\lambda f(\mathbb{X}).M \longrightarrow M$. The replacement $\texttt{theta} = \{\mathbb{X} \triangleright x, M \triangleright x\}$ does not satisfy the path condition for M in this rule.

We now define the set of "good" replacements to instantiate rewrite rules. For that we remark that given a rewrite rule $l \longrightarrow r$, the metaterm l does not contain the pattern-matching constructor, so that for any replacement `theta` the term $\texttt{theta}(l)$ is a singleton.

Definition 32 (Admissible replacement for metaterms/rules). *A replacement* `theta` *is* ***admissible for a metaterm*** *t iff*

- $\texttt{theta}(t)$ *contains only acceptable terms*
- `theta` *has the path condition for every term metavariable in t.*

A replacement `theta` *is* ***admissible for a rule*** *(l, r) iff* `theta` *is admissible for $\mapsto (l, r)$, where $\mapsto$ is* a fresh binary *function symbol.*

We remark that this definition implies that given a rule (l, r) both $\texttt{theta}(l)$ and $\texttt{theta}(r)$ are defined, so in particular all the pattern/term metavariables in l are also in $Dom(\texttt{theta})$.

Definition 33 (Admissible reduction relation). *Let $\mathcal{R}$ be a ERSP. We say that s rewrites to t, written $s \longrightarrow_{\mathcal{R}} t$ (or $s \stackrel{a}{\longrightarrow}_{\mathcal{R}} t$ when the distinction must be made), iff there exists a rule $(l, r) \in \mathcal{R}$, an admissible replacement* `theta` *for (l, r) and a context C such that $s = C[\texttt{theta}(l)]$ and $t \in C[\texttt{theta}(r)]$. This notion can also be defined by induction by the following sentences:*

$$\frac{(l, r) \in \mathcal{R}}{\texttt{theta}(l) \longrightarrow_{\mathcal{R}} t}\; t \in \texttt{theta}(r) \qquad\qquad \frac{t \longrightarrow_{\mathcal{R}} u}{C[t] \longrightarrow_{\mathcal{R}} C[u]}\; C \textit{ is a context}$$

Even if the relation $\longrightarrow_{\mathcal{R}}$ is defined on any kind of term, the reduction can only take place on acceptable subterms.

As expected, the relation reduction enjoys good preservation properties.

Lemma 5 (Preservation of free variables and acceptable terms). *Let us consider the reduction step $s \longrightarrow_{\mathcal{R}} t$. Then*

- $\forall a, \forall i, \mathcal{FV}^i_a(t) \subseteq \mathcal{FV}^i_a(s)$.
- $\mathcal{FV}(t) \subseteq \mathcal{FV}(s)$.
- *If s is an acceptable term, then t is also acceptable.*

6.3 A Subclass of Confluent ERSP

This section is devoted to the study of confluence for a certain class of ERSP which are called the *orthogonal l-constructor* ERSP, and a certain class of terms, which are called *l-constructor deterministic terms.* Intuitively, an orthogonal ERSP is *left-linear* and *not overlapping.* Sufficiency of orthogonality for confluence in first and higher-order rewrite systems is well-known [4]. An l-constructor ERSP is a system $\mathcal{R}$ where the set of function symbols is partitioned into two different subsets, namely, the set of *constructors*, which cannot be reduced, and the set of *defined symbols*, which cannot be matched. As an example, let us consider the following system which is not an l-constructor ERSP.

$$\mathcal{R}: \begin{cases} f \longrightarrow g \\ app(\mu f.h, M) \longrightarrow h\{f \texttt{ by } M\} \end{cases}$$

The term $app(\mu f.h, f)$ can be reduced to both $app(\mu f.h, g)$ and h which are not joinable since the substitution $\{\!\{f \texttt{ by } g\}\!\}$ is not defined. Thus, $\mathcal{R}$ turns out to be non confluent.

Unfortunately, *orthogonal l-constructor* ERSP do not immediately guarantee confluence as the rule $\beta_{PM} : app(\lambda \mathbb{X}.M, N) \longrightarrow M\{\mathbb{X} \texttt{ by } N\}$ shows: the term $t = app(\lambda a\langle x, y\rangle.a\langle 0, 1\rangle, 3)$ has two non-joinable reducts 0 and 1 by this unique rule. The reason is that t contains two "overlapping" patterns x and y inside the choice pattern $a\langle x, y\rangle$. The failure of the confluence property in this case is completely natural since the term t corresponds, informally, to a "non-orthogonal" first-order rewriting system. It is then clear that we have to get rid of this class of terms in order to get a confluence result, this will be done by introducing the notion of *l-constructor deterministic* terms.

We are now ready to give a formal definition of all these notions.

Definition 34 (L-constructor system). *A system $\mathcal{R}$ is* ***l-constructor*** *iff*

- *The set $\mathcal{F}$ of function symbols can be partitioned into two sets $\mathcal{F}_c$ and $\mathcal{F}_d$, called respectively* ***constructors*** *and* ***defined symbols****, such that:*
  - *Each defined symbol is the head of some left-hand side of $\mathcal{R}$.*
  - *All the function symbols in metapatterns of $\mathcal{R}$ are constructors and no constructor is the head of some left-hand side of $\mathcal{R}$.*
- *For every rule $(l, r) \in \mathcal{R}$, both l and r are pattern-linear metaterms.*

The system $\mathcal{R}_1 = \{\beta_{PM}\} \cup \{0 + N \longrightarrow N, s(M) + N \longrightarrow s(M + N)\}$ is l-constructor. The system $\mathcal{R}_2 = \{\mu f(\mathbb{X}).M \longrightarrow M, f(0) \longrightarrow 0\}$ is not l-constructor since the function symbol f appears as the head symbol of some rule and inside a metapattern of $\mathcal{R}$. The system $\mathcal{R}_3 = \{\mu f(\mathbb{X}, \mathbb{X}).0 \longrightarrow 0\}$ is not l-constructor since $\mu f(\mathbb{X}, \mathbb{X}).0$ is not pattern-linear.

Definition 35 (L-constructor metapattern and metaterms). *Given an l-constructor system $\mathcal{R}$, we say that a metapattern is* ***l-constructor*** *iff it is linear and all its function symbols are constructors of $\mathcal{R}$. A* ***l-constructor metaterm*** *contains only l-constructor metapatterns.*

As an example concerning our previous system $\mathcal{R}_1$, we can observe that the metapattern $s(\mathbb{X})$ is l-constructor but $\mathbb{X} + \mathbb{Y}$ is not since the symbol $+$ is not a constructor function symbol.

Even if Definition 34 depends on a given l-constructor system $\mathcal{R}$ we will make an abuse of notation by just writing l-constructor metapattern/metaterm instead of $\mathcal{R}$ l-constructor metapattern/metaterm.

One could be tempted to define l-constructor metasubstitutions to be those having only l-constructor patterns and terms in their image. However, an l-constructor metasubstitution `sigma` applied to an l-constructor metaterm t does not always gives an l-constructor term: indeed, $\texttt{sigma} = \{\mathbb{X} \triangleright x\}$ and $t = \mu f(x, \mathbb{X}).b$ are l-constructor but $\texttt{sigma}(t) = \mu f(x, x).b$ is not.

We thus define l-constructor metasubstitutions as follows:

Definition 36 (L-constructor metasubstitutions). *Let* `sigma` *be a metasubstitution. We say that* `sigma` *is* ***l-constructor*** *w.r.t. a metaterm t iff* $\texttt{sigma}(t)$ *is l-constructor. A metasubstitution* `sigma` *is said to be* ***l-constructor*** *w.r.t. a rule $l \longrightarrow r$ iff it is l-constructor w.r.t. the metaterm $\mapsto (l, r)$ where $\mapsto$ is a fresh function symbol.*

Thus, considering the system R_1 presented above, the metasubstitution $\texttt{sigma} = \{\mathbb{X} \triangleright 0 + 0, M \triangleright 0, N \triangleright 0\}$ is not l-constructor for the lhs of the β_{PM} rule but the metasubstitution $\texttt{theta} = \{\mathbb{X} \triangleright a\langle 0, s(x)\rangle, M \triangleright a\langle 0, x\rangle, N \triangleright 3 + 4\}$ is l-constructor for the β_{PM}-rule.

Definition 37 (L-constructor reduction relation). *If $\mathcal{R}$ is an l-constructor system, we say that s* ***(l-)constructor rewrites*** *to t (written $s \xrightarrow{c}_{\mathcal{R}} t$) iff there exists a rewrite rule $(l, r) \in \mathcal{R}$, an l-constructor and admissible metasubstitution* `theta` *for (l, r) and a context C such that $s = C[\texttt{theta}(l)]$ and $t \in C[\texttt{theta}(r)]$.*

As an example, given the previous system $\mathcal{R}_1$, we have $0 + 0 \xrightarrow{c}_{\mathcal{R}_1} 0$ but we do not have $t = app(\lambda(0 + 0).3, 0 + 0) \xrightarrow{c}_{\mathcal{R}_1} 3$ (even if we have $t \xrightarrow{a}_{\mathcal{R}_1} 3$) since the term t is not an l-constructor term.

Using Definition 37, it is easy to show by induction on the definition of $\xrightarrow{c}_{\mathcal{R}}$ the following property:

Lemma 6 (Preservation of l-constructor terms). *If $\mathcal{R}$ is l-constructor, s is l-constructor and $s \xrightarrow{c}_{\mathcal{R}} t$, then t is l-constructor.*

We now introduce **l-constructor deterministic terms** for which the class of orthogonal l-constructor ERSP will be confluent. Let us start by the following notion.

Definition 38 (Overlapping patterns). *Two patterns p and q are said to be* ***overlapping*** *iff there exists a term t s.t. both $\{\!\!\{p \texttt{ by } t\}\!\!\}$ and $\{\!\!\{q \texttt{ by } t\}\!\!\}$ are defined.*

The patterns $f(\_, x)$ and $f(y, g(0))$ are overlapping. Also $a\langle 0, s(x)\rangle$ and $b\langle s(0), s(s(\_))\rangle$ are overlapping.

Definition 39 (Deterministic patterns/preterms). *The set of* ***deterministic patterns*** *is defined to be the smallest subset of linear patterns containing wildcard and variables, closed by algebraic and layered patterns, and such that if $p_1, \ldots, p_n$ are deterministic and for all $i \neq j$ the patterns p_i and p_j are not overlapping, then $a\langle p_1, \ldots, p_n\rangle$ is deterministic. An acceptable preterm t is said to be a* ***deterministic preterm*** *iff for every pattern p appearing in t, p is deterministic.*

Thus for example, $b\langle s(0), s(s(\_))\rangle$ is deterministic but $b\langle s(0), s(\_)\rangle$ is not. We remark that if a term t is deterministic then any subterm of t is also deterministic.

The definition of deterministic pattern implies that whenever p is a deterministic pattern, then there exists *at most* one substitution θ belonging to $\{\!\!\{p \texttt{ by } t\}\!\!\}$.

When p is deterministic and $\{\!\!\{p \texttt{ by } t\}\!\!\}$ is defined, we will identify $\{\!\!\{p \texttt{ by } t\}\!\!\}$ with its single element.

Definition 40 (Deterministic metasubstitution for metaterms/rules). *A metasubstitution* `theta` *is said to be* ***deterministic*** *for a metaterm t iff*

- `theta` *is admissible for t,*
- `theta`(t) *is a deterministic term,*

Finally, `theta` *is* ***deterministic*** *for a rule (l, r) iff* `theta` *is deterministic for the metaterm $\mapsto (l, r)$, where $\mapsto$ is a fresh function symbol.*

Definition 41 (Deterministic reduction relation). *Given a system $\mathcal{R}$, we say that s* ***deterministically rewrites*** *to t (written $s \xrightarrow{d}_{\mathcal{R}} t$) iff there exists a rewrite rule $(l, r) \in \mathcal{R}$, a deterministic metasubstitution* `theta` *for (l, r) and a context C such that $s = C[\texttt{theta}(l)]$ and $t = C[\texttt{theta}(r)]$.*

From now on we use the notation $\xrightarrow{c,d}_{\mathcal{R}}$ to denote $\xrightarrow{c}_{\mathcal{R}} \cap \xrightarrow{d}_{\mathcal{R}}$.

As expected, orthogonal systems allow us to preserve deterministic terms.

Definition 42 (Left linear systems). *A rewrite rule $l \longrightarrow r$ is said to be* ***left linear*** *iff l contains at most one occurrence of any term metavariable. A system is* ***left linear*** *if all its rule are left linear.*

As an example, the rule $f(M, M) \longrightarrow 3$ is not left linear while $f(M) \longrightarrow g(M, M)$ and $\mu x.f(x, x) \longrightarrow 0$ are.

Definition 43 (Redexes and overlapping redexes). *Given an ERSP $\mathcal{R}$ and a relation $\leadsto \in \{\stackrel{a}{\longrightarrow}, \stackrel{c}{\longrightarrow}, \stackrel{d}{\longrightarrow}, \stackrel{c,d}{\longrightarrow}\}$, a term t is said to be a* ***redex*** *for $\leadsto$ if $t = \mathtt{theta}(l)$ for some rule $(l, r) \in \mathcal{R}$ and some* theta *satisfying the conditions for $\leadsto$.*

A rewrite system is said to be ***non-overlapping*** *for $\leadsto$ iff*

- *Whenever a redex $\mathtt{theta}(l_j)$ for $\leadsto$ contains (not necessarily properly) another redex $\mathtt{rho}(l_i)$ for $\leadsto$ $(i \neq j)$, then $\mathtt{rho}(l_i)$ must be contained in $\mathtt{theta}(M)$ for some term metavariable M of l_j.*
- *Likewise whenever a redex $\mathtt{theta}(l)$ for $\leadsto$ properly contains another redex $\mathtt{rho}(l)$ for $\leadsto$ of the same rule.*

From now on, we will make an abuse of notation by simply saying that a term is a redex when the considered reduction relation $\leadsto$ is clear from the context.

Definition 44 (Orthogonal systems). *A rewrite system $\mathcal{R}$ is said to be* ***orthogonal*** *(w.r.t.$\leadsto \in \{\stackrel{a}{\longrightarrow}, \stackrel{c}{\longrightarrow}, \stackrel{d}{\longrightarrow}, \stackrel{c,d}{\longrightarrow}\}$) iff $\mathcal{R}$ is left-linear and non-overlapping (w.r.t. $\leadsto$).*

As an example, the system $\{f(\mu x.x) \longrightarrow 0, \mu \mathbb{X}.y \longrightarrow 1\}$ is overlapping whatever $\leadsto$ should be: the redex $f(\mu y.y) = \mathtt{theta}(f(\mu x.x))$ contains the redex $\mu y.y = \mathtt{rho}(\mu \mathbb{X}.y)$. The system $\{f(\mu \mathbb{X}.M) \longrightarrow 0, \lambda \mathbb{Z}.N \longrightarrow g(2)\}$ is orthogonal whatever $\leadsto$ should be.

Lemma 7 (Preservation of deterministic terms). *Given a system $\mathcal{R}$, if s is deterministic and $s \stackrel{d}{\longrightarrow}_{\mathcal{R}} t$, then t is deterministic.*

Theorem 10 (Confluence). *Let $\mathcal{R}$ be an l-constructor ERSP which is orthogonal w.r.t. $\stackrel{c,d}{\longrightarrow}_{\mathcal{R}}$. Then the relation $\stackrel{c,d}{\longrightarrow}_{\mathcal{R}}$ is confluent.*

7 SERS as Particular ERSP

Simplified Expression Reduction Systems (SERS) were introduced by E. Bonelli, D. Kesner and A. Ríos in [15] as an appropriate simplification of ERS: binders are restricted to those binding only one (usual) variable and substitution is restricted to simple substitution (in contrast to simultaneous or parallel substitution).

We discuss here how SERS can be seen as particular ERSP. In the SERS formalism only normal variables are (meta)patterns so that SERS (meta)terms can be generated by the grammar:

$$
\begin{array}{lll}
t ::= & x & \text{usual variable} \\
| & M & \text{term metavariable} \\
| & f(t, \dots, t) & \text{algebraic} \\
| & \mu x.t & \text{abstraction} \\
| & t\{x \;\mathtt{by}\; t\} & \text{meta substitution}
\end{array}
$$

We define the *ordered parameter path* of a context as the list containing all the variables occurring in the parameter path from the hole $\Box$ to the root of the context. For example, the parameter path of the context $f(\lambda x.(z, \xi y.(h(y, \Box))))$ is the sequence yx.

Every SERS metaterm without free variables turns out to be linear, well-formed and acceptable. An SERS metasubstitution is a denumerable set of pairs of the form $M \triangleright t$, where M is a term metavariable and t is a term, while an SERS substitution is a denumerable set of pairs of the form $x \triangleright t$.

The application of a substitution θ to a preterm t (without term metavariables), written $\theta(t)$, can be expressed as follows:

$$
\begin{array}{lll}
\theta(x) & = \theta x & \text{if } (x \triangleright t) \in \theta \\
\theta(x) & = x & \text{if } x \notin Dom(\theta) \\
\theta(\mu x.t) & = \mu x.\theta(t) & \text{if no capture of variables occurs} \\
\theta(t\{x \text{ by } u\}) & = \{x \triangleright \theta(u)\}\theta(t) & \text{if no capture of variables occurs} \\
\theta(f(t_1, \ldots, t_n)) & = f(\theta(t_1), \ldots, \theta(t_n)) &
\end{array}
$$

The application of a metasubstitution $\mathtt{theta}$ to a preterm t, written $\mathtt{theta}(t)$, is defined by:

$$
\begin{array}{lll}
\mathtt{theta}(x) & = x & \\
\mathtt{theta}(M) & = t & \text{if } (M \triangleright t) \in \mathtt{theta} \\
\mathtt{theta}(\mu x.t) & = \mu x.\mathtt{theta}(t) & \\
\mathtt{theta}(t\{x \text{ by } u\}) & = \{x \triangleright \mathtt{theta}(u)\}\mathtt{theta}(t) & \\
\mathtt{theta}(f(t_1, \ldots, t_n)) & = f(\mathtt{theta}(t_1), \ldots, \mathtt{theta}(t_n)) &
\end{array}
$$

Example 2. The λx-calculus [11, 91] is defined by considering the signature containing the function symbols $\mathcal{F} = \{app, subs\}$ and binder symbols $\mathcal{B} = \{\lambda, \sigma\}$, together with the following *SERS*-rewrite rules:

$$
\begin{array}{lll}
app(\lambda x.M)N & \longrightarrow_{Beta} & subs(\sigma x.M, N) \\
subs(\sigma x.(app(M)N), L) & \longrightarrow_{App} & app(subs(\sigma x.M, L))subs(\sigma x.N, L) \\
subs(\sigma x.\lambda y.(M), L) & \longrightarrow_{Lam} & \lambda y.(subs(\sigma x.M, L)) \\
subs(\sigma x.x, L) & \longrightarrow_{Var} & L \\
subs(\sigma x.M, L) & \longrightarrow_{rGc} & M
\end{array}
$$

8 Simplified Expression Reduction Systems with Indices

This section gives an overview of the higher-order rewriting formalism $SERS_{dB}$, based on de Bruijn indices, which was introduced in [15] by E. Bonelli, D. Kesner and A. Ríos. A journal version with full proofs and details will appear as [13].

In order to distinguish a concept defined for the *SERS* formalism from its corresponding version (if it exists) in the $SERS_{dB}$ formalism we may prefix it using the qualifying term "de Bruijn", e.g.. "de Bruijn metaterms".

In what follows *label* means a finite sequence of symbols and *simple label* a label without repeated symbols. The notation $\mathtt{at}(l, i)$ is used to distinguish the

i-th element of l when it exists and $\mathtt{pos}(x, l)$ the position of the first occurrence of the element x in the label l if it exists.

We consider a set of *binder indicators* denoted x, y, z, a set of *t-metavariables* (t for term), denoted $M_l, N_l, L_l, \ldots$, where l ranges over the set of labels built over binder indicators, a set of *function symbols* equipped with a fixed (possibly zero) arity, denoted $f, g, h, \ldots$, a set of *binder symbols* equipped with a fixed (non-zero) arity, denoted $\lambda, \mu, \nu, \xi, \ldots$. We assume all these sets to be denumerable and disjoint.

Definition 45 (de Bruijn metaterms). *The set of* ***de Bruijn metaterms****, denoted PMT_{dB}, is defined by the following two-sorted grammar:*

$$\begin{array}{ll} \textit{metaindices} & I ::= 1 \mid \mathtt{S}(I) \\ \textit{pre-metaterms} & A ::= I \mid M_l \mid f(A, \ldots, A) \mid \xi(A, \ldots, A) \mid A[A] \end{array}$$

The operator $\bullet[\bullet]$ in a metaterm $A[A]$ is called the ***de Bruijn metasubstitution operator****. The binder symbols together with the de Bruijn metasubstitution operator are called* ***binder operators****. Thus the de Bruijn metasubstitution operator is a binder operator (since it has binding power) but is* not *a binder symbol.*

The set of *metavariables* of A is written $\mathcal{MV}(A)$. The set of *names* of free metavariables of A is the set $\mathcal{MV}(A)$ where each M_l is replaced simply by M. We also write, by abuse of notation, $\mathcal{MV}(A)$ to denote such a set of names. For example, $\mathcal{MV}(f(\lambda M_x, N_\epsilon)) = \{M, N\}$.

The set of *de Bruijn terms* (metaterms without metavariables) is denoted by T_{dB}. We reserve the name (meta)context for (meta)terms with a hole $\Box$.

We use $A, B, A_i, \ldots$ to denote de Bruijn metaterms, $a, b, a_i, b_i, \ldots$ for de Bruijn terms and $E, F, \ldots$ for de Bruijn contexts. We use the convention that $\mathtt{S}^0(1) = 1$ and $\mathtt{S}^{j+1}(n) = \mathtt{S}(\mathtt{S}^j(n))$. As is usual for indices, we shall abbreviate $\mathtt{S}^{j-1}(1)$ as j or $\underline{j}$. In particular, the notation $\underline{j}$ emphasises the fact that j could be represented by an alternative implementation.

As in the *SERS* formalism, we also need here a notion of well-formed metaterm. The first motivation is to guarantee that labels of t-metavariables are correct w.r.t. the context in which they appear, the second one is to ensure that indices like j correspond to bound variables. Indeed, the metaterms $\xi(M_{xy})$, $\xi(\xi(4))$ shall not make sense for us, and hence shall not be considered well-formed.

Definition 46 (Well-formed de Bruijn metaterms). *A metaterm $A \in \mathsf{PMT}_{dB}$ is said to be* ***well-formed*** *iff the predicate $\mathcal{WF}(A)$ holds, where $\mathcal{WF}(A)$ iff $\mathcal{WF}_\epsilon(A)$, and $\mathcal{WF}_l(A)$ is defined for any label l as follows:*

- $\mathcal{WF}_l(\mathtt{S}^j(1))$ *iff* $j + 1 \leq |l|$
- $\mathcal{WF}_l(M_k)$ *iff* $l = k$ *and* l *is a simple label*
- $\mathcal{WF}_l(f(A_1, \ldots, A_n))$ *iff for all* $1 \leq i \leq n$ *we have* $\mathcal{WF}_l(A_i)$
- $\mathcal{WF}_l(\xi(A_1, \ldots, A_n))$ *iff there exists* $x \notin l$ *such that for all* $1 \leq i \leq n$ *we have* $\mathcal{WF}_{xl}(A_i)$
- $\mathcal{WF}_l(A_1[A_2])$ *iff* $\mathcal{WF}_l(A_2)$ *and there exists* $x \notin l$ *such that* $\mathcal{WF}_{xl}(A_1)$

Therefore if $\mathcal{WF}_k(A)$, then any metavariable occurring in A must be of the form M_{lk} for some label l (moreover, lk is a simple label).

Example 3. Metaterms $\xi(M_x, \lambda(N_{yx}, 2))$ and $g(\lambda(\xi c))$ are well-formed, whereas the metaterms $\lambda(\xi(M_{xx}))$, $\lambda(f(M_x, N_y))$ are not.

We may refer to the *binder path number* of a context, which is the number of binders between the $\Box$ and the root.

Remark that de Bruijn terms are also de Bruijn metaterms, that is, $\mathsf{T}_{dB} \subset \mathsf{PMT}_{dB}$, although some de Bruijn terms may not be *well-formed* de Bruijn metaterms. Indeed, the term $\xi(\xi(4))$ is not a well-formed metaterm: if an arbitrary free variable is wished to be represented in a metaterm, then i-metavariables should be used.

Definition 47 (Free de Bruijn indices). *The set of* **free indices** *of a de Bruijn term a, written* FI(a), *is defined as follows:*

$$\begin{array}{ll} \mathrm{FI}(n) & =_{def} \{n\} \\ \mathrm{FI}(f(a_1, \dots, a_n)) & =_{def} \bigcup_{i=1}^{n} \mathrm{FI}(a_i) \\ \mathrm{FI}(\xi(a_1, \dots, a_n)) & =_{def} (\bigcup_{i=1}^{n} \mathrm{FI}(a_i)) \backslash\!\!\backslash 1 \end{array}$$

where for every set of indices S, the operation $S \backslash\!\!\backslash j$ is defined as $\{n - j \mid n \in S \text{ and } n > j\}$.

When encoding $SERS_{dB}$ systems as $SERS$ systems we shall need to speak of the free variable names associated with the free de Bruijn indices. For example, if $a = \xi(1, 2, 3)$, then $\mathrm{FI}(a) = \{1, 2\}$. The named variable we will associate with the free index 1 (resp. 2) is x_1 (resp. x_2).

Definition 48 (de Bruijn substitution and de Bruijn updating function). *The result of substituting a term b for the index $n \geq 1$ in a term a is denoted $a\{\!\{n \leftarrow b\}\!\}$ and defined as:*

$$\begin{array}{ll} f(a_1, \dots, a_n)\{\!\{n \leftarrow b\}\!\} & =_{def} f(a_1\{\!\{n \leftarrow b\}\!\}, \dots, a_n\{\!\{n \leftarrow b\}\!\}) \\ \xi(a_1, \dots, a_n)\{\!\{n \leftarrow b\}\!\} & =_{def} \xi(a_1\{\!\{n+1 \leftarrow b\}\!\}, \dots, a_n\{\!\{n+1 \leftarrow b\}\!\}) \\ m\{\!\{n \leftarrow b\}\!\} & =_{def} \begin{cases} m - 1 & \text{if } m > n \\ \mathcal{U}_0^n(b) & \text{if } m = n \\ m & \text{if } m < n \end{cases} \end{array}$$

where for $i \geq 0$ and $n \geq 1$ we define the updating functions $\mathcal{U}_i^n(\bullet)$ *as follows:*

$$\begin{array}{ll} \mathcal{U}_i^n(f(a_1, \dots, a_n)) & =_{def} f(\mathcal{U}_i^n(a_1), \dots, \mathcal{U}_i^n(a_n)) \\ \mathcal{U}_i^n(\xi(a_1, \dots, a_n)) & =_{def} \xi(\mathcal{U}_{i+1}^n(a_1), \dots, \mathcal{U}_{i+1}^n(a_n)) \\ \mathcal{U}_i^n(m) & =_{def} \begin{cases} m + n - 1 & \text{if } m > i \\ m & \text{if } m \leq i \end{cases} \end{array}$$

We now consider the rewrite rules of a $SERS_{dB}$. This includes defining valuations, their validity, and the term rewrite relation in $SERS_{dB}$. Rewrite rules are specified with de Bruijn metaterms, whereas the induced rewrite relation is on de Bruijn terms.

Definition 49 ($SERS_{dB}$). *A* ***de Bruijn rewrite rule*** *or* $SERS_{dB}$***-rewrite rule*** *is a pair of de Bruijn metaterms (L, R) (also written $L \longrightarrow R$) such that the metasubstitution operator does not occur in L, the head symbol of L is a function symbol or a binder symbol, and $\mathcal{MV}(R) \subseteq \mathcal{MV}(L)$. We shall use r to denote rewrite rules.*

Hence, we define a $SERS_{dB}$ to be a pair $(\Sigma, \mathcal{R})$ where Σ is a $SERS_{dB}$-signature and $\mathcal{R}$ is a set of $SERS_{dB}$-rewrite rules over Σ.

As in the case of *SERS*, we shall often omit Σ and write $\mathcal{R}$ instead of $(\Sigma, \mathcal{R})$, if no confusion arises.

Example 4. The λ_{dB}-calculus is defined by considering the signature containing the function symbols $\{app\}$ and binder symbols $\{\lambda\}$, together with the $SERS_{dB}$-rewrite rule: $app(\lambda M_x)N_\epsilon \longrightarrow_{\beta_{dB}} M_x[N_\epsilon]$. The $\lambda_{dB}\eta_{dB}$-calculus is obtained by adding the following $SERS_{dB}$-rewrite rule: $\lambda(app(M_x, 1)) \longrightarrow_{\eta_{dB}} M_\epsilon$.

Definition 50 (de Bruijn valuation). *A* ***de Bruijn valuation*** *κ is a (partial) function from t-metavariables to de Bruijn terms which defines a unique function $\overline{\kappa}$ from metaterms to terms as follows:*

$$\begin{array}{ll}
\overline{\kappa}(I) & =_{def} I \\
\overline{\kappa}(M_l) & =_{def} \kappa M \\
\overline{\kappa}(f(A_1, \ldots, A_n)) & =_{def} f(\overline{\kappa}(A_1), \ldots, \overline{\kappa}(A_n)) \\
\overline{\kappa}(\xi(A_1, \ldots, A_n)) & =_{def} \xi(\overline{\kappa}(A_1), \ldots, \overline{\kappa}(A_n)) \\
\overline{\kappa}(A_1[A_2]) & =_{def} \overline{\kappa}(A_1)\{\!\{1 \leftarrow \overline{\kappa}(A_2)\}\!\}
\end{array}$$

De Bruijn t-metavariables having the same name but different label cannot be instantiated arbitrarily as they have to reflect the renaming of variables which is indicated by their labels. Indeed, the goal pursued by the labels of metavariables is that of incorporating "context" information as a defining part of a metavariable. As a consequence, we must verify that the terms substituted for every occurrence of a fixed metavariable coincide "modulo" their corresponding context. Dealing with such notion of "coherence" of substitutions in a de Bruijn formalism is also present in other formalisms but in a more restricted form. Thus for example, as mentioned before, a pre-cooking function is used in [30] in order to avoid variable capture in the higher-order unification procedure. Our notion of "coherence" is implemented with *valid valuations*.

Definition 51 (Value function). *Let $a \in \mathcal{T}_{dB}$ and l be a label of binder indicators. We define the* ***value function*** *$Value(l, a)$ as $Value^0(l, a)$ where:*

$$\begin{array}{ll}
Value^i(l, n) & =_{def} \begin{cases} n & \text{if } n \leq i \\ \mathtt{at}(l, n-i) & \text{if } 0 < n - i \leq |l| \\ x_{n-i-|l|} & \text{if } n - i > |l| \end{cases} \\
Value^i(l, f(a_1, \ldots, a_n)) & =_{def} f(Value^i(l, a_1), \ldots\ Value^i(l, a_n)) \\
Value^i(l, \xi(a_1, \ldots, a_n)) & =_{def} \xi(Value^{i+1}(l, a_1), \ldots, Value^{i+1}(l, a_n))
\end{array}$$

The function $Value(l, a)$ interprets the de Bruijn term a in an l-context: bound indices are left untouched, free indices referring to the l-context are replaced by the corresponding binder indicator and the remaining free indices are replaced by their corresponding variable names. It might be observed that if repeated binder indicators are allowed in the label l of Definition 51, then this intuition would not seem to hold. Indeed, for our purposes the case of interest is when the label l is simple. Nevertheless, many auxiliary results may be proved without this requirement, thus we prefer not to restrict this definition prematurely (by requiring l to be simple). Finally, note also that $Value^i(l, n)$ may return three different kinds of results. This is just a technical resource to make easier later proofs. Indeed, we have for example $Value(xy, \xi(f(3,1))) = \xi(f(y,1)) = Value(yx, \xi(f(2,1)))$ and $Value(\epsilon, f(\xi 1, \lambda 2)) \neq Value(x, f(\xi 1, \lambda 2))$.

Definition 52 (Valid de Bruijn valuation). *A de Bruijn valuation κ is **valid for a rewrite rule** r if every metavariable in r is in $Dom(\kappa)$ and for every pair of t-metavariables X_l and $X_{l'}$ in r we have $Value(l, \kappa X_l) = Value(l', \kappa X_{l'})$.*

It is interesting to note that there is no concept analogous to safeness as used for named *SERS* due to the use of de Bruijn indices.

Example 5. In the above example we have that $\kappa = \{X_{yx}/2, X_{xy}/1\}$ is valid for the rule r_{dB} since $Value(yx, 2) = x = Value(xy, 1)$.

Another interesting example is the η-contraction rule $\lambda x.app(M, x) \longrightarrow M$ if $x \notin \mathcal{FV}(M)$ which can be expressed in the *SERS* formalism, without conditions, as the rule $\lambda x.app(M, x) \longrightarrow_{\eta} M$. In the $SERS_{dB}$ formalism it may be expressed as the rule $\lambda(app(M_x, 1)) \longrightarrow_{\eta_{dB}} M_{\epsilon}$.

Note that a valid valuation κ for η_{dB} could, for example, be a valuation $\kappa = \{M_x/m, M_{\epsilon}/n\}$ such that $Value(x, \kappa M_x) = Value(\epsilon, \kappa M_{\epsilon})$, that is, $m = 1$ is not possible, and n is necessarily $m - 1$.

To summarise, valid valuations guarantee that the unique value assigned to a t-metavariable M in the framework with names is translated accordingly in the de Bruijn framework w.r.t. the different parameter paths of all the occurrences of M in the rewrite rule. This is, in some sense, an updating of M w.r.t. the different parameter paths where it appears, and it gives us the right notion of coherence for valuations.

Definition 53 (Rewriting de Bruijn terms). *Let $\mathcal{R}$ be a set of de Bruijn rules and a, b de Bruijn terms. We say that a **$\mathcal{R}$-rewrites** or **$\mathcal{R}$-reduces to** b, written $a \longrightarrow_{\mathcal{R}} b$, iff there is a de Bruijn rule $(L, R) \in \mathcal{R}$ and a de Bruijn valuation κ valid for (L, R) such that $a = E[\kappa L]$ and $b = E[\kappa R]$, where E is a de Bruijn context.*

Thus, the term $\lambda(app(\lambda(app(1,3)), 1))$ rewrites by the η_{dB} rule to $\lambda(app(1,2))$, using the (valid) valuation $\kappa = \{M_x/\lambda(app(1,3), M_{\epsilon}/\lambda(app(1,2))\}$.

As expected, the rewrite relation on de Bruijn terms preserves free de Bruijn indices.

8.1 From Names to Indices

In this section we show how rewriting in the *SERS* formalism may be simulated in the $SERS_{dB}$ formalism. This requires two well-distinguished phases which we can refer to as the *definition phase* and the *rewrite-preservation phase*. The definition phase consists in defining appropriate translations from pre-metaterms, terms and valuations in the *SERS* setting into the corresponding notions in the $SERS_{dB}$ setting, work which is carried out in the first part of this section. The second part deals with the rewrite-preservation phase, that is, showing how *SERS* rewrite steps can be simulated via $SERS_{dB}$ rewrite steps.

Definition 54 (From metaterms to de Bruijn metaterms). *A metaterm M is translated as $T(M)$, where $T(M) = T_\epsilon(M)$ and $T_k(M)$ is defined by*

$$\begin{array}{ll} T_k(x) & =_{def} \begin{cases} \mathtt{pos}(x,k) & \text{if } x \in k \\ \mathcal{O}(x) + |k| & \text{if } x \notin k \end{cases} \\ T_k(M) & =_{def} M_k \\ T_k(f(M_1,\dots,M_n)) & =_{def} f(T_k(M_1),\dots,T_k(M_n)) \\ T_k(\xi x.(M_1,\dots,M_n)) & =_{def} \xi(T_{xk}(M_1),\dots,T_{xk}(M_n)) \\ T_k(M_1[x \leftarrow M_2]) & =_{def} T_{xk}(M_1)[T_k(M_2)] \end{array}$$

The translation of a (meta)context, denoted $T(C)$, is defined as above but adding the clause $T_k(\Box) =_{def} \Box$.

Note that if M is a well-formed metaterm, then $T(M)$ will be defined and will only have t-metavariables with simple labels. Moreover, if M is a well-formed metaterm then $T(M)$ is a well-formed de Bruijn metaterm [3].

This translation is of course compatible with α-conversion in the sense that $s =_\alpha t$ implies $T_k(s) = T_k(t)$ for any label of variables k.

Example 6. Let $M' = \xi x.(M, \lambda y.(Y, x))$ and $M'' = g(\lambda x.(\xi y.c))$. Then their respective translations are $A' = \xi(M_x, \lambda(Y_{yx}, \mathtt{S}(1)))$ and $A'' = g(\lambda(\xi c))$, which are metaterms as remarked in Example 3.

Now, given a rewrite rule (G, D) in the *SERS* formalism, its translation in $SERS_{dB}$ is given by $T(G, D) = (T(G), T(D))$. As a consequence, if (G, D) is an *SERS* rewrite rule, then $T(G, D)$ is an $SERS_{dB}$ rewrite rule.

Example 7 (λx continued). Following Example 2, the specification of λx in the $SERS_{dB}$ formalism is given below.

$$\begin{array}{lll} app(\lambda M_x, N_\epsilon) & \longrightarrow & subs(\sigma M_x, N_\epsilon) \\ subs(\sigma(app(M_x, N_x)), L_\epsilon) & \longrightarrow & app(subs(\sigma M_x, L_\epsilon))subs(\sigma N_x, L_\epsilon) \\ subs(\sigma(\lambda(M_{yx})), L_\epsilon) & \longrightarrow & \lambda(subs(\sigma(M_{xy}), L_y)) \\ subs(\sigma(1), L_\epsilon) & \longrightarrow & L_\epsilon \\ subs(\sigma(M_x), L_\epsilon) & \longrightarrow & M_\epsilon \end{array}$$

[3] This can be proved by showing a more general property, namely, for every pre-metaterm M, if $\mathcal{WF}_l(M)$, then $\mathcal{WF}_l(T_l(M))$.

The rule $subs(\sigma(\lambda M_{yx}), L_\epsilon) \longrightarrow \lambda(subs(\sigma M_{xy}, L_y))$ is interesting since it illustrates the use of binder commutation from M_{yx} to M_{xy} and shows that some index adjustment is necessary when going from L_ϵ to L_y.

Suppose some rewrite rule (L, R) is used to rewrite a term s. Then $s =_\alpha C[\theta(L)]$ for some context C and admissible valuation θ. When encoding this rewrite step in the $SERS_{dB}$ setting we have to encode not only terms and metaterms, but also the valuation θ. In particular, we need to know what the names of the variables of the binders above the $\Box$ of the context $C[\,]$ are. This is the rôle of the label k in the following definition.

Definition 55 (From valuations to de Bruijn valuations). *Let θ be a valuation and k be a label of variables. Then the translation of θ w.r.t. the label k (referred to as the context label) is defined as the de Bruijn valuation:*

$$T_k(\theta)(X_l) =_{def} T_{lk}(\theta(X)) \text{ where } X \in Dom(\theta)$$

We now arrive to the rewrite-preservation phase, that is, rewriting in the formalism with de Bruijn indices has the same semantics as the corresponding one with names. For that, we essentially need two compositionality properties: compositionality w.r.t. contexts which is given by $T_k(C[t]) = T_k(C)[T_{lk}(t)]$, and compositionality w.r.t. valuations which is given by $T_{lk}(\theta M) = T_k(\theta)(T_l(M))$. We refer the reader to [15] to full details and proofs about these properties.

Using compositionality we can finally conclude this section by stating the rewrite-preservation property:

Proposition 1 (Simulating *SERS*-rewriting via *SERS$_{dB}$*-rewriting). *Suppose $s \longrightarrow t$ in the SERS formalism using the rewrite rule (G, D). Then $T(s) \longrightarrow T(t)$ in the $SERS_{dB}$ formalism using the de Bruijn rewrite rule $T(G, D)$.*

8.2 From Indices to Names

In this section we show that *SERS* are operationally equivalent to $SERS_{dB}$. For that, we show how the notion of rewriting in the $SERS_{dB}$ formalism may be simulated in the *SERS*. As in Section 8.1 we shall develop the required results by distinguishing the *definition phase* and the *rewrite-preservation phase.*

Definition 56 (From de Bruijn (meta)terms to (meta)terms). *The translation of $a \in \mathcal{T}_{dB}$, denoted $U(a)$, is defined as $U^N_\epsilon(a)$ where, N is the set of names associated to the free indices of a and for every finite set of variables S, and every label of variables k, $U^S_k(a)$ is defined as follows:*

$$
\begin{array}{ll}
U^S_l(n) & =_{def} \begin{cases} \mathtt{at}(l,n) \text{ if } n \le |l| \\ x_{n-|l|} \quad \text{if } n > |l| \text{ and } x_{n-|l|} \in S \end{cases} \\
U^S_l(X_l) & =_{def} X \\
U^S_l(f(A_1,\dots,A_n)) & =_{def} f(U^S_l(A_1),\dots,U^S_l(A_n)) \\
U^S_l(\xi(A_1,\dots,A_n)) & =_{def} \xi x.(U^S_{xl}(A_1),\dots,U^S_{\alpha l}(A_n)), \\
 & \qquad \text{if } 1 \le i \le n\ \mathcal{WF}_{xl}(A_i) \text{ for some } x \notin (l \cup S) \\
U^S_l(A_1[A_2]) & =_{def} U^S_{xl}(A_1)[x \leftarrow U^S_l(A_2)], \\
 & \qquad \text{if } \mathcal{WF}_{xl}(A_1) \text{ for some } x \notin (l \cup S)
\end{array}
$$

The translation of a de Bruijn context E, denoted $U(E)$, is defined as above but adding the clause $U_k^S(\Box) =_{def} \Box$.

Note that $U(\bullet)$ is not a function in the sense that the choice of bound variables is non-deterministic.

Now, given a de Bruijn rewrite rule (L, R) in the $SERS_{dB}$ formalism, its translation in *SERS* is given by $U(L, R) = (U(L), U(R))$.

Note that if A is such that $\mathcal{WF}_l(A)$ holds then its translation $U_l^{\emptyset}(A)$ is also a named metaterm, that is, $\mathcal{WF}_l(U_l^{\emptyset}(A))$ also holds. Therefore, by definition, the translation of a de Bruijn rewrite rule is a rewrite rule in the *SERS* formalism.

Example 8. Consider the rule $app(\Delta M_x, N_\epsilon) \longrightarrow \Delta(M_{xy}[\lambda(app(2, app(1, N_{zy})))])$. The rule obtained by the translation introduced before is

$$app(\Delta x.M, N) \longrightarrow \Delta y.(M[x \leftarrow \lambda z.(app(y, app(z, N)))])$$

Definition 57 (From de Bruijn valuations to valuations). *Given a finite set of variables S and a label of variables k, we define the translation of κ as the valuation $U_{(S,k)}(\kappa)$, where $U_{(S,k)}(\kappa)(M) =_{def} U_{lk}^S(\kappa M_l)$ for any $M_l \in Dom(\kappa)$ such that $l \cap (S \cup k) = \emptyset$.*

Now, if κ is a valid de Bruijn valuation then this definition is *correct*, that is, the definition does not depend on the choice of the t-metavariable M_l in $Dom(\kappa)$.

We now arrive to the rewrite-preservation phase, that is, rewriting in the formalism with names has the same semantics as the corresponding one with de Bruijn indices. As before, we need some compositionality properties: compositionality w.r.t. de Bruijn contexts which is given by $U_k^S(E[a]) =_\alpha U_k^S(E)[U_{lk}^S(a)]$, and compositionality w.r.t. valuations which is given by the equation $U_{lk}^S(\kappa A) =_\alpha U_{(S,k)}(\kappa)U_l^S(A)$. We again refer the reader to [15] for full details and proofs of these properties.

Using compositionality we can now state that the *SERS* formalism preserves $SERS_{dB}$-rewriting.

Proposition 2 (Simulating $SERS_{dB}$-rewriting via *SERS*-rewriting). *Assume $a \longrightarrow b$ in the $SERS_{dB}$ formalism using rewrite rule (L, R). Then $U(a) \longrightarrow U(b)$ in the SERS formalism using rule $U(L, R)$.*

8.3 Preserving Properties

Sections 8.1 and 8.2 state that rewriting is preserved when going from names to indices and from indices to names. But the relationship between the *SERS* and $SERS_{dB}$ formalisms is even more deep because it gives rise to two results stating, respectively, that given a metaterm M then $U(T(M))$ is equal (modulo some appropriate equivalence notion) to M, and that given a de Bruijn metaterm A then $T(U(A))$ is identical to A. These results are used to show that properties such as confluence, local confluence, the diamond property and strong and weak normalisation are preserved when translating an *SERS* rewrite system into a $SERS_{dB}$ rewrite system and vice versa.

Theorem 11 (Preservation of Confluence).

- *If the SERS $\mathcal{R}$ has the confluence (resp. local confluence or diamond) property, then the $SERS_{dB}$ $T(\mathcal{R})$ has the confluence (resp. local confluence or diamond) property.*
- *If the $SERS_{dB}$ $\mathcal{R}$ has the confluence (resp. local confluence or diamond) property, then the SERS $U(\mathcal{R})$ has the confluence (resp. local confluence or diamond) property.*

Theorem 12 (Preservation of Normalisation).

1. *If $\mathcal{R}$ is a weakly (resp. strongly) normalizing SERS, then $T(\mathcal{R})$ is a weakly (resp. strongly) normalizing $SERS_{dB}$.*
2. *If $\mathcal{R}$ is a weakly (resp. strongly) normalizing $SERS_{dB}$, then $U(\mathcal{R})$ is a weakly (resp. strongly) normalizing SERS.*

9 From Indices to First-Order Systems

Section 8 reviewed the $SERS_{dB}$ formalism [15] based on de Bruijn indices which does away with α-conversion and establishes a precise correspondence with the *SERS* formalism. However, substitution remains in both formalisms as a metalevel operation. This becomes a concrete problem in real implementations where substitutions must be denoted by symbols and constructors of the language, and the computational behaviour of substitutions must be specified by reduction rules belonging to the operational rules of the language itself. Thus, all $SERS_{dB}$ can be encoded as first-order rewriting systems with the aid of explicit substitutions and the goal of this section is to give an overview of the work done by E. Bonelli, D. Kesner and A. Ríos [16] in this direction. The reader interested in full proofs and details is referred to [14].

The case of the λ-calculus is interesting but at the same time not fully representative of the problems we are faced with when encoding a higher-order system into a first-order setting. For this particular case it is enough to replace the usual variable names by de Bruijn indices and to promote metalevel substitution to the object-level in order to obtain a first-order rewrite system. However, this is not always the case for an arbitrary higher-order rewrite system. The reason is that in higher-order rewriting the left-hand side of a rewrite rule is a higher-order pattern so that we must somehow *also* encode higher-order pattern matching when encoding the higher-order system in a first-order framework. To illustrate this purpose let us consider the η_{dB}-rewrite rule:

$$\lambda(app(X_x, 1)) \longrightarrow_{\eta_{dB}} X_\epsilon$$

One may verify that the term $\lambda(app(3,1))$ rewrites to 2. In a first-order setting with explicit substitution, we have the alternative formulation:

$$\lambda(app(X[\uparrow], 1)) \longrightarrow X$$

However, in order for the term $X[\uparrow]$ to match the subterm 3 syntactic matching is no longer sufficient as we need $\mathcal{E}$-matching, that is, we would need to solve the matching equation $X[\uparrow] \stackrel{?}{=}_{\mathcal{E}} 3$ in an appropriate substitution calculus $\mathcal{E}$. This may be seen as the reason why the η_{dB}-rule has received so much attention [90, 39, 19, 48].

Another less evident example is given by the commutation rule C_{dB}:

$$imply(\exists\forall X_{yx}, \forall\exists X_{xy}) \longrightarrow_{C_{dB}} true$$

The naïve translation to first-order, namely $imply(\exists\forall X, \forall\exists X) \longrightarrow true$, is evidently not correct: in order for a term to be an instance of this rule, a term a instantiated for the leftmost X must be the one instantiated for $X[2 \cdot 1 \cdot \uparrow^2]$, say a', except that all 1-level and 2-level indices in a shall be interchanged in order to obtain a'. The following rewrite rules C_{fo} and C'_{fo} do the job:

$$\begin{array}{rcl} imply(\exists\forall X, \forall\exists X[2 \cdot 1 \cdot \uparrow^2]) & \longrightarrow & true \\ imply(\exists\forall X[2 \cdot 1 \cdot \uparrow^2], \forall\exists X) & \longrightarrow & true \end{array}$$

Now, the rules C_{fo} and C'_{fo} have exactly the same intended meaning as the original higher-order rule C. Note that both rules induce the same rewrite relation on terms.

The goal of this part is to provide a conversion algorithm for encoding higher-order rewriting systems into first-order rewriting modulo an equational theory $\mathcal{E}$. A distinctive feature of our original algorithm is that we do not attach to the encoding any particular substitution calculus. Instead, we work with an abstract formulation of substitution calculi, called *Basic Substitution Calculi* and originally used in [48, 49] to deal with confluence proofs of λ-calculi with explicit substitutions. This macro-based presentation of calculi of explicit substitutions gives us the freedom of choosing from a wide range of calculi of explicit substitution, such as σ [1], $\sigma_{\Uparrow}$ [40], υ [8], f [48], d [48], s [47], λ_ϕ [79]. For brevity we illustrate the conversion algorithm via a concrete substitution calculus, namely $\sigma_{\Uparrow}$.

The conversion procedure that we introduce in this section consists then in transforming a $SERS_{dB}$ $\mathcal{R}$ into a first-order rewrite system $fo(\mathcal{R})$. The rewrite rules produced by the conversion may or may not have occurrences of the explicit substitution operator on the *LHS*s. In the case that they do, as in the η_{dB} example, we need matching modulo the induced equational theory of the substitution calculus $\sigma_{\Uparrow}$. Otherwise, syntactic matching suffices and thus the $SERS_{dB}$ $\mathcal{R}$, called in this case an *essentially first-order* higher-order rewrite system, can be translated to a *full* first-order rewrite system, where equational reasoning is not needed at all. This is for example the case of the λ-calculus which can be translated to the (full) first order $\lambda_{\sigma_{\Uparrow}}$ calculus.

9.1 The $\sigma_{\Uparrow}$ Calculus

The $\lambda\sigma$ calculus [1] was introduced as a bridge between the classical λ-calculus and concrete implementations of functional programming. It is inspired by de

Bruijn notation [27, 28] and Categorical Combinatory Logic (CCL) [25], it is very useful for deriving machines for the λ-calculus [41] or implementing higher-order unification [30]. The $\lambda\sigma$-calculus is confluent on closed terms, and remains confluent when meta-variables for terms are added to the syntax (*i.e.*, $\lambda\sigma$ is confluent on semi-open terms), but is no longer confluent when variables for substitutions are also considered (confluence fails for open terms). To overcome this problem, Hardin and Lévy introduced in [40] the $\lambda\sigma_{\Uparrow}$-calculus which considers a new operator, written $\Uparrow$, that allows to recover confluence on the set of open terms [26].

The grammar of $\lambda\sigma_{\Uparrow}$ is given by :

$$\begin{array}{ll} (Terms) & a ::= 1 \mid app(a,b) \mid \lambda a \mid a[s] \\ (Substitutions) & s ::= \uparrow \mid id \mid \Uparrow s \mid s \circ s \mid a \cdot s \end{array}$$

We use the notation $\uparrow^k$ for $k \geq 1$ to denote the substitution defined as follows: $\uparrow^1 =_{def} \uparrow$ and $\uparrow^{n+1} =_{def} \uparrow \circ \uparrow^n$. Note that indices in $\sigma_{\Uparrow}$ are represented slightly differently than in Definition 45. Indeed, the notation $\underline{j}$ now represents the $\sigma_{\Uparrow}$-term $1[\uparrow^{j-1}]$.

The set of rewriting rules of $\lambda\sigma_{\Uparrow}$ contains the *Beta* rule, which is used to start computation :

$$(Beta) \quad app(\lambda a, b) \longrightarrow a[b \cdot id]$$

and the $\sigma_{\Uparrow}$ rules, given in Fig 1, which are used to propagate and apply substitutions :

We denote by $\sigma_{\Uparrow}(a)$ the unique $\sigma_{\Uparrow}$-normal form of the term a (this normal form exists since $\sigma_{\Uparrow}$ is a terminating and confluent system [40]).

9.2 The Conversion Procedure

We now present the *Conversion Procedure*, an algorithm to translate any higher-order rewrite system in the formalism $SERS_{dB}$ to a first-order rewrite system (eventually modulo an equational theory). The Conversion Procedure is somewhat involved since several conditions, mainly related to the labels of metavariables, must be met in order for a valuation to be admitted as *valid* (Definition 52). The idea is to replace all occurrences of metavariables X_l by a first-order variable X followed by an appropriate *index-adjusting explicit substitution* which computes valid valuations.

In order to define the conversion procedure we need two key notions that are essential to correctly manipulate all the metavariables appearing in a de Bruijn rewriting rule: binding allowance and pivot. The notion of binding allowance gives the common binder indicators appearing in all the labels of the metavariables of a rule. If this binding allowance is empty, then the conversion is trivial, otherwise, we have to take into account the position in which these binder indicators occur to correctly define the conversion. This is done via the second notion called shifting index.

(App)	$app(a,b)[s]$	$\longrightarrow$	$app(a[s],b[s])$
$(Lambda)$	$(\lambda a)[s]$	$\longrightarrow$	$\lambda(a[\Uparrow s])$
$(Clos)$	$(a[s])[t]$	$\longrightarrow$	$a[s\circ t]$
$(VarShift1)$	$n[\uparrow]$	$\longrightarrow$	$n+1$
$(VarShift2)$	$n[\uparrow \circ s]$	$\longrightarrow$	$n+1[s]$
$(FVar)$	$1[a\cdot s]$	$\longrightarrow$	a
$(FVarLift1)$	$1[\Uparrow s]$	$\longrightarrow$	1
$(FVarLift2)$	$1[(\Uparrow s)\circ t]$	$\longrightarrow$	$1[t]$
$(RVar)$	$n+1[a\cdot s]$	$\longrightarrow$	$n[s]$
$(RVarLift1)$	$n+1[\Uparrow s]$	$\longrightarrow$	$n[s\circ \uparrow]$
$(RVarLift2)$	$n+1[(\Uparrow s)\circ t]$	$\longrightarrow$	$n[s\circ(\uparrow \circ t)]$
(Ass)	$(s_1\circ s_2)\circ s_3$	$\longrightarrow$	$s_1\circ(s_2\circ s_3)$
(Map)	$(a\cdot s)\circ t$	$\longrightarrow$	$a[t]\cdot(s\circ t)$
$(Shift)$	$\uparrow \circ(a\cdot s)$	$\longrightarrow$	s
$(ShiftLift1)$	$\uparrow \circ \Uparrow s$	$\longrightarrow$	$s\circ \uparrow$
$(ShiftLift2)$	$\uparrow \circ(\Uparrow s\circ t)$	$\longrightarrow$	$s\circ(\uparrow \circ t)$
$(Lift1)$	$\Uparrow s\circ \Uparrow t$	$\longrightarrow$	$\Uparrow s\circ t$
$(Lift2)$	$\Uparrow s\circ(\Uparrow t\circ u)$	$\longrightarrow$	$\Uparrow (s\circ t\circ u$
$(LiftEnv)$	$\Uparrow s\circ(a\cdot t)$	$\longrightarrow$	$a\cdot(s\circ t)$
(IdL)	$id\circ s$	$\longrightarrow$	s
(IdR)	$s\circ id$	$\longrightarrow$	s
$(LiftId)$	$\Uparrow id$	$\longrightarrow$	id
(Id)	$a[id]$	$\longrightarrow$	a

Fig. 1. The $\sigma_{\Uparrow}$ calculus

Definition 58 (Binding allowance). *The* ***binding allowance*** *of X in the metaterm A (resp. the rule (L,R)), denoted $\mathtt{Ba}_A(X)$ (resp. $\mathtt{Ba}_{(L,R)}(X)$), is the set of binder indicators appearing at the same time in* all *the metavariables with name X in A (resp. in L and R).*

As an example, if $A = f(\xi(X_x), g(\xi(\lambda(X_{yx})), \xi(\lambda(X_{xz}))))$, then $\mathtt{Ba}_A(X) = \{x\}$.

Definition 59 (Shifting index). *We define the* ***shifting index*** *determined by the metavariable X_l at position i in l, denoted $\mathtt{Sh}(X_l, i)$, as the total number of binder indicators in l at positions $1..i-1$ that do not belong to $\mathtt{Ba}_A(X)$.*

Thus for example, if $A = f(\xi(X_x), g(\xi(\lambda(X_{yx})), \xi(\lambda(X_{xz}))))$, then $\mathtt{Sh}(X_x, 1) = \mathtt{Sh}(X_{xz}, 2) = 0$, $\mathtt{Sh}(X_{yx}, 2) = 1$. Remark that $\mathtt{Sh}(X_l, 1)$ is always 0.

Consider the rewrite rule $\lambda(\lambda(X_{xy})) \longrightarrow \lambda(\lambda(X_{yx}))$ and a valid valuation κ for this rule. If κ maps the metavariable X_{xy} to a term a, then by the condition of validity it must be the case that it maps X_{yx} to the term b resulting from a where all 1-level and 2-level indices have been interchanged. For example, if $a = \underline{1}$ then $b = \underline{2}$ and if $a = \lambda(\underline{2})$ then $b = \lambda(\underline{3})$. Therefore, the conversion of the aforementioned rule would be

$$\lambda(\lambda(X)) \longrightarrow \lambda(\lambda(X[\underline{2}\cdot\underline{1}\cdot\uparrow^2])) \quad (1)$$

In this discussion our focus was set on the metavariable X_{xy} in the sense that κ was assumed valid if the term mapped to X_{yx} was a suitable transformation of the one mapped by κ to X_{xy}. However, we may also state that κ is valid if the term it maps to X_{xy} is a suitable transformation of the one mapped by κ to X_{yx}. In this case, the conversion of the rewrite rule would be

$$\lambda(\lambda(X[\underline{2} \cdot \underline{1} \cdot \uparrow^2])) \longrightarrow \lambda(\lambda(X)) \tag{2}$$

As a consequence, for each metavariable name in a rewrite rule, the metavariable that is set into focus determines the form that the conversion of this rule shall take (see also Example 10). The metavariable that is set into focus is called the *pivot* metavariable.

Definition 60 (Pivot). *Let $\{X_{l_1}, \ldots, X_{l_n}\}$ the set of all X-based metavariables in the SERS$_{dB}$-rewrite rule (L, R). The t-metavariable X_{l_j} is called an* ***(X-based) pivot*** *for (L, R) if $|l_j| \leq |l_i|$ for all $i \in 1..n$, and*

1. *$X_{l_j} \in L$, or*
2. *$X_{l_j} \in R$ and $|l_j| < |l_i|$ for all $X_{l_i} \in \mathcal{FMV}(L)$.*

A pivot set *for a rewrite rule (L, R) is a set of pivot metavariables, one for each name X in L such that $\mathtt{Ba}_{(L,R)}(X) \neq \emptyset$. This notion extends to a set of rewrite rules as expected.*

A pivot set for (L, R) fixes a metavariable for each metavariable name having a non-empty binding allowance. Note that Definition 60 admits the existence of more than one X-based pivot metavariable. One can prove (Proposition 3), however, that the induced rewrite relation is unique, thus it is not biased by any particular choice of pivots. Nevertheless, the fact remains that the converted rewrite rule in each case differs substantially. For example, the rule (1) is a first-order rule in which syntactic matching suffices in order to apply it. However, the rule (2) requires matching modulo the equational theory of the substitution calculus. In order to favour the former over the latter in our definition of pivot we select a metavariable with shortest label on the *LHS* whenever possible. As a consequence, rule (2) is no longer obtainable since X_{yx} is not considered a valid X-based pivot according to Definition 60.

Example 9. Both metavariables X_{xy} and X_{yx} can be chosen as X-based pivot in the rewrite rule

$$\mathit{Implies}(\exists(\forall(X_{xy})), \forall(\exists(X_{yx}))) \longrightarrow \mathit{true}$$

In the rewrite rule $f(Y_\epsilon, g(\lambda(\xi(X_{xy})), \lambda(\xi(X_{yx})))) \longrightarrow \nu(X_x, Y_x)$ the metavariable X_x is the only possible X-based pivot. Also, since the binding allowance of Y in this rewrite rule is the empty set, no Y-based metavariable is declared as pivot.

As in Section 8, if l is a label of binder indicators then $\mathtt{at}(l, i)$ is used to denote the i-th element of l when it exists. Also, $\mathtt{pos}(x, l)$ is the position of the first occurrence of the element x in l when it exists.

Definition 61 (Index-Adjusting Substitutions). *Let (L,R) be a $SERS_{dB}$-rewrite rule and suppose $\mathtt{Ba}_{L,R}(X) \neq \emptyset$. Let X_l be the X-based pivot for (L,R) and let X_k be any t-metavariable X_k appearing in the rule (L,R). The* ***index-adjusting substitution*** *for X_k w.r.t. the pivot X_l is given by $b_1 \cdot \ldots \cdot b_{|l|} \cdot \uparrow^j$, where $j = |k| + |l \setminus \mathtt{Ba}_{(L,R)}(X)|$ and each b_i is defined as follows:*

1. *if X_k is the pivot (hence $l = k$), then*

$$b_i = \begin{cases} \underline{i} & \text{if } \mathtt{at}(l,i) \in \mathtt{Ba}_{(L,R)}(X) \\ \underline{|l| + 1 + \mathtt{Sh}(X_l, i)} & \text{if } \mathtt{at}(l,i) \notin \mathtt{Ba}_{(L,R)}(X) \end{cases}$$

2. *if X_k is not the pivot then*

$$b_i = \begin{cases} \underline{\mathtt{pos}(x_h, k)} & \text{if } i = \mathtt{pos}(x_h, l) \text{ for some } x_h \in \mathtt{Ba}_{(L,R)}(X) \\ \underline{|k| + 1 + \mathtt{Sh}(X_l, i)} & \text{otherwise} \end{cases}$$

Note that for an index-adjusting substitution $b_1 \cdot \ldots \cdot b_{|l|} \cdot \uparrow^j$ each b_i is a distinct de Bruijn index and less than or equal to j. Substitutions of this form, in the particular case where we fix the basic substitution calculus to σ, have been called pattern substitutions in [31], where unification of higher-order patterns via explicit substitutions is studied.

We can address the conversion of rewrite rules. Before proceeding we recall that the name of a metavariable X_l is X and that by abuse of notation we write $\mathcal{FMV}(A)$ to denote the set of all the names of the free metavariables of M.

Definition 62 (Conversion of rewrite rules). *Let r be a $SERS_{dB}$-rewrite rule and let P be a pivot set for r. The* conversion of the rewrite rule r via P *is defined as $(\mathcal{C}^r(L), \mathcal{C}^r(R))$ where $\mathcal{C}^r(A)$ is defined by induction on A, where $\mathcal{FMV}(A) \subseteq \mathcal{FMV}(L)$, as:*

$$\begin{array}{ll}
\mathcal{C}^r(\underline{n}) & =_{def} \underline{n} \\
\mathcal{C}^r(X_k) & =_{def} \begin{cases} X[\uparrow^{|k|}] & \text{if } \mathtt{Ba}_r(X) = \emptyset \text{ and } k \neq \epsilon \\ X[b_1 \cdot \ldots \cdot b_{|l|} \cdot \uparrow^j] & \text{if } \mathtt{Ba}_r(X) \neq \emptyset \text{ and} \\ & b_1 \cdot \ldots \cdot b_{|l|} \cdot \uparrow^j \neq 1 \cdot \ldots \cdot |l| \cdot \uparrow^{|l|} \\ X & \text{otherwise} \end{cases} \\
\mathcal{C}^r(f(A_1, \ldots, A_n)) & =_{def} f(\mathcal{C}^r(A_1), \ldots, \mathcal{C}^r(A_n)) \\
\mathcal{C}^r(\xi(A_1, \ldots, A_n)) & =_{def} \xi(\mathcal{C}^r(A_1), \ldots, \mathcal{C}^r(A_n)) \\
\mathcal{C}^r(A_1[A_2]) & =_{def} \mathcal{C}^r(A_1)[\mathcal{C}^r(A_2) \cdot id]
\end{array}$$

The term $X[b_1 \cdot \ldots \cdot b_{|l|} \cdot \uparrow^j]$ on the RHS of the second clause is the index-adjusting substitution for X_k w.r.t. the pivot $X_l \in P$ computed in Definition 61.

It should be noted how the de Bruijn metasubstitution operator $\bullet[\bullet]$ is converted to the term substitution operator $\bullet[\bullet]$.

Below we present some examples.

$SERS_{dB}$-rewrite rule	Converted rewrite rule
$\lambda(app(X_x, 1)) \longrightarrow X_\epsilon$	$\lambda(app(X[\uparrow], 1)) \longrightarrow X$
$\lambda(\lambda(X_{xy})) \longrightarrow \lambda(\lambda(X_{yx}))$	$\lambda(\lambda(X)) \longrightarrow \lambda(\lambda(X[2 \cdot 1 \cdot \uparrow^2]))$
$f(\lambda(\lambda(X_{xy})), \lambda(\lambda(X_{yx}))) \longrightarrow \lambda(X_z)$	$f(\lambda(\lambda(X[\uparrow^2])), \lambda(\lambda(X[\uparrow^2]))) \longrightarrow \lambda(X[\uparrow])$
$app(\lambda X_x, Y_\epsilon) \longrightarrow_{\beta_{dB}} X_x[Y_\epsilon]$	$app(\lambda X, Y) \longrightarrow X[Y \cdot id]$

Note that no pivot is selected for the first and third rows since the binding allowance of X in the respective rule is the empty set. The pivot selected in the second row is X_{xy} and in the last rule X_x (no Y-based pivot required).

In the general case the system resulting from the Conversion Procedure is coded as a first-order rewrite system where equational matching may be used. Moreover, it is possible in some cases to get a first-order system where the sets of equations needed to perform the equational matching part is empty, so that matching becomes just syntactic first-order matching, and the resulting first-order system is called a *full first-order system.*

Definition 63 (Conversion Procedure). *Let $\mathcal{R}$ be a $SERS_{dB}$. The* Conversion Procedure *consists in selecting a pivot set for each rewrite rule in $\mathcal{R}$ and converting all its rewrite rules as dictated by Definition 62. The resulting set of rewrite rules is written $fo(\mathcal{R})$ and called a* ***first order-version*** *of $\mathcal{R}$. If the LHS of each rule in $fo(\mathcal{R})$ contains no occurrences of the substitution operator $\bullet[\bullet]$, then $fo(\mathcal{R})$ is said to be a* ***full first-order system.***

Of course, we must also consider pivot selection and how it affects the conversion procedure. Assume given some rewrite rule r and different pivot sets P and Q for this rule. It is clear that conversion of the rewrite rule r via P and Q shall not be identical.

Example 10. Let us consider again the following binder-commutation rule discussed in the introduction of this section:

$$imply(\exists\forall X_{yx}, \forall\exists X_{xy}) \longrightarrow_C true$$

If we select X_{yx} as the X-based pivot we obtain the following version of C:

$$imply(\exists\forall X, \forall\exists X[\underline{2} \cdot \underline{1} \cdot \uparrow^2]) \longrightarrow_{C_{fo}} true$$

However, X_{xy} may also be selected as an X-based pivot metavariable. In this case, the resulting converted rewrite rule shall be different:

$$imply(\exists\forall X[\underline{2} \cdot \underline{1} \cdot \uparrow^2], \forall\exists X) \longrightarrow_{C'_{fo}} true$$

Nevertheless, the rewrite relation generated by both of these converted rewrite rules is identical.

Proposition 3 (Pivot Selection). *Let r be a $SERS_{dB}$-rewrite rule and let P and Q be different pivot sets for this rule. Then the rewrite relation generated by the two conversions of the rewrite rule r via P and Q are the same.*

Proposition 3 is important, for it makes clear that the Conversion Procedure is not biased by the selection of pivot sets (as regards the induced rewrite relation). Thus, with full precision, we may now speak of *the* first-order version of a $SERS_{dB}$ $\mathcal{R}$.

The main properties concerning the translation presented in this section is that the *Simulation Proposition* holds: Any higher-order rewrite step may be simulated or implemented by first-order rewriting. Also, rewrite steps in the first-order version of a higher-order system $\mathcal{R}$ can be projected in $\mathcal{R}$. Finally, we give in Section 9.3 a syntactical characterisation of higher-order rewriting systems that can be translated into first-order rewriting systems modulo an empty theory. We shall see that, for example, the λ-calculus is covered by this characterisation.

In order to introduce the simulation proposition let us recall this standard notion of reduction modulo.

Given a rewrite system S and an equational system E on a set $\mathbb{O}$, the relation S-**reduction modulo** E is defined by $a \longrightarrow_{S/E} b$ iff there exist $a', b' \in \mathbb{O}$ such that $a =_E a' \longrightarrow_S b' =_E b$. We recall also that $a \twoheadrightarrow_S b$ means that there exist a finite S-reduction sequence from a to b.

Proposition 4 (Simulation Proposition). *Let $\mathcal{R}$ be a $SERS_{dB}$ and let $fo(\mathcal{R})$ be its first-order version. Suppose $a \longrightarrow_{\mathcal{R}} b$ then*

1. *if $fo(\mathcal{R})$ is not full first-order, then $a \longrightarrow_{fo(\mathcal{R})/\sigma_{\Uparrow}} b$.*
2. *if $fo(\mathcal{R})$ is full first-order, then $a \longrightarrow_{fo(\mathcal{R})} \twoheadrightarrow_{\sigma_{\Uparrow}} b$.*

The first statement of the proposition says that any higher-order reduction step in the setting with indices R can be simulated by first order rewriting in $fo(\mathcal{R})$ but modulo the equational theory $\sigma_{\Uparrow}$ generated by the rules in Figure 1 (the rules must be taken as equations). The second statement gives an optimisation for the case where the translation of R gives a full first-order system $fo(\mathcal{R})$. Thus, any higher-order reduction step in the setting with indices R will be simulated by one step of first-order rewriting in $fo(\mathcal{R})$ followed by finitely many steps of first-order rewriting in $\sigma_{\Uparrow}$ (this time $\sigma_{\Uparrow}$ is taken as a rewrite system and not as an equational theory). This second statement corresponds to the more common higher-order languages such as λ-calculus.

We conclude this part by considering the relationship between first-order rewriting and higher-order rewriting in the setting with indices which is given by the following property, where $\sigma_{\Uparrow}(c)$ denotes the $\sigma_{\Uparrow}$-normal form of the term c.

Proposition 5 (Projection Proposition). *Let $\mathcal{R}$ be a $SERS_{dB}$ and let $fo(\mathcal{R})$ be its first-order version. If $a \longrightarrow_{fo(\mathcal{R})} b$, then $\sigma_{\Uparrow}(a) \twoheadrightarrow_{\mathcal{R}} \sigma_{\Uparrow}(b)$.*

9.3 Essentially First-Order $SERS_{dB}$

This last subsection provides a very simple syntactical criterion that can be used to decide if a given higher-order rewrite system can be translated into a full first-order rewrite system (modulo an empty equational theory). In particular, we can check that many higher-order calculi in the literature, such as the lambda calculus, verify this property.

Definition 64 (Essentially first-order $SERS_{dB}$). *A $SERS_{dB}$ $\mathcal{R}$ is called* ***essentially first-order*** *if $fo(\mathcal{R})$ is a full first-order rewrite system.*

Definition 65 (fo-condition). *A $SERS_{dB}$ $\mathcal{R}$ satisfies the* ***fo-condition*** *if every rewrite rule $(L, R) \in \mathcal{R}$ satisfies: for every name X in L such that $X_{l_1}, \ldots, X_{l_n}$ are all the X-based metavariables in L, then*

1. *all the labels $l_1 \ldots l_n$ are identical and equal to $\mathtt{Ba}_{(L,R)}(X)$, and*
2. *for all $X_k \in R$ the length of k is greater or equal to $|\mathtt{Ba}_{(L,R)}(X)|$.*

As an example, consider the rules

$$\begin{array}{rcl} app(\lambda X_x, Y_\epsilon) & \longrightarrow_{\beta_{dB}} & X_x[Y_\epsilon] \\ \lambda(app(X_x, \underline{1})) & \longrightarrow_{\eta_{db}} & X_\epsilon \end{array}$$

The β_{dB}-rule satisfies the fo-condition but the η_{db} rule does not: the label of X_x in $\lambda(app(X_x, \underline{1}))$ is different from $\mathtt{Ba}_{(\lambda(app(X_x,\underline{1})),X_\epsilon)}(X) = \emptyset$.

Proposition 6 puts forward the importance of the fo-condition. Its proof relies on a close inspection of the Conversion Procedure.

Proposition 6. *Let $\mathcal{R}$ be a $SERS_{dB}$. Then $\mathcal{R}$ satisfies the fo-condition iff $\mathcal{R}$ is essentially first-order.*

Note that many results on higher-order systems (e.g. perpetuality [63], standardisation [74]) require *left-linearity* and *fully-extendedness or locality.* The reader may find it interesting to observe that these conditions together seem to imply the fo-condition. A proof of this fact would require either developing the results of this work in the above mentioned HORS or via some suitable translation to the $SERS_{dB}$ formalism, and is left to future work.

Of course, all first-order rewriting systems are essentially first-order $SERS_{dB}$: Indeed all metavariables in first-order rewriting systems carry ϵ as label. Hence the latter systems need not be left-linear. Also, an orthogonal $SERS_{dB}$ need not be essentially first-order, the prime example of this fact being the rewrite system consisting of the sole rule η_{dB}.

10 Conclusions and Further Work

In this chapter we have given an overview of Expression Reduction Systems in its original form and also extended to express context-sensitive and pattern-directed rewriting. We have presented the major results proven in this framework.

Many future directions remain to be explored. First of all, a rich theory of strategies exists for Expression Reduction Systems, so it would be interesting to explore if the theory applies also for *ERSP*. Further, an appropriate notion of context-sensitive *ERSP* would give the possibility to enlarge further the class of programs based on pattern matching.

It would be also interesting to explore ERSP with a more expressive syntax for patterns as in [23] or [45]. Also, pioneer work on typed pattern calculi [50] was inspired by the Curry-Howard isomorphism, via a computational interpretation of Gentzen sequent calculus for intuitionistic minimal logic. As a consequence, each pattern constructor comes from the interpretation of some *left* logical rule of Gentzen calculus. It is however less evident how to associate a Curry-Howard style interpretation with the entire ERSP syntax.

The encoding of Expression Reduction Systems into First-Order Term Rewriting Systems opens up the possibility of transferring results (such as confluence, termination, completion, evaluation strategies, implementation techniques, etc.) from the first-order framework to the higher-order framework. A first step in this direction is studied in [75, 12] where the standardisation property is lifted from first-order to higher-order rewriting. Thus, the translation proposed in this paper for encoding higher-order rewriting could provide a new means for studying properties of higher-order rewriting through corresponding results in the first-order setting.

Acknowledgements

This overview is based on works of the authors, as well as on several joint works on higher-order rewriting with E. Bonelli, J. Forest, R. Kennaway, M. Ogawa, V. van Oostrom, and A. Ríos, [15, 16, 13, 14, 34, 35, 63].

References

1. Martín Abadi, Luca Cardelli, Pierre Louis Curien, and Jean-Jacques Lévy. Explicit substitutions. *Journal of Functional Programming*, 4(1):375–416, 1991.
2. The Alfa proof editor. `http://www.cs.chalmers.se/~hallgren/Alfa/`.
3. Zena Ariola, Matthias Felleisen, John Maraist, Martin Odersky, and Philip Wadler. A call-by-need lambda calculus. In *Proceedings of the 22nd Symposium on Principles of Programming Languages*, pages 233–246. ACM Press, 1995.
4. Franz Baader and Tobias Nipkow. *Term Rewriting and All That.* Cambridge University Press, 1998.
5. Henk Barendregt. *The Lambda Calculus: Its Syntax and Semantics*, volume 103 of *Studies in Logic and the Foundations of Mathematics*. North-Holland, 1984. Revised Edition.
6. Henk Barendregt, Jan A. Bergstra, Jan-Willem Klop, and Henri Volken. Some notes on lambda-reduction *in* "degrees, reductions, and representability in the lambda calculus". Technical Report 22, Department of mathematics, University of Utrecht, 1976.

7. Henk Barendregt, Richard Kennaway, Jan-Willem Klop, and M. Ronan Sleep. Needed reduction and spine strategies for the lambda calculus. *Information and Computation*, 75(3):191–231, 1987.
8. Zine-El-Abidine Benaissa, Daniel Briaud, Pierre Lescanne, and Jocelyne Rouyer-Degli. λv, a calculus of explicit substitutions which preserves strong normalisation. *Journal of Functional Programming*, 6(5):699–722, 1996.
9. Jan A. Bergstra and Jan-Willem Klop. Strong normalization and perpetual reductions in the lambda calculus. *Journal of Information Processing and Cybernetics*, 18(7/8):403–417, 1982.
10. Jan A. Bergstra and Jan-Willem Klop. Conditional rewrite rules: confluence and termination. *Journal of Computer and System Science*, 32(3):323–362, 1986.
11. Roel Bloo and Kristoffer Rose. Combinatory reduction systems with explicit substitution that preserve strong normalisation. In *7th International Conference on Rewriting Techniques and Applications*, volume 1103 of *Lecture Notes in Computer Science*, pages 169–183. Springer-Verlag, 1996.
12. Eduardo Bonelli. A normalization result for higher-order calculi with explicit substitutions. In *Foundations of Software Science and Computation Structures*, volume 2620 of *Lecture Notes in Computer Science*, pages 153–168. Springer-Verlag, 2003.
13. Eduardo Bonelli, Delia Kesner, and Alejandro Ríos. De Bruijn indices for metaterms. *Journal of Logic and Computation.* To appear.
14. Eduardo Bonelli, Delia Kesner, and Alejandro Ríos. Relating higher-order and first-order rewriting. *Journal of Logic and Computation.* To appear.
15. Eduardo Bonelli, Delia Kesner, and Alejandro Ríos. A de Bruijn notation for higher-order rewriting. In *11th International Conference on Rewriting Techniques and Applications*, volume 1833 of *Lecture Notes in Computer Science*, pages 62–79. Springer-Verlag, 2000.
16. Eduardo Bonelli, Delia Kesner, and Alejandro Ríos. From higher-order to first-order rewriting (extended abstract). In *12th International Conference on Rewriting Techniques and Applications*, volume 2051 of *Lecture Notes in Computer Science*, pages 47–62. Springer-Verlag, 2001.
17. Gérard Boudol. Computational semantics of term rewriting systems. In Maurice Nivat and John Reynolds, editors, *Algebraic methods in semantics*, pages 169–236. Cambridge University Press, 1985.
18. Nicolas Bourbaki. *Elements of Mathematics, Theory of Sets.* Addison-Wesley, 1968.
19. Daniel Briaud. An explicit *eta* rewrite rule. In *Proceedings of the 2nd International Conference of Typed Lambda Calculus and Applications*, volume 902 of *Lecture Notes in Computer Science*, pages 94–108. Springer-Verlag, 1995.
20. Rod Burstall, David MacQueen, and Donald Sanella. Hope: An experimental applicative language. In *Proceedings of the LISP Conference*, pages 136–143, Stanford University, Computer Science Department, 1980.
21. Serenella Cerrito and Delia Kesner. Pattern matching as cut elimination. In *14th Symposium on Logic in Computer Science*, pages 98–108. IEEE Computer Society Press, 1999.
22. Serenella Cerrito and Delia Kesner. Pattern matching as cut elimination. *Theoretical Computer Science*, 323:71–127, 2004.
23. Horatiu Cirstea and Claude Kirchner. ρ-calculus, the rewriting calculus. In *5th International Workshop on Constraints in Computational Logics*, 1998.
24. The Coq Proof Assistant. `http://coq.inria.fr/`.

25. Pierre-Louis Curien. *Categorical combinators, sequential algorithms and functional programming*. Progress in Theoretical Computer Science. Birkhäuser, 1986. first edition.
26. Pierre-Louis Curien, Thérèse Hardin, and Jean-Jacques Lévy. Confluence properties of weak and strong calculi of explicit substitutions. Technical Report 1617, INRIA-Rocquencourt, 1992.
27. Nicolaas G. de Bruijn. Lambda calculus notation with nameless dummies, a tool for automatic formula manipulation, with application to the Church-Rosser theorem. *Indag. Mathematicae*, 5(35):381–392, 1972.
28. Nicolaas G. de Bruijn. A namefree lambda calculus with facilities for internal definition of expressions and segments. Technical Report 78-WSK-03, Eindhoven University of Technology, 1978.
29. Nachum Dershowitz, Mitsu Okada, and G. Sivakumar. Canonical conditional rewrite systems. In *9th Conference on Automated Deduction*, volume 310 of *Lecture Notes in Computer Science*, pages 538–549. Springer-Verlag, 1988.
30. Gilles Dowek, Thérèse Hardin, and Claude Kirchner. Higher-order unification via explicit substitutions. In *10th Symposium on Logic in Computer Science*. IEEE Computer Society Press, 1995.
31. Gilles Dowek, Thérèse Hardin, and Claude Kirchner. Unification via explicit substitutions: The case of higher-order patterns. Technical Report RR3591, INRIA, 1998.
32. The ELAN system. `http://elan.loria.fr/`.
33. Julien Forest. A weak calculus with explicit operators for pattern matching and substitution. In *13th International Conference on Rewriting Techniques and Applications*, volume 2378 of *Lecture Notes in Computer Science*, pages 174–191. Springer-Verlag, 2002.
34. Julien Forest and Delia Kesner. Expression reduction systems with patterns. In *14th International Conference on Rewriting Techniques and Applications*, volume 2706 of *Lecture Notes in Computer Science*, pages 107–122. Springer-Verlag, 2003.
35. John Glauert, Richard Kennaway, and Zurab Khasidashvili. Stable results and relative normalization. *Journal of Logic and Computation*, 10(3):323–348, 2000. Special Issue: Type Theory and Term Rewriting.
36. John Glauert and Zurab Khasidashvili. An abstract Böhm-normalization. In *2nd International Workshop on Reduction Strategies in Rewriting and Programming*, volume 70(6) of *Electronic Notes in Theoretical Computer Science*. Elsevier Science, 2002.
37. Bernhard Gramlich. *Termination and confluence properties of structured rewrite systems*. PhD thesis, Universität Kaiserslautern, Germany, 1996.
38. Michael Hanus and Christian Prehofer. Higher-order narrowing with definitional trees. In *7th International Conference on Rewriting Techniques and Applications*, volume 1103 of *Lecture Notes in Computer Science*, pages 138–152. Springer-Verlag, 1996.
39. Thérèse Hardin. η-reduction for explicit substitutions. In *3rd International Conference on Proceedings of International Symposium Algebraic and Logic Programming*, volume 632 of *Lecture Notes in Computer Science*, pages 306–321. Springer-Verlag, 1992.
40. Thérèse Hardin and Jean-Jacques Lévy. A confluent calculus of substitutions. In *France-Japan Artificial Intelligence and Computer Science Symposium*, Izu (Japan), 1989.

41. Thérèse Hardin, Luc Maranget, and Bruno Pagano. Functional back-ends within the lambda-sigma calculus. In *Proceedings of the International Conference on Functional Programming*, pages 25–33. ACM Press, 1996.
42. Paul Hudak, Simon Peyton-Jones, and Philip Wadler. Report on the programming language Haskell, a non-strict, purely functional language (version 1.2). Sigplan Notices, 1992.
43. Gérard Huet and Jean-Jacques Lévy. Computations in orthogonal rewriting systems. In Jean-Louis Lassez and Gordon Plotkin, editors, *Computational Logic, Essays in Honor of Alan Robinson*, pages 394–443. MIT Press, 1991.
44. The Isabelle theorem prover. `http://isabelle.in.tum.de/`.
45. C. Barry Jay. The pattern calculus. *ACM Transactions on Programming Languages and Systems*, 26(6):911–937, 2004.
46. Jean-Pierre Jouannaud and Mitsuhiro Okada. A computation model for executable higher-order algebraic specification languages. In *6th Symposium on Logic in Computer Science*, pages 350–361. IEEE Computer Society Press, 1991.
47. Fairouz Kamareddine and Alejandro Ríos. A λ-calculus à la de Bruijn with explicit substitutions. In *Proceedings of the 7th International Symposium on Proceedings of the International Symposium on Programming Language Implementation and Logic Programming*, volume 982 of *Lecture Notes in Computer Science*, pages 45–62. Springer-Verlag, 1995.
48. Delia Kesner. Confluence properties of extensional and non-extensional λ-calculi with explicit substitutions. In *7th International Conference on Rewriting Techniques and Applications*, volume 1103 of *Lecture Notes in Computer Science*, pages 184–199. Springer-Verlag, 1996.
49. Delia Kesner. Confluence of extensional and non-extensional lambda-calculi with explicit substitutions. *Theoretical Computer Science*, 238(1-2):183–220, 2000.
50. Delia Kesner, Laurence Puel, and Val Tannen. A Typed Pattern Calculus. *Information and Computation*, 124(1):32–61, 1996.
51. Zurab Khasidashvili. β-reductions and β-developments of λ-terms with the least number of steps. In *International Conference on Computer Logic 88*, volume 417 of *Lecture Notes in Computer Science*, pages 105–111. Springer-Verlag, 1990.
52. Zurab Khasidashvili. Expression reduction systems. Technical Report 36 : 200-220, I. Vekua Institute of Applied Mathematics of Tbilisi State University, 1990.
53. Zurab Khasidashvili. The church-rosser theorem in orthogonal combinatory reduction systems. Technical Report 1825, INRIA-Rocquencourt, 1992.
54. Zurab Khasidashvili. The longest perpetual reductions in orthogonal expression reduction systems. In *Proceedings of the 3rd International Symposium on Logical Foundations of Computer Science*, volume 813 of *Lecture Notes in Computer Science*, pages 191–203. Springer-Verlag, 1994.
55. Zurab Khasidashvili. On higher order recursive program schemes. In *19th International Colloquium on Trees in Algebra and Programming*, volume 787 of *Lecture Notes in Computer Science*, pages 172–186. Springer-Verlag, 1994.
56. Zurab Khasidashvili. Perpetuality and strong normalization in orthogonal term rewriting systems. In *11th Annual Symposium on Theoretical Aspects of Computer Science*, volume 775 of *Lecture Notes in Computer Science*, pages 163–174. Springer-Verlag, 1994.
57. Zurab Khasidashvili. On the longest perpetual reductions in orthogonal expression reduction systems. *Theoretical Computer Science*, 266(1/2):737–772, 2001.
58. Zurab Khasidashvili. Optimal normalization in orthogonal term rewriting systems. In *14th International Conference on Rewriting Techniques and Applications*,

volume 2706 of *Lecture Notes in Computer Science*, pages 243–258. Springer-Verlag, 2003.
59. Zurab Khasidashvili and John Glauert. Relating conflict-free stable transition and event models via redex families. *Theoretical Computer Science*, 286(1):65–95, 2002.
60. Zurab Khasidashvili and John Glauert. An abstract concept of optimal implementation. In *3rd International Workshop on Reduction Strategies in Rewriting and Programming*, volume 84(4) of *Electronic Notes in Theoretical Computer Science*. Elsevier Science, 2003.
61. Zurab Khasidashvili and John Glauert. Stable computational semantics of conflict-free rewrite systems (partial orders with erasure). In *14th International Conference on Rewriting Techniques and Applications*, volume 2706 of *Lecture Notes in Computer Science*, pages 467–482. Springer-Verlag, 2003.
62. Zurab Khasidashvili and John Glauert. The geometry of conflict-free reduction spaces. *Theoretical Computer Science*, 2005. To appear.
63. Zurab Khasidashvili, Mizuhito Ogawa, and Vincent van Oostrom. Perpetuality and uniform normalization in orthogonal rewrite systems. *Information and Computation*, 164(1):118–151, 2001.
64. Zurab Khasidashvili and Adolfo Piperno. Perpetuality and uniform normalization. In *Join International Conference on Algebraic and Logic Programming and International Workshop on Higher-Order Algebra*, volume 1298 of *Lecture Notes in Computer Science*, pages 240–255. Springer-Verlag, 1997.
65. Zurab Khasidashvili and Vincent van Oostrom. Context-sensitive conditional expression reduction systems. In *Workshop on Graph Rewriting and Computation*, volume 2 of *Electronic Notes in Theoretical Computer Science*. Elsevier Science, 1995.
66. Zurab Khasidashvili and Vincent van Oostrom. Context-sensitive conditional rewrite systems. Technical Report SYS–C95–06, University of East Anglia, 1995.
67. Jan-Willem Klop. *Combinatory Reduction Systems*. PhD thesis, Mathematical Centre Tracts 127, CWI, Amsterdam, 1980.
68. Jan-Willem Klop. Term Rewriting Systems. In *Handbook of Logic in Computer Science*, volume 2, pages 1–116. Oxford University Press, 1992.
69. Jan-Willem Klop, Vincent van Oostrom, and Femke van Raamsdonk. Combinatory reduction systems: introduction and survey. *Theoretical Computer Science*, 121(1/2):279–308, 1993.
70. Jean-Jacques Lévy. *Réductions correctes et optimales dans le lambda-calcul*. PhD thesis, Université Paris VII, France, 1978.
71. Jean-Jacques Lévy. Optimal reductions in the lambda-calculus. In J. Roger Hindley and Jonathan P. Seldin, editors, *To H. B. Curry*, pages 159–192. Academic Press, 1980.
72. Luc Maranget. Optimal derivations in weak lambda-calculi and in orthogonal term rewriting systems. In *Proceedings of the 18th Symposium on Principles of Programming Languages*, pages 255–269. ACM Press, 1991.
73. The MAUDE System. `http://maude.cs.uiuc.edu/`.
74. Paul-André Melliès. *Description Abstraite des Systèmes de Réécriture*. PhD thesis, Université Paris VII, 1996.
75. Paul-André Melliès. Axiomatic rewriting theory II: the lambda-sigma calculus enjoys finite normalisation cones. *Journal of Logic and Computation*, 10(3):461–487, 2000.

76. Aart Middeldorp. Call by need computations to root-stable form. In *Proceedings of the 24th Symposium on Principles of Programming Languages*, pages 94–105. ACM Press, 1997.
77. Robin Milner. Functions as processes. *Mathematical Structures in Computer Science*, 2(2):119–141, 2005.
78. Robin Milner, Mads Tofte, and Robert Harper. *The definition of Standard ML*. MIT Press, 1990.
79. César Muñoz. A left-linear variant of $\lambda\sigma$. In *Join International Conference on Algebraic and Logic Programming and International Workshop on Higher-Order Algebra*, volume 1298 of *Lecture Notes in Computer Science*, pages 224–239. Springer-Verlag, 1997.
80. Robert Nederpelt. *Strong normalization for a typed lambda-calculus with lambda structured types*. PhD thesis, Technische Hogeschool Eindhoven, Netherlands, 1973.
81. Tobias Nipkow. Higher-order critical pairs. In *6th Symposium on Logic in Computer Science*, pages 342–349. IEEE Computer Society Press, 1991.
82. Tobias Nipkow. Orthogonal higher-order rewrite systems are confluent. In *Proceedings of the 1st International Conference of Typed Lambda Calculus and Applications*, volume 664 of *Lecture Notes in Computer Science*, pages 306–317. Springer-Verlag, 1993.
83. Eric Nöcker. *Efficient Functional Programming: Compilation and Programming Techniques*. PhD thesis, University of Nijmegen, Netherlands, 1994.
84. The Objective Caml language. `http://caml.inria.fr/`.
85. Michael J. O'Donnell. *Computing in Systems Described by Equations*, volume 58 of *Lecture Notes in Computer Science*. Springer-Verlag, 1977.
86. Shalva Pkhakadze. *Some problems of the notation theory*. I. Vekua Institute of Applied Mathematics of Tbilisi State University, 1977. (In Russian).
87. Shalva Pkhakadze. An n. bourbaki-type general theory and the properties of contracting symbols and corresponding contracted forms. *Georgian Mathematical Journal*, 6(2):179–190, 1999.
88. Davis Plaisted. Polynomial time termination and constraint satisfaction tests. In *14th International Conference on Rewriting Techniques and Applications*, volume 2706 of *Lecture Notes in Computer Science*, pages 405–420. Springer-Verlag, 2003.
89. The PVS system. `http://pvs.csl.sri.com/`.
90. Alejandro Ríos. *Contribution à l'étude des λ-calculs avec substitutions explicites*. Thèse de doctorat, Université de Paris VII, 1993.
91. Kristoffer Rose. Explicit cyclic substitutions. In *Proceedings of the 3rd International Workshop on Conditional Term Rewriting Systems*, volume 656 of *Lecture Notes in Computer Science*, pages 36–50. Springer-Verlag, 1992.
92. Morten Heine Sørensen. *Normalization in λ-calculus and type theory*. PhD thesis, University of Copenhagen, Denmark, 1997.
93. Morten Heine Sørensen. Properties of infinite reduction paths in untyped λ-calculus. In *1st Tbilisi Symposium on Logic, Language and Computation, Selected papers*, pages 353–367. SiLLI Publications, CSLI, Stanford, 1998.
94. Terese. *Term Rewriting Systems*, volume 55 of *Cambridge Tracts in Theoretical Computer Science*. Cambridge University Press, 2003.
95. Yoshihito Toyama. Confluent term rewriting systems with membership conditions. In *Proceedings of the 1st International Workshop on Conditional Term Rewriting Systems*, volume 308 of *Lecture Notes in Computer Science*, pages 128–141. Springer-Verlag, 1988.

96. David Turner. Miranda: A non-strict functional language with polymorphic types. In *Functional Programming Languages and Computer Architecture*, volume 201 of *Lecture Notes in Computer Science*, pages 1–16. Springer-Verlag, 1985.
97. Vincent van Oostrom. *Confluence for Abstract and Higher-order Rewriting*. PhD thesis, Vrije University, Amsterdam, Netherlands, 1994.
98. Vincent van Oostrom and Femke van Raamsdonk. Weak orthogonality implies confluence: the higher-order case. In *Proceedings of the 3rd International Symposium on Logical Foundations of Computer Science*, volume 813 of *Lecture Notes in Computer Science*, pages 379–392. Springer-Verlag, 1994.
99. Femke van Raamsdonk. Confluence and superdevelopments. In *5th International Conference on Rewriting Techniques and Applications*, volume 690 of *Lecture Notes in Computer Science*, pages 168–182. Springer-Verlag, 1993.
100. Femke van Raamsdonk, Paula Severi, Morten Heine Sørensen, and Hongwei Xi. Perpetual reductions in λ-calculus. *Information and Computation*, 149(2):173–229, 1999.
101. David Wolfram. *The Causal Theory of Types*, volume 21 of *Cambridge Tracts in Theoretical Computer Science*. Cambridge University Press, 1993.

Axiomatic Rewriting Theory I: A Diagrammatic Standardization Theorem

Paul-André Melliès

Equipe Preuves, Programmes et Systèmes,
CNRS, Université Paris 7 Denis Diderot

Abstract. By extending *nondeterministic* transition systems with *concurrency* and *copy mechanisms*, Axiomatic Rewriting Theory provides a uniform framework for a variety of rewriting systems, ranging from higher-order systems to Petri nets and process calculi. Despite its generality, the theory is surprisingly simple, based on a mild extension of transition systems with independence: an axiomatic rewriting system is defined as a 1-dimensional transition graph $\mathcal{G}$ equipped with 2-dimensional transitions describing the *redex permutations* of the system, and their orientation. In this article, we formulate a series of elementary axioms on axiomatic rewriting systems, and establish a diagrammatic standardization theorem.

Foreword by the Author

Many concepts of Rewriting Theory started in the λ-calculus — which is by far the most studied rewriting system in history. A remarkable illustration is the ***confluence theorem***. The theorem was formulated by A. Church and J.B. Rosser in the early years of the λ-calculus [7]. The theorem was then generalized and applied extensively to other rewriting systems. It became eventually an object of study in itself, in a line of research pioneered by H.-B. Curry and R. Feys in their book on Combinatory Logic (1958). This culminated in a series of beautiful papers by G. Huet, J. W. Klop, and J.-J. Lévy published at the end of the 1970s and beginning of the 1980s. Today, more than half a century after its appearance in the λ-calculus, the confluence property is universally accepted as the theoretical principle underlying *deterministic* computations.

The article is concerned with another key property of the λ-calculus: the ***standardization theorem***, which was discovered by A. Church and J.B. Rosser quite at the same time as the confluence property. We advocate in this article that, in the same way as confluence underlies deterministic computations, standardization guides *causal* computations. It is worth clarifying here what kind of causality we have in mind, since the concept has been used in so many different ways. First of all, by *computation*, we mean a rewriting path

$$M_1 \xrightarrow{u_1} M_2 \xrightarrow{u_2} M_3 \longrightarrow \cdots \longrightarrow M_{n-1} \xrightarrow{u_n} M_n$$

in which every term M_k describes a particular state of the system, and in which every redex u_k describes a particular transition on states, for $1 \leq k \leq n$. Then,

A. Middeldorp et al. (Eds.): Processes... (Klop Festschrift), LNCS 3838, pp. 554–638, 2005.

by *causal computation*, we mean a computation in which every transition u_k is enabled by a chain or cascade of previous transitions. We are particularly interested in situations where the chain of causality leading to u_k is not necessarily the whole rewriting path

$$M_1 \xrightarrow{u_1} M_2 \xrightarrow{u_2} M_3 \longrightarrow \cdots \longrightarrow M_{k-1} \xrightarrow{u_{k-1}} M_k. \tag{1}$$

At this point, we advise the reader to practice the following spiritual exercise: think of today as a particular sequence of transitions (1) starting from your bedroom (state M_1) and leading you to the current position in the day (state M_k). Then, call $v = u_k$ the transition consisting in reading this very article:

$$v = u_k \quad : \quad M_k \longrightarrow M_{k+1}.$$

You must admit that some transitions performed today among the $u_1, \ldots, u_{k-1}$ are not necessary to read this article. And that it seems particularly difficult to disentangle the necessary transitions from the unnecessary ones. This is the point of this article: we investigate how to perform this task in Rewriting Theory by *permuting* transitions — in the spirit of true concurrency and Mazurkiewicz traces. Suppose for instance that your last action $u = u_{k-1}$ today has been to drink coffee:

$$u = u_{k-1} \quad : \quad M_{k-1} \longrightarrow M_k.$$

Do you really need that coffee to read these lines? The simplest way to answer is to check whether the transition v may be permuted before the transition u. If this is the case, then coffee is not necessary. Of course, you may reply that you have already drunk your coffee ten minutes ago, and thus, that it is far too late *now* to permute the order of events! You are certainly right... but this is not what matters here: the very fact that permuting the transition v before the transition u is possible *in principle* is sufficient to establish that performing transition u is not necessary in order to perform transition v.

Suppose on the other hand that your last action u has been to fetch this article from the library. In that case, performing the transition u is absolutely necessary in order to perform the transition v. There is no way indeed (either in reality or in principle) to permute the order of the two transitions... and this is precisely the reason why you went to the library on the first hand!

Of course, separating the necessary transitions from the unnecessary ones may involve more than just one permutation. Suppose for instance that you have drunk coffee just before fetching the article from the library. In that case, it takes two permutations (permute your coffee time after your visit to the library, and then after your exploration of the article) in order to demonstrate that drinking coffee is not necessary.

Everyday life shows that chains of causality may be reconstructed by applying relevant series of permutations on transitions. Now, Rewriting Theory complicates matters by implementing a symbolic universe in which computations may be *erased* or *duplicated* at will. New situations arise, which often defy common sense! We illustrate this with a simple example, involving three transitions, a coffee machine M producing a cup of coffee C, and a duplicator replicating all that. The situation proceeds in three transitions:

1. M produces the cup of coffee C,
2. Duplicator replicates M in two exact copies M_1 and M_2, each one containing its own cup of coffee C_1 and C_2,
3. You fetch the cup of coffee C_1 from M_1, and drink it.

The situation is particularly intricate from a conceptual point of view. On one hand, producing the cup of coffee C (first transition) is necessary to fetch the cup of coffee C_1 (last transition) since the cup C_1 is just a copy of the cup C. On the other hand, the first two transitions produce the cup of coffee C_2 which is not necessary to fetch the cup of coffee C_1 in the last transition. The only way to clarify things here is to permute the duplication of the machine M (second transition) *before* the production of the cup of coffee C (first transition). From this results a series of four transitions:

1. Duplicator replicates M in two exact copies M_1 and M_2,
2. M_1 produces the cup of coffee C_1,
3. M_2 produces the cup of coffee C_2,
4. You fetch the cup of coffee C_1 from M_1, and drink it.

There is more work for everybody now (except for Duplicator possibly) since each machine M_1 and M_2 has to produce its own cup of coffee C_1 and C_2. On the other hand, starting by duplicating the machine M enables to disentangle the necessary part (producing the cup of coffee C_1) from the unnecessary part (producing the cup of coffee C_2). It then becomes possible to exhibit the chain of causality leading to the cup of coffee C_1, by permuting the two last steps in the previous sequence of transitions:

1. Duplicator replicates M in two exact copies M_1 and M_2,
2. M_1 produces the cup of coffee C_1,
3. You fetch the cup of coffee C_1 from M_1, and drink it.

This long discussion explains why standardization reorganizes computations by giving priority to duplicators and erasers: duplication and erasure are an inherent part of disentanglement. This aspect of causality is fundamental but subtle, and thus often misunderstood, even by specialists.

Technically speaking, the article is built on a seminal observation made by Jan Willem Klop in his PhD thesis, more than twenty-five years ago. The PhD thesis, published in 1980, contains two proofs of the standardization theorem for the leftmost-outermost λ-calculus. In the second proof, Jan Willem Klop reduces standardization to strong normalization and confluence of a 2-*dimensional* rewriting process on the β-rewriting paths, thus understood as 1-*dimensional* entities. The process consists in permuting the so-called *anti-standard* pairs of β-redexes u and v in the following way:

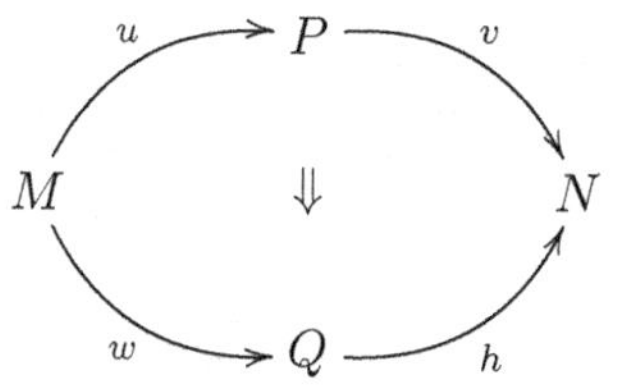

The 2-dimensional transition $f \Longrightarrow g$ transforms the β-rewriting path $f = u \cdot v$ into the β-rewriting path $g = w \cdot h$ where:

- the β-redex w is the ancestor of the β-redex v before β-reduction of the β-redex u,
- the β-rewriting path h develops the residuals of the β-redex u after β-reduction of the β-redex w.

By anti-standard pair, one means that the β-redex w lies outside or to the left of the β-redex u. Jan Willem Klop shows that the 2-dimensional procedure $\Longrightarrow$ strongly normalizes and converges on a unique normal form for every β-rewriting path. The resulting normal form is precisely the standard (that is, leftmost-outermost) β-rewriting path associated to the original β-rewriting path.

In this article, we generalize the construction to a wide class of rewriting systems, ranging from higher-order systems to Petri nets or process calculi. This provides evidence that *causality* is a general phenomenon in Rewriting Theory, and that its scope is not limited to deterministic computations. We proceed in a purely diagrammatic way: we start by formulating a series of *3-dimensional principles* which regulate the *2-dimensional permutations* acting on the *1-dimensional rewriting paths.* We then show that every Rewriting System satisfying these elementary principles (called axioms) satisfies our diagrammatic standardization theorem. The theorem states that applying 2-dimensional permutations to a rewriting path f leads eventually to a *unique* rewriting path g — modulo a fundamental notion of reversible permutation introduced in the course of the article. The standard rewriting path is finally *defined* as the unique normal form obtained at the end of the 2-dimensional procedure.

I have had several occasions to appreciate the extraordinary quality and insight of Jan Willem Klop's contribution to Rewriting Theory. It is thus a great pleasure and honour for me to dedicate today this article to Jan Willem Klop, on the occasion of his 60th birthday.

1 Standardization: From Syntax to Diagrams

1.1 Computing Leftmost Outermost is Judicious... in the λ-Calculus

The λ-calculus is the pure calculus of functions. It has a unique reduction rule, called the β-rule,

$$(\lambda x.M)P \longrightarrow M[x := P] \qquad (2)$$

which substitutes every free variable x in the λ-term M with the λ-term P. Despite its simplicity, the β-rule enables an extraordinary range of behaviours.

For instance, depending on the number of times the variable x occurs in M, the β-redex (2) duplicates its argument P, or erases it... Typically, the λ-term $\Delta = (\lambda x.xx)$ defines a duplicator, while the λ-term $K = (\lambda x.\lambda y.x)$ defines an eraser, with the following behaviours:

$$\Delta P \longrightarrow PP, \qquad\qquad KPQ \longrightarrow (\lambda y.P)Q \longrightarrow P.$$

Amusingly, the duplicator Δ applied to itself defines a λ-term $\Delta\Delta$ whose computation loops:

$$\Delta\Delta \longrightarrow \Delta\Delta \longrightarrow \cdots$$

The λ-term $Ka(\Delta\Delta)$ obtained by applying the eraser K to the variable a and to the loop $\Delta\Delta$ is particularly interesting, because its behaviour depends on the strategy chosen to compute it. When computed from left to right, the λ-term $Ka(\Delta\Delta)$ reduces in two steps to its result a:

$$Ka(\Delta\Delta) \longrightarrow (\lambda y.a)(\Delta\Delta) \longrightarrow a \tag{3}$$

On the other hand, when computed from right to left, the same λ-term $Ka(\Delta\Delta)$ loops for ever on the unnecessary computation of its subterm $\Delta\Delta$:

$$Ka(\Delta\Delta) \longrightarrow Ka(\Delta\Delta) \longrightarrow \cdots \tag{4}$$

To summarize: applying the "wrong" strategy on the λ-term $Ka(\Delta\Delta)$ computes it for ever, whereas applying the more judicious strategy (3) transforms it into its result a. This raises a very pragmatic question: does there exist a "judicious" strategy for every λ-term? This strategy would avoid useless computations, and reach the result of the λ-term, whenever this result exists. Remarkably, such a "judicious" strategy exists, and its recipe is surprisingly uniform: reduce at each step the *leftmost outermost* β-redex of the λ-term! Note that this is precisely the strategy applied successfully in (3) to compute the λ-term $Ka(\Delta\Delta)$.

We recall below the definition of the leftmost outermost strategy, formulated originally by A. Church and J. B. Rosser in the λI-calculus (the λ-calculus without erasers) then adapted to the λ-calculus by H.-B. Curry and R. Feys. A *β-redex* is a pattern $(\lambda x.P)Q$ occurring in the syntactical tree of a λ-term. The λ-terms $(\lambda x.P)$ and Q are called respectively the *function* and the *argument* of the β-redex $(\lambda x.P)Q$. A λ-term which does not contain any β-redex is called a *normal form*: it cannot be computed further. Now, consider a λ-term M containing a β-redex at least. Its *leftmost outermost* β-redex is defined by induction on the size of the λ-term M:

1. as $(\lambda x.P)Q$ when $M = \lambda x_1 ... \lambda x_k.((\lambda x.P)QR_1...R_m)$,
2. as the leftmost outermost β-redex of Q when

$$M = \lambda x_1 ... \lambda x_k.(xP_1...P_mQR_1...R_n)$$

   and every P_i is a normal form.

Theorem 1 (Curry-Feys). *Suppose that there exists a rewriting path from a λ-term M to a normal form P. The strategy consisting in rewriting at each step M_i the leftmost outermost β-redex in M_i constructs a rewriting path*

$$M = M_0 \longrightarrow M_1 \longrightarrow \cdots \longrightarrow M_{k-1} \longrightarrow M_k = P$$

from M to P.

Theorem 1 may be stated alternatively by defining $\dashrightarrow$ as the least relation between λ-terms satisfying the inductive steps of Figure 1, then by establishing that $M \twoheadrightarrow P$ is equivalent to $M \dashrightarrow P$, for every λ-term M and normal form P. We leave the reader check as exercise that the definition of $\dashrightarrow$ constructs the rewriting path (3) in the case of $M = Ka(\Delta\Delta)$.

1.2 Computing Leftmost Outermost is not Necessarily Judicious... in Other Rewriting Systems

This clarifies how a term should be computed in the λ-calculus: from left to right. It appears however that this orientation is very particular to the λ-calculus. Consider for instance the term rewriting system defined by the rules

$$\begin{array}{c} A \to A \\ B \to C \\ F(x, C) \to D \end{array} \tag{5}$$

Then, the rightmost outermost strategy (6) rewrites the term $F(A, B)$ to a result D:

$$F(A, B) \longrightarrow F(A, C) \longrightarrow D \tag{6}$$

whereas the leftmost outermost strategy loops for ever on the term $F(A, B)$:

$$F(A, B) \longrightarrow F(A, B) \longrightarrow \cdots \tag{7}$$

One must admit here that there exists no universal "syntactic orientation" in Rewriting Theory. This should not be a surprise: after all, the "syntactic orientation" of a rewriting system is extremely sensitive to its notation! Think only

(VAR)	$x \dashrightarrow x$
(BETA)	$\dfrac{M \dashrightarrow \lambda x.P \qquad P[x := N] \dashrightarrow Q}{MN \dashrightarrow Q}$
(APP)	$\dfrac{M \dashrightarrow xP_1...P_k \qquad N \dashrightarrow Q}{MN \dashrightarrow xP_1...P_kQ}$
(XI)	$\dfrac{M \dashrightarrow P}{\lambda x.M \dashrightarrow \lambda x.P}$

Fig. 1. An inductive definition of Curry and Feys' leftmost outermost strategy

of the λ-calculus written through the Looking Glass, in a reverse notation: now, the calculus is oriented right to left, instead of left to right... The general case is even worse. A rewriting system does not enjoy any uniform orientation in general, and finding the "judicious" strategy, even if we know that it exists, is a non decidable problem, see [18].

Despite the apparent mess, we will initiate in this article a *generic* theory of orientations and causality in rewriting systems. But on what foundations? Obviously, we need to abstract away from syntax in order to describe uniformly examples (3), (4), (6) and (7). We are thus compelled to reason *diagrammatically* instead of *syntactically*, and to develop a *syntax-free* Rewriting Theory, based on a 2-dimensional refinement of the traditional notion of *Abstract Rewriting System* developed in [32, 17, 21].

1.3 Forget Syntax, Think Diagrammatically!

The diagrammatic approach to Rewriting Theory which we have in mind is justified by a simple but surprising observation: despite their syntactic differences, the two terms $Ka(\Delta\Delta)$ and $F(A,B)$ define *exactly* the same transition system, which we draw below.

$$
\begin{array}{ccc}
Ka(\Delta\Delta) & \xrightarrow{\Delta_1} & Ka(\Delta\Delta) \\
{\scriptstyle K}\downarrow & & \downarrow{\scriptstyle K} \\
(\lambda y.a)(\Delta\Delta) & \xrightarrow{\Delta_2} & (\lambda y.a)(\Delta\Delta) \\
{\scriptstyle \lambda}\downarrow & & \downarrow{\scriptstyle \lambda} \\
a & \underset{id_a}{=\!=\!=} & a
\end{array}
\qquad\qquad
\begin{array}{ccc}
F(A,B) & \xrightarrow{A_1} & F(A,B) \\
{\scriptstyle B}\downarrow & & \downarrow{\scriptstyle B} \\
F(A,C) & \xrightarrow{A_2} & F(A,C) \\
{\scriptstyle F}\downarrow & & \downarrow{\scriptstyle F} \\
D & \underset{id_D}{=\!=\!=} & D
\end{array}
\qquad (8)
$$

Apparently, the *dynamical analogy* between the two terms $Ka(\Delta\Delta)$ and $F(A,B)$ goes beyond the equality of their transition systems. Observe that in the lefthand side and the righthand side of the diagram:

- the steps Δ_1 and A_1 are "unnecessary" because they may be "erased" by the paths $K \cdot \lambda$ and $B \cdot F$,
- the paths $K \cdot \lambda$ and $B \cdot F$ are more "judicious" than the paths $\Delta_1 \cdot K \cdot \lambda$ and $A_1 \cdot B \cdot F$ because they avoid computing the "unnecessary" redexes Δ_1 and A_1.

This analogy between the two terms $Ka(\Delta\Delta)$ and $F(A,B)$ is too subtle to be reflected by the transition systems of Diagram 8. However, it is possible to *refine* the notion of transition system, in order to capture the analogy. The refinement is based on the concept of *redex permutation* introduced by J.-J. Lévy in his work on the λ-calculus and on term rewriting systems, see [24, 18, 3]. Permuting redexes inside rewriting paths enables to express by local transformations that two different rewriting paths compute the same *events*, but in a different order.

Typically, the transition system of the terms $Ka(\Delta\Delta)$ and $F(A,B)$ may be equipped with two redex permutations [1] and [2] indicated below:

$$
\begin{array}{ccc}
Ka(\Delta\Delta) & \xrightarrow{\Delta_1} & Ka(\Delta\Delta) \\
{\scriptstyle K}\downarrow & [1] & \downarrow{\scriptstyle K} \\
(\lambda y.a)(\Delta\Delta) & \xrightarrow{\Delta_2} & (\lambda y.a)(\Delta\Delta) \\
{\scriptstyle \lambda}\downarrow & [2] & \downarrow{\scriptstyle \lambda} \\
a & \underset{id_a}{=\!=\!=} & a
\end{array}
\qquad\qquad
\begin{array}{ccc}
F(A,B) & \xrightarrow{A_1} & F(A,B) \\
{\scriptstyle B}\downarrow & [1] & \downarrow{\scriptstyle B} \\
F(A,C) & \xrightarrow{A_2} & F(A,C) \\
{\scriptstyle F}\downarrow & [2] & \downarrow{\scriptstyle F} \\
D & \underset{id_D}{=\!=\!=} & D
\end{array}
\qquad (9)
$$

Consider for instance the transition system of the λ-term $Ka(\Delta\Delta)$ on the left-hand side of Diagram 9:

- the two paths $\Delta_1 \cdot K \cdot \lambda$ and $K \cdot \Delta_2 \cdot \lambda$ are *equivalent* modulo permutation [1] of the β-redexes Δ_1 and K, and
- the two paths $K \cdot \Delta_2 \cdot \lambda$ and $K \cdot \lambda$ are *equivalent* modulo permutation [2] of the β-redexes Δ_2 and λ.

All put together, the two paths $f = \Delta_1 \cdot K \cdot \lambda$ and $g = K \cdot \lambda$ are equivalent modulo the two permutations [1] and [2]. In particular, they compute the same *events*, but in a different order. Note however that the redex Δ_1 has disappeared in the process of reorganizing the rewriting path f into the rewriting path g. Remarkably, the same story may be told of the term $F(A,B)$: the redex A_1 has disappeared during the process of reorganizing the rewriting path $f = A_1 \cdot B \cdot F$ into the rewriting path $g = B \cdot F$ using the two permutations [1] and [2].

The process of reorganizing a path $f : P \twoheadrightarrow Q$ into the properly oriented path $g : P \twoheadrightarrow Q$ is known as *the standardization procedure.* The rewriting path g obtained at the end of the procedure is called *the standard path* associated to the path f. J.-J. Lévy introduced the idea of an *equivalence relation* between rewriting paths *modulo redex permutation.* Here, we *orient* the redex permutations and thus refine Lévy equivalence relation into a *preorder* on rewriting paths. We call this preorder the *standardization preorder.* This enables us to describe standardization in a purely diagrammatic way, as an *extremal problem*:

$$\textit{standard paths} \quad = \quad \textit{minimal paths wrt. the standardization order.}$$

All this is explained in Sections 1.4—1.8, and illustrated by the λ-calculus in three different ways in Section 1.9. A concise and subjective history of the standardization theorem is provided in Section 1.10.

1.4 Standardization as 2-Dimensional Rewriting "Modulo"

Standardization is too often explained syntactically, and this complicates matters... In order to understand the reorganization of redexes in a simple and diagrammatic way, we decide to *orient* the permutations [1] and [2], and to define standardization as the 2-dimensional process of transforming the path $\Delta_1 \cdot K \cdot \lambda$ into the path $K \cdot \lambda$. During that transformation, each permutation [1] and [2] plays the role of a 2-dimensional rewriting step $\Longrightarrow$ reducing a rewriting path into another "more standard" rewriting path:

$$\Delta_1 \cdot K \cdot \lambda \Longrightarrow K \cdot \Delta_2 \cdot \lambda \Longrightarrow K \cdot \lambda. \tag{10}$$

The normal form of $\Delta_1 \cdot K \cdot \lambda$ is the standard path $K \cdot \lambda$. In this way, we define uniformly — for the first time — standardization for a wide class of existing rewriting system. The 2-dimensional perspective unifies already our two favourite examples: the rewriting path $A_1 \cdot B \cdot F$ is rewritten as the "rightmost outermost" rewriting path $B \cdot F$ by the same 2-dimensional procedure as example (10):

$$A_1 \cdot B \cdot F \Longrightarrow B \cdot A_2 \cdot F \Longrightarrow B \cdot F.$$

The interpretation of standardization as 2-dimensional rewriting is the author's rediscovery of an old idea published fifteen years earlier by J. W. Klop in his PhD thesis. At the time of J. W. Klop's PhD thesis (1975-80) standardization was limited to the λ-calculus and similar "leftmost-outermost" standardization theorems. J. W. Klop observed that standardization could be expressed nicely as a *plain* 2-dimensional rewriting system. Quite at the same time, G. Huet and J.-J. Lévy reshaped the field entirely by establishing a revolutionary standardization theorem for term rewriting systems, in [18]. Unfortunately, the richer standardization mechanisms disclosed by G. Huet and J.-J. Lévy cannot be expressed as a plain 2-dimensional rewriting system anymore — and J. W. Klop's elegant idea would simply not work.

It is only fifteen years later, trying to abstract away from the syntactical details of [18] that the 2-dimensional approach took shape again. This was a completely independent discovery originating from a long and obsessive reflexion on the diagrammatic presentation of [13]. Already in germ there and in the author's PhD thesis [27] the idea emerged finally that the standardization mechanism described by G. Huet and J.-J. Lévy reduces to distinguishing two classes of permutations:

- the *reversible* permutations — for instance, permutation [1] in Diagram (9),
- the *irreversible* permutations — for instance, permutation [2] in Diagram (9).

In this way, the standardization mechanisms disclosed by G. Huet and J.-J. Lévy can be reformulated as a 2-dimensional rewriting system *modulo reversible permutations* — which then specializes to a plain 2-dimensional rewriting system in the case of the "leftmost-outermost" standardization theorems studied by J. W. Klop in his PhD thesis.

At this point, it is worth explaining briefly and informally the difference between a reversible and an irreversible permutation. Permutation [1] is called *reversible* because it permutes two *disjoint* rewriting steps K and Δ_1, or B and A_1 — disjoint in the syntactic sense that no redex contains the other redex in the tree nesting order. The permutation is thus *neutral* from the point of view of standardization.

$$Ka(\Delta\Delta) \xrightarrow{\ \Delta_1\ } Ka(\Delta\Delta)$$

$$K \downarrow \qquad [1] \qquad \downarrow K$$

$$(\lambda y.a)(\Delta\Delta) \xrightarrow[\Delta_2]{} (\lambda y.a)(\Delta\Delta)$$

$$F(A,B) \xrightarrow{\ A_1\ } F(A,B)$$

$$B \downarrow \qquad [1] \qquad \downarrow B$$

$$F(A,C) \xrightarrow[A_2]{} F(A,C)$$

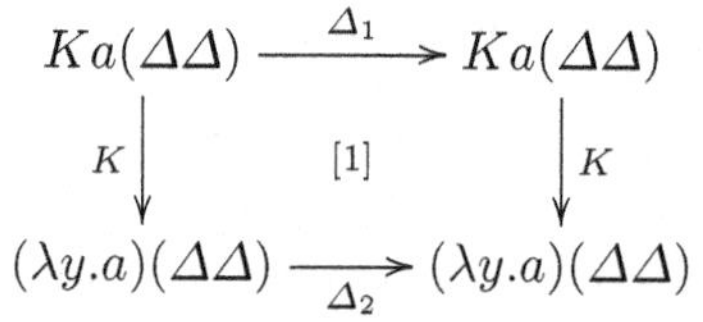

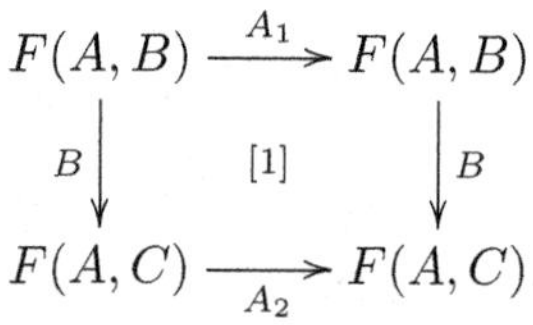

Permutation [2] is called *irreversible* because it replaces the "inside-out" computation $\Delta_2 \cdot \lambda$ or $A_2 \cdot F$ by its "outside-in" equivalent λ or F — thus *strictly improving* the computation from the point of view of standardization.

$$(\lambda y.a)(\Delta\Delta) \xrightarrow{\ \Delta_2\ } (\lambda y.a)(\Delta\Delta)$$

$$\lambda \downarrow \qquad [2] \qquad \downarrow \lambda$$

$$a \underset{\mathrm{id}_a}{=\!=\!=\!=} a$$

$$F(A,C) \xrightarrow{\ A_2\ } F(A,C)$$

$$F \downarrow \qquad [2] \qquad \downarrow F$$

$$D \underset{\mathrm{id}_D}{=\!=\!=\!=} D$$

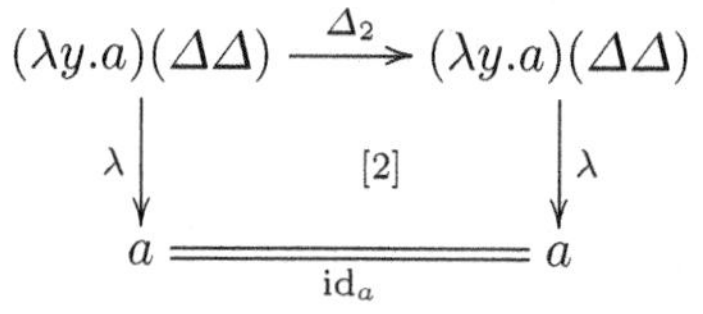

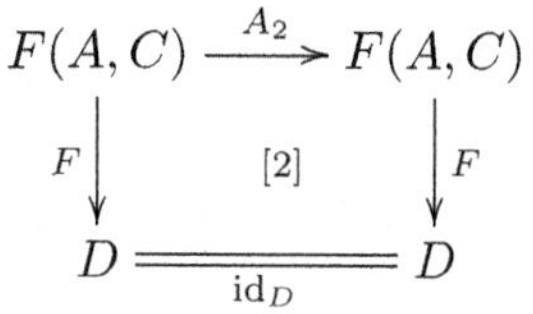

1.5   The Basic Vocabulary of Axiomatic Rewriting Theory

It is time to introduce several key definitions related to our diagrammatic theory of standardization.

Definition 1 (transition system). *A transition system (or oriented graph)* $\mathcal{G}$ *is a quadruple*

$$(\mathbf{terms}, \mathbf{redexes}, \mathbf{source}, \mathbf{target})$$

consisting of a set **terms** *of vertices (= terms), a set* **redexes** *of edges (= rewriting steps, or redexes), and two functions* **source**, **target** : **redexes** → **terms** *(= the source and target functions). We write*

$$u : M \longrightarrow N \quad \text{when } \mathbf{source}(u) = M \text{ and } \mathbf{target}(u) = N.$$

Recall that a *path* in a transition system $\mathcal{G}$ is a sequence

$$f = (M_1, u_1, M_2, ..., M_m, u_m, M_{m+1}) \tag{11}$$

where $u_i : M_i \longrightarrow M_{i+1}$ for every $i \in [1...m]$. We write $f : M_1 \longrightarrow M_{m+1}$. The *length* of f is m and f is said to be empty when $m = 0$. Two paths $f : M \longrightarrow\!\!\!\!\rightarrow N$ and $g : P \longrightarrow\!\!\!\!\rightarrow Q$ are coinitial (resp. cofinal) when $M = P$ (resp. $N = Q$). The path $f; g : M \longrightarrow\!\!\!\!\rightarrow Q$ denotes the concatenation of two paths $f : M \longrightarrow\!\!\!\!\rightarrow P$ and $g : P \longrightarrow\!\!\!\!\rightarrow Q$.

Definition 2 (2-dimensional transition system). *A 2-dimensional transition system is a pair* $(\mathcal{G}, \rhd)$ *consisting of a transition system* $\mathcal{G}$ *and a binary relation* $\rhd$ *on the paths of* $\mathcal{G}$. *The relation* $\rhd$ *is required to relate coinitial and cofinal paths:*

$$\forall f : M \longrightarrow\!\!\!\!\rightarrow N, \ g : P \longrightarrow\!\!\!\!\rightarrow Q, \qquad f \rhd g \ \Rightarrow \ (M,N) = (P,Q)$$

The idea of Axiomatic Rewriting Theory is to replace a concrete rewriting system by its 2-dimensional transition system. This has the effect of revealing unexpected similarities: typically, the two terms $Ka(\Delta\Delta)$ and $F(A, B)$ behave differently syntactically (left to right vs. right to left) but induce the same 2-dimensional transition system (drawn below) in the λ-calculus and in the term rewriting system (5).

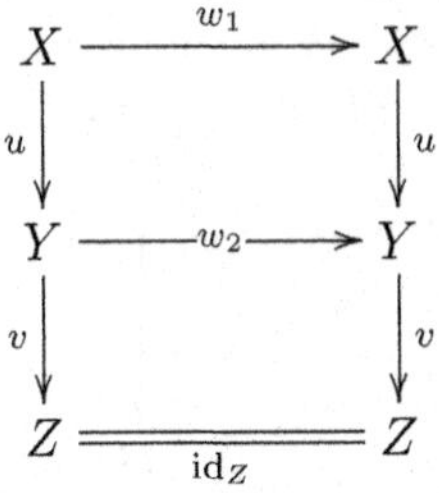

$$\begin{array}{l} w_1 \cdot u \rhd u \cdot w_2 \\ u \cdot w_2 \rhd w_1 \cdot u \\ w_2 \cdot v \rhd \quad v \end{array} \tag{12}$$

It should be obvious at this point of the exposition that the dynamical analogy observed previously between the terms $Ka(\Delta\Delta)$ and $F(A, B)$ (Section 1.3) follows from the identity of their 2-dimensional transition system.

Definition 3 (permutation). *A* permutation (f, g) *in a 2-dimensional transition system* $(\mathcal{G}, \rhd)$ *is a pair of paths such that* $f \rhd g$. *We often use the more explicit (and overloaded) notation* $f \rhd g$ *for a permutation* (f, g).

Definition 4 (standardization step, $\stackrel{1}{\Longrightarrow}$). *A* standardization step *from a path* $d : M \twoheadrightarrow N$ *to a coinitial and cofinal path* $e : M \twoheadrightarrow N$ *in a 2-dimensional transition system* $(\mathcal{G}, \rhd)$, *is a triple* $(d_1, f \rhd g, d_2)$ *consisting of a permutation* $f \rhd g$ *and two paths* d_1, d_2 *such that:*

$$d \;=\; M \stackrel{d_1}{\twoheadrightarrow} P \stackrel{f}{\twoheadrightarrow} Q \stackrel{d_2}{\twoheadrightarrow} N \qquad\qquad e \;=\; M \stackrel{d_1}{\twoheadrightarrow} P \stackrel{g}{\twoheadrightarrow} Q \stackrel{d_2}{\twoheadrightarrow} N$$

We write $d \stackrel{1}{\Longrightarrow} e$ *when there exists a standardization step from* d *to* e.

Definition 5 (standardization preorder $\Longrightarrow$, Lévy equivalence $\equiv$). *In every 2-dimensional transition system* $(\mathcal{G}, \rhd)$

- *the* standardization preorder $\Longrightarrow$ *is the least transitive reflexive relation containing* $\stackrel{1}{\Longrightarrow}$. *We say that a path* $e : M \twoheadrightarrow N$ *is* more standard *than a path* $d : M \twoheadrightarrow N$ *when* $d \Longrightarrow e$.
- *the* Lévy permutation equivalence $\equiv$ *is the least equivalence relation containing* $\Longrightarrow$. *Alternatively, the equivalence relation* $\equiv$ *is the least equivalence relation containing* $\rhd$ *and closed under composition.*

To illustrate our definitions with diagram (12), one shows that the path $u \cdot v$ is more standard than the path $w_1 \cdot u \cdot v$ by exhibiting the sequence of standardization steps:

$$w_1 \cdot u \cdot v \stackrel{1}{\Longrightarrow} u \cdot w_2 \cdot v \stackrel{1}{\Longrightarrow} u \cdot v.$$

1.6 Reversible and Irreversible Permutations

Permutations of $(\mathcal{G}, \rhd)$ are discriminated in two classes, reversible and irreversible, according to the following definition.

Definition 6 (reversible, irreversible permutation). *In every 2-dimensional transition system* $(\mathcal{G}, \rhd)$

1. *A permutation* (f, g) *is* reversible *when* $g \rhd f$. *A box* $\diamondsuit$ *signals reversible permutations* $f \diamondsuit g$ *in text and diagrams.*
2. *A permutation* (f, g) *is* irreversible *when* $\neg(g \rhd f)$. *A triangle* $\blacktriangleright$ *signals irreversible permutations* $f \blacktriangleright g$ *in text and diagrams.*

Check that the definition matches the previous qualification in Section 1.4 of permutation [1] as reversible, and permutation [2] as irreversible, in diagrams (9) and (12). We illustrate our new diagrammatic conventions on the 2-dimensional transition system (12).

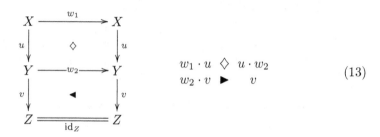

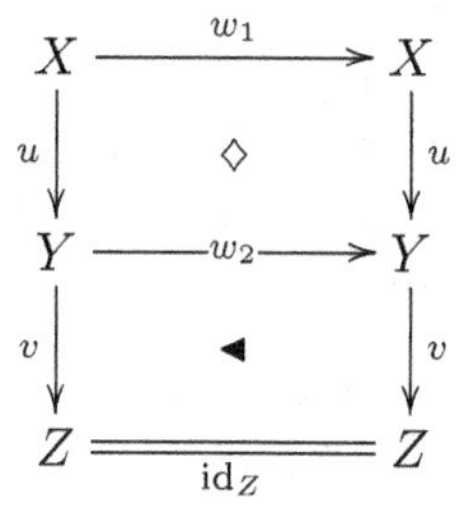

$$\begin{array}{ccc} w_1 \cdot u & \diamondsuit & u \cdot w_2 \\ w_2 \cdot v & \blacktriangleright & v \end{array} \qquad (13)$$

In the definition below, the discrimination on permutations generalizes to the obvious discrimination on standardization steps. The key concept of reversible permutation equivalence $\simeq$ is revealed, as a stronger version of usual Lévy permutation equivalence $\equiv$.

Definition 7 ($\overset{REV}{\Longrightarrow}, \overset{IRR}{\Longrightarrow}$, reversible permutation equivalence $\simeq$). *In every 2-dimensional transition system* $(\mathcal{G}, \rhd)$

– *A standardization step* $(e, f \rhd g, h)$ *is* reversible *(resp.* irreversible*) when the permutation* $f \rhd g$ *is reversible (resp. irreversible). We write*

$$d \overset{REV}{\Longrightarrow} e \qquad\qquad d \overset{IRR}{\Longrightarrow} e$$

when there exists a Reversible (resp. Irreversible) standardization step from d *to* e.

– *The* reversible permutation equivalence $\simeq$ *is the least equivalence relation containing the relation* $\overset{REV}{\Longrightarrow}$.

1.7 Standard Rewriting Paths

Definition 8 (standard path). *A rewriting path $d : M \twoheadrightarrow N$ is* standard *when there does not exist any sequence of standardization steps*

$$d \overset{REV}{\Longrightarrow} d_1 \overset{REV}{\Longrightarrow} \cdots \overset{REV}{\Longrightarrow} d_k \overset{IRR}{\Longrightarrow} d_{k+1}$$

consisting of a series of k Reversible steps followed by an Irreversible step.

So, a standard path is just a normal form of the standardization process, modulo reversible steps. So, when a rewriting path d is standard, and when $d \Longrightarrow e$, then $d \simeq e$ and the rewriting path e is standard.

For instance, the path $X \xrightarrow{w_1} X \xrightarrow{u} Y \xrightarrow{v} Z$ in diagram (12) is transformed in two steps in the standard path $X \xrightarrow{u} Y \xrightarrow{v} Z$. The rewriting path $X \xrightarrow{w_1} X \xrightarrow{u} Y$ is another example of standard path, because every standardization sequence from it to itself or to $X \xrightarrow{u} Y \xrightarrow{w_2} Y$ is reversible.

1.8 The Standardization Theorem

One main challenge of Axiomatic Rewriting Theory is to capture the diagrammatic properties of redex permutations in *syntactic* rewriting systems, in order to establish the following diagrammatic *standardization theorem*: for every rewriting path $d : M \twoheadrightarrow P$ in the transition system $\mathcal{G}$,

1. **existence:** there exists a standardization sequence
$$d \Longrightarrow e$$
transforming the rewriting path d into a standard path e,
2. **uniqueness:** every standardization sequence
$$d \Longrightarrow f$$
may be extended to a standardization sequence leading to the standard path e:
$$d \Longrightarrow f \Longrightarrow e.$$

The uniqueness property has a series of remarkable consequences. Suppose for instance that the rewriting path f is standard. In that case, the standardization sequence

$$f \Longrightarrow e$$

consists of Reversible steps. Thus,

$$f \simeq e.$$

From this follows that there exists a unique standard path e such that

$$d \Longrightarrow e$$

modulo reversible permutation equivalence. In fact, the uniqueness property ensures that there exists a unique standard path, modulo reversible permutation equivalence, in the Lévy equivalence class of the rewriting path d.

In this article, we formulate a series of nine elementary axioms on the 2-dimensional transition system $(\mathcal{G}, \rhd)$ and deduce from them the diagrammatic standardization theorem stated above. The axioms uncover a series of simple and elegant *principles* of causality in computations. They also illustrate that a purely *diagrammatic* and *syntax-free* theory of computations is possible, and useful, since it enscopes almost every existing rewriting system, from Petri nets to higher-order rewriting systems.

1.9 Illustration: The λ-Calculus and Its Three Standardization Orders

There are at least three different ways to interpret the λ-calculus as a 2-dimensional transition system, each one associated to a particular *nesting order* on the β-redexes of λ-terms. The underlying transition system $\mathcal{G}_\lambda$ is the same in the three cases. It is defined in [10, 24] as follows:

- its vertices are the λ-terms, modulo α-conversion,
- its edges are the β-redexes $u : M \longrightarrow N$.

Recall that a β-redex $u = (M, o, N)$ is a triple consisting of a λ-term M, the occurrence o of a β-pattern $(\lambda x.P)Q$ in M and the λ-term N obtained after β-reducing

$$(\lambda x.P)Q \longrightarrow P[x := Q]$$

in the λ-term M.

It is worth noting that there are two different edges $I(Ia) \longrightarrow Ia$ in the graph $\mathcal{G}_\lambda$: each edge corresponds to the reduction of a particular identity combinator $I = (\lambda x.x)$ in the λ-term $I(Ia)$.

There are at least three different ways to refine the transition system $\mathcal{G}_\lambda$ as a 2-dimensional transition system, depending on the order chosen on β-redexes:

- the *tree-order*: a β-redex u is smaller than a β-redex v when v occurs in the function or argument part of u; or equivalently, when the occurrence of u is a strict prefix of the occurrence of v. We use the notation: $u \preceq_{\mathsf{tree}} v$.
- the *left-order*: a β-redex u is smaller than a β-redex v when v occurs in the function or argument part of u, or when there exists an occurrence o of an application node PQ in the λ-term M, such that u occurs in P and v occurs in Q. We use the notation: $u \preceq_{\mathsf{left}} v$.
- the *argument-order*: a β-redex u is smaller than a β-redex v when v occurs in the argument of u. We use the notation: $u \preceq_{\mathsf{arg}} v$.

Each order induces in turn its own permutation relation $\rhd_{\mathsf{tree}}$, $\rhd_{\mathsf{left}}$ and $\rhd_{\mathsf{arg}}$ on the transition system $\mathcal{G}_\lambda$. The order considered in the literature is generally the *left-order*, see [10, 24, 20]. However, we prefer to study here the tree-order, because this seems the most natural choice after the work by G. Huet and J.-J. Lévy on term rewriting systems [18]. The two alternative orders $\preceq_{\text{left}}$ and $\preceq_{\text{arg}}$ are discussed briefly in Section 8.

We define the relation $\rhd_{\text{tree}}$ as follows. Two paths f, g are related as $f \rhd_{\text{tree}} g$ precisely when:

1. the paths f and g factor as $f = v \cdot u'$ and $g = u \cdot h$ where u, v, u' are β-redexes and h is a path,
2. the two β-redexes u and v are coinitial, and $\neg(v \preceq_{\text{tree}} u)$,
3. the β-redex u' is the (unique) residual of u after v, and the path h develops the (possibly) several residuals of v after u. [For a definition of residual and complete development, see [10, 24, 18, 3, 21, 22] or Section 6.]

Thus, every permutation $f \rhd_{\text{tree}} g$ is of the form:

$$\begin{array}{ccc} M & \xrightarrow{\quad v \quad} & Q \\ {\scriptstyle u}\big\downarrow & \Leftarrow_{\text{tree}} & \big\downarrow{\scriptstyle u'} \\ P & \xrightarrow[\quad h \quad]{} & N \end{array} \qquad \begin{array}{l} f = v \cdot u' \\ g = u \cdot h \end{array} \tag{14}$$

where u and v are different β-redexes, u' is a β-redex and h is a path. The three paradigmatic examples of β-redex permutation $f \rhd_{\text{tree}} g$ are:

$$\begin{array}{ccc} PQ & \xrightarrow{\quad v \quad} & P'Q \\ {\scriptstyle u}\big\downarrow & \Leftarrow_{\text{tree}} & \big\downarrow{\scriptstyle u'} \\ PQ' & \xrightarrow[\quad v' \quad]{} & P'Q' \end{array} \qquad \begin{array}{ccc} (\lambda x.a)P & \xrightarrow{\ v\ } & (\lambda x.a)P' \\ {\scriptstyle u}\big\downarrow & \Leftarrow_{\text{tree}} & \big\downarrow{\scriptstyle u'} \\ a & \underset{\text{id}_a}{=\!=\!=\!=} & a \end{array} \qquad \begin{array}{ccc} \Delta P & \xrightarrow{\quad v \quad} & \Delta P' \\ {\scriptstyle u}\big\downarrow & \Leftarrow_{\text{tree}} & \big\downarrow{\scriptstyle u'} \\ PP & \xrightarrow[\ v_1 \cdot v_2\ ]{} & P'P' \end{array}$$

where $P \longrightarrow P'$ and $Q \longrightarrow Q'$ are two β-redexes. The three permutations are respectively reversible, irreversible and irreversible in the 2-dimensional transition system $(\mathcal{G}_\lambda, \rhd_{\text{tree}})$.

Remark: the argument-order $\preceq_{\text{arg}}$ is included in the tree-order $\preceq_{\text{tree}}$ which is included in the left-order $\preceq_{\text{left}}$. From this follows that the permutation relation $\rhd_{\text{arg}}$ contains the permutation relation $\rhd_{\text{tree}}$ which contains in turn the permutation relation $\rhd_{\text{left}}$. It is not difficult then to establish that every rewriting path standard wrt. the left-order $\preceq_{\text{left}}$ is standard wrt. the tree-order $\preceq_{\text{tree}}$, and that every rewriting path standard wrt. the tree-order $\preceq_{\text{tree}}$ is standard wrt. the argument-order $\preceq_{\text{arg}}$. The converse is obviously false in the two cases.

1.10 A Concise History of the Standardization Theorem

Many authors have written on the standardization theorem. We do not draw below a comprehensive list, but deliver a concise history of the subject, in eight key steps.

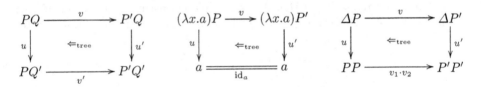

[1936]. *A. Church* and *J.B. Rosser* introduce the λI-calculus, a λ-calculus without erasure, and prove that the number of β-steps from a λI-term to its normal form is *bounded* by the length of the leftmost outermost computation. This result is the ancestor of all later standardization theorems.

[1958]. *H.B. Curry* and *R. Feys* formulate the first standardization theorem for the λ-calculus: the two authors prove that every time a λ-term P β-reduces to a λ-term Q, there exists also a *standard* way to β-reduce P to Q. The theorem extends Church and Rosser result for the λI-calculus, and plays a role in Curry and Feys' defense of their erasing combinator K.

[1978]. *J.-J. Lévy* formulates the standardization theorem in its modern algebraic form: using an equivalence relation on rewriting paths — called today Lévy permutation equivalence — Lévy proves that there exists a *unique* standard rewriting path in each equivalence class. The uniqueness result was so striking at the time that the theorem was called the *strong* standardization theorem by subsequent authors. Despite its conceptual novelty, the theorem is still limited to the λ-calculus and to its leftmost-outermost order.

[1979]. *G. Huet* and *J.-J. Lévy* formulate and establish a standardization theorem for term rewriting systems without critical pairs. This is probably the most revolutionary step in the history of standardization, the first time at least that another standardization order is considered than the "leftmost outermost" order of the λ-calculus. The theorem is still limited to term rewriting systems — because its proof relies heavily on syntactical notions like tree-occurrence — but the article delivers the message that standardization is a general property of rewriting systems, related to causality and domain-theoretic notions like stability and sequentiality.

[1980]. *J. W. Klop* introduces a 2-dimensional rewriting system on paths, consisting in permuting "anti-standard" paths of length 2 into "standard" paths of arbitrary length. In this way, Klop deduces Lévy's *strong* standardization theorem for leftmost-outermost λ-calculus, by establishing confluence and strong normalization of the 2-dimensional rewriting process: the standard path is obtained as the normal form of the procedure. Another important contribution of J. W. Klop is to stress the role of the finite development lemma in the proof of standardization, and to extend to any "left-regular" Combinatory Reduction System the standardization theorem for leftmost-outermost λ-calculus.

[Early 1980s]. *G. Boudol* extends G. Huet and J.-J. Lévy standardization theorem to term rewriting systems with critical pairs. This is another decisive step, because it extends the principle of standardization to non deterministic rewriting systems.

[1992]. *G. Gonthier* and *J.-J. Lévy* and *P-A. Melliès* deliver an axiomatic standardization theorem, where the syntactical proof of[18] is replaced by diagrammatic arguments on redexes, residuals and the nesting relation. Subsequently reworked by the author in his PhD thesis [27], the theorem extends G. Huet and J.-J. Lévy's original theorem to a great variety of rewriting systems with and without critical pairs — with the remarkable and puzzling exception (as first noted by R. Kennaway) of rewriting systems based on directed acyclic graphs.

[1996]. *D. Clark* and *R. Kennaway* adapt the syntactical works of G. Huet, J.-J. Lévy and G. Boudol and establish a standardization theorem for (possibly conflicting) rewriting systems based on directed acyclic graphs (dags).

It took the author nine years to derive the current axiomatics from [13]. One difficulty was to find the simplest possible description of rewriting systems with critical pairs. The trinity of residual, compatibility and nesting relations operating in [13] was certainly too complicated. Slowly, the 2-dimensional presentation emerged, leading the author to the elementary axiomatics of this article. Twenty-five years ago, the work of [18, 6] on term rewriting systems revealed that the "conflict-free left-regular" rewriting systems considered earlier was the emerged part of the much wider and exciting world of *causal computations.* This is that world and its boundaries which we will explore here in our 2-dimensional diagrammatic language.

1.11 Structure of the Paper

Axiomatic Rewriting Systems (AxRS) are introduced in Section 2, along with their nine standardization axioms. A less innovative but more traditional axiomatics based on residuals, critical pairs and nesting is formulated in Section 6. Standard paths are characterized in Section 3 as the paths which do not contain a particular "anti-standard" pattern, just as in [13, 27]. The standardization theorem is proved in Section 4, and reformulated 2-categorically in Section 5. An alternative axiomatization based on residuals and nesting orders is formulated in Section 6. A few additional hypotheses on axiomatic rewriting systems are discussed in Section 7. Finally, we illustrate our definition of AxRS with a series of examples in Section 8, like asynchronous transition systems, term rewriting systems, call-by-value λ-calculus, λ-calculus with explicit substitutions.

2 The Standardization Axiomatics

An *Axiomatic Rewriting System (AxRS)* is defined as a 2-dimensional transition system $(\mathcal{G}, \rhd)$ which satisfies moreover the series of nine *standardization axioms* presented in this section. Each axiom of the section is illustrated by the λ-calculus and its 2-dimensional transition system $(\mathcal{G}_\lambda, \rhd_{\mathbf{tree}})$ defined in Section 1.9.

2.1 Axiom 1: Shape

The first axiom generalizes to every AxRS the shape of permutations encountered in the λ-calculus — see Diagram (14)in Section 1.9.

Axiom 1 (Shape). *We ask that in every permutation $f \rhd g$,*

- *the path f is of length 2,*
- *the path g is of length at least 1,*
- *the initial redexes of f and g are different.*

Thus, every permutation $f \rhd g$ in the 2-dimensional transition system $(\mathcal{G}, \rhd)$ has the following shape:

$$\begin{array}{ccc} M & \xrightarrow{\;v\;} & Q \\ {\scriptstyle u}\downarrow & \Leftarrow & \downarrow{\scriptstyle u'} \\ P & \xrightarrow[\;h\;]{} & N \end{array} \qquad \begin{array}{l} f = v \cdot u' \\ g = u \cdot h \end{array} \tag{15}$$

where u and v are different redexes, u' is a redex and h is a path. In case of a *reversible* permutation $f \diamond g$, this shape specializes to a 2×2 square:

$$\begin{array}{ccc} M & \xrightarrow{\;v\;} & Q \\ {\scriptstyle u}\downarrow & \diamond & \downarrow{\scriptstyle u'} \\ P & \xrightarrow[\;v'\;]{} & N \end{array} \qquad \begin{array}{l} f = v \cdot u' \\ g = u \cdot v' \end{array}$$

where u, u', v and v' are redexes, u and v different.

2.2 Axioms 2, 3, 4, 5: Ancestor, Reversibility, Irreversibility and Cube

The standardization theorem is usually established by a fine-grained analysis of syntactic mechanisms like erasure, duplication, etc... related to Lévy theory of residuals. The fragment of Lévy theory necessary to the theorem, e.g. the finite development property, appears in our axiomatics... but reformulated, because the more geometric idea of "oriented permutation" replaces the traditional concept of "residual of a redex". The residual theory is particularly visible in the four Axioms **ancestor**, **reversibility**, **irreversibility** and **cube** introduced below, as well as in Axiom **termination** of Section 2.6.

Axiom **ancestor** incorporates two properties of the λ-calculus, traditionally called *uniqueness of ancestor* and *finite development*. The existence of a permutation $f \rhd_{\text{tree}} g$ between two β-rewriting paths:

$$f = M \xrightarrow{v} Q \xrightarrow{u'} N \qquad\qquad g = M \xrightarrow{u} P \overset{h}{\twoheadrightarrow} N$$

means that the β-redex u' is the unique residual of the β-redex u after β-reduction of the redex v, and that the path h is a complete development of the residuals of the redex v after β-reduction of the redex u. In that case, we say that the redex u is an *ancestor* of the redex u' before β-reduction of the redex v. The *uniqueness of ancestor* property states that the redex u is the unique such ancestor of the redex u'. Besides, the *finite development* property of the λ-calculus, recalled in Section 6, states that two complete developments of the same set of β-redexes, are Lévy equivalent. From this follows that any rewriting path g' involved in a permutation $f \rhd_{\mathrm{tree}} g'$ factors as $g' = u' \cdot h'$ where $u = u'$ and $h \equiv_{\mathrm{tree}} h'$. This leads us to formulate the

Axiom 2 (Ancestor). *Suppose that u, u' are redexes, that f, h, h' are rewriting paths, forming together permutations $f \rhd u \cdot h$ and $f \rhd u' \cdot h'$. We ask that $u = u'$ and $h \equiv h'$.*

Axiom **reversibility** indicates that every permutation $f \rhd g$ is either reversible, or reduces to a rewriting path g for which there exists no permutation of the form $g \rhd h$. This mirrors the following property of the λ-calculus. Suppose that $f, g, h : M \twoheadrightarrow N$ are three β-rewriting paths involved in permutations $f \rhd_{\mathrm{tree}} g$ and $g \rhd_{\mathrm{tree}} h$. The paths f and g are of length 2, the path h is of length at least 1, and the paths f, g, h decompose as

$$f \; = \; M \xrightarrow{v} Q \xrightarrow{u'} N, \qquad g \; = \; M \xrightarrow{u} P \xrightarrow{v'} N, \qquad h \; = \; M \xrightarrow{v''} O \xrightarrow{h_u} N$$

where the two redexes v and v'' are ancestor of the same redex v', and thus $v = v''$; and where the β-redex u' is the unique residual of the β-redex u, and the rewriting path h_u is a development of the residuals of u after v, and thus $h_u = u'$. It follows that $f = h$.

Axiom 3 (Reversibility). *We ask that $f = h$ when $f \rhd g$ and $g \rhd h$.*

Axiom **irreversibility** completes the two previous axioms. The axiom mirrors the fact that in the λ-calculus and in many rewriting systems, standardization preserves complete developments — see [24, 18] or Section 6 for a definition of complete developments. Let us explain briefly what we mean here. Consider any β-rewriting path $h : M \twoheadrightarrow N$ which defines a complete development of a multi-redex (M, U) in the λ-calculus, and suppose that the path h factors as

$$h \quad = \quad M \overset{h_1}{\twoheadrightarrow} M' \overset{h_2}{\twoheadrightarrow} N' \overset{h_3}{\twoheadrightarrow} N$$

where the β-rewriting path h_2 is involved in a standardization permutation

$$h_2 \rhd h_2'.$$

By definition of $\rhd_{\mathrm{tree}}$, the two β-rewriting paths h_2 and h_2' decompose as

$$h_2 = M' \xrightarrow{v} P \xrightarrow{u'} N' \qquad \text{and} \qquad h_2' = M' \xrightarrow{u} Q \overset{h''}{\twoheadrightarrow} N'.$$

We claim here that the resulting β-rewriting path

$$h' \quad = \quad M \overset{h_1}{\twoheadrightarrow} M' \overset{h_2'}{\twoheadrightarrow} N' \overset{h_3}{\twoheadrightarrow} N$$

defines a complete development of (M, U). How do we prove this? We establish first that the two redexes u and v are residual of a redex in U after the β-rewriting path h_1. The very definition of the path h as a complete development of the multi-redex (M, U) induces already that:

- the redex v is residual of a redex $v_0 \in U$ after the β-rewriting path h_1; and
- the redex u' is residual of a redex $u_0 \in U$ after the β-rewriting path $h_1 \cdot v$.

We know moreover that the β-redex u is the *unique* ancestor of the β-redex u' before reduction of the β-redex v. This uniqueness property ensures that the β-redex u is residual of the redex $u_0 \in U$ after the β-rewriting path h_1. This establishes that the two redexes u and v are residual of a redex in U after the β-rewriting path h_1. Now, we know by definition of $\triangleright_{\text{tree}}$ that the two paths h_2 and h_2' define complete developments of the multi-redex $(M', \{u, v\})$. The *finite development* property of the λ-calculus states moreover that the two β-rewriting paths h_2 and h_2' define the same residual relation. It follows quite immediately that, as we claimed, the β-rewriting path $h_1 \cdot h_2' \cdot h_3$ defines a complete development of the multi-redex (M, U). We conclude more generally that every path more standard than the path h is also a complete development of the multi-redex (M, U).

How is this result interpreted in our axiomatic setting? Consider an irreversible permutation $f \blacktriangleright_{\text{tree}} g$ between two β-rewriting paths

$$f = M \xrightarrow{v} Q \xrightarrow{u'} N \qquad g = M \xrightarrow{u} P \xrightarrow{h_v} \!\!\!\!\rightarrow N$$

and a β-rewriting path h such that

$$g \Longrightarrow h.$$

It follows from our previous argument that, just like the β-rewriting path f and g, the β-rewriting path h is a complete development of the multi-redex $(M, \{u, v\})$. Besides, the first β-redex reduced in the path h is not the β-redex v. Thus, the β-rewriting path h decomposes necessarily as

$$h = M \xrightarrow{u} P \xrightarrow{h_v'} \!\!\!\!\rightarrow N$$

where

$$h_v \Longrightarrow h_v'.$$

Here, we apply our previous argument another time, and deduce from $h_v \Longrightarrow h_v'$ that, just like the β-rewriting path h_v, the β-rewriting path h_v' is a complete development of the residuals of the β-redex v after reduction of the β-redex u. This shows in particular that $f \blacktriangleright_{\text{tree}} h$. This leads to

Axiom 4 (Irreversibility). *We ask that $f \blacktriangleright h$ when $f \blacktriangleright g$ and $g \Longrightarrow h$.*

Axiom **cube** incorporates the *cube lemma* established in [24, 18] as well as a careful analysis of nesting in the λ-calculus. Suppose that $C[-]$ is a context, see

[3] for a definition, and that a β-rewriting path $g : C[M] \longrightarrow\!\!\!\!\rightarrow C[N]$ computes only inside M, never inside $C[-]$. Then, just as the β-rewriting path g, every Lévy equivalent β-rewriting path $f : C[M] \longrightarrow\!\!\!\!\rightarrow C[N]$ computes only inside M, never inside $C[-]$. So, every β-redex w inside $C[-]$ has the same (unique) residual w'' after the β-rewriting paths f and g. Diagrammatically speaking, the property amounts to the cube property stated in the next axiom, when $f \rhd_{\text{tree}} g$ and $f = v \cdot u'$ and $g = u \cdot v_1 \cdots v_n$ and $w'' = w_{n+1}$. The axiom requires that the property holds in every AxRS.

Axiom 5 (Cube). *We ask that every diagram*

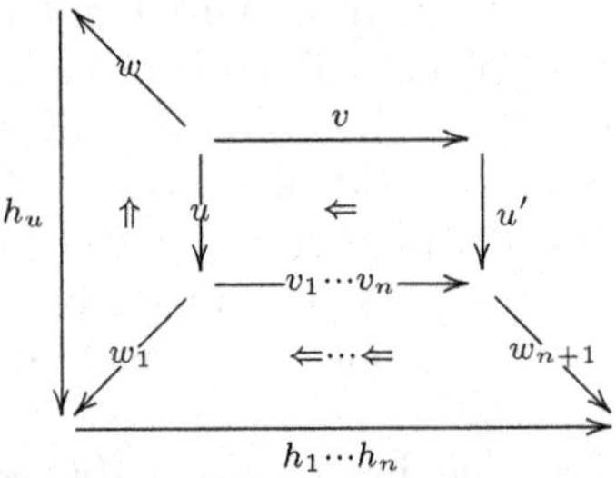

with u, u', v *and* $v_1, ..., v_n$ *and* $w, w_1, ..., w_n, w_{n+1}$ *a series of redexes and* $h_1, ..., h_n$ *a series of paths forming permutations*

$$v \cdot u' \rhd u \cdot v_1 \cdots v_n \qquad u \cdot w_1 \rhd w \cdot h_u \qquad v_i \cdot w_{i+1} \rhd w_i \cdot h_i \qquad \text{for } 1 \le i \le n$$

may be completed as a diagram:

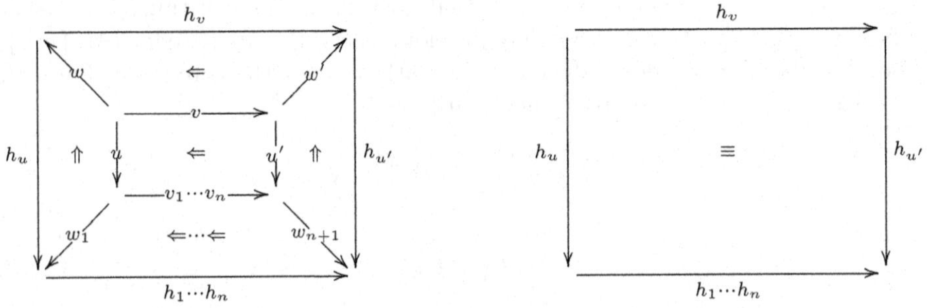

where w' *is a redex and* $h_v, h_{u'}$ *are paths which form permutations*

$$u' \cdot w_{n+1} \rhd w' \cdot h_{u'} \qquad v \cdot w' \rhd w \cdot h_v$$

and induce the equivalence

$$h_v \cdot h_{u'} \equiv h_u \cdot h_1 \cdots h_n.$$

2.3 Axiom 6: Enclave

Axiom **enclave** is based on a fundamental property of the λ-calculus, observed for the first time in the preliminary work of [13]. Suppose that a β-redex v is nested under a β-redex u — that is $u \preceq_{\mathsf{tree}} v$ — and that the β-redex v creates a β-redex w'. By *creation*, we mean that the β-redex w' has no ancestor before reduction of the β-redex v. In that case, the β-redex w' is necessarily nested under the (unique) residual u' of the β-redex u after reduction of the β-redex v.

The next axiom formulates the property as its contrapose. The existence of the permutation

$$u' \cdot w_{n+1} \rhd_{\mathsf{tree}} w' \cdot h_{u'}$$

means that the β-redex w' is *not* nested under the β-redex u'. And from this follows that the β-redex w' is not created, and thus, has an ancestor w before reduction of the β-redex v. The axiom requires that this *enclave property* holds in every AxRS.

Axiom 6 (Enclave). *We ask that every diagram*

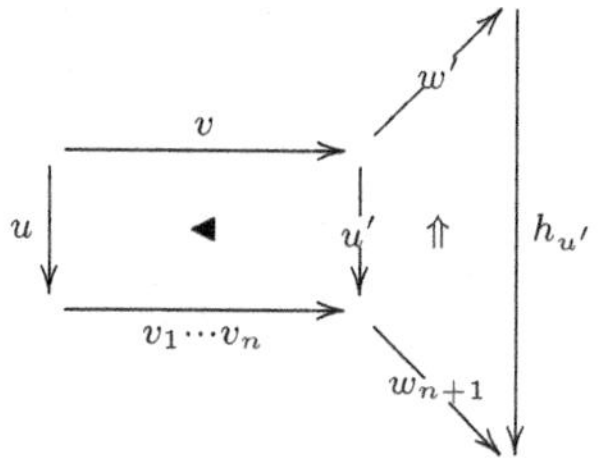

where u, v, u' and $v_1, ..., v_n$ and w', w_{n+1} are redexes, and $h_{u'}$ is a path, forming the permutations (recalling our convention, the symbol ▶ means that the permutation is irreversible)

$$v \cdot u' \blacktriangleright u \cdot v_1 \cdots v_n \qquad\qquad u' \cdot w_{n+1} \rhd w' \cdot h_{u'}$$

may be completed as a diagram:

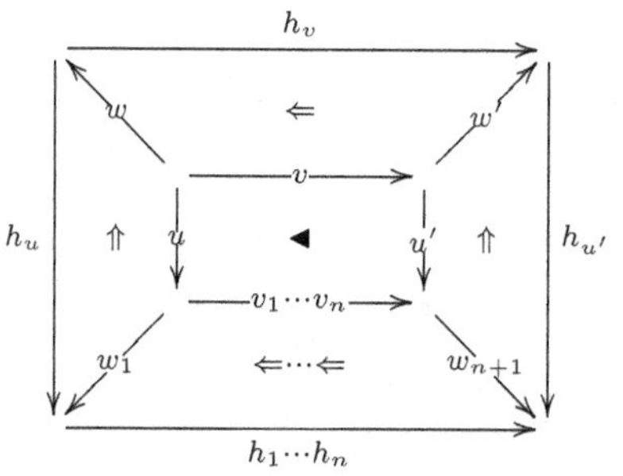

with $w, w_1, ..., w_n$ a series of redexes and h_u, h_v and $h_1, ..., h_n$ a series of paths, forming the $n+2$ permutations

$$v \cdot w' \rhd w \cdot h_v \qquad u \cdot w_1 \rhd w \cdot h_u \qquad v_i \cdot w_{i+1} \rhd w_i \cdot h_i \qquad \textit{for } 1 \leq i \leq n$$

2.4 Axioms 7 and 8: Stability and Reversible Stability

Axiom **stability** incorporates another key property of the λ-calculus, also observed for the first time in the preliminary work of [13]. Consider any reversible permutation

$$M \xrightarrow{u} P \xrightarrow{v'} N \quad \diamondsuit_{\text{tree}} \quad M \xrightarrow{v} Q \xrightarrow{u'} N$$

in which the β-redex u creates a β-redex w_1 and the β-redex v creates a β-redex w_2. It is not difficult to establish that there exists no β-redex w_{12} in the λ-term N which would be at the same time residual of the β-redex w_1 after reduction of the β-redex v', and residual of the β-redex w_2 after reduction of the β-redex u'. The property is axiomatized below as its contrapose. The axiom states that the *characteristic function* of the *event* of creating the β-redex w_{12} (or equivalently the β-redex w_1, or the β-redex w_2) is *stable* in the sense of G. Berry, see [5]. Axiom **reversible-stability** repeats the axiom in the reversible case.

Axiom 7 (Stability). *We ask that every diagram*

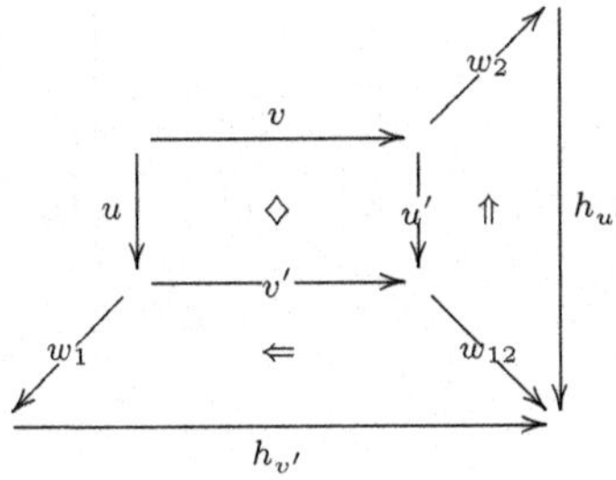

where u, v, u', v' and w_1, w_2, w_{12} are redexes and $h_{u'}, h_{v'}$ are paths, forming the permutations (recalling our convention, the symbol $\diamondsuit$ means that the permutation is reversible)

$$v \cdot u' \diamondsuit u \cdot v' \qquad u' \cdot w_{12} \rhd w_2 \cdot h_{u'} \qquad v' \cdot w_{12} \rhd w_1 \cdot h_{v'}$$

may be completed as a diagram

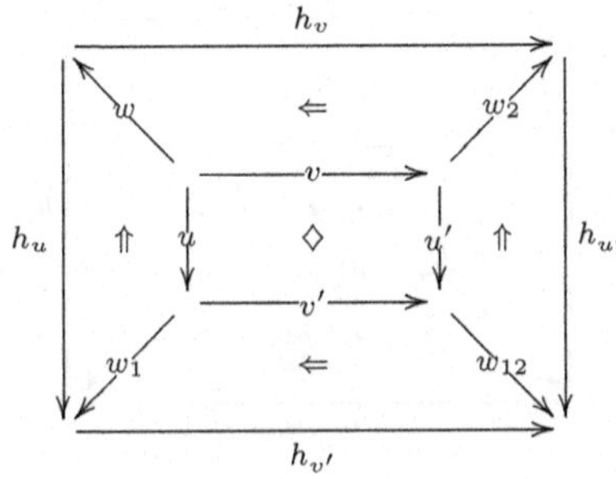

where w is a redex and h_u, h_v are two paths, forming two permutations

$$v \cdot w_2 \rhd w \cdot h_v \qquad u \cdot w_1 \rhd w \cdot h_u$$

Axiom 8 (Reversible stability). *We ask that every diagram*

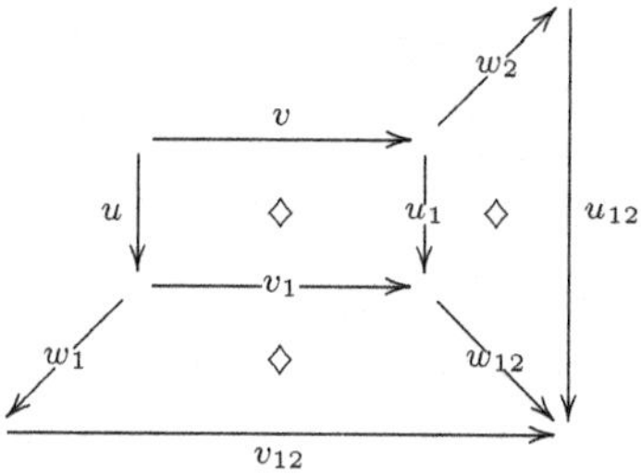

(16)

where u, v, u_1, v_1 *and* $w_1, w_2, w_{12}, u_{12}, v_{12}$ *are redexes forming the reversible permutations*

$$v \cdot u_1 \diamond u \cdot v_1 \qquad u_1 \cdot w_{12} \diamond w_2 \cdot u_{12} \qquad v_1 \cdot w_{12} \diamond w_1 \cdot v_{12}$$

may be completed as a diagram

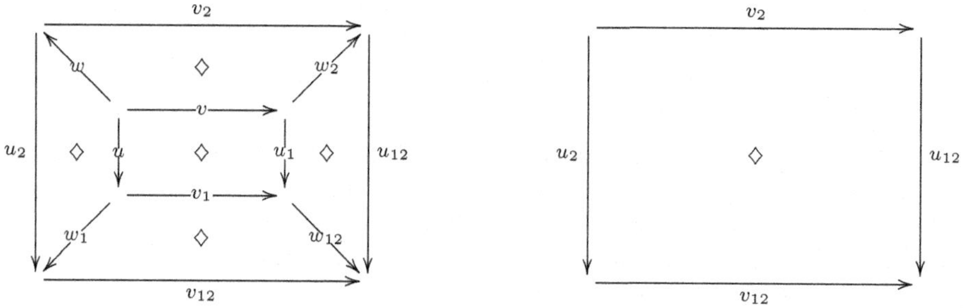

where w, u_2, v_2 *are three redexes forming the reversible permutations*

$$v \cdot w_2 \diamond w \cdot v_2 \qquad \text{and} \qquad u \cdot w_1 \diamond w \cdot u_2 \qquad \text{and} \qquad v_2 \cdot u_{12} \diamond u_2 \cdot v_{12}$$

Remark: Axiom **reversible-stability** may be understood as a converse of the reversible variant of Axiom **cube** formulated in Section 7.3. Indeed, Axiom **reversible-stability** states that every diagram

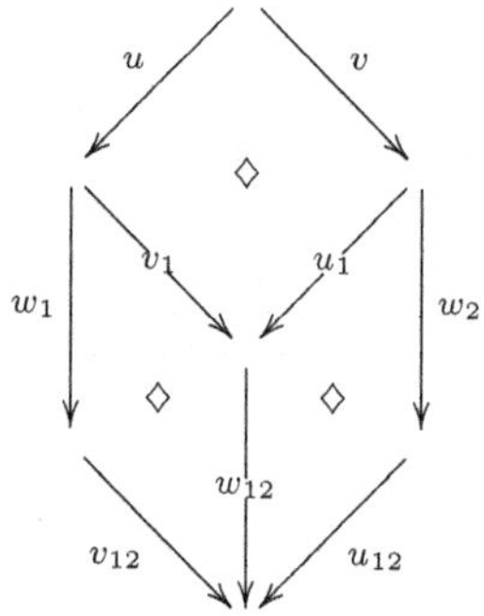

(17)

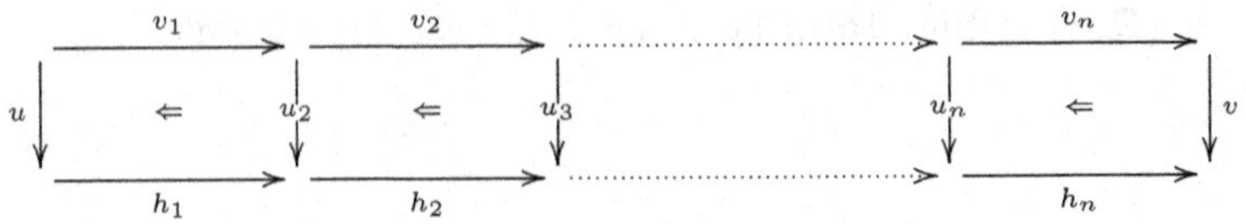

Fig. 2. The path $f = v_1 \cdots v_n$ drags the redex v to the redex u

may be completed into the diagram

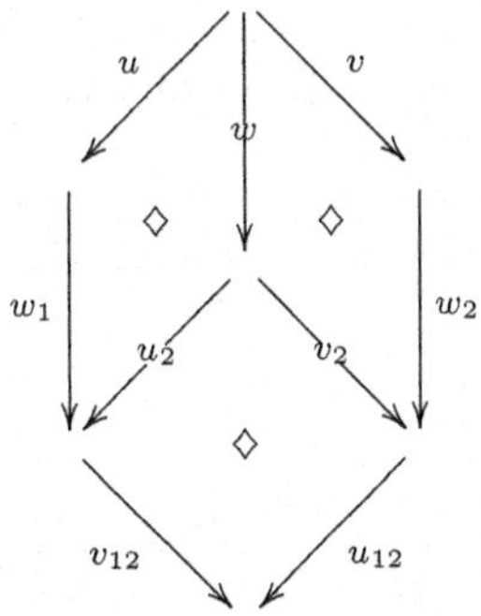

(18)

and conversely, Axiom **reversible-cube** formulated in Section 7.3 states that Diagram (18) may be completed as Diagram (17). Besides, it is remarkable that the two Axioms **reversible-stability** and **reversible-cube** are *dual* in the sense that each axiom may be obtained from the other one by *reversing* the orientations of all the arrows in diagrams.

2.5 Drag and Extraction

We need to introduce a few definitions related to standardization in order to state the last axiom of the theory (Axiom 9).

Definition 9 (drag). *A path* $f : M \twoheadrightarrow N$ drags *a redex* v *outgoing from* N *to a redex* u *outgoing from* M*, when*

- $f = \mathrm{id}_M$ *and* $v = u$,
- *or* $f = v_1 \cdots v_n$ *and there exists* $n+1$ *redexes* $u_1, ..., u_{n+1}$ *and* n *paths* $h_1, ..., h_n$ *such that:*
  - $u_1 = u$ *and* $u_{n+1} = v$,
  - *the rewriting paths* $v_i \cdot u_{i+1}$ *and* $u_i \cdot h_i$ *form a permutation* $v_i \cdot u_{i+1} \rhd u_i \cdot h_i$ *for every index* $1 \leq i \leq n$.

Notation: We write $u \overset{f}{\hookleftarrow} v$ when the rewriting path f drags the redex v to the redex u. See Figure 2.

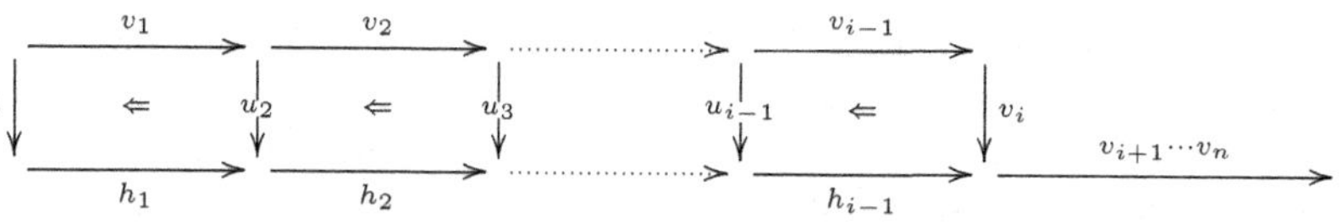

Fig. 3. The redex u is extractible from the path $f = v_1 \cdots v_n$ and the path $g = h_1 \cdots h_{i-1} \cdot v_{i+1} \cdots v_n$ is a projection of the rewriting path f by extraction of the redex u.

Lemma 10 (preservation of drag). *For every path $f : M \twoheadrightarrow N$, the relation $\overset{f}{\longleftarrow\!\!\!\shortmid}$ is a partial function, from the redexes outgoing from N to the redexes outgoing from M. Moreover, the relation is invariant by permutation on f:*

$$\forall g : M \twoheadrightarrow N, \qquad f \equiv g \;\Rightarrow\; \overset{f}{\longleftarrow\!\!\!\shortmid} \;=\; \overset{g}{\longleftarrow\!\!\!\shortmid}.$$

Proof. Suppose that $u \overset{f}{\longleftarrow\!\!\!\shortmid} v$ and $u' \overset{f}{\longleftarrow\!\!\!\shortmid} v$. Then $u = u'$ by Axiom **ancestor**, and an easy induction on the length of f. Now, by Axiom **cube**, the relation increases by *anti-standardization*: if the rewriting path g drags the redex v to the redex u, and $f \Longrightarrow g$, then the rewriting path f drags the redex v to the redex u. By Axiom **enclave**, the relation increases also by *standardization*: if the rewriting path f drags the redex v to the redex u, and $f \Longrightarrow g$, then the rewriting path g drags the redex v to the redex u as well. We conclude. □

Definition 11 (extraction, projection, $\searrow_u$). *A redex $u : M \longrightarrow P$ is* extractible *from a path $f = v_1 \cdots v_n : M \twoheadrightarrow N$ when there exists an index $1 \leq i \leq n$ such that the path $v_1 \cdots v_{i-1}$ drags the redex v_i to the redex u. In that case, we call* projection *of the rewriting path f by extraction of the redex $u : M \longrightarrow P$ any rewriting path $g : P \twoheadrightarrow N$ which decomposes as*

$$g = h_1 \cdots h_{i-1} \cdot v_{i+1} \cdots v_n$$

where there exists redexes $u_1, ..., u_i$ with $u_1 = u$ and $u_i = v_i$ and a permutation

$$v_j \cdot u_{j+1} \rhd u_j \cdot h_j$$

for every index $1 \leq j \leq i - 1$.

Notation: We write $f \searrow_u g$ when the redex u is extractible from the path f, and g is a projection of f by extraction of the redex u. See figure 3.

Lemma 12 (preservation of extraction). *Suppose that a redex u is extractible from a path $g : M \twoheadrightarrow N$ more standard than a path $f : M \twoheadrightarrow N$. Then the redex u is also extractible from the path f. Moreover, every projection of f by extraction of u and every projection of g by extraction of u are Lévy equivalent.*

Proof. Suppose that the redex u is extractible from the path $f = v_1 \cdots v_n : M \longrightarrow\!\!\!\!\rightarrow N$. By definition, there exists an index $1 \leq i \leq n$ such that the path $v_1 \cdots v_{i-1}$ drags the redex v_i to the redex u. We show that the index i is unique. Suppose that there exists another index $1 \leq j \leq n$ such that $v_1 \cdots v_{j-1}$ drags the redex v_j to the redex u. We may suppose without loss of generality that $i < j$. Let the rewriting path h be a projection of the rewriting path $v_1 \cdots v_i$ by extraction of the redex u at position i. By definition of extraction and projection, the two rewriting paths $v_1 \cdots v_i$ and $u \cdot h$ are Lévy equivalent. From this follows that the two paths

$$v_1 \cdots v_{j-1} = v_1 \cdots v_i \cdot v_{i+1} \cdots v_{j-1} \qquad \text{and} \qquad u \cdot h \cdot v_{i+1} \cdots v_{j-1}$$

are Lévy equivalent. Here comes the contradiction. By Lemma 10 (*preservation of drag*), the path $u \cdot h \cdot v_{i+1} \cdots v_{j-1}$ drags the redex v_j to the redex u. This may be decomposed in two steps: first, the path $h \cdot v_{i+1} \cdots v_{j-1}$ drags the redex v_j to a redex v, then the redex u drags the redex v to the redex u. This very last point means that there exists a permutation of the form $u \cdot v \rhd u \cdot h'$. This contradicts the Axiom **shape**. We thus conclude that the index i is unique for a given u.

We may suppose without loss of generality that there exists a unique standardization step from the rewriting path f to the rewriting path g. The remainder of the lemma follows then from Axioms **reversibility** and **cube** when the standardization step from f to g is reversible, and from Axioms **irreversibility**, **ancestor** and **cube** when the standardization step is irreversible. □

Remark: The uniqueness of the index i in the proof of Lemma 12 is not really necessary to establish the property, but it is a safeguard, since after all, we have not supposed anything like the optional hypothesis **descendant** formulated in Section 7.1.

2.6 Axiom 9: Termination

Axiom **termination** mirrors in our theory the *finite development* property of the λ-calculus, which states that every development of a set of β-redexes terminates. Jan Willem Klop uses the property in his PhD thesis to deduce that it is not possible to extract infinitely many times a β-redex from a fixed β-rewriting path, see [20] as well as Section 6.

Axiom 9 (Termination). *There exists no infinite sequence*

$$f_1 \searrow_{u_1} f_2 \searrow_{u_2} \cdots \searrow_{u_{k-1}} f_k \searrow_{u_k} \cdots$$

where f_i are paths and u_i are redexes.

3 A Direct Characterization of the Standard Paths

In this section, we establish a key preliminary step in our proof of the standardization theorem, performed in Section 4, by characterizing standard rewriting

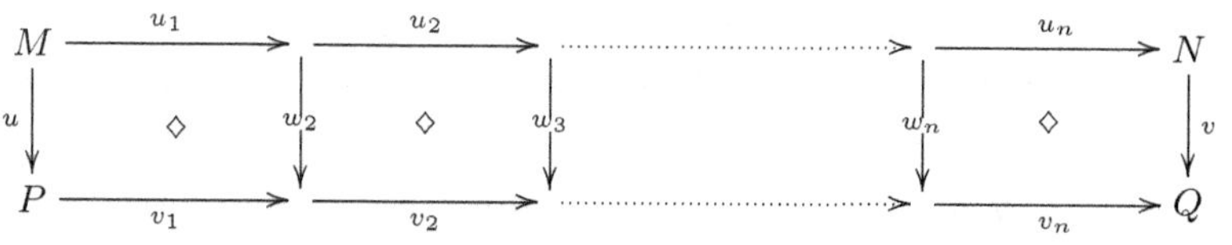

Fig. 4. The path $f = u_1 \cdots u_n : M \twoheadrightarrow N$ followed by the redex $v : N \longrightarrow Q$ permutes reversibly to the redex $u : M \longrightarrow P$ followed by the path $g = v_1 \cdots v_n : P \twoheadrightarrow Q$. Alternatively, the redex $u : M \longrightarrow P$ followed by the path $g = v_1 \cdots v_n : P \twoheadrightarrow Q$ permutes reversibly to the path $f = u_1 \cdots u_n : M \twoheadrightarrow N$ followed by the redex $v : N \longrightarrow Q$.

path in a more direct and explicit way. In Section 3.1, we introduce the notions of *starts* and *stops* of a rewriting path, and analyze their properties. From this, we deduce in Section 3.2 that every path is epi (left cancellable) with relation to the Reversible permutation relation $\simeq$. In Section 3.3, we introduce the notion of *anti-standard* path and establish that a rewriting path is standard if and only if it does not contain any occurrence of such anti-standard path.

3.1 The Structure of Starts and Stops

Definition 13 (starts and stops). *A redex $u : M \longrightarrow P$* starts *a path $f : M \twoheadrightarrow N$ when there exists a path $g : P \twoheadrightarrow N$ such that $f \simeq u \cdot g$. A redex $v : Q \longrightarrow N$* stops *a path $f : M \twoheadrightarrow N$ with* remainder *$g : M \twoheadrightarrow Q$ when $f \simeq g \cdot v$. A redex $v : Q \longrightarrow N$* stops *a path $f : M \twoheadrightarrow N$ when the redex v stops the path f with some remainder $g : M \twoheadrightarrow Q$.*

Definition 14 (reversible permutation of path and redex). *A path $f : M \twoheadrightarrow N$ followed by a redex $v : N \longrightarrow Q$* permutes reversibly *to a redex $u : M \longrightarrow P$ followed by a path $g : P \twoheadrightarrow Q$, when*

- *$f = \mathrm{id}_M$ and $g = \mathrm{id}_P$ and $v = u : M \longrightarrow P$,*
- *or $f = u_1 \cdots u_n$ and $g = v_1 \cdots v_n$ and there exists a series of $n + 1$ redexes $w_1, ..., w_{n+1}$ such that*
  - *$w_1 = u$ and $w_{n+1} = v$,*
  - *the two paths $u_i \cdot w_{i+1}$ and $w_i \cdot v_i$ form a reversible permutation $u_i \cdot w_{i+1} \diamond w_i \cdot v_i$ for every index $1 \leq i \leq n$.*

In that case, we say also that the redex $u : M \longrightarrow P$ followed by the path $g : P \twoheadrightarrow Q$ permutes reversibly *to the path $f : M \twoheadrightarrow N$ followed by the redex $v : N \longrightarrow Q$. See Figure 4.*

Remark: In Definition 14, the redex u and the rewriting path g are uniquely determined by the rewriting path f and the redex v — and conversely, the rewriting path f and the redex v are uniquely determined by the redex u and the rewriting path g. The one-to-one relationship follows from Axiom **reversibility**.

Lemma 15 (structure of stops). *A redex $v : Q \longrightarrow N$ stops a path $f = u_1 \cdots u_n : M \longrightarrow\!\!\!\!\!\rightarrow N$ with remainder $g : M \longrightarrow\!\!\!\!\!\rightarrow Q$ iff there exists an index $1 \leq i \leq n$ and a path $v_{i+1} \cdots v_n$ such that*

- *the redex u_i followed by the path $u_{i+1} \cdots u_n$ permutes reversibly to the path $v_{i+1} \cdots v_n$ followed by the redex v,*
- *the rewriting path $(u_1 \cdots u_{i-1}) \cdot (v_{i+1} \cdots v_n)$ is equivalent to the path g modulo $\simeq$.*

Proof. We declare that a redex $v : Q \longrightarrow N$ *super-stops* a path $f = u_1 \cdots u_n : M \longrightarrow\!\!\!\!\!\rightarrow N$ at position $1 \leq i \leq n$ with remainder $g : M \longrightarrow\!\!\!\!\!\rightarrow Q$ when there exists a path $v_{i+1} \cdots v_n$ such that

- the redex u_i followed by the path $u_{i+1} \cdots u_n$ permutes reversibly to the path $v_{i+1} \cdots v_n$ followed by the redex v,
- the rewriting path $(u_1 \cdots u_{i-1}) \cdot (v_{i+1} \cdots v_n)$ is equivalent to the path g modulo $\simeq$.

We declare that a redex v *super-stops* a path f with remainder g when it super-stops the path f with remainder g at some position i.

The lemma states that a redex v stops a path f with remainder a path g iff the redex v super-stops f with remainder g. Right-to-left implication ($\Leftarrow$) is immediate. The other direction ($\Rightarrow$) reduces to showing that whenever the two assertions below holds:

- a redex $v : Q \longrightarrow N$ super-stops a path $f = u_1 \cdots u_n$ with remainder g, and
- the path f' is equivalent to the path f modulo reversible permutations,

then the redex v super-stops the path f' with remainder the same rewriting path g. This elementary but fundamental preservation property is established in the following way. We may suppose without loss of generality that the two rewriting paths $f = u_1 \cdots u_n$ and $f' = u'_1 \cdots u'_n$ are related by a unique reversible permutation

$$f \overset{REV}{\Longrightarrow} f'$$

occurring at a position $1 \leq j \leq n-1$ in the rewriting path f. We thus have:

- $u'_k = u_k$ for every index $1 \leq k \leq n$ different to j and $j+1$, and
- $u_j \cdot u_{j+1} \diamondsuit u'_j \cdot u'_{j+1}$.

Now, call i any position (there exists in fact only one of these positions, $1 \leq i \leq n$, but nobody cares about that here) such that the redex $v : Q \longrightarrow N$ super-stops the path $f = u_1 \cdots u_n$ at position i with remainder g. We show by case analysis on the indices i and j that there exists an index $1 \leq k \leq n$ such that the redex $v : Q \longrightarrow N$ super-stops the path

$$f' = u'_1 \cdots u'_{k-1} \cdot u'_k \cdot u'_{k+1} \cdots u'_n$$

at position k with remainder g. To that purpose, we define a rewriting path $v'_{k+1} \cdots v'_n$ consisting of $n-k$ redexes, such that:

a. the redex u'_k followed by the path $u'_{k+1} \cdots u'_n$ permutes reversibly to the path $v'_{k+1} \cdots v'_n$ followed by the redex v,
b. the rewriting path $(u'_1 \cdots u'_{k-1}) \cdot (v'_{k+1} \cdots v'_n)$ is equivalent to the path g modulo $\simeq$.

∘ The construction is immediate when $j+1 \leq i$: simply take $k = i$ and $v'_i \cdots v'_n = v_i \cdots v_n$.
∘ The construction is also nearly immediate when $j = i$: simply take $k = i + 1$ and $v'_{i+2} \cdots v'_n = v_{i+2} \cdots v_n$, then apply Axiom **reversibility** to establish the two properties a. and b.
∘ The difficult case is the remaining case when $j > i$. In that case, let the redex x denote the unique redex such that the redex u_i followed by the path $u_{i+1} \cdots u_{j-1}$ permutes reversibly to the path $v_{i+1} \cdots v_{j-1}$ followed by the redex x. Consider the diagram below, which describes in two perspectives how the redex x followed by the path $u_j \cdot u_{j+1}$ permutes reversibly to the path $v_j \cdot v_{j+1}$ followed by the redex z:

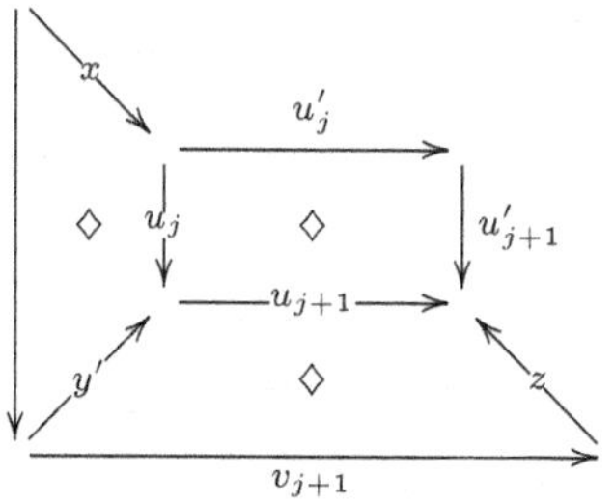

or

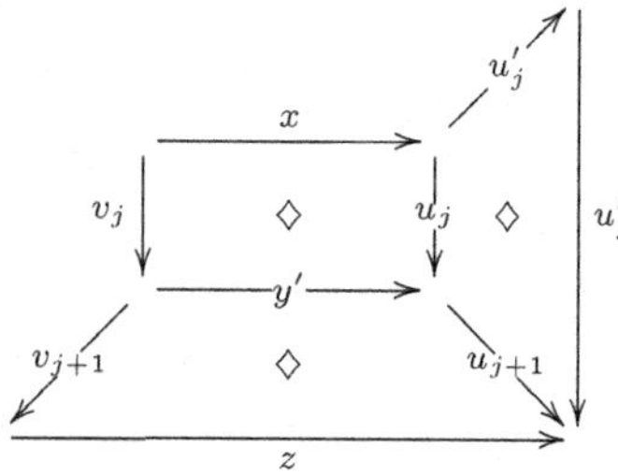

By Axiom **reversible-stability**, the diagram may be completed in the following way

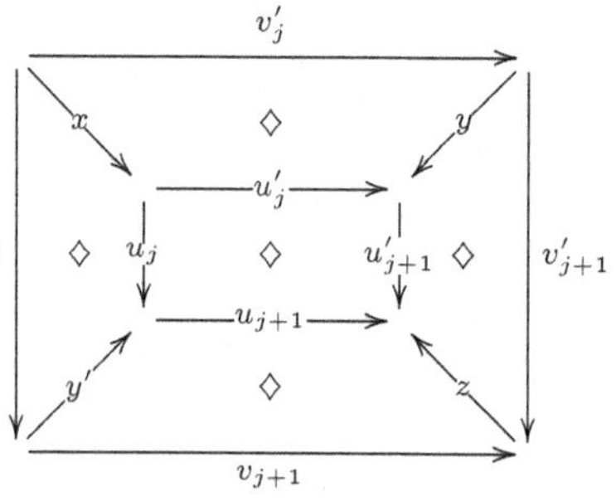

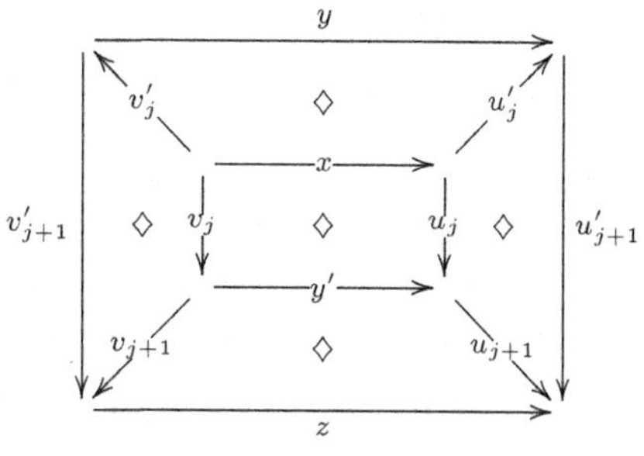

where y and v'_j and v'_{j+1} denote three redexes involved in the three reversible permutations:

$$x \cdot u'_j \diamondsuit v'_j \cdot y, \quad \text{and} \quad v_j \cdot v_{j+1} \diamondsuit v'_j \cdot v'_{j+1} \quad \text{and} \quad y \cdot u'_{j+1} \diamondsuit v'_{j+1} \cdot z.$$

The completed diagram shows (in two perspectives again) that the redex x followed by the path $u'_j \cdot u'_{j+1}$ permutes reversibly to the path $v'_j \cdot v'_{j+1}$ followed by the redex z. So, by taking $k = i$ and by defining $v'_l = v_l$ for every index $i + 1 \leq l \leq n$ different to j and $j + 1$, one obtains that:

a. the redex u_i followed by the path $u'_{i+1} \cdots u'_n$ permutes reversibly to the path $v'_i \cdots v'_n$ followed by the redex v,
b. the rewriting path $(u_1 \cdots u_{i-1}) \cdot (v'_{i+1} \cdots v'_n)$ is equivalent to the path g modulo $\simeq$. This very last point follows from the series of equivalence

$$g \simeq (u_1 \cdots u_{i-1}) \cdot (v_{i+1} \cdots v_n) \quad \text{and} \quad v_{i+1} \cdots v_n \simeq v'_{i+1} \cdots v'_n.$$

□

Unfortunately, the characterization of *starts* is not as simple as the characterization of *stops*. The main reason is that the following 2-dimensional transition system

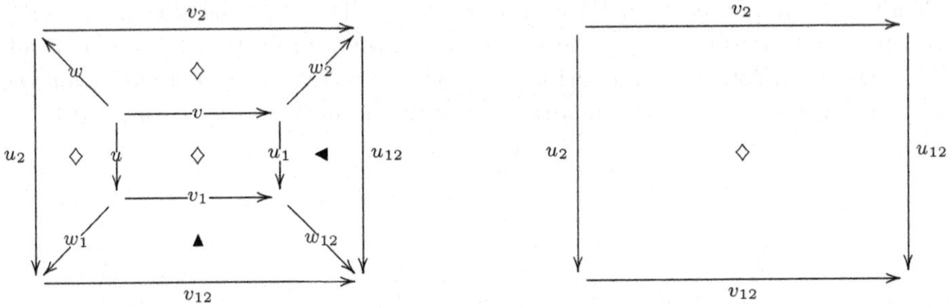

where

$$u \cdot v_1 \ \diamond \ v \cdot u_1 \qquad v \cdot w_2 \ \diamond \ w \cdot v_2 \qquad w_2 \cdot u_{12} \blacktriangleright u_1 \cdot w_{12}$$
$$u_2 \cdot v_{12} \ \diamond \ v_2 \cdot u_{12} \qquad u \cdot w_1 \ \diamond \ w \cdot u_2 \qquad w_1 \cdot v_{12} \blacktriangleright v_1 \cdot w_{12}$$

satisfies the nine properties required of an axiomatic rewriting system in Section 2. The series of equivalence

$$u \cdot w_1 \cdot v_{12} \simeq w \cdot u_2 \cdot v_{12} \simeq w \cdot v_2 \cdot u_{12} \simeq v \cdot w_2 \cdot u_{12}$$

illustrates then that a redex u may start the path $v \cdot w_2 \cdot u_{12}$ even if the path $v \cdot w_2$ followed by the redex u_{12} does not permute reversibly. However, the situation is not entirely hopeless: observe that the path $v \cdot w_2$ is $\simeq$-equivalent to the path $w \cdot v_2$ which followed by the redex u_{12} permutes reversibly to the redex u followed by the path $w_1 \cdot v_{12}$. Next lemma shows that the property characterizes *starts* in any axiomatic rewriting system.

Lemma 16 (structure of starts). *A redex $u : M \longrightarrow P$ starts a path $u_1 \cdots u_n : M \twoheadrightarrow N$ if and only there exists an index $1 \leq i \leq n$ and two paths $v_1 \cdots v_{i-1}$ and $w_1 \cdots w_{i-1}$ such that*

- *the path $v_1 \cdots v_{i-1}$ is equivalent to the path $u_1 \cdots u_{i-1}$ modulo $\simeq$,*
- *the path $v_1 \cdots v_{i-1}$ followed by the redex u_i permutes reversibly to the redex u followed by the path $w_1 \cdots w_{i-1}$.*

Proof. We declare that a redex $u : M \longrightarrow P$ *super-starts* a path $u_1 \cdots u_n : M \twoheadrightarrow N$ when there exists an index $1 \leq i \leq n$ and two paths $v_1 \cdots v_{i-1}$ and $w_1 \cdots w_{i-1}$ such that

- $u_1 \cdots u_{i-1} \simeq v_1 \cdots v_{i-1}$,
- the path $v_1 \cdots v_{i-1}$ followed by the redex u_i permutes reversibly to the redex u followed by the path $w_1 \cdots w_{i-1}$.

We prove that a redex u starts a path f iff the redex u super-starts f. Right-to-left implication ($\Leftarrow$) is immediate: the redex u super-starts the path f implies the redex u starts the path f. The converse implication ($\Rightarrow$) reduces to the following preservation property: when a redex u super-starts a path f, and when the path g is obtained from the path f by applying a reversible permutation, then the redex u super-starts also the path g.

So, consider a redex $u : M \longrightarrow P$ and a path $f = u_1 \cdots u_n : M \twoheadrightarrow N$ such that the redex u super-starts the path f. By definition, there exists an index $1 \leq i \leq n$ and two paths $v_1 \cdots v_{i-1}$ and $w_1 \cdots w_{i-1}$ such that

- $u_1 \cdots u_{i-1} \simeq v_1 \cdots v_{i-1}$,
- the redex u followed by the path $w_1 \cdots w_{i-1}$ permutes reversibly to the path $v_1 \cdots v_{i-1}$ followed by the redex u_i.

Consider any reversible standardization step

$$f \stackrel{REV}{\Longrightarrow} g$$

or equivalently, any index $1 \leq j \leq n-1$ and reversible permutation $u_j \cdot u_{j+1} \diamondsuit u'_j \cdot u'_{j+1}$. We claim that the redex u super-starts the path

$$g = (u_1 \cdots u_{j-1}) \cdot (u'_j \cdot u'_{j+1}) \cdot (u_{j+2} \cdots u_n).$$

We proceed by case analysis.

- ○ The two first cases, when $j \leq i-2$ or when $j \geq i$, are immediate.
- ○ The remaining case, when $j = i-1$, is the only difficult case. The equivalence

$$u_1 \cdots u_{i-1} \simeq v_1 \cdots v_{i-1}$$

shows that the redex u_{i-1} stops the path $v_1 \cdots v_{i-1}$ with remainder $u_1 \cdots u_{i-2}$. By Lemma 15, there exists an index $1 \leq k \leq i-1$ and a path $v'_{k+1} \cdots v'_{i-1}$ such that

- the redex v_k followed by the path $v_{k+1} \cdots v_{i-1}$ permutes reversibly to the path $v'_{k+1} \cdots v'_{i-1}$ followed by the redex u_{i-1},
- the path $(v_1 \cdots v_{k-1}) \cdot (v'_{k+1} \cdots v'_{i-1})$ is equivalent to the path $u_1 \cdots u_{i-2}$ modulo $\simeq$.

We are also in a situation where

- there exists a reversible permutation $u_{i-1} \cdot u_i \diamondsuit u'_{i-1} \cdot u'_i$
- the path $v_{k+1} \cdots v_{i-1}$ followed by the redex u_i permutes reversibly to a redex y followed by the path $w_{k+1} \cdots w_{i-1}$.

All put together, we deduce by applying Axiom **reversible-stability** $i-k-1$ times, and Axiom **reversibility** once, that there exists a redex x and path $w'_{k+1} \cdots w'_{i-1}$ such that

a. the redex u followed by the path $w_1 \cdots w_{k-1}$ permutes reversibly to the path $v_1 \cdots v_{k-1}$ followed by the redex x,
b. the redex x followed by the redex w_k permutes reversibly to the redex v_k followed by the redex y,
c. the redex y followed by the path $w_{k+1} \cdots w_{i-1}$ permutes reversibly to the path $v_{k+1} \cdots v_{i-1}$ followed by the redex u_i,
d. the redex x followed by the path $w'_{k+1} \cdots w'_{i-1}$ permutes reversibly to the path $v'_{k+1} \cdots v'_{i-1}$ followed by the redex u'_{i-1},
e. the redex w_k followed by the path $w_{k+1} \cdots w_{i-1}$ permutes reversibly to the path $w'_{k+1} \cdots w'_{i-1}$ followed by the redex u'_i.

Points a–d. are summarized in the diagram below.

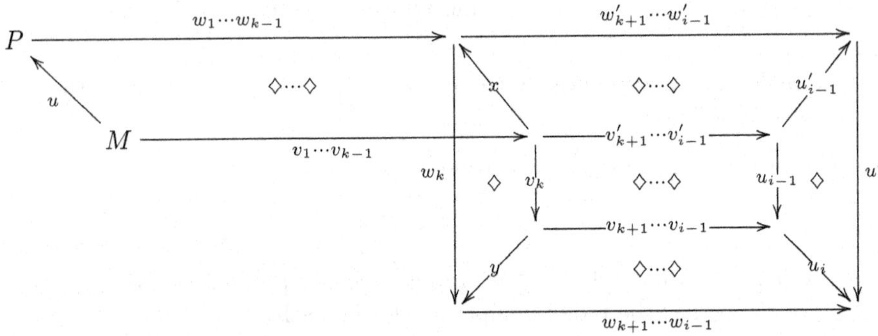

Point e. completes the diagram above by providing the front face of the cuboid generated by the redexes x and v_k and the path $v'_{k+1} \cdots v'_{i-1}$.

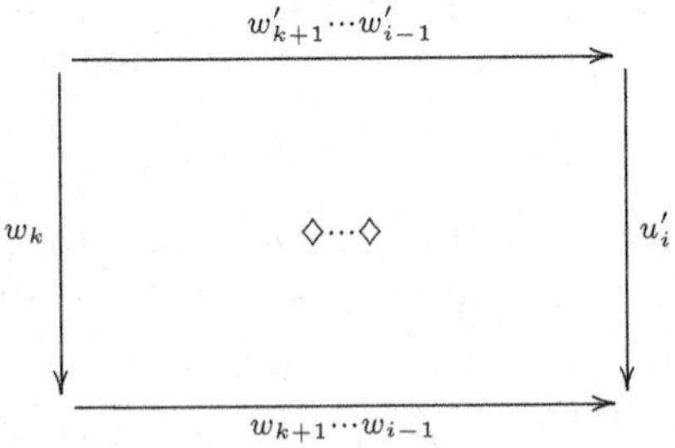

It appears now that the redex u super-starts the path

$$g = (u_1 \cdots u_{i-2}) \cdot (u'_{i-1} \cdot u'_i) \cdot (u_{i+1} \cdots u_n).$$

because

- the path $u_1 \cdots u_{i-2}$ is equivalent to the path $(v_1 \cdots v_{k-1}) \cdot (v'_{k+1} \cdots v'_{i-1})$ modulo $\simeq$,
- the path $(v_1 \cdots v_{k-1}) \cdot (v'_{k+1} \cdots v'_{i-1})$ followed by the redex u'_{i-1} permutes reversibly to the redex u followed by the path $(w_1 \cdots w_{k-1}) \cdot (w'_{k+1} \cdots w'_{i-1})$.

This establishes the equivalence between starting and super-starting a path. Since this is precisely what our lemma asserts, we conclude. □

3.2 Application: Every Rewriting Path is Epi wrt. $\simeq$

We illustrate the previous section with an application of Lemma 15.

Lemma 17 (epi wrt. $\simeq$). *If $f \cdot g_1 \simeq f \cdot g_2$ then $g_1 \simeq g_2$.*

Proof. We may suppose without loss of generality that the rewriting path f is a redex u. We prove that $u \cdot g_1 \simeq u \cdot g_2$ implies $g_1 \simeq g_2$ by induction on the length of g_1 (and of g_2). The property is immediate when g_1 (and therefore g_2) is empty. Otherwise, the path g_1 factors as $g_1 = h_1 \cdot v$ for some path h_1 and redex v. By Lemma 15, because the redex v stops the path $u \cdot g_2$ with remainder $u \cdot h_1$, one of the two following cases occurs:

- either there exists a path h_2 such that $g_2 \simeq h_2 \cdot v$ and $u \cdot h_1 \simeq u \cdot h_2$,
- or there exists a path h_2 such that the redex u followed by the path g_2 permutes reversibly to the path h_2 followed by the redex v, and such that $h_2 \simeq u \cdot h_1$.

In the first case, we deduce that $h_1 \simeq h_2$ by induction hypothesis on $u \cdot h_1 \simeq u \cdot h_2$, and conclude that $g_1 \simeq g_2$ by the series of equivalence:

$$g_1 = h_1 \cdot v \simeq h_2 \cdot v \simeq g_2$$

Now, we prove that the second case does not occur. Obviously, the path h_2 drags the redex v to the redex u. By Lemma 10 (*preservation of drag*) and equivalence $h_2 \simeq u \cdot h_1$, the path $u \cdot h_1$ drags the redex v to the redex u. In particular, there exists a redex w and a path h such that $u \cdot w \rhd u \cdot h$. This contradicts Axiom **shape**, and we conclude. □

Remark: In Section 7.2 an additional hypothesis of **reversible-shape** is required to complete the property to an epi-mono property wrt. $\simeq$.

3.3 Characterization Lemma

We introduce below the fundamental notion of *anti-standard* path. These anti-standard paths are called *conflicts* in [13, 27]. We change the terminology here because the word *conflict* is generally understood as *non determinism*, and because the notion of *anti-standard path* specializes to the notion of *anti-standard pair* introduced by J. W. Klop in the particular case of the λ-calculus equipped with the left-order $\preceq_{\text{left}}$ — see [20] and Section 1.9.

Definition 18. *A path is* anti-standard *(see Figure 5) when it factors as*

$$M \xrightarrow{u} P \xrightarrow{f} Q \xrightarrow{y} N$$

where u and y are redexes and f is a rewriting path, and

- *the redex u followed by the path f permutes reversibly to the path g followed by the redex v,*

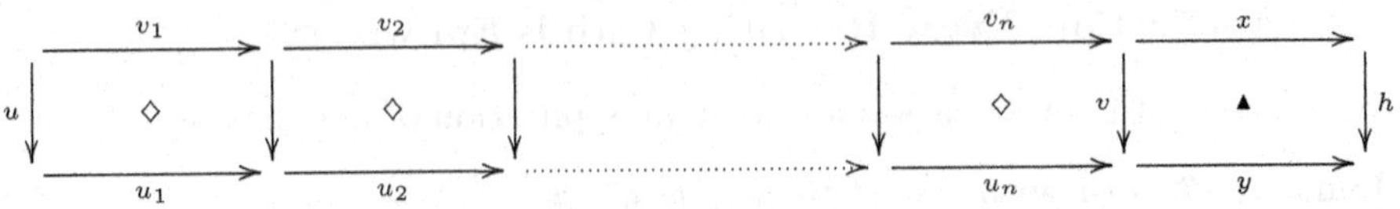

Fig. 5. The definition of an anti-standard path $u \cdot u_1 \cdots u_n \cdot y$: the redex u followed by the path $u_1 \cdots u_n$ permutes reversibly to the path $v_1 \cdots v_n$ followed by the redex v which permutes irreversibly with the redex y, as follows: $v \cdot y \blacktriangleright x \cdot h$.

– *the redex v and the redex y induce an irreversible permutation $v \cdot y \blacktriangleright x \cdot h$, for some redex x and rewriting path h.*

The β-rewriting path taken earlier as illustration

$$Ka(\Delta\Delta) \xrightarrow{\Delta_1} Ka(\Delta\Delta) \xrightarrow{K} (\lambda x.a)(\Delta\Delta) \xrightarrow{\lambda} a$$

is a typical example of anti-standard path in the axiomatic rewriting system $(\mathcal{G}_\lambda, \rhd_{\mathsf{tree}})$. Compare indeed Diagrams (9) and (13) to Figure 5.

This leads us to the main result of the section.

Lemma 19 (characterization). *A path $u_1 \cdots u_n$ is standard if and only if there exists no pair of indices $1 \leq i < j \leq n$ such that $u_i \cdots u_j$ defines an anti-standard path.*

Proof. Left-to-Right implication ($\Rightarrow$) is immediate. Proving the converse direction ($\Leftarrow$) reduces to showing that:

– when two rewriting paths f and g are equivalent modulo reversible permutations $\simeq$, and
– when the path f contains an anti-standard path,

then the path g contains also an anti-standard path.

So, consider two rewriting paths $f = u_1 \cdots u_n$ and $g = u'_1 \cdots u'_n$, and suppose that the path g is obtained after a unique reversible standardization step on the path f:

$$f \overset{REV}{\Longrightarrow} g. \tag{19}$$

Let $1 \leq k \leq n-1$ denote the index where the reversible permutation occurs in the path f. Obviously,

$$u'_1 \cdots u'_{k-1} = u_1 \cdots u_{k-1} \text{ and } u'_k \cdot u'_{k+1} \diamond u_k \cdot u_{k+1} \text{ and } u'_{k+2} \cdots u'_n = u_{k+2} \cdots u_n.$$

Now, suppose that the path f contains an anti-standard path, in the sense that there exist two indices $1 \leq i < j \leq n$ such that the path $u_i \cdots u_j$ is anti-standard. Let y denote the redex u_j. By definition of an anti-standard path, there exists a path $v_{i+1} \cdots v_{j-1}$ and redex w such that:

– the redex u_i followed by the path $u_{i+1} \cdots u_{j-1}$ permutes reversibly to the path $v_{i+1} \cdots v_{j-1}$ followed by the redex w,
– the redexes w and y form an irreversible permutation $w \cdot y \blacktriangleright x \cdot h$ for some redex x and path h.

We establish now that there exist two indices $1 \leq I < J \leq n$ such that the path $u'_I \cdots u'_J$ is anti-standard. This will show in particular that the path g contains an anti-standard path.
∘ The property is immediate when $k > j$: simply take $(I, J) = (i, j)$.
∘ The property follows from Lemma 15 when $k + 1 < j$:

– take $(I, J) = (i - 1, j)$ when $k = i - 1$,
– take $(I, J) = (i + 1, j)$ when $k = i$,
– take $(I, J) = (i, j)$ otherwise.

There remain only two difficult cases to treat: when $k = j - 1$ and when $k = j$.

∘ We treat the first case, when $k = j - 1$. The situation is summarized by the diagram:

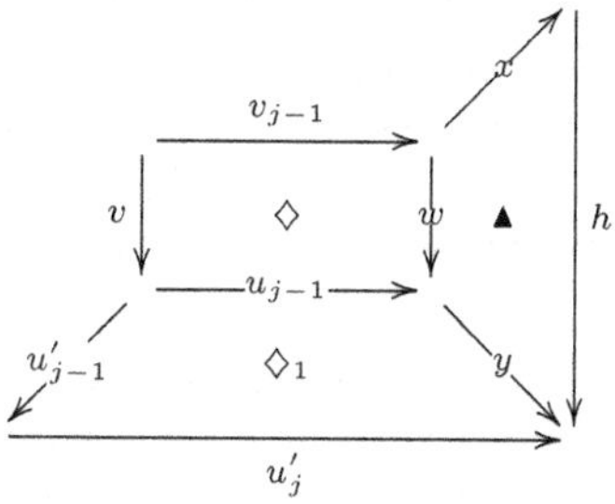

where the reversible permutation $\diamond_1$ relates the rewriting paths f and g in Equation (19) and where the irreversible permutation $w \cdot y \blacktriangleright x \cdot h$ between the redex w and the redex y witnesses the fact that the path $u_i \cdots u_{j-1} \cdot y$ (or equivalently the path $u_i \cdots u_{j-1} \cdot u_j$) is anti-standard.

The diagram may be completed by Axiom **stability** in the following way:

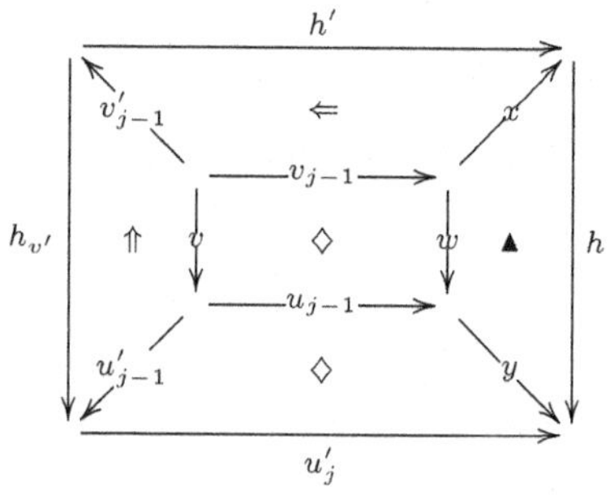

where v'_{j-1} is a redex, where h' and $h_{v'}$ are two rewriting paths, forming permutations

$$v \cdot u'_{j-1} \rhd v'_{j-1} \cdot h_{v'} \qquad \text{and} \qquad v_{j-1} \cdot x \rhd v'_{j-1} \cdot h'.$$

We proceed by case analysis on the permutation $v \cdot u'_{j-1} \rhd v'_{j-1} \cdot h_{v'}$:

— **Either the permutation is irreversible.** In that case, the path $u_i \cdots u_{j-2} \cdot u'_{j-1}$ is anti-standard, and we may thus conclude with $(I, J) = (i, j-1)$.

— **Or the permutation is reversible.** In that case, the path $h_{v'}$ is a redex; we write it v' for clarity's sake. We claim that the path $u_i \cdots u_{j-2} \cdot u'_{j-1} \cdot u'_j$ is anti-standard. Indeed, the redex u_i followed by the path $u_{i+1} \cdots u_{j-2} \cdot u'_{j-1}$ permutes reversibly to the path $v_{i+1} \cdots v_{j-2} \cdot v'_{j-1}$ followed by the redex v', and we establish now that the redexes v' and u'_j are involved in an irreversible permutation $v' \cdot u'_j \blacktriangleright v'_j \cdot h''$ for some redex v'_j and rewriting path h''. First of all, the rewriting path $v \cdot u'_{j-1}$ drags the redex u'_j to the redex v_{j-1}. So, by Lemma 10 (*preservation of drag*), the path $v'_{j-1} \cdot v'$ which is Lévy equivalent to the path $v \cdot u'_{j-1}$, drags the redex u'_j to the redex v_{j-1}. From this follows that there exists a permutation of the form $v' \cdot u'_j \rhd v'_j \cdot h''$ for some redex v'_j and rewriting path h''. There remains to show that this permutation is irreversible in order to establish our claim. We proceed by contradiction and suppose that the permutation $v' \cdot u'_j \rhd v'_j \cdot h''$ is reversible. Then, it follows from Axiom **reversible-stability** applied around the permutation $v \cdot u'_{j-1} \diamondsuit v'_{j-1} \cdot v'$ that:

- there exists a reversible permutation starting from the rewriting path $v \cdot u_{j-1}$; this permutation is necessarily the permutation $v \cdot u_{j-1} \diamondsuit v_{j-1} \cdot w$ by Axiom **reversibility**,
- there exists a reversible permutation starting from the rewriting path $w \cdot y$.

By Axiom **reversibility**, this last assertion contradicts the fact that there exists an irreversible permutation starting from the rewriting path $w \cdot y$. From this, we conclude that the permutation $v' \cdot u'_j \rhd v'_j \cdot h''$ starting from the rewriting path $v' \cdot u'_j$ is irreversible, and thus that the rewriting path $u_i \cdots u_{j-2} \cdot u'_{j-1} \cdot u'_j$ is anti-standard. We may thus take $(I, J) = (i, j)$.

∘ We treat the second case, when $k = j$, and thus, the two redexes u_j and u_{j+1} are permuted reversibly in the path f to obtain the path g. Again, we let y denote the redex u_j. So, the redex u_i followed by the path $u_{i+1} \cdots u_{j-1}$ permutes reversibly to the path $v_{i+1} \cdots v_{j-1}$ followed by the redex w; and the redex w induces the irreversible permutation $w \cdot y \blacktriangleright x \cdot h$ with the redex y, witnessing the fact that the path $u_i \cdots u_{j-1} \cdot y$ (or equivalently the path $u_i \cdots u_{j-1} \cdot u_j$) is anti-standard.

The situation is summarized in the diagram below:

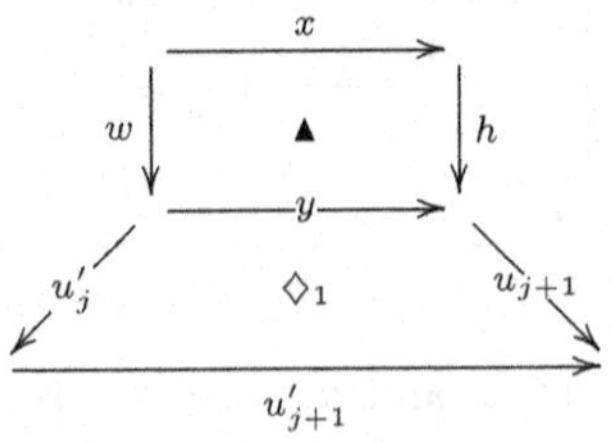

where the reversible permutation $\diamondsuit_1$ relates the rewriting paths f and g in Equation (19).

Here, we apply Axiom **enclave** and complete the diagram in the following way:

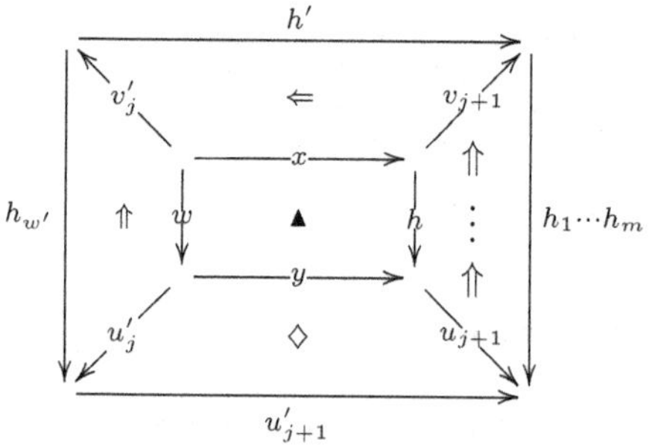

with two redex v'_j and two rewriting paths $h_{w'}$ and h' inducing permutations:

$$w \cdot u'_j \rhd v'_j \cdot h_{w'} \qquad \text{and} \qquad x \cdot v_{j+1} \rhd v'_j \cdot h'.$$

Note moreover that the path h grabs the redex u_{j+1} to a redex v_{j+1}, and that the redex x grabs the redex v_{j+1} to the redex v'_j.

We proceed by case analysis on the permutation $w \cdot u'_j \rhd v'_j \cdot h_{w'}$:

— **Either the permutation is irreversible.** In that case, the rewriting path $u_i \cdots u_{j-1} \cdot u'_j$ is anti-standard, and we may thus conclude with $(I, J) = (i, j)$.

— **Or the permutation is reversible.** In that case, the path $h_{w'}$ is a redex; we thus write it w' for clarity's sake. We claim that the rewriting path $u_i \cdots u_{j-1} \cdot u'_j \cdot u'_{j+1}$ is anti-standard. Indeed, the redex u_i followed by the path $u_{i+1} \cdots u_{j-1} \cdot u'_j$ permutes reversibly to the path $v_{i+1} \cdots v_{j-1} \cdot v'_j$ followed by the redex w', and we establish now that the redexes w and u'_{j+1} induce together an irreversible permutation starting from the path $w' \cdot u'_{j-1}$. The path $w \cdot u'_j$ grabs the redex u'_{j+1} to the redex x. By Lemma 10 (*preservation of drag*), the path $v'_j \cdot w'$ which is Lévy equivalent to the path $w \cdot u'_j$, drags the redex u'_{j+1} to the redex x. This ensures that the two redexes w' and u'_{j+1} induce together a permutation starting from the rewriting path $w' \cdot u'_{j+1}$. There remains to show that this permutation is irreversible. We proceed by contradiction and suppose that the permutation $v' \cdot u'_j \rhd v'_j \cdot h''$ is reversible. Then, it follows from Axiom **reversible-stability** applied around the permutation $w \cdot u'_j \diamond v'_j \cdot w'$ that there exists a reversible permutation starting from the rewriting path $w \cdot y$. This together with Axiom **reversibility** contradicts the existence of the irreversible permutation $w \cdot y \blacktriangleright x \cdot h$ which starts also from the rewriting path $w \cdot y$. We conclude that, as claimed, the two redexes w' and u'_{j+1} are involved in an irreversible permutation starting from the rewriting path $w' \cdot u'_{j+1}$. Thus, the rewriting path $u_i \cdots u_{j-1} \cdot u'_j \cdot u'_{j+1}$ is anti-standard. This concludes the proof, with $(I, J) = (i, j+1)$.

Conclusion: We have just established that when a path f contains an anti-standard path, then every path g equivalent to the path f modulo reversible permutations $\simeq$ contains also an anti-standard path. Lemma 19 follows immediately. □

Lemma 20 (interface). *Suppose that two paths $f : M \twoheadrightarrow P$ and $g : P \twoheadrightarrow N$ are standard. Then, the composite path $f \cdot g : M \twoheadrightarrow N$ is standard if and only if the path $u \cdot g$ is standard, for every redex u which stops f.*

Proof. Follows immediately from Lemma 19. □

4 The Standardization Theorem

All along this section, we suppose that the 2-dimensional transition system $(\mathcal{G}, \triangleright)$ defines an axiomatic rewriting system — equivalenly, that it satisfies the nine axioms formulated in Section 2. From this assumption, we deduce the diagrammatic standardization theorem (Theorem 2) evocated in the Introduction — in Section 1.8.

4.1 The Outermost Redex

For every nonempty path $f : M \twoheadrightarrow N$, we define a redex $outm(f) : M \longrightarrow P$ extractible from the path f, in these sense of Definition 11. This redex is called the *outermost* redex of the rewriting path f. We will see at the later stage of the proof that the redex $outm(f)$ is the first redex of a particular standard path g associated to the path f. The definition of the redex $outm(f)$ is by induction on the length of the path f.

Definition 21 (outermost redex). *For every non-empty path $f : M \twoheadrightarrow N$, the redex $outm(f)$ is defined as follows:*

$$\begin{array}{|ll|} outm(v) = v & \textit{for a redex } v, \\ outm(v \cdot f) = & \begin{cases} u \textit{ when the redex } v \textit{ drags the redex } outm(f) \textit{ to the redex } u, \\ v \textit{ when there is no permutation of the form } v \cdot outm(f) \triangleright h. \end{cases} \end{array}$$

Lemma 22 (preservation of outermost). *Let $f : M \twoheadrightarrow N$ be a path. Suppose that $u : M \longrightarrow P$ is a redex extractible from f, and that g is a projection of f by extraction of u. Then,*

- *either $outm(f) = u$,*
- *or the path g is nonempty, and $outm(g) \overset{u}{\longleftarrow\!\!\dashv} outm(f)$.*

Proof. By induction on the length of the path f. The property is immediate when the path f is a redex. Otherwise, suppose that the path f factors as $f = v \cdot f'$ where v is a redex and where f' is a nonempty path satisfying the property stated in the lemma. Suppose moreover that the redex u is extractible from the path f, and that $f \searrow_u g$ (see Definition 11 for a definition of the notation $\searrow_u$.)

We proceed by case analysis, depending whether the two redexes u and v coincide.
∘ Suppose that $u = v$, and thus, that the redex u is the first redex rewritten in the path f. Then, by definition of the redex $outm(-)$, either $u = outm(f)$ or $outm(f') \overset{u}{\longleftarrow\!\!\!\dashv} outm(f)$. We conclude because the equality $f' = g$ holds.
∘ Suppose now that $u \neq v$. By definition of $f \searrow_u g$, there exists a redex u' and two paths $h_{v'}$ and g' such that (1) the path g factors as $g = h_{v'} \cdot g'$, and (2) $f' \searrow_{u'} g'$ and (3) $v \cdot u' \rhd u \cdot h_{v'}$. The situation is summarized in the diagram below:

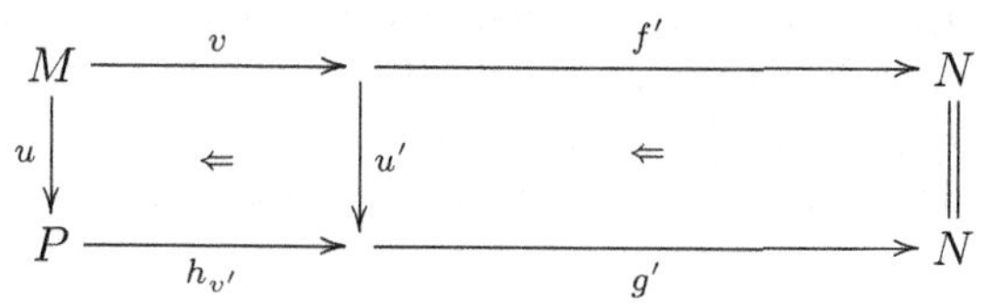

Since the proof is finished when $outm(f) = u$, we suppose from now on that $outm(f) \neq u$. From this follows that $outm(f') \neq u'$ by definition of $outm(-)$ and by Axiom **ancestor**. Here, we apply our induction hypothesis on the path f', and deduce that $outm(f') \overset{u'}{\longleftarrow\!\!\!\dashv} outm(g')$. The diagram below describes the situation:

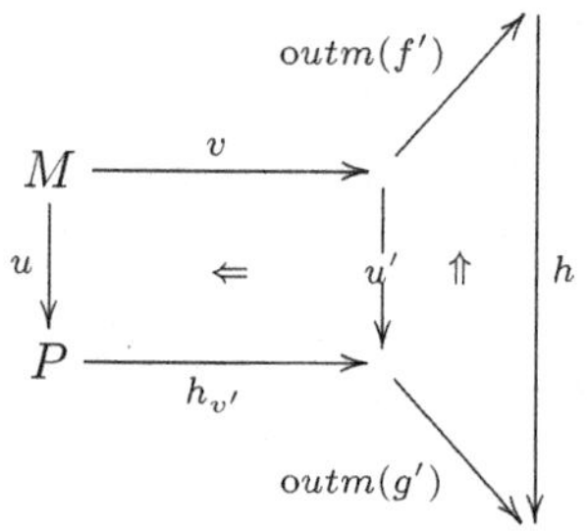

From now on, we proceed by case analysis on the permutation $v \cdot u' \rhd u \cdot h_{v'}$.

— **Either the permutation** $v \cdot u' \rhd u \cdot h_{v'}$ **is irreversible.** In that case, we apply Axiom **enclave**, and deduce that

1. the redex v drags the redex $outm(f')$ to the redex $outm(g)$, and
2. the path $h_{v'}$ drags the redex $outm(g')$ to the redex $outm(g)$, and
3. the redex u drags the redex $outm(g)$ to the redex $outm(f)$.

The third assertion concludes the proof.

— **Or the permutation** $v \cdot u' \rhd u \cdot h_{v'}$ **is reversible.** In that case, the path $h_{v'}$ is a redex. We write it v' for clarity's sake. Again, we proceed by case analysis, depending on whether the redex v coincides with the redex $outm(f)$.

1. Suppose that the redex v does not coincide with $outm(f)$. By definition of $outm(-)$, the redex v drags the redex $outm(f')$ to the redex $outm(f)$.

From this follows that the path $v \cdot u'$ drags the redex $outm(g')$ to the redex $outm(f)$. By Lemma 10 (*preservation of drag*), the path $u \cdot v'$ which is Lévy equivalent to the path $v \cdot u'$, the path $u \cdot v'$ drags the redex $outm(g')$ to the redex $outm(f)$. From this follows that the redex v' drags the redex $outm(g')$ to the redex $outm(g)$, and that the redex u drags the redex $outm(g)$ to the redex $outm(f)$. This concludes the proof.

2. Suppose that the redex v is equal to the redex $outm(f)$. In that case, we claim that the redex v' coincides with the redex $outm(g)$. We proceed by contradiction and suppose that $v' \neq outm(g)$. By definition of $outm(-)$, the redex v' drags the redex $outm(g')$ to the redex $outm(g)$. It follows from Axiom **stability** applied around the reversible permutation $v \cdot u' \diamond u \cdot v'$, that the redex v drags the redex $outm(f')$ to a redex w. This contradicts the equality $v = outm(f)$. We conclude that $v' = outm(g)$, and thus, that the redex u drags the redex $v' = outm(g)$ to the redex $v = outm(f)$. We conclude.

All this concludes our proof by induction on the length of the path f. □

Lemma 23. *Let $f : M \twoheadrightarrow N$ be a path. The redex $outm(f)$ is extractible from any path $u_1 \cdots u_n : M \twoheadrightarrow N$ obtained as follows:*

$$f \searrow_{u_1} f_2 \searrow_{u_2} \cdots f_n \searrow_{u_n} \mathrm{id}_N.$$

Proof. Immediate consequence of Lemma 22. □

4.2 Uniqueness

Lemma 24. *Suppose that $(M_1 \xrightarrow{u_1} M_2 \xrightarrow{u_2} \cdots \xrightarrow{u_{n-1}} M_n \xrightarrow{u_n} M_{n+1})$ is a standard path. Suppose moreover that, for every index $1 \leq i \leq n$, the path $u_i \cdots u_n$ is more standard than every path in its Lévy equivalence class:*

$$\forall\, 1 \leq i \leq n,\quad \forall h : M_i \twoheadrightarrow M_{n+1}, \qquad h \equiv u_i \cdots u_n \ \textit{implies}\ h \Longrightarrow u_i \cdots u_n.$$

Then, for every path $f_1 : M_1 \twoheadrightarrow M_{n+1}$ Lévy equivalent to the path $u_1 \cdots u_n$, there exists a series of rewriting paths $f_i : M_i \twoheadrightarrow M_{n+1}$ indexed by $1 \leq i \leq n$ and a sequence of extractions:

$$f_1 \searrow_{u_1} f_2 \searrow_{u_2} \cdots f_n \searrow_{u_n} \mathrm{id}_{M_{n+1}}.$$

Proof. We proceed by induction on the length n of the rewriting path $u_1 \cdots u_n$. Suppose that $f : M \twoheadrightarrow N$ is a rewriting path Lévy equivalent to the path $u_1 \cdots u_n$. Note that the redex u_1 is extractible from the path $u_1 \cdots u_n$ with resulting projection the path $u_2 \cdots u_n$. Now, by hypothesis, the path $u_1 \cdots u_n$ is more standard than the path f. From this and Lemma 12 (*preservation of extraction*) follows that the redex u_1 is extractible from the path $f_1 = f$ with projection a path f_2 Lévy equivalent to the path $u_2 \cdots u_n$. We know by induction that there exists a sequence of extractions

$$f_2 \searrow_{u_2} f_3 \searrow_{u_3} \cdots f_n \searrow_{u_n} \mathrm{id}_{M_{n+1}}.$$

We have thus established that there exists a sequence of extractions

$$f_1 \searrow_{u_1} f_2 \searrow_{u_2} \cdots f_n \searrow_{u_n} \mathrm{id}_{M_{n+1}}.$$

This concludes our proof by induction. □

Lemma 25 (uniqueness). *A standard path is more standard than every path in its Lévy equivalence class.*

Proof. We proceed by induction on the length of the standard path. Suppose from now on that the property is satisfied for every path of length $n-1$, and suppose that

$$f \; = \; (M_1 \xrightarrow{u_1} M_2 \xrightarrow{u_2} \cdots \xrightarrow{u_{n-1}} M_n \xrightarrow{u_n} M_{n+1})$$

is a standard path of length n. We establish that the path f is more standard than every path in its Lévy equivalence class.

Step 1. First of all, we claim that in order to establish that property of the path f, we only need to show that the redex u_1 is extractible from every path Lévy equivalent to the path f. Suppose indeed that this is the case, and consider a path g Lévy equivalent to the standard path f. By definition of Lévy equivalence, there exists a sequence of permutations

$$f = f_1 \stackrel{1}{\equiv} f_2 \stackrel{1}{\equiv} \cdots \stackrel{1}{\equiv} f_m \stackrel{1}{\equiv} f_{m+1} = g$$

of standardization steps $f_i \stackrel{1}{\Longrightarrow} f_{i+1}$ or $f_i \stackrel{1}{\Longleftarrow} f_{i+1}$, for every $1 \leq i \leq m$. For each such index i, the rewriting path f_i is Lévy equivalent to the path f. We have just assumed that the redex u_1 is thus extractible from each path f_i. Now, we may apply Lemma 12 (*preservation of extraction*) as many times as there are permutation steps from the path f to the path g to deduce that the two paths f and g have the same projections (modulo Lévy equivalence) after extraction of the redex u_1. Now, the path $u_2 \cdots u_n$ is the unique projection of the path f by extraction of the redex u_1. We conclude that any projection g' of the rewriting path g obtained by extraction of the redex u_1 is Lévy equivalent to the path $u_2 \cdots u_n$. By applying our induction hypothesis on the path $u_2 \cdots u_n$, we know that the path $u_2 \cdots u_n$ is more standard than the path g'. It follows that the path $f = u_1 \cdots u_n$ is more standard than the path $u_1 \cdot g'$, which is, by construction, more standard than the path g. This establishes that the path f is more standard than every path in its Lévy equivalence class.

Step 2. We have just shown in Step 1. that we only need to prove here that the redex u_1 is extractible from every path Lévy equivalent to the path $f = u_1 \cdots u_n$. We introduce the necessary notation to that purpose. The proof proceeds by contradiction. We suppose that the redex u_1 is not extractible from a particular path in the Lévy equivalence class of the path f. By definition of Lévy equivalence, there exists a sequence

$$f_1 \stackrel{1}{\equiv} f_2 \stackrel{1}{\equiv} \cdots \stackrel{1}{\equiv} f_m \stackrel{1}{\equiv} f_{m+1}$$

of standardization steps $f_i \overset{1}{\Longrightarrow} f_{i+1}$ or $f_i \overset{1}{\Longleftarrow} f_{i+1}$, for every $1 \leq i \leq m$, such that:

- $f_1 = f$,
- the redex u_1 is extractible from the path f_j, for every index $1 \leq j \leq m$,
- the redex u_1 is not extractible from the path f_{m+1}.

For each index $1 \leq i \leq m$, we define the path g_i as any projection of the path f_i by extraction of the redex u_1. So,

$$\forall 1 \leq i \leq m, \qquad f_i \searrow_{u_1} g_i.$$

Note that Lemma 12 (*preservation of extraction*) implies that all the paths $g_1 = u_2 \cdots u_n$, and $g_2, \dots, g_m$ are Lévy equivalent.

Step 3. Here, we will be slightly more explicit than in Step 2. Let p denote the length of the path f_m. Thus, the path f_m factors as

$$f_m = v_1 \cdots v_p$$

where each v_i denotes a redex, for $1 \leq i \leq p$. We know by construction that $f_m \overset{1}{\equiv} f_{m+1}$. It follows from Lemma 12 (*preservation of extraction*) that in fact

$$f_m \overset{1}{\Longrightarrow} f_{m+1}$$

because the redex u_1 is extractible from the path f_m but not from the path f_{m+1}. By definition of $\overset{1}{\Longrightarrow}$, the paths f_m and f_{m+1} factor as:

$$f_m = v_1 \cdots v_{k-1} \cdot (v_k \cdot v_{k+1}) \cdot v_{k+2} \cdots v_p \qquad f_{m+1} = v_1 \cdots v_{k-1} \cdot (w_k \cdot h) \cdot v_{k+2} \cdots v_p$$

for some index $1 \leq k \leq p-1$, where w_k is a redex and h is a path involved in a permutation $v_k \cdot v_{k+1} \rhd w_k \cdot h$. Now, it follows from Lemma 10 (*preservation of drag*) and Axiom **ancestor** that:

- the permutation $v_k \cdot v_{k+1} \blacktriangleright w_k \cdot h$ is irreversible,
- the path $v_1 \cdots v_{k-1}$ drags the redex v_k to the redex u_1.

The situation is summarized in the diagram below:

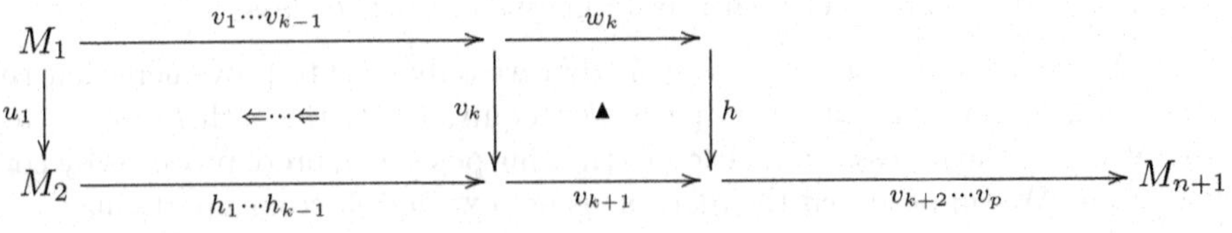

Step 4. We establish the equality $outm(f_m) = outm(f_{m+1})$. We proceed by case analysis, depending whether the redex v_{k+1} coincides with the redex $outm(v_{k+1} \cdots v_p)$.

- Suppose that the redex v_{k+1} is not equal to the redex $outm(v_{k+1} \cdots v_p)$. By Lemma 22, the path $v_{k+2} \cdots v_p$ is nonempty, and the redex v_{k+1} drags the redex $outm(v_{k+2} \cdots v_p)$ to the redex $outm(v_{k+1} \cdots v_p)$. By Axiom **enclave** applied around the irreversible permutation $v_k \cdot v_{k+1} \blacktriangleright w_k \cdot h$, the two paths $v_k \cdot v_{k+1}$ and $w_k \cdot h$ drag the redex $outm(v_{k+2} \cdots v_p)$ to the same redex $outm(v_k \cdots v_p) = outm(w_k \cdot h \cdot v_{k+2} \cdots v_p)$. The inductive definition of $outm(-)$ ensures then that $outm(f_m) = outm(f_{m+1})$. We conclude.
- Suppose now that the redex v_{k+1} coincides with the redex $outm(v_{k+1} \cdots v_p)$. In that case, $outm(v_k \cdots v_p) = w_k$ because the redex v_k drags the redex $v_{k+1} = outm(v_{k+1} \cdots v_p)$ to the redex w_k. Now, we claim that $outm(w_k \cdot h \cdot v_{k+2} \cdots v_p) = w_k$. First of all, it follows from Axioms **ancestor** and **irreversibility** and from $v_k \cdot v_{k+1} \blacktriangleright w_k \cdot h$ that the redex w_k is the only redex extractible from the path $w_k \cdot h$. So, there only remains to prove that the redex $outm(w_k \cdot h \cdot v_{k+2} \cdots v_p)$ is extractible from the path $w_k \cdot h$. Suppose that it is not. In that case, the path $w_k \cdot h$ drags the redex $outm(v_{k+2} \cdots v_p)$ to the redex $outm(w_k \cdot h \cdot v_{k+2} \cdots v_p)$. By Lemma 10 (*preservation of drag*) the path $v_k \cdot v_{k+1}$ which is Lévy equivalent to the path $w_k \cdot h$, drags the redex $outm(v_{k+2} \cdots v_p)$ to the same redex $outm(v_k \cdots v_p) = outm(w_k \cdot h \cdot v_{k+2} \cdots v_p)$. This contradicts the equality $w_k = outm(v_k \cdots v_p) = v_{k+1}$. We conclude that $outm(v_k \cdots v_p) = w_k = outm(w_k \cdot h \cdot v_{k+2} \cdots v_p)$ and thus that $outm(f_m) = outm(f_{m+1})$.

Step 5. We deduce from Step 4 that the redex u_1 drags the redex $outm(g_m)$ to the redex $outm(f_m)$. We have just proved that $outm(f_m) = outm(f_{m+1})$. From this follows that the redex $outm(f_m)$ is extractible from the path f_{m+1}. Since by construction of the path f_{m+1}, the redex u_1 is not extractible from that path, the two redexes u_1 and $outm(f_m)$ are necessarily different. We may thus apply Lemma 22 on the extraction $f_m \searrow_{u_1} g_m$. This establishes our claim: the redex u_1 drags the redex $outm(g_m)$ to the redex $outm(f_m)$.

Step 6. We prove that the redex $outm(g_m)$ is extractible from the path $g_1 = u_2 \cdots u_n$. By induction hypothesis, each path $u_i \cdots u_n$ is more standard than any of its Lévy equivalent paths, for $2 \leq i \leq n$. We may thus apply Lemma 24 to the paths g_1 and $u_2 \cdots u_n$, and deduce that there exists a series of extractions

$$g_1 \searrow_{u_2} \cdots \searrow_{u_n} \mathrm{id}_{M_{n+1}}.$$

By Lemma 23, the series implies that the redex $outm(g_m)$ is extractible from the path $u_2 \cdots u_n$.

Step 7. We deduce from Step 6 that the redex $outm(g_m)$ is extractible from all the paths $g_1, ..., g_m$. We have already noted at the end of Step 2 that all the paths $g_1 = u_2 \cdots u_n$, g_2, ..., g_m are Lévy equivalent. By induction hypothesis, the standard path $g_1 = u_2 \cdots u_n$ is more standard than every path g_i, for every index $1 \leq i \leq m$. We also know that the redex $outm(g_m)$ is extractible from the path g_1. By Lemma 12 (*preservation of extraction*), the redex $outm(g_m)$ is thus extractible from the path g_i, for every index $1 \leq i \leq m$.

Step 8. We deduce from Steps 4, 5 and 7 that the redex $outm(f_m)$ is extractible from the paths $f_1, ..., f_m, f_{m+1}$. By Step 4, the redex $outm(f_m)$ is extractible from the path f_{m+1}. So, there remains to show that the redex $outm(f_m)$ is extractible from the paths $f_1, ..., f_m$. By Step 5, the redex u_1 drags the redex $outm(g_m)$ to the redex $outm(f_m)$. By Step 7, the redex $outm(g_m)$ is extractible from all the paths $g_1, \ldots, g_m$. From this follows that the redex g_m is extractible from the paths $u_1 \cdot g_1, \ldots, u_1 \cdot g_m$. Now, for every index $1 \leq i \leq m$, the path $u_1 \cdot g_i$ is more standard than the path f_i because $f_i \searrow_{u_1} g_i$. We conclude by Lemma 12 (*preservation of extraction*) that the redex $outm(f_m)$ is extractible from the paths $f_1, ..., f_m$.

Step 9. By Step 8, we may define for every index $1 \leq i \leq m+1$ the path f'_i as an (arbitrary) projection of the path f_i by extraction of $outm(f_m)$. We thus have $f_i \searrow_{outm(f_m)} f'_i$. By Lemma 12 (*preservation of extraction*) applied m times, the rewriting paths f'_1 ,..., f'_{m+1} are Lévy equivalent.

Step 10. In order to reach a contradiction with our hypothesis, we prove that the redex u_1 is extractible from the rewriting path f_{m+1}. We have already noted in Step 9 that the paths $f'_1, ..., f'_{m+1}$ are Lévy equivalent. The path f'_1 is standard of length $n-1$ since it is defined as the projection of the standard path $f_1 = u_1 \cdots u_n$ by extraction of the redex $outm(f_m)$. By induction hypothesis, the path f'_1 is more standard than all the paths $f'_1, ..., f'_{m+1}$. Besides, the rewriting path f'_1 is not empty. We have proved indeed in Step 5 that the redexes u_1 and $outm(f_m)$ are different redexes, and more precisely, that the redex u_1 drags the redex $outm(g_m)$ to the redex $outm(f_m)$. From this follows that the extraction of the redex $outm(f_m)$ from the standard path $f_1 = u_1 \cdots u_n$ induces a reversible permutation $u_1 \cdot outm(g_m) \diamondsuit outm(f_m) \cdot u'_1$. The redex u'_1 is the first redex of the path f'_1, and the path f'_1 is more standard than all the paths $f'_1, ..., f'_{m+1}$. By Lemma 12 (*preservation of extraction*), the redex u'_1 is extractible from all the paths $f'_1, ..., f'_{m+1}$. The diagram below summarizes the situation:

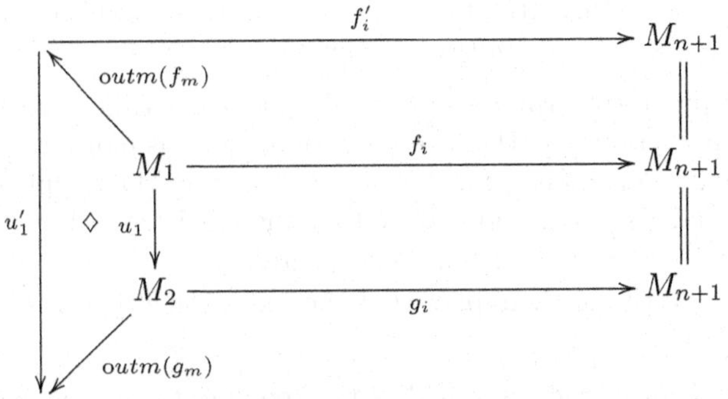

All this has the remarkable consequence that the redex u'_1 is extractible from the rewriting path f'_{m+1}. From this follows that the redex u_1 is extractible from the rewriting path $outm(f_m) \cdot f'_{m+1}$. Now, the path $outm(f_m) \cdot f'_{m+1}$ is more standard than the path f_{m+1} by definition of $f_{m+1} \searrow_{outm(f_m)} f'_{m+1}$. We conclude by Lemma 12 (*preservation of extraction*) that the redex u_1 is extractible from the rewriting path f_{m+1}.

Step 11. This is the concluding step. We deduce from the contradiction reached in Step 10 that the redex u_1 is extractible from every path Lévy equivalent to the rewriting path f. By the preliminary discussion of Step 1, this concludes our proof by induction of Lemma 25. □

4.3 Existence

Lemma 26 (towards existence). *Suppose that* $f : M_1 \twoheadrightarrow M_{n+1}$ *is a nonempty path whose projection by extraction of the redex* $outm(f) : M_1 \longrightarrow M_2$ *is Lévy equivalent to a standard path*

$$M_2 \xrightarrow{u_2} M_3 \xrightarrow{u_3} \cdots \xrightarrow{u_{n-1}} M_n \xrightarrow{u_n} M_{n+1}.$$

Then, the rewriting path

$$M_1 \overset{outm(f)}{\longrightarrow} M_2 \xrightarrow{u_2} M_3 \xrightarrow{u_3} \cdots \xrightarrow{u_{n-1}} M_n \xrightarrow{u_n} M_{n+1}$$

is standard.

Proof. By induction on n. The lemma is immediate when $n = 1$ because the path $outm(f)$ is standard, like every path of length 1. Suppose that the property is established for every standard path of length $n-2$, and consider a standard path

$$M_2 \xrightarrow{u_2} M_3 \xrightarrow{u_3} \cdots \xrightarrow{u_{n-1}} M_n \xrightarrow{u_n} M_{n+1}$$

of length $n-1$. Consider moreover a nonempty path $f : M_1 \twoheadrightarrow M_{n+1}$, and suppose that (one of) its projection g by extraction of the redex $outm(f)$: $M_1 \longrightarrow M_2$ is Lévy equivalent to the standard path $u_2 \cdots u_n$. We write u_1 for the redex $outm(f)$.

We want to prove that the path $u_1 \cdot u_2 \cdots u_n$ is standard. We proceed by contradiction, and suppose that the path $u_1 \cdot u_2 \cdots u_n$ is *not* standard. By Lemma 19 (*characterization lemma*) there exists an anti-standard path inside the rewriting path $u_1 \cdot u_2 \cdots u_n$. Since the path $u_2 \cdots u_n$ is standard, this anti-standard path is necessarily of the form $u_1 \cdots u_{k+1}$ for some index $1 \leq k \leq n-1$.

By definition of an anti-standard path, and whatever the value of the index k, there exists a redex u_2' and a path $h_{u_1'}$ forming a permutation $u_1 \cdot u_2 \triangleright u_2' \cdot h_{u_1'}$. The situation is summarized in the the diagram below:

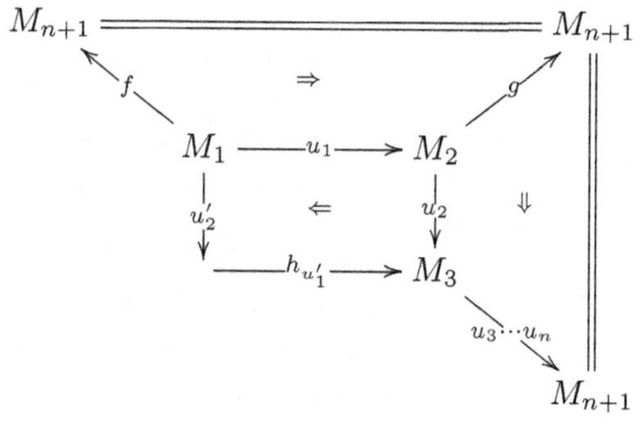

(20)

We show in Steps 2, 3, 4, 5 and 6 that the permutation $u_1 \cdot u_2 \rhd u_2' \cdot h_{u_1'}$ is reversible, or equivalently, that $k \geq 2$.

Step 2. We show that the redex u_2' is extractible from the path f. By Lemma 25 (*uniqueness*), the path $u_2 \cdots u_n$ is more standard than every Lévy equivalent path. In particular, the path $u_2 \cdots u_n$ is more standard than the path g. It follows from Lemma 12 (*preservation of extraction*) that the redex u_2 which is extractible from the path $u_2 \cdots u_n$ is also extractible from the path g. This and the existence of the permutation $u_1 \cdot u_2 \rhd u_2' \cdot h_{u_1'}$ implies that the redex u_2' is extractible from the path $u_1 \cdot g$. The path $u_1 \cdot g$ is more standard than the path f by definition of extraction $f \searrow_{u_1} g$. Thus, by applying Lemma 12 (*preservation of extraction*) again, the redex u_2' is extractible from the path f.

Step 3. Let the path f' denote an arbitrary projection of the path f by extraction of the redex u_2'. By construction, and Axiom **shape**, the redex u_2' does not coincide with the redex $outm(f) = u_1$. By Lemma 22, the path f' is non-empty and the redex u_2' drags the redex $outm(f')$ (denoted u_1' from now) to the redex $u_1 = outm(f)$. More explicitly, the two redexes u_1' and u_2' are involved in a permutation $u_2' \cdot u_1' \rhd u_1 \cdot h_{u_2}$ for some path h_{u_2}. Let the path g' denote an arbitrary projection of the path f' by extraction of the redex u_1'. The situation is summarized in the diagram below:

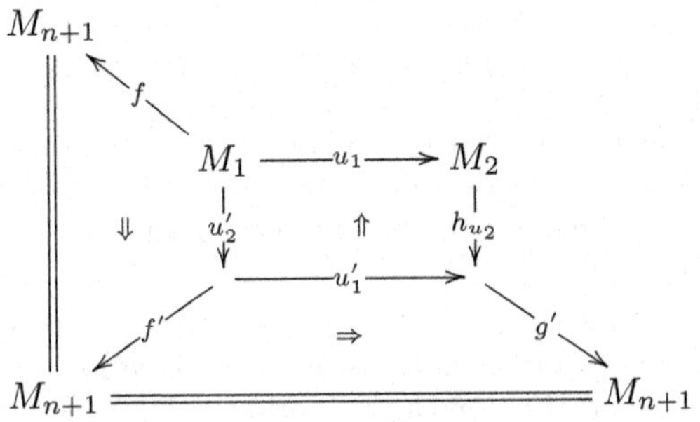

(21)

In the next Steps 4–7, we analyze the relationship between the two diagrams (20) and (21). We establish in Steps 4–6 that the paths $h_{u_1'}$ and h_{u_2} coincide respectively with the redexes u_1' and u_2, and thus, that the permutation $u_1 \cdot u_2 \rhd u_2' \cdot h_{u_1'}$ is reversible. We establish in Step 7 that the path g' is Lévy equivalent to the path $u_3 \cdots u_n$. This enables to combine the two diagrams (20) and (21) in a larger diagram.

Step 4. Here, we deduce from Lemma 25 (*uniqueness*) that the redex u_2 is extractible from the path $h_{u_2} \cdot g'$. By construction, the path $u_1 \cdot h_{u_2} \cdot g'$ is more standard than the path f. The paths $h_{u_2} \cdot g'$ and g are the projections of the paths $u_1 \cdot h_{u_2} \cdot g'$ and f by extraction of the redex u_1, respectively. By Lemma 12 (*preservation of extraction*), the two paths $h_{u_2} \cdot g'$ and g are Lévy equivalent. Now, the path g is also Lévy equivalent to the standard path $u_2 \cdots u_n$. From this and Lemma 25 (*uniqueness*) follows that the path $u_2 \cdots u_n$ is more standard than the path $h_{u_2} \cdot g'$. By Lemma 12 (*preservation of extraction*), we conclude that the redex u_2 is extractible from the path $h_{u_2} \cdot g'$.

Step 5. We deduce from Step 4 that the redex u_2 is extractible from the path h_{u_2}. We proceed by contradiction, and suppose that it is not. The redex u_2 is extractible from the path $h_{u_2} \cdot g'$. By definition of extraction, there exists a redex v extractible from the path g' such that the path h_{u_2} drags the redex v to the redex u_2. From this follows that the path $u_1 \cdot h_{u_2}$ drags the redex v to the redex u_2'. Now, the path $u_1 \cdot h_{u_2}$ is Lévy equivalent to the path $u_2' \cdot u_1'$. By Lemma 10 (*preservation of drag*), the path $u_2' \cdot u_1'$ drags the redex v to the redex u_2'. More explicitly, there exists a redex w such that: (a) the redex u_1' drags the redex v to the redex w; and (b) the redex u_2' drags the redex w to the redex u_2'. This very last statement (b) contradicts the Axiom **shape** since it implies that there exists a path h and permutation $u_2' \cdot w \rhd u_2' \cdot h$. We conclude that the redex u_2 is extractible from the path h_{u_2}.
Step 6. We deduce from Step 5 that the paths $h_{u_1'}$ and h_{u_2} coincide respectively with the redexes u_1' and u_2, and that the permutation $u_1 \cdot u_2 \rhd u_2' \cdot h_{u_1'}$ is reversible. By definition of extraction, there exists a path h such that $h_{u_2} \Longrightarrow u_2 \cdot h$. From this follows that $u_2' \cdot u_1' \rhd u_1 \cdot h_{u_2}$ and $u_1 \cdot h_{u_2} \Longrightarrow u_2' \cdot h_{u_1'} \cdot h$. Diagrammatically,

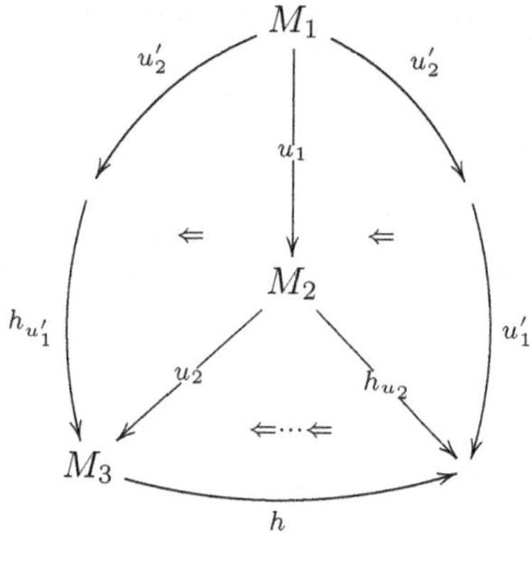

Suppose that the permutation $u_2' \cdot u_1' \rhd u_1 \cdot h_{u_2}$ is irreversible. In that case, it follows from Axiom **irreversibility** that $u_2' \cdot u_1' \rhd u_2' \cdot h_{u_1'} \cdot h$. This last statement contradicts Axiom **shape**, and we thus conclude that the permutation $u_2' \cdot u_1' \rhd u_1 \cdot h_{u_2}$ is reversible. From this follows that the path h_{u_2} is a redex. The equality $h_{u_2} = u_2$ follows immediately from the fact that the redex u_2 is extractible from the path h_{u_2}. We conclude that $u_2' \cdot u_1' \rhd u_1 \cdot u_2$. At this point, there only remains to apply Axiom **reversibility** on the permutations $u_2' \cdot u_1' \rhd u_1 \cdot u_2$. $u_1 \cdot u_2 \rhd u_2' \cdot h_{u_1'}$, from which we deduce that $h_{u_1'} = u_1'$ and that the permutation $u_1 \cdot u_2 \rhd u_2' \cdot h_{u_1'}$ is reversible.
Step 7. We have just established that the permutation $u_1 \cdot u_2 \diamondsuit u_2' \cdot u_1'$ is reversible. In Step 4, we have also proved that $u_2 \cdots u_n$ is more standard than the path $h_{u_2} \cdot g'$. We know now that the path $h_{u_2} \cdot g'$ is equal to the path $u_2 \cdot g'$. The two paths $u_3 \cdots u_n$ and g' are respectively the projections of the paths $u_2 \cdots u_n$ and $u_2 \cdot g'$ by extraction of the redex u_2. By Lemma 12 (*preservation of extraction*), the path g' is Lévy equivalent to the path $u_3 \cdots u_n$.

Step 8. We have just established in Step 7 that the projection g' of the path f' by extraction of the redex $u'_1 = outm(f')$ is Lévy equivalent to the path $u_3 \cdots u_n$. This enables to apply our induction hypothesis on the standard path $u_3 \cdots u_n$. We deduce that the path $u'_1 \cdot u_3 \cdots u_n$ is standard. In particular, the path $u'_1 \cdot u_3 \cdots u_{k+1}$ is *not* anti-standard. From this follows that the path $u_1 \cdot u_2 \cdots u_{k+1}$ is *not* anti-standard. This contradicts our original hypothesis. The path $u_1 \cdot u_2 \cdots u_n$ is thus standard. This concludes the reasoning by induction, and the proof of Lemma 4.3. □

Lemma 27 (existence). *For every path $f : M \twoheadrightarrow N$ there exists a standard path $g : M \twoheadrightarrow N$ such that $f \Longrightarrow g$.*

Proof. First, we show that every rewriting path $u_1 \cdots u_n : M \twoheadrightarrow N$ is standard when it is obtained as a sequence of extractions from a path $f_1 : M \twoheadrightarrow N$:

$$f_1 \searrow_{u_1} f_2 \searrow_{u_2} f_3 \cdots f_n \searrow_{u_n} \mathrm{id}_N \tag{22}$$

where $u_i = outm(f_i)$ for every index $1 \leq k \leq n$. The proof is nearly immediate, by induction on the length n. Suppose that the property is established for every path of length $n - 1$, and consider a path $u_1 \cdots u_n$ obtained as a series of extractions (22). By induction hypothesis, the path

$$f_2 \searrow_{u_2} f_3 \searrow_{u_3} f_4 \cdots f_n \searrow_{u_n} \mathrm{id}_N$$

is standard. By Lemma 26, the path $u_1 \cdot u_2 \cdots u_n = outm(f_1) \cdot u_2 \cdots u_n$ is also standard. We conclude.

Now, suppose that $f : M \twoheadrightarrow N$ is an arbitrary rewriting path. By Axiom **termination**, every sequence of extractions

$$f = f_1 \searrow_{outm(f_1)} f_2 \searrow_{outm(f_2)} f_3 \cdots f_n \searrow_{outm(f_n)} \cdots$$

is finite. Thus, there exists an index n such that

$$f_1 \searrow_{u_1} f_2 \searrow_{u_2} f_3 \cdots f_n \searrow_{u_n} \mathrm{id}_N$$

where $u_i = outm(f_i)$, for all $1 \leq i \leq n$. By construction, the path $u_1 \cdots u_n : M \twoheadrightarrow N$ is more standard than the path f, and it is standard by the previous argument. We conclude. □

4.4 Standardization Theorem

Theorem 2 (standardization). *Suppose that $(\mathcal{G}, \rhd)$ is an axiomatic rewriting system and that $f : M \twoheadrightarrow N$ is a path in the transition system $\mathcal{G}$. Then:*

- *there exists a standard path $g : M \twoheadrightarrow N$ more standard than f,*
- *every standard path Lévy equivalent to f is equal to g modulo reversible permutation equivalence $\simeq$.*

The standard path of any path $f : M \twoheadrightarrow N$ may be computed by extracting recursively the outermost redex $outm(f_i)$ in a sequence of rewriting paths

$$f = f_1 \searrow_{outm(f_1)} f_2 \searrow_{outm(f_2)} f_3 \searrow_{outm(f_3)} \quad \cdots \quad f_n \searrow_{outm(f_n)} \mathrm{id}_N.$$

We call this algorithm **STD** as in [13]. Note that the algorithm is non deterministic because it depends at each step f_i on the choice of the next rewriting path f_{i+1}.

Corollary 28. *The relation $\Longrightarrow$ on paths is confluent modulo $\simeq$. The $\Longrightarrow$-normal form of a path is computed by the algorithm* **STD**.

5 Standardization from the 2-Categorical Point of View

In Sections 1—4. we interpret standardization as a 2-dimensional rewriting procedure on 1-dimensional paths, and establish a confluence and normalization property for that procedure. However, we say nothing there about the 2-dimensional reductions $f \Longrightarrow g$ themselves. Intuitively, each such reduction $f \Longrightarrow g$ describes a possible way to tile the 2-dimensional *surface* lying between the two rewriting paths f and g. In this section is to show that all tilings $f \Longrightarrow g$ from a path f to its standard path g, are equivalent in an intuitive sense. We refer the reader to the last chapter of [25] (second edition) for a nice and motivated introduction to 2-categories.

5.1 Tiling Graph, Tiling Paths, and Partial Injections

To every 2-dimensional transition system $(\mathcal{G}, \rhd)$ we associate a *tiling graph* in the following way:

Definition 29 (tiling graph, path, step). *The graph* ***tiling-graph***$(\mathcal{G}, \rhd)$ *has the paths of $\mathcal{G}$ as vertices, and the standardization steps $(e, f \rhd g, h)$ as edges $e \cdot f \cdot h \Longrightarrow e \cdot g \cdot h$. The paths in* ***tiling-graph***$(\mathcal{G}, \rhd)$ *are called* tiling paths *to avoid confusion with the* rewriting paths *of the transition system $\mathcal{G}$. According to that spirit, we often call* tiling step *a standardization step. In the graph* ***tiling-graph***$(\mathcal{G}, \rhd)$, *we write $\mathrm{id}^f : f \Longrightarrow f$ for the identity of f, and $\alpha * \beta : f \Longrightarrow h$ for the composite of two paths $\alpha : f \Longrightarrow g$ and $\beta : g \Longrightarrow h$.*

Definition 30 (canonical equivalence on tiling path). *To every tiling path $\alpha : f \Longrightarrow g$, we associate a partial injection $[\alpha] : [g] \rightharpoonup [f]$ as follows.*

- *to every vertex of* ***tiling-graph***$(\mathcal{G}, \rhd)$ *we associate the finite set $[f] = \{1, ..., n\}$ of cardinal n the length of f as 1-dimensional path,*
- *to every edge $\alpha = (e, f \rhd g, h)$ of* ***tiling-graph***$(\mathcal{G}, \rhd)$ *where e, f, g and h decompose as:*

$$e = u_1 \cdots u_m \qquad f = v \cdot u' \qquad g = v_1 \cdots v_n \qquad h = w_1 \cdots w_p$$

we associate the partial injection $[\alpha] : [e \cdot g \cdot h] \rightharpoonup [e \cdot f \cdot h]$ defined as

- *when $f \Diamond g$:*

$$\begin{cases} k & \mapsto & k & \text{for every } 1 \leq k \leq m \\ m+1 & \mapsto & m+2 \\ m+2 & \mapsto & m+1 \\ m+2+k & \mapsto & m+2+k & \text{for every } 1 \leq k \leq p \end{cases}$$

- *when $f \blacktriangleright g$:*

$$\begin{cases} k & \mapsto & k & \text{for every } 1 \leq k \leq m \\ m+1 & \mapsto & m+2 \\ m+n+k & \mapsto & m+2+k & \text{for every } 1 \leq k \leq p \end{cases}$$

The partial injection $[\alpha] : \{1, ..., n\} \rightharpoonup \{1, ..., m\}$ associated to a tiling path

$$\alpha : u_1 \cdots u_m \Longrightarrow v_1 \cdots v_n$$

is defined by composing the partial injections $[\alpha_i]$'s:

$$[\alpha] = [\alpha_n] \circ \cdots \circ [\alpha_1]$$

Intuitively, the function $[\alpha]$ traces every redex v_k back to its unique "ancestor" $u_{[\alpha](k)}$ in the 1-dimensional path $u_1 \cdots u_m$, when this redex exists.

The main result of the section states that

Theorem 3. *Suppose that g is a standard rewriting path in an axiomatic rewriting system $(\mathcal{G}, \rhd)$. Then, every two tiling paths $\alpha, \beta : f \Longrightarrow g$ from a rewriting path f to the rewriting path g define the same partial injection $[\alpha] = [\beta]$.*

Reformulated 2-categorically, the theorem states that in the 2-category **2-cat**$(\mathcal{G}, \rhd)$ defined at the beginning of Section 5.3, the standard path $g : M \twoheadrightarrow N$ is terminal in its connected component in the hom-category **2-cat**$(\mathcal{G}, \rhd)(M, N)$. The standard path g is in fact strongly terminal, in the sense that in every cell $g \Longrightarrow h$, the path h is also standard, and thus terminal.

We proceed methodologically, and prove the theorem in two steps. In Section 5.2, we give a series of conditions on an equivalence relation $\cong$ on the paths of **tiling-graph**$(\mathcal{G}, \rhd)$ to ensure that every two tiling paths $\alpha, \beta : f \Longrightarrow g$ from a path f to a standard path g, are equal modulo $\cong$. In Section 5.3, we prove that the equivalence relation $\alpha \cong \beta$ induced by the equality $[\alpha] = [\beta]$ of partial injections, satisfies the formal conditions of Section 5.2.

Remark: Theorem 3 repeats in dimension 2 the observation by J.-J. Lévy in the λ-calculus, or in any conflict-free (term) rewriting system, that there exists a unique path from a term to its normal form, modulo permutation. Here, objects are 1-dimensional, paths are 2-dimensional, permutations are 3-dimensional — and the concept of a conflict-free 2-dimensional system remains to be clarified.

5.2 Standard=Strong Terminal

Definition 31 (horizontal composition). *The horizontal composite $\alpha \cdot h$ of a tiling step (=standardization step)*

$$\alpha = (e, f \rhd g, h) : e \cdot f \cdot h \Longrightarrow e \cdot g \cdot h : M \twoheadrightarrow N$$

and of a 1-dimensional path $h' : N \twoheadrightarrow P$ is defined as the tiling step:

$$\alpha \cdot h = (e, f \rhd g, h \cdot h') : e \cdot f \cdot h \cdot h' \Longrightarrow e \cdot g \cdot h : M \twoheadrightarrow P$$

The horizontal composite $\alpha \cdot h$ of a tiling path

$$\alpha = \alpha_1 * \cdots * \alpha_n : f \Longrightarrow g : M \twoheadrightarrow N$$

and a 1-dimensional path $h : N \twoheadrightarrow P$ is defined as the tiling path

$$\alpha \cdot h = (\alpha_1 \cdot h) * \cdots * (\alpha_n \cdot h) : M \twoheadrightarrow P$$

The horizontal composite $e \cdot \alpha$ of a 1-dimensional path $e : L \twoheadrightarrow M$ and a tiling path $\alpha : f \cdot g : M \twoheadrightarrow N$ is defined symmetrically.

From now on, we consider an equivalence relation $\cong$ between the tiling paths of **tiling-graph**$(\mathcal{G}, \rhd)$, satisfying the four properties below:

1. for all tiling paths $\alpha : f \Longrightarrow f'$ and $\beta : g \Longrightarrow g'$,
$$\alpha \cong \beta \ \Rightarrow \ f = g \text{ and } f' = g'$$
2. for all tiling paths $\alpha, \alpha' : f \Longrightarrow g$ and $\beta, \beta' : g \Longrightarrow h$,
$$\alpha \cong \alpha' \text{ and } \beta \cong \beta' \ \Rightarrow \ \alpha * \beta \cong \alpha' * \beta'$$
3. for all tiling paths $\alpha, \beta : g \Longrightarrow g' : M \twoheadrightarrow N$ and all 1-dimensional paths $f : L \twoheadrightarrow M$ and $h : N \twoheadrightarrow P$,
$$\alpha \cong \beta \ \Rightarrow \ f \cdot \alpha \cdot h \cong f \cdot \beta \cdot h$$
4. for all of tiling paths $\alpha : f \Longrightarrow f' : M \twoheadrightarrow N$ and $\beta : g \Longrightarrow g' : N \twoheadrightarrow P$,
$$(\alpha \cdot g) * (f' \cdot \beta) \cong (f \cdot \beta) * (\alpha \cdot g')$$

Lemma 32. *The equivalence relation $\cong$ defines a 2-category* **2-cat**$_{\cong}(\mathcal{G}, \rhd)$.

Proof. The 2-category **2-cat**$_{\cong}(\mathcal{G}, \rhd)$ has vertices and paths of $\mathcal{G}$ as objects and morphisms, and equivalence classes modulo $\cong$ of tiling paths as cells. Conditions 1–3. ensure the necessary compositionality properties of **2-cat**$_{\cong}(\mathcal{G}, \rhd)$, while condition 4. ensures the so-called interchange law of 2-categories, see [25]. □

Suppose moreover that:

5. for every path $f = u \cdot v$ where u drags the redex v to a redex v_0, and for every standard path g,
$$\forall \alpha, \beta, \qquad \alpha, \beta : f \Longrightarrow g \quad \Rightarrow \quad \alpha \cong \beta$$
6. for every path $f = u \cdot v \cdot w$ where the redex u drags the redex v to a redex v_0, and where the path $u \cdot v$ drags the redex w to a redex w_0, and for every standard path g,
$$\forall \alpha, \beta, \qquad \alpha, \beta : f \Longrightarrow g \quad \Rightarrow \quad \alpha \cong \beta$$

These two additional conditions 5 and 6 regulate the potential *critical pairs* occurring during the 2-dimensional transitions implementing standardization. The lemma below establishes that the two assumptions are sufficient to the purpose.

Lemma 33. *Suppose that the equivalence relation $\cong$ satisfies Conditions 1–6. Then, every standard path $h : M \twoheadrightarrow N$ is strongly terminal in its connected component in the hom-category $\mathbf{2\text{-}cat}_{\cong}(\mathcal{G}, \rhd)(M, N)$.*

Proof. By induction on the length of $h : M \twoheadrightarrow N$. Suppose that the property is established for every standard path of length n, and that the path $u \cdot h$ is standard of length $1 + n$. Suppose that f is a path Lévy equivalent to $u \cdot h$. We claim that for every tiling path $\gamma : f \Longrightarrow u \cdot g$ resulting of an extraction $f \searrow_u g$, and for every tiling path $\alpha : f \Longrightarrow f'$ starting from f, there exists a tiling path $\gamma' : f' \Longrightarrow u \cdot g'$ resulting of an extraction $f' \searrow_u g'$, such that

$$\gamma * (u \cdot \delta_g) \;\cong\; \alpha * \gamma' * (u \cdot \delta_{g'}) \quad : f \Longrightarrow u \cdot h \tag{23}$$

where $\delta_g : g \Longrightarrow h$ and $\delta_{g'} : g' \Longrightarrow h$ are arbitrary tiling paths to the terminal object h. To prove the claim, it is sufficient to consider the case when α is a tiling step $(f_1, f_2 \rhd f_2', f_3)$. The general case follows by a straightforward induction on the length of α. So, we want to establish that the diagram below commutes modulo $\cong$ for a tiling step $\alpha = (f_1, f_2 \rhd f_2', f_3) : f \Longrightarrow f'$ and a tiling path $\gamma : f \Longrightarrow u \cdot g$ resulting of an extraction $f \searrow_u g$.

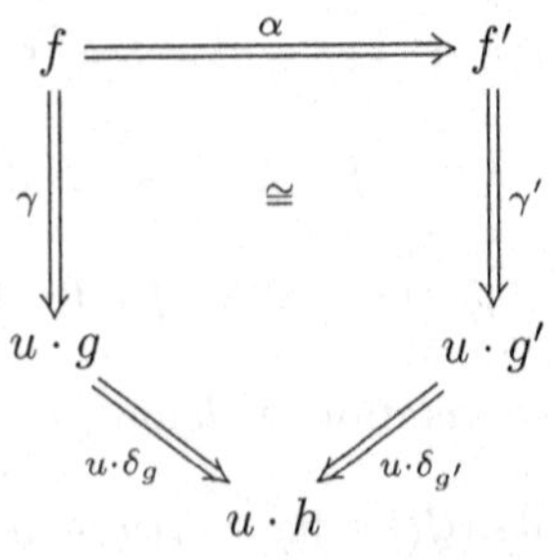

By definition of α, the paths f and f' factor as

$$f = f_1 \cdot f_2 \cdot f_3 \qquad\qquad f' = f_1 \cdot f_2' \cdot f_3.$$

The redex u is extractible from the path $f = f_1 \cdot f_2 \cdot f_3$. One of the three following situations occurs. We say that the redex u is

1. extractible from the component f_1 when the redex u is extractible from the path f_1,
2. extractible from the component f_2 when the redex u is extractible from the path $f_1 \cdot f_2$ but not from the path f_1,
3. extractible from the component f_3 when the redex u is extractible from the path $f_1 \cdot f_2 \cdot f_3$ but not from the path $f_1 \cdot f_2$.

By definition of γ as the tiling path produced by the extraction $f \searrow_u g$, the rewriting paths g and the tiling path γ factor as

$$g = g_1 \cdot g_2 \cdot g_3$$

$$\gamma = (f_1 \cdot f_2 \cdot \gamma_3) * (f_1 \cdot \gamma_2 \cdot g_3) * (\gamma_1 \cdot g_2 \cdot g_3)$$

where the definitions of g_1, g_2, g_3 and $\gamma_1, \gamma_2, \gamma_3$ depend on the component f_1 or f_2 or f_3 from which the redex u is extractible:

1. The redex u is extractible from the component f_1: in that case, $g_3 = f_3$ and $\gamma_3 = \mathrm{id}_{f_3}$, $g_2 = f_2$ and $\gamma_2 = \mathrm{id}_{f_2}$, and $\gamma_1 : f_1 \Longrightarrow u \cdot g_1$ is the result of an extraction $f_1 \searrow_u g_1$,
2. The redex u is extractible from the component f_2: in that case, $g_3 = f_3$ and $\gamma_3 = \mathrm{id}_{f_3}$, $\gamma_2 : f_2 \Longrightarrow u' \cdot g_2$ is the result of an extraction $f_2 \searrow_{u'} g_2$, the path f_1 drags u' to u and $\gamma_1 : f_1 \cdot u' \Longrightarrow u \cdot g_1$ is the result of the extraction $f_1 \cdot u' \searrow_u g_1$,
3. The redex u is extractible from the component f_3: in that case, $\gamma_3 : f_3 \Longrightarrow u'' \cdot g_3$ is the result of an extraction $f_3 \searrow_{u''} g_3$, the path f_2 drags u'' to u' and $\gamma_2 : f_2 \cdot u'' \Longrightarrow u' \cdot g_2$ is the result of the extraction $f_2 \cdot u'' \searrow_{u'} g_2$, the path f_1 drags u' to u and $\gamma_1 : f_1 \cdot u' \Longrightarrow u \cdot g_1$ is the result of the extraction $f_1 \cdot u' \searrow_u g_1$.

The tiling path γ' is defined as

$$\gamma' = (f_1 \cdot f_2' \cdot \gamma_3) * (f_1 \cdot \gamma_2' \cdot g_3) * (\gamma_1 \cdot g_2' \cdot g_3)$$

where the definition of the tiling path γ_2' is by case analysis.

1. The redex u is extractible from the component f_1: in that case, $g_2' = f_2'$ and $\gamma_2' : f_2' \Longrightarrow g_2'$ is defined as $\mathrm{id}_{f_2'}$. Equivalence (23) follows from induction hypothesis on h, as well as conditions 2, 3 and 4 on the equivalence relation $\cong$.
2. The redex u is extractible from the component f_2: in that case, the path g_2' and $\gamma_2' : f_2' \Longrightarrow u' \cdot g_2'$ are the result of an arbitrary extraction $f_2' \searrow_{u'} g_2'$.

Equivalence (23) follows from the series of equivalence:

$$
\begin{array}{ll}
\gamma * (u \cdot \delta_g) & \\
\cong \gamma * (u \cdot g_1 \cdot (g_2 \Longrightarrow g_2'') \cdot g_3) * (u \cdot \delta_{g_1 \cdot g_2'' \cdot g_3}) & \text{by ind. hyp.} \\
\cong (f_1 \cdot (\gamma_2 * \eta_2) \cdot g_3) * (\gamma_1 \cdot g_2'' \cdot g_3) * (u \cdot \delta_{g_1 \cdot g_2'' \cdot g_3}) & \text{by cond. 2, 3, 4.} \\
\cong (f_1 \cdot ((f_2 \rhd f_2') * \gamma_2' * \eta_2') \cdot g_3) * (\gamma_1 \cdot g_2'' \cdot g_3) * (u \cdot \delta_{g_1 \cdot g_2'' \cdot g_3}) & \text{by cond. 5.} \\
\cong (f_1 \cdot (f_2 \rhd f_2') \cdot f_3) * (f_1 \cdot (\gamma_2' * \eta_2') \cdot g_3) * (\gamma_1 \cdot g_2'' \cdot g_3) * (u \cdot \delta_{g_1 \cdot g_2'' \cdot g_3}) & \\
 & \text{by cond. 2, 3, 4.} \\
\cong \alpha * (f_1 \cdot \gamma_2' \cdot g_3) * (\gamma_1 \cdot g_2' \cdot g_3) * (u \cdot g_1 \cdot \eta_2' \cdot g_3) * (u \cdot \delta_{g_1 \cdot g_2'' \cdot g_3}) & \text{by cond. 2, 3, 4.} \\
\cong \alpha * \gamma' * (u \cdot \delta_{g_1 \cdot g_2' \cdot g_3}) & \text{by ind. hyp.} \\
\cong \alpha * \gamma' * (u \cdot \delta_{g'}) &
\end{array}
$$

where g_2'' is a standard path Lévy equivalent to the paths g_2 and g_2', and where

$$
\begin{array}{rcl}
\eta_2 : u \cdot g_2 \Longrightarrow u \cdot g_2'' & \text{and} & \eta_2' : u \cdot g_2' \Longrightarrow u \cdot g_2'' \\
\delta_{g_1 \cdot g_2' \cdot g_3} : g_1 \cdot g_2' \cdot g_3 \Longrightarrow h & \text{and} & \delta_{g_1 \cdot g_2'' \cdot g_3} : g_1 \cdot g_2'' \cdot g_3 \Longrightarrow h
\end{array}
$$

are arbitrary tiling paths.

3. The redex u is extractible from the component f_3: in that third case, g_2' and $\gamma_2' : f_2' \cdot u'' \Longrightarrow u' \cdot g_2'$ are the result of an arbitrary extraction $f_2' \cdot u'' \searrow_{u'} g_2'$. Equivalence (23) follows from the series of equivalence:

$$
\begin{array}{ll}
\gamma * (u \cdot \delta_g) & \\
\cong \gamma * (u \cdot g_1 \cdot (g_2 \Longrightarrow g_2'') \cdot g_3) * (u \cdot \delta_{g_1 \cdot g_2'' \cdot g_3}) & \text{by ind. hyp.} \\
\cong (f_1 \cdot f_2 \cdot \gamma_3) * (f_1 \cdot (\gamma_2 * \eta_2) \cdot g_3) * (\gamma_1 \cdot g_2'' \cdot g_3) * (u \cdot \delta_{g_1 \cdot g_2'' \cdot g_3}) & \text{by cond. 2, 3, 4.} \\
\cong (f_1 \cdot f_2 \cdot \gamma_3) * (f_1 \cdot (((f_2 \rhd f_2') \cdot u'') * \gamma_2' * \eta_2') \cdot g_3) * (\gamma_1 \cdot g_2'' \cdot g_3) * (u \cdot \delta_{g_1 \cdot g_2'' \cdot g_3}) & \\
 & \text{by cond 6.} \\
\cong (f_1 \cdot (f_2 \rhd f_2') \cdot f_3) * (f_1 \cdot f_2' \cdot \gamma_3) * (f_1 \cdot (\gamma_2' * \eta_2') \cdot g_3) * (\gamma_1 \cdot g_2'' \cdot g_3) * (u \cdot \delta_{g_1 \cdot g_2'' \cdot g_3}) & \\
 & \text{by cond. 2, 3, 4.} \\
\cong \alpha * (f_1 \cdot f_2' \cdot \gamma_3) * (f_1 \cdot \gamma_2' \cdot g_3) * (\gamma_1 \cdot g_2' \cdot g_3) * (u \cdot g_1 \cdot \eta_2' \cdot g_3) * (u \cdot \delta_{g_1 \cdot g_2'' \cdot g_3}) & \\
 & \text{by cond. 2, 3, 4.} \\
\cong \alpha * \gamma' * (u \cdot \delta_{g_1 \cdot g_2' \cdot g_3}) & \text{by ind. hyp.} \\
\cong \alpha * \gamma' * (u \cdot \delta_{g'}) &
\end{array}
$$

where g_2'' is a standard path Lévy equivalent to g_2 and g_2', and where

$$
\begin{array}{rcl}
\eta_2 : u \cdot g_2 \Longrightarrow u \cdot g_2'' & \text{and} & \eta_2' : u \cdot g_2' \Longrightarrow u \cdot g_2'' \\
\delta_{g_1 \cdot g_2' \cdot g_3} : g_1 \cdot g_2' \cdot g_3 \Longrightarrow h & \text{and} & \delta_{g_1 \cdot g_2'' \cdot g_3} : g_1 \cdot g_2'' \cdot g_3 \Longrightarrow h
\end{array}
$$

are arbitrary tiling paths.

This proves our introductory claim. Now, we prove the lemma as follows. Let $\gamma : f \Longrightarrow u \cdot g$ be the result of an arbitrary extraction $f \searrow_u g$. Consider any tiling path α from f to $u \cdot h$. By property (23) proved above, there exists a tiling path γ' such that:

$$\gamma * (u \cdot \delta_g) \cong \alpha * \gamma' * (u \cdot \delta_h) \quad : f \Longrightarrow u \cdot h$$

In that particular case, as the result of the "empty" extraction $u \cdot h \searrow_u h$, the tiling path γ' is the identity $\mathrm{id}^{u \cdot h} : u \cdot h \Longrightarrow u \cdot h$. Moreover, the tiling path δ_h is the identity $\mathrm{id}^h : h \Longrightarrow h$ by induction hypothesis. It follows that

$$\alpha \quad \cong \quad \gamma * (u \cdot \delta_g)$$

This concludes the proof. □

5.3 The 2-Category 2-cat$(\mathcal{G}, \rhd)$

Definition 34 (2-cat$(\mathcal{G}, \rhd)$). *The 2-category* **2-cat**$(\mathcal{G}, \rhd)$ *is the 2-category* **2-cat**$_{\cong}(\mathcal{G}, \rhd)$ *associated to the following equivalence relation on tiling paths:*

$$\alpha \cong \beta \qquad \Longleftrightarrow \qquad [\alpha] = [\beta]$$

The main goal of the section is to prove Theorem 4.

Lemma 35. *Suppose that $\alpha : f \Longrightarrow g : M \twoheadrightarrow N$ is a tiling path between the 1-dimensional paths $f = u_1 \cdots u_m$ and $g = v_1 \cdots v_n$. Suppose that w is a redex outgoing from M. The two following assertions are equivalent:*

1. *the path $v_1 \cdots v_{i-1}$ drags the redex v_i to the redex w,*
2. *the index $[\alpha](i) = j$ is defined and the path $u_1 \cdots u_{j-1}$ drags the redex u_j to the redex w.*

Proof. By induction on the length of α. □

Theorem 4. *In the 2-category* **2-cat**$(\mathcal{G}, \rhd)$*, every standard path is strongly terminal in its Lévy equivalence class.*

Proof. By Lemma 33, we only need to check conditions 5 and 6 on the equivalence relation $\cong$ on tiling paths $\alpha, \beta : f \Longrightarrow g$ induced by the equality $[\alpha] = [\beta]$. Consider

- a path $f = u \cdot v'$ such that u drags v' to a redex v
- or a path $f = u \cdot v' \cdot w''$ such that u drags v' to a redex v, and $u \cdot v'$ drags w'' to a redex w.

Consider two tiling paths $\alpha, \beta : f \Longrightarrow g$ standardizing f into a standard path $g = v_1 \cdots v_n$. Suppose that $[\alpha](i) = j$ for some $i \in [n]$. By Lemma 35($1 \Rightarrow 2$), the path $v_1 \cdots v_{i-1}$ drags the redex v_i to the redex $t = u$ when $j = 1$, to redex $t = v$ when $j = 2$, or to the redex $t = w$ when $j = 3$. Thus, by Lemma 35($1 \Rightarrow 2$), the index $[\beta](i) = k$ is defined and such that the path $u_1 \cdots u_{k-1}$ drags the redex u_k to the redex t. This implies that $j = k$. Applying the argument to every $i \in [n]$, and by symmetry, we deduce that $[\alpha] = [\beta]$. This proves conditions 5 and 6, and we conclude. □

Remark: In the case of the λ-calculus, and more generally in any axiomatic rewriting system derived from an axiomatic nesting system, see Section 6, the partial injection $[\alpha] : [g] \rightharpoonup [f]$ may be replaced by a *total function* $[\alpha] : [g] \twoheadrightarrow [f]$ without breaking Theorem 4. The idea is to replace the partial function $[\alpha]$ associated to an irreversible standardization step α in Definition 30 by the following total function $[\alpha]$:

$$[\alpha] : \begin{cases} k & \mapsto & k & \text{for every } 1 \le k \le m \\ m+1 & \mapsto & m+2 & \\ m+1+k & \mapsto & m+1 & \text{for every } 1 \le k \le n-1 \\ m+n+k & \mapsto & m+2+k & \text{for every } 1 \le k \le p \end{cases}$$

It is not difficult to show that conditions 5 and 6 of Section 5.2 still hold with the new definition — in the case of the λ-calculus or any axiomatic nesting systems. Theorem 4 follows. However, Theorem 4 does not generally hold with the alternative definition. The axiomatic rewriting system

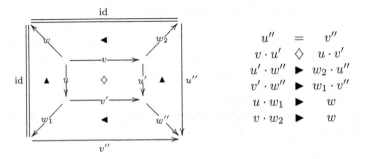

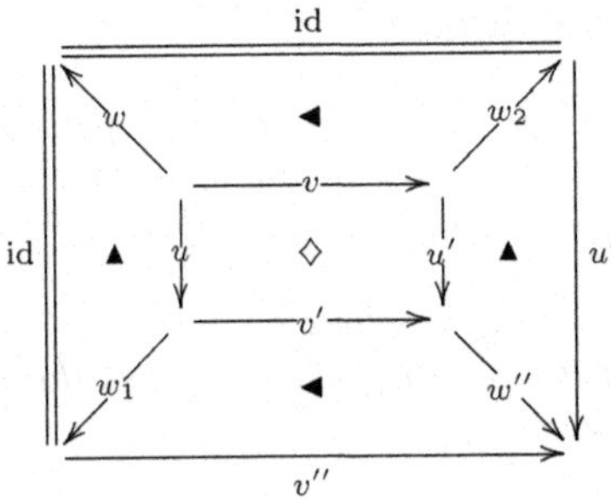

$$\begin{array}{ccc} u'' & = & v'' \\ v \cdot u' & \diamond & u \cdot v' \\ u' \cdot w'' & \blacktriangleright & w_2 \cdot u'' \\ v' \cdot w'' & \blacktriangleright & w_1 \cdot v'' \\ u \cdot w_1 & \blacktriangleright & w \\ v \cdot w_2 & \blacktriangleright & w \end{array}$$

and tiling paths

$$\alpha_1 : u \cdot v' \cdot w'' \overset{1}{\Longrightarrow} v \cdot u' \cdot w'' \overset{1}{\Longrightarrow} v \cdot w_2 \cdot u'' \overset{1}{\Longrightarrow} w \cdot u''$$

$$\alpha_2 : u \cdot v' \cdot w'' \overset{1}{\Longrightarrow} u \cdot w_1 \cdot v'' \overset{1}{\Longrightarrow} w \cdot v'' = w \cdot u''$$

illustrate this point, since both α_1 and α_2 transform the path $u \cdot v' \cdot w''$ to the standard path $w \cdot u'' = w \cdot v''$, but do not define the same *total* functions $[\alpha_1]$ and $[\alpha_2]$, since $[\alpha_1](2) = 1$ and $[\alpha_2](2) = 2$.

6 An Alternative Axiomatics Based on Residuals and Nesting

The 2-dimensional axiomatics formulated in Section 2 is particularly adapted to reason and prove diagrammatically... but it is also far away from common practice, and may be difficult to understand for someone simply interested in checking that the axioms are satisfied by his or her favorite rewriting system. For that reason, we step back (in this section only) to the axiomatics developed in [13] and [27] and based on the trinity of *residuals*, *critical pairs* and *nesting order*. Since the formulation is nearly independent of the remainder of the article, the reader may very well jump this section at a first reading.

The section is organised as follows. Axiomatic nesting system are defined in Section 6.1, and their axioms are formulated in Sections 6.2–6.5. We establish in Section 6.6 that every axiomatic nesting system $(\mathcal{G}, [\![-]\!], \preceq, \uparrow)$ defines an axiomatic rewriting system, that is, a 2-dimensional transition system $(\mathcal{G}, \triangleright)$ which satisfies the axioms of Section 2.

Remark: we provide two examples in Section 8

- the argument-nesting λ-calculus,
- the graph of sequentializations of an ordered set X.

which demonstrate that the axiomatics presented in this section is at the same time *strictly more general* than the axiomatics of [13] which inspired it, and *strictly less general* than the 2-dimensional axiomatics formulated in Section 2.

6.1 Axiomatic Nesting Systems

The main definition of the section follows.

Definition 36. *An* Axiomatic Nesting System *is a quadruple* $(\mathcal{G}, [\![-]\!], \preceq, \uparrow)$ *consisting of:*

1. *a transition system (or oriented graph)* $\mathcal{G} = (\mathbf{terms}, \mathbf{redexes}, \mathbf{source}, \mathbf{target})$,
2. *for every redex* $u : M \longrightarrow N$, *a binary relation* $[\![u]\!]$ *relating the redexes outgoing from* M *to the redexes outgoing from* N,
3. *for every vertex* M *of* $\mathcal{G}$, *a transitive reflexive antisymmetric relation* $\preceq_M$ *between the redexes outgoing from* M,
4. *for every vertex* M *of* $\mathcal{G}$, *a reflexive relation* $\uparrow_M$ *between the redexes outgoing from* M.

Every nesting system is supposed to satisfy a series of ten (4+2+4) axioms. The first four Axioms **Finite**, **Compat**, **Ancestor**, **Self** state elementary properties of residuals and compatibility. The two next Axioms **FinDev**, **Perm** enforce the well-known property of finite developments, appearing for instance in [32, 18, 20, 3, 27]. The four last Axioms **I**, **II**, **III**, **IV** regulate the properties of the nesting relation vs. the compatibility and residual relations. The ten axioms are called *N-axioms* (N stands for nesting) to distinguish them from the 2-dimensional axioms of Section 2.

6.2 The First N-axioms: Finite, Compat, Ancestor, Self

N-axiom Finite (finite residuals). We ask that a redex $v : M \longrightarrow Q$ has at most a finite number of residuals after a coinitial redex $u : M \longrightarrow P$.

$$\forall u, v \in \mathbf{redexes}, \qquad \text{the set} \quad \{v' \mid v[\![u]\!]v'\} \quad \text{is finite.}$$

N-axiom Compat (forth compatibility). We ask that two compatible redexes $u : M \longrightarrow P$ and $v : M \longrightarrow Q$ have compatible residuals u' and v' after a coinitial redex $w : M \longrightarrow N$.

$$\forall u, v, w, u', v' \in \mathbf{redexes}, \quad u[\![w]\!]u' \text{ and } v[\![w]\!]v' \text{ and } u \uparrow v \quad \Rightarrow \quad u' \uparrow v'$$

N-axiom Ancestor (unique ancestor). We ask that two different coinitial redexes $u : M \longrightarrow P$ and $v : M \longrightarrow Q$ do not have any residual in common after a coinitial redex $w : M \longrightarrow N$.

$$\forall u, v, w, u', v' \in \mathbf{redexes}, \quad u[\![w]\!]u' \text{ and } v[\![w]\!]v' \text{ and } u' = v' \quad \Rightarrow \quad u = v$$

N-axiom Self (self-destruction). We ask that a redex $v : M \longrightarrow Q$ has no residual after itself, or after an incompatible coinitial redex $u : M \longrightarrow P$.

$$\forall u, v \in \mathbf{redexes}, \quad (u = v \text{ or } \neg(u \uparrow v)) \quad \Rightarrow \quad \{v' \mid v[\![u]\!]v'\} = \emptyset$$

6.3 A Few Preliminary Definitions: Multi-redex, Development

We need a few preliminary definitions to formulate the N-axioms **FinDev** and **Perm**.

Definition 37 (residual through path). *Given a path $f : M \twoheadrightarrow N$, the relation $[\![f]\!]$ between the redexes outgoing from M and the redexes outgoing from N, is defined as follows:*

- *$[\![f]\!]$ is the identity relation when $f = \mathrm{id}_M$,*
- *$[\![f]\!]$ is the composite relation $[\![v_1]\!] \cdots [\![v_n]\!]$ when $f = v_1 \cdots v_n$.*

Explicitly, for every two redexes u and u',

$$u[\![\mathrm{id}_M]\!]u' \iff u = u'$$

$$u[\![v_1 \cdots v_n]\!]u' \iff \begin{array}{l} \exists u_2, ..., u_{n-1} \in \mathbf{redexes}, \\ u[\![v_1]\!]u_2[\![v_2]\!]u_3 \cdots u_{n-2}[\![v_{n-1}]\!]u_{n-1}[\![v_n]\!]u' \end{array}$$

Definition 38 (multi-redex). *A* multi-redex *in $(\mathcal{G}, [\![-]\!], \preceq, \uparrow)$ is a pair (M, U) consisting of a term M and a finite set U of pairwise compatible redexes of source M.*

Remark: every redex $u : M \longrightarrow N$ may be identified to the multi-redex $(M, \{u\})$.

Definition 39 (multi-residual). *Suppose that (M, U) is a multi-redex and that v is a redex compatible with every redex in U. The* multi-residual *of (M, U) after v, notation $(M, U)[\![v]\!]$, is the multi-redex (N, W) where $W = \{w \mid u[\![v]\!]w\}$.*

Remark: Definition 39 defines a multi-redex (N, W) thanks to the N-axioms **Finite** and **Compat**.

Definition 40 (development). *A* complete development *of a multi-redex (M, U) is a path f such that:*

- *$f = \mathrm{id}_M$ when U is empty,*
- *$f = u \cdot g$ when $u : M \longrightarrow N$ is a redex in U, and the path g is a complete development of the multi-redex $(M, U)[\![u]\!]$.*

A development *of (M, U) is a path $f : M \longrightarrow\!\!\!\!\!\rightarrow P$ which is prefix of a complete development $g : M \longrightarrow\!\!\!\!\!\rightarrow N$ of (M, U). Here, we call f a prefix of g when there exists a path $h : P \longrightarrow\!\!\!\!\!\rightarrow N$ such that $g = f \cdot h$.*

We define two notions mentioned informally in Sections 1 and 2, and which appear in the N-axioms **III** and **IV**.

Definition 41 (created redex). *A redex $u : M \longrightarrow P$ creates a redex $v : P \longrightarrow N$, when there does not exist any redex w outgoing from M, such that v is a residual of w after u.*

Definition 42 (disjoint). *Two redexes u and v are disjoint when $\neg(u \preceq v)$ and $\neg(v \preceq u)$.*

6.4 The N-axioms Related to Finite Developement: FinDev and Perm

N-axiom FinDev (finite developments). Let (M, U) be a multi-redex. Then, there does not exist any infinite sequence of redexes

$$M_1 \xrightarrow{u_1} M_2 \xrightarrow{u_2} \cdots \xrightarrow{u_{n-1}} M_n \xrightarrow{u_n} M_{n+1} \xrightarrow{u_{n+1}} \cdots$$

such that, for every index n, the path $u_1 \cdots u_n$ is a development of (M, U).

N-axiom Perm (compatible permutation). For every two coinitial, compatible and different redexes $u : M \longrightarrow P$ and $v : M \longrightarrow Q$, there exists a complete development h_u of $u[\![v]\!]$, and a complete development h_v of $v[\![u]\!]$, such that:

1. the paths h_u and h_v are cofinal,
2. the residual relations $[\![u \cdot h_v]\!]$ and $[\![v \cdot h_u]\!]$ are equal.

6.5 The Fundamental N-axioms: I, II, III, IV

N-axiom I (unique residual). We ask that

$$u \uparrow v \text{ and } \neg(v \preceq u) \qquad \Rightarrow \qquad \exists! u',\ u[\![v]\!]u'$$

when u and v are coinitial redexes.

N-axiom II (context-free). Suppose that u, v, w are pairwise compatible redexes, that the redex u' is residual of u after w, and the redex v' residual of v after w. We ask that,

$$\text{a.} \quad (u \preceq v \Rightarrow u' \preceq v') \quad \text{or} \quad (w \preceq u \ \text{ and } \ w \preceq v)$$
$$\text{b.} \quad (u' \preceq v' \Rightarrow u \preceq v) \quad \text{or} \quad w \preceq v$$

N-axiom III (enclave). Suppose that u and v are two compatible redexes, and that $u \prec v$. Call u' the residual of u after v. We ask that for every redex v' created by v,

$$u' \prec v' \qquad \text{or} \qquad \neg(u' \uparrow v')$$

N-axiom IV (stability). Suppose that u and v are two compatible disjoint redexes. Call u' the residual of u after v, and v' the residual of v after u. We ask that there exists no triple of redexes (w_1, w_2, w) such that w_1 is a redex created by u, w_2 is a redex created by v, and

$$w_1 [\![v']\!] w \qquad \text{and} \qquad w_2 [\![u']\!] w$$

6.6 Every Axiomatic Nesting System Defines an Axiomatic Rewriting System

Definition 43. *Every axiomatic nesting system* $(\mathcal{G}, [\![-]\!], \preceq, \uparrow)$ *defines a 2-dimensional transition system* $(\mathcal{G}, \rhd)$ *as follows:*

> $\rhd$ *is the least relation between paths of* $\mathcal{G}$ *such that* $v \cdot h_u \rhd u \cdot h_v$ *when*
>
> – *the paths* $u \cdot h_v$ *and* $v \cdot h_u$ *are cofinal, and satisfy* $[\![u \cdot h_v]\!] = [\![v \cdot h_u]\!]$,
> – u *and* v *are two coinitial redexes outgoing from a term* M,
> – $u \uparrow v$ *and* $\neg(v \preceq u)$,
> – *the path* h_u *is a complete development of* $(M, \{u\})[\![v]\!]$,
> – *the path* h_v *is a complete development of* $(M, \{v\})[\![u]\!]$.

Observe that the 2-dimensional transition system $(\mathcal{G}_\lambda, \rhd_{\text{tree}})$ of Section 1.9 is the result of applying Definition 43 to the axiomatic nesting system $(\mathcal{G}_\lambda, [\![-]\!]_\lambda, \preceq_{\text{tree}}, \uparrow_\lambda)$ below:

– $[\![-]\!]_\lambda$ is the usual residual relation between β-redexes in the λ-calculus, as defined in [10, 24, 20, 3],
– $\uparrow_\lambda$ is the compatibility relation between β-redexes, in that case the total relation, indicating that every two coinitial β-redexes are compatible,
– $\preceq_{\text{tree}}$ is the tree-nesting relation between β-redexes, defined in Section 1.9.

The main result of the section (Theorem 5) states that the 2-dimensional transition system $(\mathcal{G}, \rhd)$ of Definition 43 satisfies the standardization axiomatics of Section 2. Before proving that theorem, we start with five preliminary lemmas.

Lemma 44. *The 2-dimensional transition system $(\mathcal{G}, \triangleright)$ of Definition 43 satisfies Axiom* ***shape****.*

Proof. Suppose that $f \triangleright g$ is a permutation in $(\mathcal{G}, \triangleright)$. By definition, the two first steps of f and g are different. By the N-axioms **I** and **Self**, the length of the rewriting path f is 2. Axiom **shape** follows. □

Definition 43 exports from axiomatic rewriting systems to axiomatic nesting systems the definitions of standardization preorder $\Longrightarrow$ and Lévy equivalence relation $\equiv$ in Section 1.5, as well as (thanks to Lemma 44) the definitions of extraction and projection in Section 2.5. We prove

Lemma 45 (cube lemma). *Suppose that (M, U) is a multi-redex in an axiomatic nesting system $(\mathcal{G}, [\![-]\!], \preceq, \uparrow)$. Then, every two complete developments f and g of (M, U) are Lévy equivalent.*

Proof. By the N-axioms **Finite** and **Compat**, the complete developments of (M, U) ordered by prefix, define a finitely branching tree. The tree is thus finite by König's lemma and N-axiom **FinDev**. We proceed by induction on the length of the longest path of that tree, called the "depth" of (M, U). Suppose that the lemma is established for every multi-redex of depth less than n, and let (M, U) be a multi-redex of depth $n + 1$. Let f and g be two complete developments of (M, U). If one of the two paths f or g is empty, then the set U is empty, and thus the two complete developments f and g are empty: it follows that $f \equiv g$. Otherwise, the two paths f and g factor as $f = u \cdot f'$ and $g = v \cdot g'$ where the redexes u and v are elements of the multi-redex (M, U), the path f' is a complete development of $(M, U)[\![u]\!]$, and the path g' is a complete development of $(M, U)[\![v]\!]$. We proceed by case analysis. Either $u = v$ or $u \neq v$. In the first case, both paths f' and g' are complete developments of the multi-redex $(M, U)[\![u]\!] = (M, U)[\![v]\!]$; the equivalence $f' \equiv g'$ follows from our induction hypothesis applied to the multi-redex $(M, U)[\![u]\!]$, and we conclude that $f \equiv g$. In the second case, when $u \neq v$, it follows from N-axiom **Perm** that there exist two complete developments h_u of $u[\![v]\!]$ and h_v of $v[\![u]\!]$, such that the paths $v \cdot h_u$ and $u \cdot h_v$ are coinitial and cofinal, and induce the same residual relation $[\![u \cdot h_v]\!] = [\![v \cdot h_u]\!]$. Let h be any complete development of the multi-redex $(M, U)[\![u \cdot h_v]\!] = (M, U)[\![v \cdot h_u]\!]$. By definition of a complete development, the path $h_v \cdot h$ is a complete development of $(M, U)[\![u]\!]$, and the path $h_u \cdot h$ is a complete development of $(M, U)[\![v]\!]$. The two equivalence relations $h_v \cdot h \equiv f'$ and $h_u \cdot h \equiv g'$ follow from our induction hypothesis applied to the multi-redexes $(M, U)[\![u]\!]$ and $(M, U)[\![v]\!]$. We conclude that $f \equiv g$ by the series of equivalence:

$$f \; = \; u \cdot f' \; \equiv \; u \cdot h_v \cdot h \; \equiv \; v \cdot h_u \cdot h \; \equiv \; v \cdot g' \; = \; g$$

□

Lemma 46. *Suppose that the path f is a complete development of a multi-redex (M, U) in an axiomatic nesting system $(\mathcal{G}, [\![-]\!], \preceq, \uparrow)$. Suppose that a redex u is element of U, and satisfies $\neg(v \preceq u)$ for every redex v in the set $U - \{u\}$. Then, the redex u is extractible from the path f.* □

Proof. By induction on the length of the complete development f. The path f is not empty. It thus factors as $f = w \cdot g$, where $w : M \longrightarrow P$ is a redex of U, and g is a complete development of $(M, U)[\![w]\!]$. The lemma is obvious when $u = w$. Otherwise, by hypothesis, $u \uparrow w$ and $\neg(w \preceq u)$. By N-axiom **I**, the redex u has a unique residual residual after reduction of the redex w. Let us call this redex u'. Let v' denote any redex in $(N, U') = (M, U)[\![w]\!]$ different from the redex u'. We prove that $\neg(v' \preceq u')$. By definition of the redex v', there exists a redex v in U, such that $v[\![w]\!]v'$. Obviously, the redex v is different from the redex u because u' is the unique residual of the redex u after w. It follows from hypothesis on u that $\neg(v \preceq u)$. We apply the N-axiom **IIb**. to $\neg(v \preceq u)$ and $\neg(w \preceq u)$ to deduce that $\neg(v' \preceq u')$. We have just proved that $\neg(v' \preceq u')$ for any redex v' in $U' - \{u'\}$. Our induction hypothesis implies then that the redex u' is extractible from the complete development g of (N, U').

To summarize, we know that $u \uparrow w$, that $\neg(w \preceq u)$, and that the unique residual of u after w, denoted u', is extractible from the path g. We claim that it follows from this that the redex u is extractible from the path $w \cdot g$. Indeed, by N-axiom **Perm**, there exists a complete development h_u of the multi-redex $(M, \{u\})[\![v]\!]$ and a complete development h_w of the multi-redex $(M, \{w\})[\![u]\!]$, such that the paths $u \cdot h_w$ and $v \cdot h_u$ are coinitial, cofinal, and induce the same residual relation $[\![u \cdot h_w]\!] = [\![v \cdot h_u]\!]$. Moreover, $h_u = u'$ by N-axioms **I** and **Self**. By definition, $w \cdot u' \rhd u \cdot h_w$. It follows that the redex u is extractible from the path $f = w \cdot g$. This concludes our proof by induction. □

Lemma 47. *Suppose that $f : M \longrightarrow\!\!\!\!\!\rightarrow N$ is a complete development of a multi-redex (M, U) in an axiomatic nesting system $(\mathcal{G}, [\![-]\!], \preceq, \uparrow)$. Then, every path more standard than f is a complete development of (M, U).*

Proof. Suppose that a complete development of (M, U) factors as

$$M \xrightarrow{f_1}\!\!\!\!\!\rightarrow P \xrightarrow{f}\!\!\!\!\!\rightarrow Q \xrightarrow{f_2}\!\!\!\!\!\rightarrow N$$

and that $f \rhd g$. We show that the path $f_1 \cdot g \cdot f_2$ is also a complete development of (M, U). By definition of a complete development, we may suppose without loss of generality that the path f_1 is empty. By definition of $\rhd$, the paths f and g are two cofinal complete development of a multi-redex $(M, \{u, v\})$, and factor as $f = v \cdot u'$ and $g = u \cdot h_v$ where $\neg(v \preceq u)$, the redex u' is the unique residual of u after v and h_v is a complete development of the residuals of v after u. By definition of a complete development of (M, U), one ancestor of v' before u is element of U. By the N-axiom **Ancestor**, this ancestor is unique, and we already have one candidate: the redex v. We conclude that the redex v is element of U. By definition of $\rhd$, the rewriting paths f and g induce the same residual relation $[\![f]\!] = [\![g]\!]$. We conclude that $f_1 \cdot g \cdot f_2$ is a complete development of the multi-redex (M, U). □

Lemma 48. *Suppose that the rewriting path $f : M \longrightarrow\!\!\!\!\!\rightarrow N$ is a complete development of a multi-redex (M, U) in the axiomatic nesting system $(\mathcal{G}, [\![-]\!], \preceq, \uparrow)$. Then,*

– *every redex u extractible from the path f is element of U,*
– *every projection of the path f by extraction of a redex u is a complete development of the multi-redex $(M, U)[\![u]\!]$.*

Proof. Immediate consequence of Lemma 47. □

Theorem 5. *By Definition 43, every axiomatic nesting system $(\mathcal{G}, [\![-]\!], \preceq, \uparrow)$ defines an axiomatic rewriting system $(\mathcal{G}, \triangleright)$.*

Proof. We establish that the 2-dimensional transition system $(\mathcal{G}, \triangleright)$ satisgies the nine axioms of Section 2.

Axiom 1. Axiom **shape** is established in Lemma 44,
Axiom 2. Axiom **ancestor** follows from N-axiom **Ancestor** and Lemma 45,
Axiom 3. We prove Axiom **reversibility**. Suppose that $f \triangleright g \triangleright h$. By definition of $\triangleright$, there exists five redexes u, v, w, u', v' and a path h' such that $f = u \cdot v'$ and $g = v \cdot u'$ and $h = w \cdot h'$, and $u \uparrow v$ and $v \uparrow w$ and $\neg(u \preceq v)$ and $\neg(v \preceq w)$. By definition of $f \triangleright g$, the redex u' is the complete development of the residuals of u after v, thus a fortiori a residual of u after v. By definition of $g \triangleright h$, the redex u' is a residual of w after v. The equality $u = w$ follows from N-axiom **Ancestor**. Thus, h' is a complete development of the residuals of v after $u = w$. But, by definition of $f \triangleright g$ and N-axiom **I**, the redex v' is the unique residual of v after u. Thus, $h' = v'$ and we conclude Axiom **reversibility** with the equality $h = u \cdot h' = u \cdot v' = f$.
Axiom 4. We prove Axiom **irreversibility**. Suppose that $f \blacktriangleright g$ and $g \Longrightarrow h$. By definition of $f \triangleright g$, the paths f and g are complete developments of a multi-redex $(M, \{u, v\})$ with, say, the paths f and g starting by reducing v and u respectively. The nesting relation $u \prec v$ follows easily from $f \blacktriangleright g$. By Lemma 47, and our hypothesis that $g \triangleright h$, the path h is a complete development of $(M, \{u, v\})$. We prove that h starts by reducing the redex u. By definition of $g \Longrightarrow h$, there exists a sequence

$$g = h_1 \overset{1}{\Longrightarrow} h_2 \overset{1}{\Longrightarrow} \cdots h_n \overset{1}{\Longrightarrow} h_{n+1} = h$$

of complete development of $(M, \{u, v\})$ and an index $1 \leq i \leq n$ such that h_i starts by reducing the redex u, and h_{i+1} starts by reducing the redex v. This means that h_i and h_{i+1} factor as $h_i = u \cdot w \cdot h'$ and $h_{i+1} = v \cdot h_u \cdot h'$, where $u \cdot w \triangleright v \cdot h_u$. This contradicts $u \prec v$. We conclude that the path h starts by reducing u. Obviously, the complete developments f and h are cofinal and induce the same residual relation $[\![f]\!] = [\![h]\!]$. The relation $f \blacktriangleright h$ follows from that and $u \preceq v$. This proves Axiom **irreversibility**.
Axiom 5. We prove Axiom **cube**. Among its hypothesis, we have that $v \cdot u' \triangleright u \cdot v_1 \cdots v_n$ and that the redex w_{n+1} is residual of the redex w after the path $u \cdot v_1 \cdots v_n$. By definition of $v \cdot u' \triangleright u \cdot v_1 \cdots v_n$, the redex w_{n+1} is also residual of w after the path $v \cdot u'$. By N-axiom **Self**, the redexes u, v, w are pairwise compatible and different. Thus, the pair $(M, \{u, v, w\})$ defines a multi-redex.

We prove that $\neg(u \preceq w)$ and $\neg(v \preceq w)$. The first relation follows from the hypothesis that $u \cdot w_1 \triangleright w \cdot h_u$. The second relation is established by case analysis, depending on whether $u \preceq v$ or $\neg(u \preceq v)$. In the first case, the relation $\neg(v \preceq w)$ holds by transitivity of $\preceq$, because $\neg(u \preceq w)$. In the second case, observe that the permutation $v \cdot u' \rhd u \cdot v_1 \cdots v_n$ is reversible. We write it $v \cdot u' \rhd u \cdot v_1$. The relation $\neg(v_1 \preceq w_1)$ follows from the hypothesis that $v_1 \cdot w_2 \rhd w_1 \cdot h_1$. By $\neg(u \preceq v)$ and $\neg(u \preceq w)$, and N-axiom **II**a. the relation $\neg(v \preceq w)$ follows from $\neg(v_1 \preceq w_1)$.

We have just proved that $\neg(u \preceq w)$ and $\neg(v \preceq w)$. By Lemma 46, the redex w is extractible from the two complete developments $v \cdot u' \cdot w_{n+1}$ and $u \cdot v_1 \cdots v_n \cdot w_{n+1}$ of $(M, \{u, v, w\})$. In particular, there exists a redex w' and two paths h_v and $h_{u'}$ forming permutations $u' \cdot w_{n+1} \rhd w' \cdot h_{u'}$ and $v \cdot w' \rhd w \cdot h_v$. This proves half of Axiom **cube**.

There remains to prove that the paths $h_v \cdot h_{u'}$ and $h_u \cdot h_1 \cdots h_n$ are Lévy equivalent. The two paths are projections by extraction of w of the complete developments $v \cdot u' \cdot w_{n+1}$ and $u \cdot v_1 \cdots v_n \cdot w_{n+1}$ of $(M, \{u, v, w\})$. The Lévy equivalence follows from Lemma 48. This concludes the proof of Axiom **cube**.

Axiom 6. We prove Axiom **enclave**. We recall its hypothesis: the irreversible permutation $v \cdot u' \blacktriangleright u \cdot v_1 \cdots v_n$ and the permutation $u' \cdot w_{n+1} \rhd w' \cdot h_{u'}$. The relations $u \uparrow v$ and $u \prec v$ and $u' \uparrow w'$ and $\neg(u' \preceq w')$ follow from this. By N-axiom **III**, the redex $u : M \longrightarrow N$ does not create the redex w'. Thus, there exists a redex outgoing from M with residual w' after u. This redex is unique by N-axiom **Ancestor**. We call it w.

By definition of $v \cdot u' \rhd u \cdot v_1 \cdots v_n$, the residual relation $w[\![v \cdot u']\!]w_{n+1}$ implies that $w[\![v \cdot v_1 \cdots v_n]\!]w_{n+1}$. It follows from N-axiom **Self** that the three redexes u, v, w are pairwise different and compatible, thus define a multi-redex $(M, \{u, v, w\})$.

We prove that $\neg(u \preceq w)$ and $\neg(v \preceq w)$. The first relation follows from N-axiom **II**a. applied to the relations $\neg(v \preceq u)$ and $\neg(u' \preceq w')$. The second relation follows from transitivity of $\preceq$ and $\neg(u \preceq w)$ and $u \preceq v$.

By Lemma 46, it follows that the redex w is extractible from the complete developments $v \cdot u' \cdot w_{n+1}$ and $u \cdot v_1 \cdots v_n \cdot w_{n+1}$ of the multi-redex $(M, \{u, v, w\})$. Equivalently, both paths $v \cdot u'$ and $u \cdot v_1 \cdots v_n$ drag the redex w_{n+1} to the redex w. This concludes the proof of Axiom **enclave**.

Axiom 7. We prove Axiom **stability**. By definition of $u \cdot v' \rhd v \cdot u' \rhd u \cdot v'$, the two redexes $u : M \longrightarrow P$ and $v : M \longrightarrow Q$ are compatible, and disjoint. By N-axiom **IV**, either the redex w_1 is not created by u, or the redex w_1 is not created by v.

Suppose for instance that w_2 is not created by v. In that case, there exists a redex w such that $w[\![v]\!]w_2$. Consequently, the redex w_{12} is residual of w after the path $v \cdot u'$. By definition of $u \cdot v' \rhd v \cdot u'$, the redex w_{12} is also residual of w after $u \cdot v'$. Thus, there exists a residual w_1' of w after u, such that $w_1'[\![v']\!]w_{12}$. The equality $w_1 = w_1'$ follows from $w_1[\![v']\!]w_{12}$ and N-axiom **Ancestor**. We conclude that w_1 is not created by v, and residual of w after u. The case when w_1 is not created by u, is symmetric.

By N-axiom **II**a. and $v[\![u]\!]v'$, $w[\![u]\!]w_1$, the relation $\neg(v \preceq w)$ follows from $\neg(v' \preceq w_1)$ and $\neg(u \preceq v)$. The relation $\neg(u \preceq w)$ holds for symmetric reasons.

Axiom **stability** follows easily.

Axiom 8. We prove Axiom **reversible-stability**. By Axiom **stability**, which was established above, applied to the hypothesis of Axiom **reversible-stability**, there exists a redex w such that

- $u \uparrow w$, $\neg(u \preceq w)$, and w_1 is the unique residual of w after v,
- $v \uparrow w$, $\neg(v \preceq w)$, and w_2 is the unique residual of w after u.

We prove that $\neg(w \preceq u)$ and $\neg(w \preceq v)$. Suppose for instance that $w \preceq u$. By N-axiom **IIa**. and $u[\![v]\!]u_1$ and $w[\![v]\!]w_2$, the relation $w_2 \preceq u_1$ follows from this and $\neg(v \preceq w)$. This contradicts definition of $w_2 \cdot u_{12} \rhd u_1 \cdot w_{12}$. Thus, $\neg(w \preceq u)$, and symmetrically $\neg(w \preceq v)$. Axiom **reversible-stability** follows from Lemma 46 applied alternatively to extract the redex u from the complete development $w \cdot v_2 \cdot u_{12}$, and the redex v from the complete development $w \cdot u_2 \cdot v_{12}$.

Axiom 9. We prove Axiom **termination** using an argument found in [20]. Suppose that h_1 is a complete development of a multi-redex (M, U). By N-axiom **FinDev** and Lemma 48, there does not exist any infinite sequence of extraction:

$$h_1 \searrow_{u_1} h_2 \searrow_{u_2} \cdots \searrow_{u_{i-1}} h_i \searrow_{u_i} h_{i+1} \cdots$$

where, for every $i \geq 1$, the path h_{i+1} is a projection of the path h_i by extraction of the redex u_i. Now, we prove that there does not exist any infinite sequence

$$f_1 \searrow_{u_1} f_2 \searrow_{u_2} \cdots \searrow_{u_{i-1}} f_i \searrow_{u_i} f_{i+1} \cdots \tag{24}$$

starting from a path $f_1 : M_1 \twoheadrightarrow N$. We proceed by induction on the length of f_1. Clearly, the property holds when $f_1 = \mathrm{id}_{M_1}$. From now on, we suppose that the path f_1 factors as $f_1 = u \cdot g_1$ composed of a redex u and a path g_1 ¡ of length strictly smaller than the length of f_1. Consider any infinite sequence of the form (24). We prove that, for every index $i \geq 1$, the path f_i factors as $f_i = h_i \cdot g_{\phi(i)}$ where

- h_i is a complete development of the multi-redex (M_i, U_i) defined as:

$$(M_i, U_i) = (M_i, u[\![u_1 \cdots u_{i-1}]\!]) = (M_1, \{u\})[\![u_1]\!] \cdots [\![u_{i-1}]\!]$$

- $\phi(i)$ is an index $1 \leq \phi(i) \leq i$ defining a sequence of extraction starting from g_1:

$$g_1 \searrow_{v_1} g_2 \searrow_{v_2} \cdots \searrow_{v_{\phi(i)-2}} g_{\phi(i)-1} \searrow_{v_{\phi(i)-1}} g_{\phi(i)}$$

  for a series of redexes $v_1, \ldots, v_{\phi(i)-1}$.

Suppose that the property holds for a given index $i \geq 1$, and let us prove it for the next index $i+1$. Consider the path $f_i = h_i \cdot g_{\phi(i)}$ and the redex u_i. Either the redex u_i is extractible from h_i, or there exists a redex $v_{\phi(i)}$ extractible from g_i and dragged to u_i by the path h_i. In the first case, we define $\phi(i+1)$ as $\phi(i)$, and conclude that the path f_{i+1} factors as $f_{i+1} = h_{i+1} \cdot g_{\phi(i+1)}$, where h_{i+1} is a projection of h_i by extraction of u_i; here, by Lemma 48, the path h_{i+1} is a complete development of $(M_i, U_i)[\![u_i]\!] = (M_{i+1}, U_{i+1})$ because h_i is a complete development of (M_i, U_i). In the second case, we define $\phi(i+1)$ as $\phi(i)+1$, and observe that the path f_{i+1} factors as $f_{i+1} = h_{i+1} \cdot g_{\phi(i+1)}$, where $h_i \cdot v_{\phi(i)} \searrow_{u_i} h_{i+1}$ and $g_{\phi(i)} \searrow_{v_{\phi(i)}} g_{\phi(i+1)}$; here, by Lemma 48, the path h_{i+1} is a complete development of the multi-redex $(M_i, \{u_i\} \cup U_i)[\![u_i]\!] = (M_{i+1}, U_{i+1})$ because $h_i \cdot v_{\phi(i)}$ is a complete development of the multi-redex $(M_i, \{u_i\} \cup U_i)$. We conclude that the factorization property holds, for every index $i \geq 1$.

The end of the proof follows easily. By induction hypothesis applied to g, there exists an index $j \geq 1$ such that $\phi(j+i) = \phi(j)$, for every index $i \leq j$. Thus, the infinite sequence (24) induces an infinite sequence

$$h_j \searrow_{u_j} h_{j+1} \searrow_{u_{j+1}} \cdots \searrow_{u_{j+i-1}} h_{j+i} \searrow_{u_{j+i}} h_{j+i+1} \cdots$$

from the complete development h_j of (M_j, U_j). This contradicts a preliminary result deduced from N-axiom **FinDev**. It follows that there exists no infinite sequence of the form (24) starting from f. This concludes our reasoning by induction, and establishes Axiom **termination**. □

7 Optional Hypothesis on Standardization

7.1 Epimorphisms wrt. ≡

In Lemma 17 of Section 3.1, we establish that every path is epi (=left-cancellable) in the quotient category **2-cat**$(\mathcal{G}, \rhd)/\simeq$. The same epiness property modulo $\equiv$ instead of $\simeq$ has been established in [24, 18, 6] for the λ-calculus and any (left-linear) term rewriting system. Quite interestingly, the redex v and Lévy equivalence

$$M \xrightarrow{v} N \xrightarrow{u_1} P \quad \equiv \quad M \xrightarrow{v} N \xrightarrow{u_2} P$$

in the axiomatic rewriting system

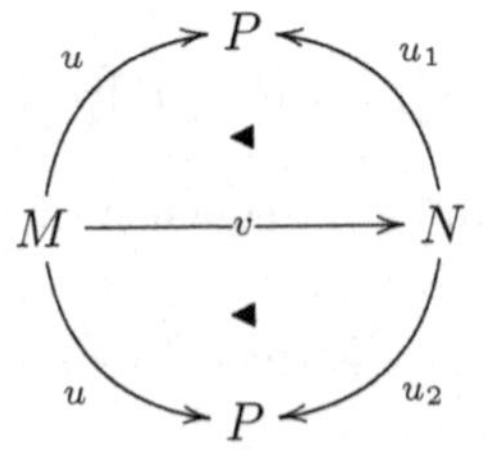

$$v \cdot u_1 \blacktriangleright u$$
$$v \cdot u_2 \blacktriangleright u$$

illustrate that the epiness property modulo $\equiv$ does not generalize to axiomatic rewriting systems. However, an additional hypothesis may be added on $(\mathcal{G}, \rhd)$ to ensure epiness of morphisms in the category **2-cat**$(\mathcal{G}, \rhd)/\equiv$.

Optional hypothesis (descendant). Two redexes u' and u'' are equal when they are involved in permutations $v \cdot u' \rhd u \cdot f$ and $v \cdot u'' \rhd u \cdot g$, where u, v are redexes and f, g are paths.

Diagrammatically,

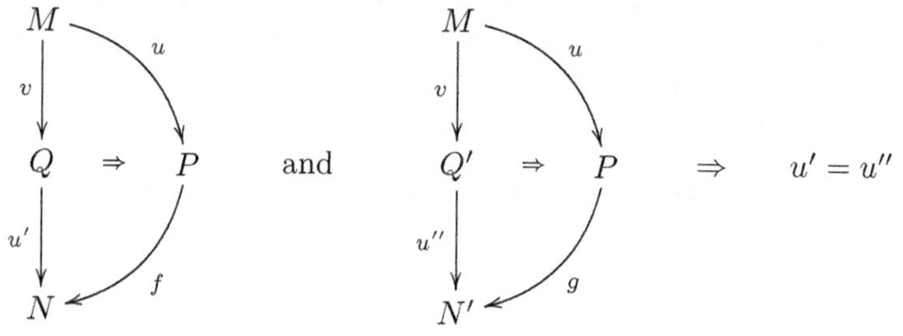

Obviously, hypothesis **descendant** holds in every axiomatic rewriting system derived from an axiomatic nesting system, see Definition 43. Thus, Lemma 49 generalizes the property of epiness modulo $\equiv$ established in [24, 18, 6] for the λ-calculus and term rewriting systems.

Lemma 49 (epi wrt. $\equiv$). *Suppose that $f : M \twoheadrightarrow P$ and $g_1, g_2 : P \twoheadrightarrow N$ are three paths in an axiomatic rewriting system $(\mathcal{G}, \rhd)$ and that $(\mathcal{G}, \rhd)$ satisfies hypothesis* ***descendant****. Then,*

$$f \cdot g_1 \equiv f \cdot g_2 \quad \Rightarrow \quad g_1 \equiv g_2$$

Proof. By induction on the length of the standard path h of $f \cdot g_1$ (and of $f \cdot g_2$.) Let u be the first redex computed in h. We conclude by induction hypothesis when u is extractible from f. Otherwise, there exist a redex v_1 extractible from g_1 and a redex v_2 extractible from g_2, such that f drags v_1 and v_2 to the redex u. By hypothesis **descendant**, the two redexes v_1 and v_2 are the same redex v. We write f', h_1 and h_2 for arbitrary results of the extractions $f \cdot v \searrow_u f'$ and $g_1 \searrow_v h_1$ and $g_2 \searrow_v h_2$. Equivalence $f' \cdot h_1 \equiv f' \cdot h_2$ follows from Lemma 12 (*preservation of extraction*), and definition of u as the first redex of a standard path of $f \cdot g_1$ and $f \cdot g_2$. Equivalence $h_1 \equiv h_2$ follows from this equivalence and our induction hypothesis. The series of equivalence

$$g_1 \equiv v \cdot h_1 \equiv v \cdot h_2 \equiv g_2$$

concludes the proof by induction. □

7.2 Monomorphisms wrt. $\simeq$

A well-known example in [24] shows that β-rewriting paths are not necessarily mono (=right-cancellable) modulo Lévy equivalence $\equiv$. The example is the β-redex w in the Lévy permutation equivalence

$$I(Ia) \xrightarrow{u} Ia \xrightarrow{w} a \qquad \equiv \qquad I(Ia) \xrightarrow{v} Ia \xrightarrow{w} a$$

The example may be adapted to show that β-rewriting paths are not necessarily mono modulo $\simeq$-equivalence in the λ-calculus equipped with the *argument-order* on β-redexes, in the following way:

$$(\lambda x.(\lambda y.y)x)a \xrightarrow{u} (\lambda y.y)a \xrightarrow{w} a \qquad \diamond \qquad (\lambda x.(\lambda y.y)x)a \xrightarrow{v} (\lambda x.x)a \xrightarrow{w} a$$

In contrast, we show that rewriting paths are mono modulo $\simeq$ in every axiomatic rewriting system satisfying the additional property **reversible-shape**. It follows that monoicity modulo $\simeq$ holds in almost every rewriting system, in particular in the λ-calculus equipped with the *tree-order* or the *left-order* on β-redexes, as well as on Petri nets and term rewriting systems.

Optional hypothesis (reversible shape). Two redexes v and v' are different when they are involved in a reversible permutation $u \cdot v \diamondsuit u' \cdot v'$.

Lemma 50 (epi-mono wrt. $\simeq$). *Suppose that $f : M \twoheadrightarrow P$ and $g_1, g_2 : P \twoheadrightarrow Q$ and $h : Q \twoheadrightarrow N$ are four paths in an axiomatic rewriting system $(\mathcal{G}, \rhd)$ satisfying hypothesis* ***reversible-shape****. Then,*

$$f \cdot g_1 \cdot h \simeq f \cdot g_2 \cdot h \quad \Rightarrow \quad g_1 \simeq g_2$$

Proof. Immediate consequence of Lemma 15 for right-cancellation and Lemma 17 for left-cancellation. □

7.3 A Simpler Structure of Starts

The *structure of starts* described in Lemma 16 (Section 3.1) appears to be surprisingly more complicated than the *structure of stops* described in Lemma 15. However, a much simpler characterization of starts is possible in any axiomatic rewriting system $(\mathcal{G}, \rhd)$ satisfying the additional hypothesis **reversible-cube** formulated below. The new characterization of starts appears in Lemma 51. Note that the property is satisfied by the λ-calculus and more generally by any axiomatic rewriting system derived from an axiomatic nesting system. On the other hand, it is not satisfied by the axiomatic rewriting system defined on order sequentializations, and defined at the endof Section 8.

Optional hypothesis (reversible cube). We ask that every diagram

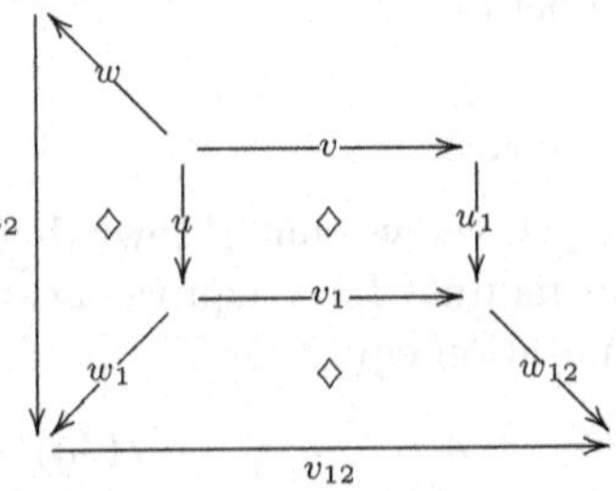

where u, v, u_1, v_1 and $w, w_1, w_{12}, u_2, v_{12}$ are redexes forming the reversible permutations

$$v \cdot u_1 \diamondsuit u \cdot v_1 \qquad u \cdot w_1 \diamondsuit w \cdot u_2 \qquad v_1 \cdot w_{12} \diamondsuit w_1 \cdot v_{12}$$

may be completed as a diagram

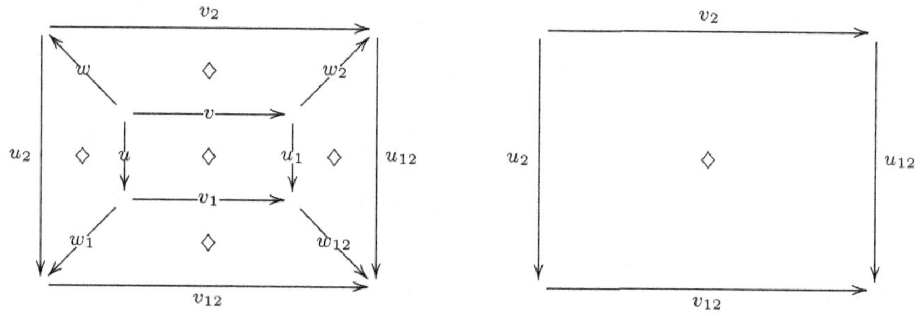

where w_2, v_2, u_{12} are three redexes forming reversible permutations

$$v \cdot w_2 \diamond w \cdot v_2 \qquad u_1 \cdot w_{12} \diamond w_2 \cdot u_{12} \qquad v_2 \cdot u_{12} \diamond u_2 \cdot v_{12}$$

Lemma 51 (simpler structure of starts). *Suppose that $u_1 \cdots u_n : M \twoheadrightarrow N$ is a path in an axiomatic rewriting system $(\mathcal{G}, \rhd)$ satisfying hypothesis* **reversible-cube***. Then, a redex $u : M \longrightarrow P$ starts the path $u_1 \cdots u_n : M \twoheadrightarrow N$ if and only there exists an index $1 \leq i \leq n$ and a path $v_1 \cdots v_{i-1}$ such that the path $u_1 \cdots u_{i-1}$ followed by the redex u_i permutes reversibly to the redex u followed by the path $v_1 \cdots v_{i-1}$.*

Proof. Suppose that a path f followed by a redex v permutes reversibly to a redex u followed by a path g. Hypothesis **reversible-cube** implies that for every path $f' \simeq f$, there exists a path $g' \simeq g$ such that the path f' followed by the redex v permutes reversibly to the redex u followed by the path g'. The lemma follows immediately from this, and Lemma 16. □

8 Examples and Open Problems

Asynchronous transition systems. Asynchronous transition systems extend both non-deterministic transition systems, and Mazurkiewicz trace languages. They were introduced independently in [4] and [39], see also [33].

An *asynchronous transition system* T is a quintuple $T = (S, i, E, I, \mathrm{Tran})$ where

- S is a set of *states* with *initial state* i,
- E is a set of *events*,
- $\mathrm{Tran} \subset S \times L \times S$ is the *transition relation*,
- $I \subset E \times E$ is an irreflexive, symmetric relation called the *independence relation.*

Every asynchronous transition system is supposed to satisfy four axioms:

1. *parsimony:* $\forall e \in E,\ \exists (s, s') \in S \times S,\ \ (s, e, s') \in \mathrm{Tran}$,
2. *determinacy:* $\ \forall (s, e, s'), (s, e, s'') \in \mathrm{Tran}, \quad s' = s''$,

3. *independence:* $\forall (s, e_1, s_1), (s, e_2, s_2) \in \text{Tran},$

$$e_1 I e_2 \;\Rightarrow\; \exists s', (s_1, e_2, s') \in \text{Tran and } (s_2, e_1, s') \in \text{Tran}$$

4. *together:* $\forall (s, e_2, s_2), (s_2, e_1, s') \in \text{Tran},$

$$e_1 I e_2 \;\Rightarrow\; \exists s_1, \quad (s, e_1, s_1) \in \text{Tran and } (s_1, e_2, s') \in \text{Tran}$$

Every asynchronous transition system T defines an axiomatic rewriting system $(\mathcal{G}_T, \rhd_T)$, as follows:

- the graph $\mathcal{G}_T$ has states as vertices and transitions (s, e, s') as arrows,
- two paths f and g are related as $f \rhd_T g$, precisely when there exist four transitions (s, e_1, s_1), (s, e_2, s_2), (s_1, e_2, s'), (s_2, e_1, s') in Tran, such that
  - $f = (s, e_2, s_2) \cdot (s_2, e_1, s')$,
  - $g = (s, e_1, s_1) \cdot (s_1, e_2, s')$,
  - the two events e_1 and e_2 are independent: $e_1 I e_2$.

We check that the standardization axioms hold in $(\mathcal{G}_T, \rhd_T)$. Axiom **shape** follows from anti-reflexivity of the independence relation. Observe that every permutation $f \rhd_T g$ is reversible: it coexists with a permutation $g \rhd_T f$. The three Axioms **irreversibility**, **enclave** and **termination** follow from this, as well as the equivalence between Axiom **stability** and Axiom **reversible-stability**. We establish now the four Axioms **ancestor**, **reversibility**, **cube** and **reversible-stability**. The property (2) of **determinacy** has two remarkable consequences in every asynchronous transition system T:

$$f \diamondsuit_T g \quad \text{and} \quad f \diamondsuit_T h \;\Rightarrow\; g = h.$$

$$f \diamondsuit_T g \diamondsuit_T h \;\Rightarrow\; f = h.$$

The two Axioms **ancestor** and **reversibility** follow from the first and second assertions, respectively. By definition of the permutation relation $\rhd_T$, the three events e_1, e_2, e_3 are pairwise independent:

$$e_1 I e_2 \qquad e_2 I e_3 \qquad e_1 I e_3.$$

in every diagram

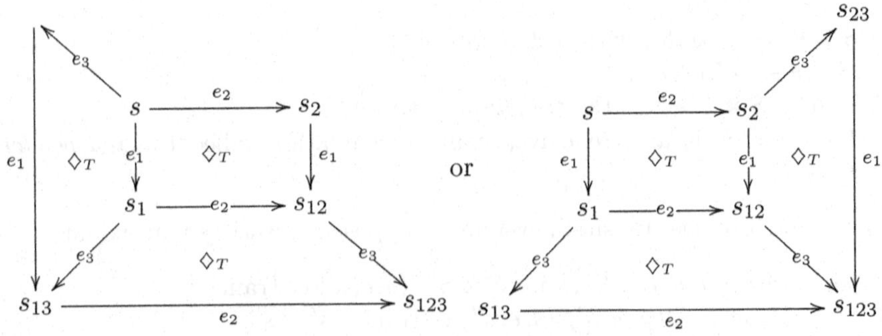

So, it follows from the properties (2) and (4) of **determinacy together** with the properties of the asynchronous transition system T, that the two diagrams above may be completed as:

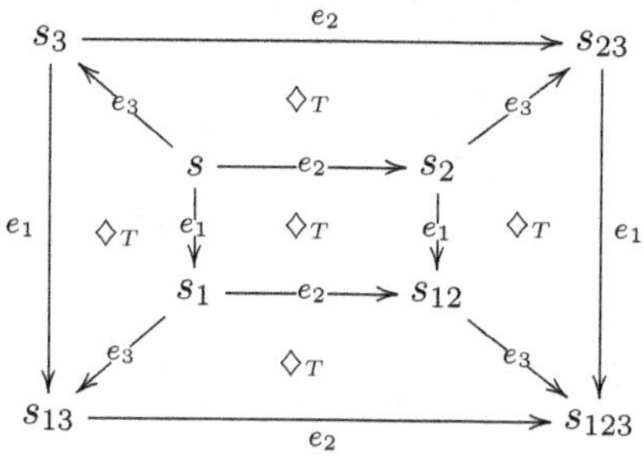

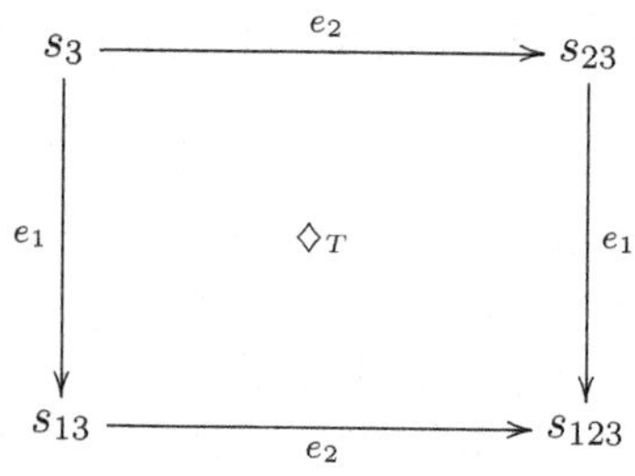

Axioms **cube** and **reversible-stability** follow immediately. It is also nearly immediate that $(\mathcal{G}_T, \rhd_T)$ enjoys the additional hypothesis **descendant**, **reversible-shape** and **reversible-cube** formulated in Section 7.

Remark: We have just proved the axiomatics (and the additional hypothesis) without ever using properties (1) and (3) of the asynchronous transition system T.

Remark: The standardization theorem is not really informative in $(\mathcal{G}_T, \rhd_T)$ because every permutation being reversible, all paths are standard. However, the axiomatics itself ensures that every asynchronous system satisfies the stability theorem stated in [30] which describes the structure of its successful runs.

PETRI NETS. The theory of Petri nets illustrates nicely the notion of asynchronous transition system. A Petri net is a quintuple $N = (C, j, F, \boldsymbol{pre}, \boldsymbol{post})$ where

- C is a set of *conditions*,
- j is a particular marking of N, called the *initial marking*, where a *marking* of N is defined as a multi-set of conditions,
- F is a set of *firings*,
- $\boldsymbol{pre}, \boldsymbol{post}$ are two functions associating to every firing $e \in F$ the nonempty markings $\boldsymbol{pre}(e)$ and $\boldsymbol{post}(e)$, called respectively the *pre-condition* and *post-condition* of e.

An asynchronous transition system $T_N = (S, i, E, I, \text{Tran})$ is associated to every Petri net N in the following way, see [33]:

- S is the set of markings of N,
- i is the marking $j \in S$,
- E is the set F of firings,
- Tran is the set of triples (p, e, q) such that $p = p_0 \uplus \boldsymbol{pre}(e)$ and $q = p_0 \uplus \boldsymbol{post}(e)$ for a marking p_0, where $\uplus$ is the multi-set addition.

- I relates two firings $e_1, e_2 \in F$ precisely when $\boldsymbol{pre}(e_1) \cap \boldsymbol{pre}(e_2)$ and $\boldsymbol{post}(e_1) \cap \boldsymbol{post}(e_2)$ are empty multi-sets.

The axiomatic rewriting system $(\mathcal{G}_N, \rhd_N)$ associated to the asynchronous transition system T_N may be described directly, as follows. Its transition system $\mathcal{G}_N$ has the markings of N as vertices, and the triples

$$(p_0 \uplus \boldsymbol{pre}(e), e, p_0 \uplus \boldsymbol{post}(e)) = (p, e, q)$$

as edges $p \longrightarrow q$. The permutation relation $\rhd_N$ relates two paths $u \cdot v' \rhd v \cdot u'$ precisely when:

1. u and v are edges $u = (p, e_1, p_1)$ and $v = (p, e_2, p_2)$,
2. u' and v' are edges $u' = (p_2, e_1, p')$ and $v' = (p_1, e_2, p')$,
3. $\boldsymbol{pre}(e_1) \cap \boldsymbol{pre}(e_2)$ and $\boldsymbol{post}(e_1) \cap \boldsymbol{post}(e_2)$ are empty multi-sets.

BUBBLE SORT. The standardization procedure may be viewed as a generalization of the bubble sort algorithm, in which the order is not given *globally* but *locally*. Define $\mathcal{G}$ as the graph with a unique vertex M and, for every natural number $i \in \mathbb{N}$, an edge $[i] : M \longrightarrow M$. Let $\rhd$ be the least relation on paths such that

$$[j] \cdot [i] \rhd [i] \cdot [j]$$

when $i < j$. All the standardization axioms introduced in Section 2 are immediate on $(\mathcal{G}, \rhd)$ — except Axiom **enclave** which follows from the transitivity of the order on natural numbers. The standardization theorem of $(\mathcal{G}, \rhd)$ states that every sequence of natural numbers $[j_1] \cdots [j_k]$ may be reordered by local permutations into an increasing sequence $[i_1] \cdots [i_k]$ — and that this reordering is unique, since all the permutations of $(\mathcal{G}, \rhd)$ are irreversible.

HIERARCHICAL TRANSITION SYSTEMS. Here, we subsume the two previous examples of asynchronous transition systems, and of bubble sort on natural numbers, into what we call a *hierarchical transition system*. The idea is to order events in an asynchronous transition system (typically firings in a Petri net) with a precedence relation $\preceq$ satisfying a *weak transitivity* condition.

A *hierarchical transition system* is a quintuple $T = (S, i, E, \preceq, \mathrm{Tran})$ where

- S is a set of *states* with *initial state* i,
- E is a set of *events*,
- $\mathrm{Tran} \subset S \times L \times S$ is a *transition relation*,
- $\preceq \subset E \times E$ is a reflexive relation called the *precedence relation*.

The independence relation I is defined as

$$eIe' \iff \neg(e \preceq e') \text{ and } \neg(e' \preceq e) \tag{25}$$

The strict precedence relation $\prec$ is defined as

$$e \prec e' \iff e \preceq e' \text{ and } \neg(e' \preceq e)$$

Every hierarchical transition system is supposed to satisfy three axioms:

1. *determinacy:* $\forall (s, e, s'), (s, e, s'') \in \text{Tran}, \quad s' = s''$,
2. *independence:* $\forall (s, e_2, s_2), (s_2, e_1, s') \in \text{Tran}$,

$$\neg(e_2 \preceq e_1) \;\Rightarrow\; \exists s_1, \quad (s, e_1, s_1) \in \text{Tran and } (s_1, e_2, s') \in \text{Tran}$$

3. *weak transitivity:* $\forall (e, e', e'') \in E \times E \times E$,

$$e \prec e' \preceq e'' \;\Rightarrow\; e \preceq e''.$$

Hierarchical transition systems extend usual asynchronous transition systems, since every asynchronous transition system $T = (S, i, E, I, \text{Tran})$ may be seen as the hierarchical transition system $V(T) = (S, i, E, \preceq_{V(T)}, \text{Tran})$ with precedence relation $\preceq_{V(T)}$ defined as:

$$\forall (e, e') \in E \times E, \qquad e \preceq_{V(T)} e' \iff \neg(eIe')$$

Here, weak transitivity of $\preceq_{V(T)}$ follows from symmetricity. Now, we associate to every hierarchical transition system $T = (S, i, E, \preceq, \text{Tran})$ the following AxRS $(\mathcal{G}_T, \rhd_T)$:

- whose transition system $\mathcal{G}_T$ has states as vertices and transitions (s, e, s') as arrows,
- whose permutation relation $\rhd_T$ relates two paths f and g as $f \rhd_T g$, precisely when $f = (s, e_2, s_2) \cdot (s_2, e_1, s')$, $g = (s, e_1, s_1) \cdot (s_1, e_2, s')$ and the two events e_1 and e_2 satisfy $\neg(e_2 \preceq e_1)$.

In particular: the permutation $f \rhd_T g$ is reversible iff $e_1 I e_2$ and irreversible iff $e_1 \prec e_2$. We claim that $(\mathcal{G}_T, \rhd_T)$ is an axiomatic rewriting system. All the standardization axioms hold in $(\mathcal{G}_T, \rhd_T)$ for the same reasons as in the case of asynchronous transition systems — except for Axiom **enclave**, which follows from the weak transitivity of the precedence relation $\preceq$.

This enables to state a standardization theorem for every hierarchical transition system T. A particularly interesting case is when the precedence relation $\preceq$ is a partial order. In that case, the standard paths of $(\mathcal{G}_T, \rhd_T)$ may be characterized as the sequences of transition:

$$s_1 \xrightarrow{e_1} \xrightarrow{e_2} \cdots \xrightarrow{e_{n-1}} s_n$$

in which there exists no pair of indices $1 \leq i < j \leq n$ such that $e_j \prec e_i$ (Hint: use the characterization lemma, Lemma 19). Thus, the standardization theorem states that every sequence of transitions in T

$$s_1 \xrightarrow{e_1} \xrightarrow{e_2} \cdots \xrightarrow{e_{n-1}} s_n$$

may be reorganised, after a series of permutations $\rhd_T$, into such an ordered sequence, and that this sequence is unique, modulo permutation of independent events.

We illustrate our point that weak transitivity of $\preceq$ is necessary to establish standardization. Consider the pseudo hierarchical transition system T with one state s, three events a, b, c, and the following precedence relation $\preceq$

$$a \preceq b \qquad b \preceq c \qquad c \preceq a$$

The relation $\preceq$ is not weakly transitive, and consequently, the uniqueness property fails: the sequence

$$s \xrightarrow{c} s \xrightarrow{b} s \xrightarrow{a} s$$

may be standardized as any of the two transition paths

$$s \xrightarrow{b} s \xrightarrow{c} s \xrightarrow{a} s \qquad \text{and} \qquad s \xrightarrow{c} s \xrightarrow{a} s \xrightarrow{b} s$$

which are not equal modulo permutation of independent events (the independence relation is empty in T.)

ERASING TRANSITION SYSTEMS. We mention only briefly that it is possible to enrich hierarchical transition systems with a notion of *erasure* between events. Start from a hierarchical transition system $(S, i, E, \preceq, \text{Tran})$ and equip it with a binary relation K on events, called the *erasing* relation, chosen among the subrelations of $\prec$. Then, replace property (2) of hierarchical transition systems, by the two axioms:

1. *K-erasure:* $\forall (s, e_2, s_2), (s_2, e_1, s') \in \text{Tran}$,

$$e_1 K e_2 \text{ and } \neg(e_2 \preceq e_1) \quad \Rightarrow \quad (s, e_1, s') \in \text{Tran}$$

2. *K-permutation:* $\forall (s, e_2, s_2), (s_2, e_1, s') \in \text{Tran}$,

$$\neg(e_1 K e_2) \text{ and } \neg(e_2 \preceq e_1) \quad \Rightarrow \quad \exists s_1, \quad (s, e_1, s_1) \in \text{Tran and } (s_1, e_2, s') \in \text{Tran}$$

This defines what we call an *erasing transition system* $T = (S, i, E, \preceq, K, \text{Tran})$. The definition of the AxRS $(\mathcal{G}_T, \rhd_T)$ associated to T proceeds as in the case of hierarchical transition system, except that permutations of the form

$$
\begin{array}{ccc}
p & \xrightarrow{\ e_2\ } & p_1 \\
{\scriptstyle e_1} \downarrow & {\scriptstyle \Leftarrow_T} & \downarrow {\scriptstyle e_1} \\
p' & \underset{\text{id}_{p'}}{=\!=\!=} & p'
\end{array}
$$

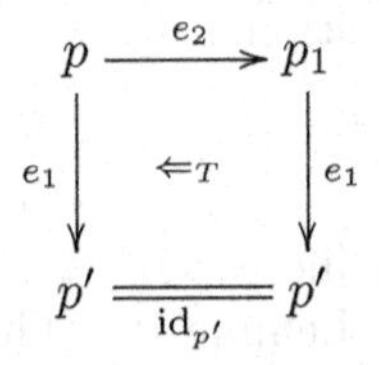

are considered when $e_1 K e_2$. The standardization axioms hold in $(\mathcal{G}_T, \rhd_T)$ for the same reasons as in the hierarchical case.

TERM REWRITING SYSTEMS. The reader interested in term rewriting systems will find an introduction to the subject in [21, 19, 2, 11] and a comprehensive study of standardization in [35]. Here, we recall only that

1. a term rewriting system is a pair $\Sigma = (\mathcal{F}, \{\rho_1, ..., \rho_n\})$ where $\mathcal{F}$ is the *signature* of an algebra and every ρ_i is a rewriting rule on this algebra.
2. a *rewriting rule* $\rho : L \to R$ is a pair of open terms of the algebra such that every variable in R also occurs in L,
3. a *redex* in Σ is a quadruple (M, o, ρ, σ) where M is a term, o is an occurrence of M, ρ is a rewriting rule $L \to R$ of the system and σ is a valuation of the variables appearing in L, such the term M decomposes as $M = C[L\sigma]_o$ for some context $C[-]_o$ with unique hole $[-]$ at occurrence o. Notation: we write $u : M \longrightarrow N$ for $N = C[R\sigma]_o$.
4. If the variable x occurs $k \geq 1$ times in L, every redex v in a term $\sigma(x)$ corresponds to k redexes $v_1, ..., v_k$ in the term $M = C[L\sigma]_o$. We say that $u = (M, o, \rho, \sigma)$ *nests* each of the redexes v_i; and that it *nests* the redex v *linearly* when $k = 1$,
5. We say that two redexes $u : M \longrightarrow P$ and $v : M \longrightarrow Q$ are *disjoint* when their occurrences in M are non comparable w.r.t the prefix order.
6. a rewriting rule $L \to R$ is *left-linear* when L does not contain two occurrences of the same variable. In that case, the only possibility for a redex to nest another redex, is to nest it linearly.

The transition system $\mathcal{G}_\Sigma$ of the rewriting system Σ has the terms M of the algebra as vertices and the redexes $u : M \longrightarrow N$ induced by the system as edges. The relation $\triangleright$ on path in $\mathcal{G}_\Sigma$ is the least relation such that:

1. $v \cdot u' \triangleright_\Sigma u \cdot v'$ when the redexes $u = (M, o_1, \rho_1, \sigma_1) : M \longrightarrow P$ and $v = (M, o_2, \rho_2, \sigma_2) : M \longrightarrow Q$ are disjoint and $u' = (Q, o_1, \rho_1, \sigma_1)$ and $v' = (P, o_2, \rho_2, \sigma_2)$,
2. $v \cdot u' \triangleright_\Sigma u \cdot f$ when $u = (M, o_1, \rho_1, \sigma_1) : M \longrightarrow P$ nests $v = (M, o_1; o, \rho_2, \sigma_2) : M \longrightarrow Q$ linearly, $u' = (Q, o_1, \rho_1, \sigma_1) : Q \longrightarrow N$ and $f : P \longrightarrow N$ is the *complete development* of the copies of v through u (see [21, 18, 27, 22] for a formal definition of complete developments and copies).

In order to prove that $(\mathcal{G}_\Sigma, \triangleright_\Sigma)$ satisfies the standardization axioms, we mediate through an axiomatic nesting system $(\mathcal{G}_\Sigma, [\![-]\!]_\Sigma, \preceq_\Sigma, \uparrow_\Sigma)$ and the ten N-axioms of Section 6. Our diagrammatic standardization theorem 2 will generalize the results of [18, 6] to possibly non-left-linear term rewriting systems.

The main point to clarify is: how shall the usual compatibility, nesting and residual relations be extended from left-linear to general term rewriting systems? There is a constraint: that the resulting axiomatic nesting system $(\mathcal{G}_\Sigma, [\![-]\!]_\Sigma, \preceq_\Sigma, \uparrow_\Sigma)$ generates the axiomatic rewriting system $(\mathcal{G}_\Sigma, \triangleright_\Sigma)$ defined hereabove. The definition follows immediately. Two coinitial redexes u and v are *compatible*, what we write $u \uparrow_\Sigma v$, when

- the redexes u and v are disjoint,
- or when the redex u nests the redex v linearly,
- or when the redex v nests the redex u linearly.

We define the relation $[\![-]\!]_\Sigma$. When u and v are not compatible, the redex u has simply no residual after v (in particular, $u[\![u]\!]_\Sigma$ is empty). When u and v are compatible, the definition of the residuals of u after v proceeds as in left-linear rewriting systems:

- when the redexes u and $v : M \longrightarrow N$ are disjoint, or when u nests v linearly, then $u = (M, o_1, \rho_1, \sigma_1)$ has the redex $u' = (N, o_1, \rho_1, \sigma_1')$ with same occurrence in N as residual.
- when the redex $v = (M, o_2, \rho_2 = L \longrightarrow R, \sigma_2)$ nests the redex u linearly, then the redex u has a residual u' after v for each occurrence of the variable x in R — where x is the variable substituted in L by the term $\sigma_2(x)$ containing the redex u.

Finally, we write $u \preceq_\Sigma v$ when the redex u nests the redex v linearly. Obviously, the axiomatic rewriting system $(\mathcal{G}_\Sigma, \rhd_\Sigma)$ derives from the resulting axiomatic rewriting system, by Definition 43. Moreover, each of the ten N-axioms are nearly immediate: N-axioms **Finite**, **Compat**, **Ancestor**, **Self** are obvious, while N-axioms **FinDev** and **Perm** generalize the well-known finite development lemma for left-linear term rewriting systems, established in [18, 20, 3, 27]. The four remaining N-axioms **I**, **II**, **III** and **IV** are also immediate.

Remark: Consider the term $F(A, A)$ in the non left-linear rewriting system Σ:

$$F(x, x) \longrightarrow G(x) \qquad\qquad A \longrightarrow B$$

Intuitively, there *should* be a permutation:

$$\begin{array}{ccccc} F(A,A) & \xrightarrow{A_1} & F(B,A) & \xrightarrow{A_2} & F(B,B) \\ {\scriptstyle F}\downarrow & & & & \downarrow{\scriptstyle F} \\ G(A) & & \xrightarrow{\quad\quad A \quad\quad} & & G(B) \end{array} \tag{26}$$

oriented as follows: $A_1 \cdot A_2 \cdot F \Longrightarrow F \cdot A$. However, in our presentation, we replace the permutation by a critical pair (= a hole) between the two redexes $F(A, A) \longrightarrow G(A)$ and $F(A, A) \longrightarrow F(B, A)$. This is one limit of our current axiomatic theory: we do not know how to integrate permutations like (26) in our standardization framework. The 2-categorical approach of Section 5 is likely to provide a solution, at least because it replaces the Axiom **shape** by the more flexible notion of partial injection $[\alpha]$.

λ-CALCULUS [TREE-NESTING ORDER]. We have already established in Section 2, at least informally, that the nine standardization axioms hold for this λ-calculus, and its associated 2-dimensional transition system $(\mathcal{G}_\lambda, \rhd_{\text{tree}})$. It is worth observing that the axiomatic nesting system $(\mathcal{G}_\lambda, [\![-]\!]_\lambda, \preceq_{\text{tree}}, \uparrow_\lambda)$ satisfies moreover the ten N-axioms of Section 6. This follows on one part from traditional results on β-redexes and residuals appearing in [24, 3], and on the other part, from elementary arguments on the dynamics of β-reduction which establish together the N-axioms **I**, **II**, **III** and **IV**. By Theorem 5, this provides another way to prove that $(\mathcal{G}_\lambda, \rhd_{\text{tree}})$ satisfies the 2-dimensional axiomatics of Section 2.

λ-CALCULUS [LEFT ORDER]. It is interesting to examine the reasons why the axiomatic nesting system associated to the λ-calculus and its left-order $\preceq_{\text{left}}$ satisfies the N-axioms formulated in Section 6. Six of the ten N-axioms do not mention the nesting order, and were thus already discussed in the previous paragraph. The four remaining axioms are N-axioms **I**, **II**, **III** and **IV**. The two N-axiom **I** and **IV** are easy to check. N-axiom **IV** for instance follows from the fact that the order $\preceq_{\text{left}}$ is *total*, and thus, that there exists no reversible permutations in the system. The two remaining N-axioms **II** and **III** are less obvious to establish. However, both of them hold inherently for the reason that in a λ-term PQ, no computation in Q may induce (by creation or residual) a β-redex above the λ-term P. This fundamental property of the λ-calculus is precisely the reason for the *left-orientation* of this calculus, discussed at length in the introduction of this article.

In that specific case, the diagrammatic standardization theorem repeats the traditional *leftmost-outermost* standardization theorem established in [24, 20, 3]. Since there exists no reversible permutation, the equivalence relation $\simeq$ modulo reversible permutation coincides with the equality. This explains why the standard path g of a path f is *unique* in that case — and not just unique *modulo.*

λ-CALCULUS [ARGUMENT ORDER]. In contrast to the two orders $\preceq_{\text{tree}}$ and $\preceq_{\text{left}}$, this particular order on β-redexes does not fall into the scope of our previous axiomatic presented in [13] for the following reason. An axiom requires that whenever two β-redexes u and v have respective residuals u' and v' after β-reduction of a coinitial β-redex w, then:

$$(u' \preceq_{\text{arg}} v' \Rightarrow u \preceq_{\text{arg}} v) \quad \text{or} \quad (w \preceq_{\text{arg}} u \text{ and } w \preceq_{\text{arg}} v). \tag{27}$$

The axiom states that a redex w may only alter the relative positions of redexes u and v when the two redexes are under the redex w. The argument-order $\preceq_{\text{arg}}$ does not satisfy this property in general, typically when the β-redex $w : (\lambda x.M)P$ substitutes its argument P containing the β-redex v inside the argument of a β-redex u in the function $(\lambda x.M)$. This is illustrated by the three coinitial β-redexes u, v and w:

$$\begin{array}{ccc}
(\lambda w.w)((\lambda v.v)a) & & (\lambda v.v)a \\
\uparrow u & & \uparrow u' \\
(\lambda w.(\lambda u.u)w)((\lambda v.v)a) & \xrightarrow{\quad w \quad} & (\lambda u.u)((\lambda v.v)a) \\
\downarrow v & & \downarrow v' \\
(\lambda w.(\lambda u.u)w)a & & (\lambda u.u)a
\end{array}$$

It is not difficult to see that Property (27) is not satisfied, since:

- the β-redexes u is not in the argument of the β-redex w: thus, $\neg(w \preceq_{\text{arg}} u)$.
- the β-redex v is not in the argument of the β-redex u: thus, $\neg(u \preceq_{\text{arg}} v)$,

– after β-contraction of the β-redex w, the residual v' of the β-redex v appears in the argument of the residual u' of the β-redex u: thus, $u' \preceq_{\mathrm{arg}} v'$.

It took us a lot of time to realize after [13] that Property (27) can be weakened and replaced by the N-axiom **IIb.** formulated in Section 6, without breaking the standardization theorem. We recall that the N-axiom **IIb.** states that in the earlier situation:

$$(u' \preceq_{\mathrm{arg}} v' \Rightarrow u \preceq_{\mathrm{arg}} v) \quad \text{or} \quad w \preceq_{\mathrm{arg}} v.$$

In other words, it is possible for a redex w above a redex v to position one of its residuals v' under a redex u not nested by the redex w. This is precisely what happens in our example. So, this weaker property and the nine other N-axioms are satisfied by the axiomatic nesting system $(\mathcal{G}_\lambda, \preceq_{\mathrm{arg}}, [\![-]\!]_\lambda, \uparrow_\lambda)$. Thus, contrary to what happened in [13], our axiomatics does not discriminate between the three different partial orders $\preceq_{\mathrm{tree}}$, $\preceq_{\mathrm{left}}$ and $\preceq_{\mathrm{arg}}$ on the β-redexes in λ-terms. Consequently, the argument-order $\preceq_{\mathrm{arg}}$ induces a well-behaved standardization theorem on the λ-calculus — just like the tree-order $\preceq_{\mathrm{tree}}$ and the left-order $\preceq_{\mathrm{left}}$.

λ-CALCULUS [CALL-BY-VALUE]. A *value* of the λ-calculus is defined either as a variable or as a λ-term of the form $\lambda x.M$. G. Plotkin introduces in [38] the call-by-value λ-calculus, whose unique β_v-reduction $(\lambda x.M)V \to M[V/x]$ is the β-rule restricted to value arguments V. It is not difficult to show that the λ_v-calculus — interpreted as an axiomatic nesting system — satisfies the ten N-axioms formulated in Section 6. The resulting standardization theorem, which is non-trivial to prove directly on the syntax, leads to Plotkin's formalization of Landin's SECD machine, see [12] for instance.

EXPLICIT SUBSTITUTIONS. The usual β-reduction $(\lambda x.M)P \longrightarrow M[P/x]$ copies its argument P as many times as the variable x occurs in M. This is fine theoretically, but inefficient if one wants to implement β-reduction in a computer. Thus, in most implementations of the λ-calculus, the argument P is not substituted, but stored in a *closure* and applied only when necessary. Unfortunately, the alternative evaluation mechanism complicates the task of checking the *correctness* of the implementation, by translating it back to the λ-calculus.

So, the $\lambda\sigma$-calculus was introduced in [1] to bridge the λ-calculus and its implementations. In the $\lambda\sigma$-calculus, substitutions are *explicit*, they can be delayed and stored just like closures. This enables to factorize many translations from abstract machines to the λ-calculus, see [15].

$$\text{Abstract Machine} \xrightarrow{\text{translation}} \lambda\sigma\text{-calculus} \xrightarrow{\text{interpretation}} \lambda\text{-calculus}$$

Formally, the $\lambda\sigma$-calculus contains two classes of objects: terms and substitutions. Terms are written in the de Bruijn notation.

$$\begin{array}{ll} \textbf{terms} & a ::= \mathbf{1} \mid ab \mid \lambda a \mid a[s] \\ \textbf{substitutions} & s ::= id \mid \uparrow \mid a \cdot s \mid s \circ t \end{array}$$

Beta	$(\lambda a)b \to a[b \cdot id]$		
App	$(ab)[s] \to a[s]b[s]$	*VarId*	$\mathbf{1}[id] \to \mathbf{1}$
Abs	$(\lambda a)[s] \to \lambda(a[\mathbf{1} \cdot (s \circ \uparrow)])$	*VarCons*	$\mathbf{1}[a.s] \to a$
Clos	$a[s][t] \to a[s \circ t]$	*IdL*	$id \circ s \to s$
Map	$(a \cdot s) \circ t \to a[t] \cdot (s \circ t)$	*ShiftId*	$\uparrow \circ\, id \to \uparrow$
Ass	$(s_1 \circ s_2) \circ s_3 \to s_1 \circ (s_2 \circ s_3)$	*ShiftCons*	$\uparrow \circ\, (a \cdot s) \to s$

Fig. 6. The 11 rules of the $\lambda\sigma$-calculus

Ten rules (called the σ-rules) describe how substitutions should be delayed, propagated, composed and performed. An eleventh rule of the calculus, the *Beta* rule, mimicks the β-rule of the λ-calculus, see Figure 6.

This makes the $\lambda\sigma$-calculus a *fibered* rewriting system with underlying *basis* the λ-calculus. The σ-calculus is strongly normalizing and confluent. Thus, every (closed) $\lambda\sigma$-term may be interpreted as the λ-term $\sigma(a)$ obtained by σ-normalization. The fiber F_M indexed by the λ-term M contains all $\lambda\sigma$-terms a interpreted as $\sigma(a) = M$. It is possible to extend the interpretation from terms to computations, and to project every $\lambda\sigma$-rewriting path $a \longrightarrow b$ to a β-rewriting path $\sigma(a) \longrightarrow \sigma(b)$ (modulo equivalence $\simeq$ though). Properties of the interpretation are studied thoroughly in [14, 9, 40, 28].

The $\lambda\sigma$-calculus is kind of hybrid between deterministic and non-deterministic rewriting systems. As a fibered system over the λ-calculus, it satisfies many properties of conflict-free rewriting systems, like confluence. At the same time, with eleven rules and eleven critical pairs (see Figure 7) the $\lambda\sigma$-calculus is an elaborate instance of a calculus with conflicts. Besides, to add some spice, its evaluation mechanism may behave counter-intuitively, as witnessed by the author's non-termination example of a simply-typed $\lambda\sigma$-term, presented in [26].

For all these reasons, the $\lambda\sigma$-calculus has been our training partner since the early days of the axiomatic theory. Many fundamental ideas of the theory (e.g. factorization, stability) originate from the meticulous analysis of its evaluation mechanism. Of course, like every term rewriting system, the $\lambda\sigma$-calculus defines an axiomatic rewriting system. As such, it satisfies the standardization theorem established in the article, as well as the factorization and stability theorems established in later articles [29, 30]. We believe that this series of structure theorems play the same regulating role for the $\lambda\sigma$-calculus as the Church-Rosser property plays traditionnaly for the λ-calculus. For instance, we were able to formulate and establish in this way a normalization theorem for the needed strategies of the $\lambda\sigma$-calculus, see [28].

DAGS. The definition of a rewriting system Σ on directed acyclic graphs (dags) may be found in [8]. We interpret any dag rewriting system Σ as the following axiomatic rewriting system $(\mathcal{G}_\Sigma, \rhd_\Sigma)$. The graph $\mathcal{G}_\Sigma$ has dags and redexes of Σ as vertices and edges. Two paths f and g are related as $f \rhd_\Sigma g$ in two cases only:

$App + Beta$	$(\lambda a)[s](b[s])$	$\overset{App}{\longleftarrow}$	$((\lambda a)b)[s]$	$\overset{Beta}{\longrightarrow}$	$a[b \cdot id][s]$
$Clos + App$	$(ab)[s \circ t]$	$\overset{Clos}{\longleftarrow}$	$(ab)[s][t]$	$\overset{App}{\longrightarrow}$	$(a[s](b[s]))[t]$
$Clos + Abs$	$(\lambda a)[s \circ t]$	$\overset{Clos}{\longleftarrow}$	$(\lambda a)[s][t]$	$\overset{Abs}{\longrightarrow}$	$(\lambda(a[\mathbf{1} \cdot s \circ \uparrow]))[t]$
$Clos + VarId$	$\mathbf{1}[id \circ s]$	$\overset{Clos}{\longleftarrow}$	$\mathbf{1}[id][s]$	$\overset{VarId}{\longrightarrow}$	$\mathbf{1}[s]$
$Clos + VarCons$	$\mathbf{1}[(a \cdot s) \circ t]$	$\overset{Clos}{\longleftarrow}$	$\mathbf{1}[a \cdot s][t]$	$\overset{VarCons}{\longrightarrow}$	$a[t]$
$Clos + Clos$	$a[s][t \circ t']$	$\overset{Clos}{\longleftarrow}$	$a[s][t][t']$	$\overset{Clos}{\longrightarrow}$	$a[s \circ t][t']$
$Ass + Map$	$(a \cdot s) \circ (t \circ t')$	$\overset{Ass}{\longleftarrow}$	$((a \cdot s) \circ t) \circ t'$	$\overset{Map}{\longrightarrow}$	$(a[t] \cdot s \circ t) \circ t'$
$Ass + IdL$	$id \circ (s \circ t)$	$\overset{Ass}{\longleftarrow}$	$(id \circ s) \circ t$	$\overset{IdL}{\longrightarrow}$	$s \circ t$
$Ass + ShiftId$	$\uparrow \circ (id \circ s)$	$\overset{Ass}{\longleftarrow}$	$(\uparrow \circ id) \circ s$	$\overset{ShiftId}{\longrightarrow}$	$\uparrow \circ s$
$Ass + ShiftCons$	$\uparrow \circ ((a \cdot s) \circ t)$	$\overset{Ass}{\longleftarrow}$	$(\uparrow \circ (a \cdot s)) \circ t$	$\overset{ShiftCons}{\longrightarrow}$	$s \circ t$
$Ass + Ass$	$(s \circ s') \circ (t \circ t')$	$\overset{Ass}{\longleftarrow}$	$((s \circ s') \circ t) \circ t'$	$\overset{Ass}{\longrightarrow}$	$(s \circ (s' \circ t)) \circ t'$

Fig. 7. The 11 critical pairs of the $\lambda\sigma$-calculus

- the reversible case: $f = v \cdot u'$ and $g = u \cdot v'$, when u and v are different compatible redexes, u' is the unique residual of u after v, and v' is the unique residual of v after u.
- the irreversible case: $f = v \cdot u'$ and $g = u$, when u and v are different compatible redexes, u' is the unique residual of u after v, and v does not have any residual after v, or equivalently, v is erased by u.

The nine standardization axioms are not too difficult to establish on $(\mathcal{G}_\Sigma, \rhd_\Sigma)$ in the same way as for erasing transition systems, considered a few paragraphs above.

Remark: In the case of a non-erasing dag rewriting system Σ, every rewriting path is standard. This indicates that our current axiomatic description of dag rewriting systems is not really satisfactory. Obviously, standardization should consider redex occurrence instead of simply redex erasure. We still do not know how to integrate such considerations in our standardization theory, see the discussion [27]. One solution may be to relax the notion of 2-dimensional normal form (=standard path) in a way similar to B. Hilken when he relaxes the definition of 1-dimensional normal form, in order to characterize the $\beta\eta$-long normal forms of simply-typed λ-calculus, see [16, 28] and the paragraph below.

λ-CALCULUS [ETA-EXPANSION]. B. Hilken considers the following permutation in simply-typed λ-calculus with β-reduction and η-expansion, see [16]:

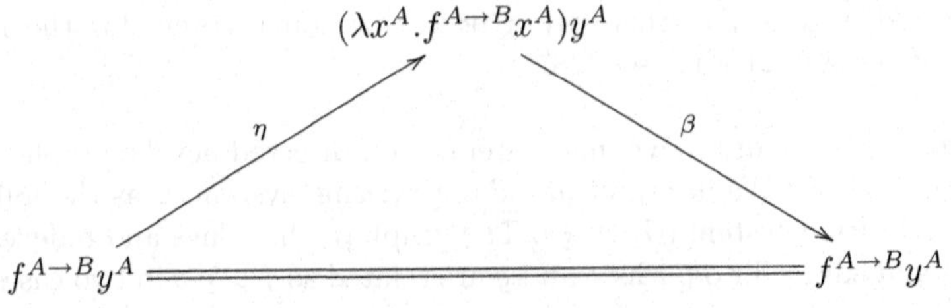

(28)

In this way, B. Hilken characterizes the $\beta\eta$-long normal forms as the λ-terms M such that, for every rewriting path $f : M \twoheadrightarrow N$, there exists a path $g : N \twoheadrightarrow M$ such that $f \cdot g : M \twoheadrightarrow M$ is equivalent to $\mathrm{id}_M : M \twoheadrightarrow M$ modulo permutation. This is one of the most interesting open problems of our Axiomatic Rewriting Theory: despite much effort, we do not know yet how permutations like (28) should be integrated in our diagrammatic theory.

Order sequentialization. Here, we illustrate the fact that axiomatic rewriting systems *strictly* generalize axiomatic nesting systems. We fix a set X, and construct the transition system $\mathcal{G}_X$ as follows:

- its vertices are the partial orders on the set X,
- its edges $\leq_1 \longrightarrow \leq_2$ are the quadruples $(\leq_1, a, b, \leq_2)$ where (a, b) is a pair of incomparable elements in the partial order $(X, \leq_1)$, and the partial order $\leq_2$ is defined as:

$$\leq_2 \quad = \quad \leq_1 \quad \cup \quad \{(x,y) \in X \times X \mid x \leq_1 a \text{ and } b \leq_1 y\}$$

The 2-dimensional transition system $(\mathcal{G}_X, \rhd_X)$ is then defined as follows. Its *irreversible* permutations $f \blacktriangleright_X g$ relate two paths

$$\begin{array}{ccc} \leq_1 & \xrightarrow{(a,b)} & \leq_2 \\ {\scriptstyle (c,d)}\downarrow & \blacktriangleright_X & \downarrow{\scriptstyle (c,d)} \\ \leq_3 & \underset{\mathrm{id}}{=\!=\!=} & \leq_3 \end{array}$$

when $c \leq_1 a$ and $b \leq_1 d$. The *reversible* permutation relation $\diamond_X$ relates two paths

$$\begin{array}{ccc} \leq_1 & \xrightarrow{(a,b)} & \leq_2 \\ {\scriptstyle (c,d)}\downarrow & \diamond_X & \downarrow{\scriptstyle (c,d)} \\ \leq_3 & \xrightarrow[(c,d)]{} & \leq_4 \end{array}$$

when neither ($c \leq_1 a$ and $b \leq_1 d$) nor ($d \leq_1 a$ and $b \leq_1 c$).

It is easy to prove that the 2-dimensional transition system $(\mathcal{G}_X, \rhd_X)$ defines an axiomatic rewriting system, for every set X. The normal forms of this system are the total orders on X. The interesting point is that the axiomatic rewriting system $(\mathcal{G}_X, \rhd_X)$ associated to $X = \{a, b, c\}$ does not satisfy Axiom **reversible-cube** formulated in Section 7.3 — and thus, cannot be expressed as an axiomatic nesting system. Indeed, $(\mathcal{G}_X, \rhd_X)$ contains the diagram

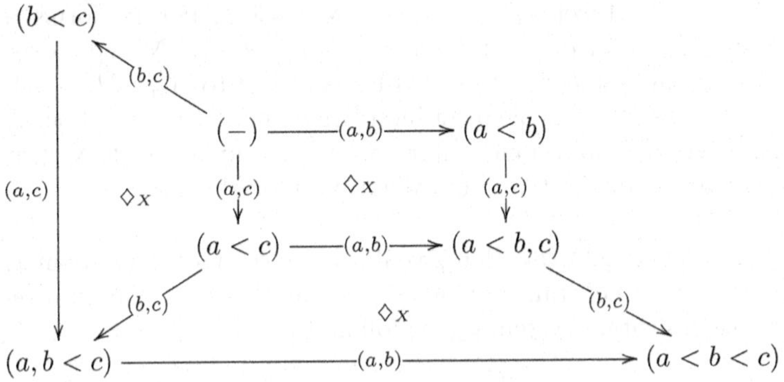

By Lemma 45 and 46 any such diagram may be completed as a reversible cube in an axiomatic rewriting system associated to an axiomatic nesting system. However, this diagram cannot be completed in $(\mathcal{G}_X, \rhd_X)$.

9 Conclusion

Axiomatic Rewriting Theory is the latest attempt since Abstract Rewriting Theory [32, 17, 21] to describe *uniformly* all existing rewriting systems — from Petri nets to higher-order rewriting systems. The theory uncovers a series of diagrammatic principles underlying the syntactic mechanisms of computation, and reduces in this way the endemic variety of syntax to a uniform geometry of causality. In about a decade, the theory has bridged the gap with category theory and denotational semantics, and solved several difficult problems of Rewriting Theory:

- a normalization theorem for needed strategies in the $\lambda\sigma$-calculus, a λ-calculus with explicit substitutions, has been formulated and established in [28],
- a factorization theorem separating functorially the useful part of a rewriting path from the junk has been established in [29],
- an algebraic characterization of head-reductions in rewriting systems with critical pairs has been formulated in [30]. A syntactic characterization of head-reductions has been also formulated in the case of the $\lambda\sigma$-calculus [28].

This series of results demonstrates that a purely diagrammatic approach to Rewriting Theory is possible and fruitful. It also opens a series of interesting research directions, at the frontier of Rewriting Theory and Higher-Dimensional Categories, see for instance [23] and [31]. More specifically, we would like to capture *properly* the causal principles underlying Rewriting Systems like the λ-calculus with β-reduction and η-expansion, the non left-linear term rewriting systems, or the directed acyclic graph rewriting systems. We are inclined to think that the diagrammatic language has something singular and innovative to articulate on these traditional topics of Rewriting Theory.

References

1. M. Abadi, L. Cardelli, P.-L. Curien, J.-J. Lévy. Explicit substitutions. *Proceedings of Principle Of Programming Languages*, 1990.
2. F. Baader, T. Nipkow. *Term rewriting and all that.* Cambridge University Press, 1998.
3. H. Barendregt. *The Lambda Calculus: Its Syntax and Semantics.* North Holland, 1985.
4. M. A. Bednarczyck, *Categories of asynchronous systems.* PhD thesis, University of Sussex, 1988.
5. G. Berry. *Modèles complètement adéquats et stables des lambda-calculs typés.* Thèse de Doctorat d'Etat, Université Paris VII, 1979.
6. G. Boudol. Computational semantics of term rewriting systems. *Algebraic methods in Semantics*, Maurice Nivat and John C. Reynolds (eds). Cambridge University Press, 1985.
7. A. Church, J.B. Rosser. Some properties of conversion. *Trans. Amer. Math. Soc. 39*, pp. 472-482, 1936.
8. D. Clark, R. Kennaway. Event structures and non-orthogonal term graph rewriting. *Mathematical Structure in Computer Science*, vol. 6, pp. 545-578, 1996.
9. P.-L. Curien, T. Hardin, A. Ríos. Strong normalization of substitutions. *Lecture Notes in Computer Science*, 629:209–217, 1992.
10. H.-B. Curry, R. Feys. *Combinatory Logic.* North Holland Volume 1, 1958.
11. N. Dershowitz, J.-P. Jouannaud. Rewrite systems. Chap. 6 of *Handbook of Theoretical Computer Science B: Formal Methods and Semantics*, J. van Leeuwen, ed., North-Holland, Amsterdam (1990) 243-320
12. M. Felleisen, R. Hieb. The revised report on the syntactic theories of sequential control and state. *Theoretical Computer Science*, 1992.
13. G. Gonthier, J.-J. Lévy, P.-A. Melliès. An abstract standardization theorem. *Proceedings of the 7th Annual IEEE Symposium on Logic In Computer Science*, Santa Cruz, 1992.
14. T. Hardin. Confluence Results for the Pure Strong Categorical Combinatory Logic. λ-calculi as subsystems of CCL. *Journal of Theoretical Computer Science*, 1989.
15. T. Hardin, L. Maranget, B. Pagano. Functional Back-Ends within the Lambda-Sigma Calculus. *Proc. of the 1996 International Conference on Functional Programming*, 1996.
16. B. P. Hilken. Towards a proof theory of rewriting: the simply typed 2λ-calculus. *Theoretical Computer Science* 170, pages 407-444, 1996.
17. G. Huet. Confluent Reductions : Abstract Properties and Applications to Term Rewriting Systems. *Journal of the Association for Computing Machinery* vol. 27, No 4 (1980), 797–821.
18. G. Huet, J.-J. Lévy. Call by Need Computations in Non-Ambiguous Linear Term Rewriting Systems. *Rapport de recherche INRIA 359*, 1979. Reprinted as: Computations in orthogonal rewriting systems. In J.-L. Lassez and G. D. Plotkin, editors, *Computational Logic; Essays in Honor of Alan Robinson*, pages 394–443. MIT Press, 1991.
19. J.-P. Jouannaud. Rewrite proofs and computations. In *Proof and Computation.* Helmut Schwichtenberg, ed. NATO series F: Computer and Systems Sciences, vol. 139, pp. 173-218, Springer Verlag, 1995.
20. J.W. Klop. *Combinatory Reduction Systems.* Thèse de l'Université d'Utrecht, Pays-Bas (1980).

21. J.W. Klop. Term Rewriting Systems. *Handbook of Logic in Computer Science*, Volume 2, in S. Abramsky, Dov M. Gabbay, T.S.E. Maibaum, editors, Oxford Science Publications, 1992.
22. J.W. Klop, V. van Oostrom, R. de Vrijer. Orthogonality. Chapter 4 in the book *Term Rewriting System*, edited by TeReSe, Cambridge Tracts in Theoretical Computer Science, Volume 55, Cambridge University Press, 2003.
23. T. Leinster. *Higher Operads, Higher Categories*. London Mathematical Society Lecture Note Series 298, Cambridge University Press, 2003.
24. J.-J. Lévy. *Réductions correctes et optimales dans le λ-calcul*. Thèse de Doctorat d'Etat, Université Paris VII, 1978.
25. S. Mac Lane. "*Categories for the working mathematician*", Second Edition, Graduate Texts in Mathematics 5, Springer-Verlag, 1998.
26. P.-A. Melliès. Typed Lambda-Calculi with Explicit Substitutions may not terminate. *Proceedings of TLCA'95*. Lecture Notes in Computer Science 902, Springer, 1995.
27. P.-A. Melliès. *Description abstraite des Systèmes de Réécriture*. Thèse de Doctorat, Université Paris VII, 1996.
28. P.-A. Melliès. Axiomatic Rewriting Theory II: The lambda-sigma-calculus enjoys finite normalisation cones. *Journal of Logic and Computation*, special issue devoted to the School on Rewriting and Type Theory, 2000.
29. P.-A. Melliès. Axiomatic Rewriting Theory III: A factorisation theorem in Rewriting Theory. *Proceedings of the 7th Conference on Category Theory and Computer Science*, Santa Margherita Ligure. Lecture Notes in Computer Science 1290, pp. 49-68, 1997.
30. P.-A. Melliès. Axiomatic Rewriting Theory IV: A stability theorem in Rewriting Theory. *Proceedings of the 14th Annual Symposium on Logic in Computer Science*, Indianapolis, 1998.
31. P.-A. Melliès. Axiomatic Rewriting Theory VI: Residual Theory Revisited. *Proceedings of Conference on Rewriting Techniques and Applications*, LNCS 2378, Springer Verlag, Kobenhavn, 2002.
32. M. H. A. Newman. On theories with a combinatorial definition of "equivalence". *Annals of Mathematics*, 43, Number 2, pages 223–243, 1942.
33. M. Nielsen, G. Winskel. Models for concurrency. In *Handbook of Logic in Computer Science*, Abramsky, Gabbay, Maibaum editors, Oxford Science, 1995.
34. V. van Oostrom. Confluence for abstract and higher-order rewriting. PhD Thesis, Vrije Universiteit, 1994.
35. V. van Oostrom, R. de Vrijer. Equivalence of reductions. Chapter 8 in the book *Term Rewriting System*, edited by TeReSe, Cambridge Tracts in Theoretical Computer Science, Volume 55, Cambridge University Press, 2003.
36. F. van Raamsdonk. *Confluence and normalization for higher-order rewriting*, PhD Thesis, Vrije Universiteit, Amsterdam, 1996.
37. P. Panangaden, V. Shanbhogue, E. W. Stark. Stability and sequentiality in data flow networks. *Proceedings of ICALP'90*, Lecture Notes in Computer Science 443, Springer, 1990.
38. G. Plotkin. Call-by-name, call-by-value, and the λ-calculus. *Theoretical Computer Science 1*, 1975.
39. M. W. Shields. Concurrent machines. *Computer Journal*, 28:449-465, 1985.
40. H. Zantema. Termination of term rewriting by interpretation. *Lecture Notes in Computer Science 656*, Springer Verlag, 1993.

Author Index

Lecture Notes in Computer Science

For information about Vols. 1–3738

please contact your bookseller or Springer

Vol. 3838: A. Middeldorp, V. van Oostrom, F. van Raamsdonk, R. de Vrijer (Eds.), Processes, Terms and Cycles: Steps on the Road to Infinity. XVIII, 639 pages. 2005.

Vol. 3837: K. Cho, P. Jacquet (Eds.), Technologies for Advanced Heterogeneous Networks. IX, 307 pages. 2005.

Vol. 3835: G. Sutcliffe, A. Voronkov (Eds.), Logic for Programming, Artificial Intelligence, and Reasoning. XIV, 744 pages. 2005. (Subseries LNAI).

Vol. 3833: K.-J. Li, C. Vangenot (Eds.), Web and Wireless Geographical Information Systems. XI, 309 pages. 2005.

Vol. 3829: P. Pettersson, W. Yi (Eds.), Formal Modeling and Analysis of Timed Systems. IX, 305 pages. 2005.

Vol. 3828: X. Deng, Y. Ye (Eds.), Internet and Network Economics. XVII, 1106 pages. 2005.

Vol. 3826: B. Benatallah, F. Casati, P. Traverso (Eds.), Service-Oriented Computing - ICSOC 2005. XVIII, 597 pages. 2005.

Vol. 3824: L.T. Yang, M. Amamiya, Z. Liu, M. Guo, F.J. Rammig (Eds.), Embedded and Ubiquitous Computing. XXIII, 1204 pages. 2005.

Vol. 3823: T. Enokido, L. Yan, B. Xiao, D. Kim, Y. Dai, L.T. Yang (Eds.), Embedded and Ubiquitous Computing. XXXII, 1317 pages. 2005.

Vol. 3822: D. Feng, D. Lin, M. Yung (Eds.), Information Security and Cryptology. XII, 420 pages. 2005.

Vol. 3821: R. Ramanujam, S. Sen (Eds.), FSTTCS 2005: Foundations of Software Technology and Theoretical Computer Science. XIV, 566 pages. 2005.

Vol. 3820: L.T. Yang, X. Zhou, W. Zhao, Z. Wu, Y. Zhu, M. Lin (Eds.), Embedded Software and Systems. XXVIII, 779 pages. 2005.

Vol. 3818: S. Grumbach, L. Sui, V. Vianu (Eds.), Advances in Computer Science – ASIAN 2005. XIII, 294 pages. 2005.

Vol. 3815: E.A. Fox, E.J. Neuhold, P. Premsmit, V. Wuwongse (Eds.), Digital Libraries: Implementing Strategies and Sharing Experiences. XVII, 529 pages. 2005.

Vol. 3814: M. Maybury, O. Stock, W. Wahlster (Eds.), Intelligent Technologies for Interactive Entertainment. XV, 342 pages. 2005. (Subseries LNAI).

Vol. 3810: Y.G. Desmedt, H. Wang, Y. Mu, Y. Li (Eds.), Cryptology and Network Security. XI, 349 pages. 2005.

Vol. 3809: S. Zhang, R. Jarvis (Eds.), AI 2005: Advances in Artificial Intelligence. XXVII, 1344 pages. 2005. (Subseries LNAI).

Vol. 3808: C. Bento, A. Cardoso, G. Dias (Eds.), Progress in Artificial Intelligence. XVIII, 704 pages. 2005. (Subseries LNAI).

Vol. 3807: M. Dean, Y. Guo, W. Jun, R. Kaschek, S. Krishnaswamy, Z. Pan, Q.Z. Sheng (Eds.), Web Information Systems Engineering – WISE 2005 Workshops. XV, 275 pages. 2005.

Vol. 3806: A.H. H. Ngu, M. Kitsuregawa, E.J. Neuhold, J.-Y. Chung, Q.Z. Sheng (Eds.), Web Information Systems Engineering – WISE 2005. XXI, 771 pages. 2005.

Vol. 3805: G. Subsol (Ed.), Virtual Storytelling. XII, 289 pages. 2005.

Vol. 3804: G. Bebis, R. Boyle, D. Koracin, B. Parvin (Eds.), Advances in Visual Computing. XX, 755 pages. 2005.

Vol. 3803: S. Jajodia, C. Mazumdar (Eds.), Information Systems Security. XI, 342 pages. 2005.

Vol. 3802: Y. Hao, J. Liu, Y. Wang, Y.-m. Cheung, H. Yin, L. Jiao, J. Ma, Y.-C. Jiao (Eds.), Computational Intelligence and Security, Part II. XLII, 1166 pages. 2005. (Subseries LNAI).

Vol. 3801: Y. Hao, J. Liu, Y. Wang, Y.-m. Cheung, H. Yin, L. Jiao, J. Ma, Y.-C. Jiao (Eds.), Computational Intelligence and Security, Part I. XLI, 1122 pages. 2005. (Subseries LNAI).

Vol. 3799: M. A. Rodríguez, I.F. Cruz, S. Levashkin, M.J. Egenhofer (Eds.), GeoSpatial Semantics. X, 259 pages. 2005.

Vol. 3798: A. Dearle, S. Eisenbach (Eds.), Component Deployment. X, 197 pages. 2005.

Vol. 3797: S. Maitra, C. E. V. Madhavan, R. Venkatesan (Eds.), Progress in Cryptology - INDOCRYPT 2005. XIV, 417 pages. 2005.

Vol. 3796: N.P. Smart (Ed.), Cryptography and Coding. XI, 461 pages. 2005.

Vol. 3795: H. Zhuge, G.C. Fox (Eds.), Grid and Cooperative Computing - GCC 2005. XXI, 1203 pages. 2005.

Vol. 3794: X. Jia, J. Wu, Y. He (Eds.), Mobile Ad-hoc and Sensor Networks. XX, 1136 pages. 2005.

Vol. 3793: T. Conte, N. Navarro, W.-m.W. Hwu, M. Valero, T. Ungerer (Eds.), High Performance Embedded Architectures and Compilers. XIII, 317 pages. 2005.

Vol. 3792: I. Richardson, P. Abrahamsson, R. Messnarz (Eds.), Software Process Improvement. VIII, 215 pages. 2005.

Vol. 3791: A. Adi, S. Stoutenburg, S. Tabet (Eds.), Rules and Rule Markup Languages for the Semantic Web. X, 225 pages. 2005.

Vol. 3790: G. Alonso (Ed.), Middleware 2005. XIII, 443 pages. 2005.

Vol. 3789: A. Gelbukh, Á. de Albornoz, H. Terashima-Marín (Eds.), MICAI 2005: Advances in Artificial Intelligence. XXVI, 1198 pages. 2005. (Subseries LNAI).

Vol. 3788: B. Roy (Ed.), Advances in Cryptology - ASIACRYPT 2005. XIV, 703 pages. 2005.

Vol. 3785: K.-K. Lau, R. Banach (Eds.), Formal Methods and Software Engineering. XIV, 496 pages. 2005.

Vol. 3784: J. Tao, T. Tan, R.W. Picard (Eds.), Affective Computing and Intelligent Interaction. XIX, 1008 pages. 2005.

Vol. 3783: S. Qing, W. Mao, J. Lopez, G. Wang (Eds.), Information and Communications Security. XIV, 492 pages. 2005.

Vol. 3781: S.Z. Li, Z. Sun, T. Tan, S. Pankanti, G. Chollet, D. Zhang (Eds.), Advances in Biometric Person Authentication. XI, 250 pages. 2005.

Vol. 3780: K. Yi (Ed.), Programming Languages and Systems. XI, 435 pages. 2005.

Vol. 3779: H. Jin, D. Reed, W. Jiang (Eds.), Network and Parallel Computing. XV, 513 pages. 2005.

Vol. 3778: C. Atkinson, C. Bunse, H.-G. Gross, C. Peper (Eds.), Component-Based Software Development for Embedded Systems. VIII, 345 pages. 2005.

Vol. 3777: O.B. Lupanov, O.M. Kasim-Zade, A.V. Chaskin, K. Steinhöfel (Eds.), Stochastic Algorithms: Foundations and Applications. VIII, 239 pages. 2005.

Vol. 3776: S.K. Pal, S. Bandyopadhyay, S. Biswas (Eds.), Pattern Recognition and Machine Intelligence. XXIV, 808 pages. 2005.

Vol. 3775: J. Schönwälder, J. Serrat (Eds.), Ambient Networks. XIII, 281 pages. 2005.

Vol. 3774: G. Bierman, C. Koch (Eds.), Database Programming Languages. X, 295 pages. 2005.

Vol. 3773: A. Sanfeliu, M.L. Cortés (Eds.), Progress in Pattern Recognition, Image Analysis and Applications. XX, 1094 pages. 2005.

Vol. 3772: M. Consens, G. Navarro (Eds.), String Processing and Information Retrieval. XIV, 406 pages. 2005.

Vol. 3771: J.M.T. Romijn, G.P. Smith, J. van de Pol (Eds.), Integrated Formal Methods. XI, 407 pages. 2005.

Vol. 3770: J. Akoka, S.W. Liddle, I.-Y. Song, M. Bertolotto, I. Comyn-Wattiau, W.-J. van den Heuvel, M. Kolp, J. Trujillo, C. Kop, H.C. Mayr (Eds.), Perspectives in Conceptual Modeling. XXII, 476 pages. 2005.

Vol. 3769: D.A. Bader, M. Parashar, V. Sridhar, V.K. Prasanna (Eds.), High Performance Computing – HiPC 2005. XXVIII, 550 pages. 2005.

Vol. 3768: Y.-S. Ho, H.J. Kim (Eds.), Advances in Multimedia Information Processing - PCM 2005, Part II. XXVIII, 1088 pages. 2005.

Vol. 3767: Y.-S. Ho, H.J. Kim (Eds.), Advances in Multimedia Information Processing - PCM 2005, Part I. XXVIII, 1022 pages. 2005.

Vol. 3766: N. Sebe, M.S. Lew, T.S. Huang (Eds.), Computer Vision in Human-Computer Interaction. X, 231 pages. 2005.

Vol. 3765: Y. Liu, T. Jiang, C. Zhang (Eds.), Computer Vision for Biomedical Image Applications. X, 563 pages. 2005.

Vol. 3764: S. Tixeuil, T. Herman (Eds.), Self-Stabilizing Systems. VIII, 229 pages. 2005.

Vol. 3762: R. Meersman, Z. Tari, P. Herrero (Eds.), On the Move to Meaningful Internet Systems 2005: OTM 2005 Workshops. XXXI, 1228 pages. 2005.

Vol. 3761: R. Meersman, Z. Tari (Eds.), On the Move to Meaningful Internet Systems 2005: CoopIS, DOA, and ODBASE, Part II. XXVII, 653 pages. 2005.

Vol. 3760: R. Meersman, Z. Tari (Eds.), On the Move to Meaningful Internet Systems 2005: CoopIS, DOA, and ODBASE, Part I. XXVII, 921 pages. 2005.

Vol. 3759: G. Chen, Y. Pan, M. Guo, J. Lu (Eds.), Parallel and Distributed Processing and Applications - ISPA 2005 Workshops. XIII, 669 pages. 2005.

Vol. 3758: Y. Pan, D.-x. Chen, M. Guo, J. Cao, J.J. Dongarra (Eds.), Parallel and Distributed Processing and Applications. XXIII, 1162 pages. 2005.

Vol. 3757: A. Rangarajan, B. Vemuri, A.L. Yuille (Eds.), Energy Minimization Methods in Computer Vision and Pattern Recognition. XII, 666 pages. 2005.

Vol. 3756: J. Cao, W. Nejdl, M. Xu (Eds.), Advanced Parallel Processing Technologies. XIV, 526 pages. 2005.

Vol. 3754: J. Dalmau Royo, G. Hasegawa (Eds.), Management of Multimedia Networks and Services. XII, 384 pages. 2005.

Vol. 3753: O.F. Olsen, L.M.J. Florack, A. Kuijper (Eds.), Deep Structure, Singularities, and Computer Vision. X, 259 pages. 2005.

Vol. 3752: N. Paragios, O. Faugeras, T. Chan, C. Schnörr (Eds.), Variational, Geometric, and Level Set Methods in Computer Vision. XI, 369 pages. 2005.

Vol. 3751: T. Magedanz, E.R. M. Madeira, P. Dini (Eds.), Operations and Management in IP-Based Networks. X, 213 pages. 2005.

Vol. 3750: J.S. Duncan, G. Gerig (Eds.), Medical Image Computing and Computer-Assisted Intervention – MICCAI 2005, Part II. XL, 1018 pages. 2005.

Vol. 3749: J.S. Duncan, G. Gerig (Eds.), Medical Image Computing and Computer-Assisted Intervention – MICCAI 2005, Part I. XXXIX, 942 pages. 2005.

Vol. 3748: A. Hartman, D. Kreische (Eds.), Model Driven Architecture – Foundations and Applications. IX, 349 pages. 2005.

Vol. 3747: C.A. Maziero, J.G. Silva, A.M.S. Andrade, F.M.d. Assis Silva (Eds.), Dependable Computing. XV, 267 pages. 2005.

Vol. 3746: P. Bozanis, E.N. Houstis (Eds.), Advances in Informatics. XIX, 879 pages. 2005.

Vol. 3745: J.L. Oliveira, V. Maojo, F. Martín-Sánchez, A.S. Pereira (Eds.), Biological and Medical Data Analysis. XII, 422 pages. 2005. (Subseries LNBI).

Vol. 3744: T. Magedanz, A. Karmouch, S. Pierre, I. Venieris (Eds.), Mobility Aware Technologies and Applications. XIV, 418 pages. 2005.

Vol. 3742: J. Akiyama, M. Kano, X. Tan (Eds.), Discrete and Computational Geometry. VIII, 213 pages. 2005.

Vol. 3740: T. Srikanthan, J. Xue, C.-H. Chang (Eds.), Advances in Computer Systems Architecture. XVII, 833 pages. 2005.

Vol. 3739: W. Fan, Z. Wu, J. Yang (Eds.), Advances in Web-Age Information Management. XXIV, 930 pages. 2005.